D0202628

INTRODUCTION TO ELECTRICAL ENGINEERING

McGRAW-HILL SERIES IN ELECTRICAL ENGINEERING

Consulting Editor

Stephen W. Director, Carnegie-Mellon University

CIRCUITS AND SYSTEMS
COMMUNICATIONS AND SIGNAL PROCESSING
CONTROL THEORY
ELECTRONICS AND ELECTRONIC CIRCUITS
POWER AND ENERGY
ELECTROMAGNETICS
COMPUTER ENGINEERING
INTRODUCTORY
RADAR AND ANTENNAS
VLSI

Previous Consulting Editors

Ronald M. Bracewell, Colin Cherry, James F. Gibbons, Willis W. Harman, Hubert Heffner, Edward W. Herold, John G. Linvill, Simon Ramo, Ronald A. Rohrer, Anthony E. Siegman, Charles Susskind, Frederick E. Terman, John G. Truxal, Ernst Weber, and John R. Whinnery

INTRODUCTORY

Consulting Editor

Stephen W. Director, Carnegie-Mellon University

Belove and Drossman
SYSTEMS AND CIRCUITS
FOR ELECTRICAL
ENGINEERING TECHNOLOGY
Belove, Schachter, and Schilling
DIGITAL AND ANALOG SYSTEMS,
CIRCUITS AND DEVICES
Brophy
BASIC ELECTRONICS FOR SCIENTISTS
Fitzgerald, Higginbotham, and Grabel
BASIC ELECTRICAL ENGINEERING
Hammond and Gehmlich
ELECTRICAL ENGINEERING
Paul, Nasar, and Unnewehr
INTRODUCTION TO ELECTRICAL
ENGINEERING
Piel and Truxal
TECHNOLOGY: HANDLE WITH CARE

INTRODUCTION TO ELECTRICAL ENGINEERING

C. R. Paul
S. A. Nasar
Department of Electrical Engineering
University of Kentucky

L. E. Unnewehr
Allied Automotive

McGraw-Hill Book Company

New York St. Louis San Francisco
Auckland Bogotá Hamburg
Johannesburg London Madrid
Mexico Montreal New Delhi
Panama Paris São Paulo
Singapore Sydney Tokyo
Toronto

To my mother,

Louise Bollinger Paul,

who showed me the meaning of love and sacrifice.

C. R. Paul

To my wife,

Sara

for her love and companionship.

S. A. Nasar

To my wife,

Jean

for her love and understanding.

L. E. Unnewehr

INTRODUCTION TO ELECTRICAL ENGINEERING

234567890 VNHVNH 89876

ISBN 0-07-045878-2

This book was set in Times Roman. The editors were Sanjeev Rao and J. W. Maisel; the design was done by INK, Graphic Design; the production supervisor was Charles Hess. The drawings were done by Wellington Studios Ltd.
Von Hoffmann Press, Inc., was printer and binder.

Library of Congress Cataloging in Publication Data
Paul, Clayton R.
Introduction to electrical engineering.

Bibliography: p.
1. Electrical engineering. I. Nasar, S. A.
II. Unnewehr, L. E., 1925– . III. Title.
TK146.P345 1986 621.3 85-9630
ISBN 0-07-045878-2
ISBN 0-07-045879-0 (solutions manual)

PART 2 ELECTRONIC CIRCUITS

PREFACE

The field of electrical engineering has developed at a dramatic rate in the last 30 years. Solid-state electronics virtually shouldered vacuum-tube electronics aside after the second world war, electrical engineering courses in automatic control soon became commonplace, and within the past 15 years integrated-circuit technology has dynamically expanded the electrical engineering curriculum. These topics have come to be integral parts of the discipline. Nevertheless, very few of such older, more traditional subjects as electric machinery, power, and instrumentation have been deemphasized. Even though their treatment may have been compacted, they continue to form an important part of the curriculum.

This text provides a survey of all topics integral to the electrical engineering discipline. Seeking to characterize the field of electrical engineering, one invariably finds four broad, central topics—electric circuits, electronics, electric machines and power systems, and systems control and instrumentation. The text has been divided along those lines.

Part 1—Linear Electric Circuits—provides the basic circuit-analysis techniques and concepts used throughout the subsequent topics. The basic circuit elements and the concepts of voltage, current, ideal sources, resistors, signal waveforms and Kirchhoff's laws are introduced in Chap. 2. This forms the basis for resistive-circuit analysis, presented in Chap. 3. We have chosen to delay the discussion of energy storage elements—the capacitor and the inductor—until Chap. 4. Circuit-analysis techniques can be illustrated with resistive circuits without the additional complication of differential equations.

We have chosen to discuss sinusoidal steady-state analysis (Chap. 5) prior to general waveform excitation (Chap. 6). We believe that this ordering of the material is easier for the student to assimilate. First-order *RL* and *RC* circuits, along with *RLC* circuits, are discussed for general excitation in Chap. 6. Simple circuits are used to illustrate the concepts of transient response. More general circuits are handled using the differential operator in the last section of Chap. 6.

Part 2—Electronic Circuits—covers the traditional material concerning the analysis of discrete electronic devices in Chaps. 7 (Diodes), 8 (Transistors and Amplifiers), and 9 (Small-Signal Analysis of Amplifiers). The operational amplifier and its uses are covered in Chap. 10. Chapter 11 provides a discussion of the use of these discrete elements to form digital circuits. The systems application of these digital circuits is reserved for discussion in Part 4.

Part 3—Electric Power and Machines—covers three-phase circuits (Chap. 12), magnetic circuits and transformers (Chap. 13), and electric machines (Chaps. 14, 15, 16). The control of electric motors and certain important topics pertinent to power systems are presented in Chaps. 17 and 18, respectively.

Part 4—Control and Instrumentation—brings together the techniques and concepts of the previous three parts in the design of systems to accomplish specific tasks. The important topics of instrumentation and transducers are covered in Chap. 20. This part closes with a discussion of the predominant method of modern system implementation—digital techniques—in Chaps. 21 and 22. Chapter 22 also contains material on data acquisition and coding.

The appendixes cover unit conversion (A), physical constants (B), Fourier series (C), and Laplace transforms (D).

The text is suitable for a one-year (two semesters or three quarters) introductory course offered to non-EE majors. We suggest coverage of Part 1 (Linear Electric Circuits) and Part 2 (Electronic Circuits) in the first half year, with Part 3 (Electric Power and Machines) and Part 4 (Control and Instrumentation) covered in the second half year.

The level of the material requires the student to have completed the basic college-level courses in algebra, trigonometry, and elementary calculus, as well as basic physics. A knowledge of differential equations will be helpful but not mandatory, since we develop those solution techniques within the text when they naturally arise.

We have endeavored in this text to provide the concepts, analytical methods, and topics basic to the field of electrical engineering. The depth of coverage has consciously been kept at the minimum level for each subject—yet at the same time we have chosen not merely to give equations and rules alone, without logical development. Rote memorization provides no long-term value to the student. We believe that students will not only be the more attracted to a subject and retain more from it when we give them simple, logical developments of the principles, but that they will also be the better able to apply these principles to future situations.

We would like to express our thanks for the many useful comments and suggestions provided by colleagues who reviewed this text during the course of its development, especially to Robert Cromwell; H. Hayre, University of Houston; Martin Kaliski, Northeastern University; Martin Kaplan, Drexel University; Douglas Miron, South Dakota State University; Louis Roemer, University of Akron; Lee Rosenthal; Martha Sloan, Michigan Technological University; William Swanson, San Jose State University; and R. P. Misra, New Jersey Institute of Technology.

C. R. Paul
S. A. Nasar
L. E. Unnewehr

Chapter

1

Introduction

Electrical engineering is basically concerned with providing efficient, convenient, and reliable means for transforming, processing, and transmitting energy and information. Hence, the study of electrical engineering involves the analysis, design, and applications of devices and systems for conversion, processing, and transmission of electrical energy and information. Owing to the enormous progress made in the field of electrical engineering during the last few decades, the scope of its applications has become virtually unlimited. Furthermore, electrical engineers encounter phenomena with a wide range of conditions. For instance, temperatures range from near absolute zero for the operation of superconductors to those approaching that of the sun for the feasibility of thermonuclear fusion. Time intervals are precisely controlled and measured from picoseconds to several hours or days. Distances range from micronmeters to light-years. Electrical engineers deal with quantities of power ranging from hundreds of gigawatts in the power industry to microwatts in radio astronomy. In short, the scope of electrical engineering is broad and we are witnessing an explosion of progress in the various branches of the profession.

In view of current progress and future prospects, it is necessary for every engineer to acquire a basic knowledge of electrical engineering. The following chapters are aimed at aiding the achievement of this goal. We begin our study with some very elementary concepts such as electric charges and certain laws governing them and we end our study with automatic control systems, instrumentation, and digital systems. Also included are chapters that relate to electric circuits, electronic circuits and circuit elements, electric machines, and power systems. The importance of electrical engineering to those in other fields of engineering stems from the fact that in almost every walk of life we are concerned with one or more aspects of electrical engineering—the conversion, transmission, processing, and storing of energy and intelligence. A broad range of electrical engineering principles and phenomena must be clearly understood to exploit present systems and to conceive and develop new ones. Instrumentation by electrical means is growing ever more important to all engineers, and no end is in sight. The digital computer, whose importance to all engineers is unquestionable, is one of the most significant elements in the fields of instrumentation and data processing.

Although various areas of electrical engineering—such as power and electronics—were once regarded as reasonably distinct, they are now very much interrelated. As engineering systems become more complex, electric elements will probably be employed more often. In order to understand such systems a thorough grasp of electrical engineering fundamentals will be absolutely essential. Our aim is to present a unified overview of the major areas of electrical engineering. Every engineer should be familiar with the following four major areas.

Electric Circuits Engineers must acquire a knowledge of electric circuit analysis, in other words, a proficiency in carrying out a mathematical study of various types of interconnections of electrical elements having paths for the flow of electric current. On the surface it may appear that such a skill is useful only to electrical engineers. However, as shown in the following chapters, electric circuit analysis is a prerequisite for the study of other major areas of electrical engineering. Electric circuit analysis not only aids us in pursuing further studies in electrical engineering but also broadens our scope and helps us to communicate with others involved in a common project (since an engineer invariably becomes a member of a team consisting of other engineers and managers). For a coordinated group effort it is important that various members of the group meaningfully communicate with each other. The concepts pertaining to electric circuits extend to problems such as traffic flow, socioeconomic systems, control and instrumentation, fluid flow, etc.

The term "electric circuit" in this text implies electric circuits consisting of only linear circuit elements and sources of electric power.

Electronic Circuits In contrast to linear electric circuits, electronic circuits invariably contain nonlinear elements such as diodes, transistors, thyristors (or SCRs), etc. These elements are introduced in electric circuits to develop technological solutions to problems resulting from social and economic needs. Examples of such problems can be seen in the areas of electrical communications systems, information processing, instrumentation, and control of all sorts of industrial processes. In fact, the range of applications of electronic circuits is enormous and is probably limited only by our imagination. Our goal in this major area is to develop analytical techniques for quantitative studies of some basic electronic circuits. Relying on the methods developed in the study of electric circuits, we study the different types of diodes and transistors and then use these electronic components to obtain electronic circuits which perform such functions as amplification, waveshaping, and switching.

A thorough understanding of the operation and analysis of electronic circuits is essential to the engineer and is of great value in choosing and applying electronic circuits to perform the desired task.

Electric Power and Machines Harnessing and utilizing energy has always been a key factor in improving the quality of life. The use of electrical energy has aided our ability to develop socially and economically and live with physical comfort. The major area of electrical engineering concerned with the generation, transmission, distribution, and utilization of electrical energy comes under the heading of "electric power and machines," which is covered in Part III of this book. The subject matter contained in this part encompasses a wide variety of disciplines of engineering. The importance of electric power and machines in almost every aspect of life can hardly be overemphasized. The number of electric motors in the average U.S. residence today is probably a minimum of 10 and can easily exceed 50. There are at least five electric machines on even the most spartan compact automobile, and this number is increasing steadily as emission and fuel economy systems are added. An aircraft contains many more electric machines. More people travel by means of electrical propulsion each day—in elevators, escalators, and horizontal people movers—than by any other mode of propulsion. Many electric machines have been on the moon and play an important role in most aerospace systems. Of course, electric machines are involved in every industrial and manufacturing process of a technological society. Electrical blackouts constantly remind us of our almost total dependence on electric power. Therefore, an understanding of the principles of energy conversion, electric machines, and power systems is important for all who wish to extend the usefulness of electrical technology in order to ameliorate the problems of energy, pollution, and poverty that presently face humanity.

Control and Instrumentation The fourth major area of electrical engineering of interest to all engineers is control and instrumentation. Control systems influence our everyday lives just as much as some of the other areas of electrical engineering. For example, household appliances, such as clothes washers and dryers, manufacturing and processing plants, and navigation and guidance systems all utilize the concepts of analysis and design of control systems. Therefore, in general, a control system is an interconnection of components—electrical, mechanical, hydraulic, thermal, etc.—to obtain a desired function in an efficient and accurate fashion. The control engineer is involved in the control of industrial processes and systems. The concepts of control engineering are not limited to any particular branch of engineering. Hence, a basic understanding of control theory is very useful to every engineer involved in the understanding of the dynamics of various types of systems.

Measurements of quantities such as voltage, current, temperature, and strain are of utmost importance to the engineer. Measurements tell us much about the universe, the environment, the economy, and ourselves. The basis of all scientific principles and engineering concepts rests, ultimately, on their verification by tests and measurements. Therefore, accurate measurement techniques in everyday engineering problems, and knowing the limitations of these techniques, is of great value to the engineer.

We feel that engineers who acquire a basic knowledge of linear electric circuits, electronic circuits, electric power and machines, and control and instrumentation will have a well-rounded background and be better suited to join a team effort.

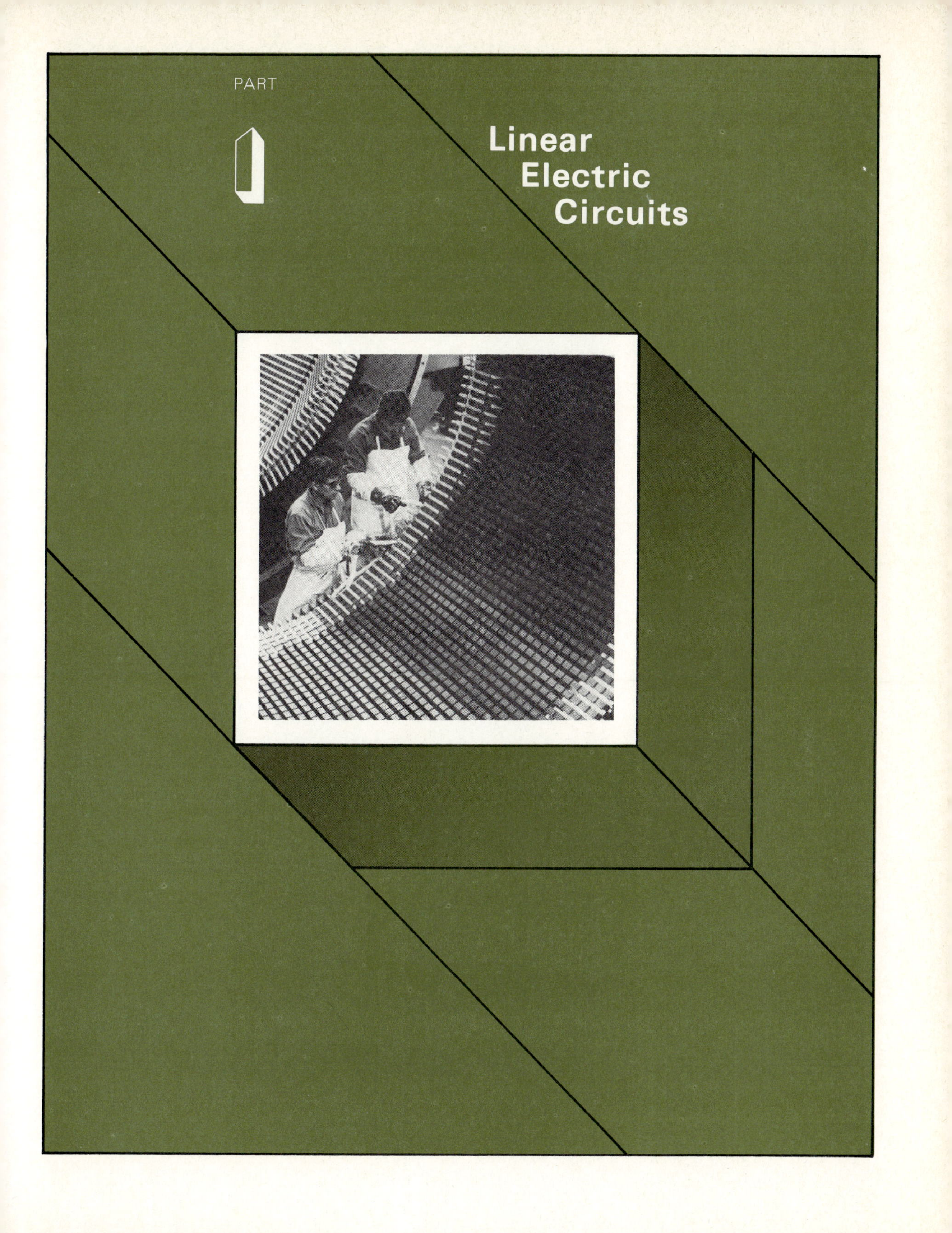

PART 1
Linear Electric Circuits

Chapter

Circuit Elements and Laws

In the chapters of Part 1 we will study electric circuits. These circuits are interconnections of certain basic electric elements. All electric devices are composed of circuit elements, and the particular interconnection of these circuit elements determines how the device performs a useful function. The basic circuit elements are the resistor, the inductor, the capacitor, and the ideal voltage and current sources. We will study how each of these circuit elements behaves individually and how, when they are interconnected, their interactions are governed by circuit laws.

There are many ways of looking at the function of an electric circuit. The view we will take is that of its function as a signal processor. There are many instances where we would like to process a signal in some fashion. For example, not only are radio signals transmitted simultaneously from numerous stations, but various types of unwanted signals, generally classified as noise, are also present; consequently, a radio receiver would be useless if it did not have the ability to "filter out" the desired signal from all the others which are present at the antenna. Understanding how an electric circuit performs this important task will be one result of studying Part 1. Yet the main objective is to master the laws governing a wide variety of signal-processing circuits. Thus, rather than limiting the discussion to specific types of circuits, we will examine the general methods for analyzing numerous types.

2.1 CHARGE AND ELECTRIC FORCES

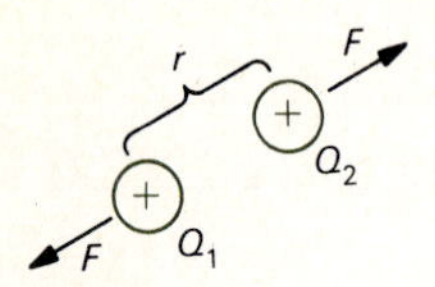

FIGURE 2.1
Illustration of Coulomb's Law.

Electric charge and its movement are the most basic items of interest in electrical engineering. We will not be concerned with the charge on a microscopic basis such as electrons but will be more concerned with large-scale (relatively speaking) charge movement.

The basic component of charge is the electron, which carries a negative charge of 1.6×10^{-19} C, where C is the unit of charge given in coulombs. One coulomb of charge therefore represents a tremendous number of electrons. The nuclei of atoms contain an equal amount of positive charge in the heavier protons.

Charges of the same sign tend to repel each other, and charges of opposite sign tend to be attracted together. Thus, charges exert forces on each other. It is this electric force that we are interested in utilizing and controlling. For example, consider two charges Q_1 and Q_2 which are separated a distance r. The force F exerted on one charge by the other varies inversely with the square of the separation between them and directly as the strengths of the charges according to Coulomb's law (see Fig. 2.1):

$$F = k\frac{Q_1 Q_2}{r^2} \qquad \text{N}$$

In the International System of Units (SI), the unit of force is the newton (N), where $1\ \text{N} = 1\ (\text{kg} \cdot \text{m})/\text{s}^2$. In free space the proportionality constant k is 9×10^9 when F is in newtons, Q_1 and Q_2 are in coulombs, and r is in meters (m). The reader should verify that the force between two 1-C charges separated a distance of 1 m in free space is approximately 1 million tons.

2.2 VOLTAGE

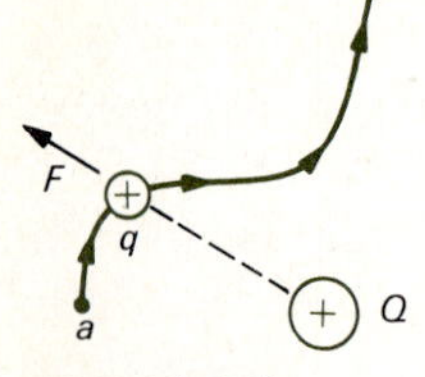

FIGURE 2.2
Voltage (potential difference) between two points in terms of the work required to move a charge between those points.

Since charges exert forces on other charges, energy must be expended in moving a charge in the vicinity of other charges. The unit of energy is the joule (J), where one joule is the energy expended in the application of one newton of force in moving an object through a distance of one meter ($\text{J} = \text{N} \cdot \text{m}$).

For example, consider Fig. 2.2. Moving a charge q from point a to point b in the presence of some other charge Q may require a net expenditure of energy. The force of Q may oppose the movement of q over certain portions of the route, while over other portions this force may be in a direction so as to aid the movement of q. Thus, it reasonably follows that if we move the charge from point a to point b and return it to point a, the net expenditure of energy will be zero. This result is explained by the fact that charge Q has a type of force field around it which tends to repel charges of like sign and attract charges of unlike sign.

The action of the electric force field is similar to that of the gravitational force field. If we raise an object above the surface of the earth, the object acquires a certain amount of potential energy which is released once the object falls back to earth. We

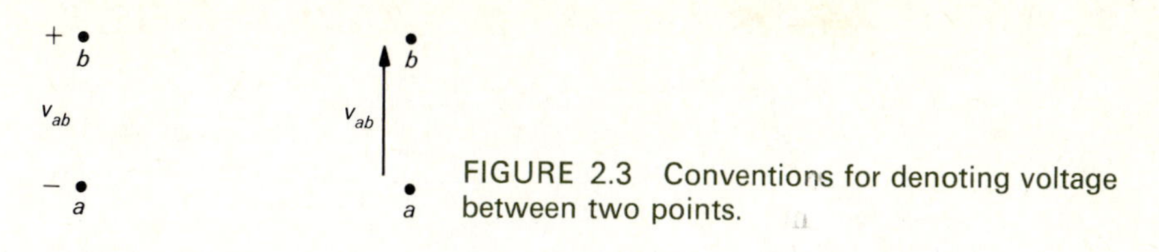

FIGURE 2.3 Conventions for denoting voltage between two points.

will characterize the potential of an electric force field for doing work by a similar concept—voltage. If we move a charge q from point a to point b and energy w_{ab} is expended, we say that the voltage v—or potential difference—between these two points is the work required per unit charge:

$$v_{ab} = \frac{w_{ab}}{q} \quad \text{V} \tag{2.1}$$

Note that, knowing the voltage between two points, we can easily compute the energy required to move some other charge Q between these two points by multiplying the voltage by the charge: $v_{ab}Q$. Therefore, knowing the voltage between two points is a convenient way of characterizing the force field. The unit of voltage is, from Eq. (2.1), joules per coulomb. This combination of units is referred to in condensed form as volts (V), where V = J/C.

Knowing the direction (polarity) of this voltage, or electric potential, is obviously as important as knowing its magnitude, since this will determine whether energy is to be expended by us or by the force field when a charge is moved between the two points. Let us reconsider the gravitational potential example. If we lift an object to a point above the surface of the earth, energy is expended by us in doing so and we say that points above the earth's surface are at a higher gravitational potential. In the same fashion, we describe point b as being at a higher electric potential than point a if net energy must be expended by us in moving a positive charge from a to b. Thus, v_{ab} in Eq. (2.1) is a "positive" number. This will be denoted in two ways, as shown in Fig. 2.3. In some situations we will find it convenient to use + and − signs to designate which point is at the *assumed* higher potential (the + point). (We use the word "assumed" because the guess may be incorrect, in which case v_{ab} will simply be a negative number such as $v_{ab} = -2$ V.) In other situations it may be convenient to use an arrow to point to the terminal of assumed higher potential. Again, it is important to understand that we need not be concerned with designating the correct point of higher potential, since this can easily be reversed by using a negative sign with the numeral value of the voltage.

2.3 CURRENT AND MAGNETIC FORCES

Current is the rate of movement of electric charge. For example, consider Fig. 2.4*a* in which we have shown charge moving along a cylinder. A certain amount of positive charge Q_R^+ and negative charge Q_R^- is moving to the right across some cross section of

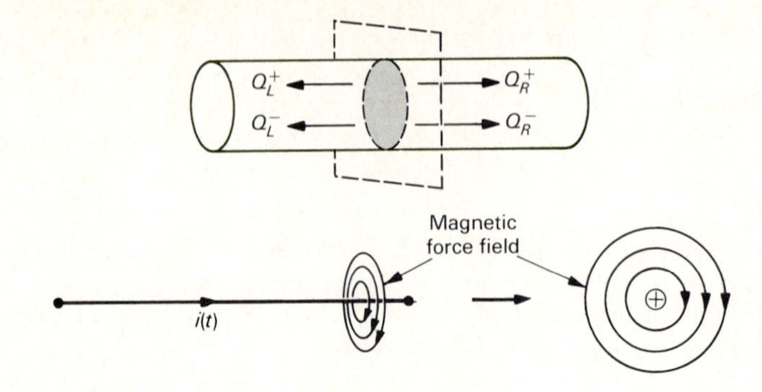

FIGURE 2.4 Electric current: (*a*) net movement of positive charge; (*b*) magnetic force field around a current-carrying wire.

the cylinder, and similar quantities Q_L^+ and Q_L^- are moving to the left. The *net* positive charge moving to the right is

$$Q = Q_R^+ - Q_R^- - Q_L^+ + Q_L^-$$

since negative charge moving to the right is equivalent to positive charge moving to the left. If we observe the charge crossing the area in certain time interval Δt, the current i directed to the right is the rate of movement of net positive charge to the right per unit of time:

$$i = \frac{Q}{\Delta t} \quad \text{A} \tag{2.2}$$

The unit of current is the ampere (A), where one ampere of current is the movement of one coulomb of net positive charge past a point in one second (s): A = C/s.

The rate of charge movement may not be constant but may vary with time. To determine the current at a specific time, we reduce the time of observation to a very small interval about that time and obtain

$$i(t) = \frac{dQ}{dt} \tag{2.3}$$

which gives the current as function of time t.

In most metallic conductors, such as wires, current is exclusively the movement of free electrons in the wire. Since electrons are negative, the charges are thus moving in a direction opposite to the direction of the current designation. The net positive charge movement is nevertheless in the direction we designate for i.

Electric currents produce forces, just as do electric charges. The forces produced by electric currents are referred to as magnetic forces, and they possess many of the properties of ordinary bar magnets. A current-carrying wire, for example, has a magnetic force field around it just as an electric charge has an electric force field around it. This magnetic force field appears as concentric circles around the wire; at a constant radius r from the wire, the magnetic force is the same (see Fig. 2.4*b*). The electric force field of a positive charge, on the other hand, always points away from the charge, and vice versa for a negative charge. If we place an ordinary bar magnet of a compass in the vicinity of the wire, the needle of the compass will align with the direction of the magnetic force field.

2.4 LUMPED-CIRCUIT ELEMENTS

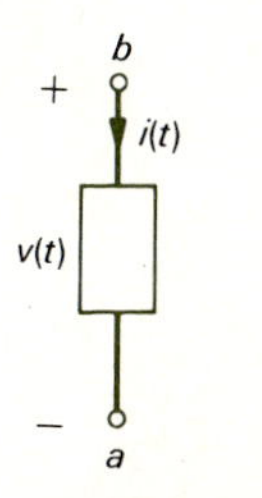

FIGURE 2.5 The passive sign convention for labeling the voltage and current of a two-terminal element.

We have seen that electric charges and currents both possess force fields and therefore the ability to do work. We have also seen that these force fields are distributed throughout space. In our study of electric circuits we will consider various electric circuit elements which utilize these force fields. In each of these elements we will consider the force fields to be confined primarily to some small region around the element. Thus, although the effects of each element are distributed throughout space, we will lump them into boxes as shown in Fig. 2.5. We refer to these as lumped elements. We will associate a voltage $v(t)$ between the terminals of the element with a current $i(t)$ entering one terminal of the element (and leaving the other terminal). In electric circuits the important quantities of interest are the voltages and currents of the circuit elements. If we can determine these in a circuit, we will have "analyzed" the circuit completely. This will be our primary goal.

If we label the element current and element voltage as shown in Fig. 2.5, where the current is assumed to enter the terminal of assumed higher voltage, then the element is said to be labeled with the *passive sign convention*. The word "passive" means that the element is assumed to absorb power. From Eqs. (2.1) and (2.3) we have

$$\begin{aligned} p(t) &= \frac{dw_{ab}}{dt} \\ &= \frac{dw_{ab}}{dq}\frac{dq}{dt} \\ &= v(t)i(t) \qquad \text{W} \end{aligned} \tag{2.4}$$

The unit of power is the watt (W), where W = J/s. Thus, the instantaneous power delivered to an element at some time t is the product of the voltage and current associated with that element at that time. If either v or i is truly negative (opposite in direction from the assumed direction), then negative power is delivered to the element or, equivalently, absorbed by the element. This is equivalent to saying that the element is delivering power. Several examples are shown in Fig. 2.6. Note that it is not important which terminal is which. It is only important to determine whether the element is labeled with the passive sign convention—the current enters the terminal of higher potential. For example, in Fig. 2.6*a*

$$\begin{aligned} P_{\text{absorbed}} &= (-v)(i) \\ &= (-3)(2) \\ &= -6 \text{ W} \end{aligned}$$

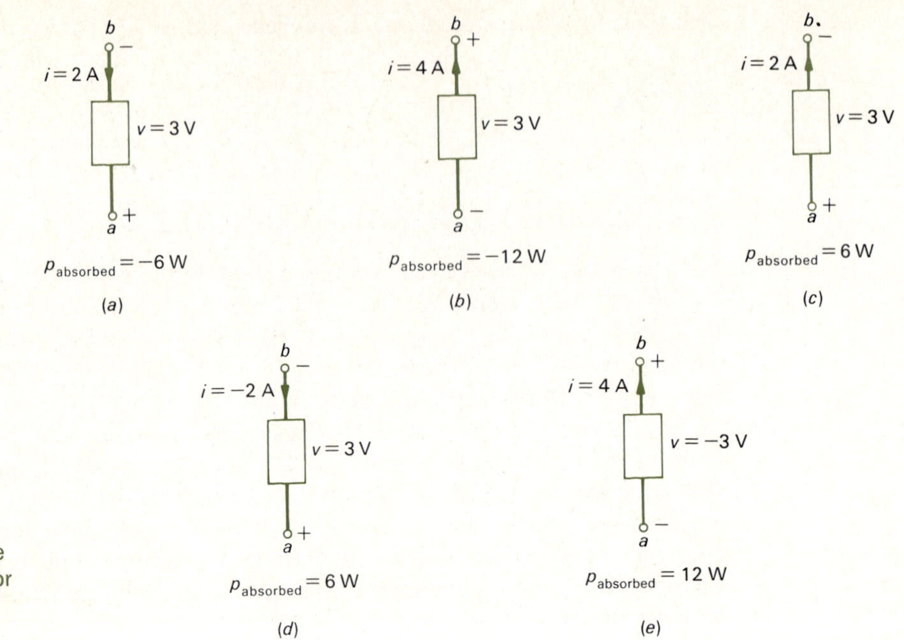

FIGURE 2.6 Illustrations of the power absorbed or delivered by an element.

and the element is delivering 6 W of power P (at that time). In Fig. 2.6*b* the element is absorbing

$$
\begin{aligned}
P_{\text{absorbed}} &= (v)(-i) \\
&= (3)(-4) \\
&= -12 \text{ W}
\end{aligned}
$$

whereas in Fig. 2.6*c*

$$
\begin{aligned}
P_{\text{absorbed}} &= (-v)(-i) \\
&= vi \\
&= 6\ W
\end{aligned}
$$

Some of the voltages or currents may have negative values. In these cases it is simpler first to write the power expression in terms of the symbols v and i and then to substitute numerical values. For example, in Fig. 2.6*e*

$$
\begin{aligned}
P_{\text{absorbed}} &= (v)(-i) \\
&= (-3)(-4) \\
&= 12 \text{ W}
\end{aligned}
$$

2.5 KIRCHHOFF'S VOLTAGE AND CURRENT LAWS

In later sections we will examine the various circuit elements—resistors, inductors, capacitors, and ideal voltage and current sources—which relate the voltage and current of the element. When these elements are interconnected to form an electric circuit, the interconnections establish relationships between the element voltages and currents. These relationships are known as Kirchhoff's laws.

The first law we will consider is Kirchhoff's current law, abbreviated KCL. The terminals of the elements will be referred to as *nodes*. In an electric circuit, these nodes are connected in some fashion. For example, consider the circuit in Fig. 2.7. This circuit consists of five elements. The elements constitute *branches* of the circuit. Kirchhoff's current law dictates that the sum of the currents entering a node must equal zero:

$$\sum_{\text{entering a node}} i = 0 \tag{2.5}$$

Note that a current i which is leaving a node is equivalent to a current $-i$ entering that node. In Fig. 2.7 we have

node a:

$$-i_1 - i_5 + i_4 = 0$$

node b:

$$i_1 - i_2 = 0$$

node c:

$$i_2 + i_5 - i_3 = 0$$

node d:

$$i_3 - i_4 = 0$$

Note that each of these equations may be multiplied by a minus sign so that KCL could have been stated as requiring the sum of the currents leaving a node to be zero:

$$\sum_{\text{leaving a node}} i = 0 \tag{2.6}$$

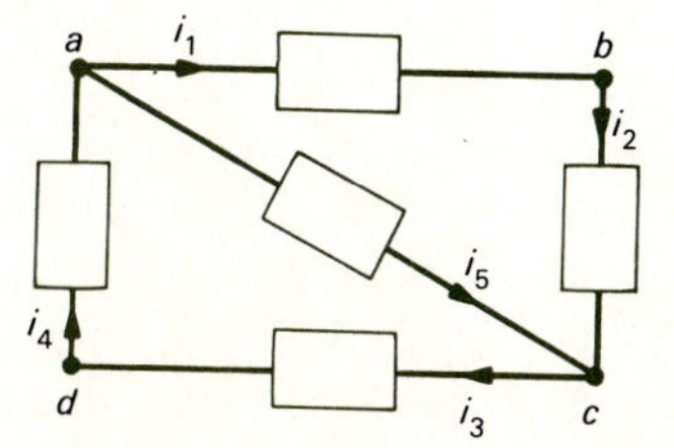

FIGURE 2.7 Example illustrating Kirchhoff's current law (KCL).

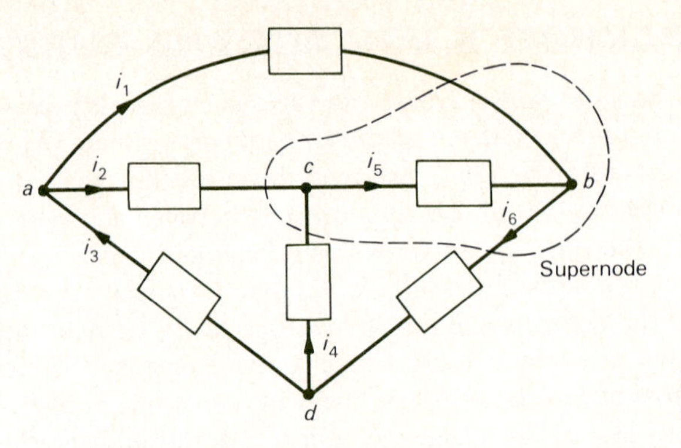

FIGURE 2.8 Example illustrating KCL for supernodes.

KCL also applies to closed regions of a circuit known as *supernodes*. For example, applying KCL at all nodes in Fig. 2.8 yields

node a:

$$-i_1 - i_2 + i_3 = 0$$

node c:

$$i_2 + i_4 - i_5 = 0$$

node b:

$$i_1 + i_5 - i_6 = 0$$

node d:

$$-i_3 - i_4 + i_6 = 0$$

Applying KCL to the supernode, we have

supernode:

$$i_2 + i_1 + i_4 - i_6 = 0$$

but this is the sum of KCL at nodes b and c.

KCL is a logical result if we recall the definition of current as the rate of flow of charge. KCL simply states that nodes cannot accumulate charge; whatever charge enters a node must leave that node.

Kirchhoff's voltage law (KVL) provides relationships between the branch (element) voltages of a circuit. KVL states that the algebraic sum of the branch voltages around a closed-circuit loop is identically zero:

$$\sum_{\text{closed loops}} v = 0 \tag{2.7}$$

Note the use of the term *algebraic sum*. This term means the following. For the element in Fig. 2.5, a positive charge proceeding through the element from terminal a to terminal b moves through a voltage *rise* of v volts. On the other hand, if the positive charge proceeds through the element from terminal b to terminal a, it moves through a voltage *drop* of v volts. A voltage rise is the negative of a voltage drop, and vice versa. KVL basically states that if we select a closed loop in a circuit, then as we proceed around the closed loop and add all the voltage rises encountered the result will be zero. Keep in mind that if we proceed through an element from the + terminal to the − terminal, we will have gone through a voltage drop of v, which is a voltage rise of $-v$. Thus, KVL is written as

$$\sum_{\text{closed loops}} \text{voltage rises} = 0 \tag{2.8}$$

or

$$\sum_{\text{closed loops}} \text{voltage drops} = 0 \tag{2.9}$$

KVL must hold for *all* loops in a circuit just as KCL must hold for *all* nodes and supernodes in a circuit. Also, KVL applies to voltages at the same instant of time in the same way that KCL applies to currents at the same instant of time.

As an example, consider the circuit in Fig. 2.9. The branch currents are labeled as in Fig. 2.8 and we have added the branch voltages labeled with the passive sign convention. There are four loops in this circuit, labeled L_1, L_2, L_3, L_4. Consider loop L_1: KVL becomes

$$\sum_{\substack{\text{voltage}\\\text{drops}\\L_1}} = v_1 - v_5 - v_2 = 0$$

or

$$\sum_{\substack{\text{voltage}\\\text{rises}\\L_1}} = -v_1 + v_5 + v_2 = 0$$

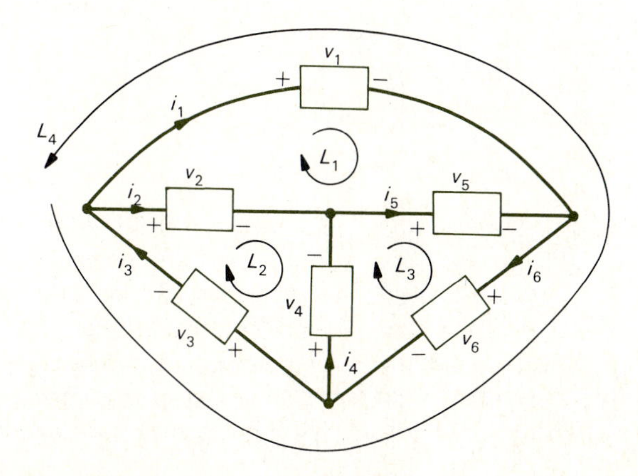

FIGURE 2.9 Example illustrating Kirchhoff's voltage law (KVL).

Similarly, for loops L_2, L_3, and L_4 (proceeding in the indicated directions around these loops)

$$\sum_{\substack{\text{voltage}\\\text{drops}\\L_2}} = v_2 - v_4 + v_3 = 0$$

$$\sum_{\substack{\text{voltage}\\\text{rises}\\L_3}} = -v_5 - v_6 - v_4 = 0$$

$$\sum_{\substack{\text{voltage}\\\text{rises}\\L_4}} = v_1 + v_6 + v_3 = 0$$

Both the starting node in applying KVL to a loop and the direction around the loop are arbitrary.

Example 2.1 For the circuit in Fig. 2.10, determine the voltages v_x and v_y and the current i_x.

Solution Applying KVL around the loop containing elements *A*, *B*, *C*, *D*, we have

$$2 - v_y + 3 - 4 = 0$$

so

$$v_y = 1 \text{ V}$$

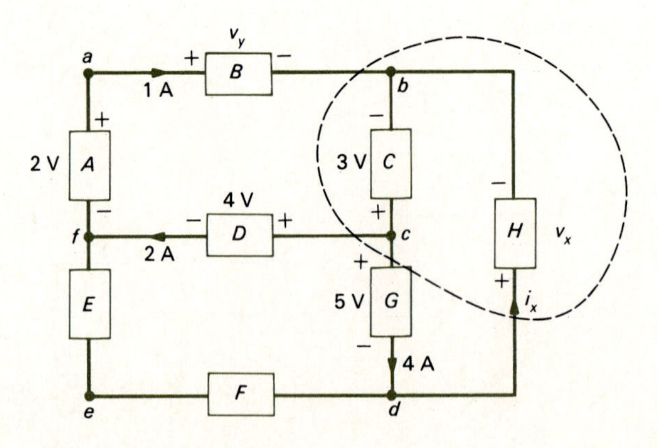

FIGURE 2.10 Example 2.1.

Now applying KVL around the loop containing elements *H, B, A, D, G*,

$$-v_x + v_y - 2 + 4 - 5 = 0$$

or

$$v_x = v_y - 3$$
$$= -2\ \text{V}$$

Applying KCL at a supernode containing nodes *b* and *c*, we have

$$1 - 2 - 4 + i_x = 0$$

or

$$i_x = 5\ \text{A}$$

2.6 IDEAL VOLTAGE AND CURRENT SOURCES

KVL provides constraints among the branch voltages and KCL provides constraints among the branch currents of a circuit. As yet we have not investigated how an element voltage and its current are related. These relationships are determined by the types of elements.

The first elements we will consider are the ideal sources. These sources do not provide a relationship between an element voltage and an element current but, instead, dictate the value of either the element voltage or the element current.

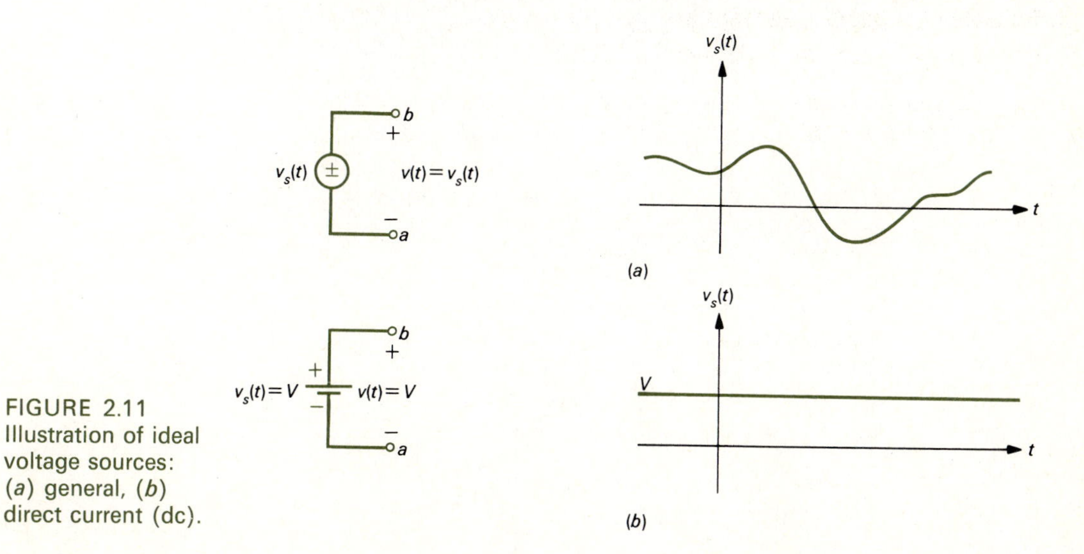

FIGURE 2.11 Illustration of ideal voltage sources: (*a*) general, (*b*) direct current (dc).

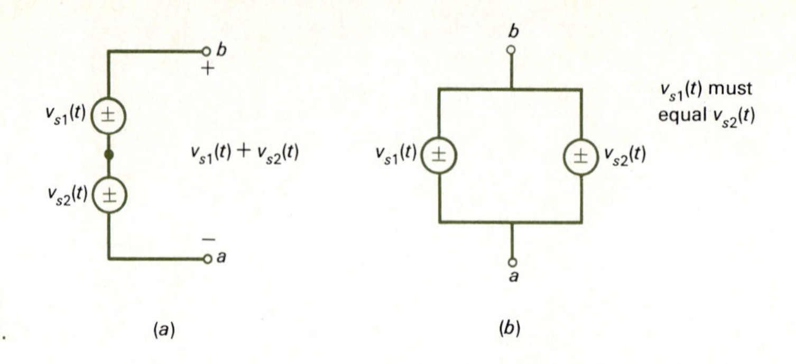

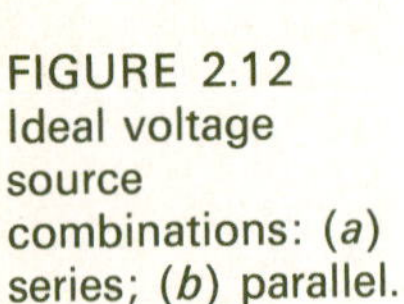
FIGURE 2.12 Ideal voltage source combinations: (*a*) series; (*b*) parallel.

The ideal voltage source dictates the value of the element voltage and is depicted as shown in Fig. 2.11. In Fig. 2.11*a* we have shown a general time-varying voltage source in which the element voltage $v(t)$ is some prescribed function of t: $v(t) = v_s(t)$. Here we have specified $v_s(t)$ graphically. In some cases we may be able to specify $v_s(t)$ as a function such as $v_s(t) = 2 \sin 3t$. In Fig. 2.11*b* we have shown a battery in which the element voltage remains constant at V for all time. Certainly this is not a good representation of typical batteries, such as are used in automobiles, since over a period of usage the available voltage decreases. This is why we have called this element an *ideal* voltage source. In later sections we will see how this ideal source can be used to represent an actual voltage source, such as an automobile battery.

We may connect ideal voltage sources end to end in a series connection, as shown in Fig. 2.12*a*. The overall terminal voltage is, according to KVL, the algebraic sum of the individual source voltages. Note that if we connect two or more ideal voltage sources in parallel, as shown in Fig. 2.12*b*, the values of the sources must be identical; otherwise, KVL would be violated.

Note that the element voltage of an ideal voltage source is a known quantity. However, the element current is still unknown. This element current will be determined by the circuit to which this source is attached. It is *not necessarily* zero.

The other type of ideal source is the ideal current source, shown in Fig. 2.13. For this source, the element current is constrained to be $i_s(t)$, the value of the current source, but the element voltage is unknown. The element voltage will be determined by the circuit to which the current source is attached and is *not necessarily* zero. Note in Fig. 2.14 that the overall terminal current resulting from the parallel connection of two or more current sources is, according to KCL, the algebraic sum of the source values, and for the series connection the values of the sources must be identical, or otherwise KCL would be violated.

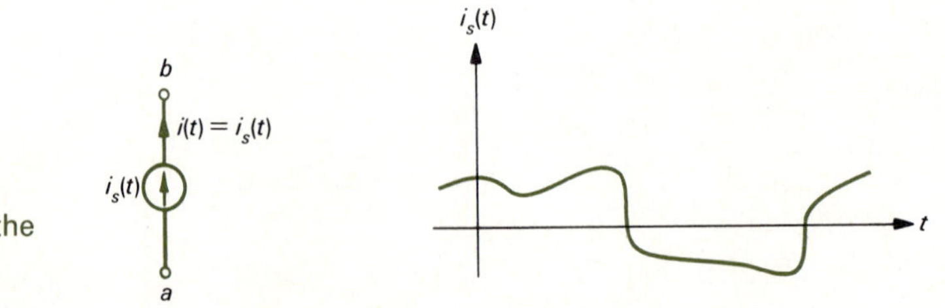

FIGURE 2.13 Illustration of the ideal current source.

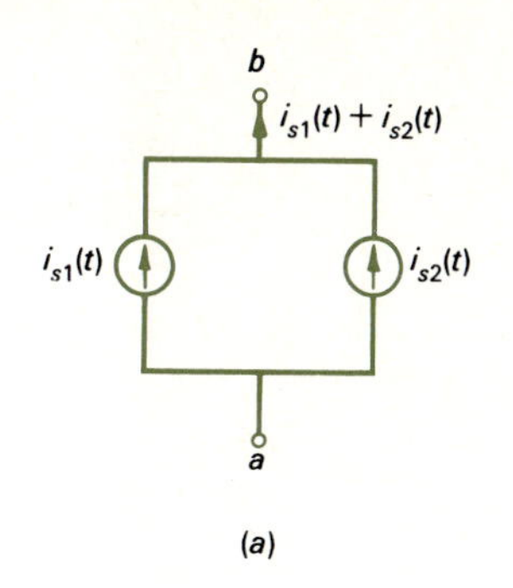

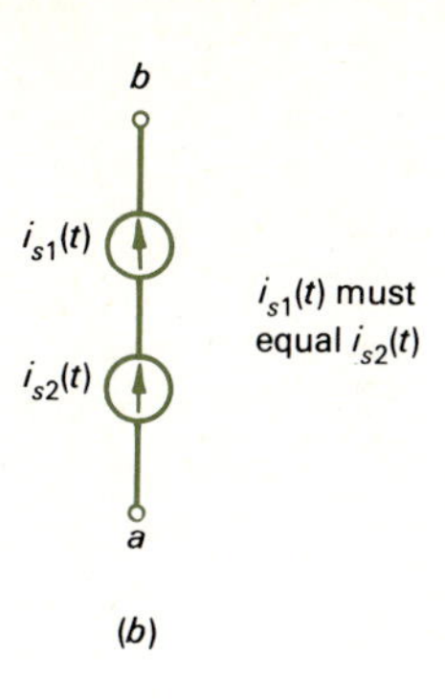

FIGURE 2.14 Ideal current source combinations: (*a*) parallel; (*b*) series.

2.7 THE RESISTOR

The ideal sources considered in the previous section constrained either the element voltage or the element current. The resistor, on the other hand, provides a relationship between the element voltage and element current.

For many conducting materials, the voltage across a block of the material is linearly related to the current passing through it. This result is credited to a German physicist, Georg Simon Ohm, and bears his name: Ohm's law. The symbol for a linear resistor is given in Fig. 2.15 along with the graphical relationship between the element voltage and current. Note that the graph is a straight line with slope R passing through the origin. If the graph were not a straight line or did not pass through the origin, the resistor would be a nonlinear one. Nonlinear resistors will be considered in Part 2. For the linear resistor, we may dispense with the graph and write Ohm's law as:

$$v(t) = Ri(t) \tag{2.10}$$

The slope R of the graph is called the resistance and its units are ohms (Ω), where $\Omega = \text{V/A}$. It is also somewhat common to write Eq. (2.10) as

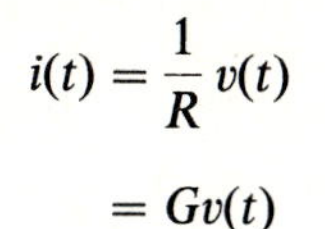

$$\begin{aligned} i(t) &= \frac{1}{R} v(t) \\ &= Gv(t) \end{aligned} \tag{2.11}$$

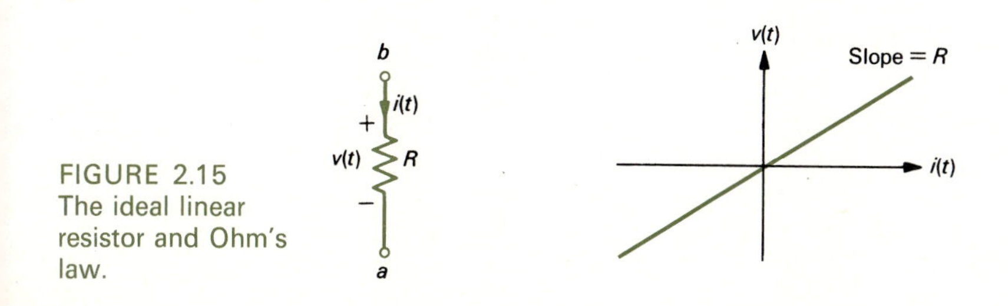

FIGURE 2.15 The ideal linear resistor and Ohm's law.

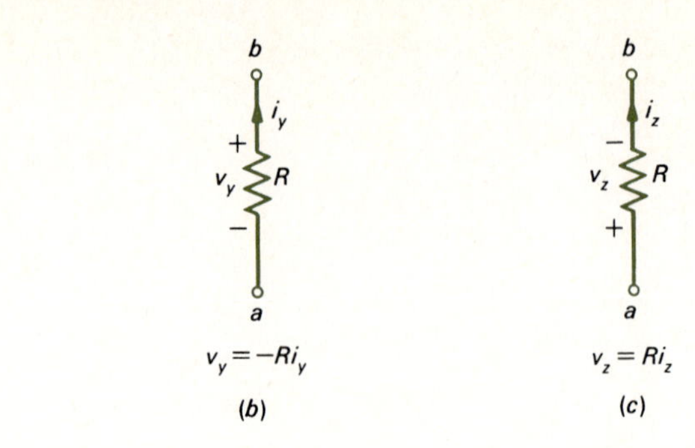

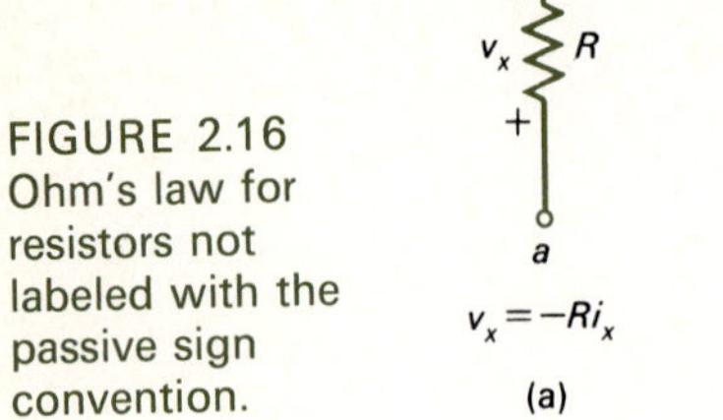

FIGURE 2.16 Ohm's law for resistors not labeled with the passive sign convention.

where

$$G = \frac{1}{R} \tag{2.12}$$

is called the conductance, whose units are siemens (S), the reciprocal of ohms. In the past, the unit and symbol of conductance were mhos (℧), mhos in a certain sense being the reciprocal of ohms (Ω). The internationally accepted units, however, are siemens (S).

Note that Ohm's law in Eq. (2.10) *presumes* that the element has been labeled with the passive sign convention. If this has not been the case, a minus sign may be needed in Eq. (2.10). Several examples are shown in Fig. 2.16; they should be studied carefully. Compare each of the labelings in Fig. 2.16 to those in Fig. 2.15. In Fig. 2.16*a*, $i_x = i$ but $v_x = -v$. Thus

$$-v_x = Ri_x$$

or

$$v_x = -Ri_x$$

In Fig. 2.16*b*, $v_y = v$ but $i_y = -i$. Thus

$$v_y = R(-i_y)$$
$$= -Ri_y$$

In Fig. 2.16*c*, $v_z = -v$ and $i_z = -i$. Thus

$$-v_z = R(-i_z)$$

or

$$v_z = Ri_z$$

TABLE 2.1 Resistivity of Common Conductors

Material	ρ, $\Omega \cdot$m
Copper	1.72×10^{-8}
Silver	1.62×10^{-8}
Gold	2.44×10^{-8}
Aluminum	2.62×10^{-8}
Brass	6.67×10^{-8}
Stainless steel	9.09×10^{-7}
Nichrome	1.0×10^{-6}
Sea water	0.25

This last result shows that for a linear resistor it doesn't matter which terminal is which, so long as the element voltage and current have been labeled with the passive sign convention.

For many types of conducting materials, the resistance of a block of material whose length l and cross-sectional area A are known can be calculated from

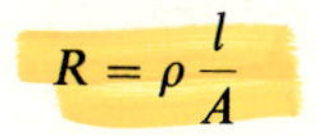

$$R = \rho \frac{l}{A}$$

where ρ is the resistivity of the material in $\Omega \cdot$m. Values for representative types of conducting materials are given in Table 2.1.

Example 2.2

Determine the resistance of a 100-ft length of No. 14 gauge copper wire (commonly used in extension cords). The diameter of No. 14 gauge wire is 64.1 mils, where 1 mil = 0.001 inch (in). If this wire is attached to a 120-V residential outlet and used to power a pair of hedge clippers which draw 1 A of current, determine the voltage at the terminals of the hedge clippers.

Solution The cross-sectional area of the wire is πr^2, where r is the wire radius. We first convert the wire radius in mils (32.05 mils) to meters:

$$32.05\ \text{mils} \times \frac{1\ \text{in}}{1000\ \text{mils}} \times \frac{2.54\ \text{cm}}{1\ \text{in}} \times \frac{1\ \text{m}}{100\ \text{cm}} = 8.14 \times 10^{-4}\ \text{m}$$

The resistivity of copper from Table 2.1 is $1.72 \times 10^{-8}\ \Omega \cdot$m. Thus, the total resistance of one wire in the cord is

$$R = \frac{1.72 \times 10^{-8} \times 30.48}{\pi(8.14 \times 10^{-4})^2}$$

$$= 0.252\ \Omega$$

where we have converted the length of the cord from feet (ft) to meters:

$$100\,\text{ft} \times \frac{12\,\text{in}}{1\,\text{ft}} \times \frac{2.54\,\text{cm}}{1\,\text{in}} \times \frac{1\,\text{m}}{100\,\text{cm}} = 30.48\,\text{m}$$

Each of the two wires of the cord has $0.252\ \Omega \times 1\ \text{A} = 0.252\ \text{V}$ dropped across it, thus the available voltage at the terminals of the hedge clippers is

$$120 - 0.252 - 0.252 = 119.5\ \text{V}$$

The linear resistor always absorbs power. This can be seen by substituting Ohm's law into the power equation:

$$\begin{aligned} P_{\text{absorbed}}(t) &= v(t)i(t) \\ &= Ri^2(t) \\ &= \frac{v^2(t)}{R} \end{aligned} \tag{2.13}$$

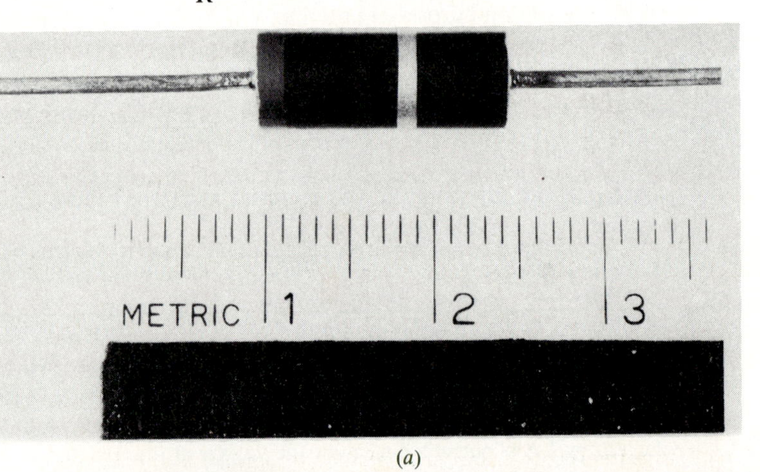

(*a*)

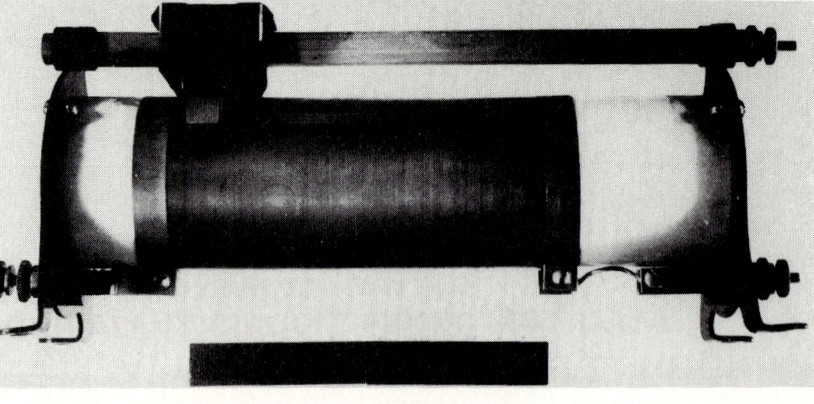

(*b*)

FIGURE 2.17 Physical resistors: (*a*) carbon composition; (*b*) wire-wound.

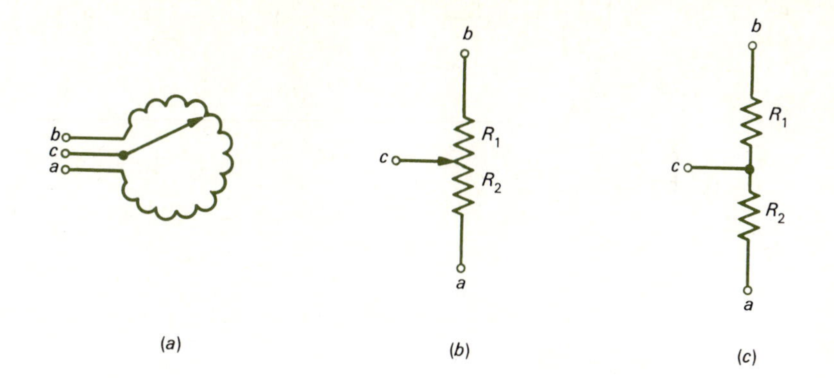

FIGURE 2.18 A variable resistor—the potentiometer.

A true negative resistor would be one in which the element voltage and current are labeled with the passive sign convention *but* in which $v(t) = -Ri(t)$. In this case the power absorbed would be $-Ri^2(t)$ or, in other words, the resistor would deliver power. The power dissipated in the extension cord of Example 2.2 is $i^2R = (1\text{ A})^2 \times 0.504\ \Omega = \frac{1}{2}$ W.

The carbon resistor shown in Fig. 2.17*a* is constructed by forming a block of carbon into a cylindrical shape and attaching leads. Wire-wound resistors such as shown in Fig. 2.17*b* are constructed of a long length of wire wound on a central axis to conserve space. In either type of resistor, the terminal voltage and current obey Ohm's law. There are also variable resistors known as potentiometers; these are usually constructed as a wire-wound resistor with a movable tap which slides along the surface of the wires, as shown in Fig. 2.18. The position of the center tap determines the portion of the total resistance between the tap and each end of the resistor.

There are many other useful types of elements which use this principle of resistance. One common such device is the strain gauge shown in Fig. 2.19. This device is constructed of a long section of wire woven on a flexible backing. When the backing is attached to a material and a force is applied to that material, the wire is stretched, which reduces the wire's cross section and increases its total resistance. This increase in resistance is a measure of the elongation of the wire and hence of the strain in the material. Ordinarily, a battery is attached to the strain gauge and an ammeter (an instrument for measuring current) is inserted to measure the change in current and hence to measure the resistance.

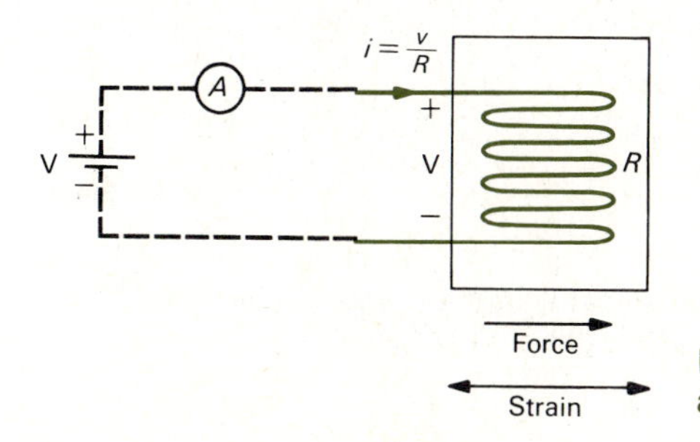

FIGURE 2.19 Illustration of a resistance application in transducers—a strain gauge.

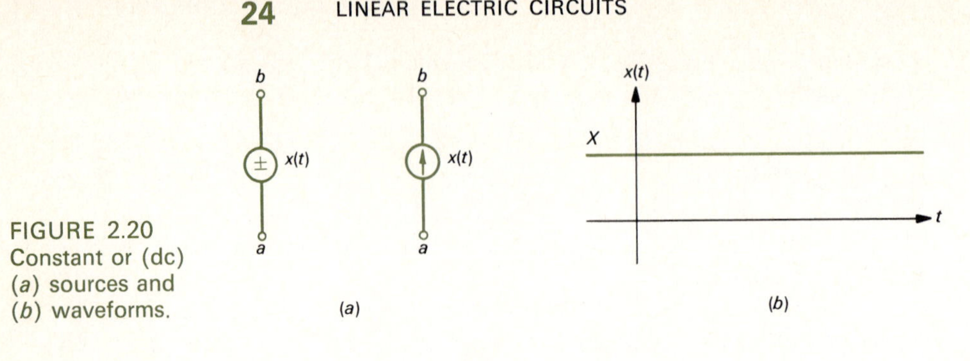

FIGURE 2.20 Constant or (dc) (*a*) sources and (*b*) waveforms.

2.8 SIGNAL WAVEFORMS

The ideal voltage and current sources provide signals which are shaped or otherwise altered by the circuit to which they are attached. It is important that we now consider how to describe, or characterize, these signals.

Perhaps the simplest type of waveform is the constant, or dc, waveform shown in Fig. 2.20*b*. In this section, we will denote the waveform as a function of time t—as $x(t)$—regardless of whether it is produced by a voltage or a current source. Characterizing this waveform is quite simple (the graph is unnecessary):

$$x(t) = X \tag{2.14}$$

These waveforms are referred to as dc since they characterize the direct current supplied by a constant voltage source or battery.

We will often be interested in waveforms which are not constant in time. Generally, we have no choice but to describe these waveforms graphically. However, there is an important class of time-varying waveform which is quite easy to describe—periodic waveforms. A periodic waveform is one which repeats itself over intervals of time of length T. In other words,

$$x(t) = x(t \pm nT)$$

where $n = 1, 2, 3, \ldots$. The length of time for repetition of the waveform, T, is called the period of the waveform. An example of a periodic waveform is shown in Fig. 2.21. Every point on the waveform is identical nT seconds later or earlier in time. Obviously, for a waveform to be periodic it must continue indefinitely in time. Note that

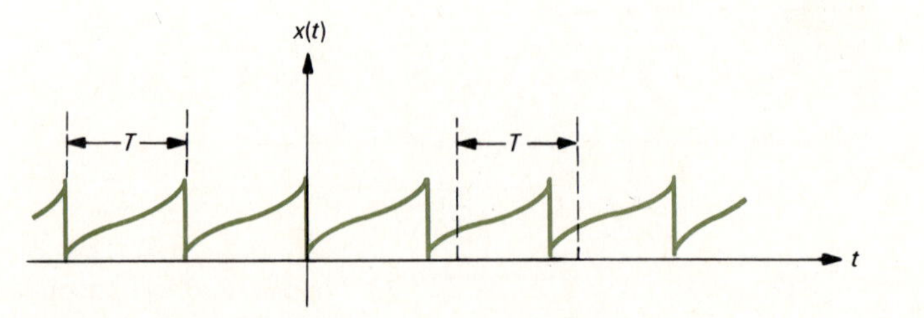

FIGURE 2.21 A general periodic waveform (signal).

the dc waveform in Fig. 2.20*b* may be considered to be periodic with an infinite period. The frequency f of a periodic waveform is the reciprocal of its period:

$$f = \frac{1}{T} \qquad \text{Hz} \tag{2.15}$$

The unit of frequency is the hertz (Hz), where one hertz represents a repetition of the waveform in one second.

Obviously, since we only need to describe the waveform over one period, periodic waveforms are simpler to characterize than general waveforms which are not periodic.

The periodic waveform to which we will devote considerable attention is the sinusoid shown in Fig. 2.22. The amplitude of the sinusoid is A, and ϕ is some phase offset. The radian frequency of the wave is $\omega = 2\pi f = 2\pi/T$. If the phase offset is zero, the wave is classified as a sine wave, and if $\phi = 90°$, the wave is classified as a cosine wave, or $A \sin(\omega t + 90°) = A \cos \omega t$.

These sinusoidal waveforms have a practical utility, and we will concentrate our study of electric circuits on those containing ideal voltage and current sources which are either dc or sinusoidal. Electric power is generated, distributed, and utilized as sinusoidal voltage or current at a frequency of 60 Hz (50 Hz in Europe). Sinusoidal waveforms are commonly classified as ac waveforms since they are characteristic of the alternating current produced by electric power generators. Radio and television transmission makes use of sinusoids which have much higher frequencies than 60 Hz. For example, standard AM transmission utilizes frequencies in the range of 550 kHz (550×10^3 Hz) to 1600 kHz (1.6×10^6 Hz). These radio transmissions utilize a single frequency in this range which is called a *carrier*. The music or other information to be transmitted is used to vary the amplitude of this carrier (amplitude modulation) so that a receiver tuned to the carrier frequency may extract these variations and hence the information. In FM, or frequency modulation, the information is used to vary the carrier frequency.

Periodic waveforms may also be classified according to their average value. The average value (AV) of a periodic waveform is the net positive area under the curve for one period divided by the period:

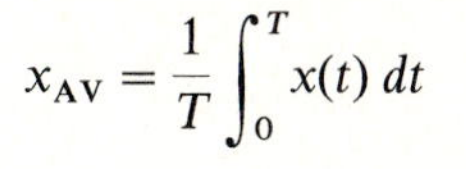

$$x_{AV} = \frac{1}{T}\int_0^T x(t)\, dt \tag{2.16}$$

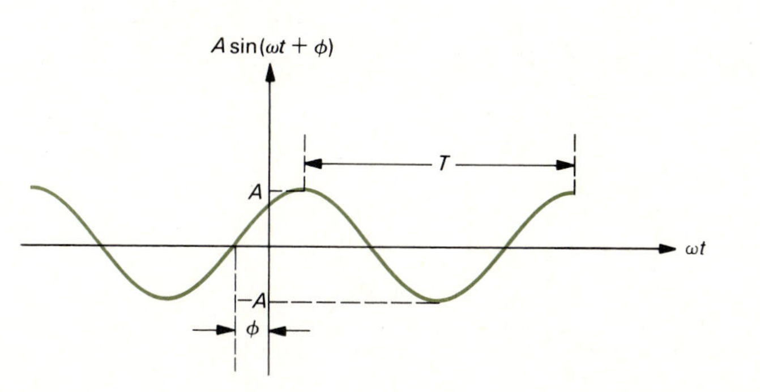

FIGURE 2.22
A sinusoidal waveform (signal).

The effective, or root-mean-square (RMS), value is the square root of the average of $x^2(t)$:

$$x_{\text{RMS}} = \sqrt{\frac{1}{T}\int_0^T x^2(t)\,dt} \tag{2.17}$$

This definition of the RMS, or effective, value of a periodic waveform arises for the following reason. Consider a resistor R through which a periodic current $i_R(t)$ passes. The instantaneous power dissipated in the resistor is

$$\begin{aligned} p_R(t) &= v_R(t)i_R(t) \\ &= Ri_R^2(t) \end{aligned} \tag{2.18}$$

Since $i_R(t)$ is periodic, $p_R(t)$ will be periodic with the same period. We may compute an average power dissipated in the resistor (converted to heat) as

$$\begin{aligned} P_{\text{AV}} &= \frac{1}{T}\int_0^T p_R(t)\,dt \\ &= R\left[\frac{1}{T}\int_0^T i_R^2(t)\,dt\right] \end{aligned} \tag{2.19}$$

But this may be written as

$$P_{\text{AV}} = RI_{\text{RMS}}^2 \tag{2.20}$$

where, comparing Eqs. (2.19) and (2.20),

$$I_{\text{RMS}} = \sqrt{\frac{1}{T}\int_0^T i_R^2(t)\,dt} \tag{2.21}$$

For a particular periodic current waveform $i_R(t)$, it is somewhat inconvenient to compute Eq. (2.19). However, if we know the RMS value of the waveform we may easily compute the average power dissipated in a resistance via Eq. (2.20) as though the waveform were dc with a value of I_{RMS}. Similarly, we may compute

$$P_{\text{AV}} = \frac{v_{\text{RMS}}^2}{R} \tag{2.22}$$

where

$$V_{\text{RMS}} = \sqrt{\frac{1}{T}\int_0^T v^2(t)\,dt} \tag{2.23}$$

Example 2.3 Compute the power dissipated in a 5-Ω resistor which has a current with a waveform given in Fig. 2.23 passing through it.

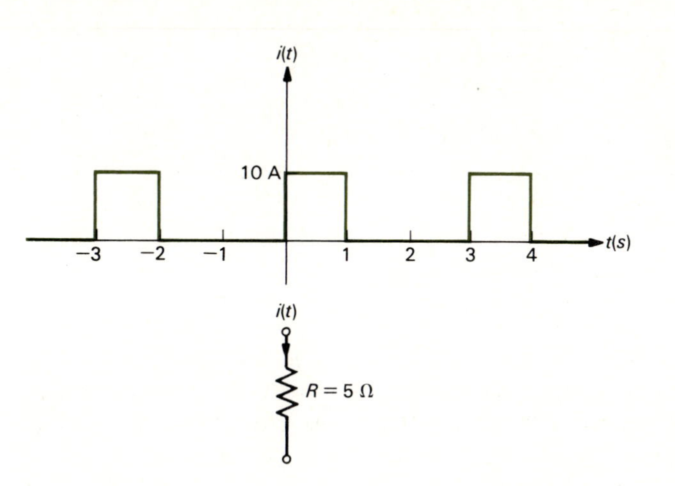

FIGURE 2.23
Example 2.3.

Solution The period is 3 s and the RMS value is

$$
\begin{aligned}
i_{\text{RMS}} &= \sqrt{\frac{1}{3}\int_0^3 i^2(t)\,dt} \\
&= \sqrt{\frac{1}{3}\int_0^1 (10)^2\,dt} \\
&= \sqrt{\frac{100}{3}} \\
&= 5.77 \text{ A}
\end{aligned}
$$

Thus

$$
\begin{aligned}
P_{\text{AV}} &= 5\ \Omega \times (5.77)^2 \\
&= 166.7 \text{ W}
\end{aligned}
$$

For the special case of a dc waveform shown in Fig. 2.20, the average and RMS values are

$$
\begin{aligned}
x_{\text{AV}} &= x_{\text{RMS}} \\
&= X
\end{aligned}
\tag{2.24}
$$

For the sinusoid

$$x(t) = A \sin \omega t$$

the average value is zero

$$x_{\mathrm{AV}} = 0$$

but the RMS value is

$$\begin{aligned} x_{\mathrm{RMS}} &= \sqrt{\frac{1}{T}\int_0^T A^2 \sin^2\left(\frac{2\pi}{T}t\right) dt} \\ &= \frac{A}{\sqrt{2}} \\ &\doteq 0.707\ \mathrm{A} \end{aligned} \tag{2.25}$$

where we have used the trigonometric identity $\sin^2 \theta = \frac{1}{2}(1 - \cos \theta)$. The same result would be obtained for the cosine wave $x(t) = A \cos \omega t$ or for any general sinusoid $x(t) = A \sin (\omega t + \phi)$. Thus, if a sinusoidal current of amplitude A passes through a resistor R, the average power dissipated in that resistor is

$$\begin{aligned} P_{\mathrm{AV}} &= RI_{\mathrm{RMS}}^2 \\ &= \tfrac{1}{2}A^2 R \end{aligned} \tag{2.26}$$

PROBLEMS

2.1 Determine the approximate number of electrons required for 1 C of charge.

2.2 Three charges in microcoulombs (μC) are placed in a line along the x axis: $Q_1 = +10\ \mu$C at $x =$ 1 m, $Q_2 = -7\ \mu$C at$x = 6$ m, and $Q_3 = +8\ \mu$C at $x =$ 10 m. A $+5$-μC charge is placed at $x = 8$ m. Determine the force (magnitude and direction) exerted on that $+5$-μC charge.

2.3 Three $+1$-μC charges are located on the corners of an equilateral triangle having side lengths of 1 m. Determine the force (magnitude and direction) exerted by these charges on a $+5$-μC charge which is placed (a) at the midpoint of one of the sides and (b) at the centroid of the triangle.

2.4 A $+10$-μC charge is held fixed. Determine the energy required to move a $+5$-μC charge from a distance of 1 m away from the $+10$-μC charge to a distance of 6 m. Determine the voltage between the beginning and end points, assuming that $x = 6$ m is at the higher potential. Repeat for movement of a -7-μC charge.

2.5 Suppose the current passing a point in a wire is constant at 2 A. Determine the total amount of net positive charge which has moved in the direction of the current over an interval of 1 hour (h).

2.6 Suppose the current passing a point along a wire is as sketched in Fig. P2.6. Determine the total net positive charge which has passed that point from $t = 1$ s to $t = 6$ s.

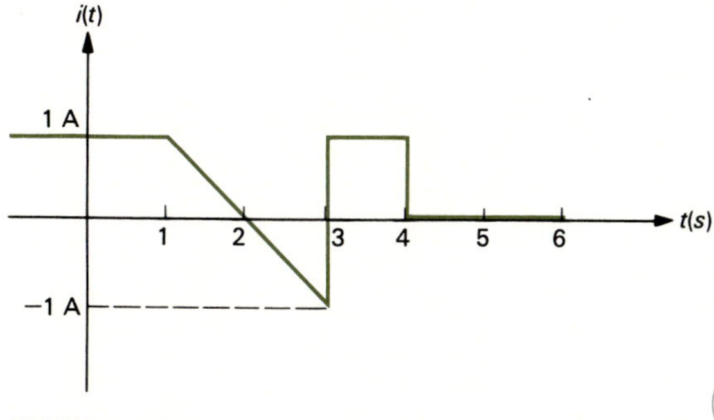

FIGURE P2.6

2.7 Suppose the net positive charge passing a point along a wire is as shown in Fig. P2.7. Sketch the current from $t = 1$ s to $t = 5$ s.

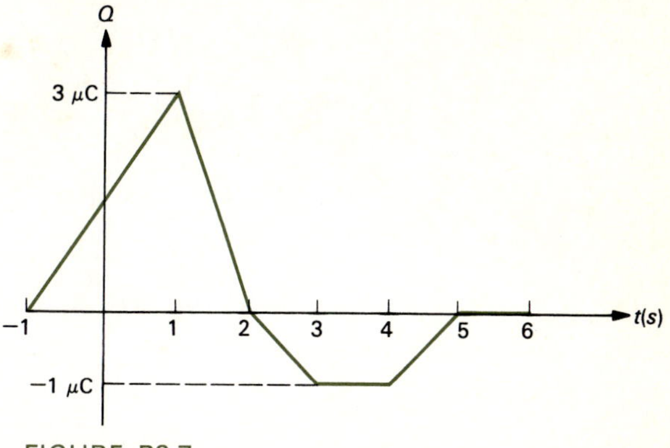

FIGURE P2.7

2.8 In the circuit shown in Fig. P2.8, determine v_x, v_y, i_a, and i_b.

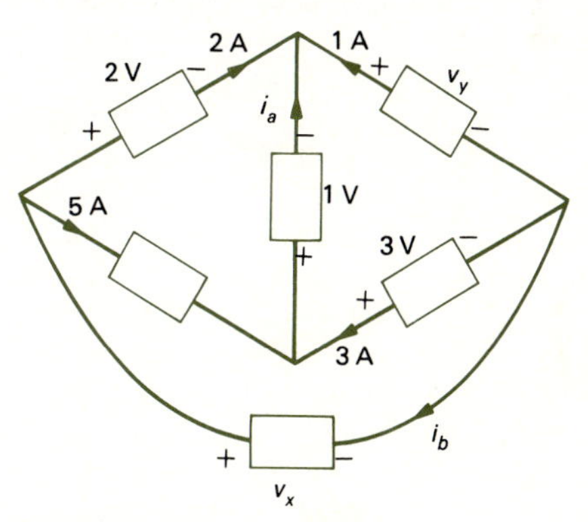

FIGURE P2.8

2.9 In the circuit shown in Fig. P2.9, write KVL around all loops and KCL at all nodes and supernodes.

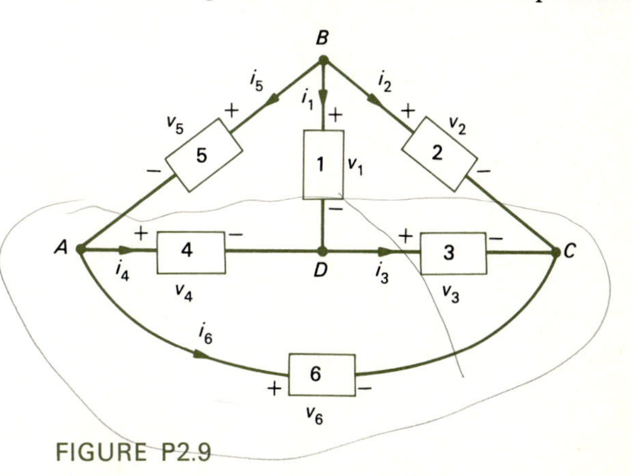

FIGURE P2.9

2.10 In the circuit shown in Fig. P2.10, determine v_x, v_y, and i_z.

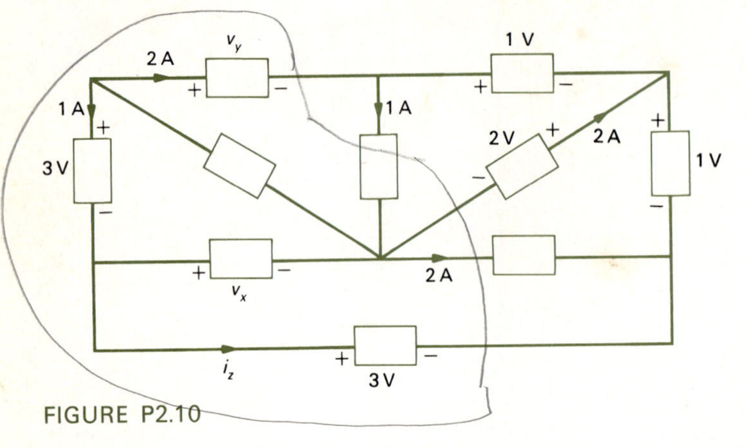

FIGURE P2.10

2.11 Automobile storage batteries are rated in terms of their terminal voltage (12-V) and their ampere-hour (Ah) capacities. For a typical 12-V battery having a 115-Ah capacity, determine the length of time this battery will light a 6-W bulb. (Assume that the battery voltage is constant at 12 V, even though this is not true for real batteries.) Determine the total energy stored in the battery before it is connected to the bulb.

2.12 In the circuit shown in Fig. P2.12, apply KVL, KCL, and Ohm's law to determine the currents and voltages associated with all the elements. Compute the power absorbed by the resistors and the power delivered by the ideal sources. Does conservation of energy result?

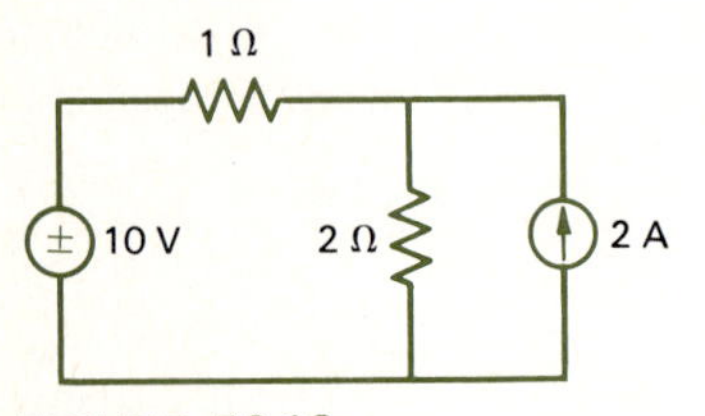

FIGURE P2.12

2.13 In the circuit in Fig. P2.13, calculate i_x and v_x. (Label the branch voltages and currents and apply KVL, KCL, and Ohm's law.)

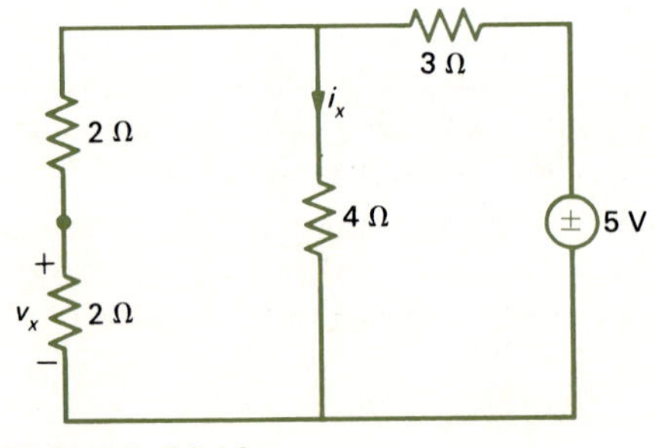

FIGURE P2.13

2.14 For the resistors shown in Fig. P2.14, determine the unknown items. In each case determine the power absorbed by each resistor.

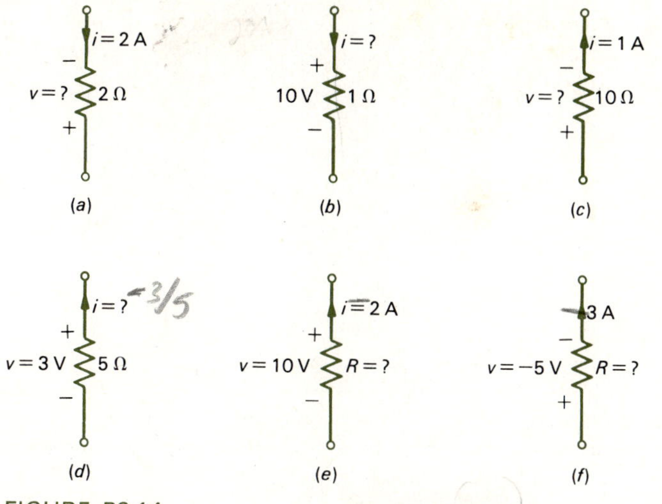

FIGURE P2.14

2.15 A 5-Ω resistor has the periodic voltage waveform shown in Fig. P2.15 applied across it. Determine the average power dissipated by the resistor.

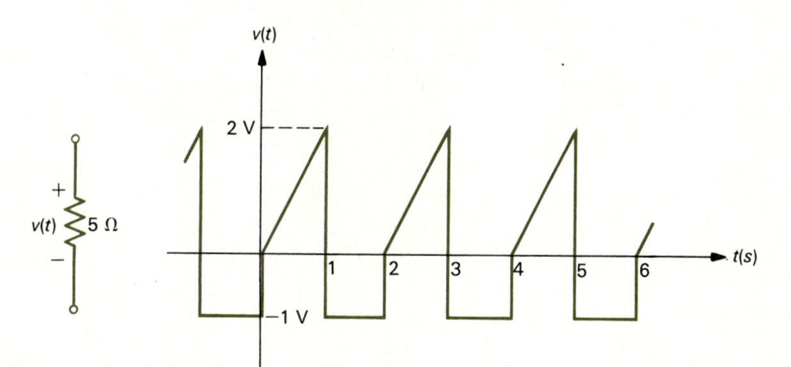

FIGURE P2.15

2.16 A 3-Ω resistor has the periodic current waveform shown in Fig. P2.16*a* passing through it. Compute the average power dissipated in the resistor. Repeat for the waveform shown in Fig. P2.16*b*.

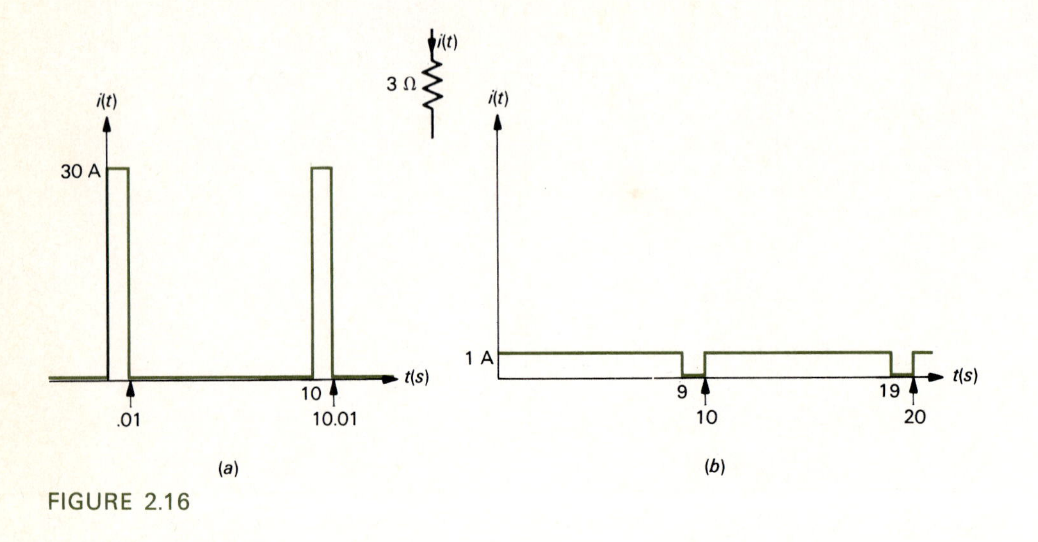

FIGURE 2.16

REFERENCES

2.1 A. B. Carlson and D. G. Gisser, *Electrical Engineering: Concepts and Applications*, Addison-Wesley, Reading, Mass., 1981.

2.2 Ralph J. Smith, *Circuits, Devices and Systems: A First Course in Electrical Engineering*, 4th ed., Wiley, New York, 1984.

2.3 A. E. Fitzgerald, D. Higginbotham, and A. Grabel, *Basic Electrical Engineering*, 5th ed., McGraw-Hill, New York, 1981.

2.4 S. E. Schwartz and W. G. Oldham, *Electrical Engineering: An Introduction*, Holt, Rinehart & Winston, New York, 1984.

2.5 R. L. Boylestad, *Introductory Circuit Analysis*, 4th ed., Charles E. Merrill, Columbus, Ohio, 1982.

Chapter

Analysis Methods for Resistive Circuits

In the previous chapter we studied various types of electric circuit elements—the resistor and the ideal voltage and current sources—and the laws which govern their voltages and currents when they are interconnected to form an electric circuit—KVL and KCL. Our ultimate objective in the analysis of an electric circuit is to determine some or all of the branch (element) voltages and currents. These branch voltages and currents are the result of the excitation of the circuit by its ideal sources. The particular interconnections of these circuit elements determine the values of these branch voltages and currents.

In this chapter we will study various techniques for solving a circuit for its branch voltages and currents. We will concentrate on dc sources, although the techniques are equally valid for other time-varying sources.

3.1 SERIES AND PARALLEL CONNECTIONS OF RESISTORS

A *series* connection of resistors is shown in Fig. 3.1. Note that the current through each resistor must be the same:

$$i = i_1 = i_2 = \cdots = i_n \tag{3.1}$$

but that the overall voltage is the sum of the resistor voltages:

$$v = v_1 + v_2 + \cdots + v_n \tag{3.2}$$

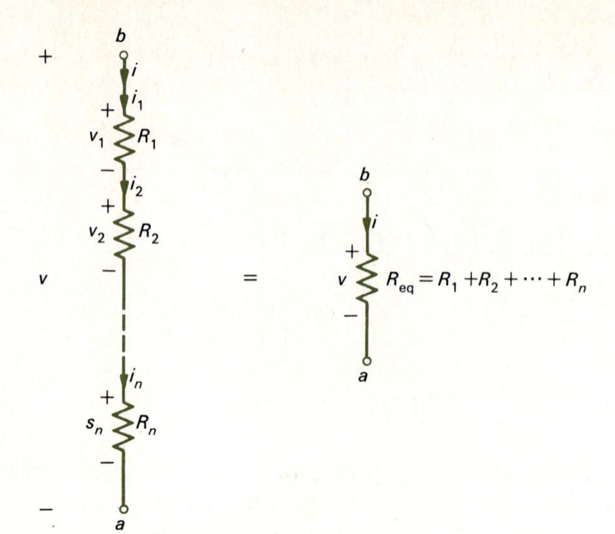

FIGURE 3.1 The equivalent resistance of resistors in series.

Substituting Ohm's law for each resistor into Eq. (3.2), we obtain

$$\begin{aligned} v &= R_1 i + R_2 i + \cdots + R_n i \\ &= (R_1 + R_2 + \cdots + R_n)i \end{aligned} \tag{3.3}$$

Thus, from the standpoint of terminals *a-b*, the series combination is equivalent to a single equivalent resistor whose value is the sum of the values of the resistors:

$$R_{eq} = R_1 + R_2 + \cdots + R_n \tag{3.4}$$

Now let us consider the *parallel* combination of resistors, as shown in Fig. 3.2.

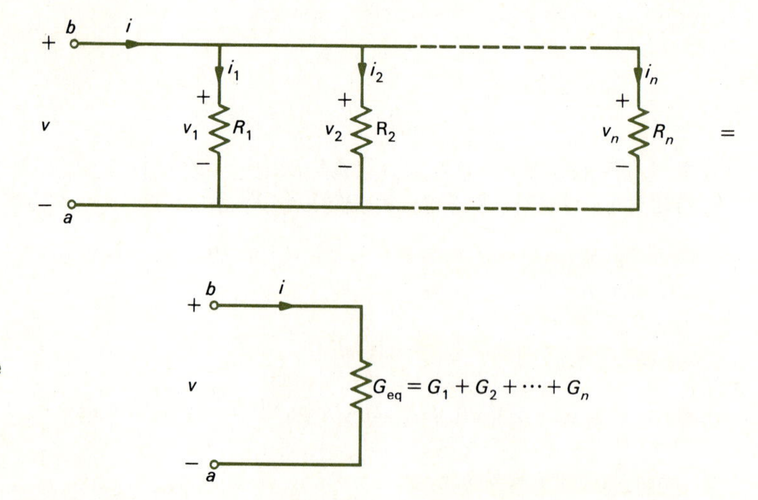

FIGURE 3.2 The equivalent resistance of resistors in parallel.

Note that for this case, as opposed to the series connection, the overall current entering the terminals of the combination is the sum of the individual currents:

$$i = i_1 + i_2 + \cdots + i_n \tag{3.5}$$

but the individual resistor voltages must all be equal:

$$v = v_1 = v_2 = \cdots = v_n \tag{3.6}$$

Substituting Ohm's law in terms of conductance into Eq. (3.5) yields

$$\begin{aligned} i &= G_1 v + G_2 v + \cdots + G_n v \\ &= (G_1 + G_2 + \cdots + G_n)v \end{aligned} \tag{3.7}$$

Thus, a parallel combination of resistors is equivalent (at the terminals) to a single resistor whose conductance is the sum of the conductances of each of the resistors in the parallel connection. Note in both the series and parallel reductions that once we replace the combination with an equivalent resistor, we no longer have retained the individual voltages and currents of the resistors in the combination.

One very common case which we will frequently use is the parallel combination of two resistors shown in Fig. 3.3. The equivalent conductance is the sum of the conductances of the two resistors, and the equivalent resistance is the reciprocal of this:

$$\begin{aligned} R_{eq} &= \frac{1}{G_{eq}} \\ &= \frac{1}{G_1 + G_2} \\ &= \frac{1}{1/R_1 + 1/R_2} \\ &= \frac{R_1 R_2}{R_1 + R_2} \\ &\triangleq R_1 || R_2 \end{aligned} \tag{3.8}$$

This will be denoted as $R_1 || R_2$.

In a parallel combination of resistors, it should be noted that the equivalent resistance is smaller than the smallest resistance in the parallel combination and that

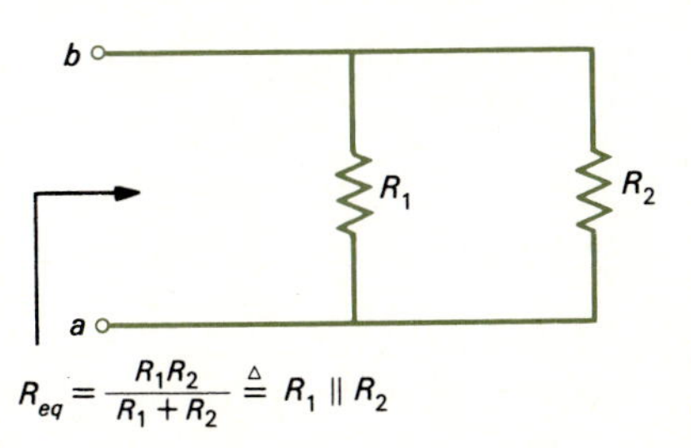

FIGURE 3.3 A very common case of equivalent resistance—two resistors in parallel.

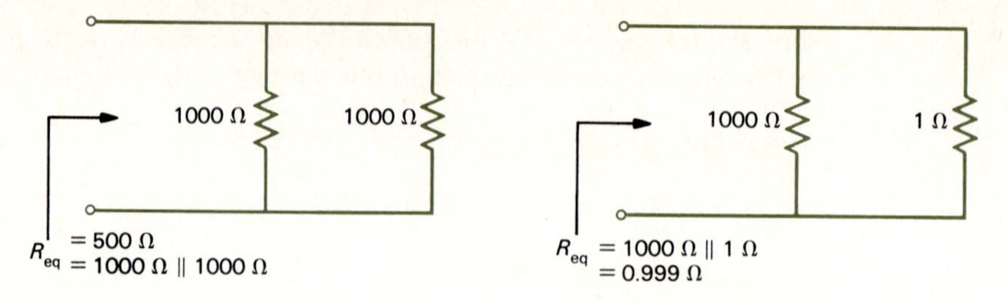

FIGURE 3.4 Illustration of frequently encountered cases: (*a*) two resistors of equal value in parallel; (*b*) two resistors of widely dissimilar values in parallel (the equivalent resistance is smaller than and approximately equal to the smaller resistance).

the parallel combination of two identical resistors yields an equivalent resistance equal to one-half the value of one of the resistors:

$$R||R = \tfrac{1}{2}R \tag{3.9}$$

These points are illustrated in Fig. 3.4.

Example 3.1 Consider the parallel-series combination of resistors shown in Fig. 3.5. Determine an equivalent resistance at the terminals *a-b*.

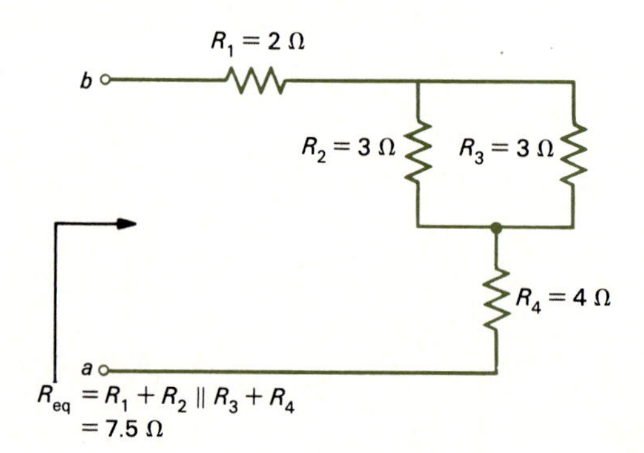

FIGURE 3.5 Example 3.1: reduction of series-parallel combinations.

Solution First replace the parallel combination of R_2 and R_3 with $R_2||R_3 = 1.5\ \Omega$. Then R_{eq} becomes the series combination of R_1, $R_2||R_3$, and R_4, so that

$$R_{eq} = R_1 + R_2||R_3 + R_4$$
$$= 7.5\ \Omega$$

These series-parallel reduction rules can be quite useful in determining the voltages and currents of a circuit, as the following examples show.

Example 3.2 Determine the current *i* delivered by the 10-V battery in Fig. 3.6*a*.

Solution First reduce the parallel combination of 2-Ω resistors to an equivalent 2-Ω||2-Ω = 1-Ω combination at terminal *c*. Then add this to the 1-Ω resistor in series with it. Then reduce the parallel combination of $1\ \Omega + 2\ \Omega||2\ \Omega = 2\ \Omega$ and $3\ \Omega$, which is equivalent to $2\ \Omega||3\ \Omega = \frac{6}{5}\ \Omega$. Add this to the 2-Ω resistor which is in series with the battery so that the battery "sees" a resistance of $2\ \Omega + \frac{6}{5}\ \Omega = 2\frac{6}{5}\ \Omega$. Replacing the circuit attached to the 10-V battery with this equivalent resistance at terminals *a-b*, we then compute

$$i = \frac{10\text{ V}}{2\frac{6}{5}\ \Omega} = 3.13\text{ A}$$

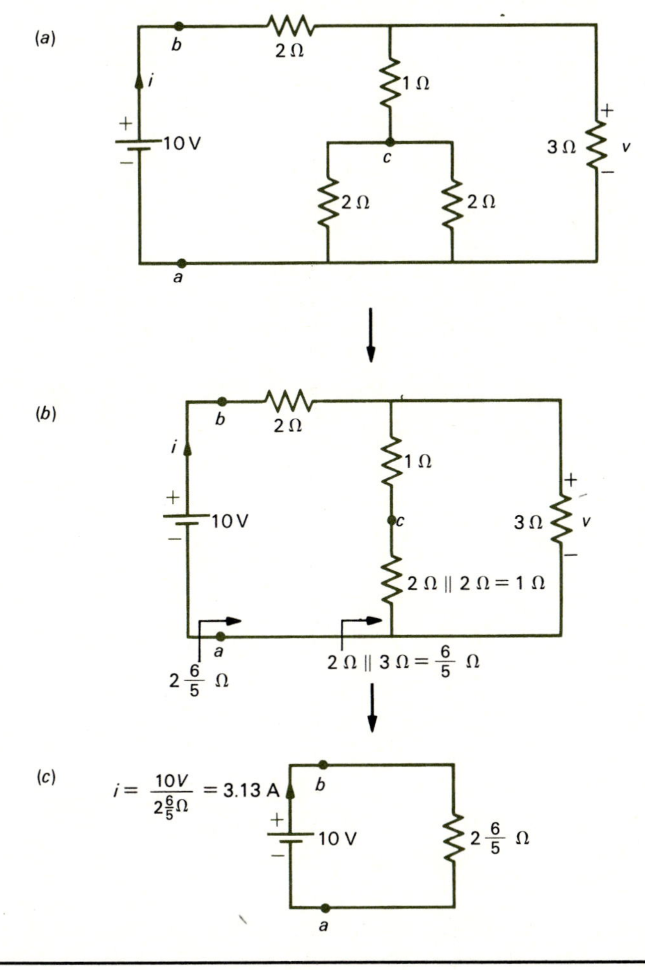

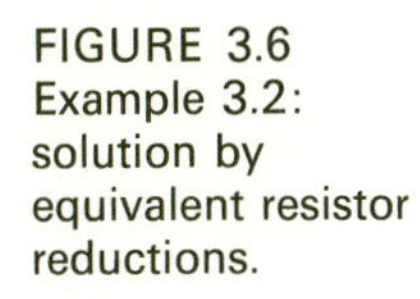
FIGURE 3.6 Example 3.2: solution by equivalent resistor reductions.

This example illustrates an important point. When using these series-parallel reduction rules, it is important to label points in the circuit at which a replacement is being made. Note in the last step in Fig. 3.6*c* that node *c* has disappeared. However, as far as computing *i* is concerned, this is unimportant since the effect of the resistors attached to node *c* has been included in the $2\frac{6}{5}$-Ω equivalent resistance. Note also that the voltage *v* across the 3-Ω resistor has disappeard in the last step (Fig. 3.6*c*). In order to compute *v*, we must return to Fig. 3.6*b*. Note in Fig. 3.6*b* that the current *i* passes through the 2-Ω resistor and produces a voltage across it of $i \times 2\,\Omega = 6.25$ V. The unknown voltage v can be determined by writing KVL around the outside loop:

$$10 - 2i - v = 0$$

or

$$\begin{aligned} v &= 10 - 2i \\ &= 3.75 \text{ V} \end{aligned}$$

Alternatively, we may compute v by multiplying i by the equivalent resistance across which v is located, which is $\frac{6}{5}\,\Omega$:

$$\begin{aligned} v &= \tfrac{6}{5}\,\Omega \times i \\ &= 3.75 \text{ V} \end{aligned}$$

Example 3.3 Determine voltages v, v_x, and v_y and current i_y in Fig. 3.7*a*.

Solution First reduce the parallel combination of 2-Ω and 1-Ω resistors to an equivalent 2-Ω||1-Ω $= \frac{2}{3}$-Ω resistor, as shown in Fig. 3.7*b*. Now it is clear that the 1-A current passes through each resistor in Fig. 3.7*b* so that

$$\begin{aligned} v_x &= -2\,\Omega \times 1 \text{ A} \\ &= -2 \text{ V} \\ v_y &= -\tfrac{2}{3}\,\Omega \times 1 \text{ A} \\ &= -\tfrac{2}{3} \text{ V} \end{aligned}$$

(Note the signs.) Next, v can be found in two ways. One way is to note that the 1-A current source sees an equivalent resistance of $2\,\Omega + \frac{2}{3}\,\Omega = \frac{8}{3}\,\Omega$ and that v is across this equivalent resistance. Thus

$$\begin{aligned} v &= -\tfrac{8}{3}\,\Omega \times 1 \text{ A} \\ &= -\tfrac{8}{3} \text{ V} \end{aligned}$$

Alternatively, the other way is to write KVL around the outside loop to obtain

$$v - v_y - v_x = 0$$

or

$$\begin{aligned} v &= v_y + v_x \\ &= -2 - \tfrac{2}{3} \\ &= -\tfrac{8}{3} \text{ V} \end{aligned}$$

Knowing v_y, we may now determine i_y from Fig. 3.7*a* as

$$i_y = \frac{v_y}{2\,\Omega} = \frac{-\frac{2}{3}}{2} = -\tfrac{1}{3}\text{ A}$$

Note in Fig. 3.7*b* that i_y has disappeared at this reduction stage.

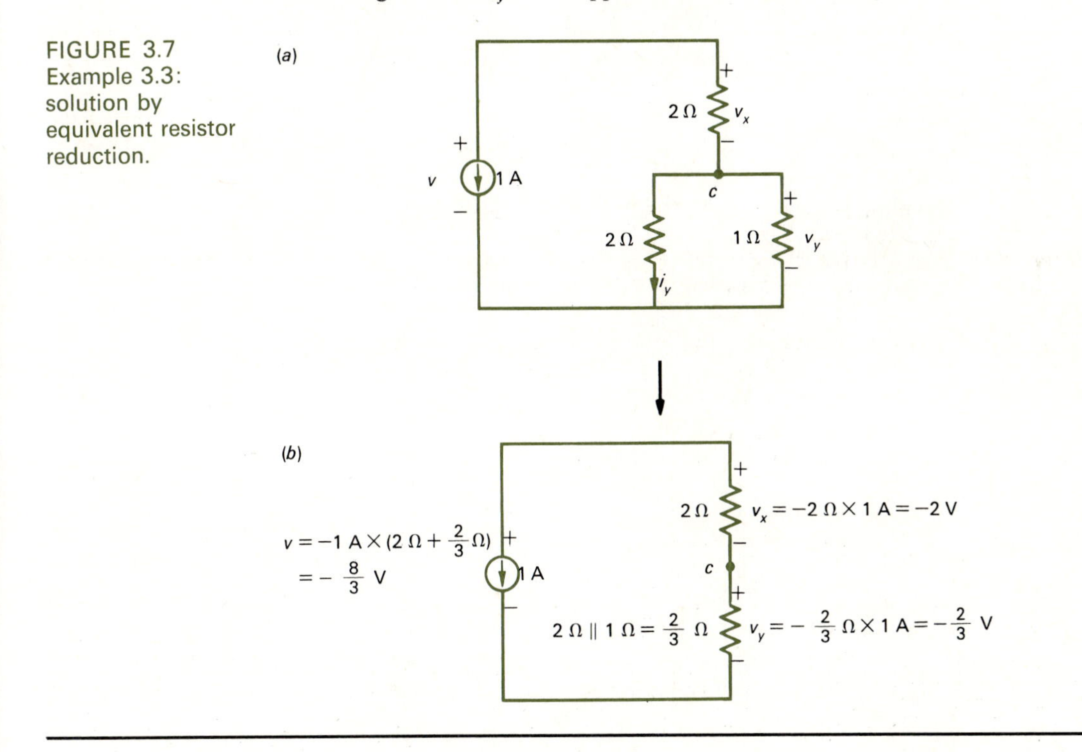

FIGURE 3.7 Example 3.3: solution by equivalent resistor reduction.

3.2 VOLTAGE AND CURRENT DIVISION

There are two additional reduction techniques which, like series and parallel combinations of resistors, are useful in solving for the voltages and currents in a circuit. These are the rules of voltage and current division.

First let us consider voltage division. Suppose that a known voltage is applied across a series combination of resistors as shown in Fig. 3.8. The current i through the series combination is

$$i = \frac{v_s}{R_1 + R_2 + \cdots + R_n} \tag{3.10}$$

since the voltage source sees an equivalent resistance $R_{eq} = R_1 + R_2 + \cdots + R_n$. The

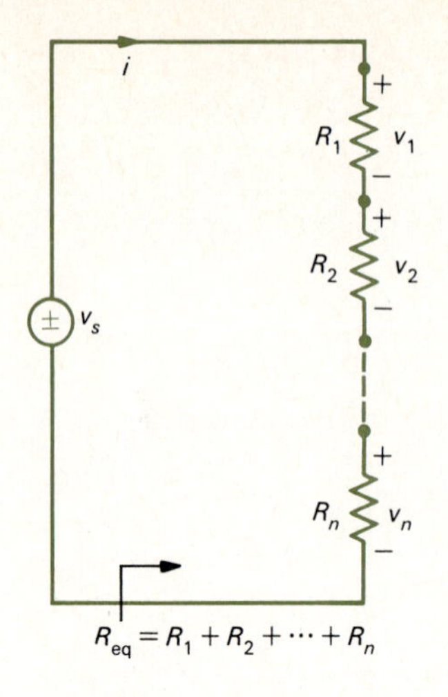

FIGURE 3.8 Illustration of the voltage-division rule.

individual voltages across the individual resistors are

$$
\begin{aligned}
v_1 &= R_1 i \\
v_2 &= R_2 i \\
&\vdots \\
v_n &= R_n i
\end{aligned}
\tag{3.11}
$$

Substituting Eq. (3.10), we find that

$$v_i = \frac{R_i}{R_1 + R_2 + \cdots + R_n} v_s \tag{3.12}$$

Thus, the voltage divides according to the ratio of the resistor in question to the sum of the resistors in the series combination. This is called the voltage-division rule.

Example 3.4 Consider the example shown in Fig. 3.9. Determine voltage v.

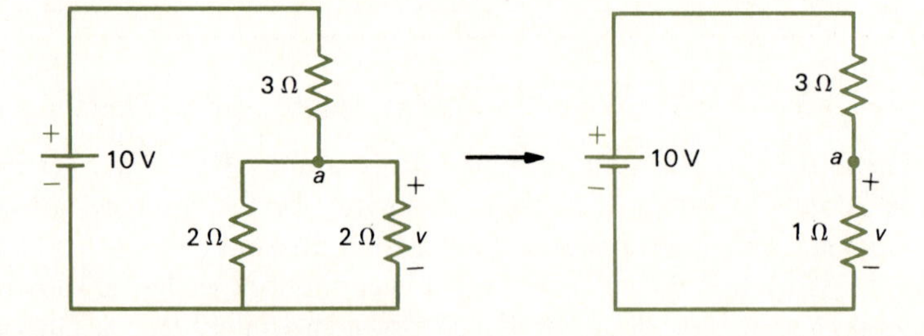

FIGURE 3.9 Example 3.4: illustration of voltage division.

Solution First combine the two 2-Ω resistors into an equivalent 1-Ω resistor. Then the voltage v is, by voltage division,

$$
\begin{aligned}
v &= \frac{1\ \Omega}{1\ \Omega + 3\ \Omega} 10\ \text{V} \\
&= 2.5\ \text{V}
\end{aligned}
$$

The voltage across the series combination need not be provided by an ideal source. It may be a result of some other portion of the circuit.

Example 3.5 Consider the circuit in Fig. 3.10. Determine voltages v_x and v_y.

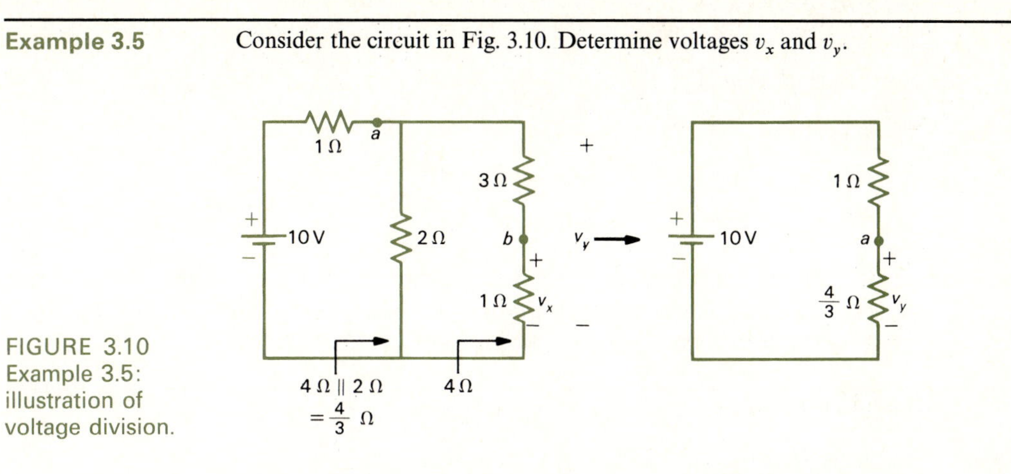

FIGURE 3.10 Example 3.5: illustration of voltage division.

Solution First reduce the circuit connected to node a to an equivalent resistance of $\frac{4}{3}\,\Omega$ and redraw the circuit slightly. Note that node b and v_x have disappeared in this reduction. Nevertheless, by voltage division we may obtain

$$\begin{aligned} v_y &= \frac{\frac{4}{3}\,\Omega}{\frac{4}{3}\,\Omega + 1\,\Omega}\,10\text{ V} \\ &= \tfrac{40}{7}\text{ V} \end{aligned}$$

Now from the original circuit we have, by voltage division,

$$\begin{aligned} v_x &= \frac{1\,\Omega}{1\,\Omega + 3\,\Omega}\,v_y \\ &= \tfrac{10}{7}\text{ V} \end{aligned}$$

The next rule, current division, is similar to the voltage-division rule. Consider the parallel combination of resistors shown in Fig. 3.11. The equivalent conductance seen by the current source is

$$G_{eq} = G_1 + G_2 + \cdots + G_n \tag{3.13}$$

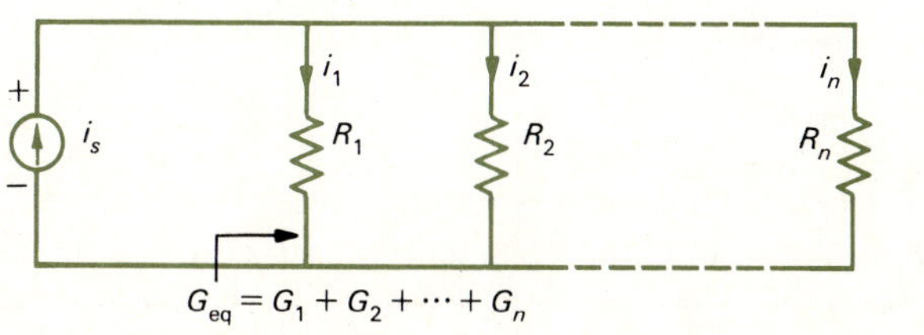

FIGURE 3.11 Illustration of the current-division rule.

and the resulting voltage across the current source and parallel combination is

$$v = \frac{1}{G_{eq}} i_s \tag{3.14}$$

Now, each current is given by

$$\begin{aligned} i_i &= G_i v \\ &= \frac{G_i}{G_1 + G_2 + \cdots + G_n} i_s \end{aligned} \tag{3.15}$$

This is the current-division rule. The current entering a parallel combination of resistors divides according to the ratio of the conductance in question to the sum of the conductances in the parallel combination.

Example 3.6

Consider the circuit shown in Fig. 3.12. Determine i_x.

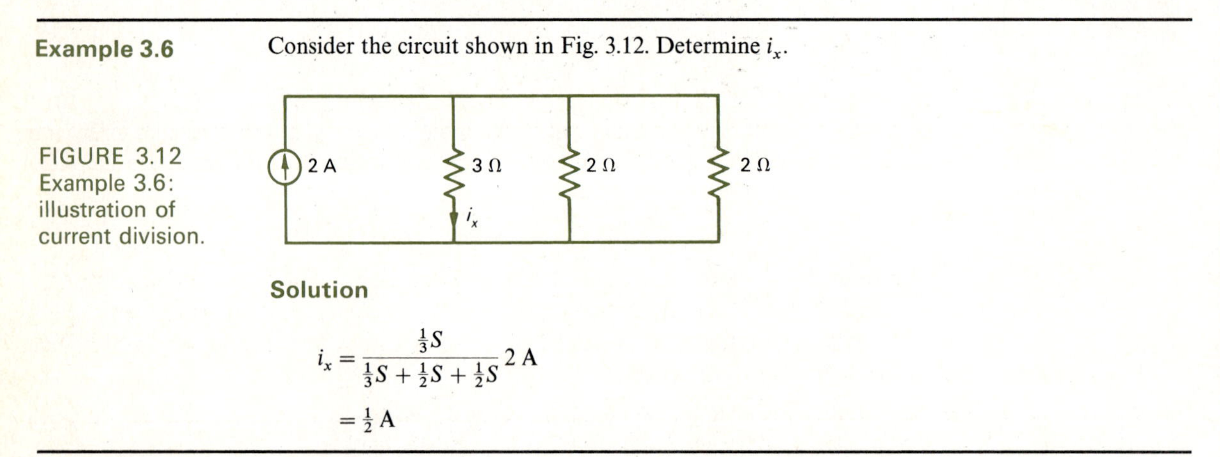

FIGURE 3.12 Example 3.6: illustration of current division.

Solution

$$\begin{aligned} i_x &= \frac{\frac{1}{3}S}{\frac{1}{3}S + \frac{1}{2}S + \frac{1}{2}S} 2\text{ A} \\ &= \tfrac{1}{2}\text{ A} \end{aligned}$$

Quite often we tend to deal in resistance rather than conductance. A common case where this can be easily done when using current division is the case of two parallel resistances, shown in Fig. 3.13. The currents are, by current division,

$$\begin{aligned} i_1 &= \frac{G_1}{G_1 + G_2} i_s \\ &= \frac{R_2}{R_1 + R_2} i_s \\ i_2 &= \frac{G_2}{G_1 + G_2} i_s \\ &= \frac{R_1}{R_1 + R_2} i_s \end{aligned} \tag{3.16}$$

Note that when dealing with resistance instead of conductance, the current divides according to the ratio of the resistance *opposite* the one in question to the sum of the

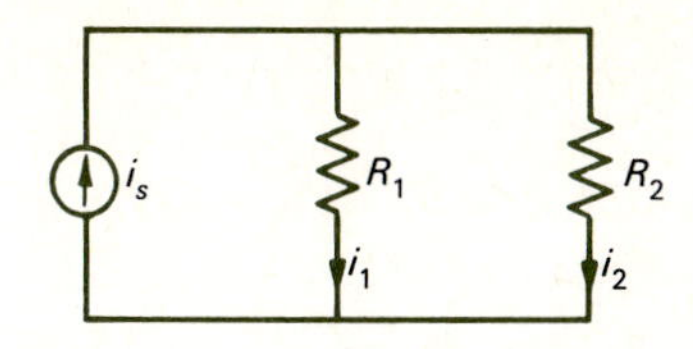

FIGURE 3.13 Illustration of current division for an important case—two resistors in parallel.

resistances in the combination. This technique can be adapted to cases involving more than two parallel resistors, as the following example shows.

Example 3.7 Reconsider the circuit in Fig. 3.12. Determine i_x.

Solution The two 2-Ω resistors are in parallel with each other and we may combine them into an equivalent 1-Ω resistor. Then we have 1 Ω in parallel with 3 Ω and wish to find the current through the 3-Ω resistor:

$$i_x = \frac{1\ \Omega}{3\ \Omega + 1\ \Omega} 2\ \text{A}$$

$$= \tfrac{1}{2}\ \text{A}$$

3.3 THE STRETCH-AND-BEND PRINCIPLE

Quite often it is helpful to redraw a circuit in an equivalent form. We may stretch, bend, and slide wires along other wires to form circuits which are equivalent to the original ciruits. We may *not*, however, cut any wires or attach them at any other place than their original connection.

Example 3.8 Consider the circuit shown in Fig. 3.14. Draw several equivalent circuits.

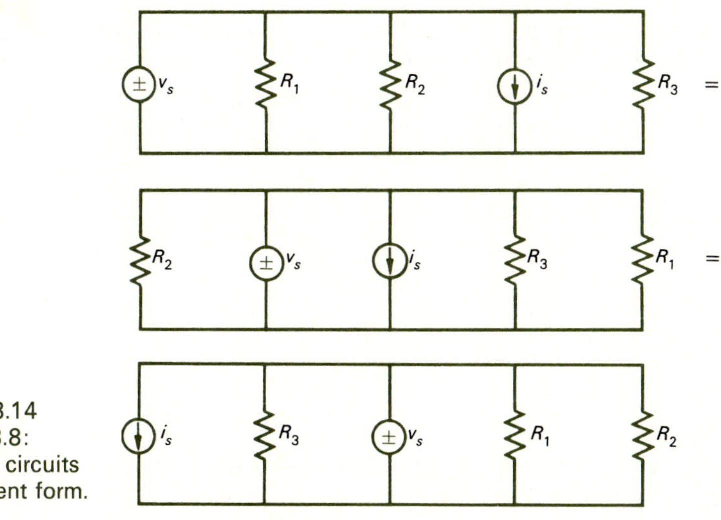

FIGURE 3.14 Example 3.8: redrawing circuits in equivalent form.

Solution Several equivalent circuits are shown. In certain cases these other circuits may be easier to solve than the original, but equivalent, circuit.

Another example is shown in Fig. 3.15*a*. The reader should verify that the circuit in Fig. 3.15*b* is an equivalent one, whereas, the circuit in Fig. 3.15*c* is not.

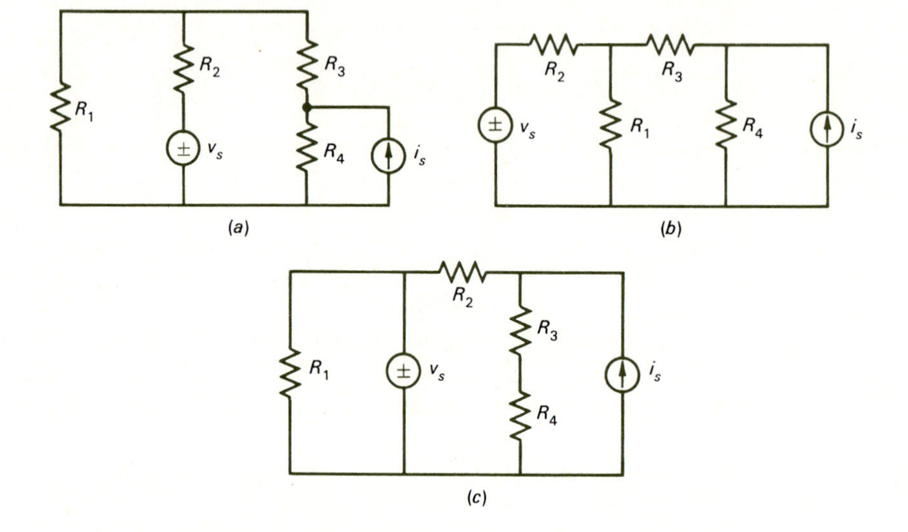

FIGURE 3.15 Illustration of correct and incorrect redrawing of circuits.

3.4 SUPERPOSITION

Perhaps the most powerful principle in analyzing *linear* circuits is the principle of *superposition.* This principle does *not* apply to *nonlinear* circuits. It is therefore important to determine what constitutes a linear circuit: *A linear circuit is one which contains only linear circuit elements.* The circuit elements which we have considered—the resistor and the ideal voltage and current source—are linear elements. All of the ideal sources are linear, so we need to focus on the resistors to determine whether a circuit is nonlinear. An example of a nonlinear resistor is one in which the terminal voltage-current characteristic is not a straight line or does not pass through the origin, such as is shown in Fig. 3.16. For nonlinear resistors, the terminal voltage and current are not linearly related. For example, $v = Ri^2$ is a nonlinear relationship since the voltage does not depend directly on the current but depends on the square of the current.

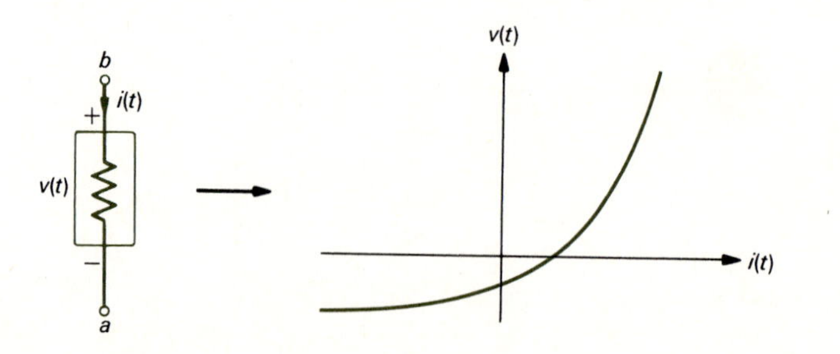

FIGURE 3.16 A general nonlinear resistor.

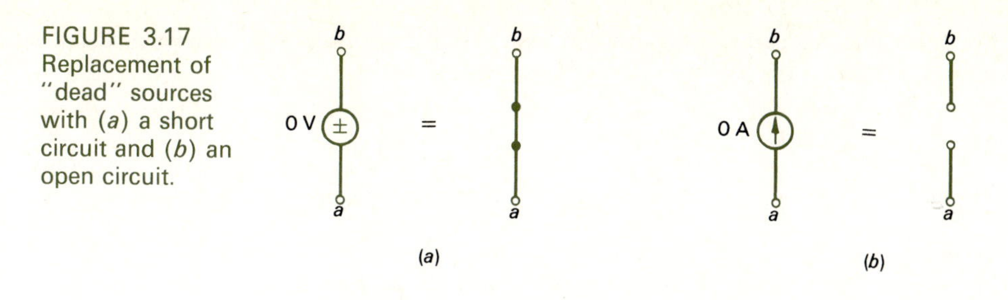

FIGURE 3.17 Replacement of "dead" sources with (*a*) a short circuit and (*b*) an open circuit.

The principle of superposition is quite simple: *If a linear circuit has* N *ideal sources present, any branch voltage or current is composed of the sum of* N *contributions, each of which is due to each source acting individually when all others are set equal to zero (inactivated).* When we set the value of an ideal *voltage source* to zero (kill it), the voltage of that branch becomes zero. Thus, we replace the voltage source with a *short circuit,* as shown in Fig. 3.17*a*. Conversely, when we set an ideal *current source* to zero (kill it), the current of that branch becomes zero. Thus, we replace it with an *open circuit,* as shown in Fig. 3.17*b*.

To illustrate the principle of superposition, let us consider the (linear) circuit in Fig. 3.18*a*. The voltage across R_1 consists of two components, v' due to the ideal voltage source and v'' due to the ideal current source; similarly, the current i through resistor R_2 consists of two components. Each of these contributions is computed from the circuits in Fig. 3.18*b* and *c*. The contributions due to the voltage source, with the current source killed, are computed from Fig. 3.18*b* as

$$v' = \frac{R_1}{R_1 + R_2} v_s$$

$$i' = \frac{v_s}{R_1 + R_2}$$

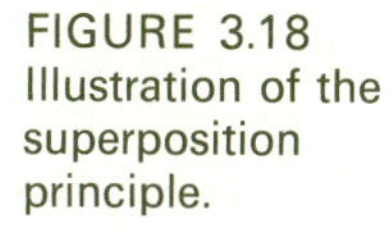

FIGURE 3.18 Illustration of the superposition principle.

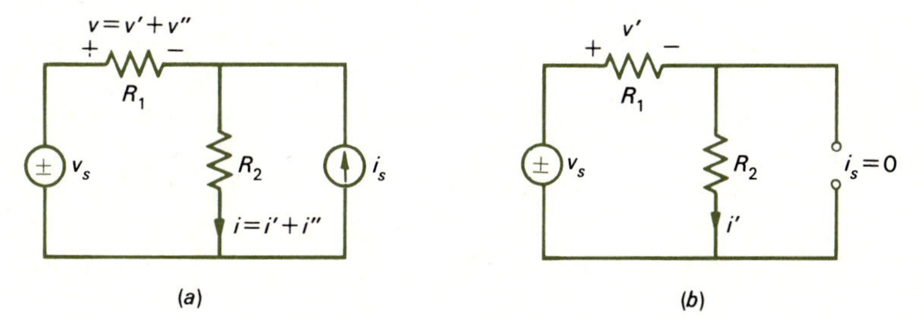

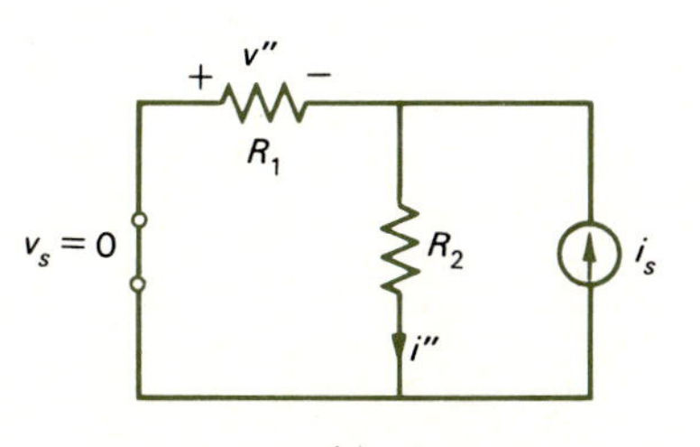

The contributions due to the current source, with the voltage source killed, are computed from Fig. 3.18*c* as

$$v'' = -R_1||R_2 i_s$$

$$i'' = \frac{R_1}{R_1 + R_2} i_s$$

The reader should verify these results, using voltage and current division and equivalent resistances. The total voltage v and current i due to both sources activated becomes

$$v = v' + v''$$

$$i = i' + i''$$

Example 3.9 Using superposition, compute the current i and voltage v in the circuit of Fig. 3.19*a*. Show that these results satisfy KVL and KCL.

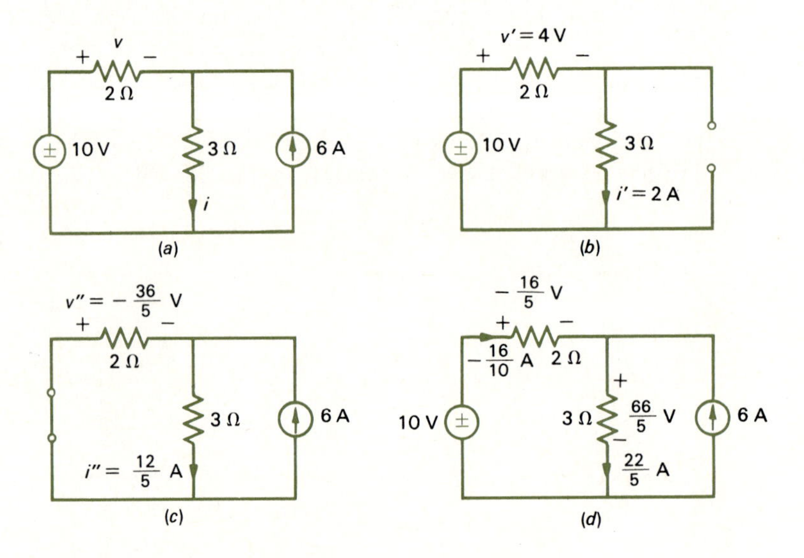

FIGURE 3.19 Example 3.9: illustration of superposition.

Solution We determine from Fig. 3.19*b* and *c* that

$$v' = 4\text{ V} \qquad v'' = -\tfrac{36}{5}\text{ V}$$

$$i' = 2\text{ A} \qquad i'' = \tfrac{12}{5}\text{ A}$$

so that

$$v = v' + v''$$

$$= -\tfrac{16}{5}\text{ V}$$

$$i = i' + i''$$

$$= \tfrac{22}{5}\text{ A}$$

To verify that these solutions satisfy KVL and KCL, we must compute all element voltages and currents. This is done in Fig. 3.19*d* using only Ohm's law. Applying KVL around the loop containing the two resistors and the voltage source, we have

$$10 - (-\tfrac{16}{5}) - \tfrac{66}{5} \stackrel{?}{=} 0$$

which is true. Applying KCL at the upper node to which the current source is attached, we have

$$6 - \tfrac{22}{5} + (-\tfrac{16}{10}) \stackrel{?}{=} 0$$

which is true. Since the branch voltages and currents satisfy KVL and KCL for the circuit, we have found the solution.

In applying superposition, it is not necessary to consider each source individually. We may group several sources and determine the contribution due to this group of sources.

Example 3.10 Determine voltage v in the circuit of Fig. 3.20*a*.

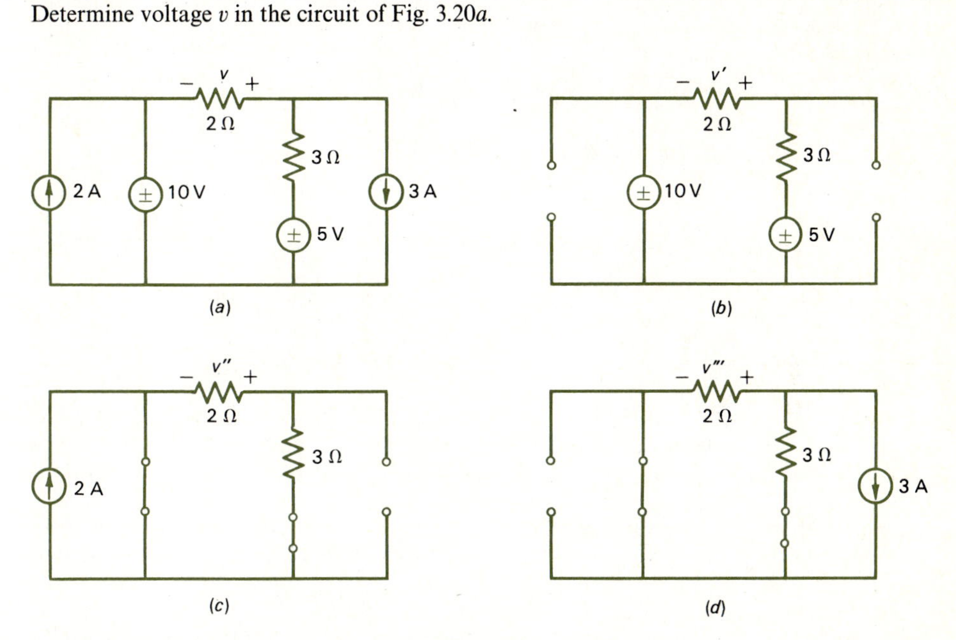

FIGURE 3.20 Example 3.10: illustration of superposition by combining groups of sources.

Solution To illustrate this concept, we will consider the two voltage sources as jointly contributing v' and the current sources as individually contributing v'' and v'''. These solutions are shown in Fig. 3.20*b*, *c*, and *d*:

$$v' = \frac{2}{2+3}(5 - 10)$$

$$= -2 \text{ V}$$

$$v'' = 0$$

$$v''' = -\tfrac{18}{5} \text{ V}$$

(Note that, by current division, all of the 2-A current source goes through the short circuit caused by killing the 10-V voltage source.) Thus

$$\begin{aligned} v &= v' + v'' + v''' \\ &= -2 + 0 - \tfrac{18}{5} \\ &= -\tfrac{28}{5}\ \text{V} \end{aligned}$$

It is worthwhile at this stage to point out that superposition does *not* apply to the computation of power.

Example 3.11 Determine the power delivered to the 3-Ω resistor in Fig. 3.19.

Solution Since $i = \frac{22}{5}$ A (from Example 3.9), we find that

$$\begin{aligned} p_{3\,\Omega} &= 3i^2 \\ &= 58.08\ \text{W} \end{aligned}$$

If we try to compute this by superposition, we find that

$$\begin{aligned} p_{3\,\Omega} &\stackrel{?}{=} 3(i')^2 + 3(i'')^2 \\ &= 3(2)^2 + 3(\tfrac{12}{5})^2 \\ &= 12 + 17.28 \\ &= 29.28 \\ &\neq 58.08\ \text{W} \end{aligned}$$

It is also important to note that an ideal voltage or current source may actually be absorbing power (having power delivered to it). Whether or not the element absorbs or delivers power depends solely on the circuit to which the source is connected.

Example 3.12 Determine the power delivered by the 2- and 3-A current sources in Fig. 3.20*a*.

Solution In order to do this we need only determine the terminal voltage of each current source and its polarity. The voltage across the 2-A current source is simply the value of the 10-V voltage source. Because of the polarity of this voltage and the current of the current source, the 2-A current source is delivering 2 A × 10 V = 20 W. The terminal voltage across the 3-A current source is, by KVL, the sum $10\ \text{V} + v = 10 - \frac{28}{5} = \frac{22}{5}$ V. But this voltage has the polarity + at the top of the current source and − at the bottom. Because of this terminal voltage polarity and the direction of the current source, it is delivering $-(\frac{22}{5})(3) = -\frac{66}{5}$ W. Thus, the 3-A current source is actually absorbing $\frac{66}{5}$ W of power!

Now we will show, very simply, why superposition works. The branch voltages and currents are governed by (must simultaneously satisfy) KVL, KCL, and the ele-

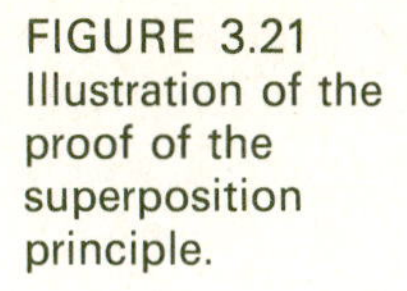

FIGURE 3.21 Illustration of the proof of the superposition principle.

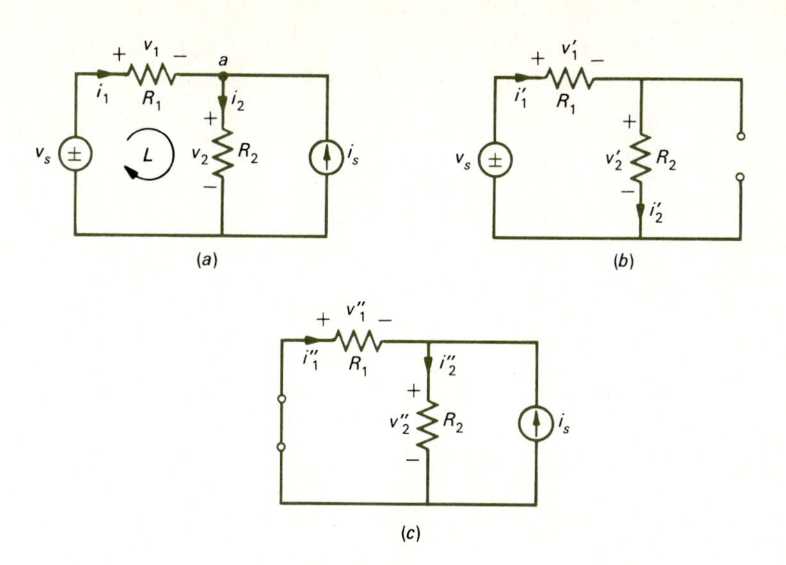

ment relations. For example, consider the circuit in Fig. 3.21*a*. KVL only needs to be written around one loop, *L*, and KCL needs to be written at only one node, *a*. These two equations, along with the resistor relations, become

KVL loop *L*:

$$v_1 + v_2 = v_s$$

KCL node *a*:

$$i_2 - i_1 = i_s$$

Ohm's law for R_1: (3.17)

$$v_1 - R_1 i_1 = 0$$

Ohm's law for R_2:

$$v_2 - R_2 i_2 = 0$$

Note that we have placed the values of the ideal sources (the known quantities in these equations) on the right-hand side of these equations and all other quantities on the left. We now have four equations which can be solved for the four unknowns: v_1, i_1, v_2, i_2. We would not, of course, resort to solving these simultaneously, since simpler methods of analysis will do (as illustrated in previous examples); nevertheless, they serve to illustrate why superposition works. The solutions, by superposition, for the contributions of the two sources to these four branch voltages and currents are shown in Fig. 3.21*b* and *c*. The principle of superposition states that if these contributions satisfy the circuit equations individually, then their sum will satisfy the circuit equations. From Fig. 3.21*b*, the contributions due to the voltage source satisfy [set $i_s = 0$ in the circuit equations of Eq. (3.17)]

$$\begin{aligned} v_1' + v_2' &= v_s \\ i_2' - i_1' &= 0 \\ v_1' - R_1 i_1' &= 0 \\ v_2' - R_2 i_2' &= 0 \end{aligned} \tag{3.18}$$

Similarly, from Fig. 3.21*c*, the contributions due to the current source satisfy [set $v_s = 0$ in the circuit equations of Eq. (3.17)]

$$\begin{aligned} v_1'' + v_2'' &= 0 \\ i_2'' - i_1'' &= i_s \\ v_1'' - R_1 i_1'' &= 0 \\ v_2'' - R_2 i_2'' &= 0 \end{aligned} \tag{3.19}$$

The question now is, given that v_1', v_2', i_1', i_2' satisfy Eq. (3.18) and that $v_1'', v_2'', i_1'', i_2''$ satisfy Eq. (3.19), is it true that $v_1' + v_1''$, $v_2' + v_2''$, $i_1'' + i_1''$, $i_2' + i_2''$ *automatically* satisfy Eq. (3.17)? Let us see. Substitute the sum variables into Eq. (3.17) to yield

$$\begin{aligned} (v_1' + v_1'') + (v_2' + v_2'') &= v_s + 0 \\ (i_2' + i_2'') - (i_1' + i_1'') &= 0 + i_s \\ (v_1' + v_1'') - R_1(i_1' + i_1'') &= 0 + 0 \\ (v_2' + v_2'') - R_2(i_2' + i_2'') &= 0 + 0 \end{aligned} \tag{3.20}$$

Note that we have added a zero to the right-hand side of each of the Eq. (3.17) equations. Now it is easy to see that Eq. (3.20) can be factored as the sum of Eqs. (3.18) and (3.19), which were assumed to be true, so therefore Eq. (3.20) is true.

Now it is easy to see why superposition does not work for nonlinear circuits. Suppose that resistor R_1 in the above circuit were nonlinear and characterized by

$$v_1 = R_1 i_1^2 \tag{3.21}$$

The prime variables due to the voltage source satisfy

$$v_1' - R_1 i_1'^2 = 0 \tag{3.22}$$

and the double-prime variables due to the current source satisfy

$$v_1'' - R_1 i_1''^2 = 0 \tag{3.23}$$

Now can

$$(v_1' + v_1'') - R_1(i_1' + i_1'')^2 = 0 \tag{3.24}$$

be factored into the sum of Eqs. (3.22) and (3.23)? Let us try:

$$v_1' + v_1' - R_1 i_1^{2\prime} - 2R_1 i_1' i_1'' - R_1 i_1^{2\prime\prime} = 0 \tag{3.25}$$

In Eq. (3.25) we identify Eqs. (3.22) and (3.23), but there is an additional term, $-2R_1 i_1' i_1''$, which is not accounted for. Thus, we cannot say that Eq. (3.25) is satisfied solely because Eqs. (3.22) and (3.23) were presumed to be satisfied. Now we clearly see why superposition works for linear circuits and does not work for nonlinear circuits. All it takes to make an entire circuit a nonlinear one is for one of the resistors to be nonlinear! In our future use of superposition we will begin to see its extraordinary usefulness in solving linear circuits.

3.5 THEVENIN AND NORTON EQUIVALENT CIRCUITS

The Thevenin and Norton equivalent circuits allow us to replace a portion of a circuit which is connected to two terminals with another circuit which has the same effect on the rest of the circuit. The principle is quite simple and, again, is valid only for a linear portion of a circuit. In other words, a circuit may be nonlinear due to the presence of a nonlinear resistor—but we may *replace* that linear portion of the circuit which is attached to the terminals of the nonlinear resistor with an equivalent, and often simpler, circuit.

To illustrate this principle and show how sensible it is, let us consider an electric circuit which has been partitioned into parts connected by wires as shown in Fig. 3.22*a*. Now let us ask the question "How are the voltage v and current i at the terminals of the linear portion related"? The obvious answer is that they are related as a linear equation such as

$$v = Ai + B \tag{3.26}$$

where A and B are two constants.

How else would they be related? A relation such as $v = Ai^2 + B$ would not be logical, since the portion of the circuit was stipulated to be linear. Now all we need to do to obtain an equivalent circuit with which to replace this linear portion at terminals *a-b* is to find one which has the same terminal voltage and current relation as in Eq. (3.26).

The Thevenin equivalent circuit is shown in Fig. 3.22*b*. Writing KVL around the loop containing V_{OC} and R_{TH}, we obtain

$$v = R_{TH} i + V_{OC} \tag{3.27}$$

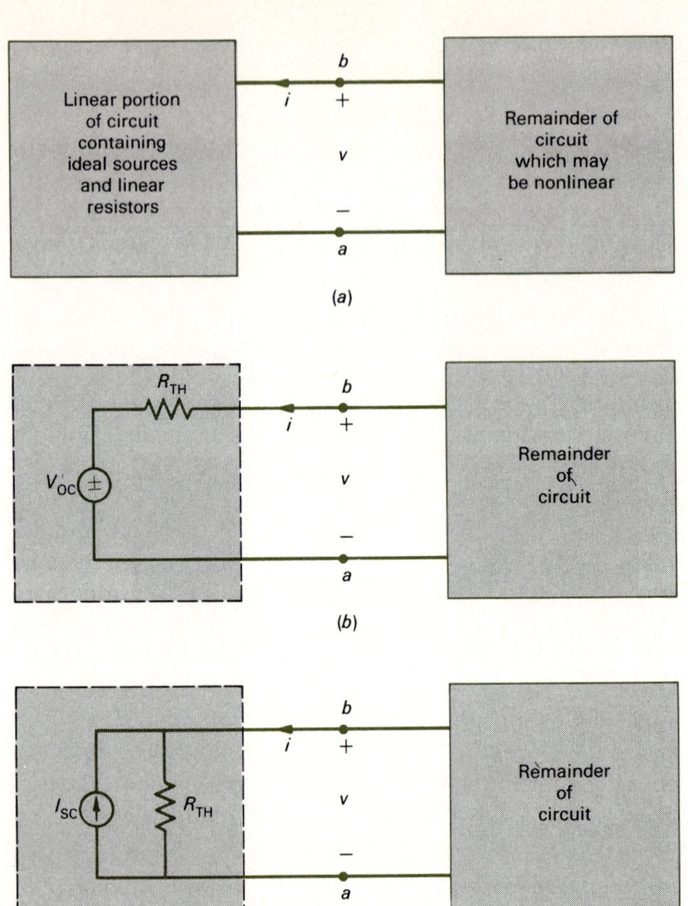

FIGURE 3.22 Equivalent representation of two-terminal circuits: (*a*) two-part circuit connected by wires; (*b*) Thevenin equivalent; (*c*) Norton equivalent.

Thus, if we know A and B in Eq. (3.26), perhaps from some measurements of the circuit, we identify $V_{OC} = B$ and $R_{TH} = A$. The quantity V_{OC} is called the open-circuit voltage, because when we open-circuit the terminals, $i = 0$, we obtain from Eq. (3.27)

$$V_{OC} = v|_{i=0} \tag{3.28}$$

Obviously, V_{OC} is due to the ideal sources present in this portion of the circuit; if there were no ideal sources present (some may be present in the remainder circuit), then V_{OC} would be zero. Similarly, R_{TH} is called the Thevenin equivalent resistance, because when we set $V_{OC} = 0$ (kill all ideal sources in this portion of the circuit), then from Eq. (3.27)

$$R_{TH} = \frac{v}{i}\bigg|_{V_{OC}=0} \tag{3.29}$$

These concepts are illustrated in Fig. 3.23*a* and*b*.

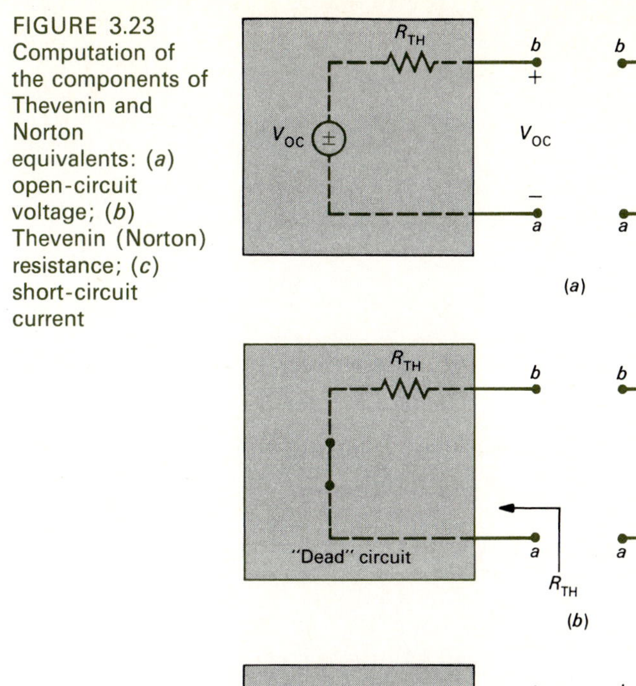
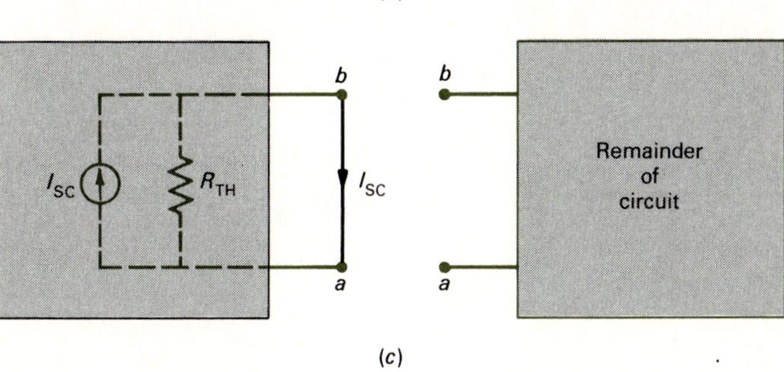

FIGURE 3.23 Computation of the components of Thevenin and Norton equivalents: (*a*) open-circuit voltage; (*b*) Thevenin (Norton) resistance; (*c*) short-circuit current

---

Example 3.13 Consider the circuit shown in Fig. 3.24*a*. Determine the current *i*.

Solution Reduce the portion of the circuit to the left of terminals *a*-*b* to a Thevenin equivalent. First, remove the 5-Ω resistor (the remainder circuit) and compute the open-circuit voltage, as shown in Fig. 3.24*b*; this can be computed by superposition to be

$$\begin{aligned} V_{OC} &= 2\text{ V} - 10\text{ V} \\ &= -8\text{ V} \end{aligned}$$

Note that the 3-Ω resistor has no effect on this, since no current flows through it when terminals *a*-*b* are open-circuited. Second, in order to obtain the Thevenin resistance, kill all independent sources in this portion of the circuit—replace ideal voltage sources with short circuits and ideal current sources with open circuits. This is shown in Fig. 3.24*c*. Then R_{TH} is the resistance, seen

FIGURE 3.24
Example 3.13: illustrating use of a Thevenin equivalent.

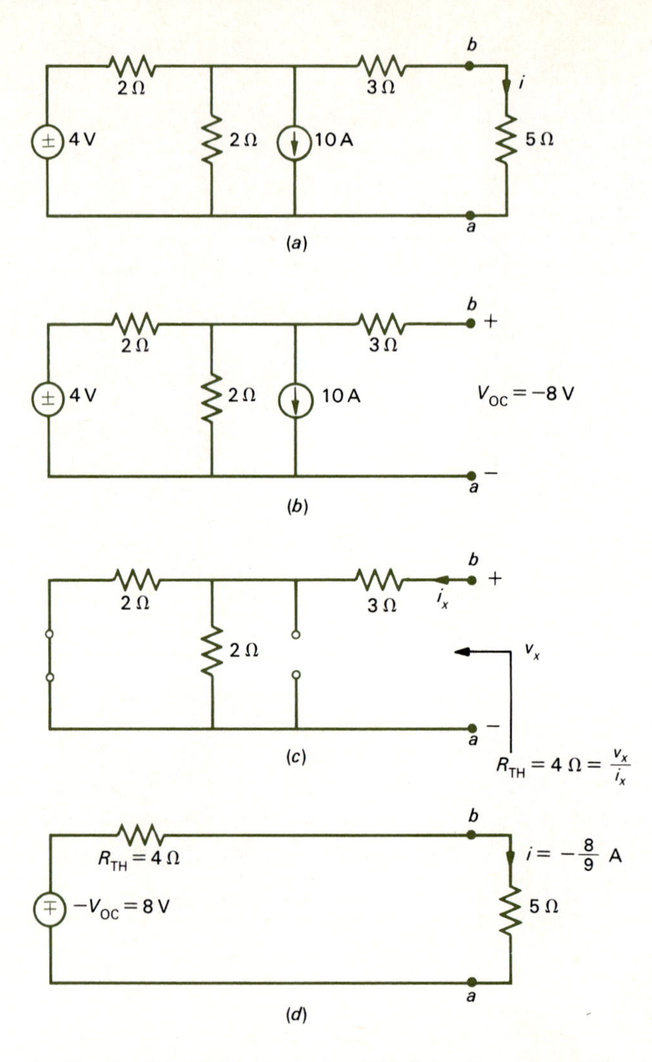

looking back into the terminals of this "dead" portion of the circuit:

$$
\begin{aligned}
R_{TH} &= 2\ \Omega \| 2\ \Omega + 3\ \Omega \\
&= 4\ \Omega
\end{aligned}
$$

It is also the ratio of

$$R_{TH} = \frac{v_x}{i_x}$$

in this circuit. Now attach the Thevenin equivalent circuit to the remainder of the circuit (the 5-Ω resistor), taking care to observe the proper polarity of V_{OC} and the terminal labels. This is shown in Fig. 3.24*d*. From this we obtain

$$i = -\tfrac{8}{9}\,\text{A}$$

Note in the previous example that it is vitally important to label the terminals at which we wish to make a replacement, to define V_{OC} with respect to these, and to attach the Thevenin equivalent circuit in such a way that the polarity of V_{OC} conforms to this chosen designation.

An alternative form is the Norton equivalent circuit shown in Fig. 3.22*c*. Writing KCL at node *b* in Fig. 3.22c, we obtain

$$i = \frac{1}{R_{TH}} v - I_{SC} \tag{3.30}$$

If we rewrite Eq. (3.27) in this form, we have

$$i = \frac{1}{R_{TH}} v - \frac{V_{OC}}{R_{TH}} \tag{3.31}$$

Comparing Eqs. (3.30) and (3.31), we find that

$$\begin{aligned} I_{SC} &= \frac{V_{OC}}{R_{TH}} \\ &= -i|_{v=0} \end{aligned} \tag{3.32}$$

From Eq. (3.32) it is clear that I_{SC} is the current through a short circuit placed across terminals *a-b* after the portion of the circuit is disconnected from the remainder circuit. As in the case of V_{OC}, I_{SC} is caused by ideal sources; if no ideal sources are present in this portion of the circuit, I_{SC} will be zero. This is illustrated in Fig. 3.23*c*.

Example 3.14 Compute *i* in Example 3.13 by using a Norton equivalent.

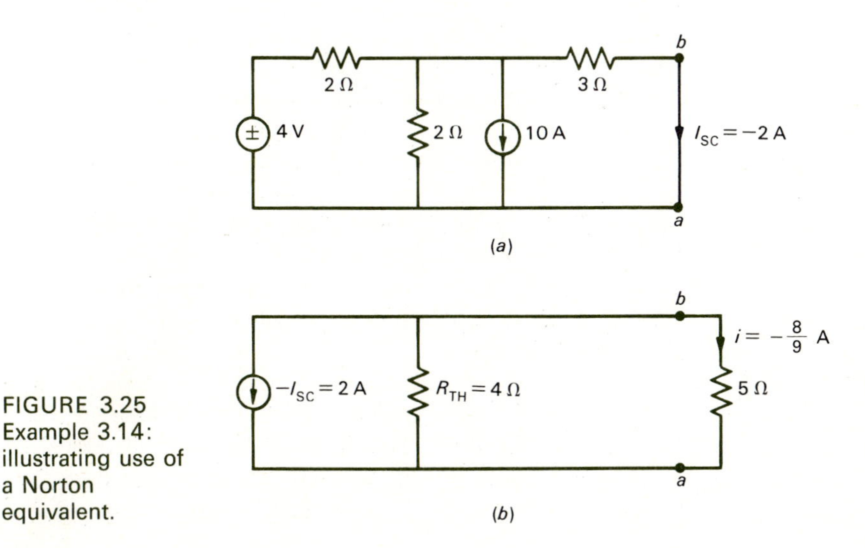

FIGURE 3.25 Example 3.14: illustrating use of a Norton equivalent.

Solution R_{TH} was computed Example 3.13 to be $R_{TH} = 4\ \Omega$. Thus, we only need to find I_{SC}. Removing the remainder circuit (the 5-Ω resistor) and placing a short circuit across terminals *a-b* as shown in Fig. 3.25*a* we obtain, by superposition,

$$I_{SC} = \tfrac{1}{2}\ \text{A} - \tfrac{5}{2}\ \text{A}$$

$$= -2\ \text{A}$$

Note that

$$I_{SC} = \frac{V_{OC}}{R_{TH}}$$

Now reattaching the Norton equivalent circuit to the remainder circuit as shown in Fig. 3.25*b*, we find (by current division) that, just as in Example 3.13,

$$i = -\frac{4\ \Omega}{4\ \Omega + 5\ \Omega} 2\ \text{A}$$

$$= -\tfrac{8}{9}\ \text{A}$$

Note that in the examples of this section we have had numerous occasions to use the analysis techniques of the previous sections. It is very important that the reader understand those techniques, since they will be used on numerous occasions throughout this text.

3.6 NODE VOLTAGE AND MESH CURRENT ANALYSIS

The analysis techniques of the previous sections are the primary techniques we will use for the analysis of linear electric circuits. Virtually all of the circuits which we will analyze throughout the remainder of this text are capable of being analyzed with those techniques and nothing more. Thus, when the reader is confronted with a circuit to be analyzed, the first impulse should be to use one or more of those techniques. They are simple and involve no solution of simultaneous equations.

Perhaps the most important thing that must be done to avoid errors in using those reduction techniques properly is to label the element voltages and currents *as well as* the nodes of the circuit. Thus, when using the circuit reduction techniques, we will not be tempted to think that an element voltage or current has been solved for when, in fact, it has disappeared in the reduction. Furthermore, when we apply the stretch-and-bend principle to redraw a circuit in an equivalent but simpler form, the labeling not only of the element voltages and currents but also of the nodes in the original circuit will help to show when we have obtained an entirely different circuit, rather than one equivalent to the original circuit. If we should make a mistake at any stage of the reduction, such as redrawing a circuit in a form which we think is equivalent but is not, then we would be wasting our time to analyze this erroneous circuit

—for the task was not to analyze this new circuit but to analyze the original one or one equivalent to it. If we are careful to ensure that circuits equivalent to the original one are used, the previous techniques can be very powerful, simple, and useful.

There are, however, circuits which cannot be analyzed with the simple reduction techniques of the previous section. For these types of circuits, we have no recourse but to use the analysis techniques of this section—node voltages or mesh currents. Yet these two techniques involve the simultaneous solution of sets of algebraic equations and as such are prone to numerical errors. Consequently, if we adopt the attitude that we shall always attempt to use the reduction techniques of the previous sections and shall use the node voltage and mesh current techniques of this section *only as a last resort*, we will minimize our numerical errors.

3.6.1 Solution of Linear Simultaneous Equations—Cramer's Rule

Since the node voltage and mesh current methods involve writing (and solving) linear simultaneous algebraic equations, it is appropriate to examine methods for solving these types of equations.

A set of two linear simultaneous algebraic equations in two unknowns is written in the form

$$\begin{aligned} a_{11}x_1 + a_{12}x_2 &= b_1 \\ a_{21}x_1 + a_{22}x_2 &= b_2 \end{aligned} \tag{3.33}$$

Here x_1 and x_2 are the two unknowns to be solved for. The coefficients a_{11}, a_{12}, a_{21}, and a_{22} are known quantities. The first subscript on these quantities denotes the equation and the second subscript denotes the unknown which it multiplies. The two quantities on the right-hand side, b_1 and b_2, are also known quantities. A set of three linear simultaneous algebraic equations in the unknowns x_1, x_2, x_3 would appear as

$$\begin{aligned} a_{11}x_1 + a_{12}x_2 + a_{13}x_3 &= b_1 \\ a_{21}x_1 + a_{22}x_2 + a_{23}x_3 &= b_2 \\ a_{31}x_1 + a_{32}x_2 + a_{33}x_3 &= b_3 \end{aligned} \tag{3.34}$$

These equations are said to be linear since none of the unknowns x_1, x_2, x_3 appears in the equations as products of each other or raised to a power other than unity. If any equation contained, for example, x_1x_3 or x_2^3, then the entire set would be classified as a nonlinear set. Linear equations are considerably easier to solve than nonlinear ones. The equations are said to be algebraic since they do not involve derivatives of the unknowns, such as $dx_2(t)/dt$. Algebraic equations are also easier to solve than differential equations, which involve derivatives of the unknowns. Differential equations will result from adding the two final circuit elements, the inductor and capacitor, to our list of basic circuit elements. These will be considered in Chaps. 4, 5, and 6.

Probably the simplest technique for solving linear simultaneous algebraic equations is the method of Gauss elimination. Gauss elimination is also the technique which is used exclusively to solve these equations with a high-speed digital computer. The basic idea is to reduce the equations to an equivalent form which is triangular. For example, we would try to reduce the set of three equations in Eq. (3.34) to the equivalent, triangular form:

$$\begin{aligned} c_{11}x_1 + c_{12}x_2 + c_{13}x_3 &= d_1 \\ c_{22}x_2 + c_{23}x_3 &= d_2 \\ c_{33}x_3 &= d_3 \end{aligned} \tag{3.35}$$

If the solutions for x_1, x_2, x_3 of Eqs. (3.35) and (3.34) are the same, the equations are said to be equivalent. However, Eq. (3.35) is much easier to solve than is Eq. (3.34). Once the set is in triangular form, as in Eq. (3.35), we may use the "back substitution" technique. First solve the last equation for x_3:

$$x_3 = \frac{d_3}{c_{33}} \tag{3.36}$$

Then substitute into the second equation

$$c_{22}x_2 + c_{23}\frac{d_3}{c_{33}} = d_2 \tag{3.37}$$

and solve:

$$x_2 = \frac{1}{c_{22}}\left[-c_{23}\left(\frac{d_3}{c_{33}}\right) + d_2\right] \tag{3.38}$$

Now substitute x_3 from Eq. (3.36) and x_2 from Eq. (3.38) into the first equation of Eq. (3.35) and solve for x_1.

The final point we have to address is how to reduce the set in Eq. (3.34) to the equivalent triangular form in Eq. (3.35). We can multiply any equation in the set by a nonzero number without changing the solution; we can also add or subtract two equations and replace one of the two with the result. These are the only two techniques needed to reduce a set to equivalent triangular form. Once triangular form is achieved, we can solve for the unknowns by back substitution.

Example 3.15 Consider the set of linear simultaneous algebraic equations

$$\begin{aligned} x + 3y - z &= 4 \\ 2x - 4y + 2z &= 3 \\ 3x + y + z &= 2 \end{aligned}$$

Multiply the first equation by -2 and add the result to the second equation;

$$x + 3y - z = 4$$
$$0 - 10y + 4z = -5$$
$$3x + y + z = 2$$

Note that this has removed x from the second equation. Now multiply the first equation by -3 and add the result to the third equation:

$$x + 3y - z = 4$$
$$0 - 10y + 4z = -5$$
$$0 - 8y + 4z = -10$$

Thus, we have eliminated x from the third equation. Now let us eliminate y from the third equation and we will have achieved the triangular form. Multiply the second equation by $-\frac{8}{10}$ and add the result to the third equation:

$$x + 3y - z = 4$$
$$0 - 10y + 4z = -5$$
$$0 + 0 + \tfrac{4}{5}z = -6$$

Now solve by back substitution for the variables. The third equation yields

$$z = \frac{-6}{\frac{4}{5}}$$
$$= -\tfrac{15}{2}$$

Substituting this result into the second equation gives

$$-10y + 4(-\tfrac{15}{2}) = -5$$

from which we obtain

$$y = -\tfrac{5}{2}$$

Substituting these results into the first equation gives

$$x + 3(-\tfrac{5}{2}) - (-\tfrac{15}{2}) = 4$$

or

$$x = 4$$

We should substitute these supposed solutions into the original set to see if any numerical errors were made:

$$(4) + 3(-\tfrac{5}{2}) - (-\tfrac{15}{2}) \stackrel{?}{=} 4$$
$$2(4) - 4(-\tfrac{5}{2}) + 2(-\tfrac{15}{2}) \stackrel{?}{=} 3$$
$$3(4) + (-\tfrac{5}{2}) + (-\tfrac{15}{2}) \stackrel{?}{=} 2$$

This shows that our "solutions" do indeed satisfy the original equations. It is a good idea always to substitute back into the original equations to see whether we have obtained the solution.

A second technique which is useful for hand calculations is Cramer's rule. It is generally not used for digital computer calculations, since Gauss elimination is much faster. Cramer's rule requires the use of the concept of a determinant. To illustrate why this occurs, let us solve the set of two equations in Eq. (3.33) in general form by Gauss elimination. We obtain (the reader should verify this)

$$x_1 = \frac{a_{22}b_1 - a_{12}b_2}{a_{11}a_{22} - a_{12}a_{21}}$$
$$x_2 = \frac{a_{11}b_2 - a_{21}b_1}{a_{11}a_{22} - a_{12}a_{21}} \tag{3.39}$$

Note that the denominator of each solution involves

$$a_{11}a_{22} - a_{12}a_{21} \tag{3.40}$$

If we arrange the coefficients of Eq. (3.33) in the following array

$$\begin{bmatrix} a_{11} & a_{12} \\ a_{21} & a_{22} \end{bmatrix} \tag{3.41}$$

then Eq. (3.40) is said to be the determinant of the 2 × 2 array. The terms a_{11} and a_{22} are said to be the main diagonal terms of the array, whereas the terms a_{12} and a_{21} are said to be the off-diagonal terms. The determinant of the 2 × 2 array is denoted with vertical bars and is the product of the main diagonal terms minus the product of the off-diagonal terms:

$$\begin{vmatrix} a_{11} & a_{12} \\ a_{21} & a_{22} \end{vmatrix} = a_{11}a_{22} - a_{12}a_{21} \tag{3.42}$$

Now we see a simple rule for computing the solutions in Eq. (3.39):

$$x_1 = \frac{\begin{vmatrix} b_1 & a_{12} \\ b_2 & a_{22} \end{vmatrix}}{\begin{vmatrix} a_{11} & a_{12} \\ a_{21} & a_{22} \end{vmatrix}}$$

$$x_2 = \frac{\begin{vmatrix} a_{11} & b_1 \\ a_{21} & b_2 \end{vmatrix}}{\begin{vmatrix} a_{11} & a_{12} \\ a_{21} & a_{22} \end{vmatrix}} \tag{3.43}$$

This result is called Cramer's rule and we will generalize it to larger numbers of equations. Note that the solution for an unknown is the ratio of two determinants. The denominator determinant is the determinant of the array. The numerator determinant is formed from the coefficient array by replacing the column of the array corresponding to the unknown to be solved for with the right-hand side known quantities of the equations.

The extension of Cramer's rule to more than two equations is identical to the result for two equations but is slightly more involved in the evaluation of the resulting determinants. For example, Cramer's rule for the set of three equations in Eq. (3.34) gives the solutions as

$$x_1 = \frac{\begin{vmatrix} b_1 & a_{12} & a_{13} \\ b_2 & a_{22} & a_{23} \\ b_3 & a_{32} & a_{33} \end{vmatrix}}{\Delta}$$

$$x_2 = \frac{\begin{vmatrix} a_{11} & b_1 & a_{13} \\ a_{21} & b_2 & a_{23} \\ a_{31} & b_3 & a_{33} \end{vmatrix}}{\Delta} \tag{3.44}$$

$$x_3 = \frac{\begin{vmatrix} a_{11} & a_{12} & b_1 \\ a_{21} & a_{22} & b_2 \\ a_{31} & a_{32} & b_3 \end{vmatrix}}{\Delta}$$

where Δ is the 3×3 determinant of the array of coefficients:

$$\Delta = \begin{vmatrix} a_{11} & a_{12} & a_{13} \\ a_{21} & a_{22} & a_{23} \\ a_{31} & a_{32} & a_{33} \end{vmatrix} \tag{3.45}$$

Evaluation of each of these determinants is done in terms of the 2×2 determinants each contains. Select any row or column and multiply each element in that selected row or column by its cofactor. The sum of these products is the value of the determinant of the 3×3 array. A cofactor of a particular element appearing in the ith row and jth column is the product of $(-1)^{i+j}$ and the determinant formed by removing the ith row and jth column. For example, to evaluate the 3×3 determinant in Eq. (3.45) we may (arbitrarily) select the second column and the determinant is

$$\begin{aligned}\Delta = a_{12}(-1)^{1+2}\begin{vmatrix} a_{21} & a_{23} \\ a_{31} & a_{33} \end{vmatrix} \\ + a_{22}(-1)^{2+2}\begin{vmatrix} a_{11} & a_{13} \\ a_{31} & a_{33} \end{vmatrix} \\ + a_{32}(-1)^{3+2}\begin{vmatrix} a_{11} & a_{13} \\ a_{21} & a_{23} \end{vmatrix}\end{aligned} \tag{3.46}$$

The remaining 2×2 determinants are evaluated as before. We will consider determinants of no higher order than 3×3, so the above discussion suffices for Cramer's rule.

The sign of each cofactor can be found by using the simple "checkerboard sign pattern"

$$\begin{vmatrix} + & - & + \\ - & + & - \\ + & - & + \end{vmatrix} \tag{3.47}$$

Example 3.16 Consider the set of three equations in Example 3.15:

$$\begin{aligned} x + 3y - z &= 4 \\ 2x - 4y + 2z &= 3 \\ 3x + y + z &= 2 \end{aligned}$$

Solve for x, y, and z using Cramer's rule.

Solution The array of coefficients is

$$\begin{bmatrix} 1 & 3 & -1 \\ 2 & -4 & 2 \\ 3 & 1 & 1 \end{bmatrix}$$

The determinant of this array is found by expanding along the (arbitrarily selected) third row:

$$\begin{aligned}\Delta &= 3(-1)^{3+1}\begin{vmatrix} 3 & -1 \\ -4 & 2 \end{vmatrix} + 1(-1)^{3+2}\begin{vmatrix} 1 & -1 \\ 2 & 2 \end{vmatrix} + 1(-1)^{3+3}\begin{vmatrix} 1 & 3 \\ 2 & -4 \end{vmatrix} \\ &= 3(6-4) - 1(2+2) + 1(-4-6) \\ &= -8\end{aligned}$$

The solutions are therefore

$$x = \frac{\begin{vmatrix} 4 & 3 & -1 \\ 3 & -4 & 2 \\ 2 & 1 & 1 \end{vmatrix}}{-8} = \frac{-32}{-8} = 4$$

$$y = \frac{\begin{vmatrix} 1 & 4 & -1 \\ 2 & 3 & 2 \\ 3 & 2 & 1 \end{vmatrix}}{-8} = \frac{20}{-8} = -\tfrac{5}{2}$$

$$z = \frac{\begin{vmatrix} 1 & 3 & 4 \\ 2 & -4 & 3 \\ 3 & 1 & 2 \end{vmatrix}}{-8} = \frac{60}{-8} = -\tfrac{15}{2}$$

as were obtained by Gauss elimination in Example 3.15.

3.6.2 Node Voltage Analysis

Node voltage analysis is a method for obtaining a set of equations to be solved for a set of circuit voltages—the node voltages. Once these node voltages are obtained, we can easily obtain all branch voltages and currents from them.

For example, consider the circuit shown in Fig. 3.26*a*, which contains three nodes. Select one of the three nodes as a reference node. (The particular choice is arbitrary.) Next define the node voltages as the voltages of each of the remaining nodes *with respect to the reference node*, as shown in Fig. 3.26*b*. Next write KCL at each node in terms of these node voltages. For example, at node a in Fig. 3.26*a* we have

$$\frac{1}{R_1} V_a + \frac{1}{R_2}(V_a - V_b) = i_{s1} - i_{s2} \tag{3.48}$$

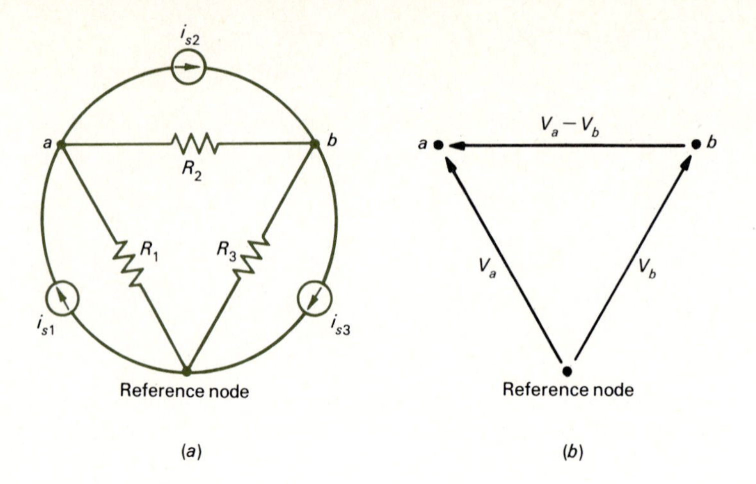

FIGURE 3.26 Definition of node voltages.

and at node b.

$$\frac{1}{R_3} V_b + \frac{1}{R_2}(V_b - V_a) = i_{s2} - i_{s3} \tag{3.49}$$

Rewriting Eqs. (3.48) and (3.49) by grouping terms multiplying each node voltage, we have

node a:

$$\left(\frac{1}{R_1} + \frac{1}{R_2}\right)V_a - \frac{1}{R_2} V_b = i_{s1} - i_{s2} \tag{3.50}$$

node b:

$$-\frac{1}{R_2} V_a + \left(\frac{1}{R_3} + \frac{1}{R_2}\right)V_b = i_{s2} - i_{s3}$$

These equations can be solved by Gauss elimination or Cramer's rule for the node voltages V_a and V_b. Once these are obtained, all other branch voltages may be obtained in terms of the node voltages. For example, the branch voltage across R_1 is V_a and across R_3 is V_b. The branch voltage across R_2 is the difference $V_a - V_b$. Once all branch voltages are determined, the branch currents are obtained by Ohm's law.

We may set up the node voltage equations for a circuit by inspection. Note the pattern in Eq. (3.50). When we write the equation at a particular node, the coefficient of that node voltage is the sum of the conductances connected to that node. The coefficient of one of the other node voltages in that equation is negative and is equal to the conductance connected between the two nodes. The right-hand side of that equation is the sum of the values of the ideal current sources attached to that node. They appear as positive in this sum if they are pointing to the node and as negative if they are pointing away from this node.

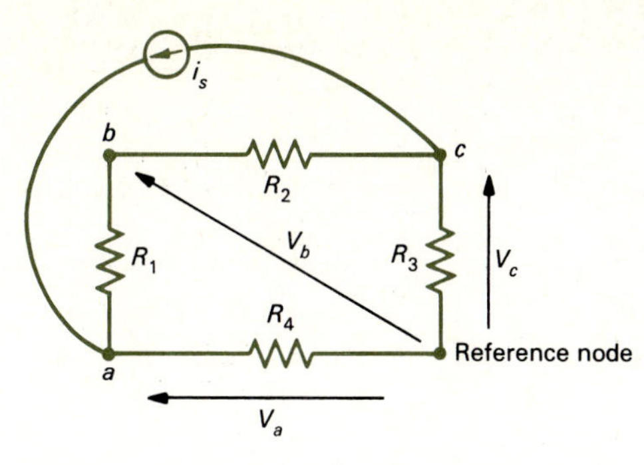

FIGURE 3.27 Illustration of writing node voltage equations.

This result can be generalized to larger circuits. For example, consider the circuit in Fig. 3.27. In terms of the node voltages shown, we write, by inspection,

node a:

$$\left(\frac{1}{R_1}+\frac{1}{R_4}\right)V_a-\frac{1}{R_1}V_b=i_s$$

node b:

$$-\frac{1}{R_1}V_a+\left(\frac{1}{R_1}+\frac{1}{R_2}\right)V_b-\frac{1}{R_2}V_c=0 \tag{3.51}$$

node c:

$$-\frac{1}{R_2}V_b+\left(\frac{1}{R_2}+\frac{1}{R_3}\right)V_c=-i_s$$

Apply the general rule developed above to obtain Eq. (3.51).

If the ideal sources in a circuit are not all current sources, we may convert voltage sources to current sources with a Thevenin-to-Norton transformation. For example, consider the circuit shown in Fig. 3.28*a*. First convert the voltage source v_{s1} in series with R_1 to a current source $i_{s1}=v_{s1}/R_1$ in parallel with R_1, as shown in Fig. 3.28*b*.

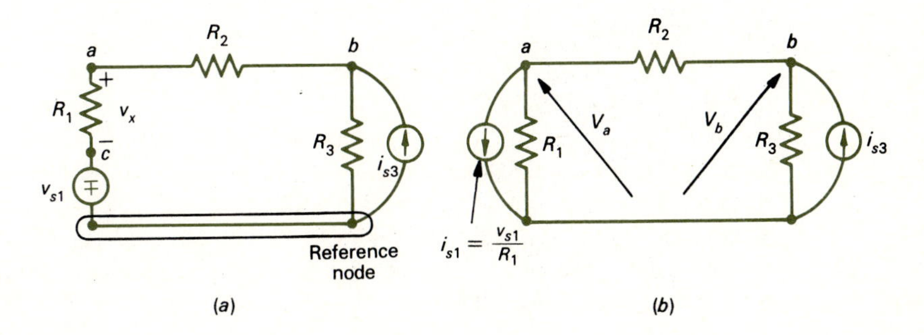

FIGURE 3.28 Illustration of writing node voltage equations for circuits containing voltage sources.

Note that node c has disappeared in this reduction. Writing the node voltage equations for this equivalent circuit, we obtain

node a:

$$\left(\frac{1}{R_1}+\frac{1}{R_2}\right)V_a-\frac{1}{R_2}V_b=-i_{s1}=-\frac{v_{s1}}{R_1}$$

node b: (3.52)

$$-\frac{1}{R_2}V_a+\left(\frac{1}{R_2}+\frac{1}{R_3}\right)V_b=i_{s3}$$

Once these are solved for V_a and V_b, we can calculate voltages v_x across R_1 from Fig. 3.28a as

$$v_x = V_a + v_{s1} \tag{3.53}$$

Example 3.17 Determine the current i in Fig. 3.29a.

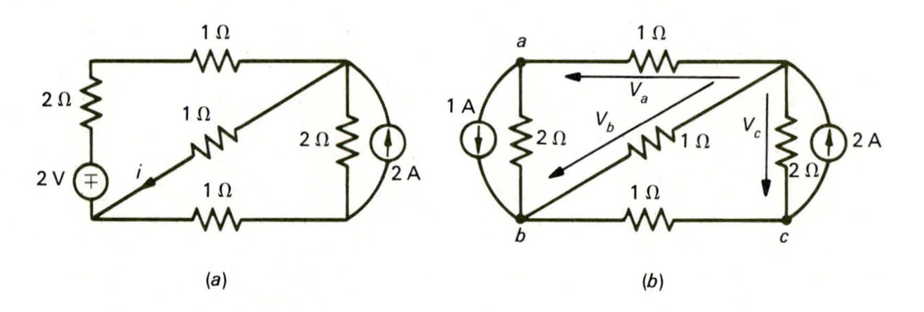

FIGURE 3.29 Example 3.17: writing node voltage equations for circuits with mixed sources.

Solution The voltage source is converted to a current source and the node voltages are labeled as shown in Fig. 3.29b. The node voltage equations become

node a:

$$(\tfrac{1}{2}+1)V_a-\tfrac{1}{2}V_b=-1$$

node b:

$$-\tfrac{1}{2}V_a+(\tfrac{1}{2}+1+1)V_b-1V_c=1$$

node c:

$$-1V_b+(1+\tfrac{1}{2})V_c=-2$$

We only need to determine V_b since

$$i = -\frac{V_b}{1\ \Omega}$$

By Cramer's rule

$$V_b = \frac{\begin{vmatrix} \frac{3}{2} & -1 & 0 \\ -\frac{1}{2} & 1 & -1 \\ 0 & -2 & \frac{3}{2} \end{vmatrix}}{\begin{vmatrix} \frac{3}{2} & -\frac{1}{2} & 0 \\ -\frac{1}{2} & \frac{5}{2} & -1 \\ 0 & -1 & \frac{3}{2} \end{vmatrix}}$$

$$= \frac{\frac{3}{2}\begin{vmatrix} 1 & -1 \\ -2 & \frac{3}{2} \end{vmatrix} - (-1)\begin{vmatrix} -\frac{1}{2} & -1 \\ 0 & \frac{3}{2} \end{vmatrix}}{\frac{3}{2}\begin{vmatrix} \frac{5}{2} & -1 \\ -1 & \frac{3}{2} \end{vmatrix} - (-\frac{1}{2})\begin{vmatrix} -\frac{1}{2} & -1 \\ 0 & \frac{3}{2} \end{vmatrix}}$$

$$= \frac{\frac{3}{2}(\frac{3}{2} - 2) + 1(-\frac{3}{4})}{\frac{3}{2}(\frac{15}{4} - 1) + \frac{1}{2}(-\frac{3}{4})}$$

$$= \frac{-\frac{6}{4}}{\frac{30}{8}}$$

$$= -\tfrac{2}{5}\ \text{V}$$

where we have chosen (arbitrarily) to expand both determinants along the first row. Thus

$$i = -\frac{V_b}{1\ \Omega}$$

$$= \tfrac{2}{5}\ \text{A}$$

3.6.3 Mesh Current Analysis

The mesh current analysis method, like the node voltage analysis method, generates a set of simultaneous equations. The solution to this set of equations can be used to find the solutions for all branch voltages and currents of the circuit. In the mesh current method we use a set of mesh currents, whereas in the node voltage method we used a set of node voltages.

A mesh current is a fictitious current which is defined to circulate around a mesh of the circuit. A circuit mesh can be thought of as panes in a window. If we draw the circuit on a piece of paper with no branches crossing each other, the meshes are circuit loops which do not encircle other circuit elements; examples are given in Fig. 3.30*a*. Not all circuits can be laid out to contain only meshes. Those which cannot are called nonplanar circuits; an example is shown in Fig. 3.30*b*. For nonplanar circuits, the mesh current analysis method which we will discuss cannot be used, but the node voltage method discussed previously can be used.

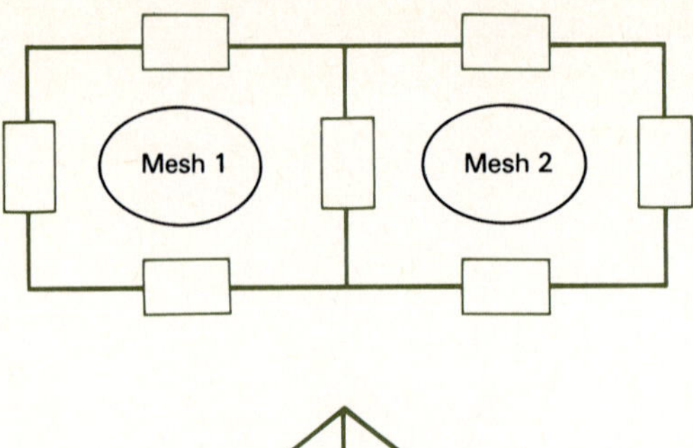

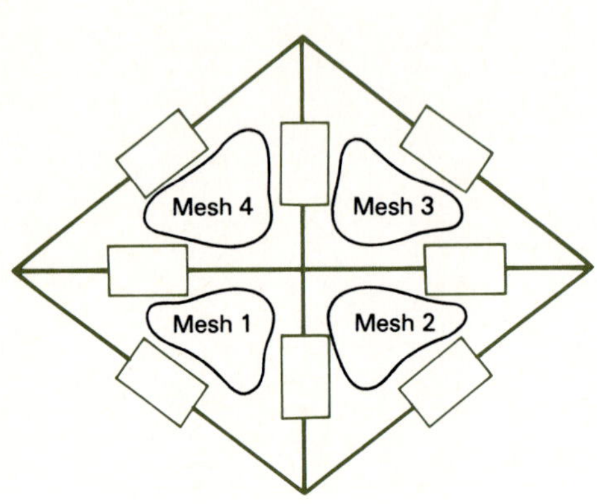

(*a*) Meshes of a circuit

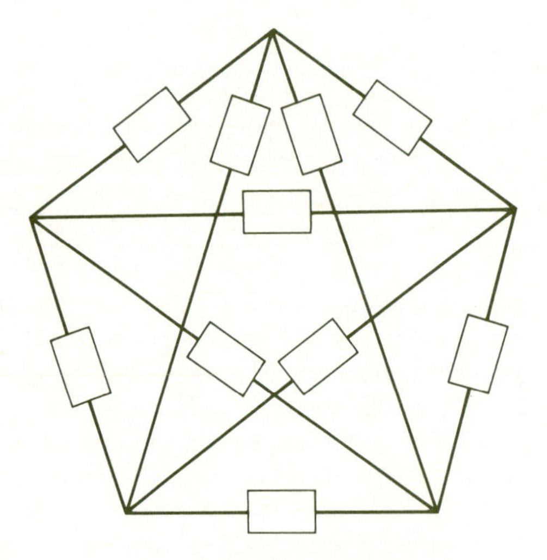

(*b*) Nonplanar circuit

FIGURE 3.30 Defining meshes of a circuit: (*a*) examples of mesh definition; (*b*) a nonplanar circuit for which meshes cannot be defined.

A mesh current is a fictitious current defined to circulate clockwise around only that mesh. Counterclockwise circulation could have been used, but we will always assume clockwise circulation for the mesh currents. An example is shown in Fig. 3.31. Note that the current through R_2 is the difference of the two mesh currents, since both mesh currents flow through this branch but in opposite directions. Now write KVL around each mesh:

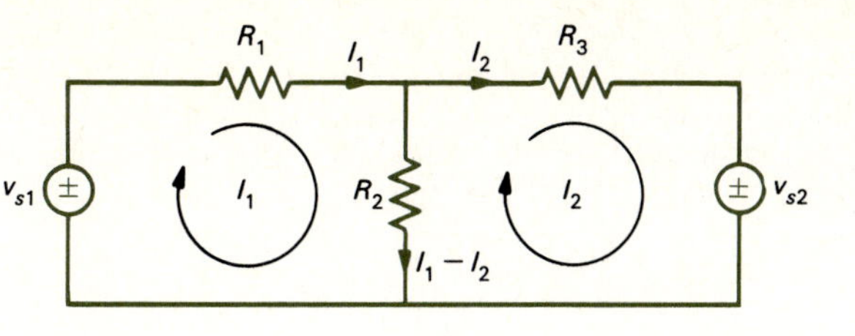

FIGURE 3.31 Illustration of writing mesh current equations.

mesh 1:

$$R_1I_1 + R_2(I_1 - I_2) = v_{s1} \tag{3.54}$$

mesh 2:

$$R_3I_2 + R_2(I_2 - I_1) = -v_{s2}$$

Group coefficients of I_1 and I_2 and write Eq. (3.54) as

mesh 1:

$$(R_1 + R_2)I_1 - R_2I_2 = v_{s1} \tag{3.55}$$

mesh 2:

$$-R_2I_1 + (R_2 + R_3)I_2 = -v_{s2}$$

Note that each equation for a mesh has, on the left-hand side of the equation, the sum of the resistances common to that mesh multiplying that mesh current minus the product of the adjacent mesh current and the resistance common to both meshes —and on the right-hand side the sum of any voltage sources encountered in traversing the mesh. If the voltage source tends to "push" in the direction of the mesh current, it appears in the sum on the right-hand side as a positive quantity, and as a negative quantity if it tends to oppose the mesh current. We will find this to be a general rule.

A slightly more complicated example is shown in Fig. 3.32. For this circuit, with the mesh currents as indicated, we obtain

mesh 1:

$$(R_1 + R_2 + R_4)I_1 - R_2I_2 - R_4I_3 = v_{s1} - v_{s3}$$

mesh 2:

$$-R_2I_1 + (R_2 + R_3 + R_5)I_2 - R_5I_3 = v_{s2} + v_{s3} \tag{3.56}$$

mesh 3:

$$-R_4I_1 - R_5I_2 + (R_4 + R_5 + R_6)I_3 = -v_{s2}$$

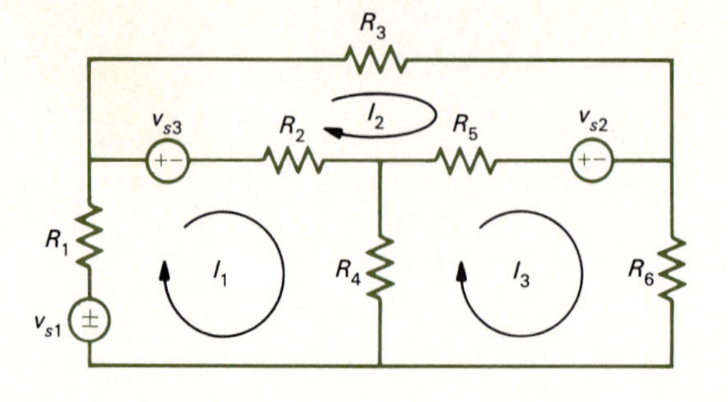

FIGURE 3.32 Illustration of writing mesh current equations for more complicated circuits.

Note that the rule for writing the mesh equations by inspection, as developed previously for the two-mesh circuit in Fig. 3.31, works here. For example, in writing the mesh equation for mesh 3, $R_4 + R_5 + R_6$ is the common resistance around this mesh, while R_4 is shared with mesh 1 and R_5 is shared with mesh 2. Also note that v_{s2} is encountered in traversing the mesh but that it tends to oppose I_3; it is therefore entered as a minus quantity on the right-hand side of that mesh equation.

In order to write mesh equations, all sources must be voltage sources. If some are current sources, they can be converted to voltage sources with a Norton-to-Thevenin transformation.

Example 3.18 Using mesh currents, determine the current i in the circuit of Example 3.17 (Fig. 3.29*a*).

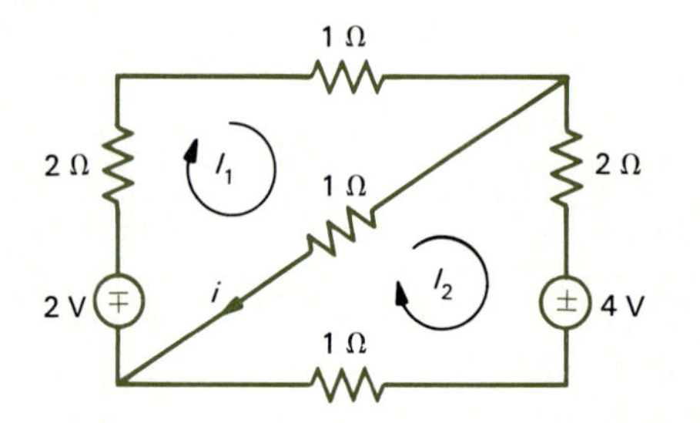

FIGURE 3.33 Example 3.18.

Solution The 2-A current source is converted to a voltage source as shown in Fig. 3.33, and mesh currents are defined as shown. The mesh current equations become

mesh 1:

$$(2 + 1 + 1)I_1 - 1I_2 = -2$$

mesh 2:

$$-1I_1 + (2 + 1 + 1)I_2 = -4$$

Here we must solve for both $I_1 - I_2$ since

$$i = I_1 - I_2$$

By Cramer's rule we obtain

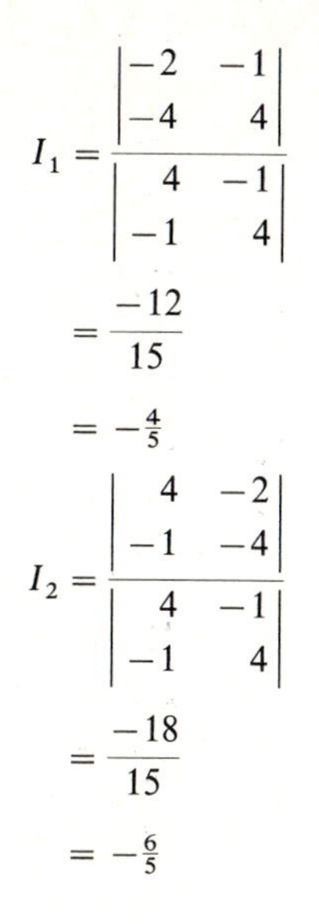

$$I_1 = \frac{\begin{vmatrix} -2 & -1 \\ -4 & 4 \end{vmatrix}}{\begin{vmatrix} 4 & -1 \\ -1 & 4 \end{vmatrix}} = \frac{-12}{15} = -\tfrac{4}{5}$$

$$I_2 = \frac{\begin{vmatrix} 4 & -2 \\ -1 & -4 \end{vmatrix}}{\begin{vmatrix} 4 & -1 \\ -1 & 4 \end{vmatrix}} = \frac{-18}{15} = -\tfrac{6}{5}$$

Thus

$$i = I_1 - I_2 = \tfrac{2}{5}\ \text{A}$$

as was obtained in Example 3.17 by the node voltage method.

3.6.4 Comparison of Solution Difficulty

When deciding whether to use node voltage or mesh current analysis for a circuit, it is important to determine which method will result in the smallest number of simultaneous equations. To do so, first visualize converting all voltage sources to current sources and count the number of existing nodes N. The number of node voltage equations will be $N - 1$. Then visualize converting all current sources to voltage sources and count the number of meshes M. The number of mesh current equations will then be M. If $M < N - 1$, use mesh current analysis; if $M > N - 1$, use node voltage analysis. For instance, for the circuit of Fig. 3.29 used in Examples 3.17 and 3.18 the number of node voltage equations was three and the number of mesh current equations was two. Thus, it was simpler to use mesh current analysis even though we had to solve for two mesh currents and only one node voltage. However, some thought will reveal that we need not have solved for two mesh currents if we had redrawn the circuit in Fig. 3.33 using the stretch and bend principle so that the branch containing i was on the outside of a mesh, such that only one mesh current passed through it.

It is worthwhile to close this section by reiterating that *whenever possible* avoid the use of node voltage or mesh current analysis—since the numerical errors one is likely to make in solving large systems of simultaneous equations by Cramer's rule will be eliminated. The circuit considered in Examples 3.17 and 3.18 can be solved by circuit reduction techniques.

Example 3.19 Solve the circuit in Fig. 3.29*a* (considered in Examples 3.17 and 3.18) by the use of circuit-reduction techniques.

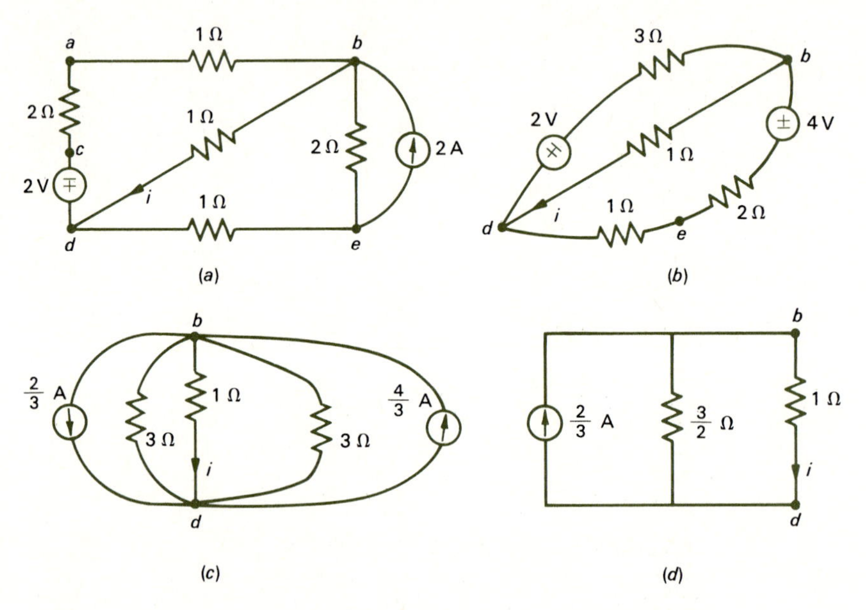

FIGURE 3.34 Example 3.19: circuit reduction as an alternative to writing mesh current or node voltage equations.

Solution The circuit is shown with the nodes labeled (an important first step in using reduction techniques) in Fig. 3.34*a*. First convert the circuit as shown in Fig. 3.34*b* through *d*. Then in Fig. 3.34*d* use current division to obtain

$$
\begin{aligned}
i &= \frac{\frac{3}{2}\,\Omega}{\frac{3}{2}\,\Omega + 1\,\Omega}\,\tfrac{2}{3}\,\text{A} \\
&= \tfrac{2}{5}\,\text{A}
\end{aligned}
$$

as was obtained with mesh current and node voltage analysis. But here we did not need to solve any simultaneous equations; we simply drew several equivalent circuits.

Alternatively, let us convert the circuit attached to the branch containing i to a Thevenin equivalent. This is shown in Fig. 3.35. V_{OC} is found in Fig. 3.35*b* by superposition and by voltage and current division to be

$$
\begin{aligned}
V_{OC} &= -\frac{3\,\Omega}{3\,\Omega + 3\,\Omega}\,2\,\text{V} + \left(\frac{2\,\Omega}{2\,\Omega + 4\,\Omega}\,2\,\text{A}\right)3\,\Omega \\
&= 1\,\text{V}
\end{aligned}
$$

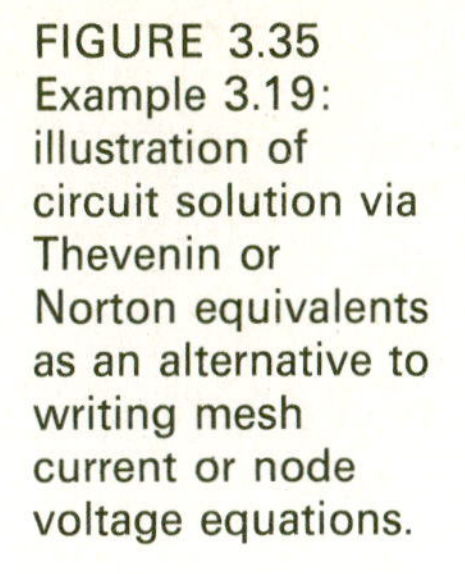

FIGURE 3.35 Example 3.19: illustration of circuit solution via Thevenin or Norton equivalents as an alternative to writing mesh current or node voltage equations.

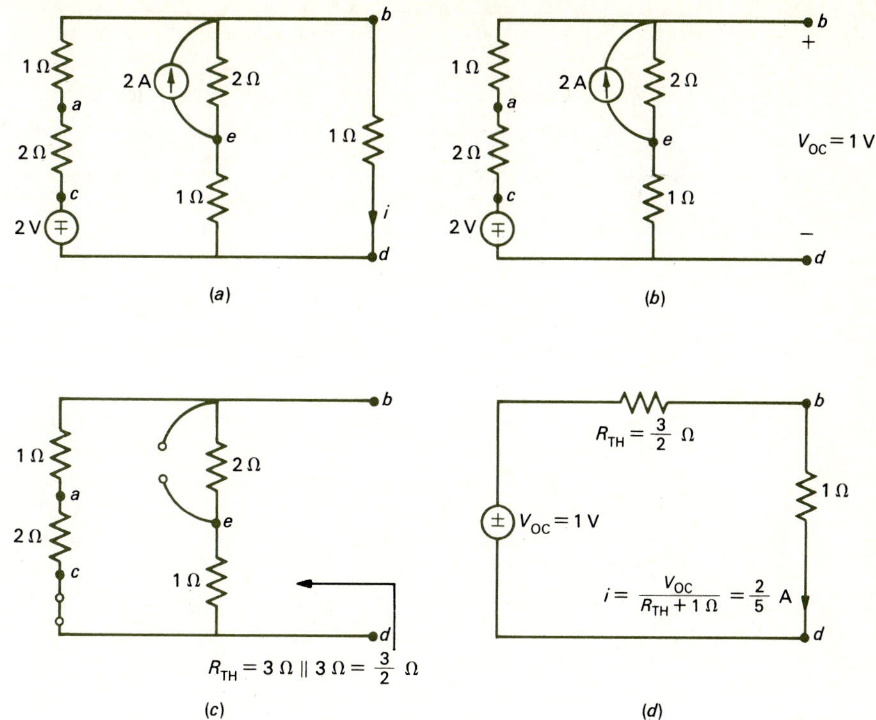

and R_{TH} is found from Fig. 3.35*c* to be

$$R_{TH} = \tfrac{3}{2}\,\Omega$$

The Thevenin equivalent is reattached to terminals *b*-*d* in Fig. 3.35*d* and *i* is easily determined to be

$$\begin{aligned} i &= \frac{V_{OC}}{R_{TH} + 1\,\Omega} \\ &= \tfrac{2}{5}\,\text{A} \end{aligned}$$

as before.

This last example clearly illustrates that there are many ways to solve a circuit. It is important to understand all of the techniques of this chapter so that the simplest ones for a particular circuit can be used.

One final example will be given to illustrate that mesh current and node voltage analysis should be used as a last resort. Of course, there are circuits for which one has no other recourse than to use those techniques.

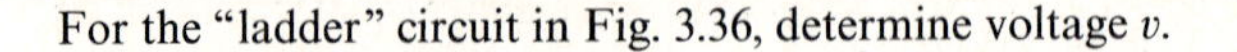

Example 3.20 For the "ladder" circuit in Fig. 3.36, determine voltage v.

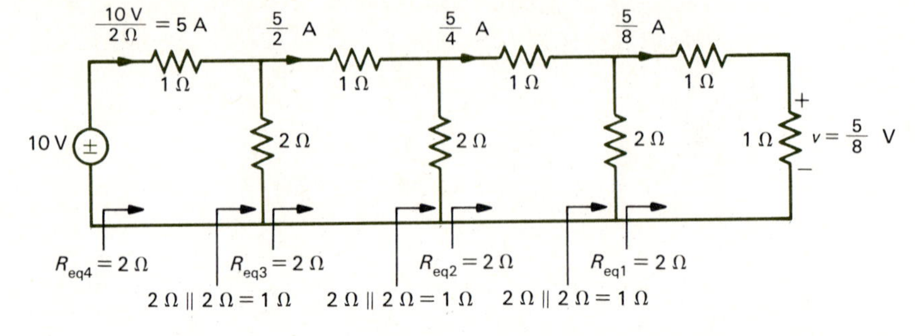

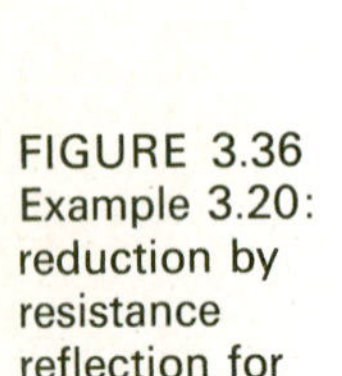

FIGURE 3.36 Example 3.20: reduction by resistance reflection for ladder circuits.

Solution Here, for example, one may be tempted to use mesh current analysis, but this would involve the solution of four simultaneous mesh current equations. A simpler method is to work back to the source by using series-parallel resistor reduction to obtain the resistance seen by the source, which is $R_{eq\,4} = 2\,\Omega$. Thus, the current leaving the 10-V is 5 A. This current divides equally between $2\,\Omega$ and $R_{eq\,3} = 2\,\Omega$ as $\frac{5}{2}$, and this divides between $2\,\Omega$ and $R_{eq\,2} = 2\,\Omega$ as $\frac{5}{4}$, and so on, such that the current in the final branch is $\frac{5}{8}$ A and the required voltage is $v = \frac{5}{8} \times 1\,\Omega = \frac{5}{8}$ V. This is a much simpler method and less prone to numerical errors than mesh current or node voltage would have been.

3.7 CONSERVATION OF POWER IN RESISTIVE CIRCUITS

At any instant of time the power delivered *by* the ideal sources must equal the power delivered *to* the resistors of a circuit. This is a rather obvious but important result. Thus, KVL and KCL must be such that this is satisfied, and they are therefore not independent laws but are somehow related.

Example 3.21 Show that conservation of power is satisfied for the circuit in Fig. 3.37*a*.

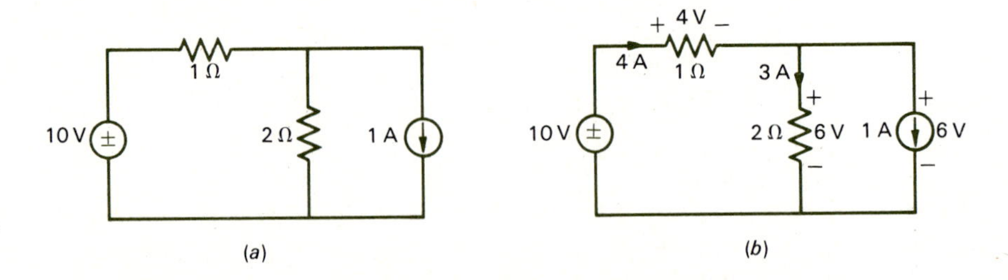

FIGURE 3.37 Example 3.21: illustration of conservation of power in resistive circuits.

Solution The resistor voltages and currents can be obtained by using superposition, Ohm's law, and current division, as shown in Fig. 3.37*b*. The 10-V voltage source delivers 10 V × 4 A = 40 W to the circuit, whereas the 1-A current source delivers −6 V × 1 A = −6 W to the circuit and is actually absorbing 6 W. The two resistors absorb $(4\text{ A})^2 \times 1\,\Omega + (3\text{ A})^2 \times 2\,\Omega =$ 34 W of power and conservation of power is achieved.

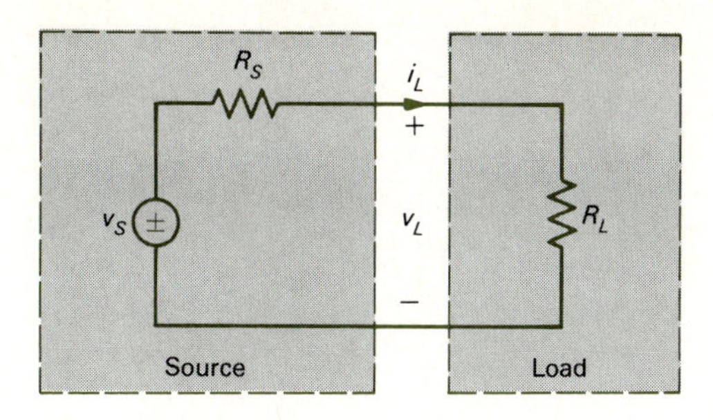

FIGURE 3.38 Maximum power transfer from a source to a load.

With regard to power, one final point should be mentioned. All practical sources have internal resistance, represented as R_S in Fig. 3.38. Therefore, when this source is connected to a load, represented by R_L, all of the open-circuit voltage of the source, v_S, will not appear across R_L. Some of v_S will be dropped across R_S, $i_L R_S$. The question now arises as to what would be the optimum choice for R_L (if we have a choice) such that maximum power will be delivered from the source to the load. Ordinarily, we do not have any choice over R_S since it is the equivalent internal resistance of the source. Let us determine the optimum value of R_L to achieve maximum power transfer from the source to the load. The load current is

$$i_L = \frac{v_S}{R_S + R_L} \tag{3.57}$$

and the power delivered to the load is

$$\begin{aligned} p_L &= i_L^2 R_L \\ &= \frac{R_L}{(R_S + R_L)^2} v_S^2 \end{aligned} \tag{3.58}$$

The important question here is what value of R_L will maximize this expression. This is found by differentiating p_L with respect to R_L and setting the result equal to zero:

$$\begin{aligned} \frac{dp_L}{dR_L} &= 0 \\ &= \frac{(R_S + R_L)^2 - 2R_L(R_S + R_L)}{(R_S + R_L)^4} v_S^2 \end{aligned} \tag{3.59}$$

To find the value of R_L which satisfies this, we simply set the numerator equal to zero and obtain

$$R_L = R_S \tag{3.60}$$

Thus, for maximum power transfer to a load, we should choose the load resistor to be equal to the internal resistance of the source.

PROBLEMS

3.1 In the circuits shown in Fig. P3.1, determine the equivalent resistance at terminals *a-b*.

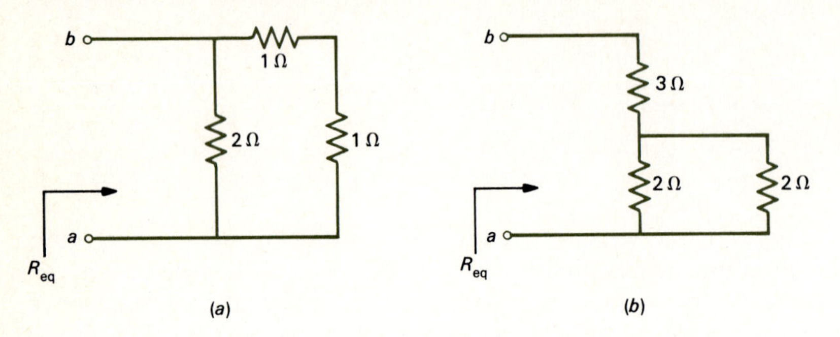

(*a*) (*b*)

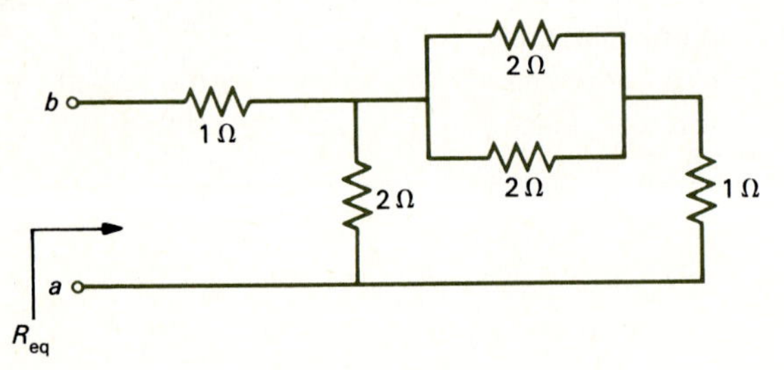

(*c*)

FIGURE P3.1

3.2 In the circuits of Fig. P3.2, determine the equivalent resistances seen at terminals *a-b*.

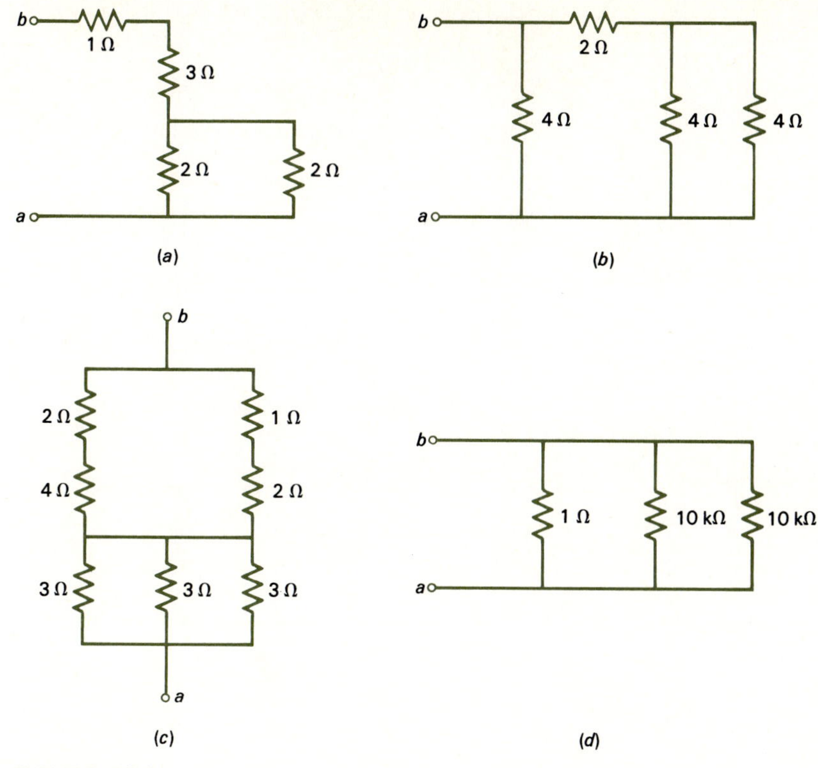

FIGURE P3.2

3.3 In the circuit shown in Fig. P3.3, determine the current i_x and voltage v_x.

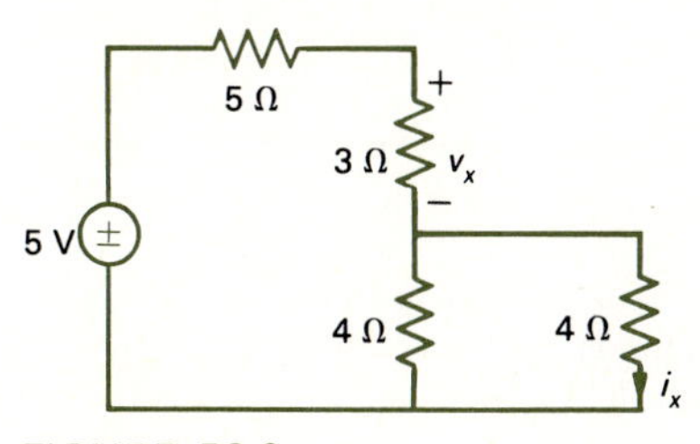

FIGURE P3.3

3.4 In the circuit in Fig. P3.4, determine current i_x and voltage v_x.

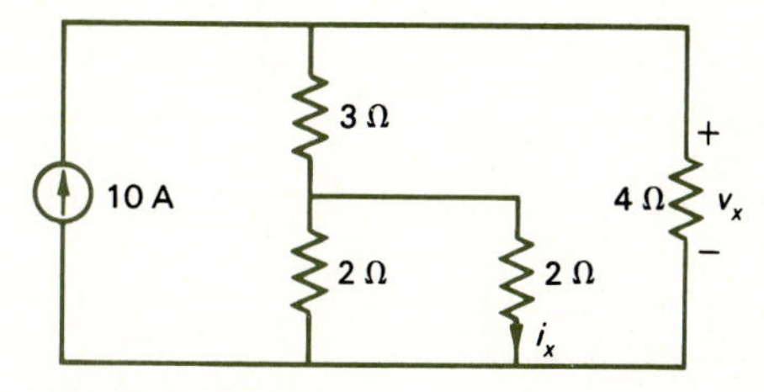

FIGURE P3.4

3.5 A 10-A current source is in parallel with 10 identical 1-Ω resistors. Determine the current in each resistor.

3.6 Repeat Prob. 3.5 if the resistors are all 2-Ω.

3.7 A 10-V voltage source is in series with 20 identical 1-Ω resistors. Determine the voltages across each resistor.

3.8 Determine current i_x in the circuit in Fig. P3.8 by superposition.

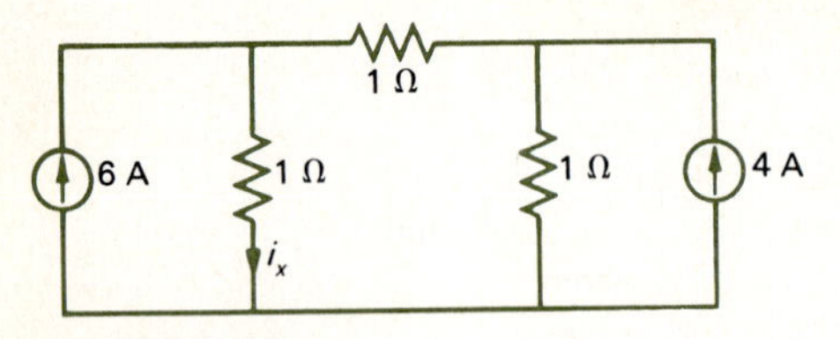

FIGURE P3.8

3.9 Replace the 4-A current source in Fig. P3.8 with a 4-V voltage source and determine i_x by superposition.

3.10 Determine voltage v_x in the circuit shown in Fig. P3.10.

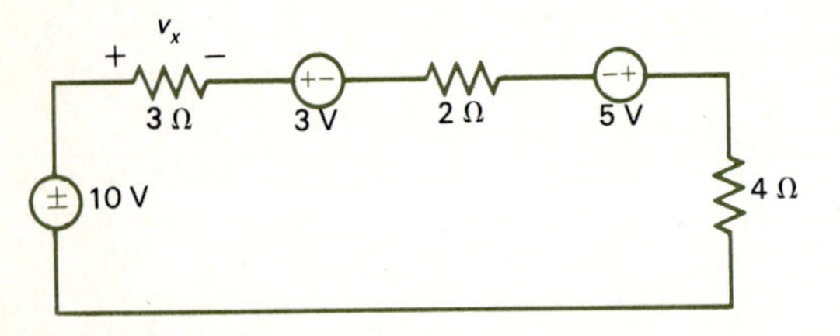

FIGURE P3.10

3.11 Determine current i_x in the circuit shown in Fig. P3.11.

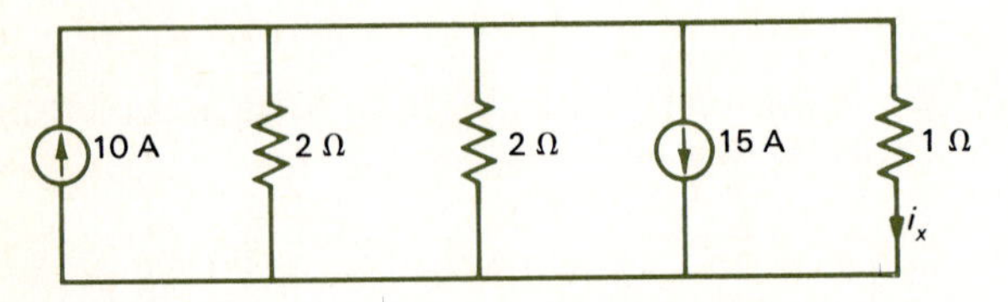

FIGURE P3.11

3.12 Determine current i_x in the circuit of Fig. P3.12.

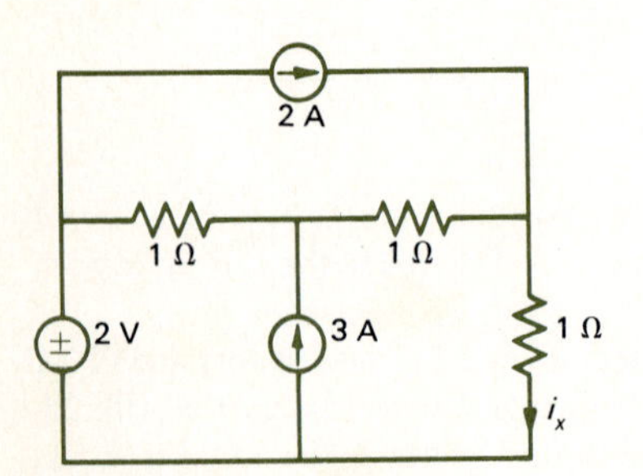

FIGURE P3.12

3.13 Determine a Thevenin equivalent circuit at terminals *a-b* for the circuit in Fig. P3.13.

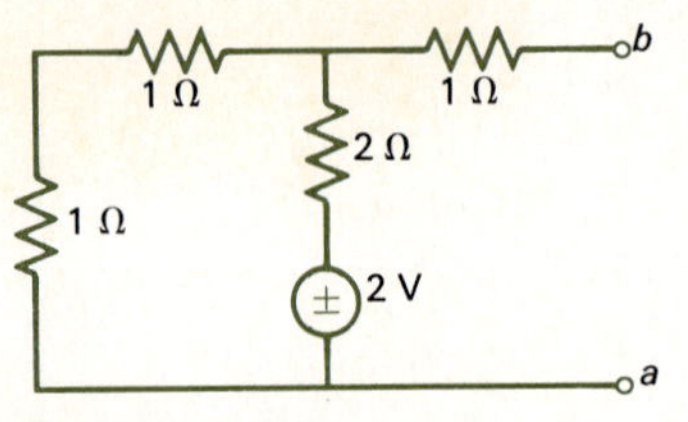

FIGURE P3.13

3.14 Determine a Norton equivalent circuit at terminals *a-b* in the circuit of Fig. P3.13. *Note*: Compute I_{SC} directly and verify that $I_{SC} = V_{OC}/R_{TH}$.

3.15 Reduce the circuit in Fig. P3.15 to the left of terminals *a-b* to a Thevenin equivalent and compute i_x. Compute i_x directly, using superposition.

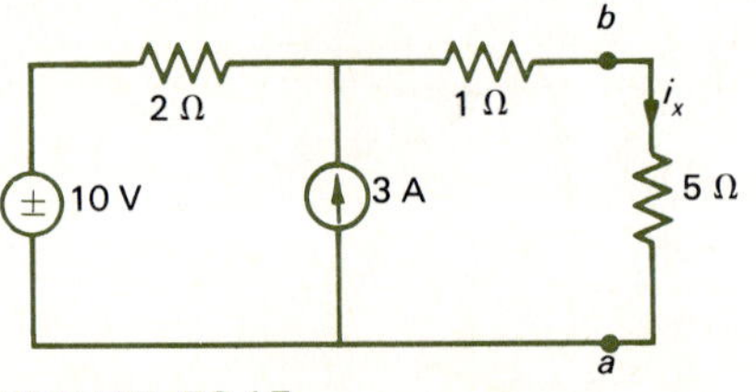

FIGURE P3.15

3.16 In the circuit shown in Fig. P3.16, determine a Thevenin equivalent circuit to the left of terminals *a-b* and compute v_x.

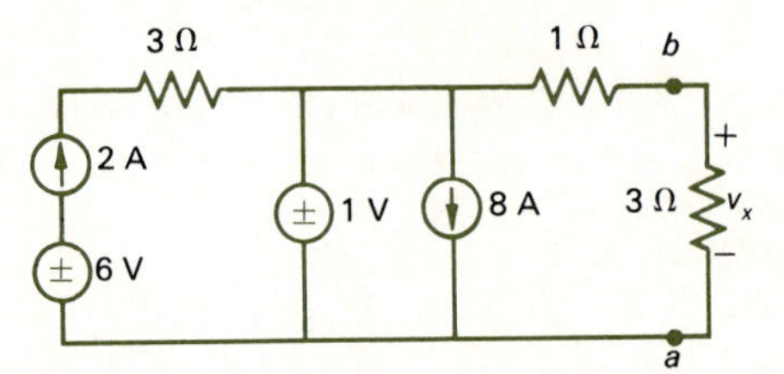

FIGURE P3.16

3.17 Solve for voltage v_x in the circuit of Fig. P3.17, using the node voltage method. Verify your result using superposition.

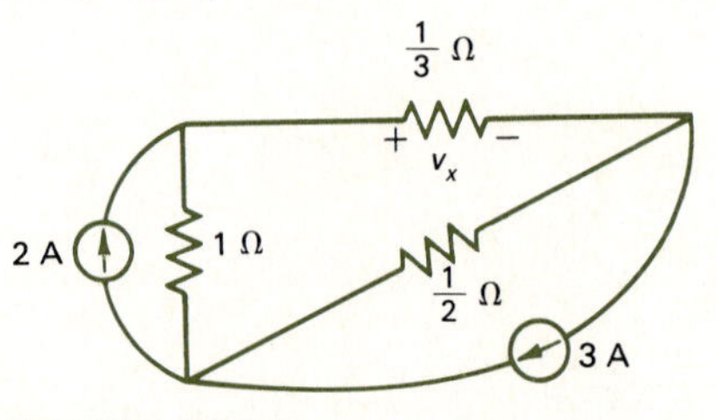

FIGURE P3.17

3.18 Determine voltage v_x in the circuit shown in Fig. P3.18 by the mesh current method. Verify your result using superposition.

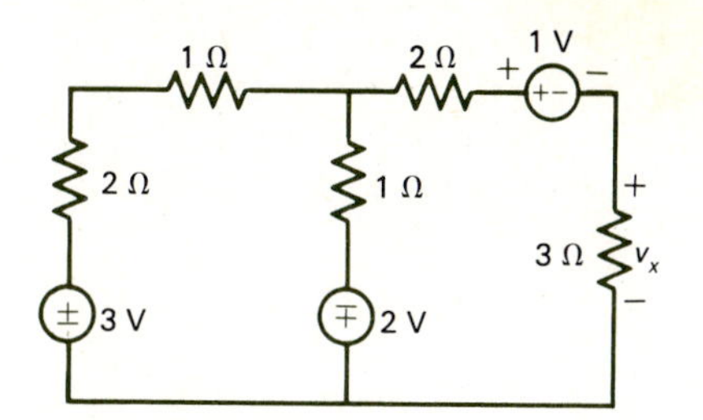

FIGURE P3.18

3.19 Solve for current i_x in the circuit of Fig. P3.19 by using (*a*) the node voltage method and (*b*) the mesh current method. Check your result with superposition.

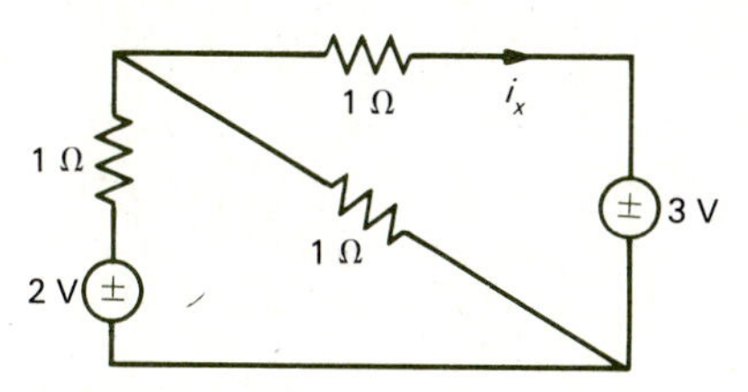

FIGURE P3.19

3.20 In the circuit in Fig. P3.3, verify that the power delivered by the voltage source equals the power delivered to the resistors.

3.21 In the circuit of Fig. P3.8, verify that the power delivered by the current sources equals the power delivered to the resistors.

3.22 In the circuit of Fig. P3.12, verify that the power delivered by the ideal sources equals the power delivered to the resistors.

3.23 For the circuit shown in Fig. P.3.23, determine the value of R such that maximum power is dissipated in it. Determine that maximum power.

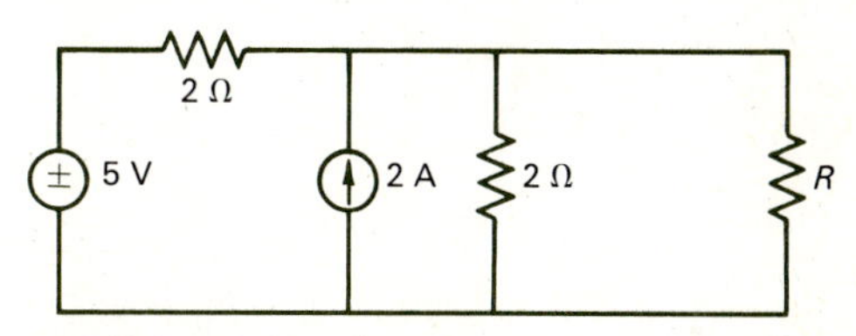

FIGURE P3.23

3.24 In the circuit shown in Fig. 3.38, suppose v_S and R_L to be fixed. What value of R_S would result in maximum power being dissipated in R_L?

REFERENCES

3.1 A. B. Carlson and D. G. Gisser, *Electrical Engineering: Concepts and Applications*, Addison-Wesley, Reading, Mass., 1981.

3.2 W. H. Hayt, Jr., and J. E. Kemmerly, *Engineering Circuit Analysis*, 3d ed., McGraw-Hill, New York, 1978.

3.3 J. A. Edminister, *Schaum's Outline of Electric Circuits*, 2d ed., McGraw-Hill, New York, 1983.

3.4 J. D. Irwin, *Basic Engineering Circuit Analysis*, Macmillan, New York, 1983.

3.5 R. L. Boylestad, *Introductory Circuit Analysis*, 4th ed., Charles E. Merrill, Columbus, Ohio, 1982.

Chapter

4

The Energy Storage Elements

In the previous two chapters we studied circuits composed only of ideal voltage and current sources and resistors. The circuit equations governing the branch voltages and currents—KVL, KCL, and the element relations—were algebraic equations primarily because the resistor voltage and current were related by an algebraic equation: $v = Ri$. Recall that the resistor constantly absorbed power: $p = i^2R = v^2/R$.

In this chapter we will study the two remaining basic circuit elements—the capacitor and the inductor. Unlike the resistor, these elements store energy and their voltages and currents are related by differential equations. Therefore, whenever we include one of these elements in a circuit, the circuit equations governing the branch voltages and currents become differential equations, which are more difficult to solve than the algebraic circuit equations encountered with resistive circuits.

Although circuits containing these energy storage elements are more difficult to analyze than resistive circuits, they allow us to build circuits which provide a greater variety of signal processing than is attainable with resistive circuits. An example of this benefit is in the construction of electric filters, which are considered in the next chapter.

4.1 THE CAPACITOR

The first energy storage element which we will consider is the capacitor. Suppose we place two rectangular metal plates in close proximity and, via wires, attach a battery to the plates as shown in Fig. 4.1. When the battery is connected to the plates, charge Q is transferred from the battery to the plates; a positive amount is present on the plate attached to the positive battery, and an equal but negative amount is placed on the other plate. Thus, the plates separate charge. This charge and voltage between the

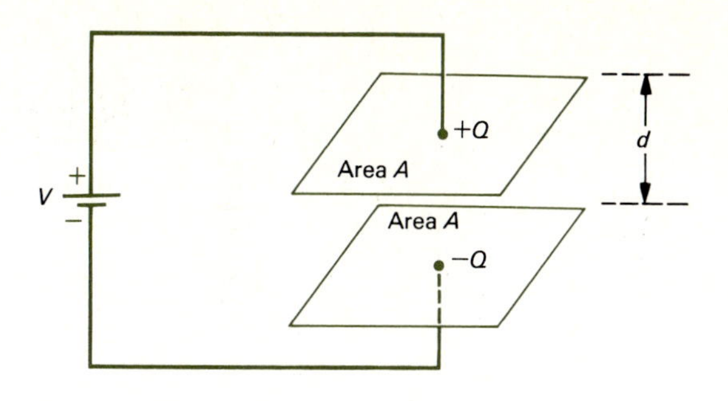

FIGURE 4.1
A capacitor.

plates (provided here by the battery) are related by the capacitance of this capacitor:

$$Q = CV \tag{4.1}$$

The quantity C is the capacitance of the capacitor whose units are, according to Eq. (4.1), coulombs per volt. This combination of units is given the name of farads (F), where F = C/V.

For the parallel-plate capacitor shown in Fig. 4.1 with plates of area A and separation D in air, the capacitance may be computed from

$$C = 8.84 \times 10^{-12} \frac{A}{d} \tag{4.2}$$

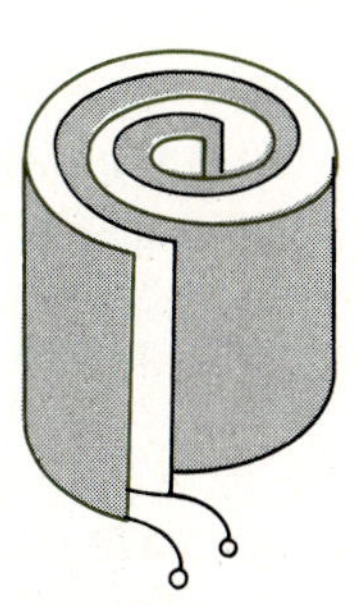

FIGURE 4.2
Physical construction of a capacitor to occupy a small space.

The units of capacitance for practical capacitors are usually given in microfarads (μF $= 10^{-6}$ F) or picofarads (pF $= 10^{-12}$ F) since a 1-F capacitor would be equivalent to a separation of 1 m between two plates, each having an area of approximately 44,000 mi^2!

Capacitors are constructed in several ways. One common method is to roll two sheets of metal foil which are separated by a dielectric such as paper into a tubular shape to conserve space, as shown in Fig. 4.2. Leads (wires) are attached to each sheet and form the terminals of the capacitor. Very large values of capacitance, such as 100 μF, are constructed as electrolytic capacitors. Modern integrated-circuit technology allows the construction of a wide variety of capacitance values in an extremely small space.

The element symbol for a capacitor is shown in Fig. 4.3. The current entering one terminal of the capacitor is equal to the rate of buildup of charge on the plate attached to this terminal:

$$i(t) = \frac{dQ}{dt} \tag{4.3}$$

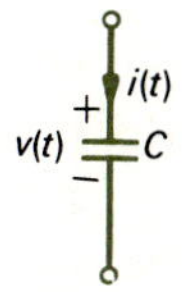

$i(t)$, $v(t)$, $+$, $-$, C

$$i(t) = C\frac{dv(t)}{dt}$$

$$v(t) = \frac{1}{C}\int_{-\infty}^{t} i(\tau)\, d\tau$$

FIGURE 4.3 The terminal v–i relationship for a capacitor.

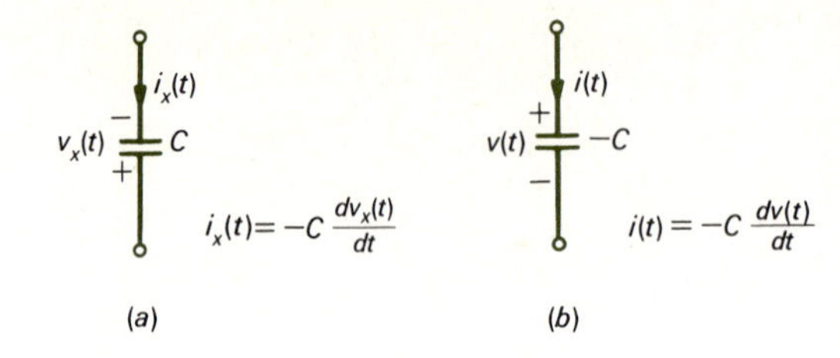

FIGURE 4.4 Terminal relations for capacitors, and the passive sign convention: (*a*) element not labeled with the passive sign convention; (*b*) a negative capacitor (note that the terminal voltage and current are labeled with the passive sign convention).

If we differentiate Eq. (4.1) with respect to t and substitute the result into Eq. (4.3) we obtain the terminal voltage-current relationship for the capacitor:

$$i(t) = C \frac{dv(t)}{dt} \tag{4.4}$$

Here we have assumed that the capacitance is fixed. If the separation distance between the plates changed with time, then we would write $C(t)$, since the capacitance would vary with time and we would have to take this into account when differentiating Eq. (4.1) with respect to time. Note also that in order for Eq. (4.4) to be true, $v(t)$ and $i(t)$ must be labeled on the element with the passive sign convention. If they were not labeled with the passive sign convention, a negative sign would be needed in Eq. (4.4). If, on the other hand, $v(t)$ and $i(t)$ were labeled on the element with the passive sign convention and we found that $i(t) = -C[dv(t)/dt]$, then this would be a true negative capacitor. (We will only consider positive capacitors.) Examples are shown in Fig. 4.4. This proper labeling of the element voltage and current to conform to the passive sign convention in order that Eq. (4.4) is the correct terminal relation should be carefully studied. If the terminal voltage and current of a positive capacitor are not labeled with the passive sign convention but Eq. (4.4) is used to relate them nevertheless, the result is nonsense.

The terminal relationship in Eq. (4.1) is said to describe a linear capacitor, since the terminal voltage and charge on the plates are linearly related. If, on the other hand, the charge and terminal voltage of a capacitor were related by a nonlinear equation such as $Q = CV^2$, then the capacitor would be classified as nonlinear. We will only consider linear capacitors.

The terminal voltage-current relationship in Eq. (4.4) can be written in an alternative form by integrating both sides of the equation to yield

$$v(t) = \frac{1}{C} \int_{-\infty}^{t} i(\tau)\, d\tau \tag{4.5}$$

where τ is the dummy variable of integration. This result shows that the terminal voltage depends on the past values of the terminal current. This makes sense because the integral of the current over all previous time yields the present value of charge on the capacitor plates.

Usually, we will designate some origin in time for our investigations as $t = 0$. If we separate the integral in Eq. (4.5) into a portion for $t < 0$ and a portion for $t > 0$, we have

$$v(t) = \frac{1}{C}\int_0^t i(\tau)\,d\tau + \frac{1}{C}\int_{-\infty}^0 i(\tau)\,d\tau \tag{4.6}$$

Note that the portion for $t < 0$ is the value of the capacitor voltage at $t = 0$:

$$v(0) = \frac{1}{C}\int_{-\infty}^0 i(\tau)\,d\tau \tag{4.7}$$

Thus, Eq. (4.6) can be written in terms of this "initial" capacitor voltage as

$$v(t) = \frac{1}{C}\int_0^t i(\tau)\,d\tau + v(0) \tag{4.8}$$

The energy stored in a capacitor at a particular time can be found by integrating the expression for the instantaneous power delivered to the capacitor

$$\begin{aligned} p(t) &= v(t)i(t) \\ &= Cv(t)\frac{dv(t)}{dt} \end{aligned} \tag{4.9}$$

to give

$$\begin{aligned} w(t) &= \int_{-\infty}^t p(\tau)\,d\tau \\ &= C\int_{-\infty}^t v(\tau)\frac{dv(\tau)}{d\tau}\,d\tau \\ &= \tfrac{1}{2}Cv^2(t) - \tfrac{1}{2}Cv^2(-\infty) \end{aligned} \tag{4.10}$$

Now if we assume that the capacitor voltage was zero at $t = -\infty$, the energy stored in the capacitor at some time t depends only on the voltage of the capacitor at that time:

$$w(t) = \tfrac{1}{2}Cv^2(t) \tag{4.11}$$

an interesting result.

This energy is stored in the electric force field between the plates by the separation of charge, as was discussed in Chap. 2. We will find that the next energy storage element which we will consider, the inductor, stores energy in the dual force field—the magnetic force field. These duality concepts will occur quite frequently and the reader should be alert to find them.

Example 4.1 Consider the voltage waveform $v_s(t)$ applied to the 5-F capacitor in Fig. 4.5*a*. Sketch the capacitor current and stored energy as a function of time.

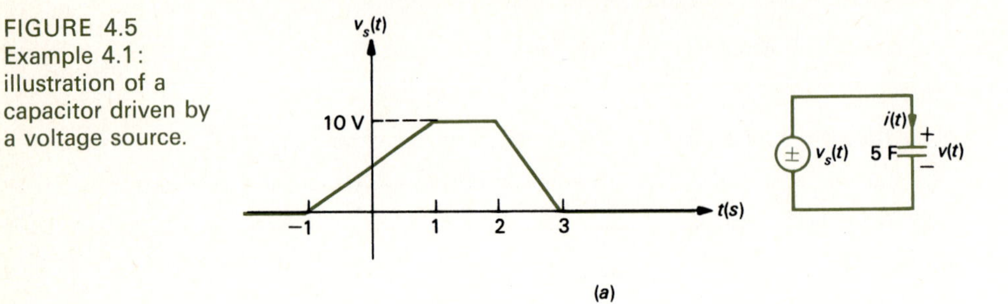

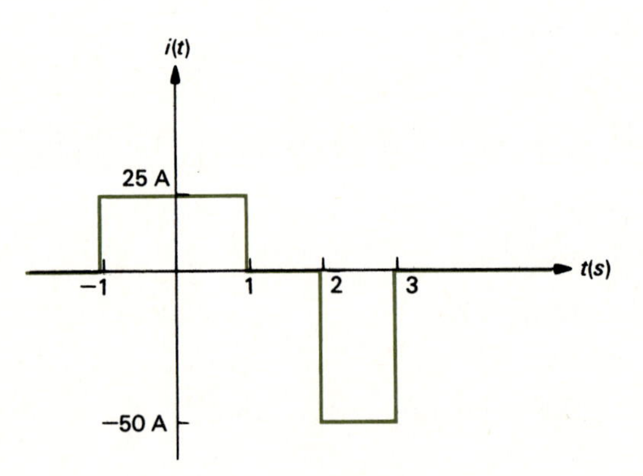

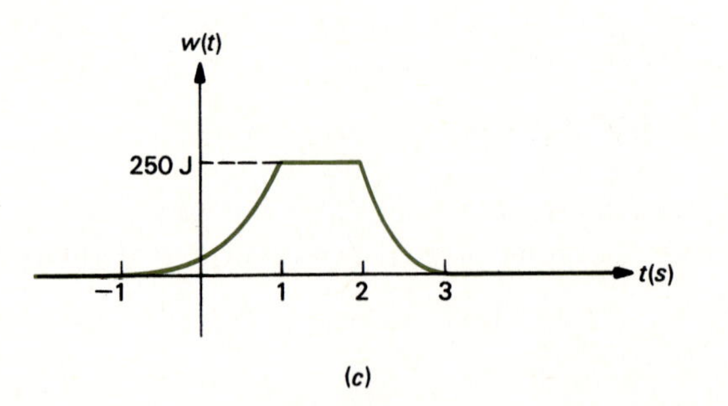

FIGURE 4.5 Example 4.1: illustration of a capacitor driven by a voltage source.

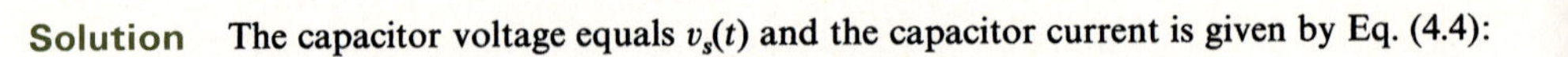

Solution The capacitor voltage equals $v_s(t)$ and the capacitor current is given by Eq. (4.4):

$$i(t) = 5\,\frac{dv_s(t)}{dt}$$

Now $v_s(t)$ is characterized by

$$
\begin{aligned}
v_s(t) &= 0 && t \leq -1 \\
&= 5(t+1) \quad \text{V} && -1 \leq t \leq 1 \\
&= 10 \text{ V} && 1 \leq t \leq 2 \\
&= -10(t-3) \quad \text{V} && 2 \leq t \leq 3 \\
&= 0 && 0 \leq t
\end{aligned}
$$

Thus, the capacitor current is found from Eq. (4.4) by differentiating $v_s(t)$ and multiplying the result by $C = 5$ F:

$$
\begin{aligned}
i(t) &= 0 && t \leq -1 \\
&= 25 \quad \text{A} && -1 \leq t \leq 1 \\
&= 0 && 1 \leq t \leq 2 \\
&= -50 \quad \text{A} && 2 \leq t \leq 3 \\
&= 0 && 0 \leq t
\end{aligned}
$$

which is sketched in Fig. 4.5*b*. The energy stored at any instant of time is given by Eq. (4.11):

$$
\begin{aligned}
w(t) &= \tfrac{1}{2}(5)v_s^2(t) \\
&= 0 && t \leq -1 \\
&= 62.5(t^2 + 2t + 1) \quad \text{J} && -1 \leq t \leq 1 \\
&= 250 \quad \text{J} && 1 \leq t \leq 2 \\
&= 250(t^2 - 6t + 9) \quad \text{J} && 2 \leq t \leq 3 \\
&= 0 && 0 \leq t
\end{aligned}
$$

which is sketched in Fig. 4.5*c*.

Example 4.2 Suppose a current source is attached to a 5-F capacitor as shown in Fig. 4.6*a*. Sketch the waveform of the capacitor voltage.

Solution The capacitor voltage is given by Eq. (4.5):

$$
v(t) = \tfrac{1}{5} \int_{-\infty}^{t} i_s(\tau)\, d\tau
$$

The waveform of the current source is the same shape as the waveform of the voltage source in the previous example:

$$
\begin{aligned}
i_s(t) &= 0 && t \leq -1 \\
&= 5(t+1) \quad \text{A} && -1 \leq t \leq 1 \\
&= 10 \text{ A} && 1 \leq t \leq 2 \\
&= -10(t-3) \quad \text{A} && 2 \leq t \leq 3 \\
&= 0 && 3 \leq t
\end{aligned}
$$

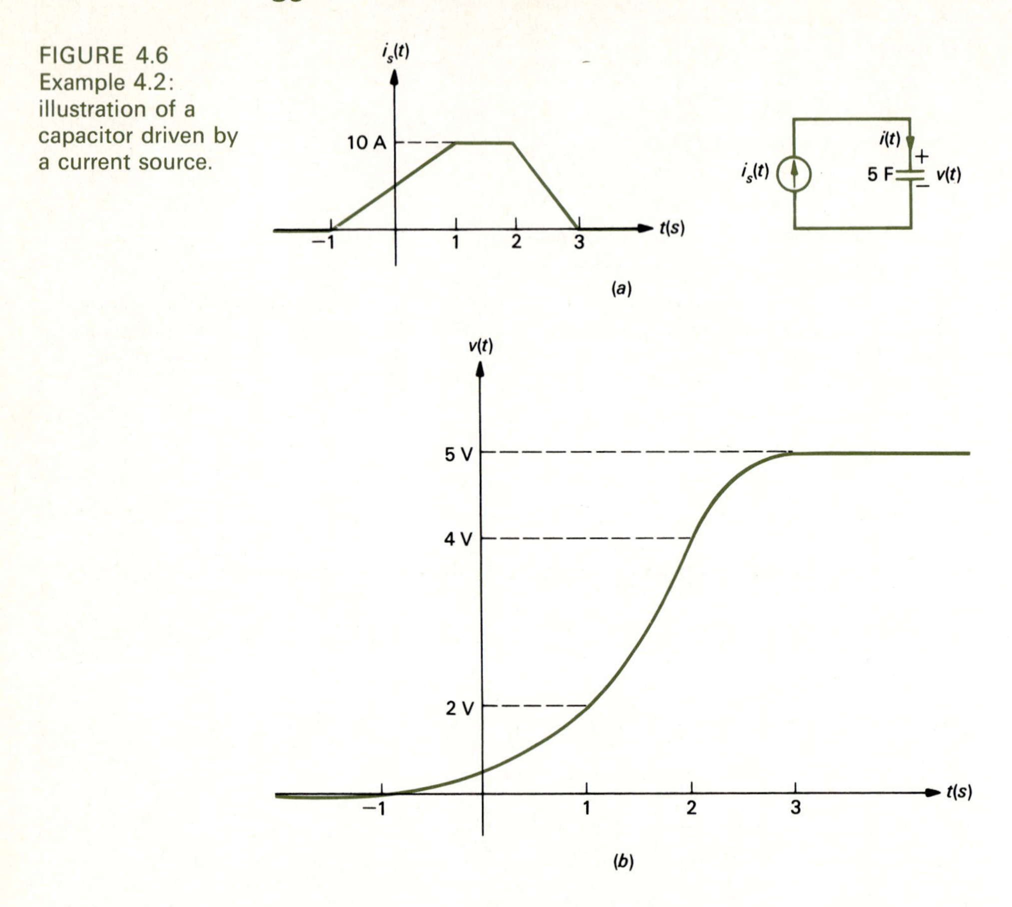

FIGURE 4.6 Example 4.2: illustration of a capacitor driven by a current source.

Thus, the capacitor voltage is found by integrating the waveform of the current source and multiplying the result by $\frac{1}{5}$ F. The result is

$$
\begin{aligned}
v(t) &= 0 && t \leq -1 \\
&= \frac{t^2}{2} + t + \frac{1}{2} \quad \text{V} && -1 \leq t \leq 1 \\
&= 2t \quad \text{V} && 1 \leq t \leq 2 \\
&= -t^2 + 6t - 4 \quad \text{V} && 2 \leq t \leq 3 \\
&= 5 \quad \text{V} && 3 \leq t
\end{aligned}
$$

which is sketched in Fig. 4.6*b*. For example, for the interval $2 \leq t \leq 3$ we integrate Eq. (4.5) to obtain

$$
\begin{aligned}
v(t) &= \tfrac{1}{5} \int_{-\infty}^{t} v(t)\, d\tau \\
&= \tfrac{1}{5} \int_{2}^{t} [-10(\tau - 3)]\, d\tau + v(2)
\end{aligned}
$$

$$= -2\left(\frac{\tau^2}{2} - 3\tau\right)\Bigg|_2^t + v(2)$$

$$= -t^2 + 6t - 8 + 4$$

$$= -t^2 + 6t - 4$$

The energy stored in the capacitor can be obtained from Eq. (4.11) by squaring the waveform for $v(t)$ determined above and multiplying the result by $C/2 = 2.5$.

Several capacitors in series or parallel may be combined into an equivalent capacitance. First consider the parallel connection of capacitors shown in Fig. 4.7. The terminal voltage of the combination is equal to each of the capacitor terminal voltages:

$$v(t) = v_1(t) = v_2(t) = \cdots = v_n(t) \tag{4.12}$$

and the terminal current of the combination is the sum of the terminal currents of each one:

$$i(t) = i_1(t) + i_2(t) + \cdots + i_n(t) \tag{4.13}$$

Substituting the terminal relation for each capacitor given in Eq. (4.4) into Eq. (4.13) gives

$$i(t) = C_1 \frac{dv(t)}{dt} + C_2 \frac{dv(t)}{dt} + \cdots + C_n \frac{dv(t)}{dt}$$

$$= (C_1 + C_2 + \cdots + C_n) \frac{dv(t)}{dt} \tag{4.14}$$

Thus, an equivalent capacitor at the terminals *a*-*b* is

$$C_{eq} = C_1 + C_2 + \cdots + C_n \tag{4.15}$$

and capacitors in parallel combine like resistors in series.

Next consider the series connection of capacitors shown in Fig. 4.8. Here the terminal current of the combination is equal to each of the capacitor currents:

$$i(t) = i_1(t) = i_2(t) = \cdots = i_n(t) \tag{4.16}$$

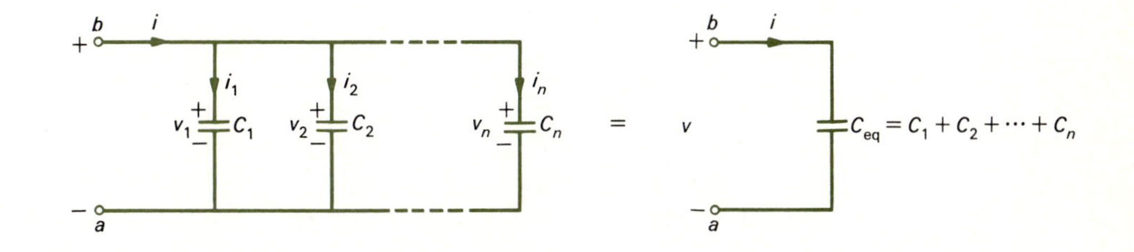

FIGURE 4.7 The equivalent capacitance of capacitors in parallel.

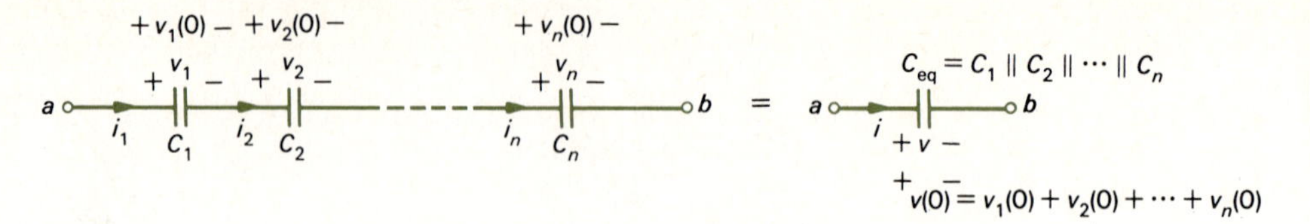

FIGURE 4.8 The equivalent capacitance of capacitors in series.

and the terminal voltage of the combination is the sum of the capacitor voltages:

$$v(t) = v_1(t) + v_2(t) + \cdots + v_n(t) \tag{4.17}$$

Substituting the terminal relation for each capacitor in the form given in Eq. (4.8) into Eq. (4.17) yields

$$\begin{aligned} v(t) &= \frac{1}{C_1}\int_0^t i(\tau)\,d\tau + v_1(0) + \frac{1}{C_2}\int_0^t i(\tau)\,d\tau + v_2(0) \\ &\quad + \cdots + \frac{1}{C_n}\int_0^t i(\tau)\,d\tau + v_n(0) \\ &= \left(\frac{1}{C_1} + \frac{1}{C_2} + \cdots + \frac{1}{C_n}\right)\int_0^t i(\tau)\,d\tau \\ &\quad + [v_1(0) + v_2(0) + \cdots + v_n(0)] \end{aligned} \tag{4.18}$$

Thus, the equivalent capacitor at terminals *a-b* has the value

$$\frac{1}{C_{\text{eq}}} = \frac{1}{C_1} + \frac{1}{C_2} + \cdots + \frac{1}{C_n} \tag{4.19}$$

or

$$C_{\text{eq}} = C_1||C_2||\cdots||C_n \tag{4.20}$$

with an initial voltage which is the sum of the initial voltages of each capacitor:

$$v(0) = v_1(0) + v_2(0) + \cdots + v_n(0) \tag{4.21}$$

and capacitors in series combine like resistors in parallel.

4.2 THE INDUCTOR

The second energy storage element we will consider is the inductor. Inductors store energy in a magnetic field just as capacitors store energy in the dual electric force field.

Suppose we form a length of wire into a loop and pass a current through this wire as shown in Fig. 4.9*a*. A magnetic force field is generated around the wire in the form of magnetic flux ψ. This magnetic force field has the capability to attract ferrous

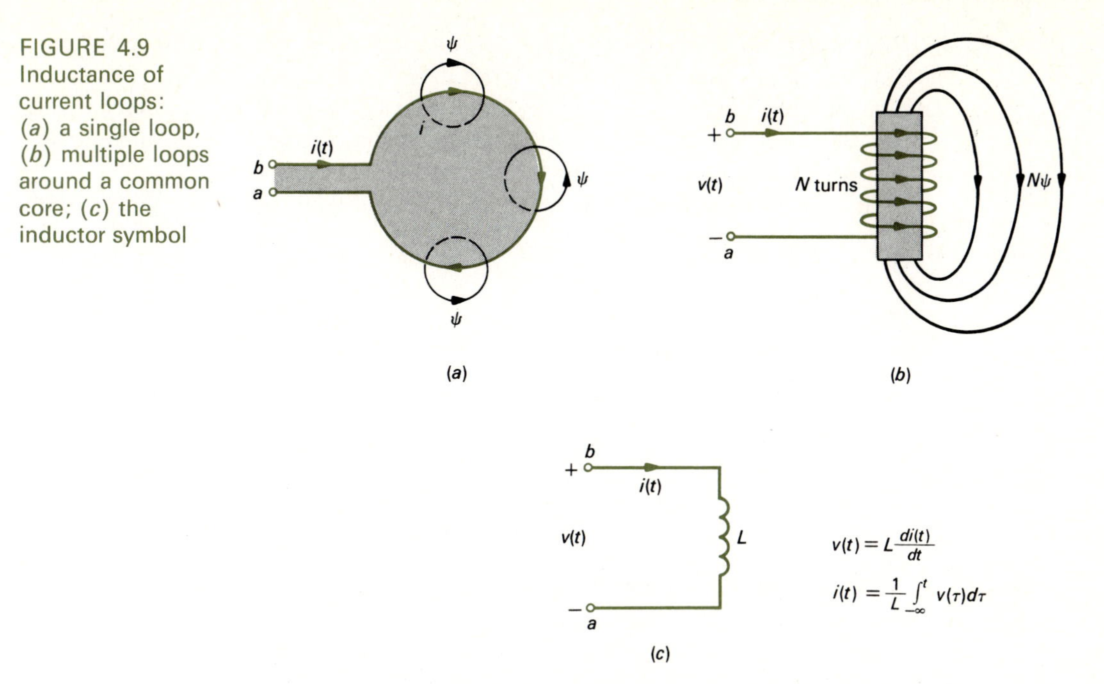

FIGURE 4.9 Inductance of current loops: (*a*) a single loop, (*b*) multiple loops around a common core; (*c*) the inductor symbol

metals, such as iron. The magnetic flux ψ and the current i producing it are linearly related by a quantity called the inductance of the loop, L:

$$\psi = Li \tag{4.22}$$

If N of these loops are wound about some central core as shown in Fig. 4.9*b*, the total magnetic flux encircling, or "linking," current i is N times Eq. (4.22). Thus, a current produces magnetic flux which links that current and the current and resulting flux are related by the inductance L. The unit of inductance is the henry (H), where H = Wb/A and Wb denotes webers, the unit of magnetic flux.

Faraday's law states that the time-varying magnetic flux in the above examples will induce a voltage in the loops which tends to produce a current in the loop. This induced current and associated magnetic flux tend to oppose the change in the original current and its associated magnetic flux. This induced voltage is related to the time rate of change of the magnetic flux:

$$v(t) = \frac{d\psi}{dt} \tag{4.23}$$

Substituting Eq. (4.22), we obtain, for an inductance which does not change with time (i.e., the physical shape of the loop is constant),

$$v(t) = L\frac{di(t)}{dt} \tag{4.24}$$

This is illustrated in Fig. 4.9*c*, where the circuit symbol for the inductor is shown.

Mathematically, the inductor is the dual of the capacitor. For example, compare the terminal relationship for the inductor in Eq. (4.24) to the terminal relationship for the capacitor given in Eq. (4.4). One can be obtained from the other by interchanging v and i and interchanging L and C. We will find many other relationships for the inductor which are duals of the corresponding capacitor relationships.

The alternative characterization of the inductor is obtained by integrating both sides of Eq. (4.24) to yield

$$i(t) = \frac{1}{L}\int_{-\infty}^{t} v(\tau)\, d\tau = \frac{1}{L}\int_{0}^{t} v(\tau)\, d\tau + i(0) \tag{4.25}$$

where $i(0)$ is the inductor current at $t = 0$:

$$i(0) = \frac{1}{L}\int_{-\infty}^{0} v(\tau)\, d\tau \tag{4.26}$$

Note that the current through an inductor at a particular time depends on the values of the inductor voltage at all prior instants of time.

It is once again important that the terminal voltage and current for the inductor are labeled in Fig. 4.9*c* with the passive sign convention. If they are not, a minus sign is required in the terminal relationships in Eqs. (4.24) and (4.25).

If the inductor, or coil, in Fig. 4.9*b* is wound about a ferrous core such as iron, the relationship between the magnetic flux ψ and current i will be a nonlinear one, as shown in Fig. 4.10, and the inductor is said to be nonlinear. We will only consider linear inductors in this text.

Typical values for inductors are in the millihenry (mH $= 10^{-3}$ H) or microhenry (μH $= 10^{-6}$ H) range. An approximate value for an air-core inductor consisting of N turns each having an area of A which is wound on a toroid with mean circumference l is

$$L \doteq 4\pi \times 10^{-7} \frac{N^2 A}{l} \tag{4.27}$$

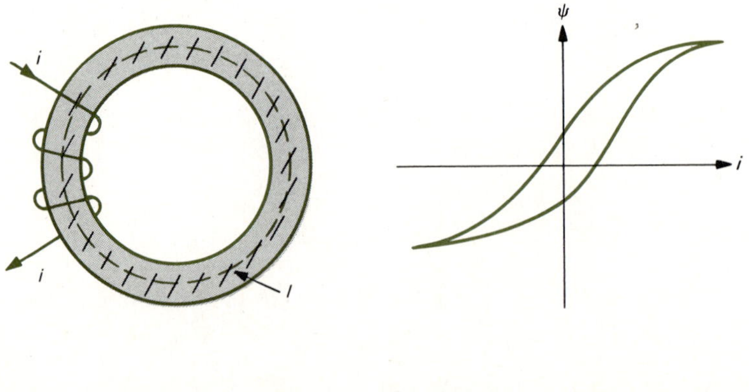

FIGURE 4.10
A nonlinear inductor—a toroid with a ferrous core.

The energy stored in an inductor (in its magnetic field) is obtained by integrating the instantaneous power

$$\begin{aligned} p(t) &= v(t)i(t) \\ &= Li(t)\frac{di(t)}{dt} \end{aligned} \tag{4.28}$$

to yield

$$\begin{aligned} w(t) &= \int_{-\infty}^{t} p(\tau)\, d\tau \\ &= \tfrac{1}{2}Li^2(t) - \tfrac{1}{2}Li^2(-\infty) \\ &= \tfrac{1}{2}Li^2(t) \end{aligned} \tag{4.29}$$

where we have assumed that the inductor current at $t = -\infty$ is zero. Thus, the energy stored in an inductor at an instant of time depends only on the inductor current at that time. This is the dual of the result for the energy stored in a capacitor, which depended only on the capacitor voltage.

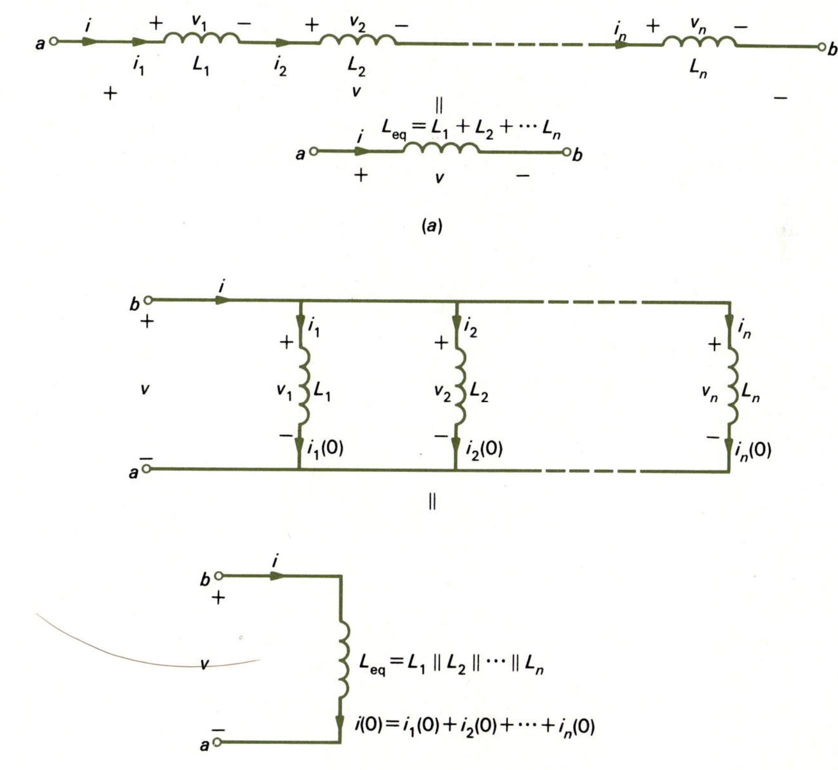

FIGURE 4.11 The equivalent inductance of inductors in series and parallel: (*a*) series combination; (*b*) parallel combination.

The other results which are duals of corresponding results for the capacitor are the series and parallel reductions of inductors shown in Fig. 4.11. For these we can easily show, by substituting the terminal relations for the individual inductors given in Eqs. (4.24) or (4.25) into

$$\begin{aligned} v &= v_1 + v_2 + \cdots + v_n \\ i &= i_1 = i_2 = \cdots = i_n \end{aligned} \tag{4.30}$$

for the series combination or

$$\begin{aligned} v &= v_1 = v_2 = \cdots = v_n \\ i &= i_1 + i_2 + \cdots + i_n \end{aligned} \tag{4.31}$$

for the parallel combination, that inductors in series combine like resistors in series:

$$L_{eq} = L_1 + L_2 + \cdots + L_n \tag{4.32}$$

and that inductors in parallel combine like resistors in parallel:

$$L_{eq} = L_1 || L_2 || \cdots || L_n \tag{4.33}$$

with an initial current equal to the sum of the initial currents of each inductor:

$$i(0) = i_1(0) + i_2(0) + \cdots + i_n(0) \tag{4.34}$$

Example 4.3 Consider the current source applied to the 5-H inductor shown in Fig. 4.12*a*. Sketch the inductor voltage and energy stored in the inductor.

Solution The inductor voltage is found from Eq. (4.24) as the product of the inductance and the slope of the current waveform of the inductor:

$$v(t) = 5 \frac{di_s(t)}{dt}$$

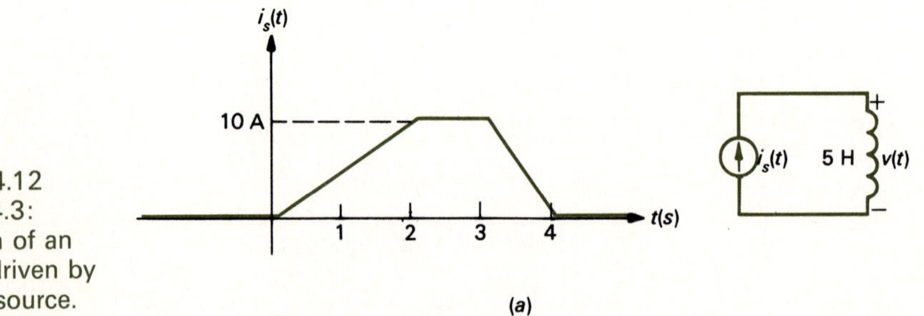

FIGURE 4.12 Example 4.3: illustration of an inductor driven by a current source.

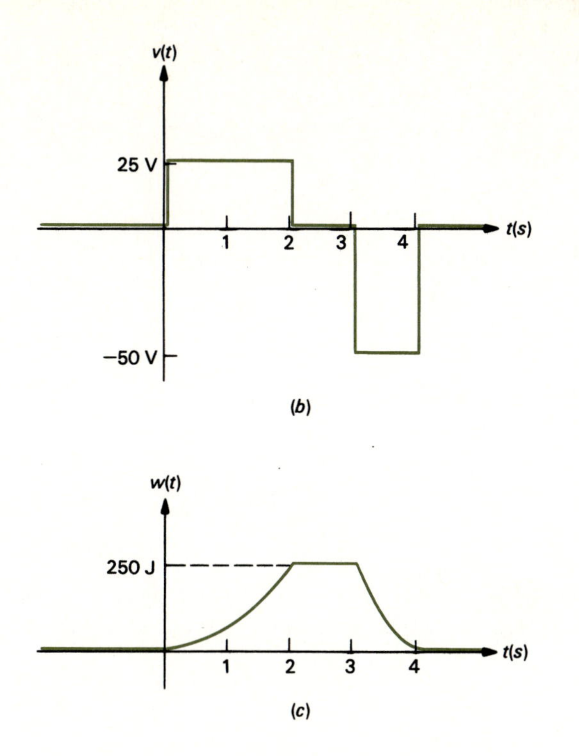

which is sketched in Fig. 4.12*b*. The stored energy at any instant of time is related to the square of the inductor current at that time by Eq. (4.29):

$$w(t) = \tfrac{1}{2}5i_s^2(t)$$

which is sketched in Fig. 4.12*c*.

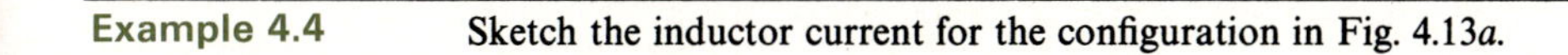

Example 4.4 Sketch the inductor current for the configuration in Fig. 4.13*a*.

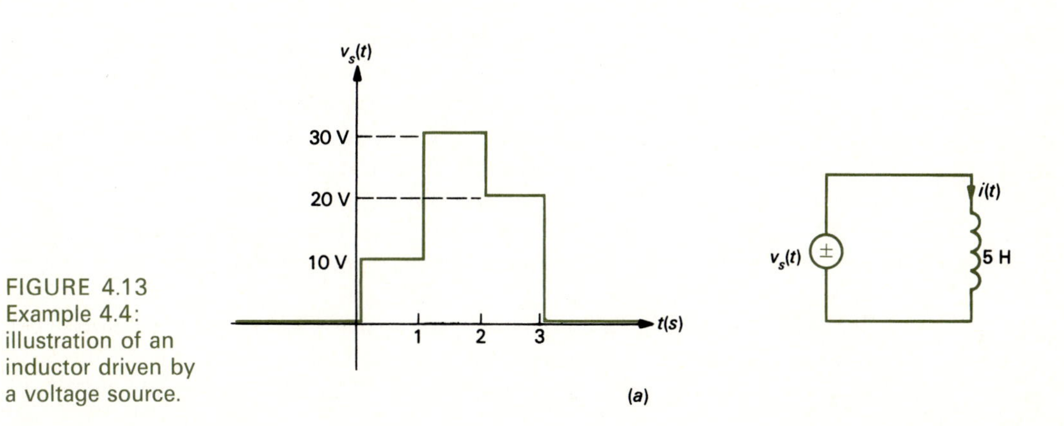

FIGURE 4.13 Example 4.4: illustration of an inductor driven by a voltage source.

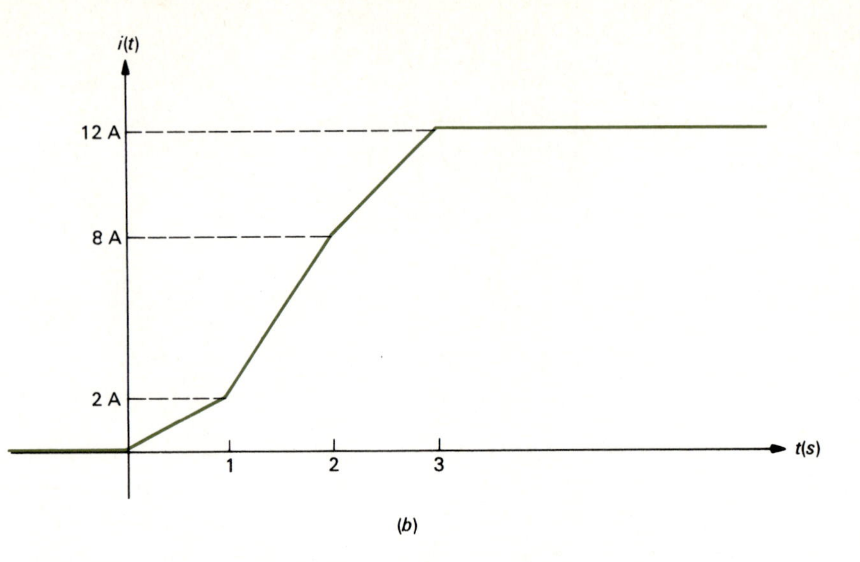

FIGURE 4.13 (*Continued*)

Solution The inductor voltage waveform is $v_s(t)$ and the inductor current is related to the integral of this current as the ratio of the area under the curve up to this t and the inductance, as in Eq. (4.25):

$$i(t) = \tfrac{1}{5} \int_{-\infty}^{t} v_s(\tau)\, d\tau$$

This is sketched in Fig. 4.13*b*.

4.3 CONTINUITY OF CAPACITOR VOLTAGES AND INDUCTOR CURRENTS

The voltage across a capacitor cannot change value instantaneously. This can be seen in several ways. From the relation between the terminal voltage and current given in Eq. (4.4), we see that if the terminal voltage were to change value instantaneously the capacitor current would be infinite: an unrealistic situation. Alternatively, from the integral relation in Eq. (4.5) we see that the terminal voltage is related to the integral of the terminal current. Realistic currents $i(t)$, when integrated, will yield a continuous function. Thus, the voltage cannot change value instantaneously.

Another way of seeing this result is to examine the expression for stored energy in Eq. (4.11). If the capacitor voltage changes value instantaneously, the stored energy would also change instantaneously: an unrealistic situation.

However, there is no reason to rule out an instantaneous change in the capacitor current. This was illustrated in Example 4.1, where the capacitor current shown in Fig. 4.5*b* changes value instantaneously at $t = -1, 1, 2, 3$.

Similarly, *the current through an inductor cannot change value instantaneously.* This is easily seen from the terminal relations for the inductor in Eqs. (4.24) and (4.25), as well as from the relation for the stored energy in Eq. (4.29). However, there is no

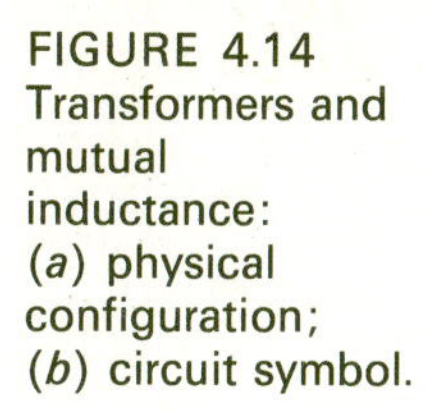

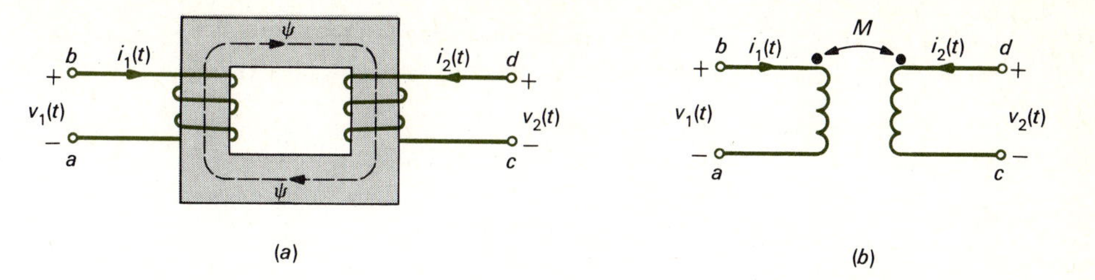

FIGURE 4.14 Transformers and mutual inductance: (*a*) physical configuration; (*b*) circuit symbol.

reason to rule out an instantaneous change in the value of the inductor voltage (see Fig. 4.12*b*).

This latter property of inductors is used in generating large voltages in an automobile ignition system to fire the spark plugs. A current is passed through an inductor, the "ignition coil." A cam opens and closes a switch connected in series with this ignition coil, thus reducing the current through the coil to zero in a very short period of time. From the terminal equation for the ignition coil in Eq. (4.24), we see that changing the current in the coil in a short period of time can generate large voltages dependent on di/dt and the inductance of the coil.

4.4 TRANSFORMERS AND COUPLED COILS

We have seen how a time-varying current passing through the windings of an inductor can produce a time-varying magnetic flux which links this current and produces an induced voltage at the terminals of the inductor; this is given by Eq. (4.24). If some of this time-varying magnetic flux also links the turns of some other inductor, it will also induce a voltage at the terminals of that inductor. We then say that there is mutual inductance between the two inductors and refer to them as being coupled.

For example, consider the two inductors which are wound on a common core (such as iron) in Fig. 4.14*a*. A current $i_1(t)$ entering the terminals *a-b* of one inductor will produce a magnetic flux ψ which will link not only its own turns but also those of the second inductor. Thus, this time-varying current $i_1(t)$ produces an induced voltage at terminals *a-b* of

$$v_1'(t) = L_1 \frac{di_1(t)}{dt} \tag{4.35}$$

The resulting magnetic flux also produces an induced voltage at terminals *c-d* of the second inductor of

$$v_2'(t) = M \frac{di_1(t)}{dt} \tag{4.36}$$

where M is said to be the mutual inductance between the two inductors. Because of the way in which the inductors have been wound on the common core, the flux ψ induces at terminals *c-d* a voltage [Eq. (4.36)] which has the same polarity as shown for $v_2(t)$. This is the meaning of the dots on the circuit symbol in Fig. 4.14*b*. A current entering a dotted terminal produces an induced voltage which is positive at the dotted terminals. Similarly, a time-varying current $i_2(t)$ entering terminal d will induce a voltage at the terminals of the first inductor of

$$v_1''(t) = M \frac{di_2(t)}{dt} \tag{4.37}$$

(positive at terminal b) and at the terminals of the second inductor of

$$v_2''(t) = L_2 \frac{di_2(t)}{dt} \tag{4.38}$$

(positive at terminal d). Combining Eqs. (4.35) and (4.37) and also combining Eqs. (4.36) and (4.38), we have

$$\begin{aligned} v_1(t) &= v_1'(t) + v_1''(t) \\ &= L_1 \frac{di_1(t)}{dt} + M \frac{di_2(t)}{dt} \\ v_2(t) &= v_2'(t) + v_2''(t) \\ &= M \frac{di_1(t)}{dt} + L_2 \frac{di_2(t)}{dt} \end{aligned} \tag{4.39}$$

Equations (4.39) are the circuit equations for two coupled inductors. It is very important to observe the dot notation when writing the terminal equations. This result can be extended to any number of coupled coils. An example of three coupled coils is shown in Fig. 4.15. The circuit equations become

$$\begin{aligned} v_1 &= L_1 \frac{di_1}{dt} + M_{12} \frac{di_2}{dt} + M_{13} \frac{di_3}{dt} \\ v_2 &= M_{12} \frac{di_2}{dt} + L_2 \frac{di_2}{dt} + M_{23} \frac{di_3}{dt} \\ v_3 &= M_{13} \frac{di_1}{dt} + M_{23} \frac{di_2}{dt} + L_3 \frac{di_3}{dt} \end{aligned} \tag{4.40}$$

Note the dot notation and resulting polarity of each component of induced voltage in Eq. (4.40).

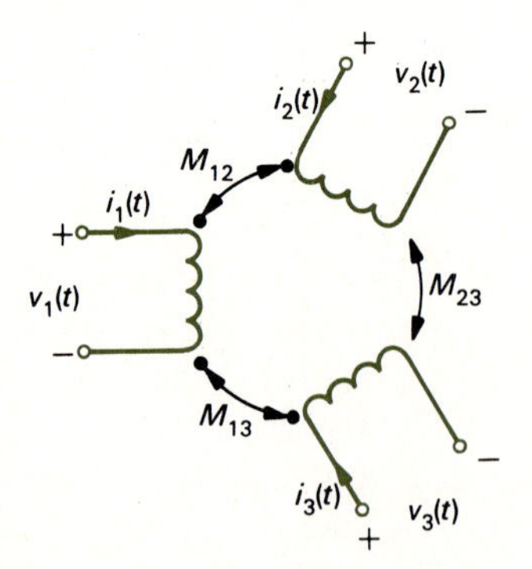

FIGURE 4.15 Coupled coils.

Example 4.5

Consider the set of two coupled inductors shown in Fig. 4.16*a*. Write the terminal equations in terms of the voltages and currents defined on the circuit. Repeat this for the coupled set of three inductors shown in Fig. 4.16*b*.

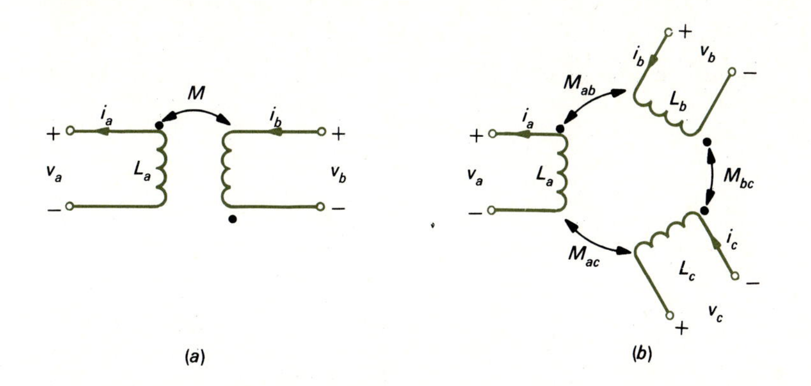

FIGURE 4.16 Example 4.5: writing terminal equations for coupled coils.

Solution If (1) the defined voltages and currents at terminal pair do not conform to the passive sign convention and/or (2) the positive voltage terminal is not at the dot, the simplest way to write *correct* terminal equations is first to reverse the necessary currents and voltages to (1) conform to the passive sign convention and (2) have the positive terminal of the voltage at the dot. For the two coupled coils in Fig. 4.16*a*, for example, we must reverse i_a, call it i_a', reverse i_b, call it i_b', and reverse v_b and call it v_b'. In terms of the usual polarities, we have

$$v_a = L_1 \frac{di_a'}{dt} + M \frac{di_b'}{dt}$$

$$v_b' = M \frac{di_a'}{dt} + L_2 \frac{di_b'}{dt}$$

but $v_b' = -v_b$, $i_a' = -i_a$, and $i_b' = -i_b$. Substituting these, we finally have

$$v_a = -L_1 \frac{di_a}{dt} - M \frac{di_b}{dt}$$

$$v_b = M \frac{di_a}{dt} + L_2 \frac{di_b}{dt}$$

Similarly, for the three coupled coils in Fig. 4.16*b* we must reverse i_a, i_b, v_b, and v_c. Thus

$$v_a = L_a \frac{di_a'}{dt} + M_{ab} \frac{di_b'}{dt} + M_{ac} \frac{di_c}{dt}$$

$$v_b' = M_{ab} \frac{di_a'}{dt} + L_b \frac{di_b'}{dt} + M_{bc} \frac{di_c}{dt}$$

$$v_c' = M_{ac} \frac{di_a'}{dt} + M_{bc} \frac{di_b'}{dt} + L_c \frac{di_c}{dt}$$

Now $i_a' = -i_a$, $i_b' = -i_b$, $v_b' = -v_b$, and $v_c' = -v_c$, so that in terms of the terminal voltages and currents defined in Fig. 4.16*b* we have

$$v_a = -L_a \frac{di_a}{dt} - M_{ab} \frac{di_b}{dt} + M_{ac} \frac{di_c}{dt}$$

$$v_b = M_{ab} \frac{di_a}{dt} + L_b \frac{di_b}{dt} - M_{bc} \frac{di_c}{dt}$$

$$v_c = M_{ac} \frac{di_a}{dt} + M_{bc} \frac{di_b}{dt} - L_c \frac{di_c}{dt}$$

Coupled coils are often referred to as transformers for the following reason. Consider the pair of coupled coils shown in Fig. 4.17. The first coil has N_1 turns and the second has N_2 turns. A current i_1 in the first coil produces magnetic flux ψ which links the N_2 turns of the second coil. The voltage induced in the first coil by i_1 is equal to

$$\begin{aligned} v_1' &= \frac{d}{dt}(N_1\psi) \\ &= N_1 \frac{d\psi}{dt} \end{aligned} \tag{4.41}$$

while the voltage induced in the second coil by i_1 is equal to

$$\begin{aligned} v_2' &= \frac{d}{dt}(N_2\psi) \\ &= N_2 \frac{d\psi}{dt} \end{aligned} \tag{4.42}$$

[Each turn in coil 2 has a voltage $(d/dt)(\psi)$ induced in it since flux ψ links each turn.] Now the ratio of these two voltages is

$$\frac{v_2'}{v_1'} = \frac{N_2}{N_1} \tag{4.43}$$

Thus, suppose a voltage v_1 is applied to coil 1; then a voltage

$$v_2 = \frac{N_2}{N_1} v_1 \tag{4.44}$$

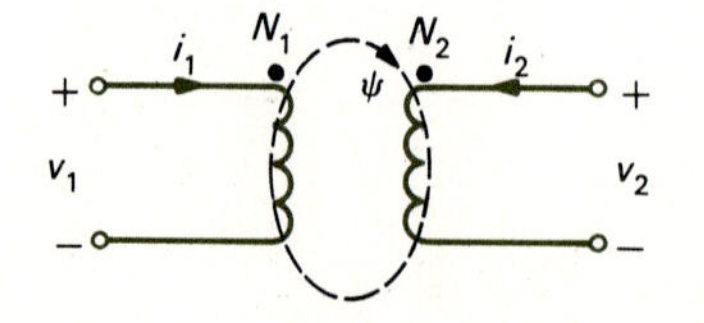

FIGURE 4.17 The ideal transformer.

will be induced at the terminals of coil 2. If $N_2 > N_1$, v_2 will be larger than v_1 and we say that the voltage v_1 has been "stepped up" by the "turns ratio" N_2/N_1. Of course, if the voltage had been applied to the terminals of coil 2, the induced voltage at the terminals of coil 1 would have been "stepped down" by the ratio N_1/N_2 if $N_1 < N_2$. Step-up transformers are used to obtain from a much lower voltage the very high voltage (as high as 365,000 V) used in power transmission lines.

Note that the instantaneous power delivered to the first coil is

$$p_1 = v_1 i_1 \tag{4.45}$$

and that the power delivered to the second coil at that time is

$$p_2 = v_2 i_2 \tag{4.46}$$

If we assume that the transformer is lossless, then the total power delivered to the transformer must be zero:

$$p_1 + p_2 = 0 \tag{4.47}$$

or

$$v_1 i_1 + v_2 i_2 = 0 \tag{4.48}$$

Substituting Eq. (4.44) into Eq. (4.48) yields

$$v_1 i_1 + \frac{N_2}{N_1} v_1 i_2 = 0 \tag{4.49}$$

or

$$i_2 = -\frac{N_1}{N_2} i_1 \tag{4.50}$$

Thus, for current i_1 entering the dotted terminal of coil 1, current i_2 must leave the dotted terminal of coil 2 and is $(N_1/N_2)i_1$. The current ratio in Eq. (4.50) is the

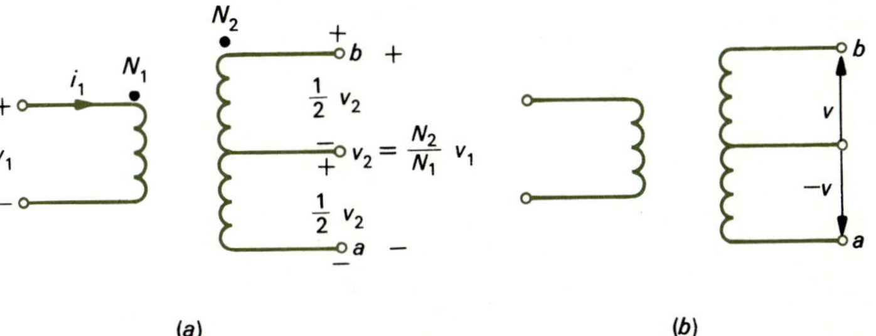

FIGURE 4.18 Center-tapped transformers.

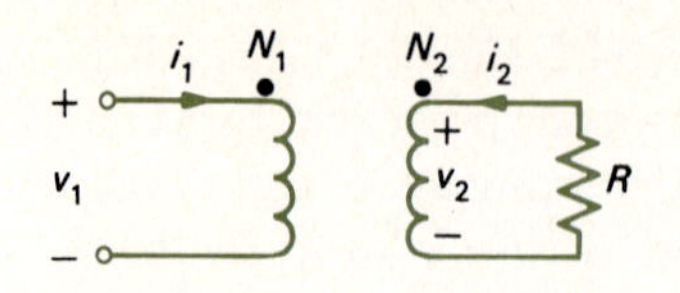

FIGURE 4.19 Illustration of resistance reflection through a transformer.

reciprocal of the voltage ratio in Eq. (4.44). Thus, if a transformer steps up a voltage applied to coil 1, it steps down the current entering coil 1 by the same ratio so that the transformer remains lossless. Such a transformer is said to be *ideal.*

Quite often the center-tapped ideal transformers illustrated in Fig. 4.18*a* are used. The second coil has a tap placed at its center so that the voltage induced across the second coil, $(N_2/N_1)v_1$, is divided equally between both halves of that coil. Note that this transformer can be used to provide two voltages which are opposite to each other in polarity, as shown in Fig. 4.18*b*. The voltage of terminal *a* with respect to the center tap is opposite in polarity to the voltage of terminal *b* with respect to the center tap.

One final point should be mentioned—reflection of resistance through a transformer. Consider the two-coil transformer in Fig. 4.19 which has the second coil attached to a resistor. The resistance seen, looking into the terminals of the first coil, is

$$R_{eq} = \frac{v_1}{i_1} \tag{4.51}$$

Substituting

$$v_1 = \frac{N_1}{N_2} v_2$$
$$i_1 = -\frac{N_2}{N_1} i_2 \tag{4.52}$$

and

$$v_2 = -Ri_2 \tag{4.53}$$

we obtain

$$R_{eq} = \left(\frac{N_1}{N_2}\right)^2 R \tag{4.54}$$

Thus, the resistor R appears at the terminals of the first coil larger or smaller by the ratio $(N_1/N_2)^2$, depending on whether $N_1 > N_2$ or $N_1 < N_2$. This can be useful in matching for maximum power transfer, as the following example shows.

Example 4.6 Consider the source and load shown in Fig. 4.20*a*. Design a circuit to provide maximum power transfer from the source to the load. Determine the average power delivered to the load.

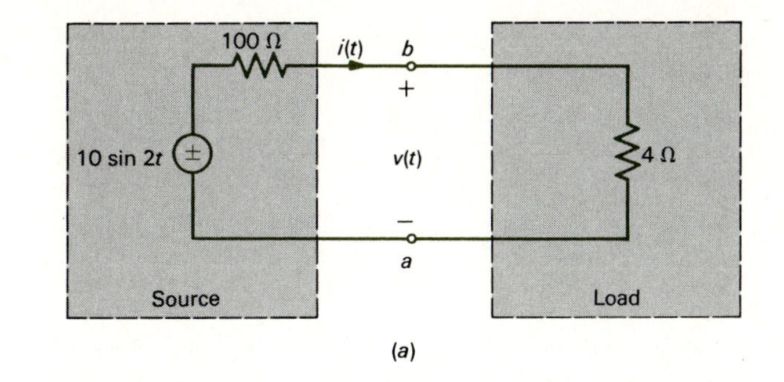

(*a*)

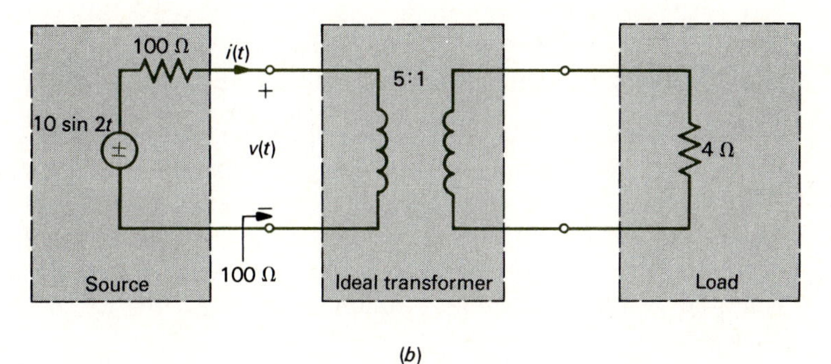

(*b*)

FIGURE 4.20 Example 4.6: maximum power transfer, utilizing a matching transformer.

Solution If the load resistance had been 100 Ω, maximum power transfer to the load would take place. If we place an ideal transformer with a turns ratio of 5:1 between the source and load, the source will see a resistance of

$$\left(\tfrac{5}{1}\right)^2 \times 4\ \Omega = 100\ \Omega$$

Thus, the source will deliver maximum power to the transformer and the 4-Ω load. But the transformer is assumed to be lossless (ideal), so all of this power is delivered to the 4-Ω resistor. From Chap. 2, the average power delivered to the equivalent 100-Ω resistor seen by the source is

$$P_{AV} = \frac{V_{RMS}^2}{100\ \Omega}$$

where V_{RMS} is the RMS value of the sinusoidal voltage $v(t)$. But, by voltage division,

$$\begin{aligned} v(t) &= \frac{100\ \Omega}{100\ \Omega + 100\ \Omega}\, 10 \sin 2t \\ &= 5 \sin 2t \qquad \text{V} \end{aligned}$$

which has an RMS value of

$$V_{\text{RMS}} = \frac{5}{\sqrt{2}} \text{ V}$$

Thus, the average power delivered to the 4-Ω load in Fig. 4.20*b* is

$$P_{\text{AV}} = \frac{(5/\sqrt{2} \text{ V})^2}{100 \ \Omega}$$

$$= 0.125 \text{ W}$$

4.5 THE CIRCUIT EQUATIONS FOR CIRCUITS CONTAINING ENERGY STORAGE ELEMENTS

The equations governing the branch voltages and currents consists of KVL, KCL, and the element relations; including energy storage elements does not change this fact. However, if the circuit contains at least one energy storage element, then these circuit equations will become differential equations, which are more difficult to solve than the algebraic equations encountered with the purely resistive circuits in Chap. 3. The solution of these differential equations will be considered in the following two chapters.

Example 4.7 Write the circuit equations for the circuit shown in Fig. 4.21.

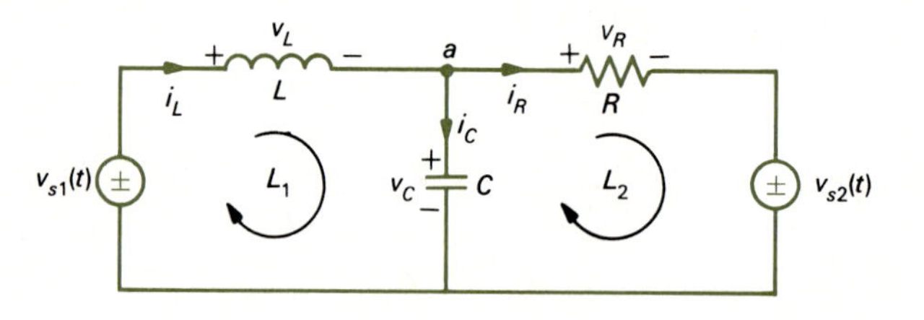

FIGURE 4.21 Example 4.7: writing circuit equations for circuits containing energy storage elements.

Solution Using the labeled element voltages and currents, we need six simultaneous equations in the six unknown branch voltages and currents: $v_L, i_L, v_C, i_C, v_R, i_R$. Writing KCL at node *a*, KVL around loops L_1 and L_2, and the three element equations yields those six equations:

KCL node *a*:

$$i_L - i_C - i_R = 0$$

KVL loop L_1:

$$v_L + v_C = v_{s1}(t)$$

KVL loop L_2:

$$v_C - v_R = v_{s2}(t)$$

resistor:

$$v_R - Ri_R = 0$$

inductor:

$$v_L - L\frac{di_L}{dt} = 0$$

capacitor:

$$i_C - C\frac{dv_C}{dt} = 0$$

Note that the two energy storage elements have introduced derivatives of two of the element unknowns, i_L and v_C. Thus, the set of six equations have become differential equations.

Nevertheless, superposition still holds for these "dynamic" circuits containing energy storage elements so long as the energy storage elements (and all resistors in the circuit) are linear. This can easily be demonstrated for the circuit in Fig. 4.21. Suppose that v_L', i_L', v_C', i_C', v_R', i_R' are due to $v_{s1}(t)$ with $v_{s2}(t) = 0$ and that v_L'', i_L'', v_C'', i_C'', v_R'', i_R'' are due to $v_{s2}(t)$ with $v_{s2}(t) = 0$. KVL and KCL are linear and are satisfied by the sums of the prime and double-prime solutions by superposition. Now if the prime variables satisfy the element relations:

$$\begin{aligned} v_R' - Ri_R' &= 0 \\ v_L' - L\frac{di_L'}{dt} &= 0 \\ i_C' - C\frac{dv_C'}{dt} &= 0 \end{aligned} \tag{4.55}$$

and the double-prime variables satisfy the element relations, too:

$$\begin{aligned} v_R'' - Ri_R'' &= 0 \\ v_L'' - L\frac{di_L''}{dt} &= 0 \\ i_C'' - C\frac{dv_C''}{dt} &= 0 \end{aligned} \tag{4.56}$$

then the sums of the prime and double-prime variables satisfy these element relations, since

$$\begin{aligned}(v_R' + v_R'') - R(i_R' + i_R'') &= 0 \\ (v_L' + v_L'') - L\frac{d}{dt}(i_L' + i_L'') &= 0 \\ (i_C' + i_C'') - C\frac{d}{dt}(v_C' + v_C'') &= 0\end{aligned} \tag{4.57}$$

can be factored into the sum of Eqs. (4.55) and (4.56), which were presumed to be true.

4.6 THE RESPONSE OF THE ENERGY STORAGE ELEMENTS TO DC SOURCES

We have seen in the previous section that the principle of superposition applies to linear circuits which contain linear resistors, inductors, and capacitors. In this section we will investigate how to determine the portions of the branch voltages and currents of a circuit which are due to the dc, or constant-value, sources in the circuit. We will find a simple technique for doing so—*replace all inductors with short circuits and all capacitors with open circuits* and solve the resulting resistive circuit. For example, consider the circuit shown in Fig. 4.22*a* which contains a dc voltage source V_s and a dc current source I_s, a resistor R, an inductor L, and a capacitor C. The solution for each branch voltage and current in the circuit—for example, v_R, i_R, v_L, i_L, v_C, and i_C—consists of two components, and each component is due to each source acting individually, as shown in Fig. 4.22*b*. In each of the circuits in Fig. 4.22*b* there is only one source and it is dc. The important question is "What would we reasonably expect the form of the branch currents and voltages due to the dc sources to be?" The obvious answer is that they will be dc (constant), too.

FIGURE 4.22 Illustration of the principle of superposition for circuits containing energy storage elements.

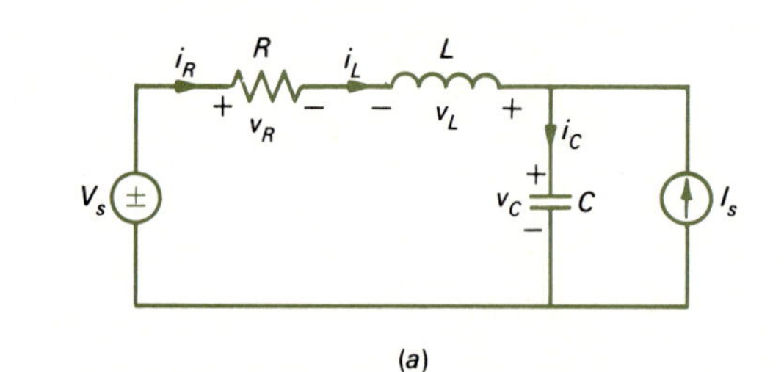

(a)

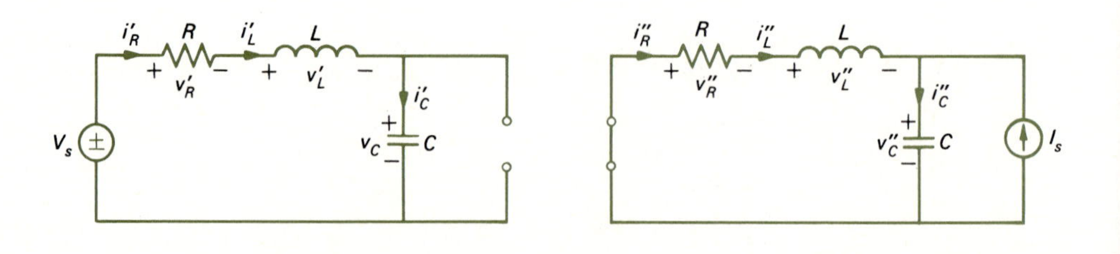

(b)

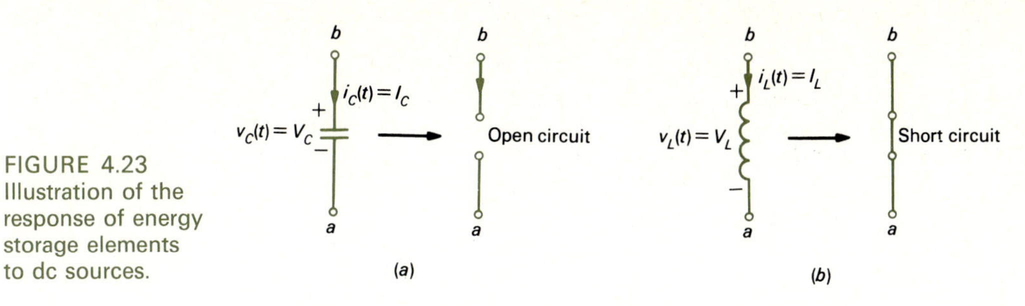

FIGURE 4.23 Illustration of the response of energy storage elements to dc sources.

Now let us examine the implication of this observation. Consider the voltage-current relationship for the capacitor shown in Fig. 4.23*a*:

$$i_C(t) = C\frac{dv_C}{dt} \tag{4.58}$$

Now suppose that both the current and the voltage are dc (constant): $i_C(t) = I_C$ and $v_C(t) = V_C$. Then Eq. (4.58) becomes

$$\begin{aligned} I_C &= C\frac{dV_C}{dt} \\ &= 0 \end{aligned} \tag{4.59}$$

since the derivative of a constant is zero. Thus, the dc value of the capacitor current is zero, which is equivalent to replacing the capacitor with an open circuit!

Similarly, consider the inductor shown in Fig. 4.23*b*. If the current and voltage are dc, $i_L(t) = I_L$ and $v_L(t) = V_L$. Then the terminal relation of the inductor,

$$v_L(t) = L\frac{di_L(t)}{dt} \tag{4.60}$$

shows that

$$\begin{aligned} V_L &= L\frac{dI_L}{dt} \\ &= 0 \end{aligned} \tag{4.61}$$

Thus, the dc value of the inductor voltage is zero, which is equivalent to replacing the inductor with a short circuit.

Now let us return to the example in Fig. 4.22 and apply these results. Since the sources are dc, we replace inductors with short circuits and capacitors with open circuits and thereby get the circuits in Fig. 4.24. Thus, we easily solve these for

$$\begin{aligned} i_C' &= 0 \\ i_R' &= i_L' = 0 \\ v_R' &= v_L' = 0 \\ v_C' &= V_s \end{aligned}$$

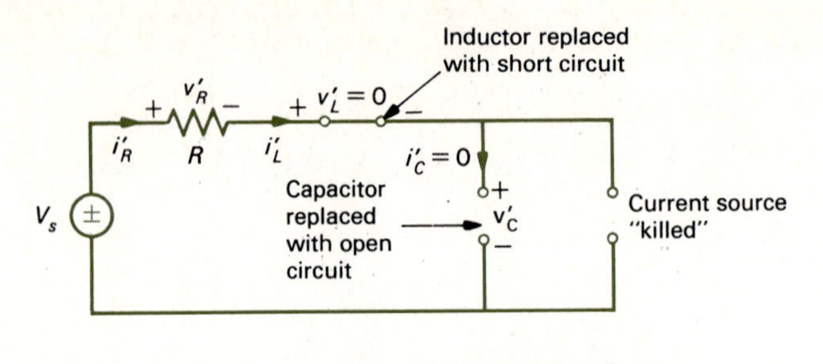

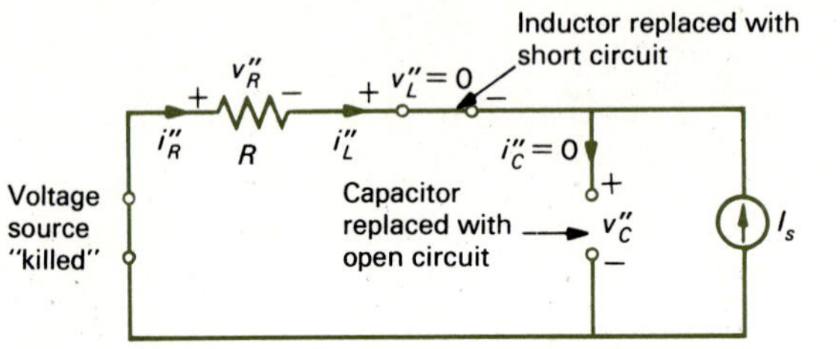

FIGURE 4.24
Illustration of the application of the principle of superposition for circuits containing energy storage elements and dc sources.

and

$$i_C'' = 0$$
$$i_R'' = i_L'' = -I_s$$
$$v_R'' = -RI_s$$
$$v_C'' = RI_s$$
$$v_L'' = 0$$

Therefore

$$v_R = v_R' + v_R''$$
$$= -RI_s$$
$$i_R = i_R' + i_R''$$
$$= -I_s$$
$$v_L = v_L' + v_L''$$
$$= 0$$
$$i_L = i_L' + i_L''$$
$$= -I_s$$
$$v_C = v_C' + v_C''$$
$$= V_s + RI_s$$
$$i_C = i_C' + i_C''$$
$$= 0$$

Example 4.8 Determine voltage v_x and current i_x in the circuit of Fig. 4.25*a*.

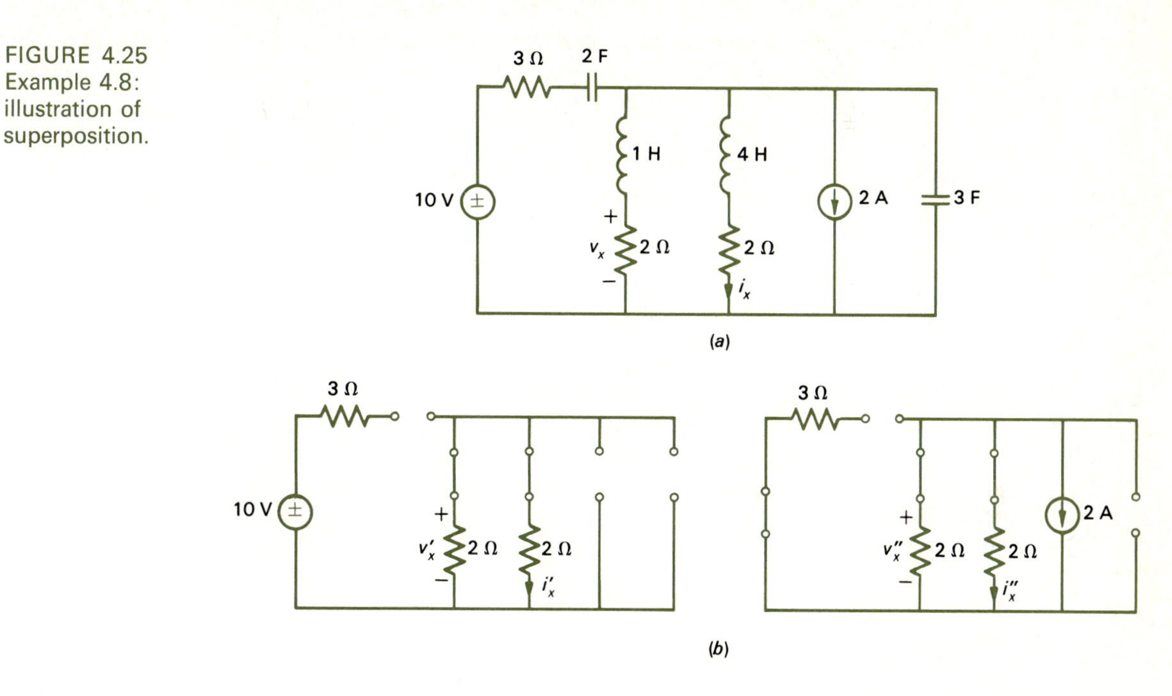

FIGURE 4.25 Example 4.8: illustration of superposition.

Solution By superposition, the solution is shown in Fig. 4.25*b* where, because both sources are dc, we have replaced the inductors with short circuits and the capacitors with open circuits. We can easily solve these, using our resistive-circuit-analysis techniques developed in Chap. 3, to yield

$$v_x' = i_x' = 0$$

$$v_x'' = -2 \text{ V}$$

$$i_x'' = -1 \text{ A}$$

as the reader should verify. Thus

$$v_x = -2 \text{ V}$$

$$i_x = -1 \text{ A}$$

Note that the 2-F capacitor in series with the 10-V voltage source effectively blocks any effect of this source on the circuit.

We have seen in this section that computing the contribution of a dc source to a branch voltage or current in a circuit is very simple—replace inductors with short circuits and capacitors with open circuits and solve the resulting resistive circuit. The two forms of ideal sources which we will study exclusively in this text are the dc source and the ac, or sinusoidal, source. In the next chapter we will determine a method of solving for that contribution to a branch voltage or current of a circuit which is due to a sinusoidal source in the circuit. We will find that a simple method evolves there which also makes use of the resistive-circuit-analysis techniques we developed in Chap. 3. Now we are beginning to see why a thorough understanding of those resistive-circuit-analysis methods is so important—we will utilize them over and over again.

PROBLEMS

4.1 A 10-μF capacitor is to be constructed by rolling two strips of aluminum foil into a tubular shape. If the strips are 1 in wide and separated by a sheet of paper which is 1 mil thick, determine the length of each sheet in inches.

4.2 A 10-μF capacitor has a voltage applied across it as shown in Fig. P4.2. Sketch the capacitor current and charge delivered to the capacitor.

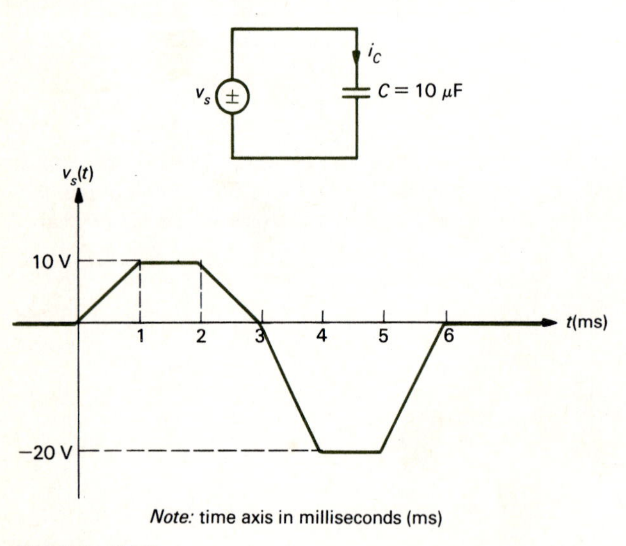

FIGURE P4.2

4.3 A 20-pF capacitor has the current waveform shown in Fig. P4.3 applied to it. Sketch the voltage across the capacitor.

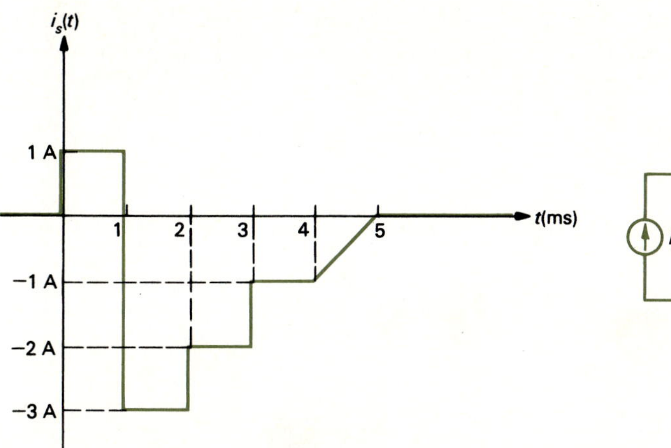

FIGURE P4.3

4.4 In Prob. 4.3, determine the total charge transferred to the capacitor at $t = 4$ milliseconds (ms).

4.5 Replace the capacitors in Fig. P4.5 at terminals *a-b* with an equivalent capacitance.

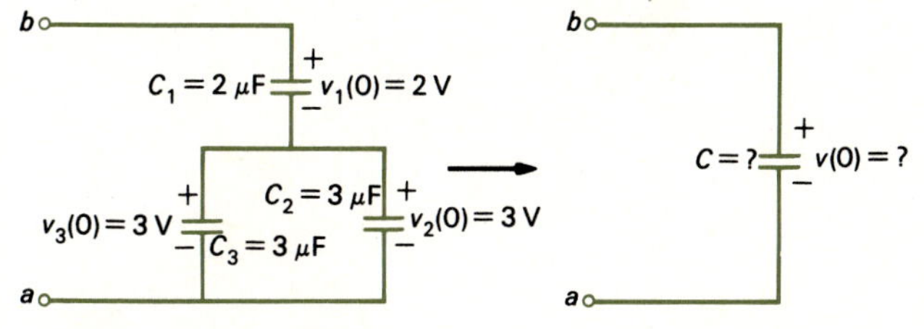

FIGURE P4.5

4.6 A 5-mH inductor has the voltage waveform shown in Fig. P4.6 applied across it. Sketch the inductor current.

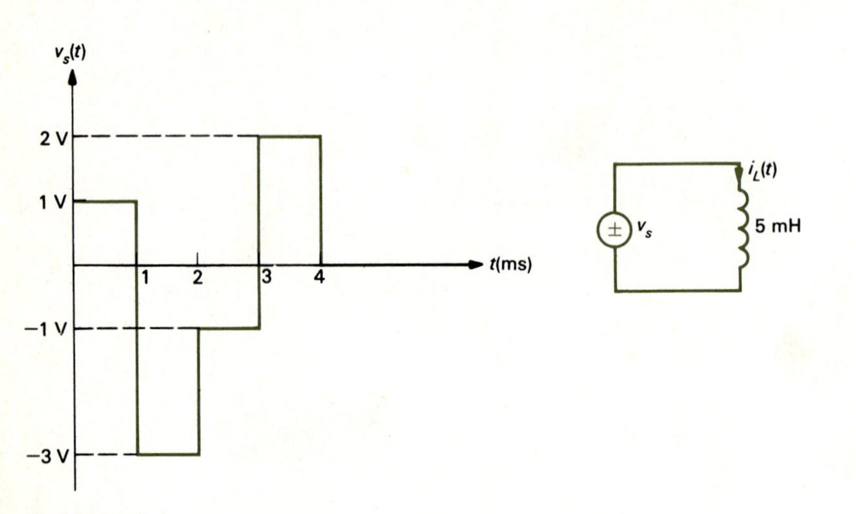

FIGURE P4.6

4.7 A 10-mH inductor has the current waveform shown in Fig. P4.7 applied to it. Sketch the inductor voltage.

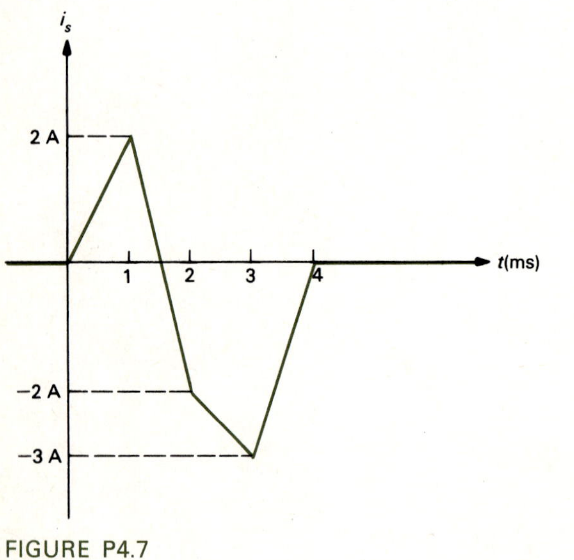

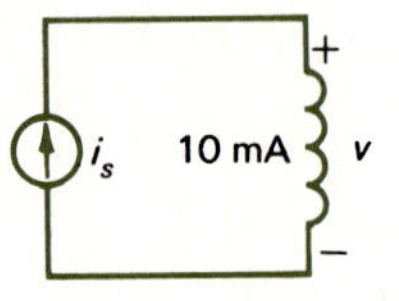

FIGURE P4.7

4.8 Replace the inductors shown in Fig. P4.8 with an equivalent inductance at terminals *a-b*.

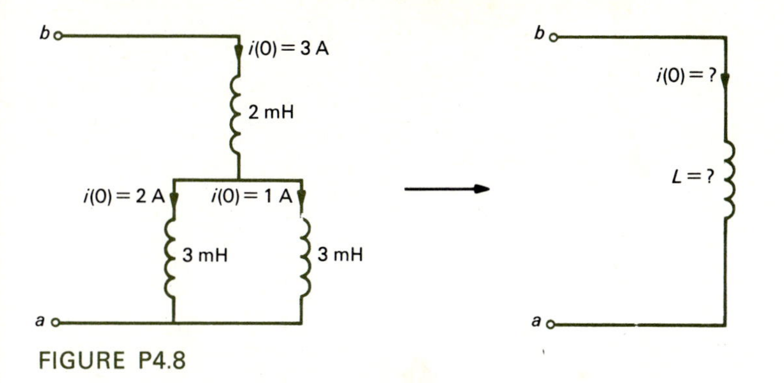

FIGURE P4.8

4.9 The voltage waveform shown in Fig. P4.9 is applied across a 5-mH inductor and a 2-μF capacitor. Sketch the current through each element.

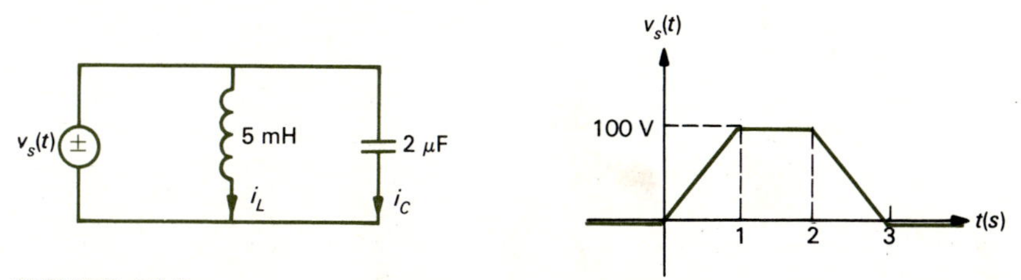

FIGURE P4.9

4.10 Sketch the energy stored in the capacitor of Fig. P4.2.

4.11 In Fig. P4.7, determine the energy stored in the inductor at $t = 3$ ms.

4.12 In Fig. P4.9, determine the total energy stored in the inductor and capacitor at $t = 1$ s.

4.13 In Fig. P4.13, determine v_x, v_z, and i_y.

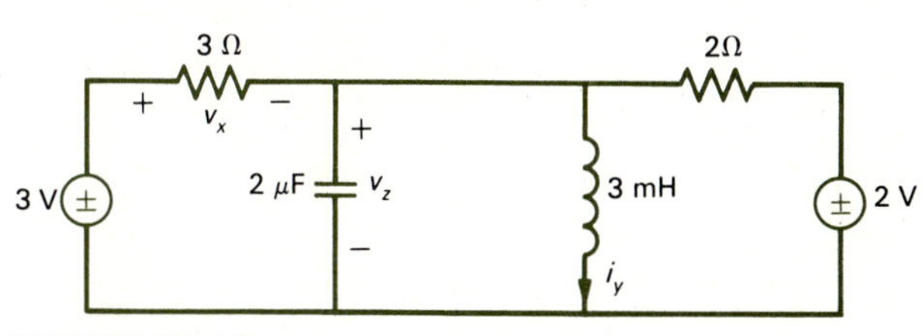

FIGURE P4.13

4.14 In the circuit shown in Fig. P4.14, determine v_x, i_y, v_z, and i_w.

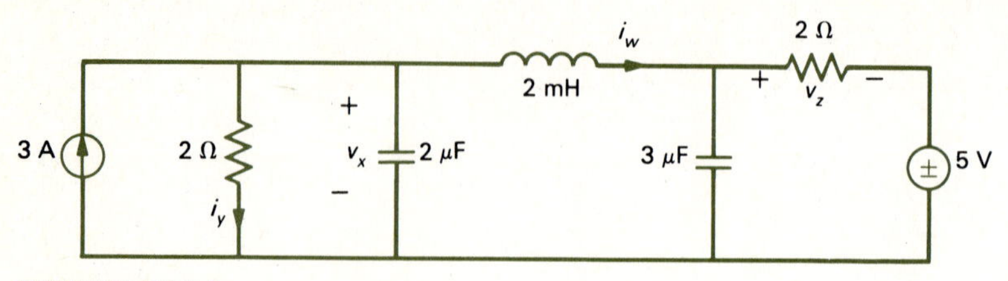

FIGURE P4.14

4.15 Determine the current i in Fig. P4.15.

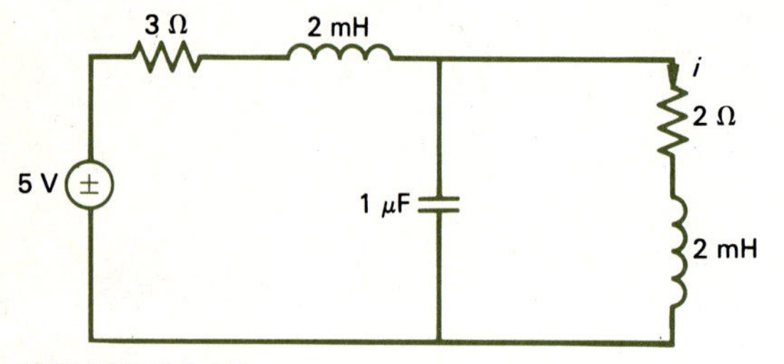

FIGURE P4.15

REFERENCES

4.1 A. B. Carlson and D. G. Gisser, *Electrical Engineering: Concepts and Applications*, Addison-Wesley, Reading, Mass., 1981.

4.2 Ralph J. Smith, *Circuits, Devices and Systems: A First Course in Electrical Engineering*, 4th ed., Wiley, New York, 1984.

4.3 A. E. Fitzgerald, D. Higginbotham, and A. Grabel, *Basic Electrical Engineering*, 5th ed., McGraw-Hill, New York, 1981.

4.4 J. D. Irwin, *Basic Engineering Circuit Analysis*, Macmillan, New York, 1983.

Chapter

5

AC Circuits

In a circuit containing several ideal current and voltage sources, we may determine a particular branch voltage or current by superposition—that is, by computing the contributions to this branch voltage or current of each source acting individually, with the other sources killed. In the last section of the previous chapter we determined a simple method for computing the contribution to a branch voltage or current of a dc (constant-value) source—replace all inductors with short circuits and all capacitors with open circuits and solve the resulting resistive circuit.

In this chapter we will examine this problem for ac, or sinusoidal, sources. We will find that the contribution to a particular branch voltage or branch current due to a sinusoidal source either of value $A \sin(\omega t + \phi)$ or $A \cos(\omega t + \phi)$ is of the same form and frequency. We will also determine a simple technique for computing this contribution; the method makes use of our resistive-circuit-analysis techniques developed in Chap. 3 with only one small, additional complication—the method requires the manipulation of complex numbers rather than real numbers. However, we will find that this requirement to deal with complex numbers is a small price to pay for the tremendous solution advantages of the method.

Most of our later work will require the analysis of circuits containing sinusoidal sources. The results of this chapter are therefore very important, since we will be using them frequently throughout the remainder of this text. Hence, they should be studied very carefully.

Sinusoidal sources are of considerable importance for a number of reasons. One reason is that electric power is transmitted as a sinusoidal voltage and current at a frequency of 60 Hz (50 Hz in Europe). But perhaps the most important reason for their study is that any periodic function of time can be represented as an infinite sum of sinusoids with a Fourier series representation, as is shown in App. C; therefore, any ideal voltage or current source whose value is a periodic function of time may be represented as a sum of sinusoidal sources, and superposition can be employed to find the contribution of each of these sinusoidal components to a particular branch voltage or current of the circuit. Furthermore, nonperiodic functions of time can also be represented as an infinite sum of sinusoids with a Fourier integral; any arbitrary function of time can thus be viewed as being composed of sinusoidal components. Therefore, we do not need to investigate methods of finding the solutions for branch voltages or currents of a circuit due to sources which are arbitrary functions of time; we only need to investigate methods of finding the solution for sinusoidal sources. This chapter will be devoted exclusively to that problem.

5.1 AN EXAMPLE

Consider the circuit shown in Fig. 5.1. The ideal voltage source is sinusoidal at a radian frequency of $\omega = 3$ radians per second (rad/s). Let us examine the determination of the solution for the inductor current $i(t)$. First we will derive an equation relating $i(t)$ and the source. To do so, we note that $i(t)$ also passes through the 3-Ω resistor, developing a voltage across it of $Ri(t) = 3i(t)$. The voltage across the inductor is $L\,di(t)/dt = 2di(t)/dt$. Applying KVL around the loop, we have

$$2\frac{di(t)}{dt} + 3i(t) = 10 \sin 3t \tag{5.1}$$

Now let us find the solution to this equation. Let us guess the form of the solution. If we can obtain a guess which will satisfy Eq. (5.1), then we need look no further. A reasonable choice would be to assume $i(t)$ in the form of another sinusoid at a radian frequency of 3 rad/s also:

$$i(t) = I \sin 3t \tag{5.2}$$

Substituting into Eq. (5.1), we have

$$6I \cos 3t + 3I \sin 3t = 10 \sin 3t \tag{5.3}$$

But this equation cannot be satisfied by any choice of I. So try

$$i(t) = I_s \sin 3t + I_c \cos 3t \tag{5.4}$$

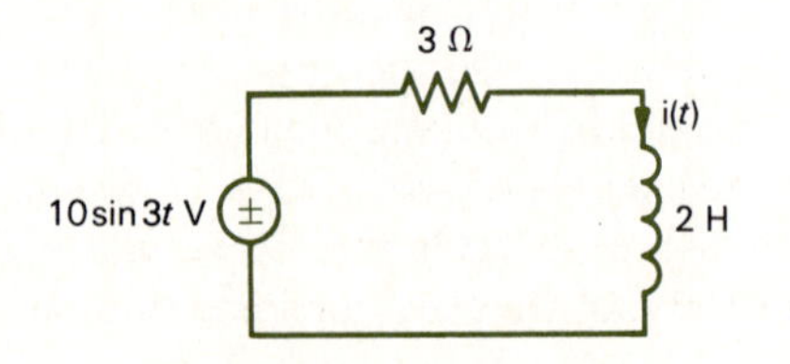

FIGURE 5.1 Illustration of the response of circuits to sinusoidal sources.

Substituting into Eq. (5.1) gives

$$6I_s \cos 3t - 6I_c \sin 3t + 3I_s \sin 3t + 3I_c \cos 3t = 10 \sin 3t \tag{5.5}$$

Matching coefficients of sin $3t$ and cos $3t$, we have

$$\begin{aligned} 3I_s - 6I_c &= 10 \\ 6I_s + 3I_c &= 0 \end{aligned} \tag{5.6}$$

By Cramer's rule we obtain

$$\begin{aligned} I_s &= \frac{\begin{vmatrix} 10 & -6 \\ 0 & 3 \end{vmatrix}}{\begin{vmatrix} 3 & -6 \\ 6 & 3 \end{vmatrix}} \\ &= \tfrac{30}{45} \\ &= 0.67 \\ I_c &= \frac{\begin{vmatrix} 3 & 10 \\ 6 & 0 \end{vmatrix}}{45} \\ &= -\tfrac{60}{45} \\ &= -1.33 \end{aligned} \tag{5.7}$$

and the solution has been found. Substituting Eq. (5.7) into (5.4) yields the solution

$$i(t) = 0.67 \sin 3t - 1.33 \cos 3t \tag{5.8}$$

This may be placed in an alternative form with the trigonometric identity

$$A \sin \theta + B \cos \theta = M \sin (\theta + \phi)$$

where $M = \sqrt{A^2 + B^2}$

$$\phi = \tan^{-1} \frac{B}{A}$$

so that

$$i(t) = 1.49 \sin (3t - 63.43°) \tag{5.9}$$

For this rather simple circuit we needed to (1) derive the differential equation relating the unknown branch current to the source and (2) solve that differential equation. Step (2) required the solution of two simultaneous equations. In the next sections we will develop a much simpler method for finding this solution which avoids both of these steps. The tradeoff is that we must deal with complex numbers in that method. However, we will see that this is a small price to pay for the method's considerable computational advantage.

5.2 COMPLEX NUMBERS AND COMPLEX ALGEBRA

As indicated in the previous section, the method which we will develop will involve the manipulation of complex numbers. We therefore need to become thoroughly familiar with complex numbers and the rules for their manipulation (complex algebra).

A complex number consists of two parts: a real part and an imaginary part. Complex numbers will be indicated by boldface throughout this text. For example, the complex number **C** may be represented as

$$\mathbf{C} = a + jb \tag{5.10}$$

where a and b are real numbers and j symbolizes the square root of -1:

$$j = \sqrt{-1} \tag{5.11}$$

The real part of **C** is a and is denoted by

$$a = \text{Re}\,\mathbf{C} \tag{5.12}$$

Similarly, the imaginary part of **C** is b and is denoted by

$$b = \text{Im}\,\mathbf{C} \tag{5.13}$$

The complex number **C** can be represented as a vector in the complex plane by plotting the real part on the horizontal axis and the imaginary part on the vertical axis, as shown in Fig. 5.2. The length of this vector is referred to as the magnitude of the complex number and is denoted as $|\mathbf{C}|$ or C, where

$$C = \sqrt{a^2 + b^2} \tag{5.14}$$

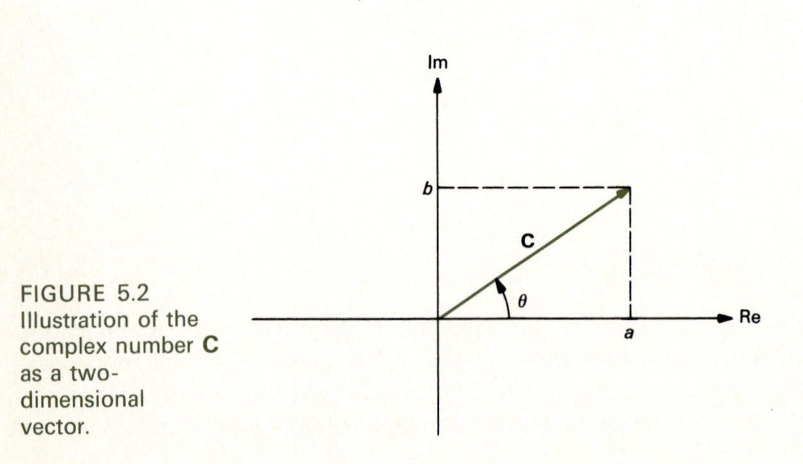

FIGURE 5.2 Illustration of the complex number **C** as a two-dimensional vector.

The angle θ, measured counterclockwise from the positive, horizontal axis, is

$$\theta = \tan^{-1} \frac{b}{a} \tag{5.15}$$

The representation of the complex number in terms of its real and imaginary parts as in Eq. (5.10) is referred to as the *rectangular form*. Alternatively, the complex number may be represented in terms of its magnitude and angle in *polar form*:

$$\mathbf{C} = C\underline{/\theta} \tag{5.16}$$

From Fig. 5.2

$$\begin{aligned} a &= C \cos \theta \\ b &= C \sin \theta \end{aligned} \tag{5.17}$$

Thus

$$\begin{aligned} \mathbf{C} &= C \cos \theta + jC \sin \theta \\ &= C (\cos \theta + j \sin \theta) \end{aligned} \tag{5.18}$$

The last portion of Eq. (5.18) is written as

$$e^{j\theta} \triangleq \cos \theta + j \sin \theta \tag{5.19}$$

This is referred to as Euler's (pronounced "oiler's") identity. Thus

$$\mathbf{C} = Ce^{j\theta} \tag{5.20}$$

Note that the magnitude of $e^{j\theta}$ is unity, since

$$\begin{aligned} |e^{j\theta}| &= \sqrt{\cos^2 \theta + \sin^2 \theta} \\ &= 1 \end{aligned} \tag{5.21}$$

The angle of $e^{j\theta}$ is simply

$$\begin{aligned} \underline{/e^{j\theta}} &= \tan^{-1} \frac{\sin \theta}{\cos \theta} \\ &= \theta \end{aligned} \tag{5.22}$$

Thus, the polar form of $e^{j\theta}$ is

$$e^{j\theta} = 1\underline{/\theta} \tag{5.23}$$

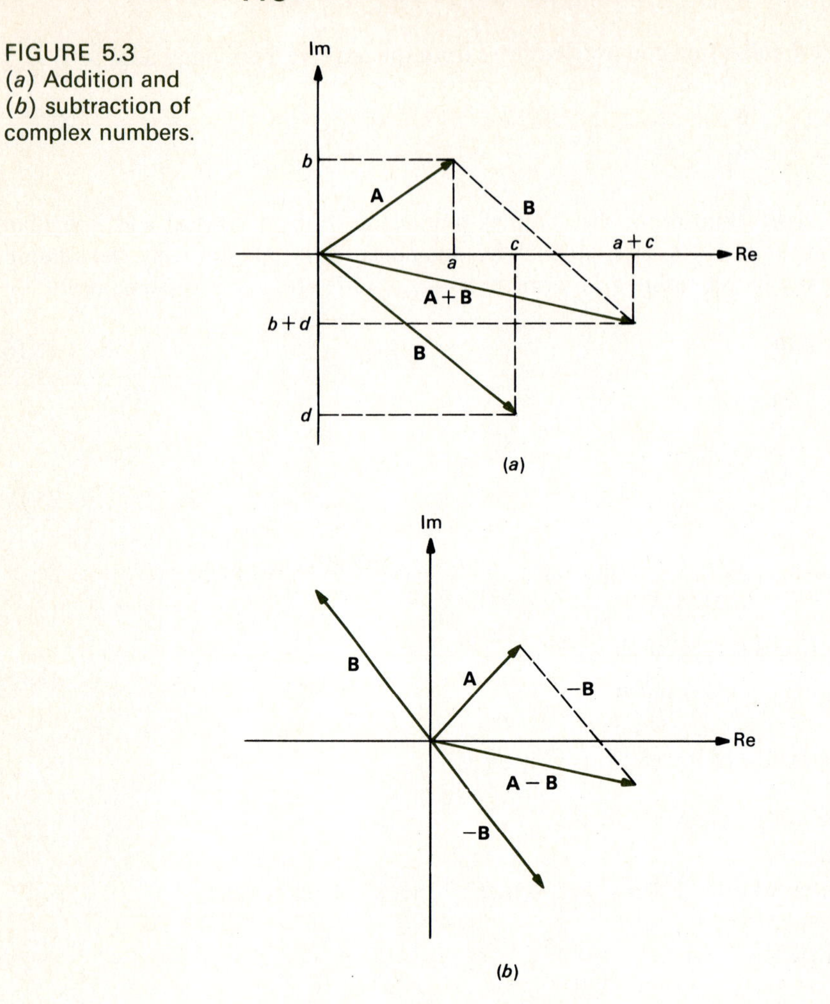

FIGURE 5.3
(*a*) Addition and (*b*) subtraction of complex numbers.

Now consider the addition, subtraction, multiplication, and division of two complex numbers. If

$$\begin{aligned} \mathbf{A} &= a + jb \\ \mathbf{B} &= c + jd \end{aligned} \tag{5.24}$$

then the sum of these two is simply the vector addition of the two complex vectors, as shown in Fig. 5.3*a*. The result is

$$\begin{aligned} A + B &= (a + jb) + (c + jd) \\ &= (a + c) + j(b + d) \end{aligned} \tag{5.25}$$

Thus

$$\begin{aligned} \text{Re}\,(\mathbf{A} + \mathbf{B}) &= \text{Re}\,\mathbf{A} + \text{Re}\,\mathbf{B} \\ \text{Im}\,(\mathbf{A} + \mathbf{B}) &= \text{Im}\,\mathbf{A} + \text{Im}\,\mathbf{B} \end{aligned} \tag{5.26}$$

The subtraction of two complex numbers is defined as

$$\begin{aligned}\mathbf{A} - \mathbf{B} &= \mathbf{A} + (-\mathbf{B}) \\ &= (a - c) + j(b - d)\end{aligned} \tag{5.27}$$

as shown in Fig. 5.3*b*.

The multiplication of two complex numbers is defined by the following. First we represent the two numbers in their polar form:

$$\begin{aligned}\mathbf{A} &= A\underline{/\theta_A} \\ \mathbf{B} &= B\underline{/\theta_B}\end{aligned} \tag{5.28}$$

The product is defined by

$$\mathbf{AB} = AB\underline{/\theta_A + \theta_B} \tag{5.29}$$

Thus, the magnitude of the product is the product of the magnitudes, and the angle of the product is the sum of the angles. Note that

$$\begin{aligned}j^2 &= \sqrt{-1}\sqrt{-1} \\ &= -1 \\ j^3 &= -j \\ j^4 &= 1 \\ j^5 &= j \\ &\vdots\end{aligned} \tag{5.30}$$

Thus, the product can be obtained in rectangular form as

$$\begin{aligned}\mathbf{AB} &= (a + jb)(c + jd) \\ &= ac + jad + jbc + j^2bd \\ &= (ac - bd) + j(ad + bc)\end{aligned} \tag{5.31}$$

Division is defined as

$$\frac{\mathbf{A}}{\mathbf{B}} = \frac{A}{B}\underline{/\theta_A - \theta_B} \tag{5.32}$$

so that the magnitude is the magnitude of the numerator A divided by the magnitude of the denominator B, and the angle is obtained by subtracting the angle of the denominator from the angle of the numerator.

The conjugate of a complex number **C** is denoted with a star (*) as **C*** and is obtained by changing the sign of the imaginary part. For example, if

$$\mathbf{C} = a + jb \tag{5.33}$$

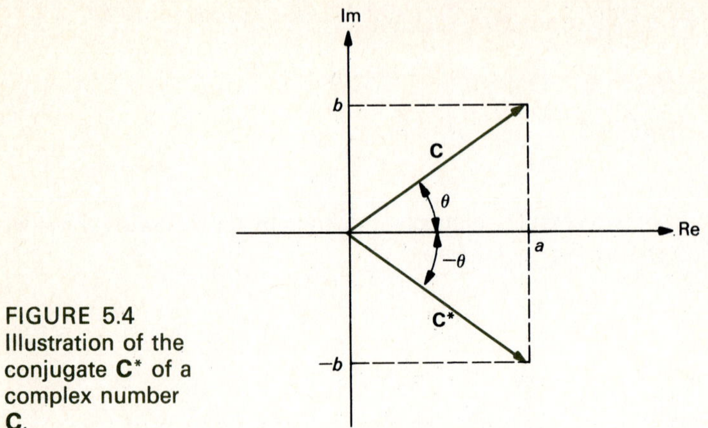

FIGURE 5.4
Illustration of the conjugate **C*** of a complex number **C**.

then

$$\mathbf{C}^* = a - jb \tag{5.34}$$

In terms of the polar form,

$$\begin{aligned} \mathbf{C} &= C\underline{/\theta} \\ \mathbf{C}^* &= C\underline{/-\theta} \end{aligned} \tag{5.35}$$

as is easily seen from Fig. 5.4. In terms of the conjugate we may obtain some other interesting results. For example, for the complex numbers **A** and **B** defined in Eq. (5.24), we may write

$$\begin{aligned} \frac{\mathbf{A}}{\mathbf{B}} &= \frac{a+jb}{c+jd} \\ &= \frac{a+jb}{c+jd}\frac{c-jd}{c-jd} \\ &= \frac{ac+bd}{c^2+d^2} + j\frac{bc-ad}{c^2+d^2} \\ &= \frac{\mathbf{A}}{\mathbf{B}}\cdot\frac{\mathbf{B}^*}{\mathbf{B}^*} \end{aligned} \tag{5.36}$$

Note that

$$\begin{aligned} \mathbf{BB}^* &= (c+jd)(c-jd) \\ &= c^2+d^2 \\ &= |\mathbf{B}|^2 \end{aligned} \tag{5.37}$$

Thus, the square of the magnitude of a complex number is the product of itself and its conjugate. This is a very important result and will be used on numerous occasions.

Several quite useful identities may now be obtained. For example,

$$
\begin{aligned}
(\mathbf{A}^*)^* &= \mathbf{A} \\
(\mathbf{AB})^* &= \mathbf{A}^*\mathbf{B}^* \\
\left(\frac{\mathbf{A}}{\mathbf{B}}\right)^* &= \frac{\mathbf{A}^*}{\mathbf{B}^*} \\
(\mathbf{A} + \mathbf{B})^* &= \mathbf{A}^* + \mathbf{B}^*
\end{aligned}
\tag{5.38}
$$

The reader should study these identities and prove each one.

Example 5.1 If

$$\mathbf{A} = 2 + j3 = 3.61\underline{/56.31^\circ}$$

$$\mathbf{B} = 3 + j5 = 5.83\underline{/59.04^\circ}$$

find **C** in rectangular and polar form when

(a) $\mathbf{C} = \mathbf{A} + \mathbf{B}$

(b) $\mathbf{C} = \mathbf{A} - \mathbf{B}$

(c) $\mathbf{C} = \mathbf{AB}$

(d) $\mathbf{C} = \dfrac{\mathbf{A}}{\mathbf{B}}$

(e) $\mathbf{C} = \mathbf{A} + \mathbf{B}^* = (\mathbf{A}^* + \mathbf{B})^*$

(f) $\mathbf{C} = \mathbf{A}^*\mathbf{B} = (\mathbf{AB}^*)^*$

(g) $\mathbf{C} = \dfrac{\mathbf{A}}{\mathbf{B}^*} = \left(\dfrac{\mathbf{A}^*}{\mathbf{B}}\right)^*$

Solution

(a)
$$
\begin{aligned}
\mathbf{C} &= \mathbf{A} + \mathbf{B} \\
&= (2 + j3) + (3 + j5) \\
&= 5 + j8 \\
&= 9.43\underline{/58^\circ}
\end{aligned}
$$

(b)
$$
\begin{aligned}
\mathbf{C} &= \mathbf{A} - \mathbf{B} \\
&= (2 + j3) - (3 + j5) \\
&= -1 - j2 \\
&= 2.24\underline{/-116.57^\circ}
\end{aligned}
$$

(c) $\mathbf{C} = \mathbf{AB}$

$$= (2 + j3)(3 + j5)$$

$$= -9 + j19$$

$$\mathbf{C} = 3.61\angle 56.31^\circ \; 5.83\angle 59.04^\circ$$

$$= 21.02\angle 115.35^\circ$$

Note: $-9 + j19 = 21.02\angle 115.35^\circ$

(d) $\mathbf{C} = \dfrac{\mathbf{A}}{\mathbf{B}}$

$$= \frac{2 + j3}{3 + j5}$$

$$= \frac{2 + j3}{3 + j5}\,\frac{3 - j5}{3 - j5}$$

$$= 0.62 - j0.03$$

$$\mathbf{C} = \frac{3.61\angle 56.31^\circ}{5.83\angle 59.04^\circ}$$

$$= 0.62\angle -2.73^\circ$$

Note: $0.62 - j0.03 = 0.62\angle -2.73^\circ$

(e) $\mathbf{C} = \mathbf{A} + \mathbf{B}^*$

$$= (2 + j3) + (3 - j5)$$

$$= 5 - j2$$

Note: $\mathbf{C} = \mathbf{A} + \mathbf{B}^*$

$$= (\mathbf{A}^* + \mathbf{B})^*$$

$$= (2 - j3 + 3 + j5)^*$$

$$= (5 + j2)^*$$

$$= 5 - j2$$

(f) $\mathbf{C} = \mathbf{A}^*\mathbf{B}$

$$= (2 - j3)(3 + j5)$$

$$= 21 + j1$$

$$= 21.02\angle 2.73^\circ$$

$$= 3.61\angle -56.31^\circ \; 5.83\angle 59.04^\circ$$

$$= 21.02\angle 2.73^\circ$$

$$= (\mathbf{AB}^*)^*$$

$$= (3.61\angle 56.31^\circ \; 5.83\angle -59.04^\circ)^*$$

$$= (21.02\angle -2.73^\circ)^*$$

$$= 21.02\angle 2.73^\circ$$

(g) $\mathbf{C} = \dfrac{\mathbf{A}}{\mathbf{B}^*}$

$$= \frac{2 + j3}{3 - j5}$$

$$= \frac{2 + j3}{3 - j5} \frac{3 + j5}{3 + j5}$$

$$= \frac{-9 + j19}{34}$$

$$= -0.26 + j0.56$$

$$= 0.62\underline{/115.35^\circ}$$

$$= \frac{3.61\underline{/56.31^\circ}}{5.83\underline{/-59.04^\circ}}$$

$$= 0.62\underline{/115.34}$$

$$= \left(\frac{\mathbf{A}^*}{\mathbf{B}}\right)^*$$

$$= \left(\frac{3.61\underline{/-56.31^\circ}}{5.83\underline{/59.04^\circ}}\right)^*$$

$$= (0.62\underline{/-115.35^\circ})^*$$

$$= 0.62\underline{/115.35^\circ}$$

The above complex operations can be used repeatedly to reduce a complex expression to a real and imaginary part or, as the following example shows, to a magnitude and angle.

Example 5.2 Suppose

$$\mathbf{A} = 1 + j2$$

$$\mathbf{B} = 3 - j5$$

$$\mathbf{C} = 2 + j4$$

$$\mathbf{D} = 1 - j3$$

$$\mathbf{E} = 4 + j1$$

Evaluate

$$\mathbf{F} = \frac{\mathbf{A} + \mathbf{B}/\mathbf{C}}{\mathbf{DE}}$$

Solution First we evaluate **B**/**C**. Writing in polar form,

$$\frac{\mathbf{B}}{\mathbf{C}} = \frac{5.83\angle -59.04^\circ}{4.47\angle 63.43^\circ}$$

$$= 1.30\angle -122.47^\circ$$

Converting to rectangular form,

$$\frac{\mathbf{B}}{\mathbf{C}} = -0.70 - j1.10$$

Now

$$\mathbf{A} + \frac{\mathbf{B}}{\mathbf{C}} = (1 + j2) + (-0.7 - j1.1)$$

$$= 0.3 + j0.9$$

$$= 0.95\angle 71.57^\circ$$

$$\mathbf{F} = \frac{\mathbf{A} + \mathbf{B}/\mathbf{C}}{\mathbf{DE}}$$

$$= \frac{0.95\angle 71.57^\circ}{(1 - j3)(4 + j1)}$$

$$= \frac{0.95\angle 71.57^\circ}{3.16\angle -71.57^\circ \; 4.12\angle 14.04^\circ}$$

$$= \frac{0.95\angle 71.57^\circ}{13.02\angle -57.53^\circ}$$

$$= 0.07\angle 129.10^\circ$$

The manipulation of complex-number expressions, as in the preceding example, generally requires several conversions of complex numbers between rectangular and polar forms. An electronic calculator having the ability to convert between these forms is particularly advantageous in reducing these expressions, and we will have numerous occasions in the remainder of this chapter to reduce such expressions.

5.3 BRUTE-FORCE SOLUTION OF THE AC CIRCUIT

With this knowledge of complex numbers and algebra, let us reexamine the solution of the circuit in Fig. 5.1 which we considered in Sec. 5.1. The differential equation relating $i(t)$ to the source was

$$2\frac{di(t)}{dt} + 3i(t) = 10 \sin 3t \tag{5.39}$$

Instead of solving Eq. (5.39), let us replace the sinusoidal source with

$$10 \sin 3t \rightarrow 10e^{j3t} \tag{5.40}$$

as shown in Fig. 5.5*a*. Thus, Eq. (5.39) becomes

$$2\frac{dI}{dt} + 3I = 10e^{j3t} \tag{5.41}$$

A reasonable guess for the form of the solution to this equation would be to assume I to be of the same form as the right-hand side:

$$I = \mathbf{I}e^{j3t} \tag{5.42}$$

Substitute Eq. (5.42) into Eq. (5.41)

$$2(j3)\mathbf{I}e^{j3t} + 3\mathbf{I}e^{j3t} = 10e^{j3t} \tag{5.43}$$

Canceling e^{j3t}, which is common to both sides, we obtain the solution for $\mathbf{I}$ as

$$\begin{aligned} \mathbf{I} &= \frac{10}{3 + j6} \\ &= 1.49\underline{/-63.43^\circ} \end{aligned} \tag{5.44}$$

Thus, the solution to Eq. (5.41) is

$$\begin{aligned} I &= \mathbf{I}e^{j3t} \\ &= 1.49\underline{/-63.43^\circ}\; e^{j3t} \\ &= 1.49e^{j(3t - 63.43^\circ)} \end{aligned} \tag{5.45}$$

Now let us examine how the solution, I, to Eq. (5.41) is related to the solution, $i(t)$, to the original equation in Eq. (5.39). The solution for I in Eq. (5.45) can be written, with Euler's identity, as

$$\begin{aligned} I &= 1.49e^{j(3t - 63.43^\circ)} \\ &= 1.49 \cos (3t - 63.43^\circ) + j1.49 \sin (3t - 63.43^\circ) \end{aligned} \tag{5.46}$$

Also, the right-hand side of Eq. (5.41) can be written, using Euler's identity, as

$$10e^{j3t} = 10 \cos 3t + j10 \sin 3t \tag{5.47}$$

Thus, Eq. (5.41) can be written as

$$\begin{aligned} &2\frac{d}{dt}[1.49 \cos (3t - 63.43^\circ) + j1.49 \sin (3t - 63.43^\circ)] \\ &\quad + 3[1.49 \cos 3t - 63.43^\circ) + j1.49 \sin (3t - 63.43^\circ)] \\ &= 10 \cos 3t + j10 \sin 3t \end{aligned} \tag{5.48}$$

Equating the real and imaginary parts of this equation yields two equations:

$$2\frac{d}{dt}[1.49\cos(3t - 63.43°)] + 3[1.49\cos(3t - 63.43°)] = 10\cos 3t \tag{5.49a}$$

$$2\frac{d}{dt}[1.49\sin(3t - 63.43°)] + 3[1.49\sin(3t - 63.43°)] = 10\sin 3t \tag{5.49b}$$

Now compare the original equation, Eq. (5.39), with Eq. (5.49*b*), and we see that the solution to this original equation in Eq. (5.39) is

$$i(t) = 1.49\sin(3t - 63.43°) \tag{5.50}$$

as was obtained in Sec. 5.1 by guessing a form of solution and substituting that guess into the equation to find the values of the unknowns in the assumed form of the solution.

Right now, in order to perceive a method in this, it is important to review what we have done. Note that the right-hand side of the original equation, $10\sin 3t$ (the value of the source), is the imaginary part of the right-hand side of Eq. (5.41), $10e^{j3t}$ (the new value of the source):

$$10\sin 3t = \text{Im } 10e^{j3t} \tag{5.51}$$

Note, too, that the solution to the original equation in Eq. (5.39) is also the imaginary part of the solution to the new equation in Eq. (5.41):

$$i(t) = \text{Im } \mathbf{I}e^{j3t} \tag{5.52}$$

At this point it is easy to see why this happens. Since

$$10e^{j3t} = 10\cos 3t + j10\sin 3t \tag{5.53}$$

we see that we may represent the 10^{j3t} source as the series combination of two sources, $10\cos 3t$ and $j10\sin 3t$, as shown in Fig. 5.5*b*, when we replace the original sinusoidal source in Fig. 5.1 with this new source, as in Fig. 5.5*a*. Thus, by superposition, the solution for I will also consist of two components, one due to the real part of $10e^{j3t}$,

$$10\cos 3t \rightarrow I_R \tag{5.54}$$

and one due to the imaginary part of $10e^{j3t}$,

$$j10\sin 3t \rightarrow jI_I \tag{5.55}$$

Therefore, if we replace the original source with $10e^{j3t}$, the solution for the current with the original source present is the imaginary part of the solution due to the source

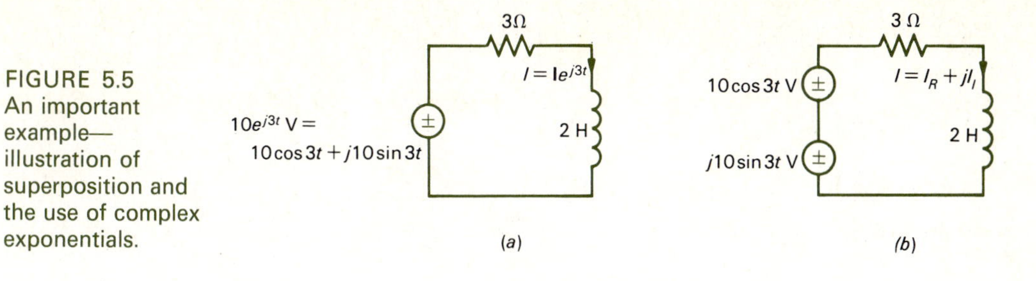

FIGURE 5.5 An important example—illustration of superposition and the use of complex exponentials.

$10e^{j3t}$. Superposition has come into play once again. It is also clear that if the circuit had been a nonlinear one (at least one circuit element had been nonlinear), then this method (which relied on the validity of superposition) *would not work.*

Suppose that the ideal voltage source had been cosinusoidal with a value of $10 \cos 3t$. How would we find the solution for $i(t)$? The answer is now quite clear; the solution would be the real part of the solution due to the $10e^{j3t}$ source, I_R, since $10 \cos 3t$ is the real part of $10e^{j3t}$.

Now it is easy to see that this method will work for any linear circuit. To compute the contribution to a branch current or voltage due to a sinusoidal source $A \sin (\omega t + \phi)$ or $A \cos (\omega t + \phi)$, replace the source with $Ae^{j\phi}e^{j\omega t}$ and solve the resulting circuit for the complex branch current or voltage of interest:

$$\begin{aligned} I &= \mathbf{I}e^{j\omega t} \\ V &= \mathbf{V}e^{j\omega t} \end{aligned} \tag{5.56}$$

where

$$\begin{aligned} \mathbf{I} &= I\underline{/\theta_I} \\ \mathbf{V} &= V\underline{/\theta_V} \end{aligned} \tag{5.57}$$

Now if the source was a sine source, the solution is the imaginary part of Eq. (5.56),

$$A \sin (\omega t + \phi) \rightarrow \begin{matrix} I \sin (\omega t + \theta_I) \\ V \sin (\omega t + \theta_V) \end{matrix} \tag{5.58}$$

and if the source was a cosine source, the solution is the real part of Eq. (5.56),

$$A \cos (\omega t + \phi) \rightarrow \begin{matrix} I \cos (\omega t + \theta_I) \\ V \cos (\omega t + \theta_V) \end{matrix} \tag{5.59}$$

One final point should be noted. Observe that in this example e^{j3t} was common to *all* currents and voltages in the circuit. Therefore, we may drop it from all assumed solutions and solve directly for $\mathbf{I}$, as shown in Fig. 5.6. Once this is found, we simply multiply $\mathbf{I}$ by e^{j3t}, as shown in Eq. (5.42).

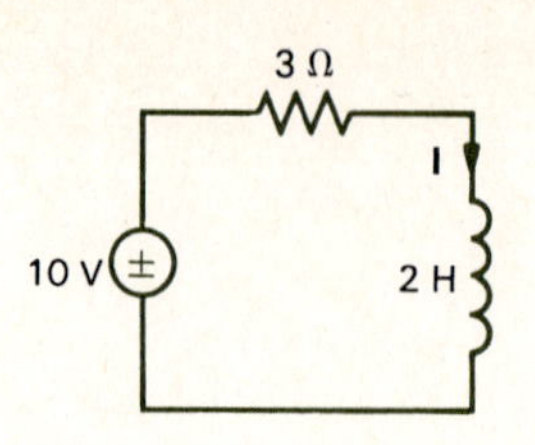

FIGURE 5.6 A simplification of the circuit in Fig. 5.5.

Example 5.3

Consider the circuit in Fig. 5.7*a*. Determine $v(t)$.

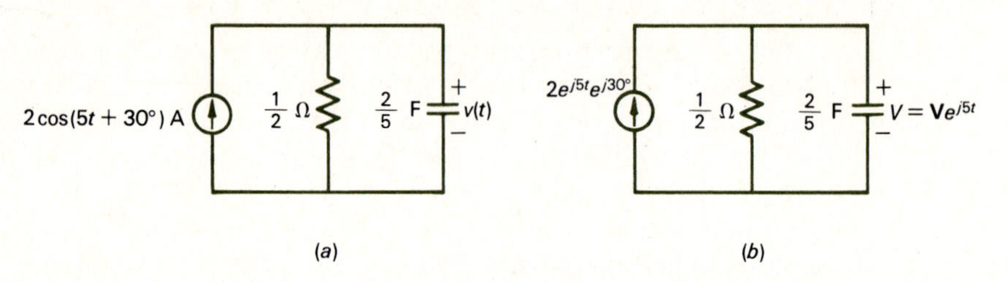

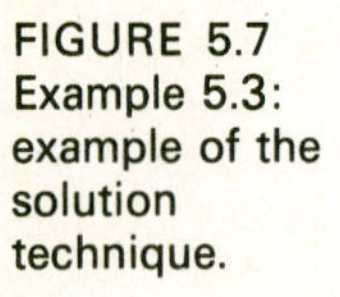

FIGURE 5.7 Example 5.3: example of the solution technique.

Solution The differential equation relating $v(t)$ to the source is obtained by applying KCL at the upper node. The current passing through the $\frac{1}{2}$-Ω resistor is $2v(t)$ and the current passing through the 2/5-F capacitor is $\frac{2}{5}[dv(t)/dt]$, so that

$$\frac{2}{5}\frac{dv(t)}{dt} + 2v(t) = 2\cos(5t + 30°)$$

$$= \text{Re } 2e^{j5t}e^{j30°}$$

Thus, replacing the source with $2e^{j5t}e^{j30°}$, as in Fig. 5.7*b*, and substituting

$$V = \mathbf{V}e^{j5t}$$

into the above equation yields

$$\tfrac{2}{5}j5\mathbf{V}e^{j5t} + 2\mathbf{V}e^{j5t} = 2e^{j5t}e^{j30°}$$

Solving for **V** gives

$$\mathbf{V} = \frac{2\underline{/30°}}{\frac{2}{5}j5 + 2}$$

$$= \frac{2\underline{/30°}}{2\sqrt{2}\underline{/45°}}$$

$$= 0.707\underline{/-15°}$$

Thus

$$V = \mathbf{V}e^{j5t}$$

$$= 0.707e^{-j15°}e^{j5t}$$

$$= 0.707e^{j(5t-15°)}$$

Since the original source was the real part of the replacement source,

$$2\cos(5t + 30°) = \text{Re } 2\underline{/30°}e^{j5t}$$

then

$$v(t) = \text{Re}\,[V = \mathbf{V}e^{j5t}]$$
$$= 0.707\cos(5t - 15°)$$

The reader should check, by direct substitution into the original differential equation, that this is, indeed, the solution.

5.4 THE PHASOR CIRCUIT

From the previous example, it is clear that a considerable simplification in the solution of an ac circuit has been obtained. In finding the component of a branch voltage or current due to an ideal voltage or current source which has the form

$$A\cos(\omega t + \phi) \tag{5.60a}$$

or

$$A\sin(\omega t + \phi) \tag{5.60b}$$

we replace that source with

$$Ae^{j(\omega t + \phi)} = A\underline{/\phi}e^{j\omega t} \tag{5.61}$$

We next assume that the unknown branch voltages and currents are of the form

$$Ie^{j(\omega t + \theta_I)} = I\underline{/\theta_I}e^{j\omega t} \tag{5.62a}$$

or

$$Ve^{j(\omega t + \theta_V)} = V\underline{/\theta_V}e^{j\omega t} \tag{5.62b}$$

where the ω in the assumed forms of the solutions for the branch currents and voltages in Eq. (5.62) is the same as the ω of the source in Eq. (5.60). Since $e^{j\omega t}$ is common to all quantities in the circuit, we may discard it and solve the resulting phasor circuit for $I\underline{/\theta_I}$ or $V\underline{/\theta_V}$. If the source was cosinusoidal, as shown in Eq. (5.60*a*), then the solution is the real part of the solution in Eq. (5.62):

$$\begin{aligned} &I\cos(\omega t + \theta_I) \\ &V\cos(\omega t + \theta_V) \end{aligned} \tag{5.63a}$$

and if the source was sinusoidal, as in Eq. (5.60*b*), then the solution is the imaginary part of the solution in Eq. (5.62):

$$I \sin(\omega t + \theta_I)$$
$$V \sin(\omega t + \theta_V) \tag{5.63b}$$

When we replace the value of the actual source in Eq. (5.60) with the value in Eq. (5.61), replace the actual branch voltages and currents with their assumed forms in this circuit, as given in Eq. (5.62), and remove the common $e^{j\omega t}$ from all quantities, then the resulting circuit is referred to as the *phasor circuit*. The assumed complex solutions in Eq. (5.62) without the common $e^{j\omega t}$ terms, $\mathbf{I} = I\underline{/\theta_I}$ and $\mathbf{V} = V\underline{/\theta_V}$, are referred to as the *phasor* currents and voltages of the circuit. The original circuit is referred to as the *time-domain circuit*.

Now let us examine the solution of this phasor circuit *without* deriving the differential equation relating the branch variable of the interest to the source. First let us examine how the voltage and current for each type of element are related. For the resistor,

$$v_R(t) = Ri_R(t) \tag{5.64}$$

Replacing $v_R(t)$ and $i_R(t)$ with their phasor equivalents, we have

$$\mathbf{V}_R e^{j\omega t} = R\mathbf{I}_R e^{j\omega t} \tag{5.65}$$

or

$$\mathbf{V}_R = R\mathbf{I}_R \tag{5.66}$$

Thus, the phasor voltage and current for a resistor are related again by Ohm's law. Next consider the inductor:

$$v_L(t) = L\frac{di_L(t)}{dt} \tag{5.67}$$

Replacing with phasor equivalents, we have

$$\mathbf{V}_L e^{j\omega t} = L\frac{d}{dt}\mathbf{I}_L e^{j\omega t} \tag{5.68}$$

or

$$\mathbf{V}_L = j\omega L\mathbf{I}_L \tag{5.69}$$

Note that the phasor voltage and phasor current for an inductor are also related in a form similar to Ohm's law as

$$\mathbf{V}_L = \mathbf{Z}_L\mathbf{I}_L \tag{5.70}$$

FIGURE 5.8 Circuit components in the phasor circuit.

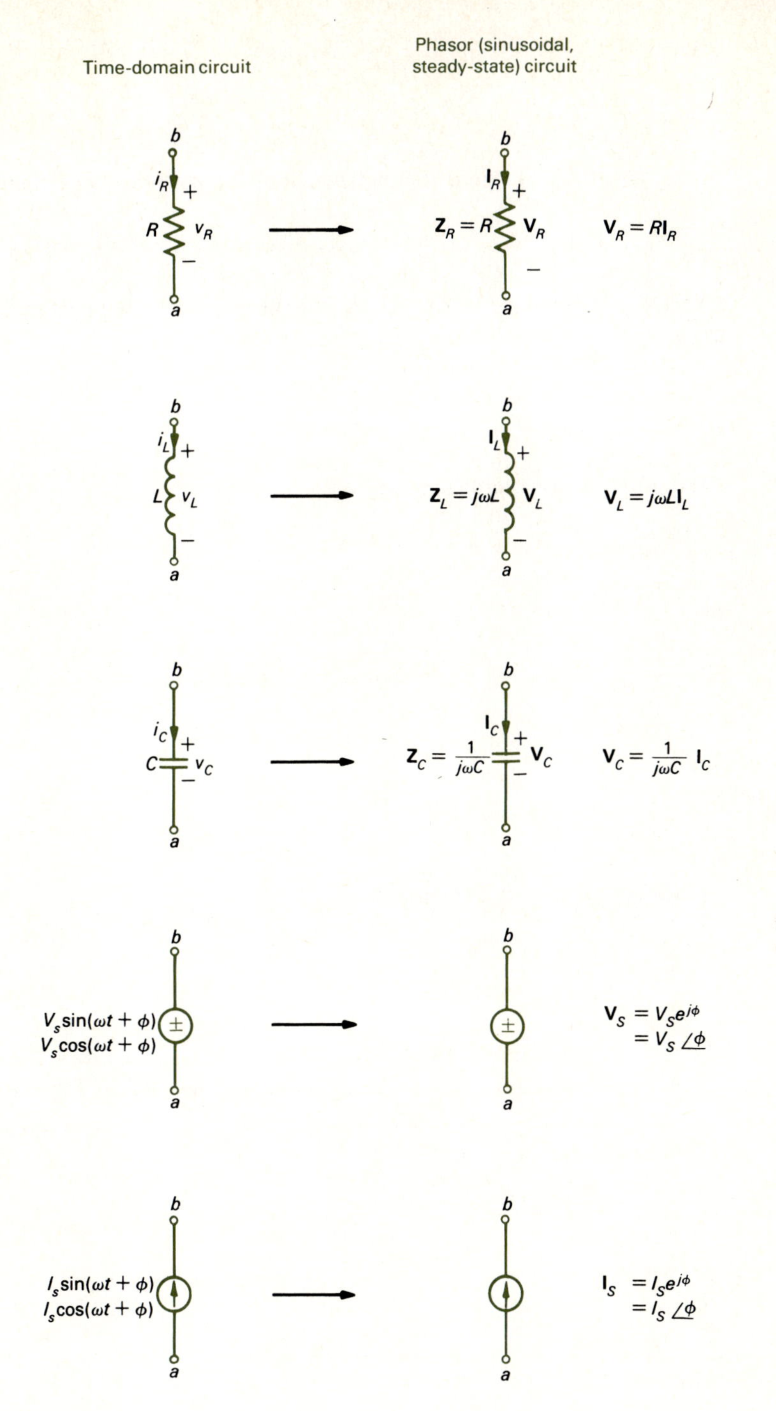

where

$$\mathbf{Z}_L = j\omega L \tag{5.71}$$

The quantity $\mathbf{Z}_L$ is called the *impedance* of the inductor. For the resistor,

$$\mathbf{V}_R = \mathbf{Z}_R \mathbf{I}_R \tag{5.72}$$

where the impedance of the resistor is the value of the resistance. Similarly, for the capacitor,

$$i_C(t) = C\frac{dv_C(t)}{dt} \tag{5.73}$$

Assuming phasor quantities, we have

$$\mathbf{I}_C e^{j\omega t} = j\omega C \mathbf{V}_C e^{j\omega t} \tag{5.74}$$

or

$$\mathbf{V}_C = \frac{1}{j\omega C}\mathbf{I}_C \tag{5.75}$$

Thus

$$\mathbf{V}_C = \mathbf{Z}_C \mathbf{I}_C \tag{5.76}$$

and the impedance of the capacitor is

$$\mathbf{Z}_C = \frac{1}{j\omega C} \tag{5.77}$$

The reciprocals of these impedances are referred to as the *admittances* of the elements and are denoted by **Y**. These concepts are summarized in Fig. 5.8.

Example 5.4 Determine the current $i(t)$ in Fig. 5.9a.

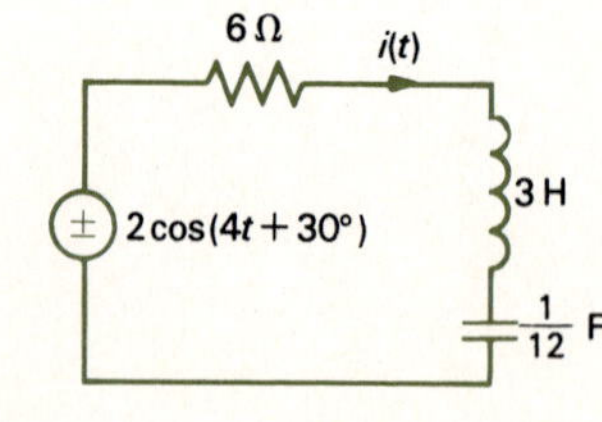

(*a*) Time-domain circuit

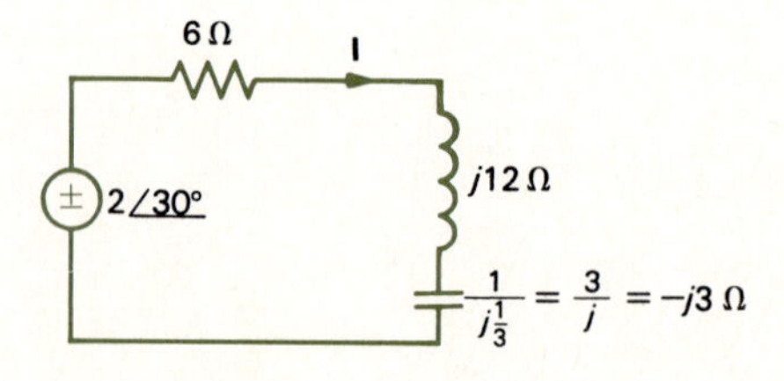

(*b*) Phasor circuit

FIGURE 5.9 Example 5.4: an example of the use of the phasor circuit.

Solution The phasor circuit is shown in Fig. 5.9*b* ($\omega = 4$). Solving, we obtain

$$\mathbf{I} = \frac{2\underline{/30^\circ}}{6 + j12 + 3/j}$$

$$= \frac{2\underline{/30^\circ}}{6 + j12 - j3}$$

$$= \frac{2\underline{/30^\circ}}{6 + j9}$$

$$= 0.18\underline{/-26.31^\circ}$$

Thus, the current is

$$i(t) = \text{Re } \mathbf{I}e^{j4t}$$

$$= 0.18 \cos (4t - 26.31^\circ)$$

because the original source was the real part of $2e^{j(4t+30^\circ)}$:

$$2 \cos (4t + 30^\circ) = \text{Re } 2e^{j(4t+30^\circ)}$$

If the circuit contains several sinusoidal sources, not all of which are of the same frequency, the result can be obtained by superposition in the time domain.

Example 5.5 Determine the current $i(t)$ in the circuit of Fig. 5.10*a*.

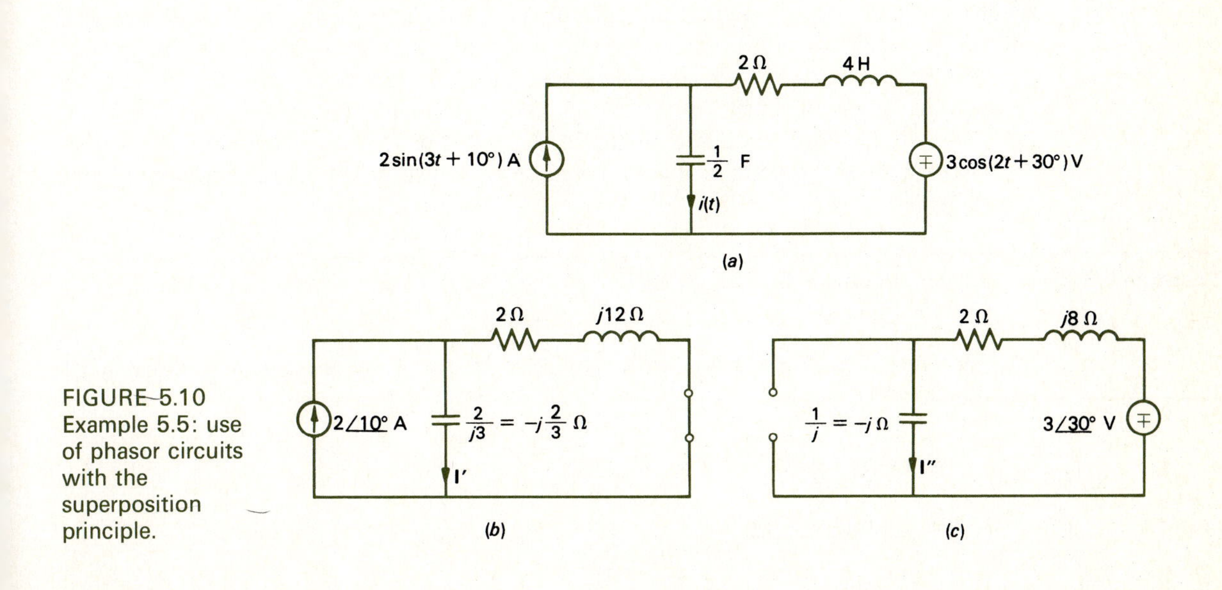

FIGURE 5.10 Example 5.5: use of phasor circuits with the superposition principle.

Solution The phasor circuit diagram for the current source ($\omega = 3$) is shown in Fig. 5.10*b*. We obtain the phasor current due to this source by current division as

$$\mathbf{I}' = \frac{2 + j12}{(2 + j12) + 2/j3} 2\underline{/10^\circ}$$
$$= 2.11\underline{/10.55^\circ}$$

and

$$i'(t) = 2.11 \sin (3t + 10.55^\circ)$$

The phasor circuit for the voltage source ($\omega = 2$) is shown in Fig. 5.10*c*. The phasor current due to this source is

$$\mathbf{I}'' = \frac{3\underline{/30^\circ}}{2 + j8 + 1/j}$$
$$= -0.41\underline{/-44.05^\circ}$$

and

$$i''(t) = -0.41 \cos (2t - 44.05^\circ)$$

Thus, the current due to both sources is

$$i(t) = i'(t) + i''(t)$$
$$= 2.11 \sin (3t + 10.55^\circ) - 0.41 \cos (2t - 44.05^\circ)$$

It should be clear that the phasor circuit can be treated as a resistive circuit having "complex-valued resistors" when solving for the phasor branch variable of interest. All of the various techniques which we obtained in Chap. 3 for solving resistive circuits—voltage division, current division, Thevenin and Norton equivalents (using a complex Thevenin "impedance" $\mathbf{Z}_{TH}$ with a phasor open-circuit voltage $\mathbf{V}_{OC}$ or short-circuit current $\mathbf{I}_{SC}$), node voltage analysis and mesh current analysis—can be used in solving the phasor circuit for the phasor branch variable of interest. Thus, the analysis of phasor circuits is no more difficult, theoretically, than that of resistive circuits; the manipulation of complex numbers, however, adds to the computational difficulty.

Example 5.6 Solve for the voltage $v(t)$ in the circuit of Fig. 5.11*a* by reducing the circuit at terminals *a-b* to a Thevenin equivalent.

Solution The phasor circuit is shown in Fig. 5.11*b*. The phasor open-circuit voltage is found from Fig. 5.11*c* as

$$\mathbf{V}_{OC} = \frac{2}{2 + j9} 2\underline{/10^\circ}$$
$$= 0.43\underline{/-67.67^\circ}$$

FIGURE 5.11 Example 5.6: Thevenin equivalent reduction in phasor circuits.

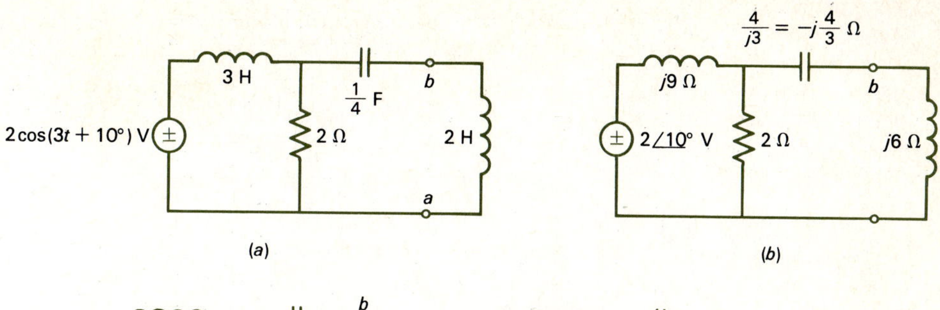

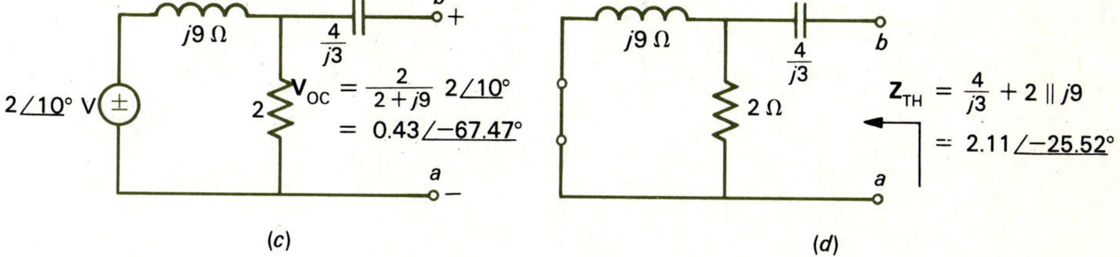

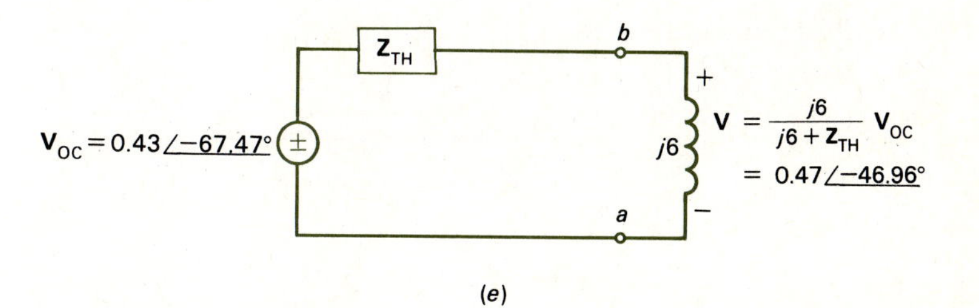

The Thevenin impedance is found from Fig. 5.11*d* as

$$\mathbf{Z}_{\text{TH}} = \frac{4}{j3} + 2\,||\,j9$$

$$= 2.11\underline{/-25.52^\circ}$$

Reattaching, the complete circuit is shown in Fig. 5.11*e*, from which we obtain, by voltage division,

$$\mathbf{V} = \frac{j6}{j6 + \mathbf{Z}_{\text{TH}}}\mathbf{V}_{\text{OC}}$$

$$= 1.10\underline{/20.53^\circ}\ 0.43\underline{/-67.47^\circ}$$

$$= 0.47\underline{/-46.94^\circ}$$

so that

$$v(t) = \text{Re}\ \mathbf{V}e^{j3t}$$

$$= 0.47\cos(3t - 46.94^\circ)$$

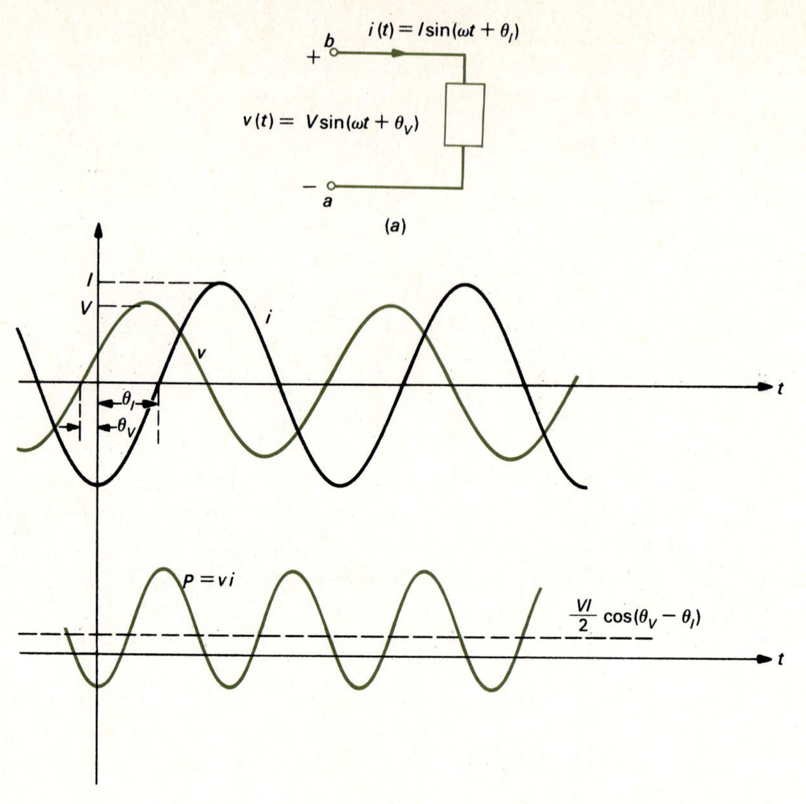

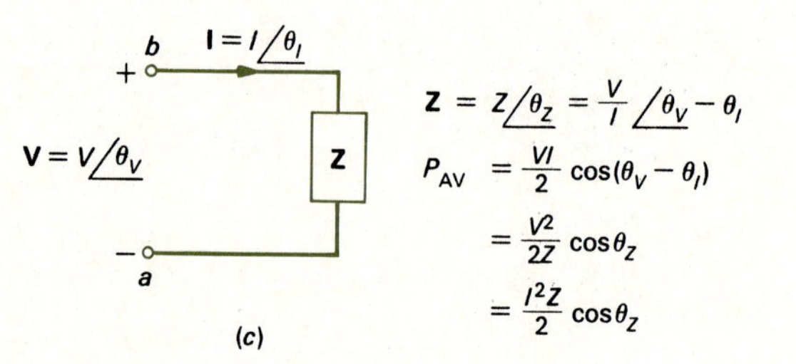

FIGURE 5.12 Average power dissipation in (absorbed by) an element because of sinusoidal excitation.

5.5 AVERAGE POWER, REACTIVE POWER, AND POWER FACTOR

Now that we have considered how to obtain the solution for any branch voltage or current of a circuit due to a sinusoidal source, let us consider how to calculate the power resulting from sinusoidal sources. The instantaneous power delivered to the element shown in Fig. 5.12*a* is

$$p(t) = v(t)i(t) \tag{5.78}$$

Since each variable will be sinusoidal, we may write

$$\begin{aligned} p(t) &= V \sin(\omega t + \theta_V) I \sin(\omega t + \theta_I) \\ &= VI \sin(\omega t + \theta_V) \sin(\omega t + \theta_I) \end{aligned} \tag{5.79}$$

Denoting the difference in the phase angles of the voltage and current as

$$\theta = \theta_V - \theta_I \tag{5.80}$$

we may write, using trigonometric identities,

$$p(t) = \frac{VI}{2} \cos \theta - \frac{VI}{2} \cos (2\omega t + \theta_V + \theta_I) \tag{5.81}$$

which is plotted in Fig. 5.12*b*. Thus, the instantaneous power consists of a dc component, $VI/2 \cos \theta$, and of a sinusoidal component, $VI/2 \cos (2\omega t + \theta_V + \theta_I)$, which varies at twice the frequency f of the voltage or current. The average power is the integral of this instantaneous power over a period of $p(t)$, $T = 1/2f$, averaged over that period:

$$\begin{aligned} P_{AV} &= \frac{1}{T} \int_0^T p(t)\, dt \\ &= \frac{VI}{2} \cos (\theta_V - \theta_I) \end{aligned} \tag{5.82}$$

which becomes obvious from the plot in Fig. 5.12*b*. The phasor representation is shown in Fig. 5.12*c*. Note that

$$\begin{aligned} \mathbf{Z} &= Z\underline{/\theta_Z} \\ &= \frac{\mathbf{V}}{\mathbf{I}} \\ &= \frac{V\underline{/\theta_V}}{I\underline{/\theta_I}} \\ &= \frac{V}{I} \underline{/\theta_V - \theta_I} \end{aligned} \tag{5.83}$$

where $\mathbf{Z}$ is the phasor impedance of the element. Therefore,

$$Z = \frac{V}{I} \tag{5.84}$$

and

$$\theta_Z = \theta_V - \theta_I \tag{5.85}$$

Thus, we may write the average power in Eq. (5.82) as

$$P_{AV} = \frac{V^2}{2Z}\cos\theta_Z$$
$$= \tfrac{1}{2} I^2 Z \cos\theta_Z \tag{5.86}$$

In Chap. 2 we define the effective, or RMS, value of a sinusoidal voltage or current as

$$V_{RMS} = \frac{V}{\sqrt{2}}$$
$$I_{RMS} = \frac{I}{\sqrt{2}} \tag{5.87}$$

Therefore Eq. (5.86) can be written as

$$P_{AV} = \frac{V_{RMS}^2}{Z}\cos\theta_Z$$
$$= I_{RMS}^2 Z \cos\theta_Z \tag{5.88}$$

which provides a convenient method for computing the average power dissipated in a circuit element.

The average power dissipated in an element can be written as a more general result:

$$P_{AV} = \tfrac{1}{2}\,\text{Re}\,\mathbf{VI}^*$$
$$= \tfrac{1}{2}\,\text{Re}\,\mathbf{V}^*\mathbf{I} \tag{5.89}$$

where $\mathbf{I}^*$ is the conjugate of the phasor $\mathbf{I}$ and Re $(\cdot)$ denotes "real part." To show that Eq. (5.89) reduces to Eq. (5.82), we write

$$\mathbf{V} = V\underline{/\theta_V}$$
$$\mathbf{I} = I\underline{/\theta_I} \tag{5.90}$$

Substituting Eq. (5.90) into Eq. (5.89), we obtain

$$P_{AV} = \tfrac{1}{2}\,\text{Re}\,(V\underline{/\theta_V}\,I\underline{/-\theta_I})$$
$$= \tfrac{1}{2}\,\text{Re}\,(VI\underline{/\theta_V - \theta_I})$$
$$= \frac{VI}{2}\,\text{Re}\,[\cos(\theta_V - \theta_I) + j\sin(\theta_V - \theta_I)]$$
$$= \frac{VI}{2}\cos(\theta_V - \theta_I) \tag{5.91}$$

Equation (5.89) is a very important result and will be used in numerous instances. The phasor power is

$$\begin{aligned}\mathbf{P} &= \tfrac{1}{2}\mathbf{VI}^* \\ &= V_{\text{RMS}}\mathbf{I}^*_{\text{RMS}} \\ &= \frac{VI}{2}\cos(\theta_V - \theta_I) + j\frac{VI}{2}\sin(\theta_V - \theta_I)\end{aligned} \tag{5.92}$$

The real part of Eq. (5.92) is, of course, the average power dissipated in the element. The imaginary part is referred to as the reactive power and is denoted as

$$\begin{aligned}Q &= \tfrac{1}{2}\operatorname{Im}\mathbf{VI}^* \\ &= \operatorname{Im}\mathbf{V}_{\text{RMS}}\mathbf{I}^*_{\text{RMS}} \\ &= \frac{VI}{2}\sin(\theta_V - \theta_I)\end{aligned} \tag{5.93}$$

so that

$$\mathbf{P} = P_{\text{AV}} + jQ \tag{5.94}$$

The reactive power does not represent real power loss in the element and only indicates temporary energy storage, as we shall see.

Now let us investigate the application of these concepts to the R, L, and C elements. For the resistor shown in Fig. 5.13, the phasor voltage and current are in phase:

$$\mathbf{V} = R\mathbf{I} \tag{5.95}$$

so that

$$\begin{aligned}P_{\text{AV}} &= \frac{V^2}{2R} \\ &= \frac{I^2R}{2}\end{aligned} \tag{5.96}$$

The reactive power Q for this element is zero since $\theta_V = \theta_I$ and $\sin(\theta_V - \theta_I) = 0$. Therefore, no reactive power is associated with the resistor.

For the inductor shown in Fig. 5.14, the phasor voltage and current are related by

$$\begin{aligned}\mathbf{V}_L &= j\omega L\mathbf{I}_L \\ &= \omega L\angle 90^\circ\, \mathbf{I}_L\end{aligned} \tag{5.97}$$

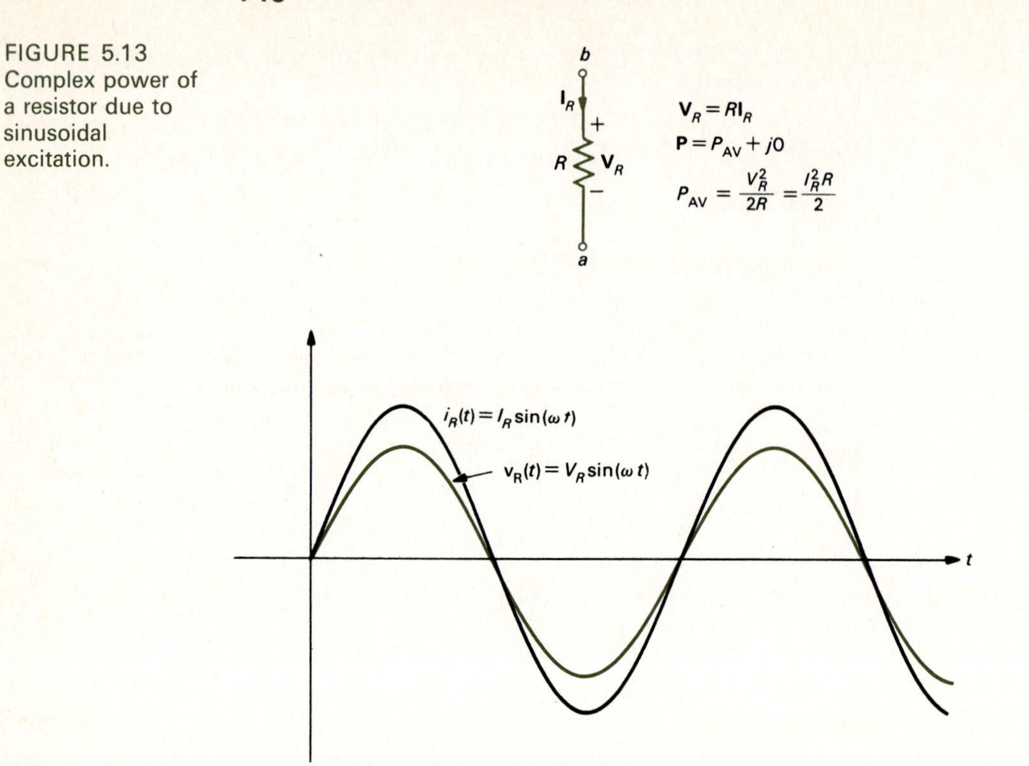

FIGURE 5.13
Complex power of a resistor due to sinusoidal excitation.

and the phasor voltage leads the current by 90°. Thus

$$P_{AV} = \frac{V_L I_L}{2} \cos 90^\circ$$

$$= 0 \tag{5.98a}$$

$$Q = \frac{V_L I_L}{2} \sin 90^\circ$$

$$= \frac{V_L I_L}{2} \tag{5.98b}$$

The inductor current absorbs no (real) power but has a reactive power associated with it. These results are shown in Fig. 5.14.

For the capacitor shown in Fig. 5.15, the phasor voltage and current are related by

$$\mathbf{V}_C = \frac{1}{j\omega C} \mathbf{I}_C \tag{5.99}$$

or

$$\mathbf{I}_C = j\omega C \mathbf{V}_C$$

$$= \omega C \underline{/90^\circ}\, \mathbf{V}_C \tag{5.100}$$

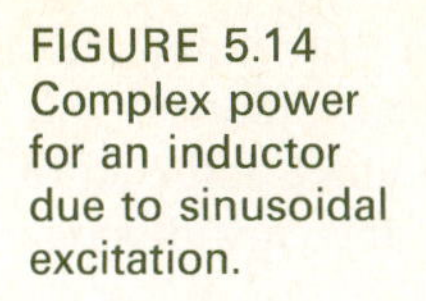

FIGURE 5.14
Complex power for an inductor due to sinusoidal excitation.

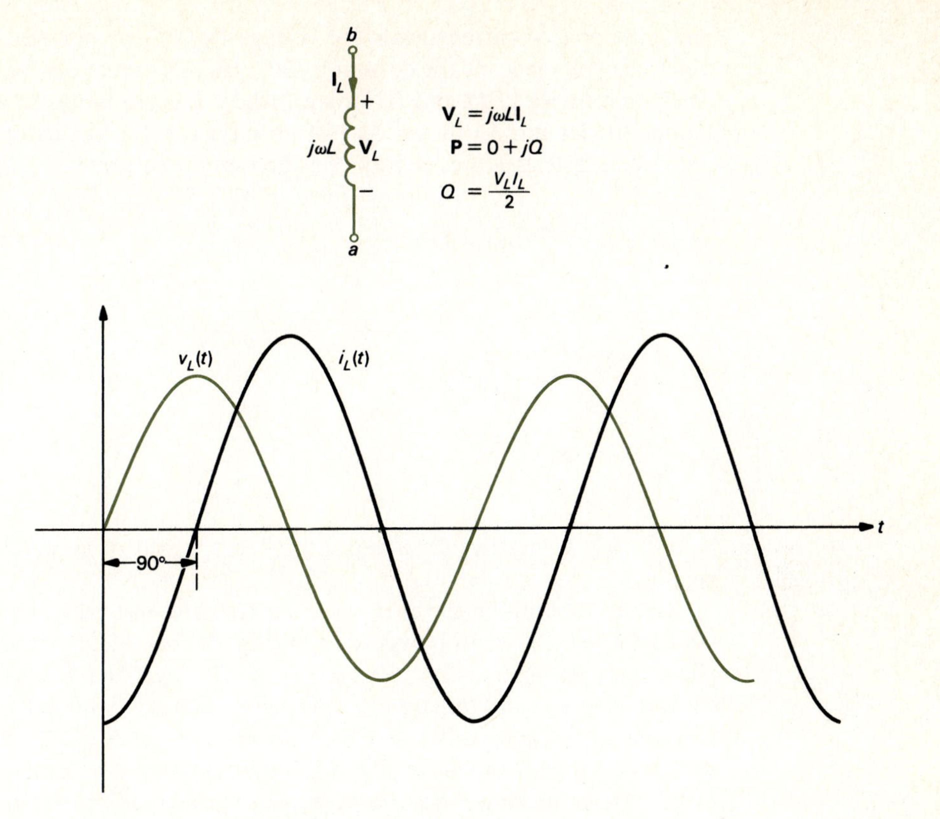

FIGURE 5.15
Complex power for a capacitor due to sinusoidal excitation.

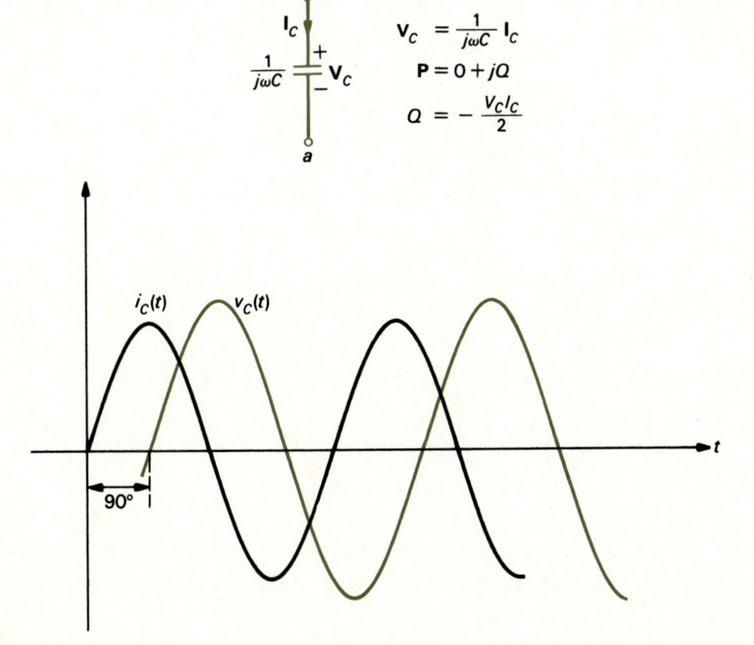

Thus, the phasor current leads the voltage by 90°—as opposed to the inductor, in which the voltage leads the current by 90°. This difference can be remembered by the mnemonic device "*ELI* and *ICE* man," where *E* corresponds to voltage and *I* corresponds to current. As in the case of an inductor, the capacitor does not dissipate power but, instead, stores energy. This becomes clear since

$$P_{AV} = \frac{V_C I_C}{2} \cos(-90°)$$

$$= 0 \tag{5.101a}$$

$$Q = \frac{V_C I_C}{2} \sin(-90°)$$

$$= -\frac{V_C I_C}{2} \tag{5.101b}$$

Note that Q is negative for a capacitor—as opposed to an inductor, for which Q is positive.

We have seen that the energy storage elements dissipate no power ($P_{AV} = 0$) but store energy ($Q \neq 0$). In power distribution, the net energy transfer from the power plant to the consumer is the average power. The reactive power is not power which is converted to a useful function and therefore represents no net energy transfer. It is desirable to minimize this reactive power in a load, as the following shows. Consider the series *RL* load shown in Fig. 5.16*a*, representing a consumer load. Motors and other typical consumer loads are typically represented in this fashion. The source is represented as an ideal voltage source v_S in series with a source resistance R_S. The transmission line is represented as a simple resistance, R_{line}, corresponding to the resistance of the line conductors. The phasor circuit is shown in Fig. 5.16*b* and the phasor load current $\mathbf{I}_L$ is

$$\mathbf{I}_L = \frac{\mathbf{V}_S}{R_S + R_{line} + R + j\omega L} \tag{5.102}$$

The total average power disspated in the load is the power dissipated in the load resistance R;

$$P_{AV} = \frac{I_L^2 R}{2} \tag{5.103}$$

The magnitude of the phasor current is, from Eq. (5.102),

$$I_L = \frac{V_S}{\sqrt{(R_S + R_{line} + R)^2 + (\omega L)^2}} \tag{5.104}$$

For a fixed value of source voltage V_S, the magnitude of the load current and consequently the average power delivered to the load is obviously smaller than if there were

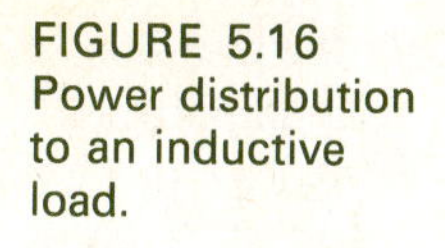
FIGURE 5.16
Power distribution to an inductive load.

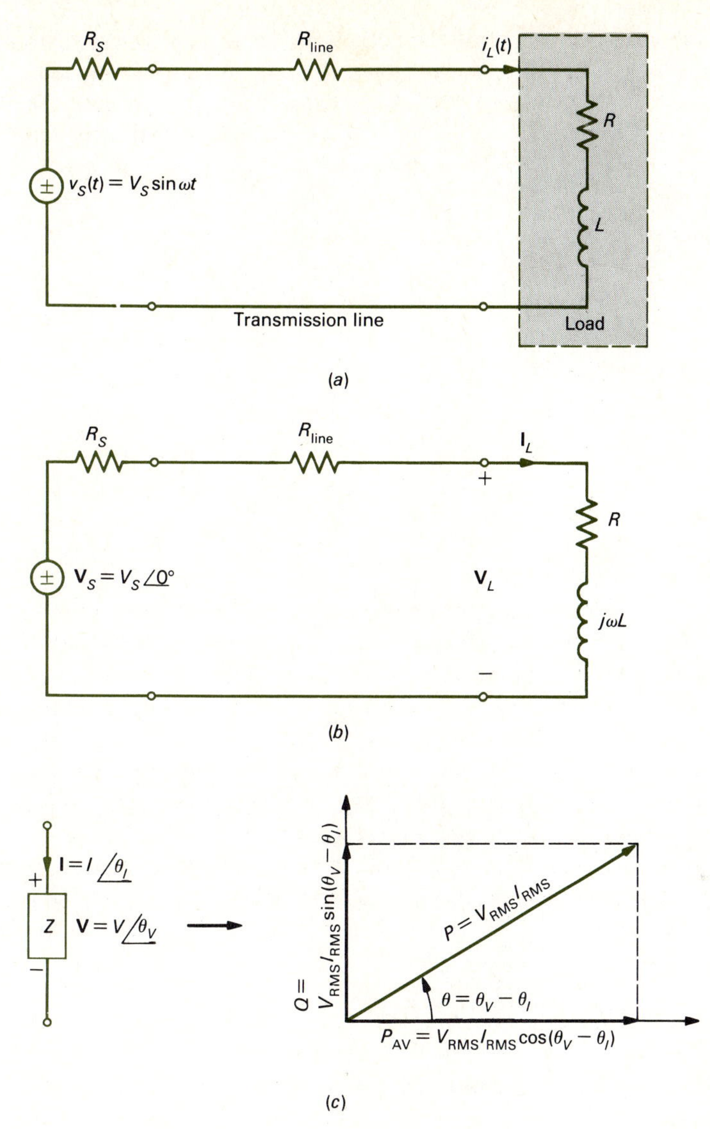

no inductance in the load ($L = 0$). Having an inductive component of the load increases the required current and source voltage for a given average power consumption. But a higher current increases the line losses ($\frac{1}{2}I_L^2 R_{\text{line}}$) and generator losses ($\frac{1}{2}I_L^2 R_S$) which the power company and not the consumer must bear.

This suitability of a particular impedance for absorption of average power is given by its power factor:

$$\cos(\theta_V - \theta_I) = \frac{P_{\text{AV}}}{V_{\text{RMS}} I_{\text{RMS}}} \tag{5.105}$$

where V_{RMS} and I_{RMS} are the RMS voltage and current associated with the terminals of the impedance. The phasor power associated with the impedance has a magnitude

$$\begin{aligned}|\mathbf{P}| &= P \\ &= V_{\text{RMS}} I_{\text{RMS}} \end{aligned} \tag{5.106}$$

and

$$P_{\text{AV}} = P\cos(\theta_V - \theta_I) \tag{5.107a}$$

$$Q = P\sin(\theta_V - \theta_I) \tag{5.107b}$$

as shown in Fig. 5.16*c*. Thus, the power factor

$$\begin{aligned}\text{pf} &= \cos(\theta_V - \theta_I) \\ &= \frac{P_{\text{AV}}}{P}\end{aligned} \tag{5.108}$$

gives a measure of the element's ability to absorb power. For the resistor, pf = 1, and for the inductor and capacitor, pf = 0. For any other element which is a combination of R, L, or C elements,

$$0 \le \text{pf} \le 1 \tag{5.109}$$

Example 5.7 For the circuit shown in Fig. 5.17*a*, find the average and reactive power associated with the load and compute the power factor of the load.

Solution The phasor circuit is shown in Fig. 5.17*b* ($\omega = 3$). The phasor voltage and current of the load are

$$\begin{aligned}\mathbf{V}_L &= \frac{5 + j9 - j2}{4 + j6 + 5 + j9 - j2}\, 100\underline{/0^\circ} \\ &= \frac{5 + j7}{9 + j13}\, 100\underline{/0^\circ} \\ &= 54.41\underline{/-0.84^\circ}\end{aligned}$$

$$\begin{aligned}\mathbf{I}_L &= \frac{100\underline{/0^\circ}}{4 + j6 + 5 + j9 - j2} \\ &= 6.32\underline{/-55.3}\end{aligned}$$

The average power delivered to the load can be found by simply computing the average power dissipated in the 5-Ω resistor of the load:

$$\begin{aligned}P_{\text{AV}} &= \tfrac{1}{2} I_L^2 5 \\ &= \tfrac{1}{2}(6.32)^2 5 \\ &= 100 \text{ W}\end{aligned}$$

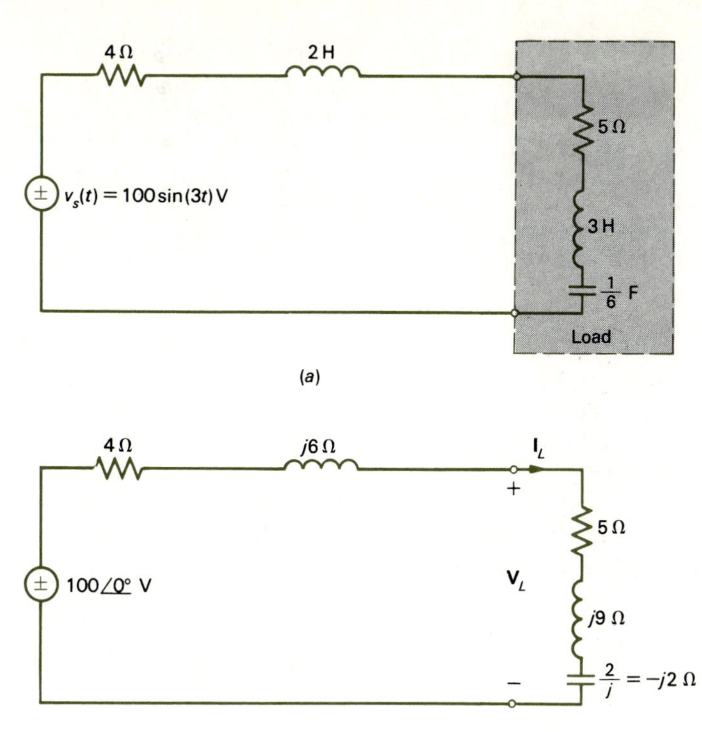

FIGURE 5.17 Example 5.7: illustration of average and reactive power delivered to a complex load.

Both the average and reactive power can be found from the phasor power of the load:

$$\begin{aligned}\mathbf{P} &= \tfrac{1}{2}\mathbf{V}_L\mathbf{I}_L^* \\ &= \tfrac{1}{2}54.41\underline{/-0.84°}\ 6.32\underline{/55.3°} \\ &= 172.05\underline{/54.46°} \\ &= 100 + j140\end{aligned}$$

as

$$\begin{aligned}P_{\text{AV}} &= \text{Re}\,\mathbf{P} \\ &= 100\text{ W} \\ Q &= \text{Im}\,\mathbf{P} \\ &= 140\end{aligned}$$

Thus, the load appears to have a net inductive component (since Q is nonzero and positive), as is clear from the phasor circuit. The power factor of the load is

$$\begin{aligned}\text{pf} &= \cos\theta_Z \\ &= \cos(\theta_V - \theta_I) \\ &= \cos 54.46° \\ &= 0.58\end{aligned}$$

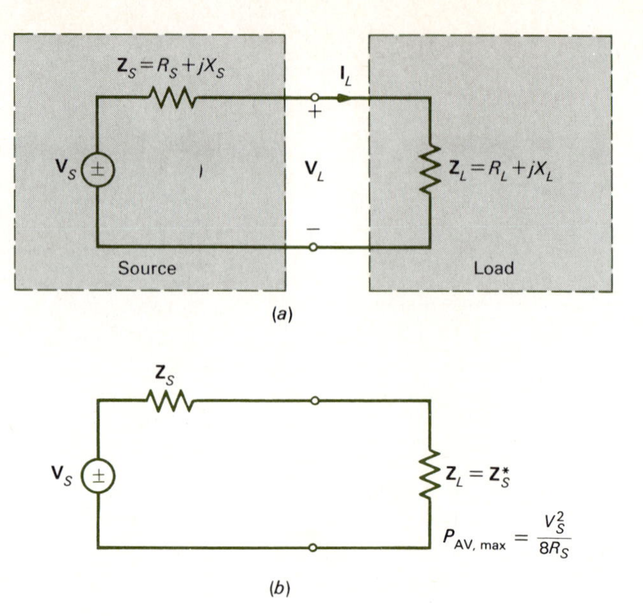

FIGURE 5.18 Illustration of maximum power transfer for complex source and load impedances.

5.6 MAXIMUM POWER TRANSFER

Consider the circuit shown in Fig. 5.18*a*. A source having phasor voltage $\mathbf{V}_S$ and source impedance $\mathbf{Z}_S$ is connected to a load impedance $\mathbf{Z}_L$. Suppose that the source voltage and impedance, $\mathbf{V}_S$ and $\mathbf{Z}_S$, are fixed, but that we have a choice as to the value of the load impedance $\mathbf{Z}_L$. We wish to find the value of $\mathbf{Z}_L$ which will result in the maximum average power being transferred from the source to the load. If we represent each impedance with a real and imaginary part,

$$\begin{aligned} \mathbf{Z}_S &= R_S + jX_S \\ \mathbf{Z}_L &= R_L + jK_L \end{aligned} \tag{5.110}$$

then the phasor load current $\mathbf{I}_L$ is

$$\mathbf{I}_L = \frac{\mathbf{V}_S}{\mathbf{Z}_S + \mathbf{Z}_L} \tag{5.111}$$

and its magnitude becomes

$$I_L = \frac{V_S}{\sqrt{(R_S + R_L)^2 + (X_S + X_L)^2}} \tag{5.112}$$

The average power delivered to the load is

$$P_{AV} = \tfrac{1}{2} I_L^2 R_L \tag{5.113}$$

which could also be determined from

$$\begin{aligned} P_{AV} &= \tfrac{1}{2} \operatorname{Re} \mathbf{V}_L \mathbf{I}_L^* \\ &= \tfrac{1}{2} \operatorname{Re} \mathbf{I}_L \mathbf{I}_L^* \mathbf{Z}_L \\ &= \tfrac{1}{2} \operatorname{Re} I_L^2 \mathbf{Z}_L \\ &= \frac{I_L^2}{2} \operatorname{Re} \mathbf{Z}_L \\ &= \frac{I_L^2}{2} R_L \end{aligned} \tag{5.114}$$

Substituting Eq. (5.112) into Eq. (5.113), we obtain

$$P_{AV} = \frac{V_S^2}{2} \frac{R_L}{(R_S + R_L)^2 + (X_S + X_L)^2} \tag{5.115}$$

If we can only vary X_L, then maximum power is delivered to the load if

$$X_L = -X_S \tag{5.116}$$

If only R_L can be varied, the result is a bit more difficult to see. If we form dP_{AV}/dR_L and set this equal to zero, we find that maximum power transfer occurs for

$$R_L = \sqrt{R_S^2 + (X_S + X_L)^2} \tag{5.117}$$

Now if both R_L and X_L can be varied, we obtain maximum power transfer to the load by substituting Eq. (5.116) into Eq. (5.117) to yield

$$\begin{aligned} X_L &= -X_S \\ R_L &= R_S \end{aligned} \tag{5.118}$$

or

$$\begin{aligned} \mathbf{Z}_L &= R_S - jX_S \\ &= \mathbf{Z}_S^* \end{aligned} \tag{5.119}$$

Therefore, maximum power is delivered from a source to a load when the load impedance is the conjugate of the source, as shown in Fig. 5.18*b*. If the source and load are purely resistive, $\mathbf{Z}_S = R_S$ and $\mathbf{Z}_L = R_L$, the maximum power is delivered to the load when $R_L = R_S$ (as we found in Chap. 3) and, once again, $\mathbf{Z}_L = \mathbf{Z}_S^*$.

When the load impedance is the conjugate of the source impedance, the load and source are said to be matched. Under matched conditions, the total average power delivered to $\mathbf{Z}_S$ and $\mathbf{Z}_L$ are equally divided. Denoting $\mathbf{Z}_S = R + jX$ and $\mathbf{Z}_L = \mathbf{Z}_S^* = R - jX$, the average power delivered to the load is, from Eq. (5.115),

$$P_{\mathrm{AV,\,load}} = \frac{V_S^2}{8R} \qquad \text{matched} \tag{5.120}$$

5.7 RESONANCE AND AN INTRODUCTION TO FILTERS

Periodic functions of time of period T can be represented with the Fourier series as an infinite sum of sinusoids having frequencies, or harmonics, which are multiples of the fundamental frequency $f = 1/T$ (see App. C):

$$x(t) = x_{\mathrm{AV}} + a_1 \sin \omega t + a_2 \sin 2\omega t + a_3 \sin 3\omega t + \cdots$$
$$+ b_1 \cos \omega t + b_2 \cos 2\omega t + b_3 \cos 3\omega t + \cdots \tag{5.121}$$

where

$$\omega = 2\pi f$$
$$= \frac{2\pi}{T} \tag{5.122}$$

The frequency components of $x(t)$ are separated in frequency by multiples of $1/T$. If an ideal voltage source has periodic voltage $x(t)$, then it can alternatively be considered to be the series connection of voltage sources each having a voltage corresponding to a term in Eq. (5.121), as shown in Fig. 5.19*a*. Similarly, if $x(t)$ is the value of a periodic current source, it can alternatively be represented as the parallel connection of current sources each having a value in Eq. (5.121), as shown in 5.19*b*.

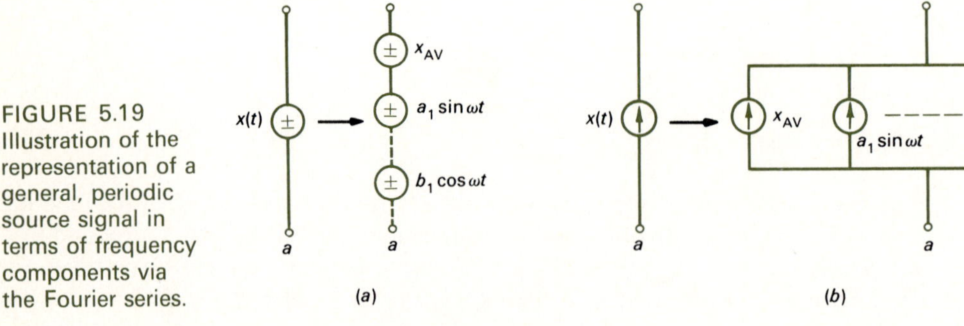

FIGURE 5.19 Illustration of the representation of a general, periodic source signal in terms of frequency components via the Fourier series.

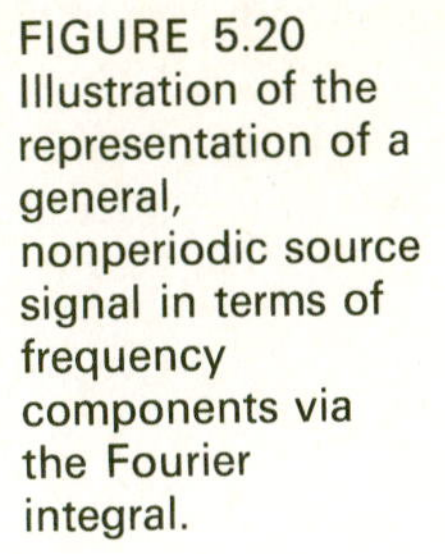
FIGURE 5.20 Illustration of the representation of a general, nonperiodic source signal in terms of frequency components via the Fourier integral.

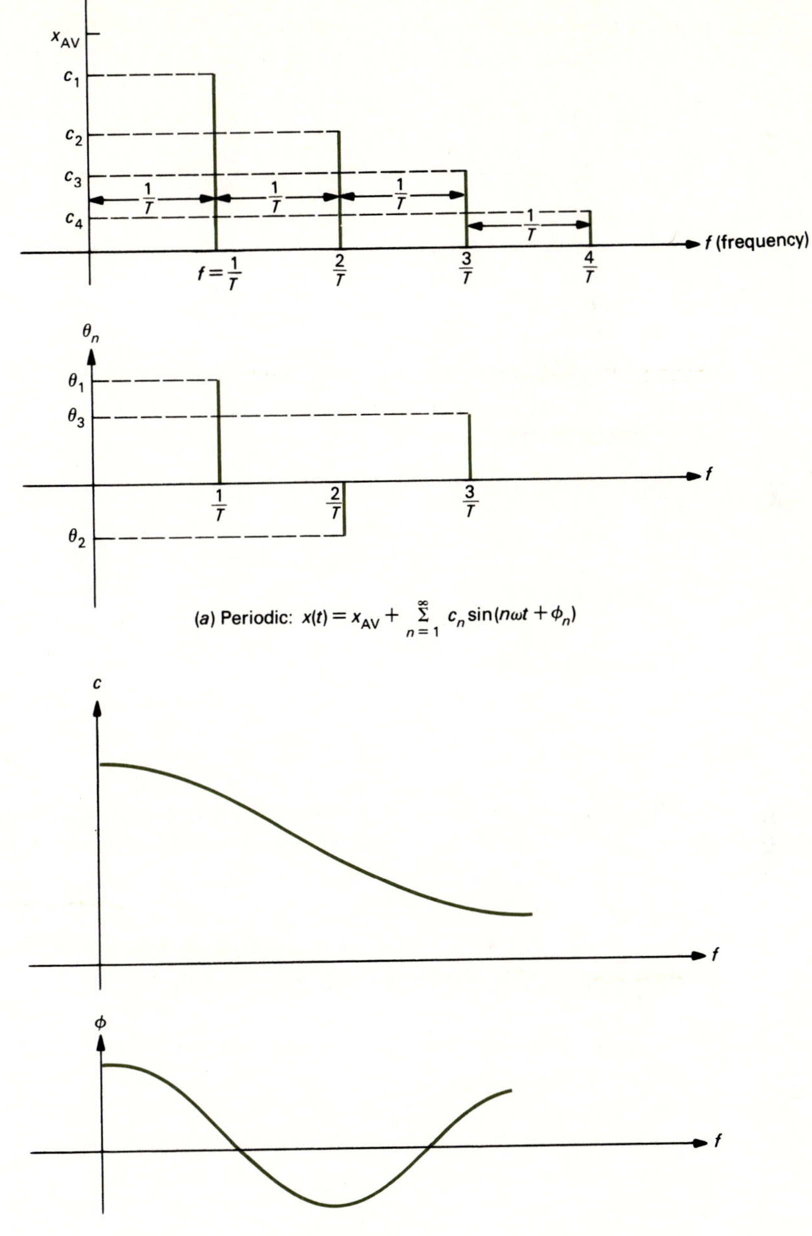

(a) Periodic: $x(t) = x_{AV} + \sum_{n=1}^{\infty} c_n \sin(n\omega t + \phi_n)$

(b) Nonperiodic

The Fourier series in Eq. (5.121) can be written in an alternative form by combining corresponding sine and cosine harmonic components as

$$a_n \sin n\omega t + b_n \cos n\omega t = c_n \sin (n\omega t + \phi_n) \tag{5.123}$$

With the identity

$$\sin (A + B) = \sin A \cos B + \cos A \sin B$$

we can write Eq. (5.123) as

$$c_n \sin (n\omega t + \theta_n) = c_n \cos \theta_n \sin n\omega t + c_n \sin \theta_n \cos n\omega t \tag{5.124}$$

and identify by comparing Eqs. (5.123) and (5.124):

$$a_n = c_n \cos \theta_n \tag{5.125a}$$

$$b_n = c_n \sin \theta_n \tag{5.125b}$$

or

$$c_n = \sqrt{a_n^2 + b_n^2} \tag{5.126a}$$

$$\phi_n = \tan^{-1} \frac{b_n}{a_n} \tag{5.126b}$$

Thus, we may write

$$x(t) = x_{AV} + c_1 \sin (\omega t + \phi_1) + c_2 \sin (2\omega t + \phi_2) + \cdots \tag{5.127}$$

The Fourier series in Eq. (5.127) can alternatively be represented as a plot of the coefficients c_n and angles ϕ_n as a function of frequency, as shown in Fig. 5.20*a*. This plot gives a convenient, graphical portrayal of the relative importance of each harmonic in the overall makeup of the signal, since the height of each bar in the plot of the c_n coefficients is the relative amplitude of that sinusoidal component.

Nonperiodic functions of time, such as voice, can also be represented as an infinite sum of sinusoids via the Fourier integral. However, as opposed to periodic functions, the frequency components of the nonperiodic wave are not separated by some constant $(1/T)$ but are continuously distributed throughout the frequency spectrum, as shown in Fig. 5.20*b*.

Nevertheless, for either periodic or nonperiodic functions we may view the time-domain wave as a combination of sinusoids. The effect of a circuit on $x(t)$ can, by superposition, be described by its effect on each frequency component of the wave. There exist numerous instances—radio communication, noise suppression, and many others—in which we would like to eliminate certain frequency components from the waveform. Devices which do this are called filters. In this section, we will investigate some rather simple examples of filters to illustrate the point.

As an example, suppose that an ideal voltage source $x(t)$ with source resistance R_S is connected to a resistive load R_L as shown in Fig. 5.21*a*. Suppose that we wish to eliminate certain frequency components of $x(t)$ by inserting a filter between the source and load as shown in Fig. 5.21*b*. Let us now investigate some rather simple filters for doing this.

The first filter will be known as a low-pass filter which is to pass only certain low frequencies and reject higher ones. What we therefore need is a device which provides a direct connection between the source and load at low frequencies and provides isolation between the source and load at higher frequencies. A possibility consisting simply of a series inductor is shown in Fig. 5.22*a*. We have replaced $x(t)$ with a general sinusoid, $V \sin(\omega t + \phi)$, representing one of its frequency components. Note that at dc ($\omega = 0$) the inductor is a short circuit ($Z_L = j\omega L$), but that as frequency ω is increased its impedance increases, thus isolating the source and load at higher frequencies. The phasor circuit is shown in Fig. 5.22*b* and the phasor load current is

$$\mathbf{I}_L = \frac{\mathrm{V} \angle \phi}{R_S + R_L + j\omega L} \tag{5.128}$$

The magnitude is

$$I_L = \frac{V}{\sqrt{(R_S + R_L)^2 + (\omega L)^2}} \tag{5.129}$$

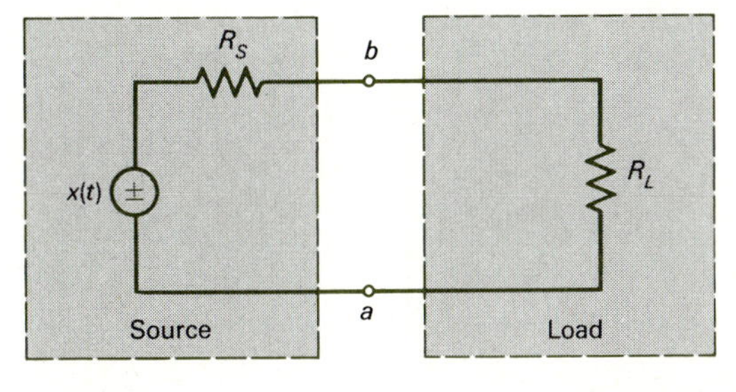

(a)

(b)

FIGURE 5.21 Illustration of the use of a filter to eliminate certain frequency components of a signal.

which is plotted in Fig. 5.22*c* in terms of the ratio

$$\frac{I_L}{V} = \frac{1}{\sqrt{(R_S + R_L)^2 + (\omega L)^2}}$$

$$= \frac{1}{R_S + R_L} \frac{1}{\sqrt{1 + [\omega L/(R_S + R_L)]^2}} \tag{5.130}$$

At dc this ratio is $1/(R_S + R_L)$, as is clear from the circuit diagram. As the frequency increases, this ratio eventually decreases, asymptotically, to zero. At a frequency ω_c such that

$$\omega_c = \frac{R_S + R_L}{L} \tag{5.131}$$

Eq. (5.130) has been reduced by a factor of $1/\sqrt{2}$ of its value at dc. The average power delivered to R_L is

$$P_{AV} = \tfrac{1}{2} I_L^2 R_L \tag{5.132}$$

and at ω_c the average power delivered to R_L is one-half of its value at dc. Thus, ω_c is referred to as the half-power point of the low-pass filter.

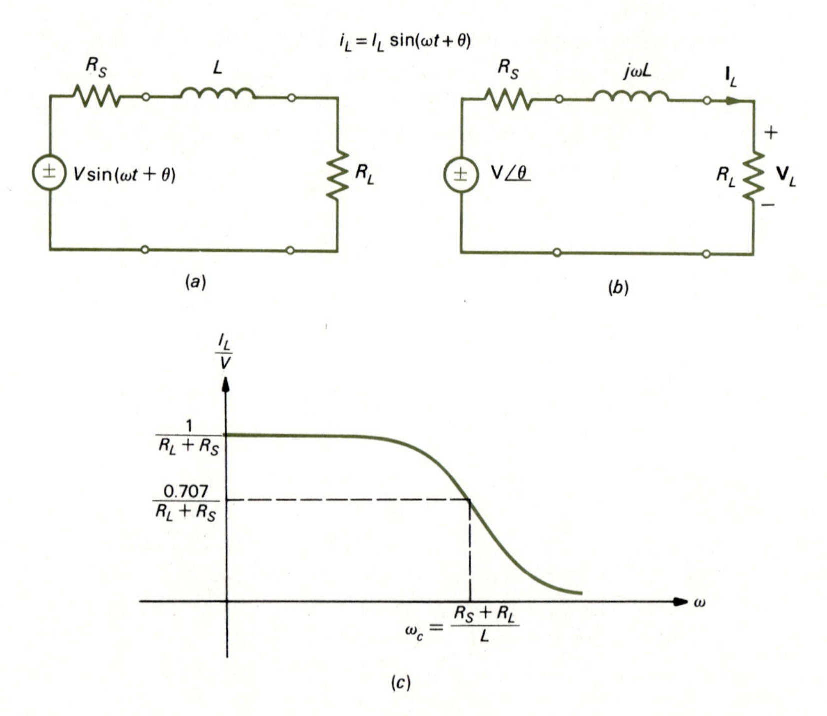

FIGURE 5.22
A low-pass filter.

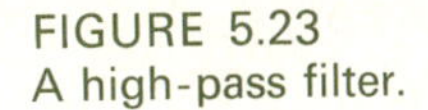
FIGURE 5.23
A high-pass filter.

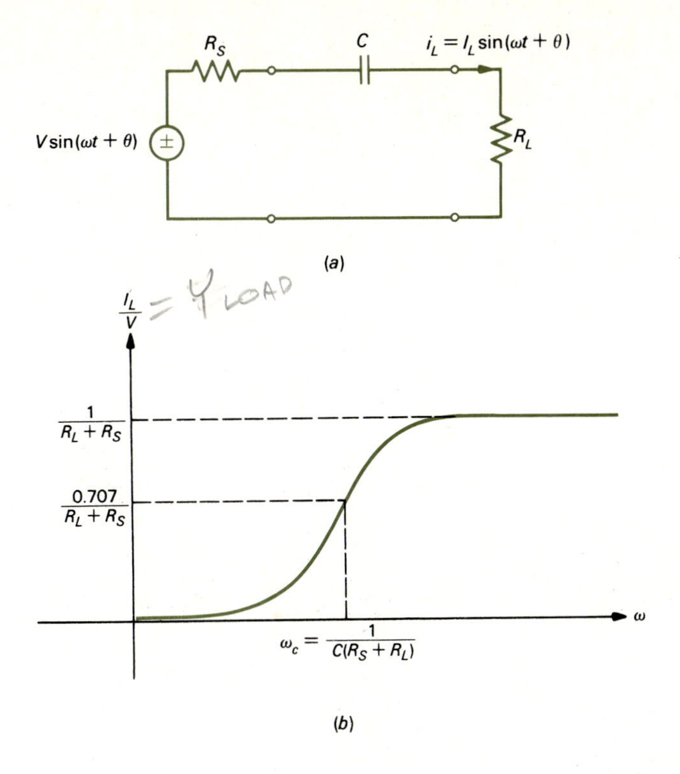

Now let us consider the design of a high-pass filter, one which passes high-frequency components of the signal and rejects low-frequency ones. The obvious choice is to substitute a capacitor for the inductor of the low-pass filter, since the impedance of a capacitor, $Z_C = 1/j\omega C$, is infinite at dc but is reduced as the frequency is increased. The circuit and frequency response are shown in Fig. 5.23. The ratio of load current to source voltage is

$$\frac{I_L}{V} = \frac{1}{R_S + R_L} \frac{1}{\sqrt{1 + \left[\frac{1}{\omega C(R_S + R_L)}\right]^2}} \tag{5.133}$$

At a frequency

$$\omega_c = \frac{1}{C(R_S + R_L)} \tag{5.134}$$

the ratio in Eq. (5.133) has been reduced from its high-frequency value by $1/\sqrt{2}$ and the average power delivered to R_L has been reduced by 1/2.

The low-pass (high-pass) filter passes low frequencies (high frequencies) and rejects high frequencies (low frequencies). Filters can also be designed to pass or reject certain bands of frequencies. These filters make use of more than one energy storage

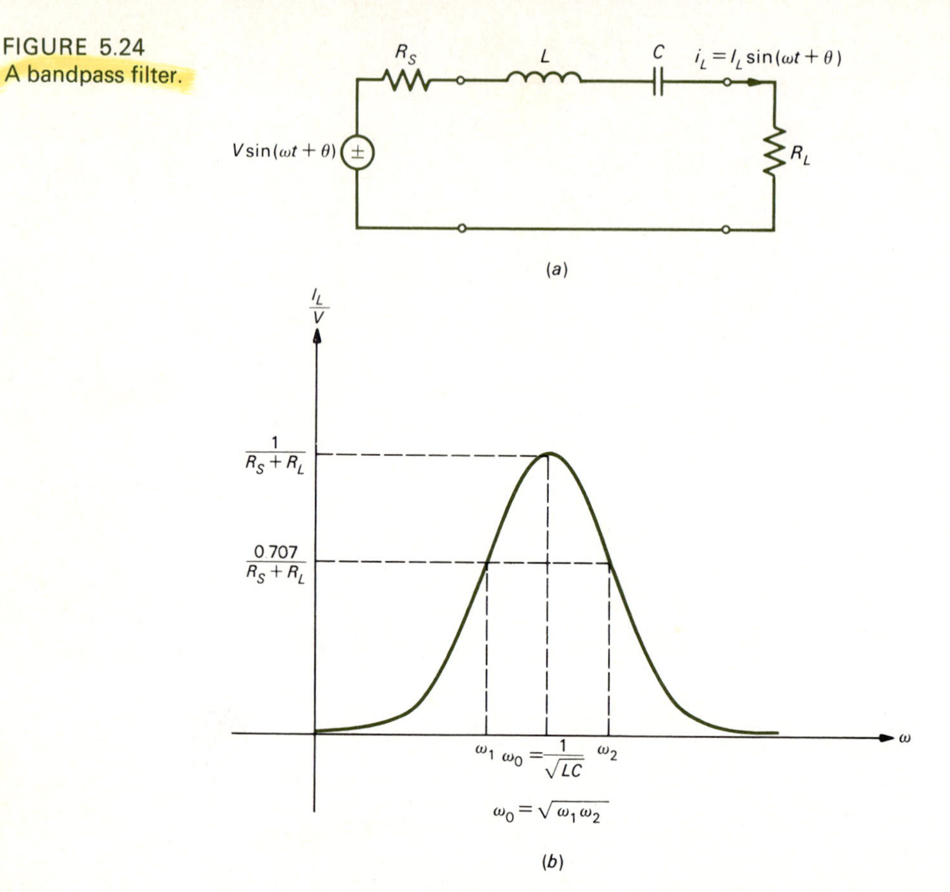

FIGURE 5.24
A bandpass filter.

element and utilize the phenomenon of resonance. A bandpass filter is shown in Fig. 5.24*a*. At dc the capacitor presents an infinite impedance, and at $\omega = \infty$ the inductor presents an infinite impedance. At intermediate frequencies, the impedance of the series L–C combination

$$\mathbf{Z} = j\omega L + \frac{1}{j\omega C}$$

$$= \frac{1 - \omega^2 LC}{j\omega C} \tag{5.135}$$

is less than infinite. At a frequency of

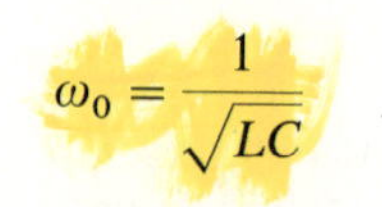

$$\omega_0 = \frac{1}{\sqrt{LC}} \tag{5.136}$$

the impedance is zero and the source and load are directly connected. The ratio of the load current and source voltage can easily be obtained from the phasor circuit and becomes

$$\frac{I_L}{V} = \frac{1}{\sqrt{(R_S + R_L)^2 + (\omega L - 1/\omega C)^2}}$$
$$= \frac{1}{R_S + R_L} \frac{1}{\sqrt{1 + \left[\dfrac{1 - \omega^2 LC}{\omega C(R_S + R_L)}\right]^2}} \tag{5.137}$$

At frequency

$$\omega_0 = \frac{1}{\sqrt{LC}} \tag{5.138}$$

Eq. (5.137) becomes

$$\frac{I_L}{V} = \frac{1}{R_S + R_L} \qquad \omega = \omega_0 \tag{5.139}$$

which is the largest value this ratio can attain at any frequency. As the frequency is lowered from ω_0, the ratio eventually decreases to zero at dc, as is clear from the circuit, since at dc the capacitor presents an open circuit. As the frequency is raised above ω_0, the ratio also eventually decreases to zero at $\omega = \infty$, as is clear from the circuit, since at $\omega = \infty$ the inductor presents an open circuit. At $\omega = \omega_0$, the inductor and capacitor are said to resonate such that their series impedance $\omega_0 L - 1/\omega_0 C$ is zero.

The half-power points are frequencies such that the ratio I_L/V is reduced to $1/\sqrt{2}$ of its value at ω_0. From Eq. (5.137) this occurs when

$$\frac{1 - \omega^2 LC}{\omega C(R_S + R_L)} = 1 \tag{5.140}$$

or

$$\omega^2 + \frac{R_S + R_L}{L}\omega - \frac{1}{LC} = 1 \tag{5.141}$$

There are two values of frequency which satisfy this equation:

$$\omega^+, \omega^- = -\frac{R_S + R_L}{2L} \pm \tfrac{1}{2}\sqrt{\frac{(R_S + R_L)^2}{L^2} + \frac{4}{LC}} \tag{5.142}$$

Defining

$$\alpha = \frac{R_S + R_L}{2L} \tag{5.143}$$

we obtain

$$\omega^+, \omega^- = -\alpha \pm \sqrt{\alpha^2 + \omega_0^2} \tag{5.144}$$

Although two solutions are obtained, the negative frequency case is meaningless and we obtain

$$\omega_1 = -\alpha + \sqrt{\alpha^2 + \omega_0^2} \tag{5.145}$$

Returning to Eq. (5.137), we see that there is another possibility:

$$\frac{1 - \omega^2 LC}{\omega C(R_S + R_L)} = -1 \tag{5.146}$$

Again we obtain

$$\omega^+, \omega^- = \alpha \pm \sqrt{\alpha^2 + \omega_0^2} \tag{5.147}$$

and, selecting the positive frequency, we have

$$\omega_2 = \alpha + \sqrt{\alpha^2 + \omega_0^2} \tag{5.148}$$

The bandwidth B is the frequency separation between the half-power points

$$\begin{aligned} B &= \omega_2 - \omega_1 \\ &= 2\alpha \\ &= \frac{R_S + R_L}{L} \end{aligned} \tag{5.149}$$

and represents the effective passband of the filter. It can readily be shown that ω_0, ω_1, and ω_2 are related such that

$$\omega_0 = \sqrt{\omega_1 \omega_2} \tag{5.150}$$

[Substitute Eqs. (5.148) and (5.145) into Eq. (5.150).]

The ratio of the center frequency ω_0 to the bandwidth B of the filter is a measure of the frequency discrimination of the filter; the narrower the bandwidth, the greater the ability of the filter to isolate or separate out a single frequency component of the signal. This ratio is defined as the Q of the filter:

$$\begin{aligned} Q &= \frac{\omega_0}{B} \\ &= \frac{\omega_0 L}{R_S + R_L} \end{aligned} \tag{5.151}$$

FIGURE 5.25
A band-reject filter.

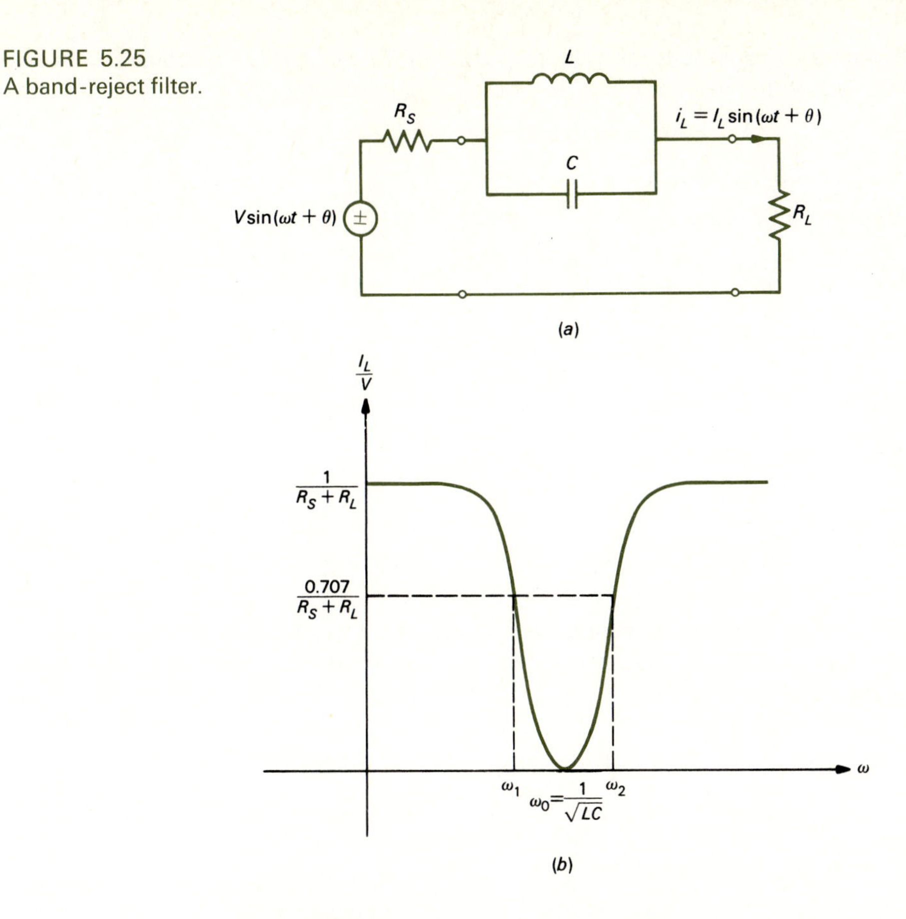

The smaller the source and load resistances, the larger the Q and resulting sharpness of the filter.

A band-reject filter can similarly be constructed to reject a narrow band of frequencies and pass the remaining ones. This type of filter is useful in eliminating unwanted frequency components from interfering with transmissions in radio circuits. Figure 5.25 shows a simple example and its frequency response. At dc the inductor acts as a short-circuit, and at $\omega = \infty$ the capacitor acts as a short circuit. However, at intermediate frequencies the impedance of the parallel LC combination

$$\mathbf{Z} = \frac{1}{j\omega C + 1/j\omega L}$$

$$= \frac{j\omega L}{1 - \omega^2 LC} \tag{5.152}$$

is larger than zero. In particular, at

$$\omega_0 = \frac{1}{\sqrt{LC}} \tag{5.153}$$

the impedance is infinite, as seen from Eq. (5.152), and the LC combination isolates the source from the load. From the phasor circuit we can obtain

$$\frac{I_L}{V} = \frac{1}{\sqrt{(R_S + R_L)^2 + \left(\dfrac{\omega L}{1 - \omega^2 LC}\right)^2}}$$

$$= \frac{1}{R_S + R_L} \frac{1}{\sqrt{1 \mp \left[\dfrac{\omega L/(R_S + R_L)}{1 - \omega^2 LC}\right]^2}} \tag{5.154}$$

At a frequency such that

$$\frac{\omega L/(R_S + R_L)}{1 - \omega^2 LC} = \pm 1 \tag{5.155}$$

the frequency response is reduced by $1/\sqrt{2}$ from its maximum value and the average power delivered to the load is reduced by 1/2 from its maximum value. Solving Eq. (5.155), we obtain

$$\omega^2 + \frac{1}{C(R_S + R_L)}\omega \pm \frac{1}{LC} = 0 \tag{5.156}$$

Solving Eq. (5.156), we obtain

$$\omega_1 = -\alpha + \sqrt{\alpha^2 + \omega_0^2} \tag{5.157a}$$

$$\omega_2 = \alpha + \sqrt{\alpha^2 + \omega_0^2} \tag{5.157b}$$

where

$$\alpha = \frac{1}{2C(R_S + R_L)} \tag{5.158}$$

The bandwidth becomes

$$B = \omega_2 - \omega_1$$

$$= 2\alpha$$

$$= \frac{1}{C(R_S + R_L)} \tag{5.159}$$

and the Q of the filter becomes

$$Q = \frac{\omega_0}{B} = \omega_0 C(R_S + R_L) \quad (5.160)$$

Obviously, the response characteristics and the other properties of these filters depend on the source and load. Therefore, we cannot speak of the properties of a filter without considering the particular source and load with which it will be used. The simple filters considered in this section show via their frequency-response characteristics how a particular frequency component of the source waveform will be affected. Other, more elaborate filters have corresponding properties but are not as easily obtained.

Example 5.8 A square-wave voltage source having an amplitude of 5 V, a frequency of 1 kHz, a pulse width of 0.5 ms, and internal source resistance of 50 Ω is applied to a load which is represented at its input terminals as a 100-Ω resistor; this is shown in Fig. 5.26*a*. The source may represent the output of a digital device, such as a digital computer, and the 100-Ω load may represent a unit designed to process that data. This representation of the output of the digital device as a known sequence of square pulses is not a valid representation of a digital device since digital data would have pulses occurring randomly in time, as we will see later. However, for illustration we will use this model.

The problem is that the abrupt rise and fall of the leading- and trailing edges are troublesome for the receiver. What we would like to do is to smooth these pulses somewhat to provide a less abrupt transition. The abruptness of the transitions is primarily due to the high-frequency components of the wave; to slow these transitions, we will insert a filter between the source and load to remove some of these high-frequency components. Suppose that we arbitrarily attempt to reduce all components above 5 kHz.

Design a low-pass filter to accomplish this. Determine the resulting voltage $v_L(t)$ across the load and show its amplitude spectrum.

Solution The Fourier series of the square wave is (see App. C)

$$v_S(t) = 2.5 + \frac{10}{\pi}\sin\omega_0 t + \frac{10}{3\pi}\sin 3\omega_0 t + \frac{10}{5\pi}\sin 5\omega_0 t + \cdots$$

where $\omega_0 = 2\pi/f_0$ and $f_0 = 1/T = 1$ kHz. The lowest frequency component, $f_0 = 1$ kHz, is called the fundamental frequency of the wave; the next frequency, $3f_0 = 3$ kHz, is called the third harmonic of the wave; etc. The term 2.5 is the average value (dc component) of the

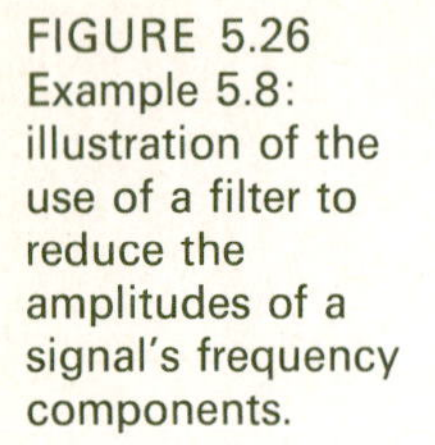
FIGURE 5.26 Example 5.8: illustration of the use of a filter to reduce the amplitudes of a signal's frequency components.

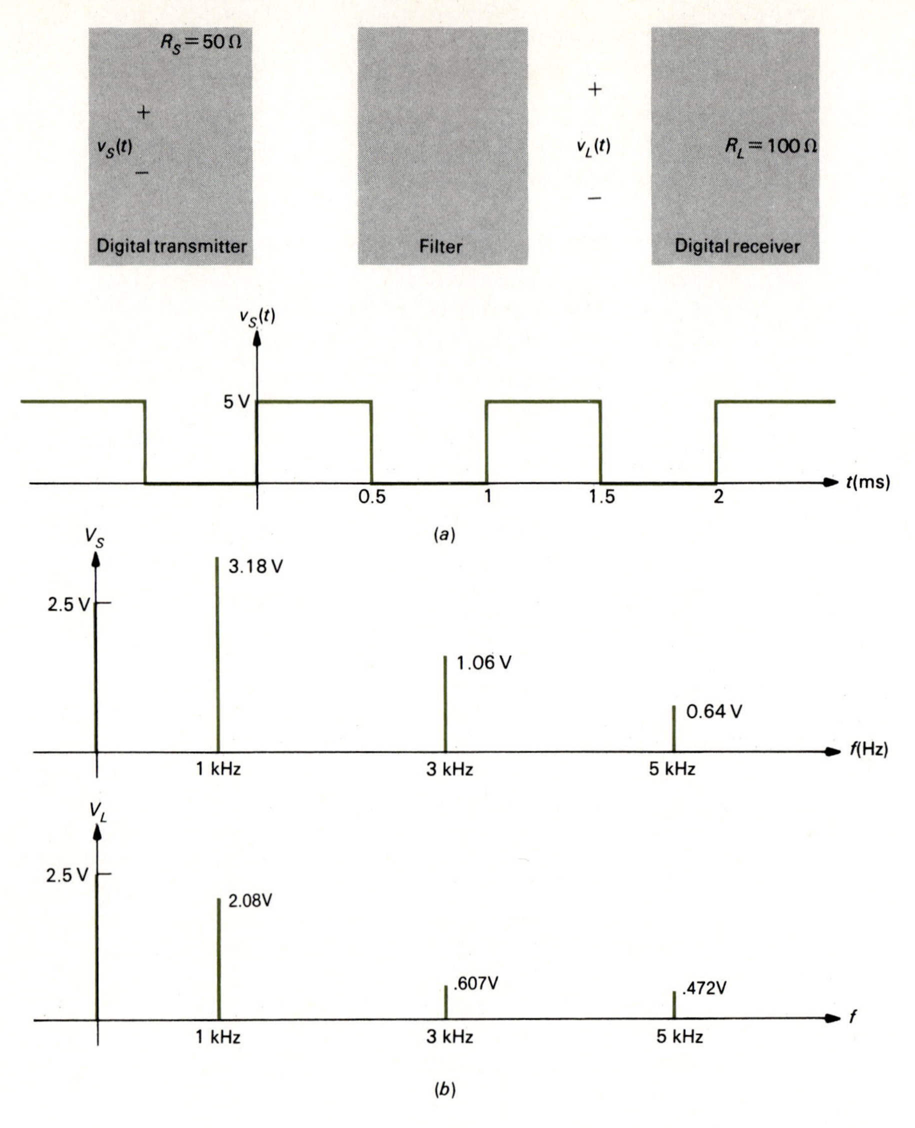

waveform, which is obvious from the sketch of $v_S(t)$. Suppose we select the half-power point of the filter to be 5 kHz. Thus, from Eq. (5.131)

$$2\pi f_c = 2\pi \times 5 \times 10^3$$

$$= \frac{R_S + R_L}{L}$$

$$= \frac{150}{L}$$

Solving for L gives

$$L = \frac{150}{5 \times 10^3 \times 2\pi}$$

$$= 4.8 \text{ mH}$$

The ratio of the load voltage to $V_S(t)$ is from Eq. (5.130):

$$\frac{V_L}{V_S} = \frac{R_L}{R_S + R_L} \frac{1}{\sqrt{1 + \left[\frac{\omega L}{(R_S + R_L)}\right]^2}}$$

$$= \frac{6.67 \times 10^{-1}}{\sqrt{1 + (2 \times 10^{-4} f)^2}}$$

The dc component ($f = 0$) is passed unaltered. The fundamental frequency component ($f =$ 1 kHz) of $10/\pi = 3.18$ V becomes

$$V_L = 6.54 \times 10^{-1} V_S$$

$$= 6.54 \times 10^{-1} \times \frac{10}{\pi}$$

$$= 2.08 \text{ V}$$

The third harmonic ($f = 3$ kHz) becomes

$$V_L = 5.72 \times 10^{-1} V_S$$

$$= 5.72 \times 10^{-1} \times \frac{10}{3\pi}$$

$$= .607 \text{ V}$$

while the fifth harmonic becomes

$$V_L = .472 \text{ V}$$

The seventh harmonic is

$$V_L = .388 \text{ V}$$

The amplitude spectra of $v_S(t)$ and $v_L(t)$ are sketched in Fig. 5.26*b*. To see how effective the filtering has been, let us compute the ratio of the amplitude of the fundamental and the seventh harmonic ($f = 7$ kHz). For $v_S(t)$ this is

$$\frac{10/\pi}{10/7\pi} = 7$$

Thus, the fundamental component is 7 times larger in amplitude than the seventh harmonic. Now for $V_L(t)$ this ratio is

$$\frac{2.08}{.388} = 5.36$$

Thus, the amplitude of the seventh harmonic has been reduced in comparison to the amplitude of the fundamental frequency.

A sketch of $v_L(t)$ can be obtained from the Fourier series of $v_L(t)$ found by combining the above amplitudes:

$$\begin{aligned} V_L(t) = 2.5 &+ 2.08 \text{ V} \sin(2\pi \times 10^3 t) \\ &+ .607 \text{ V} \sin(6\pi \times 10^3 t) \\ &+ .472 \text{ V} \sin(10\pi \times 10^3 t) \\ &+ .388 \text{ V} \sin(14\pi \times 10^3 t) \\ &+ \cdots \end{aligned}$$

5.8 SUPERPOSITION OF AVERAGE POWER IN AC CIRCUITS

In Chap. 1 we found that, although we may apply superposition to a branch voltage or current, we may generally not apply superposition to the branch power. For example, suppose a branch voltage has two components, $v'(t)$ and $v''(t)$, each due to one of the ideal sources in the circuit. Similarly, the branch current has two components, $i'(t)$ and $i''(t)$, each due to one of the two sources. The total branch voltage and current are

$$v(t) = v'(t) + v''(t) \tag{5.161a}$$

$$i(t) = i'(t) + i''(t) \tag{5.161b}$$

The instantaneous power associated with the branch is

$$\begin{aligned} p(t) &= v(t)i(t) \\ &= v'i' + v''i'' + v''i' + v'i'' \end{aligned} \tag{5.162}$$

If we compute the branch powers due to each source acting alone as

$$\begin{aligned} p' &= v'i' \\ p'' &= v''i'' \end{aligned} \tag{5.163}$$

we see from comparing Eqs. (5.162) and (5.163) that

$$p \neq p' + p'' \tag{5.164}$$

Thus, we cannot apply superposition to instantaneous power.

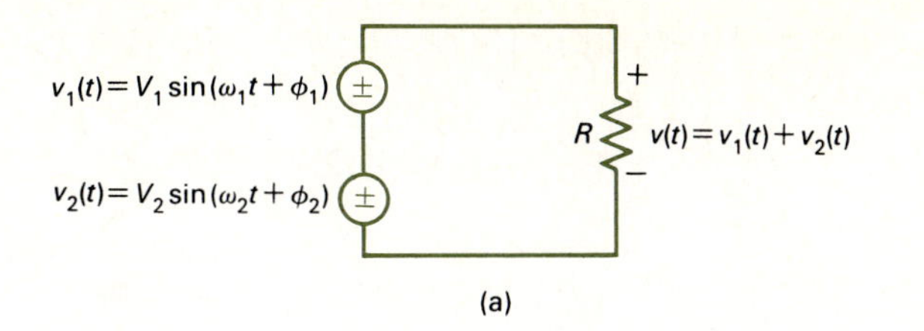

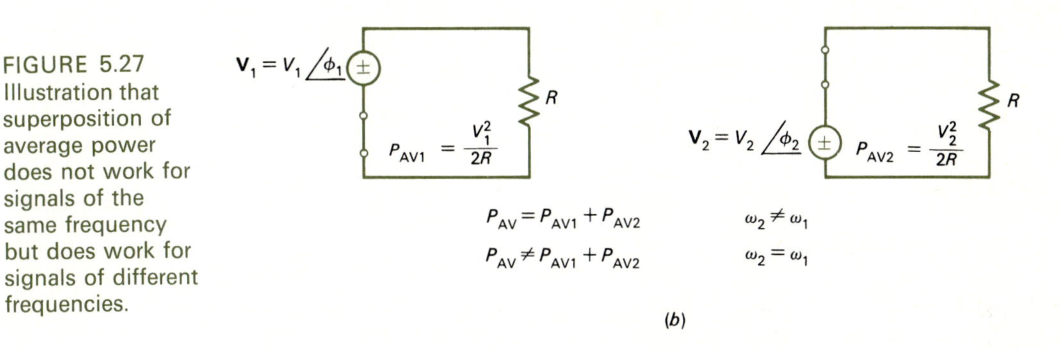

FIGURE 5.27 Illustration that superposition of average power does not work for signals of the same frequency but does work for signals of different frequencies.

Although we may not apply superposition to instantaneous power, we will find that, with certain simple restrictions, we can apply superposition to average power calculations in ac circuits. To begin, let us consider a simple case of two sinusoidal voltage sources attached to a resistor R, as shown in Fig. 5.27*a*. The voltage across the resistor is

$$\begin{aligned} v(t) &= v_1(t) + v_2(t) \\ &= V_1 \sin(\omega_1 t + \phi_1) + V_2 \sin(\omega_2 t + \phi_2) \end{aligned} \tag{5.165}$$

The instantaneous power is

$$\begin{aligned} p(t) &= \frac{v^2(t)}{R} \\ &= \frac{1}{R}[V_1 \sin(\omega_1 t + \phi_1) + V_2 \sin(\omega_2 t + \phi_2)]^2 \\ &= \frac{1}{R}[V_1^2 \sin^2(\omega_1 t + \phi_1) + V_2^2 \sin^2(\omega_2 t + \phi_2) \\ &\quad + 2V_1 V_2 \sin(\omega_1 t + \phi_1)\sin(\omega_2 t + \phi_2)] \end{aligned} \tag{5.166}$$

Noting that

$$\sin^2 A = \tfrac{1}{2}(1 - \cos 2A)$$

$$\sin A \sin B = \tfrac{1}{2}[\cos(A - B) - \cos(A + B)]$$

we may write Eq. (5.166) as

$$p(t) = \frac{V_1^2}{2R} + \frac{V_2^2}{2R} - \frac{V_1^2}{2R}\cos(2\omega_1 t + 2\phi_1) - \frac{V_2^2}{2R}\cos(2\omega_2 t + 2\phi_2)$$
$$+ \frac{V_1V_2}{R}\{\cos[(\omega_2 - \omega_1)t + \phi_2 - \phi_1] - \cos[(\omega_2 + \omega_1)t + \phi_2 + \phi_1]\} \tag{5.167}$$

An interesting observation is apparent from Eq. (5.167). The instantaneous power is the sum of six terms

$$p(t) = T_1 + T_2 + T_3 + T_4 + T_5 + T_6 \tag{5.168}$$

where $T_1 = \frac{V_1^2}{2R}$

$$T_2 = \frac{V_2^2}{2R}$$

$$T_3 = -\frac{V_1^2}{2R}\cos(2\omega_1 t + 2\phi_1)$$

$$T_4 = -\frac{V_2^2}{2R}\cos(2\omega_2 t + 2\phi_2)$$

$$T_5 = \frac{V_1V_2}{R}\cos[(\omega_2 - \omega_1)t + \phi_2 - \phi_1]$$

$$T_6 = -\frac{V_1V_2}{R}\cos[(\omega_2 + \omega_1)t + \phi_2 + \phi_1]$$

Note that T_1 and T_2 are constants but that T_3, T_4, T_5, T_6 are all sinusoidal functions with zero average value. Thus, it is clear that the average power is

$$P_{\text{AV}} = \frac{V_1^2}{2R} + \frac{V_2^2}{2R} \tag{5.169}$$

However, Eq. (5.169) is the sum of the average powers dissipated in R by each source individually, as shown in Fig. 5.27*b*:

$$P_{\text{AV}} = P_{\text{AV1}} + P_{\text{AV2}} \qquad \omega_1 \neq \omega_2 \tag{5.170}$$

However, if both sources are of the same frequency, $\omega_2 = \omega_1$, then terms T_3, T_4, and T_6 remain sinusoidal functions with zero average value *but* T_5 becomes

$$T_5 = \frac{V_1V_2}{R}\cos(\phi_2 - \phi_1) \qquad \omega_2 = \omega_1 \tag{5.171}$$

which is a constant. Then

$$P_{AV} = \frac{V_1^2}{2R} + \frac{V_2^2}{2R} + \frac{V_1V_2}{R}\cos(\phi_2 - \phi_1) \qquad \omega_2 = \omega_1 \tag{5.172}$$

and

$$P_{AV} \neq P_{AV1} + P_{AV2} \qquad \omega_2 = \omega_1 \tag{5.173}$$

In general, we can show by similar methods that *we may superposition the average powers from sinusoidal sources of different frequencies* but may not superposition average powers from sinusoidal sources of the same frequency. Note that a dc source can be considered to be a sinusoid at zero frequency ($\omega = 0$) since

$$A\cos(0t + \phi) = A\cos\phi$$

This result provides a considerable simplification in the calculation of average power, as shown by the following example.

Example 5.9 For the circuit of Fig. 5.28*a*, compute the average power delivered to the load.

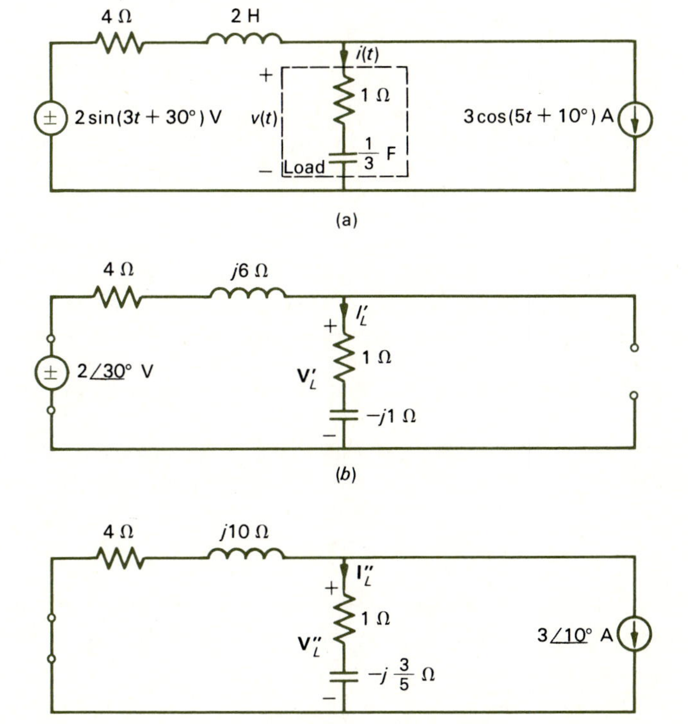

FIGURE 5.28 Example 5.9: illustration of superposition of average power.

Solution The average power due to the voltage source can be computed from the phasor circuit in Fig. 5.28*b*. The phasor current through the load is

$$\mathbf{I}_L' = \frac{2\angle 30^\circ}{4 + j6 + 1 - j1}$$

$$= \frac{2\angle 30^\circ}{5 + j5}$$

$$= 0.28\angle -15^\circ \text{ A}$$

The average power due to this source is

$$P_{AV}' = \tfrac{1}{2}(0.28)^2 1$$

$$= 0.0393 \text{ W}$$

The phasor circuit for the current source is shown in Fig. 5.28*c*. From this we compute

$$\mathbf{I}_L'' = -\frac{4 + j10}{4 + j10 + 1 - j3/5} 3\angle 10^\circ$$

$$= -3.03\angle 16.21^\circ \text{ A}$$

and

$$P_{AV}'' = \tfrac{1}{2}(3.03)^2 1$$

$$= 4.59 \text{ W}$$

Since the two sources are of different frequencies, the average power delivered to the load with both sources functioning is

$$P_{AV} = P_{AV}' + P_{AV}''$$

$$= 4.63 \text{ W}$$

Example 5.10 Consider the circuit in Fig. 5.28 but with the current source having the same radian frequency ($\omega = 3$) as the voltage source. Show that we cannot superposition average power in this case.

Solution The phasor load current due to the voltage source remains unchanged from the previous example:

$$I_L' = 0.28\angle -15^\circ \text{ A}$$

The phasor load current due to the current source becomes

$$I_L'' = -\frac{4 + j6}{4 + j6 + 1 - j1} 3\angle 10^\circ$$

$$= -3.06\angle 21.31^\circ$$

We cannot add these two phasors to find the total phasor current, since $\mathbf{I}_L'$ was from a sinusoidal source and $\mathbf{I}_L''$was from a cosinusoidal source. If we did, how would we choose to convert to the time-domain form:

$$i_L(t) \stackrel{?}{=} \text{Re}\,(\mathbf{I}_L' + \mathbf{I}_L'')e^{j\omega t}$$

or

$$i_L(t) \stackrel{?}{=} \text{Im}\,(\mathbf{I}_L' + \mathbf{I}_L'')e^{j\omega t}$$

However, no problem would have arisen if we had converted both sources to either sines or cosines. Therefore, let us convert the current source to a sinusoidal form:

$$\begin{aligned} 3\cos(3t + 10°) &= 3\sin(3t + 10° + 90°) \\ &= 3\sin(3t + 100°) \end{aligned}$$

Thus, $\mathbf{I}''$ for this form is the previous result with 90° added, or

$$\mathbf{I}_L'' = -3.06\underline{/111.31°}$$

and

$$\mathbf{I}_L = \mathbf{I}_L' + \mathbf{I}_L''$$

Then the time-domain current is

$$\begin{aligned} i_L(t) &= \text{Im}\,(\mathbf{I}_L' + \mathbf{I}_L'')e^{j\omega t} \\ &= 3.23\sin(3t - 64.69°) \end{aligned}$$

and the total average power delivered to the load is

$$\begin{aligned} P_{\text{AV}} &= \tfrac{1}{2}I_L^2 \cdot 1 \\ &= 5.23\ \text{W} \end{aligned}$$

Note that

$$\begin{aligned} P_{\text{AV}}' &= \tfrac{1}{2}(I_L)^2 \cdot 1 \\ &= 0.04 \end{aligned}$$

and

$$\begin{aligned} P_{\text{AV}}'' &= \tfrac{1}{2}(I_L'')^2 \cdot 1 \\ &= 4.68 \end{aligned}$$

and

$$\begin{aligned} P_{\text{AV}}' + P_{\text{AV}}'' &= 4.72\ \text{W} \\ &\neq P_{\text{AV}} = 5.23\ \text{W} \end{aligned}$$

One final point which is appropriate to this section should be made. If a current in an element is composed, by superposition, of several parts due to sinusoidal sources of different frequencies

$$i(t) = I_1 \sin(\omega_1 t + \theta_1) + I_2 \sin(\omega_2 t + \theta_2) + \cdots + I_N \sin(\omega_N t + \theta_N) \tag{5.174}$$

then the RMS, or effective, value of $i(t)$ is *not* the sum of the RMS values of all components but is

$$I_{\text{RMS}} = \sqrt{I_{\text{RMS1}}^2 + I_{\text{RMS2}}^2 + \cdots + I_{\text{RMS}N}^2} \tag{5.175}$$

This can be shown quite easily by recalling that if this current is applied to a resistor R, the average power dissipated in the resistor is

$$\begin{aligned} P_{\text{AV}} &= \frac{I_1^2 R}{2} + \frac{I_2^2 R}{2} + \cdots + \frac{I_N^2 R}{2} \\ &= (I_{\text{RMS1}}^2 + I_{\text{RMS2}}^2 + \cdots + I_{\text{RMS}N}^2) R \end{aligned} \tag{5.176}$$

But, in terms of the RMS value of $i(t)$,

$$\begin{aligned} P_{\text{AV}} &= \frac{I^2}{2} R \\ &= I_{\text{RMS}}^2 R \end{aligned} \tag{5.177}$$

Comparing Eqs. (5.177) and (5.176), we obtain Eq. (5.175). The result similarly applies to a voltage

$$v(t) = V_1 \sin(\omega_1 t + \theta_1) + V_2 \sin(\omega_2 t + \theta_2) + \cdots + V_N \sin(\omega_N t + \theta_N) \tag{5.178}$$

as

$$V_{\text{RMS}} = \sqrt{V_{\text{RMS1}}^2 + V_{\text{RMS2}}^2 + V_{\text{RMS3}}^2 + \cdots + V_{\text{RMS}N}^2} \tag{5.179}$$

PROBLEMS

5.1 For the sinusoidal waveform shown in Fig. P5.1, suppose we write $x(t) = A \sin(\omega t + \theta)$. Determine A, ω, θ. Repeat if we write $x(t) = A \cos(\omega t + \theta)$.

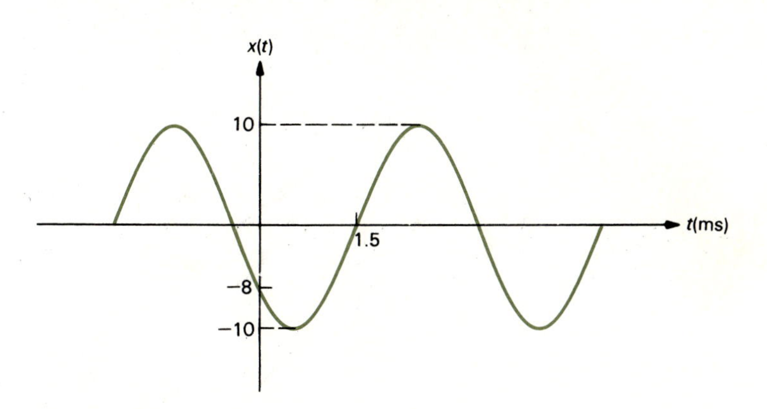

FIGURE P5.1

5.2 For the circuit shown in Fig. P5.2, write a differential equation relating $v(t)$ to $i_S(t)$. Determine a solution to that equation. Write your solution in the form $A \sin(2t + \theta)$.

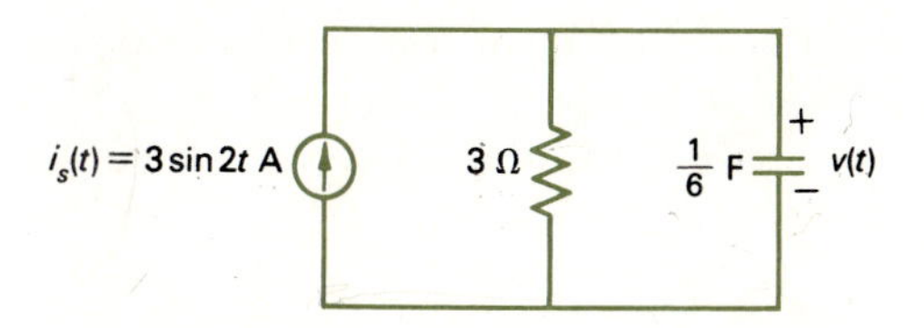

FIGURE P5.2

5.3 For the circuit shown in Fig. P5.3, write a differential equation relating $i(t)$ to $v_S(t)$. Determine a solution to that equation.

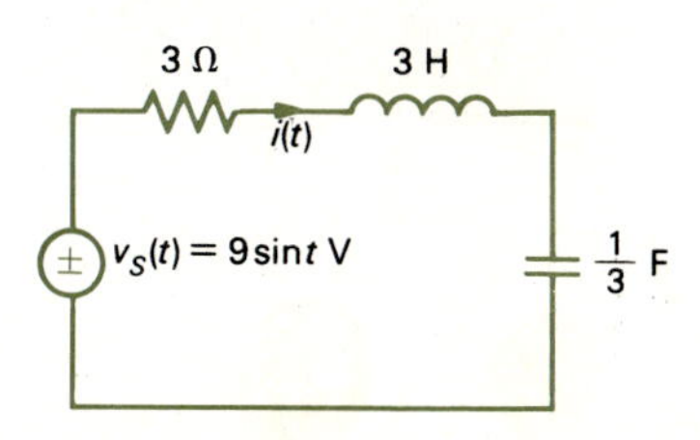

FIGURE P5.3

5.4 For the complex numbers $\mathbf{A} = 2 + j2$ and $\mathbf{B} = 1 - j1$, determine the following in both rectangular and polar form:

(*a*) $\mathbf{A} + \mathbf{B}$ (*d*) $\mathbf{AB}$
(*b*) $\mathbf{A} - \mathbf{B}$ (*e*) $\mathbf{A}^* + \mathbf{B}$
(*c*) $\mathbf{A}/\mathbf{B}$

Verify your results in (*a*), (*b*), and (*e*) by plotting in the complex plane.

5.5 Verify to your satisfaction that $A\underline{/\theta} = Ae^{j\theta} = A\cos\theta + jA\sin\theta$.

5.6 Show that $\mathbf{A} + \mathbf{B} + \mathbf{C} = 0$ when

$$\mathbf{A} = 1\underline{/0^\circ}$$

$$\mathbf{B} = 1\underline{/120^\circ} = e^{j(2\pi/3)}$$

$$\mathbf{C} = 1\underline{/240^\circ} = 1\underline{/-120^\circ} = e^{j(4\pi/3)} = e^{-j(2\pi/3)}$$

5.7 For the complex numbers $\mathbf{A} = 1 + j2$ and $\mathbf{B} = 2 - j1$, compute:

(*a*) $\mathbf{A}$ and $\mathbf{B}$ in polar form (*d*) $\mathbf{A}^* - \mathbf{B}$
(*b*) $\mathbf{A} + \mathbf{B}$ (*e*) $\mathbf{B}/\mathbf{A}$
(*c*) $\mathbf{A} - \mathbf{B}$ (*f*) $\mathbf{AB}$

5.8 For the complex numbers $\mathbf{A} = 2\underline{/-45^\circ}$, $\mathbf{B} = 1 - j3$, and $\mathbf{C} = 2 + j1$, compute:

(*a*) $\mathbf{A} + \mathbf{B}$ (*d*) $\mathbf{A} + \mathbf{B}/\mathbf{C}$
(*b*) $\mathbf{A}/\mathbf{B}$ (*e*) $2\mathbf{C} - 3\mathbf{A}/\mathbf{B}$
(*c*) $\mathbf{AC}$

5.9 Given $\mathbf{A} = 1 + j3$, $\mathbf{B} = 2\underline{/30^\circ}$, and $\mathbf{C} = 2 - j3$, determine in polar and rectangular form

(*a*) $j\mathbf{A}(\mathbf{B} + \mathbf{C})^2$ (*b*) $\mathbf{B}\ \text{Re}\ \mathbf{C} - j\mathbf{A}\ \text{Im}\ \mathbf{B}$

5.10 For the circuit in Fig. P5.2, replace $i_S(t)$ with $3e^{j2t}$ and solve the resulting differential equation. From that solution determine $v(t)$.

5.11 In the circuit shown in Fig. P5.3, replace $v_S(t)$ with $9e^{jt}$ and solve the resulting differential equation. From that solution determine $i(t)$.

5.12 For the circuit shown in Fig. P5.12, write the differential equation relating $i(t)$ to $v_S(t)$. (*Hint*; Use a Thevenin equivalent.) Solve for $i(t)$.

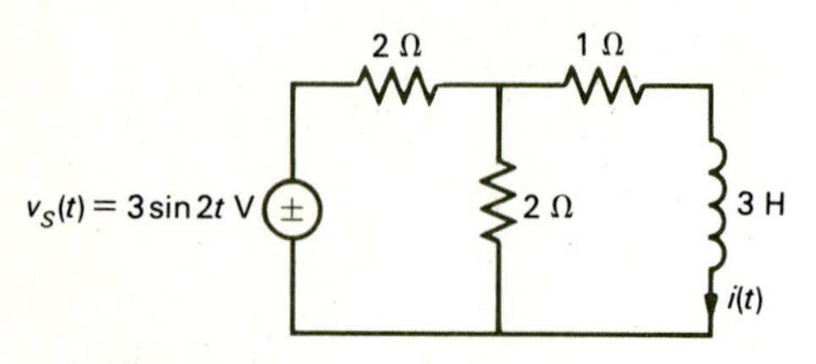

FIGURE P5.12

5.13 Draw the phasor circuit for Fig. P5.2 and solve for $v(t)$.

5.14 For the circuit in Fig. P5.3, draw the phasor circuit and solve for $i(t)$.

5.15 For the circuit in Fig. P5.12, draw the phasor circuit and solve for $i(t)$.

5.16 For the circuit shown in Fig. P5.16, draw the phasor circuit and solve for $i(t)$.

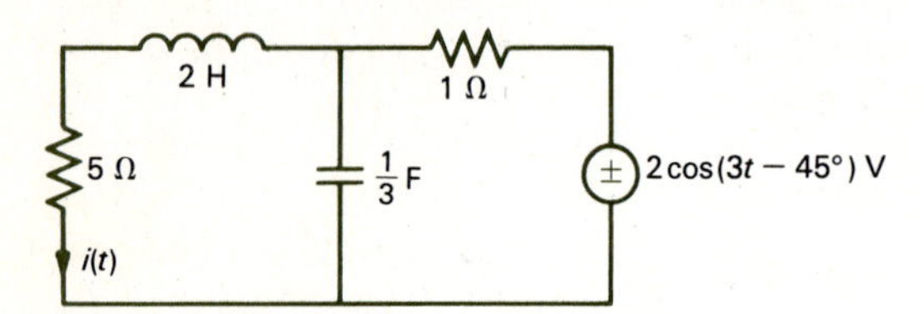

FIGURE P5.16

5.17 For the circuit shown in Fig. P5.17, draw the phasor circuit and solve for $i(t)$ by obtaining a Thevenin equivalent at terminals *a-b*.

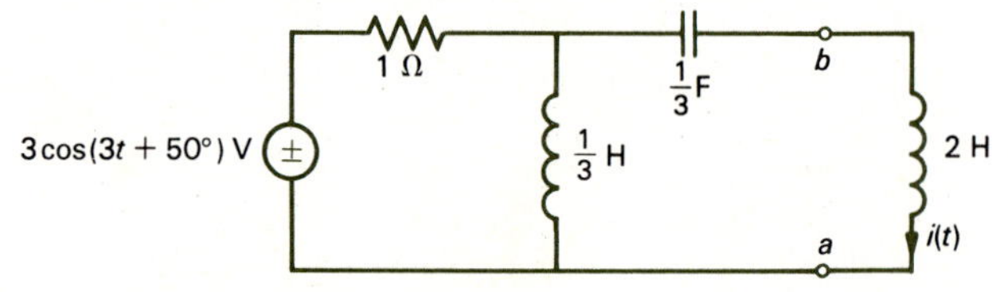

FIGURE P5.17

5.18 For the circuit shown in Fig. P5.18, calculate $i(t)$.

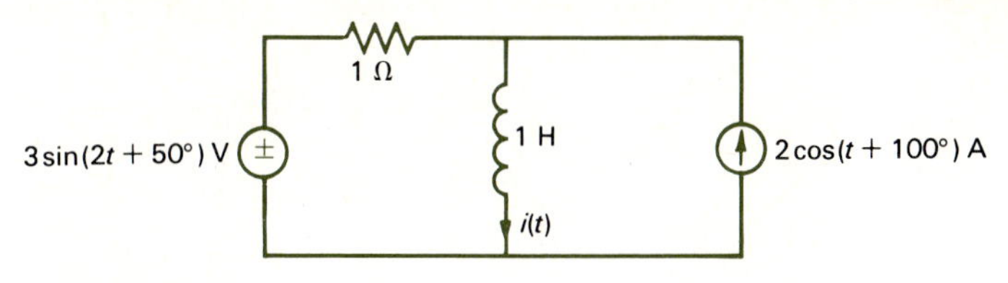

FIGURE P5.18

5.19 Change the frequency of the voltage source in Fig. P5.18 from 2 rad/s to 1 rad/s and repeat Prob. 5-18. Obtain your solution two ways. First use superposition. Next obtain the solution with both sources acting simultaneously. Show that the two solutions are equivalent. Could this be done with the circuit in Fig. P5.18? Why not?

5.20 In the circuit of Fig. P5.20, determine the average power delivered by the two sources. Show that this is the average power consumed by the resistor.

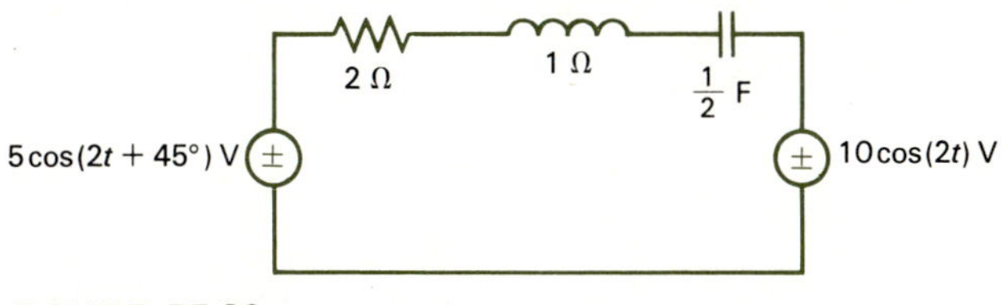

FIGURE P5.20

5.21 In the circuit shown in Fig. P5.21, determine the complex power delivered by the source. Show that the total average power delivered to the circuit by the source is zero. Why is this to be expected?

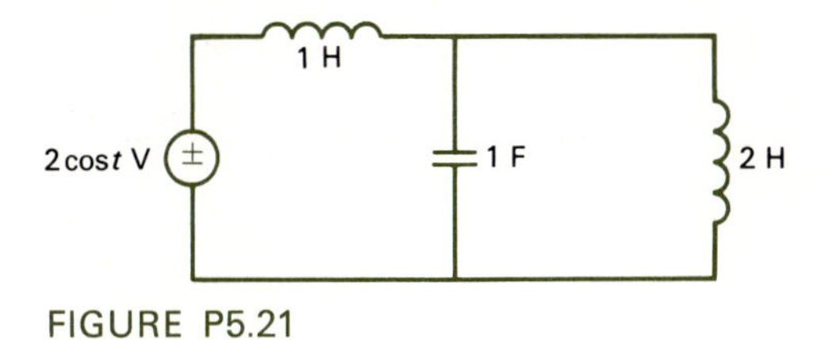

FIGURE P5.21

5.22 Determine the power factor of the load represented in Fig. P5.22 if the radian frequency of excitation is $\omega = 2$ rad/s. Can you replace this load with (*a*) a resistor in series with an inductor or (*b*) a resistor in series with a capacitor? If so, give the values of these elements.

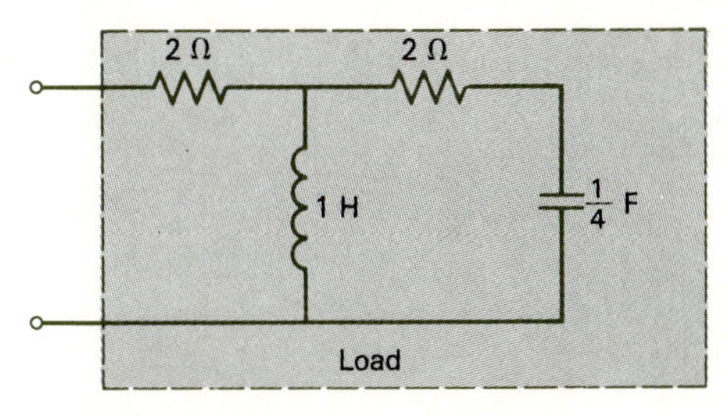

FIGURE P5.22

5.23 An industrial plant utilizes several electric motors which are highly inductive. The equivalent circuit of the plant as seen by the 60-Hz commercial power company is shown in Fig. P5.23. Determine the power factor of the plant. It is desired to change the power factor to unity. What value of capacitor C may be placed across the plant input terminals to achieve this?

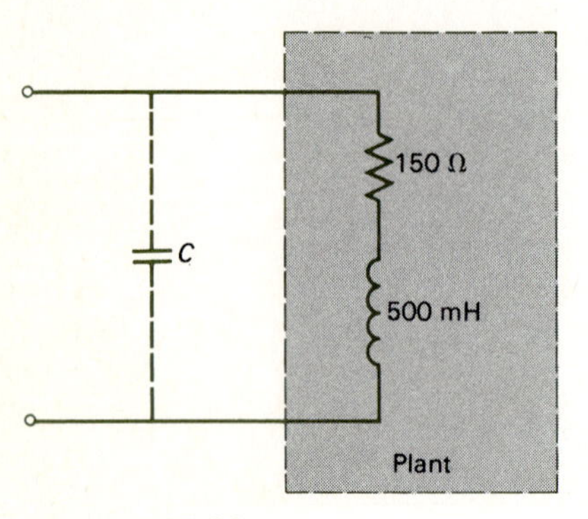

FIGURE P5.23

5.24 In the circuit shown in Fig. P5.24, determine the value of Z_L such that maximum average power is delivered to it. Determine that average power.

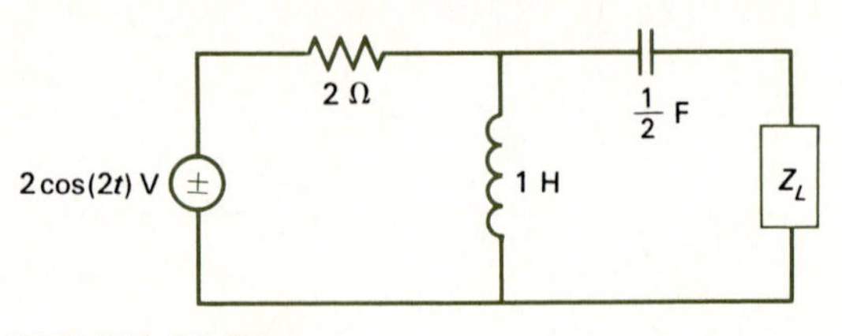

FIGURE P5.24

5.25 A low-pass filter is inserted between a sinusoidal source and a load as shown in Fig. P5.25. Determine the frequency of the source at which the magnitude of the voltage across R_L is reduced to one-half its value when the source is dc ($\omega = 0$).

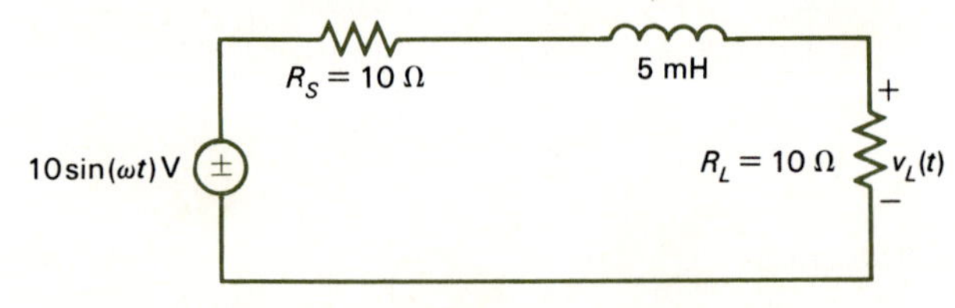

FIGURE P5.25

5.26 Determine the center frequency, bandwidth, and Q of the bandpass filter shown in Fig. P5.26.

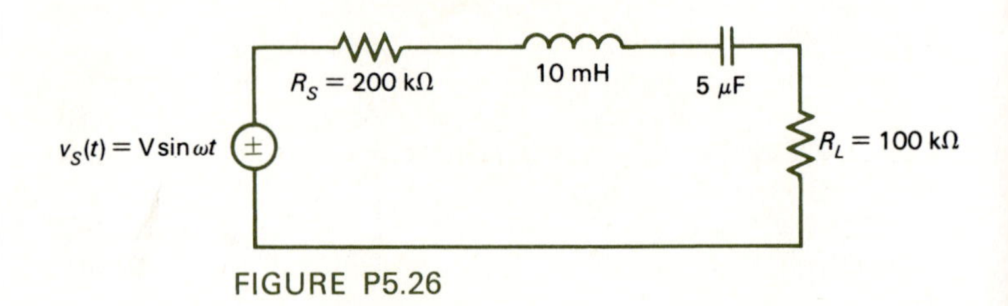

FIGURE P5.26

5.27 A signal $v_S(t)$ with source resistance $R_S = 50\ \Omega$ is applied to the input of a band-reject filter having $L = 2.533\ \mu H$ and $C = 1\ \mu F$. The output of the filter is applied across a load $R_L = 50\ \Omega$. If $v_S(t)$ can be represented through the fourth harmonic as

$$v_S(t) = 10 + 2\sin(2\pi \times 50 \times 10^3 t) + 1.5\sin(2\pi \times 10^5 t) + 1\sin(2\pi \times 1.5 \times 10^5 t) + 0.5\sin(2\pi \times 2 \times 10^5 t)$$

determine the resulting $v_L(t)$ as well as the center frequency, bandwidth, and Q of the filter. Are any frequency components of $v_S(t)$ effectively eliminated?

5.28 In the circuit shown in Fig. P5.28, compute the average power dissipated in the resistor. Show that this average power may not be computed by superimposing the average powers dissipated by each source. (Why?)

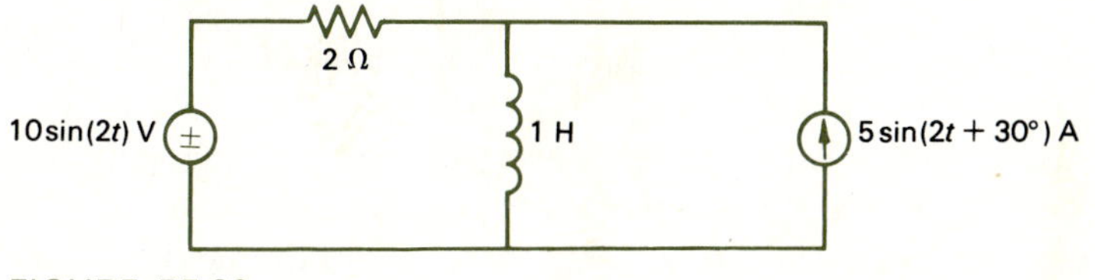

FIGURE P5.28

5.29 Change the frequency of the current source in Fig. P5.28 to 3 rad/s. Show that the average power dissipated in the 2-Ω resistor may be computed by superimposing the average powers dissipated by each source.

5.30 Sketch the resulting $v_L(t)$ in Example 5.8.

REFERENCES

5.1 A. B. Carlson and D. G. Gisser, *Electrical Engineering: Concepts and Applications.* Addison-Wesley, Reading, Mass., 1981.

5.2 Ralph J. Smith, *Circuits, Devices and Systems: A First Course in Electrical Engineering*, 4th ed., Wiley, New York, 1984.

5.3 S. E. Schwartz and W. G. Oldham, *Electrical Engineering: An Introduction*, Holt, Rinehart & Winston, New York, 1984.

5.4 W. H. Hayt, Jr., and J. E. Kemmerly, *Engineering Circuit Analysis*, 3d ed., McGraw-Hill, New York, 1978.

5.5 J. A. Edminister, *Schaum's Outline of Electric Circuits*, 2d ed., McGraw-Hill, New York, 1983.

Chapter

6

Transients

In the previous chapters we investigated the primary circuit elements—the resistor R, the capacitor C, the inductor L, and the ideal voltage and current sources. We determined the circuit element voltages and currents when the circuits contained dc (constant value) or ac (sinusoidal waveform) ideal voltage and current sources. These solutions are said to be the steady-state values of the element voltages and currents.

In this chapter we will investigate the transient values of the element voltages and currents. In circuits containing energy storage elements, the inductor current $i_L(t)$ and capacitor voltage $v_C(t)$ cannot change value instantaneously; otherwise, their stored energies $w_L(t) = \frac{1}{2}Li_L^2(t)$ and $w_C = \frac{1}{2}Cv_C^2(t)$ would change instantaneously. Thus, there is a certain amount of "inertia" associated with these elements. Suppose, for example, that we find the solutions for the element voltages and currents for a particular circuit and then, at some later time, open or close a switch located somewhere in the circuit; activating the switch *changes the circuit*, and now we must solve this new circuit. The element voltages and currents will have changed due to the fact that we have a new circuit caused by the switching operation. The object of this chapter is to determine the behavior of the element voltages and currents in the intermediate, or transient, time interval while they are adjusting to their new values.

This idea is illustrated in Fig. 6.1. The original circuit consists of A and B connected together. Suppose we solve for the element voltages and currents in the B part of the circuit. At $t = 0$ a switch opens, disconnecting A from B. Now the element currents and voltages in the B part which we solved for previously must have new values since the A part has been removed by the switch. There must be a time interval when the element currents and voltages are adjusting to their new values. We call this the transient period, and the solution is called the transient (exists only momentarily) solution.

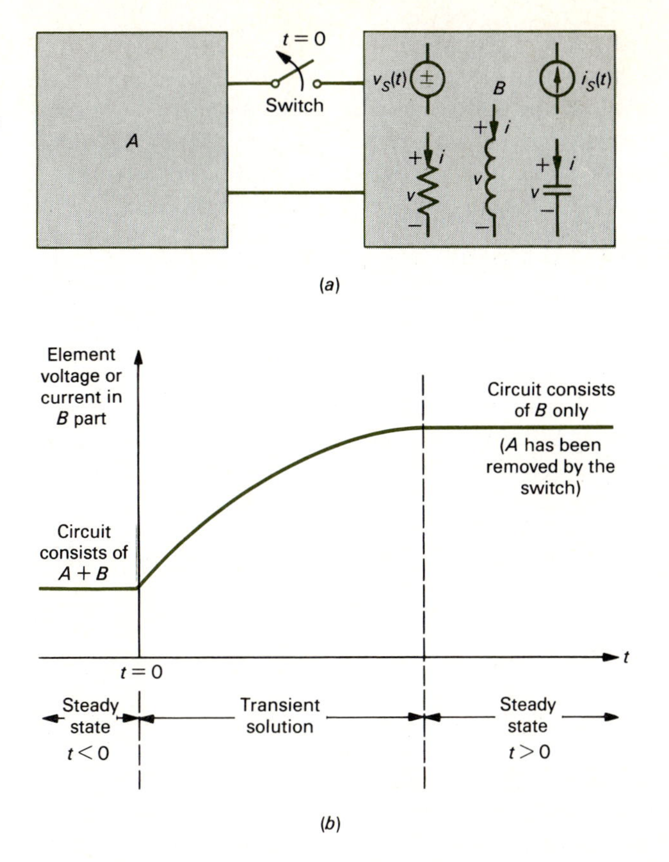

FIGURE 6.1 Illustration of the transient period of the circuit solution following a switching operation.

The final solution after this transient period is called the steady-state solution for the new circuit. The portion of this steady-state solution due to dc sources in B is found by replacing all inductors with short circuits and all capacitors with open circuits. This is due to the fact that all currents and voltages in the circuit due to a constant-value source must also be constant in value. Since

$$v_L = L\frac{di_L}{dt} \tag{6.1}$$

if the inductor currents i_L are constant, then $v_L = 0$, which is a short circuit. Similarly, if the capacitor voltages v_C are constant, then

$$i_C = C\frac{dv_C}{dt} \tag{6.2}$$

and $i_C = 0$, which is equivalent to an open circuit. If the source is sinusoidal (ac), we may use the phasor principles developed in the previous chapter in order to find the portion of the steady-state solution due to this type of source.

These steady-state computational methods also apply to the circuit $A + B$ before the switch is closed at $t = 0$. In other words, we assume that for $t < 0$, A and B are connected together and no switching operation or other disturbance has occurred in either part A or part B for a very long time prior to the switching operation at $t = 0$. Thus, all currents and voltages in $A + B$ have reached steady-state at $t = 0^-$; that is, they have reached steady-state immediately prior to the switching operation.

6.1 THE *RL* CIRCUIT

Consider the circuit shown in Fig. 6.2*a* consisting of a dc voltage source, a resistor, and an inductor. Before $t = 0$, the switch is open. The value of the inductor current at $t = 0^-$ (immediately prior to the closing of the switch) is obviously zero since no sources are attached to it for $t < 0$, as shown in Fig. 6.2*b*. At $t = 0$ the switch closes, thus attaching the voltage source and resistor to the inductor, as shown in Fig. 6.2*c*. The steady-state (ss) value of the inductor current for $t > 0$ is found in Fig. 6.2*d* by replacing the inductor with a short circuit since the source is dc. Therefore,

$$\begin{aligned} i_{L,\mathrm{ss}} &= \frac{V_S}{R} \\ &= 5 \text{ A} \end{aligned} \tag{6.3}$$

We found in Chap. 4 that this inductor current cannot change instantaneously from its value at $t = 0^-$, 0 A, to its steady-state value for $t > 0$ of 5 A or else the inductor voltage given by Eq. (6.1) would become infinite. In order to examine this transitional behavior of the inductor current, let us obtain an equation relating the

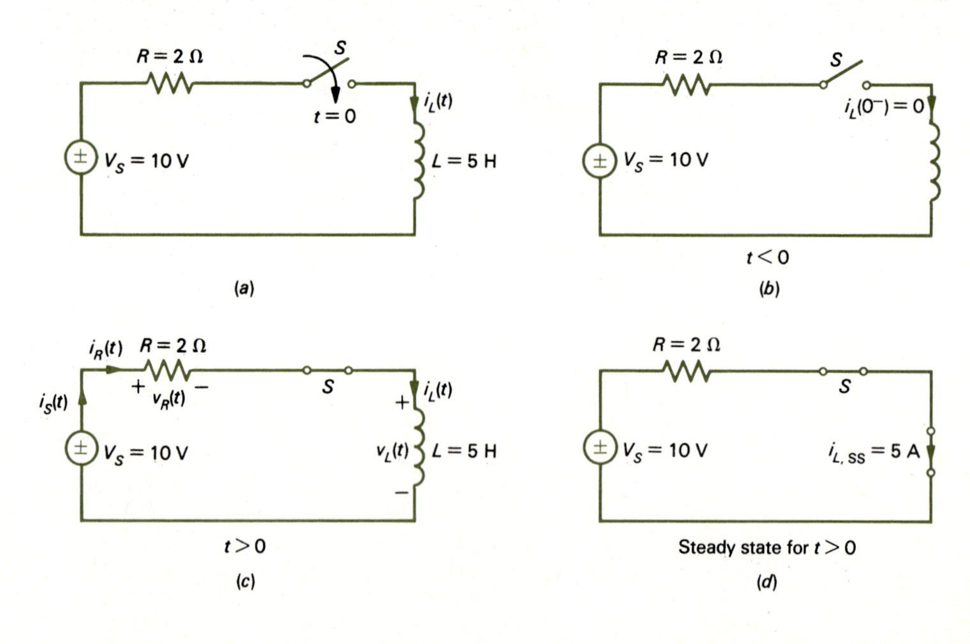

FIGURE 6.2 The *RL* circuit and its complete solution.

inductor current i_L and the voltage source V_S. Applying KVL around the loop in the $t > 0$ circuit of Fig. 6.2*c*, we obtain

$$v_L(t) + v_R(t) = V_S \tag{6.4a}$$

or

$$L\frac{di_L}{dt} + Ri_L = V_S \tag{6.4b}$$

Rewriting Eq. (6.4) by dividing both sides by L, we have

$$\frac{di_L}{dt} + \frac{R}{L}i_L = \frac{V_S}{L} \tag{6.5}$$

This type of equation is a differential equation (DE) since it involves a derivative of the unknown variable, i_L. It is the simplest possible type of differential equation and has a very simple solution; other types of differential equations are more difficult to solve. It is said to be an ordinary DE since only ordinary derivatives are involved; a partial differential equation would involve partial derivatives. Note that since the unknown, $i_L(t)$, is a function of only one variable, time t, no partial derivatives are necessary. It is also said to be a first-order differential equation since the highest derivative of the unknown appearing in the equation is the first derivative; a second-order DE would involve the second derivative of the unknown, i_L, as the highest derivative. The equation is a constant coefficient one since the coefficients of the unknown, 1 and R/L, are independent of t. If the resistor has been a time-varying one such as $R(t) = 2t$, then the equation would have been a nonconstant coefficient one, which would be very difficult to solve. The final and most important property of the DE is that it is a linear one. A DE is linear if the unknown, i_L, and any of its derivatives are not raised to a power other than unity or do not appear as products of each other. For example, terms such as $i_L^2(t)$ and $i_L di_L/dt$ would make the equation nonlinear. Having a term such as $d^2 i_L(t)/dt^2$ makes the equation second order (if it is the highest derivative), but a term such as $[di_L(t)/dt]^2$ makes the equation nonlinear.

We will encounter this type of differential equation in numerous places, so it is important to outline a simple method of solution. To illustrate this method, let us consider a general form of a linear first-order ordinary constant coefficient DE:

$$\frac{dx(t)}{dt} + ax(t) = f(t) \tag{6.6}$$

Here $x(t)$ denotes the unknown, $f(t)$ denotes the known right-hand side "forcing function," and a, which is a constant, is the coefficient of $x(t)$. Note in our example in Eq. (6.5) that $x(t) = i_L(t)$, $f(t) = V_S/L$, and $a = R/L$. If the source is not dc, $f(t)$ will be a function of time. The solution to Eq. (6.6) is easily obtained if we do something to Eq. (6.6) which may appear to be inconsequential—add a zero to the right-hand side:

$$\frac{dx(t)}{dt} + ax(t) = 0 + f(t) \tag{6.7}$$

The addition of this zero to the right-hand side, along with the important property of linearity, allows us to see that the solution for $x(t)$ will be the sum of two parts. The transient or natural solution, $x_T(t)$, is the portion of the solution to Eq. (6.7) due to the zero part of the right-hand side:

$$\frac{dx_T(t)}{dt} + ax_T(t) = 0 \tag{6.8}$$

The steady-state or forced solution, $x_{ss}(t)$, is the portion of the solution to Eq. (6.7) due to $f(t)$:

$$\frac{dx_{ss}(t)}{dt} + ax_{ss}(t) = f(t) \tag{6.9}$$

In terms of these two solutions, the solution to Eq. (6.7) is

$$x(t) = x_T(t) + x_{ss}(t) \tag{6.10}$$

This is easy to see by substituting Eq. (6.10) into Eq. (6.7) and factoring as

$$\frac{d}{dt}(x_T + x_{ss}) + a(x_T + x_{ss}) = 0 + f(t) \tag{6.11}$$

or

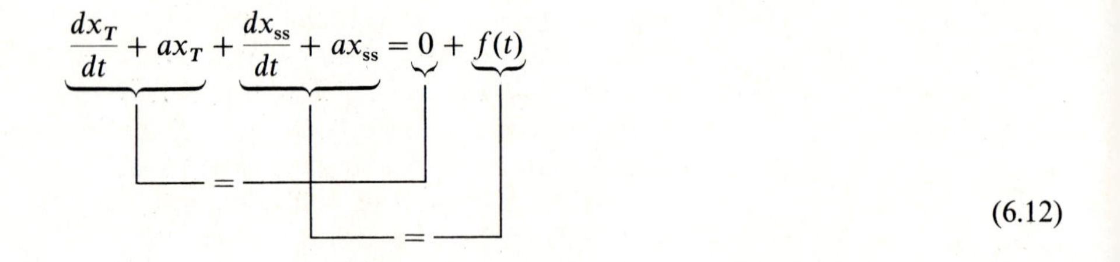

$$\underbrace{\frac{dx_T}{dt} + ax_T}_{} + \underbrace{\frac{dx_{ss}}{dt} + ax_{ss}}_{} = \underbrace{0}_{} + \underbrace{f(t)}_{} \tag{6.12}$$

The factorization in Eq. (6.12) shows that the result is the sum of two equalities, Eqs. (6.8) and (6.9), which is an equality. Superposition and linearity come into play once again.

First consider the transient solution to Eq. (6.8). In order to satisfy Eq. (6.8), $x_T(t)$ must be such that when differentiated with respect to t the result must be of the same form as $x_T(t)$ if the sum of dx_T/dt and ax_T is to equal zero. A possible form for $x_T(t)$ which achieves this objective is the exponential e^{pt}. If we differentiate e^{pt} we obtain pe^{pt} which is of the same form as e^{pt}. Thus, let us try a solution of the form

$$x_T(t) = Ae^{pt} \tag{6.13}$$

where A and p are, as yet, undetermined constants. Substituting Eq. (6.13) into Eq. (6.8), we obtain

$$pAe^{pt} + aAe^{pt} = 0 \tag{6.14}$$

or

$$(p + a)Ae^{pt} = 0 \tag{6.15}$$

Now A cannot be zero or x_T would be zero for all t: a trivial solution. Also, e^{pt} cannot be zero for all t regardless of p. Thus, we conclude that

$$p + a = 0 \tag{6.16}$$

or

$$p = -a \tag{6.17}$$

Thus, the transient solution is

$$x_T(t) = Ae^{-at} \tag{6.18}$$

where A is, as yet, some undetermined constant. We will evaluate A later.

We have already determined how to find the steady-state or forced solution to Eq. (6.9) without, perhaps, knowing that we have done so. The steady-state solution is any function which satisfies Eq. (6.9). Since Eq. (6.9) is simply a mathematical statement of the circuit constraints, we may as well find $x_{ss}(t)$ directly from the circuit. Each ideal source will contribute a distinct portion to $f(t)$. For example, suppose a circuit has N ideal sources. Then $f(t)$ will consist of N pieces with each piece due to only one of the sources:

$$f(t) = f_1(t) + f_2(t) + \cdots + f_N(t) \tag{6.19}$$

Here $f_i(t)$ is due to the ith ideal source with all other sources killed. Since the DE in Eq. (6.9) is linear, there will also be N contributions to $x_{ss}(t)$:

$$x_{ss}(t) = x_{ss1}(t) + x_{ss2}(t) + \cdots + x_{ssN}(t) \tag{6.20}$$

where $x_{ssi}(t)$ is due to $f_i(t)$ with the other sources contributing to the other $f(t)$'s killed. Thus, we may find the steady-state solution directly from the circuit by superposition. To find the portion of the steady-state solution due to a dc source, we first kill all the other sources and replace inductors with short circuits and capacitors with open circuits and solve the resulting circuit for the element voltage or current of interest, $x_{ss,i}(t)$. To find the portion of the steady-state solution due to an ac source, we kill all the other sources and use the phasor method of the previous chapter to obtain $x_{ss,i}(t)$ directly from the circuit. Therefore, we already know how to find the steady-state solution for dc or ac sources.

The solution to Eq. (6.7) is the sum of the transient and steady-state solutions:

$$\begin{aligned} x(t) &= x_T(t) + x_{ss}(t) \\ &= Ae^{-at} + x_{ss}(t) \end{aligned} \tag{6.21}$$

Two points about the total solution in Eq. (6.21) should be made. The constant A is, as yet, unknown. To evaluate it we need some additional information. Suppose we wish to obtain the solution for $x(t)$ for all times greater than some initial time t_0; that is, we wish to find $x(t)$ such that Eq. (6.21) is correct for all $t \geq t_0$. Without any loss of generality, we may select this initial time as $t_0 = 0$. Evaluating Eq. (6.21) immediately after $t = 0$, or at $t = 0^+$, we have

$$x(0^+) = A + x_{ss}(0^+) \tag{6.22}$$

so that

$$A = x(0^+) - x_{ss}(0^+) \tag{6.23}$$

The value of $x(t)$ at the initial time, $x(0^+)$, is called the initial condition on $x(t)$. We will have found $x_{ss}(t)$ for a particular $f(t)$, so it is simple to obtain the constant $x_{ss}(0^+)$ by substituting $t = 0$ into $x_{ss}(t)$. Substituting Eq. (6.23) into Eq. (6.21), we have the solution for $x(t)$ for all $t \geq 0$ in terms of the initial value of $x(t)$:

$$x(t) = \underbrace{[x(0^+) - x_{ss}(0^+)]e^{-at}}_{\text{transient}} + \underbrace{x_{ss}(t)}_{\substack{\text{steady} \\ \text{state}}} \tag{6.24}$$

As time progresses, e^{-at} decreases (assuming a is positive) so that the transient solution eventually goes to zero. Thus, the name *transient solution* is quite appropriate: it is only present momentarily. At a time $t = 1/a$, the transient solution has decayed to $1/e$ or 37 percent of its initial value at $t = 0^+$, which is $x(0^+) - x_{ss}(0^+)$. Clearly, the constant a must have the dimension of reciprocal time; that is, the units of $1/a$ are seconds. Thus, the reciprocal of a is known as the time constant and is denoted by

$$T = \frac{1}{a} \quad \text{s} \tag{6.25}$$

so that Eq. (6.24) may be written as

$$x(t) = [x(0^+) - x_{ss}(0^+)]e^{-t/T} + x_{ss}(t) \tag{6.26}$$

This final result is very important and should be committed to memory. We will no longer need to repeat the solution process for a first-order DE; simply put it in the form of Eq. (6.6) where the coefficient of dx/dt is unity, identify a, determine $x_{ss}(t)$ and $x(0^+)$ from the circuit, and write the solution in the form of Eq. (6.26).

With this solution to a general linear first-order constant coefficient DE, let us return to the example shown in Fig. 6.2. The DE relating the inductor current $i_L(t)$ to the dc source V_S is

$$\frac{di_L(t)}{dt} + \frac{R}{L} i_L(t) = \frac{V_S}{L} \tag{6.27}$$

Comparing Eqs. (6.27) and (6.6), we identify $a = R/L$ so that the time constant is

$$\begin{aligned} T &= \frac{1}{a} \\ &= \frac{L}{R} \quad \text{s} \end{aligned} \tag{6.28}$$

The steady-state solution can be found directly from Eq. (6.27) by substituting an assumed solution which is a constant [since the right-hand side of Eq. (6.27) is constant] and obtaining

$$\begin{aligned} i_{L,\text{ss}} &= \frac{V_S}{R} \\ &= 5 \text{ A} \end{aligned} \tag{6.29}$$

Note that this can also be obtained from the $t > 0$ circuit by replacing the inductor with a short circuit (since the source is dc), as shown in Fig. 6.2*d*. Therefore, the solution for the inductor current is of the form of Eq. (6.26):

$$i_L(t) = [i_L(0^+) - i_{L,\text{ss}}]e^{-t/T} + i_{L,\text{ss}} \tag{6.30}$$

where $T = L/R = \frac{5}{2}$. Therefore,

$$i_L(t) = [i_L(0^+) - 5]e^{-2t/5} + 5 \tag{6.31}$$

To finish the solution we need $i_L(0^+)$; that is, we need the value of the inductor current immediately after the switch closes. We know that the inductor current cannot change value instantaneously; otherwise, the inductor voltage

$$v_L(t) = L\frac{di_L}{dt}$$

would become infinite. Therefore, the inductor current immediately before the switch closes, $i_L(0^-)$, and the inductor current immediately after the switch closes, $i_L(0^+)$, must be the same:

$$\begin{aligned} i_L(0^+) &= i_L(0^-) \\ &= 0 \end{aligned} \tag{6.32}$$

Substituting into Eq. (6.31), we have the total solution for the inductor current for $t > 0$:

$$i_L(t) = (0 - 5)e^{-2t/5} + 5$$
$$= 5(1 - e^{-2t/5}) \quad \text{A} \quad t > 0 \tag{6.33}$$

The total solution in Eq. (6.33) is plotted in Fig. 6.3*a*. Note that the solution never actually reaches the steady-state value of 5 A but that it comes very close to 5 A after a short period of time. At $t = T = L/R = \frac{5}{2}$ s, the inductor current is $5(1 - e^{-1}) = 5(0.63) = 3.16$ A; thus, after one time constant has elapsed, the current has risen to 63 percent of its final, or steady-state, value. After two time constants it has risen to 86 percent of its steady-state value, and after five time constants have elapsed, $t = 5T = 12.5$ s, the inductor current is 99 percent of its final value; thus, after five time constants have elapsed, the inductor current has essentially reached its steady-state value. This "long" time (12.5 s) is a result of the values chosen for our circuit elements. Suppose we had chosen more typical values such as $R = 200\ \Omega$ and $L = 5$ mH; then the time constant would have been $L/R = 2.5 \times 10^{-5}$, or 25 μs. If the source voltage had been chosen as $V_S = 100$ V, then the inductor current would have risen to 99 percent of its steady-state value of 0.5 A in about 125 μs, or 0.125 ms.

The inductor voltage can be obtained by simply differentiating the solution for the inductor current in Eq. (6.33):

$$v_L = L\frac{di_L}{dt}$$
$$= 10e^{-\frac{2}{5}t} \quad t > 0 \tag{6.34}$$

This is shown in Fig. 6.3*b*. Note that at $t = 0^+$, $v_L(0^+) = 10$ V and the inductor voltage has *changed instantaneously* from its value at $t = 0^-$ of $v_L(0^-) = 0$ V. Therefore, *an inductor voltage may change value instantaneously but the inductor current immediately prior to and immediately after a switching operation must be the same*:

$$i_L(0^+) = i_L(0^-) \tag{6.35}$$

Also note that the inductor voltage eventually decays to zero. This is obviously correct, since in the steady state the inductor behaves as a short circuit for dc sources and $v_{L,\text{ss}} = 0$. After one time constant the inductor voltage has decayed to $10(e^{-1}) = 10(0.37) = 3.7$, or 37 percent of its initial value.

Once we have found the solution for the inductor voltage and current, we may work back into the $t > 0$ circuit to find any other element voltages and currents of interest for $t > 0$. For example, from Fig. 6.2*c* the current associated with the voltage source is

$$i_S(t) = i_L(t)$$
$$= 5(1 - e^{-\frac{2}{5}t}) \quad t > 0 \tag{6.36}$$

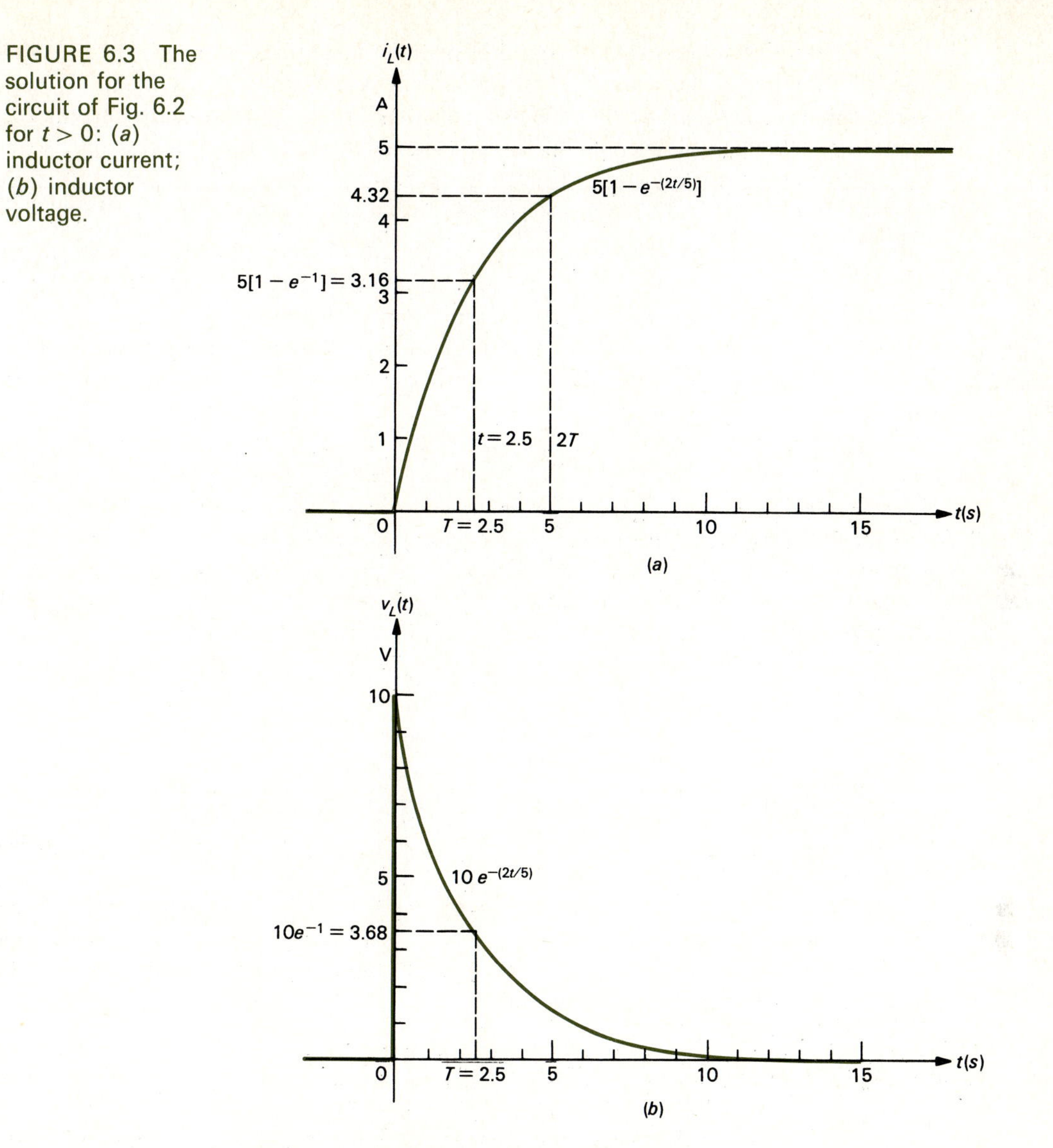

FIGURE 6.3 The solution for the circuit of Fig. 6.2 for $t > 0$: (*a*) inductor current; (*b*) inductor voltage.

The resistor voltage and current are

$$
\begin{aligned}
i_R(t) &= i_L(t) \\
&= 5(1 - e^{-\frac{2}{5}t}) \qquad t > 0
\end{aligned}
\tag{6.37}
$$

$$
\begin{aligned}
v_R(t) &= Ri_R(t) \\
&= 10(1 - e^{-\frac{2}{5}t}) \qquad t > 0
\end{aligned}
\tag{6.38}
$$

Thus, finding the solution for the inductor current is the crucial part of the process. Once that is done, all other element voltages and currents in the circuit can be obtained in terms of the inductor current.

It is worthwhile to explain what is happening in qualitative terms. The inductor stores $w_L(t) = \frac{1}{2}Li_L^2(t)$ energy in its magnetic field. At $t = 0^+$, $i_L(0^+) = 0$ and no magnetic field is established by the inductor current. For $t > 0$, the inductor current and resulting magnetic field are building up to their final (steady-state) values.

Example 6.1

For the circuit in Fig. 6.4a, determine the solution or the current in resistor R_1, $i_{R1}(t)$, for $t > 0$. The switch opens at $t = 0$.

FIGURE 6.4 Example 6.1: illustration of the complete solution for an RL circuit with nonzero initial conditions.

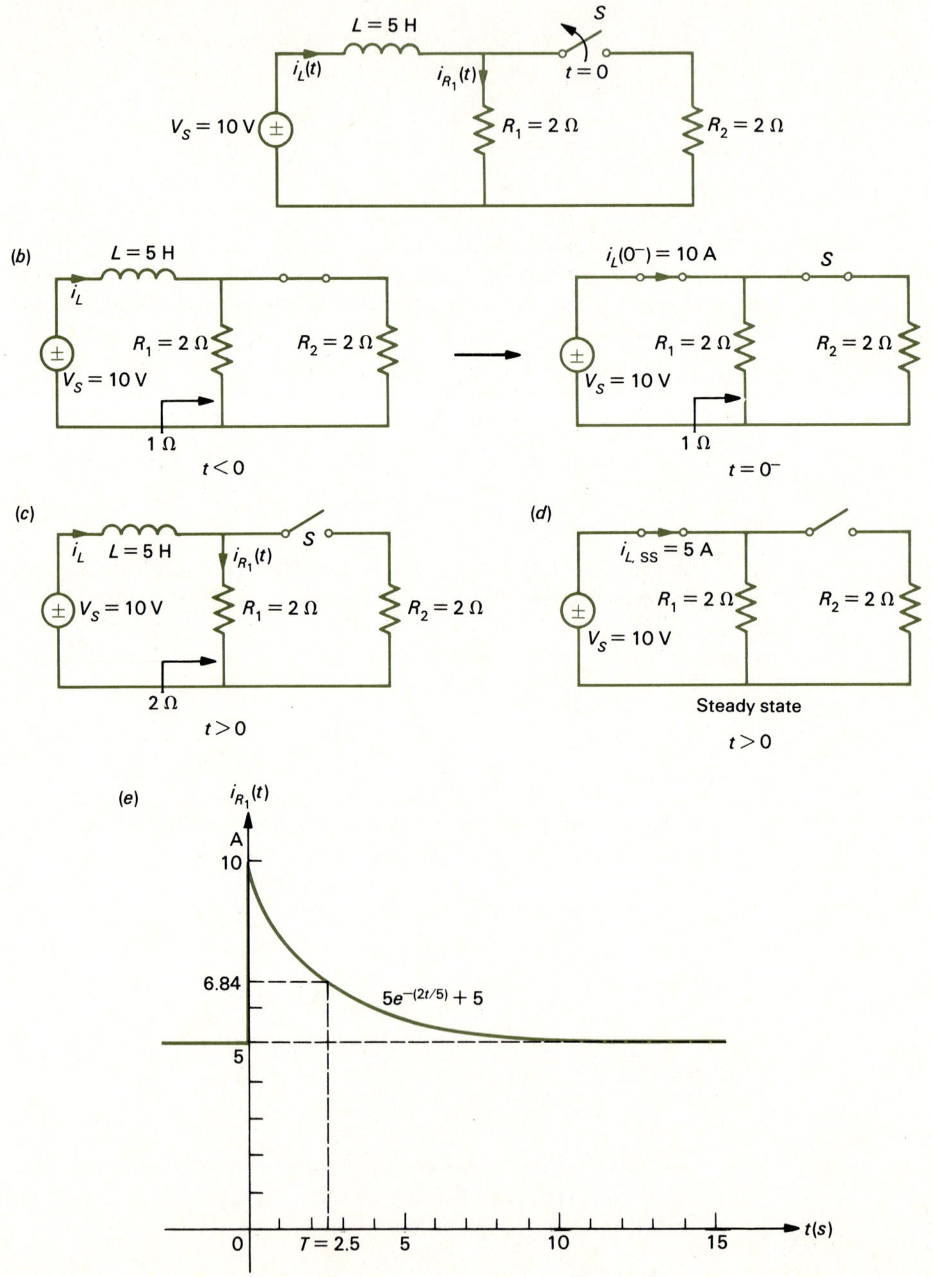

Solution Our basic method will be to solve for the inductor current for $t > 0$ and then find all other element voltages and currents from that. We again choose to solve for the inductor current, rather than its voltage, for the important reason that we will need to know the initial condition at $t = 0^+$ on the variable we are solving for in terms of its value at $t = 0^-$. Since the inductor current cannot change value instantaneously,

$$i_L(0^+) = i_L(0^-)$$

we can easily obtain the initial condition $i_L(0^+)$. If we had tried to solve instead for the inductor voltage, we would need its initial value at $t = 0^+$, $v_L(0^+)$. However,

$$v_L(0^+) \neq v_L(0^-)$$

necessarily, so we could not assume that the inductor voltage was continuous in order to establish that initial condition.

Since we will eventually need $i_L(0^+)$, let us solve for it now. The $t < 0$ circuit is shown in Fig. 6.4*b*. Since the source is dc, we replace the inductor with a short circuit and obtain

$$\begin{aligned} i_L(0^-) &= \frac{V_S}{R_1 || R_2} \\ &= 10 \text{ A} \\ &= i_L(0^+) \end{aligned}$$

The next step is to find the time constant for the inductor current for $t > 0$. The $t > 0$ circuit is shown in Fig. 6.4*c*. From this we identify the time constant as

$$\begin{aligned} T &= \frac{L}{R_1} \\ &= 2.5 \text{ s} \end{aligned}$$

The steady-state solution is found from the $t > 0$ circuit by replacing the inductor with a short circuit (since the source is dc), as shown in Fig. 6.4*d*:

$$\begin{aligned} i_{L,\text{ss}} &= \frac{V_S}{R_1} \\ &= 5 \text{ A} \qquad t > 0 \end{aligned}$$

Therefore, the total solution is

$$\begin{aligned} i_L(t) &= [i_L(0^+) - i_{L,\text{ss}}]e^{-R_1 t/L} + i_{L,\text{ss}} \\ &= (10 - 5)e^{-2t/5} + 5 \\ &= 5e^{-t/2.5} + 5 \qquad \text{A} \qquad t > 0 \end{aligned}$$

Now in order to find the resistor current $i_{R1}(t)$ for $t > 0$, we merely need to note that, in the $t > 0$ circuit in Fig. 6.4c,

$$i_{R1}(t) = i_L(t)$$
$$= 5e^{-t/2.5} + 5 \quad \text{A}$$

The result is plotted in Fig. 6.4e. Note that the value of the resistor current at $t = 0^-$ is found, by current division, from Fig. 6.4b to be

$$i_{R1}(0^-) = \frac{R_2}{R_1 + R_2} i_L(0^-)$$
$$= 5 \text{ A}$$

Thus, the resistor current changes value instantaneously at $t = 0$.

6.2 THE *RC* CIRCUIT

As a second example, consider a similar circuit shown in Fig. 6.5a containing a dc current source, a resistor, and a capacitor. At $t = 0$ the switch closes, connecting the capacitor to the current source and resistor. For $t < 0$, the value of the capacitor voltage is zero since no sources are attached to it, as shown in Fig. 6.5b. The steady-state value for the capacitor voltage for $t > 0$ is found by replacing the capacitor with an open circuit (since the source is dc), as shown in Fig. 6.4d:

$$v_{C,\text{ss}} = 20 \text{ V} \qquad t > 0 \tag{6.39}$$

(Since the source is dc, the capacitor voltage as well as all other steady-state element voltages and currents are constant. Thus, the capacitor current is zero, $i_C = C dv_C/dt = 0$, an open circuit.) The complete solution for the capacitor voltage can be found by deriving the differential equation relating v_C and the source I_S. Applying KCL at the upper node in the $t > 0$ circuit in Fig. 6.5c, we obtain

$$i_C + i_R = I_S \tag{6.40a}$$

or

$$C\frac{dv_C}{dt} + \frac{v_C}{R} = I_S \tag{6.40b}$$

Dividing both sides by C gives

$$\frac{dv_C(t)}{dt} + \frac{1}{RC} v_C(t) = \frac{I_S}{C} \tag{6.41}$$

FIGURE 6.5 The *RC* circuit and its complete solution.

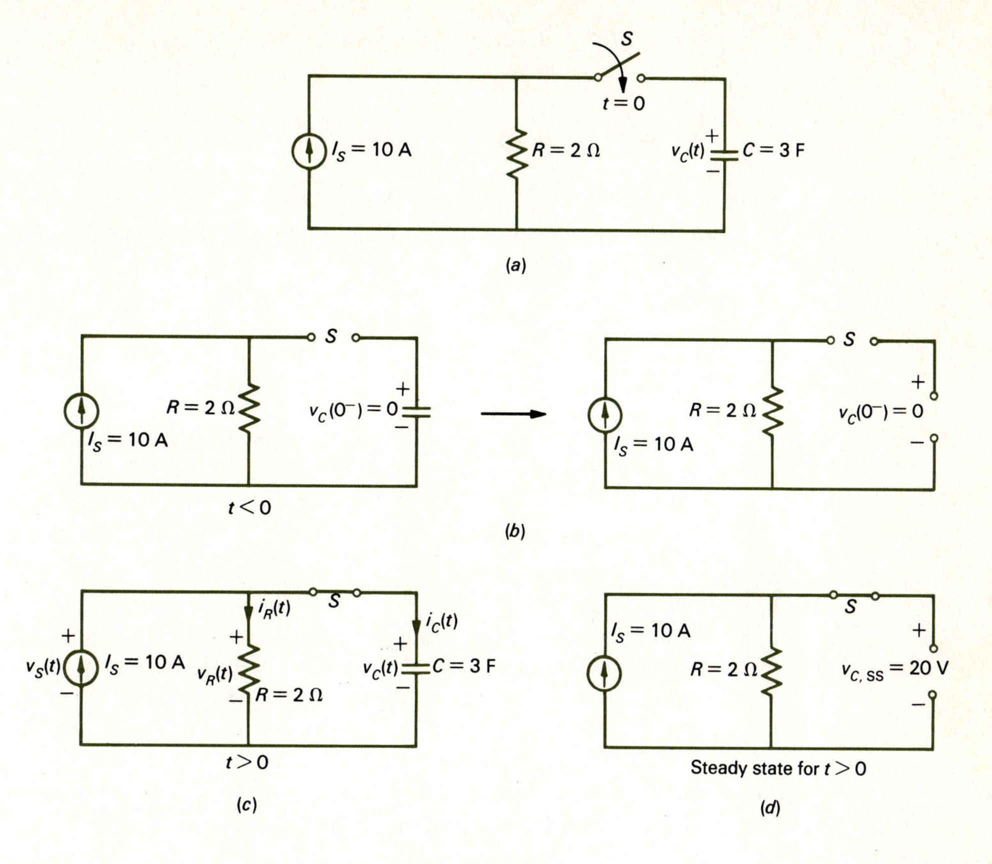

This equation is identical in form to the one derived for the inductor current in the previous example. Comparing Eq. (6.41) to Eq. (6.6), we identify $a = 1/RC$, $f(t) = I_S/C$. Thus, the total solution is of the form of Eq. (6.26):

$$v_C(t) = [v_C(0^+) - v_{C,ss}]e^{-t/RC} + v_{C,ss} \tag{6.42}$$

where we identify the time constant as

$$\begin{aligned} T &= \frac{1}{a} \\ &= RC \\ &= 6 \text{ s} \end{aligned} \tag{6.43}$$

The steady-state solution was found previously as

$$\begin{aligned} v_{C,ss}(t) &= RI_S \\ &= 20 \text{ V} \end{aligned} \tag{6.44}$$

Note that Eq. (6.44) satisfies Eq. (6.41) since I_S is constant (dc). Thus, the total solution is

$$v_C(t) = [v_C(0^+) - 20]e^{-t/6} + 20 \qquad \text{V} \qquad t > 0 \tag{6.45}$$

The value of the capacitor voltage immediately prior to the closure of the switch, $t = 0^-$, is zero. This must also be the value of the capacitor voltage immediately after the switch closure

$$\begin{aligned} v_C(0^+) &= v_C(0^-) \\ &= 0 \text{ V} \end{aligned} \tag{6.46}$$

since the capacitor voltage cannot change value instantaneously; otherwise, the capacitor current

$$i_C = C\frac{dv_C}{dt}$$

would become infinite. Also, the energy stored in the capacitor, $w_C = \frac{1}{2}Cv_C^2$, would change instantaneously if the capacitor voltage did so. Applying Eq. (6.46) to Eq. (6.45) gives

$$\begin{aligned} v_C(t) &= (0 - 20)e^{-t/6} + 20 \\ &= 20(1 - e^{-t/6}) \qquad t > 0 \end{aligned} \tag{6.47}$$

This solution is plotted in Fig. 6.6*a*. The capacitor voltage has reached 63 percent of its final value at $t = T = 6$ s and 86 percent after two time constants have elapsed.

The capacitor current can be obtained by differentiating the solution for the capacitor voltage in Eq. (6.47):

$$\begin{aligned} i_C &= C\frac{dv_C}{dt} \\ &= 10e^{-t/6} \qquad t > 0 \end{aligned} \tag{6.48}$$

This is plotted in Fig. 6.6*b*. Note that the capacitor current eventually decays to zero, which is a sensible result since the capacitor appears as an open circuit to a dc source in the steady state. Note also that the capacitor current has changed value instantaneously at $t = 0$ from its value at $t = 0^-$ of 0 A to its value at $t = 0^+$ of 10 A. Therefore, *a capacitor current may change value instantaneously but the capacitor voltage immediately prior to and immediately after a switching operation must be the same*:

$$v_C(0^+) = v_C(0^-) \tag{6.49}$$

Again, as was the case for the inductor, once we find the solution for the voltage and current of the energy storage element, we may work back into the remainder of

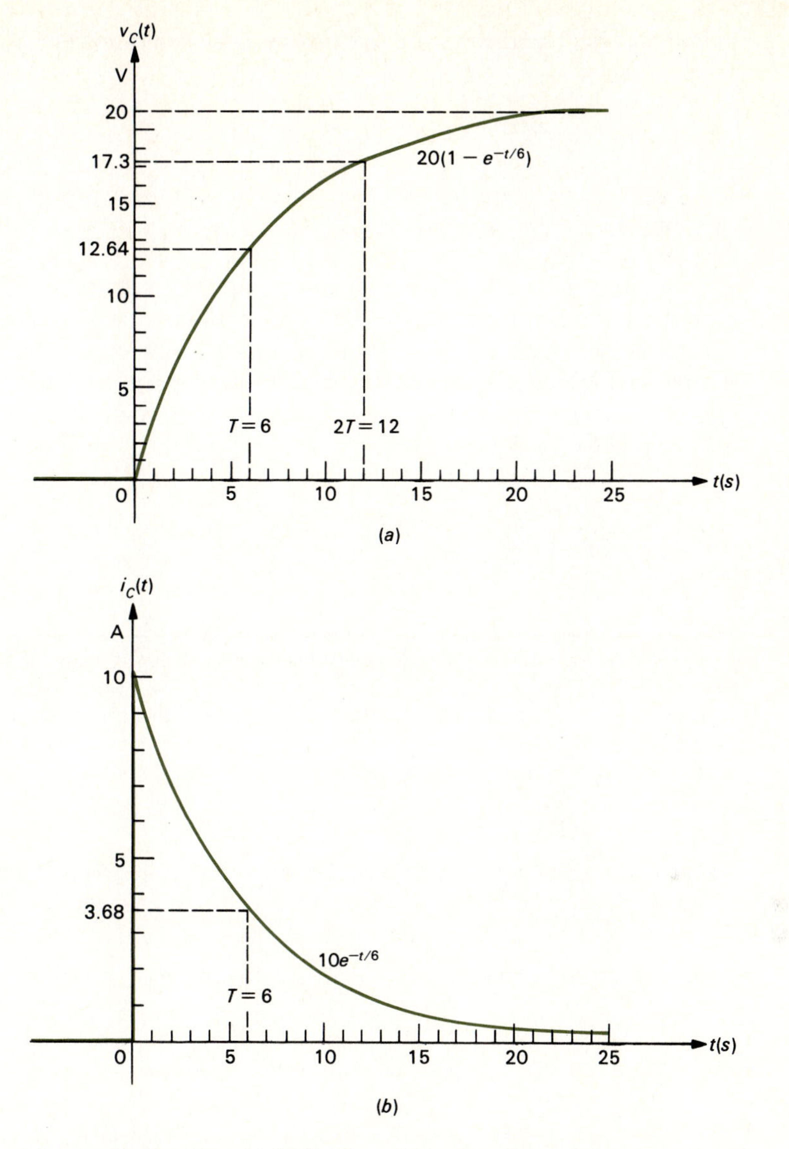

FIGURE 6.6 The solution for the circuit of Fig. 6.5 for $t \geqq 0$: (*a*) capacitor voltage; (*b*) capacitor current.

the $t > 0$ circuit to find all other element voltages and currents for $t > 0$:

$$v_S(t) = v_C(t)$$

$$= 20(1 - e^{-t/6}) \qquad t > 0 \tag{6.50}$$

$$v_R(t) = v_C(t)$$

$$= 20(1 - e^{-t/6}) \qquad t > 0 \tag{6.51}$$

$$i_R(t) = \frac{v_R(t)}{R}$$

$$= 10(1 - e^{-t/6}) \qquad t > 0 \tag{6.52}$$

Note that KVL and KCL are satisfied by these solutions. For example, KCL at the upper node is

$$i_C(t) + i_R(t) = I_S \tag{6.53}$$

or

$$10e^{-t/6} + 10(1 - e^{-t/6}) = 10 \tag{6.54}$$

Again it is worthwhile to explain in qualitative terms what is happening. The capacitor stores energy $w_C(t) = \frac{1}{2}Cv_C^2(t)$ in its electric field established by the electric charge on the capacitor plates. At $t = 0^+$, $v_C(0^+) = 0$ and no electric field exists. As t increases, the charge is being transferred to the capacitor plates by the capacitor current "charging up" the capacitor to its steady-state value. At steady-state, the charge transferred to the capacitor is

$$\begin{aligned} Q_{ss} &= Cv_{C,ss} \\ &= 3\text{F } (20 \text{ V}) \\ &= 60 \text{ C} \end{aligned} \tag{6.55}$$

Example 6.2 For the circuit shown in Fig. 6.7*a*, determine $i(t)$ for $t > 0$. The switch opens at $t = 0$.

Solution Our method will again be to solve for the capacitor voltage $v_C(t)$ for $t > 0$ and then to find $i(t)$ from that. The $t < 0$ circuit is shown in Fig. 6.7*b*. Since the source is dc, we replace the capacitor with an open circuit and obtain

$$\begin{aligned} v_C(0^-) &= (R_1||R_2)I_S \\ &= 5 \text{ V} \\ &= v_C(0^+) \end{aligned}$$

The $t > 0$ circuit is shown in Fig. 6.7*c*. We see that the time constant is

$$\begin{aligned} T &= R_1C \\ &= 6 \text{ s} \end{aligned}$$

The steady-state solution is found from the $t > 0$ circuit by replacing the capacitor with an open circuit, as shown in Fig. 6.7*d*:

$$\begin{aligned} v_{C,ss} &= I_SR_1 \\ &= 10 \text{ V} \end{aligned}$$

Thus, for $t > 0$,

$$\begin{aligned} v_C(t) &= [v_C(0^+) - v_{C,ss}]e^{-t/RC} + v_{C,ss} \\ &= (5 - 10)e^{-t/6} + 10 \\ &= -5e^{-t/6} + 10 \quad \text{V} \quad t > 0 \end{aligned}$$

FIGURE 6.7 Example 6.2: illustration of the complete solution for an *RC* circuit with nonzero initial conditions.

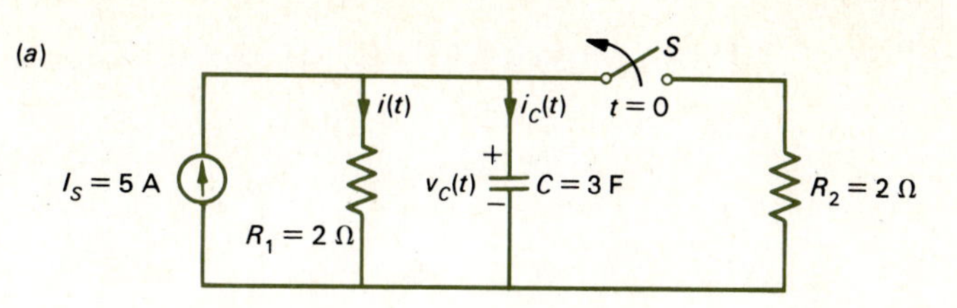

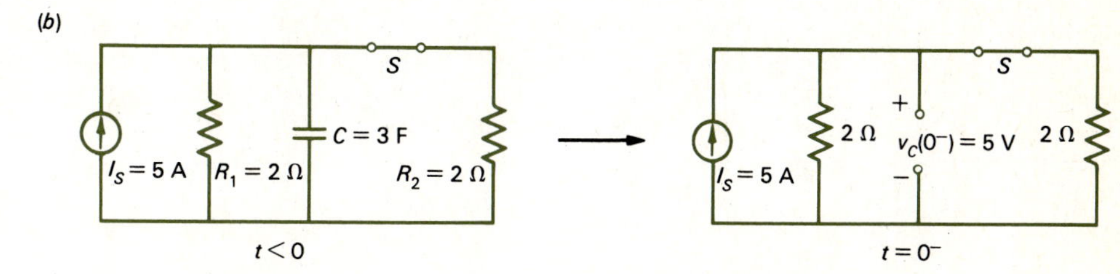

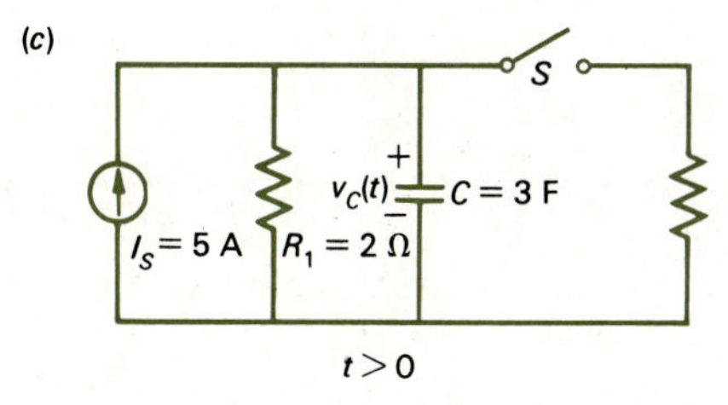

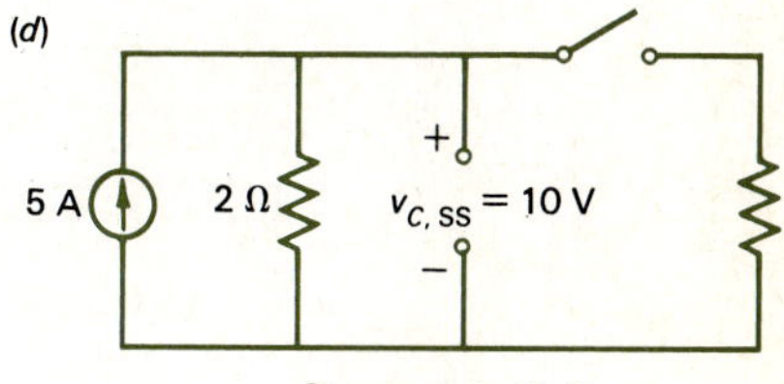

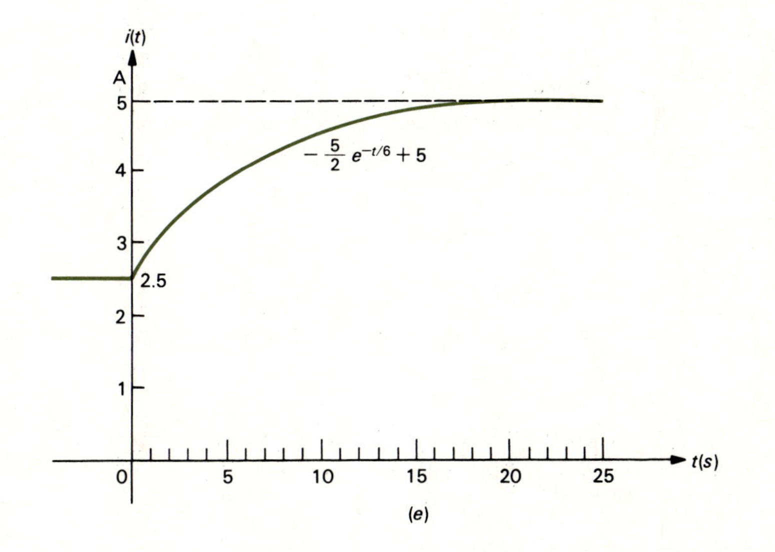

In the $t > 0$ circuit of Fig. 6.7c we see that

$$i(t) = \frac{v_C(t)}{R_1}$$
$$= -\tfrac{5}{2}e^{-t/6} + 5 \qquad t > 0$$

This is plotted in Fig. 6.7e. [Note that $i(0^-) = 2.5$ A, which can also be obtained from Fig. 6.7b.] The resistor current rises to its steady-state value of $i(\infty) = 5$ A after approximately five time constants have elapsed. This steady-state value for the resistor current can also be obtained from Fig. 6.7d. This type of final check on the answer using the steady-state circuits for $t < 0$ and for $t > 0$ should always be made.

6.3 WRITING THE SOLUTION BY INSPECTION FOR CIRCUITS CONTAINING ONE ENERGY STORAGE ELEMENT

We now have the necessary tools to solve any circuit which contains only one energy storage element—an inductor or a capacitor—in which a switching operation takes place. The key is to put the $t > 0$ circuit in the form of the basic circuits of the previous sections—*RL* or *RC*.

For example, suppose the energy storage element is an inductor. Draw the $t > 0$ circuit (after the switching operation has taken place). Then reduce the remainder of the $t > 0$ circuit (which contains only ideal sources and resistors) attached to the inductor terminals to a Thevenin equivalent, as shown in Fig. 6.8. From this reduction it is clear that the time constant of the inductor current is

$$T = L/R_{TH} \tag{6.56}$$

Then from the $t > 0$ circuit find the steady-state value of the inductor current $i_{L,ss}$. Next draw the $t < 0$ circuit as it appears prior to the switching operation. Then find in

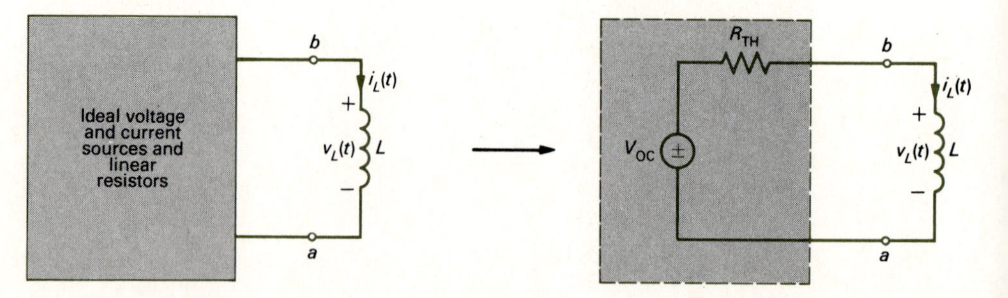

FIGURE 6.8 Solution of a circuit containing one inductor by use of a Thevenin equivalent.

this $t < 0$ ciruit the steady-state value of the inductor current $i_L(0^-)$. By continuity of inductor currents,

$$i_L(0^+) = i_L(0^-) \tag{6.57}$$

Then write the total solution for the inductor current for $t > 0$:

$$i_L(t) = [i_L(0^+) - i_{L,\mathrm{ss}}]e^{-R_{\mathrm{TH}}t/L} + i_{L,\mathrm{ss}} \tag{6.58}$$

The process is very straightforward and simple since the only circuit analysis required is in the determination of the steady-state value of the inductor current in the $t > 0$ circuit, $i_{L,\mathrm{ss}}$, and the steady-state value of the inductor current in the $t < 0$ circuit, $i_L(0^-)$. In previous chapters we have examined methods for obtaining the steady-state solutions due to dc sources and due to ac sources.

Note that we really don't need to find V_{OC} in the Thevenin equivalent. The time constant $T = L/R_{\mathrm{TH}}$ and the form of the transient solution depend only on the equivalent Thevenin resistance seen by the inductor. The steady-state solution may be found directly from the circuit without the need to compute V_{OC}.

To reiterate, the solution steps are:

1. Draw the $t > 0$ circuit.
2. Find the Thevenin resistance seen by the inductor, R_{TH}, in this $t > 0$ circuit.
3. Find the steady-state value of the inductor current in the $t > 0$ circuit, $i_{L,\mathrm{ss}}$.
4. Draw the $t < 0$ circuit and find the steady-state value of the inductor current in this circuit, $i_L(0^-)$.
5. By continuity of inductor currents, obtain

$$i_L(0^+) = i_L(0^-) \tag{6.59}$$

6. Write the solution for the inductor current for $t > 0$:

$$i_L(t) = [i_L(0^+) - i_{L,\mathrm{ss}}]e^{-R_{\mathrm{TH}}t/L} + i_{L,\mathrm{ss}} \qquad t > 0 \tag{6.60}$$

7. Work back into the circuit attached to the inductor terminals to find any other variable of interest.

Example 6.3 Find the complete solution for the voltage $v_x(t)$ for $t > 0$ for the circuit in Fig. 6.9.

Solution The Thevenin resistance seen by the inductor for $t > 0$ can be found in the usual fashion by drawing the $t > 0$ circuit and setting all ideal sources to zero, as shown in Fig. 6.9*b*, with the result

$$R_{\mathrm{TH}} = 3\ \Omega$$

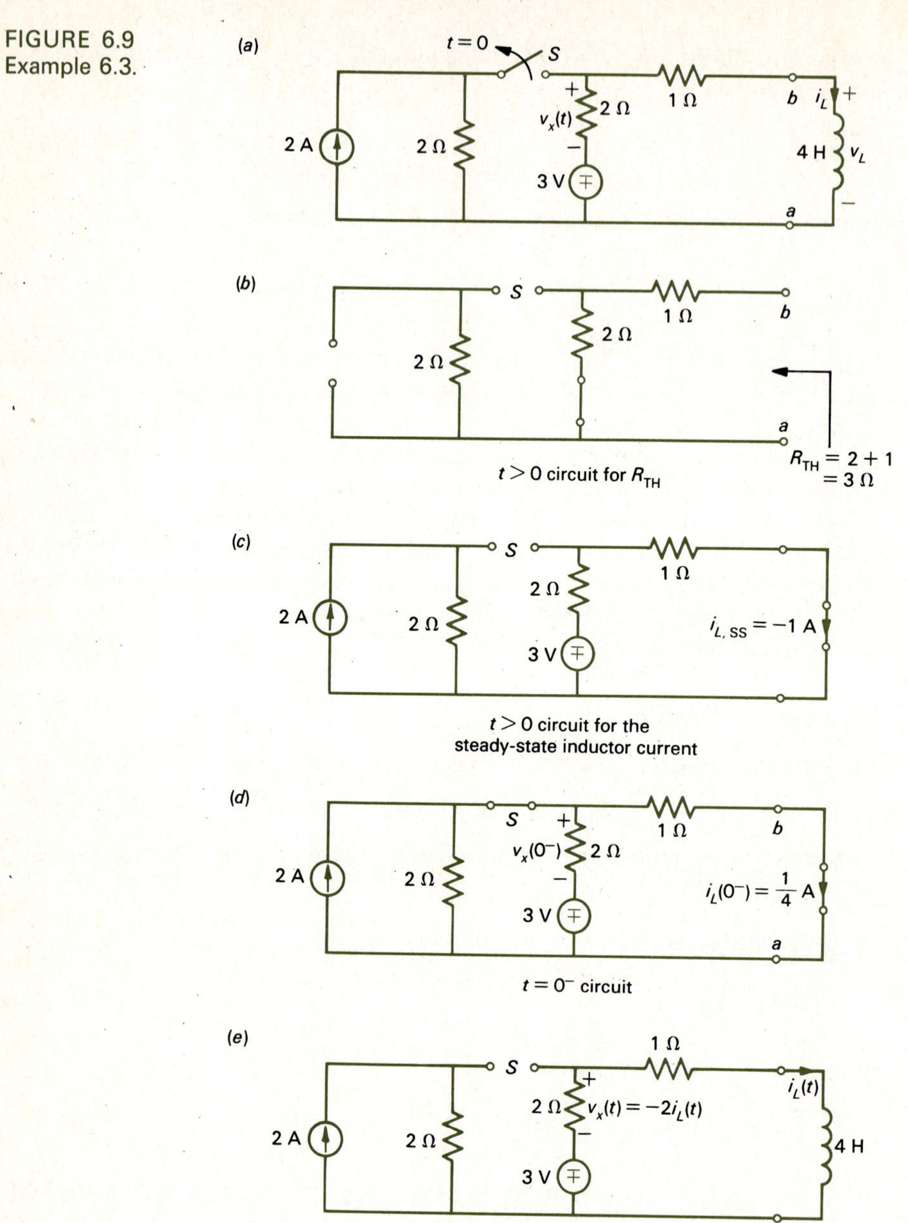

FIGURE 6.9 Example 6.3.

The time constant of the inductor current is

$$T = L/R_{TH}$$
$$= \tfrac{4}{3}\ \text{s}$$

The steady-state value of the inductor current for $t > 0$ is found by replacing the inductor with a short circuit in the $t > 0$ circuit, as shown in Fig. 6.9*c*, since the sources are both dc. The result, as determined by any of the previous analysis techniques for resistive circuits, is

$$i_{L,\text{ss}} = -1 \text{ A}$$

The initial current at $t = 0^-$ is found from the $t < 0$ circuit as the steady-state value of the inductor current for $t < 0$. Since both sources are dc, we may replace the inductor with a short circuit in the $t < 0$ circuit, as shown in Fig. 6.9*d*, and obtain

$$\begin{aligned} i_L(0^-) &= \tfrac{1}{4} \text{ A} \\ &= i_L(0^+) \end{aligned}$$

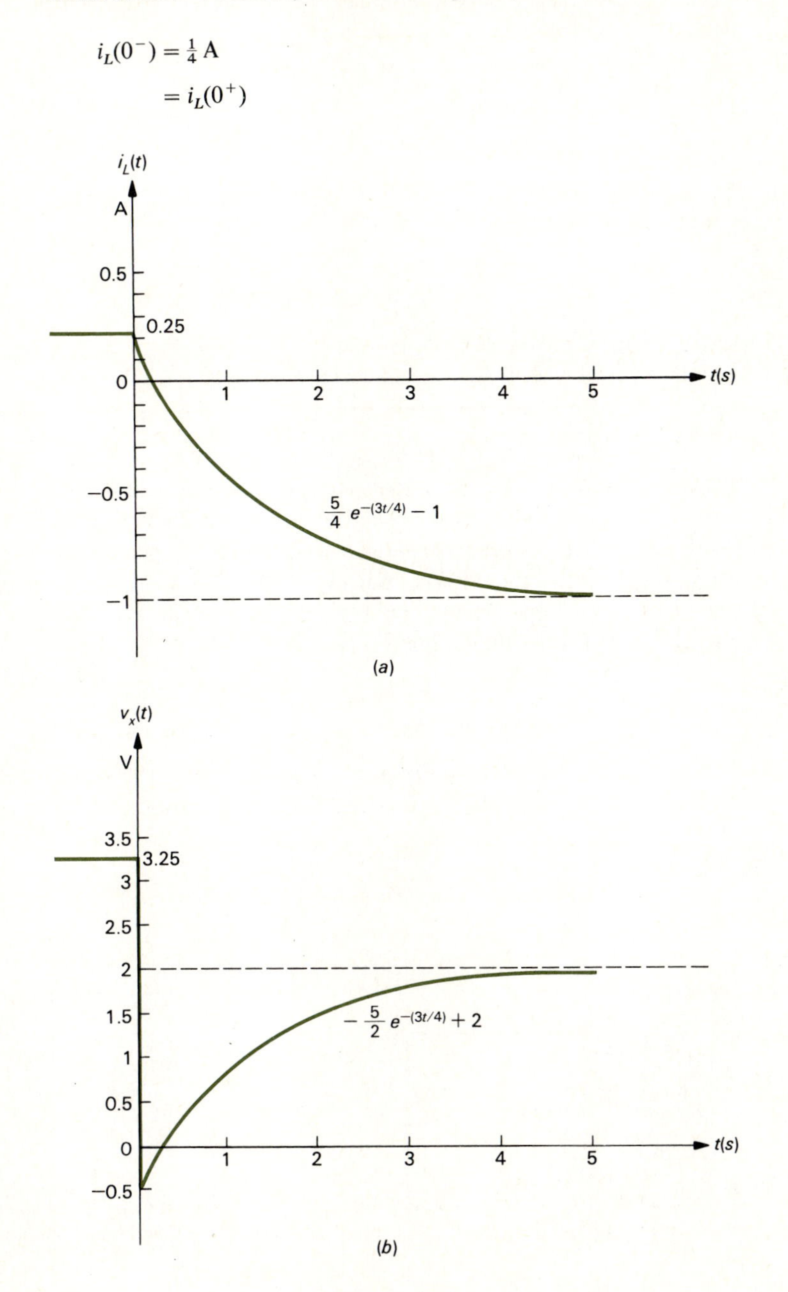

FIGURE 6.10 Example 6.3 solution.

Thus, the solution for $i_L(t)$ for $t > 0$ is

$$i_L(t) = (\tfrac{1}{4} + 1)e^{-3t/4} - 1$$
$$= \tfrac{5}{4}e^{-3t/4} - 1$$

The unknown voltage $v_x(t)$ can be found as usual by working back in the $t > 0$ circuit shown in Fig. 6.9*e*. Note that the current through the 2 Ω resistor across which $v_x(t)$ is defined is equal to $i_L(t)$ for $t > 0$. Thus

$$v_x(t) = -2i_L(t)$$
$$= -\tfrac{5}{2}e^{-3t/4} + 2$$

The inductor current and voltage v_x are sketched in Fig. 6.10. Note that $v_x(0^-)$ can be found from the $t = 0^-$ circuit in Fig. 6.9*d* to be

$$v_x(0^-) = \tfrac{13}{4} \text{ V}$$

and v_x changes instantaneously at $t = 0$, as shown in Fig. 6.10*b*. Note that both solutions asymptotically approach their steady-state values: $i_{L,\text{ss}} = -1$ A and $v_{x,\text{ss}} = 2$ V.

If the energy storage element is a capacitor, we reduce the remainder of the circuit attached to the capacitor terminals to a Norton equivalent, as shown in Fig. 6.11, and essentially repeat the process used above for the inductor problem. Here we will solve for the capacitor voltage $v_C(t)$. We do so since we will need to use continuity of capacitor voltages to find the initial condition, $v_C(0^+) = v_C(0^-)$. The only difference between this problem and the inductor problem is (1) in finding the steady-state values, $v_C(0^-)$ and $v_{C,\text{ss}}$, and (2) the time constant is

$$T = R_{\text{TH}}C \tag{6.61}$$

Once again, we do not actually need to find I_{SC} since the steady-state value $v_{C,\text{ss}}$ may be found directly from the $t > 0$ circuit. The total solution for the capacitor voltage for $t > 0$ is written

$$v_C(t) = [v_C(0^+) - v_{C,\text{ss}}]e^{-t/R_{\text{TH}}C} + v_{C,\text{ss}} \qquad t > 0 \tag{6.62}$$

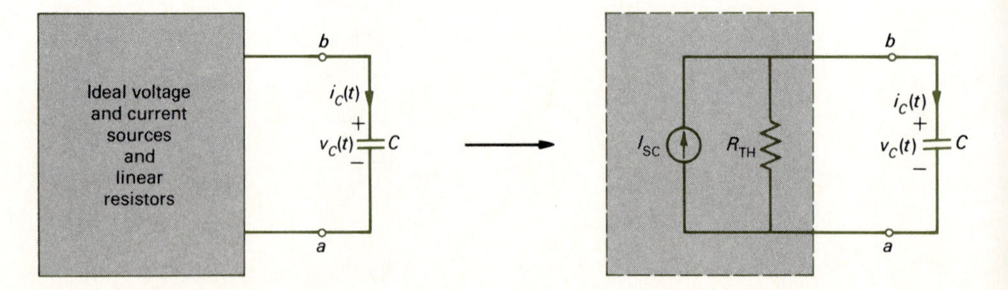

FIGURE 6.11 Solution of a circuit containing one capacitor by use of a Norton equivalent.

Again we solve for the capacitor voltage rather than the capacitor current since we can use continuity of capacitor voltages:

$$v_C(0^+) = v_C(0^-) \tag{6.63}$$

Example 6.4 Determine the solution for the voltage $v_x(t)$ across the 2-Ω resistor in Fig. 6.12*a*.

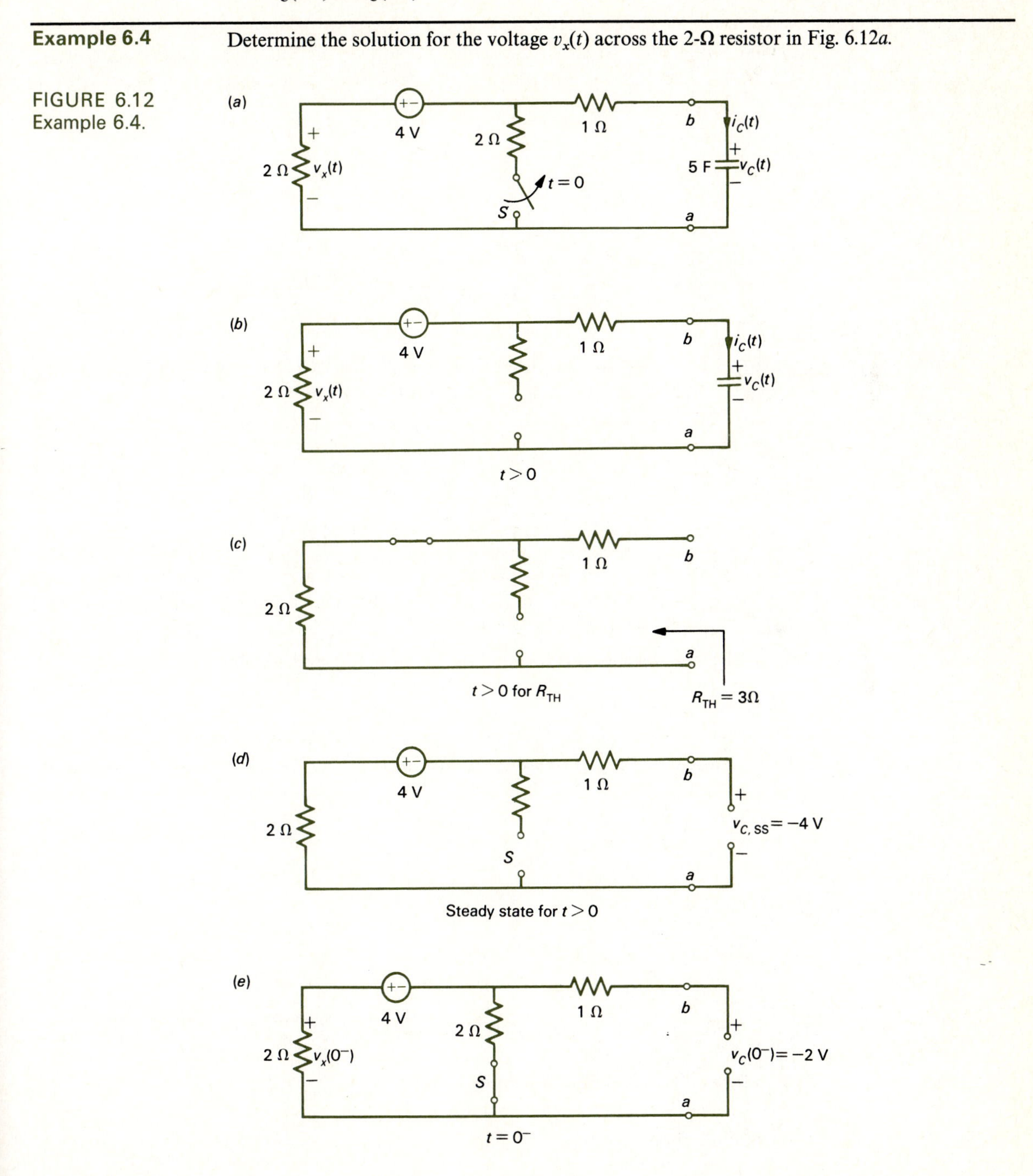

FIGURE 6.12 Example 6.4.

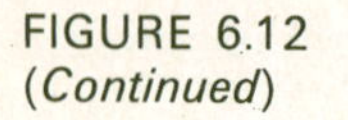

FIGURE 6.12 (*Continued*)

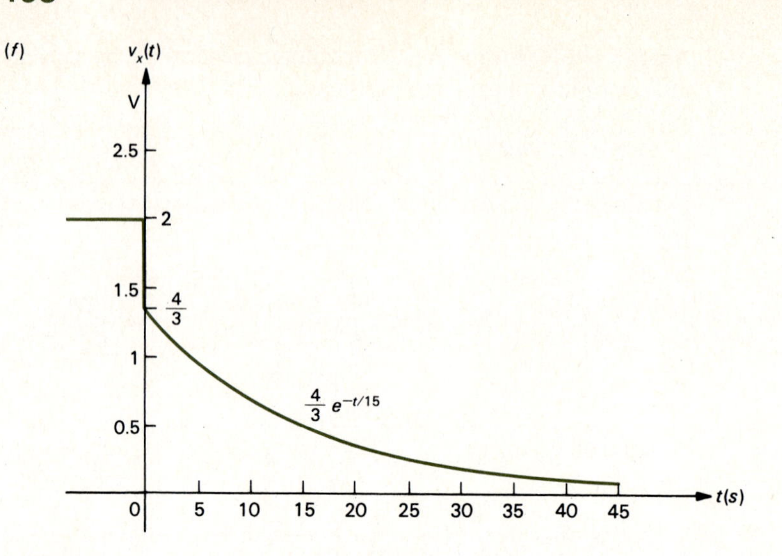

Solution The $t > 0$ circuit is shown in Fig. 6.12*b*. Killing the source in this circuit, we find that $R_{TH} = 3\ \Omega$, as shown in Fig. 6.12*c*. Thus, the time constant is

$$\begin{aligned} T &= R_{TH}C \\ &= 15\ \text{s} \end{aligned}$$

The steady-state circuit for $t > 0$ is shown in Fig. 6.12*d*, where we have replaced the capacitor with an open circuit since the source in the $t > 0$ circuit is dc. From this we obtain

$$v_{C,ss} = -4\ \text{V}$$

Drawing the steady-state $t < 0$ circuit at $t = 0^-$, as in Fig. 6.4*e*, we obtain

$$v_C(0^-) = -2\ \text{V}$$

Thus, by continuity of capacitor voltage, we obtain

$$\begin{aligned} v_C(0^+) &= v_C(0^-) \\ &= -2\ \text{V} \end{aligned}$$

Thus, for $t > 0$

$$\begin{aligned} v_C(t) &= [v_C(0^+) - v_{C,ss}]e^{-t/R_{TH}C} + v_{C,ss} \\ &= (-2 + 4)e^{-t/15} - 4 \\ &= 2e^{-t/15} - 4 \qquad t > 0 \end{aligned}$$

Now in the $t > 0$ circuit in Fig. 6.12*b* we observe that

$$v_x(t) = -2i_C(t)$$

So we need to find the capacitor current:

$$i_C(t) = C\frac{dv_C(t)}{dt}$$

$$= -\tfrac{2}{3}e^{-t/15}$$

and

$$v_x(t) = \tfrac{4}{3}e^{-t/15} \qquad t > 0$$

This is plotted in Fig. 6.12*f*. Note in the $t = 0^-$ circuit in Fig. 6.12*e* that

$$v_x(0^-) = 2 \text{ V}$$

From the above solution

$$v_x(0^+) = \tfrac{4}{3}$$

Thus, the resistor voltage has changed value instantaneously at $t = 0$. The solution for $v_x(t)$ asymptotically approaches zero. This can be seen to be the correct result from the steady-state $t > 0$ circuit in Fig. 6.12*d*.

6.4 CIRCUITS INVOLVING TWO ENERGY STORAGE ELEMENTS

Our previous results on circuits containing one energy storage element are of more benefit than we perhaps realize. Most of these results and techniques may be easily extended to handle a large number of circuits which contain two energy storage elements. We will only consider the cases in which one energy storage element is an inductor and the other is a capacitor and these are either in series with each other or in parallel. Other cases can be handled by similar methods.

First consider the series *LC* case shown in Fig. 6.13*a*. The remainder of the network containing only resistors and ideal voltage and current sources can be represented as a Thevenin equivalent in Fig. 6.13*b*. In Fig. 6.13*b*, we may write a differential equation in terms of the inductor current by applying KVL around the loop to yield

$$v_L(t) + v_C(t) = V_{OC}(t) - R_{TH}i_L(t) \tag{6.64}$$

Note that the current around the loop is the same as $i_L(t)$ so that

$$v_L(t) = L\frac{di_L(t)}{dt} \tag{6.65}$$

$$v_C(t) = \frac{1}{C}\int_{-\infty}^{t} i_C(\tau)\,d\tau$$

$$= \frac{1}{C}\int_{-\infty}^{t} i_L(\tau)\,d\tau \tag{6.66}$$

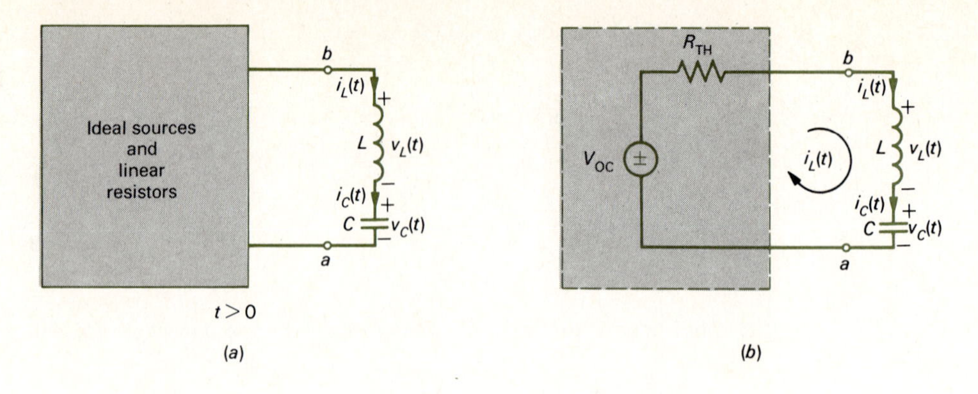

FIGURE 6.13 A series LC circuit.

Substituting Eqs. (6.66) and (6.65) into Eq. (6.64), we obtain

$$L\frac{di_L(t)}{dt}+\frac{1}{C}\int_{-\infty}^{t} i_L(\tau)\,d\tau = V_{OC}(t) - R_{TH}i_L(t) \tag{6.67}$$

Differentiating Eq. (6.67) and dividing both sides by L, we obtain

$$\frac{d^2i_L(t)}{dt^2}+\frac{R_{TH}}{L}\frac{di_L(t)}{dt}+\frac{1}{LC}i_L(t)=\frac{1}{L}\frac{dV_{OC}(t)}{dt} \tag{6.68}$$

This equation is a second-order constant coefficient linear ordinary differential equation. Once we solve Eq. (6.68) for $i_L(t)$, we may work back into the remainder circuit to find any other variables of interest. The inductor voltage $v_L(t)$ is found from $i_L(t)$ via Eq. (6.65). The total voltage across the terminals *a-b*, $v_L + v_C$, can be found from

$$v_L(t) + v_C(t) = V_{OC}(t) - R_{TH}i_L(t) \tag{6.69}$$

and therefore the capacitor voltage is, from Eq. (6.69),

$$v_C(t) = V_{OC}(t) - R_{TH}i_L(t) - v_L(t) \tag{6.70}$$

The capacitor current equals the inductor current:

$$\begin{aligned} i_C(t) &= C\frac{dv_C(t)}{dt} \\ &= i_L(t) \end{aligned} \tag{6.71}$$

Therefore, it is sufficient to find the solution for the inductor current from Eq. (6.68) and then use this solution to find any other variable of interest.

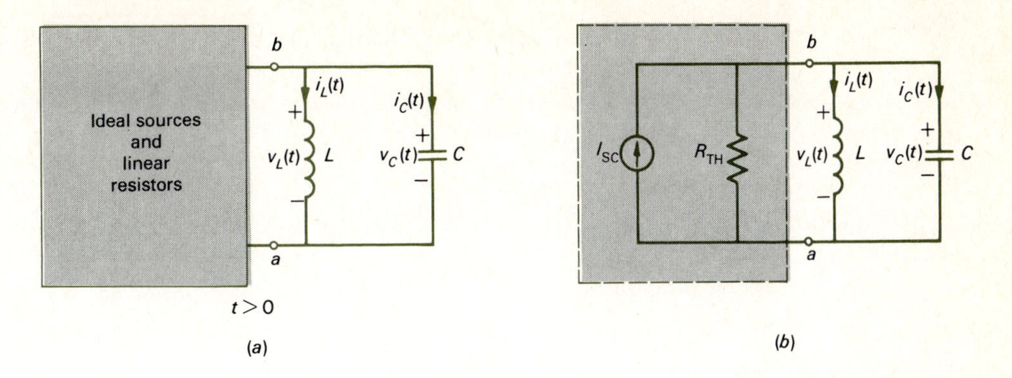

FIGURE 6.14 A parallel *LC* circuit.

For the parallel LC circuit shown in Fig. 6.14*a*, we may obtain a Norton equivalent representation, as shown in Fig. 6.14*b*. From this circuit, we may write a differential equation in terms of the capacitor voltage $v_C(t)$ by writing KCL at node b:

$$i_L(t) + i_C(t) = I_{SC}(t) - \frac{v_C(t)}{R_{TH}} \tag{6.72}$$

Substituting

$$i_C(t) = C\frac{dv_C(t)}{dt} \tag{6.73}$$

and

$$i_L(t) = \frac{1}{L}\int_{-\infty}^{t} v_C(\tau)\,d\tau \tag{6.74}$$

[since $v_L(t) = v_C(t)$] into Eq. (6.72) and rearranging, we obtain

$$\frac{d^2v_C(t)}{dt} + \frac{1}{CR_{TH}}\frac{dv_C(t)}{dt} + \frac{1}{LC}v_C(t) = \frac{1}{C}\frac{dI_{SC}(t)}{dt} \tag{6.75}$$

This is also a second-order constant coefficient linear ordinary differential equation of the same form as the one obtained for the series LC problem in Eq. (6.68). Once we solve Eq. (6.75) for $v_C(t)$, we may obtain any other branch variable in the remainder network by working back into this remainder network since the voltage of terminals *a-b* is $v_C(t)$. The capacitor current is found from Eq. (6.73). The inductor current is then found from Eq. (6.72). The inductor voltage equals the capacitor voltage:

$$\begin{aligned} v_L(t) &= L\frac{di_L(t)}{dt} \\ &= v_C(t) \end{aligned} \tag{6.76}$$

Thus, it is seen that the solution for $v_C(t)$ is the key to finding the solution for the other variables of interest.

For either case, we must solve a second-order constant coefficient linear ordinary differential equation of the form

$$\frac{d^2x(t)}{dt^2} + a\frac{dx(t)}{dt} + bx(t) = f(t) \tag{6.77}$$

[Compare Eq. (6.77) to Eqs. (6.68) and (6.75).] For the series LC problem

$$a = \frac{R_{\text{TH}}}{\text{L}}$$

$$b = \frac{1}{LC} \tag{6.78}$$

For the parallel LC problem

$$a = \frac{1}{R_{\text{TH}}C}$$

$$b = \frac{1}{LC} \tag{6.79}$$

Note that for both problems b is the same. For the series LC circuit, $1/a$ is the time constant of the RL problem of the previous sections. For the parallel LC circuit, $1/a$ is the time constant of the RC problem. Again, because of the linearity of the equation, the solution will consist of the sum of a transient solution $x_T(t)$ and a steady-state solution $x_{ss}(t)$:

$$x(t) = x_T(t) + x_{ss}(t) \tag{6.80}$$

where the transient solution satisfies Eq. (6.77) with $f(t) = 0$:

$$\frac{d^2x_T(t)}{dt^2} + a\frac{dx_T(t)}{dt} + bx_T(t) = 0 \tag{6.81}$$

and the steady-state solution satisfies Eq. (6.77) with $f(t) \neq 0$:

$$\frac{d^2x_{ss}(t)}{dt^2} + a\frac{dx_{ss}(t)}{dt} + bx_{ss}(t) = f(t) \tag{6.82}$$

First let us discuss the steady-state solution. Once again, $x_{ss}(t)$ is any function which, when substituted into Eq. (6.82), identically satisfies it. In Chap. 4 we considered a simple method for obtaining the steady-state solution due to a dc source (replace inductors with short circuits and capacitors with open circuits), and in Chap. 5 we considered simple ways of obtaining the steady-state solution due to a sinusoidal source. In this chapter we will only consider circuits containing dc ideal sources, so

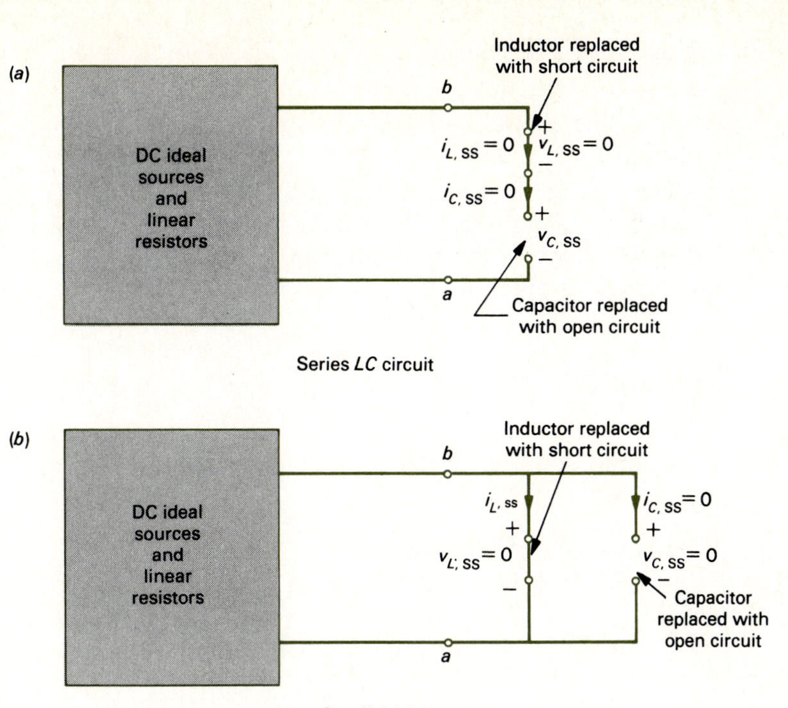

FIGURE 6.15 Determining steady-state solutions for dc sources: (*a*) series *LC* circuit; (*b*) parallel *LC* circuit.

that V_{OC} and I_{SC} will be constant. Therefore, again, *to find the steady-state solution for dc sources for any branch voltage of any energy storage element, we replace the inductor with a short circuit and the capacitor with an open circuit*, as shown in Fig. 6.15. Note that in the series *LC* problem only the steady-state capacitor voltage is nonzero and is determined by the remainder circuit. Similarly, for the parallel *LC* problem, only the steady-state inductor current will be nonzero. Note that in the series *LC* circuit Eq. (6.68) shows that since V_{OC} will be dc if the circuit contains only dc sources, then $i_{L,ss} = 0$. Similarly, for the parallel *LC* circuit containing only dc sources, Eq. (6.75) shows that $v_{C,ss} = 0$. These facts need not be memorized since they will become evident when one draws the steady-state circuit for a particular problem, as shown in Fig. 6.15.

Now for the transient solution. The solution to Eq. (6.81) must again be of the form

$$x_T(t) = Ae^{pt} \tag{6.83}$$

in order that it have the possibility of satisfying Eq. (6.81). Substituting Eq. (6.83) into Eq. (6.81), we obtain

$$p^2Ae^{pt} + apAe^{pt} + bAe^{pt} = 0 \tag{6.84}$$

or

$$(p^2 + ap + b)Ae^{pt} = 0 \tag{6.85}$$

This can only be true if

$$p^2 + ap + b = 0 \tag{6.86}$$

which is called the characteristic equation for the differential equation.

Depending on the values of the circuit elements R_{TH}, L, and C, we have three possibilities for the two roots p_1 and p_2 of the characteristic equation—Eq. (6.86). The two roots are given by

$$p_1, p_2 = -\frac{a}{2} \pm \frac{1}{2}\sqrt{a^2 - 4b} \tag{6.87}$$

which can be written, by factoring out $-a/2$, as

$$p_1, p_2 = -\frac{a}{2}\left(1 \pm \sqrt{1 - \frac{4b}{a^2}}\right) \tag{6.88}$$

The two roots will be real and unequal if

$$\frac{4b}{a^2} < 1$$

and will be equal if

$$\frac{4b}{a^2} = 1$$

The two roots will be complex if

$$\frac{4b}{a^2} > 1$$

Now let us investiage each of these three cases in more detail.

Case 1: Roots Real and Distinct For $4b/a^2 < 1$, the roots of the characteristic equation will be real and unequal (distinct). From Eq. (6.88) it is clear that if a is positive, both of these roots will be negative. For circuits containing positive resistors, inductors, and capacitors, R_{TH}, L, and C will be positive and so will a. Therefore, for realistic circuits, both roots will be negative and we will designate them as

$$p_1 = -\alpha_1$$

$$p_2 = -\alpha_2 \tag{6.89}$$

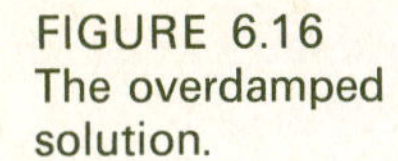
FIGURE 6.16
The overdamped solution.

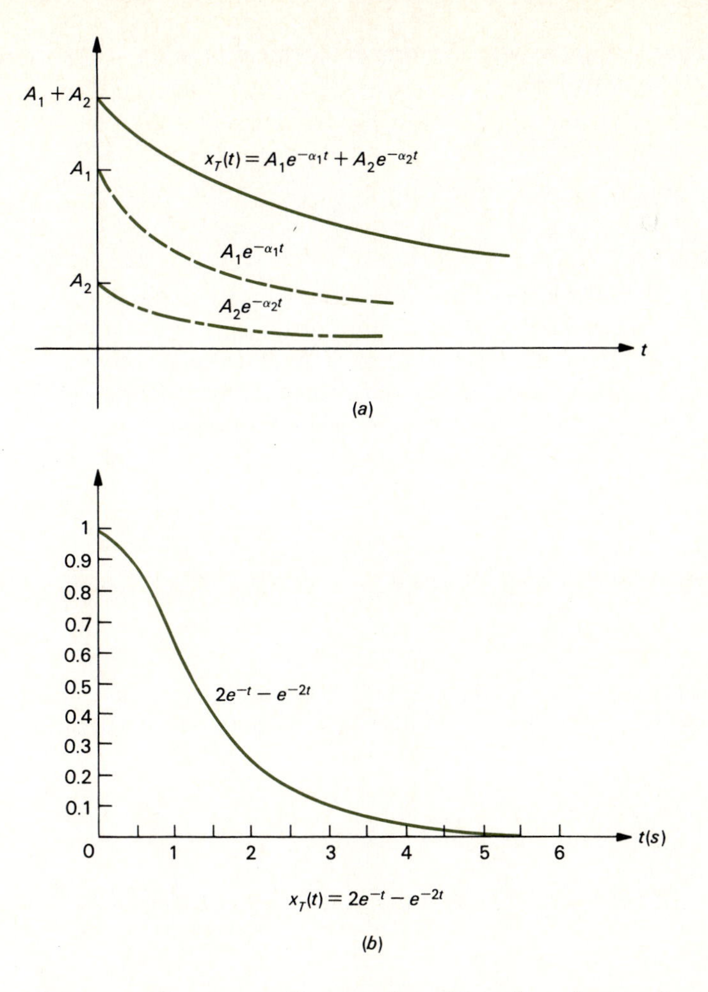

where α_1 and α_2 are positive numbers. Therefore, the transient solution will be of the form

$$x_T(t) = A_1e^{-\alpha_1 t} + A_2e^{-\alpha_2 t} \tag{6.90}$$

where A_1 and A_2 are, as yet, undetermined constants. Thus, the transient solution consists of the sum of two decaying exponentials, as shown in Fig. 6.16*a*. The solution for the specific case $A_1 = 2$, $\alpha_1 = 1$, $A_2 = -1$, $\alpha_2 = 2$ is shown in Fig. 6.16*b*. For this case, the transient solution is said to be overdamped.

Case 2: Roots Real and Equal If $4b = a^2$, the radical in Eq. (6.88) will be zero and the roots will be equal:

$$\begin{aligned} p_1, p_2 &= -\frac{a}{2} \\ &= -\alpha \end{aligned} \tag{6.91}$$

If we attempt to write the form of the transient solution as previously done for distinct roots, we find that

$$\begin{aligned} x_T(t) &= A_1 e^{-\alpha t} + A_2 e^{-\alpha t} \\ &= (A_1 + A_2)e^{-\alpha t} \\ &= Ae^{-\alpha t} \end{aligned} \tag{6.92}$$

In other words, this form of the transient solution reduces to the same as for a first-order differential equation. Thus, for the case of equal roots, we would suspect that Eq. (6.92) is not the proper form of the transient solution.

Let us now reexamine our procedure for determining the form of the transient solution. Recall that in order to find a solution of

$$\frac{d^2 x_T(t)}{dt^2} + a\frac{dx_T(t)}{dt^2} + bx_T(t) = 0 \tag{6.93}$$

we assumed a form for $x_T(t)$ as

$$x_T(t) = Ae^{pt} \tag{6.94}$$

substituted into Eq. (6.93), and determined p such that this form would satisfy Eq. (6.93). For equal roots, another possible form which will satisfy Eq. (6.93) is

$$x_T(t) = Ate^{pt} \tag{6.95}$$

To see this, we substitute Eq. (6.95) into Eq. (6.93) to obtain

$$(pAe^{pt} + pAe^{pt} + p^2 Ate^{pt}) + a(Ae^{pt} + pAte^{pt}) + b(Ate^{pt}) = 0 \tag{6.96}$$

Grouping terms, we obtain

$$(2p + a)Ae^{pt} + \underbrace{(p^2 + ap + b)}_{0}Ate^{pt} = 0 \tag{6.97}$$

The coefficient of Ate^{pt} is the characteristic equation and thus is automatically equal to zero. The question remains as to whether

$$2p + a \stackrel{?}{=} 0 \tag{6.98}$$

for these real, repeated roots. But this real, repeated root given in Eq. (6.91) exactly satisfies Eq. (6.98) so that Eq. (6.95) is a valid form of the transient solution. By superposition, we may combine Eqs. (6.94) and (6.95) to obtain the form of the transient solution which satisfies Eq. (6.93):

$$\begin{aligned} x_T(t) &= A_1 e^{-\alpha t} + A_2 te^{-\alpha t} \\ &= (A_1 + A_2 t)e^{-\alpha t} \end{aligned} \tag{6.99}$$

Note that this solution cannot be reduced any further and that it contains exactly two undetermined constants. For this case, the transient solution is said to be critically damped, and Eq. (6.99) is plotted in Fig. 6.17*a*. The result for $A_1 = 2$, $A_2 = -1$, and $\alpha = 2$ is shown in Fig. 6.17*b*.

Case 3: Complex Roots If $4b/a^2 > 1$ in Eq. (6.88), the roots will be complex. Denoting

$$j = \sqrt{-1}$$

we may write Eq. (6.88) as

$$p_1, p_2 = -\frac{a}{2}\left(1 \pm j\sqrt{\frac{4b^2}{a} - 1}\right) \tag{6.100}$$

so that we will denote the roots as

$$p_1, p_2 = -\alpha \pm j\beta \tag{6.101}$$

The two roots are conjugates. As we found in the previous chapter, the conjugate of a complex number $\mathbf{c}$ is denoted by $\mathbf{c}^*$ and is obtained by replacing all imaginary terms in $\mathbf{c}$ with their negatives. For example, if $\mathbf{c} = 2 + j3$, then $\mathbf{c}^* = 2 - j3$. The transient solution becomes

$$\begin{aligned} x_T(t) &= A_1 e^{(-\alpha + j\beta)t} + A_2 e^{(-\alpha - j\beta)t} \\ &= e^{-\alpha t}(A_1 e^{j\beta t} + A_2 e^{-j\beta t}) \end{aligned} \tag{6.102}$$

Although this is a valid transient solution, we will find it convenient to write Eq. (6.102) in an alternative form:

$$x_T(t) = e^{-\alpha t}(C_1 \cos \beta t + C_2 \sin \beta t) \tag{6.103}$$

This can be easily shown by substituting Euler's identity, which we studied in the previous chapter,

$$e^{j\beta t} = \cos \beta t + j \sin \beta t$$

into Eq. (6.102) to yield

$$\begin{aligned} x_T(t) &= e^{-\alpha t}[A_1(\cos \beta t + j \sin \beta t) + A_2(\cos \beta t - j \sin \beta t)] \\ &= e^{-\alpha t}[(A_1 + A_2) \cos \beta t + j(A_1 - A_2) \sin \beta t] \end{aligned} \tag{6.104}$$

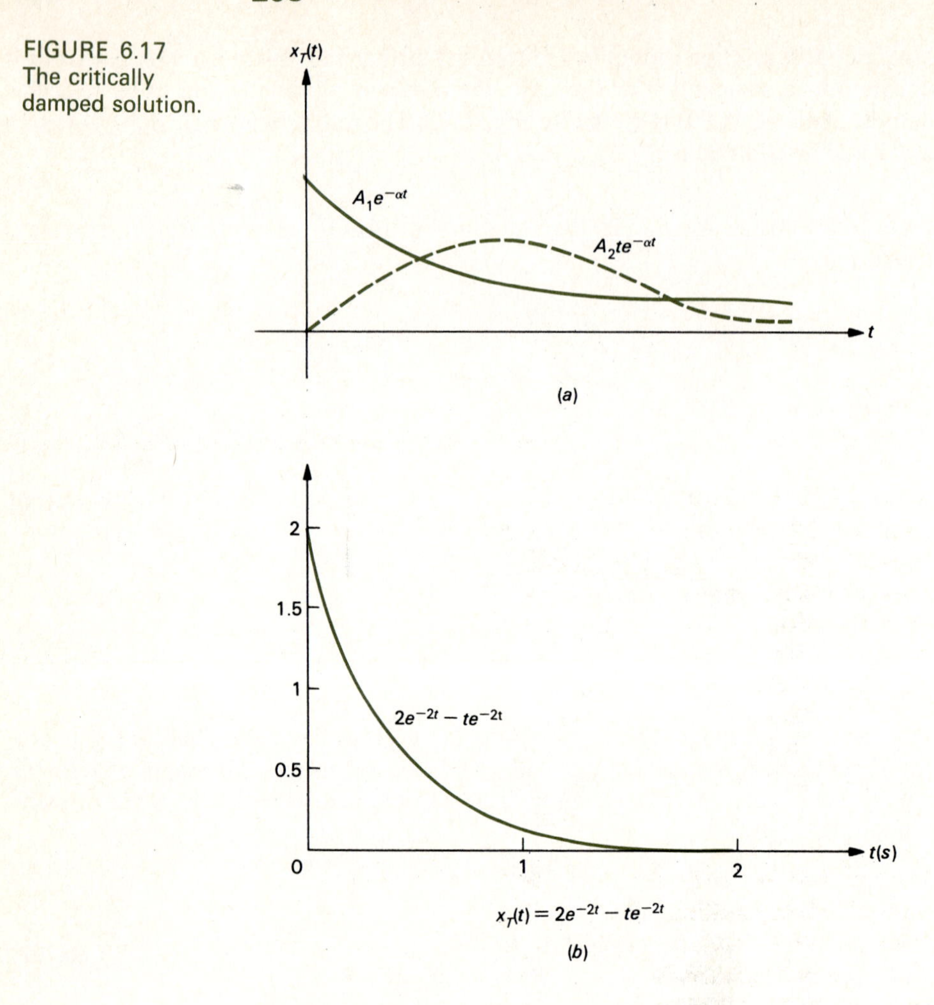

FIGURE 6.17 The critically damped solution.

Comparing Eqs. (6.103) and (6.104), we identify

$$C_1 = A_1 + A_2 \tag{6.105a}$$

$$C_2 = j(A_1 - A_2) \tag{6.105b}$$

Since A_1 and A_2 are (as yet) undetermined constants, we may as well use the form given in Eq. (6.103). Since the transient solution given in Eqs. (6.102) or (6.103) must be real, the undetermined constants in Eq. (6.103) must be real. However, the constants in Eq. (6.103), C_1 and C_2, are related to the constants in Eq. (6.102), A_1 and A_2, by Eq. (6.105). But C_2 which is real is $j(A_1 - A_2)$. In order that C_2 can be real, A_1 and A_2 must be complex and conjugates of each other. For example, if we write

$$\begin{aligned} A_1 &= A + jB \\ A_2 &= A - jB \end{aligned} \tag{6.106}$$

substitution in Eq. (6.105) yields

$$\begin{aligned} C_1 &= A_1 + A_2 \\ &= 2A \\ C_2 &= j(A_1 - A_2) \\ &= j(j2B) \\ &= -2B \end{aligned} \tag{6.107}$$

Therefore, the undetermined constants A_1 and A_2 in Eq. (6.102) are complex and conjugates.

An alternative form of the solution in Eq. (6.103) can be written as

$$x_T(t) = Ce^{-\alpha t} \sin(\beta t + \phi) \tag{6.108}$$

To show this, we expand $C \sin(\beta t + \phi)$ in Eq. (6.108) as

$$C \sin(\beta t + \phi) = C \sin \beta t \cos \phi + C \cos \beta t \sin \phi \tag{6.109}$$

Thus, we identify

$$\begin{aligned} C_1 &= C \cos \phi \\ C_2 &= C \sin \phi \end{aligned} \tag{6.110}$$

and

$$\begin{aligned} \frac{\sin \phi}{\cos \phi} &= \tan \phi \\ &= \frac{C_2}{C_1} \end{aligned} \tag{6.111}$$

or

$$\phi = \tan^{-1} \frac{C_2}{C_1} \tag{6.112}$$

The transient solution for the case of complex roots is said to be underdamped, and Eq. (6.108) is sketched in Fig. 6.18*a*. The specific case for $C = 1$, $\alpha = 2$, $\beta = 4$, $\phi = 30°$ is sketched in Fig. 6.18*b*.

Note that in all three cases the transient solution will eventually decay to zero, leaving only the steady-state solution. For the overdamped case,

$$x_T(t) = A_1 e^{-\alpha_1 t} + A_2 e^{-\alpha_2 t} \tag{6.113}$$

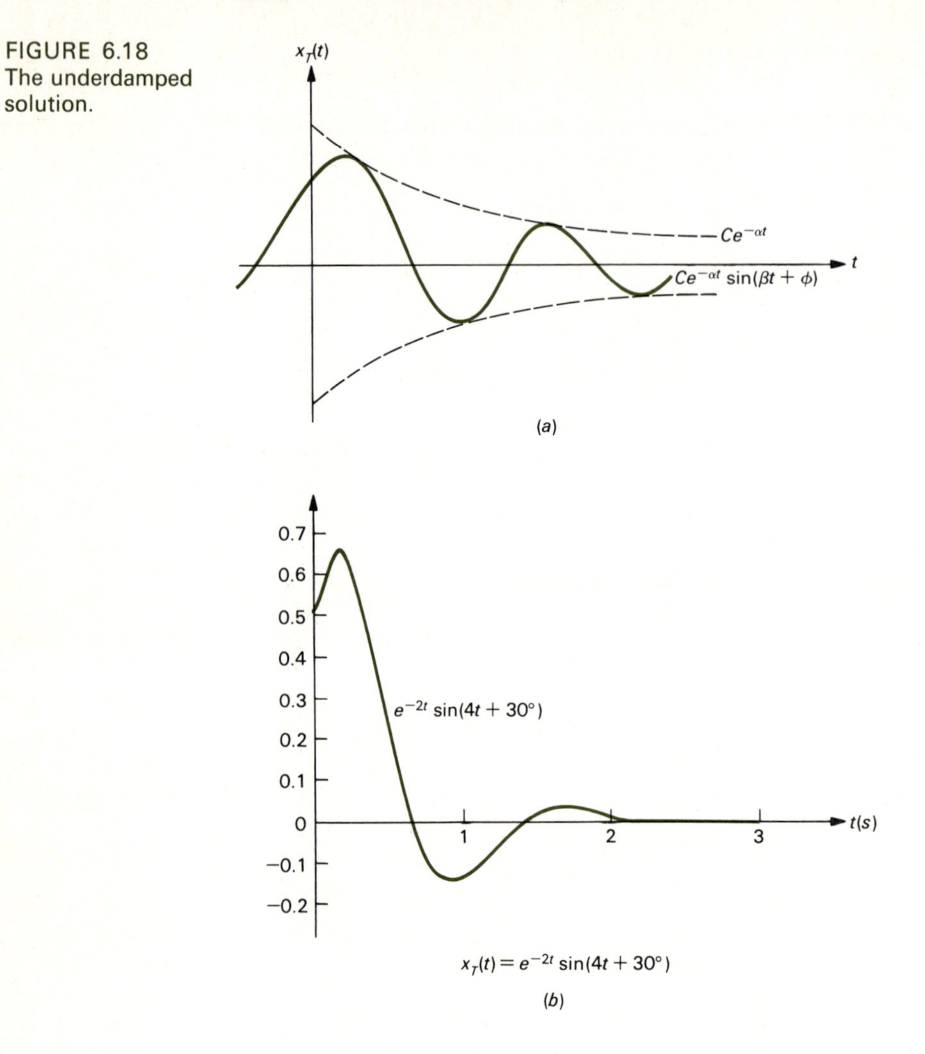

FIGURE 6.18 The underdamped solution.

will go to zero as time progresses. For the critically damped case,

$$x_T(t) = A_1 e^{-\alpha t} + A_2 t e^{-\alpha t} \tag{6.114}$$

will also go to zero since $e^{-\alpha t}$ goes to zero faster than t goes to infinity as time progresses. For the underdamped case,

$$x_T(t) = Ce^{-\alpha t} \sin(\beta t + \phi) \tag{6.115}$$

also goes to zero as time progresses. We have assumed, however, that the real parts of the roots α_1 and α_2 for the underdamped case, or α for the critically damped and underdamped cases, are nonzero. From the form of the roots in Eq. (6.88) it is easy to see that the real parts of these roots will not be zero if $R_{TH} \neq 0$. Therefore, if the circuit contains some resistance, the transient solution will eventually decay to zero. This is a realistic case, since all physical circuits contain some resistance even though it may be only the resistance of the connecting wires. Thus, we will only consider circuits for which the transient solution decays to zero.

6.5 SWITCHING OPERATIONS AND INITIAL CONDITIONS

Again we wish to obtain complete solutions for circuit voltages or circuits after some switching operation has taken place at $t = 0$. The procedure is almost identical to the case of one energy storage element, except that now we must obtain two initial conditions (at $t = 0^+$) on the variable of interest (i_L, v_L, i_C, or v_C) in order to evaluate the two undetermined constants in the transient solution.

Again we consider circuits containing only dc ideal voltage and current sources. We first draw the $t < 0$ circuit—the circuit as it exists prior to the switching operator. Next we assume that it has been in this state for a very long time so that it has reached steady-state and all transients have essentially vanished. *Only the inductor current and capacitor voltage are continuous at* $t = 0$*: the same at* $t = 0^-$ *and* $t = 0^+$. Thus, since the $t < 0$ circuit contains only dc sources and is in the steady state, we replace the inductor with a short circuit and the capacitor with an open circuit to obtain $i_L(0^-)$ and $v_C(0^-)$. We thus have

$$i_L(0^+) = i_L(0^-) \tag{6.116}$$

$$v_C(0^+) = v_C(0^-) \tag{6.117}$$

Next we draw the $t > 0$ circuit and write the form of the solution for either the inductor current or the capacitor voltage by inspection. If L and C are in series,

$$a = \frac{R_{TH}}{L} \qquad b = \frac{1}{LC} \tag{6.118}$$

If L and C are in parallel,

$$a = \frac{1}{CR_{TH}} \qquad b = \frac{1}{LC} \tag{6.119}$$

Next we find the roots of

$$p^2 + ap + b = 0 \tag{6.120}$$

and write the form of the transient solution by inspection, depending on whether the roots are real and distinct, real and repeated, or complex conjugates.

Next we find the steady-state solution by drawing the $t > 0$ circuit (which contains only dc sources) and replace the inductor with a short circuit and the capacitor with an open circuit. We then form the total solution for $t > 0$ by adding this steady-state solution to the transient solution:

$$i_L(t) = i_{L,T}(t) + i_{L,\text{ss}}(t) \qquad t > 0 \tag{6.121}$$

$$v_C(t) = v_{C,T}(t) + v_{C,\text{ss}}(t) \qquad t > 0 \tag{6.122}$$

There are two unknowns in the transient solution regardless of the form of the roots. If we wish to find the solution for the inductor current, we have one initial condition $i_L(0^+)$ to be used, but we need one more condition on $i_L(0^+)$ to evaluate the two constants. If we knew the derivative of i_L at $t = 0^+$, $di_L/dt|_{t=0^+}$, we could evaluate the two constants since we could evaluate Eq. (6.121) at $t = 0^+$ as

$$i_L(0^+) = i_{L,T}(0^+) + i_{L,\text{ss}}(0^+) \tag{6.123}$$

and then we could differentiate Eq. (6.121) and evaluate the result at $t = 0^+$:

$$\left.\frac{di_L}{dt}\right|_{t=0^+} = \left.\frac{di_{L,T}}{dt}\right|_{t=0^+} + \left.\frac{di_{L,\text{ss}}}{dt}\right|_{t=0^+} \tag{6.124}$$

Thus, Eqs. (6.123) and (6.124) would allow us to evaluate the two undetermined constants in the transient solution. Since we only know $i_L(0^+)$ and $v_C(0^+)$, how do we obtain $di_L/dt|_{t=0^+}$? Note that the inductor voltage is

$$v_L(t) = L\frac{di_L(t)}{dt} \tag{6.125}$$

Therefore,

$$\left.\frac{di_L}{dt}\right|_{t=0^+} = \frac{1}{L}v_L(0^+) \tag{6.126}$$

If we can find the inductor voltage at $t = 0^+$, we can find the derivative of the inductor current at $t = 0^+$ with Eq. (6.126). To find the inductor voltage (which is *not* necessarily continuous at $t = 0$), we draw the $t = 0^+$ circuit. Knowing $i_L(0^+)$ and $v_C(0^+)$ in this circuit, we may use KVL, KCL, and Ohm's law to find $v_L(0^+)$. A similar process would apply if we had chosen to obtain the capacitor voltage equation in Eq. (6.122); that is, find $v_C(0^+)$ and $dv_C/dt|_{t=0^+} = (1/C)i_C(0^+)$.

Example 6.5 Consider the circuit in Fig. 6.19*a*. Find the form of the inductor current.

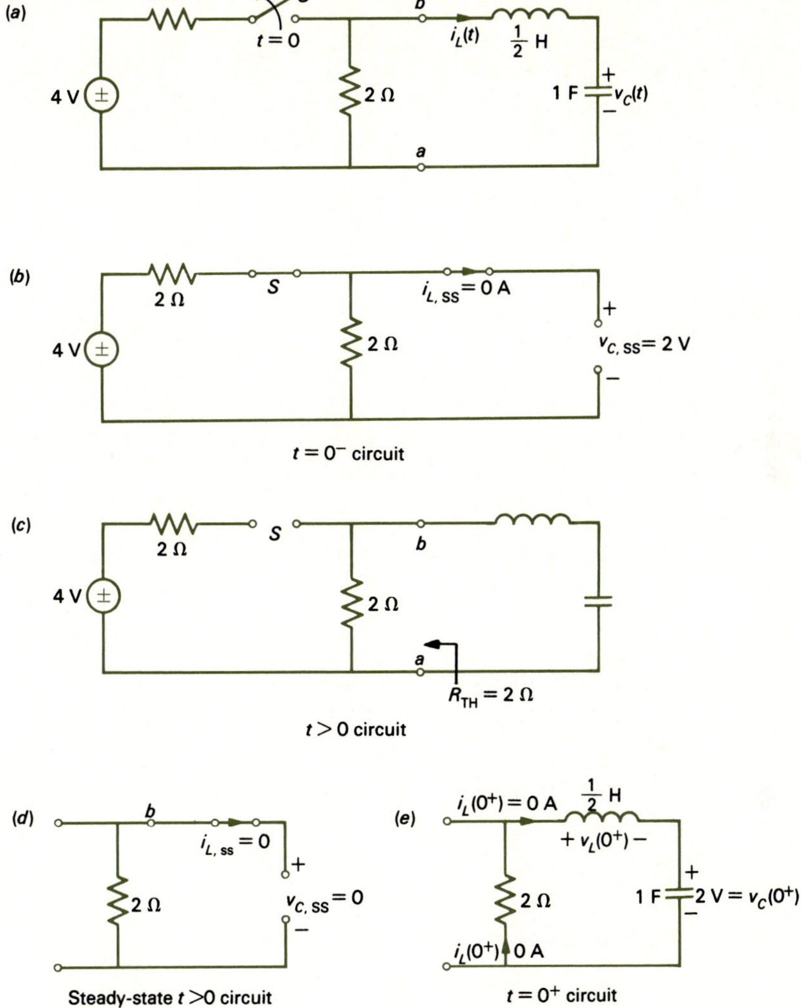

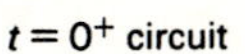

FIGURE 6.19 Example 6.5: illustration of determining initial conditions and the complete solution for $t \geqq 0$.

Solution The $t = 0^-$ circuit is shown in Fig. 6.19*b* and

$$v_C(0^+) = v_C(0^-) = 2\text{ V}$$

$$i_L(0^+) = i_L(0^-) = 0\text{ A}$$

The $t > 0$ circuit is shown in Fig. 6.19*c*. From this we obtain

$$R_{TH} = 2\ \Omega$$

and since it is a series LC circuit

$$a = \frac{R_{\text{TH}}}{L}$$

$$= 4$$

$$b = \frac{1}{LC}$$

$$= 2$$

so that the characteristic equation is

$$p^2 + 4p + 2 = 0$$

The roots are

$$p_1, p_2 = -2 \pm \tfrac{1}{2}\sqrt{16 - 8}$$

$$= -2 \pm \sqrt{2}$$

$$= -3.414, -0.586$$

Thus, the transient solution is

$$i_{L,T}(t) = A_1 e^{-3.414t} + A_2 e^{-0.586t}$$

The steady-state circuit for $t > 0$ is shown in Fig. 6.19*d*. From it we obtain

$$i_{L,\text{ss}} = 0 \qquad t > 0$$

Thus, the total solution is

$$i_L(t) = A_1 e^{-3.414t} + A_2 e^{-0.586t}$$

The $t = 0^+$ circuit is shown in Fig. 6.19*e*. From it we obtain, by writing a KVL equation around the loop,

$$v_L(0^+) = -2i_L(0^+) - v_C(0^+)$$

$$= -2 \text{ V}$$

But

$$v_L(0^+) = L \left. \frac{di_L}{dt} \right|_{t=0^+}$$

so

$$\left. \frac{di_L}{dt} \right|_{t=0^+} = \frac{1}{L} v_L(0^+)$$

$$= -4 \text{ A/s}$$

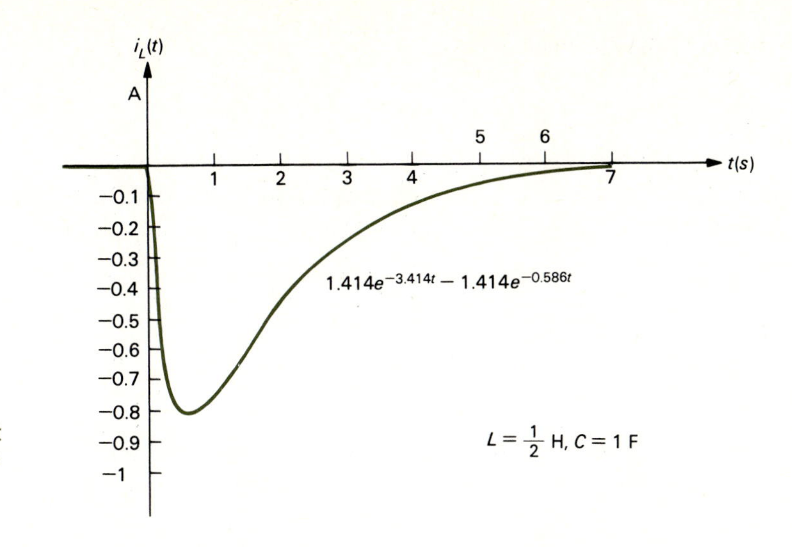

FIGURE 6.20 Plot of the inductor current for $t \geqq 0$ for the circuit of Fig. 6.19.

The total solution at $t = 0^+$ is

$$\begin{aligned} i_L(0^+) &= A_1 + A_2 \\ &= 0\,\text{A} \end{aligned}$$

Differentiating the total solution and evaluating at $t = 0^+$, we obtain

$$\begin{aligned} \left.\frac{di_L}{dt}\right|_{t=0^+} &= -3.414A_1 - 0.586A_2 \\ &= -4 \end{aligned}$$

or

$$\begin{aligned} A_1 + A_2 &= 0 \\ -3.414A_1 - 0.586A_2 &= -4 \end{aligned}$$

Thus, we solve for A_1 and A_2 by Cramer's rule as

$$\begin{aligned} A_1 &= \frac{\begin{vmatrix} 0 & 1 \\ -4 & -5.86 \end{vmatrix}}{\begin{vmatrix} 1 & 1 \\ -3.414 & -0.586 \end{vmatrix}} \\ &= 1.414 \end{aligned}$$

$$\begin{aligned} A_2 &= \frac{\begin{vmatrix} 1 & 0 \\ -3.414 & -4 \end{vmatrix}}{\begin{vmatrix} 1 & 1 \\ -3.414 & -0.586 \end{vmatrix}} \\ &= -1.414 \end{aligned}$$

Thus, the total solution is

$$i_L(t) = 1.414e^{-3.414t} - 1.414e^{-0.586t} \qquad t > 0$$

This is plotted in Fig. 6.20.

Example 6.6 Repeat Example 6.5 but with $L = 1$ H, $C = 1$ F.

Solution Several of the previous example results remain the same. Changing the values of L and C only affects (1) the roots and (2)

$$\left.\frac{di_L}{dt}\right|_{t=0^+} = \frac{1}{L} v_L(0^+)$$

Note that the initial values of $i_L(0^+)$ and $v_C(0^+)$ remain unchanged. Also, $v_L(0^+)$ remains unchanged since this depended on $i_L(0^+)$ and the circuit structure at $t = 0^+$. For this case

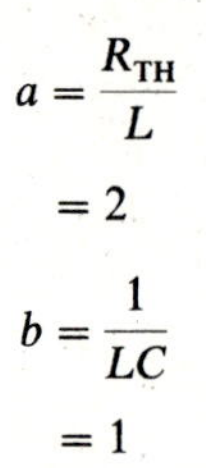

$$a = \frac{R_{TH}}{L}$$
$$= 2$$
$$b = \frac{1}{LC}$$
$$= 1$$

and

$$p^2 + 2p + 1 = 0$$

The roots are repeated,

$$p_1, p_2 = -1$$

and the solution is

$$i_L(t) = A_1 e^{-t} + A_2 t e^{-t}$$

Now $i_L(0^+)$ again is 0 A. Evaluating i_L and $t = 0^+$, we obtain

$$i_L(0^+) = A_1$$
$$= 0 \text{ A}$$

Thus, $A_1 = 0$. Now differentiating the solution, we obtain

$$\frac{di_L}{dt} = -A_1 e^{-t} - A_2 t e^{-t} + A_2 e^{-t}$$

and

$$\left.\frac{di_L}{dt}\right|_{t=0^+} = -A_1 + A_2$$
$$= A_2$$

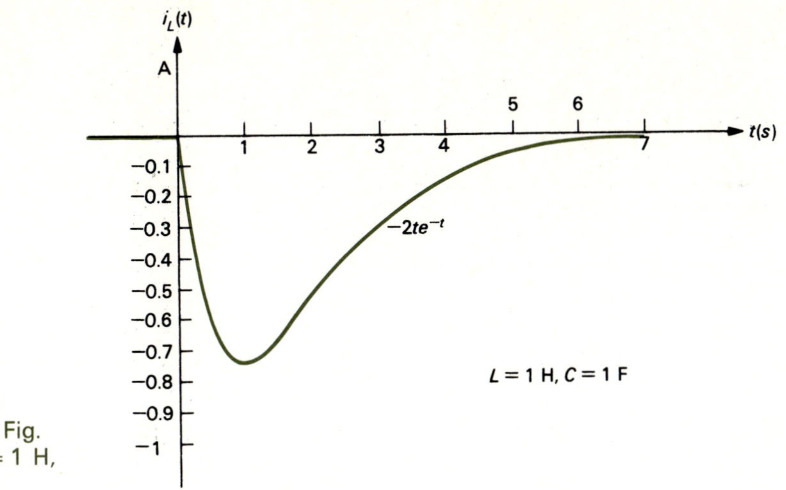

FIGURE 6.21 Example 6.6: Fig. 6.19 with $L = 1$ H, $C = 1$ F.

But

$$\left.\frac{di_L}{dt}\right|_{t=0^+} = \frac{1}{L} v_L(0^+)$$

$$= -2 \text{ V}$$

So

$$A_2 = -2$$

and

$$i_L(t) = -2te^{-t}$$

This is plotted in Fig. 6.21.

Example 6.7 Repeat Example 6.5 with $L = 5$ H, $C = 1$ F.

Solution Again, only certain parts of the solution remain unchanged. For this case

$$a = \frac{R_{TH}}{L}$$

$$= \tfrac{2}{5}$$

$$b = \frac{1}{LC}$$

$$= \tfrac{1}{5}$$

and

$$p^2 + \tfrac{2}{5}p + \tfrac{1}{5} = 0$$

The roots are

$$p_1, p_2 = -\tfrac{1}{5} \pm \tfrac{1}{2}\sqrt{\tfrac{4}{25} - \tfrac{4}{5}}$$
$$= -\tfrac{1}{5} \pm j\tfrac{2}{5}$$

Thus, the solution is

$$i_L(t) = e^{-\frac{1}{5}t}(C_1 \cos \tfrac{2}{5}t + C_2 \sin \tfrac{2}{5}t)$$

Again

$$i_L(0^+) = 0 \text{ A}$$

but

$$\left.\frac{di_L}{dt}\right|_{t=0^+} = \frac{1}{L}\,v_L(0^+)$$
$$= -\tfrac{2}{5}$$

since $v_L(0^+) = -2$ V again.

Evaluating the solution, we obtain

$$i_L(0^+) = C_1$$
$$= 0$$

or $C_1 = 0$. The derivative of i_L at $t = 0^+$ is a bit more complicated:

$$\frac{di_L}{dt} = -\tfrac{1}{5}C_1 e^{-\frac{1}{5}t}\cos(\tfrac{2}{5}t) - \tfrac{2}{5}C_1 e^{-\frac{1}{5}t}\sin \tfrac{2}{5}t$$
$$-\tfrac{1}{5}C_2 e^{-\frac{1}{5}t}\sin(\tfrac{2}{5}t) + \tfrac{2}{5}C_2 e^{-\frac{1}{5}t}\cos \tfrac{2}{5}t$$

so that

$$\left.\frac{di_L}{dt}\right|_{t=0^+} = -\tfrac{1}{5}C_1 + \tfrac{2}{5}C_2$$
$$= -\tfrac{2}{5}$$

or

$$C_2 = -1$$

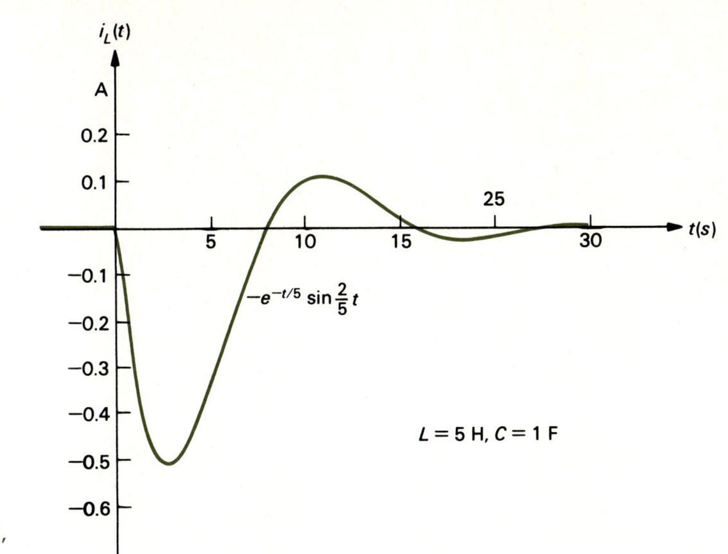

FIGURE 6.22 Example 6.7: Fig. 6.19 with $L = 5$ H, $C = 1$ F.

The solution thus becomes

$$i_L(t) = -e^{-t/5}\sin\tfrac{2}{5}t \qquad t > 0$$

This is plotted in Fig. 6.22.

Example 6.8 For the parallel LC problem in Fig. 6.23*a*, solve for the current in the 1-Ω resistor, $i_x(t)$, for $t > 0$.

Solution The $t = 0^-$ circuit is shown in Fig. 6.23*b*. From it we obtain

$$\begin{aligned} i_L(0^-) &= -\tfrac{3}{4}\text{ A} \\ &= i_L(0^+) \\ v_C(0^-) &= 0 \\ &= v_C(0^+) \end{aligned}$$

The $t > 0$ circuit is shown in Fig. 6.23*c*. From it we obtain $R_{TH} = 3\ \Omega$ and

$$\begin{aligned} a &= \frac{1}{R_{TH}C} \\ &= \tfrac{4}{15} \\ b &= \frac{1}{LC} \\ &= \tfrac{4}{45} \end{aligned}$$

So

$$p^2 + \tfrac{4}{15}p + \tfrac{4}{45} = 0$$

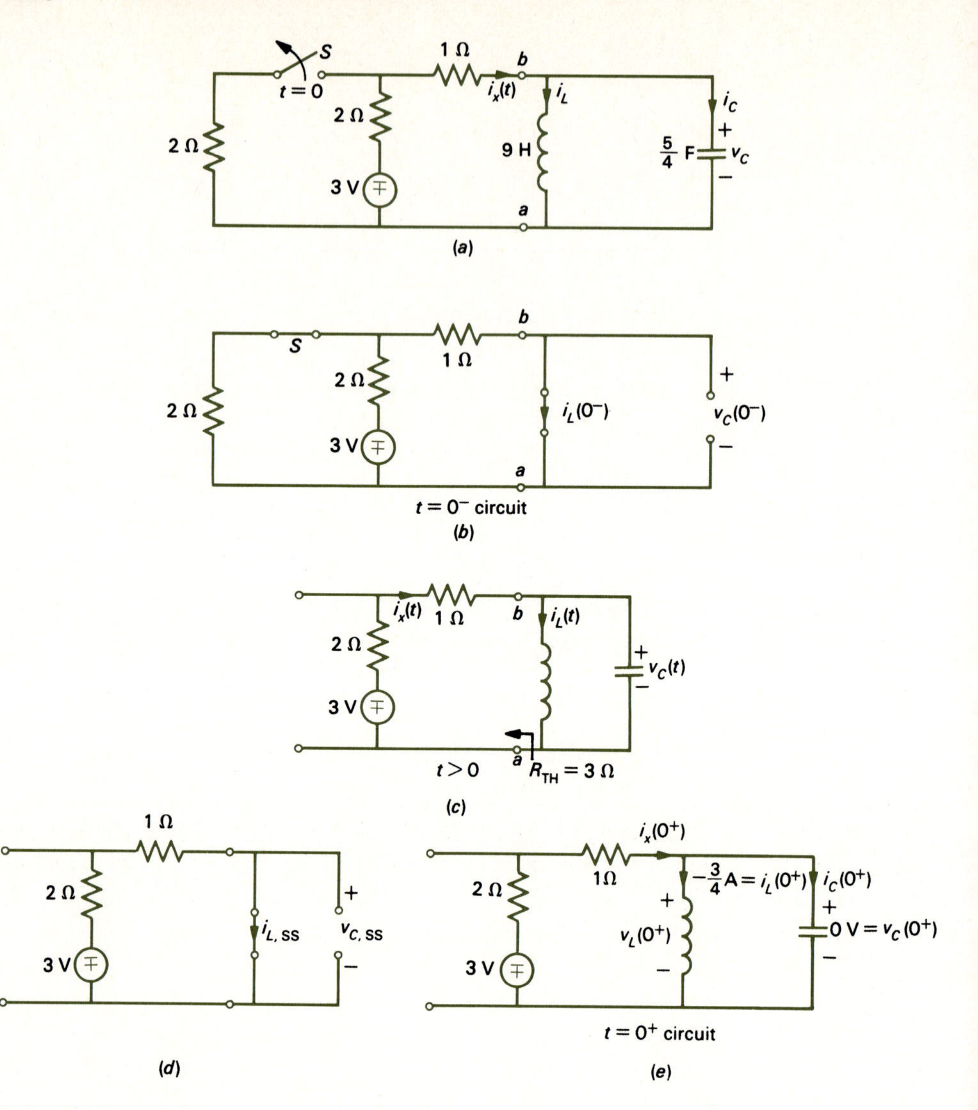

FIGURE 6.23 Example 6.8: a parallel LC circuit.

and the roots become

$$p_1, p_2 = -\tfrac{2}{15} \pm \tfrac{1}{2}\sqrt{(\tfrac{4}{15})^2 - \tfrac{16}{45}}$$
$$= -\tfrac{2}{15} \pm j\tfrac{4}{15}$$

The transient solution for the capacitor voltage is

$$v_{C,T}(t) = e^{-\frac{2}{15}t}[C_1 \cos(\tfrac{4}{15}t) + C_2 \sin \tfrac{4}{15}t]$$

The steady-state solution $v_{C,ss}$ is found from Fig. 6.23*d* as

$$v_{C,ss} = 0$$

so that

$$v_C(t) = e^{-\frac{2}{15}t}[C_1 \cos \tfrac{4}{15}t + C_2 \sin \tfrac{4}{15}t]$$

Since $v_C(0^+) = 0$, we find that

$$C_1 = 0$$

Now differentiating $v_C(t)$, we obtain

$$\frac{dv_C}{dt} = -\tfrac{2}{15}C_1 e^{-\frac{2}{15}t} \cos \tfrac{4}{15}t - \tfrac{4}{15}C_1 e^{-\frac{2}{15}t} \sin \tfrac{4}{15}t$$
$$- \tfrac{2}{15}C_2 e^{-\frac{2}{15}t} \sin \tfrac{4}{15}t + \tfrac{4}{15}C_2 e^{-\frac{2}{15}t} \cos \tfrac{4}{15}t$$

or

$$\left.\frac{dv_C}{dt}\right|_{t=0^+} = -\tfrac{2}{15}\cancelto{0}{C_1} + \tfrac{4}{15}C_2$$

or

$$C_2 = \tfrac{15}{4} \left.\frac{dv_C}{dt}\right|_{t=0^+}$$

How do we obtain $dv_C/dt|_{t=0^+}$? First draw the $t = 0^+$ circuit and observe that the only variables we know in this circuit are $i_L(0^+)$ and $v_C(0^+)$, as shown in Fig. 6.23*e*. Observe that

$$i_C = C \frac{dv_C}{dt}$$

so that

$$\left.\frac{dv_C}{dt}\right|_{t=0^+} = \frac{1}{C} i_C(0^+)$$

Therefore, we only need to find $i_C(0^+)$. Note that

$$i_x(0^+) = i_L(0^+) + i_C(0^+)$$

But by writing KVL around the outside loop we obtain

$$i_x(0^+) = \frac{-3\text{ V} - \cancelto{0}{v_C}(0^+)}{3\ \Omega}$$
$$= -1\text{ A}$$

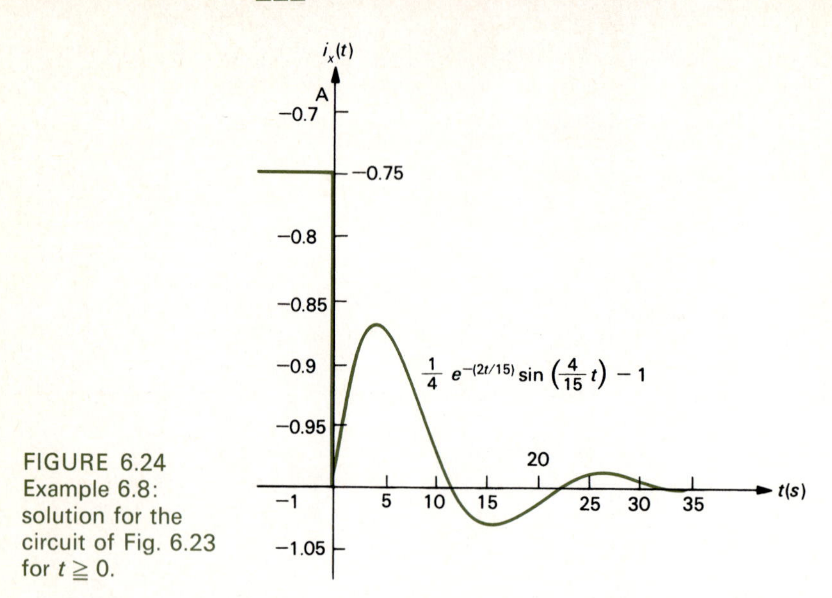

FIGURE 6.24 Example 6.8: solution for the circuit of Fig. 6.23 for $t \geqq 0$.

Therefore,

$$
\begin{aligned}
i_C(0^+) &= i_x(0^+) - i_L(0^+) \\
&= -1 + \tfrac{3}{4} \\
&= -\tfrac{1}{4}\,\text{A}
\end{aligned}
$$

and

$$
\begin{aligned}
\left.\frac{dv_C}{dt}\right|_{t=0^+} &= \tfrac{4}{5}(-\tfrac{1}{4}) \\
&= -\tfrac{1}{5}
\end{aligned}
$$

Therefore,

$$
\begin{aligned}
C_2 &= \tfrac{15}{4} \left.\frac{dv_C}{dt}\right|_{t=0^+} \\
&= -\tfrac{3}{4}
\end{aligned}
$$

and the solution for $v_C(t)$ is

$$v_C(t) = -\tfrac{3}{4}e^{-\frac{2}{15}t} \sin \tfrac{4}{15}t$$

Now we need to obtain the current $i_x(t)$. In the $t > 0$ circuit in Fig. 6.23*c* we obtain, by applying KVL around the outside loop,

$$\begin{aligned} i_x(t) &= \frac{-3\text{ V} - v_C(t)}{3\ \Omega} \\ &= -1 - \tfrac{1}{3}v_C(t) \\ &= -1 + \tfrac{1}{4}e^{-\frac{2}{15}t}\sin\tfrac{4}{15}t \end{aligned}$$

Note that as $t \to \infty$, $i_x \to -1$. Thus, $i_x = -1$ must be its steady-state value. This is seen to be true from Fig. 6.23*d*. The solution is plotted in Fig. 6.24.

6.6 TRANSFER FUNCTIONS

One of the difficult parts of the analysis in the previous sections is determining the differential equation relating the desired element voltage or current to the ideal sources. Once this is obtained, however, we only need to solve that differential equation for the transient and steady-state solutions, add these two parts of the solution, and apply the initial conditions to this total solution to evaluate the undetermined constants in the transient portion of the total solution.

In the previous sections, we avoided finding the required differential equation for every new circuit by putting the remainder of the circuit (containing only resistors and ideal sources) attached to the energy storage element(s) into a Thevenin or Norton equivalent form. Once this was done, we could then immediately write the form of the transient solution for the current or voltage of the energy storage element(s), attach the steady-state solution, and apply the initial conditions to this total solution. Once the solution for the current or voltage of the energy storage element(s) was obtained, we could then work back into the remainder circuit to find the solution for the element voltage or current of interest.

Although this method was somewhat cumbersome, it avoided finding the appropriate differential equation for the variable of interest. However, the energy storage elements must be in a particular form—single L, single C, series LC, or parallel LC—in order for us to apply this method. Obviously, there are many circuits which do not fit this mold. In this section we will obtain a very simple technique for writing the differential equation relating any element voltage or current to each of the ideal sources. The method only requires the resistive-circuit-analysis techniques of Chap. 3. Once the differential equation is obtained, we can obtain directly from it without using the circuit diagram not only the transient solution but also the steady-state solution. It is a very powerful and simple technique and deserves serious study.

The technique makes use of the differential operator D:

$$D \triangleq \frac{d}{dt} \tag{6.127}$$

Thus, when we multiply a function of t by D we intend this to mean that we "operate on it" as

$$Dx(t) \triangleq \frac{dx(t)}{dt} \tag{6.128}$$

Similarly, operating on $x(t)$ with a power of D means to differentiate $x(t)$ that many times:

$$D^n x(t) \triangleq \frac{d^n x(t)}{dt^n} \tag{6.129}$$

Dividing $x(t)$ by D symbolizes integration of $x(t)$:

$$\begin{aligned} \frac{1}{D}x(t) &\triangleq \int_{-\infty}^{t} x(\tau)\,d\tau \\ \frac{1}{D^n}x(t) &\triangleq \underbrace{\int \cdots \int}_{n} x(\tau)\,d\tau \end{aligned} \tag{6.130}$$

Note that

$$D^n \frac{1}{D^n} x(t) = x(t) \tag{6.131}$$

and

$$\begin{aligned} D[x(t) + y(t)] &= Dx(t) + Dy(t) \\ &= \frac{dx(t)}{dt} + \frac{dy(t)}{dt} \end{aligned} \tag{6.132}$$

These differential operators obey many of the rules of ordinary algebra and may be manipulated as though they were ordinary algebraic quantities. For example,

$$\begin{aligned} (D + a)(D + b)x(t) &= (D + b)(D + a)x(t) \\ &= [D^2 + (a + b)D + ab]x(t) \\ &= \frac{d^2x(t)}{dt} + (a + b)\frac{dx(t)}{dt} + abx(t) \end{aligned} \tag{6.133}$$

As an example, consider the circuit shown in Fig. 6.25. Suppose we wish to obtain the differential equation relating current $i(t)$ to each of the ideal voltage sources $v_{S1}(t)$ and $v_{S2}(t)$. Let us use mesh equations. We need only obtain a differential equation in

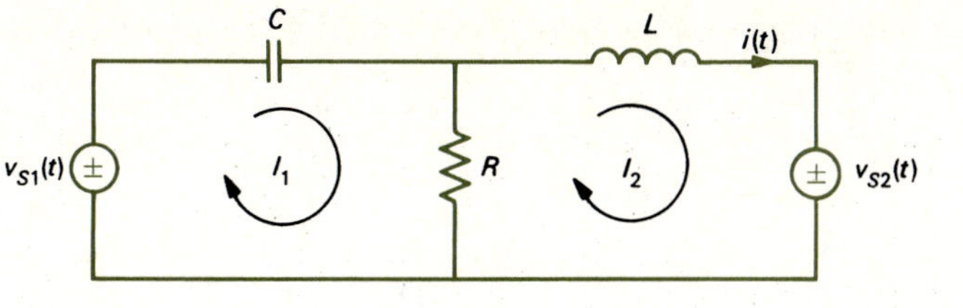

FIGURE 6.25 Illustration of writing circuit equations.

terms of mesh current I_2 since $i(t) = I_2$. The mesh equations become:

$$\text{mesh 1:} \quad \frac{1}{C}\int_{-\infty}^{t} I_1\,d\tau + RI_1 - RI_2 = v_{S1}(t)$$

$$\text{mesh 2:} \quad -RI_1 + RI_2 + L\frac{dI_2}{dt} = -v_{S2}(t)$$

Using the differential operator D, these become

$$\left(\frac{1}{CD} + R\right)I_1 - RI_2 = v_{S1}(t)$$

$$-RI_1 + (R + LD)I_2 = -v_{S2}(t)$$

Now mesh current I_1 can be eliminated from these by using any algebraic method. For example, using Cramer's rule we have

$$I_2 = \frac{\begin{vmatrix} \left(\frac{1}{CD} + R\right) & v_{S1}(t) \\ -R & -v_{S2}(t) \end{vmatrix}}{\begin{vmatrix} \left(\frac{1}{CD} + R\right) & -R \\ -R & (R + LD) \end{vmatrix}}$$

$$= \frac{Rv_{S1}(t) - (1/CD + R)v_{S2}(t)}{(1/CD + R)(R + LD) - R^2}$$

Treating D as an ordinary algebraic variable, we obtain

$$I_2 = \frac{[(1/L)D]v_{S1} - [1/RLC + (1/L)D]v_{S2}}{D^2 + (1/RC)D + 1/LC}$$

Multiplying both sides by the denominator yields

$$\left(D^2 + \frac{1}{RC}D + \frac{1}{LC}\right)I_2 = \left(\frac{1}{L}D\right)v_{S1} - \left(\frac{1}{RLC} + \frac{1}{L}D\right)v_{S2}$$

Operating on the appropriate quantities and substituting $i(t) = I_2$ gives

$$\frac{d^2i(t)}{dt^2} + \frac{1}{RC}\frac{di(t)}{dt} + \frac{1}{LC}i(t) = \frac{1}{L}\frac{dv_{S1}(t)}{dt} - \frac{1}{RLC}v_{S2}(t) - \frac{1}{L}\frac{dv_{S2}(t)}{dt}$$

and we have obtained the differential equation relating $i(t)$ to the two ideal sources. Knowing the form of $v_{S1}(t)$ and $v_{S2}(t)$, we may operate on them as indicated, giving a second-order differential equation with a known right-hand side. Thus, the operator D allows us to easily obtain the DE of interest.

There is one final simplification which we shall make in this procedure. The terminal relations of the resistor, inductor, and capacitor can be written as

$$v_R(t) = Ri_R(t) \tag{6.134a}$$

$$v_L(t) = L\frac{di_L(t)}{dt}$$

$$= LDi_L(t) \tag{6.134b}$$

$$v_C(t) = \frac{1}{C}\int_{-\infty}^{t} i_C(\tau)\,d\tau$$

$$= \frac{1}{CD}i_C(t) \tag{6.134c}$$

Note that the inductor may be considered to have a "resistance" of LD and the capacitance a "resistance" of $1/CD$. These will be referred to as *impedances*, as was done in the previous chapter, and denoted as Z:

$$Z_R = R \tag{6.135a}$$

$$Z_L = LD \tag{6.135b}$$

$$Z_C = \frac{1}{CD} \tag{6.135c}$$

so that

$$v(t) = Zi(t) \tag{6.136}$$

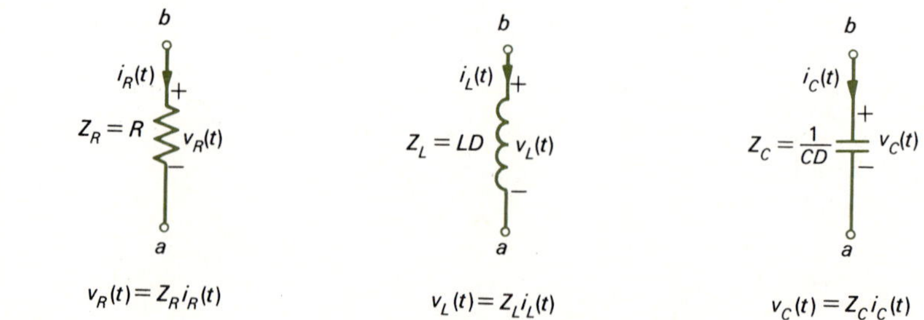

FIGURE 6.26 Circuit element impedances in terms of the differential operator D.

as shown in Fig. 6.26. Thus, the terminal equations in terms of the operator impedance for all elements are similar to Ohm's law for the resistor:

$$v_R(t) = Z_R i_R(t) \tag{6.137a}$$

$$v_L(t) = Z_L i_L(t) \tag{6.137b}$$

$$v_C(t) = Z_C i_C(t) \tag{6.137c}$$

Alternatively, we may consider the dynamic elements as having an *admittance* similar to the conductance of a resistor. This admittance is simply the reciprocal of the element impedance and is denoted by Y:

$$Y_R = \frac{1}{Z_R} = \frac{1}{R} = G \tag{6.138a}$$

$$Y_L = \frac{1}{Z_L} = \frac{1}{LD} \tag{6.138b}$$

$$Y_C = \frac{1}{Z_C} = CD \tag{6.138c}$$

and for each element $i(t) = Yv(t)$.

Now in order to write the differential equation relating any particular circuit variable to any of the ideal sources, we "pretend" that we have a resistive circuit with the elements replaced by "resistors" having the above impedances. Then we derive, *using any of our resistive circuit techniques of Chap. 3*, an equation in D relating the variable of interest and each of the ideal sources.

Example 6.9 For the circuit shown in Fig. 6.27*a*, derive differential equations relating $i_L(t)$ and $v_x(t)$ to the ideal source $v_S(t)$.

Solution The circuit labeled with impedance values is shown in Fig. 6.27*b*. The impedance seen by $v_S(t)$ is

$$Z(D) = R + \frac{1}{CD} + LD$$

$$= \frac{LCD^2 + RCD + 1}{CD}$$

Thus

$$i_L(t) = \frac{v_S(t)}{Z(D)}$$

$$= \frac{CD}{LCD^2 + RCD + 1} v_S(t)$$

FIGURE 6.27 Example 6.9: use of operator impedances to derive circuit differential equations.

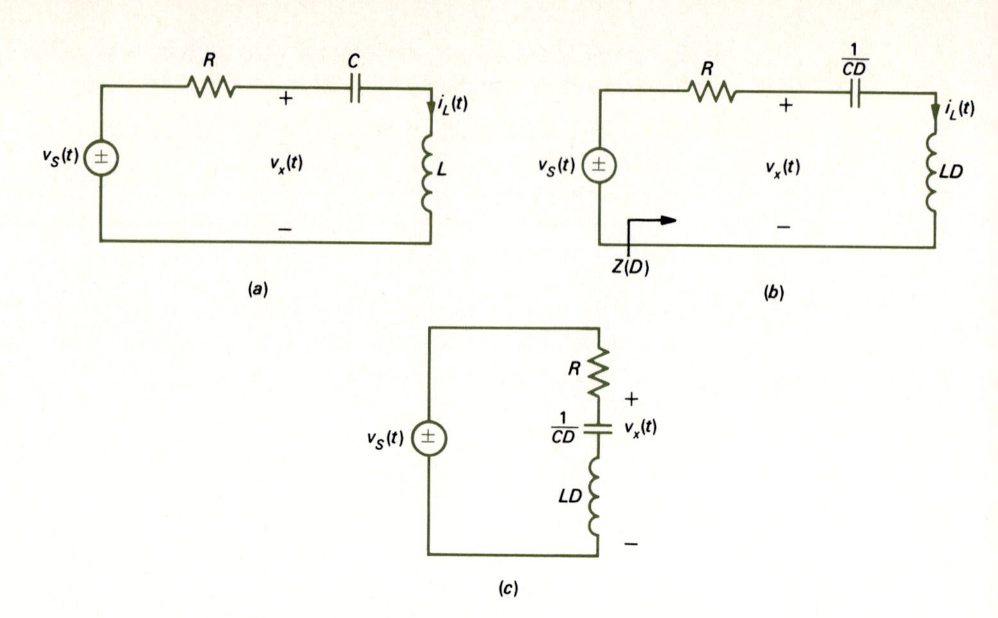

Multiplying both sides by the denominator $LCD^2 + RCD + 1$ yields

$$(LCD^2 + RCD + 1)i_L(t) = (CD)v_S(t)$$

Dividing both sides by the coefficient of D^2, LC, yields

$$\left(D^2 + \frac{R}{L}D + \frac{1}{LC}\right)i_L(t) = \left(\frac{1}{L}D\right)v_S(t)$$

Operating on the variables yields

$$\frac{d^2i_L(t)}{dt^2} + \frac{R}{L}\frac{di_L(t)}{dt} + \frac{1}{LC}i_L(t) = \frac{1}{L}\frac{dv_S(t)}{dt}$$

Compare this result to Eq. (6.68), which was derived from a similar circuit. This should serve to illustrate the power and simplification afforded by this operator technique.

In order to derive the differential equation for $v_x(t)$, we redraw the circuit as shown in Fig. 6.27*c*. Using voltage division, we obtain

$$\begin{aligned} v_x(t) &= \frac{1/CD + LD}{R + 1/CD + LD}v_S(t) \\ &= \frac{LCD^2 + 1}{LCD^2 + RCD + 1}v_S(t) \end{aligned}$$

Thus, the differential equation becomes

$$\left(D^2 + \frac{R}{L}D + \frac{1}{LC}\right)v_x(t) = \left(D^2 + \frac{1}{LC}\right)v_S(t)$$

or

$$\frac{d^2v_x(t)}{dt^2} + \frac{R}{L}\frac{dv_x(t)}{dt} + \frac{1}{LC}v_x(t) = \frac{d^2v_S(t)}{dt^2} + \frac{1}{LC}v_S(t)$$

Any of the resistive-circuit-analysis techniques of the previous chapter may be used with this technique. Consequently, this operator technique builds on our previous work.

Example 6.10 Derive a differential equation relating $i(t)$ to the voltage source $v_S(t)$ in the circuit of Fig. 6.28*a*.

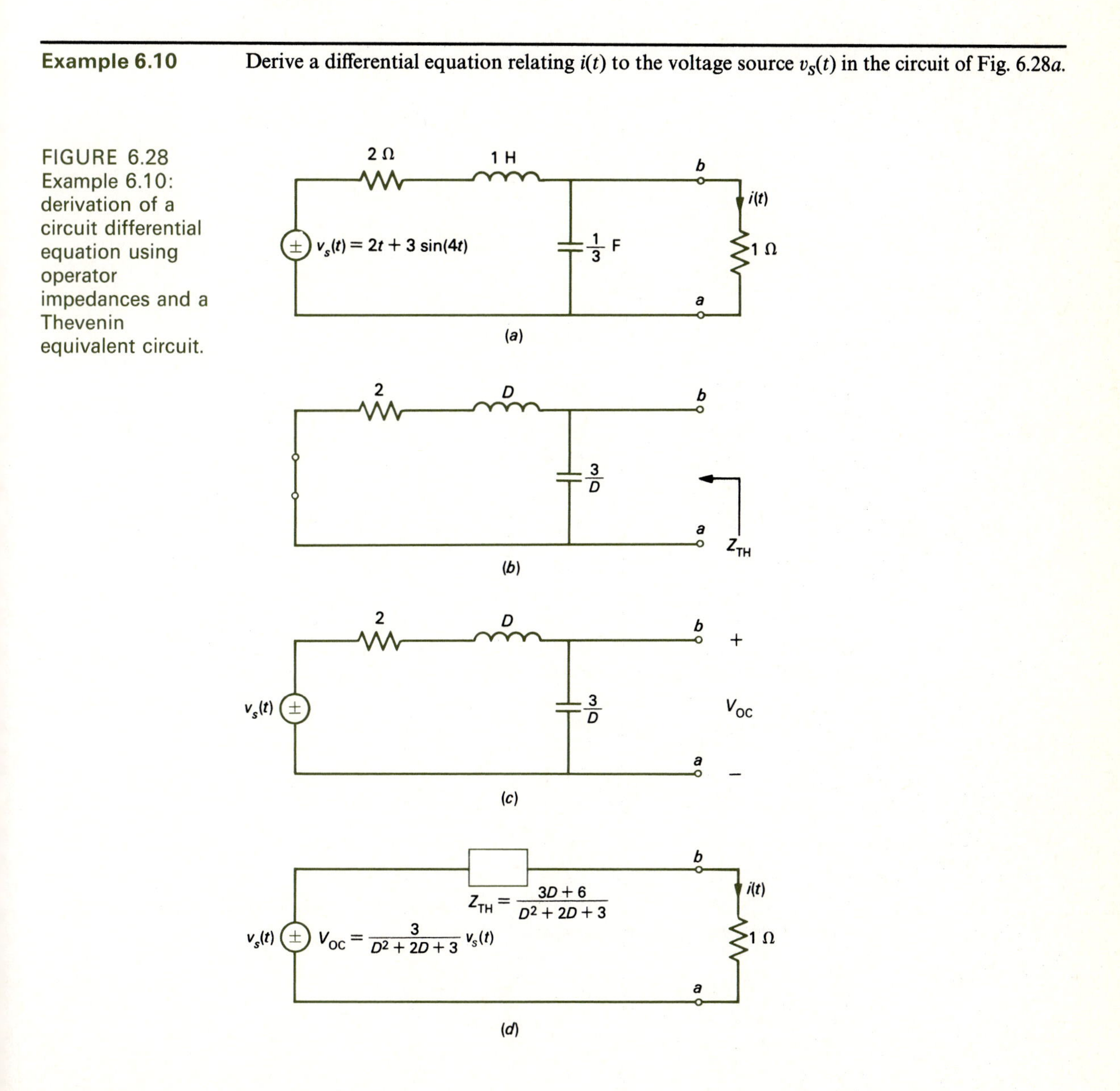

FIGURE 6.28 Example 6.10: derivation of a circuit differential equation using operator impedances and a Thevenin equivalent circuit.

Solution We will solve this problem using a Thevenin equivalent at terminals *a-b*. The Thevenin impedance Z_{TH} is computed from the dead circuit, as shown in Fig. 6.28*b*, as

$$\begin{aligned} Z_{TH} &= \left(\frac{3}{D}\right) \| (2 + D) \\ &= \frac{(3/D)(2 + D)}{3/D + 2 + D} \\ &= \frac{3D + 6}{D^2 + 2D + 3} \end{aligned}$$

The Thevenin open-circuit voltage is determined from Fig. 6.28*c* (by voltage division) as

$$\begin{aligned} V_{OC} &= \frac{3/D}{3/D + 2 + D} v_S \\ &= \frac{3v_S}{D^2 + 2D + 3} \end{aligned}$$

From Fig. 6.28*d* we obtain

$$\begin{aligned} i(t) &= \frac{V_{OC}}{Z_{TH} + 1} \\ &= \frac{[3/(D^2 + 2D + 3)]v_S}{(3D + 6)/(D^2 + 2D + 3) + 1} \\ &= \frac{3}{D^2 + 5D + 9} v_S \end{aligned}$$

or

$$(D^2 + 5D + 9)i(t) = 3v_S$$

and the differential equation becomes

$$\begin{aligned} \frac{d^2i(t)}{dt^2} + 5\frac{di(t)}{dt} + 9i(t) &= 3v_S(t) \\ &= 6t + 9 \sin 4t \end{aligned}$$

This use of the operator D and the element "impedances" simplifies the derivation of the desired differential equation. In each case, the desired circuit variable $x(t)$ is related to the values of the N ideal sources as

$$x(t) = \frac{F_1(D)f_1(t) + F_2(D)f_2(t) + \cdots + F_N(D)f_N(t)}{G(D)} \tag{6.139}$$

where $f_1(t), f_2(t), \ldots$ are the values of the ideal sources and $F_1(D), F_2(D), \ldots$ and $G(D)$ are polynomials in the operator D; for example, $G(D) = 2D^2 + 3D + 5$. (See Examples 6.9 and 6.10.) Multiplying both sides of Eq. (6.139) by $G(D)$, we obtain

$$G(D)x(t) = F_1(D)f_1(t) + F_2(D)f_2(t) + \cdots \tag{6.140}$$

Operating on the (known) $f_1(t), f_2(t), \ldots$ functions with the associated $F(D)$ polynomial, the right-hand side of Eq. (6.140) becomes a known function. Operating on $x(t)$ with $G(D)$, we obtain the various derivatives of $x(t)$ and the resulting differential equation for $x(t)$ is found.

In this technique, we do not need to memorize any of the forms of the terms in the differential equation in terms of R_{TH}, L, or C as we did for the special cases considered previously. We don't even need to find R_{TH} or determine, for example, whether we have a series LC problem or a parallel LC problem (if that is possible). The results are automatic. We only need to know the resistive-circuit techniques of the previous chapter.

The various ratios

$$T_1(D) = \frac{F_1(D)}{G(D)}$$

$$T_2(D) = \frac{F_2(D)}{G(D)}$$

$$\vdots$$

$$T_N(D) = \frac{F_N(D)}{G(D)} \tag{6.141}$$

are ratios of polynomials in the operator D and are known as the "transfer functions" between the variable of interest $x(t)$ and each of the ideal sources $f_1(t), f_2(t), \ldots, f_N(t)$. If we find, by the simple resistive-circuit techniques of Chap. 3, the transfer function

$$T_i(D) = \frac{F_i(D)}{G(D)} \tag{6.142}$$

we can immediately write the differential equation relating the variable $x(t)$ and the source $f_i(t)$. Thus, the transfer functions for a circuit contain a considerable amount of information about the circuit, and we will have numerous occasions to study not only circuits but other systems (mechanical, chemical, etc.) by similarly obtaining the transfer functions of the system. This concept will be used in later sections of this book.

The steady-state solution can easily be found from the transfer function without any further reference to the circuit. For example, suppose the transfer function relating the variable of interest $x(t)$ to an ideal source $f(t)$ is

$$x(t) = \frac{F(D)}{G(D)} f(t) \tag{6.143}$$

If $f(t)$ is dc, $f(t) = K$, simply replace D with zero since the solution will be constant also and the derivative of a constant is zero:

$$x_{ss}(t) = \frac{F(0)}{G(0)} K \tag{6.144}$$

If $f(t)$ is an ac source such as

$$f(t) = A \sin(\omega t + \phi) \tag{6.145a}$$

or

$$f(t) = A \cos(\omega t + \phi) \tag{6.145b}$$

replace it with its exponential, phasor equivalent:

$$f(t) = A\underline{/\phi} e^{j\omega t} \tag{6.146}$$

The steady-state solution will also be of this form:

$$x_{ss}(t) = X\underline{/\theta} e^{j\omega t} \tag{6.147}$$

Since $De^{j\omega t} = j\omega e^{j\omega t}$, we replace D with $j\omega$ in the transfer function and find the phasor solution:

$$X\underline{/\theta} = \frac{F(j\omega)}{G(j\omega)} A\underline{/\theta} \tag{6.148}$$

We then convert back to the time domain. If $f(t)$ is a sinusoid as in Eq. (6.145a), we have

$$\begin{aligned} x(t) &= \text{Im } X\underline{/\theta} e^{j\omega t} \\ &= X \sin(\omega t + \theta) \end{aligned} \tag{6.149}$$

If $f(t)$ is a cosine as in Eq. (6.145b), we have

$$\begin{aligned} x(t) &= \text{Re } X\underline{/\theta} e^{j\omega t} \\ &= X \cos(\omega t + \theta) \end{aligned} \tag{6.150}$$

Example 6.11 For the circuit in Fig. 6.29a, find the steady-state solution for $i(t)$.

Solution We first derive the transfer function relating $i(t)$ to the two ideal sources. Replacing the elements with their operator equivalents, we obtain Fig. 6.29b. Now $i(t)$ can be found by superposition:

$$i(t) = i'(t) + i''(t)$$

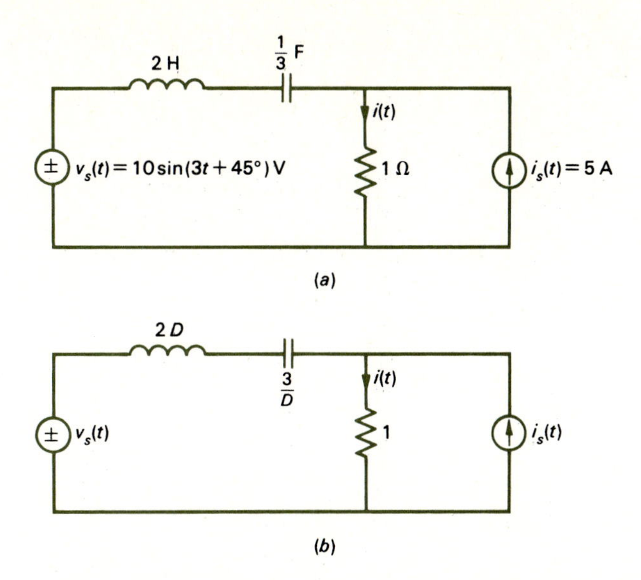

FIGURE 6.29 Example 6.11: obtaining the steady-state solution using operator impedances.

where $i'(t)$ is due to the dc current source and $i''(t)$ is due to the sinusoidal source. In terms of transfer functions, we obtain

$$i'(t) = T'(D)i_S$$

where

$$T'(D) = \frac{D^2 + \frac{3}{2}}{D^2 + \frac{1}{2}D + \frac{3}{2}}$$

and

$$i''(t) = T''(D)v_S(t)$$

where

$$T''(D) = \frac{\frac{1}{2}D}{D^2 + \frac{1}{2}D + \frac{3}{2}}$$

The steady-state solution due to the dc current source is found by substituting $D = 0$ in $T'(D)$:

$$T'(0) = 1$$

Thus

$$\begin{aligned} i_{ss}' &= T'(0)i_S \\ &= 5 \text{ A} \end{aligned}$$

This can easily be seen from the circuit by killing $v_S(t)$ and replacing the inductor with a short circuit and the capacitor with an open circuit (since i_S is dc).

To find the steady-state solution for $v_S(t)$, we replace D with $j\omega = j3$ in $T''(D)$:

$$T''(j3) = \frac{\frac{1}{2}(j3)}{(j3)^2 + \frac{1}{2}(j3) + \frac{3}{2}}$$

$$= \frac{j\frac{3}{2}}{-9 + j\frac{3}{2} + \frac{3}{2}}$$

$$= 0.196\underline{/-78.69^\circ}$$

Thus, the phasor solution for $i''(t)$ is

$$I'' = T''(j3)10\underline{/45^\circ}$$

$$= 1.96\underline{/-33.69^\circ}$$

In the time domain [since $v_S(t)$ was a sine function]

$$i''_{ss}(t) = \text{Im } I''e^{j3t}$$

$$= 1.96 \sin(3t - 33.69^\circ)$$

Thus, the complete steady-state solution is

$$i_{ss}(t) = i'_{ss}(t) + i''_{ss}(t)$$

$$= 5 + 1.96 \sin(3t - 33.69^\circ)$$

The complete solution (including the transient solution) can also be found from the transfer function. For example, in the previous example

$$i(t) = T'(D)i_s + T''(D)v_s(t)$$

$$= \frac{D^2 + \frac{3}{2}}{D^2 + \frac{1}{2}D + \frac{3}{2}} i_S + \frac{\frac{1}{2}D}{D^2 + \frac{1}{2}D + \frac{3}{2}} v_S(t)$$

Multiplying both sides by $D^2 + \frac{1}{2}D + \frac{3}{2}$ gives

$$(D^2 + \tfrac{1}{2}D + \tfrac{3}{2})i(t) = (D^2 + \tfrac{3}{2})i_S + (\tfrac{1}{2}D)v_S(t)$$

or

$$\frac{d^2i(t)}{dt^2} + \frac{1}{2}\frac{di(t)}{dt} + \frac{3}{2}i(t) = \frac{d^2i_S}{dt^2} + \frac{3}{2}i_S + \frac{1}{2}\frac{dv_S(t)}{dt}$$

and we have obtained the differential equation relating $i(t)$ to each of the two sources. Now, assuming that the circuit in Fig. 6.29a is valid for $t > 0$, the transient solution is

$$i_T(t) = e^{-\frac{1}{4}t}\left(A_1 \cos\frac{\sqrt{23}}{4}t + A_2 \sin\frac{\sqrt{23}}{4}t\right)$$

since the characteristic equation is

$$p^2 + \tfrac{1}{2}p + \tfrac{3}{2} = 0$$

with roots

$$p_1, p_2 = -\tfrac{1}{4} \pm j\frac{\sqrt{23}}{4}$$

Adding the steady-state solution, we have the complete solution:

$$\begin{aligned} i(t) &= i_T(t) + i_{ss}(t) \\ &= e^{-\frac{1}{4}t}\left(A_1 \cos\frac{\sqrt{23}}{4}t + A_2 \sin\frac{\sqrt{23}}{4}t\right) \\ &\quad + 5 + 1.96 \sin(3t - 33.69^\circ) \end{aligned}$$

In order to evaluate the two undetermined constants A_1 and A_2, we need two initial conditions $i(0^+)$ and $di(t)/dt|_{t=0^+}$ to be applied to this complete solution. These can be obtained from the $t = 0^+$ circuit by using continuity of inductor currents $i_L(0^+) = i_L(0^-)$ and capacitor voltages $v_C(0^+) = v_C(0^-)$ in the usual fashion.

Example 6.12 For the circuit shown in Fig. 6.30*a*, determine $i_x(t)$.

Solution First we determine the initial conditions on the capacitor voltage and inductor current from the $t = 0^-$ circuit shown in Fig. 6.30*b*. Note that in the $t < 0$ circuit there is only one source (the 5-A current source) which is dc. Thus, to find the steady-state values of v_C and i_L at $t = 0^-$, we replace the capacitor with an open circuit and the inductor with a short circuit. We obtain

$$\begin{aligned} v_C(0^-) &= v_C(0^+) \\ &= 0\ \text{V} \\ i_L(0^-) &= i_L(0^+) \\ &= 5\ \text{A} \end{aligned}$$

Next we draw the $t > 0$ circuit, replacing the inductor and capacitor with their operator impedances (LD and $1/CD$) as shown in Fig. 6.30*c*. From this we obtain (for example, by superposition)

$$i_x = \frac{1}{3D^2 + 5D + 4}10\cos(3t) - \frac{3D(D+1)}{3D^2 + 5D + 4}5$$

From this we obtain the steady-state value of i_x as

$$\begin{aligned} i_{x,ss} &= \text{Re}\left[\frac{1}{3(j3)^2 + 5(j3) + 4}10e^{j3t}\right] - \frac{3(0)(0+1)}{3(0)^2 + 5(0) + 4}5 \\ &= 0.36\cos(3t - 146.89^\circ) \end{aligned}$$

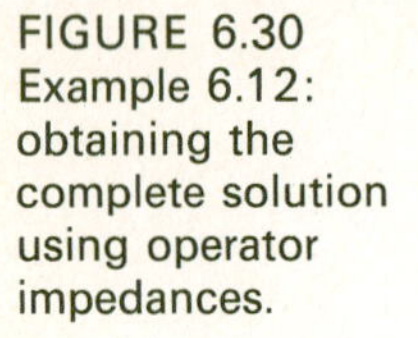

FIGURE 6.30 Example 6.12: obtaining the complete solution using operator impedances.

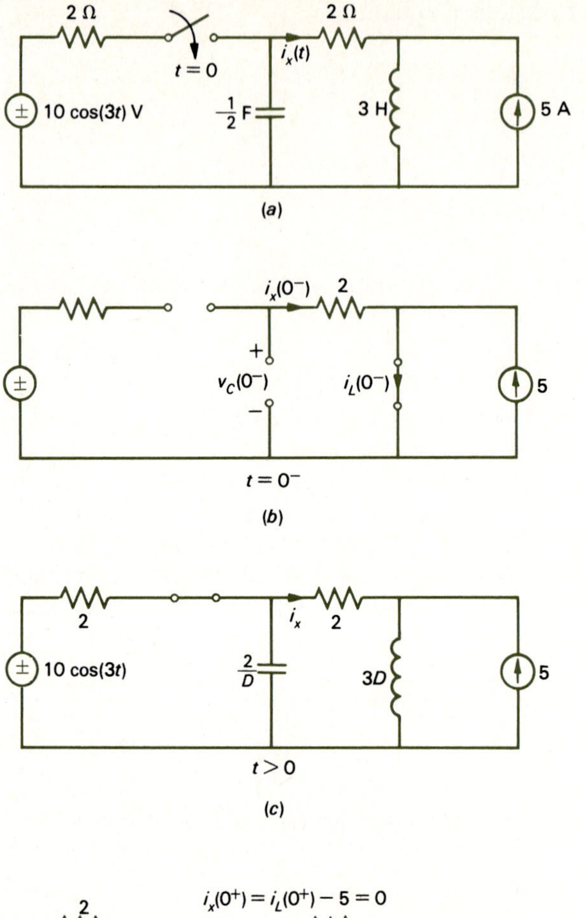

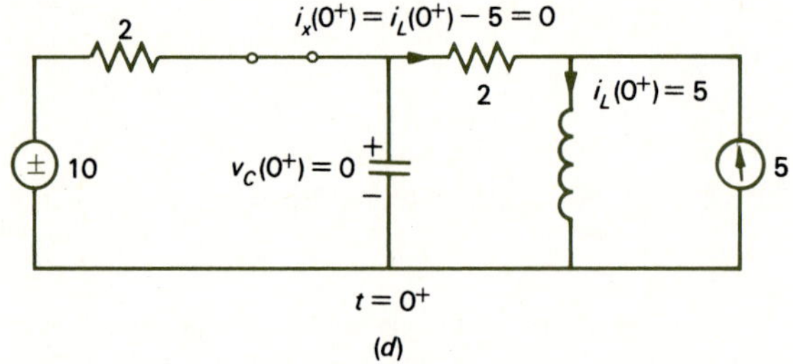

Note that the steady-state value of i_x due to the 5-A dc current source is zero. This can easily be seen by replacing the inductor with a short circuit and a capacitor with an open circuit in the $t > 0$ circuit.

Now let us determine the transient solution. From the above work we see that the differential equation relating $i_x(t)$ to the two sources is

$$(3D^2 + 5D + 4)i_x(t) = 10 \cos(3t) - 3D(D + 1)5$$

$$= 10 \cos(3t)$$

or

$$\frac{d^2 i_x(t)}{dt^2} + \frac{5}{3}\frac{di_x(t)}{dt} + \frac{4}{3}i_x(t) = \frac{10}{3}\cos 3t$$

Thus, the characteristic equation is

$$p^2 + \frac{5}{3}p + \frac{4}{3} = 0$$

with roots

$$p_1, p_2 = -\frac{5}{6} \pm j\frac{\sqrt{23}}{6}$$

Thus, the transient solution is of the form

$$i_{x,T}(t) = e^{-\frac{5}{6}t}\left[C_1 \cos\frac{\sqrt{23}}{6}t + C_2 \sin\frac{\sqrt{23}}{6}t\right]$$

and the total solution is

$$\begin{aligned} i_x(t) &= i_{x,T}(t) + i_{x,ss}(t) \\ &= e^{-\frac{5}{6}t}\left[C_1 \cos\frac{\sqrt{23}}{6}t + C_2 \sin\frac{\sqrt{23}}{6}t\right] + 0.36\cos(3t - 146.89°) \end{aligned}$$

We now need to obtain the initial conditions on $i_x(t)$:

$$i_x(0^+)$$

$$\left.\frac{di_x}{dt}\right|_{t=0^+}$$

From the $t = 0^+$ circuit shown in Fig. 6.30*d* we observe that

$$\begin{aligned} i_x(0^+) &= i_L(0^+) - 5 \\ &= 0 \end{aligned}$$

Similarly,

$$\begin{aligned} \left.\frac{di_x}{dt}\right|_{t=0^+} &= \left.\frac{di_L}{dt}\right|_{t=0^+} - \frac{d}{dt}(5) \\ &= \left.\frac{di_L}{dt}\right|_{t=0^+} \\ &= \frac{1}{L}v_L(0^+) \end{aligned}$$

Thus, we need to find $v_L(0^+)$. Writing KVL around a loop involving v_L, v_x, v_C, we have

$$\begin{aligned} v_L(0^+) &= v_C(0^+) - v_x(0^+) \\ &= v_C(0^+) - 2i_x(0^+) \\ &= 0 \end{aligned}$$

Thus

$$\left.\frac{di_x}{dt}\right|_{t=0^+} = 0$$

Applying these initial conditions to the total solution yields

$$\begin{aligned} i_x(0^+) &= 0 \\ &= C_1 + 0.36 \cos(-146.89°) \end{aligned}$$

Thus

$$\begin{aligned} C_1 &= -0.36 \cos(-146.89°) \\ &= 0.30 \end{aligned}$$

Next

$$\begin{aligned} \frac{di_x}{dt} &= -\tfrac{5}{6}e^{-\frac{5}{6}t}\left[C_1 \cos \frac{\sqrt{23}}{6} t + C_2 \sin \frac{\sqrt{23}}{6} t\right] \\ &\quad + e^{-\frac{5}{6}t}\left[-\frac{\sqrt{23}}{6} C_1 \sin \frac{\sqrt{23}}{6} t + \frac{\sqrt{23}}{6} C_2 \cos \frac{\sqrt{23}}{6} t\right] \\ &\quad - 1.08 \sin(3t - 146.89°) \end{aligned}$$

Evaluating this at $t = 0^+$ gives

$$\begin{aligned} \left.\frac{di_x}{dt}\right|_{t=0^+} &= 0 \\ &= -\tfrac{5}{6}(C_1) + \frac{\sqrt{23}}{6} C_2 - 1.08 \sin(-146.89°) \end{aligned}$$

or

$$C_2 = -.43$$

Thus

$$i_x(t) = \underbrace{e^{-\frac{5}{6}t}\left[0.3 \cos\left(\frac{\sqrt{23}}{6} t\right) - .43 \sin \frac{\sqrt{23}}{6} t\right]}_{\text{transient}} + \underbrace{0.36 \cos(3t - 146.89°)}_{\text{steady state}}$$

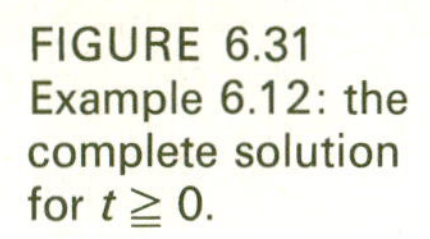

FIGURE 6.31
Example 6.12: the complete solution for $t \geqq 0$.

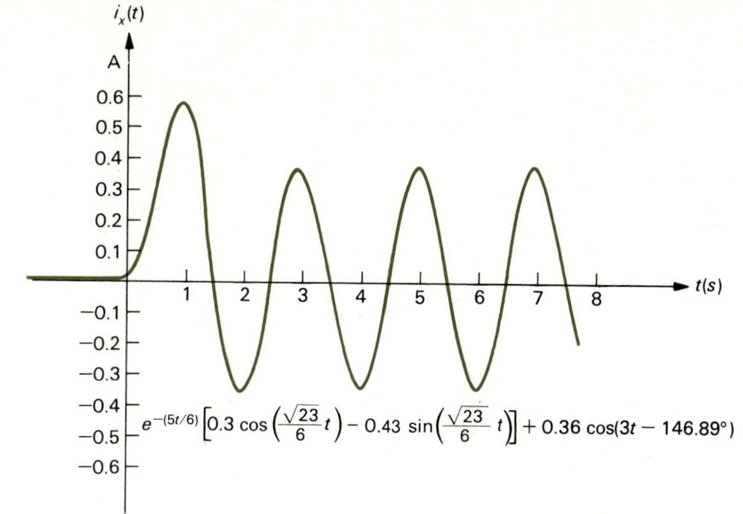

This solution is plotted in Fig. 6.31. Notice that there is evident a transient period when the current oscillation achieves a maximum of 0.58 A above that of the final steady-state peak of 0.36 A.

PROBLEMS

6.1 In the circuit of Fig. P6.1, sketch the current $i(t)$ for $0 \leq t \leq 10$ ms.

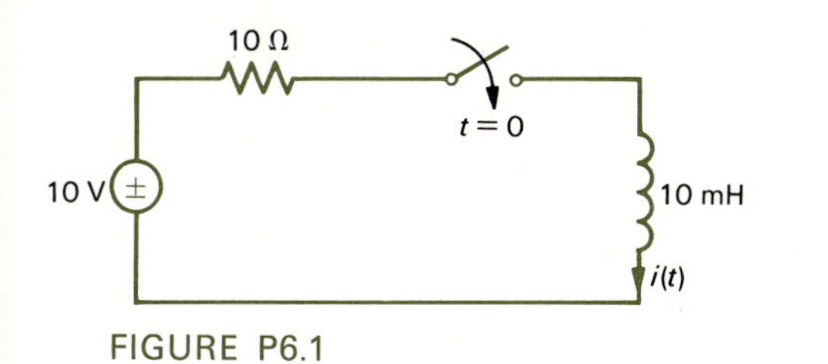

FIGURE P6.1

6.2 In the circuit shown in Fig. P6.2, sketch $i(t)$ for $0 \leq t \leq 100$ ms.

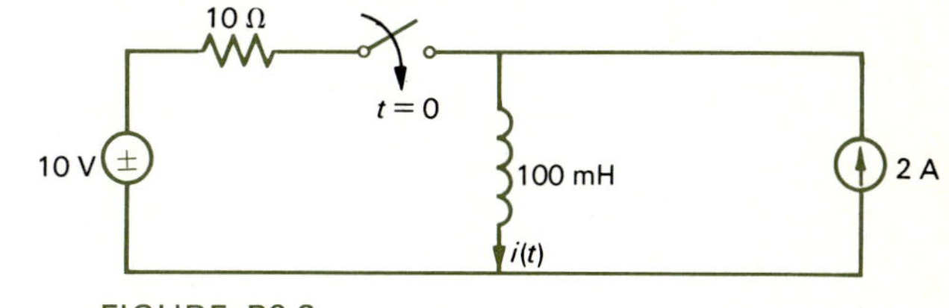

FIGURE P6.2

(Continued)

6.3 Sketch the voltage $v(t)$ in the circuit of Fig. P6.3 for $0 \leq t \leq 30\ \mu s$.

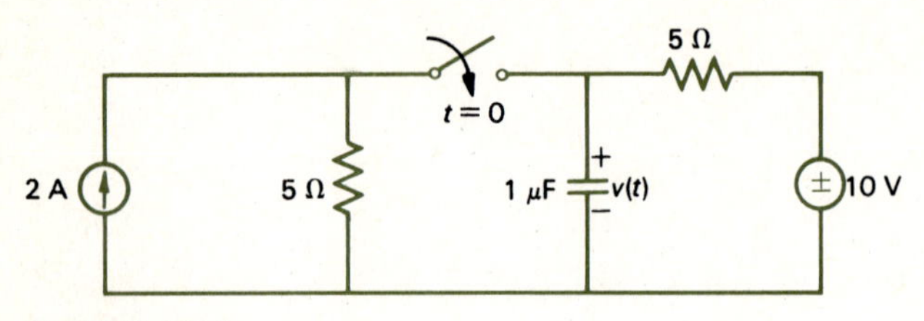

FIGURE P6.3

6.4 Change the value of the voltage source in Fig. P6.3 from 10 to 20 V and repeat Prob. 6.3.

6.5 For the circuit shown in Fig. P6.5, sketch $v(t)$ for $0 \leq t \leq 5$ ms.

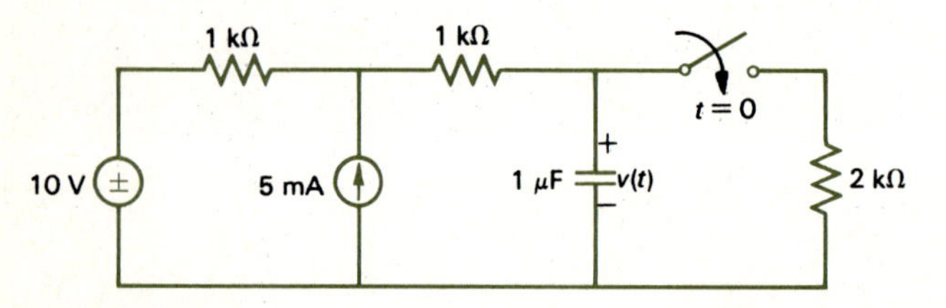

FIGURE P6.5

6.6 In the circuit of Fig. P6.6, a switch which is initially closed at $t = 0$ opens and remains open for $0 < t < 10$ ms. At $t = 10$ ms it closes once again and remains closed for $10\text{ ms} < t < 20$ ms. At $t = 20$ ms it opens once again and closes at $t = 30$ ms. Sketch $v(t)$ for $0 \leq t < 30$ ms.

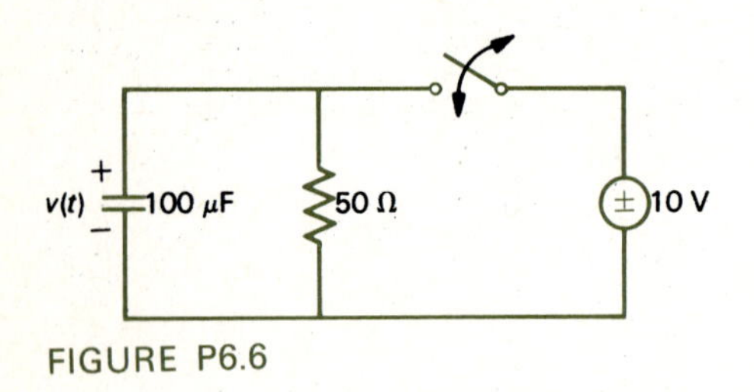

FIGURE P6.6

6.7 In the circuit of Fig. P6.7, sketch $v(t)$ for $0 \le t \le 6$ s.

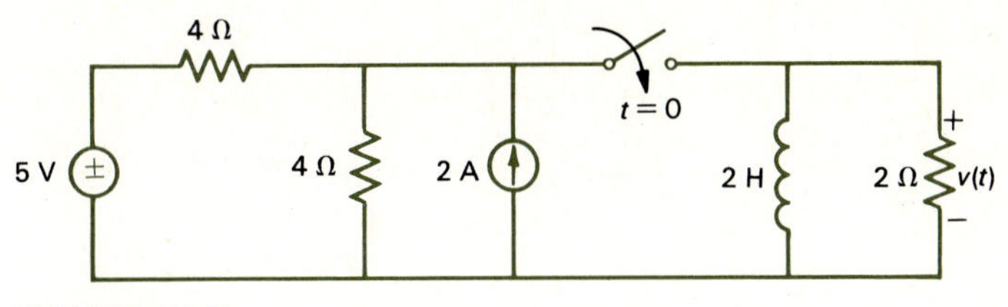

FIGURE P6.7

6.8 For the circuit shown in Fig. P6.8, sketch $i(t)$ for $0 \le t \le 5$ ms.

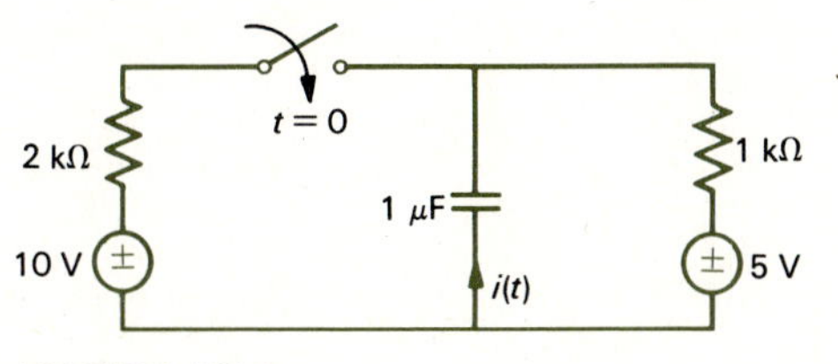

FIGURE P6.8

6.9 A 1-kΩ resistor in series with a 1-μF capacitor is plugged into a residential power outlet. Assume the outlet to be represented by an ideal voltage source having a voltage of $120 \sin (2\pi \times 60t)$ V. Determine an expression for the current through the resistor.

6.10 For the circuit shown in Fig. P6.10, determine $v(0^+)$, $i(0^+)$, and an expression for $i(t)$ for $t \ge 0$.

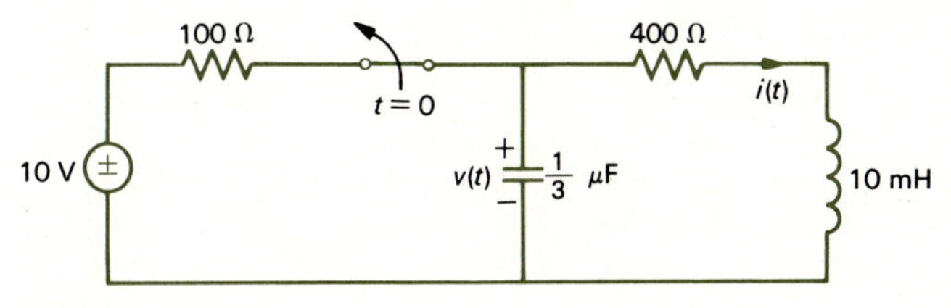

FIGURE P6.10

6.11 In the circuit of Fig. P6.10, change the capacitance value to $\frac{1}{4}$ μF and repeat Prob. 6.10.

6.12 In the circuit of Fig. P6.10, change the capacitance value to $\frac{1}{5}$ μF and repeat Prob. 6.10.

6.13 For the circuit shown in Fig. P6.13, sketch $i(t)$ for $0 \le t \le 5$ s.

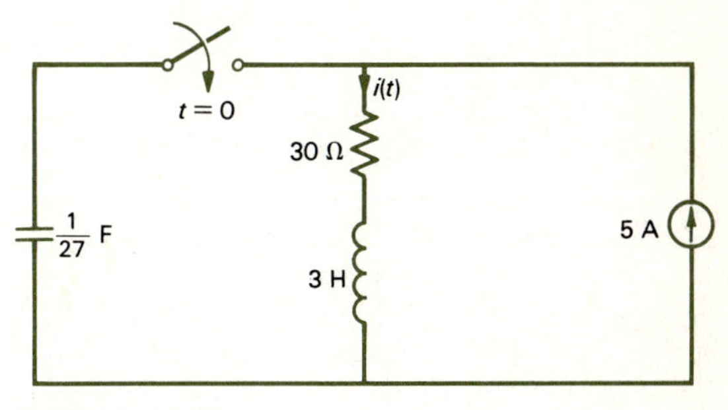

FIGURE P6.13

6.14 In the circuit shown in Fig. P6.14, determine $i(t)$ for $t > 0$.

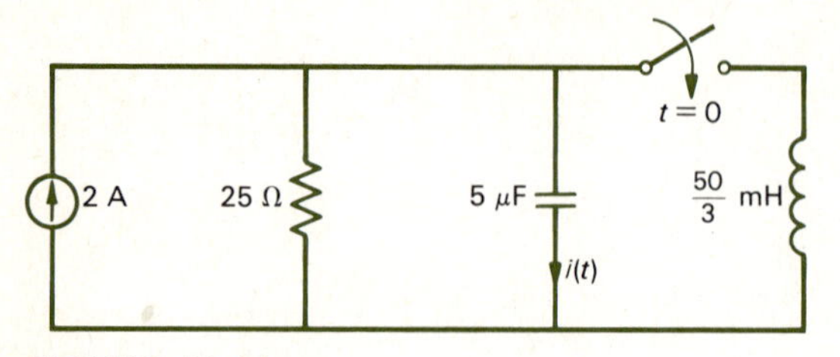

FIGURE P6.14

6.15 For the circuit shown in Fig. P6.2, determine the differential equation relating $i(t)$ and the two sources for $t > 0$. Determine the steady-state solution from this equation.

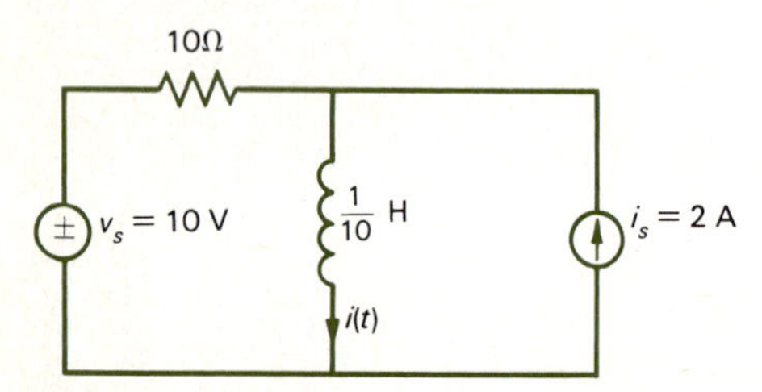

FIGURE P6.15

6.16 For the circuit shown in Fig. P6.3, determine the differential equation relating $v(t)$ to the two sources for $t > 0$. From this equation determine the steady-state solution for $v(t)$.

6.17 For the circuit shown in Fig. P6.5, determine the differential equation relating $v(t)$ to the two sources for $t > 0$. From that equation determine the steady-state solution and time constant for $v(t)$.

6.18 In the circuit of Fig. P6.7, write the differential equation relating $v(t)$ and the two sources for $t > 0$. From that equation determine the time constant and steady-state solution for $v(t)$.

6.19 For the circuit shown in Fig. P6.8, determine the differential equation relating $i(t)$ and the two sources for $t > 0$. From that equation write the form of the solution for $i(t)$ for $t > 0$.

6.20 For the circuit shown in Fig. P6.13, derive the differential equation relating $i(t)$ to the source for $t > 0$. From this equation write the form of $i(t)$ for $t > 0$.

6.21 For the circuit shown in Fig. P6.21, write the differential equation relating $i(t)$ to the two sources for $t > 0$. From that equation write the form of $i(t)$ for $t > 0$.

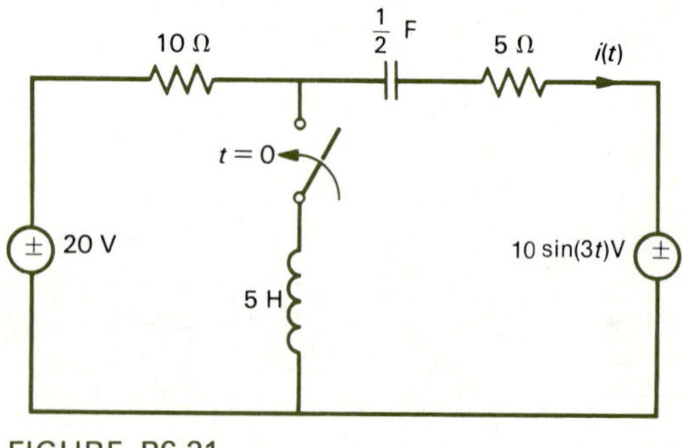

FIGURE P6.21

REFERENCES

6.1 A. B. Carlson and D. G. Gisser, *Electrical Engineering: Concepts and Applications*, Addison-Wesley, Reading, Mass., 1981.

6.2 Ralph J. Smith, *Circuits, Devices and Systems: A First Course in Electrical Engineering*, 4th ed., Wiley, New York, 1984.

6.3 A. E. Fitzgerald, D. Higginbotham, and A. Grabel, *Basic Electrical Engineering*, 5th ed., McGraw-Hill, New York, 1981.

6.4 S. E. Schwartz and W. G. Oldham, *Electrical Engineering: An Introduction*, Holt, Rinehart & Winston, New York, 1984.

6.5 R. L. Boylestad, *Introductory Circuit Analysis*, 4th ed., Charles E. Merrill, Columbus, Ohio, 1982.

6.6 W. W. Hayt, Jr., and J. E. Kemmerly, *Engineering Circuit Analysis*, 3d ed., McGraw-Hill, New York, 1978.

6.7 J. A. Edminister, *Schaum's Outline of Electric Circuits*, 2d ed., McGraw-Hill, New York, 1983.

PART
2
Electronic
Circuits

Chapter

7

Diodes

In Part 1 we have considered circuits composed of linear two-terminal elements: the resistor, the inductor, and the capacitor. In Part 2 we will consider circuits which may contain—in addition to linear resistors, inductors, and capacitors—certain nonlinear resistors. The nonlinear resistors which we will consider are commonly referred to as electronic elements, and these nonlinear circuits are referred to as electronic circuits. Of course, the introduction of these nonlinear resistors into an otherwise linear circuit means that the resulting circuit becomes a nonlinear one. Consequently, the powerful circuit-analysis techniques for linear circuits such as superposition no longer hold, and the analysis of these nonlinear circuits becomes considerably more difficult than the analysis of linear circuits.

We will study two such nonlinear resistors in this chapter—the semiconductor diode and the zener diode. We shall also study a three-terminal device commonly used in power supplies—the silicon-controlled rectifier (SCR). The primary use of these devices is in the construction of dc power supplies. A dc power supply converts an ac, or sinusoidal, waveform (typically the 60-Hz commercial signal) to a constant, or dc, voltage. We will find in later chapters of Part 2 that amplifiers which increase the magnitude of some information-bearing signal require a dc voltage source for proper operation; batteries are clumsy and require frequent maintenance or replacement. Most conventional radios (except portable ones) use a power supply to provide the required dc voltage, thus eliminating the need for batteries. There are many other uses for diodes which we shall also study.

7.1 THE SEMICONDUCTOR DIODE

The most common two-terminal nonlinear resistor is the semiconductor diode, whose symbol is shown in Fig. 7.1*a*. The terminal voltage and current are denoted as v_d and i_d, respectively.

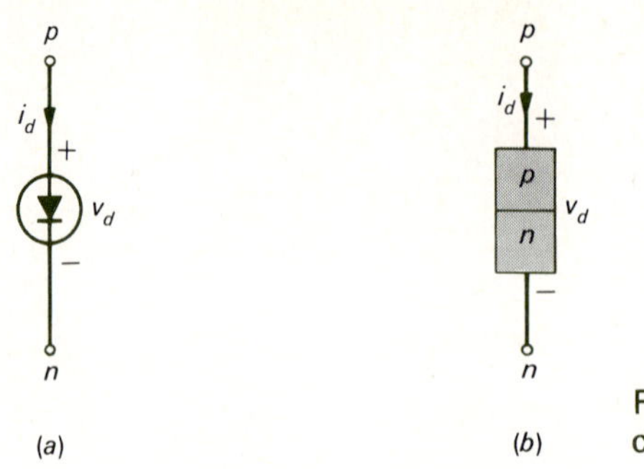

FIGURE 7.1 The semiconductor diode: (*a*) circuit symbol; (*b*) physical construction.

The semiconductor diode is constructed by "growing" two pieces of semiconductor material—a *p*-type and an *n*-type—together to form a *pn* junction, as shown in Fig. 7.1*b*. Typical types of semiconductor materials used are silicon (Si) and germanium (Ge). Silicon has recently been the predominant material, especially in integrated circuits.

Both silicon and germanium have four valence electrons in their outer atomic orbits. The atoms of pure silicon form covalent bonds, as illustrated in Fig. 7.2; each neighboring atom shares an electron in the bond. Whereas the valence electrons in a metal are free to move about the material to participate in the conduction process, the valence electrons of a pure, or intrinsic, semiconductor (silicon) are so tightly bound that no free electrons are liberated to participate in conduction at a temperature of 0 kelvin (K). At room temperature (293 K), an insignificant number of the valence electrons are liberated.

The energy band description of a material is shown in Fig. 7.3, in which the bands of the atoms' allowed energy levels are separated by gaps or forbidden levels. For a metal, shown in Fig. 7.3*a*, the partially filled upper level is referred to as the conduction, or valence, band; the addition of small amounts of energy raises the energy levels of the electrons in this band quite easily. In the filled band, an increase in the gap energy E_g is required in order to raise an electron in a filled band to the next higher unfilled band. For a semiconductor material, shown in Fig. 7.3*b*, the conduction band is empty (at 0 K) and the next lower (valence) band is filled. In order for an electron to

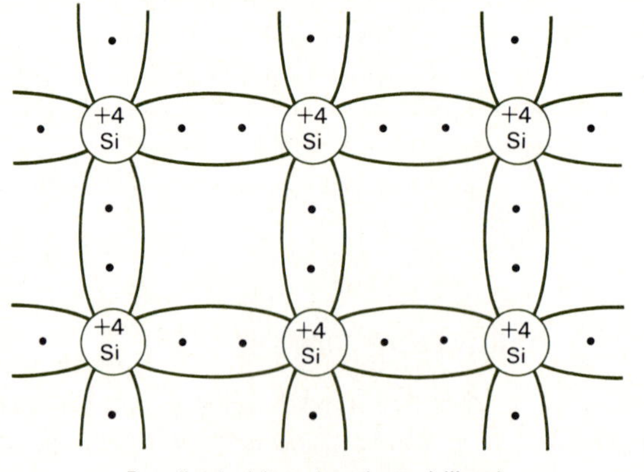

FIGURE 7.2
Silicon and covalent bonds.

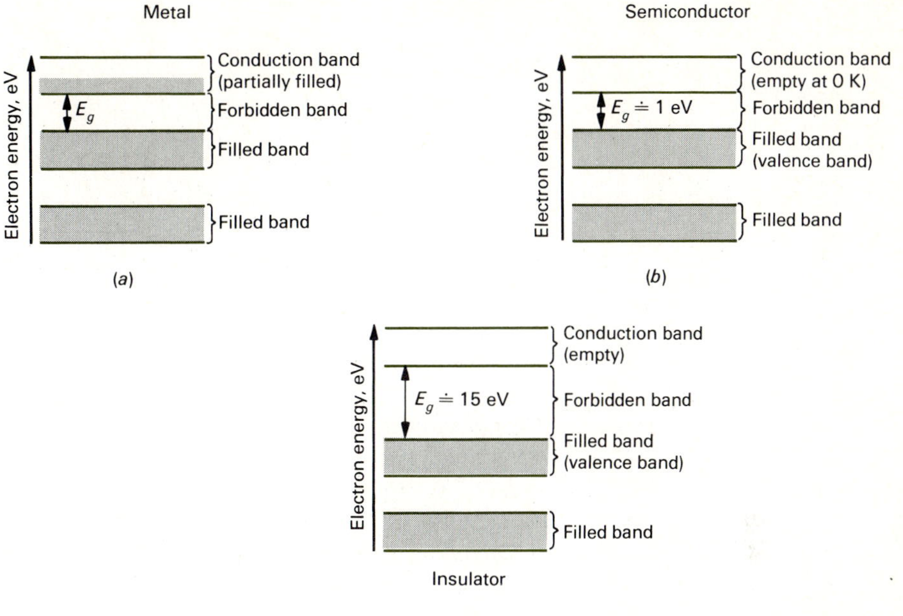

FIGURE 7.3 Energy band characterization of (*a*) metals, (*b*) semiconductors, and (*c*) insulators.

be elevated into the unfilled conduction band, an energy of approximately 1 electron-volt (eV) must be added. At room temperature, an occasional electron is liberated by thermal energy to the conduction band; ordinarily, however, there are very few such electrons and the semiconductor material behaves more like an insulator than a metal. The energy bands for an insulator are shown in Fig. 7.3*c*. To move an electron into the insulator's unfilled conduction band requires a much greater energy than for a semiconductor, on the order of 15 eV; hence, the insulator does not conduct except under the application of rather large voltages.

In order for an intrinsic semiconductor such as silicon to conduct current, certain additional sources of conduction electrons must be provided. Impurities are added to the intrinsic semiconductor for this purpose. Note that silicon and germanium are both tetravalent, which means that they have four electrons in the outer valence band; in a pure crystal, these valence electrons are shared with the other atoms in covalent bonds. Typical impurities are the pentavalent ones such as arsenic (As) and antimony (Sb), which have five valence electrons, or the trivalent ones such as gallium (Ga) and boron (B), which have only three valence electrons.

If a pentavalent impurity is added to an intrinsic semiconductor such as silicon, the atom replaces one of the semiconductor atoms and forms a covalent bond. The extra valence electron of the impurity does not participate in a bond and is available for conduction, as shown in Fig. 7.4*a*. The impurity is referred to as a donor, and the resulting material is referred to as *n*-type material, since electrons have been liberated. The addition of only very small amounts of these impurities causes a dramatic change in the conductivity of silicon. For example, when one atom in 10 million is replaced by one of these donor impurity atoms, the conductivity of silicon at room temperature increases by a factor of 10^5!

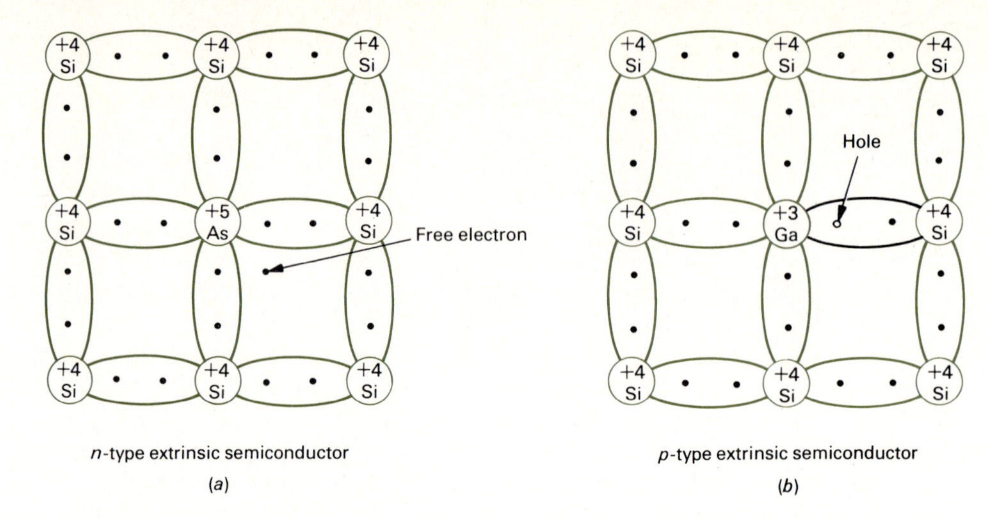

FIGURE 7.4 Addition of impurities to form extrinsic semiconductors.

If a trivalent impurity is added to an intrinsic semiconductor, the atom also replaces a host atom and forms a covalent bond. But there are only three valence electrons so that a hole or absence of an electron in a bond is created, as shown in Fig. 7.4*b*. This impurity atom may then accept an electron to complete the bond. When this happens, the hole appears to move to the position from which the electron came and can thus be thought of as a mobile charge carrier. A trivalent impurity is thus called an acceptor, and the resulting material is called a *p*-type material in reference to the holes.

Intrinsic semiconductors also have electron-hole pairs generated by thermal energy; however, at room temperature their numbers are quite small. The addition of an impurity to an intrinsic (pure) semiconductor results in an extrinsic semiconductor. Although hole-electron pairs are present via thermal agitation in both *n*-type and *p*-type materials, the dominant charge carrier (or majority carrier) is the one produced by the impurity. In *n*-type extrinsic semiconductors, for example, the electrons

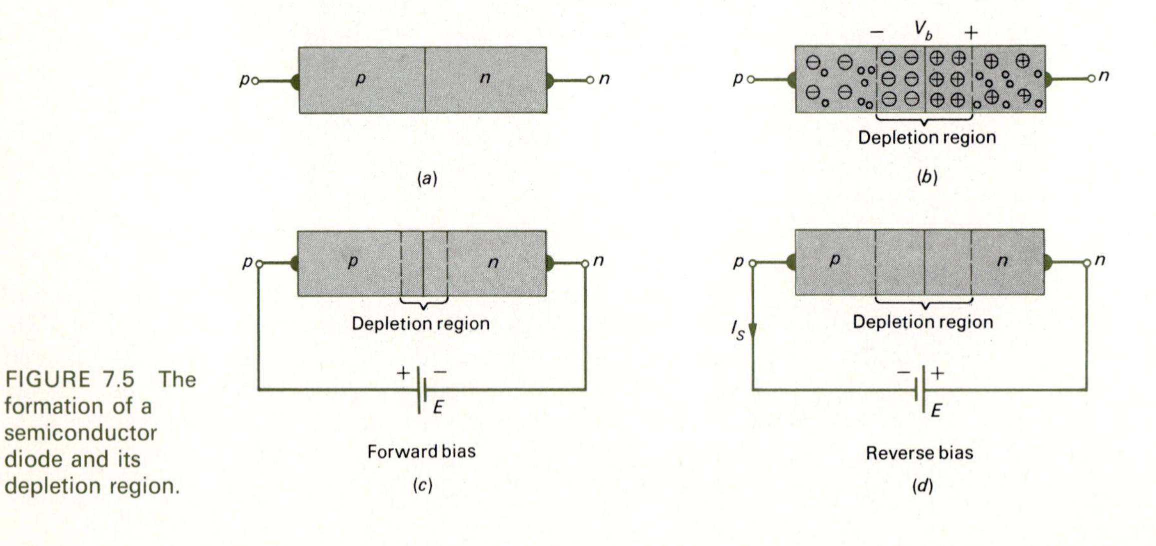

FIGURE 7.5 The formation of a semiconductor diode and its depletion region.

liberated by the impurity far exceed the holes produced by thermal agitation. Thus, for *n*-type materials, electrons are the majority carriers and holes are the minority carriers. For *p*-type materials, holes are the majority carriers and electrons are the minority carriers.

A semiconductor diode is formed as shown in Fig. 7.5 by growing a *p*-type material and an *n*-type material to form a *pn* junction diode. Leads (wires) attached to the ends of the material form the terminals of the device. The *n*-type has a high density of electrons while the *p*-type has a high density of holes, and there is a tendency for these majority carriers to diffuse across the junction. This diffusion continues until a sufficient number of immobile donor and acceptor atoms have been exposed to provide an attractive force (electric field) or potential barrier, V_b, which prevents any further diffusion, as shown in Fig. 7.5*b*. This region of immobile charges is called the depletion region, and the magnitude of the barrier voltage V_b is a few tenths of a volt.

When a battery E is attached to the terminals of the diode with the positive terminal attached to the *p*-type material and the negative terminal attached to the *n*-type material, as shown in Fig. 7.5*c* the diode is said to be forward-biased. With

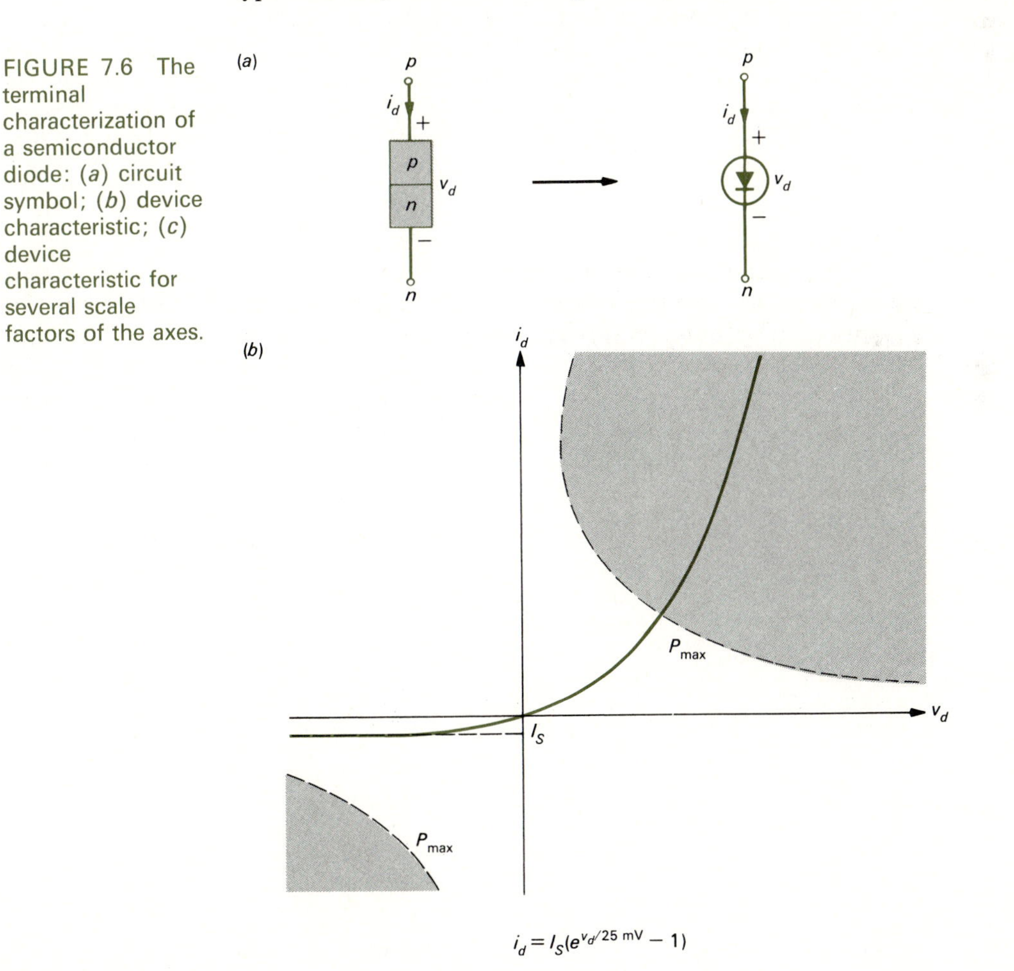

FIGURE 7.6 The terminal characterization of a semiconductor diode: (*a*) circuit symbol; (*b*) device characteristic; (*c*) device characteristic for several scale factors of the axes.

(Figure 7.6 continued)

FIGURE 7.6 (*Continued*)

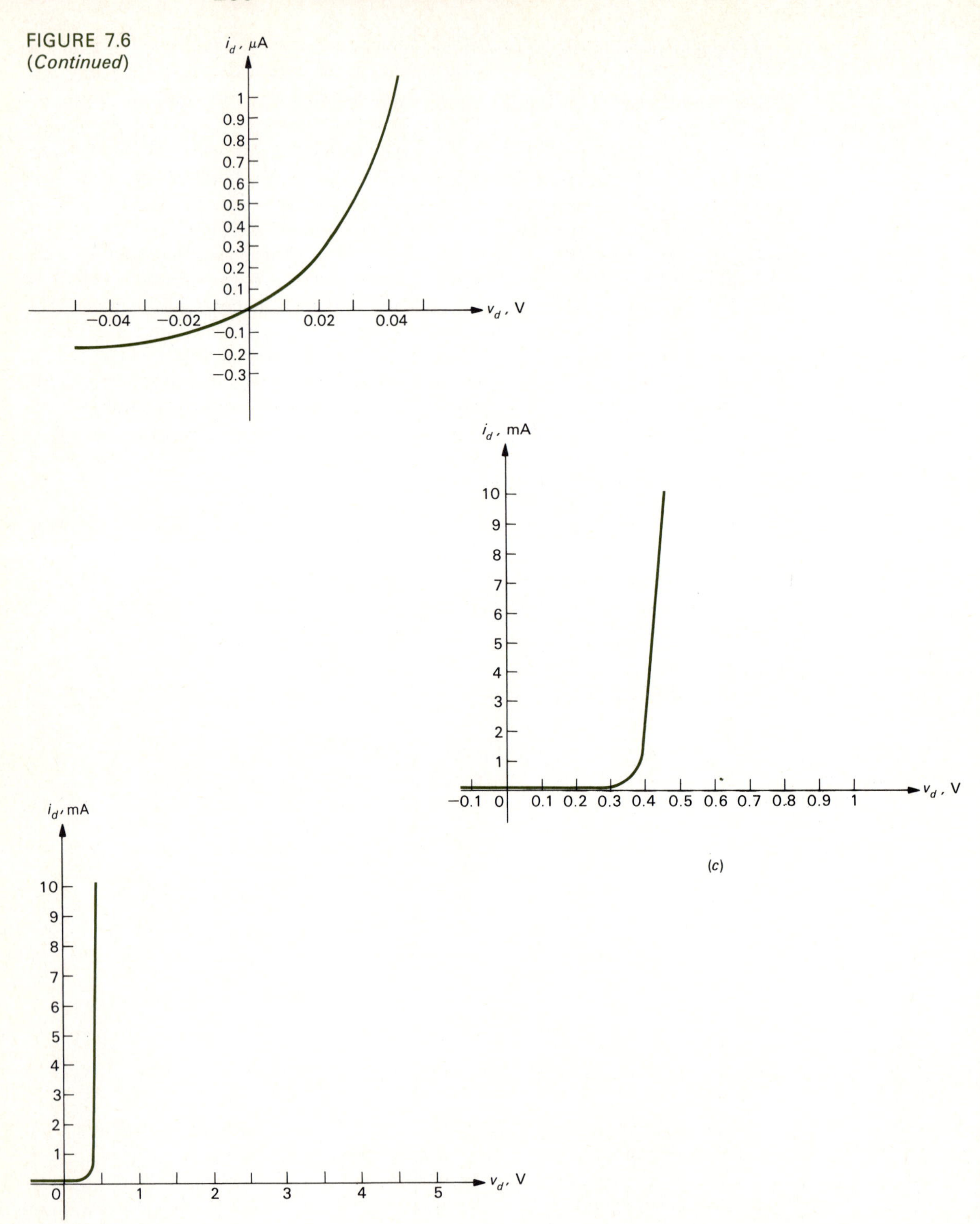

(*c*)

forward bias, more holes in the *p* region move into the *n* region. As these holes enter the *n* region, which has a larger number of electrons, recombination takes place. The electrons eliminated by recombination are replaced by electrons flowing from the negative terminal of the battery into the *n* region: a current flow. A current passes around the circuit through the battery and the diode. If the battery is reversed, as shown in Fig. 7.5*d*, the diode is said to be reverse-biased. More holes diffuse from the *n* region (where they are scarce) into the *p* region, and electrons diffuse from the *p* region into the *n* region. Since a very small number of charge carriers are involved, the current is also quite small. This reverse current is called a saturation current I_s, and it eventually reaches a constant value which is quite small, on the order of 10^{-9} A. Thus, for practical purposes, we may say that the semiconductor diode conducts no current (i.e., is an open circuit) under reverse-biased conditions.

The symbol for a semiconductor diode is shown in Fig. 7.6*a* along with the device terminal characteristic. The direction of the arrow in the symbol serves as a convenient reminder of the direction of current flow through the device occurring for forward-biased conditions (*p* terminal positive and *n* terminal negative).

The behavior of the semiconductor diode is found to be characterized by the Shockley diode equation

$$i_d = I_s(e^{qv_d/kT} - 1) \tag{7.1}$$

where q is the electron charge (1.6×10^{-19} C), k is Boltzmann's constant ($k = 1.38 \times 10^{-23}$ J/K), and T is the junction temperature in degrees kelvin. At room temperature, 293 K, we find that

$$\begin{aligned} \frac{kT}{q} &= 0.025 \text{ V} \\ &= 25 \text{ mV} \end{aligned} \tag{7.2}$$

and Eq. (7.1) can be written as

$$i_d = I_s(e^{v_d/25\,\text{mV}} - 1) \tag{7.3}$$

The diode equation in Eq. (7.3) is plotted in Fig. 7.6*b*. Note that for $v_d \gg 25$ mV, Eq. (7.3) simplifies to

$$i_d \approx I_s e^{v_d/25\,\text{mV}} \qquad v_d \gg 25 \text{ mV} \tag{7.4}$$

For $v_d \ll 25$ mV, it simplifies to

$$i_d \approx -I_s \qquad v_d \ll 25 \text{ mV} \tag{7.5}$$

The diode characteristic is plotted in Fig. 7.6*c* for several scale factors of the horizontal v_d axis. Note that as the v_d scale is increased, the characteristic approaches that of a vertical line for $v_d > 0.5$ V. This observation of the behavior for large v_d will be used in our modeling of the device.

Example 7.1 A semiconductor diode at room temperature is found to have a diode current of $i_d = 100$ milliamperes (mA) when the diode voltage is $v_d = 0.5$ V. Determine the reverse saturation current I_s.

Solution From Eq. (7.3) we have

$$I_s = \frac{i_d}{e^{v_d/25\,\mathrm{mV}} - 1}$$

and

$$I_s = \frac{100\ \mathrm{mA}}{e^{0.5/25\,\mathrm{mV}} - 1}$$

$$= 0.206 \times 10^{-9}\ \mathrm{A}$$

As current passes through the diode, average power is dissipated and converted to heat. This heat must be dissipated into the surroundings or it will accumulate and eventually destroy the diode. Semiconductor diodes are rated in terms of the maximum average power which they may safely dissipate; this is shown on the device characteristic in Fig. 7.6*b* as the hyperbola P_{max}. Ordinarily, P_{max} is an average power quantity since it relates to the ability of the device to dissipate the accumulated heat. Consequently, it is a function of the waveshape of $v_d(t)$ and $i_d(t)$. For dc operation at some operating point $i_d = I_d$, $v_d = V_d$, the requirement for safe operation of the device is that $P_{max} > V_d I_d$; thus, the operating point must not lie in the maximum average power dissipation region. For time-varying operation, the average power dissipated by the device is a function of the waveform of $v_d(t)$ and $i_d(t)$. For certain waveforms we may operate in the maximum power region such that $v_d i_d$ exceeds P_{max} if we do so only momentarily. A safe criterion is to ensure that at no time does the instantaneous product $v_d i_d$ exceed P_{max}.

Heat sinks are metal plates or flanges which are attached to the diode to facilitate the heat dissipation. Heat sinks effectively increase the P_{max} rating. These heat sinks allow the heat generated in the diode to spread out rapidly and they increase the surface area for heat dissipation.

7.2 LOAD LINES

The analysis of a circuit containing a diode is quite simple using the concept of a load line. For example, consider the circuit shown in Fig. 7.7*a* consisting of a diode, a resistor, and a battery. The diode characteristic in Fig. 7.6*b* relates the diode voltage v_d and diode current i_d. This may be thought of as an "equation" in the two unknowns v_d and i_d. Now if we are to solve for v_d and i_d, we need another equation containing these variables. This can be obtained by writing KVL around the loop in Fig. 7.7*a* to give

$$v_d = V_S - R i_d \tag{7.6}$$

FIGURE 7.7
Using a load line to solve a circuit containing a diode.

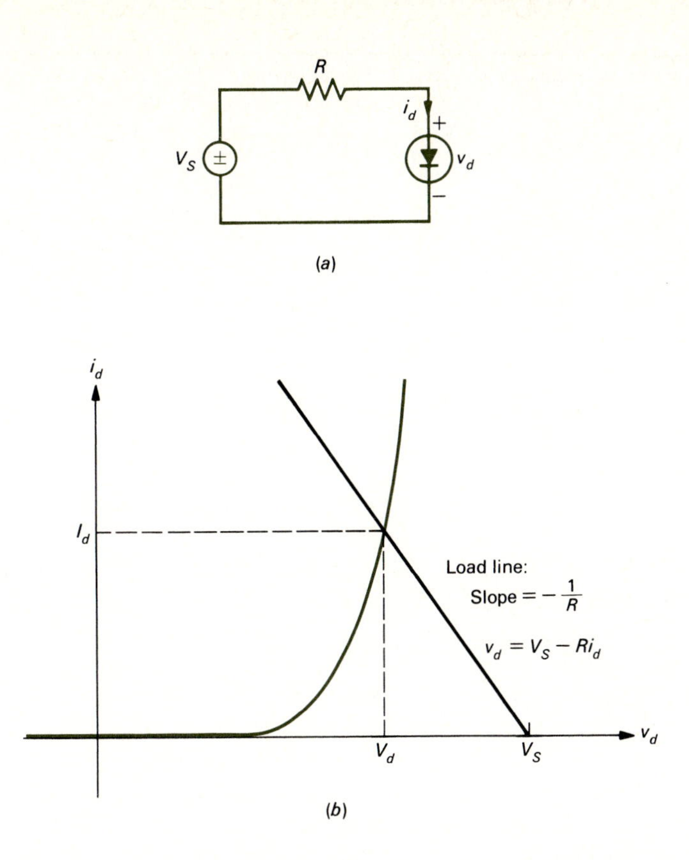

Now we have two equations in the two unknowns v_d and i_d. If we plot Eq. (7.6) on the diode characteristic and locate the point of intersection of the two curves, we will have found the simultaneous solution of these two equations. This point of intersection of the two curves is called the *operating point*. The values of diode current and voltage are read off the graph and denoted by I_d and V_d, as shown in Fig. 7.7*b*.

In plotting the load line from Eq. (7.6) on the graph, we must be careful to note that its slope is negative because of the negative sign in front of Ri_d in Eq. (7.6). Also, since the diode characteristic is plotted with i_d on the vertical axis and v_d on the horizontal axis, the slope of the load line is $-1/R$, not $-R$. To fix the actual position of the load line, we note in Eq. (7.6) that when $i_d = 0$, $v_d = V_S$. The horizontal axis of the graph is the location of all points where $i_d = 0$. Therefore, the load line must cross the v_d axis at V_S.

This concept of load lines can be easily extended to more complicated circuits so long as they contain ideal sources, linear resistors, and only one diode. To do this, we reduce the circuit attached to the diode terminals to a Thevenin equivalent so that it appears in the form of Fig. 7.7*a*. In this case R becomes R_{TH} and V_S becomes V_{OC}. Then we draw the load line on the diode characteristic and solve for the diode voltage and current at the operating point, V_d and I_d. With a knowledge of the diode voltage and current, we work back into the circuit to obtain any other element voltage or current of interest.

Example 7.2 In the circuit of Fig. 7.8*a*, determine the current *i*.

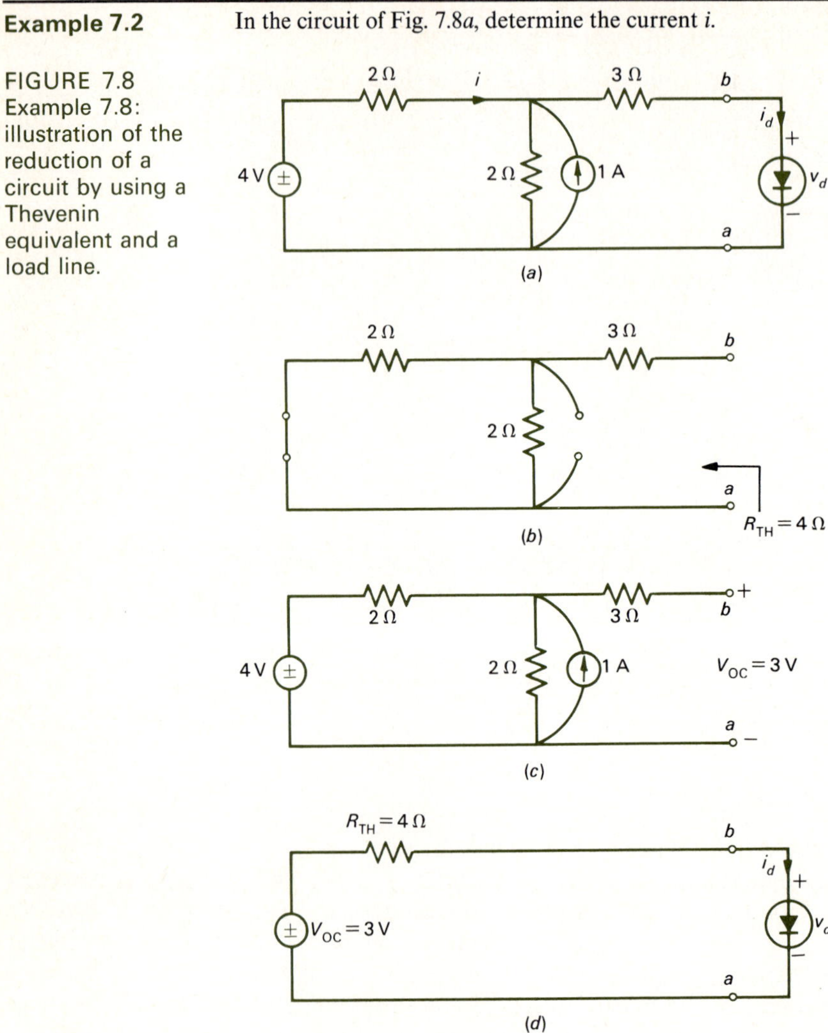

FIGURE 7.8 Example 7.8: illustration of the reduction of a circuit by using a Thevenin equivalent and a load line.

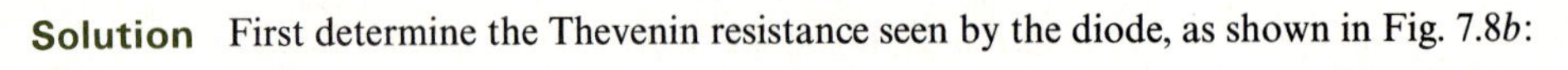

Solution First determine the Thevenin resistance seen by the diode, as shown in Fig. 7.8*b*:

$$R_{TH} = 4\ \Omega$$

Next determine the open-circuit voltage at terminals *a*-*b* (by superposition), as shown in Fig. 7.8*c*:

$$V_{OC} = 3\ \text{V}$$

Attach the Thevenin equivalent circuit to the diode, as shown in Fig. 7.8*d*. Plot the load line on the diode characteristic, as in Fig. 7.8*e*, and obtain

$$V_d = 0.38\ \text{V}$$

$$I_d = 0.66\ \text{mA}$$

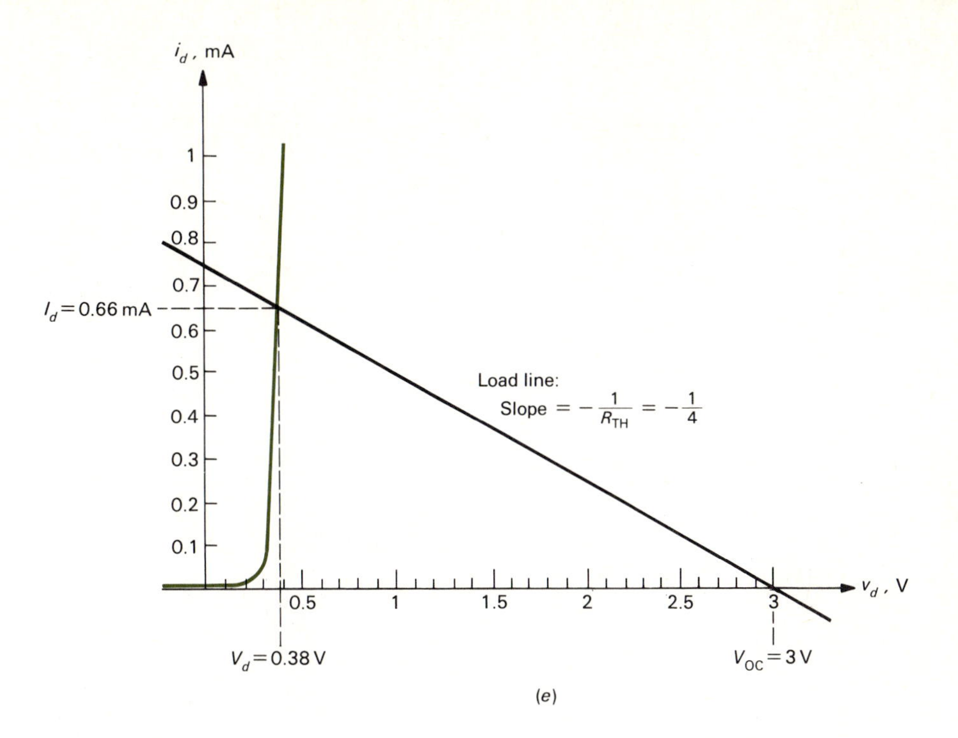

(e)

Now to determine i, we simply apply KCL at the node to which i enters to obtain

$$i = I_d - 1 - \frac{V_d - 3I_d}{2}$$

$$= -1.19 \text{ A}$$

7.3 PIECEWISE-LINEAR APPROXIMATION AND SMALL-SIGNAL ANALYSIS

Although graphical analysis of resistive circuits containing one semiconductor diode is relatively straightforward when using load lines, as we have seen in the previous section, there is nevertheless the problem of drawing the graphical characteristic of the diode, accurately plotting the load line, and reading off the values of the operation point. We will have numerous occasions to consider other approximate methods of analysis.

The primary approximate method which we will employ is the piecewise-linear approximation, shown as a dashed line in Fig. 7.9*a*. The piecewise-linear approximation consists of two straight-line segments. At a particular operating point (I_d, V_d), we construct a tangent to the curve having slope $1/R_f$. The reciprocal of this slope, R_f, is called the forward resistance of the diode *at this operating point* (I_d, V_d). The intersection of this line and the v_d axis ($i_d = 0$) is denoted by E_f, which is referred to as the forward voltage of the diode for this operating point. For $v_d < E_f$, another straight line segment with zero slope along the v_d axis completes the characterization. This

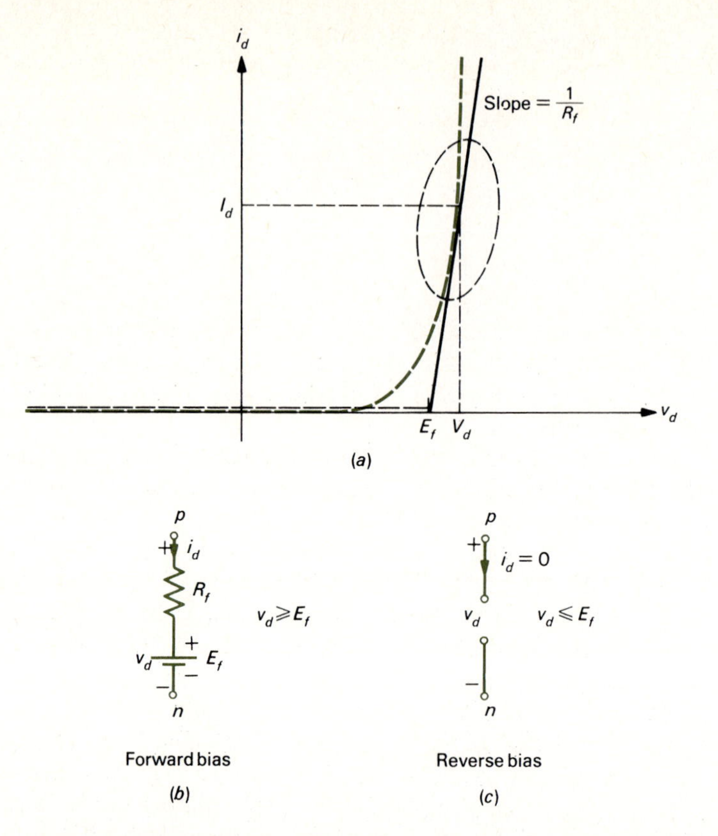

FIGURE 7.9 The piecewise-linear approximation of a diode.

latter segment corresponds to the condition that the diode is assumed to conduct no current for reverse-biased conditions ($v_d < 0$). For $v_d > E_f$, the diode approximation appears as a resistor R_f, in series with a battery E_f, as shown in Fig. 7.9*b*. For $v_d < E_f$, the diode is approximated as an open circuit conducting no current, as shown in Fig. 7.9*c*.

The value of the forward resistance R_f for a particular operating point (V_d, I_d) is found from Eq. (7.3):

$$R_f = \left.\frac{dv_d}{di_d}\right|_{V_d, I_d}$$

$$= \frac{25 \text{ mV}}{I_s e^{V_d/25 \text{ mV}}}$$

$$= \frac{25 \text{ mV}}{I_d + I_s} \tag{7.7}$$

Assuming I_s to be much smaller in magnitude than i_d, we have

$$R_f \approx \frac{25}{I_d \text{ (mA)}} \quad \Omega \tag{7.8}$$

For example, at an operating point of $I_d = 1$ mA, $R_f = 25\ \Omega$, and at $I_d = 10$ mA, $R_f = 2.5\ \Omega$. The forward voltage E_f is found from the equation of the tangent:

$$v_d = E_f + R_f i_d \tag{7.9}$$

For an operating point (I_d, V_d) we obtain

$$\begin{aligned} E_f &= V_d - R_f I_d \\ &= V_d - 25 \quad \text{mV} \end{aligned} \tag{7.10}$$

The representation of the diode with the piecewise-linear equivalent circuit in Fig. 7.9*b* only describes the characteristic in a small region about the operating point, as circled in Fig. 7.9*a*. This representation, however, is useful for what is known as small-signal analysis.

The use of this piecewise-linear approximation for small-signal analysis is illustrated in Fig. 7.10*a*. The ideal voltage source consists of a dc voltage E_S in series with a sinusoidal voltage $V_S \sin\omega t$. The load line, plotted on the characteristic in Fig. 7.10*b*, consists of a line of slope $-1/R_S$ intersecting the $i_d = 0$ axis at $v_d = E_S + V_S \sin\omega t$. As the sinusoid varies, the load line moves back and forth and the operating point moves along the characteristic. The actual values of i_d and v_d are plotted versus time as solid lines, whereas the approximate values using the tangent to the characteristic are plotted as dashed lines. Note that since the diode characteristic is nonlinear, the diode voltage and current only approximately resemble sinusoids; distortion of the input sinusoid $v_S(t)$ is said to have taken place. However, if we restrict the magnitude of the source sinusoid V_S to a small value, the nonlinearity of the device characteristic becomes less important and the diode voltage and current more closely resemble sinusoids. In this case, we may simply replace the diode with its (linear) forward-biased circuit shown in Fig. 7.9*b*.

This is an important concept which will be used in numerous instances in Part 2. Note that so long as $v_S(t)$ is restricted to being "small enough" so that excursions of the instantaneous operating point are restricted to an approximately linear region of the diode characteristic, we may replace the nonlinear diode with a linear circuit, as shown in Fig. 7.11*a*. So long as $v_S(t)$ is small enough, it doesn't matter that the diode characteristic is truly nonlinear; $v_S(t)$ never forces v_d and i_d outside of the approximately linear, circled region. Thus, the remainder of the characteristic may be approximated as shown. Under this restriction of small-signal approximately linear operation, we may replace the nonlinear diode with its small-signal linear equivalent circuit, as shown in Fig. 7.11*b*. Now, we have a strictly *linear problem* to which we may apply all of our previous linear resistive-circuit-analysis techniques, including superposition.

Granted this is an approximation but is nevertheless a rather considerable simplification. We first find the operating point V_d and I_d by plotting the load line on the diode characteristic with $v_S(t) = 0$. We then find the small-signal forward resistance R_f at the particular operating point from Eq. (7.8). The forward voltage E_f is found from Eqs. (7.8) and (7.10) (or is read off the characteristic) as

$$E_f = V_d - R_f I_d \tag{7.11}$$

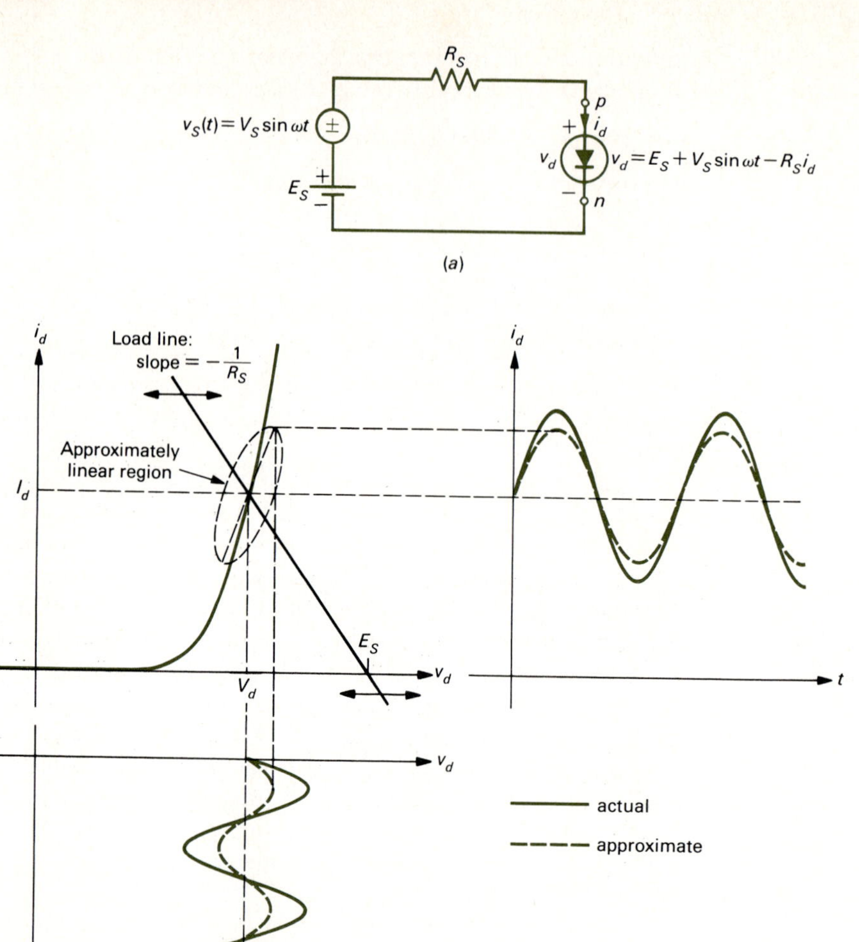

FIGURE 7.10 Small-signal operation.

Substituting Eq. (7.8) gives

$$E_f = V_d - 0.025 \tag{7.12}$$

Then we replace the diode with its small-signal linear equivalent circuit, as shown in Fig. 7.11*b*. Now, with simple linear circuit-analysis principles, we may solve for the diode voltage and current. Note that the circuit is, with this approximation, a linear one so that, by superposition, the diode current and voltage will consist of two pieces:

$$i_d = I_d + \Delta i_d \tag{7.13a}$$

$$v_d = V_d + \Delta v_d \tag{7.13b}$$

where

$$\left.\begin{matrix} I_d \\ V_d \end{matrix}\right\} \text{due to } E_S \text{ and } E_f \tag{7.14a}$$

$$\left.\begin{matrix} \Delta i_d \\ \Delta v_d \end{matrix}\right\} \text{due to } v_S(t) \tag{7.14b}$$

If we are only interested in the portion due to $v_S(t)$, we may set E_S and E_f equal to zero, as shown in Fig. 7.11*c*, and solve for Δi_d and Δv_d as

$$\Delta i_d = \frac{v_S(t)}{R_S + R_f} \tag{7.15a}$$

$$\Delta v_d = \frac{R_f}{R_f + R_S} v_S(t) \tag{7.15b}$$

This principle of small-signal approximately linear analysis will be used on numerous occasions in Part 2 and should be studied carefully.

Quite often, for practical purposes, we may neglect R_f in the small-signal equivalent circuit of a diode. For typical semiconductor diodes, the value of R_f is quite small: between 1 and 100 Ω. Note that the diode characteristic for $v_d \gg 25$ mV is exponential, as shown in Eq. (7.4). Consequently, for increasing v_d, the characteristic asymptotically approaches a vertical line quite rapidly and a simpler approximate representation of the semiconductor diode is to neglect R_f, as shown in Fig. 7.12*a*. So long as the

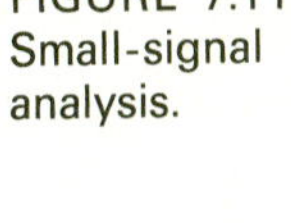

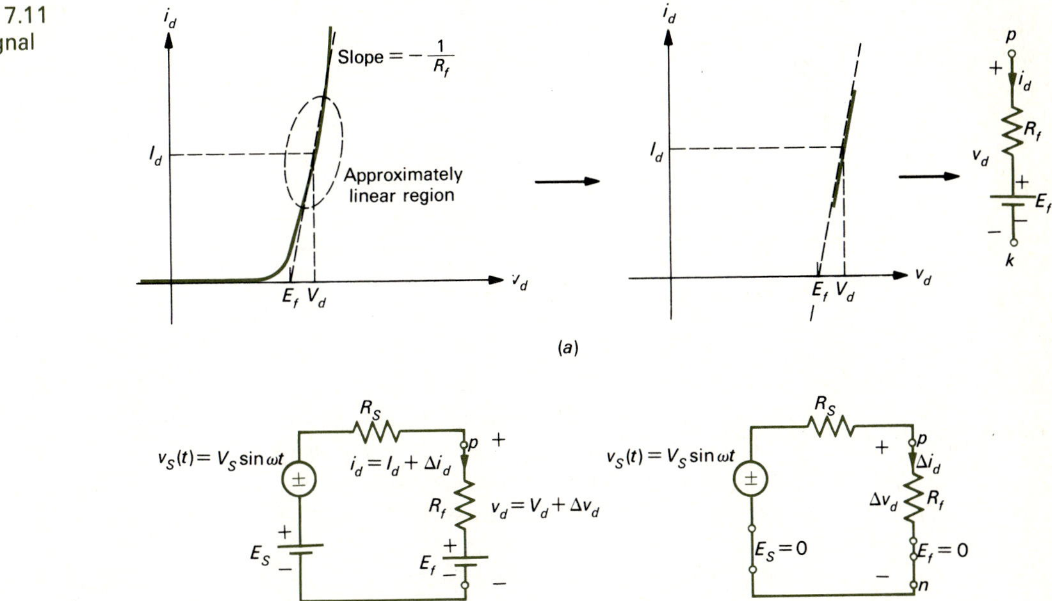

FIGURE 7.11 Small-signal analysis.

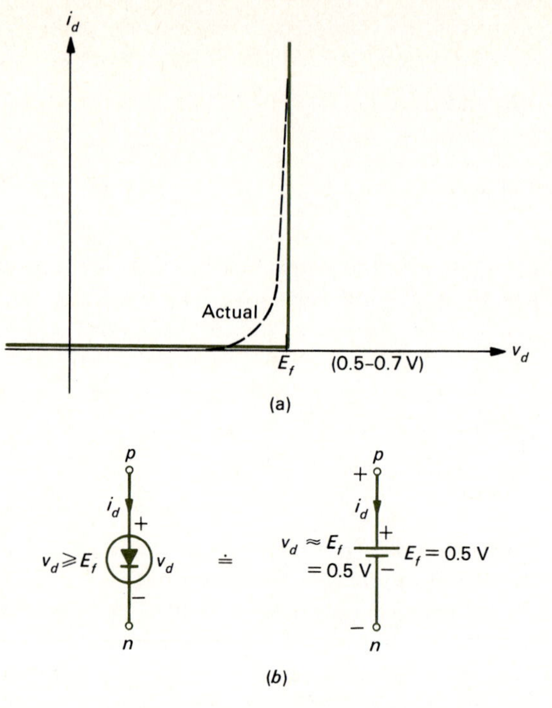

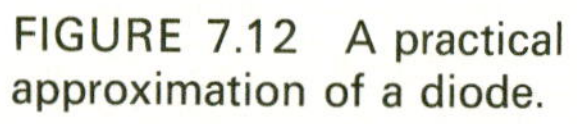
FIGURE 7.12 A practical approximation of a diode.

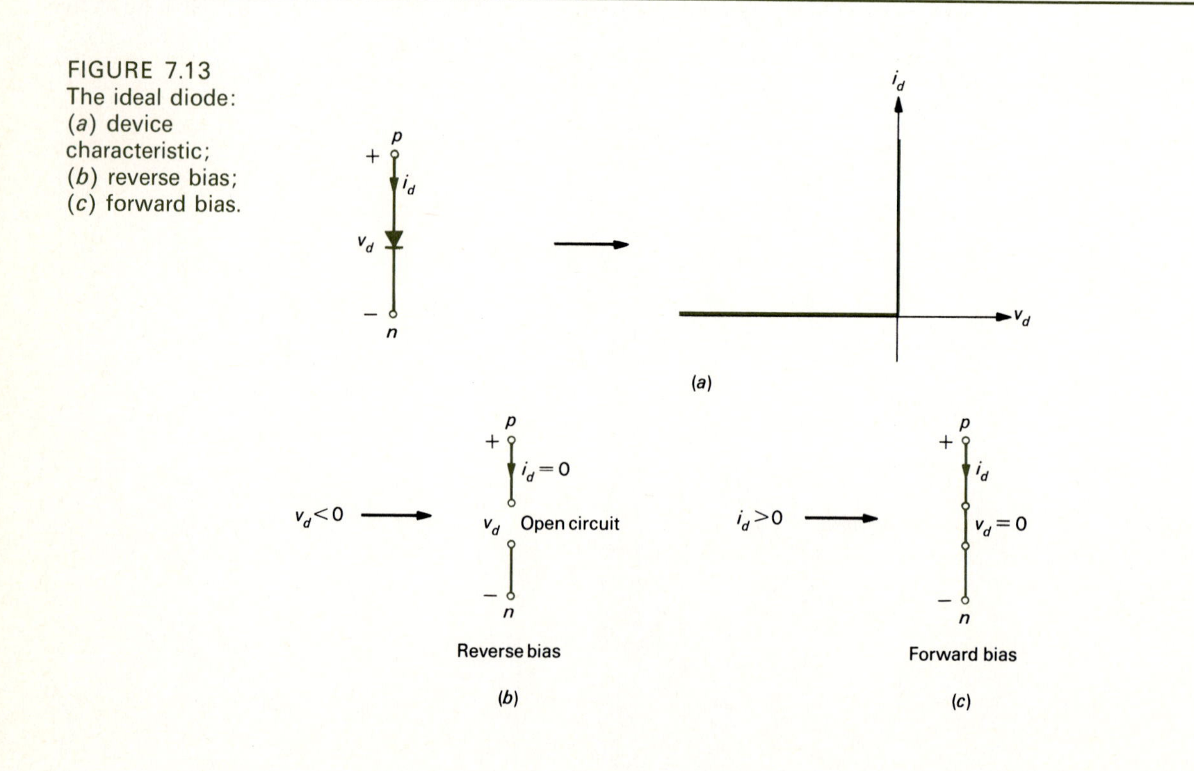

FIGURE 7.13
The ideal diode:
(*a*) device characteristic;
(*b*) reverse bias;
(*c*) forward bias.

operating point is not in the "knee" of the characteristic, this approximation is usually sufficient. For silicon diodes, typical values of E_f range from 0.5 to 0.8 V. We will use a value of 0.5 V in our analyses, as shown in Fig. 7.12*b*.

7.4 THE IDEAL DIODE

The semiconductor diode behaves in a fashion similar to a switch: current flows in only one direction. For reverse-biased diodes ($v_d < 0$), the diode appears as an open circuit. This important feature of the diode permits the construction of several useful circuits which we will discuss.

To obtain preliminary estimates of the circuit performance, we will replace the actual diode with an ideal diode, as shown in Fig. 7.13*a*. The ideal diode is effectively

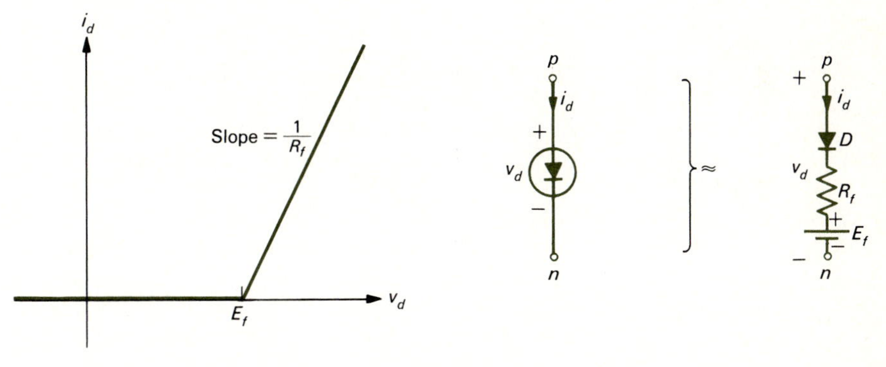

(*a*)

+
p
i_d
D closed
$v_d \geq E_f$
v_d
R_f
+
E_f
−
n
+
p
$i_d = 0$
D open
$v_d \leq E_f$
v_d
R_f
+
E_f
−
n

(*b*)

FIGURE 7.14 The piecewise-linear model of a diode, using an ideal diode.

an ideal switch. When the diode is reverse-biased ($v_d < 0$) it appears as an open circuit, as is the case for an actual diode, as shown in Fig. 7.13*b*. Under forward-biased conditions ($i_d > 0$) it closes and appears as a short circuit, as shown in Fig. 7.13*c*.

The ideal diode is a further approximation to the piecewise-linear approximation in Fig. 7.9, in that for the ideal diode $R_f = 0$ and $E_f = 0$. For a semiconductor diode, R_f is quite small (typically only a few ohms) and E_f is on the order of 0.5 V.

Once a preliminary estimate of a circuit's performance is obtained by replacing the actual diodes with ideal diodes, a more refined estimate can be obtained by the piecewise-linear models in series with the ideal diode, as shown in Fig. 7.14*a*. The ideal diode is closed in this model for $v_d > E_f$ and open for $v_d < E_f$ as shown in Fig. 7.14*b*.

Example 7.3 Nonlinear resistors with a wide range of characteristics can be obtained, approximately, with circuits containing diodes. For example, a square-law device is a two-terminal nonlinear resistor whose terminal voltage-current characteristic obeys

$$i = kv^2$$

where k is a normalization constant. The ideal characteristic is shown in Fig. 7.15*a*; this device may be used in modulators, for example, to attach a voice signal to a higher frequency carrier wave, as is done in amplitude modulation (AM) radio transmission. Design a square-law device to approximate the ideal characteristic for $0 \leq v \leq 5$ V with a normalization constant of $k = 0.001$.

Solution A circuit using ideal diodes which will accomplish the task is shown in Fig. 7.15*b*. First we describe the operation of this circuit. Slowly increase v from zero and observe the behavior of i. Initially, diodes D_1 and D_2 will be reverse-biased and open. Once v reaches the smaller of the two battery voltages E_1 or E_2, the diode associated with that branch will close. Let us assume for illustration that $E_1 < E_2$. Thus, until v reaches E_1 both diodes will be open and only resistor R_3 will be in the circuit; for $0 \leq v \leq E_1$, the curve will thus have slope $1/R_3$. For $E_1 \leq v < E_2$, diode D_1 will be closed but D_2 will be open. The input resistance at the terminals will be $R_1||R_3$. For $E_2 \leq v \leq V$, diode D_2 will also be closed and the input resistance at the terminals will be $R_1||R_2||R_3$. Thus, the input characteristic will have the piecewise-linear characteristic shown in Fig. 7.15*c*. We would choose $V = 5$ V and choose the battery voltages E_1 and E_2 arbitrarily. Suppose we choose the battery voltages as

$$E_1 = 2.0 \text{ V}$$

$$E_2 = 3.5 \text{ V}$$

The resistors are chosen such that the breakpoints at E_1 and E_2 and the point at $V = 5$ V fall on the desired ideal curve. Thus, we compute

$$\begin{aligned} I_1 &= kE_1^2 \\ &= 4 \text{ mA} \end{aligned}$$

$$\begin{aligned} I_2 &= kE_2^2 \\ &= 12.25 \text{ mA} \end{aligned}$$

$$\begin{aligned} I &= kV^2 \\ &= 25 \text{ mA} \end{aligned}$$

FIGURE 7.15
Example 7.3: synthesis of a square-law resistor, using diodes.

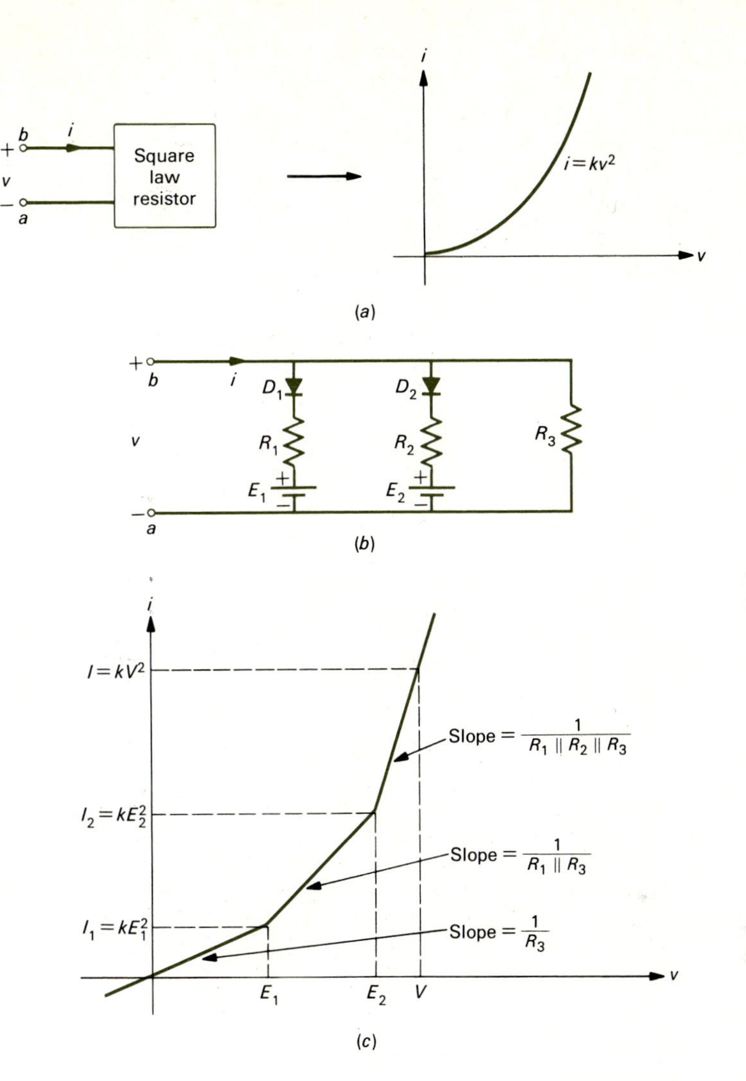

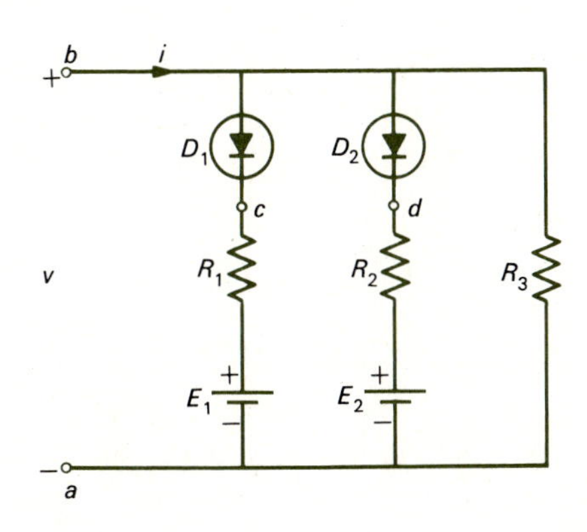

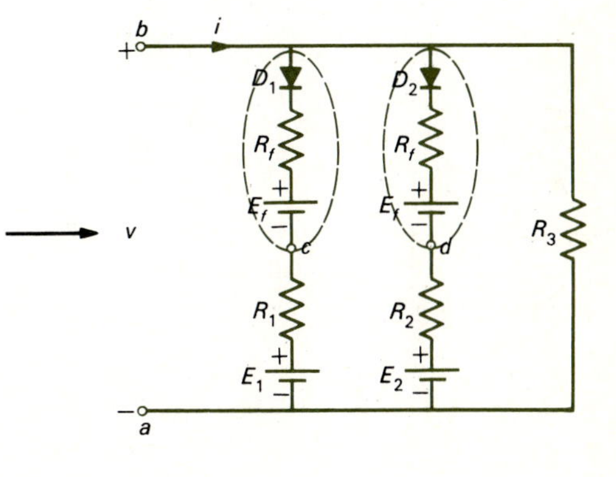

(d)

Noting the slopes of each portion, we obtain

$$R_3 = \frac{E_1}{I_1}$$

$$= 500\ \Omega$$

$$R_1 \| R_3 = \frac{E_2 - E_1}{I_2 - I_1}$$

$$= 182\ \Omega$$

so that

$$R_1 = 286\ \Omega$$

Finally,

$$R_1 \| R_2 \| R_3 = \frac{V - E_2}{I - I_2}$$

$$= 118\ \Omega$$

so that

$$R_2 = 333\ \Omega$$

Now that we have designed the circuit for ideal diodes, let us determine the actual values of R_1, R_2, R_3, E_1, and E_2 which should be used if we substitute actual diodes, as shown in Fig. 7.15*d*. Replacing the actual diodes with their piecewise-linear approximations, we see that we must subtract R_f from the above values of R_1 and R_2 and subtract E_f from the above values of E_1 and E_2. If we assume that both diodes have $R_f = 10\ \Omega$ and $E_f = 0.5$ V, then the actual circuit elements should be

$$R_1 = 276\ \Omega$$

$$R_2 = 323\ \Omega$$

$$R_3 = 500\ \Omega$$

$$E_1 = 1.5\ \text{V}$$

$$E_2 = 3.0\ \text{V}$$

In this case we may use ordinary flashlight batteries for E_1 and E_2 (a result of some preliminary planning).

In the following sections, we will investigate some useful circuits containing semiconductor diodes. These will be analyzed by replacing the actual diodes with ideal diodes. The modification of this behavior when piecewise-linear approximations are used will be indicated.

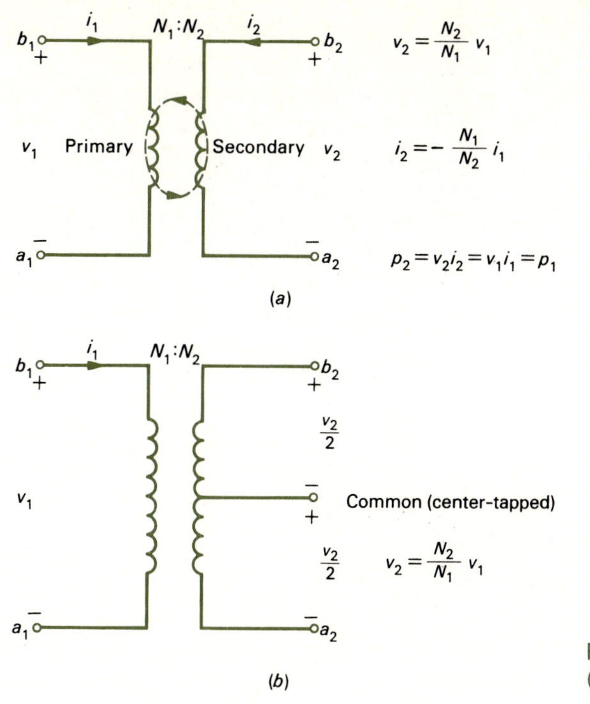

FIGURE 7.16 Transformers: (a) ideal; (b) center-tapped.

We will find it necessary to use the ideal transformers shown in Fig. 7.16; these were discussed in Chap. 4. Here we will provide a brief review of their properties. It is sufficient to describe the ideal transformer with turns ratio N_1/N_2 as a device such that the voltages v_2 and v_1 are related by the turns ratio:

$$v_2 = \frac{N_2}{N_1} v_1 \tag{7.16}$$

If $N_2 > N_1$, the transformer can be used to step up or step down a voltage, depending on which side the voltage source is attached. The currents are related by

$$i_2 = -\frac{N_1}{N_2} i_1 \tag{7.17}$$

Note that if the voltage v_2 is stepped up by ratio N_2/N_1, the current is stepped down by N_1/N_2. The power delivered to the transformer is

$$\begin{aligned} P &= P_2 + P_1 \\ &= \left(\frac{N_2}{N_1} v_1\right)\left(-\frac{N_1}{N_2} i_1\right) + v_1 i_1 \\ &= 0 \end{aligned} \tag{7.18}$$

Thus, the ideal transformer consumes no power.

Practical transformers consist of turns of wire around a ferromagnetic metal. The magnetic flux generated in one winding is linked with the turns of the other winding, inducing a voltage in that winding. For this to occur, the voltages and currents must be varying with time so that transformers are commonly used to step up or step down ac (sinusoidal) voltages, such as commercial 60-Hz power.

If a terminal (common terminal) is attached to the center of the secondary winding, as shown in Fig. 7.16*b*, the transformer is said to be center-tapped. The total voltage across the secondary, $v_2 = (N_2/N_1)v_1$, is now equally divided between the two halves. As terminal b_2 is increasing with respect to the common terminal, terminal a_2 is decreasing with respect to the common terminal. This arrangement is sometimes referred to as a phase inverter since it produces two voltages with respect to the common terminal which are 180° out of phase with each other.

7.5 RECTIFIERS

Rectifiers are circuits which contain semiconductor diodes and produce a dc voltage or current from an ac source. As we will see in subsequent chapters, numerous electronic circuits require a dc voltage for operation. Batteries are cumbersome and require frequent maintenance and replacement. A more practical method of providing a dc voltage is to rectify an ac voltage; ac voltages are somewhat easier to generate, and transmission of ac power is, at present, more efficient than transmission of dc power, as evidenced by the numerous high-voltage ac transmission lines across the countryside. The generation and transmission of ac power is discussed in Part 3.

7.5.1 The Half-Wave Rectifier

A half-wave rectifier using an ideal diode is shown in Fig. 7.17*a*. The resistor R_L represents some load to which we would like to supply a dc voltage. A 1:1 ideal transformer isolates the load from the source. The transformer is useful in providing a larger or smaller apparent value of source voltage. If instead of a 1:1 turns ratio we had used a 1:2 turns ratio, the secondary voltage would be $2v_S(t)$. We may replace the source voltage and transformer with an equivalent source voltage as shown in Fig. 7.17*b*.

During the positive half cycle of the source, the ideal diode is forward-biased and closed so that the source voltage is connected directly across the load. During the negative half cycle of the source, the ideal diode is reversed-biased so that the source voltage is disconnected from the load and the load voltage is zero. Thus, the load voltage is said to be rectified and is of one polarity, as shown in Fig. 7.17*c*. It may be described by

$$v_L(t) = V_S \sin \omega t \qquad 0 \leq \omega t \leq \pi \tag{7.19a}$$

$$v_L(t) = 0 \qquad \pi \leq \omega t \leq 2\pi \tag{7.19b}$$

Although the resulting load voltage is not truly dc as produced by a battery, it nevertheless has an average, or dc, component. The average (dc) value of v_L is

$$V_L = \frac{1}{2\pi}\int_0^{\pi} V_S \sin(\omega t) d(\omega t)$$

$$= \frac{V_S}{\pi} \tag{7.20}$$

Diodes are often rated in terms of their peak inverse voltage (PIV) and their maximum average power dissipation, $P_{\max}$, discussed previously. The PIV rating refers to the maximum reverse-biased voltage the diode can withstand. For the half-wave rectifier, this maximum reverse voltage occurs when $v_S(t)$ reaches its peak on the

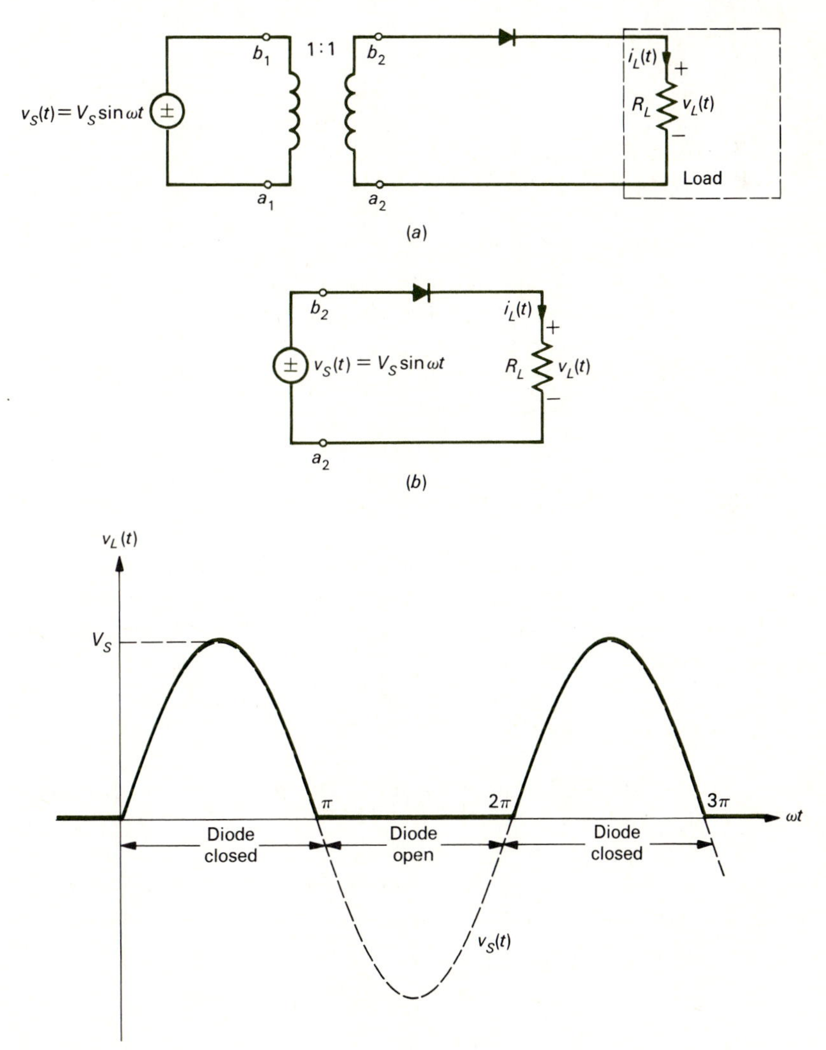

FIGURE 7.17 The half-wave rectifier.

negative half cycle $-V_S$. Thus, for a 1:1 transformer, PIV $= V_S$. For a 1:5 transformer, PIV $= 5V_S$. Ideal diodes obviously do not dissipate or consume power; thus, there is no power restriction, P_{max}. Practical diodes do, and for this case we would have to find the diode voltage waveform $v_d(t)$ and diode current waveform $i_d(t)$ and determine whether

$$P_{AV} = \frac{1}{T}\int_0^T v_d(t) i_d(t)\, dt \tag{7.21}$$

is less than P_{max}.

7.5.2 Filtering the Half-Wave Rectifier Voltage

The pulsating dc voltage produced by the half-wave rectifier is periodic and thus has a Fourier series representation. (See App. C for a discussion of the Fourier series.) The average value was determined in the previous section. Denoting this representation of the load voltage as

$$\begin{aligned} v_L(t) = V_L &+ a_1 \sin(\omega t) + a_2 \sin(2\omega t) + \cdots \\ &+ b_1 \cos(\omega t) + b_2 \cos(2\omega) + \cdots \end{aligned} \tag{7.22}$$

The Fourier coefficients are determined from App. C as

$$a_n = \frac{2}{T}\int_0^T v_L(t) \sin(n\omega t)\, dt \tag{7.23a}$$

$$b_n = \frac{2}{T}\int_0^T v_L(t) \cos(n\omega t)\, dt \tag{7.23b}$$

For the half-wave rectified voltage,

$$\begin{aligned} a_1 &= \frac{2}{T}\int_0^T v_L(t) \sin(\omega t)\, dt \\ &= \frac{1}{\pi}\int_0^\pi V_S \sin \omega t \sin(\omega t)\, d\omega t \\ &= \frac{V_S}{\pi}\int_0^\pi \frac{1}{2}[1 - \cos(2\omega t)]\, d\omega t \\ &= \frac{V_S}{2} \end{aligned}$$

Similarly, we may obtain

$$b_2 = -\frac{2V_S}{3\pi} \qquad b_3 = 0 \qquad b_4 = -\frac{2V_S}{15\pi} \qquad a_5 = 0$$

The coefficients of the sine terms a_n are zero for $n > 1$ since

$$a_n = \frac{2}{T}\int_0^T v_L(t)\sin(n\omega t)\,dt$$

$$= \frac{1}{\pi}\int_0^\pi V_s \sin(\omega t)\sin(n\omega t)\,d\omega t$$

$$= \frac{V_S}{\pi}\int_0^\pi \frac{1}{2}[\cos(n-1)\omega t - \cos(n+1)\omega t]\,d\omega t$$

$$= 0$$

Thus, the Fourier series for the half-wave rectified signal is

$$v_L(t) = \frac{V_S}{\pi} + \frac{V_S}{2}\sin(\omega t) - \frac{2V_S}{3\pi}\cos(2\omega t) - \frac{2V_S}{15\pi}\cos(4\omega t) + \cdots \tag{7.24}$$

This shows that, in addition to the average (dc) value, higher frequency harmonics are also present. This suggests the use of a filter to eliminate these unwanted sinusoidal components. One such satisfactory arrangement is obtained by placing a capacitor across the load resistor, as shown in Fig. 7.18*a*. We observe that the capacitor has a lower impedance to higher frequencies and is an open circuit to dc; thus, the capacitor should tend to short out the high-frequency components of the wave. As the source voltage initially increases positively, the diode is forward-biased since the load voltage is zero and the source is directly connected to the load over interval T_1. Once the source reaches its maximum value V_S and begins to decrease, the load voltage and consequently the capacitor voltage is maintained momentarily at V_S and the diode becomes reverse-biased and consequently open-circuited. The capacitor then discharges over interval T_2 through R_L until the source voltage $v_S(t)$ has increased to a value equal to the load voltage. At this point, the source voltage exceeds the capacitor voltage and the diode is once again forward-biased and closed. The capacitor once again charges to V_S. These points are illustrated in Fig. 7.18*b*. Thus, the load voltage is smoothed somewhat and more closely resembles a true dc voltage. The smoothing of the filter can be improved by increasing the CR_L time constant so that the discharge rate is slowed. If R_L is fixed (as it usually is), this can be accomplished by using a larger capacitor, as shown in Fig. 7.18*c*.

7.5.3 Effects of Actual Diodes

If the piecewise-linear model of an actual diode replaces the ideal diode in the half-wave rectifier, the circuit becomes as shown in Fig. 7.19*a*. The ideal diode is open for $v_s(t) < E_f$ and is closed for $v_S(t) > E_f$, as shown in Fig. 7.19*b*. The phase displacement $\omega t_1 = \theta_1$ at which the ideal diode closes is found by setting $v_S(t)$ equal to E_f:

$$V_S \sin\theta_1 = E_f \tag{7.25}$$

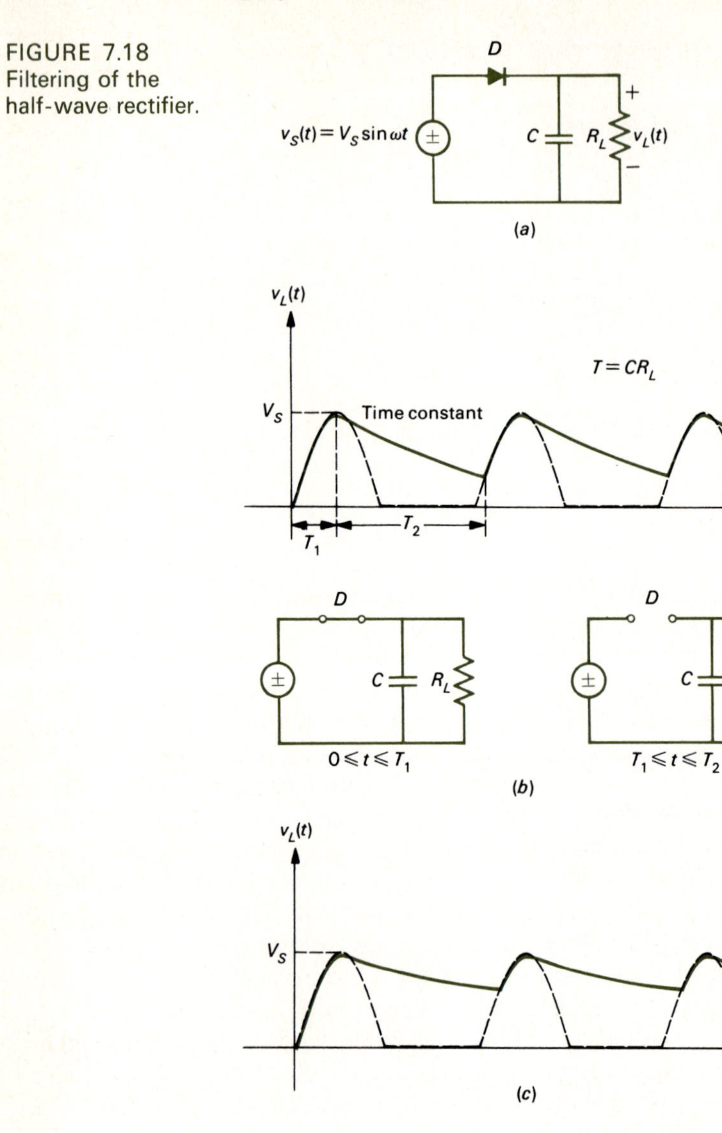

FIGURE 7.18
Filtering of the half-wave rectifier.

or

$$\theta_1 = \sin^{-1} \frac{E_f}{V_S} \tag{7.26}$$

The diode closes once again at $\theta_2 = \pi - \theta_1$. The load voltage when the ideal diode is closed is given by voltage division as

$$v_L(t) = \frac{R_L}{R_L + R_f} [v_S(t) - E_f] \tag{7.27}$$

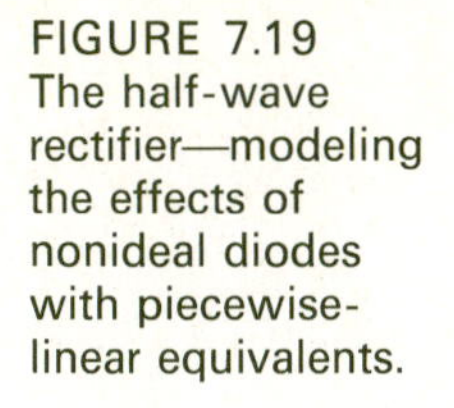
FIGURE 7.19
The half-wave rectifier—modeling the effects of nonideal diodes with piecewise-linear equivalents.

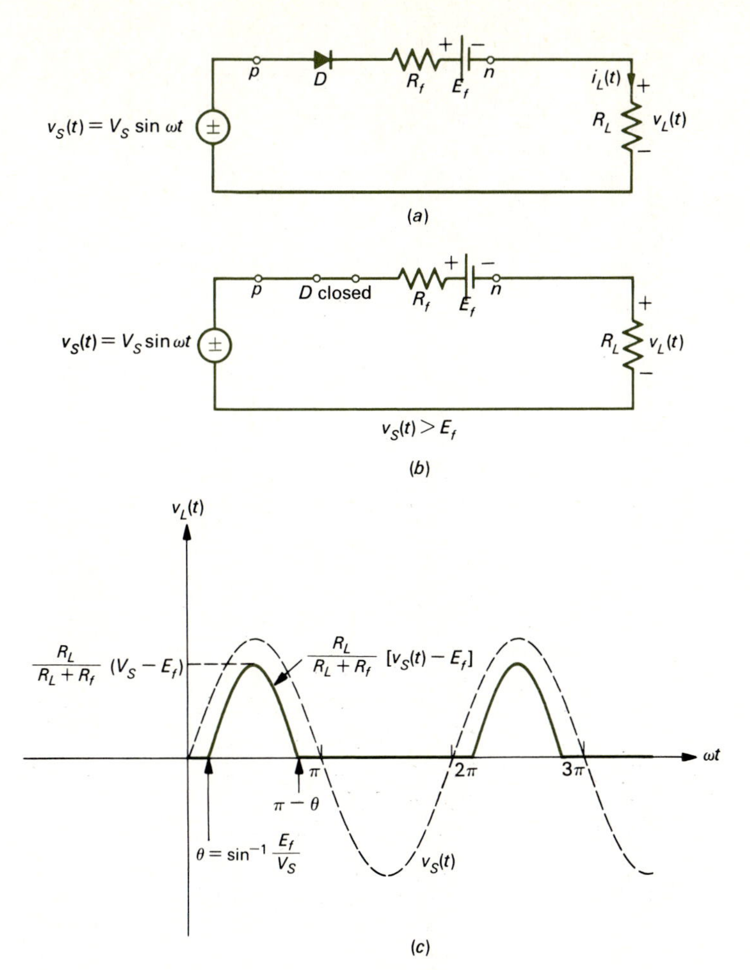

The peak is reduced from the ideal case, as shown in Fig. 7.19*c*. One of the major effects is to lower the dc component of the voltage, since the average value of $v_L(t)$ has been lowered. This average value is determined from

$$v_L = \frac{1}{2\pi}\int_{\theta_1}^{\theta_2} \frac{R_L}{R_L + R_f}[V_S \sin(\omega t) - E_f]d(\omega t)$$

$$= \frac{1}{2\pi}\frac{R_L}{R_L + R_f}(-V_S \cos\theta_2 + V_S \cos\theta_1 - E_f\theta_2 + E_f\theta_1) \tag{7.28}$$

Since $\theta_2 = \pi - \theta_1$, we obtain

$$v_L = \frac{1}{2\pi}\frac{R_L}{R_L + R_f}(2V_S \cos\theta_1 - E_f\pi + 2E_f\theta_1) \tag{7.29}$$

by using the trigonometric identity $\cos(\pi - \theta) = -\cos\theta$.

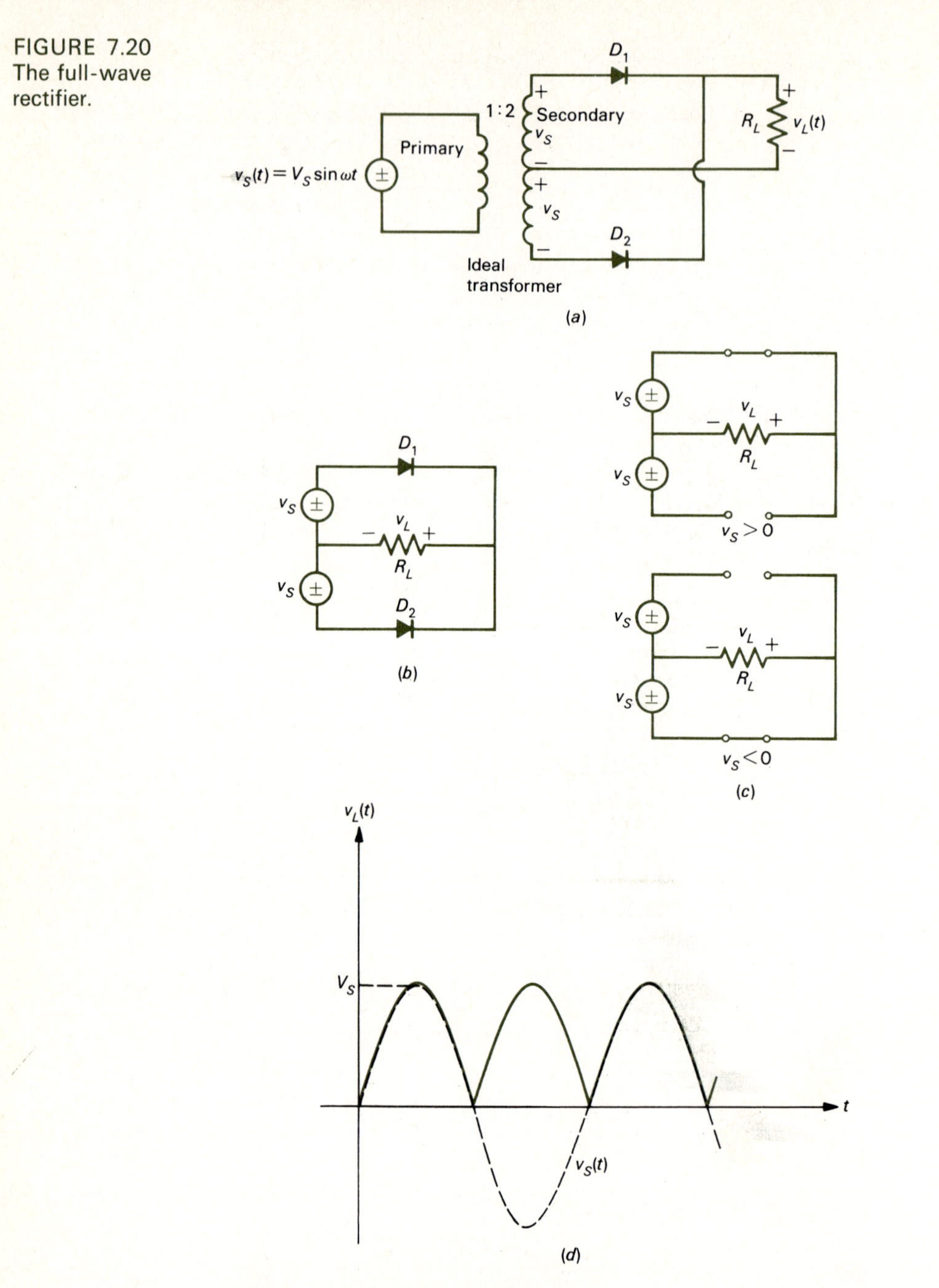

FIGURE 7.20 The full-wave rectifier.

7.5.4 The Full-Wave Rectifier

The half-wave rectifier produced a pulsating dc yet appeared to use only the positive half-cycle of the source voltage. The full-wave rectifier shown in Fig. 7.20*a* uses both half cycles of $v_S(t)$. We have used an ideal transformer with a center-tapped secondary having a 1:2 turns ratio. Therefore, the voltage across each of the halves of the secondary is $v_s(t)$, but the voltages are out of phase with respect to the common, center-tapped terminal. Redrawing this circuit as shown in Fig. 7.20*b*, we see that for the positive half cycle of $v_S(t)$, diode D_1 is forward-biased and closed and diode D_2 is reverse-biased and open (as shown in Fig. 7.20*c*) and $v_L(t) = v_S(t)$. Over the negative half cycle of $v_S(t)$, D_1 is open and D_2 is closed and $v_L(t) = -v_S(t)$. This has the net effect of reversing the polarity of the negative half cycle of $v_S(t)$ and applying this directly across the load.

The resulting load voltage waveform is shown in Fig. 7.20*d*. The average value is

$$\begin{aligned} v_L &= \frac{1}{\pi}\int_0^{\pi} V_S \sin \omega t d(\omega t) \\ &= \frac{2V_S}{\pi} \end{aligned} \tag{7.30}$$

which is twice the dc component of the half-wave rectifier. The PIV of each diode is, from Fig. 7.20*c*, twice that of the half-wave rectifier: $2V_S$. The Fourier series of this waveform is

$$v_L(t) = \frac{2V_S}{\pi} - \frac{4V_S}{3\pi}\cos(2\omega t) - \frac{4V_S}{15\pi}\cos(4\omega t) + \cdots \tag{7.31}$$

FIGURE 7.21 The full-wave rectifier—modeling the effects of nonideal diodes with piecewise-linear equivalents.

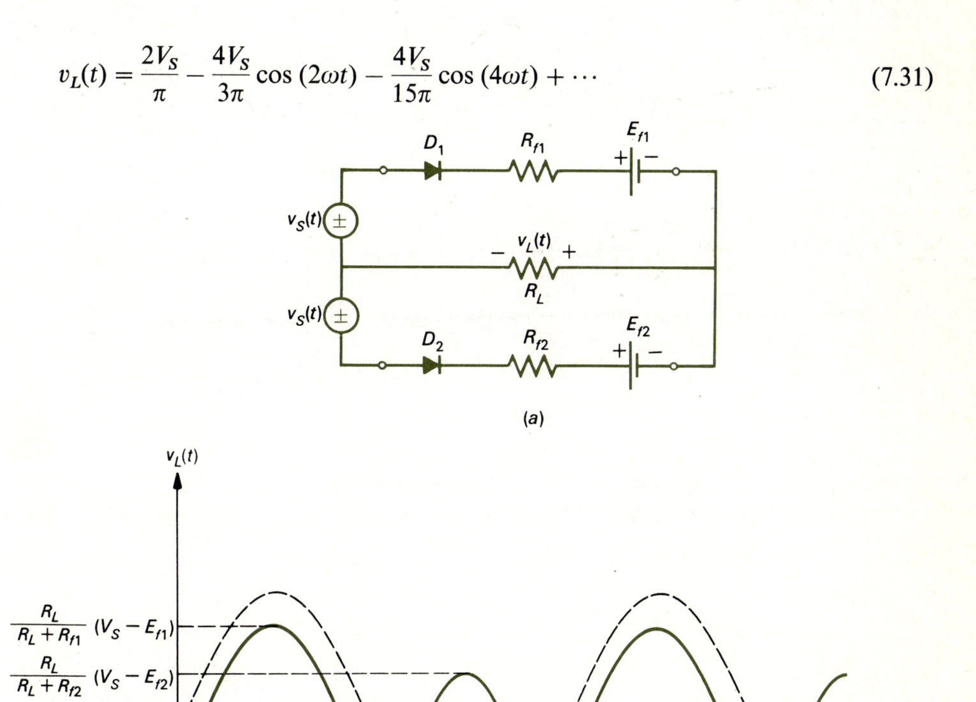

Comparing Eqs. (7.31) and (7.24), we note an advantage of the full-wave rectifier. First of all, the fundamental harmonic (sin ωt) present in the half-wave rectifier result is eliminated from the full-wave rectifier result. This indicates that the undesired (non-constant) portion of $v_L(t)$ is reduced, as is clear from the waveform.

Replacing each diode with its piecewise-linear approximation, we obtain the circuit and resulting waveform shown in Fig. 7.21. We have shown the waveform for the case where the two diodes have different values of R_f and E_f. The different values of forward voltage E_f causes each ideal diode to delay its turn-on time by different amounts.

7.5.5 The Full-Wave Bridge Rectifier

The full-wave rectifier considered in the previous section must employ a center-tapped transformer for proper operation. A full-wave rectifier can be constructed without the

FIGURE 7.22
The bridge rectifier.

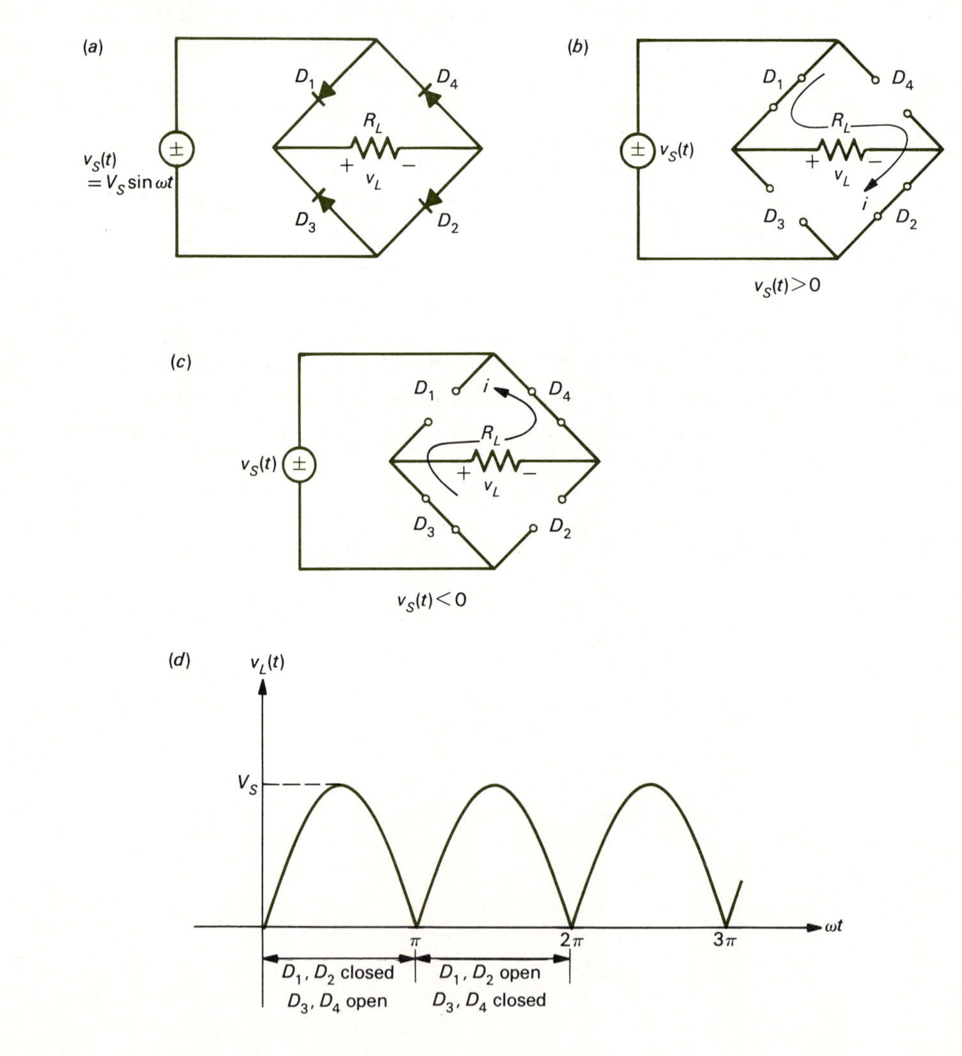

need for a transformer, but we must use four diodes instead of two. This type of rectifier is called the bridge rectifier and is shown in Fig. 7.22*a*. When the source voltage is positive, diodes D_1 and D_2 are forward-biased and closed but D_3 and D_4 are reverse-biased and open, resulting in $v_L(t) = v_S(t)$, as shown in Fig. 7.22*b*. When source voltage is negative, diodes D_1 and D_2 are reverse-biased and open and D_3 and D_4 forward-biased and closed, resulting in $v_L(t) = -v_S(t)$, as shown in Fig. 7.22*c*. The resulting waveform is a full-wave rectified wave having the same dc component as for the full-wave rectifier which used a 1:2 turns-ratio transformer. Of course, a transformer could be used to increase (or decrease) $v_S(t)$.

7.6 THE LIMITER

Quite often we need to restrict the variation of some voltage within certain limits to prevent an excessive voltage from appearing across the terminals of a device. A limiter, constructed from diodes, is a device which performs this function. For example, suppose we wish to restrict some voltage $v_L(t)$ across a load between the limits V_1 and $-V_2$, as shown in Fig. 7.23*a*. Inserting a limiter between the source and load will perform this function, as shown in Fig. 7.23*b*.

A simple limiter consisting of two ideal diodes and two batteries is shown in Fig. 7.24*a*. We may redraw the circuit, using the stretch-and-bend principle, by moving the load resistor to the input of the limiter, as shown in Fig. 7.24*b*: the load voltage is unchanged. This may be further reduced for analysis by converting the resulting circuit at the input terminals of the limiter to a Thevenin equivalent. If the load voltage attempts to exceed the battery voltage V_1, diode D_1 closes at $v_L(t) = V_1$. Similarly, if $v_L(t)$ attempts to go more negative than $-V_2$, diode D_2 closes at $v_L(t) = -V_2$. The resulting circuit for each of these cases is shown in Fig. 7.24*c*. The transfer characteristic of the limiter in terms of $v_S(t)$ and $v_L(t)$ is shown in Fig. 7.24*d* along with a typical waveform showing the performance of the limiter.

Replacing the ideal diodes with piecewise-linear ones, we obtain the circuit in Fig. 7.25*a*, which may be reduced to the circuit in Fig. 7.25*b* for the purpose of analysis. Note that when $v_L(t) > (E_{f1} + V_1)$, diode D_1 closes and D_2 is open so that

$$v_L(t) = \frac{R_{f1}}{R_{f1} + R_S||R_L}\left[\frac{R_L}{R_S + R_L} v_S(t)\right] + \frac{R_L||R_S}{R_L||R_S + R_{f1}}[E_{f1} + V_1] \tag{7.32}$$

as shown in Fig. 7.25*c*. Thus, $v_L(t)$ and $v_S(t)$ are related on the transfer characteristic shown in Fig. 7.25*d* by a line of slope

$$\frac{R_{f1}}{R_{f1} + R_S||R_L}\,\frac{R_L}{R_S + R_L}$$

When $v_L(t)$ is less than $E_{f1} + V_1$ but greater than $-(E_{f2} + V_2)$, both diodes are open and $v_S(t)$ and $v_L(t)$ are related by

$$v_L(t) = \frac{R_L}{R_L + R_S} v_S(t) \tag{7.33}$$

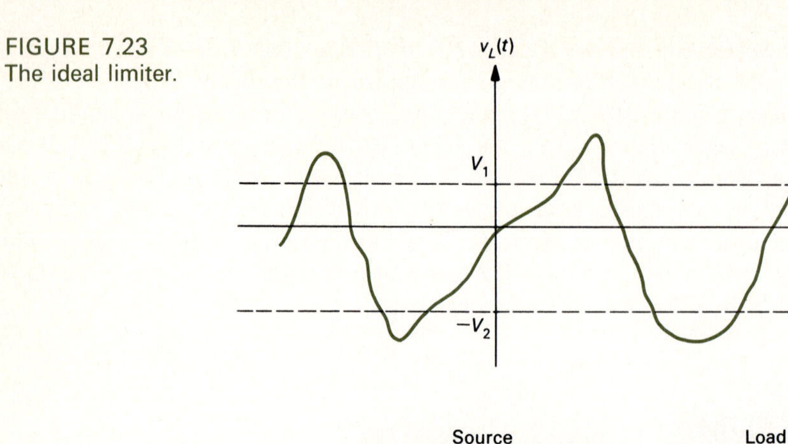

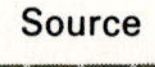

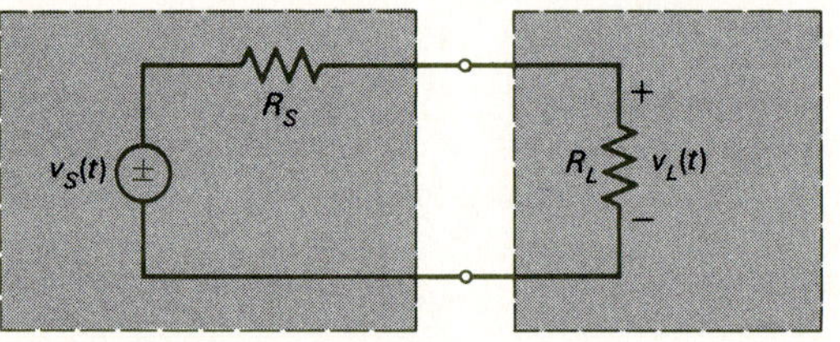

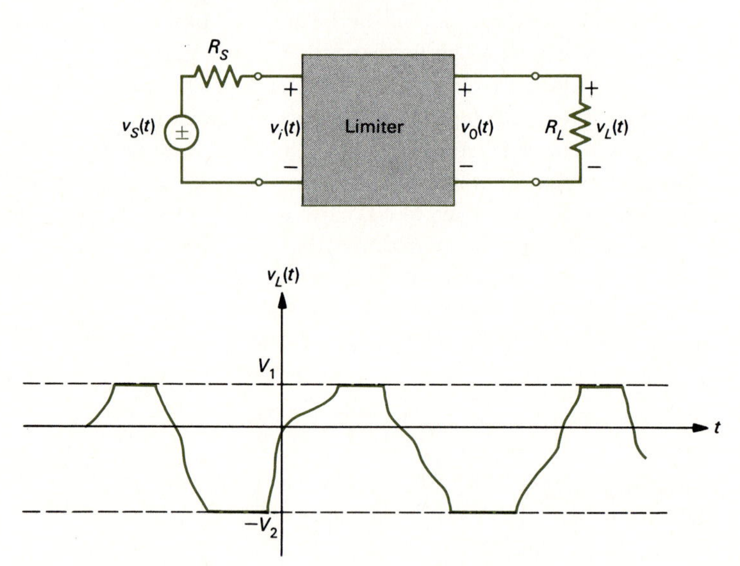

FIGURE 7.23
The ideal limiter.

The transfer characteristic and a typical waveform are shown in Fig. 7.25*d*. Note that the effect of the forward voltages E_{f1} and E_{f2} is to widen the limiting range of the limiter. The effect of the forward resistances R_{f1} and R_{f2} is to prevent perfect limiting. For typical semiconductor diodes, the forward resistances are quite small; thus their presence affects the limiting action only slightly.

FIGURE 7.24
Synthesis of a limiter, using ideal diodes.

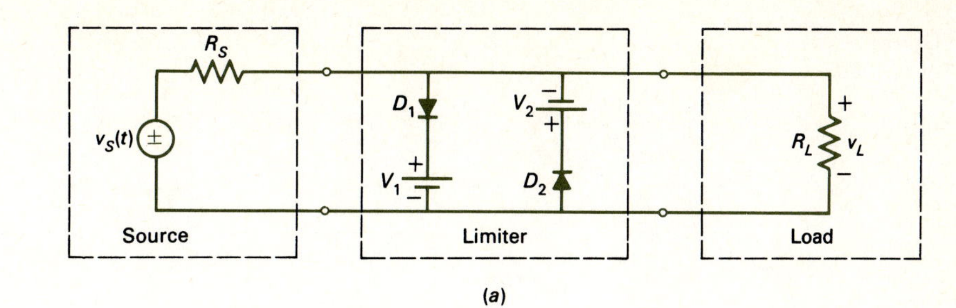

(a)

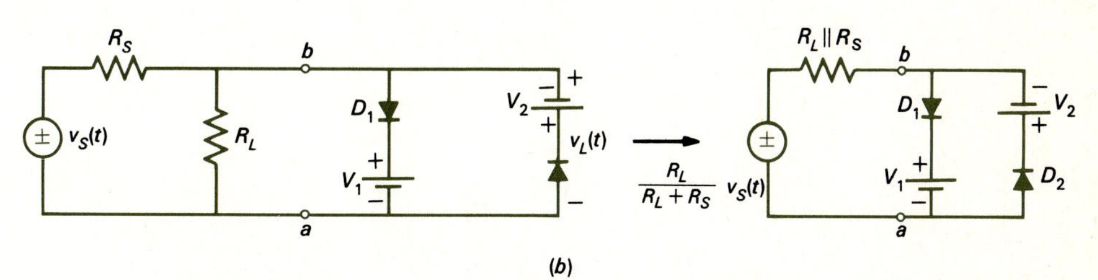

(b)

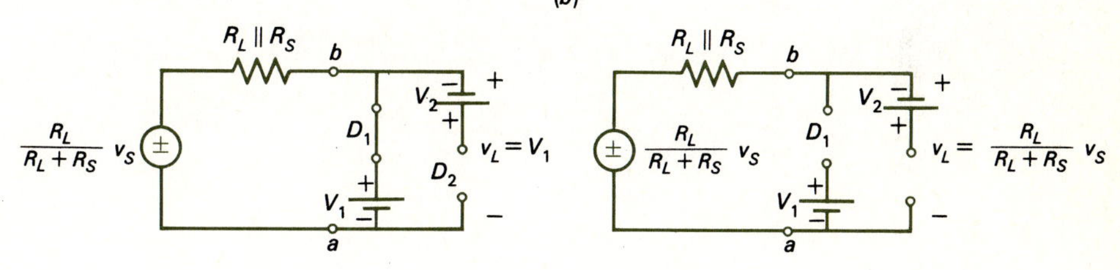

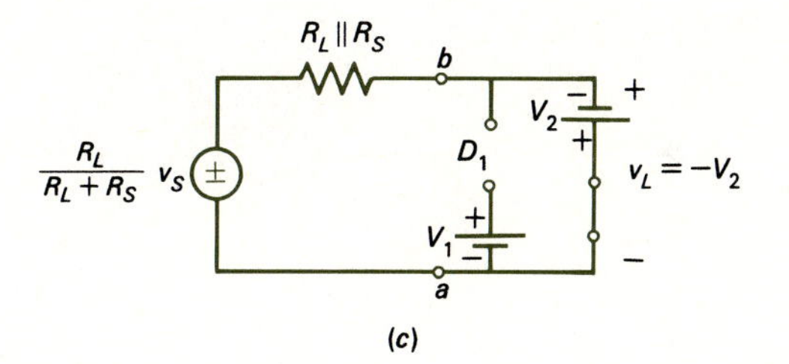

(c)

FIGURE 7.24
(*Continued*)

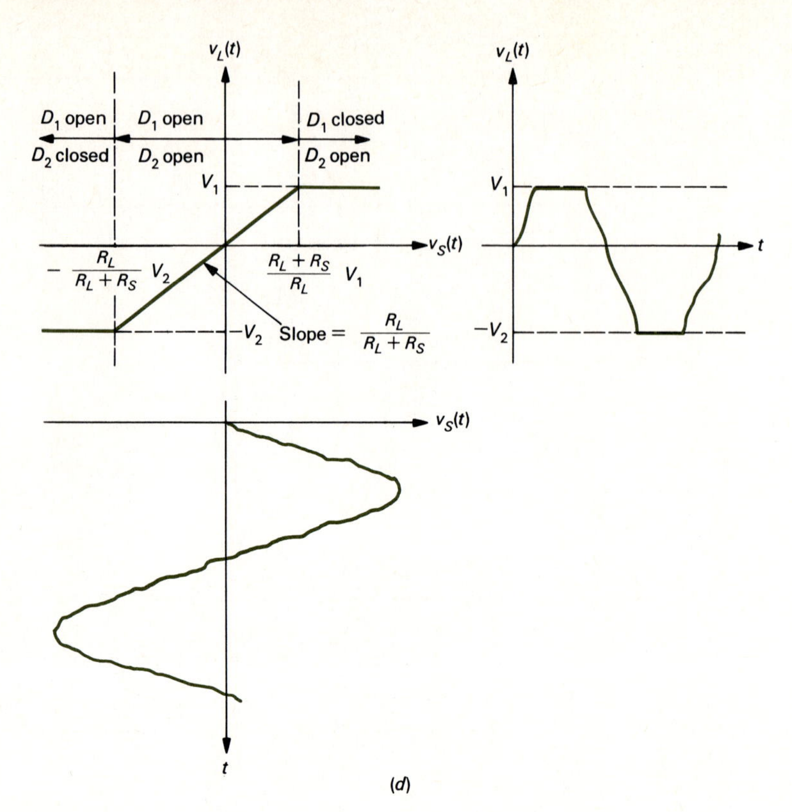

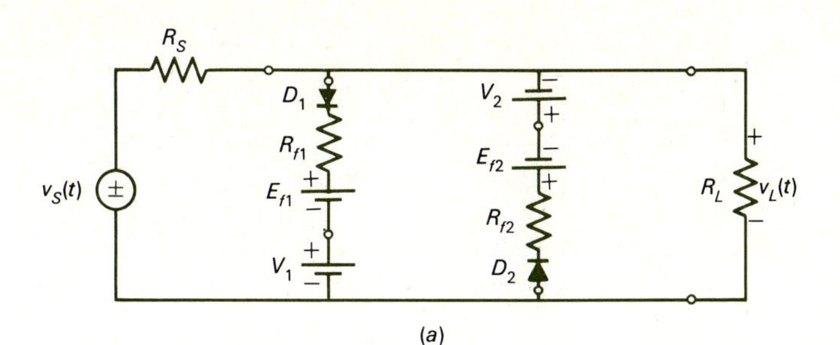

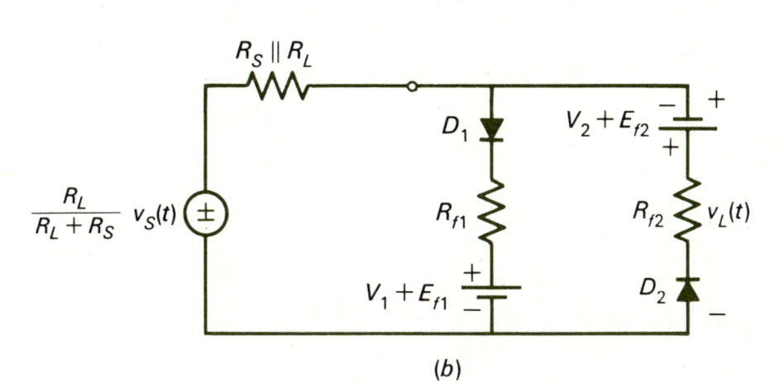

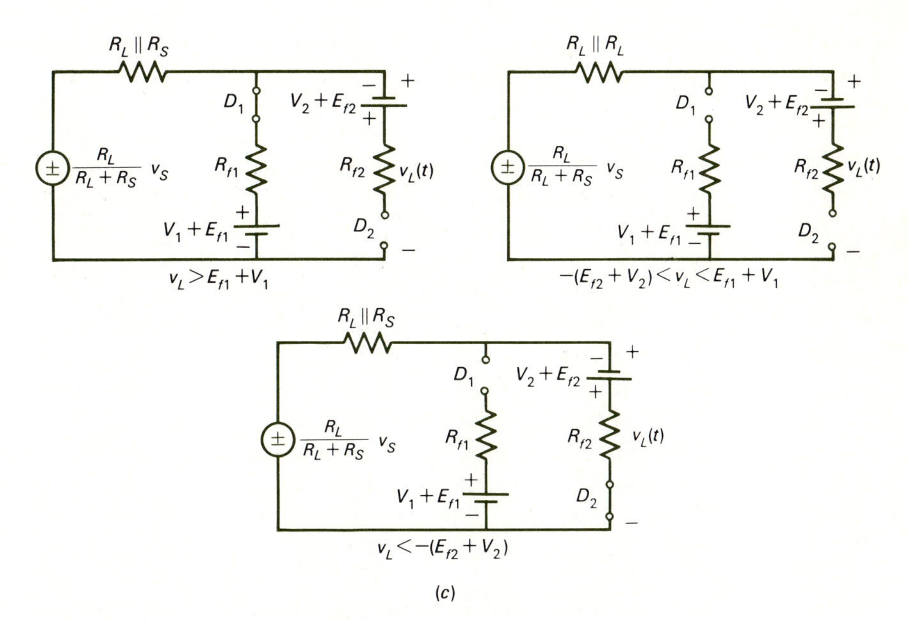

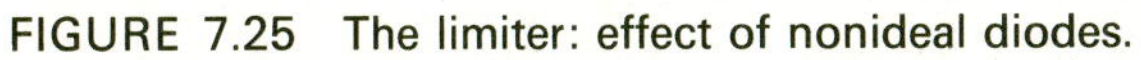
FIGURE 7.25 The limiter: effect of nonideal diodes.

FIGURE 7.25
(*Continued*)

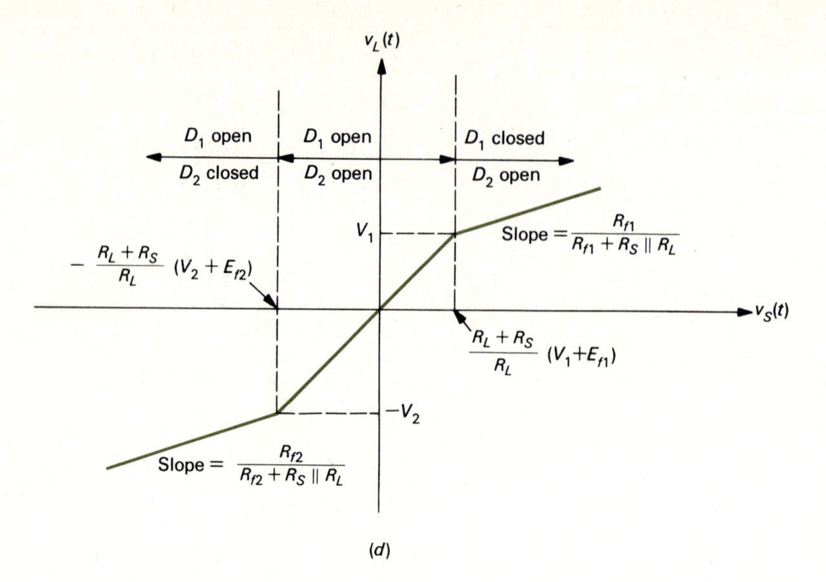

(*d*)

7.7 ZENER DIODES

It was pointed out in the discussion of semiconductor diodes that the diode conducts only a small reverse saturation current I_S under reverse-biased conditions. However, if the reverse-biased voltage is increased to a large value, the semiconductor diode will "break down" and conduct a significant current in the reverse direction. This phenomenon is referred to as avalanche breakdown; it results from the charge carriers acquiring sufficient energy between collisions to dislodge electrons from some of the covalent bonds. These dislodged electrons can cause other electrons to dislodge, which results in a cumulative, or avalanche, effect. A similar effect, the zener effect, occurs at lower reverse voltage. In this effect, the electric field becomes sufficiently intense to pull electrons from the covalent bonds, resulting in a large reverse current. Certain semiconductor diodes are commercially available with precisely controlled zener breakdown voltages V_z, and these are called zener diodes.

A typical characteristic and device symbol for a zener diode is shown in Fig. 7.26*a*. In the forward-biased direction ($v_d > 0$) the zener diode behaves in a manner which is very similar to the semiconductor diode. However, in the reverse-biased region the curve drops abruptly at $v_d = -V_z$ with a slope $1/R_z$. R_z is called the zener resistance; typical values range from 0.1 to as much as 200 Ω, depending on the zener voltage V_z and the maximum average power the device can dissipate, P_{max}. We will assume in our analyses that $R_z = 0$. Thus, in the reverse-biased region the zener voltage is $v_d = -V_z$ for all $i_d < 0$. A piecewise- linear approximation of the characteristic is shown in Fig. 7.26*b*, and the equivalent circuit using two ideal diodes is shown in Fig. 7.26*c*.

Zener diodes as well as semiconductor diodes are rated in terms of the maximum average power which they may safely dissipate. This is shown on the device character-

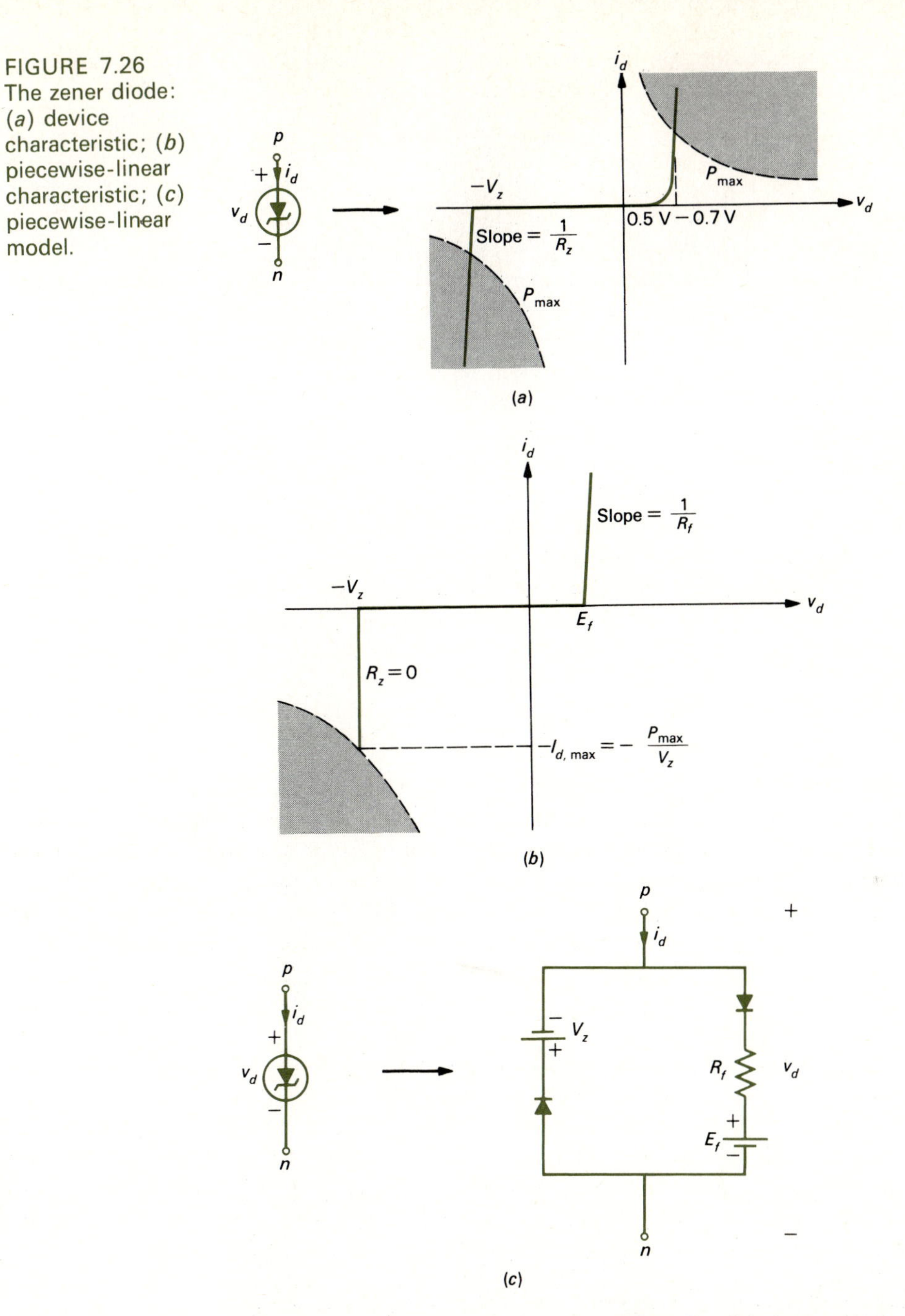

FIGURE 7.26
The zener diode: (a) device characteristic; (b) piecewise-linear characteristic; (c) piecewise-linear model.

istic as a hyperbola: $P_{max} = v_d i_d$. Ordinarily, this is an average power rating since it relates to the ability of the device to dissipate accumulated heat; consequently, it is a function of the waveshape of v_d and i_d. For certain waveforms we may operate outside the P_{max} hyperbola if we do so for a short time. A safe criterion, however, is to ensure that at no time does the product of the *instantaneous* values of v_d and i_d exceed P_{max}. In the reverse-biased region, we assume that $v_d = -V_z$ ($R_z = 0$); thus, the maximum value of reverse current which the diode may conduct is

$$I_{d,\max} = \frac{P_{\max}}{V_z} \qquad v_d < 0 \tag{7.34}$$

as shown in Fig. 7.26*b*.

Zener diodes are commonly operated in the reverse region such that the operating point lies on the reverse breakdown region of the zener curve and the reverse current does not exceed $I_{d,\max}$. In this mode of operation, the device functions as a regulator to maintain a constant voltage level of V_z across some load.

For example, typical dc voltage sources such as batteries are not ideal and have a source resistance R_S and an open-circuit voltage V_S, as shown in Fig. 7.27*a*. If this dc source is attached to a load R_L, the load voltage V_L will be dependent on the value of the source resistance since some of the open-circuit voltage $I_L R_S$ is dropped across the source resistance. Changing values of R_L will result in changing values of load voltage since the load current will change.

To maintain a constant value of load voltage when R_L changes value, we may insert a regulator between the source and load, as shown in Fig. 7.27*b*. The regulator consists of a resistor R and a zener diode. Note the connection of the zener diode. The load voltage V_L is the negative of the diode voltage; $V_L = -v_d$. The battery voltage V_S is assumed known, as is the source resistance R_S. Given a range of possible values of R_L, we would like to determine a value of R such that the load voltage is maintained at the zener voltage $V_L = V_z$ for these various values of the load resistance.

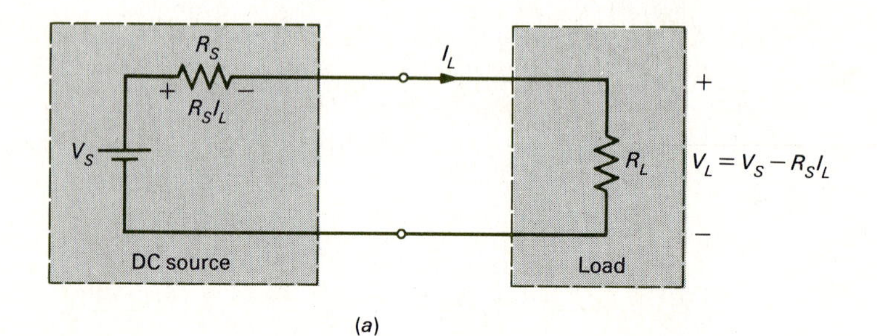

(*a*)

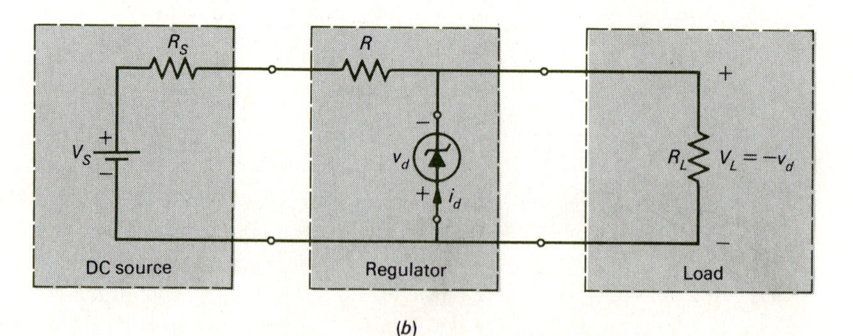

(*b*)

FIGURE 7.27 A zener diode-regulated power supply.

Another application is maintenance of a constant voltage across the load when the load resistance is fixed but the source voltage V_S varies. For example, the rectifiers discussed in a preceding section produce a pulsating dc voltage. However, we wish to obtain a constant value of voltage. Filters can be used to smooth this voltage somewhat (see Fig. 7.18, for example). A zener diode regulator may also be used for these purposes.

The operation of the regulator circuit can be understood if we redraw the circuit, using the stretch-and-bend principle, as shown in Fig. 7.28*a*, by moving the load resistor to the left of the zener diode. This circuit and the original circuit in Fig. 7.27*b* are *identical.* Next we reduce the remainder of the circuit attached to the terminals of the zener to a Thevenin equivalent, as shown in Fig. 7.28*b*, for the purposes of drawing a load line on the characteristic and determining the zener's operating point. From Fig. 7.28*b* we obtain

$$\begin{aligned} v_d &= -V_{OC} - R_{TH} i_d \\ &= -\frac{R_L}{R_L + R_S + R} V_S - [(R_S + R)||R_L] i_d \end{aligned} \tag{7.35}$$

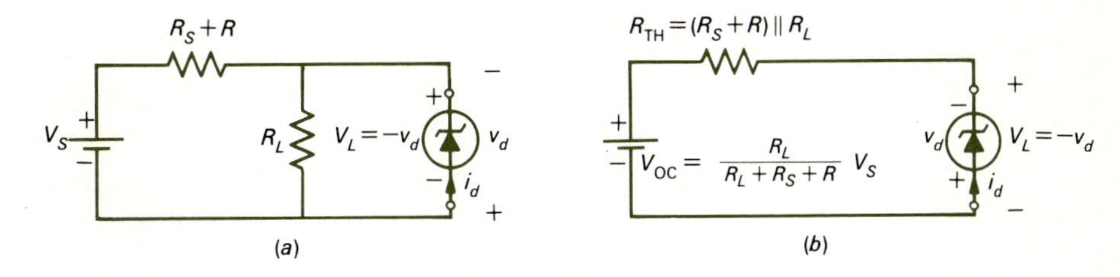

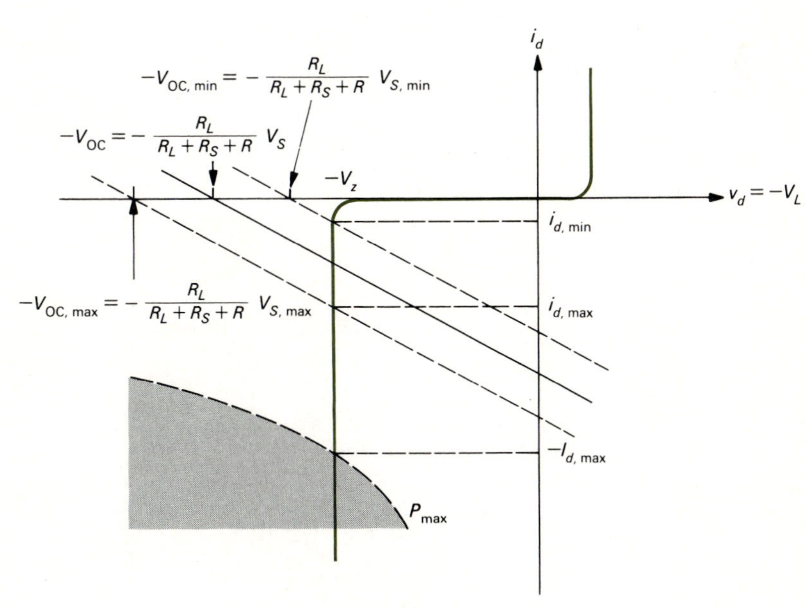

FIGURE 7.28 Analysis of the zener diode-regulated power supply.

which is plotted on the zener characteristic in Fig. 7.28*c*. The intersection of the load line on the v_d axis ($i_d = 0$) is found from Eq. (7.35) by substituting $i_d = 0$ as

$$v_d = -V_{OC}$$

$$= -\frac{R_L}{R_L + R_S + R} V_S \qquad i_d = 0 \tag{7.36}$$

The slope of the load line is $-1/R_{TH}$, where

$$R_{TH} = (R_S + R)||R_L \tag{7.37}$$

Note from Fig. 7.27 that the load voltage V_L is the negative of the diode voltage. So long as the load line remains on the reverse zener characteristic, $V_L = -v_d = V_z$.

In the case of a fixed load resistance and a source voltage V_S which varies between $V_{S,\max} > V_S > V_{S,\min}$, we may select a value of R such that for $V_S = V_{S,\min}$ the intersection of the load line on the v_d axis is to the left of $-V_z$, as shown in Fig. 7.28*c*. Thus, the value of R is such that

$$\frac{R_L}{R_L + R_S + R} V_{S,\min} \geq V_z \tag{7.38}$$

Similarly, the power rating of the zener must be such that for $V_S = V_{S,\max}$ the operating point does not fall within the maximum power hyperbola defined by $P_{\max}$.

A simple way of determining R and the maximum required power rating of the zener is from the circuit shown in Fig. 7.29. Note that for either $V_{S,\min}$ or $V_{S,\max}$ we presume that $v_d = -V_z$; thus $V_L = V_z$. The current I through $R_S + R$ is then

$$I = \frac{V_S - V_z}{R_S + R} \tag{7.39}$$

The load current is

$$I_L = \frac{V_L}{R_L}$$

$$= \frac{V_z}{R_L} \tag{7.40}$$

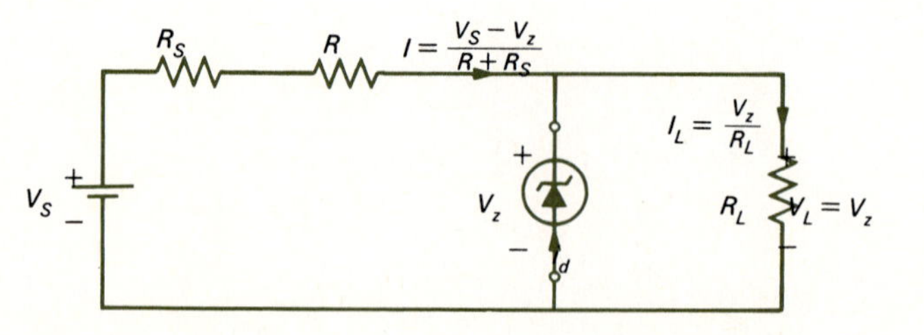

FIGURE 7.29 Simplified analysis of the zener diode-regulated power supply.

The zener current is thus

$$i_d = I_L - I$$
$$= \frac{V_z}{R_L} - \frac{V_S - V_z}{R_S + R} \quad (7.41)$$

For $V_S = V_{S,\text{min}}$, i_d must be negative and nonzero (the zener must be conducting in its reverse region). Thus, from Eq. (7.41)

$$I \geq I_L \quad (7.42)$$

or

$$\frac{V_{S,\text{min}} - V_z}{R_S + R} > \frac{V_z}{R_L} \quad (7.43)$$

which may be solved for R as

$$R \leq R_L \frac{V_{S,\text{min}} - V_z}{V_z} - R_S \quad (7.44)$$

For $V_S = V_{S,\text{max}}$, the maximum zener current is, from Eq. (7.41),

$$i_{d,\text{max}} = \frac{V_z}{R_L} - \frac{V_{S,\text{max}} - V_z}{R_S + R} \quad (7.45)$$

from which the required maximum power rating is

$$P_z \geq -V_z i_{d,\text{max}} \quad (7.46)$$

Example 7.4 A source voltage varies between 120 and 75 V. The source resistance is zero ($R_S = 0$) and the load resistance is 1000 Ω. It is desired to maintain the load voltage at 60 V. Determine R and the required power rating of the zener.

Solution We select a zener with zener voltage of 60 V. The required resistance is obtained from Eq. (7.44) as

$$R \leq \frac{R_L(V_{S,\text{min}} - V_z)}{V_z}$$
$$= \frac{1k(75 - 60)}{60}$$
$$= 250\ \Omega$$

The maximum zener current is computed from Eq. (7.45) as

$$i_{d,\,\max} = \frac{60}{1k} - \frac{120 - 60}{250}$$

$$= 0.06 - 0.240$$

$$= -0.180 \text{ A}$$

so that the required power rating of the zener is *at least*

$$P_z \geq -V_z i_{d,\,\max}$$

$$= 60(0.180)$$

$$= 10.8 \text{ W}$$

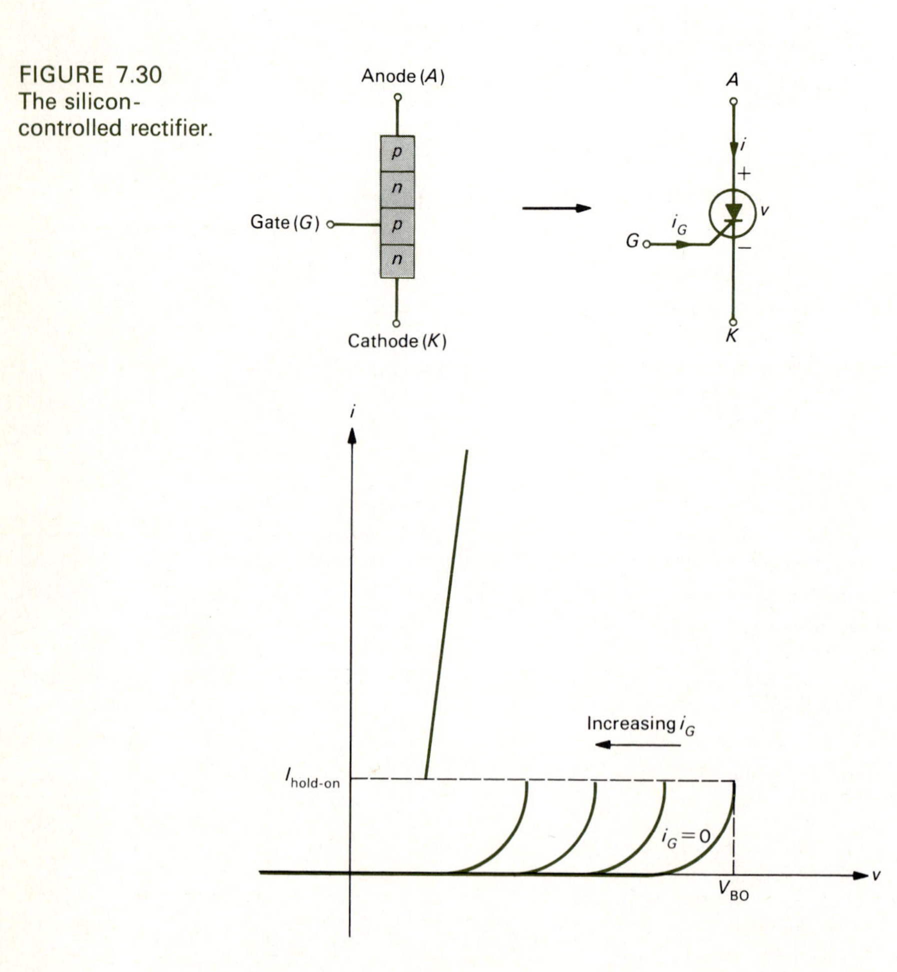

FIGURE 7.30 The silicon-controlled rectifier.

7.8 SILICON-CONTROLLED RECTIFIERS (SCRs)

The silicon-controlled rectifier (SCR) is a three-terminal device consisting of four layers of alternating *p*- and *n*-type semiconductor, as shown in Fig. 7.30. The three terminals are labeled as anode (A), cathode (K), and gate (G). Except for values of i between 0 and $I_{\text{hold-on}}$, the SCR behaves like a semiconductor diode. It does not conduct if the anode-cathode voltage v is negative. Note that for small positive values of i, the characteristic is determined by the value of the gate current i_G. As current i is increased from zero, the voltage v across the anode-cathode increases along the curve determined by i_G until it reaches the breakover voltage V_{BO}. At that point, further increases in i above $I_{\text{hold-on}}$ cause the SCR to "fire," and v drops to a much lower voltage (typically 0.5 V, similar to that of a diode). Once the SCR has fired, the gate loses control until the current i is reduced below $I_{\text{hold-on}}$.

It is possible to switch large currents i with small currents i_G. For example, consider the circuit of Fig. 7.31, where V and R are fixed. The load line is plotted on the characteristic from

$$v = V - Ri \tag{7.47}$$

which is obtained by writing KVL around the anode-cathode loop. If the gate current switches from zero to I_G', the curve changes from the value for $i_G = 0$ to $i_G = I_G'$ and the operating point switches from one to two.

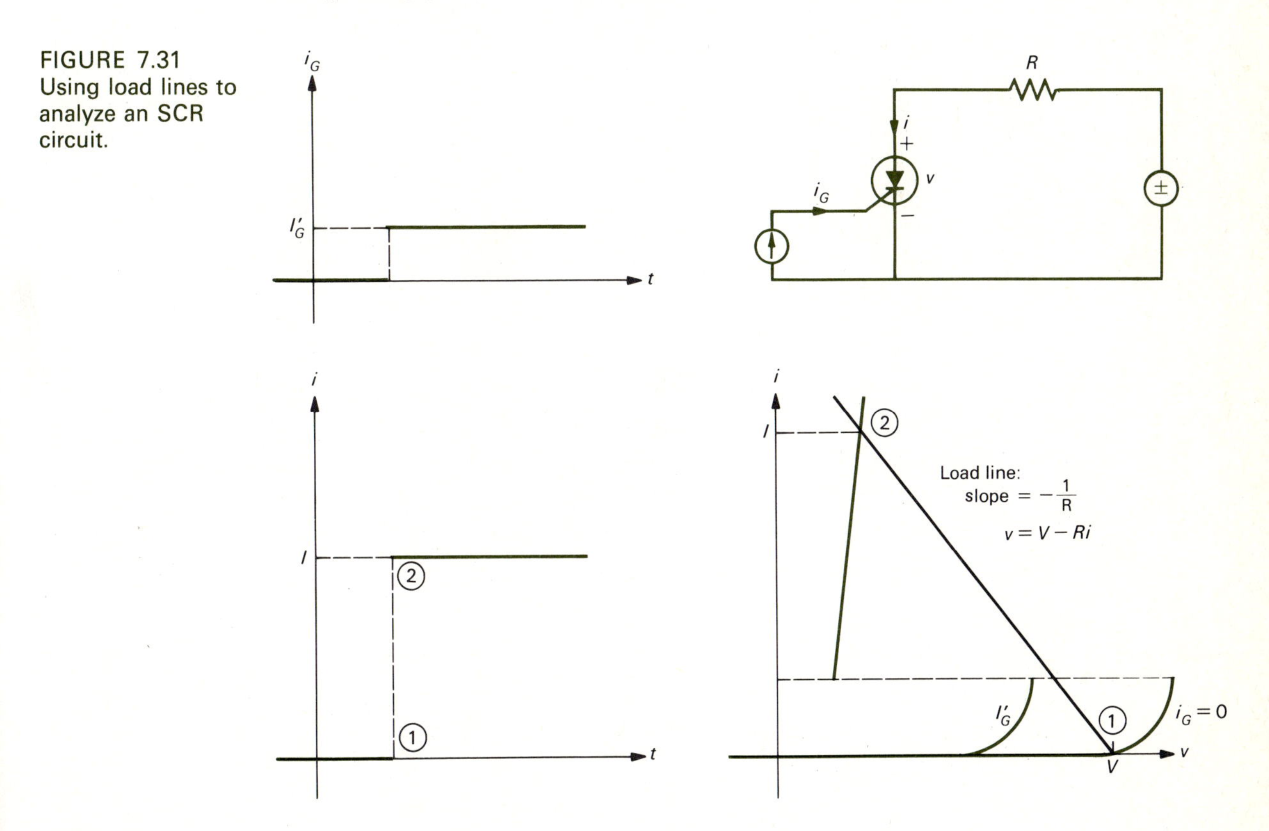

FIGURE 7.31
Using load lines to analyze an SCR circuit.

FIGURE 7.32
An SCR light dimmer circuit.

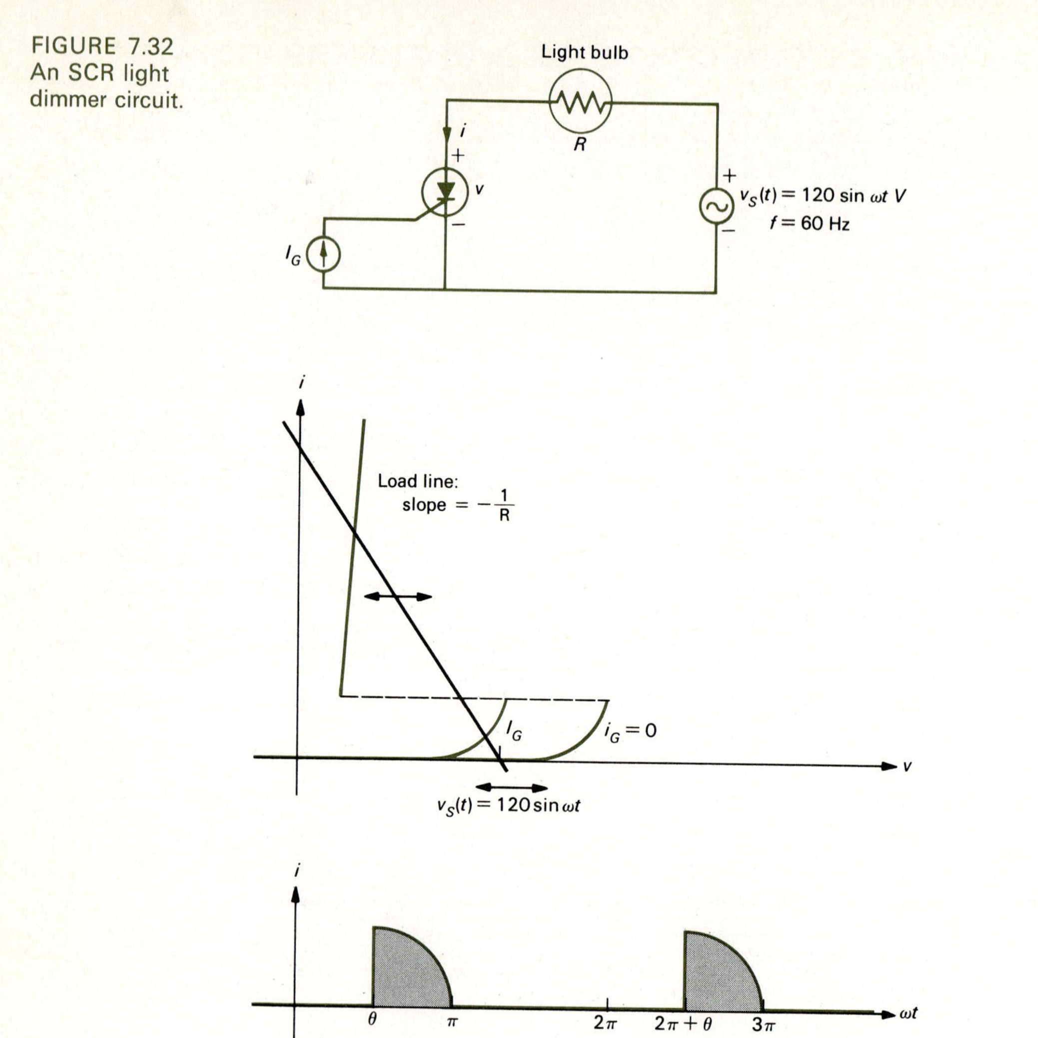

The SCR can be used as a controllable rectifier. For example, the circuit in Fig. 7.32 will control the power to the light bulb. The average power delivered to the light bulb (and consequently the light output) is a function of the power dissipated by the bulb, which is a function of the shaded area under the current waveform:

$$P_{\text{AV}} = \frac{1}{2\pi} \int_0^{2\pi} i^2 R \, d(\omega t) \tag{7.48}$$

As the sinusoidal voltage $v_S(t) = 120 \sin \omega t$ varies, the load line moves right and left on the characteristic. When $v_S(t)$ increases from zero, essentially no anode current is drawn until $v_S(t)$ exceeds V_{BO} for the particular i_G used. When that occurs, the SCR fires and becomes essentially a short circuit. (There is approximately 0.5 V dropped across the SCR, as with other semiconductor diodes.) After the SCR fires at $\omega t = \theta$, the voltage across the bulb is essentially $v_S(t)$. When $v_S(t)$ is reduced towards zero, the anode current drops below $I_{\text{hold-on}}$ and the SCR opens. For the negative half cycle of $v_S(t)$, the SCR remains open. The light output can be adjusted by adjusting the firing angle θ, which is controlled by the value of i_G.

SCRs are also used in speed control circuits for electric motors, which will be discussed in Part 3. There are other types of the thyristor class of devices of which the SCR is a member. Another common thyristor is the triac, which is very similar to the SCR except that it conducts in either direction; the curve has the mirror image of the SCR forward-biased characteristic in the reverse-biased quadrant $v < 0, i < 0$.

PROBLEMS

7.1 A silicon diode is operated in two different environments. In one, the temperature is 15°C. In the other, the temperature is 35°C. Sketch the $v - i$ characteristics for these two environments. Assume that the saturation current is the same in the two environments and is 1 nanoampere (nA). Determine the diode current when the diode voltage is 0.5 and 0.2 V.

7.2 For the circuit shown in Fig. P7.2, determine the diode current and the average power dissipated by the diode.

7.3 Repeat Prob. 7.2 when the resistor is changed from 500 to 250 Ω.

7.4 Suppose the voltage and resistance in Fig. P7.2 are changed to 10 V and 2 kΩ. Determine the diode current and voltage and the power dissipated by the diode.

7.5 In the circuit shown in Fig. P7.5, determine the diode current and voltage and the power delivered by the source. The diode charcteristic is given in Fig. P7.2.

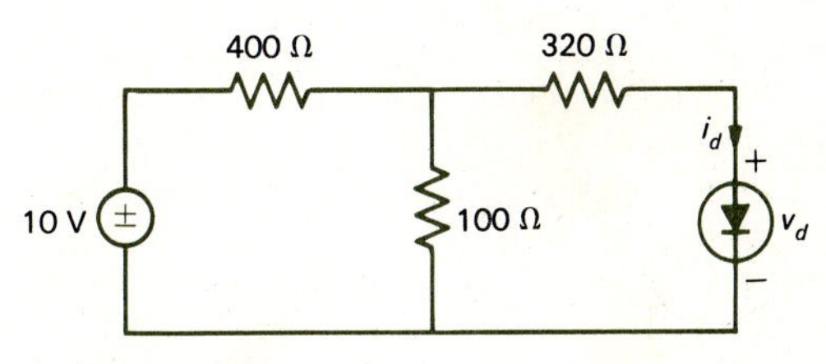

FIGURE P7.5

7.6 For the circuit shown in Fig. P7.6, determine the diode current and voltage and the power dissipated by the diode. The diode characteristic is given in Fig. P7.2.

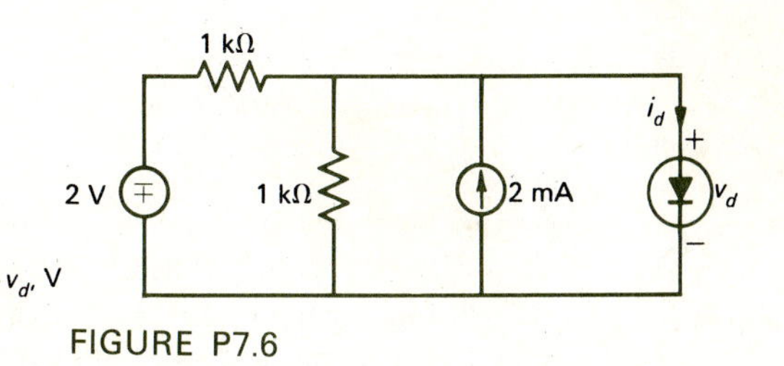

FIGURE P7.6

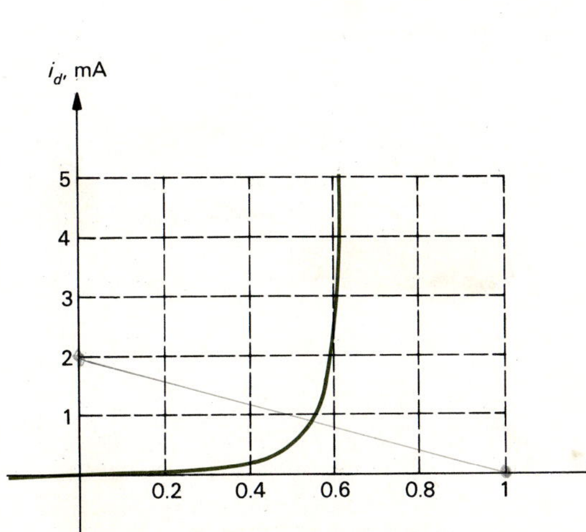

FIGURE P7.2

7.7 The diode in Fig. P7.2 is to be approximated by an ideal diode in series with a resistor R_f and voltage source E_f. Determine R_f and E_f if the operating point of the diode is $V_d = 0.6$ V, $I_d = 3$ mA. Repeat if the operating point is $V_d = 0.5$ V, $I_d = 0.5$ mA.

7.8 Repeat Prob. 7.7 using the diode equation. Assume the temperature to be 37°C.

7.9 For the circuit shown in Fig. 7.10*a*, with $V_S = 0.01$ V, $E_S = 5$ V, $R_S = 1$ kΩ, and using the diode shown in Fig. P7.2, write an approximate equation for the diode current.

7.10 For the half-wave rectifier shown in Fig. 7.17*a*, assume an actual diode whose characteristic is given in Fig. P7.2 and sketch $v_L(t)$ for $V_S = 2$ V and $R_L = 500$ Ω.

7.11 For the circuit shown in Fig. 7.18*a*, assume an ideal diode and $V_S = 10$ V, $\omega = 2\pi \times 10^3$, $C = 1$ μF, $R = 1$ kΩ. Sketch $v_L(t)$. What is the minimum value of $v_L(t)$ at any time after steady-state operation has been obtained?

7.12 For the full-wave bridge rectifier shown in Fig. 7.22*a*, assume identical piecewise-linear diodes with $R_f = 20$ Ω, $E_f = 0.5$ V, $R_L = 100$ Ω, and $V_S = 100$ V. Sketch $v_L(t)$.

7.13 For the full-wave rectifier with ideal diodes shown in Fig. 7.20*a*, assume that diode D_2 has been inserted opposite the direction shown. Sketch $v_L(t)$.

7.14 For the limiter shown in Fig. 7.24*a*, assume identical piecewise-linear diodes with $R_f = 100$ Ω, $E_f = 0.5$ V, $V_1 = V_2 = 10$ V, $R_L = 100$ Ω, $R_S = 100$ Ω, and $v_S(t) = 5 \sin \omega t$ V. Sketch $v_L(t)$.

7.15 For the zener diode regulator shown in Fig. 7.27*b*, assume that V_S varies between 40 and 60 V, $R_S = 100$ Ω, $R_L = 1$ kΩ. Choose a zener diode and R such that V_L is maintained at 30 V.

7.16 For the zener diode regulator shown in Fig. 7.27*b*, assume that $V_S = 50$ V, $R_S = 100$ Ω, and $R_L = 100$ Ω. Suppose a 20-V 1-W zener is to be used so that $V_L = 20$ V with R_L attached (full load) and with R_L removed (no load). Determine a range of values for R which will accomplish this.

7.17 In the SCR light control circuit shown in Fig. 7.32, assume that $R = 10$ Ω and $I_G = 10$ mA. Assume an ideal SCR having the characteristic shown in Fig. P7.17, and assume that when the SCR fires it is a short circuit. Sketch the current waveform of the light bulb.

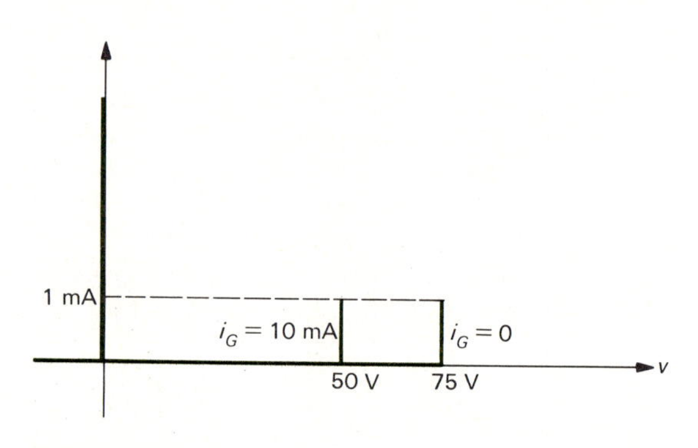

FIGURE P7.17

7.18 Repeat Prob. 7.17 if $R = 10$ kΩ.

REFERENCES

7.1 A. B. Carlson and D. G. Gisser, *Electrical Engineering: Concepts and Applications*, Addison-Wesley, Reading, Mass., 1981.

7.2 Ralph J. Smith, *Circuits, Devices and Systems: A First Course in Electrical Engineering*, 4th ed., Wiley, New York, 1984.

7.3 A. E. Fitzgerald, D. Higginbotham, and A. Grabel, *Basic Electrical Engineering*, 5th ed., McGraw-Hill, New York, 1981.

7.4 D. L. Schilling and C. Belove, *Electronic Circuits: Discrete and Integrated*, 2d ed., McGraw-Hill, New York, 1979.

7.5 R. L. Boylestad and L. Nashelsky, *Electronic Devices and Circuit Theory*, 3d ed., Prentice-Hall, Englewood Cliffs, N.J., 1982.

7.6 E. C. Lowenberg, *Electronic Circuits*, McGraw-Hill, New York, 1967.

Chapter

Transistors and Amplifiers

In the previous chapter we studied certain two terminal nonlinear resistors—which were referred to as diodes. A three terminal nonlinear resistor, the SCR, was also studied since its main use is in the construction of dc power supplies, which is one of the main uses of diodes.

In this chapter we will consider certain three terminal nonlinear resistors—which are called transistors. The two transistors which we will study are the field-effect transistor (FET) and the bipolar junction transistor (BJT). One of the main uses of these transistors is in the construction of amplifiers. We will study this use of these transistors in this chapter and in Chap. 9. Another important use of these devices is in the construction of digital circuits, such as computers. This use will be studied in Chap. 11.

As was the case for circuits containing diodes, circuits containing transistors are nonlinear. Consequently, such powerful techniques of linear circuit analysis as superposition do not generally apply. Moreover, the only feasible way of specifying these nonlinear devices is with graphical techniques; the devices cannot generally be characterized with equations. Therefore, the analysis of circuits containing these nonlinear devices is generally more difficult than for linear circuits, and graphical techniques (such as load lines) must be employed.

Quite often we will use certain approximations in order to simplify the analysis. The basis for these approximations is the small-signal linear operation of the device. Under this assumption of linear operation, the technique of superposition may be used to simplify the analysis. The use of these approximate techniques is important in gaining an understanding of the device performance.

8.1 PRINCIPLES OF LINEAR AMPLIFIERS

One of the primary uses of field-effect and bipolar junction transistors is in the construction of linear amplifiers. Essentially, a linear amplifier is a circuit which enlarges a signal (amplifier) and preserves its shape (linear). For example, consider Fig. 8.1*a*. A sinusoidal voltage source

$$v_i(t) = V_i \sin \omega t \tag{8.1}$$

is applied to the input terminals of a linear amplifier. We have shown this source to be, for illustration, a single-frequency sinusoid. However, no information is contained in a single-frequency sinusoid, so a more practical input signal would be some complicated waveform representing, for example, music from a phonograph cartridge. Since any

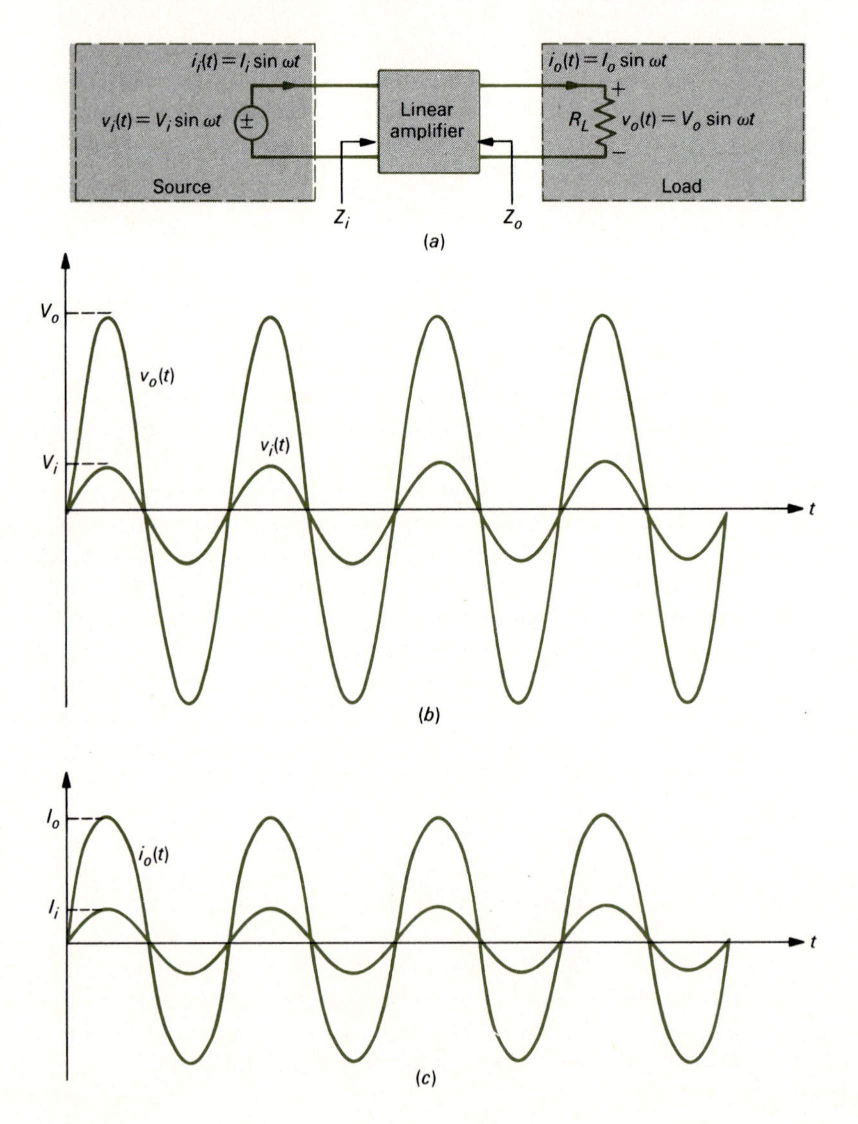

FIGURE 8.1 Illustration of general properties of a linear amplifier.

signal can alternatively be represented with Fourier methods as a sum of sinusoids at different frequencies and the amplifier is presumed to be linear, we may decompose the signal into its sinusoidal components and analyze the behavior of the amplifier for these sinusoidal components individually, and the sum of the responses to each will be, by superposition, the response to the more complicated signal. Thus, a single-frequency sinusoid can represent a rather general input signal.

Now suppose we wish to enlarge this sinusoidal signal and apply it to some load represented by R_L. This load may represent, for example, a loudspeaker. In order for the listener to hear the music, the signal applied to the loudspeaker

$$v_o(t) = V_o \sin \omega t \tag{8.2}$$

must be made much larger than the signal from the phonograph cartridge, as shown in Fig. 8.1*b*. If it is, then the amplifier is said to have *voltage gain*

$$\begin{aligned} A_v &= \frac{v_o(t)}{v_i(t)} \\ &= \frac{V_o}{V_i} > 1 \end{aligned} \tag{8.3}$$

For example, if the phonograph cartridge produces a sinusoid with a peak voltage of 1 mV and the peak voltage across the load is 100 mV, then the amplifier has a voltage gain of $A_V = 100$.

We also do not want the amplifier to distort the signal as it is being amplified. If the input is a pure 1-kHz sinusoid, we want the output to be a pure 1-kHz sinusoid. This is what is meant by a *linear* amplifier. If the music is distorted by the amplifier, a large voltage gain is unimportant.

At this point, we might wonder if there is a simple way to produce a device which has voltage gain. The answer to this is simple: If we want a linear amplifier with a voltage gain of 100, we simply use an ideal transformer with a turns ratio of 1:100! So there must be other criteria for a circuit to be an amplifier other than simply possessing a voltage gain. Another "figure of merit" of an amplifier is its current gain. For example, if the input current to the amplifier is

$$i_i(t) = I_i \sin \omega t \tag{8.4}$$

and the output current delivered to the load is

$$i_o(t) = I_o \sin \omega t \tag{8.5}$$

as shown in Fig. 8.1*c* then the amplifier is said to have a *current gain* of

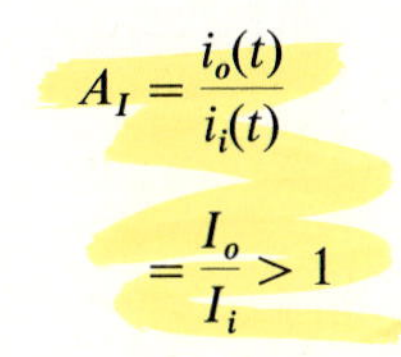

$$\begin{aligned} A_I &= \frac{i_o(t)}{i_i(t)} \\ &= \frac{I_o}{I_i} > 1 \end{aligned} \tag{8.6}$$

An ideal transformer with a turns ratio of 1:100 has a voltage gain of $A_V = 100$ but a current gain of $A_I = 1/100$ since, if the voltage is stepped up by a factor of 100, the current is stepped down by a factor of 1/100 and thus the ideal transformer has no power gain. Therefore, another figure of merit of an amplifier is its power gain:

$$\begin{aligned} A_P &= \frac{P_o}{P_i} \\ &= \frac{V_o I_o}{V_i I_i} \\ &= A_V A_I \end{aligned} \tag{8.7}$$

If the load represents a loudspeaker, it would be of little value to amplify the voltage if the power to drive the loudspeaker were not also amplified. Surely, the simple phonograph cartridge will not provide the power to drive a large loudspeaker. The additional power comes from the voltage sources within the amplifier which are used to "bias" it. This will be discussed in later sections of this chapter.

Two final figures of merit of linear amplifiers should be discussed. First, the ratio of input voltage to input current to the linear amplifier is known as the *input* impedance of the amplifier:

$$Z_i = \frac{v_i(t)}{i_i(t)} \tag{8.8}$$

It is an important quantity since, if the source has a source resistance r_s, we would like to have $Z_i = r_s$ so that maximum power transfer will take place from the source to the amplifier; in other words, the source and the amplifier would be matched.

Second, the source and amplifier can be replaced, as far as the load is concerned, with a Thevenin equivalent. The Thevenin impedance seen by the load is called the *output* impedance of the amplifier and is denoted by Z_o. For maximum power transfer from the amplifier to the load, we should have $Z_o = R_L$. Matching of the amplifier to the load is also important in the overall efficiency of the system.

In this chapter we will study linear amplifiers which are constructed from FETs and BJTs and analyze their performance. Our method of analysis will be a graphical one so that we can understand how these devices operate. In the next chapter we will investigate simpler ways of calculating A_V, A_I, A_P, Z_i, and Z_o for a particular amplifier.

8.2 THEORY OF OPERATION OF THE JUNCTION FIELD-EFFECT TRANSISTOR

There are several types of field-effect transistors. The primary ones are the junction field-effect transistor (JFET) and the metal-oxide-semiconductor field-effect transistor (MOSFET). The MOSFET is used primarily in integrated circuits and will be

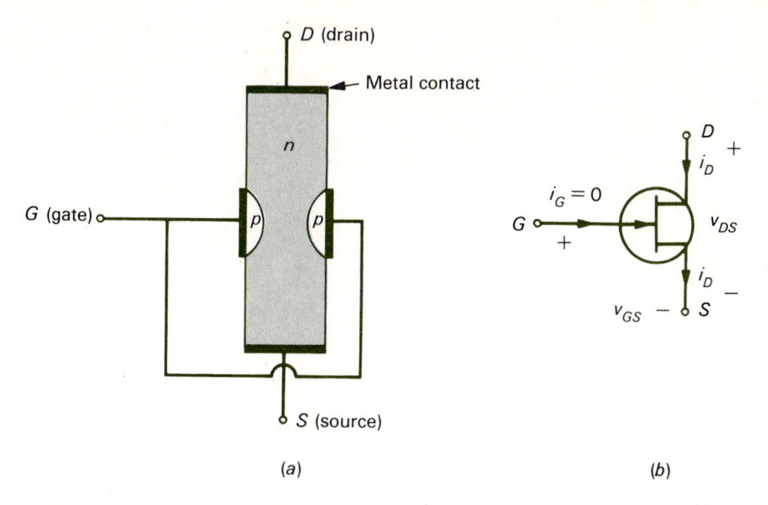

FIGURE 8.2 The junction field-effect transistor (JFET): (*a*) physical construction; (*b*) device symbol.

discussed in a later section. The analysis techniques for MOSFET amplifiers are virtually identical to those for JFET amplifiers.

The junction FET (JFET) is constructed of a piece of *n*-type material (called the channel) with two *p*-type regions imbedded in it, as shown in Fig. 8.2*a*. Two terminals, *D* (Drain) and *S* (Source), are attached to the two ends of the channel. A third terminal, *G* (Gate), is attached to the two *p* regions which are imbedded in ("grown" into) the channel. The device symbol with terminal currents and voltages is shown in Fig. 8.2*b*. We will show why the gate current i_G is essentially zero [typically 0.01 microamperes (μA)].

A *p*-channel JFET is also available and it consists of a *p*-type channel with *n*-type regions imbedded in it. The *n*-channel JFET requires positive power supplies (with respect to ground) in the amplifier biasing circuits and are thus more common than *p*-channel JFET amplifiers, which require negative power supplies. Our discussions will therefore concentrate on the *n*-channel JFET. Once we design an amplifier with an *n*-channel JFET, we may substitute a *p*-channel JFET (of the same type number) if we simply reverse the polarity of the dc power supplies.

To determine the characteristic of the JFET, let us apply a voltage V_{DS} between the drain and the source with the positive terminal on the drain and the negative terminal on the source, as shown in Fig. 8.3*a*. Let us also attach the gate to the source so that $v_{GS} = 0$. If we increase the battery voltage v_{DS}, we observe a linear increase in drain current, as shown in Fig. 8.3*b*. As the battery voltage is increased further, we reach a point $v_{DS} = V_p$ where further increases in v_{DS} result in only very small increases in drain current. Increasing v_{DS} further, we reach a point $v_{DS} = BV_{DSS}$ where the drain current increases dramatically. The current I_{DSS} is normally specified for the device by the manufacturer and is the drain-source current i_D with the gate short-circuited to the source ($v_{GS} = 0$).

The operation and resulting terminal characteristics of an *n*-channel JFET can be easily understood if we reconsider the semiconductor diode shown in Fig. 8.4. A depletion layer devoid of mobile charges is established at the *pn* junction by diffusion of the majority carriers of each material across the junction. This exposes the immobile donor and acceptor ions in a region of length d_o. If, as shown in Fig. 8.4*b*, a forward

FIGURE 8.3
Terminal properties of a JFET.

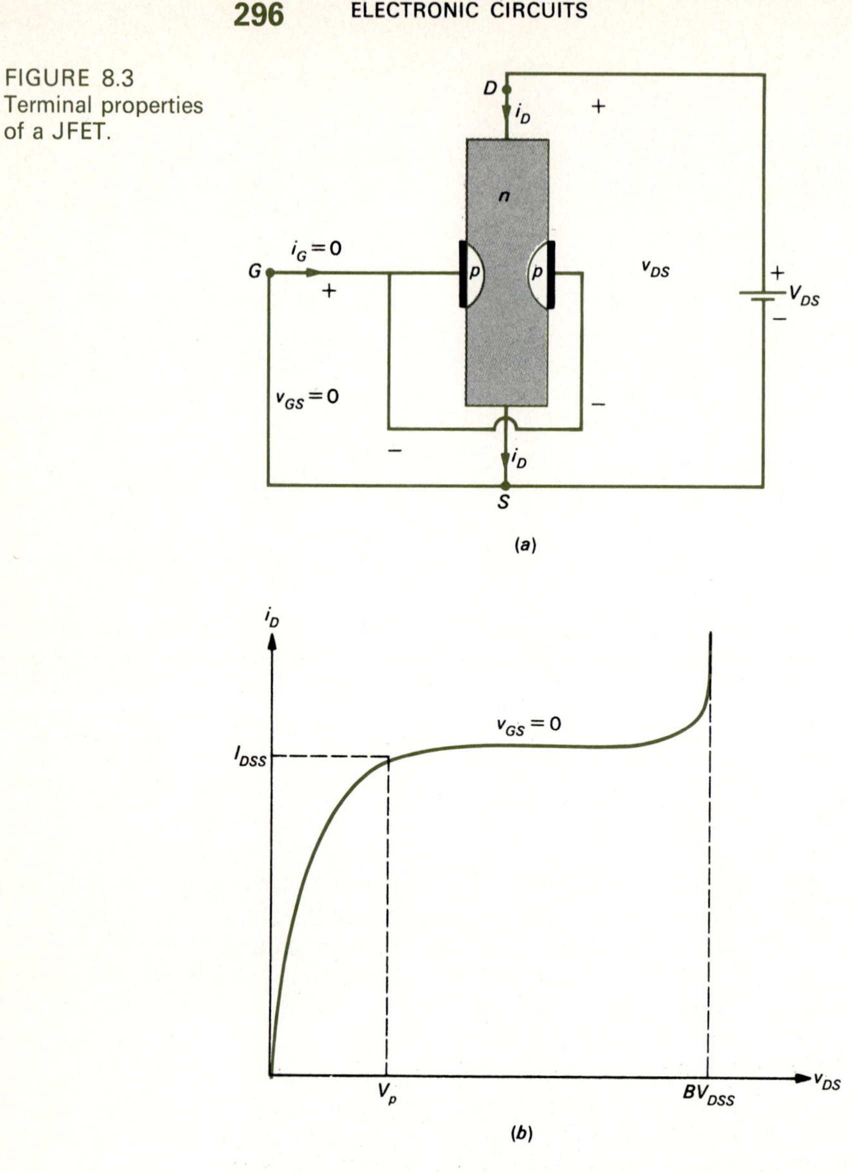

bias is applied by attaching the positive terminal of a battery V to the p region and the negative terminal to the n region, the depletion layer is reduced. If reverse bias is applied, as shown in Fig. 8.4c, the depletion layer is increased. Remember that the depletion layer is devoid of mobile charge carriers and thus may be thought of as having a large resistance.

This characteristic of an increased depletion region for a reverse-biased *pn* junction helps to explain the JFET characteristic in Fig. 8.3. Note that there are *pn* junctions formed between gate and source and between gate and drain. For $v_{GS} = 0$ in Fig. 8.3*a*, shown in Fig. 8.5*a*, a depletion region devoid of mobile charge carriers is established at the *pn* junctions. Since the gate is connected to the source, the gate-drain side of the *pn* junction will be more reverse-biased than the gate-source side since $v_{DG} > 0$ while $v_{GS} = 0$. This results in an asymmetrical depletion region. In any

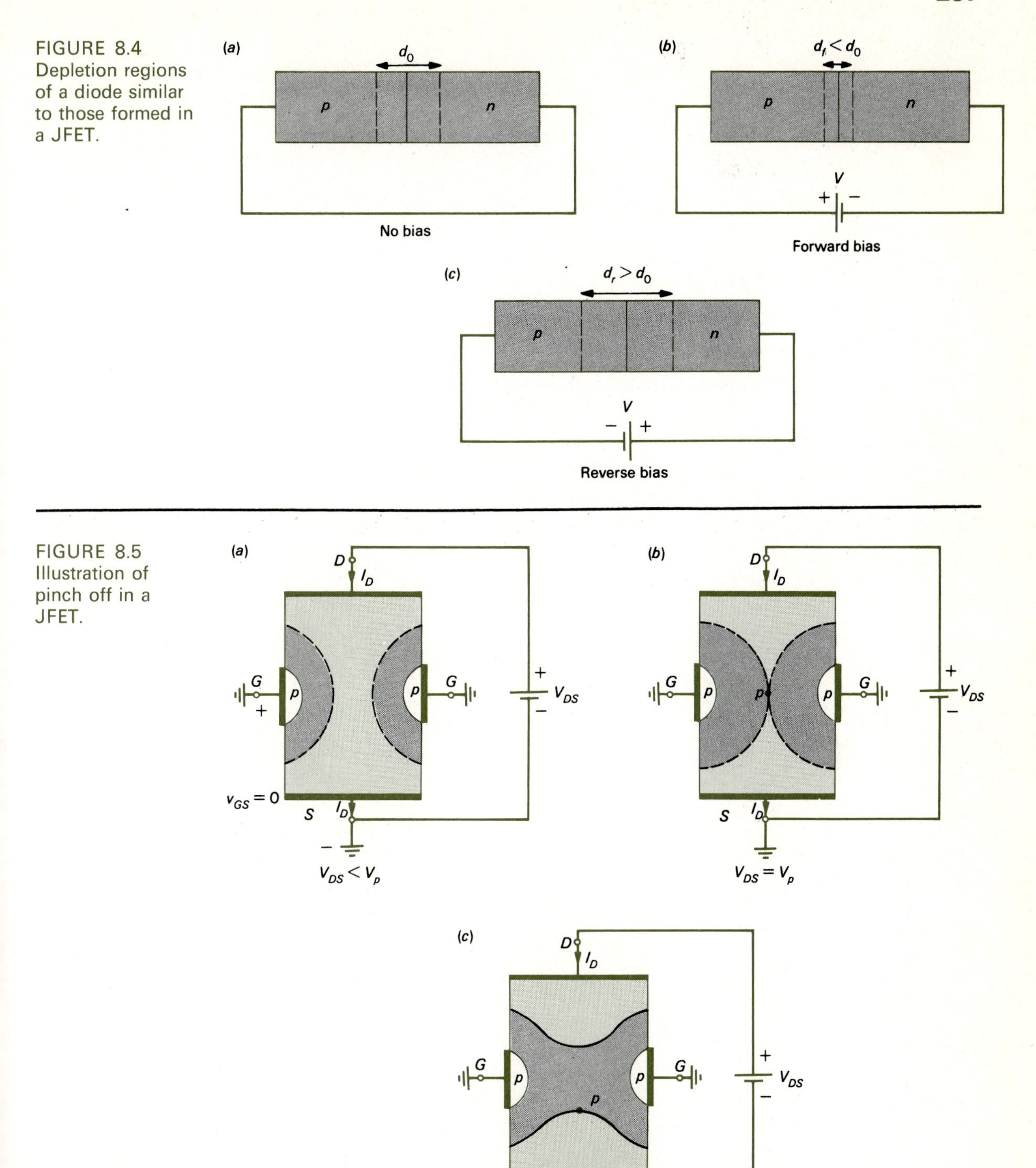

FIGURE 8.4 Depletion regions of a diode similar to those formed in a JFET.

FIGURE 8.5 Illustration of pinch off in a JFET.

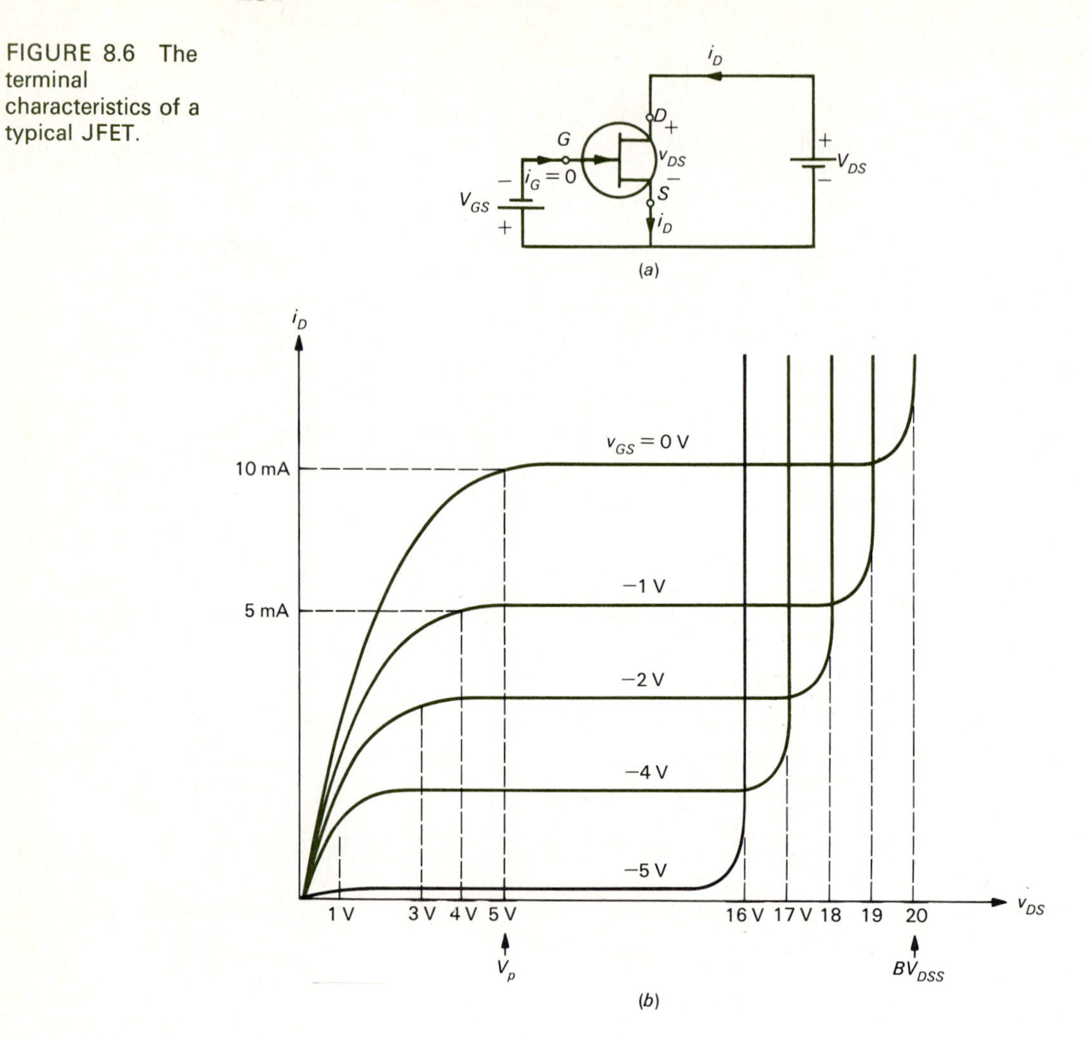

FIGURE 8.6 The terminal characteristics of a typical JFET.

case, negligible gate current flows since the *pn* junction is reverse-biased. This is why we assume that $i_G = 0$. If the gate to source is not reverse-biased ($v_{GS} > 0$), then gate current will flow.

The drain current flows through the *n*-channel and out of the source terminal and is therefore determined by the resistance of the channel. As v_{DS} is increased, i_D increases, but the depletion region and the resulting resistance of the channel become larger since the gate-drain *pn* junction becomes more reverse-biased. We eventually reach a point where the two depletion regions merge and the channel is "pinched off," as shown in Fig. 8.5*b*. The value of drain-source voltage is denoted as $v_{DS} = V_p$ and is referred to as the pinch-off voltage. Further increases in v_{DS} greater than V_p cause a further increase in the depletion region, as shown in Fig. 8.5*c*, but the voltage of a point *P* in this depletion region remains essentially constant at the pinch-off voltage. Thus, further increases in v_{DS} bring only minor increases in drain current, as shown in

Fig. 8.3*b*. If we further increase v_{DS}, we reach a point at which avalanche breakdown occurs, resulting in a rapid rise in drain current similar to that of semiconductor diodes. This voltage is denoted as BV_{DSS}, or breakdown voltage: drain to source with gate shorted (to source).

Now let us consider the resulting characteristic for other gate-source voltages. If we attach a battery V_{GS} between gate and source with the negative terminal attached to the gate, as shown in Fig. 8.6*a*, the *pn* junction will be reverse-biased and no gate current will flow. This is the typical mode of operation for the JFET. If we perform our previous experiment by increasing v_{DS} and recording the resulting drain current, we obtain the typical characteristic shown in Fig. 8.6*b*. For negative V_{GS}, the depletion region will be larger than for $V_{GS} = 0$ since the negative gate voltage adds a reverse bias to the *pn* junction in addition to the effect of v_{DS}. Thus, for the same value of v_{DS}, the drain current with $V_{GS} < 0$ will be less than with $V_{GS} = 0$ and the curves for increasingly negative V_{GS} will move downward. Note that the pinch-off and breakdown voltages become reduced for increasingly negative V_{GS}.

8.3 THE JFET AMPLIFIER

A simple JFET amplifier is shown in Fig. 8.7*a*. A battery V_{GG} is inserted in series with the voltage source $v_S(t) = V_S \sin \omega t$, with $V_{GG} > V_S$ so that the gate voltage always remains negative with respect to the source terminal. The output voltage is equal to the drain-source voltage $v_o = v_{DS}$. Here we have assumed that the load resistance in Fig. 8.1 is infinite. Writing KVL around the drain-source loop, we obtain

$$v_{DS} = V_{DD} - R_D i_D \tag{8.9}$$

which is plotted on the characteristic as a load line in Fig. 8.7*b*. Writing KVL around the gate-source loop, we obtain

$$v_{GS} = -V_{GG} + V_S \sin \omega t \tag{8.10}$$

When $v_S(t) = 0$, $v_{GS} = -V_{GG}$, which locates the dc operating point (V_{DS}, I_D), as shown in Fig. 8.7*b*. As $v_S(t)$ varies, the gate-source voltage varies sinusoidally between $-V_{GG} - V_S$ and $-V_{GG} + V_S$, and the operating point moves along the load line. The resulting drain-source voltage v_{DS} is sketched versus time. Note that as $V_S(t)$ increases, $v_{DS} = v_o$ decreases; thus, the output voltage is 180° out of phase with the source voltage. If the source voltage is made larger in magnitude (V_S is increased), the excursions along the load line are larger and we observe a distortion of the output-voltage waveform, as shown in Fig. 8.8.

If $v_S(t)$ is kept small enough so that distortion does not occur, we will have constructed a linear amplifier. Note that it is important for us to have chosen the battery V_{GG} such that the dc operating point with $v_S(t) = 0$ is in a reasonably linear region of the characteristic. If we had chosen $V_{GG} = 0$, a positive excursion would have caused the gate-source voltage to become positive and gate current would flow, which could damage the JFET. If V_{GG} had been chosen too large, the dc operating point would

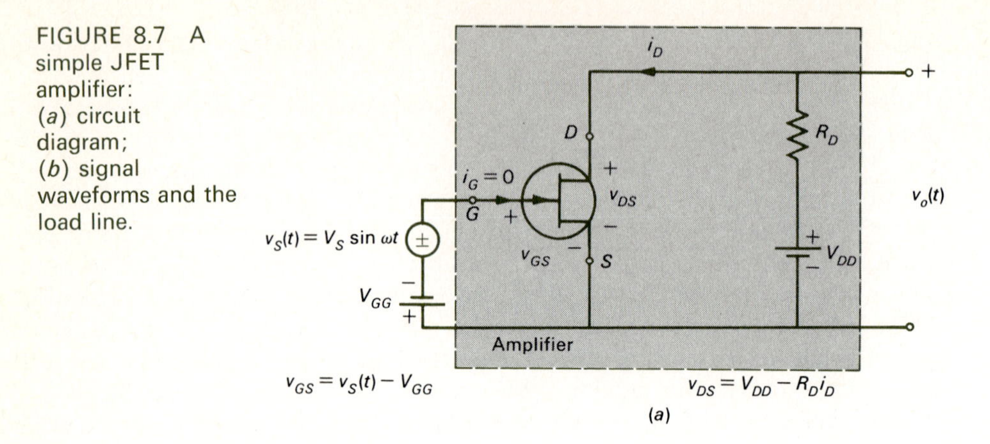

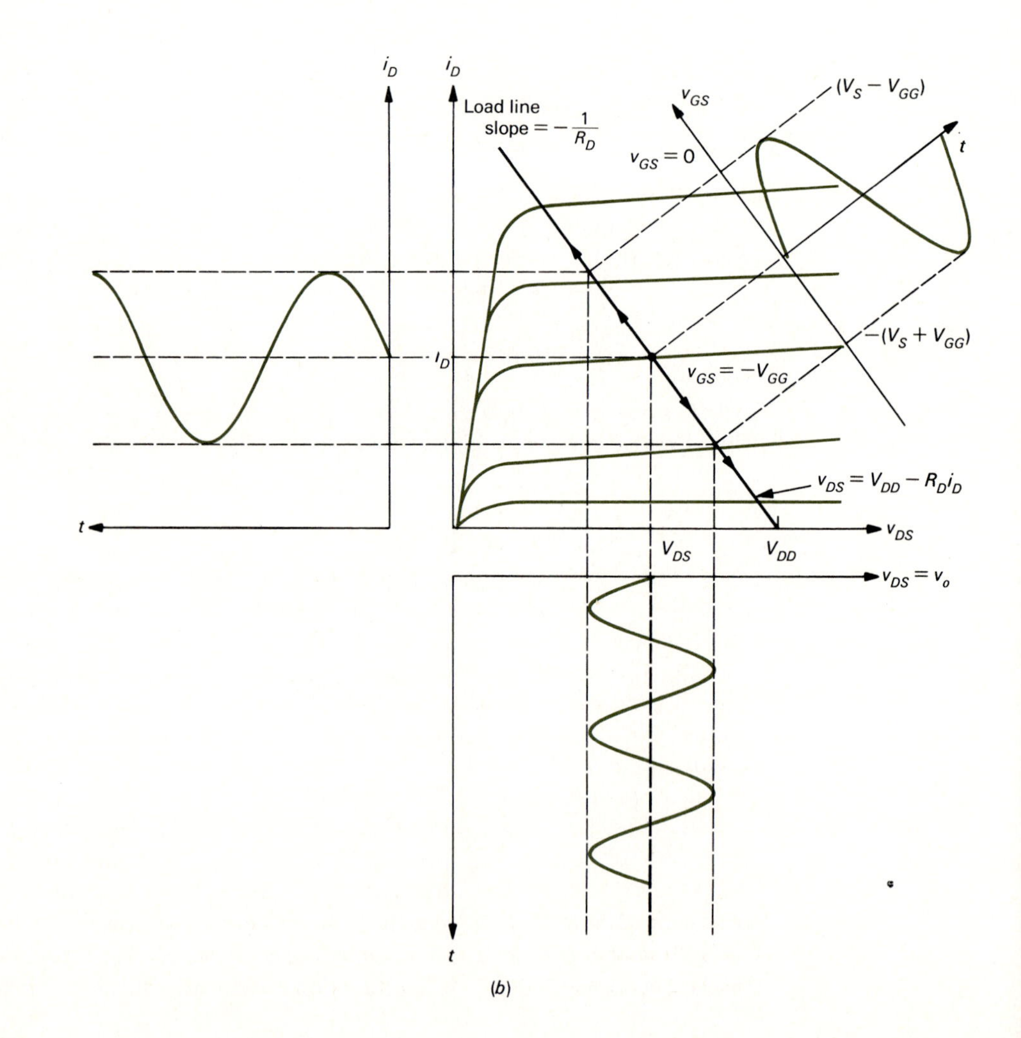

FIGURE 8.7 A simple JFET amplifier: (*a*) circuit diagram; (*b*) signal waveforms and the load line.

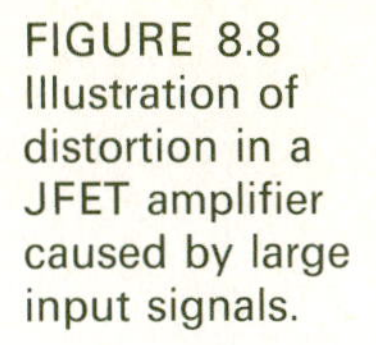

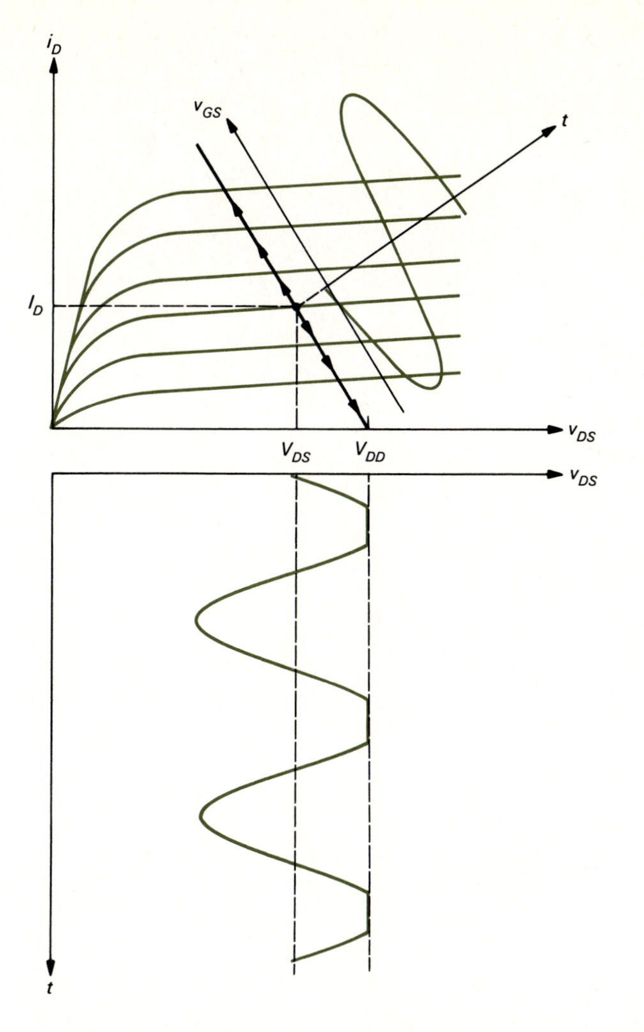

FIGURE 8.8 Illustration of distortion in a JFET amplifier caused by large input signals.

have been located near the bottom of the load line and distortion would occur with only a small $v_S(t)$, as shown in Fig. 8.8. Thus, we need to locate the dc operating point somewhere in the middle of the linear region in order to linearly amplify the largest possible $v_S(t)$.

Note also that the output voltage consists of a sinusoidal voltage riding on a dc level v_{DS}. It is only the ac portion of this output signal that we are interested in. If the output voltage swings from $V_{DS,\min}$ to $V_{DS,\max}$ symmetrically about V_{DS}, then the (ac) voltage gain is

$$A_V = \frac{V_{DS,\max} - V_{DS,\min}}{V_S} \tag{8.11}$$

Since the load is infinite, $i_o(t) = 0$ and current gain is meaningless. However, since the input current is i_G and $i_G = 0$ (for negative v_{GS}), the input impedance to the amplifier is infinite. This is one of the advantages of the FET over the BJT, as we will see.

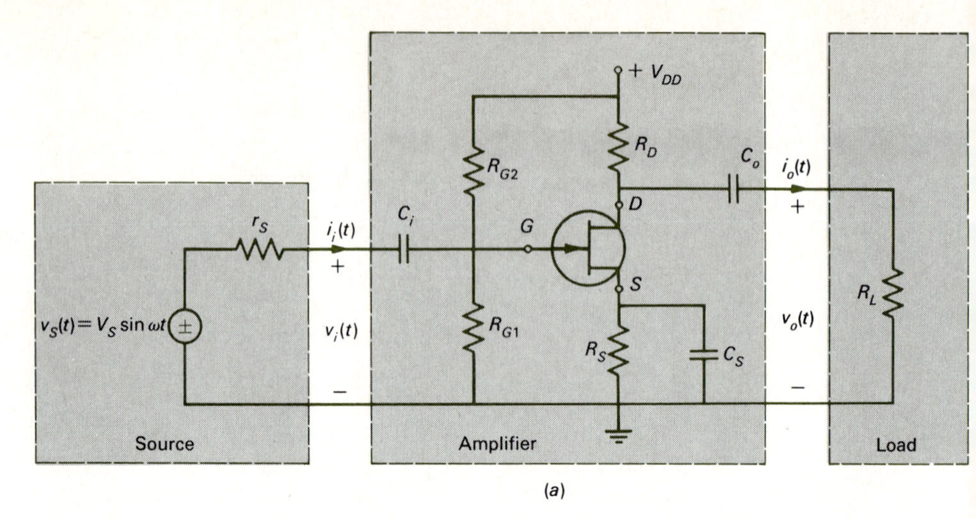

(a)

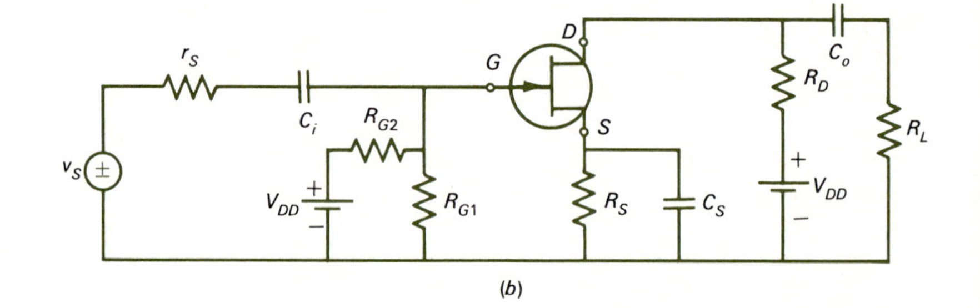

(b)

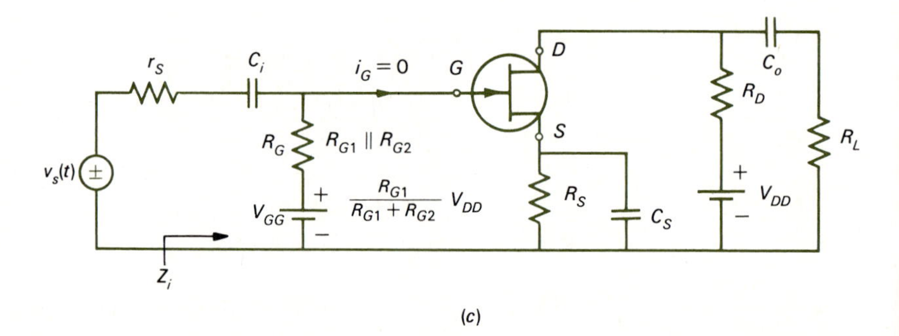

(c)

FIGURE 8.9 The dc analysis of a typical JFET amplifier in determining the operating point.

There are a few problems with this simple amplifier. One is that it requires two batteries (or dc power supplies), V_{GG} and V_{DD}. Another is that all semiconductor devices exhibit large unit-to-unit variations. For example, several 2N3819 JFETs purchased the same day from the same manufacturer will have different characteristic curves—the shapes will all be the same, as in Fig. 8.6*b*, but the numbers on these characteristics will differ between devices of the same type number. We must design our amplifiers with this variation in mind. If we wish to mass-produce a large number of amplifiers having approximately the same performance, the circuit should be relatively insensitive to these unit-to-unit variations. This is an important design criterion.

A more practical JFET amplifier which overcomes these problems is shown in Fig. 8.9*a*. The capacitors C_i and C_o are included to block any dc from getting to the source and load, respectively. We will essentially assume that they have negligible impedance (for example, 1 Ω) at the frequency of $v_S(t)$ and that they can thus be assumed to be short circuits to the ac signal.

The simple amplifier of Fig. 8.7*a* required two batteries, V_{GG} and V_{DD}. The function of the battery V_{GG} is to set a constant, negative bias voltage of the gate to place the dc operating point in a linear region of the characteristic. This is performed by the resistor R_S and the combination of R_{G1} and R_{G2}. A capacitor C_S is placed across R_S so that ac voltages are not developed across R_S. The capacitor value is chosen such that its impedance at the frequency $v_S(t)$ is negligible. Thus, C_S essentially shorts out R_S for ac, but dc current passes through R_S, developing a constant voltage across it.

Only one V_{DD} battery or dc power supply is used in this circuit connected between $+V_{DD}$ and ground (⏚). We may reduce the combination of R_{G1} and R_{G2} and its associated battery V_{DD} to a Thevenin equivalent, as shown in Fig. 8.9*b* and *c*, where

$$R_G = R_{G1} || R_{G2} \tag{8.12}$$

and

$$V_{GG} = \frac{R_{G1}}{R_{G1} + R_{G2}} V_{DD} \tag{8.13}$$

Note that since $i_G = 0$, the input impedance to this amplifier is

$$\begin{aligned} Z_i &= R_G \\ &= R_{G1} || R_{G2} \end{aligned} \tag{8.14}$$

Removing the signal source open-circuits all capacitors and also essentially removes the load (since C_o is an open circuit at dc). Thus, we are left with the dc circuit shown in Fig. 8.10*a*. Writing KVL around the drain-source loop, we obtain

$$v_{DS} = V_{DD} - (R_D + R_S) i_D \tag{8.15}$$

which may be plotted as a load line on the characteristic, as shown in Fig. 8.10*b*. From this characteristic we may determine a suitable operating point (V_{DS}, I_D) in the linear region. Reading off V_{GS} from the characteristic and writing KVL around the gate-source loop, we obtain

$$V_{GS} = V_{GG} - R_S i_D \tag{8.16}$$

(Note that since $i_G = 0$, there is no voltage drop across R_G.) This V_{GG} "battery" is opposite in polarity to the V_{GG} battery in the elementary amplifier of Fig. 8.7*a*. Thus, in order to maintain v_{GS} negative, $R_S i_D$ must be greater in magnitude than V_{GG}, as Eq. (8.16) shows. With $i_D = I_D$ and $v_{GS} = V_{GS}$ determined from the characteristic, we

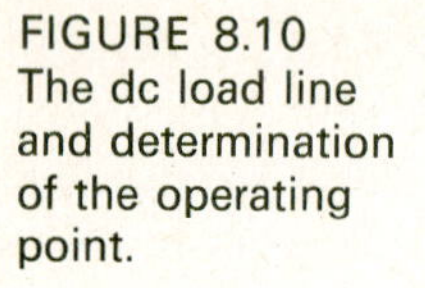

FIGURE 8.10
The dc load line and determination of the operating point.

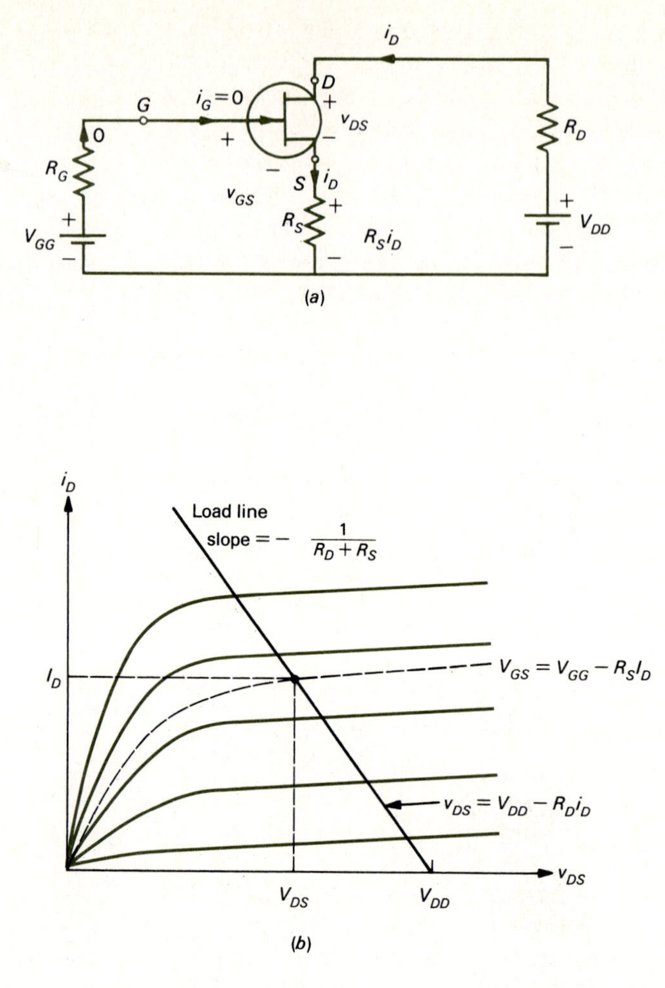

would like to select our resistor values to minimize any effects in unit-to-unit variations. The operating-point drain current and V_{GS} are related through the external circuitry via Eq. (8.16):

$$I_D = \frac{V_{GG} - V_{GS}}{R_S} \tag{8.17}$$

In order that I_D be insensitive to changes in V_{GS}, we select $V_{GG} \gg V_{GS}$. Selecting V_{GG} determines R_S from Eq. (8.17), and R_D can be determined from Eq. (8.15).

There is one final calculation: determining R_{G1} and R_{G2}. These are related to V_{GG} by Eq. (8.13). Having selected V_{DD} and V_{GG}, we have

$$\frac{R_{G1}}{R_{G1} + R_{G2}} = \frac{V_{GG}}{V_{DD}} \tag{8.18}$$

But we still haven't determined R_{G1} and R_{G2} explicitly. We need another relation among them. Note that the input impedance to the amplifier in Fig. 8.9*c* is

$$Z_i = R_G = \frac{R_{G1}R_{G2}}{R_{G1} + R_{G2}} \tag{8.19}$$

Thus, we might specify the input impedance to the amplifier and then determine R_{G1} and R_{G2} from Eqs. (8.18) and (8.19).

Example 8.1 From the manufacturer's supplied characteristics, the JFET amplifier of Fig. 8.9 is to be designed for a dc operating point of $I_D = 2$ mA, $V_{DS} = 5$ V, and $V_{GS} = -2$ V. Determine R_{G1}, R_{G2}, R_D, and R_S to bias the device at this operating point with $V_{DD} = 20$ V and $Z_i = 100$ kilohms (kΩ).

Solution In order that the design be relatively insensitive to unit variations, we would choose V_{GG} to be much larger than V_{GS}. If we choose $V_{GG} = 10$ V, then we may determine R_S from Eq. (8.17) as

$$R_S = \frac{V_{GG} - V_{GS}}{I_D}$$

or

$$R_S = 6 \text{ k}\Omega$$

From Eq. (8.15) we obtain

$$R_D = \frac{V_{DD} - V_{DS}}{I_D} - R_S = \frac{20 - 5}{2 \text{ mA}} - 6 \text{ k}\Omega = 1.5 \text{ k}\Omega$$

Finally, we must determine R_{G1} and R_{G2}. V_{GG} is related to R_{G1}, R_{G2}, and V_{DD} by

$$V_{GG} = \frac{R_{G1}}{R_{G1} + R_{G2}} V_{DD}$$

or

$$10 = \frac{R_{G1}}{R_{G1} + R_{G2}} 20$$

Therefore,

$$\frac{R_{G1}}{R_{G1} + R_{G2}} = \frac{1}{2}$$

Therefore, $R_{G1} = R_{G2}$, but we need another relation between R_{G1} and R_{G2} to determine them explicitly. This is provided by the input impedance specification:

$$\begin{aligned} Z_i &= 100\ \text{k}\Omega \\ &= R_{G1}||R_{G2} \\ &= \left(\frac{R_{G1}}{R_{G1} + R_{G2}}\right)R_{G2} \end{aligned}$$

Thus

$$R_{G2} = 200\ \text{k}\Omega$$

and

$$R_{G1} = 200\ \text{k}\Omega$$

Note that from Eq. (8.17) a 50 percent change in V_{GS} only causes an 8 percent change in I_D.

We always assume linear operation. In other words, we assume that the sinusoidal source is small enough so that at no time does the operating point move outside the linear region of the characteristic. With this assumption, we essentially have a linear circuit to deal with. The device is linear over a large portion of its characteristic. Even though the characteristic is nonlinear in certain regions, we will never move into those regions under linear operation.

Under the assumption that $v_S(t)$ is "small enough" so that the nonlinear device in the amplifier is being operated only in a linear region of its characteristic, we may apply superposition, and each circuit variable (branch voltage or branch current) will consist of two parts: one due to the dc source V_{DD} in the amplifier and one due to the sinusoidal source $v_S(t)$. We will denote the dc component of a circuit voltage or current $x(t)$ with a capital letter X, and the sinusoidal component will be denoted as a Δ quantity, Δx, so that

$$x(t) = X + \Delta x(t) \tag{8.20}$$

It is important to remember that $\Delta x(t)$ is time-varying (sinusoidal) and that X is constant (dc). For brevity, we will write Eq. (8.20) as

$$x = X + \Delta x \tag{8.21}$$

When we reattach the signal source and load, as shown in Fig. 8.11*a*, each voltage and current in the circuit will be composed of a dc component due to V_{DD} (and V_{GG}) and an ac component due to $v_S(t)$. The dc components are governed by the dc load line determined previously with slope

$$R_{\text{dc}} = -\frac{1}{R_D + R_S} \tag{8.22}$$

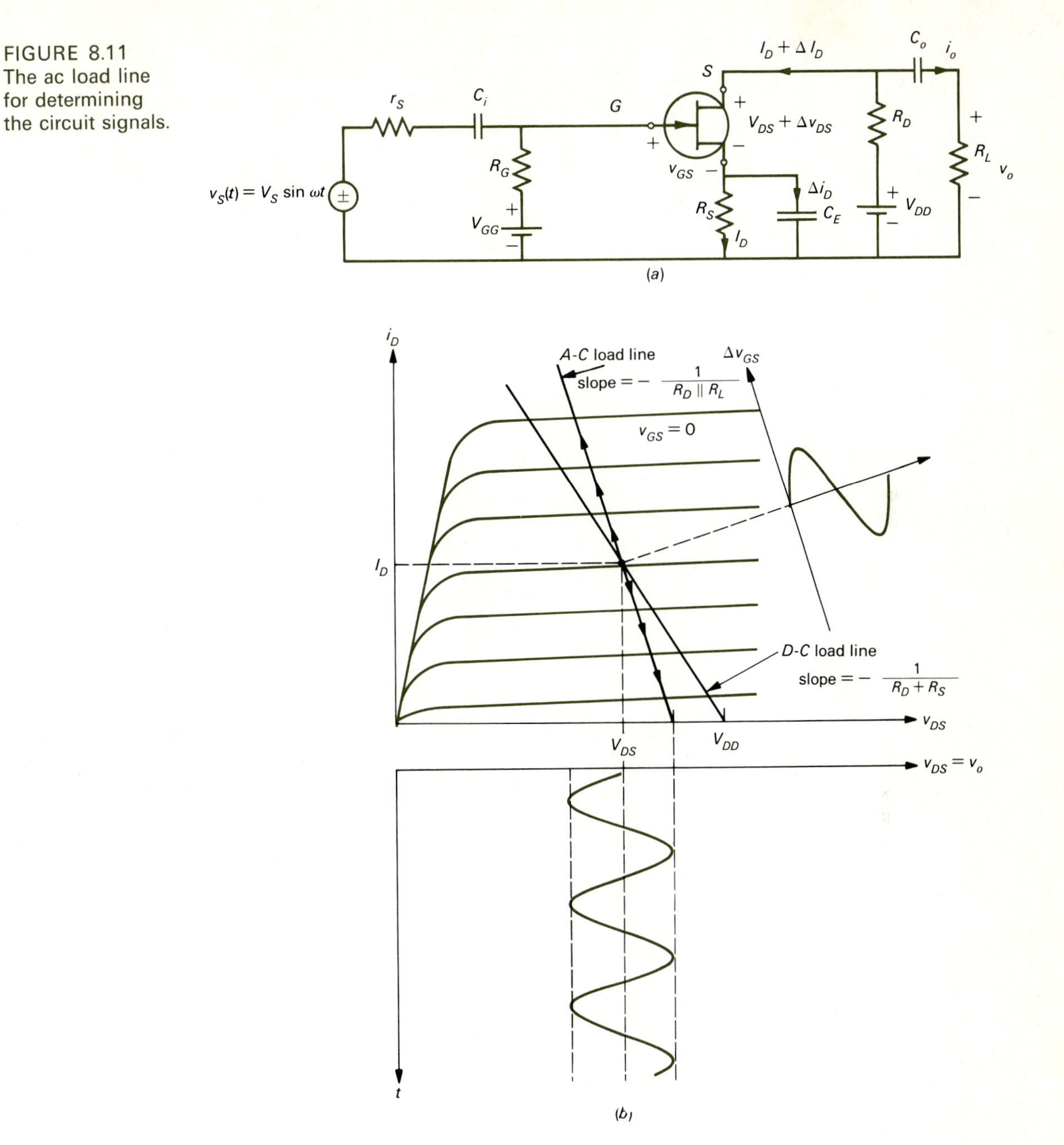

FIGURE 8.11
The ac load line for determining the circuit signals.

but the ac components are governed by an ac load line having a different slope. To determine the slope of this ac load line, we short all capacitors (assumed to have negligible impedance to the ac component) and kill the dc sources. This leaves only the delta (Δ), or ac, components of the voltages and currents, as shown in Fig. 8.12. Writing KVL around the drain-source loop for these conditions gives

$$\Delta v_{DS} = -(R_D || R_L)\, \Delta i_D \tag{8.23}$$

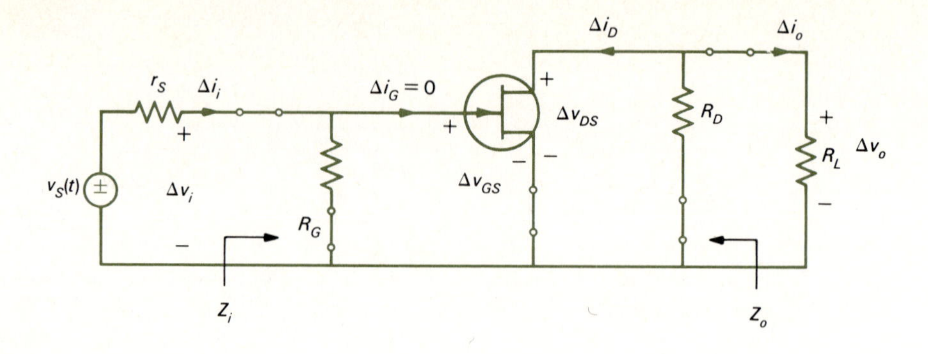

FIGURE 8.12 The small-signal ac circuit.

Thus, the ac load line has slope

$$R_{ac} = -\frac{1}{R_D||R_L} \tag{8.24}$$

The ac variations must move along the ac load line. We reason that this ac load line must pass through the dc operating point since, as we reduce the magnitude of $v_S(t)$, the variations along this ac load line must reduce to zero, which is the dc operating point. The ac variations of V_{GS} are seen to be from Fig. 8.12 (since $\Delta i_G = 0$)

$$\Delta v_{GS} = \frac{R_G}{R_G + r_S} v_S(t) \tag{8.25}$$

Thus, the gate voltage varies sinusoidally and the resulting operating point moves along the ac load line, as shown in Fig. 8.11*b*.

The output voltage $v_o = v_{DS}$ also varies sinusoidally. If the magnitude of this output-voltage sinusoid is larger than the magnitude of the sinusoidal component of the input voltage to the amplifier, voltage gain will be achieved. The voltage gain is the ratio of the ac output voltage across R_L, Δv_o, to the ac input voltage to the amplifier, Δv_i:

$$A_V = \frac{\Delta v_o}{\Delta v_i} \tag{8.26}$$

The current gain is similarly defined:

$$A_I = \frac{\Delta i_o}{\Delta i_i} \tag{8.27}$$

Also, the input and output impedances Z_i and Z_o of the amplifier are ac quantities and are determined from the ac circuit in Fig. 8.12. In the next chapter we will investigate the calculation of these ac figures of merit for an amplifier.

8.4 THEORY OF OPERATION OF THE BIPOLAR JUNCTION TRANSISTOR

An *npn* BJT consists of two *n*-type semiconductor regions with a very thin *p*-type region sandwiched between the two *n*-type regions, as shown in Fig. 8.13*a*. There exist *pnp*-type BJTs where an *n*-type region is sandwiched between two *p*-type regions, but we will only consider *npn* BJTs for the same reason that we only considered *n*-channel JFETs: Amplifiers constructed from *npn* BJTs require power supplies which have a positive voltage with respect to ground, whereas *pnp* BJT amplifiers require negative voltage power supplies. In fact, many types of BJTs (for example, a 2N718) exist in both *npn* and *pnp* construction. The only difference between these two is that the terminal currents and voltages of the *pnp* BJT are opposite in sign to those of the *npn* BJT. Thus, if we design an *npn* BJT amplifier, we may simply substitute a *pnp* BJT (of the same type number) if we reverse the polarity of the batteries (or dc power supplies) in the biasing circuitry.

Terminals are attached to each of the three regions as shown in Fig. 8.13*a*. These are designated *E* for emitter, *B* for base, and *C* for collector. The device symbol for an *npn* BJT is shown in Fig. 8.13*b*. Note that the base and collector currents i_B and i_C are defined as entering the terminals, whereas the emitter current i_E is defined as leaving the emitter terminal. Note also the (arbitrary) directions of the terminal voltages. There is a small arrow in the device symbol in the emitter lead which is in the direction of the (assumed) emitter current. We will find that this is the actual direction of the dc emitter current for an *npn* BJT under normal biasing conditions. A *pnp* BJT would have a similar device symbol, but the small arrow in the emitter lead would be reversed from that of the *npn* direction.

The operation and amplification ability of a BJT can be explained from Fig. 8.14*a*. We have supplied two batteries. One battery, V_{BE}, has its positive terminal connected to the *p*-type base region terminal and its negative terminal connected to the *n*-type emitter region terminal. The other battery, V_{CE}, has its positive terminal connected to the *n*-type collector region terminal and its negative terminal connected

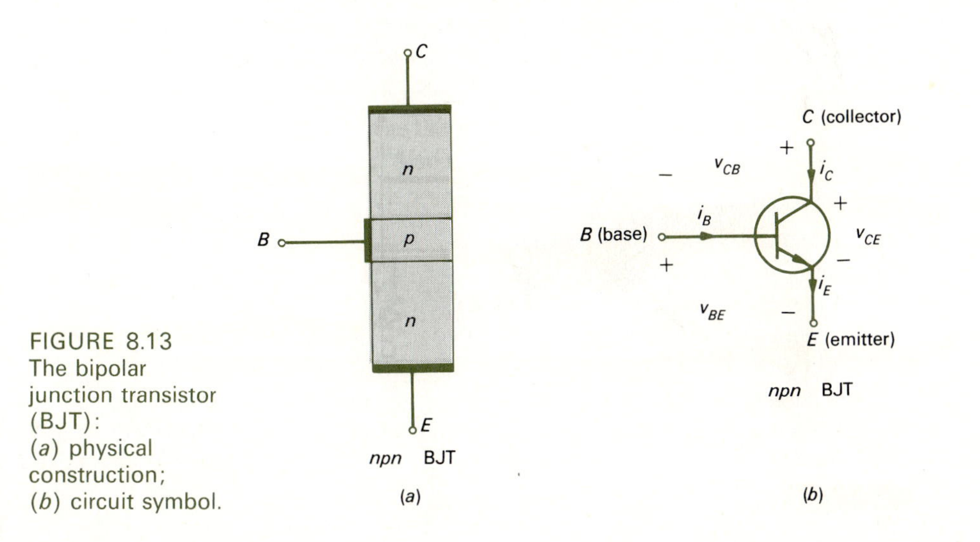

FIGURE 8.13 The bipolar junction transistor (BJT): (*a*) physical construction; (*b*) circuit symbol.

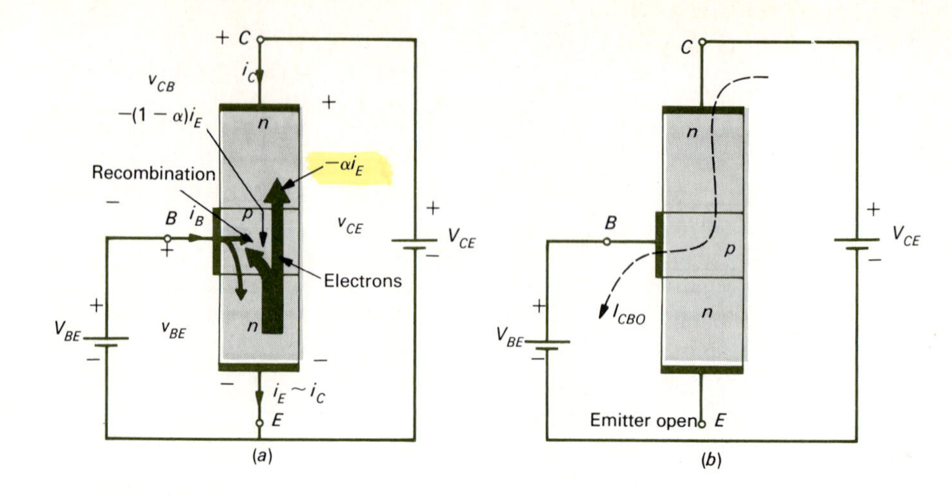

FIGURE 8.14 Charge flow in a BJT: (*a*) current amplification; (*b*) leakage current.

to the *n*-type emitter region terminal. This is the normal polarity for dc biasing, although we will not use only batteries but will also include series resistors with each battery; otherwise, the terminal voltages would be fixed at the battery voltages and no variations (amplification) would be possible. However, note that

$$v_{BE} = V_{BE} \tag{8.28}$$

and that the *pn* base-emitter junction is forward-biased. Note also that

$$v_{CB} = V_{CE} - V_{BE} \tag{8.29}$$

and that the base-collector junction is reverse-biased if $V_{CE} > V_{BE}$. In normal BJT operation, the junctions are always biased in this fashion. Each junction can be thought of as a *pn* semiconductor diode. The base-emitter junction being forward-biased permits electrons (majority carriers) in the *n*-type emitter region to be injected into the *p*-type base region. Consequently, the characteristic for this pair of terminals resembles that of a forward-biased *pn* semiconductor diode, as shown in Fig. 8.15*a*. For $V_{CE} > 1$ V, which is the usual case, the characteristic is independent of changes in $v_{CE} = V_{CE}$. For BJTs constructed of silicon, the curve breaks at approximately 0.5 to 0.7 V, as we saw for silicon *pn* diodes.

The BJT base region is very thin in order to minimize the recombination of the electrons injected from the emitter with holes (the majority carriers) in the *p*-type base region. Also, the emitter region is much more heavily doped than is the base region so that the hole current from base to emitter is much less than the electron current from emitter to base. Since the collector-base junction is reverse-biased, these injected electrons are swept into the collector region. However, a small amount of the electrons injected into the base region are "siphoned off" through the base terminal. Typically, only a few percent are collected by the base so that the collector current ranges from 0.98 to 0.99 of the emitter current. Thus, we may write

$$i_C = \alpha i_E \tag{8.30}$$

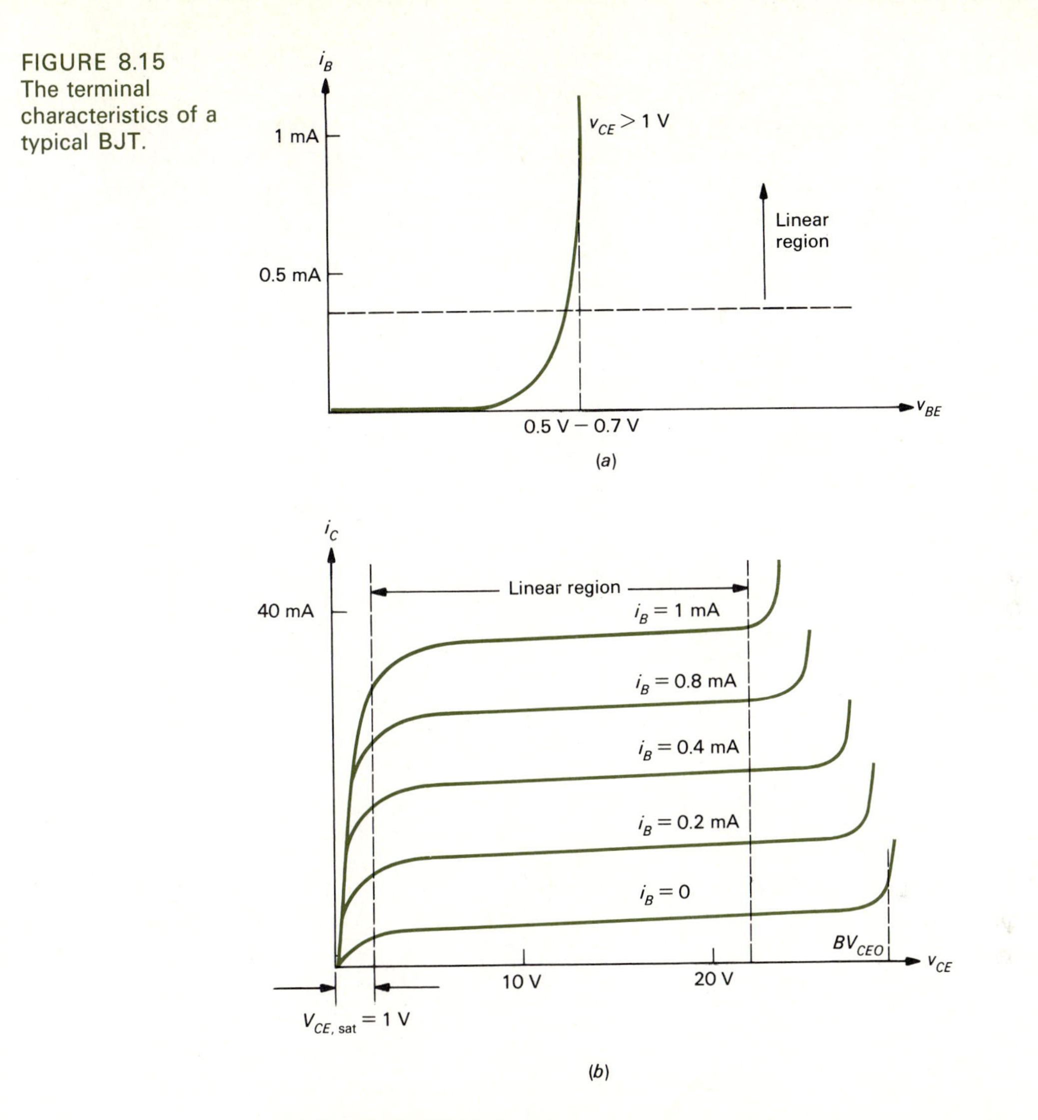

FIGURE 8.15 The terminal characteristics of a typical BJT.

where α (alpha) is called the forward-current transfer ratio whose value typically ranges from 0.98 to 0.99.

Now suppose that we open-circuit the emitter, as shown in Fig. 8.14*b*, thus reducing i_E to zero. Note that the base-collector junction resembles a reverse-biased *pn* diode; consequently we expect to find a leakage current which was designated as I_S for the diode. For the BJT, this is designated as I_{CBO}, where the subscript denotes collector to base with the emitter open-circuited. This reverse leakage current will always be present, regardless of whether or not the emitter is open, since the batteries reverse-bias the collector-base junction. Consequently, superimposing this current onto Eq. (8.20), we obtain

$$i_C = \alpha i_E + I_{CBO} \tag{8.31}$$

One important point about I_{CBO} should be mentioned—it exhibits a strong dependence on the temperature of the collector-base junction. If this temperature increases, I_{CBO} also increases. This can lead to an uncontrolled increase in collector temperature (known as thermal runaway) unless the biasing circuit configuration is chosen to minimize this effect. For example, if an initial heating of the BJT occurs (owing, say, to heat generated by neighboring components or to other ambient temperature variations), this will result in an increase in I_{CBO}. However, this current increase causes an increase in junction temperature due to $I_{CBO}^2 R_{CB}$ losses, where R_{CB} is the junction resistance. These losses therefore increase the junction temperature, which increases I_{CBO}. The process may continue, with the result that the dc collector current $I_C = \alpha I_E + I_{CBO}$ increases. This may result in the operating point moving into a nonlinear region of the characteristic, as was the result for unit variations of the JFET. Consequently, our biasing circuitry should tend to minimize this effect.

We may rewrite Eq. (8.31) in terms of i_B and i_C, since by KCL

$$i_E = i_B + i_C \tag{8.32}$$

or

$$i_B = i_E - i_C \tag{8.33}$$

Equation (8.31) may be written as

$$i_E = \frac{1}{\alpha} i_C - \frac{1}{\alpha} I_{CBO} \tag{8.34}$$

Substituting Eq. (8.34) into Eq. (8.33), we obtain

$$\begin{aligned} i_B &= \frac{1}{\alpha} i_C - \frac{1}{\alpha} I_{CBO} - i_C \\ &= \frac{1-\alpha}{\alpha} i_C - \frac{1}{\alpha} I_{CBO} \end{aligned} \tag{8.35}$$

The symbol β (beta) is used to represent the ratio $\alpha/(1 - \alpha)$; thus

$$\beta = \frac{\alpha}{1 - \alpha} \tag{8.36}$$

and

$$\alpha = \frac{\beta}{\beta + 1} \tag{8.37}$$

Typical values of β range from $\beta = 49$ ($\alpha = 0.98$) to $\beta = 99$ ($\alpha = 0.99$). Writing Eq. (8.35) in terms of β, we obtain

$$i_C = \beta i_B + \frac{I_{CBO}}{1 - \alpha}$$

$$= \beta i_B + (\beta + 1) I_{CBO} \tag{8.38}$$

The quantity $(\beta + 1)I_{CBO}$ is, from Eq. (8.38), the collector current when the base is open-circuited ($i_B = 0$) and is designated as

$$I_{CEO} = (\beta + 1) I_{CBO} \tag{8.39}$$

where the subscript CEO denotes collector-emitter current with the base open-circuited. In terms of I_{CEO}, Eq. (8.38) becomes

$$i_C = \beta i_B + I_{CEO} \tag{8.40}$$

The plot of v_{CE} and i_C versus i_B for a typical BJT is shown in Fig. 8.15*b*. For v_{CE} less than approximately 1 V, the collector-base junction is forward-biased [see Fig. 8.14*a* and Eq. (8.29)] and normal BJT operation does not occur. The collector current increases linearly with an increase in v_{CE} up to a point at which the v_{CE} is sufficiently large (> 1 V) that the collector-base junction becomes reverse-biased and normal BJT operation occurs. In this case, Eq. (8.40) applies. Actually, Eq. (8.40) does not strictly apply since there is a slight increase in i_C with increasing v_{CE}, as evidenced by the small (but nonzero) slope of the curves in Fig. 8.15*b*. Further increases in v_{CE} result in only small increases in i_C, and the curves are relatively flat up to the point at which v_{CE} is sufficiently large that avalanche breakdown occurs. At this point, which is designated as BV_{CEO} for $i_B = 0$ (breakdown voltage collector to emitter with base open-circuited), the collector current increases rapidly with a small increase in v_{CE}.

8.5 THE BJT AMPLIFIER

Obviously, for linear operation we will design the biasing circuitry so that we operate in the linear regions of the characteristics. A dc-biasing network which will accomplish this is shown in Fig. 8.16*a*. Writing KVL around the base-emitter loop and the collector-emitter loop, we obtain

$$v_{BE} = V_{BB} - R_B i_B \tag{8.41}$$

and

$$v_{CE} = V_{CC} - R_C i_C \tag{8.42}$$

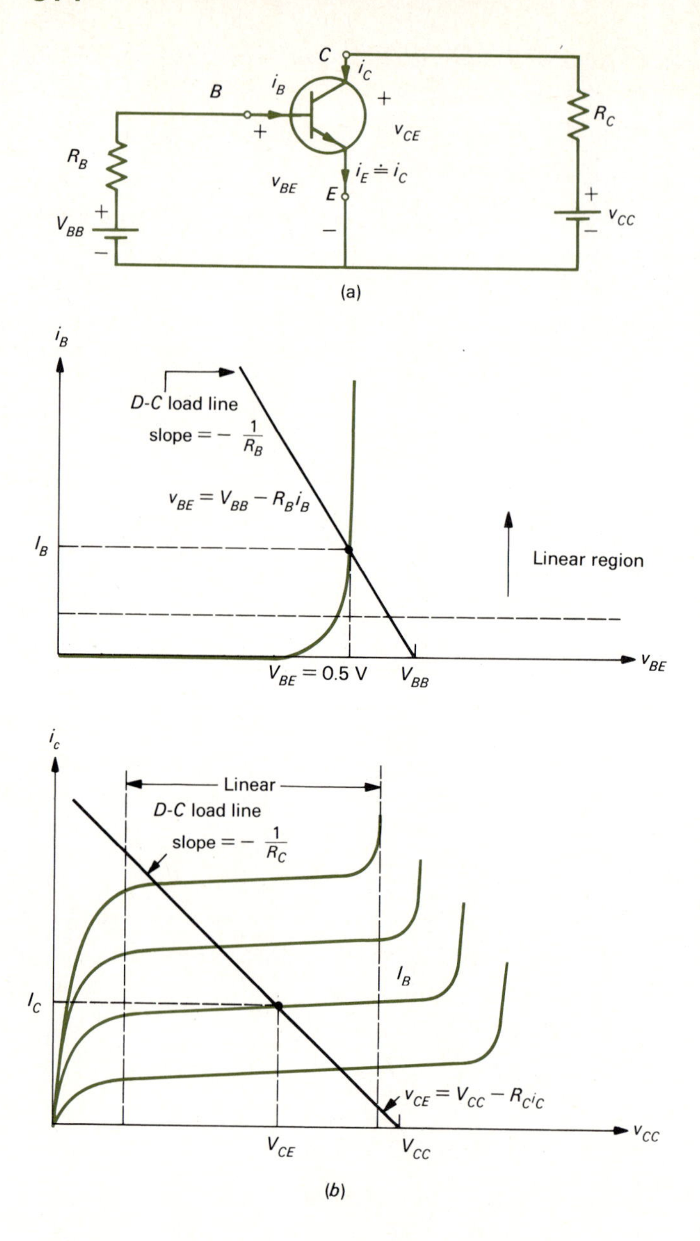

FIGURE 8.16 DC load-line analysis of a BJT amplifier.

Plotting Eq. (8.41) as a dc load line on the base-emitter characteristic and plotting Eq. (8.42) on the collector-emitter characteristics, we find the dc operating points (I_B, $V_{BE} \doteq 0.5$ V, I_C, V_{CE}), as shown in Fig. 8.16*b*.

Now let us add an ac signal source and load, as shown in Fig. 8.17*a*. Once again we assume that $v_S(t)$ is small enough so that linear operation is preserved. In this case each voltage and current in the circuit will consist, by superposition, of a dc quantity due to V_{BB} and V_{CC} and of a delta (Δ) ac quantity due to the sinusoidal source $v_S(t)$. Removing the ac source gives the dc circuit shown in Fig. 8.16*a*. The ac circuit is

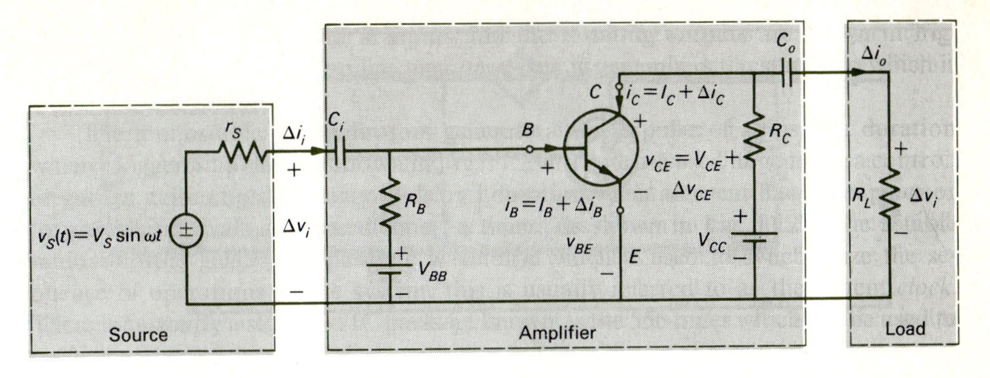

(a)

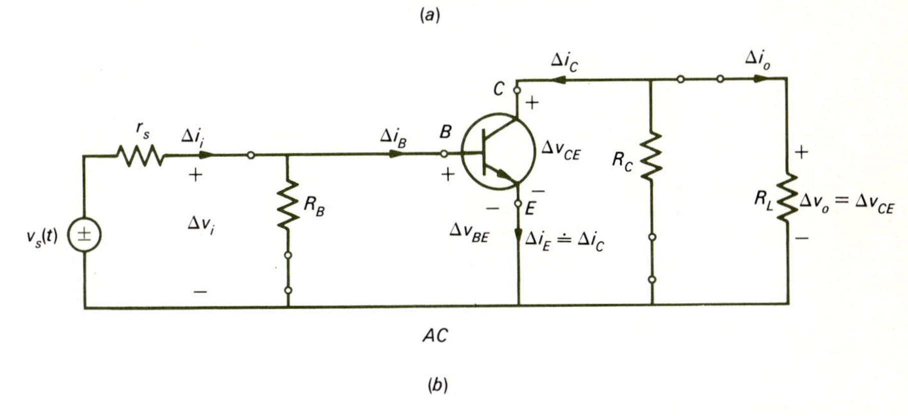

(b)

FIGURE 8.17 AC small-signal analysis of a BJT amplifier.

obtained by shorting all capacitors and killing the two dc sources V_{BB} and V_{CC}, as shown in Fig. 8.17*b*. From this we see that the ac quantities Δv_{CE} and Δi_C are related by

$$\Delta v_{CE} = -(R_C || R_L)\, \Delta i_C \tag{8.43}$$

Thus, the ac signal variations follow an ac load line with slope

$$R_{\text{ac}} = -\frac{1}{R_C || R_L} \tag{8.44}$$

as shown in Fig. 8.18.

8.6 SELECTION OF THE OPERATING POINT

At this point we may wish to determine a criterion for choosing the operating point for the BJT amplifier. Since the linear region of the collector-emitter characteristic is rather large (between $v_{CE} \doteq 1$ V and $v_{CE} \doteq BV_{CEO}$), it would appear that any combina-

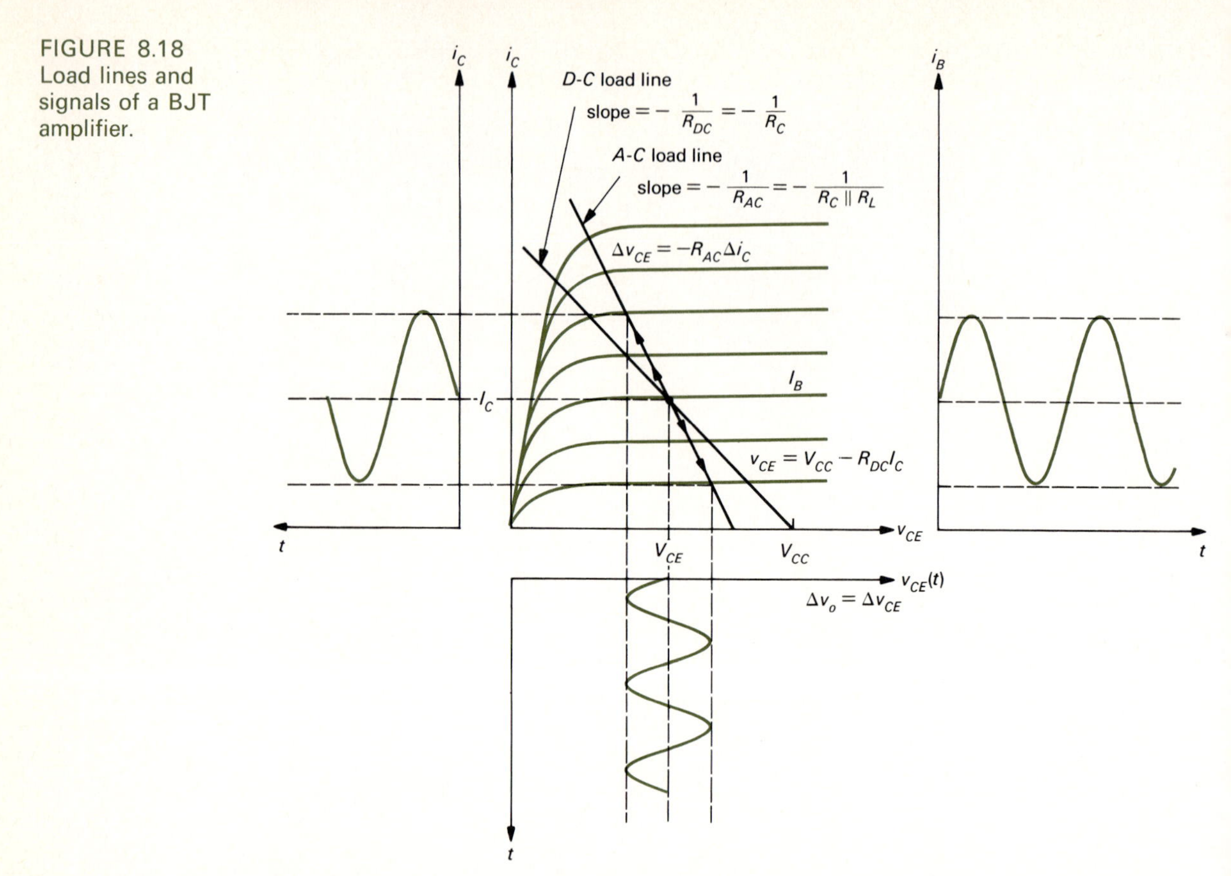

FIGURE 8.18 Load lines and signals of a BJT amplifier.

tion of I_C, V_{CE}, I_B in this region will be acceptable. Actually, we only need to specify I_C since the resulting V_{CE} can be determined from the values of R_C and V_{CC} from

$$V_{CE} = V_{CC} - R_C I_C \tag{8.45}$$

and I_B can be determined approximately from Eq. (8.40) (I_{CEO} is typically much less than βI_B):

$$I_B \doteq \frac{I_C}{\beta} \tag{8.46}$$

The base-emitter operating-point voltage V_{BE} is rather constant at around 0.5 to 0.7 V. The load lines are plotted on the collector-emitter characteristic in Fig. 8.18. Note that the ac load line must pass through the dc operating point since, if we reduce $v_S(t)$ to zero, the movement along the ac load line must reduce to zero and therefore converge on the dc operating point.

Now consider the variation of i_B. From the ac circuit, Δi_B is sinusoidal at the frequency of $v_S(t)$. A typical variation is shown on the characteristic in Fig. 8.18. As $v_S(t)$ varies, so does i_B, and the instantaneous operating point moves along the ac load

FIGURE 8.19 Illustration of operating-point selection for maximum symmetrical swing.

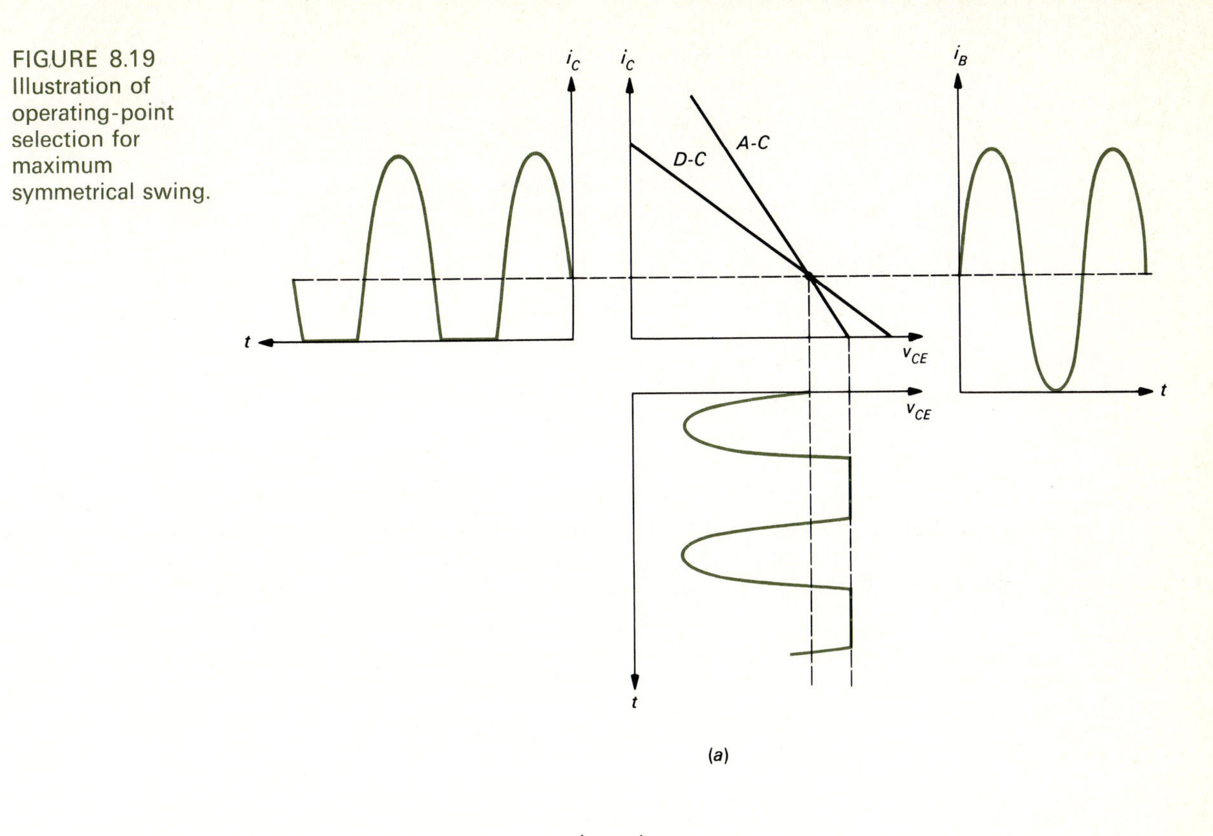

(a)

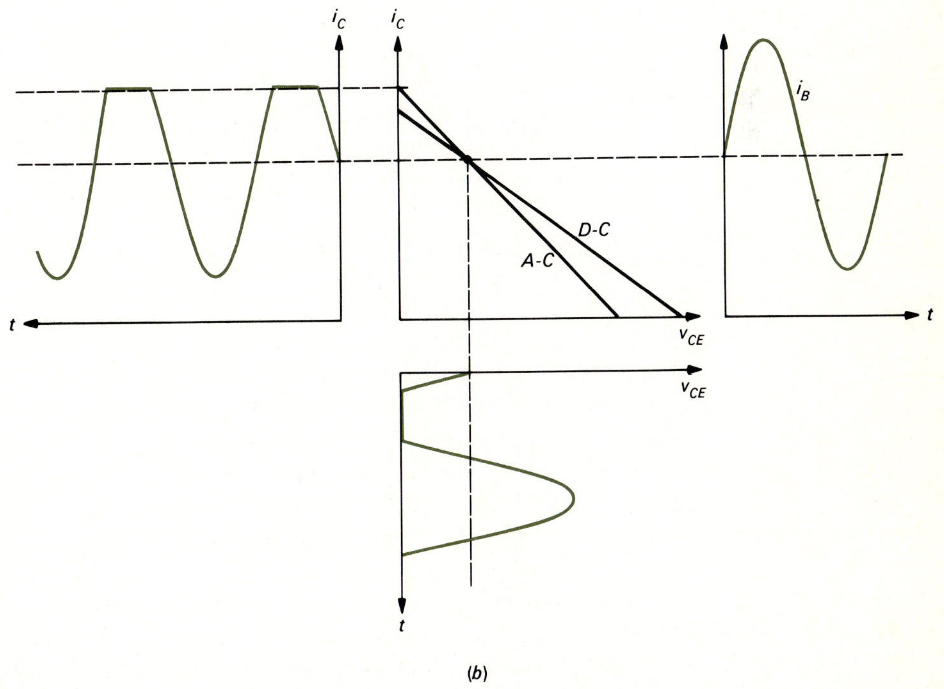

(b)

line as shown with the resulting variation of v_{CE} and i_C. Observe that as $v_S(t)$ is increased in magnitude, the magnitude of the v_{CE} change increases until it is limited to the intersection of the ac load line with either the i_C axis or the v_{CE} axis, as shown in Fig. 8.19. Further increases in the magnitude of $v_S(t)$ increase one-half of v_{CE} and i_C, but the other half is "mashed," resulting in a highly nonlinear operation of the amplifier. In these characteristics, we have neglected the nonlinearity of the characteristic for $v_{CE} < 1$ V.

Note that

$$\Delta v_o = \Delta v_{CE} \tag{8.47}$$

$$\Delta i_o = \frac{R_C}{R_C + R_L} \Delta i_C \tag{8.48}$$

so that the shape $v_o(i_o)$ is the same as the shape of $v_{CE}(i_C)$. Obviously, it does us no good to have an output voltage with one half cycle larger and resembling the half cycle of a sinusoid but with the other half cycle mashed; the output voltage does not resemble $v_S(t)$ and the amplifier is being operated nonlinearly. Distortion is said to have occurred. However, it is still desirable to obtain as large an output-voltage and output-current swing as possible while maintaining linear operation. For example, a voltage gain $A_V = 100$ merely means that if a ± 1-mV input-voltage swing is applied, a ± 100-mV output-voltage swing is obtained (assuming that 1 mV is small enough for linear operation). Also, if a ± 10-mV input voltage is applied (and is small enough for linear operation), the output-voltage swing will be ± 1 V.

Thus, a large A_V or A_I is not the total picture. We would like to obtain as large a swing of output voltage or output current as possible. For linear operation it must be a symmetrical waveform (swing). In the design of an amplifier, R_C, R_L, and V_{CC} fix the slopes of the dc and ac load lines as well as the intersection of the dc load line with the v_{CE} axis (V_{CC}). What is not fixed until we know I_B is the intersection of these two load lines I_C. But I_B is determined through V_{BB} and R_B and can be obtained by writing KVL around the base-emitter loop in the dc circuit in Fig. 8.16*a*:

$$I_B = \frac{V_{BB} - V_{BE} \doteq 0.5 \text{ V}}{R_B} \tag{8.49}$$

Once V_{BB} and R_B are fixed, so are I_B and $I_C \doteq \beta I_B$. Thus, we will determine the operating point (and the intersection of the ac and dc load lines) once we fix V_{BB} and R_B. This allows some flexibility in the design. For maximum symmetrical output-voltage and output-current swing, we would like to adjust the operating point I_C such that $X+ = X-$ and $Y+ = Y-$ in Fig. 8.20. Simple trigonometry will show that the optimal value of I_C for which this occurs is

$$I_{C,\text{opt}} = \frac{V_{CC}}{R_{\text{dc}} + R_{\text{ac}}} \tag{8.50}$$

and we may set $I_{B,\text{opt}} = I_{C,\text{opt}}/\beta$ by adjusting R_{BB} and V_B. In this case, we will obtain maximum symmetrical v_{CE} and i_C swing and therefore, from Eqs. (8.47) and (8.48),

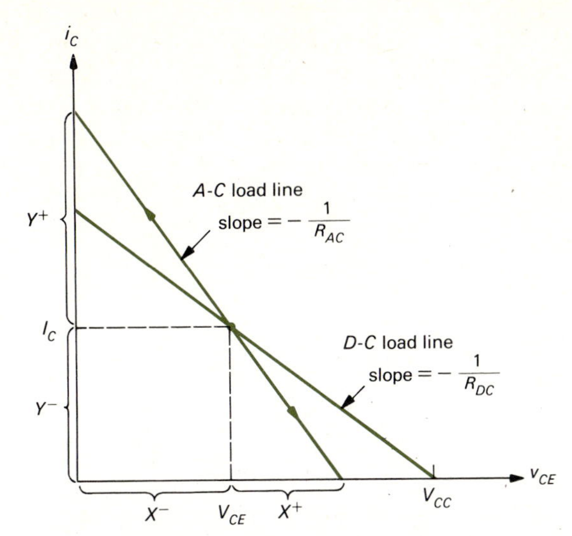

FIGURE 8.20 Selection of the optimum operating point for maximum symmetrical swing.

maximum symmetrical output-voltage (v_o) and output-current (i_o) swing for the amplifier. With R_C, R_L, and V_{CC} fixed, this is "the best we can do" in terms of obtaining the largest output signal for linear operation.

Example 8.2

Verify that the circuit in Fig. 8.21*a* is designed for maximum symmetrical swing. Here we have used one battery to provide V_{BB} and V_{CC}. It is connected between $+V_{CC}$ and ground (⏚). Assume $\beta = 100$.

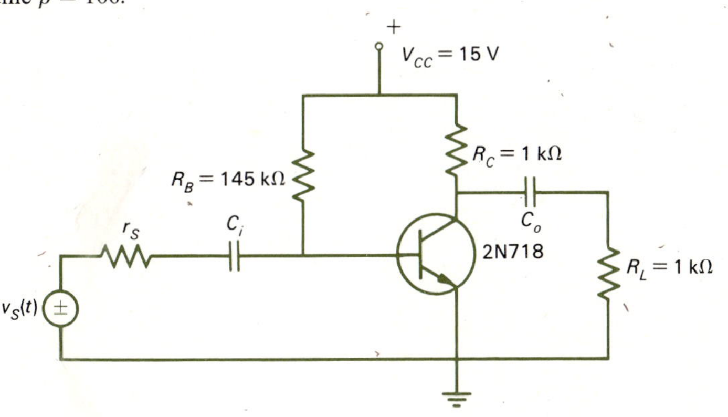

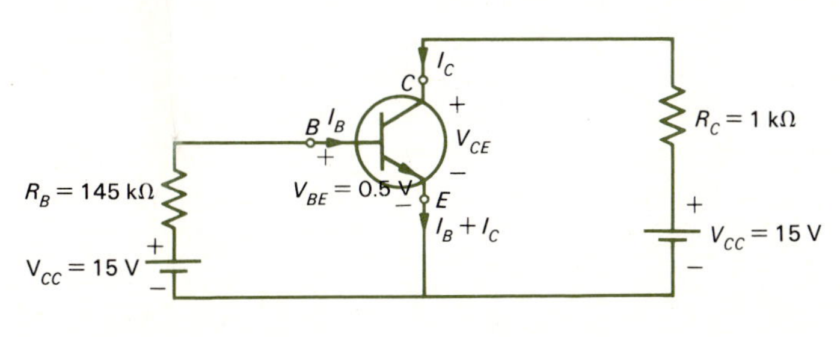

FIGURE 8.21 Example 8.2: design of the operating point for maximum symmetrical swing.

Solution

$$I_{C,\text{opt}} = \frac{V_{CC}}{R_{\text{dc}} + R_{\text{ac}}}$$

$$= \frac{V_{CC}}{R_C + R_C||R_L}$$

$$= \frac{15\text{ V}}{1\text{ k}\Omega + 500\,\Omega}$$

$$= 10\text{ mA}$$

and

$$I_{B,\text{opt}} = \frac{I_{C,\text{opt}}}{\beta}$$

$$= \frac{10\text{ mA}}{100}$$

$$= 0.1\text{ mA}$$

From Fig. 8.21*b*

$$R_{B,\text{opt}} = \frac{V_{BB}(= V_{CC}) - V_{BE}(\doteq 0.5\text{ V})}{I_{B,\text{opt}}}$$

$$= \frac{15\text{ V} - 0.5\text{ V}}{0.1\text{ mA}}$$

$$= 145\text{ k}\Omega$$

8.7 STABILITY OF THE OPERATING POINT

We have found that

$$i_C = \beta i_B + I_{CEO} \tag{8.51}$$

and that I_{CEO} is a leakage current which is strongly dependent on the base-collector junction temperature. It typically doubles for every 10°C increase in temperature. As pointed out previously, if the ambient temperature of the device increases, so does I_{CEO}, and this produces an additional increase in junction temperature due to increased i^2R losses. Thus, I_{CEO} increases once again. The process may continue, with the result being an increase in collector current such that the operating point moves into a nonlinear region of the characteristic. This phenomenon is called thermal runaway and is clearly undesirable; once we design for a certain $I_{C,\text{opt}}$, we want I_C to remain there.

Other parameters are also sensitive to temperature. For example, the base-emitter voltage V_{BE} decreases approximately 2.5 mV for every 1°C increase in temperature. Another parameter which varies with temperature is β; however, this variation is nonlinear with temperature and is difficult to estimate.

We will design our biasing circuitry so that changes in the amplifier operating point due to changes in these parameters will be minimized. For silicon transistors, the major variations to be guarded against are changes in V_{BE} and β. For germanium transistors, the major variation to guard against is a temperature change resulting in an I_{CBO} change.

An amplifier circuit which has the ability to limit changes in the operating point due to these variations is shown in Fig. 8.22*a*. A resistor R_E bypassed for ac by a capacitor C_E is placed in the emitter lead. Two base-biasing resistors R_{B1} and R_{B2}

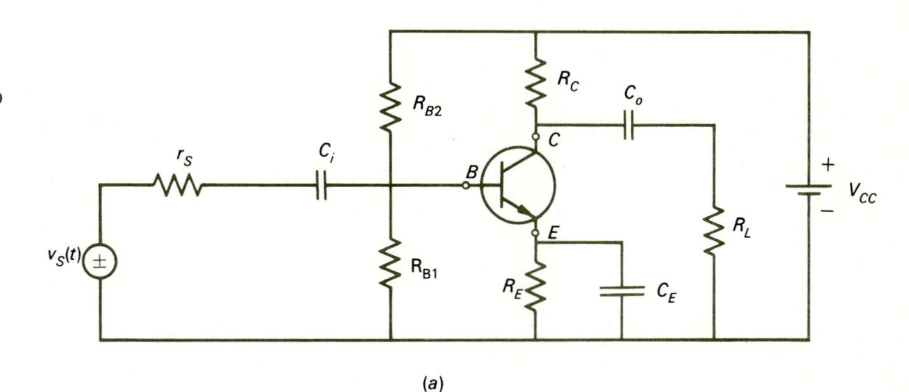

(*a*)

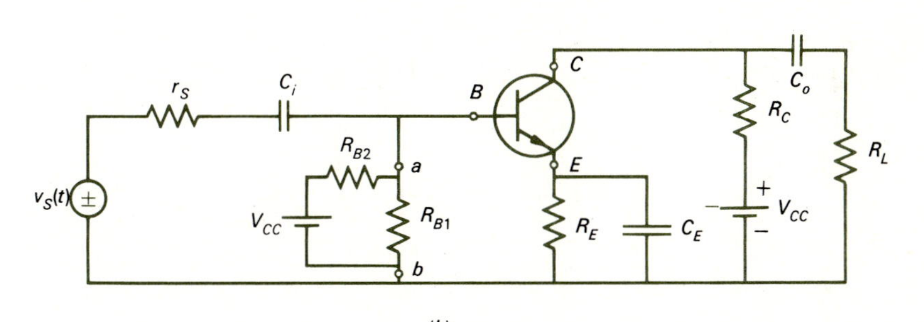

(*b*)

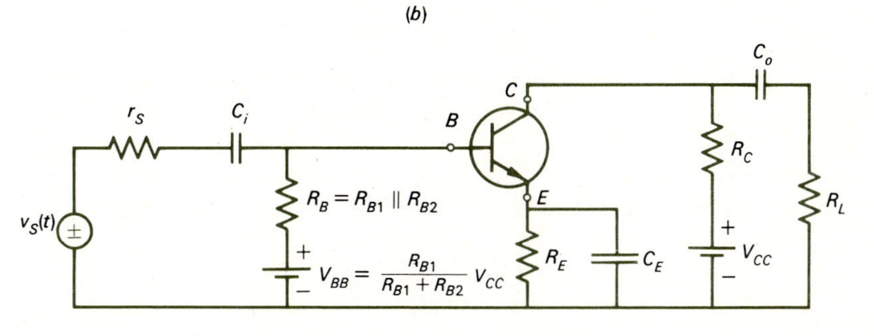

(*c*)

FIGURE 8.22 A practical BJT amplifier circuit to stabilize against operating-point changes.

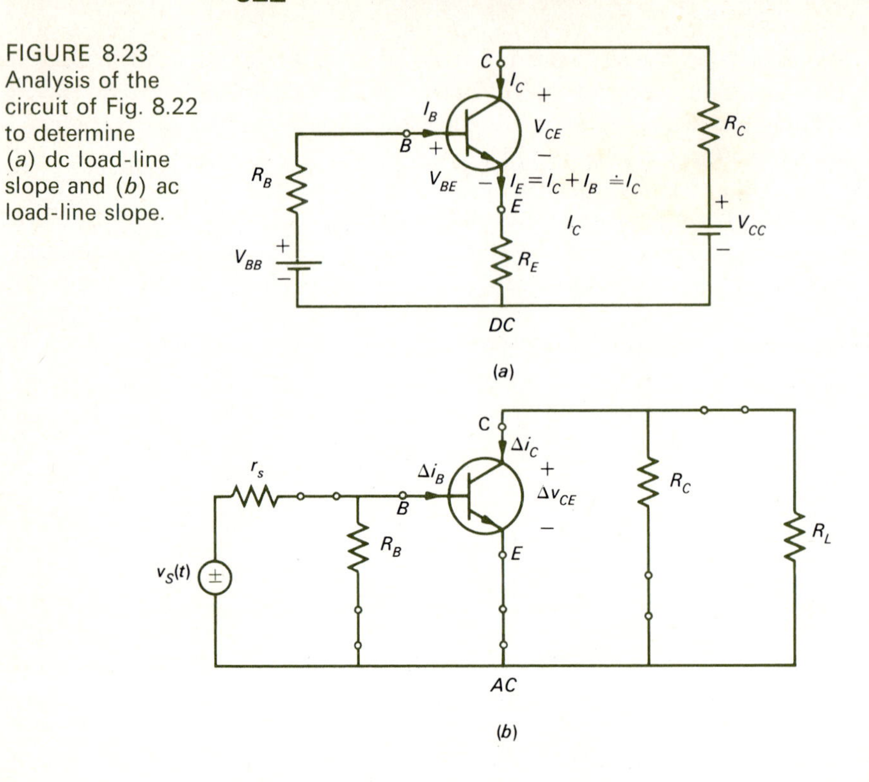

FIGURE 8.23 Analysis of the circuit of Fig. 8.22 to determine (*a*) dc load-line slope and (*b*) ac load-line slope.

allow the use of one battery. These two resistors in combination with V_{CC} can be reduced to one resistor

$$R_B = R_{B1} || R_{B2} \tag{8.52}$$

and a voltage source

$$V_{BB} = \frac{R_{B1}}{R_{B1} + R_{B2}} V_{CC} \tag{8.53}$$

with a Thevenin equivalent, as shown in Fig. 8.22*b* and *c*. With the exception of R_E and C_E, this is identical to the basic amplifier in Fig. 8.17*a*. In fact, this amplifier and the final JFET amplifier in Fig. 8.9 are identical in form.

This circuit tends to stabilize against changes in I_{CEO}, V_{BE}, and β due to temperature variations, and it also tends to stabilize changes in operating point due to unit-to-unit variations in β, as we will see. For example, the manufacturer of the 2N718 BJT specifies that $\beta_{\min} = 50$, $\beta_{\max} = 150$. Thus, a group of supposedly identical 2N718 BJTs can have β's (at the same operating points) which differ by as much as 3:1. For the original circuit which we have considered (Fig. 8.17), note that I_B is fixed by V_{BB} and R_B:

$$I_B = \frac{V_{BB} - V_{BE}(\doteq 0.5\text{ V})}{R_B} \tag{8.54}$$

and

$$I_C = \beta I_B \tag{8.55}$$

Thus, for two identical amplifiers having 2N718 BJTs, the operating points I_C may differ by as much as 3:1.

To show that the circuit in Fig. 8.22 stabilizes against temperature variations in I_{CBO} and V_{BE} as well as against unit variations in β, consider the dc circuit shown in Fig. 8.23*a*. Writing KVL around the base-emitter loop, we obtain

$$V_{BB} - R_B I_B - V_{BE} - (I_C + I_B)R_E = 0 \tag{8.56}$$

From Eq. (8.38) we obtain

$$I_B = \frac{I_C}{\beta} - \frac{\beta + 1}{\beta} I_{CBO} \tag{8.57}$$

Substituting Eq. (8.57) into Eq. (8.56) gives

$$I_C = \frac{V_{BB} - V_{BE} + [(\beta + 1)/\beta](R_B + R_E)I_{CBO}}{[(\beta + 1)/\beta]R_E + R_B/\beta} \tag{8.58}$$

For practical values of $\beta \gg 1$, $(\beta + 1)/\beta \doteq 1$ so that Eq. (8.58) becomes

$$I_C \doteq \frac{V_{BB} - V_{BE} + (R_B + R_E)I_{CBO}}{R_E + R_B/\beta} \tag{8.59}$$

In order to make I_C insensitive to changes in V_{BE}, we might select $V_{BB} \gg V_{BE} \doteq 0.5$ V. So a value of V_{BB} greater than, for example, 5 V would suffice. To make I_C insensitive to changes in β, we would select R_B and R_E such that

$$R_E \gg \frac{R_B}{\beta} \tag{8.60}$$

One of the other important changes we would like to stabilize against is temperature-induced variations in I_{CBO}. The effect of changes in I_{CBO} on I_C can be obtained by taking the partial derivative of I_C with respect to I_{CBO} in Eq. (8.59):

$$\begin{aligned} \frac{\partial I_C}{\partial I_{CBO}} &= \frac{R_B + R_E}{R_E + R_B/\beta} \\ &\doteq \frac{R_B}{R_E} + 1 \end{aligned} \tag{8.61}$$

[We have used the criterion in Eq. (8.60).] Thus, we would want the ratio of R_B/R_E to be as small as possible. Reducing R_B to satisfy Eq. (8.60) will also reduce the ratio R_B/R_E in Eq. (8.61). But we can reduce R_B only so far until we are shunting a major portion of the input current to the amplifier through R_B and causing deterioration of the current gain.

Example 8.3

An *npn* BJT is to be used to construct the amplifier in Fig. 8.24. The manufacturer specifies for this device that β will be between 50 and 150, I_{CBO} will be 1 nanoampere (nA) at room temperature (25°C), and V_{BE} will be 0.5 V at room temperature. Determine the change in operating point due to these parameter changes.

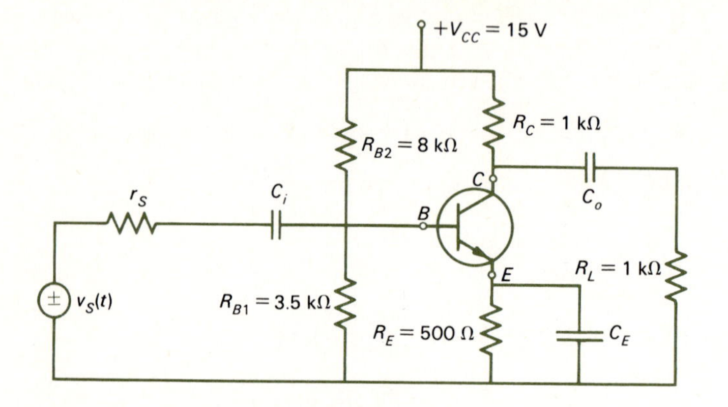

FIGURE 8.24
Example 8.3.

Solution From Eq. (8.59) we have

$$I_C = \frac{V_{BB} - V_{BE} + (R_B + R_E)I_{CBO}}{R_E + R_B/\beta}$$

But

$$R_B = R_{B1}||R_{B2}$$
$$= 2435\ \Omega$$

and

$$V_{BB} = \frac{R_{B1}}{R_{B1} + R_{B2}} V_{CC}$$
$$= 4.57\ \text{V}$$

At room temperature the changes in I_C due to changes in β are

$$I_{C,\text{min}} = \frac{4.57 - 0.5 + (2435 \times 10^{-9})}{500 + 2435/50}$$
$$= \frac{4.07}{500 + 48.7}$$
$$= 7.4\ \text{mA}$$

$$I_{C,\text{max}} = \frac{4.07}{500 + 2435/150}$$
$$= 7.88\ \text{mA}$$

Note that I_{CBO} may typically be neglected in calculating I_C. So a change in β of a factor of 3 only causes a 6 percent change in I_C. This small sensitivity to changes in β is to be expected since

$$R_E = 500\,\Omega$$

and

$$\frac{R_B}{\beta_{\min}} = 48.7$$

As for the change in I_C due to changes in V_{BE} due to temperature variations, suppose that the temperature increases from 25 to 50°C. Recall that V_{BE} changes 2.5 mV/°C (decreasing with increasing temperature). Thus, from 25 to 50°C V_{BE} will change by

$$\Delta V_{BE} = \frac{-2.5\,\text{mV}}{°\text{C}}\,25°\text{C}$$

$$= -0.0625\,\text{V}$$

so that at 50°C

$$V_{BE} = 0.4375\,\text{V}$$

Also, I_{CBO} approximately doubles for every 10°C increase in temperature. Thus, at 50°C I_{CBO} has changed by a factor of $2^{\Delta T/10} = 6.25$. Therefore, at 50°C I_{CBO} becomes 6.25 nA. The change in I_C due to these two effects (assuming an average BJT with $\beta = 100$) is thus

$$I_C = \frac{4.57 - 0.4375 + (2435)(6.25 \times 10^{-9})}{500 + 2435/100}$$

$$= \frac{4.13}{524.35}$$

$$= 7.9\,\text{mA}$$

For 25°C and $\beta = 100$, we find that $I_C = 7.4$ mA. Thus, a rather extreme temperature change causes only a 6 percent change in I_C. This is also to be expected since

$$\frac{R_B}{R_E} = 4.87$$

and $V_{BB} = 4.57 \gg V_{BE}$.

8.8 INTEGRATED CIRCUITS

In this chapter and in Chap. 7, we have discussed various semiconductor devices —nonlinear resistors—along with useful circuits which utilize these devices. All of these devices may be obtained in discrete form; that is, each device may be obtained as an individual unit with the necessary device terminals protruding from the unit in the form of wires. A single bipolar junction transistor is shown in Fig. 8.25*a*. Several such devices may also be manufactured as a single unit in the form of an integrated circuit, as shown in Fig. 8.25*b*. Integrated circuits (ICs) consisting of several hundred transistors, resistors, capacitors, and diodes can be fabricated on a single wafer which may be no larger than a pencil eraser. Thus, an entire amplifier or other special purpose device can be obtained in an extremely miniaturized form. An operational amplifier, discussed in Chap. 10, is shown in Fig. 8.25*b*. It is commonly supplied in an 8-pin dual in-line package (DIP) for insertion into some larger circuit.

There are currently two types of integrated circuits, monolithic and hybrid. The monolithic IC is one in which all of the parts (transistors, resistors, capacitors, and diodes) necessary for a complete circuit, such as an amplifier, are constructed at the same time from one wafer. In the hybrid IC, the various circuit components constructed on individual chips are connected by wire bonds or other suitable techniques so that the hybrid IC resembles a discrete circuit packaged into a single, small case.

One of the obvious advantages of IC fabrication is a tremendous reduction in the space occupied by the circuit. A second advantage is an increase in the useful frequency range of a circuit. As frequency is increased so that the circuit dimensions become on the order of a wavelength, lumped-circuit notions discussed previously no longer hold and distributed parameter effects become important. For example, in discussing the interconnection of lumped elements via wires, it was assumed that signal propagation between the lumped elements occurred instantaneously, but there is actually a certain amount of propagation delay introduced by the wire lengths. At frequencies where the circuit dimensions are electrically small, these propagation delays are inconsequential. As the frequency is increased, however, the connecting wires may introduce important propagation delays as well as other associated phenomena not present at the lower frequencies. The ability to place circuit elements closer on an IC chip thus acts to extend the frequency range of the devices.

A third and equally important advantage of IC technology is the increased reliability of circuits. All circuit elements, as well as their interconnections, can be fabricated in the initial manufacturing process. Thus, the reliability of all components in a circuit can be maintained to the same degree.

In this section we will briefly discuss the construction of ICs. Our intent is not to give an exhaustive coverage, but instead to give the reader the essential "flavor" of the topic.

8.8.1 The General Fabrication Technique

An IC is commonly fabricated by "planar processing," which starts with a *p*-type silicon wafer that is typically 5 mils (0.005 in) thick. This silicon wafer is called the substrate. The wafer is placed in an oxygen atmosphere in an oven and heated to over 1000°C. This oxidizes the silicon surface and forms a silicon dioxide (SiO_2) layer. A photosensitive emulsion is then placed on the oxidized surface, and a mask having the required area to be etched as an opaque coating is placed over the emulsion, as shown in Fig. 8.26*a*. Ultraviolet light is shown on the mask. Those portions of the mask (to be etched) which are not exposed to the light are removed along with the underlying silicon dioxide layer, as shown in Fig. 8.26*b*.

FIGURE 8.25
Packaging of transistors: (*a*) discrete transistor; (*b*) integrated circuit; (*c*) dual in-line package (DIP).

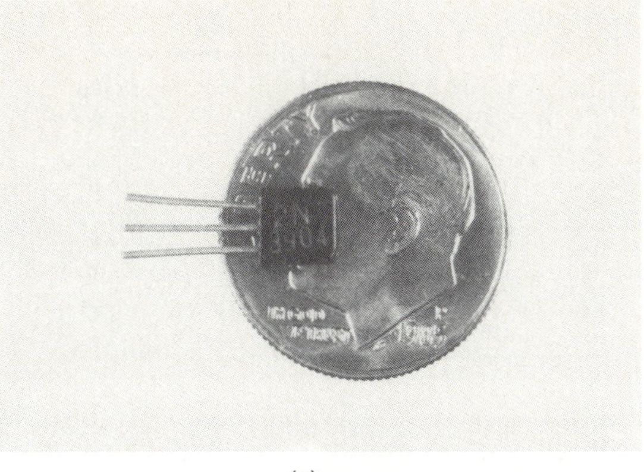
(*a*)

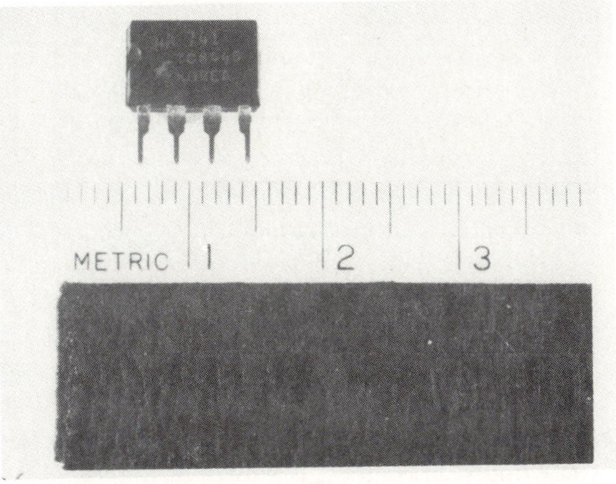

(*b*)

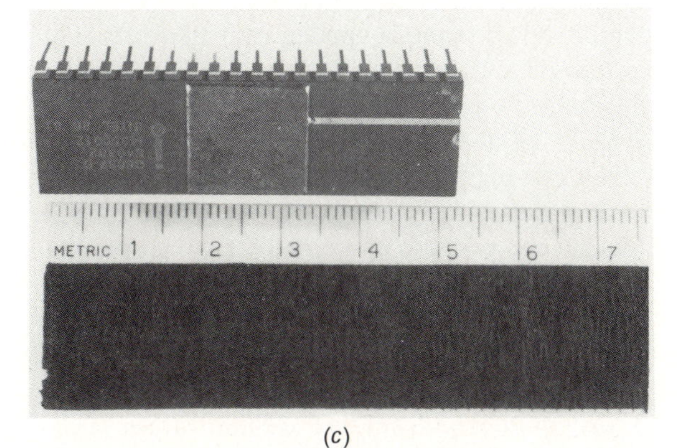

(*c*)

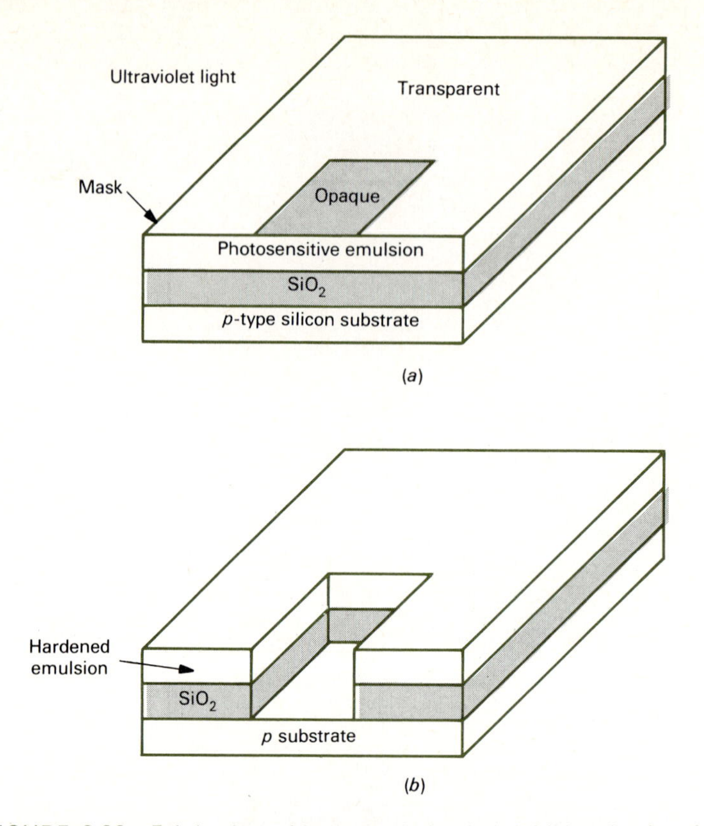

FIGURE 8.26 Fabrication of integrated circuits—initial etch of wafer.

The next step in the process is to place the etched wafer in an oven and expose it to an n-type dopant such as arsenic. The dopant diffuses into the silicon substrate, forming a heavily doped n^+ region, as shown in Fig. 8.27*a*. Passing a gas containing an n-type impurity over the wafer forms an n-type region over the n^+ region, as shown in Fig. 8.27*b*. This n-type layer will be the collector of a bipolar junction transistor. A p-type layer is diffused into this n-type layer to form the base, and an additional n^+-type layer is diffused into this p-type layer to form the emitter, as shown in Fig. 8.27*c*. A silicon dioxide coating is applied over the surface of the wafer and metallic attachments are made to the various regions. Small n^+ regions are diffused to provide good electrical connections to each region.

A silicon wafer of about 1 in in diameter is divided into squares called chips, which are on the order of 50 to 100 mils (0.05 to 0.1 in) wide. Each element of a circuit (including resistors, capacitors, and diodes) may be formed in the above manner on each chip. Thus, all components of a complete circuit may be fabricated at the same time from the same silicon wafer. This increases the reliability of the resulting circuit and also tends to ensure that the values of the circuit elements will be relatively constant between similar elements. The values of two resistors constructed from the same wafer can be controlled to within a few percent of each other. The absolute values, however, may vary quite a bit from their desired values.

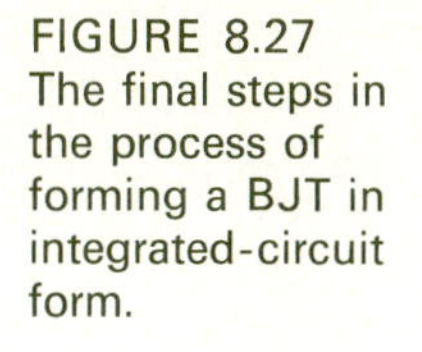
FIGURE 8.27 The final steps in the process of forming a BJT in integrated-circuit form.

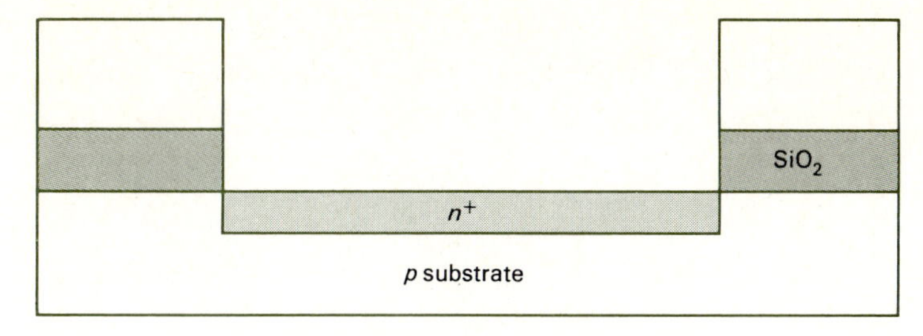

(a)

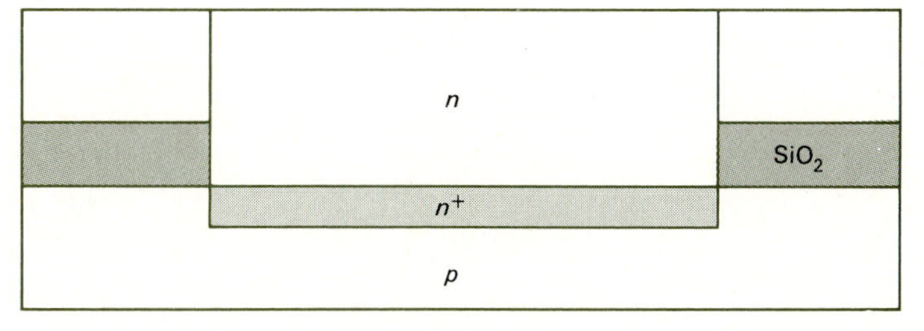

(b)

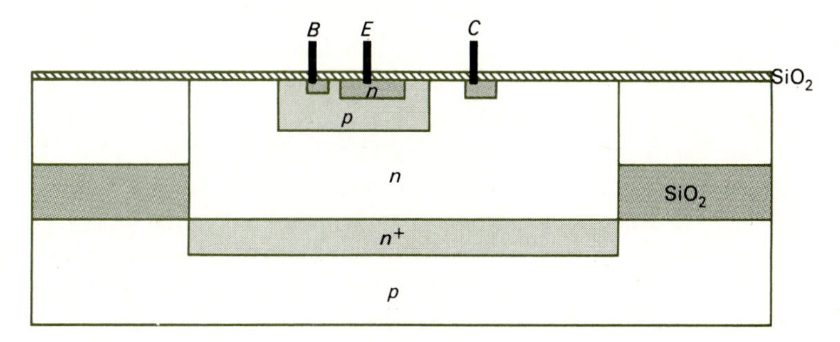

(c)

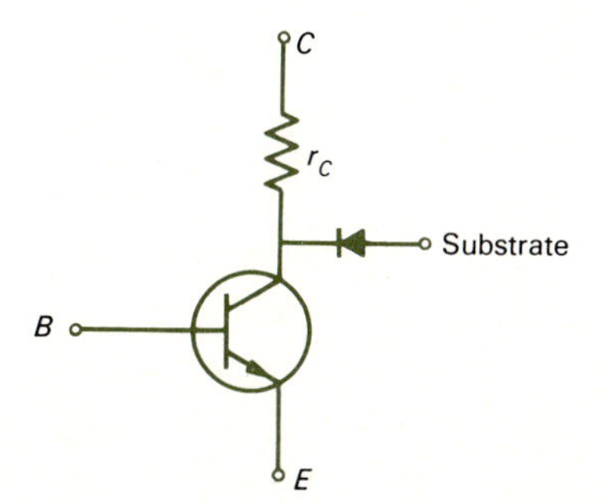

FIGURE 8.28 The IC BJT.

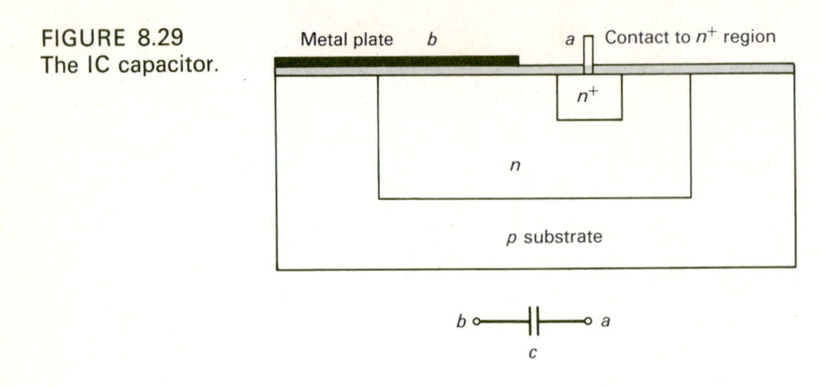

FIGURE 8.29
The IC capacitor.

8.8.2 The IC Bipolar Junction Transistor and Other Components

The basic process described in the preceding section has formed a bipolar junction transistor, whose equivalent circuit is given in Fig. 8.28. Resistor r_C represents the resistance of the relatively large n region connected to the collector. The n^+ heavily doped layer has a much lower resistance and is essentially in parallel with this n region, and it thus lowers r_C from what it would be without the n^+ region. A typical value for r_C is less than 1 Ω.

The diode shown in Fig. 8.28 represents the *pn* junction formed between the collector and the substrate. In most applications, the substrate is connected to the most negative part of the circuit or power supply to reverse-bias this diode.

An IC diode may be obtained by constructing a transistor in the above manner (Sec. 8.8.1) and using only the base-emitter leads. Recall that in normal transistor operation the base-emitter junction is essentially a forward-biased *pn* diode. Quite often the base and the collector leads are connected together.

IC capacitors may also be fabricated as shown in Fig. 8.29. A metal plate forms one plate of the capacitor and the *n* material forms the other plate.

Recall that under normal operating conditions the collector-base junction of a transistor appears as a reverse-biased *pn* junction. Also recall that a reverse-biased *pn* junction has a depletion region devoid of mobile charge which separates charges on either side of the junction. Thus, a reverse-biased *pn* junction displays a capacitance. The collector-base terminals of the transistor can also be used for this component.

The IC resistor is obtained by stopping the diffusion process after the *p*-type base region has been deposited. The result is shown in Fig. 8.30. Both contacts are applied to the *p* region (via small n^+ regions) and the protective silicon dioxide layer is applied to the surface of the wafer.

A complete circuit can be fabricated in one step. Such an amplifier circuit and its cross-sectional realization are shown in Fig. 8.31. The usual n^+ contact regions are omitted.

8.8.3 The IC Field-Effect Transistor

The junction FET, or JFET, was considered in previous sections. Another type of FET which is used in ICs is the insulated-gate FET, or IGFET. Two types of IGFETs may be obtained, the depletion-mode IGFET and the enhancement-mode IGFET.

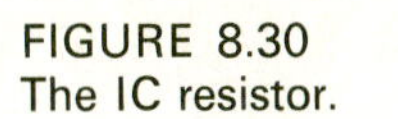

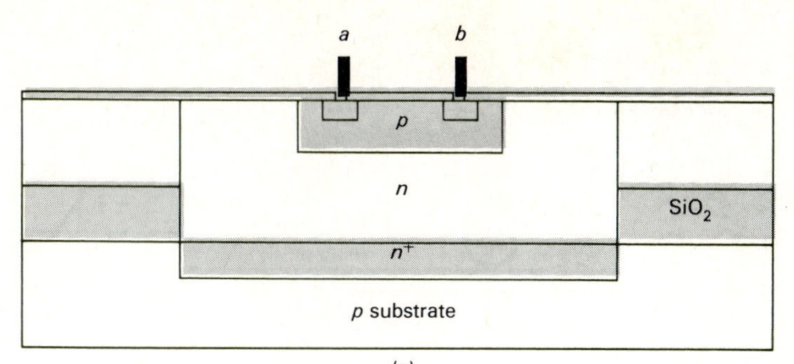

(a)

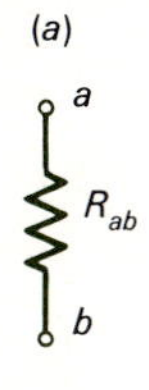

(b)

FIGURE 8.30
The IC resistor.

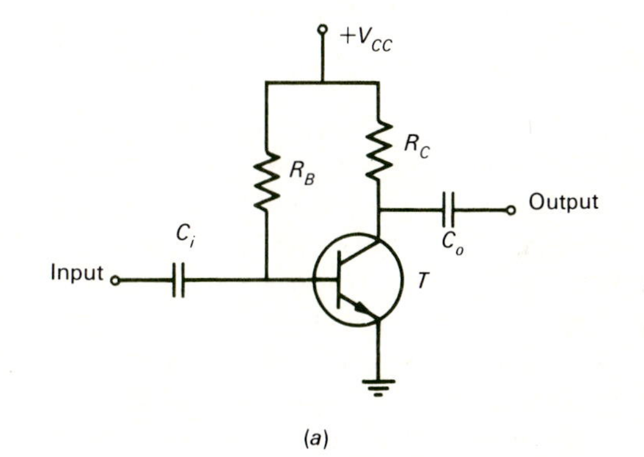

(a)

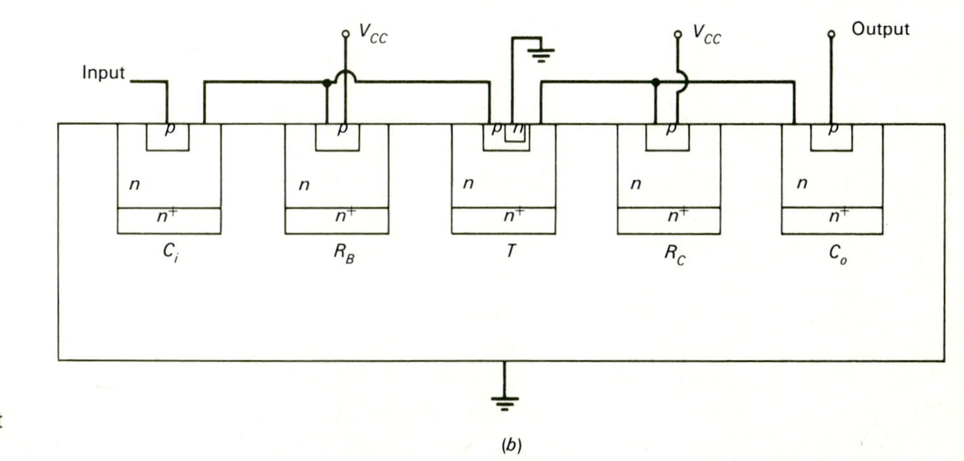

(b)

FIGURE 8.31
A complete integrated-circuit amplifier.

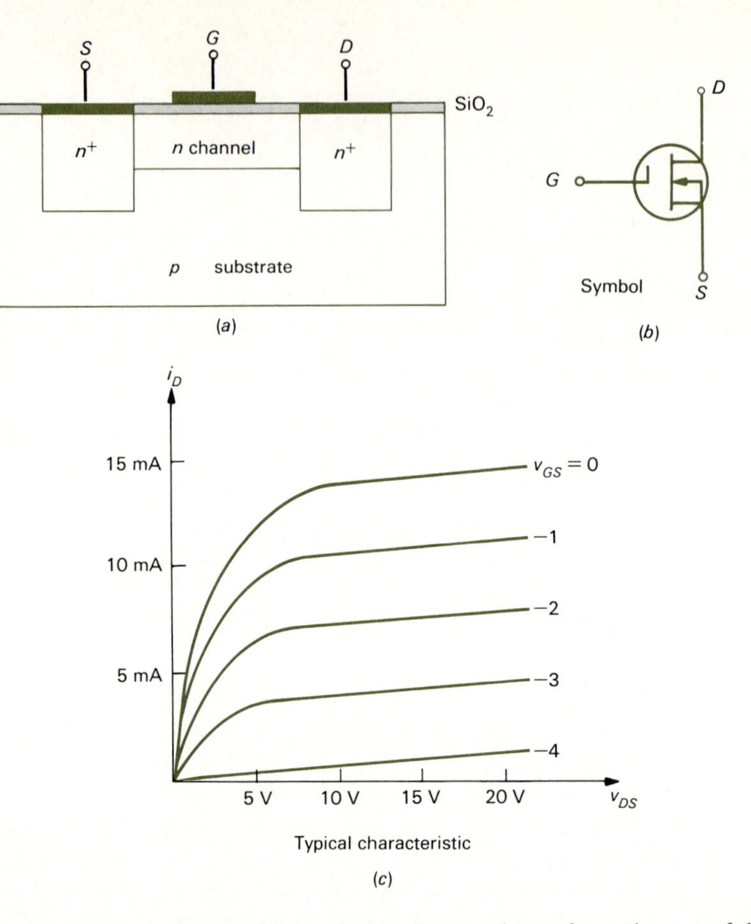

FIGURE 8.32 The depletion-mode n-channel MOSFET: (a) physical construction; (b) device symbol; (c) typical device characteristic.

Both of these are formed by insulating the metal gate from the rest of the device with a thin, silicon dioxide film; thus, these IGFETs are commonly referred to as metal-oxide-semiconductor FETs, or MOSFETs. In these types of FETs, the resistance of the oxide layer is so large ($10^{12}\ \Omega$) that the gate current is entirely negligible, regardless of the polarity of the gate-source voltage.

In the depletion-mode MOSFET, shown in Fig. 8.32a, two highly doped n^+ regions along with an n channel are diffused into the p substrate. Current flow is possible from drain to source for $v_{GS} = 0$, just as for the JFET. Applying a negative voltage to the gate such that $v_{GS} < 0$ produces an electric field which attracts holes toward the gate and pushes electrons away resulting in a depletion region being formed in the n channel near the gate. The width of the remaining portion of the channel determines its resistance and the resulting drain-source current, as with the JFET. Eventually, the channel is pinched at a value of $v_{GS} = -V_p$. The resulting characteristic in Fig. 8.32c is very similar to that of the JFET.

A transfer characteristic is shown in Fig. 8.33. Since the curves in the linear region of the characteristic are almost flat, the device is relatively independent of v_{DS} in the linear region. Thus, we may select a value of v_{GS} and determine the corresponding value of i_D. Doing this for several values of v_{DS} gives the transfer characteristic in Fig.

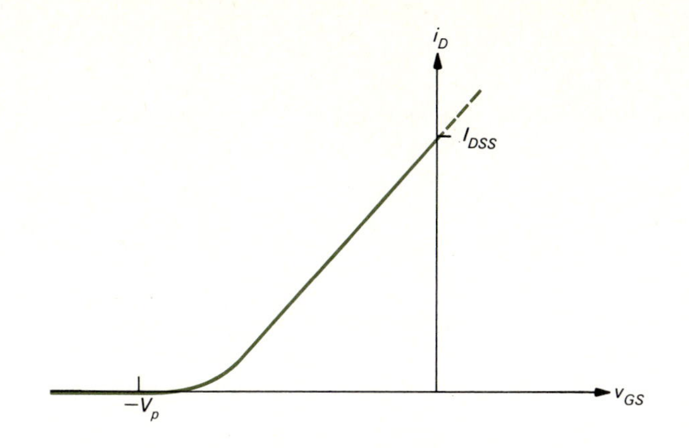

FIGURE 8.33
The transfer characteristic of a depletion-mode n-channel MOSFET.

8.33 (independent of v_{DS}). Note that for large negative values of v_{GS}, the transfer characteristic becomes nonlinear (around $v_{GS} = -V_p$). This represents a crowding of the curves on the characteristic for large negative values of v_{GS}. This relation between drain current and drain-source voltage is

$$i_D = I_{DSS}\left(1 - \frac{v_{GS}}{V_p}\right)^2 \tag{8.62}$$

where I_{DSS} is the drain-source current with the gate shorted to the source, $v_{GS} = 0$. Note that, as opposed to the JFET, the device may be operated with positive gate-source voltage. Since the gate is insulated, no gate current flows.

The n-channel enhancement-mode MOSFET is shown in Fig. 8.34. As opposed to the n-channel JFET and depletion-mode MOSFET, no n-channel between drain and source is present. Thus, for $v_{DS} > 0$ the pn junction at the drain end is reverse-biased and no drain current flows. This is indicated on the symbol in Fig. 8.34b by the broken lines. However, applying a positive gate-source voltage establishes an electric field which pulls electrons toward the gate and pushes holes away. Because of these mobile electrons being attracted to the gate, an n-type channel is formed adjacent to the gate which makes it similar to the n-channel depletion-mode MOSFET.

Above a certain value of v_{GS} known as the threshold voltage V_T, the channel forms along with a depletion region between this n channel and the p substrate. Thus, for positive v_{GS} the transfer characteristic of the n-channel enhancement-mode MOSFET shown in Fig. 8.34c is similar to the JFET and to the depletion-mode MOSFET, with the exception that positive v_{GS} voltages are used.

The transfer characteristic is shown in Fig. 8.35. This curve obeys the relation

$$i_D = k\left(\frac{v_{GS}}{V_T} - 1\right)^2 \tag{8.63}$$

where V_T is the threshold voltage and $k = i_D$ when $v_{GS} = 2V_T$. The depletion-mode MOSFET discussed previously can also be operated in the enhancement mode by applying positive gate-source voltage.

Although not specifically discussed, p-channel MOSFETs in either the depletion-mode or the enhancement-mode configuration (or both) may be obtained, as was the case for p-channel JFETs. The characteristics are similar to those for the n-channel MOSFETs except that the currents and voltages are reversed in polarity.

The majority of our design procedures developed in this chapter for the JFET hold true for the MOSFETs. The only exception is that amplifiers using enhancement-mode n-channel MOSFETs are designed to have a positive gate-source voltage.

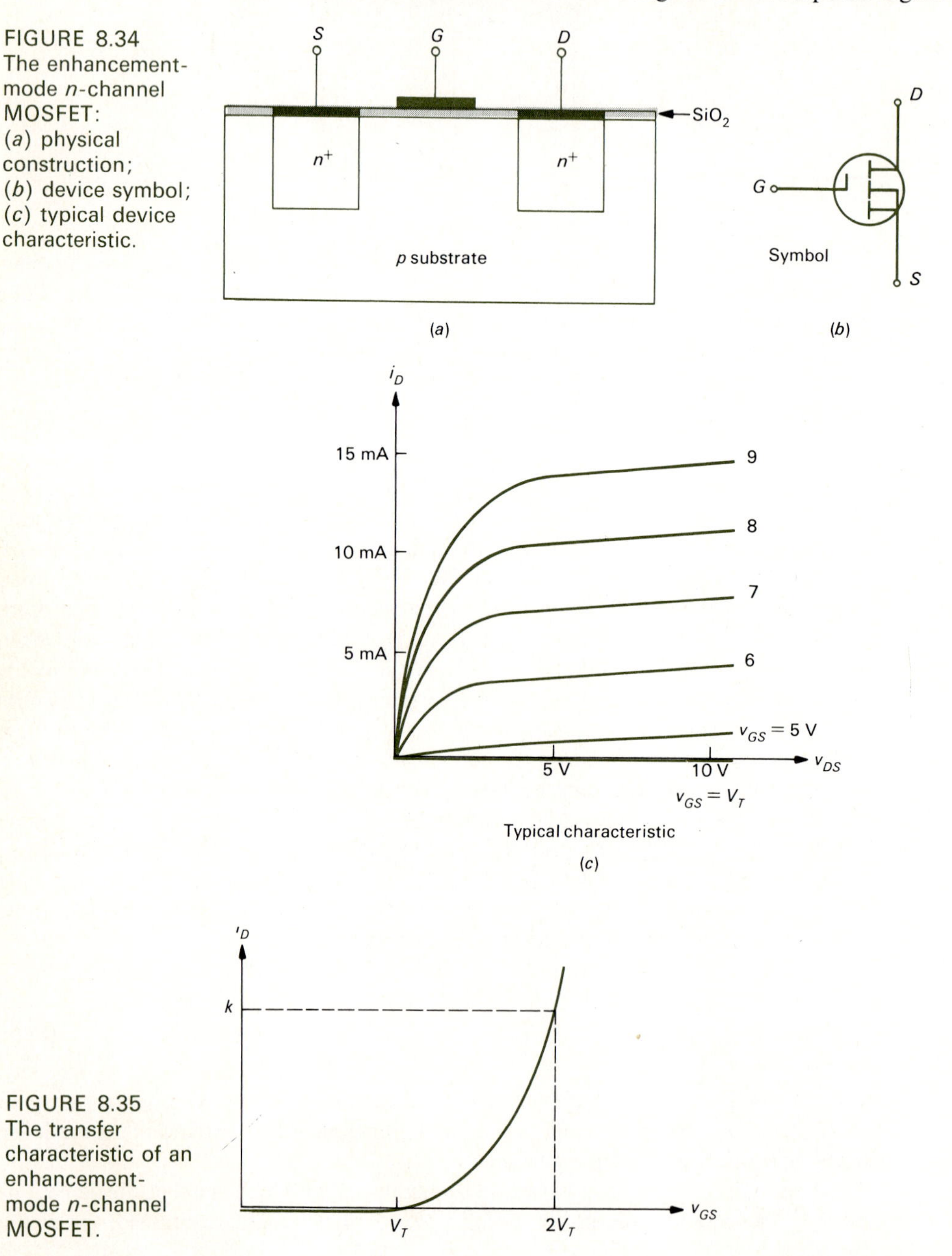

FIGURE 8.34 The enhancement-mode n-channel MOSFET: (a) physical construction; (b) device symbol; (c) typical device characteristic.

FIGURE 8.35 The transfer characteristic of an enhancement-mode n-channel MOSFET.

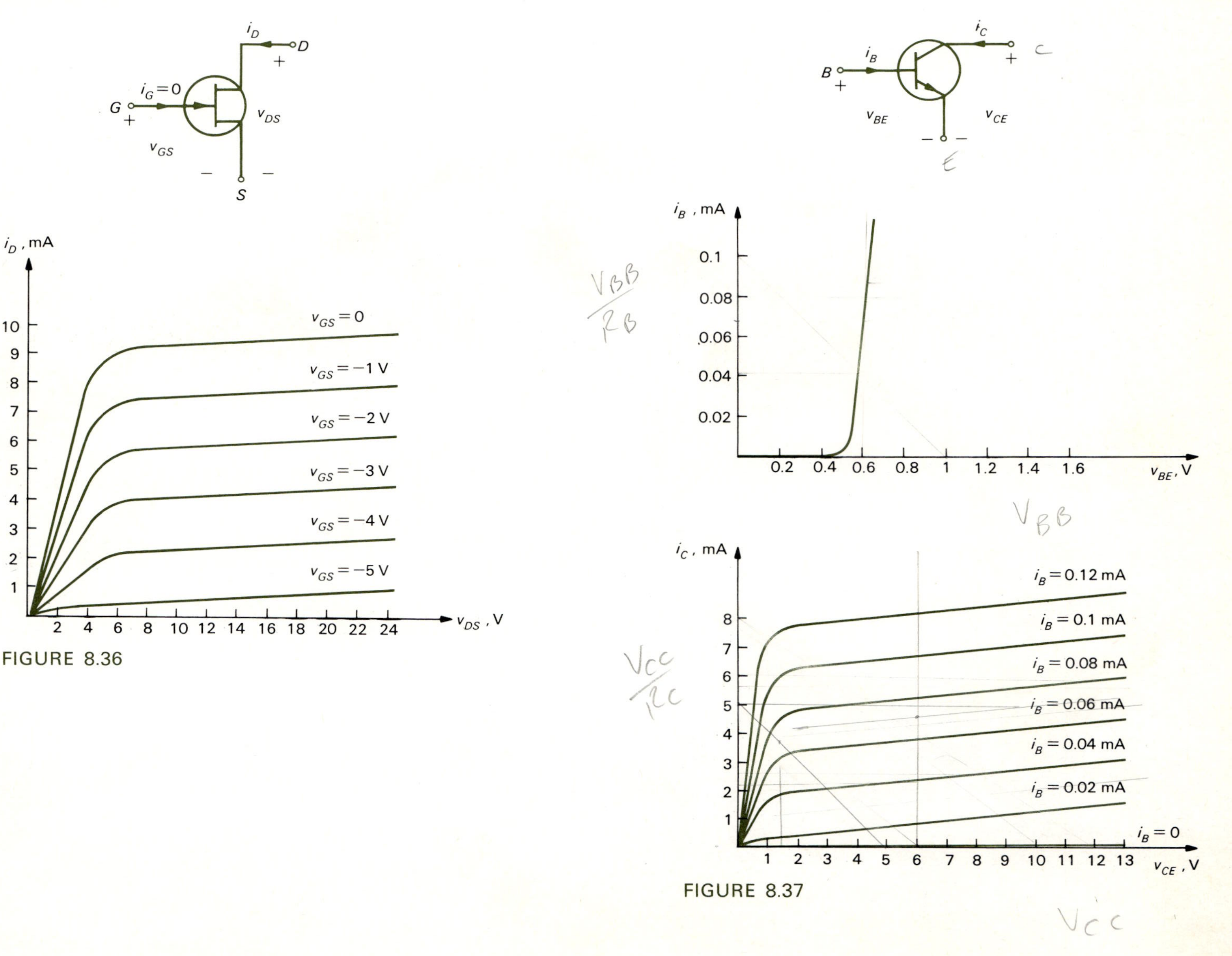

FIGURE 8.36

FIGURE 8.37

PROBLEMS

8.1 For the linear amplifier shown in Fig. 8.1, suppose $V_o = 1$ V, $V_i = 1$ mV, $Z_i = 100\,\Omega$, $R_L = 1$ kΩ. Determine the voltage and current gains.

8.2 Show that for a linear amplifier the current gain A_I can be calculated from the voltage gain A_V, input impedance Z_i, and load impedance R_L. Give a formula for A_I.

8.3 The circuit shown in Fig. P8.3 employs a JFET whose characteristic is given in Fig. 8.36. Determine the operating-point dc values for V_{GS}, I_D, V_{DS}. Sketch $v_o(t)$ and determine the voltage gain A_V.

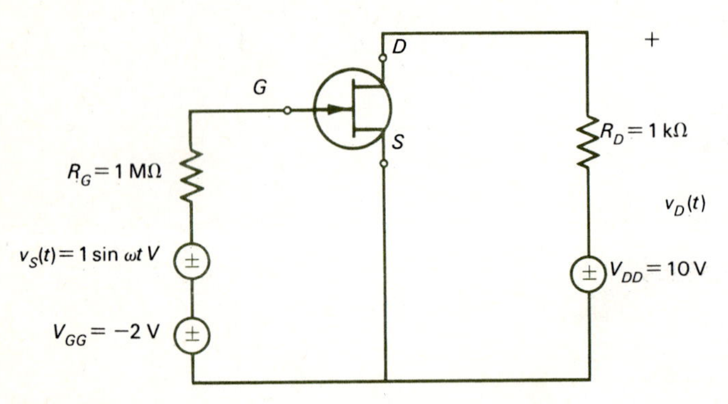

FIGURE P8.3

8.4 Repeat Prob. 8.3 but change R_D to 1.5 kΩ.

8.5 Suppose in Prob. 8.3 that V_{GG} changes from -2 to -2.5 V. Recompute the operating point and the voltage gain.

8.6 The circuit shown in Fig. 8.9 uses a JFET which has the characteristic shown in Fig. 8.36. Design the amplifier for an operating point of $I_D = 3.4$ mA, $V_{GS} = -3$ V, $V_{DS} = 12$ V, an input impedance of 50 kΩ, and $V_{DD} = 30$ V.

8.7 For the design in Prob. 8.6, calculate the voltage gain if $R_L = 1500\,\Omega$, $V_S = 100\,\Omega$, $V_S = 1$ V.

8.8 For the BJT whose characteristics are given in Fig. 8.37, calculate β and α.

8.9 A BJT is biased using the circuit shown in Fig. 8.16, with $R_B = 10$ kΩ, $V_{BB} = 1$ V, $R_C = 1.5$ kΩ, $V_{CC} = 12$ V. Determine the operating point.

8.10 Change R_B to 15 kΩ, and repeat Prob. 8.9.

8.11 The amplifier shown in Fig. 8.17*a* uses a BJT having the characteristics shown in Fig. 8.37. Suppose $V_{CC} = 12$ V, $R_C = 1500\,\Omega$, $R_B = 150$ kΩ, and $V_{BB} = 10$ V; determine the operating point.

8.12 The amplifier shown in Fig. P8.12 uses a BJT whose characteristics are shown in Fig. 8.37. For $R_B = 15\,\text{k}\Omega$,, $V_{BB} = 1.2\,\text{V}$, $V_S = 0.2\,\text{V}$, $R_C = 1.25\,\text{k}\Omega$, $V_{CC} = 10\,\text{V}$, and $R_L = 1500\,\Omega$, determine the operating point, sketch the output voltage, and determine the voltage gain $A_V = V_o/V_S$.

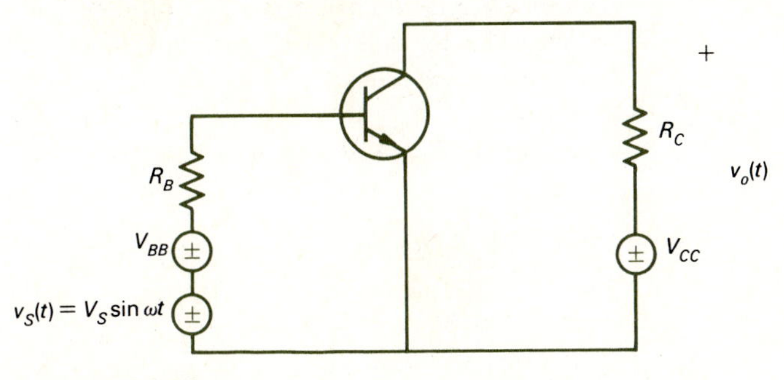

FIGURE P8.12

8.13 The circuit shown in Fig. P8.13 uses an *npn* BJT with the characteristics shown in Fig. 8.37. Determine the operating point.

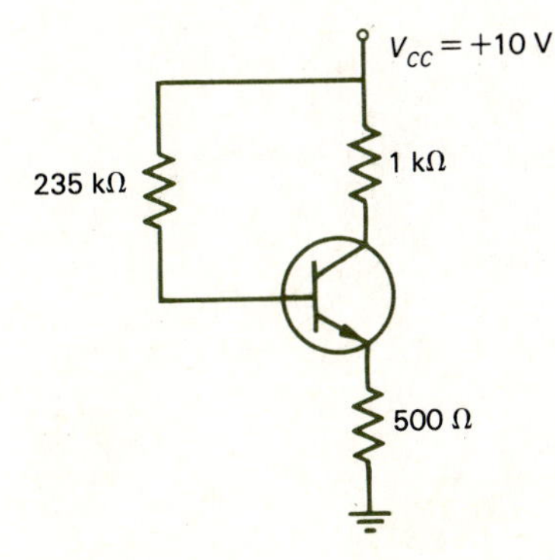

FIGURE P8.13

8.14 For the circuit shown in Fig. P8.14, the operating point is $V_{CE} = 6\,\text{V}$, $I_C = 2.3\,\text{mA}$, and the BJT whose characteristic is given in Fig. 8.37 is used. If $V_{CC} = 10\,\text{V}$, determine R_C and R_B.

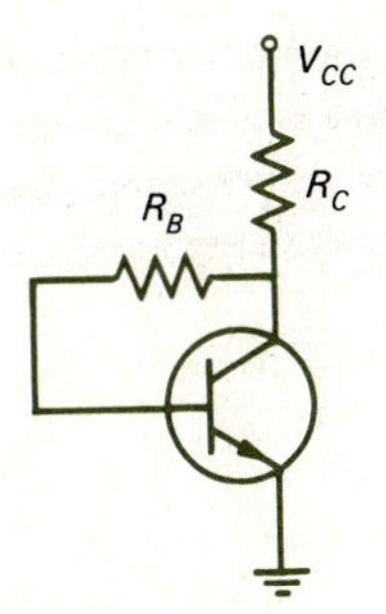

FIGURE P8.14

8.15 The Darlington compound transistor shown in Fig. P8.15 uses identical BJTs having $\beta = 100$ and $I_{CEO} = 1\ \mu A$. Determine an expression relating i_o and i_i. From this, can you deduce the β of the overall structure?

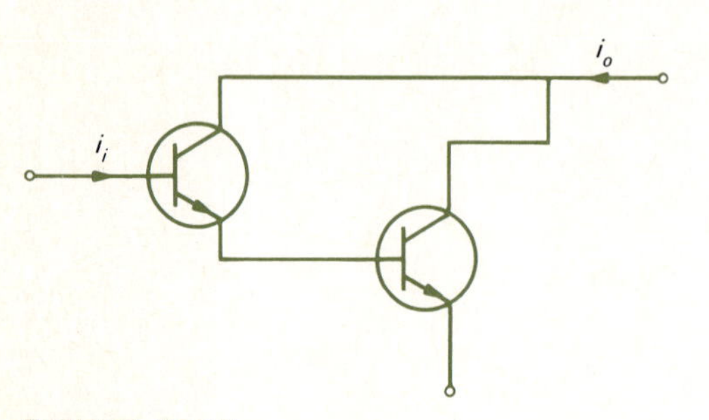

FIGURE P8.15

8.16 The circuit shown in Fig. 8.22 is to be designed for maximum symmetrical swing. If $R_C = 1\ k\Omega$, $R_L = 1\ k\Omega$,, $V_{CC} = 20$ V, $R_E = 500\ \Omega$, determine R_C, R_{B1}, and R_{B2} such that $R_{B1}||R_{B2} = 10\ k\Omega$. Assume $\beta = 100$ and $V_{BE} = 0.5$ V.

8.17 For the circuit shown in Fig. P8.14, the transistor has $\beta = 100$, $I_{CBO} = 1$ mA, $V_{BE} = 0.6$ V at room temperature (20°C). If the temperature changes to 40°C, determine the change in the operating point if β remains constant. Assume $V_{CC} = 10$ V, $R_C = 1\ k\Omega$, and $R_B = 150\ k\Omega$.

REFERENCES

8.1 A. B. Carlson and D. G. Gisser, *Electrical Engineering: Concepts and Applications*, Addison-Wesley, Reading, Mass., 1981.

8.2 Ralph J. Smith, *Circuits, Devices and Systems: A First Course in Electrical Engineering*, 4th ed., Wiley, New York, 1984.

8.3 A. E. Fitzgerald, D. Higginbotham, and A. Grabel, *Basic Electrical Engineering*, 5th ed., McGraw-Hill, New York, 1981.

8.4 D. L. Schilling and C. Belove, *Electronic Circuits: Discrete and Integrated*, 2d ed., McGraw-Hill, New York, 1979.

8.5 R. L. Boylestad and L. Nashelsky, *Electronic Devices and Circuit Theory*, 3d ed., Prentice-Hall, Englewood Cliffs, New Jersey, 1982.

8.6 E. C. Lowenberg, *Electronic Circuits*, McGraw-Hill, New York, 1967.

Chapter

Small-Signal Analysis of Amplifiers

In the previous chapter we studied the field-effect transistor (FET), the bipolar junction transistor (BJT), and their use in constructing linear signal amplifiers. One key assumption in the design of these amplifiers was that the input signal was small enough so that these nonlinear devices were being operated only in their linear regions. The reason for this restriction on the operation is that we did not want distortion of the signal. If the nonlinear devices were operated at any instant of time in the nonlinear region of their characteristics, the output signal of the amplifier would not be a faithful reproduction of the input signal. Amplification (enlargement) of a signal such as music from a phonograph cartridge would be of no value if it were altered or distorted.

In this chapter we will continue to make this important assumption and will determine approximate methods for calculating the various figures of merit of the amplifier—such as voltage gain A_V, current gain A_I, power gain $A_P = A_V A_I$, input impedance Z_i, and output impedance Z_o. Our method of approach will be to determine a linear equivalent circuit of the transistor which describes (approximately) the linear portion of its characteristic, to substitute this linear circuit model for the transistor, and to compute, with linear circuit-analysis techniques, the figures of merit for the system. Thus, once we have determined the small-signal linear equivalent circuit of the transistor, the remaining task is a simple application of circuit-analysis techniques which we studied in Part 1.

9.1 REVIEW OF AMPLIFIER FUNDAMENTALS

All of the JFET and BJT amplifiers studied in the previous chapter were designed using essentially the same procedure:

1 First we draw the dc circuit and determine the resistors and dc voltage sources in the biasing circuitry to locate the dc operating point at a desirable position in the linear portion of the device characteristic; these are also chosen such that this dc operating point is insensitive to unit-to-unit parameter variations, temperature effects, etc. The dc circuit is obtained by removing the ac source and load (caused by the open-circuit nature of the input and output capacitors to dc).

2 Then we reattach the ac signal source, short all capacitors, and kill the dc sources to obtain the ac circuit—from which we calculate the ac quantities A_V, A_I, A_P, Z_i, and Z_o for the amplifier.

FIGURE 9.1 Illustration of the use of the superposition principle in the analyis of linear amplifiers.

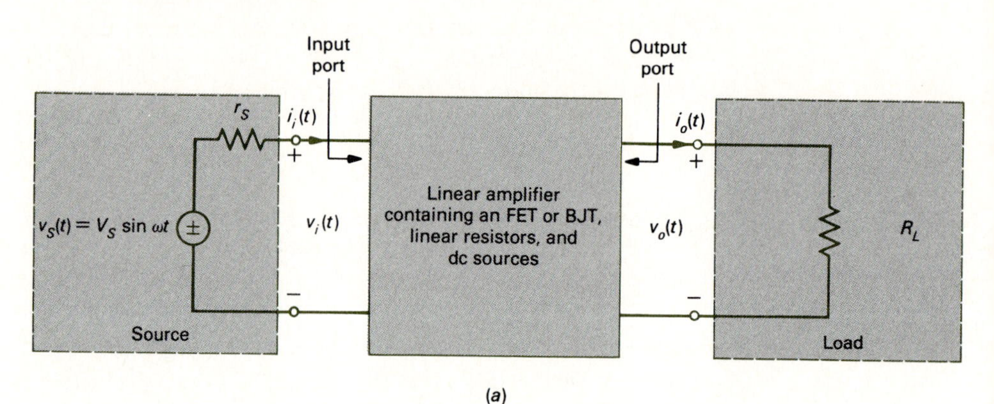

(a)

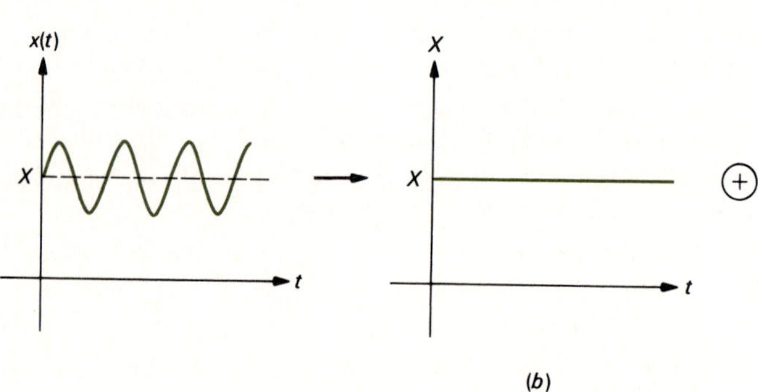

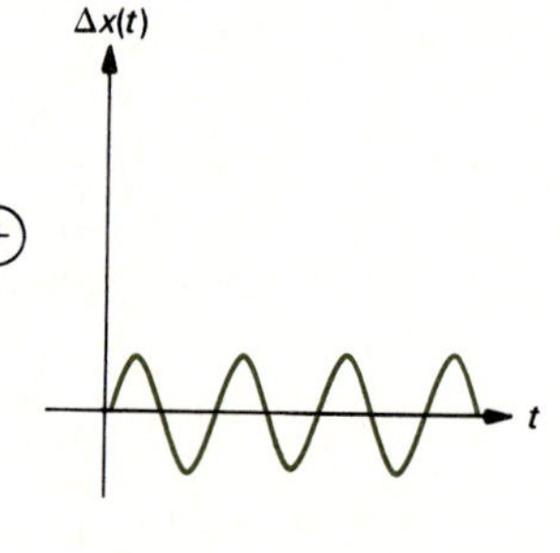

(b)

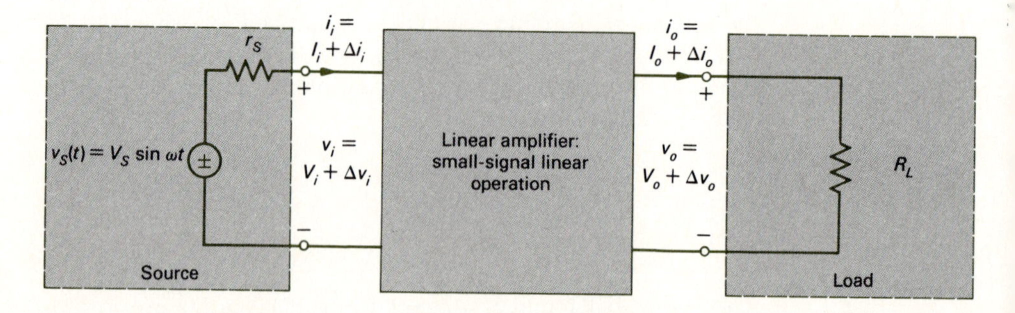

(c)

The justification for this procedure of treating dc and ac quantities separately is simply that we assume linear operation of the device so that superposition may be applied. Thus, each voltage and current in the circuit will consist of two parts. One part is dc and is due to the dc batteries in the biasing circuitry; this is denoted with a capital letter X. The other part is ac and is due to the ac source $v_S(t)$; this is denoted as a delta (Δ) quantity, $\Delta x(t)$. The total value of the particular voltage or current is, by superposition, the sum of these two parts:

$$x(t) = X + \Delta x(t) \tag{9.1}$$

where $x(t)$ represents any voltage or currrent in the circuit.

For example, for the linear amplifier circuit in Fig. 9.1*a*, the input voltage to the amplifier is, assuming small-signal linear operation,

$$v_i = V_i + \Delta v_i \tag{9.2}$$

as shown in Fig. 9.1*c*. The input current to the amplifier is

$$i_i = I_i + \Delta i_i \tag{9.3}$$

and the output voltage and current of the amplifier are

$$v_o = V_o + \Delta v_o \tag{9.4}$$

$$i_o = I_o + \Delta i_o \tag{9.5}$$

The ac circuit is shown in Fig. 9.2*a*. It is obtained (by superposition) by killing the dc batteries in the amplifier. In addition, any capacitors in the amplifier are replaced by short circuits, assuming that they have negligible impedance at the frequency of the ac source. For the ac amplifier circuit in Fig. 9.2*a*, the voltage gain of the amplifier is the ratio of the ac component of the output voltage Δv_o and the ac component of the input voltage Δv_i:

$$A_V = \frac{\Delta v_o}{\Delta v_i} \tag{9.6}$$

The current gain of the amplifier is similarly defined:

$$\begin{aligned} A_I &= \frac{\Delta i_o}{\Delta i_i} \\ &= \frac{\Delta v_o/R_L}{\Delta v_i/Z_i} \\ &= A_V \frac{Z_i}{R_L} \end{aligned} \tag{9.7}$$

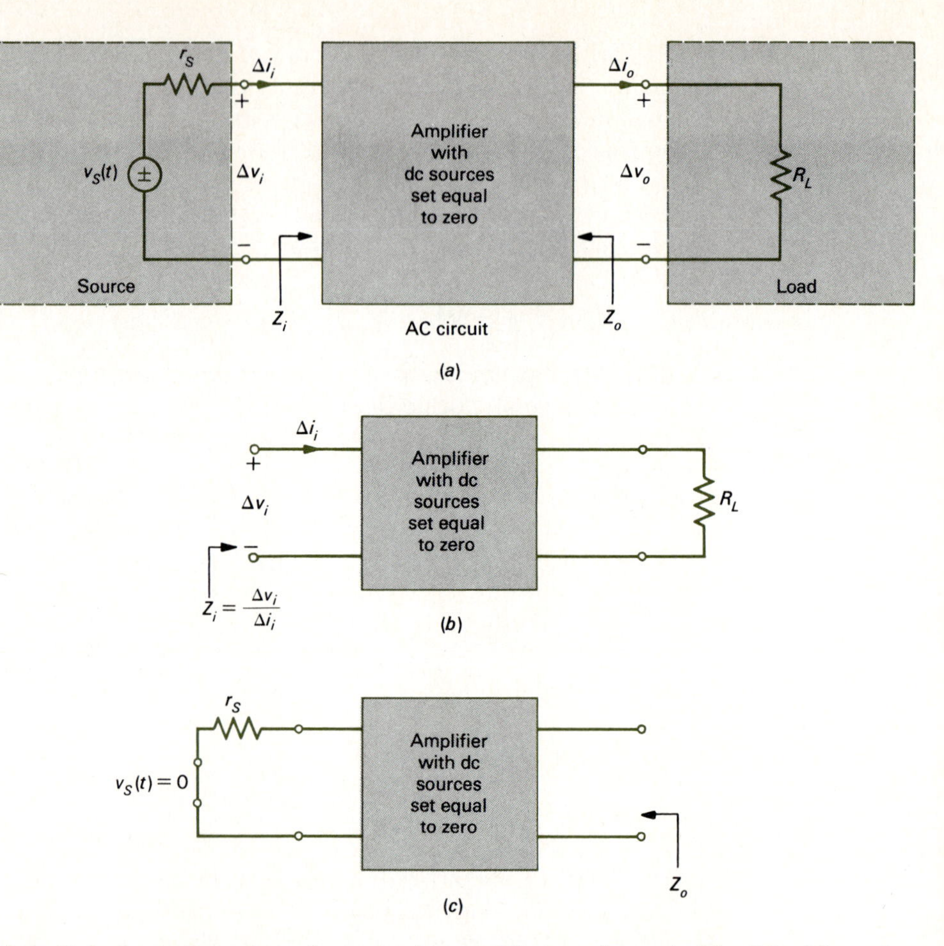

FIGURE 9.2 Illustration of ac properties of linear amplifiers: (*a*) ac analysis; (*b*) input impedance; (*c*) output impedance.

where Z_i is the ac input impedance to the amplifier, as shown in Fig. 9.2*a*. Thus, the current gain can be found from the voltage gain. The power gain is the ratio of the output ac power and the input ac power:

$$\begin{aligned} A_P &= \frac{\Delta v_o \, \Delta i_o}{\Delta v_i \, \Delta i_i} \\ &= A_V A_I \\ &= A_V^2 \frac{Z_i}{R_L} \end{aligned} \tag{9.8}$$

Certain other figures of merit are also of interest. If we set the dc sources in the amplifier to zero in applying superposition, as shown in Fig. 9.2*a*, then all variables are Δ (sinusoidal) quantities. The input impedance Z_i to the amplifier is the ac Thevenin impedance seen looking into the input terminals, as shown in Fig. 9.2*b*. Similarly, the output impedance Z_o is the Thevenin impedance seen by R_L [set $v_S(t) = 0$], as shown in Fig. 9.2*c*. One of the reasons for our interest in Z_i and Z_o relates to matching. For example, if in Fig. 9.2*a* $Z_i = r_S$, maximum power transfer will take place from the source to the amplifier. Similarly, if $Z_o = R_L$, maximum power transfer will take place from the amplifier to the load. Quite often, amplifiers are not used specifically to provide voltage or current gain but are used instead for matching from a source to a load. For example, suppose $r_S = 200\ \Omega$ and $R_L = 1000\ \Omega$. If the source and load were connected directly, maximum power transfer from the source to the load would not take place. On the other hand, if an amplifier having $Z_i = 200\ \Omega$ and $Z_o = 1000\ \Omega$ is inserted between the source and load, maximum power transfer from the source to the load will be achieved. If the amplifier possesses a power gain ($A_P >$ 1), then we have an additional benefit.

9.2 THE COMMON-SOURCE JFET AMPLIFIER

Consider the JFET amplifier discussed in the previous chapter; it is shown in Fig. 9.3*a*. Once the dc analysis is completed and the operating point is determined, the signal variations move along the ac load line, as shown in Fig. 9.4*a*. Note from the ac circuit in Fig. 9.3*b* that

$$\Delta v_o = \Delta v_{DS} \tag{9.9a}$$

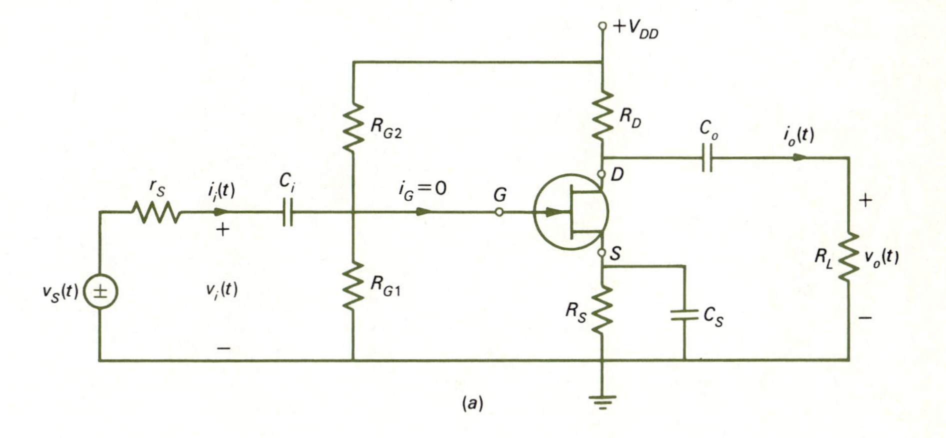

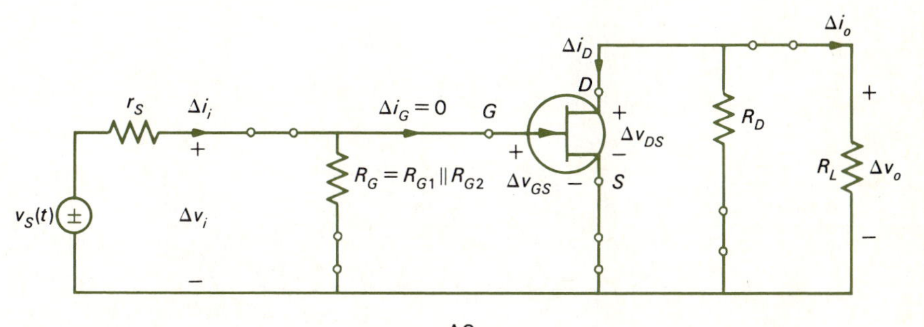

FIGURE 9.3 The ac circuit of the common-source JFET amplifier.

and

$$\Delta i_o = -\frac{R_D}{R_D + R_S}\Delta i_D \tag{9.9b}$$

Thus, the ac output-voltage and output-current variations for the amplifier may be found from the characteristic, as shown in Fig. 9.4*a*.

Once again we observe an important point. If the excursions along the ac load line are restricted to a linear portion of the characteristic (circled in Fig. 9.4*a*), then the fact that the device characteristic is nonlinear over certain portions of the characteristic outside this linear region is immaterial; we never operate in those nonlinear regions. Therefore, for small-signal linear operation, we are free to redraw the device characteristic outside this linear region in any form we choose. Why not choose a simple representation: simply extend the linear portion throughout the rest of the characteristic, as shown in Fig. 9.4*b*. We may write a linear equation describing this new characteristic. Note that three variables are related on the characteristic: i_D, v_{DS}, and v_{GS}. So our linear equation must relate these three variables. Let us presume a form of the equation to be

$$i_D = y_{fs}v_{GS} + y_{os}v_{DS} + I_{DSS} \tag{9.10}$$

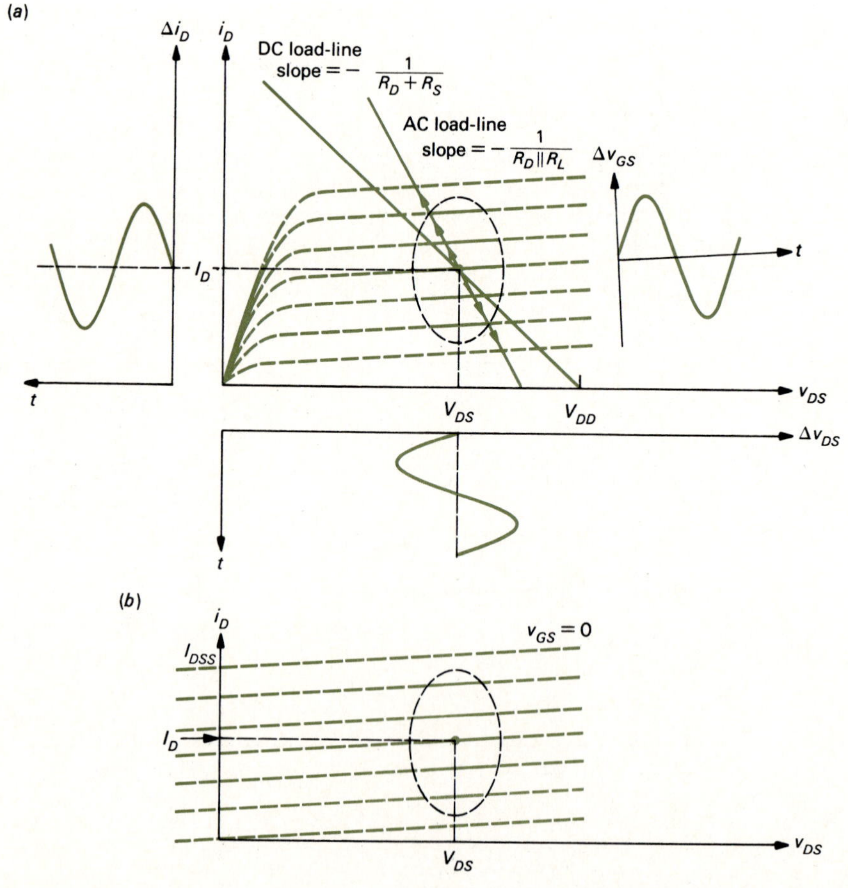

FIGURE 9.4 Small-signal linearization of JFET amplifier operation.

where I_{DSS} is the drain-source current with the gate short-circuited to the source, $v_{GS} = 0$. Obviously, this equation is a linear one, and once we determine the constants y_{fs}, y_{os}, and I_{DSS} from the characteristic at this dc operating point, we will have an alternative but equally valid representation of the characteristic.

The constant y_{fs} in Eq. (9.10) is calculated by taking a small change in v_{GS} (Δv_{GS}) and finding the resulting change in i_D (Δi_D) when no change in v_{DS} occurs ($\Delta v_{DS} = 0$), as shown in Fig. 9.5a:

$$y_{fs} = \left.\frac{\Delta i_D}{\Delta v_{GS}}\right|_{\Delta v_{DS}=0} \qquad \text{S} \tag{9.11}$$

FIGURE 9.5 Calculation of small-signal JFET parameters.

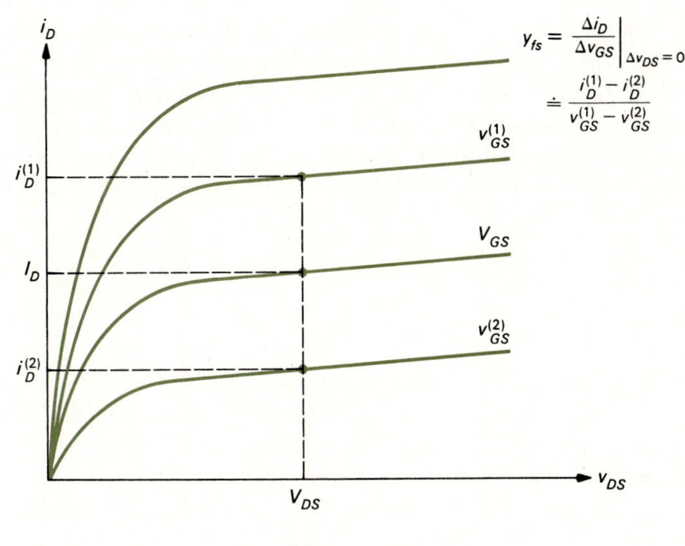

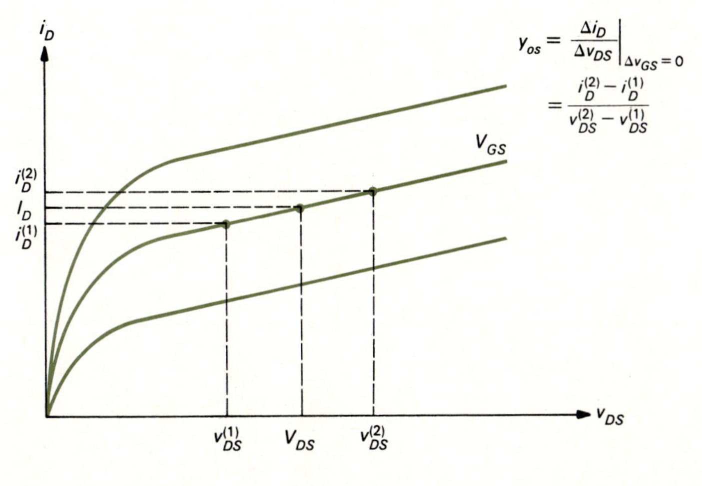

The units of y_{fs} are the units of conductance (siemens) since y_{fs} is a ratio of a current to a voltage. Typical values range from 1000 μS (10^{-3} S) to 6000 μS (6×10^{-3} S). The parameter y_{os} is the reciprocal of the slope of the characteristic line passing through the dc operating point, since from Eq. (9.10)

$$y_{os} = \left.\frac{\Delta i_D}{\Delta v_{DS}}\right|_{\Delta v_{GS}=0} \quad \text{S} \tag{9.12}$$

Thus, the units of y_{os} are also siemens. This parameter may be calculated at the operating point from the JFET characteristic, as shown in Fig. 9.5*b*. Typical values of $1/y_{os}$ range from 10 kΩ to as much as 1 megohm (MΩ). Of course it cannot be stated that a JFET has unique values of y_{fs} and y_{os} since the values depend on where the parameters are being calculated on the characteristic—namely the operating point. The parameter I_{DSS} is included in the equation since the approximation line in Fig. 9.4*b* for $v_{GS} = 0$ may not pass through $i_D = 0$ for $v_{DS} = 0$ (and generally doesn't).

Thus, the y symbols for these parameters indicate an admittance (or conductance). The origin of the subscripts is the following. The s subscript refers to the fact that the FET source terminal is common to the input and output; v_{GS} is input to one "port," and v_{DS} and i_D are output at the second port. The f subscript in y_{fs} symbolizes that this is a forward transfer conductance parameter since it yields the change in i_{DS} in the second port with a change in v_{GS} in the first port (with no change in v_{DS}). Similarly, the o subscript in y_{os} symbolizes that this is a conductance in the output port of the device.

An equivalent circuit for small-signal linear operation which represents Eq. (9.10) can be obtained as shown in Fig. 9.6. [The reader should verify that this circuit is equivalent to Eq. (9.10).] The diamond-shaped element is referred to as a controlled source. It has one of the characteristics of a current source since it maintains a current $y_{fs}v_{GS}$ through its terminals while its terminal voltage is not, as yet, known. It is quite different from an ideal current source for the following two reasons. First, the output current of this controlled source is not known but is determined by the value of some other branch variable in the circuit (in this case v_{GS}); thus, it is similar to a resistor in which the resistor current is "determined" by the resistor voltage $i = v/R$, and it is often referred to as a "controlled resistor." The second important difference between a controlled source and an ideal source is that *superposition cannot be applied to controlled sources*. This should be clear since the output current of the controlled source is

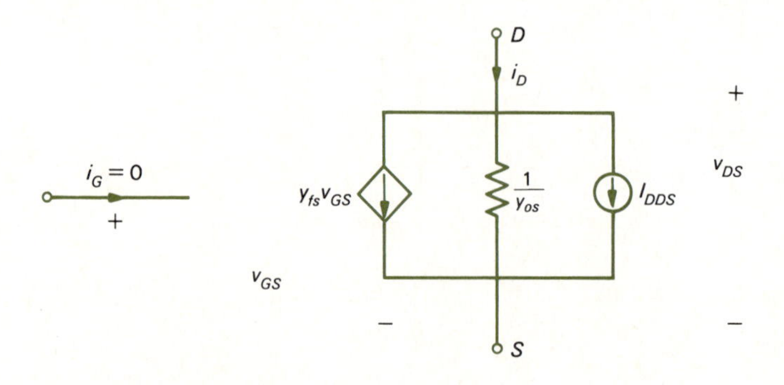

FIGURE 9.6 The small-signal linear equivalent circuit of the common-source JFET amplifier.

not fixed, as is the case for an ideal current source, but is dependent on another branch variable. Thus, its value will not appear on the right-hand side of the circuit equations but will instead be incorporated into a coefficient of v_{GS}.

Substituting this small-signal equivalent circuit for the JFET into the JFET amplifier circuit of Fig. 9.3*a*, we obtain the circuit in Fig. 9.7*a*. For ac analysis we set all dc

FIGURE 9.7 Small-signal analysis of the common-source JFET amplifier.

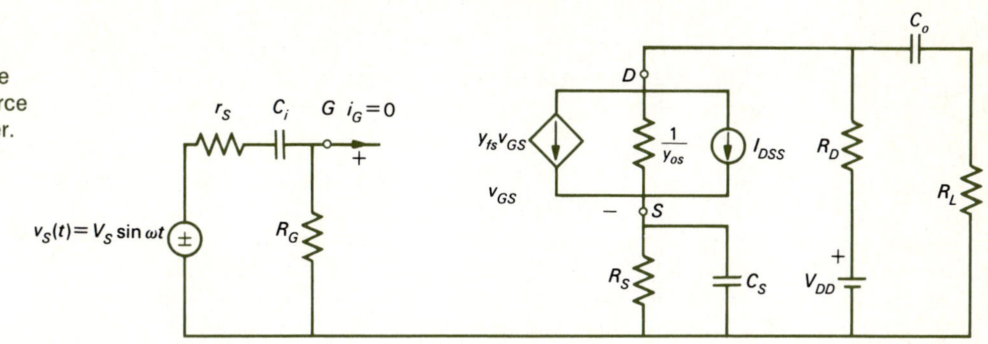

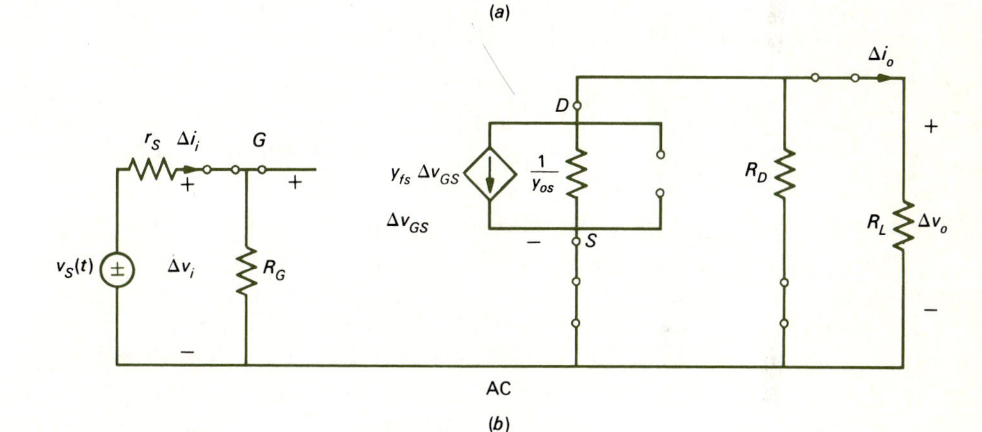

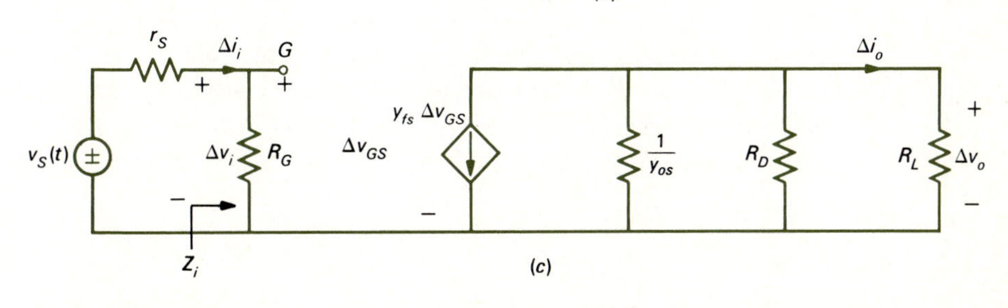

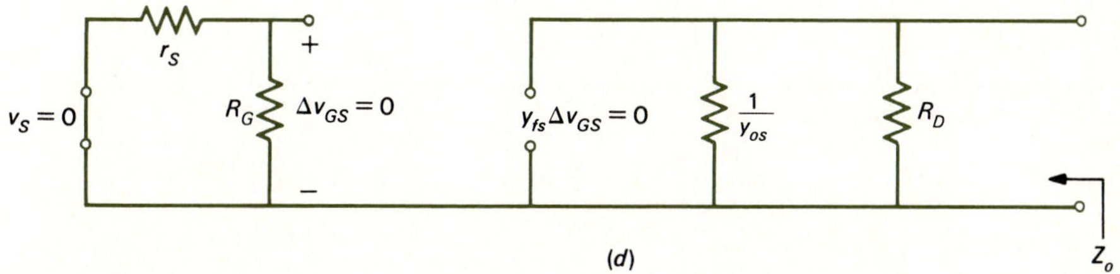

sources equal to zero and replace (as approximations) the capacitors with short circuits, as shown in Fig. 9.7*b*. The resulting circuit is simplified in Fig. 9.7*c*. From this circuit we obtain the ac output voltage as

$$\Delta v_o = -y_{fs}\,\Delta v_{GS}\left(\frac{1}{y_{os}}\,||R_D||R_L\right) \tag{9.13}$$

and the ac input voltage to the amplifier is

$$\Delta v_i = \Delta v_{GS} \tag{9.14}$$

The ratio of Eqs. (9.13) and (9.14) yields the voltage gain of the amplifier:

$$A_v = \frac{\Delta v_o}{\Delta v_i}$$

$$= -y_{fs}\left(\frac{1}{y_{os}}\,||R_D||R_L\right) \tag{9.15}$$

The negative sign in Eq. (9.15) shows that the output voltage $\Delta v_o = \Delta v_{DS}$ and the input voltage $\Delta v_i = \Delta v_{GS}$ are 180° out of phase; as $\Delta v_i = \Delta_{GS}$ increases, Δv_o decreases, which is also evident from Fig. 9.4*a*.

The input impedance is easily seen from Fig. 9.7*c* to be

$$Z_i = R_G \tag{9.16}$$

which can be made quite large (R_G is typically chosen to be 1 MΩ). The output (Thevenin) impedance seen by R_L is found by setting $v_S(t) = 0$ in the ac circuit, as shown in Fig. 9.7*d*. Since $\Delta v_{GS} = 0$, the controlled-source output current is zero and is therefore replaced by an open circuit. Note that we did *not* kill the controlled source; killing $v_S(t)$ made $\Delta v_{GS} = 0$, which killed the controlled source. In this sense, a controlled source is treated no differently than a resistor in obtaining the Thevenin impedance. The output impedance then becomes

$$Z_o = \frac{1}{y_{os}}\,||R_D \tag{9.17}$$

The input current to the amplifier is

$$\Delta i_i = \frac{\Delta v_i}{R_G}$$

$$= \frac{\Delta v_i}{Z_i} \tag{9.18}$$

The output current is

$$\Delta i_o = \frac{\Delta v_o}{R_L} \tag{9.19}$$

and the current gain is

$$A_I = \frac{\Delta i_o}{\Delta i_i}$$

$$= \frac{\Delta v_o}{\Delta v_i}\frac{Z_i}{R_L}$$

$$= A_V \frac{R_G}{R_L} \tag{9.20}$$

Example 9.1

The manufacturer lists nominal parameters for the 2N3819 JFET as $y_{os} = 50\ \mu S$, $y_{fs}(\text{min}) = 2000\ \mu S$, and $y_{fs}(\text{max}) = 6500\ \mu S$. Determine A_V, Z_i, and Z_o for this amplifier, assuming that the dc design yields $R_S = 5\ k\Omega$, $R_D = 1\ k\Omega$, $R_G = 100\ k\Omega$, and that the load is $R_L = 1\ k\Omega$. Assume a maximum JFET.

Solution With $R_G = 100\ k\Omega$ and $R_L = 1\ k\Omega$, we find that

$$A_V = -y_{fs}\left(\frac{1}{y_{os}}||R_D||R_L\right)$$

$$= -3.17$$

$$Z_o = \frac{1}{y_{os}}||R_D$$

$$= 952.38\ \Omega$$

$$Z_i = R_G$$

$$= 100\ k\Omega$$

The current gain is

$$A_I = A_V \frac{Z_i}{R_L}$$

$$= -317$$

Thus, the power gain is

$$A_P = A_V A_I$$

$$= 1004.9$$

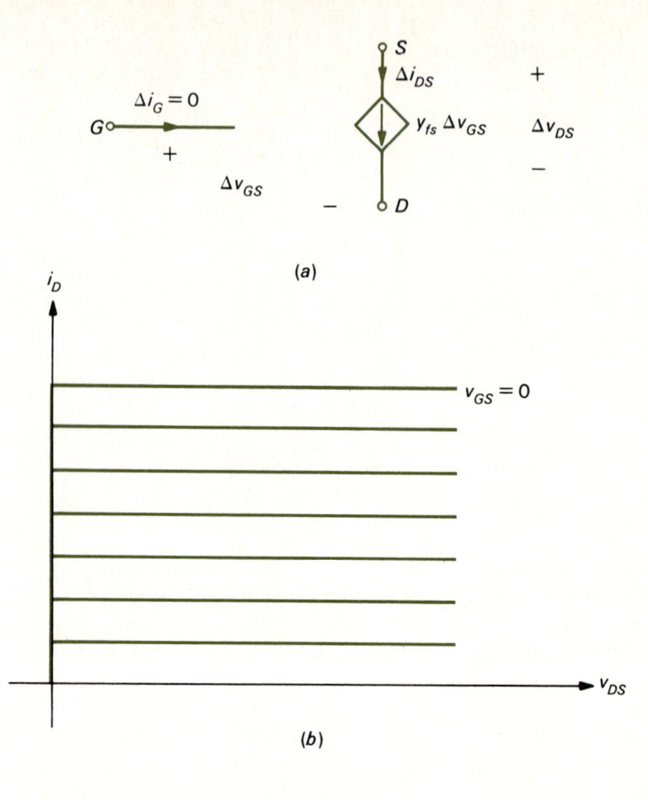

FIGURE 9.8 A simplified small-signal equivalent circuit of a JFET.

Note that since $1/y_{os} = 20\ \mathrm{k\Omega}$, it has little effect on any of the ac parameters. For example, in the voltage-gain expression in Eq. (9.15), $1/y_{os}$ is in parallel with $R_D||R_L$. But since $R_D||R_L = 500\ \Omega$, $1/y_{os}||R_D||R_L = 487.8\ \Omega$. Neglecting $1/y_{os}$ gives $A_V = -3.25$, which is reasonably close to the exact value of $A_V = -3.17$. Similarly, $1/y_{os}$ entered into the Z_o calculation in parallel with R_D, but since $R_D = 1\ \mathrm{k\Omega}$, $1/y_{os}$ may be neglected, yielding $Z_o \doteq R_D = 1\ \mathrm{k\Omega}$ as compared to the exact value of $Z_o = 952.38\ \Omega$.

Therefore, for simplification of our future calculations with this model, we will omit $1/y_{os}$ from the small-signal linear equivalent circuit. Thus, the circuit which we will use is shown in Fig. 9.8*a*. Omitting $1/y_{os}$ means that we assume that the lines on the characteristic are flat with no slope (Fig. 9.8*b*). It should be remembered that this will not always be a valid approximation. For example, if R_D and R_L in the previous example had both been 100 kΩ, then $1/y_{os} = 20\ \mathrm{k\Omega}$ would have had an effect and could not be removed from the linear model. However, to simplify our calculations with this model, we will always omit $1/y_{os}$ and use the model in Fig. 9.8*a*.

If the capacitor C_S which bypasses most of the ac signal around R_S had been removed, then R_S would appear in the ac circuit and would affect some of our ac figures of merit. To determine the effect of removing C_S, let us reexamine our calculations. The ac circuit is shown in Fig. 9.9*a*. We now replace the device symbol with the small-signal linear equivalent circuit of Fig. 9.8*a*, which results in the circuit in Fig. 9.9*b*. We must be careful to attach the correct terminals and identify the location of the

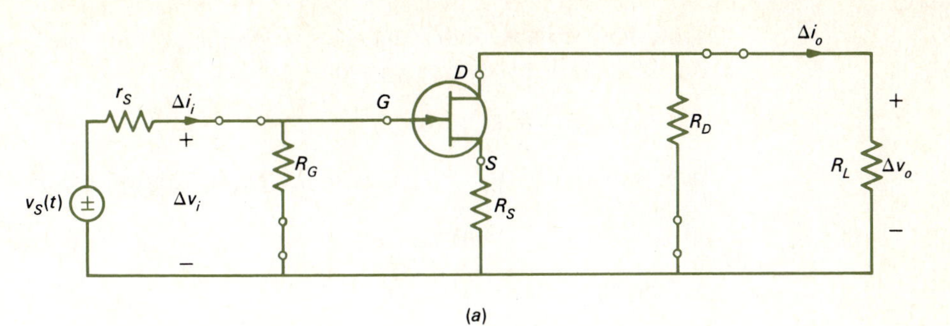

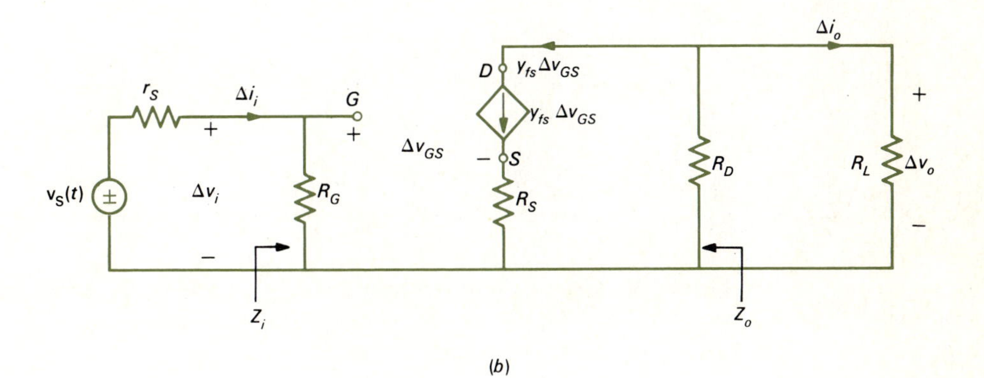

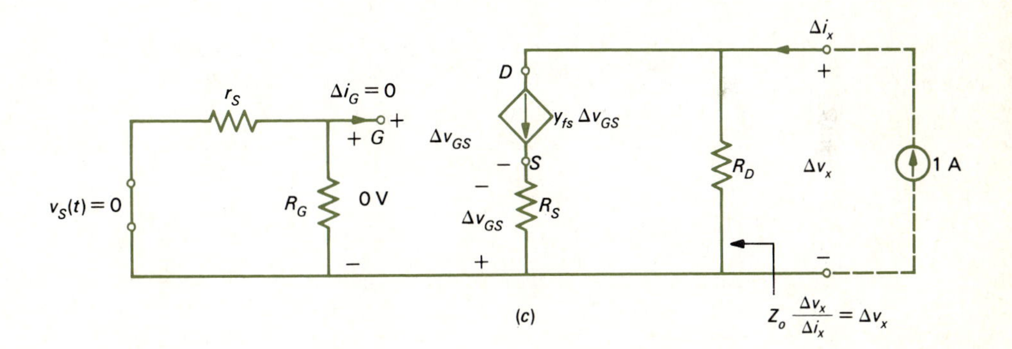

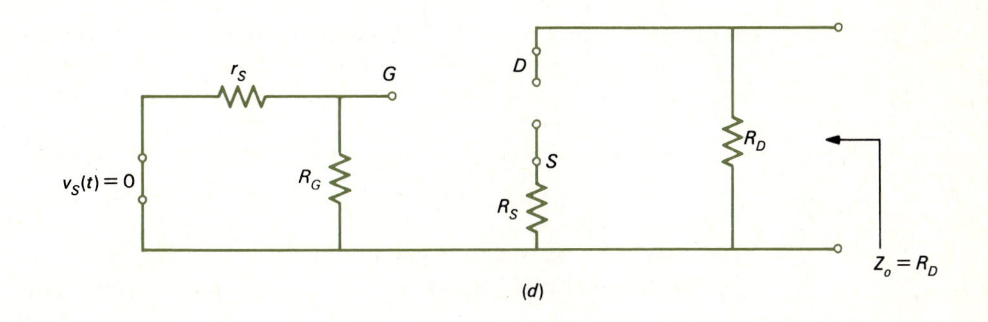

FIGURE 9.9 AC analysis of the common-source JFET amplifier, using the simplified small-signal equivalent circuit of Fig 9.8.

controlling variable of the controlled source Δv_{GS}. Note that Δv_{GS} is now not equal to Δv_i. But with KVL we may obtain

$$\Delta v_i = \Delta v_{GS} + R_S(y_{fs}\,\Delta v_{GS}) \tag{9.21}$$

(since current through R_S is determined by the controlled current source: $y_{fs}\Delta v_{GS}$) or

$$\Delta v_i = (1 + R_S y_{fs})\Delta v_{GS} \tag{9.22}$$

The output voltage is, once again,

$$\Delta v_o = -y_{fs}\,\Delta v_{GS}(R_D||R_L) \tag{9.23}$$

Thus, the voltage gain is

$$A_V = \frac{\Delta v_o}{\Delta v_i}$$

$$= \frac{y_{fs}(R_D||R_L)}{1 + R_S y_{fs}} \tag{9.24}$$

Comparing Eq. (9.24) to the result when C_S bypassed R_S (removing R_S from the ac circuit) in Eq. (9.13), we see that the voltage gain has been reduced by a factor of $1 + R_S y_{fs}$. This represents what is known as feedback; a portion of the effect of the current in the output circuit, $y_{fs}\,\Delta v_{GS}$, has been "fed back" to affect the input circuit via the voltage drop across R_S, $y_{fs} R_S\,\Delta v_{GS}$. We will see this effect many times throughout this chapter.

The input impedance is again

$$Z_i = \frac{\Delta v_i}{\Delta i_i}$$

$$= R_G \tag{9.25}$$

and R_S has no effect on this. The current gain is

$$A_I = A_V \frac{Z_i}{R_L} \tag{9.26}$$

and since A_V has been reduced by the presence of R_S in the ac circuit, A_I is also reduced. The output impedance is found by setting $v_S(t) = 0$ and finding the Thevenin impedance seen by R_L, as shown in Fig. 9.9*c*. In this circuit

$$Z_o = \frac{\Delta v_x}{\Delta i_x} \tag{9.27}$$

This may be found by constraining either Δv_x or Δi_x and finding the other. For example, if we constrain Δi_x to 1 A by applying a current source, then $Z_o = \Delta v_x$. But the voltage across R_G is zero since $\Delta i_G = 0$. Therefore, all of the controlling voltage for the controlled source Δv_{GS} appears across R_S:

$$\Delta v_{GS} = -R_S y_{fs} \Delta v_{GS} \tag{9.28}$$

from which we conclude that $\Delta v_{GS} = 0$. Thus, the controlled current source is replaced with an open circuit, as shown in Fig. 9.9*d*, and

$$Z_o = R_D \tag{9.29}$$

Example 9.2 Compute A_V, A_I, Z_i, Z_o for the JFET amplifier of Example 9.1 with C_S removed.

Solution The circuit element values were given as $R_S = 5\text{ k}\Omega$, $R_D = 1\text{ k}\Omega$, $R_L = 1\text{ k}\Omega$, $R_G = 100\text{ k}\Omega$. Assume a maximum JFET with $y_{fs}(\text{max}) = 6500\ \mu\text{S}$. Thus

$$A_V = -\frac{y_{fs}(R_D||R_L)}{1 + R_S y_{fs}}$$

$$= -0.097$$

$$Z_i = R_G$$

$$= 100\text{ k}\Omega$$

$$A_I = A_V \frac{Z_i}{R_L}$$

$$= -9.7$$

$$Z_o = R_D$$

$$= 1\text{ k}\Omega$$

Note that removing C_S dramatically reduces the voltage gain from 3.25 to 0.097 as a result of the feedback through the unbypassed R_S.

9.3 THE COMMON-DRAIN JFET AMPLIFIER

The common-drain amplifier, often called the source follower for reasons soon to become apparent, is shown in Fig. 9.10*a*. The input signal is again applied to the gate, but the output is taken off the source terminal. The dc circuit and bias design are identical to the common-source amplifier discussed in the preceding chapter.

The ac circuit is shown in Fig. 9.10*b*. Consequently, substituting the small-signal linear equivalent circuit of the JFET from Fig. 9.8*a*, we obtain the circuit of Fig. 9.11*a*. Note that, as opposed to the common-source amplifier of Example 9.1, $\Delta v_{GS} \neq \Delta v_i$. By writing KVL around the outside loop, we obtain

$$\Delta v_i = \Delta v_{GS} + \Delta v_o \tag{9.30}$$

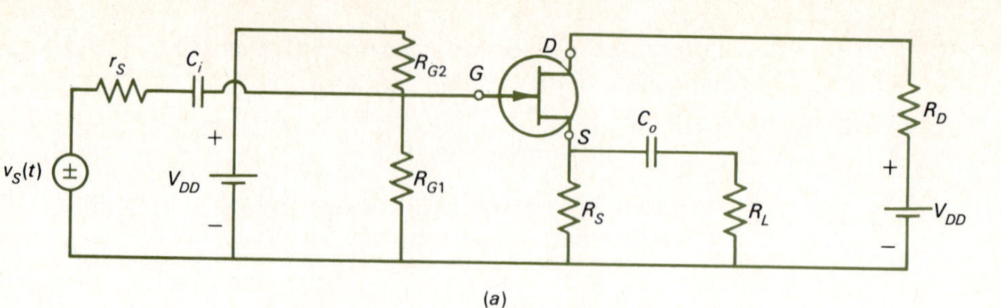

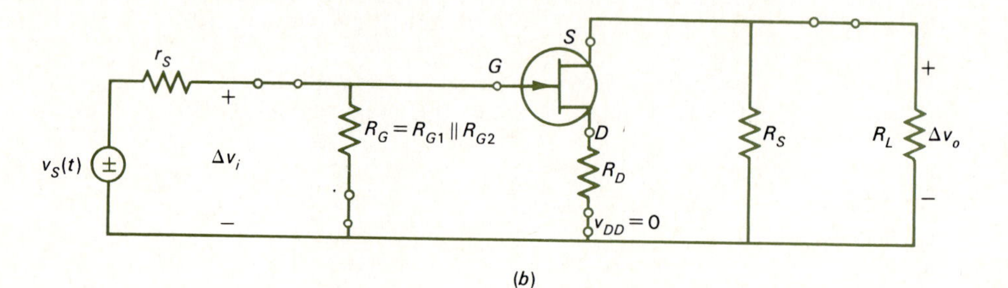

FIGURE 9.10 The common-drain JFET amplifier: (*a*) complete circuit; (*b*) ac circuit.

From this circuit

$$\Delta v_o = y_{fs}\, \Delta v_{GS}(R_S||R_L) \tag{9.31}$$

Substituting Eq. (9.31) into Eq. (9.30) yields

$$\Delta v_i = [1 + y_{fs}(R_S||R_L)]\, \Delta v_{GS} \tag{9.32}$$

so that

$$A_V = \frac{\Delta v_o}{\Delta v_i}$$

$$= \frac{y_{fs}(R_S||R_L)}{1 + y_{fs}(R_S||R_L)} \tag{9.33}$$

Note that the voltage gain is less than unity and is positive. The positive sign on the voltage gain indicates that the input and output voltages are in phase. The positive sign of A_V and the fact that the voltage gain is less than (but close to) unity mean that the output (source terminal) voltage "follows" the input voltage; hence, the name "source follower."

The input impedance is the ratio of

$$Z_i = \frac{\Delta v_i}{\Delta i_i} \tag{9.34}$$

From Fig. 9.11a, $Z_i = R_G$. The current gain is

$$A_I = A_V \frac{Z_i}{R_L} \tag{9.35}$$

The output impedance is found from Fig. 9.11b as

$$Z_o = \frac{\Delta v_x}{\Delta i_x} \tag{9.36}$$

But

$$Z_o = \hat{Z}_o || R_S \tag{9.37}$$

FIGURE 9.11 AC analysis of the common-drain JFET amplifier, using the simplified small-signal equivalent circuit of Fig 9.8.

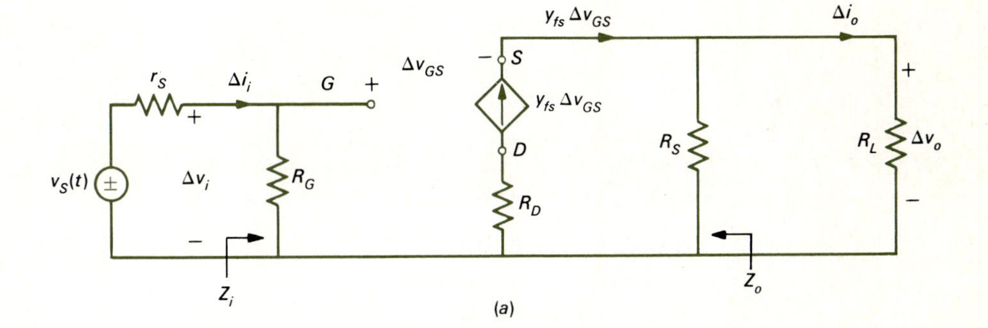

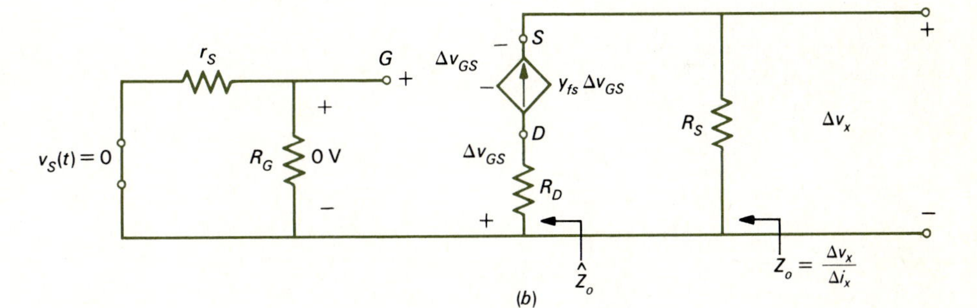

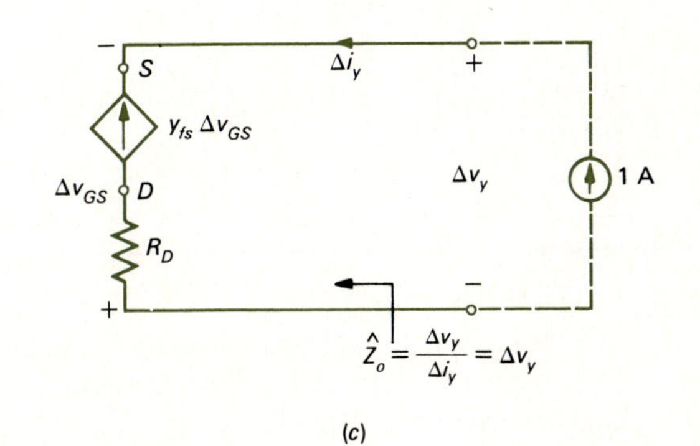

where $\hat{Z}_o$ is the impedance seen to the left of R_S. From Fig. 9.11*c* we obtain

$$\hat{Z}_o = \frac{\Delta v_y}{\Delta i_y} \tag{9.38}$$

If we set $\Delta i_y = 1$, then $\hat{Z}_o = \Delta v_y$. Note that $\Delta v_y = -\Delta v_{GS}$ and $y_{fs}\,\Delta v_{GS} = -\Delta i_y = -1$. Thus, $\Delta v_{GS} = -1/y_{fs}$ and $\Delta v_y = 1/y_{fs}$. Thus

$$\hat{Z}_o = \frac{1}{y_{fs}} \tag{9.39}$$

Therefore,

$$Z_o = \frac{1}{y_{fs}} || R_S \tag{9.40}$$

Example 9.3 For a 2N3819 JFET common-drain amplifier, compute A_V, A_I, Z_i, and Z_o if $y_{fs} = 6500\ \mu\text{S}$, $R_S = 10\ \text{k}\Omega$, $R_L = 10\ \text{k}\Omega$, $R_D = 10\ \text{k}\Omega$, and $R_G = 1\ \text{M}\Omega$.

Solution From the above results

$$A_V = \frac{y_{fs}(R_S||R_L)}{1 + y_{fs}(R_S||R_L)}$$

$$= 0.97$$

$$Z_i = R_G$$

$$= 1\ \text{M}\Omega$$

$$A_I = A_V \frac{Z_i}{R_L}$$

$$= 97.01$$

$$Z_o = \frac{1}{y_{fs}} || R_S$$

$$= 153.85 || 10\ \text{k}\Omega$$

$$= 151.5\ \Omega$$

Note that in addition to having a voltage gain less than unity, the output impedance is also quite small. Thus, the common-drain amplifier is often used to match a high-impedance source (choose $R_G = r_S$) to a low-impedance load.

9.4 THE COMMON-GATE JFET AMPLIFIER

The common-gate JFET amplifier is shown in Fig. 9.12*a*. For this amplifier, the input is applied to the source and the output is taken off the drain. The dc circuit and dc analysis are identical to those of the common-source and common-drain amplifiers considered previously.

The ac circuit is shown in Fig. 9.12*b*. Substituting the small-signal linear equivalent circuit, we obtain the circuit shown in Fig. 9.13*a*. Note that since $\Delta i_G = 0$, $\Delta v_{GS} = -v_i$. The output voltage becomes

$$\Delta v_o = -y_{fs}\,\Delta v_{GS}(R_D||R_L) \tag{9.41}$$

and

$$A_V = \frac{\Delta v_o}{\Delta v_i}$$

$$= y_{fs}(R_D||R_L) \tag{9.42}$$

Note that this is identical in magnitude to the common-source amplifier but that the output and input voltages are in phase because of the positive sign of A_V.

The input impedance is a bit more involved than usual. If we constrain $\Delta v_i = 1$ V, as shown in Fig. 9.13*b*, then $\Delta v_{GS} = -1$ V and

$$\Delta i_i = \frac{1}{R_S} + y_{fs} \tag{9.43}$$

FIGURE 9.12 The common-gate JFET amplifier: (*a*) complete circuit; (*b*) ac circuit.

(*a*)

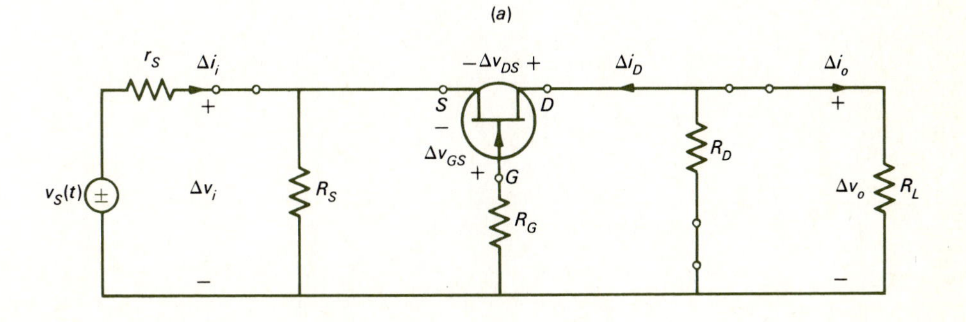

(*b*)

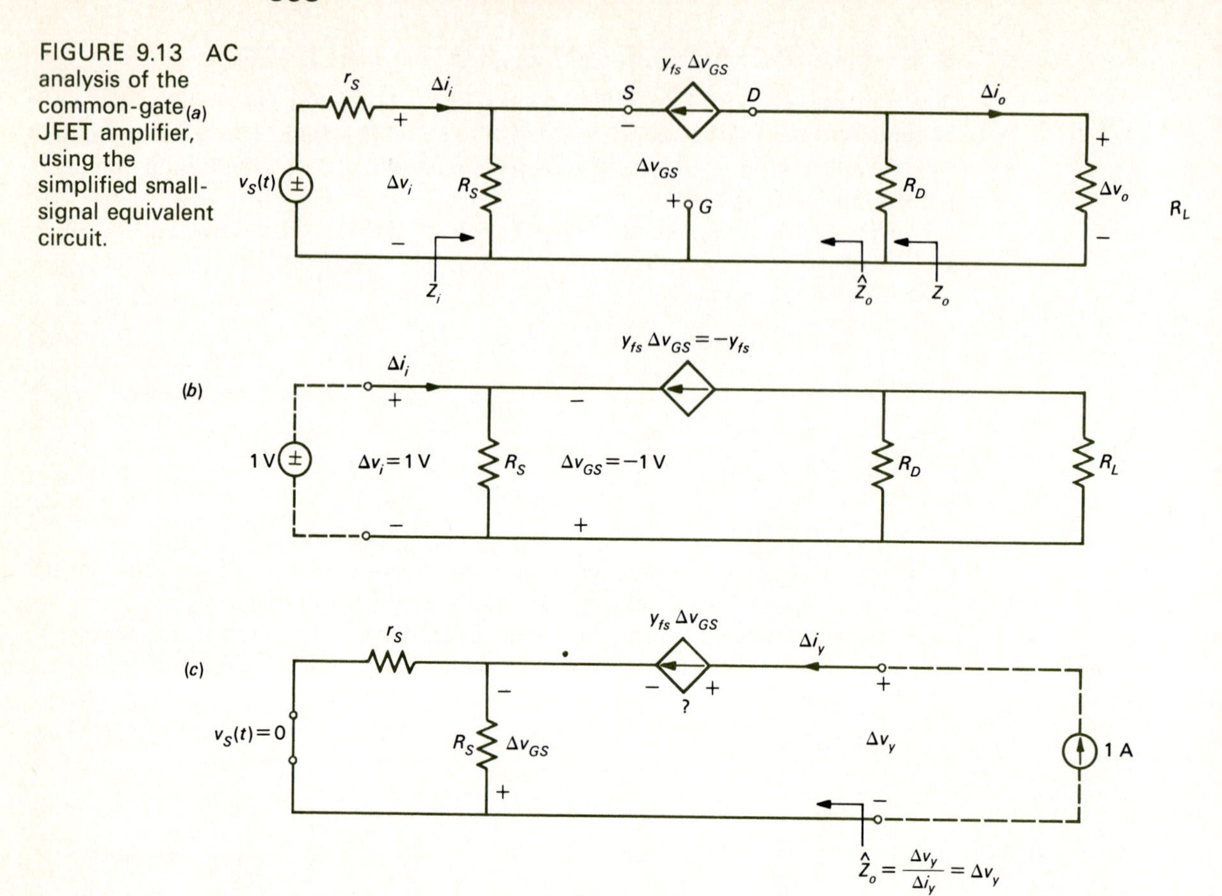

FIGURE 9.13 AC analysis of the common-gate JFET amplifier, using the simplified small-signal equivalent circuit.

Thus

$$Z_i = \frac{\Delta v_i}{\Delta i_i}$$

$$= \frac{1}{\Delta i_i}$$

$$= \frac{1}{1/R_S + y_{fs}}$$

$$= R_S || \frac{1}{y_{fs}} \tag{9.44}$$

The current gain is again

$$A_I = A_V \frac{Z_i}{R_L} \tag{9.45}$$

The output impedance is the parallel combination of R_D and $\hat{Z}_o$. Thus, if we constrain $\Delta i_y = 1$ A as in Fig. 9.13*c*, we need to find the resultant Δv_y. But in order to do this we need to know the voltage across the controlled current source in order to apply KVL around the loop. We have no way of finding this voltage, and this ambiguity is a result of our neglecting $1/y_{os}$, which is a resistor across the controlled current source. If we reinsert $1/y_{os}$, compute the result, and take the limit as $y_{os} \rightarrow 0$, then we will find that

$$\hat{Z}_o = \infty \tag{9.46}$$

Thus

$$Z_o = R_D \tag{9.47}$$

Example 9.4 For a common-gate amplifier using a 2N3819 JFET having $y_{fs} = 6500\ \mu$S and $R_L = R_D = R_S = 10$ kΩ, with $R_G = 100$ kΩ, compute A_V, A_I, Z_i, and Z_o.

Solution From the above results we have

$$\begin{aligned} A_V &= y_{fs}(R_D || R_L) \\ &= 32.5 \end{aligned}$$

$$\begin{aligned} Z_i &= R_S || \frac{1}{y_{fs}} \\ &= 151.5\ \Omega \end{aligned}$$

$$\begin{aligned} A_I &= A_V \frac{Z_i}{R_L} \\ &= 0.49 \end{aligned}$$

$$\begin{aligned} Z_o &= R_D \\ &= 10\ \text{k}\Omega \end{aligned}$$

Note that the input impedance to the common-gate amplifier is quite small. Thus, a common-gate amplifier is often used to match a low-impedance source to a high-impedance load and in this respect is the dual of the common-drain amplifier, which had a high input impedance but a low output impedance.

Note also in this example that $A_I < 1$. This is a general result. Substitute Eqs. (9.44) and (9.42) into

$$A_I = A_V \frac{Z_i}{R_L} \tag{9.48}$$

and we obtain

$$A_I = \frac{R_D}{R_D + R_L} \frac{R_S}{R_S + 1/y_{fs}}$$
$$< 1 \tag{9.49}$$

which is always less than unity. Again this supports the notion of the common-gate amplifier as the dual of the common-drain amplifier, which had a voltage gain less than unity.

9.5 FREQUENCY RESPONSE OF JFET AMPLIFIERS

The frequency response of the voltage gain of the common-source JFET amplifier is shown in Fig. 9.14. The low-frequency deterioration of the voltage gain is due to the increasing impedance of the coupling capacitors C_o and C_i and of the bypass capacitor C_S at these lower frequencies. Recall that the coupling capacitors C_i and C_o of the amplifiers were chosen such that their impedances were small (negligible) at the operating frequency of the sinusoidal source $v_S(t)$. Similarly, the capacitor C_S which bypasses ac signals around the resistor R_S in the common-source amplifier shown in Fig. 9.3*a* was chosen such that its impedance was also negligible at this operating frequency. As the frequency is lowered, these capacitors present a larger impedance and tend to "open up" as the frequency is reduced to zero. If C_i and/or C_o are replaced by open circuits, then clearly the voltage gain is zero. Also, in Example 9.2 we found that removing the capacitor C_S causes the voltage gain of the common-source amplifier to drop dramatically. Consequently, increasing the impedances of either of these capacitances (as is the case when the frequency is reduced) will cause the voltage gain to drop.

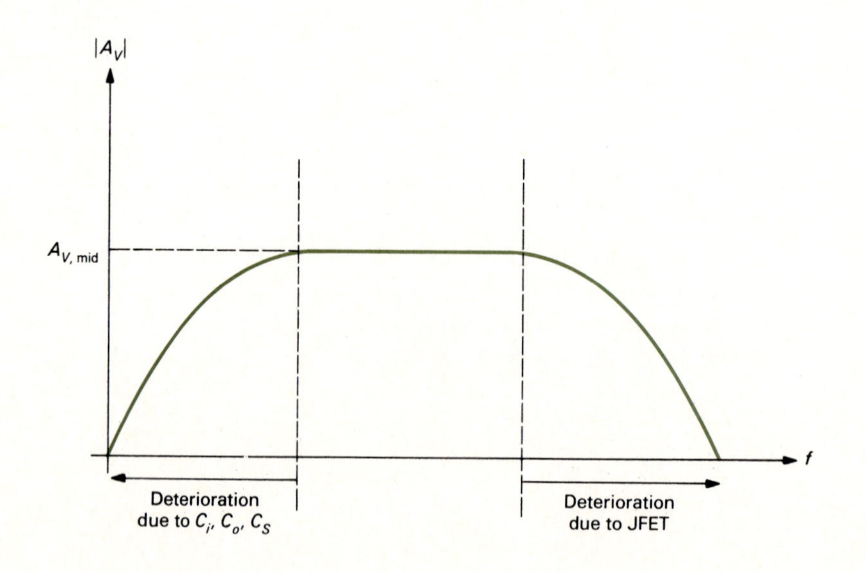

FIGURE 9.14 Frequency response of the voltage gain of a common-source JFET amplifier.

Substituting Eqs. (9.50) and (9.55) into Eq. (9.54), we obtain

$$A_V = \frac{\Delta v_o}{\Delta v_i}$$

$$= A_{V,\text{mid}} + (R_D||R_L)\frac{1 - A_{V,\text{mid}}}{R_D||R_L + 1/j\omega C_{GD}}$$

$$= A_{V,\text{mid}} + \frac{1 - A_{V,\text{mid}}}{1 + 1/j\omega C_{GD}(R_D||R_L)}$$

$$\doteq A_{V,\text{mid}}\frac{1}{1 + j\omega C_{GD}(R_D||R_L)} \tag{9.56}$$

Note that the high-frequency voltage gain rolls off like a low-pass filter with a time constant of $C_{GD}(R_D||R_L)$.

Example 9.5 For a common-source JFET amplifier with $R_L = R_D = R_S = 10\ \text{k}\Omega$ and $R_G = 100\ \text{k}\Omega$, compute the frequency at which the magnitude of the input impedance is reduced to 1 kΩ if a 2N3819 with $y_{fs} = 6500\ \mu\text{S}$, $C_{GS} = C_{GD} = 4$ pF, and $C_{DS} = 0$ is used.

Solution The midband voltage gain is

$$A_{V,\text{mid}} = -y_{fs}(R_D||R_L)$$

$$= -32.5$$

so that C_{GD} appears in the input circuit as $(1 - A_{V,\text{mid}})C_{GD} = (33.5)4\ \text{pF} = 134$ pF. Thus, the input impedance is

$$Z_i = R_G||\frac{1}{j\omega[C_{GS} + (1 - A_{V,\text{mid}})C_{GD}]}$$

$$= 1\ \text{k}\Omega$$

and we have neglected $R_D||R_L/(1 - A_{V,\text{mid}}) = 149.25\ \Omega$. Since $C_{GS} + (1 - A_{V,\text{mid}})C_{GD} = 138$ pF, for $Z_i = 1\ \text{k}\Omega$ and $R_G = 100\ \text{k}\Omega$ we have

$$1\ \text{k}\Omega = \left|\frac{100\ \text{k}\Omega\ (1/j\omega 138 \times 10^{-12})}{100\ \text{k}\Omega + 1/j\omega 138 \times 10^{-12}}\right|$$

$$= \left|\frac{100k}{1 + j\omega 1.38 \times 10^{-5}}\right|$$

or

$$1k = \frac{100k}{\sqrt{1 + (\omega 1.38 \times 10^{-5})^2}}$$

and the frequency is 1.153 megahertz (MHz). This is a rather "low" frequency to have the input impedance deteriorate by such a large factor. It illustrates how dramatic the effect of these device capacitances can be.

9.6 THE COMMON-EMITTER BJT AMPLIFIER

The analysis of the JFET amplifiers consisted of two parts: (1) a dc analysis to place the dc operating point in a linear region of the characteristic and (2) an ac analysis to determine the resulting voltage gain of the amplifier, as well as to determine the input and output impedances. For small-signal linear operation, we obtained a linear equivalent circuit representing small variations about the dc operating point. We assumed small-signal linear operation of this device such that the device terminal currents and voltages stayed in a locally linear region of the characteristic about the dc operating point. This assumption of small-signal linear operation allowed us to apply superposition even though the circuit contained a basically nonlinear element; since linear operation was preserved, it did not matter that the device was nonlinear outside the linear region.

Thus, we may separate the analysis of JFET amplifiers into two parts. One part, the dc-biasing problem, is due only to dc sources in the circuit and uses the concept of selecting an operating point and determining the required values of the biasing resistors and batteries. The second part, the ac analysis problem, consists of setting all dc sources to zero, substituting the small-signal linear equivalent circuit of the device, and determining the ac quantities of interest, such as voltage gain and the input and output impedances of the amplifier.

The analysis of amplifiers containing bipolar junction transistors follows much the same procedure. The only differences in the two procedures are that (1) we use a different design philosophy for locating the dc operating point in determining the biasing resistors and (2) the small-signal linear ac model of the BJT is different than for the JFET. But the general procedure remains the same.

Consider the typical set of charcteristics of the BJT shown in Fig. 8.15. Note the location of the linear regions. These characteristics are linearized in Fig. 9.16*a*. Now since these characteristics are linear ones, we may write linear equations describing them. For example, for the base-emitter characteristic we may write

$$v_{BE} = h_{ie} i_B + V_{BEO} \tag{9.57}$$

where h_{ie} is a constant to be determined; it represents the slope of the characteristic at the operating point, since from Eq. (9.57)

$$h_{ie} = \frac{\Delta v_{BE}}{\Delta i_B} \quad \Omega \tag{9.58}$$

where the Δ quantities represent small changes in the variables about the operating point. Since the base-emitter junction behaves as a forward-biased *pn* diode, we would

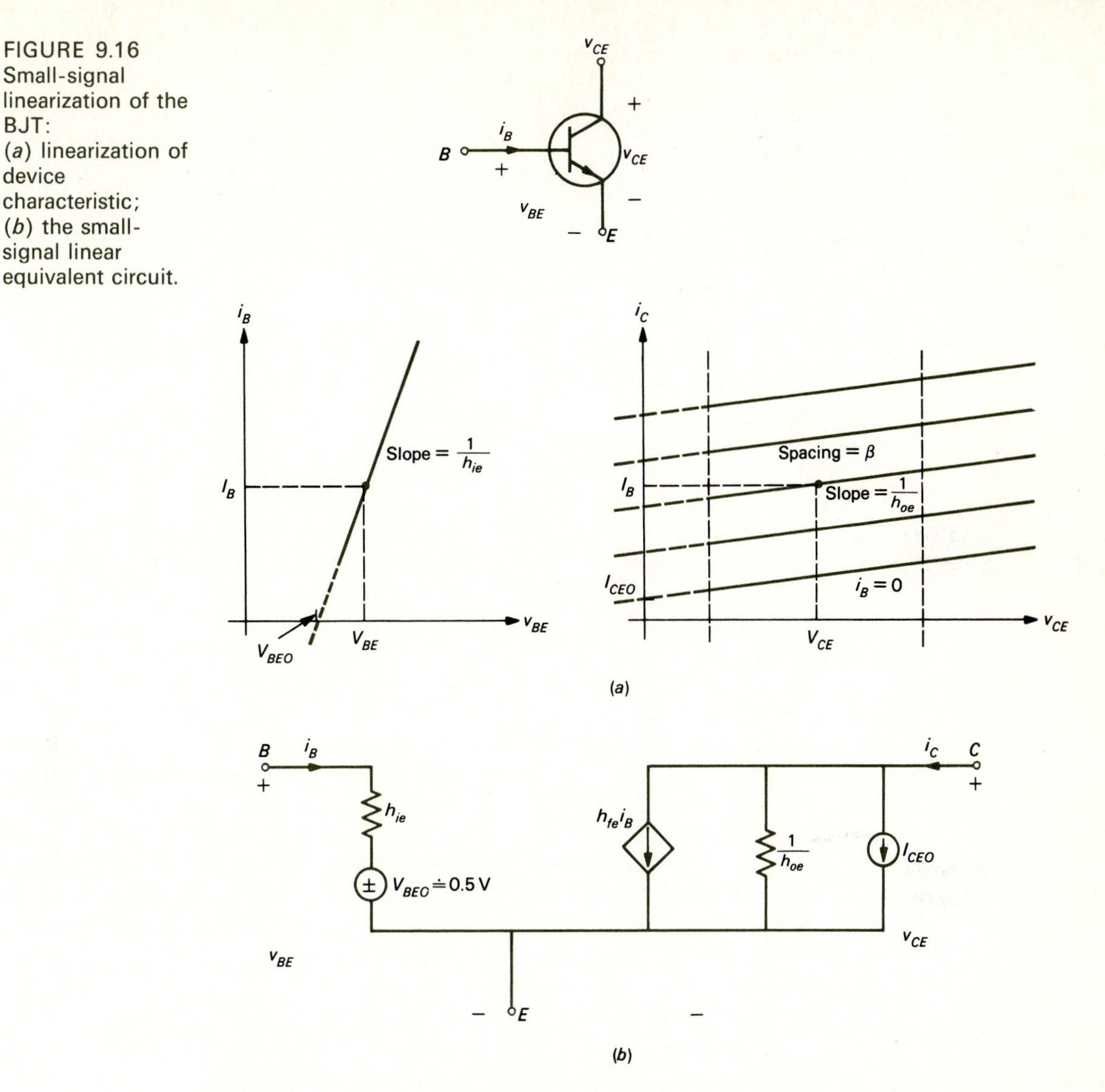

FIGURE 9.16 Small-signal linearization of the BJT: (*a*) linearization of device characteristic; (*b*) the small-signal linear equivalent circuit.

expect h_{ie} to be approximately given by the small-signal incremental resistance of a *pn* diode, as determined in Chap. 7. For example, by analogy to the *pn* diode, we may write

$$i_B = I_{BEO}(e^{qv_{BE}/kT} - 1) \tag{9.59}$$

where I_{BEO} is a reverse current for the base-emitter junction reverse-biased and the collector open-circuited. Once again, q is an electron charge, k is Boltzmann's constant, and T is the junction temperature in degrees kelvin. At room temperature we have $kT/q \doteq 25$ mV. For $v_{BE} \gg 25 \text{ mV} = 0.025$ V, Eq. (9.59) becomes

$$i_B \doteq I_{BEO}e^{v_{BE}/25\,\text{mV}} \tag{9.60}$$

Differentiating this, we obtain

$$\frac{di_B}{dv_{BE}} = \frac{1}{25\text{ mV}} I_{BEO} e^{v_{BE}/25\text{ mV}} \tag{9.61}$$

so that

$$h_{ie} \doteq \frac{25\text{ mV}}{I_{BEO} e^{v_{BE}/25\text{ mV}}} \tag{9.62}$$

Substituting Eq. (9.60), we have

$$h_{ie} = \frac{25\text{ mV}}{i_B} \tag{9.63}$$

But

$$i_B \doteq \frac{i_C}{\beta} \tag{9.64}$$

so that at the particular operating point $I_B \doteq \beta I_C$ we have

$$\begin{aligned} h_{ie} &= \frac{25\text{ mV}}{I_B} \\ &\doteq \beta \frac{25\text{ mV}}{I_C} \end{aligned} \tag{9.65}$$

For a typical operating point of $I_C = 10$ mA and $\beta = 100$, we obtain $h_{ie} = 250\ \Omega$.

Note that the intersection of the curve with the v_{BE} axis ($i_B = 0$) occurs typically in the range of 0.5 to 0.7 V for silicon semiconductor materials. This will be denoted as V_{BEO}, where from Eq. (9.57)

$$V_{BEO} = v_{BE}|_{i_B=0} \tag{9.66}$$

For simplification, we will assume that

$$V_{BEO} \doteq 0.5\text{ V} \tag{9.67}$$

Approximating the collector-emitter characteristic similarly, as shown in Fig. 9.16*a*, we may write

$$i_C = h_{fe} i_B + h_{oe} v_{CE} + I_{CEO} \tag{9.68}$$

which describes this characteristic. Note in Eq. (9.68) that

$$I_{CEO} = i_C \Big|_{\substack{i_B=0 \\ v_{CE}=0}} \tag{9.69}$$

Thus, I_{CEO} is the intersection of the $i_B = 0$ line with the i_C axis ($v_{CE} = 0$). Once again, this is the collector-emitter leakage current with the base open-circuited ($i_B = 0$), as was discussed previously. From Eq. (9.68) we obtain

$$h_{fe} = \left.\frac{\Delta i_C}{\Delta i_B}\right|_{\Delta v_{CE}=0} \qquad \text{dimensionless} \tag{9.70}$$

$$h_{oe} = \left.\frac{\Delta i_C}{\Delta v_{CE}}\right|_{\Delta i_B=0} \qquad \text{S} \tag{9.71}$$

By comparing Eq. (9.70) to Eq. (8.40) of Chap. 8, we see that†

$$h_{fe} \doteq \beta \tag{9.72}$$

and is related to the spacing between the lines on the characteristic; on the collector-emitter characteristic for $v_{CE} = V_{CE}$, take a small change in i_B (Δi_B) and find the resulting change in i_C (Δi_C) for no change in v_{CE}. Similarly, Eq. (9.71) shows that h_{oe} is the slope of the $i_B = I_B$ curve through the operating point. Thus, the BJT may be approximated with the linear equivalent circuit shown in Fig. 9.16*b*, which represents Eqs. (9.57) and (9.68), as the reader should verify. Note that a current-controlled current source is used instead of the voltage-controlled current source used for FETs since, unlike the FET, the BJT second-port characteristic is a function of the first-port current i_B.

The origin of the symbols h_{ie}, h_{fe}, and h_{oe} is as follows. They are hybrid parameters in terms of their dimensions since h_{ie} has the dimensions of resistance, h_{fe} is dimensionless, and h_{oe} has the dimensions of conductance. The e subscript denotes that the emitter terminal is common to both ports. The subscript i on h_{ie} designates that the parameter appears as a resistance in the BJT input, as shown by Eq. (9.58). The f subscript on h_{fe} designates that this is a forward-current transfer ratio since it relates the change in a current in the output circuit Δi_C to a change in a current in the input circuit Δi_B, as shown by Eq. (9.70). The subscript o on h_{oe} designates that this parameter appears as a conductance in the BJT output, as shown by Eq. (9.71).

The designation of one port as "input" and the other as "output" is not intended to mean that these are the only possibilities for inputting a signal to or outputting it from a BJT, as we will soon see. The amplifier in Fig. 9.17*a* is referred to as a common-emitter amplifier since the emitter is common to both input and output ports of the amplifier in the ac circuit of Fig. 9.17*b*. We will also consider common-base and common-collector amplifiers in later sections.

Now for small-signal linear operation, we may substitute the equivalent circuit in Fig. 9.16*b* for the BJT, resulting in Fig. 9.18*a*. Once again, since linear operation is assumed, each current and voltage in the circuit will be composed of two components. One component is due to the dc sources in the circuit [with $v_S(t) = 0$] and is therefore dc. This dc component is denoted with a capital letter. The other component is due to

† The parameters h_{fe} and β are not identical; however, the symbols are often used interchangeably. We will use the symbol β instead of the more correct symbol of h_{fe}.

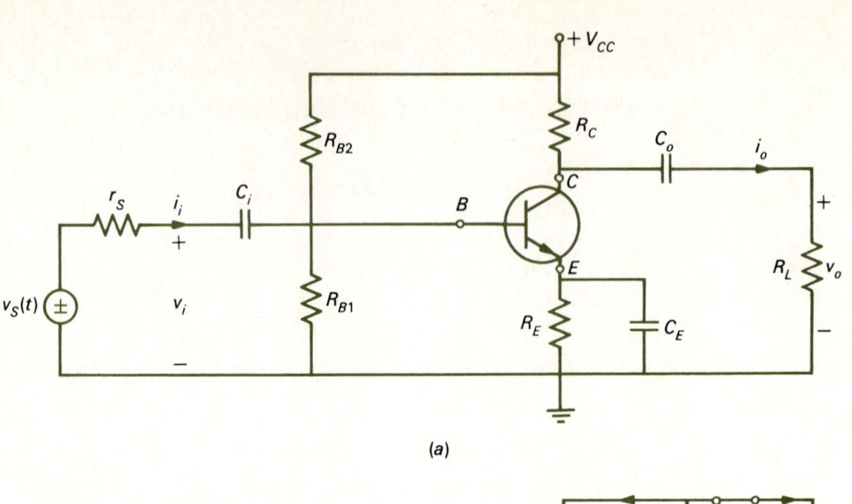

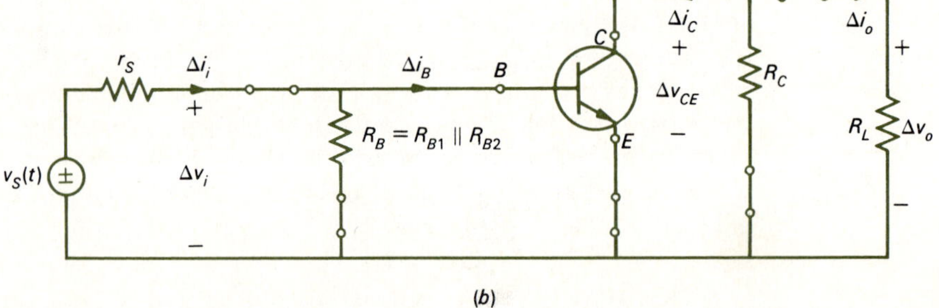

FIGURE 9.17 The common-emitter BJT amplifier: (a) complete circuit; (b) ac circuit.

$v_S(t)$ (with the dc sources set equal to zero) and is a sinusoidal quantity since $v_S(t)$ is sinusoidal. This sinusoidal component is denoted in the usual fashion as a Δ quantity. By superposition the total current is the sum of these two contributions, i.e., $i_C = I_C + \Delta i_C$. Now for computing the ac quantities of interest for the amplifier, such as

$$A_V = \frac{\Delta v_o}{\Delta v_i} \tag{9.73}$$

$$A_I = \frac{\Delta i_o}{\Delta i_i} \tag{9.74}$$

$$A_P = A_V A_I \tag{9.75}$$

and Z_i and Z_o, we may set all dc sources to zero, as shown in Fig. 9.18b, and all currents and voltages become only sinusoidal Δ quantities. To simplify the analysis, we will assume that all capacitors are chosen such that their impedances are negligible at the frequency of $v_S(t)$ so that they may be replaced by short circuits. This circuit is further simplified in Fig. 9.18c.

From this resulting circuit we see that

$$\Delta v_o = -h_{fe}\,\Delta i_B\left(\frac{1}{h_{oe}}\,||R_C||R_L\right) \tag{9.76}$$

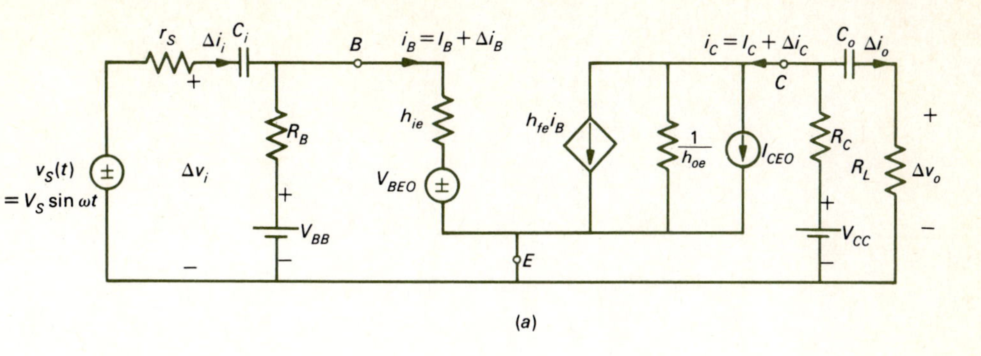

(a)

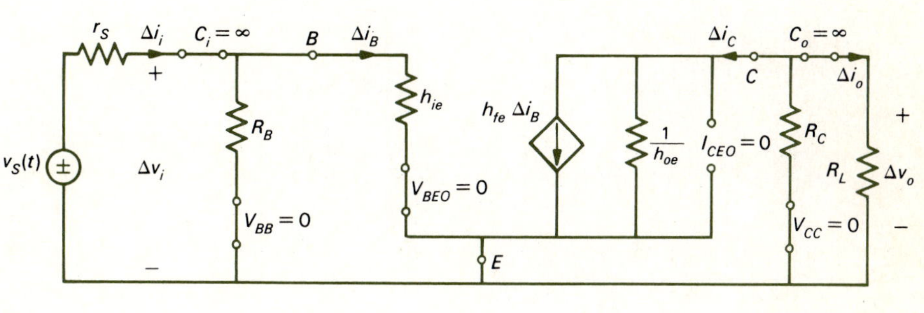

(b)

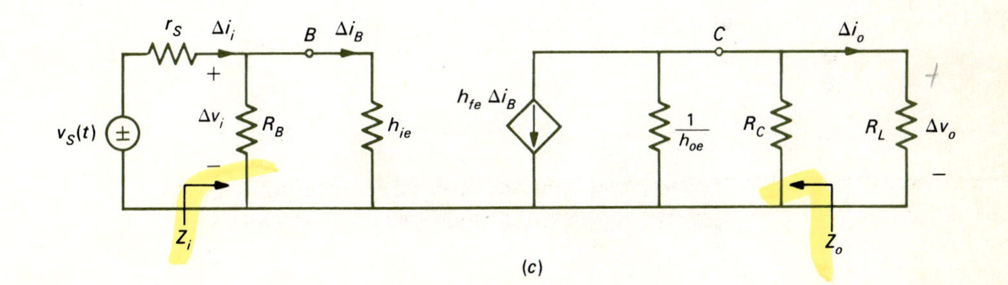

(c)

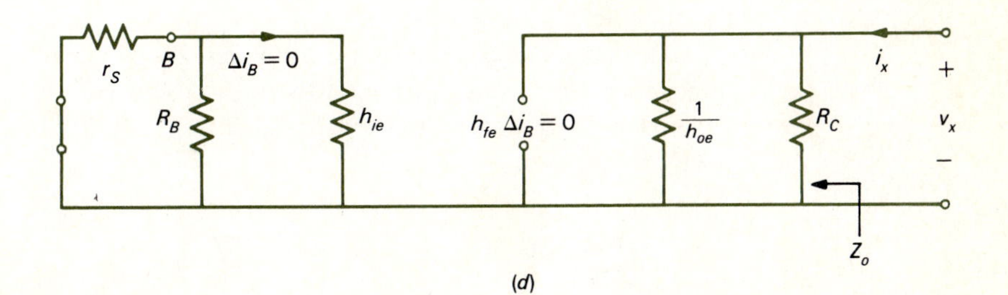

(d)

FIGURE 9.18 Use of the small-signal equivalent circuit in the ac analysis of the common-emitter BJT amplifier: (a) complete circuit; (b) ac circuit; (c) redrawn ac circuit; (d) determining the amplifier output impedance.

and

$$\Delta v_i = h_{ie}\,\Delta i_B \tag{9.77}$$

so that the voltage gain is

$$A_V = -\frac{h_{fe}}{h_{ie}}\left(\frac{1}{h_{oe}}||R_C||R_L\right) \tag{9.78}$$

The input impedance is, from Fig. 9.18*c*,

$$Z_i = \frac{\Delta v_i}{\Delta i_i}$$

$$= R_B||h_{ie} \tag{9.79}$$

From the circuit in Fig. 9.18*c* we see that, by current division,

$$\Delta i_o = -h_{fe}\,\Delta i_B \frac{R_C||1/h_{oe}}{R_C||1/h_{oe} + R_L} \tag{9.80}$$

and

$$\Delta i_B = \frac{R_B}{R_B + h_{ie}}\,\Delta i_i \tag{9.81}$$

Thus, the current gain becomes

$$A_I = \frac{\Delta i_o}{\Delta i_i}$$

$$= -\frac{R_B}{R_B + h_{ie}}\,h_{fe}\,\frac{R_C||1/h_{oe}}{R_C||1/h_{oe} + R_L} \tag{9.82}$$

This is equivalent (as the reader should verify) to

$$A_I = A_V\frac{Z_i}{R_L} \tag{9.83}$$

Note the presence of the two current-division quantities in Eq. (9.82) which are both less than unity; thus $|A_I| < h_{fe}$.

The output impedance of the amplifier, Z_o, can be found by finding the Thevenin impedance seen by R_L [set $v_S(t)$ to zero, of course], as shown in Fig. 9.18*d*. Since $\Delta i_B = 0$ in this situation, the controlled current source is zero and it is replaced by an open circuit. Thus

$$Z_o = \frac{1}{h_{oe}}||R_C \tag{9.84}$$

Example 9.6 For the 2N718 common-emitter amplifier shown in Fig. 9.19*a*, with $\beta = 100$, $h_{oe} = 10^{-5}$, compute the voltage gain, the current gain, the power gain, and the input and output impedances.

Solution The dc circuit becomes as shown in Fig. 9.19*b*. The operating point for the base current is found by summing KVL around the base-emitter loop to yield

$$I_B = \frac{V_{BB} - V_{BE}}{R_B}$$

$$= \frac{3.75 - 0.5\ \text{V}}{75\ \text{k}\Omega}$$

$$= 43\ \mu\text{A}$$

Using $\beta = 100$, we obtain

$$I_C \doteq \beta I_B$$

$$= 100 I_B$$

$$= 4.3\ \text{mA}$$

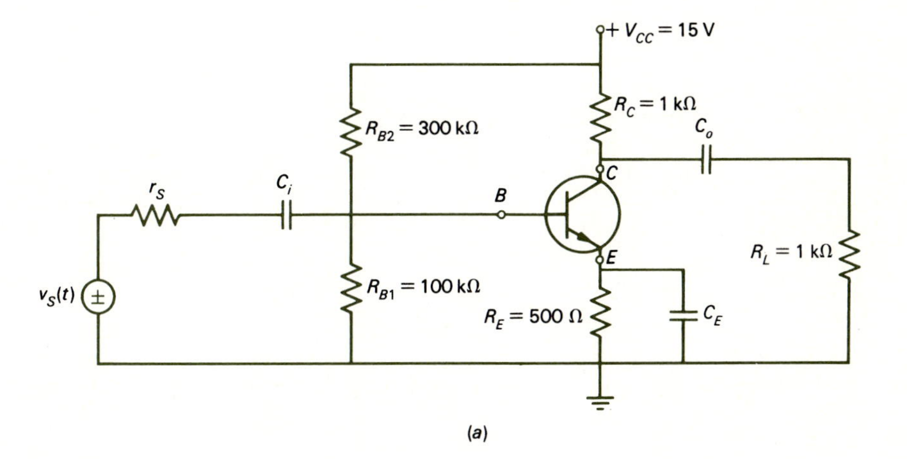

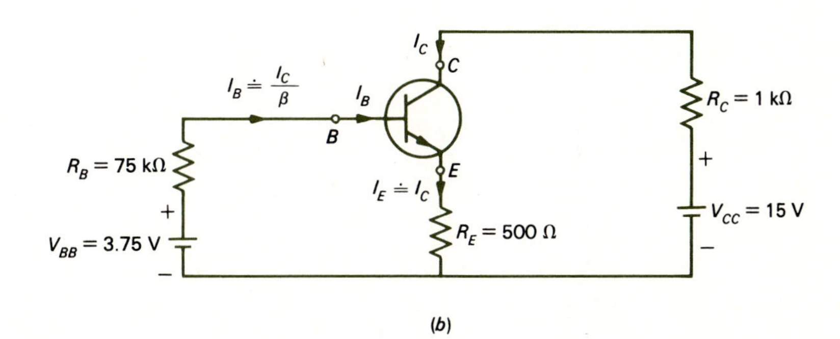

FIGURE 9.19 Example 9.6: analysis of a common-emitter BJT: (*a*) complete circuit; (*b*) dc circuit.

Thus, V_{CE} can be found by writing KVL around the collector-emitter loop as

$$\begin{aligned} V_{CE} &= V_{CC} - R_C I_C - R_E I_E \\ &= 15\ \text{V} - (1\ \text{k}\Omega)(4.3\ \text{mA}) - (500)(4.3\ \text{mA}) \\ &= 8.55\ \text{V} \end{aligned}$$

where we have used the approximation that

$$\begin{aligned} I_E &= I_C + I_B \\ &\doteq I_C \end{aligned}$$

The parameter h_{ie} is not specified but may be found from Eq. (9.65) as

$$\begin{aligned} h_{ie} &= \beta \frac{25\ \text{mV}}{I_C} \\ &= 100 \frac{25 \times 10^{-3}}{4.3 \times 10^{-3}} \\ &= 581.4\ \Omega \end{aligned}$$

Note that if I_C had been 1 mA, h_{ie} would have increased to 2500 Ω. The voltage gain is

$$\begin{aligned} A_V &= -\frac{h_{fe}}{h_{ie}}\left(\frac{1}{h_{oe}} || R_C || R_L\right) \\ &\doteq -86 \end{aligned}$$

since $1/h_{oe} = 100\ \text{k}\Omega$, which is much larger than $R_C || R_L = 500\ \Omega$. The current gain is

$$\begin{aligned} A_I &= -\underbrace{\frac{R_B}{R_B + h_{ie}}}_{1}\ \underbrace{h_{fe}}_{100}\ \underbrace{\frac{R_C || 1/h_{oe}}{R_C || 1/h_{oe} + R_L}}_{\frac{1}{2}} \\ &\doteq -50 \end{aligned}$$

since $1/h_{oe} \gg R_C$; thus $R_C || 1/h_{oe} \doteq R_C$ and $R_L = R_C$. The input impedance is

$$\begin{aligned} Z_i &= R_B || h_{ie} \\ &\doteq h_{ie} \\ &= 581\ \Omega \end{aligned}$$

and the output impedance is

$$\begin{aligned} Z_o &= \frac{1}{h_{oe}} || R_C \\ &\doteq R_C \\ &= 1\ \text{k}\Omega \end{aligned}$$

Note that the current gain can also be computed more easily from

$$A_I = A_V \frac{Z_i}{R_L}$$

$$= -86 \frac{581}{1\ \text{k}\Omega}$$

$$= -50$$

The power gain is

$$A_P = A_V A_I$$

$$= 4300$$

Note that for this rather typical range of parameters, the voltage gain is much larger than for the typical common-source JFET amplifiers. Also, the input impedance of this common-emitter BJT amplifier is much lower than for typical common-source JFET amplifiers, which ranges from 100 kΩ to 1 MΩ, and it may easily be obtained simply by selecting R_G. For this BJT amplifier, however, Z_i is no larger than h_{ie} no matter what value of R_B occurs. Thus, the input impedance of this amplifier is limited by the value of h_{ie}.

Note in the previous example that the resistor $1/h_{oe}$ in the small-signal linear model which is in parallel with the controlled current source is $1/h_{oe} = 100$ kΩ and is quite large. Consequently, in the previous example it has little effect on A_V since it is in parallel with $R_D || R_L = 500\ \Omega$. Similarly, it is in parallel with R_C in Z_o, and $R_C = 1$ kΩ. Thus, $1/h_{oe}$ has no effect and may be removed from the circuit.

Although this is not always possible (suppose $R_D = R_L = 100$ kΩ), we will omit $1/h_{oe}$ from the small-signal linear equivalent circuit of the BJT for the purposes of simplifying our future calculations. This approximation has the effect of assuming that the curves in the second-port (collector-emitter) characteristic of the BJT are flat —which is again similar to the effect of neglecting $1/y_{os}$ in the JFET model. The resulting circuit is shown in Fig. 9.20 where we will assume that $h_{fe} = \beta$. Compare this resulting circuit and the one for the JFET in Fig. 9.8*a*. We see two major differences. The controlled current source for the JFET depended on the gate-source voltage Δv_{GS}, whereas the controlled current source for the BJT depends on a current, the base current Δi_B. Also note that a resistance h_{ie} appears in the base lead of the BJT, whereas

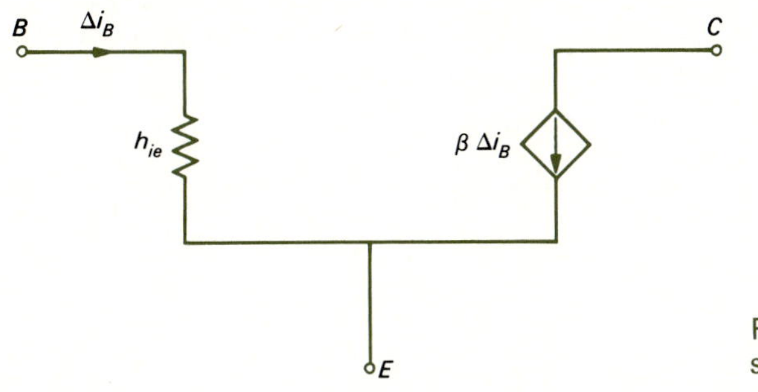

FIGURE 9.20 A simplified small signal equivalent circuit of a BJT.

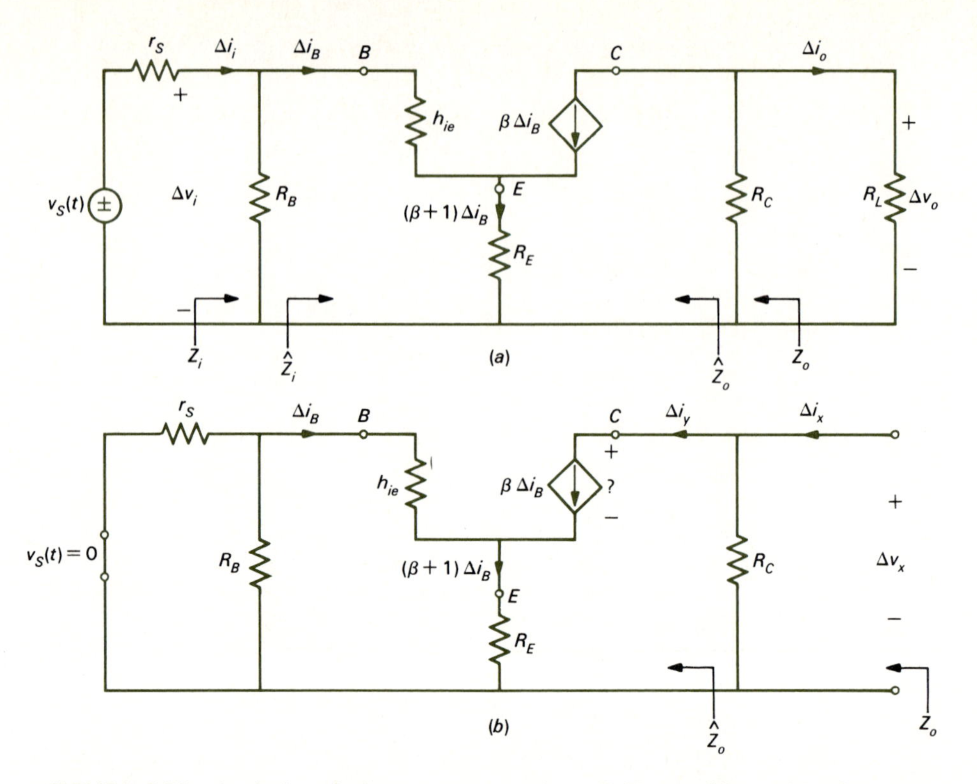

FIGURE 9.21 Analysis of the common-emitter BJT amplifier with C_E removed: (*a*) complete ac circuit; (*b*) determination of the amplifier output impedance.

the gate lead of the JFET was an open circuit. This will have the effect of giving much smaller input impedances for BJT amplifiers than was the case for JFET amplifiers.

Removal of the capacitor C_E, which bypasses ac signals around R_E, will have an effect similar to the removal of C_S in the common-source JFET amplifier: voltage and current gain will drop. It also has another important effect: increasing the input impedance to the amplifier. Note in the previous example and in Fig. 9.18*c* that the input impedance is no larger than h_{ie}, and h_{ie} can be quite small (250 Ω). Removal of C_E will effectively increase Z_i, as we now show.

The ac circuit for the common-emitter amplifier in Fig. 9.17*a* with C_E removed will be similar to Fig. 9.17*b*, but with one major difference. Because C_E has been removed, R_E will appear in the ac circuit in the emitter lead. Substituting the small-signal linear equivalent circuit gives the circuit in Fig. 9.21*a*. Note that with KCL, the current through R_E is $(\beta + 1)\, \Delta i_B$. Thus

$$\Delta v_i = h_{ie}\, \Delta i_B + R_E(\beta + 1)\, \Delta i_B \tag{9.85}$$

Also

$$\Delta v_o = -\beta\, \Delta i_B(R_C || R_L) \tag{9.86}$$

so that the voltage gain is

$$A_V = \frac{\Delta v_o}{\Delta v_i}$$

$$= -\frac{\beta(R_C||R_L)}{h_{ie} + (\beta + 1)R_E} \tag{9.87}$$

Comparing Eq. (9.87) to the corresponding expression with C_E present, given by Eq. (9.78), we see that A_V is reduced when C_E is removed.

The input impedance is

$$Z_i = \hat{Z}_i||R_B \tag{9.88}$$

Note that

$$\hat{Z}_i = \frac{\Delta v_i}{\Delta i_B}$$

$$= h_{ie} + (\beta + 1)R_E \tag{9.89}$$

and

$$Z_i = R_B||[h_{ie} + (\beta + 1)R_E] \tag{9.90}$$

Comparing this to the corresponding expression with C_E present, given by Eq. (9.79), we see that $\hat{Z}_i$ can be much larger than h_{ie}. Thus, the overall input impedance to the amplifier Z_i is not "loaded down" as much by h_{ie}. Here we see that the effect of the controlled source is to make R_E appear larger to the input by a factor of $(\beta + 1)$. The output impedance is obtained from Fig. 9.21*b* as

$$Z_o = \frac{\Delta v_x}{\Delta i_x} \tag{9.91}$$

and

$$Z_o = \hat{Z}_o||R_C \tag{9.92}$$

To find $\hat{Z}_o = \Delta v_x/\Delta i_y$, we may constrain $\Delta i_y = 1$ A and find Δv_x. But if we do, we need to find the voltage across the controlled current source. We have no way of determining this. The dilemma is caused by our having neglected $1/h_{oe}$ in the model. If we reinsert $1/h_{oe}$, find $\hat{Z}_i$, and take the limit as $h_{oe} \to 0$, we find that

$$\hat{Z}_0 = \infty \tag{9.93}$$

so that

$$Z_o = R_C \tag{9.94}$$

The current gain is

$$A_I = A_V \frac{Z_i}{R_L} \tag{9.95}$$

Example 9.7

For the common-emitter BJT amplifier of Example 9.6, recompute A_V, A_I, Z_i, Z_o, and A_P with C_E removed.

Solution

$$A_V = -\frac{\beta(R_C||R_L)}{h_{ie} + (\beta + 1)R_E}$$

$$= -0.98$$

$$\hat{Z}_i = h_{ie} + (\beta + 1)R_E$$

$$= 51{,}081\ \Omega$$

$$Z_i = \hat{Z}_i||R_B$$

$$= 30{,}386\ \Omega$$

$$Z_o = R_C$$

$$= 1\ \text{k}\Omega$$

$$A_I = A_V \frac{Z_i}{R_L}$$

$$= -30$$

$$A_P = A_V A_I$$

$$= 29$$

Note that the voltage gain has been drastically reduced by the removal of C_E. The current gain has been reduced somewhat, while the input impedance has been increased by a factor of 50!

The amplifier considered in the previous results is referred to as a common-emitter amplifier since the emitter terminal of the BJT amplifier is common to both input and output. We begin to notice an analogy in name and function between the three terminals of the BJT and the FET:

BJT		FET
Emitter	⟶	Source
Base	⟶	Gate
Collector	⟶	Drain

For example, the BJT collector "collects" electrons, as does the FET drain; similarly, the BJT emitter "emits" electrons, as does the FET source; and the BJT base acts to control this flow, as does the FET gate. Thus, we would expect to investigate other configurations of BJT amplifiers—common-base and common-collector—having properties similar to those of the corresponding FET amplifiers. The common-base BJT amplifier, being analogous to the common-gate FET amplifier, should have a low input impedance, a high output impedance, and $A_I < 1$. The common-collector BJT amplifier, being analogous to the common-drain FET amplifier, should have $A_V < 1$, a low output impedance, and a high input impedance. These analogies will be confirmed in the following sections.

9.7 THE COMMON-BASE BJT AMPLIFIER

A typical common-base BJT amplifier is shown in Fig. 9.22*a*. The input is placed on the emitter and the output is taken off the collector. The ac circuit is shown in Fig. 9.22*b*; capacitor C_x shorts out R_{B1} and R_{B2} and removes them from the ac circuit. It is therefore clear that, for ac, the base is common to both input and output.

The ac parameters can be obtained by substituting the small-signal linear equivalent circuit in Fig. 9.20 into the ac circuit of Fig. 9.22*b*, resulting in the circuit of Fig. 9.23*a*. In this circuit the current leaving the emitter terminal is $(\beta + 1)\,\Delta i_B$ since $\Delta i_E = \Delta i_C + \Delta i_B \doteq \beta\Delta i_B + \Delta i_B$. From this circuit

$$\Delta v_i = -h_{ie}\,\Delta i_B \tag{9.96}$$

and

$$\Delta v_o = -\beta\Delta i_B(R_C||R_L) \tag{9.97}$$

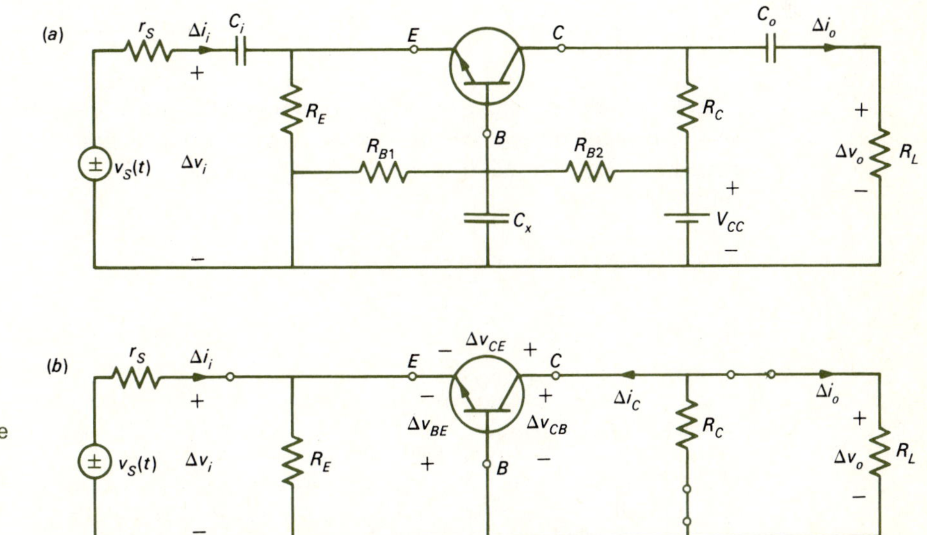

FIGURE 9.22 The common-base BJT amplifier: (*a*) complete circuit; (*b*) ac circuit.

so that

$$A_V = \frac{\Delta v_o}{\Delta v_i}$$

$$= \frac{\beta}{h_{ie}} (R_C || R_L) \tag{9.98}$$

Thus, the voltage gain of the common-base amplifier has approximately the same magnitude as the common-emitter amplifier. The sign of A_V is positive, however, which symbolizes that the input and output voltages of the amplifier are in phase.

The input impedance is

$$Z_i = R_E || \hat{Z}_i \tag{9.99}$$

where $\hat{Z}_i$ is the impedance seen to the right of R_E, as shown in Fig. 9.23*a*:

$$\hat{Z}_i = \frac{\Delta v_i}{-(\beta + 1)\, \Delta i_B}$$

$$= \frac{h_{ie}}{\beta + 1} \tag{9.100}$$

From Eq. (9.99) the input impedance is less than the smaller of R_E and $h_{ie}/(\beta + 1)$. The latter is usually the smallest, so that $Z_i \doteq h_{ie}/(\beta + 1)$, which is ordinarily quite small. For example, if $h_{ie} = 250\ \Omega$ and $\beta = 100$, $Z_i \doteq 2.5$.

The output impedance is found from the circuit in Fig. 9.23*b* as

$$Z_o = \frac{\Delta v_x}{\Delta i_x}$$

$$= \hat{Z}_o || R_C \tag{9.101}$$

We encounter a problem, however, since we do not know the voltage across the controlled current source. This is a consequence of our assumption that h_{oe} was small enough to be neglected. If we insert h_{oe}, calculate Z_o, and take the limit as $h_{oe} \to 0$, we find that

$$\hat{Z}_o = \infty \tag{9.102}$$

so that

$$Z_o = R_C \tag{9.103}$$

The current gain is

$$A_I = A_V \frac{Z_i}{R_L} \tag{9.104}$$

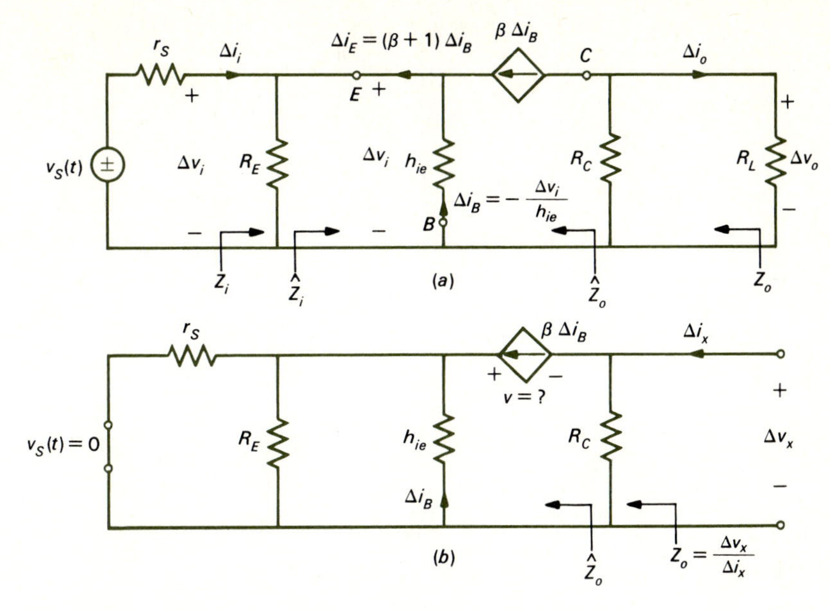

FIGURE 9.23 AC analysis of the common-base BJT amplifier: (*a*) substitution of the ac equivalent circuit; (*b*) determination of the amplifier output impedance.

Substituting Eqs. (9.98), (9.99), and (9.100), we find that

$$A_I = \frac{\beta}{\beta + 1}\frac{R_C}{R_C + R_L}\frac{R_E}{R_E + \hat{Z}_i}$$

$$< 1 \tag{9.105}$$

which is always less than unity.

Example 9.8 For the common-base amplifier with $R_L = R_C = 10\ \text{k}\Omega$, $R_E = 1\ \text{k}\Omega$, $r_S = 600\ \Omega$, and $50 < \beta < 150$, determine A_V, A_I, A_P, Z_i, and Z_o. The operating point is approximately $I_C = 1$ mA.

Solution Assuming $\beta = \beta_{\text{min}} = 50$, we may calculate h_{ie} as

$$h_{ie} = \beta_{\text{min}}\frac{25\ \text{mV}}{I_C}$$

$$= \frac{50(25 \times 10^{-3})}{1 \times 10^{-3}}$$

$$= 1250\ \Omega$$

Thus

$$A_V = \frac{\beta}{h_{ie}}(R_C||R_L)$$

$$= \frac{50}{1250}(10k||10k)$$

$$= 200$$

$$\hat{Z}_i = \frac{h_{ie}}{\beta + 1}$$

$$= \frac{1250}{51}$$

$$= 24.5\ \Omega$$

$$Z_i = R_E || \hat{Z}_i$$

$$= 1k || 24.5$$

$$= 24\ \Omega$$

$$A_I = \frac{R_E}{R_E + \hat{Z}_i} \frac{\beta}{\beta + 1} \frac{R_C}{R_C + R_L}$$

$$= \frac{1k}{1k + 24.5} \frac{50}{51} \frac{10k}{10k + 10k}$$

$$= 0.48$$

$$= A_V \frac{Z_i}{R_L}$$

$$Z_o = R_C$$

$$= 10k$$

Thus, the common-base amplifier has a very low input impedance and a high output impedance (depending on R_C), and $A_I < 1$. Quite often it is used to match from a low impedance source to a high impedance load.

9.8 THE COMMON-COLLECTOR BJT AMPLIFIER

A typical common-collector BJT amplifier is shown in Fig. 9.24*a*. It is virtually identical to the common-emitter BJT amplifier except that the output (and consequently R_L) is at the emitter terminal. C_E is not present; otherwise, it would short out the ac output.

The ac analysis can be accomplished by substituting the small-signal linear equivalent circuit into the ac circuit of Fig. 9.24*b*. This results in the circuit of Fig. 9.25*a*. Note once again that the emitter current is

$$\Delta i_E \doteq (\beta + 1)\, \Delta i_B \tag{9.106}$$

Thus

$$\Delta v_o = (\beta + 1)\, \Delta i_B (R_E || R_L) \tag{9.107}$$

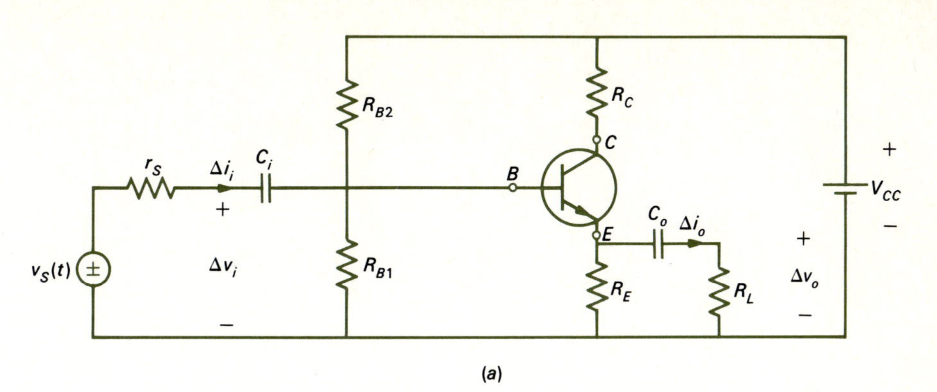

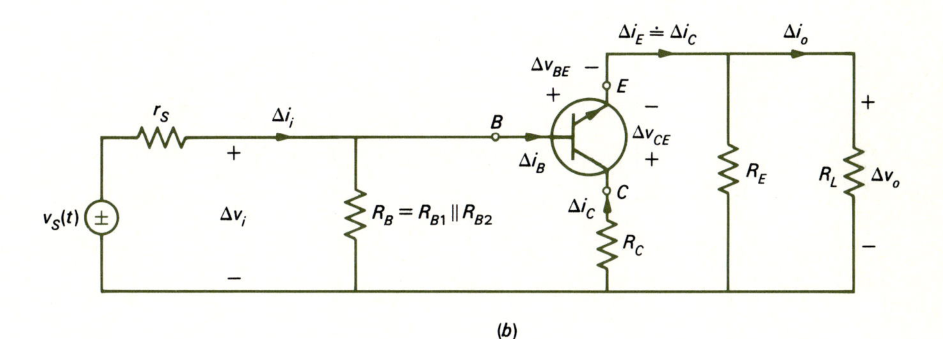

FIGURE 9.24 The common-collector BJT amplifier: (*a*) complete circuit; (*b*) ac circuit.

and, by applying KVL around the outside loop,

$$\Delta v_i = h_{ie}\,\Delta i_B + \Delta v_o \tag{9.108}$$

Solving Eqs. (9.107) and (9.108), we obtain

$$\begin{aligned} A_V &= \frac{\Delta v_o}{\Delta v_i} \\ &= \frac{(\beta + 1)(R_E||R_L)}{h_{ie} + (\beta + 1)(R_E||R_L)} \\ &< 1 \end{aligned} \tag{9.109}$$

Note that the output voltage is less than, and in phase with, the input voltage. This is the dual of the current gain for the common-base amplifier, which was $A_I < 1$. Actually, the result in Eq. (9.109) is easy to see since $\Delta v_i = \Delta v_{BE} + \Delta v_o$, or $\Delta v_o = \Delta v_i - \Delta v_{BE}$, and Δv_{BE} is quite small.

The input impedance is

$$Z_i = \hat{Z}_i||R_B \tag{9.110}$$

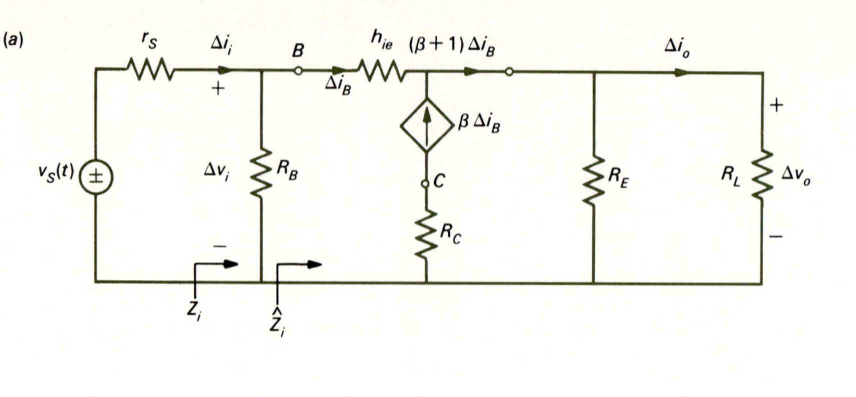

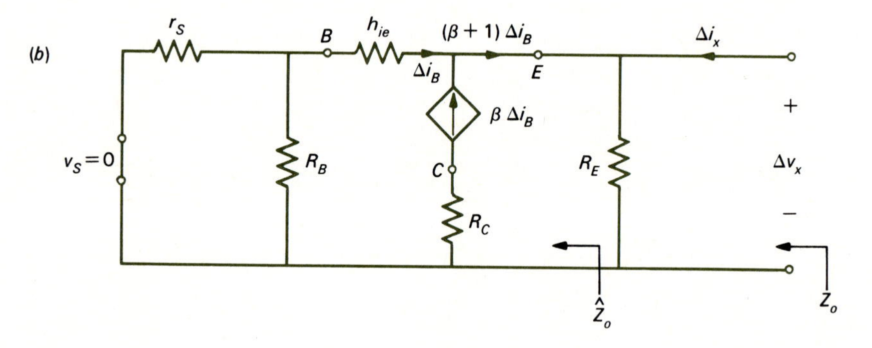

FIGURE 9.25 AC analysis of the common-collector BJT amplifier: (*a*) substitution of the ac equivalent circuit; (*b*) determination of the amplifier output impedance.

where $\hat{Z}_i$ is the impedance seen to the right of R_B, as shown in Fig. 9.25*a*. From the circuit, writing KVL around the outside loop yields

$$\Delta v_i = h_{ie}\,\Delta i_B + (\beta + 1)\,\Delta i_B(R_E||R_L) \tag{9.111}$$

so that

$$\begin{aligned}\hat{Z}_i &= \frac{\Delta v_i}{\Delta i_B}\\ &= h_{ie} + (\beta + 1)(R_E||R_L)\end{aligned} \tag{9.112}$$

Note that the parallel combination of $R_E||R_L$ in the emitter lead appears to the input larger by a factor of $(\beta + 1)$. Thus, the input impedance seen by R_B is no longer h_{ie} but is much larger. Thus, the input impedance to the amplifier Z_i is larger than for the common-emitter amplifier, which was limited to less than h_{ie}. This is the dual to the common-base amplifier, which had a very small input impedance.

The current gain can be found by substituting Eqs. (9.109), (9.110), and (9.112) into

$$A_I = A_V \frac{Z_i}{R_L} \tag{9.113}$$

to yield

$$A_I = (\beta + 1) \frac{R_E}{R_E + R_L} \frac{R_B}{R_B + \hat{Z}_i} \tag{9.114}$$

The output impedance can be found from Fig. 9.25*b*. Note that $\hat{Z}_o$ seen to the left of R_E is related to Z_o by

$$Z_o = R_E || \hat{Z}_o \tag{9.115}$$

and

$$\hat{Z}_o = \frac{\Delta v_x}{-(\beta + 1)\, \Delta i_B} \tag{9.116}$$

But writing KVL around the outside loop, we obtain

$$\Delta v_x = -[(r_S || R_B) + h_{ie}]\, \Delta i_B \tag{9.117}$$

so that

$$\hat{Z}_o = \frac{r_S || R_B + h_{ie}}{\beta + 1} \tag{9.118}$$

Therefore,

$$Z_o = \hat{Z}_o || R_E \tag{9.119}$$

which is usually *small*, as shown by the following example.

Example 9.9 Determine A_V, A_I, Z_i, and Z_o for the common-collector amplifier which has $R_C = R_E = 1\ \text{k}\Omega$, $R_L = 50\ \Omega$, $r_S = 600\ \Omega$, and $R_B = 5\ \text{k}\Omega$. Assume $\beta = 50$ and $h_{ie} = 250\ \Omega$.

Solution

$$A_V = \frac{(\beta + 1)(R_E || R_L)}{h_{ie} + (\beta + 1)(R_E || R_L)}$$

$$= 0.91$$

$$\hat{Z}_i = h_{ie} + (\beta + 1)(R_E || R_L)$$

$$= 2678.6\ \Omega$$

and

$$Z_i = \hat{Z}_i || R_B$$
$$= 1744.2\ \Omega$$

$$A_I = A_V \frac{Z_i}{R_L}$$
$$= 31.7$$

$$\hat{Z}_o = \frac{r_S || R_B + h_{ie}}{\beta + 1}$$
$$= 15.41\ \Omega$$

and

$$Z_o = \hat{Z}_o || R_E$$
$$= 15.17\ \Omega$$

9.9 FREQUENCY RESPONSE OF BJT AMPLIFIERS

Consider the common-emitter BJT amplifier in Fig. 9.17*a*. As the frequency of $v_S(t)$ is reduced, the impedances of the coupling capacitors C_i and C_o increase, resulting in a decrease in voltage gain. Also, as the frequency is decreased, the impedance of C_E increases, thus no longer removing R_E from the ac circuit. This increasing presence of an impedance (R_E) in the emitter lead forms a "feedback" path between the input and output which also reduces the voltage gain for the JFET amplifier. Thus, the magnitude of A_V decreases with decreasing frequency, which results in a response similar to the JFET shown in Fig. 9.14.

If we increase the frequency above its midband value at which the capacitors were chosen to have negligible impedance, we find that the magnitude of A_V also decreases, as was the case for the JFET amplifier. This high-frequency deterioration is once again due to the deterioration of the device performance at these higher frequencies.

The small-signal linear equivalent circuit which we have been using to model the BJT for ac analysis is valid for midband frequencies and below. For higher frequencies, we must include the inherent capacitances of the device. The depletion layers at the base-emitter and base-collector junctions give rise to capacitances between the terminals. A useful high-frequency modification of the small-signal linear equivalent circuit is shown in Fig. 9.26*a*. Capacitances C_{BE} and C_{BC} have been added to the midband model. Note that Δi_B is still the current through h_{ie}. Ordinarily, C_{BC} is given by the manufacturer as C_{ob}. C_{BE} can be calculated from

$$C_{BE} = \frac{\beta}{\omega_T h_{ie}} \tag{9.120}$$

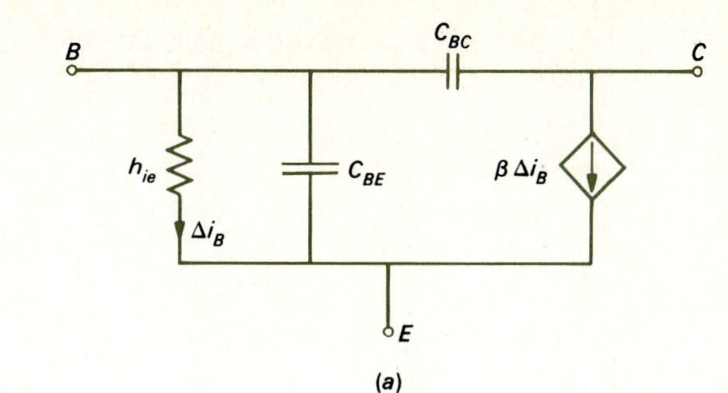

(a)

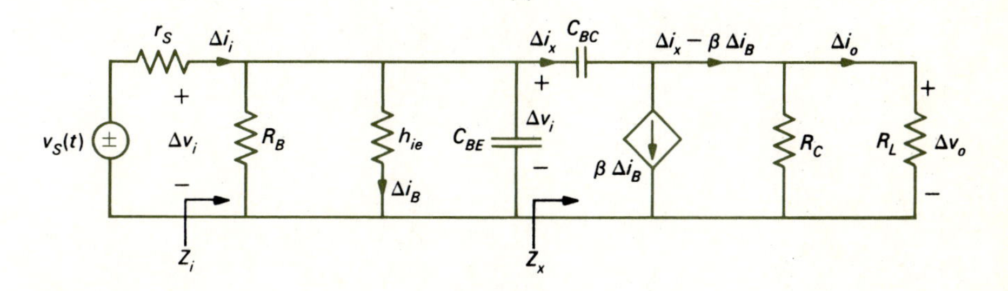

(b)

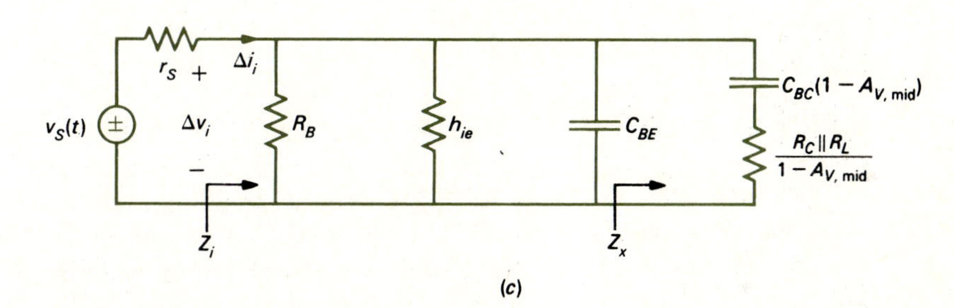

(c)

FIGURE 9.26 High-frequency model of the BJT: (*a*) circuit parasitic capacitances; (*b*) use in the common-emitter BJT amplifier; (*c*) determining the amplifier input impedance and the Miller effect.

where

$$\omega_T = 2\pi f_T \tag{9.121}$$

and f_T is given by the manufacturer on design specification sheets and is usually denoted as the current-gain-bandwidth product. It is the frequency at which the value of β drops to unity. Typical values for the 2N718 BJT are $C_{BC} = C_{ob} = 20$ pF and $f_T = 75$ MHz.

Replacing the BJT in the ac circuit for the amplifier circuit of Fig. 9.17*a* with the equivalent circuit of Fig. 9.26*a*, we obtain the circuit of Fig. 9.26*b*. (We are interested in high-frequency response, so we replace C_o, C_i, and C_E with short circuits.) For the purposes of calculating Δi_B as well as Z_i, we may replace the amplifier at the input terminals with the circuit of Fig. 9.26*c*. This may be determined from the following. Note in Fig. 9.26*b* that the voltage across h_{ie} as well as C_{BE} is Δv_i. Thus, if we find the current Δi_x entering C_{BC}, we can replace the circuit to the right of C_{BE} with an equivalent impedance $Z_x = \Delta v_i/\Delta i_x$. From Fig. 9.26*b*, we obtain

$$\Delta i_B = \frac{\Delta v_i}{h_{ie}} \tag{9.122}$$

and

$$\Delta v_i = \Delta i_x \frac{1}{j\omega C_{BC}} + (\Delta i_x - \beta \Delta i_B)(R_C||R_L) \tag{9.123}$$

Substituting Eq. (9.122) into Eq. (9.123), we obtain

$$Z_x = \frac{\Delta v_i}{\Delta i_x}$$

$$= \frac{1}{j\omega C_{BC}(1 - A_{V,\mathrm{mid}})} + \frac{R_C||R_L}{1 - A_{V,\mathrm{mid}}} \tag{9.124}$$

where the midband voltage gain is used:

$$A_{V,\mathrm{mid}} = -\frac{\beta}{h_{ie}}(R_C||R_L) \tag{9.125}$$

Thus, C_{BC} appears to the input as though it has been increased by a factor of $(1 - A_{V,\mathrm{mid}})$ and the parallel combination of R_C and R_L appears as though it is reduced by a factor of $(1 - A_{V,\mathrm{mid}})$. This corresponds almost identically to the result for JFET amplifiers and is called the Miller effect. It has the effect of reducing the amplifier's input impedance (and also its voltage and current gain) at high frequencies.

Example 9.10 For the common-emitter circuit of Example 9.6 (Fig. 9.19*a*), $I_C = 4.3$ mA, $\beta = 100$, $R_B = 75$ kΩ, $R_L = 1$ kΩ. A 2N718 with $C_{ob} = 20$ pF and $f_T = 75$ MHz is used. Compute the frequency at which the input impedance is reduced from its midband value of 581.4 to 50 Ω.

Solution

$$C_{BC} = C_{ob}$$

$$= 20 \text{ pF}$$

$$C_{BE} = \frac{\beta}{\omega_T h_{ie}}$$

$$= \frac{100}{2\pi(75 \times 10^6)581.4}$$

$$= 365 \text{ pF}$$

and

$$A_{V,\mathrm{mid}} = -86$$

From Fig. 9.26*c*

$$\frac{R_C||R_L}{1 - A_{V,\text{mid}}} = \frac{500}{87}$$

$$= 5.75\ \Omega$$

which may be neglected in this analysis. Thus

$$Z_i = R_B||h_{ie}||\frac{1}{j\omega[C_{BE} + C_{BC}(1 - A_{V,\text{mid}})]}$$

$$= 577||\frac{1}{j\omega(2105\text{ pF})}$$

We wish to find f such that

$$50 = \left|\frac{577(1/j2\pi f\ 2105\text{ pF})}{577 + 1/j2\pi\ 2105\text{ pF}}\right|$$

$$= \left|\frac{577}{1 + j7.63 \times 10^{-6}f}\right|$$

or

$$11.54 = \sqrt{1 + (7.63 \times 10^{-6}f)^2}$$

Solving, we obtain

$$f = 1.507\text{ MHz}$$

9.10 Impact of Integrated-Circuit Technology on Electronic Design

In this chapter and the previous chapter we have discussed analysis and design techniques for linear amplifiers. In the next chapter we will study a very useful linear amplifier, the operational amplifier (op amp). As opposed to the FET and BJT amplifiers which we have discussed, the operational amplifier requires very little dc-biasing design or detailed ac analysis. The entire amplifier is contained in one integrated circuit which only requires that one attach a dc voltage source (power supply). This amplifier contains numerous FET or BJT amplifiers, but the design of their biasing circuitry has already been done. Consequently, one can "build" an entire linear amplifier with virtually none of the detailed knowledge of biasing studied previously. The ac analysis is also quite simple.

This is representative of the trend in integrated-circuit technology—one can purchase entire electronic circuits in one small DIP which will perform many tasks. Consequently, the detailed analyses studied in this and the previous chapter are being done by those who design the overall IC. This trend will no doubt continue, making the application of electronics available to those who do not wish to become deeply involved in the amplifier design.

PROBLEMS

9.1 Consider the linear amplifier shown in Fig. 9.1. Suppose

$$v_i(t) = 3\text{ V} + 2\text{m V} \sin(2\pi \times 10^3 t)$$

$$i_i(t) = 1\text{ mA} + 0.01\text{ mA} \sin(2\pi \times 10^3 t)$$

$$v_o(t) = 4.5\text{ V} - 0.75\text{ V} \sin(2\pi \times 10^3 t)$$

$$i_o(t) = 6\text{ mA} - 1\text{ mA} \sin(2\pi \times 10^3 t)$$

For each of the above quantities expressed as $x(t) = X + \Delta x(t)$, identify X and $\Delta x(t)$. Compute the voltage gain, current gain, input impedance, and R_L. Can you determine the output impedance of the amplifier from the information given?

9.2 For the linear amplifier shown in Fig. 9.1, when the load resistor R_L is removed, $v_o = 2\text{ V} + 3\text{ mV} \sin(2\pi \times 10^3\, t)$, and when R_L is replaced by a short circuit the output current is $i_o = 5\text{ mA} + 0.01\text{ mA} \sin(2\pi \times 10^3\, t)$. Determine the output impedance of the amplifier.

9.3 The JFET whose characteristic is shown in Fig. 8.36 is modeled as shown in Fig. 9.6. Determine y_{fs} and y_{os} about an operating point of $V_{DS} = 14$ V, $V_{GS} = -3$ V. Calculate I_{DSS}.

9.4 Determine y_{fs} and y_{os} for the JFET whose characteristic is shown in Fig. 8.36 about an operating point of $V_{DS} = 20$ V, $V_{GS} = -4$ V.

9.5 For the JFET amplifier shown in Fig. 9.3*a*, assume $R_{G1} = R_{G2} = 100\text{ k}\Omega$, $R_S = 1.5\text{ k}\Omega$, $R_D = 10\text{ k}\Omega$, and $R_L = 10\text{ k}\Omega$. Calculate A_V, A_I, Z_i, and Z_o. Assume that at the operating point $y_{fs} = 2 \times 10^{-3}$ and $y_{os} = 0$. Assume that all capacitors are short circuits at the operating frequency.

9.6 Repeat Prob. 9.5 with C_S removed.

9.7 For the common-drain amplifier shown in Fig. 9.10*a*, assume $R_{G1} = R_{G2} = 100\text{ k}\Omega$, $R_S = R_L = 10\text{ k}\Omega$, $R_D = 5\text{ k}\Omega$, $y_{fs} = 2 \times 10^{-3}$, and $y_{os} = 0$. Calculate A_V, A_I, Z_i, and Z_o.

9.8 For the common-gate amplifier shown in Fig. 9.12, assume $R_D = R_L = 10\text{ k}\Omega$, $R_{G1} = R_{G2} = 100\text{ k}\Omega$, $R_S = 5\text{ k}\Omega$, $y_{fs} = 2 \times 10^{-3}$, $y_{os} = 0$. Calculate A_V, A_I, Z_i, and Z_o.

9.9 A JFET amplifier is to be inserted between a source and a load for impedance-matching purposes, as shown in Fig. P9.9. What type of amplifier would you choose if $r_S = 500\ \Omega$ and $R_L = 10\text{ k}\Omega$? If $y_{fs} = 10^{-3}$ and $y_{os} = 0$, calculate R_D for a power gain of 2.

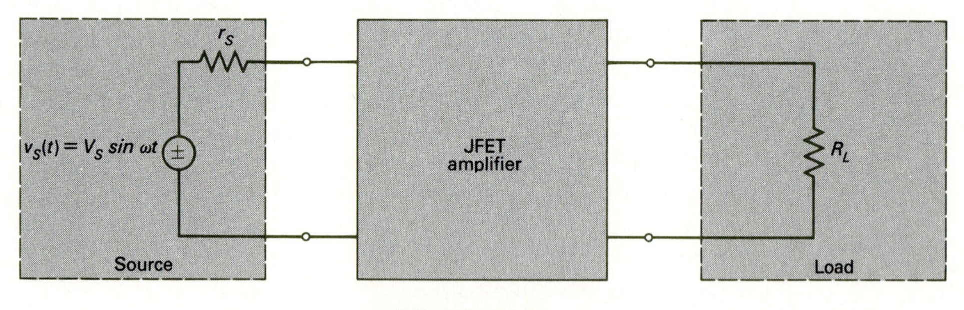

FIGURE P9.9

9.10 Repeat Prob. 9.9 if $r_S = 50\ \text{k}\Omega$ and $R_L = 500\ \Omega$.

9.11 Calculate the voltage gain of the amplifier shown in Fig. P9.11. Assume that both JFETs are identical, with $y_{fs} = 10^{-3}$ and $y_{os} = 0$.

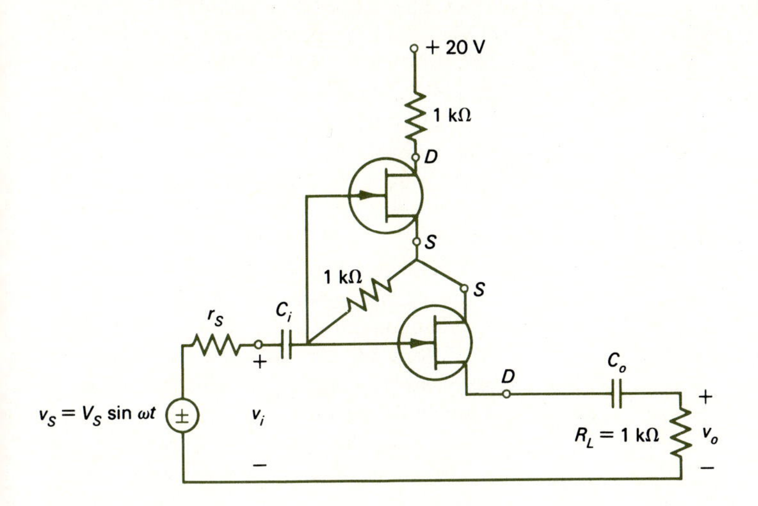

FIGURE P9.11

9.12 Two common-source JFET amplifiers are cascaded as shown in Fig. P9.12 to provide a larger overall voltage gain than could be provided by one stage. Calculate A_V for the overall amplifier, assuming that both JFETs have $y_{fs} = 10^{-3}$ and $y_{os} = 0$.

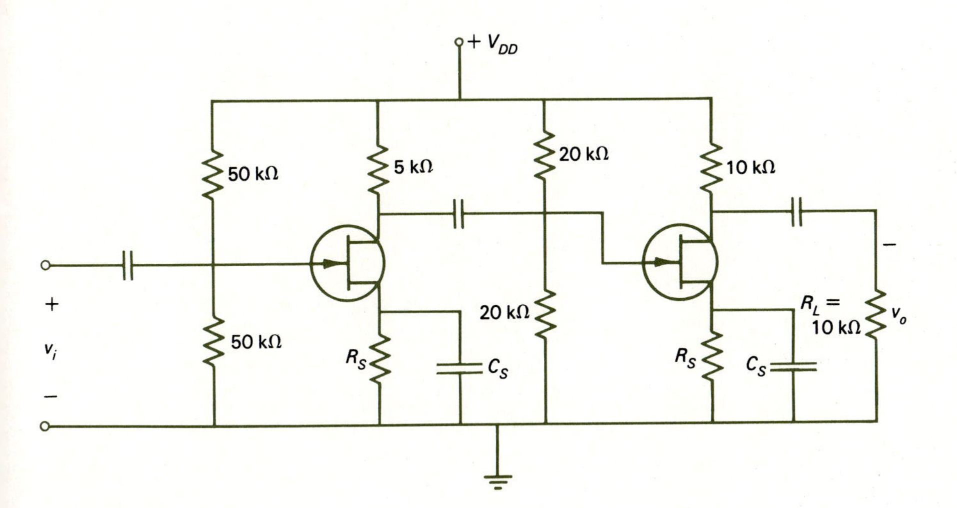

FIGURE P9.12

9.13 Determine the frequency at which the input impedance of the common-drain amplifier in Fig. 9.10 has been reduced to 1 kΩ if $R_{G1} = R_{G2} = 100$ kΩ, $R_S = 10$ kΩ, $R_L = 10$ kΩ, $R_D = 10$ kΩ, $y_{fs} = 10^{-3}$, $y_{os} = 0$, $C_{GS} = C_{GD} = 4$ pF, and $C_{DS} = 0$. Assume that C_i and C_o are short circuits. (Use reasonable approximations.)

9.14 The BJT whose characteristics are shown in Fig. 8.37 is modeled as shown in Fig. 9.16*b*. Calculate h_{ie}, h_{fe}, h_{oe} at an operating point of $I_B = 0.06$ mA, $V_{CE} = 7$ V, $I_C = 3.8$ mA.

9.15 For the common-emitter amplifier in Fig. 9.17*a*, compute A_V, A_I, Z_i, Z_o if $\beta = 100$, $h_{oe} = 0$, $h_{ie} = 500$, $R_{B1} = R_{B2} = 20$ kΩ, $R_C = 1$ kΩ, $R_L = 1$ kΩ.

9.16 Repeat Prob. 9.15 with C_E removed and $R_E = 1$ kΩ.

9.17 For the common-base amplifier shown in Fig. 9.22*a*, calculate A_V, A_I, Z_i, and Z_o if $R_C = R_L = 10$ kΩ, $R_E = 1$ kΩ, and the BJT has $h_{ie} = 500$ Ω, $\beta = 100$, $h_{oe} = 0$.

9.18 For the common-collector amplifier shown in Fig. 9.24*a*, calculate A_V, A_I, Z_i, and Z_o if $R_L = 500$ Ω, $R_E = 1$ kΩ, $R_C = 500$ Ω, $R_{B1} = R_{B2} = 100$ kΩ, $r_S = 100$ Ω, and the BJT has $h_{ie} = 300$ Ω, $h_{oe} = 0$, $h_{fe} = 50$.

9.19 A BJT amplifier is to be inserted between a source and a load as shown in Fig. P9.9 for matching purposes. If $r_S = 10$ Ω and $R_L = 1$ kΩ, what type of amplifier would you select? Design the amplifier for maximum power transfer from the source to the load if $h_{ie} = 1$ kΩ, $\beta = 80$.

9.20 As shown in Fig. P9.20, a two-stage amplifier consisting of a common-emitter amplifier followed by a common-collector amplifier is used to match to a 50-Ω load and also provide voltage gain. Calculate the overall A_V, A_I, Z_i, Z_o for this two-stage amplifier. Assume that both BJTs have $h_{ie} = 500$ Ω, $\beta = 10$, $h_{oe} = 0$.

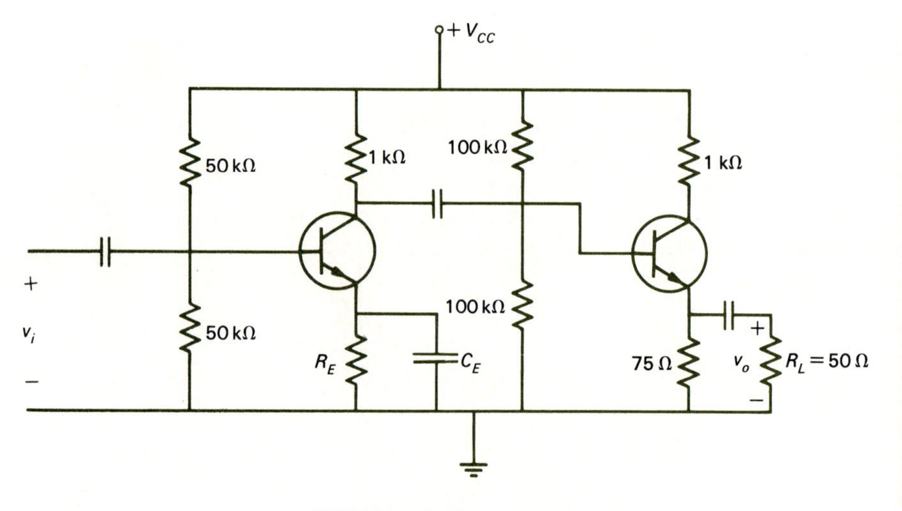

FIGURE P9.20

9.21 Calculate the frequency at which the input impedance to the common-collector amplifier shown in Fig. 9.24*a* is reduced to 50 Ω. Assume $R_{B1} = R_{B2} = 20$ kΩ, $R_C = 0$, $R_E R_L = 100$ Ω, $h_{ie} = 500$ Ω, $\beta = 100$, $h_{oe} = 0$, $C_{ob} = 0$, $f_T = 5$ MHz.

REFERENCES

9.1 A. B. Carlson and D. G. Gisser, *Electrical Engineering: Concepts and Applications*, Addison-Wesley, Reading, Mass., 1981.

9.2 Ralph J. Smith, *Circuits, Devices and Systems: A First Course in Electrical Engineering*, 4th ed., Wiley, New York, 1984.

9.3 A. E. Fitzgerald, D. Higginbotham, and A. Grabel, *Basic Electrical Engineering*, 5th ed., McGraw-Hill, New York, 1981.

9.4 D. L. Schilling and C. Belove, *Electronic Circuits: Discrete and Integrated*, 2d ed., McGraw-Hill, New York, 1979.

9.5 R. L. Boylestad and L. Nashelsky, *Electronic Devices and Circuit Theory*, 3d, ed., Prentice-Hall, Englewood Cliffs, New Jersey, 1982.

9.6 E. C. Lowenberg, *Electronic Circuits*, McGraw-Hill, New York, 1967.

Chapter

The Operational Amplifier

In Chaps. 8 and 9 we discussed the construction of linear signal amplifiers. The complete design consisted of determining a biasing circuit to place the operating point in a linear region of the nonlinear device characteristic and to stabilize against unit and temperature variations, and then of determining the ac properties of interest, such as voltage gain, current gain, power gain, input impedance, and output impedance. All of these stages of the design required considerable analysis effort. Furthermore, it was not known until the dc-biasing design was completed what values of the ac parameters would result.

In this chapter we will discuss an important type of amplifier that essentially removes all of these steps so that the design of an amplifier becomes an almost trivial process. This is the operational amplifier, which we will refer to as the op amp. The op amp consists of several transistors (BJTs and/or FETs), diodes, capacitors, and resistors and is available in integrated-circuit form in a small DIP with external leads, as shown in Fig. 8.25*b*. These op amps can presently be purchased for less than one U.S. dollar!

10.1 THE IDEAL OPERATIONAL AMPLIFIER

We will first discuss the ideal op amp shown in Fig. 10.1, along with some useful circuits, and in the next section we will discuss practical op amps. A useful preliminary analysis of circuits containing op amps can be made by replacing the actual op amps with ideal ones. Once the general characteristics of the circuit are obtained, the limitations of practical op amps can be factored into the analysis.

The symbol for the op amp is shown in Fig. 10.1. Two terminals labeled + and − are available for inputs as is an output terminal. The assumed voltages of these terminals are labeled (as shown in Fig. 10.1) with respect to the common terminal de-

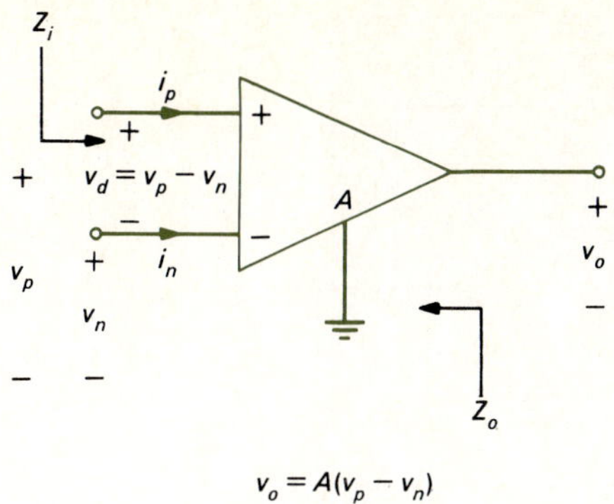

FIGURE 10.1 The operational amplifier.

noted with a ground symbol (⏚). The output voltage is related to the difference of the two input voltages as

$$v_o = A(v_p - v_n) \tag{10.1}$$

Thus, the op amp is basically a form of differential, or difference, amplifier. Practical op amps have gains A which are on the order of 10^5; thus, for practical op amps where the output voltage may be practically restricted to something on the order of 30 V, the difference voltage $v_d = v_p - v_n$ must be less than 0.3 mV, or 300 μV. The input impedance Z_i between the + and − terminals is on the order of 1 MΩ, while the output impedance, Z_0, is on the order of 100 Ω to 1 kΩ.

These practical op-amp characteristics are approximated in the ideal op amp shown in Fig. 10.2. By virtue of the high input impedance, the very large gain, and the resulting small difference voltage v_d in practical op amps, the ideal op amp is approximated by the following two important properties:

1 The input currents i_p and i_n are zero:

$$i_p = i_n = 0 \tag{10.2a}$$

2 The difference voltage v_d is zero:

$$v_d = 0 \tag{10.2b}$$

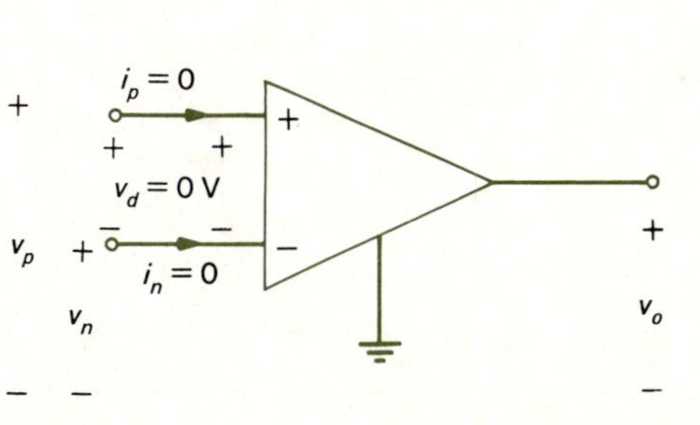

FIGURE 10.2 The ideal operational amplifier.

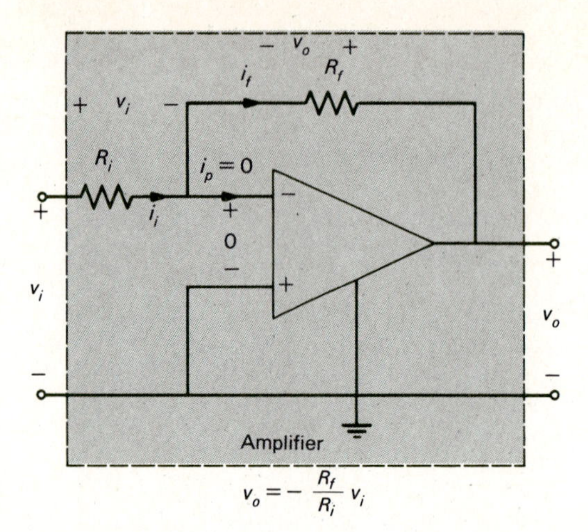

$$v_o = -\frac{R_f}{R_i} v_i$$

FIGURE 10.3
The inverting amplifier.

Note that although no current flows into the + and − terminals, the voltage between them v_d is assumed to be zero. This is referred to as the principle of "virtual short circuit" and will be crucial to our ability to analyze circuits containing ideal op amps.

As an example, consider the circuit shown in Fig. 10.3. A resistor R_i is placed in the − lead while the + lead is connected to the common terminal (grounded). Also, a resistor R_f connects the − lead and the output lead; this feedback resistor R_f is always connected to the − terminal for stability purposes, which will be discussed later. Note that because $i_p = 0$, $i_i = i_f$. Summing KVL around the outside loop, we obtain

$$v_i = R_i i_i + R_f i_f + v_o \tag{10.3}$$

Since $v_d = 0$, the input voltage and current are related by (sum KVL through R_i and v_d)

$$i_i = \frac{v_i}{R_i} \tag{10.4}$$

Substituting Eq. (10.4) and $i_i = i_f$ into Eq. (10.3) yields

$$v_i = (R_i + R_f)\frac{v_i}{R_i} + v_o \tag{10.5}$$

Solving Eq. (10.5) for the ratio of v_o/v_i yields

$$\frac{v_o}{v_i} = -\frac{R_f}{R_i} \tag{10.6}$$

Thus, the voltage gain of this overall amplifier (containing R_i, R_f, and the op amp) is simply the ratio of R_f to R_i! The negative sign symbolizes that this is an "inverting

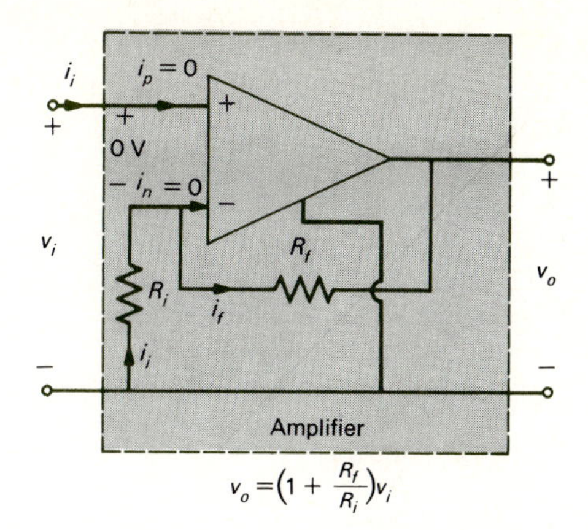

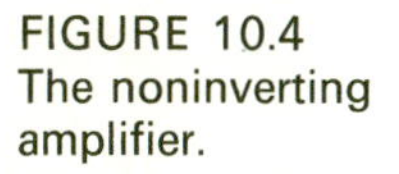
FIGURE 10.4
The noninverting amplifier.

amplifier"; that is, the output voltage is 180° out of phase with the input voltage. Nevertheless, constructing a linear amplifier with a prescribed voltage gain is exceedingly simple: select the resistors to achieve the desired ratio.

Assuming that the ideal op amp has zero output impedance, one can show that the amplifier in Fig. 10.3 also has $Z_o = 0$. However, the input impedance is the ratio $Z_i = v_i/i_i$. From Eq. (10.4) we see that $Z_i = R_i$. Thus, to construct an amplifier with a voltage gain of 10 and an input impedance of 100 kΩ, we select $R_i = 100$ kΩ and $R_f = 1$ MΩ.

Another useful amplifier is shown in Fig. 10.4. Here the input is directly connected to the + terminal. Resistors R_i and R_f are connected as in the previous amplifier. Note that since the input current i_p is zero, the input impedance to this amplifier is infinite. Once again, because $i_n = 0$, we see that $i_i = i_f$. Summing KVL around the loop containing R_i and R_f, we obtain

$$v_o = -(R_i + R_f)i_i \tag{10.7}$$

But because of the virtual short circuit

$$v_i = -R_i i_i \tag{10.8}$$

so that

$$\frac{v_o}{v_i} = 1 + \frac{R_f}{R_i} \tag{10.9}$$

Thus, we have a simple noninverting amplifier.

The op amp can be used to construct numerous other useful devices. Some of these will be discussed later in this chapter.

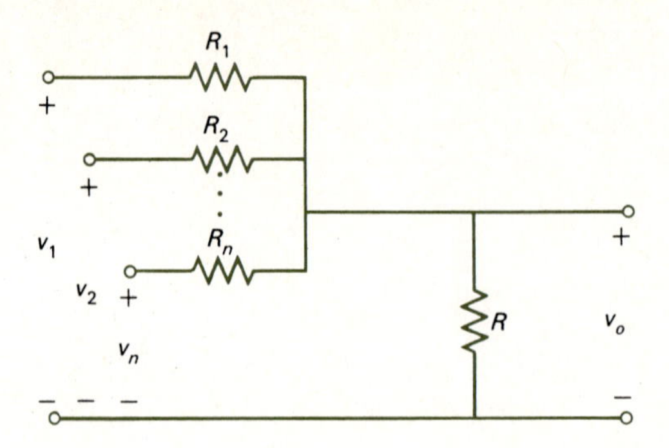

FIGURE 10.5
The resistor summer.

A summer is a device which produces as an output the sum of several input signals. A simple summer using only resistors is shown in Fig. 10.5. With $v_2 = v_3 = \cdots = v_n = 0$, we obtain

$$v_o = \left(\frac{R||R_2||\cdots||R_n}{R_1 + R||R_2||\cdots||R_n}\right)v_1 \tag{10.10}$$

If we choose R to be much smaller than $R_1, R_2, \ldots, R_n$, then

$$v_o \doteq \frac{R}{R_1 + R} v_1 \tag{10.11}$$

Similarly, we obtain by superposition

$$v_o \doteq \frac{R}{R_1 + R} v_1 + \frac{R}{R_2 + R} v_2 + \cdots + \frac{R}{R_n + R} v_n \tag{10.12}$$

so that v_o is the weighted sum of the n input signals.

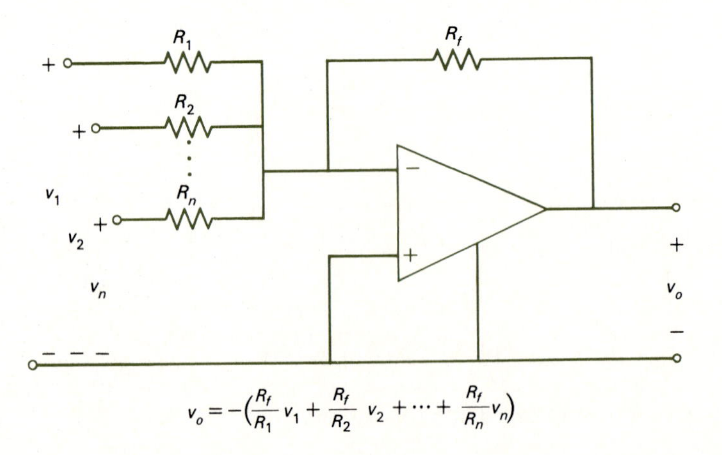

FIGURE 10.6
The operational-amplifier summer.

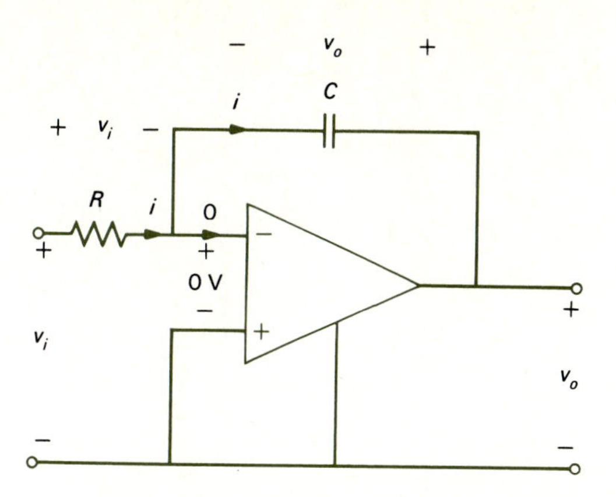

FIGURE 10.7
The integrator.

A much more useful summer can be constructed with an op amp, as shown in Fig. 10.6. With a simple extension of the analysis for Fig. 10.3, we obtain

$$v_o = -\left(\frac{R_f}{R_1}v_1 + \frac{R_f}{R_2}v_2 + \cdots + \frac{R_f}{R_n}v_n\right) \tag{10.13}$$

Another useful circuit is the integrator, shown in Fig. 10.7. Here a capacitor is used in place of the usual feedback resistor. Note that, once again, due to the principle of virtual short circuit,

$$i = \frac{v_i}{R} \tag{10.14}$$

and v_o is directly across C. Thus

$$i = -C\frac{dv_o}{dt} \tag{10.15}$$

Equating Eqs. (10.14) and (10.15), we obtain

$$\frac{dv_o}{dt} = -\frac{1}{RC}v_i \tag{10.16}$$

or

$$v_o = -\frac{1}{RC}\int v_i(\tau)\, d\tau \tag{10.17}$$

Example 10.1 Determine the input impedance to the circuit in Fig. 10.8.

Solution The input impedance to the circuit is the ratio of v_i to i_i. Utilizing the principle of virtual short circuit, we label the various voltages and currents as shown. Summing KVL around the loop containing v_i, Z, and the two resistors, we obtain

$$v_i = Zi_i + 2v_i \tag{10.18}$$

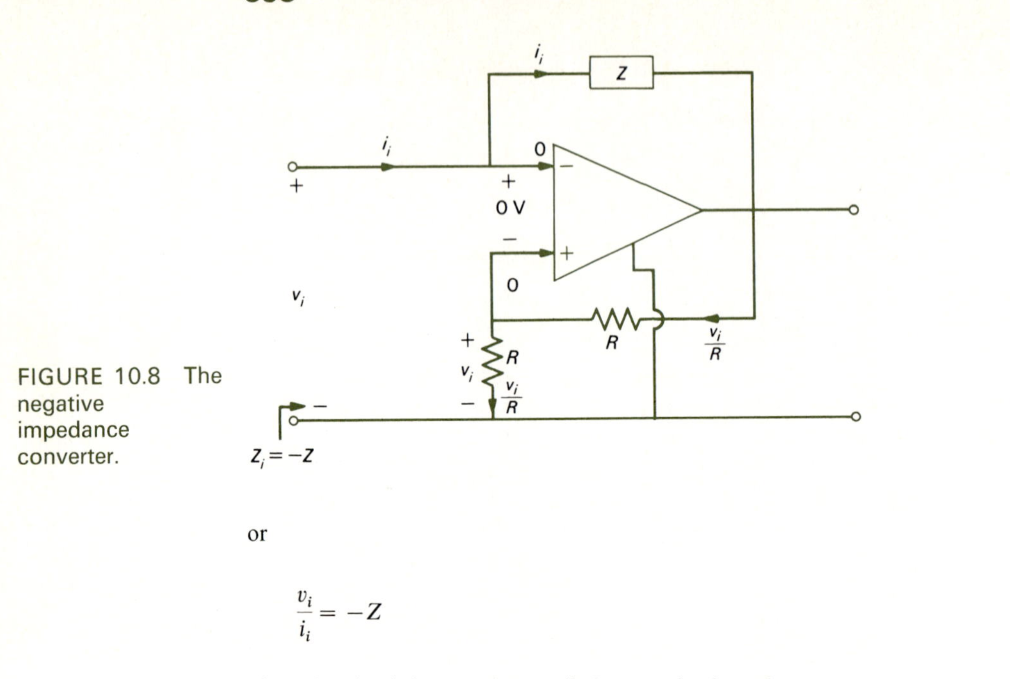

FIGURE 10.8 The negative impedance converter.

or

$$\frac{v_i}{i_i} = -Z \tag{10.19}$$

Thus, the circuit is sometimes called a negative impedance converter.

10.2 THE PRACTICAL OPERATIONAL AMPLIFIER

The ideal op amp in the previous section obtained its analysis simplification primarily by virtue of the principle of virtual short circuit; that is, the currents into the + and − input terminals were assumed to be zero, and the voltage between these two terminals was assumed to be zero. Practical op amps, even though they do not have these ideal characteristics, approximate them very closely.

A practical op amp has the equivalent circuit shown in Fig. 10.9*a*. Typical values of Z_i and Z_o are 1 MΩ and 100 to 1 kΩ, respectively, while A is typically 10^5 or larger. Also shown are the usual $+V_{CC}$ and $-V_{CC}$ dc power supplies required to bias the transistors in the op amp. The transfer characteristics for open-circuited output is shown in Fig. 10.9*b*; it exhibits a linear region with slope A. Also shown is a saturation region so that v_o cannot exceed V_{CC}, the values of the power supplies; typically, V_{CC} is on the order of 10 V. Thus, for $A = 10^5$ we find a maximum differential voltage which may be applied such that linear operation is maintained as

$$\begin{aligned} v_{d,\max} &= \frac{V_{CC}}{A} \\ &= \frac{10}{10^5} \\ &= 0.1 \text{ mV} \end{aligned} \tag{10.20}$$

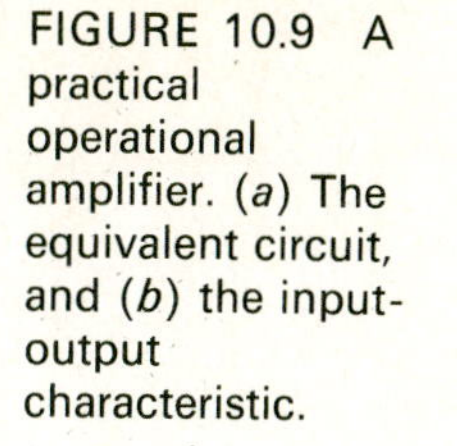
FIGURE 10.9 A practical operational amplifier. (*a*) The equivalent circuit, and (*b*) the input-output characteristic.

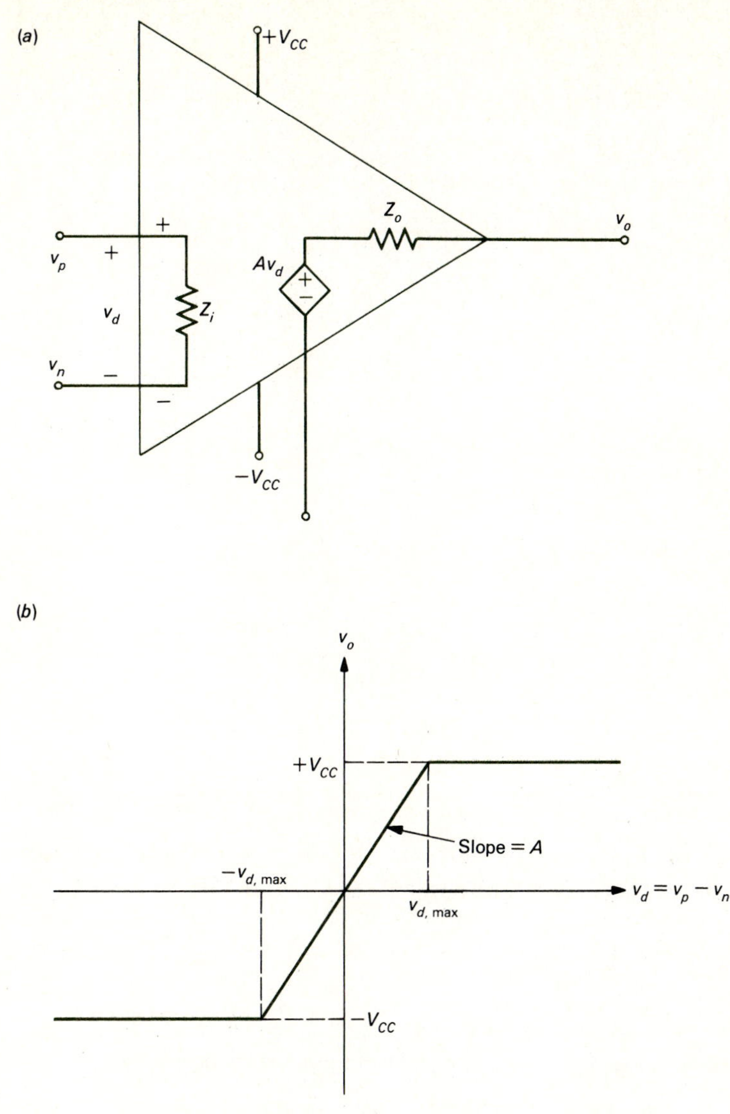

Even though the input currents are not zero, if $Z_i = 1\ \text{M}\Omega$ they are on the order of $i = v_{d,\max}/Z_i = 10^{-10}\ A$ and are essentially zero.

The restriction imposed by Eq. (10.20) that v_d must not exceed 0.1 mV to maintain linear operation seems to rule out the use of this device as a practical amplifier. But recall that in the circuits considered previously, the difference voltage v_d is *not* the input voltage to the circuit. For example, consider the circuit in Fig. 10.3. Inserting the equivalent circuit of Fig. 10.9*a*, we obtain Fig. 10.10. Let us presume that $Z_i = \infty$ and $Z_o = 0$ to simplify the analysis. In this case we obtain

$$v_o = Av_d \tag{10.21}$$

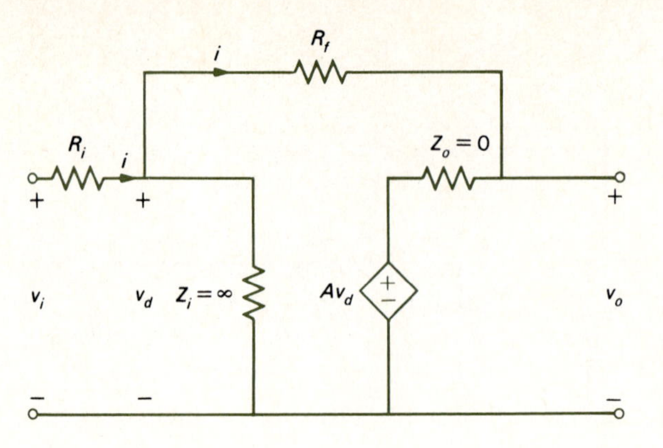

FIGURE 10.10 The inverting amplifier with operational-amplifier circuit substituted.

but by voltage division

$$v_d = \frac{R_f}{R_f + R_i}(v_i - v_o) + v_o \tag{10.22}$$

Thus

$$v_o = \frac{AR_f}{R_f + R_i}(v_i - v_o) + Av_o \tag{10.23}$$

or

$$\frac{v_o}{v_i} = \frac{AR_f}{R_f + (1 - A)R_i} \tag{10.24}$$

For A very large, this simplifies to

$$\frac{v_o}{v_i} = -\frac{R_f}{R_i} \tag{10.25}$$

which was obtained using the principle of virtual short circuit, which essentially assumed that $A \to \infty$. The ratio of v_d to v_i can then be found from

$$\frac{v_d}{v_i} = \frac{v_d}{v_o}\frac{v_o}{v_i} \tag{10.26}$$

Substituting Eqs. (10.21) and (10.24), we obtain

$$\frac{v_d}{v_i} = \frac{1}{A}\frac{AR_f}{R_f + (1 - A)R_i}$$

$$= \frac{1}{1 + (1 - A)R_i/R_f} \tag{10.27}$$

Since A is very large, the ratio in Eq. (10.27) is quite small. (For example, if $A = 10^5$ and $R_f/R_i = 10$, then $v_d/v_i \doteq 10^{-4}$.) Thus, v_d can be very small and yet v_i can be considerably larger. The feedback configuration thus allows the construction of practical amplifiers which are not limited to small input voltages.

Example 10.2 Construct an amplifier with the configuration shown in Fig. 10.3 to amplify a 200-mV (peak) sinusoid. Assume that $A = 10^5$ and $V_{CC} = 15$ V.

Solution For linear operation, v_o must not exceed 15 V. Thus, the maximum overall gain is $15/0.2 = 75$. Rather than working so close to this margin, we might select an overall gain of 68. Selecting $R_i = 10$ kΩ and $R_f = 680$ kΩ, we see that the output-voltage sinusoid will have a peak value of $68 \times 0.2 = 13.6$ V and linear operation will be maintained.

Another practical aspect of op amps is their frequency response. The open-loop gain is the gain

$$A_o = \frac{v_o}{v_d} \tag{10.28}$$

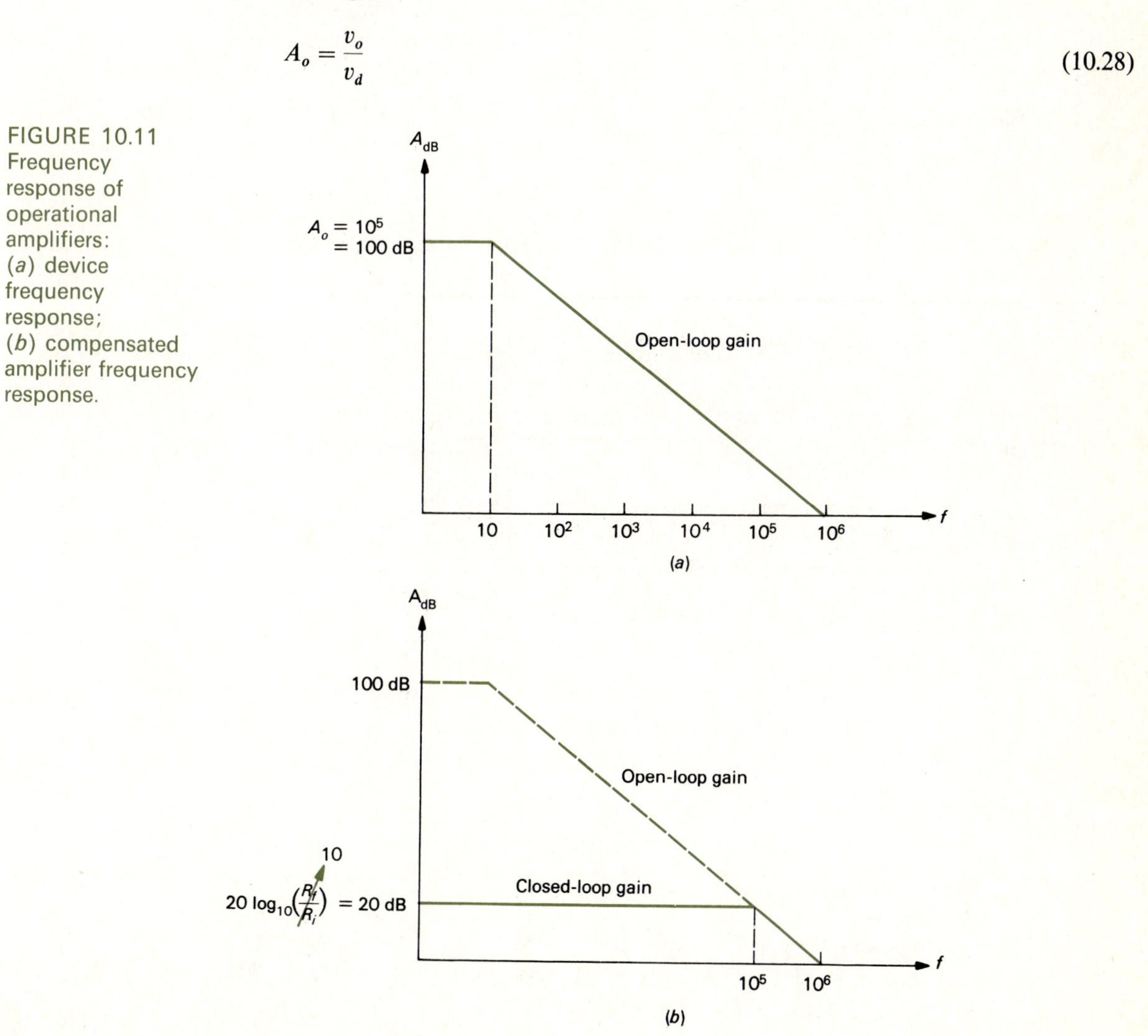

FIGURE 10.11 Frequency response of operational amplifiers: (*a*) device frequency response; (*b*) compensated amplifier frequency response.

of the op amp without any external feedback circuitry. A plot of this gain as a function of frequency for a typical op amp is shown in Fig. 10.11*a*. The frequency axis is plotted logarithmically and the gain is plotted in decibels (dB), where

$$A_{dB} = 20 \log A \tag{10.29}$$

If $A_o = 10^5$, then in decibels $A_o = 100$ dB. Note that this large open-loop gain is constant only up to around 10 Hz and drops at a rate of 20 dB per decade to a value of 0 dB (or absolute value of unity gain) at 1 MHz. Although these numbers differ among op amps, they are fairly typical. The important point to note here is that the bandwidth is very narrow, approximately 10 Hz. The frequency at which the open-loop gain (in dB) drops to zero is the gain-bandwidth product. (A gain of 0 dB corresponds to unity gain.)

When the op amp is inserted into the circuit of Fig. 10.3, the gain-versus-frequency curve becomes as shown in Fig. 10.11*b*. Note that the maximum (dc) closed-loop gain is $-R_f/R_i$ so that for a gain of 10 or 20 dB the frequency response is reduced in magnitude but the bandwidth is increased dramatically to 10^5 Hz. What we observe is that the gain-bandwidth product is the same for both amplifiers, a property which is true for other feedback amplifiers.

10.3 INDUCTORLESS (ACTIVE) FILTERS

In Chap. 5 we considered filters—low-pass, high-pass, bandpass and band-reject—which were used to pass or eliminate certain frequency components of a signal. Some of these required the use of inductors. For example, the bandpass and band-reject filters required an inductor in series or in parallel with a capacitor. The resonance phenomenon associated with the series or parallel connection of these two elements allowed the elimination or passage of certain frequency components.

In Chap. 8 we discussed the small-scale integrated-circuit fabrication of various transistors, resistors, diodes, and capacitors. Inductors are obviously not suited to IC fabrication. Thus, in order to construct filters in the form of an integrated circuit, we need some alternative to the inductor.

Filters which are suitable for IC fabrication but which do not contain inductors are referred to as active filters. One common component of active filters is the operational amplifier.

Consider the basic op-amp circuit with frequency-dependent impedances shown in Fig. 10.12. The voltage-gain or voltage-transfer function of this device is

$$\frac{v_o}{v_i} = -\frac{Z_f}{Z_i} \tag{10.30}$$

A low-pass filter can be constructed as shown in Fig. 10.13*a*. In this case

$$Z_i = R_i \tag{10.31a}$$

$$Z_f = R_f || \frac{1}{j\omega C_f}$$

$$= \frac{R_f}{1 + j\omega R_f C_f} \tag{10.31b}$$

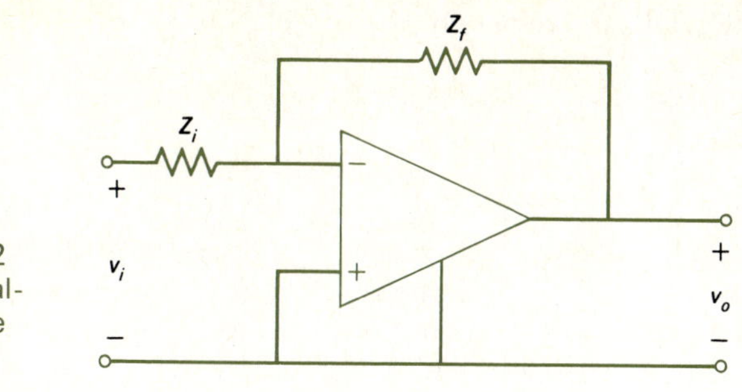

FIGURE 10.12 The operational-amplifier active filter.

FIGURE 10.13 A low-pass active filter.

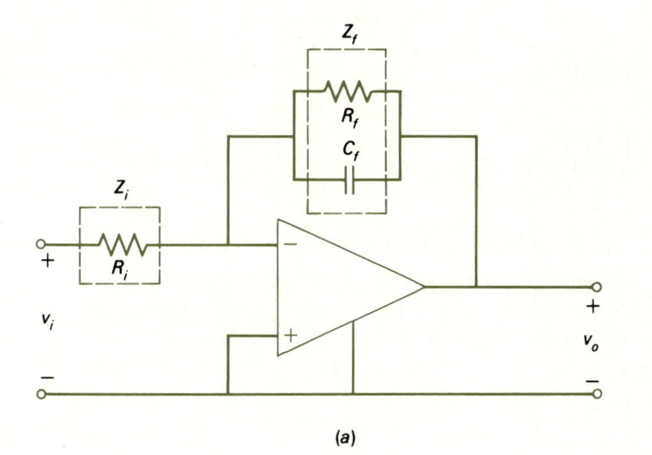

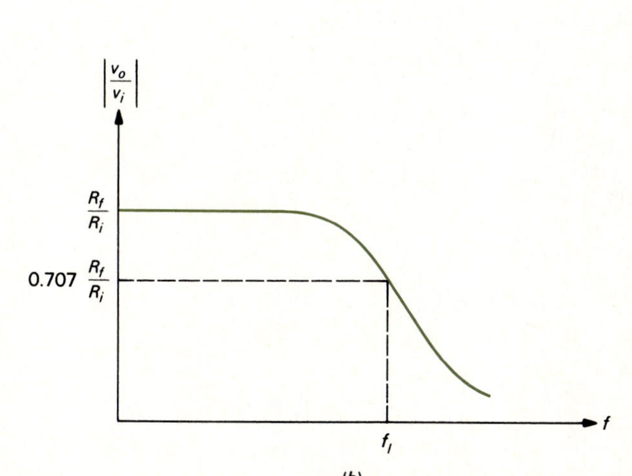

Substituting into Eq. (10.30), we obtain

$$\frac{v_o}{v_i} = -\frac{R_f}{R_i}\frac{1}{1 + j\omega R_f C_f} \tag{10.32}$$

Denoting

$$f_l = \frac{1}{2\pi R_f C_f} \tag{10.33}$$

Eq. (10.32) may be written as

$$\frac{v_o}{v_i} = -\frac{R_f}{R_i}\frac{1}{1 + jf/f_l} \tag{10.34}$$

The magnitude is

$$\left|\frac{v_o}{v_i}\right| = \frac{R_f}{R_i}\frac{1}{\sqrt{1 + (f/f_l)^2}} \tag{10.35}$$

which is sketched in Fig. 10.13*b*. The dc value is R_f/R_i, which is down by a factor of $1/\sqrt{2}$ at the half-power, or 3-dB, point of f_l.

Similarly, a high-pass filter is shown in Fig. 10.14*a*. Here

$$Z_i = R_i + \frac{1}{j\omega C_i} \tag{10.36a}$$

$$Z_f = R_f \tag{10.36b}$$

Substituting into Eq. (10.30) and defining

$$f_h = \frac{1}{2\pi R_i C_i} \tag{10.37}$$

we obtain

$$\frac{v_o}{v_i} = -\frac{R_f}{R_i}\frac{jf/f_h}{1 + jf/f_h} \tag{10.38}$$

The magnitude is

$$\left|\frac{v_o}{v_i}\right| = \frac{R_f}{R_i}\frac{1}{\sqrt{1 + (f_h/f)^2}} \tag{10.39}$$

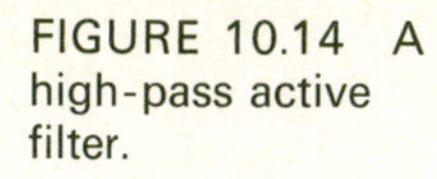

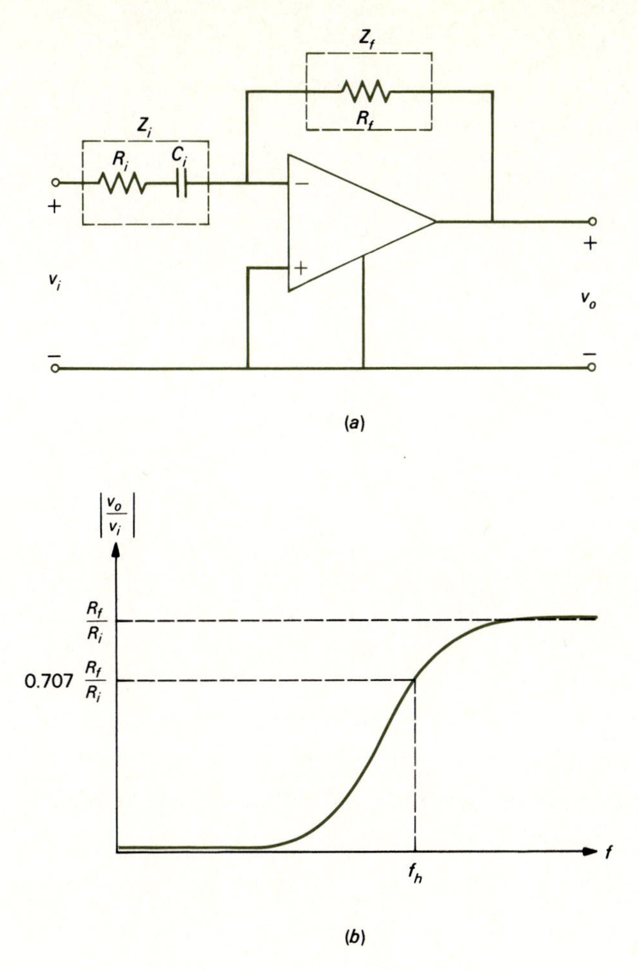

FIGURE 10.14 A high-pass active filter.

which is sketched in Fig. 10.14*b*. Note that the high-frequency gain is again R_f/R_i and that the half-power, or 3-dB, point is f_h.

In order to construct a bandpass filter, we might try using a low-pass and a high-pass filter and overlapping their transfer functions, as shown in Fig. 10.15. In order to do this we need to isolate the two filters, as shown in Fig. 10.16*a*. Here

$$Z_f = R_f || \frac{1}{j\omega C_f}$$

$$= \frac{R_f}{1 + j\omega R_f C_f} \tag{10.40a}$$

$$Z_i = R_i + \frac{1}{j\omega C_i} \tag{10.40b}$$

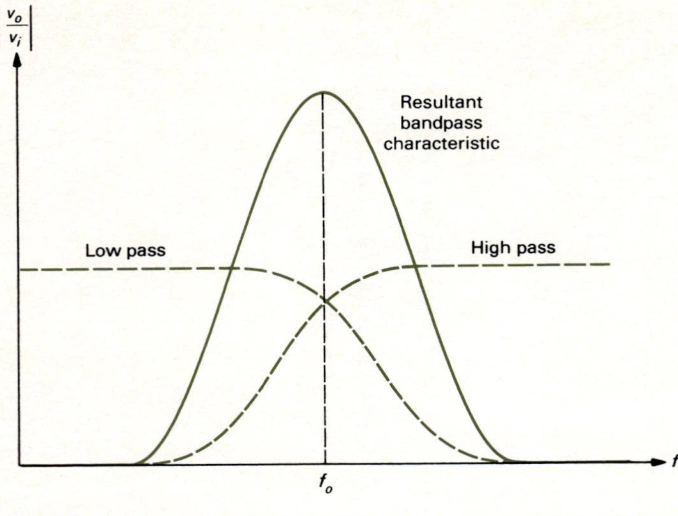

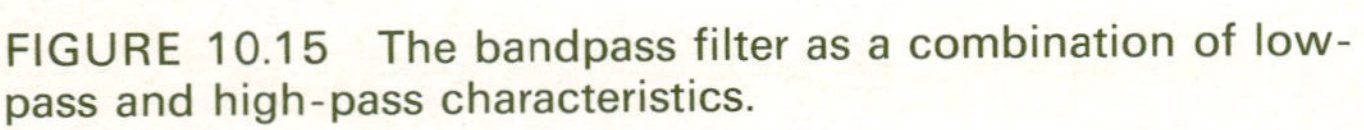
FIGURE 10.15 The bandpass filter as a combination of low-pass and high-pass characteristics.

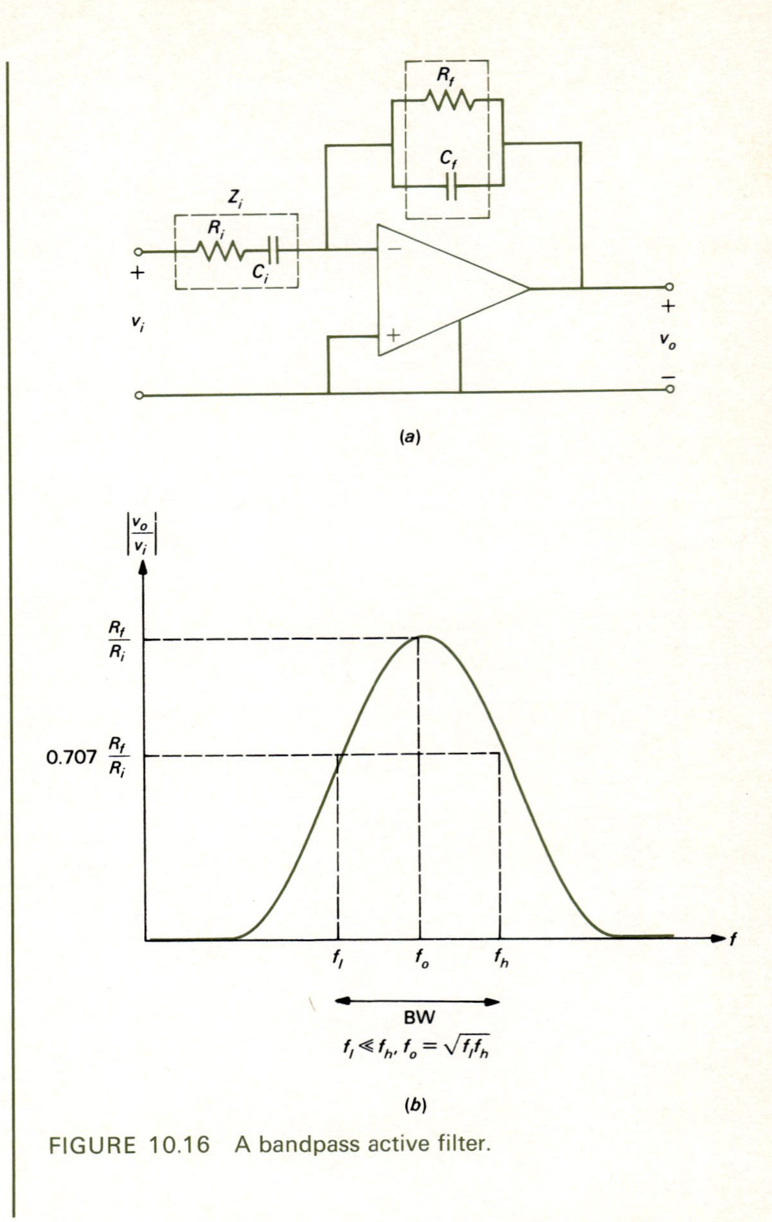

FIGURE 10.16 A bandpass active filter.

and

$$\frac{v_o}{v_i} = -\frac{Z_f}{Z_i}$$

$$= -\frac{R_f}{R_i}\frac{1}{(1 + j\omega R_f C_f)(1 + 1/j\omega R_i C_i)} \tag{10.41}$$

Denoting

$$f_l = \frac{1}{2\pi R_i C_i} \tag{10.42a}$$

$$f_h = \frac{1}{2\pi R_f C_f} \tag{10.42b}$$

Eq. (10.41) can be written as

$$\frac{v_o}{v_i} = -\frac{R_f}{R_i}\frac{1}{(1 + jf/f_h)(1 - jf_l/f)}$$

$$= -\frac{R_f}{R_i}\frac{1}{(1 + f_l/f_h) + j(f/f_h - f_l/f)}$$

$$= -\frac{R_f}{R_i}\frac{1}{[(f_h + f_l)/f_h] + j(f^2 - f_l f_h)/ff_h}$$

$$= -\frac{R_f}{R_i}\frac{f_h/(f_h + f_l)}{1 + j(f^2 - f_l f_h)/f(f_l + f_h)} \tag{10.43}$$

Note that the response is a maximum at

$$f_o = \sqrt{f_l f_h} \tag{10.44}$$

If $f_l \ll f_h$, then the magnitude of Eq. (10.43) becomes $0.707R_f/R_i$ at f_l and f_h as shown in Fig. 10.16*b*. In this case we may define the bandwidth (BW) as

$$\text{BW} = f_h - f_l \qquad f_l \ll f_h \tag{10.45}$$

10.4 ANALOG COMPUTERS

Presently (and in the conceivable future) the digital computer forms the basic numerical analysis tool for the solution of equations (algebraic as well as differential). In the past, another type of computer—the analog computer—occupied an equally

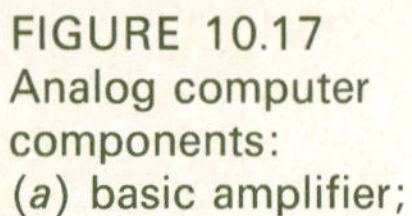

FIGURE 10.17
Analog computer components:
(*a*) basic amplifier;
(*b*) potentiometer;
(*c*) summer
(*d*) integrator.

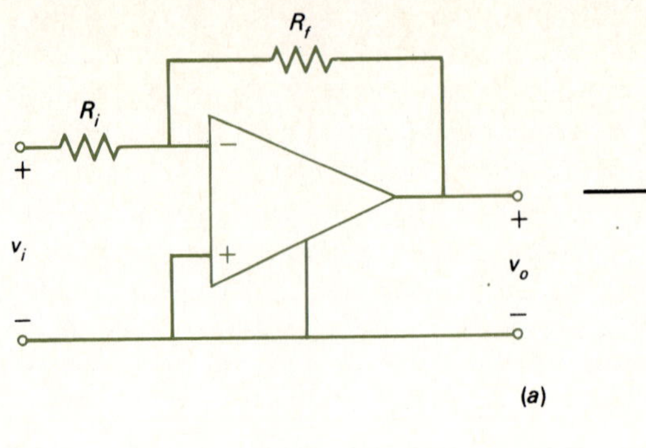

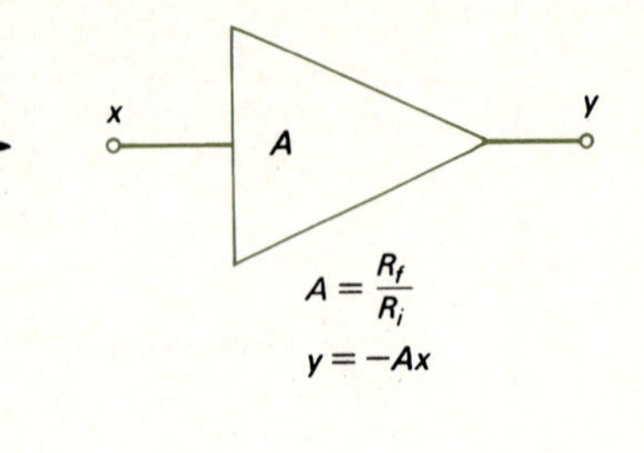

(*a*)

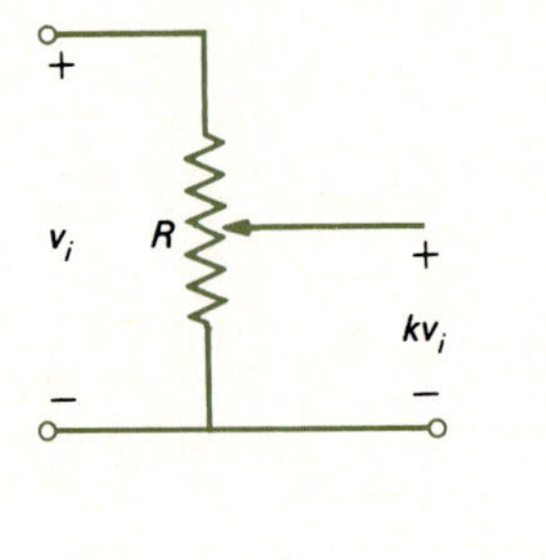

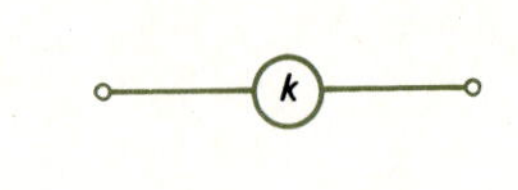

(*b*)

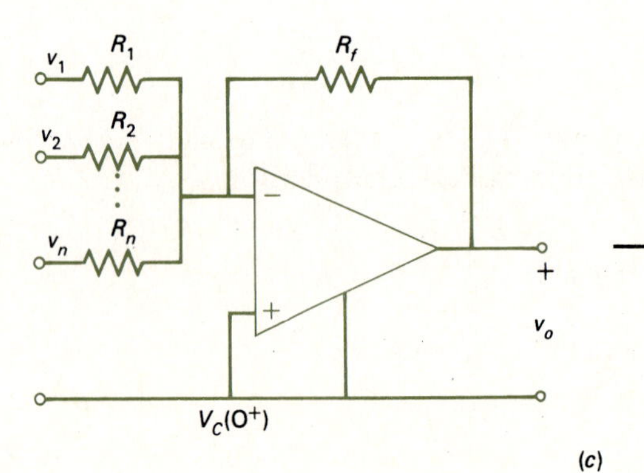

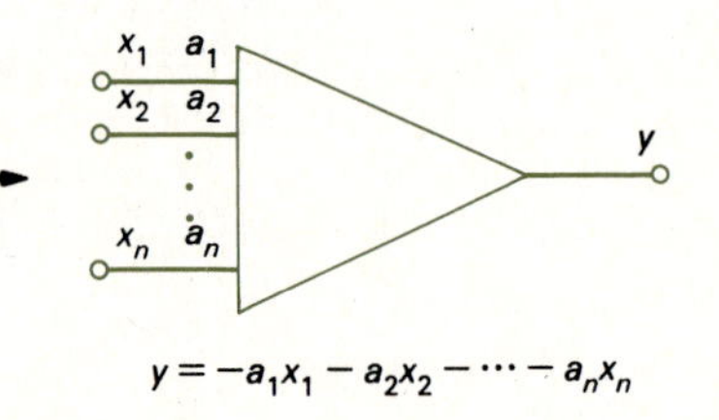

(*c*)

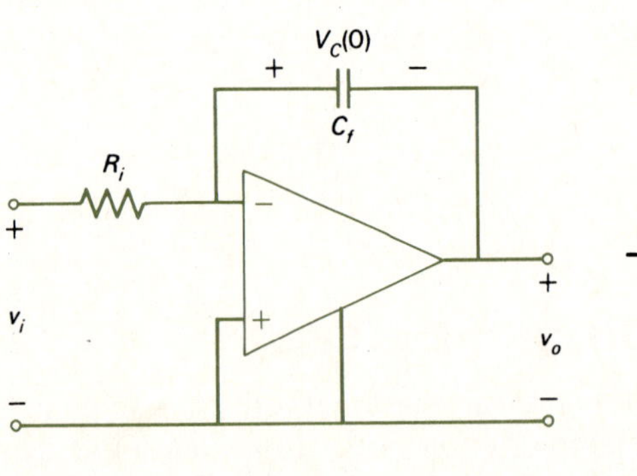

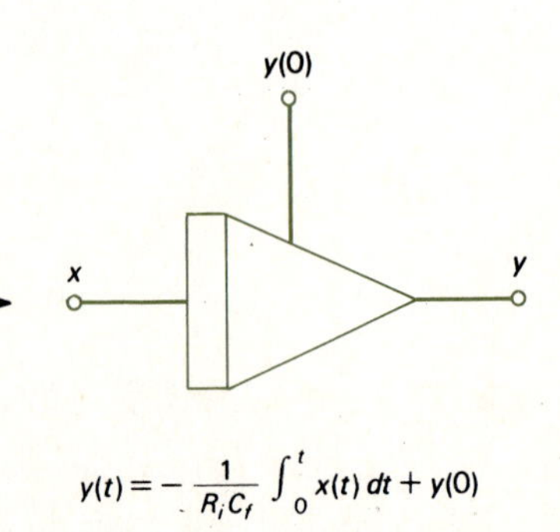

(*d*)

important role. Although not used as much as the digital computer, the analog computer still retains some important advantages over the digital computer.

The majority of the basic elements of an analog computer, shown in Fig. 10.17, are constructed with operational amplifiers. The basic amplifier in Fig. 10.17*a* is used to provide integer multiplication (often by a factor of 10), whereas the potentiometer in Fig. 10.17*b* is used for providing noninteger gains < 1. The summer in Fig. 10.17*c* and the integrator in Fig. 10.17*d* were discussed previously. In the integrator, C_f is charged to an initial voltage $V_C(0)$ to provide the initial condition on y. In the summer of Fig. 10.17*c*, if we choose $R_f/R_1 = R_f/R_2 = \cdots R_f/R_n = 1$, then $y = x_1 + x_2 + \cdots + x_n$. Similarly, in Fig. 10.17*d*, if we choose $R_i C_f = 1$, then $y = -\int_0^t x(\tau)\, d\tau + y(0)$. These cases will be indicated by 1s in the unit boxes.

For example, suppose we wish to solve the ordinary differential equation

$$\frac{d^2 y(t)}{dt} + a_1 \frac{dy(t)}{dt} + a_2 y(t) = f(t) \tag{10.46}$$

subject to the initial conditions

$$\begin{aligned} y(0) &= y_0 \\ \left.\frac{dy}{dt}\right|_{t=0} &= y_1 \end{aligned} \tag{10.47}$$

The connection shown in Fig. 10.18 will accomplish the *simulation* of the solution. We write Eq. (10.46) by isolating the highest derivative as

$$\ddot{y} = -a_1 \dot{y} - a_2 y + f \tag{10.48}$$

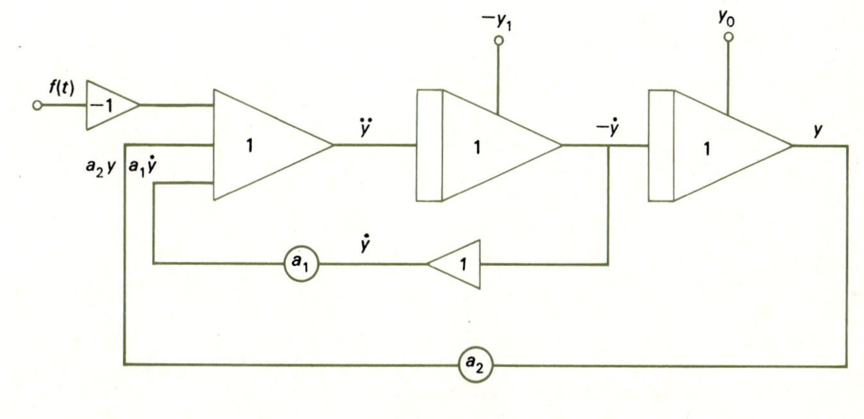

FIGURE 10.18 Analog computer implementation of a second-order differential equation.

where the dots are used to denote the various derivatives with respect to time t. Thus, we need to sum $-a_1\dot{y}$, $-a_2 y$, and f. The summer does this. The first integrator integrates $\ddot{y}$ to give $\dot{y}$:

$$\dot{y} = -\int_0^t \ddot{y}\,d\tau - \underbrace{\dot{y}(0)}_{y_1} \tag{10.49}$$

The second integrator integrates this result:

$$y = -\int_0^t \dot{y}\,d\tau - \underbrace{y(0)}_{y_0} \tag{10.50}$$

These variables are multiplied by constants a_1 and a_2 and added, along with $-f$, at the summer.

One problem with this arrangement is that if we need to investigate the actual solution for a very long time, say 1 h, we must observe the output of this analog computer for that actual time interval. Similarly, if the important part of the response lasts a very short time, say 1 μs, then the observation time may be too short for a recording instrument such as a chart recorder. These problems may be overcome by time scaling. Redefine t as

$$t = \alpha\tau \tag{10.51}$$

Thus

$$\begin{aligned} \frac{dy}{dt} &= \frac{dy}{d\tau}\frac{d\tau}{dt} \\ &= \frac{1}{\alpha}\frac{dy}{d\tau} \end{aligned} \tag{10.52}$$

Thus, the differential equation Eq. (10.46) becomes, in terms of the new time variable,

$$\frac{1}{\alpha^2}\frac{d^2y(\tau)}{d\tau^2} + \frac{a_1}{\alpha}\frac{dy(\tau)}{d\tau} + a_2 y(\tau) = f(\tau) \tag{10.53}$$

or

$$\frac{d^2y(\tau)}{d\tau^2} = -\alpha a_1 \frac{dy(\tau)}{d\tau} - \alpha^2 a_2 y(\tau) + \alpha^2 f(\tau) \tag{10.54}$$

Consequently, the coefficients may simply be scaled by the appropriate factors to change the solution time of the analog computer. For example, if $\alpha = 0.1$, then $t = 10$ s corresponds to $\tau = 1$ s and thus the solution time may be reduced.

PROBLEMS

10.1 For the circuit shown in Fig. P10.1, assume an ideal op amp and calculate v_o.

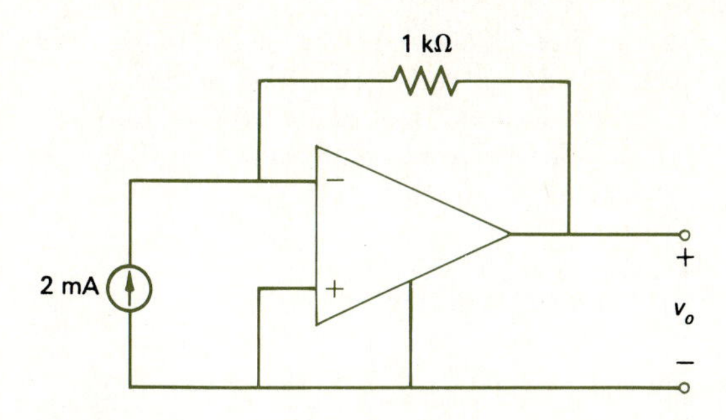

P10.1

10.2 For the circuit shown in Fig. P10.2, assume an ideal op amp and calculate v_o.

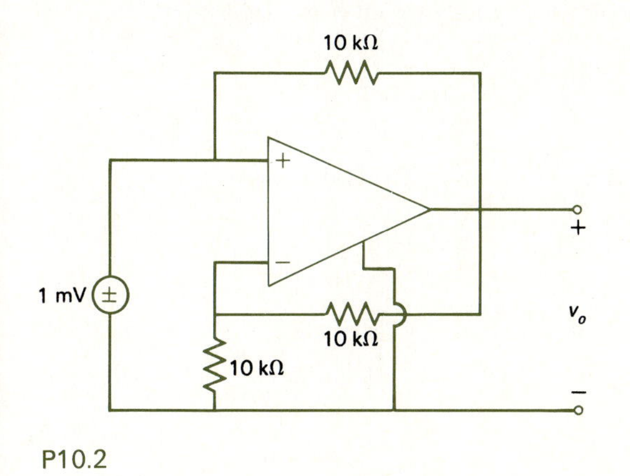

P10.2

10.3 For the circuit shown in Fig. P10.3, assume an ideal op amp and calculate v_o.

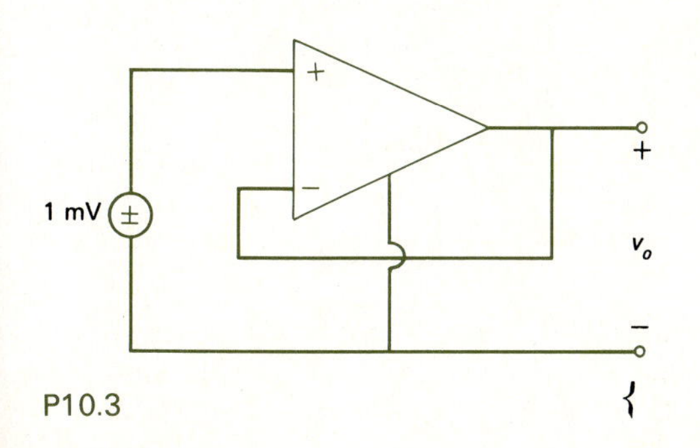

P10.3

10.4 The circuit shown in Fig. P10.4 is used as an electronic ohmmeter for measuring an unknown resistance. Assume an ideal op amp and determine the voltmeter reading V in terms of the unknown value of R.

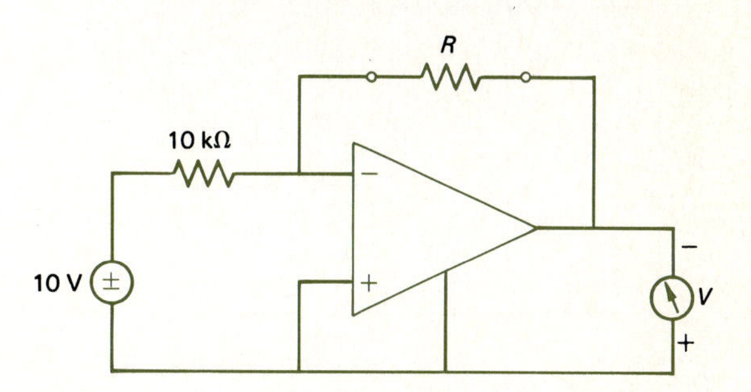

P10.4

10.5 The circuit shown in Fig. P10.5 will serve as a "difference amplifier" to measure the difference between two voltages v_1 and v_2. Write v_o in terms of v_1 and v_2, assuming an ideal op amp.

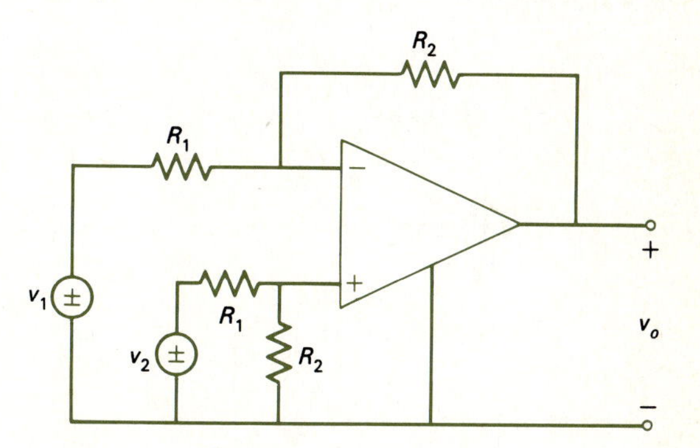

P10.5

10.6 For the circuit shown in Fig. P10.6, calculate v_o/v_i. Assume an ideal op amp. What effect do R_L and R have? Compare this circuit with that of Fig. 10.4.

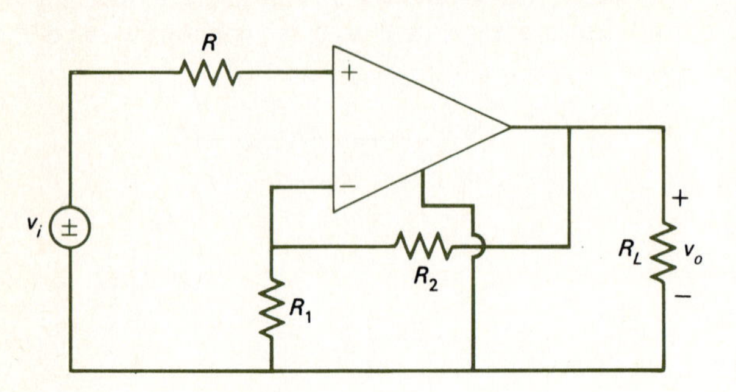

P10.6

10.7 For the op-amp circuit shown in Fig. 10.4, calculate v_o/v_i. Assume that the op amp has finite A, $Z_i = \infty$, and $Z_o = 0$. Show that as $A \to \infty$, $v_o/v_i \to (R_f + R_i)/R_i$.

10.8 For the circuit shown in Fig. P10.8, assume an ideal op amp and calculate v_o/v_i. Determine the input impedance seen by v_i.

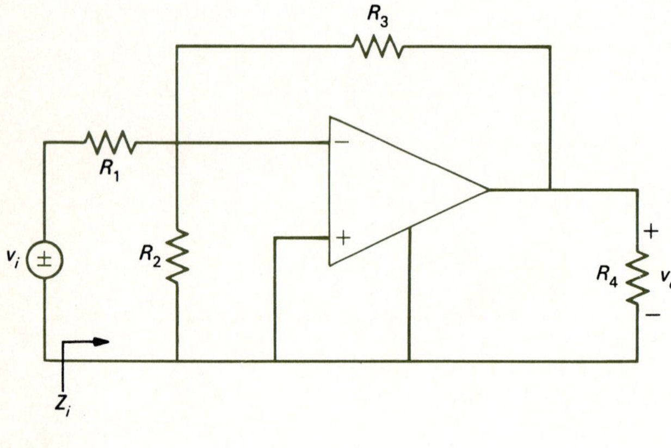

P10.8

10.9 For the circuit shown in Fig. P10.2, calculate the input impedance seen by the 1-mV source.

10.10 For the low-pass filter shown in Fig. 10.13, determine C_f such that the 3-dB point is at 1 kHz. Assume $R_i = R_f = 1\ \text{M}\Omega$.

10.11 For the high-pass filter shown in Fig. 10.14, calculate R_f and R_i to give a 3-dB point at 1 MHz. Assume $v_o/v_i = 2$ at 10 MHz and $C_i = 100$ pF.

10.12 Construct an analog computer diagram to solve the differential equation

$$\frac{d^2y(t)}{dt^2} + 10\frac{dy(t)}{dt} + 4y(t) = 5$$

with $y(0) = 2$, $\dot{y}(0) = 0$.

10.13 If the solution to the differential equation in Prob. 10.12 is to be obtained over $0 \le t \le 1$ ms but we wish to expand this over an interval of 1 s, redraw the analog computer diagram.

10.14 An integrator shown in Fig. 10.17*d* is to be constructed to solve the differential equation

$$\frac{dy(t)}{dt} + 10^3 y(t) = 0$$

with $y(0) = 2$. If $R_i = 10\ \text{k}\Omega$, determine C_i.

REFERENCES

10.1 A. B. Carlson and D. G. Gisser, *Electrical Engineering: Concepts and Applications*, Addison-Wesley, Reading, Mass., 1981.

10.2 Ralph J. Smith, *Circuits, Devices and Systems: A First Course in Electrical Engineering*, 4th ed., Wiley, New York, 1984.

10.3 A. E. Fitzgerald, D. Higginbotham, and A. Grabel, *Basic Electrical Engineering*, 5th ed., McGraw-Hill, New York, 1981.

10.4 R. L. Boylestad and L. Nashelsky, *Electronic Devices and Circuit Theory*, 3d ed., Prentice-Hall, Englewood Cliffs, New Jersey, 1982.

10.5 W. G. Oldham and S. E. Schwartz, *An Introduction to Electronics*, Holt, Rinehart & Winston, New York, 1972.

10.6 J. D. Irwin, *Basic Engineering Circuit Analysis*, Macmillan, New York, 1983.

Chapter

Digital Electronic Circuits

In previous chapters we studied the use of nonlinear devices (the BJT and the FET) in constructing linear amplifiers. Although these devices were inherently nonlinear, we confined the operation to the linear portions of the characteristics in order to produce linear amplification of a signal.

In this chapter we will be more interested in the operation of these devices in the nonlinear regions of their characteristics. The primary use of these devices will be in constructing electronic switches for use in computers and other digital devices. Digital electronic circuits are becoming of increasing importance for several reasons. Current integrated-circuit technology allows the construction of an enormous number of transistors and diodes, as well as resistors and capacitors, on a very small chip no larger than a pencil eraser. The present technology with regard to density of components on a chip is:

1. Small-scale integration (SSI) containing fewer than 100 components
2. Medium-scale integration (MSI) containing 100 to 1000 components
3. Large-scale integration (LSI) containing 1000 to 10,000 components
4. Very large-scale integration (VLSI) containing over 10,000 components on a single chip

The large number of switches (gates) required by digital systems can now be placed on a chip of very small size. Prior to the revolutionary advances in IC technology, such systems would occupy a much larger space and analog systems tended to be predominant.

Another important advantage of digital systems over analog ones is their inherent noise and interference immunity. Digital systems operate on discrete voltage levels; for typical BJT gates, these levels are 0 and 5 V.† The data bits are logical 0s (0 V) and 1s (5 V). Strings of these logical 0s and 1s can be used to represent numbers, as will be discussed in Part 4. These numbers can be manipulated—such as in addition, subtraction, multiplication, and division—by operating on the bits. Since the bit levels are widely separated (on the order of 5 V for BJT devices), any noise or other random voltage introduced into the system will have to be of sufficient magnitude to cause a "bit error" to result. The noise tolerance level tends to be larger for digital systems than for the comparable analog versions.

In this chapter we will study the construction of the basic digital circuit elements. These are composed of gates (AND, OR, NAND, NOR, NOT) for bit stream manipulation, memory circuits (flip-flops) for storage of data, and timing circuits for synchronization of the data manipulation. We will take a brief journey through this field and highlight the important points. There is much more to study in the design of these circuits, and this will be reserved for later courses.

11.1 DIODE GATES

The diode AND gate is shown in Fig. 11.1*a*. The symbol is shown in Fig.11.1*b*. Suppose that the input voltages to the gate V_A and V_B consist of pulses which are either 0 or +5 V, with 0 V representing a logical 0 and +5 V representing a logical 1. If either V_A or V_B is 0 V, the associated diode is forward-biased (by the +5-V battery) and is closed. Thus, the output voltage V_o is zero. If both V_A and V_B are +5 V, the output is also +5 V. (If V_A and V_B are both +5 V, then the diode is either open or closed; the precise state is hard to determine, for when the voltage across an ideal diode is 0 V, is it open or closed? Nevertheless, for either condition $V_o = +5$ V.) Thus, this gate performs the logical AND function:

A	B	A·B
0	0	0
0	1	0
1	0	0
1	1	1

† The actual voltage levels vary somewhat. We will use 0 and 5 V for illustration.

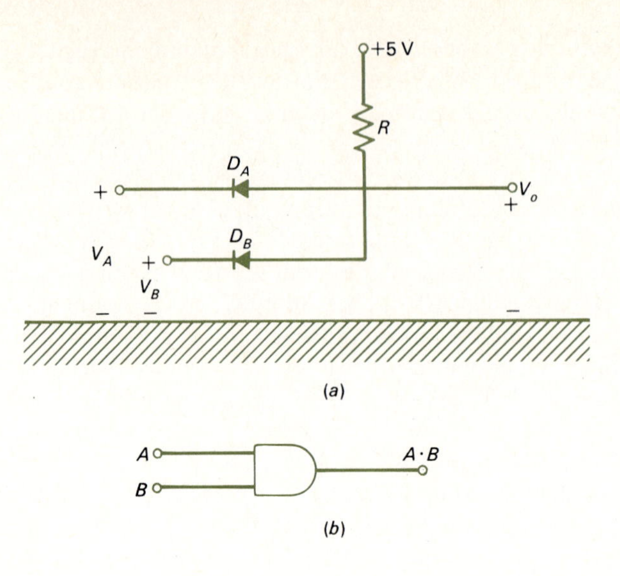

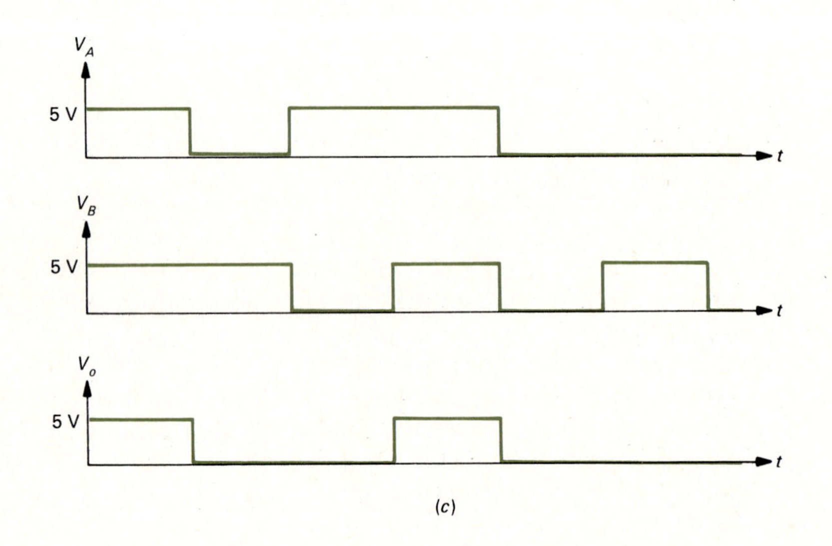

FIGURE 11.1 The AND gate. (*a*) Diode implementation, (*b*) symbol, (*c*) digital operation.

The AND operation is denoted with a dot (·): $A \cdot B$. Typical bit stream inputs to the AND gate are shown in Fig. 11.1*c* along with the resulting output stream.

The diode OR gate is shown in Fig. 11.2*a*, and the symbol is shown in Fig. 11.2*b*. If either V_A or V_B is +5 V, the associated diode will be forward-biased and closed and the output will be $V_o = 5$ V. If both V_A and V_B are 0 V, both diodes are reverse-biased and open and the output voltage will be 0 V. This gate performs the logical OR function:

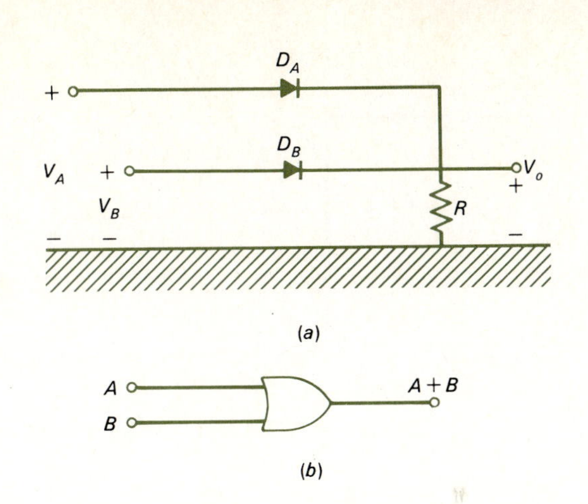

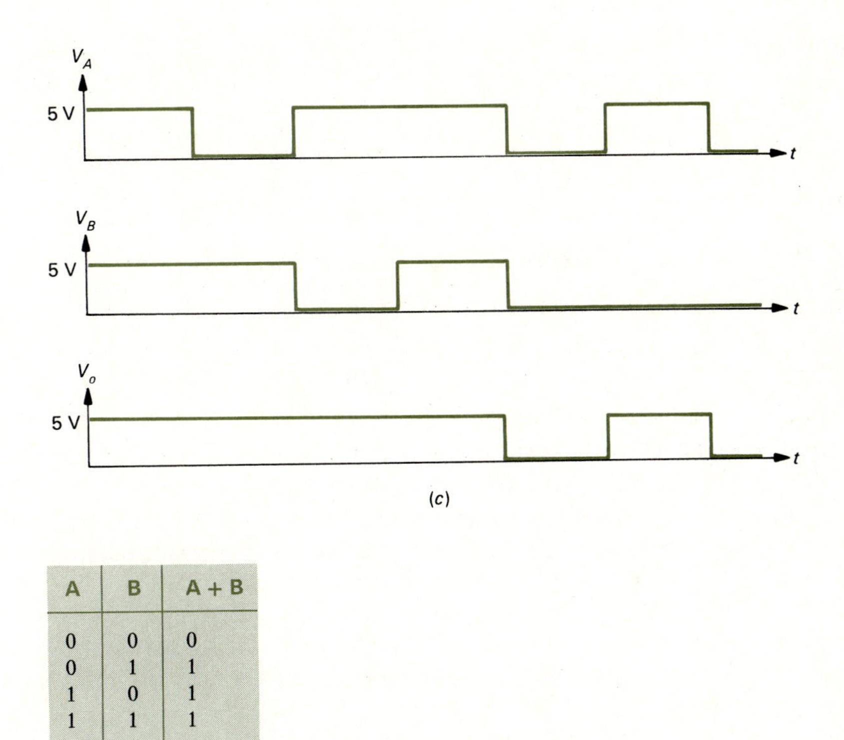

FIGURE 11.2 The OR gate. (*a*) Diode implementation, (*b*) symbol, (*c*) digital operation.

A	B	A + B
0	0	0
0	1	1
1	0	1
1	1	1

The OR function is denoted with a plus (+): $A + B$. Typical input bit streams to the OR gate are shown in Fig. 11.2*c* along with the resulting output.

Although these gates are conceptually simple, they have an important problem. In the illustration we have used ideal diodes, but when we construct these gates with

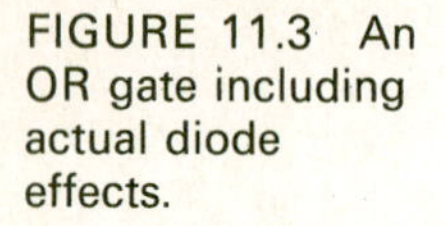
FIGURE 11.3 An OR gate including actual diode effects.

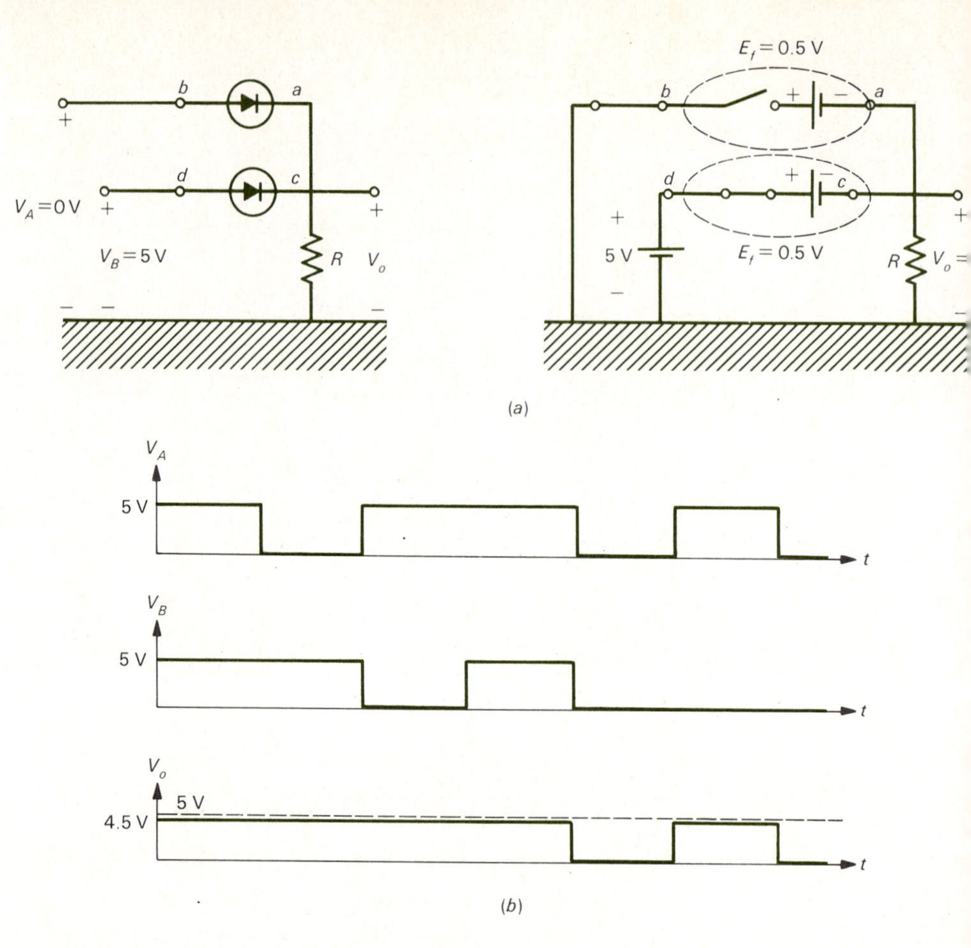

actual diodes, a drop of approximately 0.5-V (actually closer to 0.7-V) will appear across a diode when it is closed. Thus, when the output of the OR gate in Fig. 11.2*c* is supposed to be 5 V, it will instead be 4.5 V, as shown in Fig. 11.3. When these gates are cascaded, the error eventually reaches a point where the output level has dropped such that a logical 1 is interpreted as a logical 0. The AND gate has the reverse problem; namely, 0.5 V is added to the input voltage, so that a 0-V logical 0 at the input becomes a 0.5-V level at the output of the gate. In a later section we will see a way of overcoming this propagating error by using transistors to construct these gates.

A NOT gate can be constructed by using the inherent inversion property of a transistor, as shown in Fig. 11.4*a*; the symbol is shown in Fig. 11.4*b*. If V_A is 0 V (*A* is grounded), the resistors R_{B1} and R_{B2} and the battery V_{BB} produce a negative base voltage on the transistor, so that the base-emitter junction is reverse-biased (for this *npn* BJT). Thus, the base current is approximately zero, as is the collector current ($i_C = \beta i_B$), and thus $V_o = 5$ V. If $V_A = 5$ V and R_{B1}, R_{B2}, and V_{BB} are properly chosen, the base voltage will be greater than the turn-on voltage of the base-emitter junction

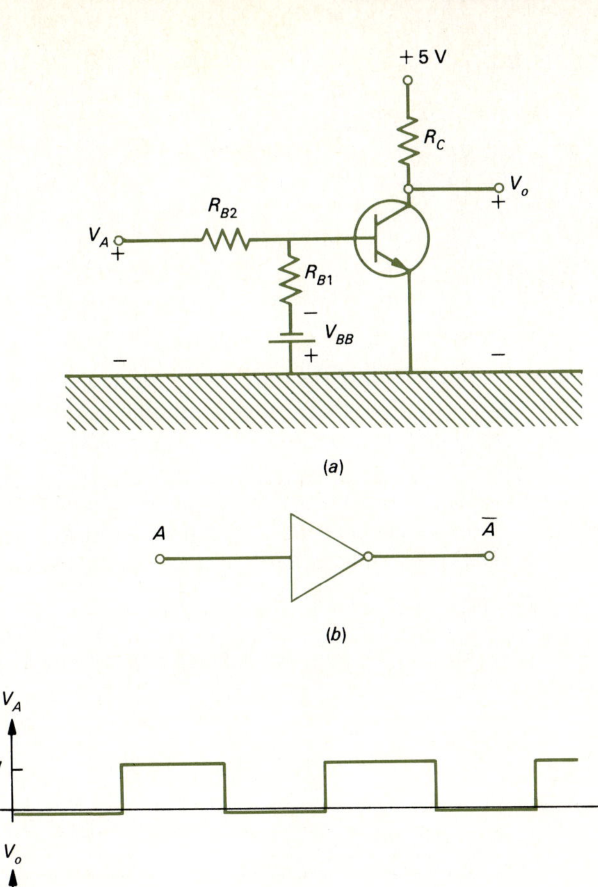

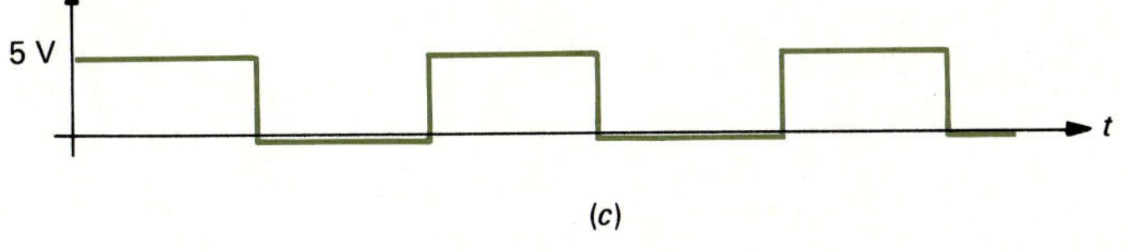

FIGURE 11.4 The NOT gate. (*a*) BJT implementation, (*b*) symbol, (*c*) digital operation.

(approximately 0.5 V) and base current will flow. Thus, collector current will flow and $V_o = 5\text{ V} - R_C i_C$. If enough collector current is caused to flow, $V_o = 0$ V and the NOT operation results. The NOT function is denoted with a bar over the operation. Typical bit streams are shown in Fig. 11.4*c*.

The NAND gate

A	B	$\overline{A \cdot B}$
0	0	1
0	1	1
1	0	1
1	1	0

and the NOR gate

A	B	$\overline{A+B}$
0	0	1
0	1	0
1	0	0
1	1	0

can be constructed by feeding the output of a diode AND or OR gate to the transistor NOT gate. The symbol for these gates is similar to the AND and OR gate symbols except that a small circle is placed on the output, as was the case for the NOT gate.

Example 11.1

The diode AND gate can be used as a pulse-code modulator. In this application the signal is sampled over equally spaced intervals of time and the samples are then processed by a digital circuit. For the AND gate, suppose that V_A is a 3-V 1-kHz sinusoid

$$V_A = 3 \sin 2000\pi t \qquad \text{V}$$

and that V_B is a sequence of 6-V pulses with duration 0.1 ms and separated by 0.1 ms. Sketch the output of the AND gate.

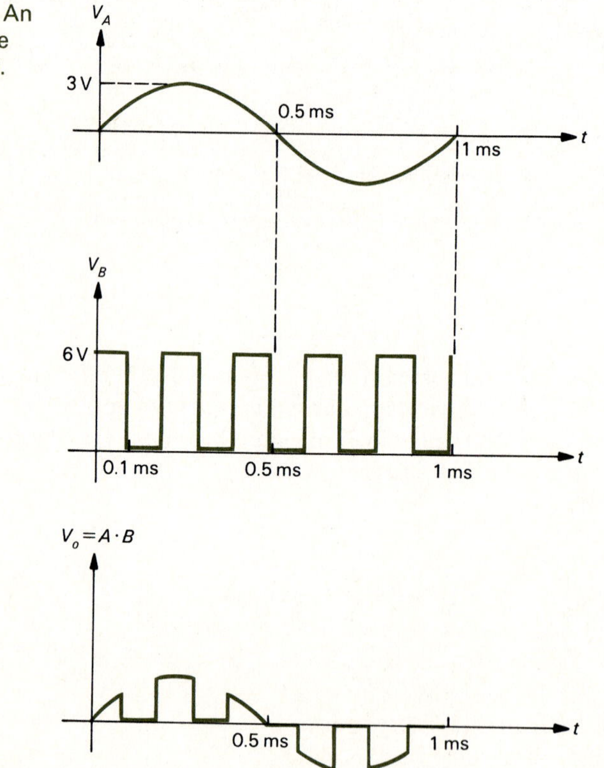

FIGURE 11.5 An AND gate, pulse code modulator.

Solution If either input to the AND gate in Fig. 11.1*a* is less than the battery voltage, 5 V, then the associated diode will be forward-biased and closed. Thus, the output will be connected to that input. When a 6-V pulse is present at *B*, D_B is open, and when the 6-V pulse is absent at *B*, D_B is closed. By comparing the two signals, we obtain the sampled output shown in Fig. 11.5.

Suppose the sinusoid has a peak value of 10 V. What changes would occur?

11.2 THE BJT SWITCH

Transistors (BJT or FET) can also be used to construct gates. They have the ability to restore the proper logic levels in a cascade of diode gates which has been degraded by the accumulation of 0.5-V drops across those diodes which are closed. Cascading a set of diode gates has another problem—impedance loading. If a diode gate is to derive successive diode gates, the input impedances to those successive gates must be much larger than *R* in this first gate in order that the next gate will not affect the operation of the first. Transistor gates also tend to remedy this loading problem to some degree.

The operation of a BJT switch is summarized in Fig. 11.6. Writing KVL around the collector-emitter loop yields $v_{CE} = V_{CC} - R_C i_C$, from which the load line may be drawn on the collector-emitter characteristic. Similarly, writing KVL around the base-emitter loop yields $v_{BE} = V_i - R_B i_B$, and the corresponding load line may be drawn on the base-emitter characteristic. Suppose the input is at 0 V (logical 0); then the base current is zero and the operating point is at ①. At this point the transistor is said to be *cut off*, or simply *off*, and only a small value of collector current flows ($I_{C,\text{cutoff}} \doteq I_{CEO}$). Thus, in the cutoff state

$$\begin{aligned} V_o &= V_{CE,\text{cutoff}} \\ &\doteq V_{CC} - R_C I_{CEO} \\ &\doteq V_{CC} \\ &= 5\text{ V} \end{aligned} \tag{11.1}$$

and the output is at the logical 1 level.

Now suppose that V_i changes to +5 V (logical level). Base current flows which is determined from

$$I_{B,\text{sat}} = \frac{V_i - v_{BE}}{R_B} \tag{11.2}$$

where $v_{BE} \doteq 0.7$ V. Note that although the base-emitter characteristic breaks at about 0.5 V, the actual value of v_{BE} is more like 0.7 V. We will refer to this as the threshold voltage $V_T = 0.7$ V. Previously, we ignored this practicality since this base-emitter voltage turned out to be negligible in our dc amplifier calculations. But here the voltage levels are much lower and the true value of v_{BE} (0.7 V) should be used. This applies to the diode gates, too. Now suppose that R_B is chosen such that $I_{B,\text{sat}} = 50\ \mu$A. The

FIGURE 11.6
The BJT switch.
(*a*) Circuit,
(*b*) typical operation using load lines.

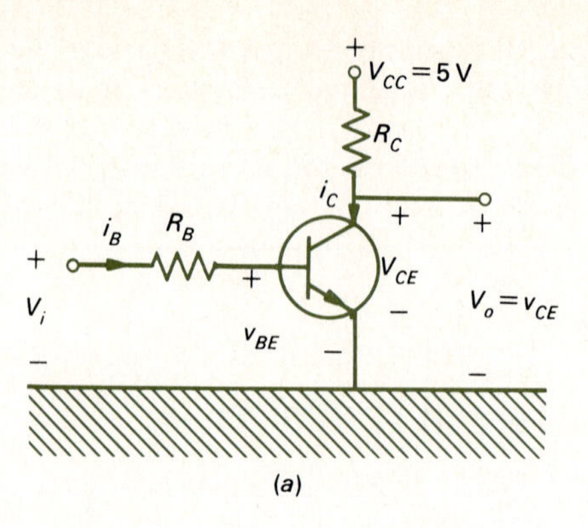

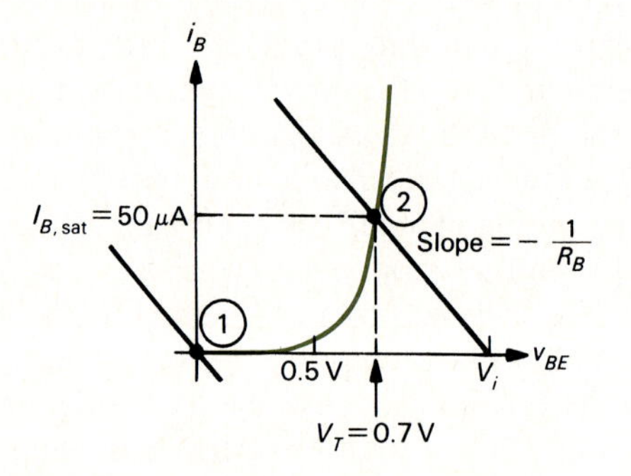

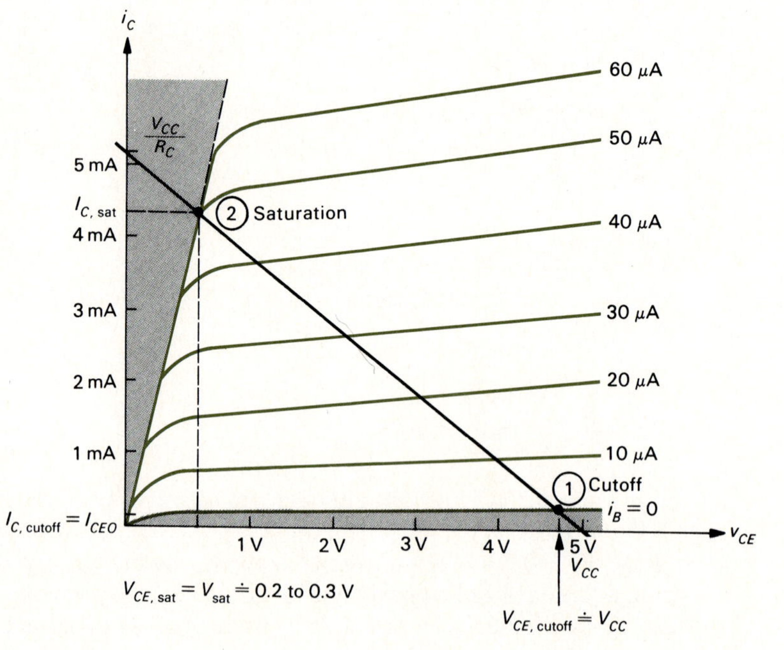

(*b*)

FIGURE 11.7
Modeling the BJT switch: (*a*) cutoff (off), (*b*) saturation (on).

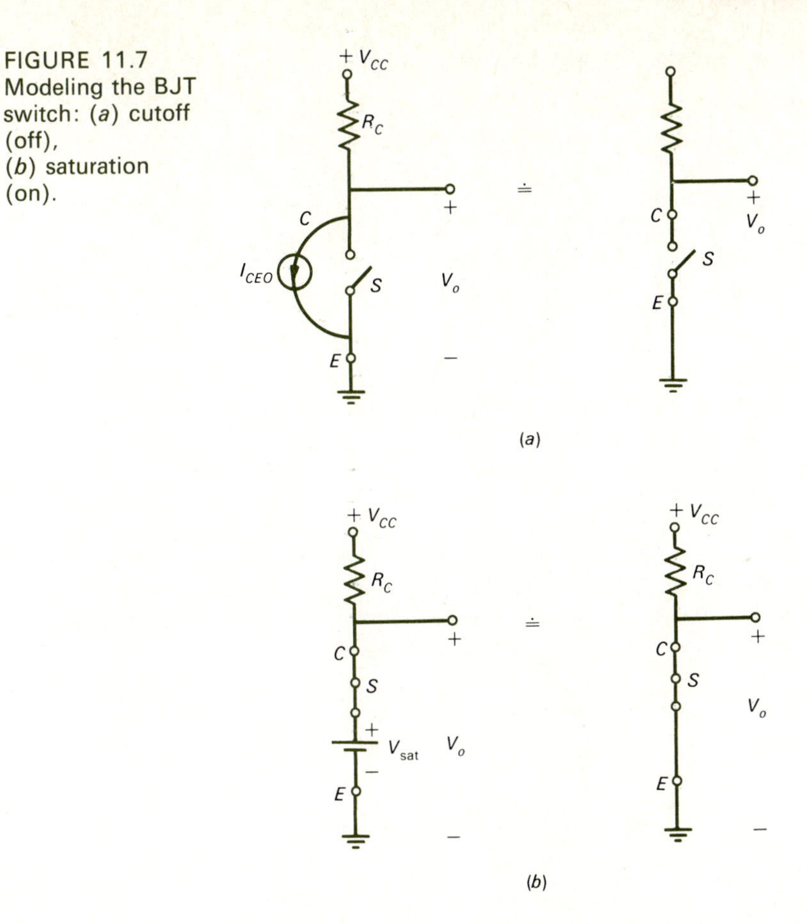

operating point "switches" to point ② on the collector-emitter characteristic and the transistor is said to be *saturated*, or simply *on*. In this saturated state $V_{CE,\text{sat}} = V_{\text{sat}}$, which is typically 0.2 to 0.3 V, depending on i_B. Note that increasing i_B further by reducing R_B causes no additional change. The collector current in saturation is

$$I_{C,\text{sat}} = \frac{V_{CC} - V_{\text{sat}}}{R_C}$$

$$\doteq \frac{V_{CC}}{R_C} \tag{11.3}$$

Thus, the transistor behaves like an ideal switch, as shown in Fig. 11.7.

In the above analysis, we have assumed that R_B was small enough such that $i_B \geq 50\ \mu\text{A}$ so as to drive the transistor into saturation. Increasing i_B further gains nothing; the BJT operating point remains at ②. But we must be sure that i_B is sufficiently large. Since $i_C \doteq \beta i_B + I_{CEO}$, we thus must have (neglecting I_{CEO}) $I_{B,\text{sat}} > I_{C,\text{sat}}/\beta$. But i_C at saturation is $I_{C,\text{sat}} = (V_{CC} - V_{\text{sat}})/R_C$. Thus, $I_{B,\text{sat}} > (V_{CC} - V_{\text{sat}})/\beta R_C$.

Also, $I_{B,\text{sat}} = (V_i - V_T)/R_B$. Therefore, saturation will occur when

$$\frac{V_i - V_T}{R_B} > \frac{V_{CC} - V_{\text{sat}}}{\beta R_C} \tag{11.4}$$

or

$$V_i > (V_{CC} - V_{\text{sat}})\frac{R_B}{\beta R_C} + V_T \tag{11.5}$$

Note that the power dissipated in the transistor, $p = v_{CE}i_C + v_{BE}i_B \doteq v_{CE}i_C$, is approximately zero (or at least very small) in either cutoff or saturation. However, power is expended in switching through the linear, or active, region from one state to the other.

Example 11.2 The transistor switch in Fig. 11.6*a* is to be designed to operate in saturation and in cutoff when a pulse signal which varies between 0 and 5 V is applied to the input. The width of the pulses is 5 μs, and 5 μs separates the pulses. The supply voltage is $V_{CC} = 5$ V and $R_C = 500\ \Omega$. Determine the minimum value of R_B and sketch the output-voltage waveform. Assume an ideal transistor with $\beta = 100$, $V_T = 0.7$ V, $V_{\text{sat}} = 0.2$ V, and $I_{CEO} = 0.1$ mA.

Solution When the signal is zero, the transistor is cut off and $I_{C,\text{cutoff}} = I_{CEO} = 0.1$ mA. Thus, $V_o = V_{CC} - R_C I_{CEO} = 4.95$ V. From Fig. 11.6 and with the transistor in saturation

$$I_{C,\text{sat}} = \frac{V_{CC} - V_{\text{sat}}}{R_C} = \frac{5 - 0.2}{500} = 9.6 \text{ mA}$$

But

$$i_C \doteq \beta i_B + I_{CEO}$$

so that

$$I_{B,\text{sat}} \doteq \frac{I_{C,\text{sat}} - I_{CEO}}{\beta} = 95\ \mu\text{A}$$

For the transistor to be in saturation, we must have $I_B > I_{B,\text{sat}}$. R_B can be found from

$$I_B = \frac{V_i - V_T}{R_B} \geq I_{B,\text{sat}}$$

or

$$R_B \le \frac{V_i - V_T}{I_{B,\text{sat}}}$$

$$= 45.26\ \text{k}\Omega$$

The resulting output voltage is sketched in Fig. 11.8.

The transfer characteristic relating V_i and V_o is shown in Fig. 11.9. A plot of a typical input waveform and the resulting output-voltage waveform is shown in Fig. 11.10. Note the inherent inversion property of the switch. Also note the important distortion of the waveform. The collector-current waveform is shown with important time parameters defined on it. With $V_i = 0$ V initially, the transistor is off, $V_O \doteq 5$ V, and $I_{C,\text{cutoff}} = I_{CEO} \doteq 0$. When V_i abruptly changes from 0 to 5 V, the output voltage and collector current do not initially react; there is a certain amount of delay t_d for a

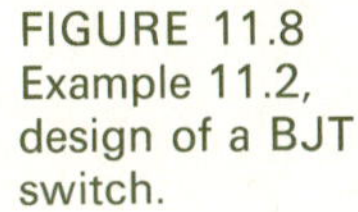

FIGURE 11.8 Example 11.2, design of a BJT switch.

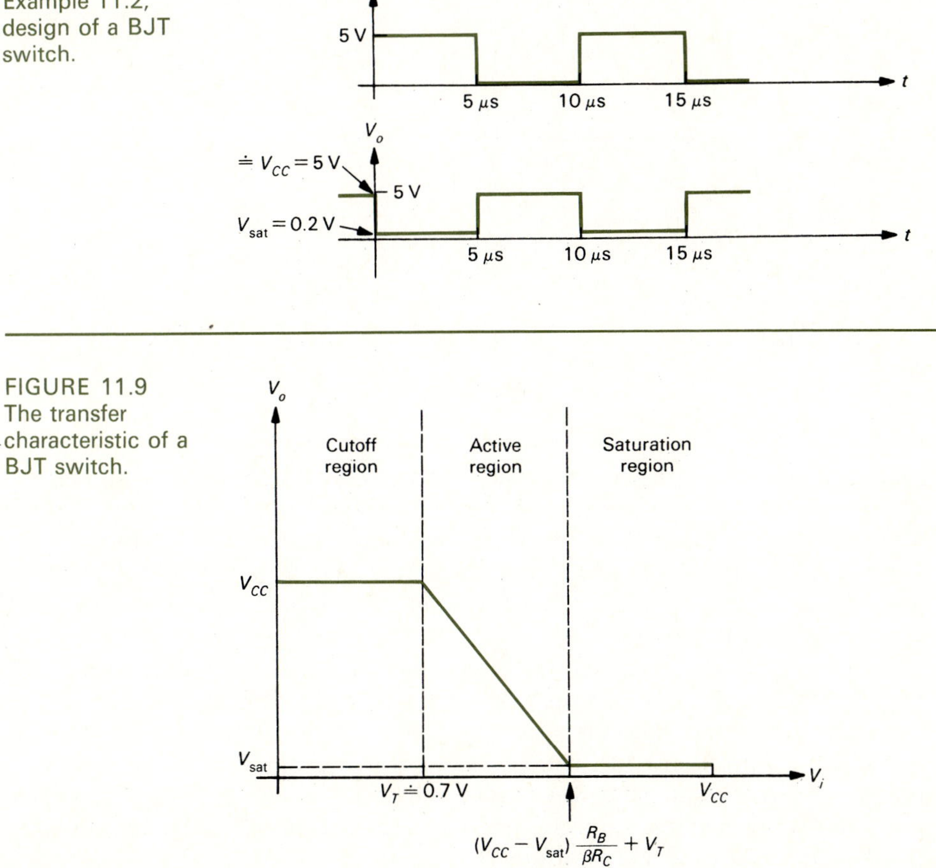

FIGURE 11.9 The transfer characteristic of a BJT switch.

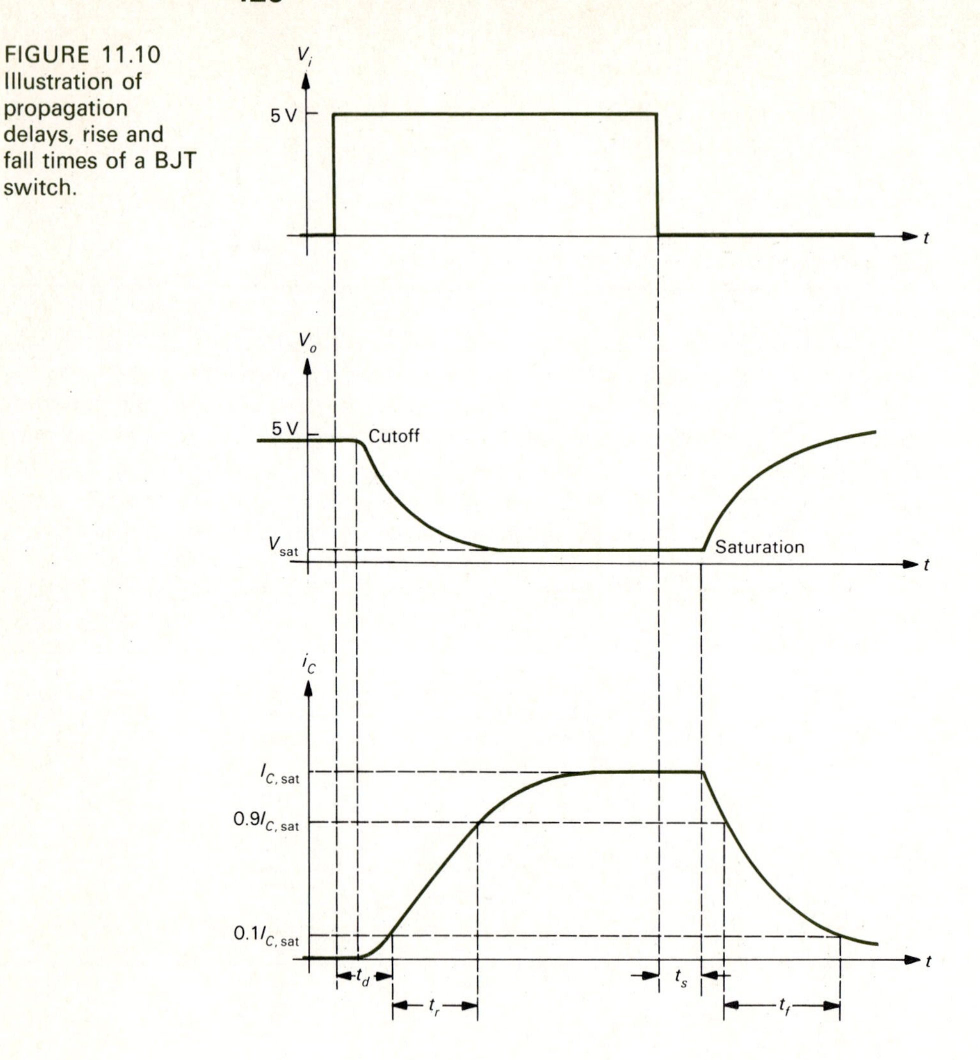

FIGURE 11.10 Illustration of propagation delays, rise and fall times of a BJT switch.

change to occur. This delay is referred to as propagation delay and is the time to change from zero to 10 percent of the final value. (Sometimes propagation delay is defined as the time between the 50 percent level of V_i and the 50 percent level of V_o.) The rise time t_r is the time required to change from 10 to 90 percent of the final level. The various capacitances inherent in the transistor, as well as other stray capacitances, influence these times.

Once the transistor is established in saturation and the input pulse returns to 0 V (logic level 0), there is again a propagation delay t_s as well as a fall time t_f required for the output voltage and collector current to change state. The propagation delay t_s is a result of the time required to remove charge stored in the base region before the transistor begins to switch out of saturation and is usually longer than t_d. The fall time is the time required to switch through the active region from saturation to cutoff; this, too, is influenced by the transistor capacitances.

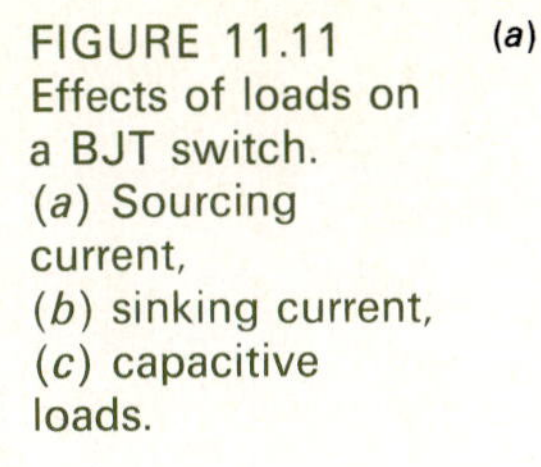

FIGURE 11.11
Effects of loads on a BJT switch.
(*a*) Sourcing current,
(*b*) sinking current,
(*c*) capacitive loads.

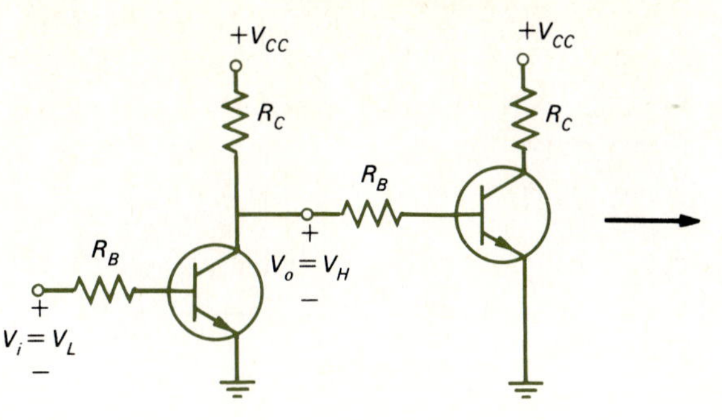

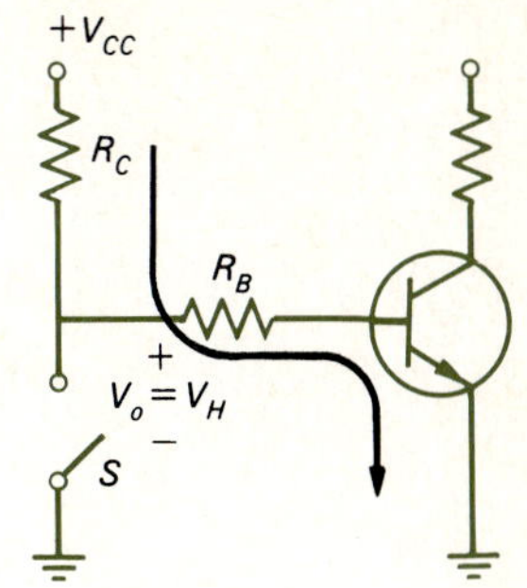

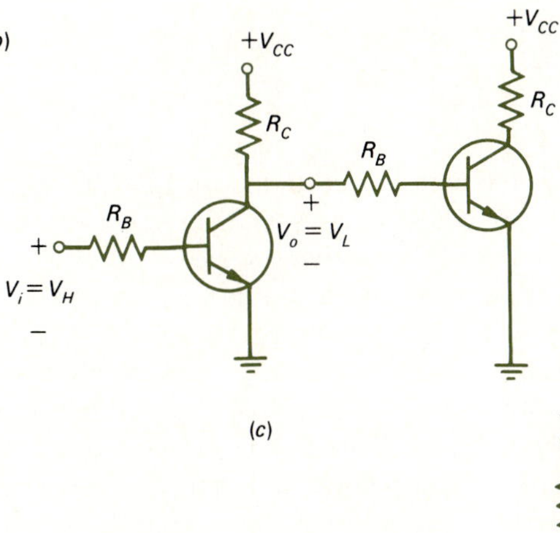

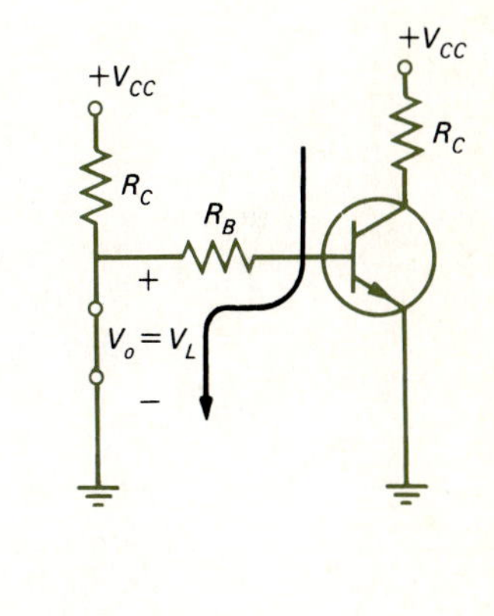

(*c*)

These propagation delays and the rise and fall times are important parameters influencing the design of a digital circuit. They are grouped under the category of switching speed.

Another important performance parameter is fan-out, which refers to the maximum number of switches that may be *driven* by a switch. The importance of this fan-out restriction is illustrated in Fig. 11.11. When the input is low ($V_i = V_L$) at the logic level 0, the output is high ($V_o = V_H$); in this state the transistor is supplying, or sourcing, current to the next switch. When the state is reversed so that the output is in the low state, $V_o = 0$ and the switch is drawing, or sinking, current from the next switch. The maximum number of switches which can *precede* a switch is referred to as fan-in. Other factors affecting fan-out are the input capacitances of the driven stages, as

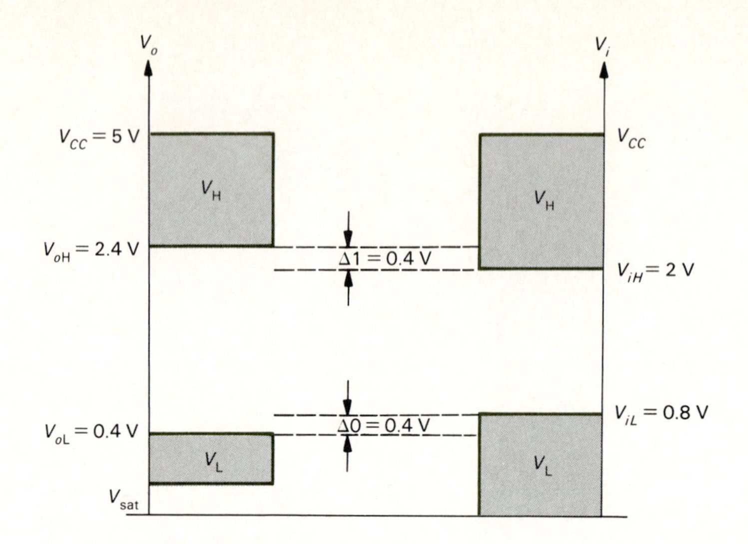

FIGURE 11.12 Illustration of noise margins.

shown in Fig. 11.11*c*. If one stage has an input capacitance of 8 pF, for instance, then 10 stages connected in parallel will have a net input capacitance of 80 pF, and this will affect the rise and fall times of the driving stage.

The third important performance parameter is noise margin. This is illustrated in Fig. 11.12. The logic levels are not defined precisely at 0 and 5 V but occupy acceptable ranges. For example, the manufacturer may specify that the output of a switch which is not driving others will not fall below 2.4 V in the high state or rise above 0.4 V in the low state. Similarly, the driven stage is specified as interpreting a level as a logical 0 if the input voltage is below 0.8 V and as a logical 1 if the input voltage is above 2 V. Thus, if the output of one stage is at 0.4 V, the next stage would interpret that as low, since it is less than 0.8 V. Suppose that noise added 0.4 V to the output of the first stage; then the next stage would still interpret this as low, but the addition of any more noise could result in an incorrect interpretation of the level by the second stage. These differences in manufacturer specified logic levels

$$\Delta 1 = V_{oH} - V_{iH}$$

$$\Delta 0 = V_{iL} - V_{oL}$$

are called noise margins.

11.3 LOGIC FAMILIES

Logic gates to perform the basic logic functions (AND, OR, NAND, NOR, NOT) can be constructed with the transistor switch. To some degree, these gates overcome the disadvantages of the simple diode gates. In designing a digital system it is generally not necessary to "design" the individual gates; these have already been designed and are available in the form of small packages—such as the dual in-line package, or DIP, shown in Fig. 11.13*a*.

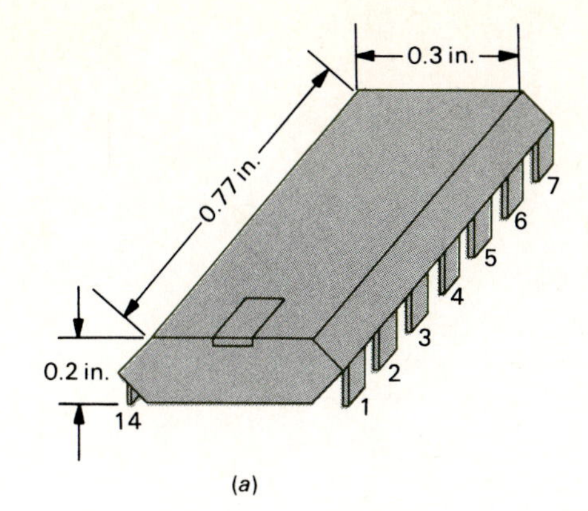

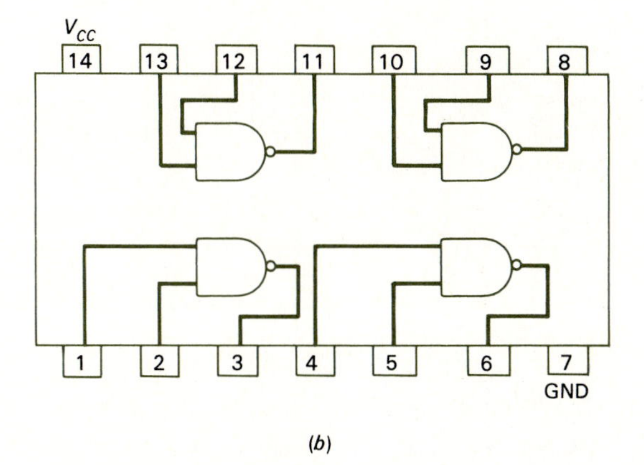

FIGURE 11.13 Packaging of digital circuits. (*a*) dual inline package (DIP) (*b*) schematic of quad NAND gate package.

A typical schematic of a unit containing several gates is shown in Fig. 11.13*b*. The "designer" merely needs to connect the proper supply voltage to the unit, observe fan-out restrictions (the maximum number of gates which this gate may drive), fan-in restrictions (the maximum number of gates which may drive this device), and propagation delays, and ensure that the gates are properly connected so that the intended logic function will be performed—a topic covered in Part 4. Gates of the same logic family can be interconnected in this fashion since they have the same logic voltage levels, impedance characteristics, and switching times. Several logic families are discussed in this section.

A gate using resistor-transistor logic (RTL) is shown in Fig. 11.14. Suppose V_A is high but V_B and V_C are low. In this case, T_1 is on and T_2 and T_3 are off, so that V_o is low. If both V_A and V_B are high, V_o is also low. V_o is high only if V_A, V_B, and V_C are all low. Thus, this gate performs the NOR function. The noise margins as well as the switching speeds of the RTL family are low. The fan-out is usually limited to about five gates.

Diode transistor logic (DTL) NOR and NAND gates are shown in Fig. 11.15. These are obtained by connecting the appropriate OR or AND diode gate to a transistor switch (inverter). The noise margins and fan-out of DTL are generally better than RTL, but the switching speeds are about the same.

FIGURE 11.14 A resistor transistor logic (RTL) gate.

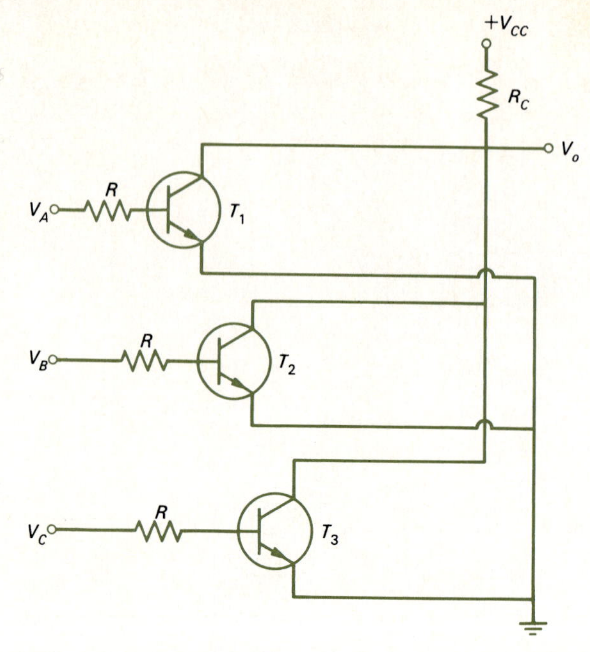

FIGURE 11.15 Diode transistor logic (DTL) gates. (*a*) NOR, (*b*) NAND.

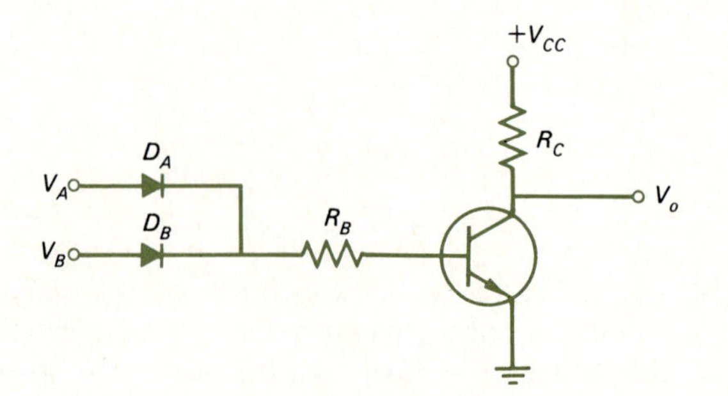

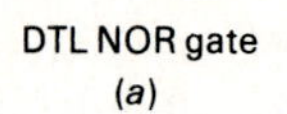

DTL NOR gate
(*a*)

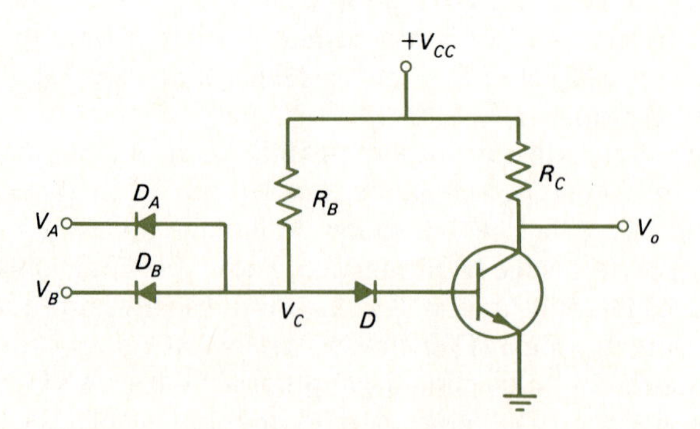

DTL NAND gate
(*b*)

Example 11.3 For the DTL NOR gate in Fig. 11.15*a*, assume diodes with $V_T = 0.7$ V, $V_{CC} = 5$ V, $R_B = 12$ kΩ, and $R_C = 500$ Ω. For the transistor, assume $\beta = 35$, $V_T = 0.7$ V, $I_{CEO} = 0$, and $V_{sat} = 0.2$ V. If $V_A = 5$ V and $V_B = 0$ V, determine V_o. Repeat these calculations if $V_A = V_B = 0$ V.

Solution With $V_A = 5$ V and $V_B = 0$ V, D_A will be closed and D_B will be open. Thus, the base current is

$$i_B = \frac{V_A - 0.7 - 0.7}{R_B}$$

$$= 0.3 \text{ mA}$$

For the transistor to be in saturation, i_C must be greater than

$$I_{C,\text{sat}} = \frac{V_{CC} - V_{\text{sat}}}{R_C}$$

$$= 9.6 \text{ mA}$$

But $i_C \doteq \beta i_B$, so

$$i_C = 35 i_B$$

$$= 10.5 \text{ mA}$$

and the transistor is in saturation, with

$$V_o = V_{\text{sat}}$$

$$= 0.2 \text{ V}$$

If $V_A = V_B = 0$, both diodes will be open and $i_B = 0$. Thus, the transistor is cut off and $V_o \doteq V_{CC}$. Thus, this functions as a NOR gate.

In recent years, the most popular BJT logic family has been transistor-transistor logic (TTL). A basic TTL NAND gate is shown in Fig. 11.16*a*. Transistor T_1 is a multiple-emitter *npn* BJT, which acts as an AND gate. Replacing the base-emitter and base-collector junctions with diodes, we arrive at Fig. 11.16*b*, With V_A, V_B, and V_C high, all three diodes D_A, D_B, and D_C are reverse-biased and open. Current flows through R_B and D to saturate T_2, which saturates T_3. Adding the base-emitter drops of T_2 and T_3, the base voltage of T_2 becomes $2V_T = 1.4$ V. Adding the 0.7-V drop of T_1 across the base-collector diode D gives the base current to T_2 of $(V_{CC} - 2.1 \text{ V})/R_B$. With T_2 on, the base voltage of T_4 is too low to cause it to saturate, and T_4 is off. Thus, with all inputs in the high state, V_o will be in the low state.

Suppose that at least one input, say V_A, is in the low state. The associated base-emitter diode will be closed and the base voltage of T_1 will be $V_A + 0.7 \text{ V} = 0.7$ V. Combined with the 0.7-V drop across D, the base voltage of T_2 will be $-0.7 + 0.7 = 0$ V, so that T_2 is cutoff. This serves to cut off T_3 and the output is high. Since T_2 is cut off, its collector voltage rises, turning T_4 on.

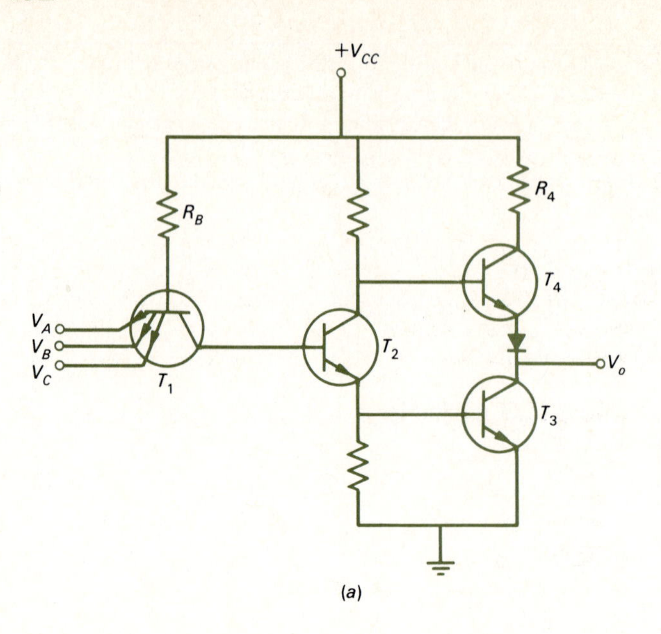

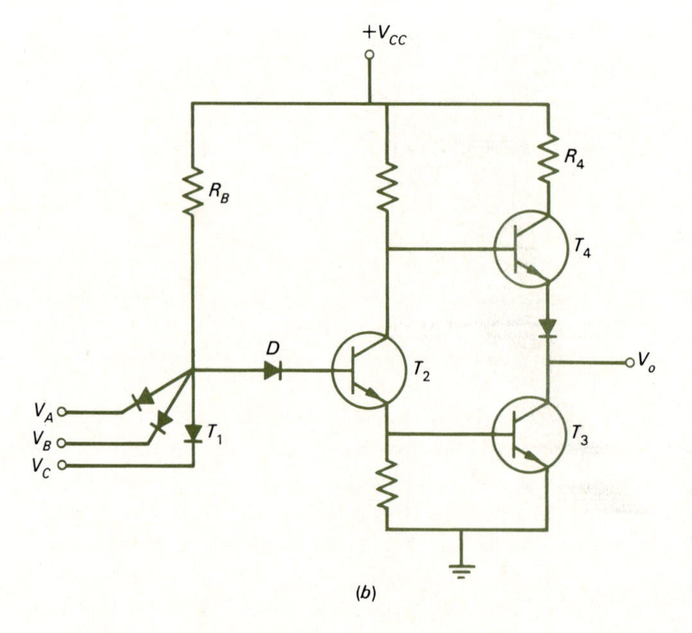

FIGURE 11.16 The transistor transistor logic (TTL) gate. (*a*) Physical schematic, (*b*) diode replacement of three-input transistor.

The combination of R_4 and T_4 acts as a variable resistance. When T_3 is on, T_4 is off, lowering the power consumption; but when T_3 is off, T_4 is on, which provides a low resistance seen by the succeeding gate, thus improving the switching time (see Fig. 11.11). The diode between T_4 and T_3 serves to add a voltage drop to help ensure that T_4 is off when T_3 is on.

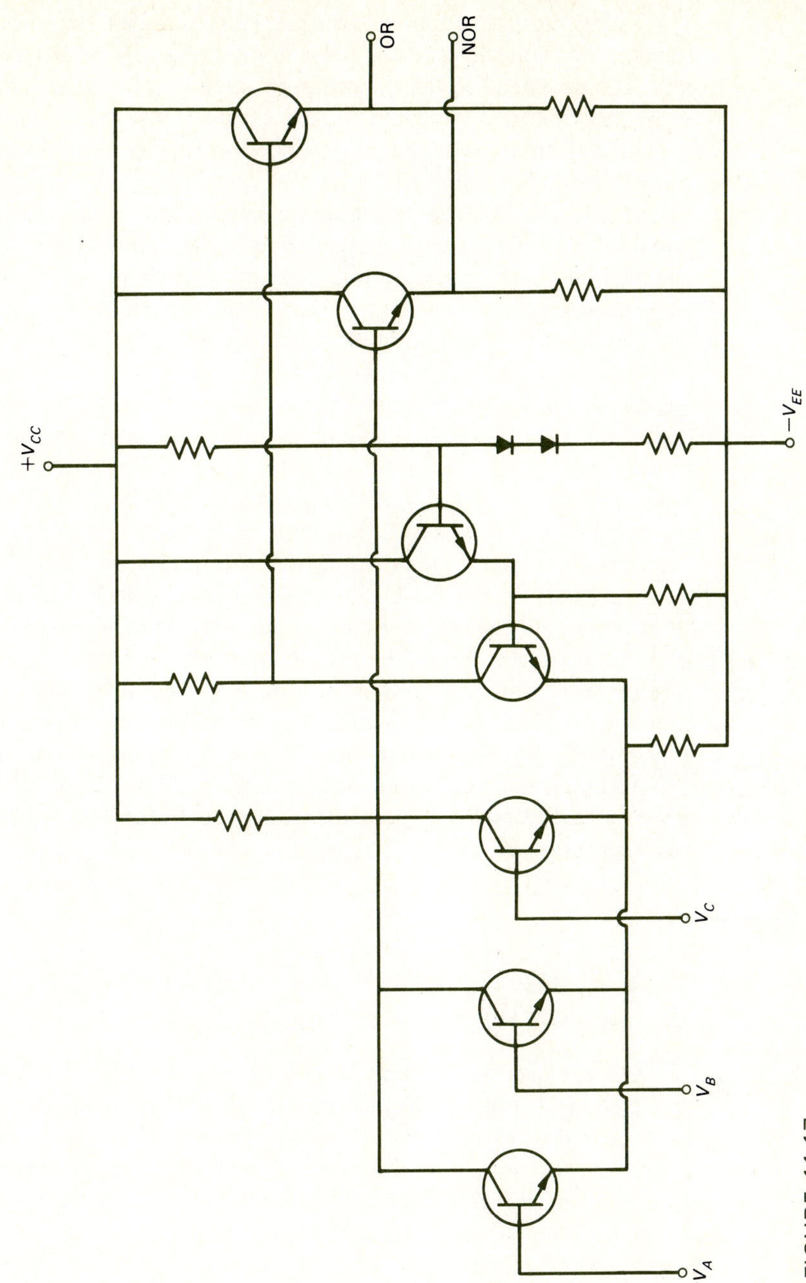

FIGURE 11.17 The emitter coupled logic (ECL) gate.

TTL is the most popular of the members of the bipolar logic families. The fan-out is quite large (on the order of 10 or more gates may be driven by one TTL NAND gate), the propagation delays are quite small [on the order of 2 to 10 nanoseconds (ns)], and the power consumption is typically 2 mW. By contrast, DTL has a typical fan-out of 8 to 10, a propagation delay on the order of 30 to 90 ns, and a power consumption of around 15 mW.

One of the reasons for the popularity of TTL over DTL is its higher speed. A primary reason for the speed restrictions in DTL and TTL is that the transistors are switched into saturation and time is required to remove the stored charge. (Driving the transistor into saturation with more than the minimum required collector current $I_{C,\text{sat}}$ accomplishes nothing, but it increases the required time to switch back out of saturation.) Part of this problem of switching speed could be eliminated if, in switching, we stayed only in the active region. DTL and TTL could not be reliably operated in this mode since the range of base-emitter voltage required to switch from cutoff to saturation is only a few tenths of a volt (from about 0.4 to 0.7 V). Temperature variations, manufacturer variability, etc., would not allow us to ensure that we would not roam out of cutoff or saturation with an "ideal" design. Furthermore, driving the transistors completely into saturation or cutoff gives reliable logic voltage levels ($V_{CC} = 5$ V or $V_{\text{sat}} \doteq 0.2$ V). If we tried to switch in the active region only, minor variations could cause rather large changes in these logic levels.

The emitter-coupled logic (ECL) gate shown in Fig. 11.17 provides a way of reliably doing this and thus increasing switching times. The name arises from the common attachment of the emitters of the input transistors. The propagation delays are on the order of 1 ns, but the power consumption is quite high (on the order of 25 mW per gate). This latter disadvantage of ECL, combined with relatively small noise margins (<0.3 V), has tended to make TTL—and particularly the high-speed Schottky diode TTL gates—the popular choice.

11.4 COMPLEMENTARY METAL-OXIDE SEMICONDUCTOR (CMOS) GATES

Switches can also be constructed with FETs. A typical FET switch constructed from an n-channel JFET (or a depletion-mode MOSFET) is shown in Fig. 11.18. Voltage V_T is referred to as the threshold voltage (previously called the pinch-off voltage). For $v_{GS} < -V_T$, no drain current flows and the device is cut off. With $V_i = 0$ V, the FET is in saturation, and with V_i more negative than the threshold voltage V_T, the FET is cut off. Here, $V_i = 0$ V may represent a logical 1 and $V_i = -V_p = -V_T$ may represent a logical 0. Thus, the most positive of the two levels represents a logical 1. This is referred to as positive logic. If the most positive level represented a logical 0, this would be referred to as negative logic. The depletion-mode MOSFET in Fig. 11.18 can also be operated in the enhancement mode in which case V_i may go positive.

FET logic circuits can also be used with positive pulses by using enhancement-mode n-channel MOSFETs as switches, as shown in Fig. 11.19. A transfer characteristic for this n-channel enhancement-mode MOSFET is shown; note that the threshold

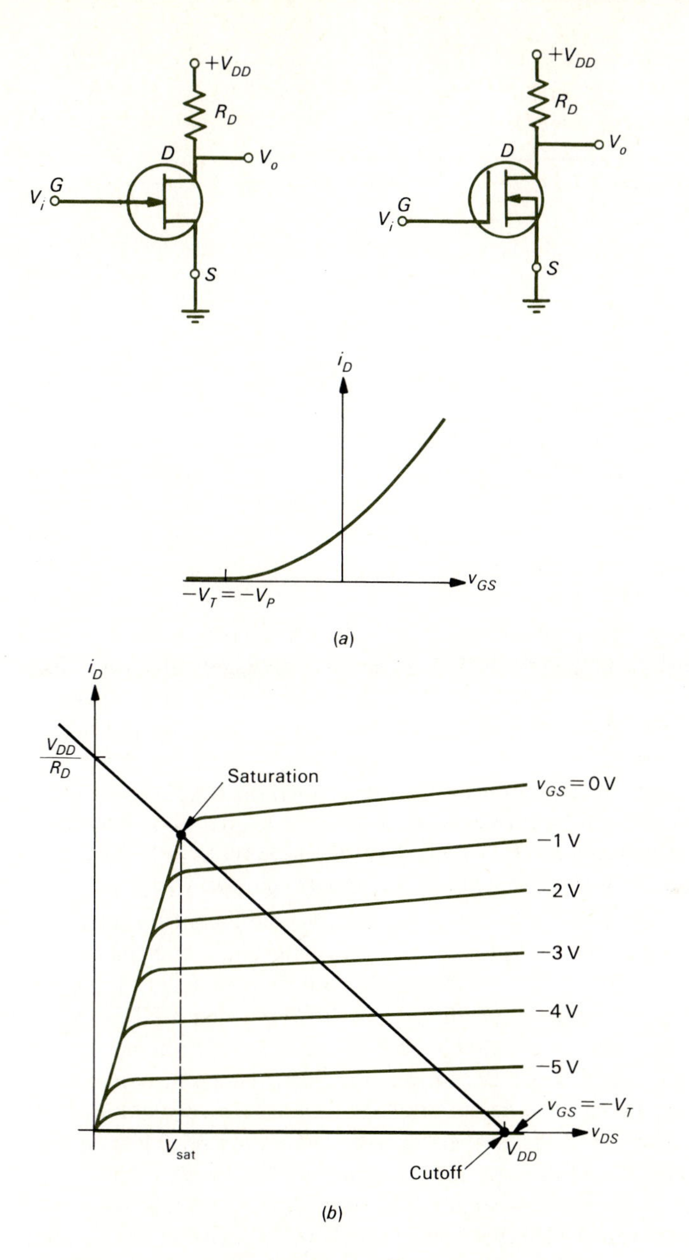

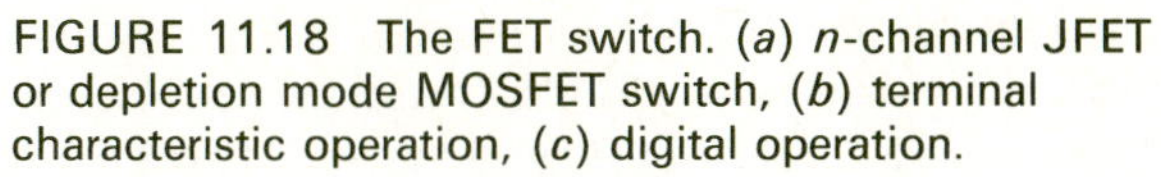

FIGURE 11.18 The FET switch. (*a*) *n*-channel JFET or depletion mode MOSFET switch, (*b*) terminal characteristic operation, (*c*) digital operation.

FIGURE 11.18 (*Continued*)

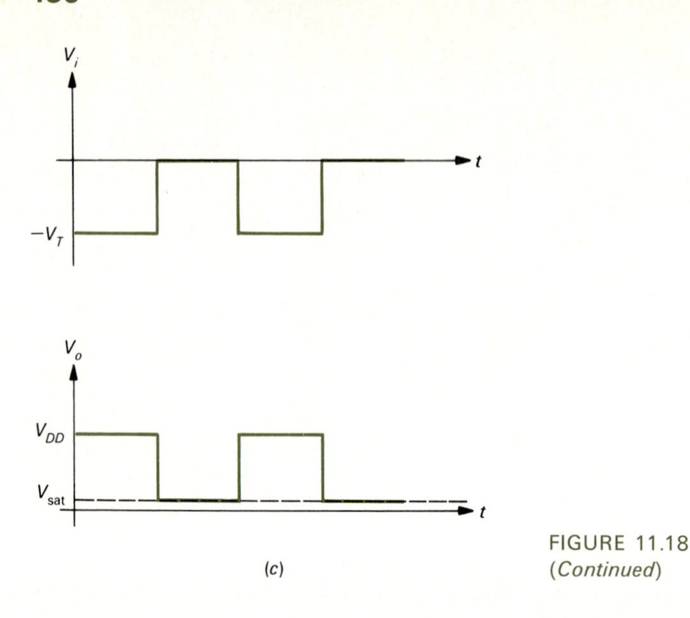

FIGURE 11.18
(*Continued*)

voltage V_T is positive and is typically on the order of a few volts (how many depends on the particular device). A *p*-channel enhancement-mode MOSFET has the characteristics shown in Fig. 11.20. Note that the characteristics of the *p*-channel and *n*-channel devices differ only in that the directions of i_D and the polarities of v_{DS} and v_{GS} are reversed. For the *p*-channel MOSFET, we must make the gate more *negative* than the source terminal by V_T in order to turn the device on. Making the gate more negative causes drain current to flow *from* the source *to* the drain.

FET switches offer important advantages over BJT switches. Since the gate current of a FET is, for all practical purposes, zero, a FET switch does not draw current from a previous stage; consequently, no significant power-loading effects are present. In contrast, BJTs do load down previous stages, and this must be taken into account in the design. Another advantage of FETs is that their logic voltage levels tend to be somewhat higher (typically, $V_{DD} = 15$ V) than with BJTs (typically, $V_{CC} = 5$ V); thus, logic circuits constructed from FETs tend to tolerate more noise than do comparable circuits using BJTs. Because of the larger inherent capacitances of FETs, however, their switching speeds tend to be somewhat slower than for BJTs.

MOSFETs are preferred over JFETs for IC digital circuits. They can be constructed as either a *p*-channel metal-oxide semiconductor (*p*MOS) or an *n*-channel metal-oxide semiconductor (*n*MOS). MOSFETs offer tremendous packing densities, since a typical MOSFET requires about 15 percent of the chip area of a BJT.

The most popular of the MOSFET family (for reasons soon to become apparent) is the complementary MOS, or CMOS (pronounced "see-moss"). A basic CMOS inverter is shown in Fig. 11.21. A *p*MOS and an *n*MOS (both are enhancement-mode) are used. When V_i is low, the gate-source voltage of the *n*MOS is less than the threshold voltage (see Fig. 11.19) and is cut off. The voltage from gate to source of the *p*MOS, however, is $-V_{SS}$, where $V_{SS} > V_T$ is the supply voltage. Thus, the *p*MOS is on (see Fig. 11.20) and the supply voltage appears at the output. When V_i goes to the high state ($V_i = V_{SS}$) the *p*MOS turns off and the *n*MOS turns on, whereupon V_{SS} appears across the drain-source terminals of the *p*MOS and V_o drops to zero.

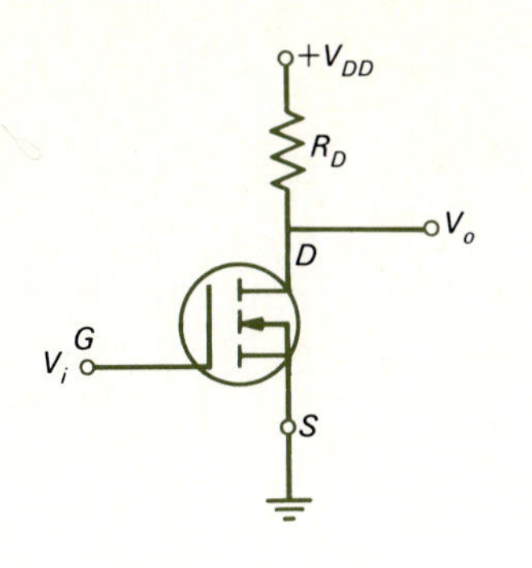

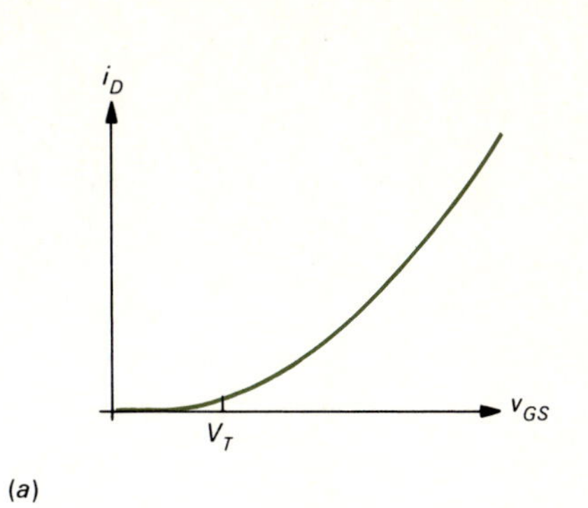

(a)

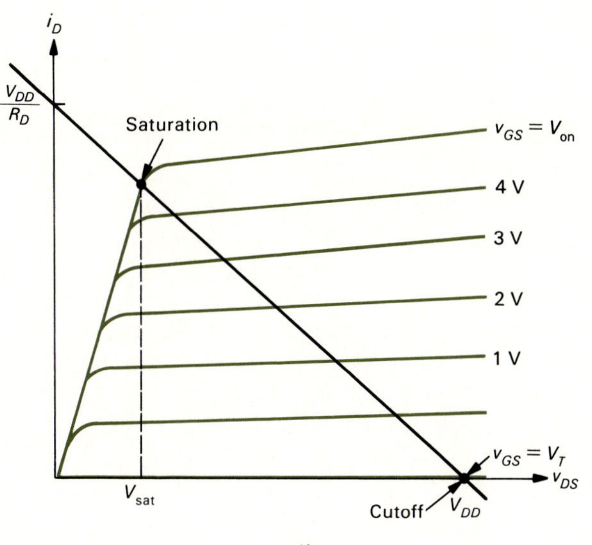

(b)

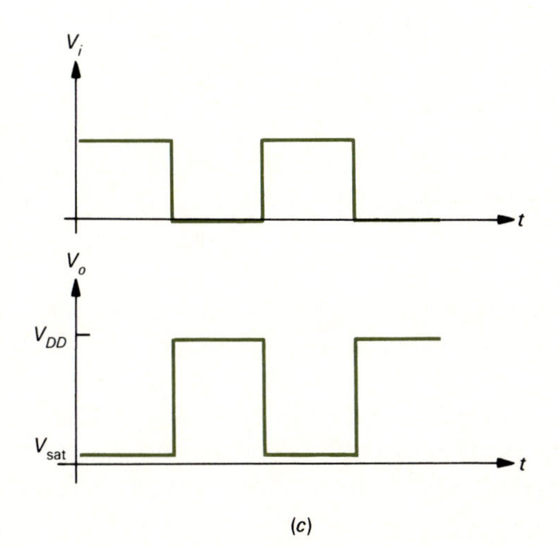

(c)

FIGURE 11.19
The *n*-channel enhancement-mode MOSFET switch.
(*a*) Circuit diagram and transfer characteristic,
(*b*) terminal characteristic and load line,
(*c*) digital operation.

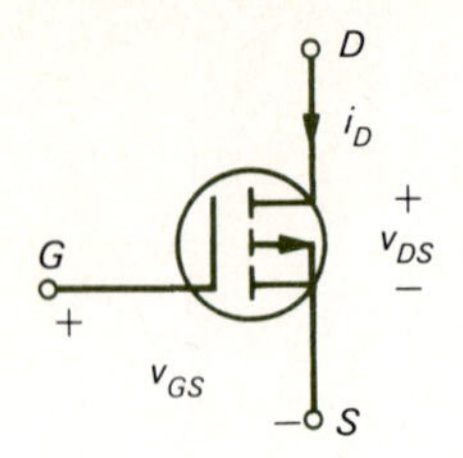

FIGURE 11.20 The p-channel enhancement-mode MOSFET switch.

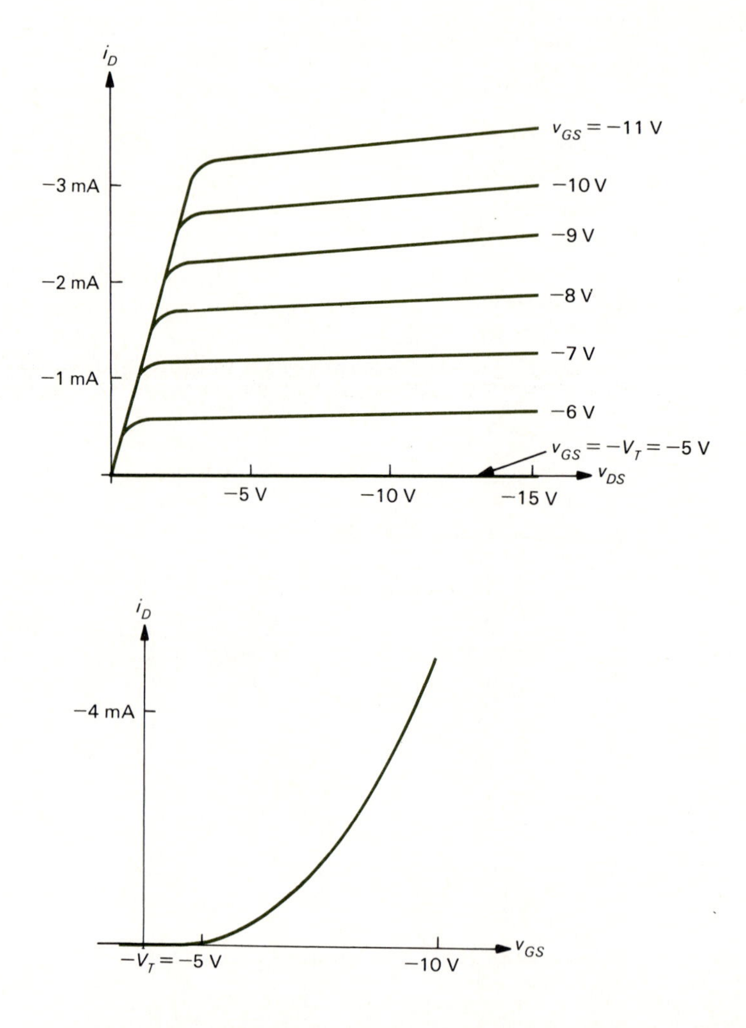

Thus, the circuit functions as an inverter with an important property. Note that when the output is in the low state the nMOS is on but the pMOS is off and virtually no current is drawn from the power supply. On the other hand, when the output is in the high state the pMOS is on but the nMOS is off and once again no power-supply current is drawn. This property of virtually no power consumption, coupled with the small consumption of chip area, makes the CMOS very attractive for such miniature,

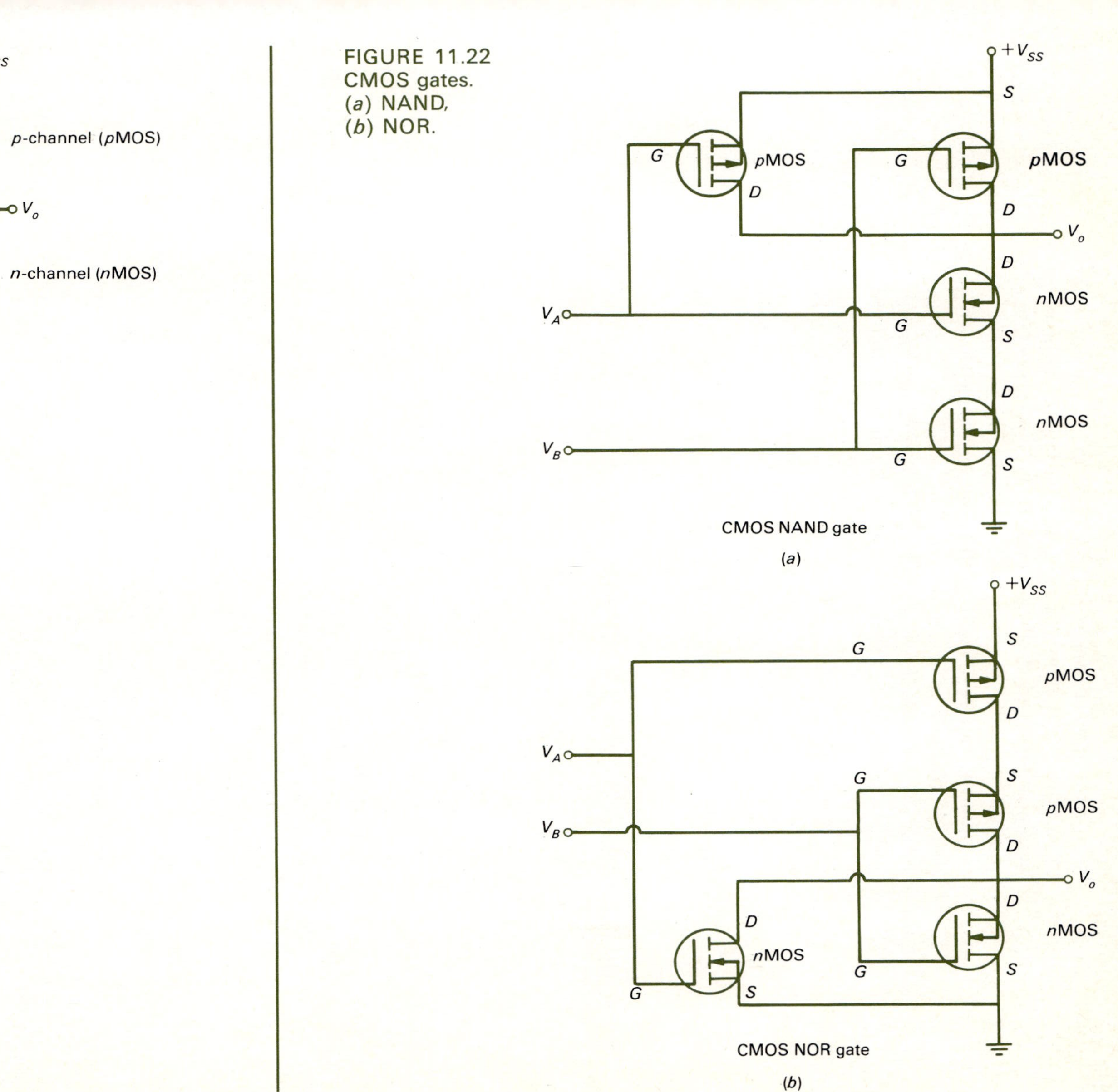

FIGURE 11.21 The complementary symmetry MOSFET (CMOS) switch.

FIGURE 11.22 CMOS gates. (a) NAND, (b) NOR.

low-power applications as wristwatches and calculators. The poor switching speeds (relative to TTL), however, relegate it to low-to-medium speed devices. The power-supply voltage and the logic level in the high state (5 to 15 V) can provide noise margins larger than for comparable TTL gates.

Two input NAND and NOR CMOS gates are shown in Fig. 11.22. For the NOR gate in Fig. 11.22*b*, if V_A and V_B are low, both *n*MOS devices are off but both *p*MOS devices are on and V_o is around V_{SS}. If one of the inputs goes high, the associated *p*MOS device turns off and the corresponding *n*MOS device turns on, and thus V_o drops to the low state. Thus, the function of a NOR gate is produced. For the NAND gate in Fig. 11.22*a*, if at least one input is in the low state the associated *p*MOS device will be on and the *n*MOS device will be off, giving a high output state. If both V_A and V_B are high, both *p*MOS devices will be off while both *n*MOS devices will be on, and V_o will be low. Thus, the device functions as a NAND gate. Neither of these gates draws virtually any power-supply current, so there is virtually no power consumption.

11.5 MULTIVIBRATORS

The above gates can be used to construct *combinational* logic circuits, in which a gate output at a particular time depends only on the gate inputs at that time. *Sequential* logic circuits have memory, in that an output at a particular time depends not only on the inputs at that time but also on inputs at previous times. Both of these types of logic circuits will be discussed in Part 4. Here we will discuss the basic construction of these types of logic devices.

The output of the BJT and FET switches discussed previously has two stable states: with the input in one state (0 or 1), the output is in a well-defined state (1 or 0). Multivibrators are devices in which, as illustrated in Fig. 11.23, (*a*) the output may appear in either of two states for the same input (bistable), (*b*) only one state is stable (monostable), or (*c*) no output state is stable (astable). For the bistable flip-flop, a spike at t_1 causes the output to switch from its present stable state to the other stable state. The device remains in that stable state until another spike is applied at t_2 and the device switches states again. Note that we do not need a long-duration pulse to switch the device; a short-duration spike is sufficient. The monostable multivibrator switches from its present stable state with the application of a spike, but after Δt (an adjustable parameter) it switches states again. The astable multivibrator switches continuously from one state to the other at well-defined times without the application of an input spike. Astables are often called oscillators.

A BJT flip-flop is shown in Fig. 11.24*a*. Resistor R_C is usually chosen much less than R_A. Assume for the moment that T_1 is off and T_2 is on, as shown in Fig. 11.24*b*. The collector C_2 of T_2 is at zero potential (neglecting $V_{\text{sat}} \doteq 0.2\text{ V}$ for T_2). The current entering the base of T_2

$$i_{B2} = \frac{V_{CC} - 0.7\text{ V}}{R_C + R_A}$$

is such that T_2 is kept on. The collector current for T_2 is

$$i_{C2} = \frac{V_{CC}}{R_C}$$

Since C_2 is low, virtually no base current flows into T_1, ensuring its off condition. Assuming T_1 on and T_2 off, one can show that this device can indeed be designed to have two stable states at 0 (and their opposite at $\bar{0}$). Now in order to change states, we simply apply a spike to S (the SET input); this causes base current to flow into T_1, turning it on, and C_1 thus drops, turning T_2 off. Note that additional spikes to S cause no change. To change states, we apply a spike to R (the RESET input), causing T_2 (which was turned off by S) to turn on, which causes T_1 to turn off. Thus, the SET (S) input causes the device to change from its present state ($\bar{0}$ high, 0 low) to its other state ($\bar{0}$ low, 0 high). A pulse must be applied to the RESET (R) input in order to change the device from its present state ($\bar{0}$ low, 0 high) to the other state ($\bar{0}$ high, 0 low). A typical

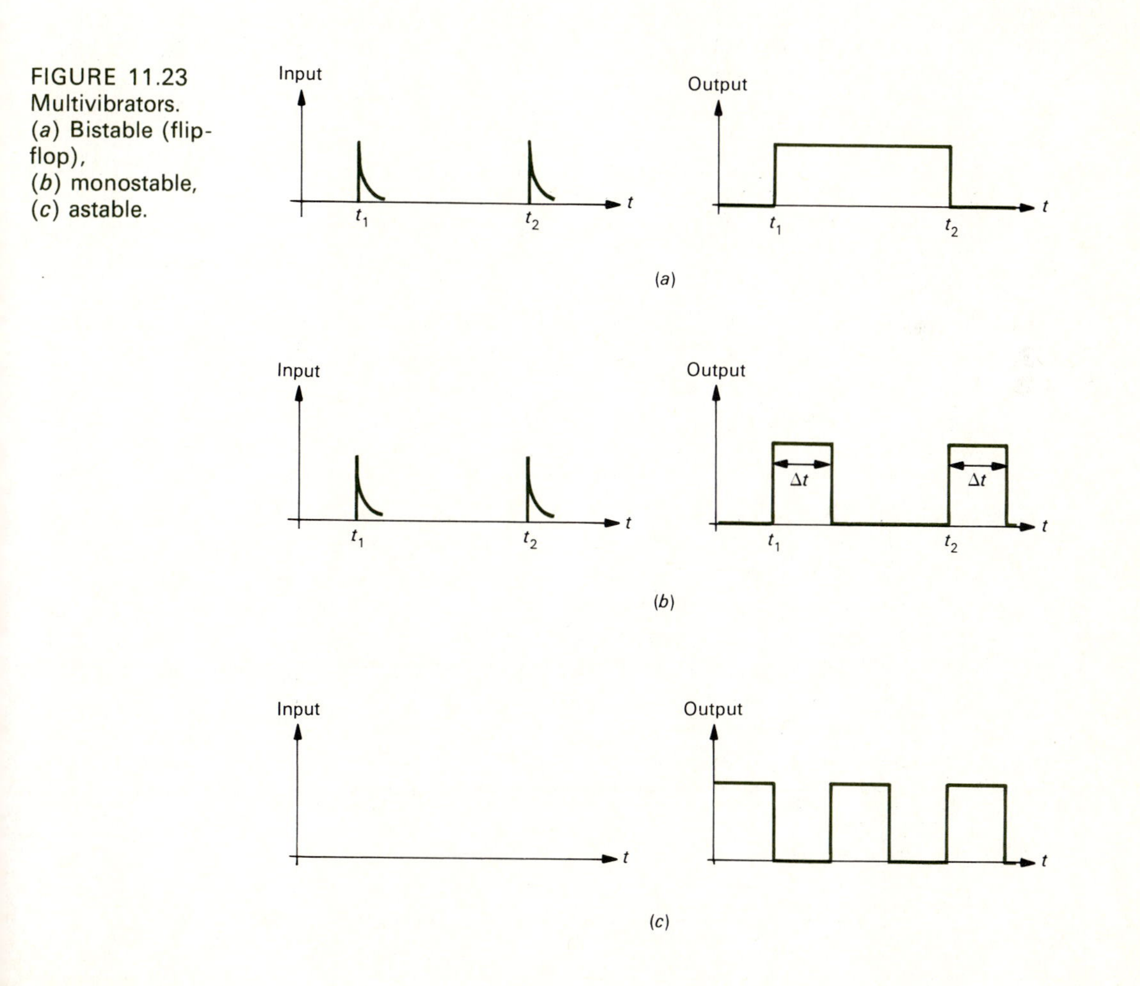

FIGURE 11.23 Multivibrators. (*a*) Bistable (flip-flop), (*b*) monostable, (*c*) astable.

FIGURE 11.24 Operation of the BJT flip-flop. (*a*) Circuit, (*b*) equivalent circuit, (*c*) digital operation.

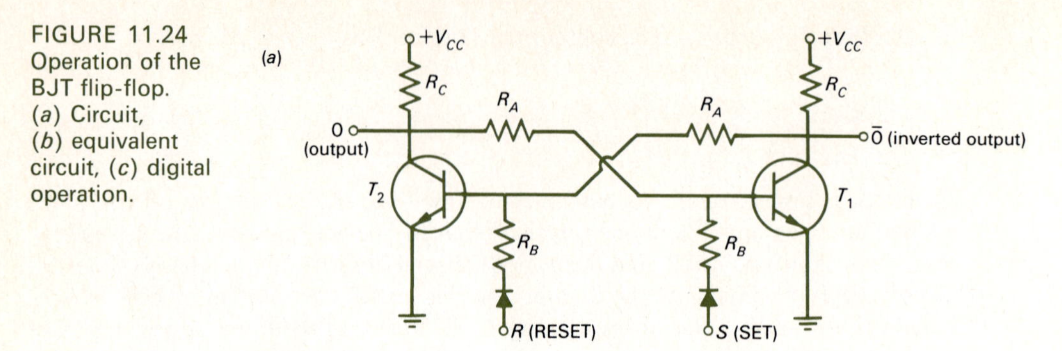

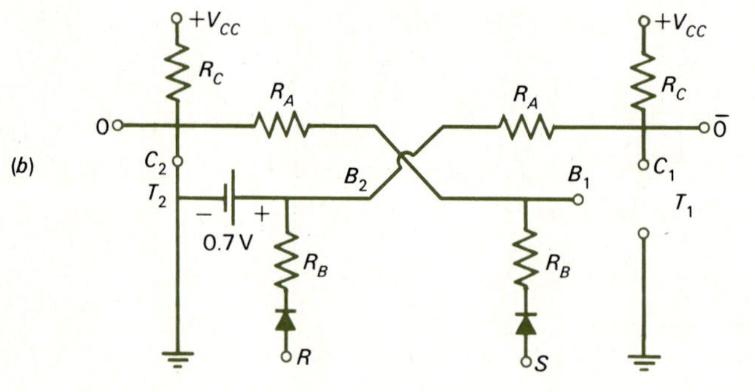

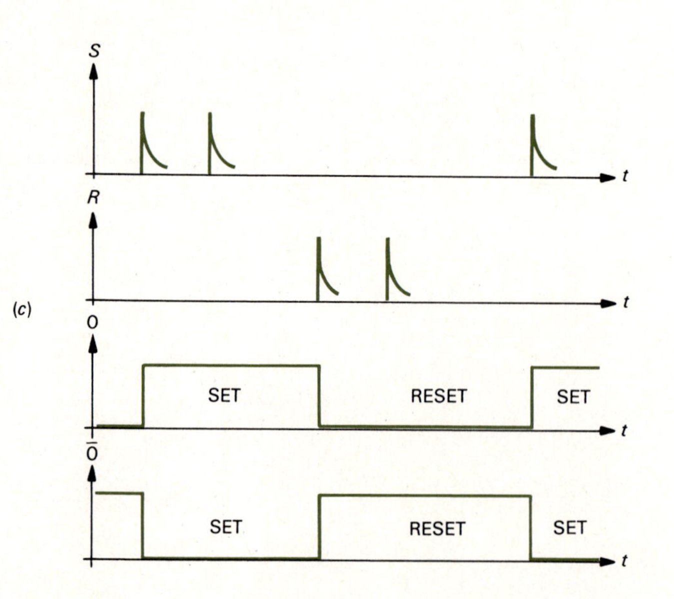

pulse sequence to the S and R inputs and the resulting outputs are shown in Fig. 11.24*c*. Note that the flip-flop has memory, since it remembers the state into which it has been triggered.

The monostable multivibrators generate a single pulse of adjustable duration when a trigger is applied, as shown in Fig. 11.23*b*. These are used to generate a control, or gating, pulse of proper magnitude and duration when an event has taken place or to provide intervals of elapsed time—a timer. As shown in Fig. 11.23*c*, the astable multivibrators generate a periodic waveform which is used to synchronize the sequence of operations of the system; this is usually referred to as the system *clock*. There is currently a standard IC package known as the 555 timer which can be used to perform either of these functions with the simple addition of a capacitor and one or two resistors to its external terminals. The pulse duration for the monostable output can be easily adjusted by proper selection of these external elements; in fact, rules for selection of the values of these external elements are given in the handbooks provided by the manufacturer. The 555 timer can also produce an astable multivibrator with a variable frequency (period of oscillation) adjustable by proper selection of the external resistors and capacitor. A functional block diagram of the 555 timer is given in Fig. 11.25. C_1 and C_2 are comparators (to be described) which provide SET and RESET signals to a flip-flop. The output of the flip-flop drives a buffer, or power, stage which may drive devices requiring larger power, such as electromechanical relays.

Before describing how the 555 timer functions as a monostable or astable multivibrator, we first describe the comparator. Comparators are essentially high-gain difference amplifiers, much like the op amp considered in the previous chapter. The characteristic of an ideal comparator is shown in Fig. 11.26*a*. When $V_i = 0$, $V_o = V_H$. As V_i increases, V_o remains at V_H until V_i reaches V_2, at which point V_o drops to V_L. (Note

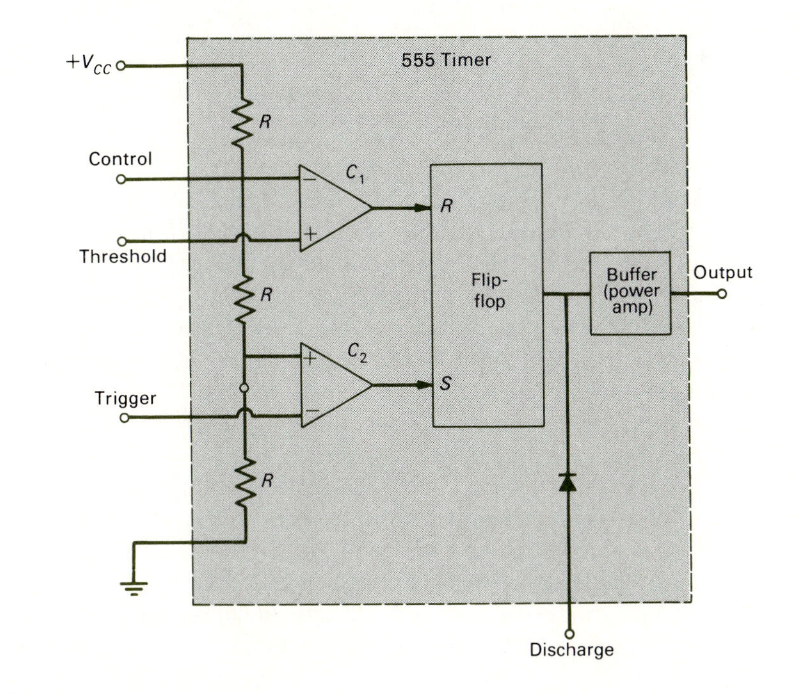

FIGURE 11.25
The 555 timer.

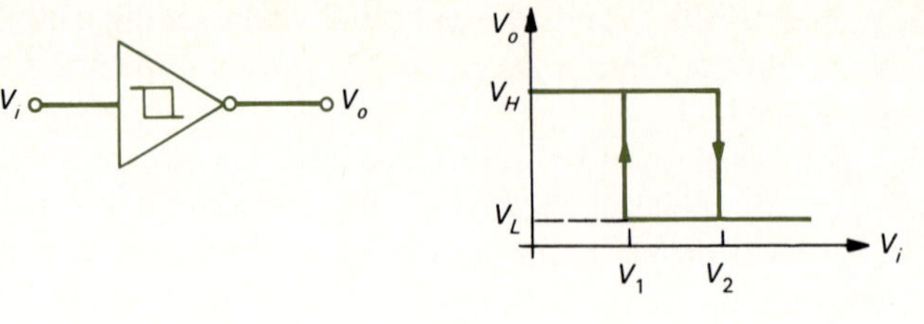

(a)

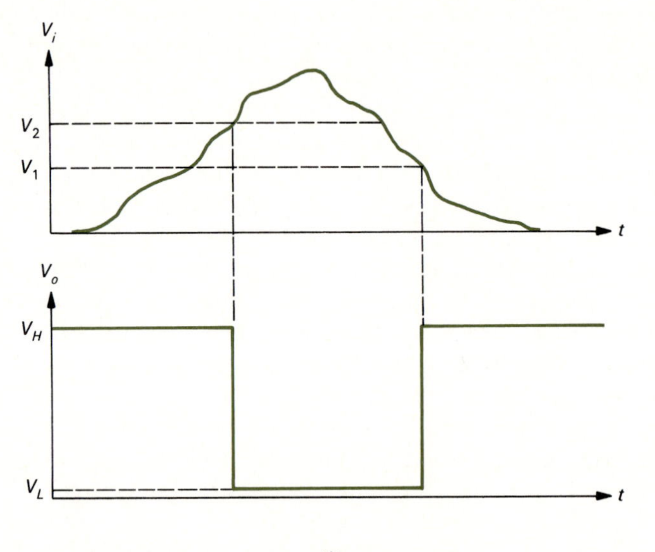

(b)

FIGURE 11.26 The ideal comparator. (a) Device symbol and transfer characteristic, (b) device operation.

the arrows on the characteristic.) The output remains at V_L until V_i is reduced to V_1, at which point the output switches back to V_H. Input-output sample waveforms are shown in Fig. 11.26*b*. The comparator shown in Fig. 11.26*a* is said to have the property of hysteresis in that the switching levels V_1 and V_2 depend upon whether the input is going from low to high or high to low. A comparator with hysteresis is referred to as a Schmitt trigger. Hysteresis reduces the possibility of random switching due to noise or other voltage variations.

A monostable multivibrator can be constructed by adding a capacitor and resistors to the input of a comparator, as shown in Fig. 11.27*a*. For the input waveform shown in Fig. 11.27*b*, when $V_i = 0$ the capacitor voltage is zero and the voltage across R, v, is zero. When the input abruptly rises to $V_{CC} = 15$ V, the voltage across the capacitor remains zero and v jumps to $V_{CC} = 15$ V. Eventually, the capacitor charges up and v drops to zero. When v passes $V_2 = 5$ V, the comparator output switches from V_H to V_L and remains there until the RC circuit charges up sufficiently that v decays to $V_1 = 3$ V. At this point the comparator switches back to V_H. The time required for this

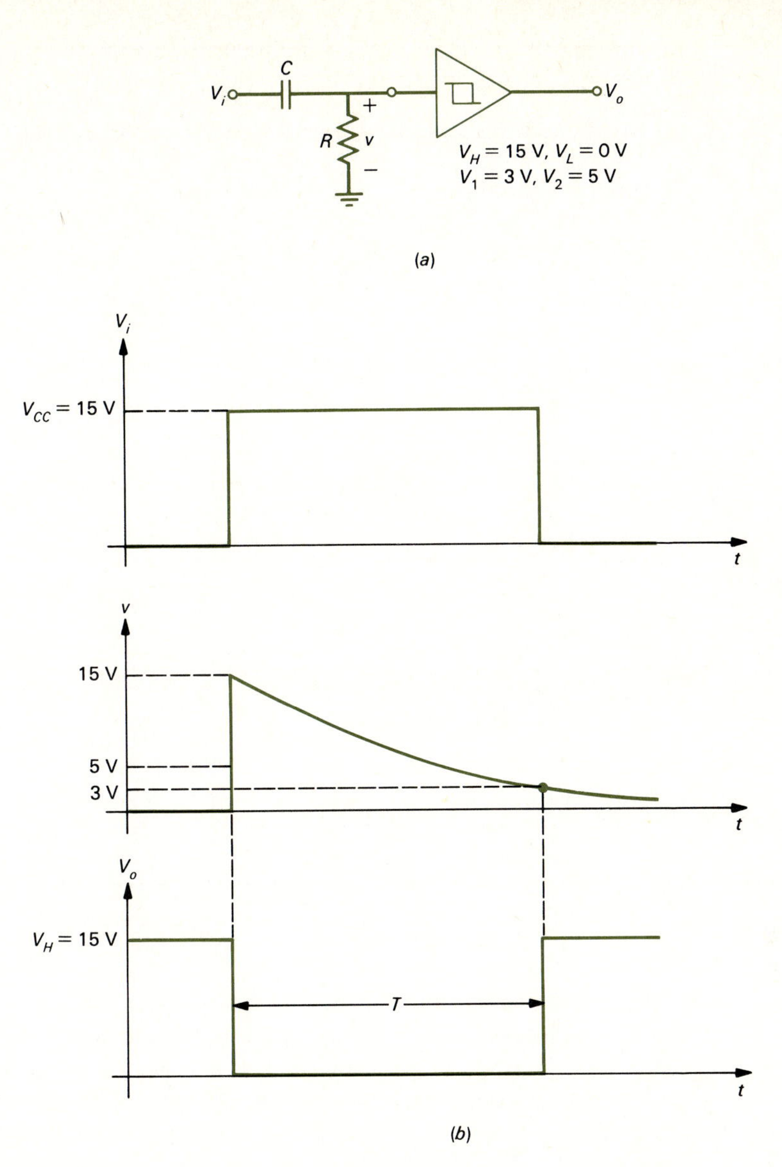

FIGURE 11.27 A monostable multivibrator using a comparator.

to happen is governed by the charge time of the *RC* circuit. This circuit can be solved (neglecting any loading of the comparator, which is typically true for CMOS comparators) to yield

$$T = RC \ln \frac{V_H}{V_1} \tag{11.6}$$

Thus, the duration of the output pulse can be adjusted by selecting *R* and *C*. The polarity of the output can be inverted by feeding it to a NOT gate.

Example 11.4 Show that Eq. (11.6) is true.

Solution Considering V_i, R, and C alone, we may write the equation for v as

$$v(t) = Ae^{-t/RC} + v_{SS}$$

For V_i constant at V_H, $v_{SS} = 0$; thus

$$v(t) = Ae^{-t/RC}$$

The initial condition on $v(t)$ is $v(0^+) = V_i - v_C(0^+) = V_i$; thus

$$v(t) = V_H e^{-t/RC}$$

The time required for v to decay to V_1 is

$$V_1 = V_H e^{-T/RC}$$

or

$$\ln \frac{V_1}{V_H} = -\frac{T}{RC}$$

Thus

$$T = RC \ln \frac{V_H}{V_1}$$

The astable multivibrator, too, can be constructed from comparators, as shown in Fig. 11.28*a*. Suppose that both comparators have identical characteristics, as shown in Fig. 11.28*b*. To begin, suppose that V_o is high and V_a is low. Since V_a (the output of C_1) is low, V_b (the input to C_1) must be greater than V_1. The capacitor is in the process of charging up, and V_b is decreasing with time constant $T = RC$ (see Fig. 11.27). When V_b decreases to V_1, C_1 switches to the high state and V_a goes high, and C_2 switches to the low state and V_o goes low. Immediately prior to this change, the voltage across C is $V_b - V_o = V_1 - V_H$, negative at b. This voltage cannot charge the capacitor instantaneously, so when V_o drops to low (0 V), V_b drops to $V_1 - V_H$. Since $V_o = 0$ V and $V_a = V_H$ the voltage across the RC circuit is $V_a - V_o = V_H$ and the capacitor charges up towards V_H. When V_b reaches V_2, C_1 switches to the low state, causing V_a to be zero, which causes C_2 to switch to the high state—and $V_o = V_H$, and the process repeats.

Example 11.5 For the astable multivibrator in Fig. 11.28, assume that $R = 500\,\Omega$, $C = 100$ pF, $V_L = 0$ V, $V_H = 15$ V, $V_1 = 3$ V, and $V_2 = 5$ V. Calculate the frequency of this clock.

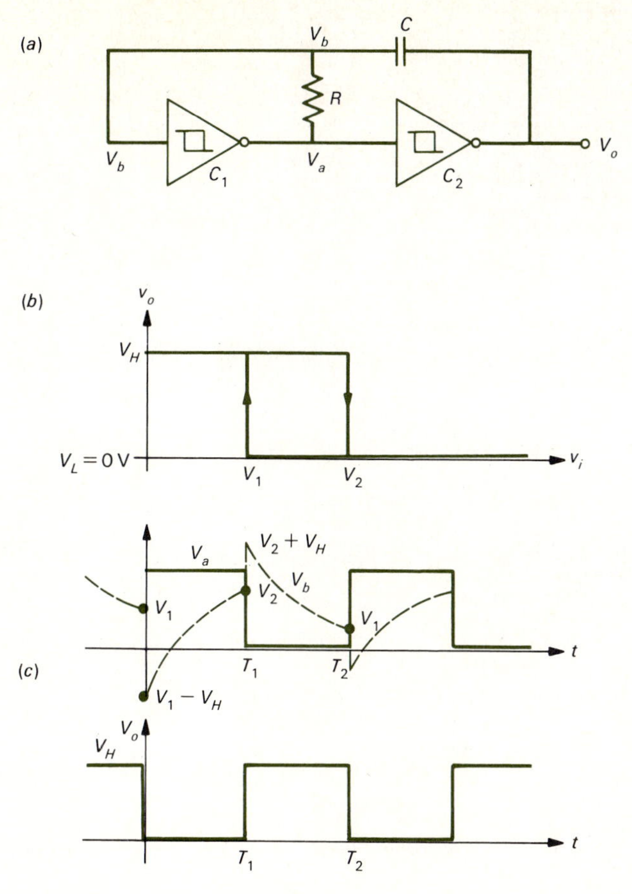

FIGURE 11.28 An astable multivibrator using comparators. (*a*) Circuit, (*b*) transfer characteristic, (*c*) circuit operation.

Solution The time required for the waveform to increase from $V_1 - V_H$ to V_2 at T_1 is determined by writing the solution of the voltage across the capacitor. See Fig. 11.29*a*. Essentially, a dc voltage of $V_a - V_o = V_H - 0 = V_H$ is applied across the *RC* combination for $0 < t < T_1$. The initial capacitor voltage is $V_b - V_o = V_1 - V_H$, so the form of the solution for $V_b - V_o = V_b$ is

$$V_b = (V_1 - 2V_H)e^{-t/RC} + V_H$$

(Check it at $t = 0$ and steady state.) The time T_1 required to increase to V_2 is found from

$$V_2 = (V_1 - 2V_H)e^{-T_1/RC} + V_H$$

or

$$T_1 = RC \ln \frac{V_1 - 2V_H}{V_2 - V_H}$$

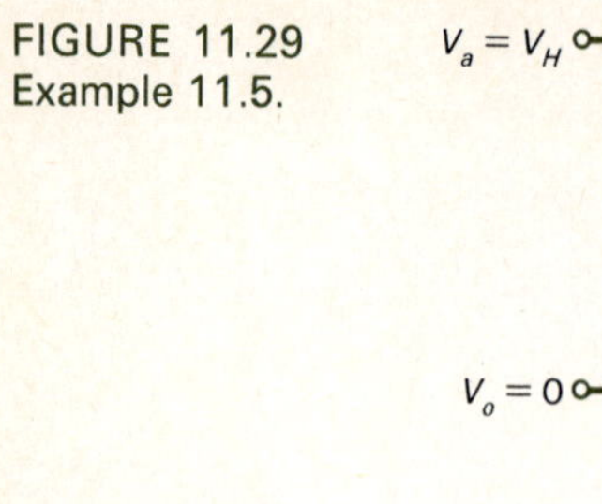
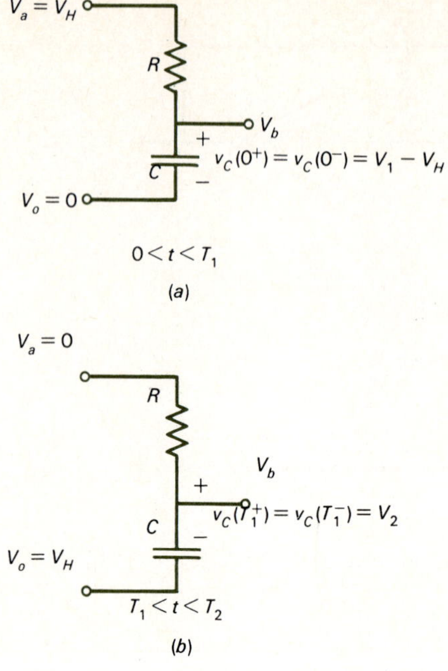

FIGURE 11.29
Example 11.5.

This is repeated for the interval $T_1 < t < T_2$ from the circuit shown in Fig. 11.29*b* with the result

$$V_b = (V_2 + V_H)e^{-(t-T_1)/RC}$$

so that at T_2

$$V_1 = (V_2 + V_H)e^{-(T_2-T_1)/RC}$$

or

$$T_2 = RC \ln\left(\frac{V_2 + V_H}{V_1}\right) + T_1$$

Thus, the period of the waveform is $T_1 + T_2$, or

$$T = RC\left[2 \ln\left(\frac{V_1 - 2V_H}{V_2 - V_H}\right) + \ln\left(\frac{V_2 + V_H}{V_1}\right)\right]$$

with the prescribed values

$$T = 1.45 \times 10^{-7}\,\text{s}$$

and the clock frequency is

$$f = \frac{1}{T}$$

$$= 6.92\ \text{MHz}$$

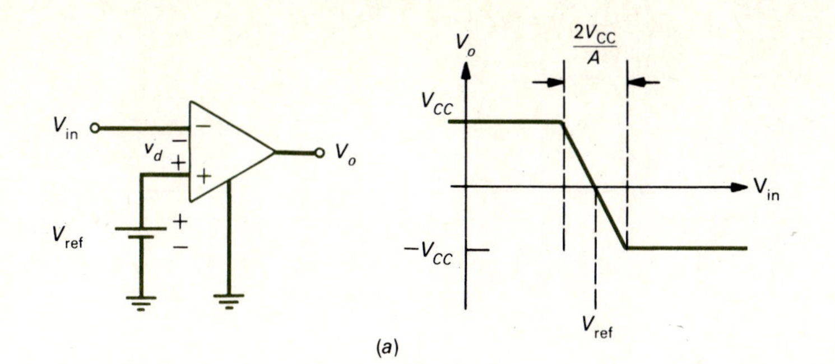

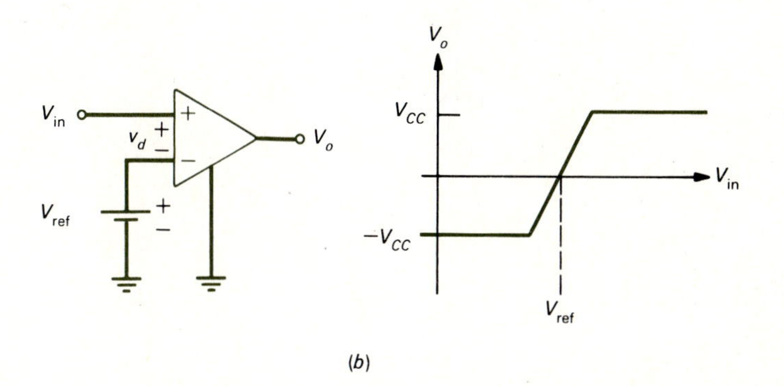

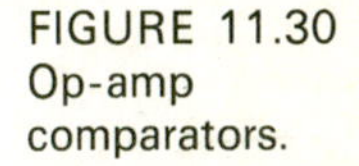
FIGURE 11.30 Op-amp comparators.

Comparators can be constructed from op amps, as illustrated in Fig. 11.30. Recall from Chap. 10 that for negative v_d (between the + − terminals of the op amp) that the op-amp output will be saturated at its negative supply voltage, $-V_{CC}$. Similarly for positive v_d, the op-amp output voltage will be at the positive supply voltage, V_{CC}. (See Fig. 10.9.) With these properties in mind, we see that v_d in Fig. 11.30*a* will be $v_d = V_{ref} - V_{in}$. So long as $V_{in} < V_{ref}$, v_d will be positive and the op-amp output will be at $+V_{CC}$. Recall that a small linear region determined by the gain of the amplifier A exists. For realistic op-amp gains, this region is quite narrow, e.g., $V_{CC} = 15$ V, $A = 10^5$, $2V_{CC}/A = 300\ \mu$V. Hysteresis can be incorporated into this basic comparator with the addition of resistors in a feedback arrangement.

The 555 timer can be used to construct an astable multivibrator with the addition of resistors R_1 and R_2 and capacitor C, as shown in Fig. 11.31*a*. The three resistors R internal to the timer divide the supply voltage such that the voltage at the negative terminal of C_1 is $\frac{2}{3}V_{CC}$ and the voltage at the positive terminal of C_2 is $\frac{1}{3}V_{CC}$. The threshold and trigger terminals are connected to the upper terminal of the capacitor so that their voltages are the capacitor voltage v_C. The capacitor charges through R_1 and R_2 until $v_C = \frac{2}{3}V_{CC}$, at which time C_1 resets the flip-flop output state to zero. The capacitor discharges through R_2 until its voltage is reduced to $\frac{1}{3}V_{CC}$. At this point C_2 sets the flip-flop. The cycle continues over and over to produce the periodic waveform shown in Fig. 11.31*b*.

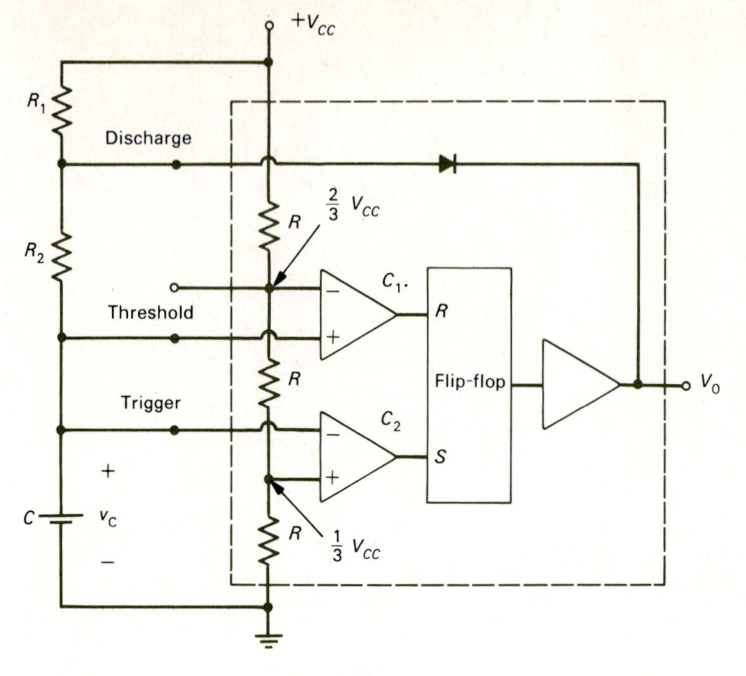

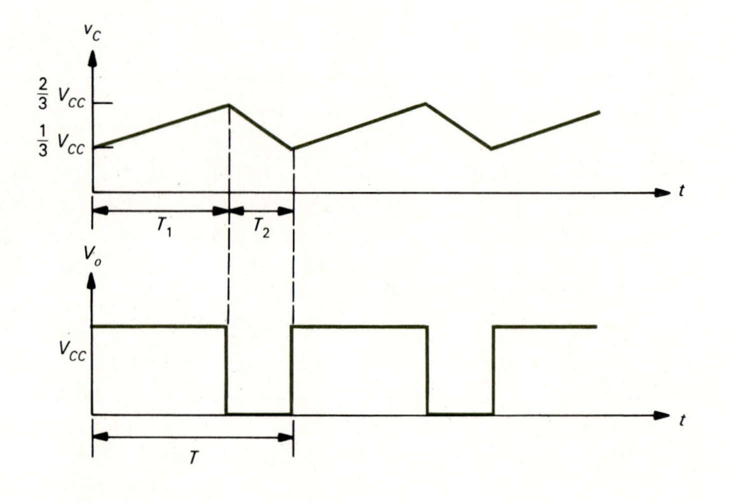

FIGURE 11.31
The 555 timer connected as an astable multivibrator.

Example 11.6 Determine equations for times T_1 and T_2 in Fig. 11.31*b*. Determine values of R_1, R_2 and C to produce a 100-kHz square wave having a 50 percent duty cycle.

Solution The various times can be calculated with the appropriate RC time constants of the charge-discharge circuits. Once v_C has charged to a value of $\frac{2}{3}V_{CC}$ through $R_1 + R_2$, the time required to discharge through R_2 to a value of $\frac{1}{3}V_{CC}$ is calculated from

$$\tfrac{2}{3}V_{CC}e^{-T_2/R_2C} = \tfrac{1}{3}V_{CC}$$

giving

$$T_2 = R_2 C \ln 2 \tag{11.6}$$

The time required to charge up to $v_C = \frac{2}{3}V_{CC}$ can similarly be obtained as

$$T_1 = (R_1 + R_2)C \ln 2 \tag{11.7}$$

If $R_1 = R_2$, $T_1 = T_2$, and the waveform is said to have a 50 percent duty cycle. The frequency of the output waveform is

$$f = \frac{1}{T_1 + T_2} \tag{11.8}$$

$$= \frac{1.44}{(R_1 + 2R_2)C}$$

To obtain a square-wave oscillator at a frequency of 100 kHz having a 50 percent duty cycle, we may choose

$$R_1 = R_2$$
$$= 480\ \Omega$$
$$C = 0.01\ \mu\text{F}$$

Choosing standard resistor values of 470 Ω would yield a frequency of 102.128 kHz.

11.6 A PRACTICAL NOTE

Although we have considered in some detail the design and operation of a large class of digital circuit "building blocks," it is not necessary to perform these laborious and detailed analysis tasks when interconnecting such devices to build a digital system, as will be described in Part 4. The numerous handbooks furnished by the device manufacturers contain more than enough design aids, simple equations, nomographs, etc., to interconnect these devices properly. In fact, many of these handbooks could (and often do) serve as textbooks for courses in logic circuit design and make enjoyable and, now that we have discussed the fundamentals, easy reading.

PROBLEMS

11.1 For the diode AND gate shown in Fig. 11.1, determine V_o if the two signals shown in Fig. P11.1 are applied. Assume ideal diodes.

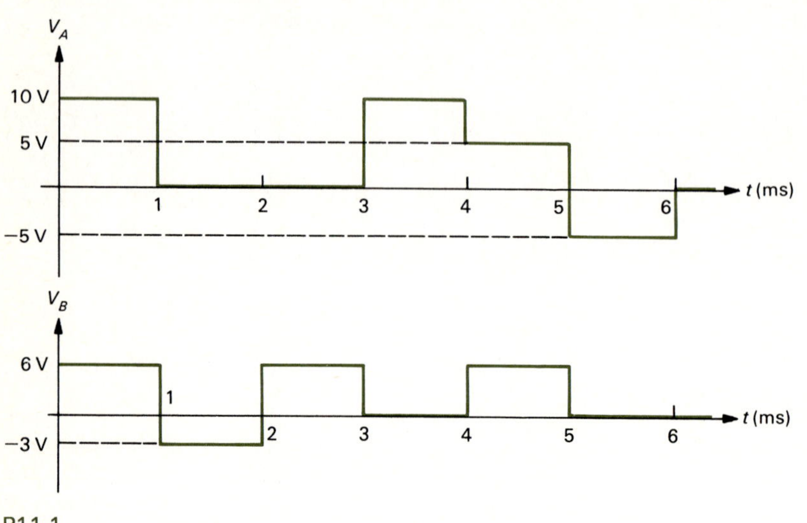

P11.1

11.2 Repeat Prob. 11.1 if piecewise-linear diodes having $E_f = 0.5$ V and $R_f = 0$ are used.

11.3 For the diode OR gate shown in Fig. 11.2*a*, sketch V_o for the two input signals shown in Fig. P11.1.

11.4 Repeat Prob. 11.3 if piecewise-linear diodes having $E_f = 0.5$ V and $R_f = 0$ are used.

11.5 The NOT gate shown in Fig. 11.4 uses a BJT having characteristics given in Fig. 8.37. For the input signal shown in Fig. P11.5, sketch V_o if $R_{B1} = R_{B2} = 40\ \text{k}\Omega$, $V_{BB} = 6$ V, $R_C = 1\ \text{k}\Omega$.

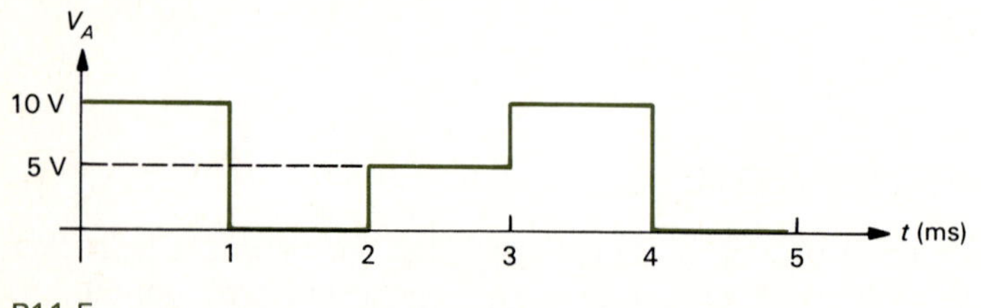

P11.5

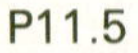

11.6 Change V_{BB} to 2 V and repeat Prob. 11.5.

11.7 For the BJT switch shown in Fig. 11.6*a*, $V_{CC} = 5$ V, $V_T = 0.7$ V, $V_{sat} = 0.2$ V, and $\beta = 20$. If V_i switches between 0 and 5 V and $i_B \leq 0.1$ mA, determine the minimum values of R_B and R_C for proper operation.

11.8 Repeat Prob. 11.7 if the input signal switches between 0 and 1 V.

11.9 Plot the transfer characteristic for the BJT switch in Fig. 11.6*a* with $V_{CC} = 5\,\text{V}$, $V_{\text{sat}} = 0.2\,\text{V}$, $V_T = 0.7\,\text{V}$, $R_C = 1\,\text{k}\Omega$, $R_B = 10\,\text{k}\Omega$, $\beta = 50$.

11.10 Repeat Prob. 11.9 with R_C changed to 500 Ω.

11.11 The transistor switch shown in Fig. 11.6 uses a BJT which has the characteristics shown in Fig. 8.37. For the input signal shown in Fig. P11.11, sketch V_o if $R_B = 10\,\text{k}\Omega$, $R_C = 750\,\Omega$.

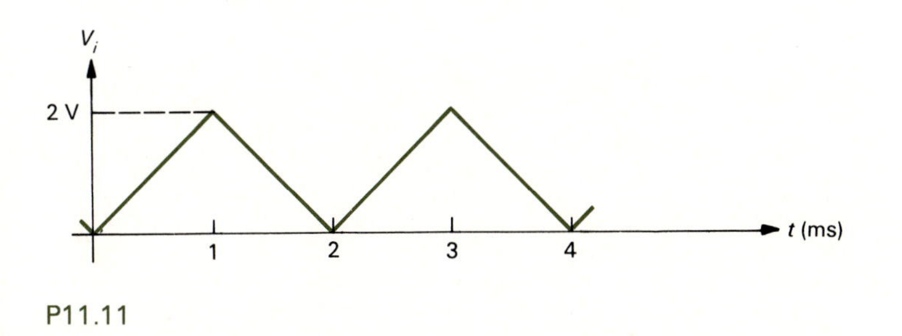

P11.11

11.12 For the DTL NOR gate shown in Fig. 11.15*a*, assume ideal diodes, $V_T = 0.7\,\text{V}$, $V_{\text{sat}} = 0.2\,\text{V}$, $R_B = 10\,\text{k}\Omega$, $R_C = 500\,\Omega$, and $V_{CC} = 5\,\text{V}$. Sketch V_o for the input signals shown in Fig. P11.1. Use reasonable approximations and assume $\beta = 20$ for the BJT.

11.13 For the TTL gate shown in Fig. 11.16, assume that the inputs vary between 0 and 5 V and that $V_{CC} = 5\,\text{V}$. Determine the maximum value of R_B to saturate T_2 if $i_{C,\text{sat}} = 3.8\,\text{mA}$.

11.14 The switch shown in Fig. 11.18, with $R_D = 3\,\text{k}\Omega$ and $V_{DD} = 12\,\text{V}$, uses a JFET whose characteristic is given in Fig. 8.36. If the input voltage is as shown in Fig. P11.14, sketch the output voltage.

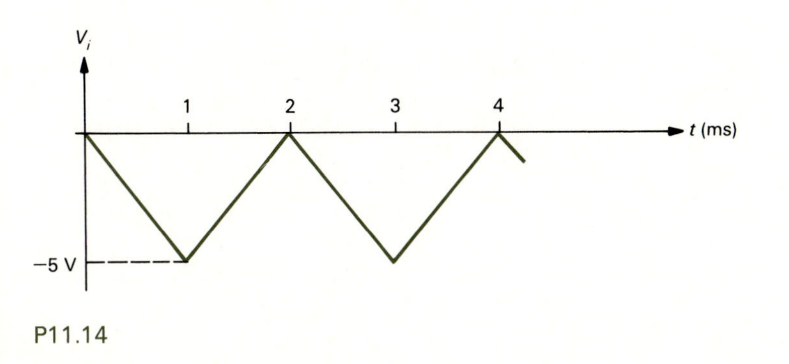

P11.14

11.15 The MOSFETs in the CMOS gate shown in Fig. 11.21 have $V_T = 5\text{ V}$. If $V_{SS} = 20\text{ V}$ and V_i is as shown in Fig. P11.15, sketch V_o. Assume that $V_{\text{sat}} = 1\text{ V}$ for both MOSFETs.

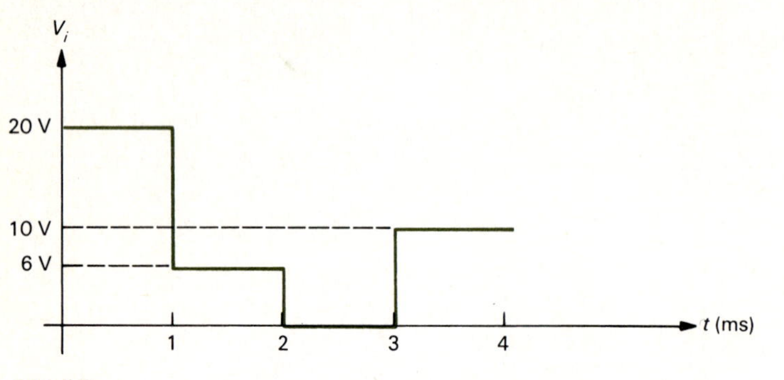

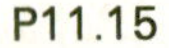
P11.15

11.16 The monostable multivibrator shown in Fig. 11.27*a* has $V_H = 10\text{ V}$, $V_L = 1\text{ V}$, $V_1 = 3\text{ V}$, and $V_2 = 5\text{ V}$. For the input voltage shown in Fig. P11.16, sketch the output voltage if $R = 1\text{ k}\Omega$ and $C = 1\ \mu\text{F}$.

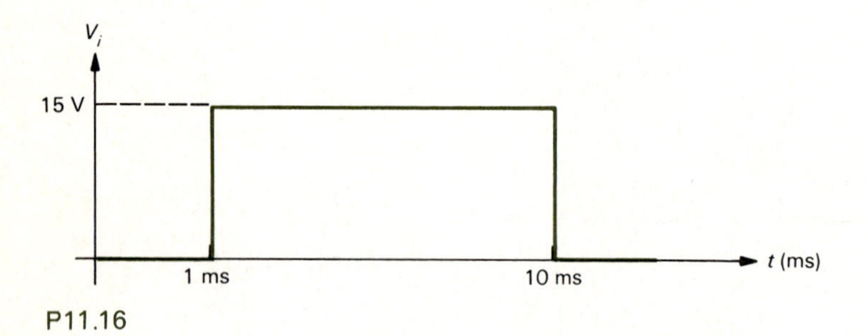

P11.16

11.17 Repeat Prob. 11.16 if R is changed to $100\text{ k}\Omega$.

REFERENCES

11.1 A. B. Carlson and D. G. Gisser, *Electrical Engineering: Concepts and Applications*, Addison-Wesley, Reading, Mass., 1981.

11.2 Ralph J. Smith, *Circuits, Devices and Systems: A First Course in Electrical Engineering*, 4th ed., Wiley, New York, 1984.

11.3 A. E. Fitzgerald, D. Higginbotham, and A. Grabel, *Basic Electrical Engineering*, 5th ed., McGraw-Hill, New York, 1981.

11.4 W. G. Oldham and S. E. Schwartz, *An Introduction to Electronics*, Holt, Rinehart & Winston, New York, 1972.

11.5 R. L. Boylestad and L. Nashelsky, *Electronic Devices and Circuit Theory*, 3d ed., Prentice-Hall, Englewood Cliffs, N.J. 1982.

11.6 R. J. Tocci, *Digital Systems, Principles and Applications*, 3d ed. Prentice-Hall, Englewood Cliffs, N.J., 1985.

PART 3

Electric Power and Machines

Chapter

12

Polyphase Circuits

Polyphase circuits—three-phase circuits in particular—are used in the majority of transmission, distribution, and energy conversion systems where the power or voltampere (VA) levels are above 10 to 20 kilowatts (kW) or kilovoltamperes (kVA). The transmission and distribution of electric power to residences, industries, and business is accomplished almost entirely by means of three-phase networks consisting of transmission and distribution lines, transformers, and associated protective devices such as circuit breakers, fuses, relays, and lightning protectors. Electromagnetic and electric energy conversion devices are likewise found in polyphase configurations for most applications at power or voltampere levels greater than those listed above.

The basic reasons for using polyphase systems are related to *power density*, which is defined as the ratio of either power to weight or power to volume of the device. Thus, an electric machine of a given weight is capable of delivering more power in polyphase than in single-phase designs. In some electrical applications the voltampere rating is often substituted for the power rating. For example, the specific weight of an ac motor is generally less for a three-phase motor than for a single-phase motor at output power ratings above 1 horsepower (hp). Above approximately 5 hp, a three- or two-phase motor will always have a lower specific weight than a single-phase motor. Because three-phase systems are by far the most common, we will restrict our discussion to them.

12.1 THREE-PHASE SYSTEMS

Suppose we have a system of three ac voltages of a certain frequency such that their amplitudes are equal but these voltages are displaced from one another by 120° in time. We may mathematically express this system of voltages as

$$v_{a'a} = V_m \sin \omega t \tag{12.1}$$

$$v_{b'b} = V_m \sin (\omega t - 120°) \tag{12.2}$$

$$v_{c'c} = V_m \sin (\omega t - 240°) \tag{12.3}$$

These voltages are graphically depicted in Fig. 12.1. In terms of their RMS values, these voltages may be written in phasor notation as

$$\begin{aligned} V_{a'a} &= V\underline{/0°} \\ &= V(1 + j0) \end{aligned} \tag{12.4}$$

$$\begin{aligned} V_{b'b} &= V\underline{/-120°} \\ &= V(-0.5 - j0.866) \end{aligned} \tag{12.5}$$

$$\begin{aligned} V_{c'c} &= V\underline{/-240°} \\ &= V(-0.5 + j0.866) \end{aligned} \tag{12.6}$$

Figure 12.2*a* shows the phasor representation of Eqs. (12.4) to (12.6). Notice that in Fig. 12.2*b* we have also shown (hypothetically) three voltage sources corresponding to Eqs. (12.4) to (12.6). Consequently, we may define a three-phase (voltage) source having three equal voltages which are 120° out of phase with one another. In particular, we call this system a *three-phase balanced system*—in contrast to an unbalanced system, in which the magnitudes may be unequal and/or the phase displacements may not by 120°. For a balanced three-phase system, it follows from Eqs. (12.1) to (12.3), as well as from Eqs. (12.4) to (12.6), that the phasor sum of the three voltages is zero.

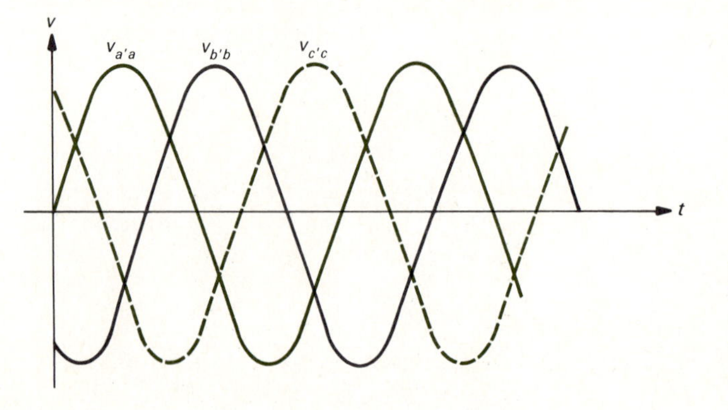

FIGURE 12.1 A system of three voltages of equal magnitude but displaced from each other by 120° in phase.

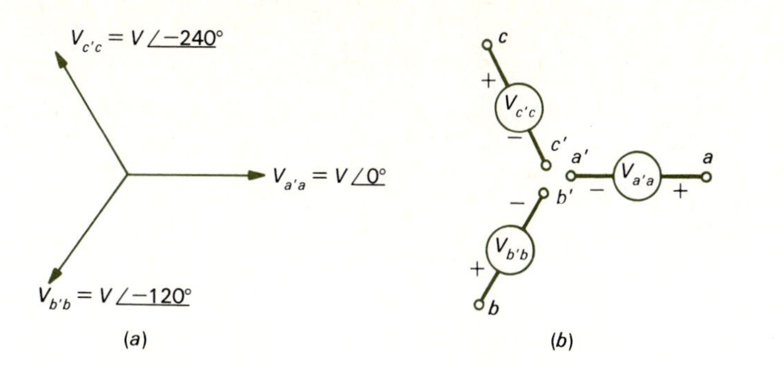

FIGURE 12.2
(*a*) Balanced three-phase phasor representation.
(*b*) Three-phase voltage source.

We abbreviate $v_{a'a}$, $v_{b'b}$, and $v_{c'c}$ as v_a, v_b, and v_c respectively. Now referring to Figs. 12.1 and 12.2*a*, we observe that the voltages attain their maximum values in the order v_a, v_b, and v_c. This order is known as the phase *sequence abc*. A reverse phase sequence will be *acb*, in which case the voltages v_c and v_b lag v_a by 120° and 240°, respectively.

12.2 THREE-PHASE CONNECTIONS

There are two common and practical ways to interconnect the three voltage sources shown in Fig. 12.2*b*. These two forms of interconnection are illustrated in Figs. 12.3*a* and *b*, respectively labeled the *wye connection* and the *delta connection*. The result, for either type of connection, is a three-terminal (*ABC*) ac source of power supplying a balanced set of three voltages to a load. In the wye connection, the terminals a', b', and c' are joined together to form the *neutral* point *o*. If a lead is brought out from the neutral point *o*, the system becomes a *four-wire three-phase* system. The terminals *a* and b', *b* and c', and *c* and a' are joined individually to form the delta connection.

Notice from Fig. 12.3*a* that we can identify two types of voltages: voltages $V_{a'a}$, $V_{b'b}$, and $V_{c'c}$ across the three individual phases, known as the *phase voltages*; and voltages V_{ab}, V_{bc}, and V_{ca} across the lines *a*, *b*, and *c* (or, A, B, and C), known as the *line voltages*. The line voltages are related to the phase voltages such that

$$V_{oa} + V_{ab} = V_{ob}$$

or

$$V_{ab} = V_{ob} - V_{oa} \tag{12.7}$$

Similarly,

$$V_{bc} = V_{oc} - V_{ob} \tag{12.8}$$

and

$$V_{ca} = V_{oa} - V_{oc} \tag{12.9}$$

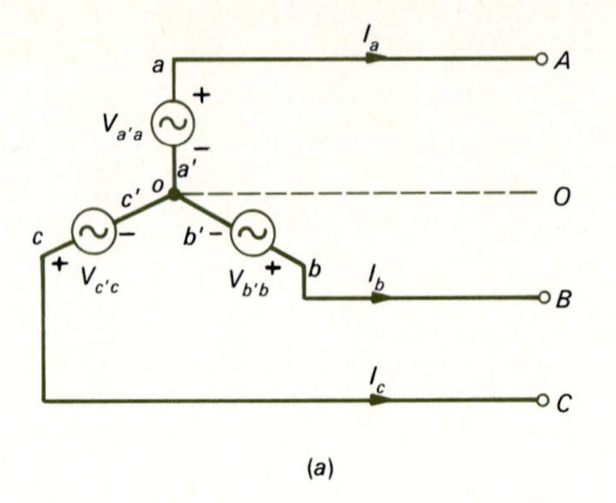

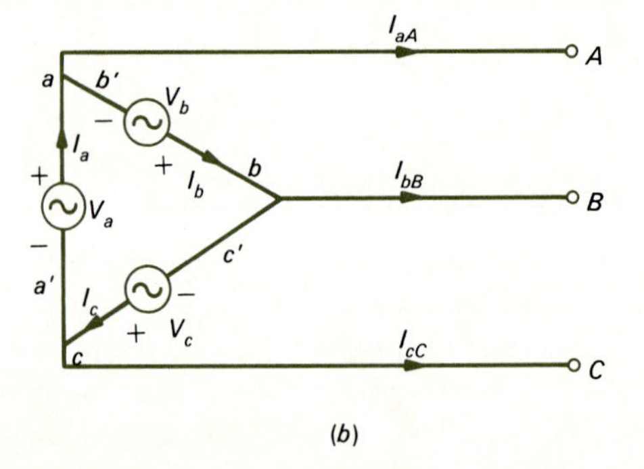

FIGURE 12.3 (*a*) Wye connection. (*b*) Delta connection.

These relationships of the phase voltages and line voltages are illustrated in the phasor diagram of Fig. 12.4. Furthermore, we may combine Eqs. (12.4), (12.5), and (12.7) to obtain

$$V_{ab} = V(-0.5 - j0.866) - V(1 + j0)$$
$$= V(-1.5 - j0.866)$$

or

$$V_{ab} = \sqrt{3}V\underline{/-120^\circ} \tag{12.10}$$

which is consistent with the phasor V_{ab} of Fig. 12.4. Relationships similar to Eq. (12.10) are valid for the phasors V_{bc} and V_{ca}. Because V_{ab} is the voltage across the lines a and b and V is the magnitude of the voltage across the phase, we may generalize Eq. (12.10) to

$$V_l = \sqrt{3}V_p \tag{12.11}$$

where V_l is the voltage across any two lines and V_p is the phase voltage.

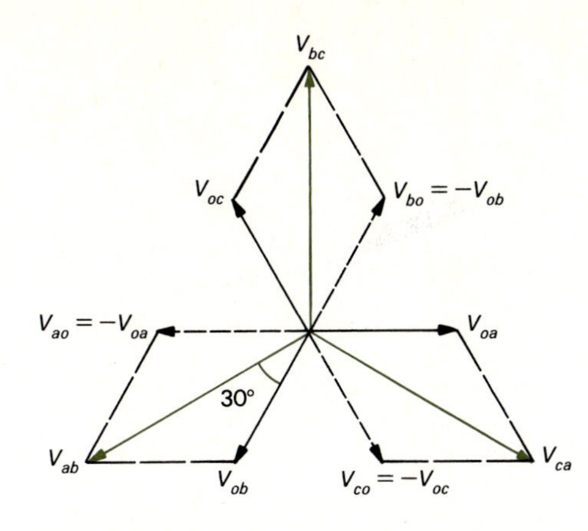

FIGURE 12.4 Voltage phasors for Y connection.

For the wye connection, it is clear from Fig. 12.3*a* that the line currents I_l and phase currents I_p are the same. Thus, we may write:

$$I_l = I_p \tag{12.12}$$

The mutual phase relationships of the currents are given in Fig. 12.5.

Turning now to the delta connection, we can verify from Fig. 12.3*b* that the line voltages V_l are the same as the phase voltages V_p. Hence,

$$V_l = V_p \tag{12.13}$$

The phase relationships of the three-phase (and line) voltages are shown in Fig. 12.6. Next, we show in Fig. 12.7 the phase currents and line currents for the delta-connected system of Fig. 12.3*b*. The phase currents and line currents are related to each other by

$$I_{aA} = I_{ca} - I_{ab}$$

$$= \sqrt{3} I_{ca} \underline{/30^\circ} \tag{12.14}$$

$$I_{bB} = \sqrt{3} I_{ab} \underline{/30^\circ} \tag{12.15}$$

and

$$I_{cC} = \sqrt{3} I_{bc} \underline{/30^\circ} \tag{12.16}$$

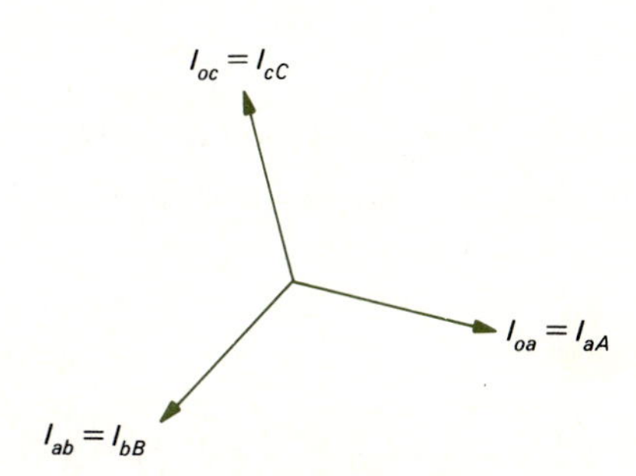

FIGURE 12.5 Current phasors for Y connection.

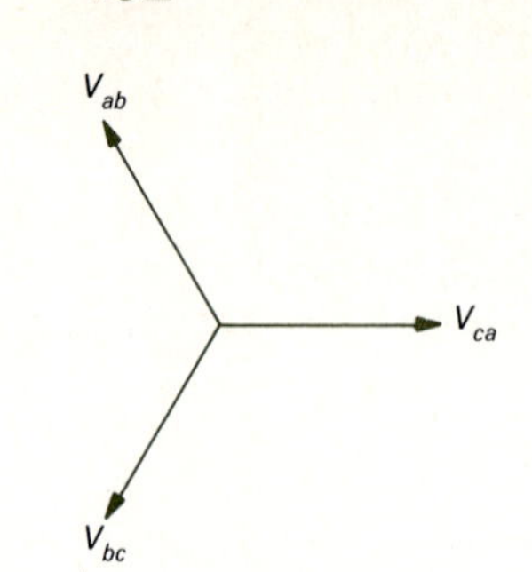

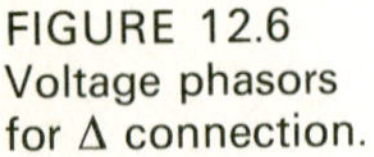
FIGURE 12.6
Voltage phasors for Δ connection.

FIGURE 12.7
Current phasors for Δ connection.

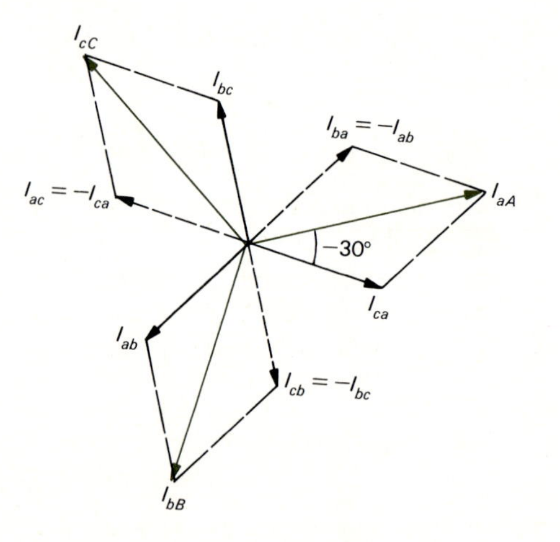

These relationships are shown in Fig. 12.7. In general, in a delta connection the line current $I_l\,(= |I_{aA}| = |I_{bB}| = |I_{cC}|)$ is related to the phase current $I_p\,(= I_{ab} = I_{bc} = I_{ca})$ by

$$I_l = \sqrt{3}I_p \tag{12.17}$$

12.3 POWER IN THREE-PHASE SYSTEMS

We know from the study of single-phase ac circuits that the instantaneous power in the circuit is pulsating in nature. For instance, if $v = V_m \sin \omega t$ and $i = I_m \sin \omega t$ are the voltage across and the current through a resistor, then the instantaneous power p is given by

$$p = V_m I_m \sin^2 \omega t = \tfrac{1}{2} V_m I_m (1 - \cos 2\omega t)$$

which has a pulsating component of double frequency. Of course, the time-average value of the power in $P = (V_m/\sqrt{2})(I_m/\sqrt{2}) = VI$, where V and I are the RMS values.

It is interesting to note that the instantaneous power in a balanced three-phase circuit is a constant. This fact can be demonstrated as follows, where we consider a purely resistive circuit for the sake of simplicity. The instantaneous power may be written as

$$\begin{aligned} p &= v_a i_a + v_b i_b + v_c i_c \\ &= (V_m \sin \omega t)(I_m \sin \omega t) \\ &\quad + [V_m \sin (\omega t - 120°)][I_m \sin (\omega t - 120°)] \\ &\quad + [V_m \sin (\omega t + 120°)][I_m \sin (\omega t + 120°)] \\ &= V_m I_m (\sin^2 \omega t + \tfrac{1}{4} \sin^2 \omega t + \tfrac{3}{4} \cos^2 \omega t \\ &\quad + \tfrac{1}{4} \sin^2 \omega t + \tfrac{3}{4} \cos^2 \omega t) \\ &= \tfrac{3}{2} V_m I_m (\sin^2 \omega t + \cos^2 \omega t) \\ &= \tfrac{3}{2} V_m I_m \\ &= 3 V_p I_p \end{aligned} \tag{12.18}$$

where V_p and I_p are RMS values of phase voltage and current. Since the instantaneous power is constant, the average power is the same as the total instantaneous power; that is,

$$P_T = 3 V_p I_p \tag{12.19}$$

If the circuit is not purely resistive and has a power factor angle θ_p, then the average power delivered by each phase is given by $V_p I_p \cos \theta_p$. So the total power delivered to the load is given (for this *balanced* system) by

$$P_T = 3(V_p I_p \cos \theta_p) \tag{12.20}$$

But for the wye connection, the phase current is the same as the current in the line I_l, and we showed earlier that $V_l = \sqrt{3}\, V_p$; thus, we can express the total power in terms of line voltages and currents as follows:

$$P_T = 3 \frac{V_l}{\sqrt{3}} I_l \cos \theta_p = \sqrt{3}\, V_l I_l \cos \theta_p \tag{12.21}$$

It should be noted that the angle θ_p is still the angle between the phase voltage and phase current.

The expression for the total power in a delta-connected system is the very same. Whereas the line voltage equals the phase voltage for a delta-connected system, the line current is $\sqrt{3}$ times greater than the phase current, as we could easily show.

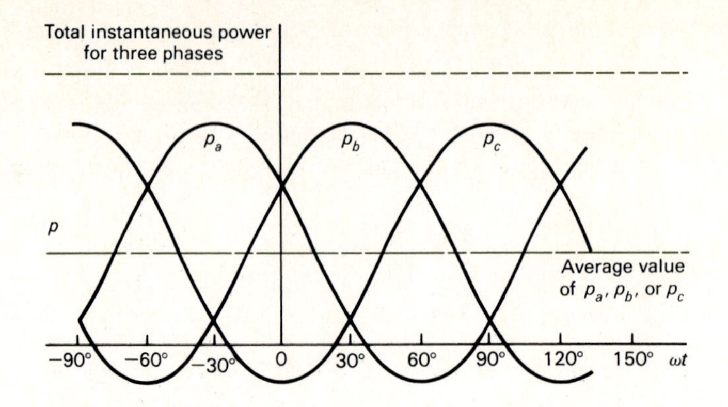

FIGURE 12.8 Power in a three-phase system.

Hence, if Eq. (12.20) is converted to line quantitites for the delta-connected system, we obtain

$$P_T = 3V_l \frac{I_l}{\sqrt{3}} \cos \theta_p = \sqrt{3}\, V_l I_l \cos \theta_p \tag{12.22}$$

A graphical representation of the instantaneous power in a three-phase system is given in Fig. 12.8. It is seen that the total instantaneous power is constant and is equal to 3 times the average power. This feature is of great value in the operation of three-phase motors where the constant instantaneous power implies an absence of torque pulsations and consequent vibrations.

Example 12.1 A three-phase delta-connected load having a $(3 + j4)$-Ω impedance per phase is connected across a 220-V three-phase source. Calculate the line current and the total power supplied to the load.

Solution

$$V_{\text{phase}} = 220 \text{ V}$$

$$Z_{\text{phase}} = 3 + j4 = 5\underline{/53.2}\ \Omega$$

$$I_{\text{phase}} = \frac{220}{5} = 44 \text{ A}$$

$$I_{\text{line}} = \sqrt{3} \times 44 = 76.21 \text{ A}$$

$$\text{Power} = \sqrt{3} \times 220 \times 76.21 \times \cos 53.2 = 17.4 \text{ kW}$$

Example 12.2 A 220-V three-phase source supplies a three-phase wye-connected load having an impedance of $(3 + j4)\ \Omega$ per phase. Calculate the phase voltage across each phase of the load and the total power consumed by the load.

Solution

$$V_{\text{phase}} = \frac{220}{\sqrt{3}}$$

$$= 127 \text{ V}$$

$$Z_{\text{phase}} = 3 + j4$$

$$= 5\underline{/53.2} \ \Omega$$

$$I_{\text{phase}} = \tfrac{127}{5}$$

$$= 25.4 \text{ A} = I_{\text{line}}$$

$$\text{Power} = \sqrt{3} \times 220 \times 25.4 \cos 53.2$$

$$= 5.8 \text{ kW}$$

Example 12.3 An unbalanced three-phase load supplied by a three-phase four-wire system is shown in Fig. 12.9*a*. The currents in phases *A* and *B* are 10 and 8 A, respectively, and phase *C* is open. The load power-factor angle for phase *A* is 30° and for phase *B* is 60°, lagging in both cases. Draw the phasor diagram, and determine the current in the neutral.

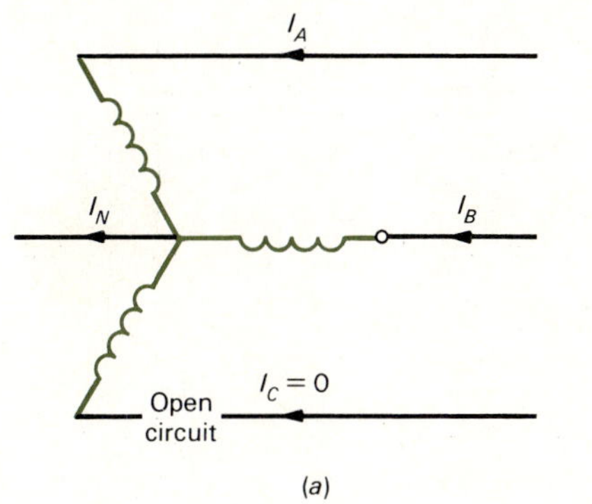

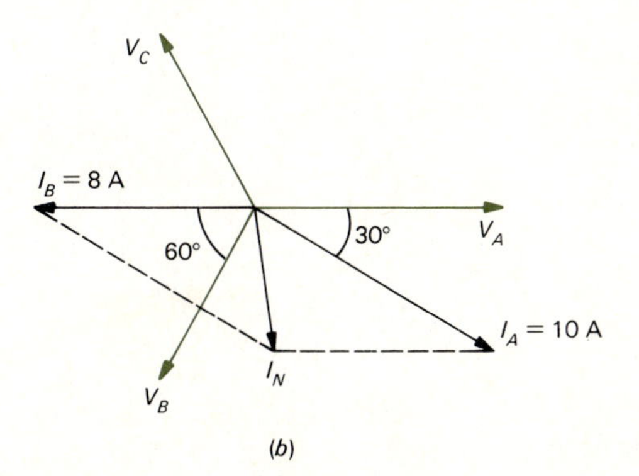

FIGURE 12.9 Example 12.3.

Solution The phasor diagram is shown in Fig. 12.9*b*. With V_A as the reference phasor,

$$I_A = 10\underline{/-30^\circ}$$
$$= 8.66 - j5 \quad \text{A}$$
$$I_B = 8\underline{/-180^\circ}$$
$$= -8.0 + j0 \quad \text{A}$$

Hence,

$$I_N = I_A + I_B$$
$$= -0.66 - j5$$
$$= 5.04\underline{/-97.5^\circ}$$

Example 12.4 A three-phase balanced load has a 10-Ω reistance in each of its phases. The load is supplied by a 220-V three-phase source. Calculate the power absorbed by the load if it is connected in wye; calculate the same if it is connected in delta.

Solution In the wye connection

$$V_p = \frac{220}{\sqrt{3}}$$
$$= 127 \text{ V}$$
$$I_p = \tfrac{127}{10}$$
$$= 12.7 \text{ A} = I_l$$
$$\cos\theta_p = 1 \qquad \text{load being purely resistive}$$

Hence,

$$P = \sqrt{3}V_l I_l \cos\theta_p$$
$$= \sqrt{3} \times 220 \times 12.7 \times 1$$
$$= 4.84 \text{ kW}$$

In the delta connection

$$V_l = 220 \text{ V}$$
$$I_p = \tfrac{220}{10}$$
$$= 22 \text{ A}$$
$$I_l = \sqrt{3} \times 22 = 38.1 \text{ A}$$

Hence,

$$P = \sqrt{3}V_l I_l \cos\theta_p$$
$$= \sqrt{3} \times 220 \times 38.1 \times 1$$
$$= 14.52 \text{ kW}$$

Notice that the power consumed in the delta connection is 3 times that of the wye connection.

Example 12.5 For the balanced load shown in Fig. 12.10*a*, draw a phasor diagram showing all currents and voltages and determine the power consumed in the load.

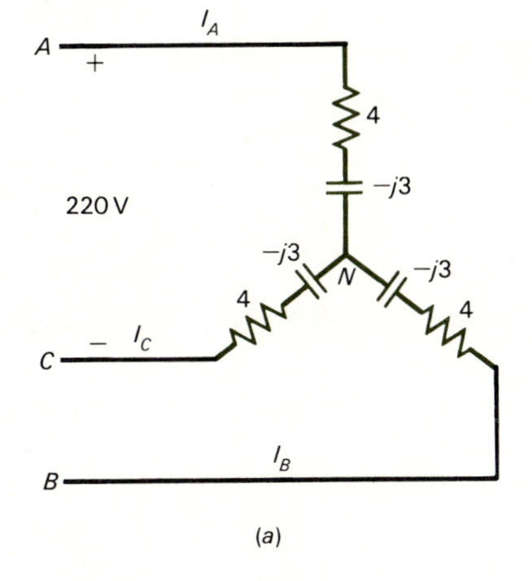

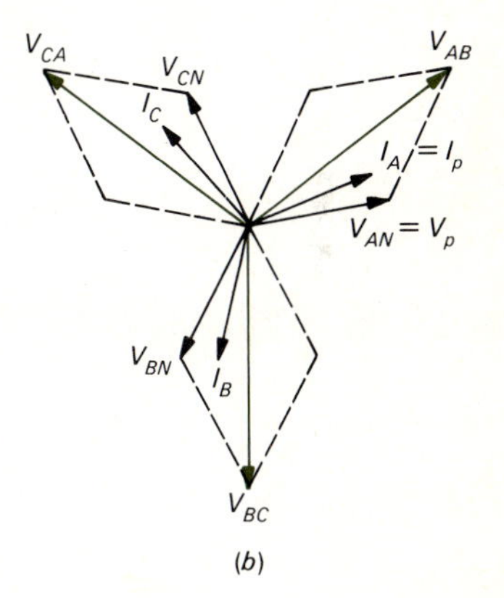

FIGURE 12.10
Example 12.5.

Solution The phasor diagram is shown in Fig. 12.10*b*.

$$V_p = \frac{220}{\sqrt{3}}$$
$$= 127 \text{ V}$$
$$Z_p = 4 - j3$$
$$= 5\underline{/-36.87} \ \Omega$$
$$I_l = I_p$$
$$= \frac{127}{5\underline{/-36.87}}$$
$$= 25.4\underline{/36.87} \text{ A}$$

Hence,

$$P = \sqrt{3} \times 220 \times 25.4 \cos 36.87$$
$$= 7.74 \text{ kW}$$

12.4 WYE-DELTA EQUIVALENCE

In working with *balanced* three-phase systems, computations are usually made on a *per-phase* basis, and then the total results for the entire circuit are obtained on the basis of the symmetry and other factors which must apply. Thus, if a set of three identical wye-connected impedances is connected to a wye-connected three-phase source (as shown in Fig. 12.11), then because of the symmetry, point O' will prove to have the same potential as point O and could therefore be connected to it. Thus, each generator seems to supply only its own phase, and the computations on a per-phase basis are therefore legitimate.

Sometimes it is necessary, or desirable, to mathematically convert a set of identical impedances connected in wye to an equivalent set connected in delta, or vice versa (Fig. 12.12). In order for this transformation to be valid, the impedances seen between any two of the three terminals must be the same. Thus, getting the impedance Z_{ab} for both the wye and the delta, and equating them, we get

$$2Z_w = \frac{Z_d(Z_d + Z_d)}{Z_d + Z_d + Z_d} = \tfrac{2}{3}Z_d \tag{12.23}$$

From this we can solve for the wye impedances in terms of the delta impedances, or vice versa, such that

$$Z_w = \frac{Z_d}{3} \tag{12.24}$$

and

$$Z_d = 3Z_w \tag{12.25}$$

Hence, we can switch back and forth between balanced wye and delta as desired.

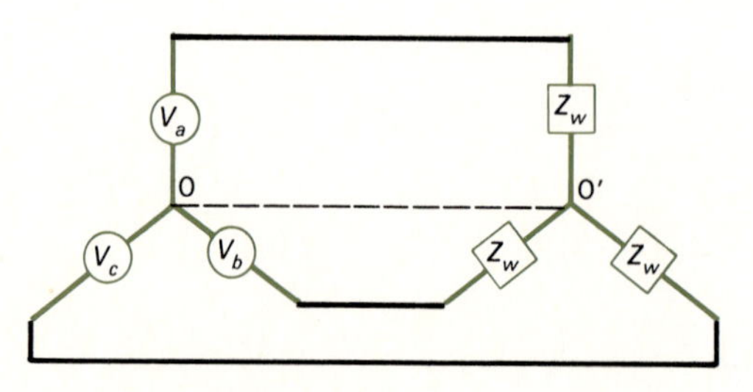

FIGURE 12.11 Reduction of a three-phase system to three single-phase systems.

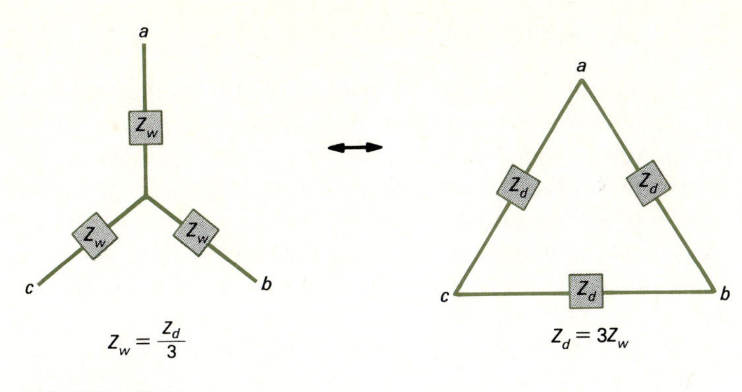

FIGURE 12.12
Equivalence of YΔ impedances.

Example 12.6 The load on a three-phase wye-connected 220-V system consists of three 6-Ω resistances connected in wye and in parallel with three 9-Ω resistances connected in delta. Calculate the magnitude of the line current.

Solution Converting the delta load to a wye, from Eq. (12.24) we have

$$R_w = \tfrac{1}{3} \times 9$$
$$= 3\ \Omega$$

This resistance combined with the wye-connected load of 6 Ω per phase gives a per-phase resistance R_p as

$$R_p = \frac{6 \times 3}{6 + 3}$$
$$= 2\ \Omega$$

The phase voltage is

$$\frac{220}{\sqrt{3}} = 127\ \text{V}$$

Hence,

$$I_p = I_l$$
$$= \tfrac{127}{2}$$
$$= 63.5\ \text{A}$$

12.5 HARMONICS IN THREE-PHASE SYSTEMS

The term *harmonics* in power circuitry refers to components of current or voltage, the frequency of which is greater than the *fundamental*, or *source*, frequency. Voltage and current components, the frequency of which is *less* than the fundamental, or applied frequency, are known as *subharmonics*. Voltage and current harmonics, as generally used in power circuit theory and in this text, are found by means of a Fourier series analysis of the voltage or current waveforms or, experimentally, by waveform analyzers or similar instrumentation. In power circuits, the source harmonic current is due to the nonlinear characteristic of the magnetic circuit of an electromechanical device, such as a transformer or alternator. Both current and voltage harmonics also result from the application of nonsinusoidal voltage or current sources. In general, sources involving power semiconductors are nonsinusoidal.

The role of harmonic voltages and currents in polyphase power systems is a fascinating subject and involves far more theory, analysis, and experimentation than could possibly be covered in this text. There are, however, several simple concepts that are worth noting. In a three-phase system, the phase shift of the third harmonic of the second phase is $3 \times 120°$, or $360°$; likewise, in the third phase the phase shift is $3 \times 240°$, or $720°$ or $360°$. This tells us that in a three-phase system the third harmonics are all shifted in such a manner as to be in phase with each other. This has profound implications concerning the manner in which the phases are connected with respect to each other. For example, the third (and all odd multiples thereof) harmonic components of current in the three windings are in phase with each other. In a wye connection, this may add up to considerable current flowing in the neutral wire; in a delta, this amounts to a "circulating" current around the delta. The delta connection is often used for just this purpose—to "trap" the third harmonic currents—that is, to contain third harmonic currents as circulating currents around the delta windings. The "price" for trapping third harmonic currents in a delta winding is, of course, increased ohmic losses—and hence increased temperature rise—in the windings. The benefits are improved voltage and current waveforms on the three-phase lines. Tradeoffs such as this are usually made in power systems on the basis of cost, efficiency, and complexity versus the penalties for nonsinusoidal waveforms.

PROBLEMS

12.1 Three identical impedances, each having a value $Z = 12\underline{/-15°}\ \Omega$, are connected in delta to a balanced three-phase source of 100 V (line-to-line). Determine the line and phase currents, three-phase power, and three-phase voltamperes, and draw the phasor diagram.

12.2 Three identical impedances, each having a value $Z = (3 + j4)\ \Omega$, are connected in wye and supplied by a 400-V three-phase source. Determine the quantities listed in Prob. 12.1 and draw the phasor diagram.

12.3 The three impedances $Z_A = 20\underline{/0°}\ \Omega$, $Z_B = 10\underline{/30°}\ \Omega$, and $Z_C = 25\underline{/0°}\ \Omega$ are connected in delta to a 100-V source. Calculate the line currents.

12.4 Repeat Prob. 12.3 if the impedances are connected in wye, the supply voltage remaining unchanged.

12.5 Repeat Prob. 12.3 if the sequence of the source voltage is reversed.

12.6 Determine the power supplied to the load of Probs. 12.3 and 12.5.

12.7 A neutral is provided to the three-phase system of Prob. 12.4. What is the neutral current?

12.8 A balanced three-phase wye-connected load having an *RC* parallel circuit in each phase is supplied by a 220-V 60-Hz source. If $R = 80\ \Omega$ and $C = 50\ \mu F$, determine the total power and voltamperage taken by the load.

12.9 The load on a 400-V three-phase system consists of a delta-connected load which has a 10-Ω resistance in each phase and operates in parallel with a wye-connected load also having a 10-Ω resistance in each phase. Calculate the line current and the power supplied by the source.

12.10 A balanced delta-connected inductive load takes 4 kW of power at 220 V while drawing a line current of 15 A. What is the impedance in each phase?

12.11 If the impedances determined in Prob. 12.10 are connected in wye and supplied by a 220-V source, determine the power and voltamperage supplied to the load.

REFERENCES

12.1 C. I. Hubert, *Electric Circuits AC-DC: An Integrated Approach*, McGraw-Hill, New York, 1982.

12.2 J. D. Irwin, *Basic Engineering Circuit Analysis*, Macmillan, New York, 1983.

12.3 D. F. Mix and N. M. Schmitt, *Circuit Analysis for Engineers*, Wiley, New York, 1985.

Chapter

Magnetic Circuits and Transformers

Magnetic materials are used in many electronic and electromechanical systems. In a sense, *all* materials are magnetic materials at room temperature in that a magnetic field can exist within the material. However, the term *magnetic material* is usually reserved for those materials which exhibit certain magnetic properties (discussed in Sec. 13.1) or which exhibit permanent magnetism. In general, the term *magnetic circuit* applies to any closed path in space, but in the analysis of electromechanical and electronic systems this term is usually used in reference to a specific class of circuits which contain considerable portions composed of magnetic materials. This class of magnetic circuit includes most rotating machines, transformers, saturable reactors, electromagnetic relays, solenoids, many types of actuators, doorbells, and mechanical choppers. An understanding of magnetic circuit concepts is useful in the design, analysis, and applications of most of these electromagnetic devices.

13.1 MAGNETIC CIRCUITS

A magnetic circuit provides a path for magnetic flux, just as an electric circuit provides a path for the flow of electric current. Magnetic circuits are an integral part of transformers and electric machines. The analysis of magnetic circuits depends on the following basic concepts.

We define the magnetic flux density B by the force equation

$$F = BlI \tag{13.1}$$

where F is the force [in newtons (N)] experienced by a straight conductor of length l [in meters (m)] carrying a current I [in amperes (A)] and oriented at right angles to a magnetic field of flux density B [in tesla (T)]. In other words, if a conductor is 1 m long, carries a 1-A current, and experiences a 1-N force when located at right angles to certain magnetic flux lines, then the flux density is 1 T. The force, the conductor, and the flux density are mutually perpendicular to each other.

The magnetic flux ϕ through a given surface is defined by

$$\phi = \int_s \mathbf{B} \cdot d\mathbf{s} \tag{13.2}$$

If B is uniform over an area A and is perpendicular to A, then

$$\phi = BA \tag{13.3}$$

from which

$$B = \frac{\phi}{A} \tag{13.4}$$

The unit of magnetic flux is the weber (Wb), and B may be expressed in $\text{Wb/m}^2 = 1$ T.

The source of magnetic flux is either a permanent magnet or an electric current. To measure the effectiveness of electric current in producing a magnetic field (or flux), we introduce the concept of magnetomotive force (or mmf) $\mathscr{F}$, defined as

$$\mathscr{F} \equiv NI \tag{13.5}$$

where I is the current (A) flowing in an N-turn coil. The unit of mmf is the ampere-turn (At). Schematically, a magnetic circuit with an mmf and magnetic flux are shown in Fig. 13.1.

Although we have defined mmf by Eq. (13.5), the mutual relationship between an electric current I and the corresponding magnetic field intensity H is given by *Ampère's circuital law*, expressed as

$$\oint \mathbf{H} \cdot d\mathbf{l} = I \tag{13.6}$$

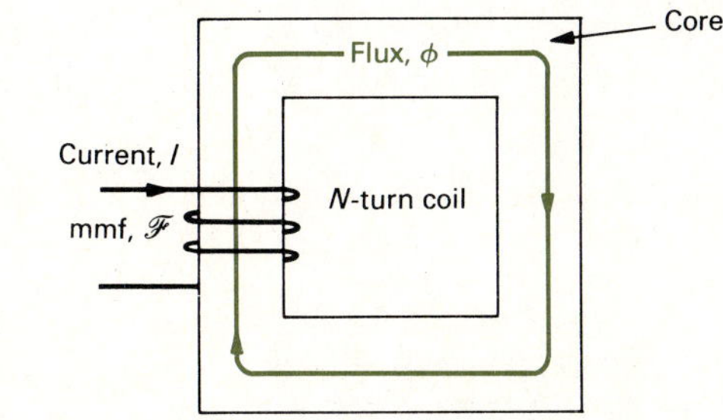

FIGURE 13.1
A magnetic circuit showing mmf and flux.

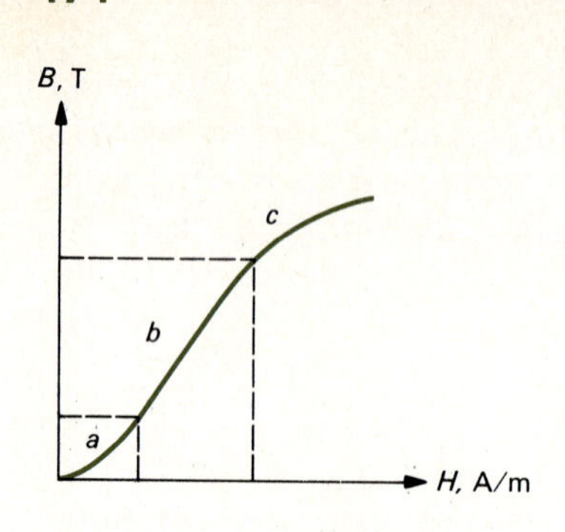

FIGURE 13.2
A B-H curve.

When the closed path is threaded by the current N times, as in Fig. 13.1, Eq. (13.6) becomes

$$\oint \mathbf{H} \cdot d\mathbf{l} = NI \equiv \mathcal{F} \tag{13.7}$$

The core material of a magnetic circuit (constituting a transformer or an electric machine) is generally of such a material having the variation of B with H as depicted by the saturation curve of Fig. 13.2. The slope of the curve depends upon the operating flux density, as classified in regions a, b, and c. For region b, which is of a constant slope, we may write

$$B = \mu H \tag{13.8}$$

where μ is defined as the permeability of the material and is measured in henrys per meter (H/m). For free space (or air), we have $\mu = \mu_o = 4\pi \times 10^{-7}$ H/m. In terms of μ_o, Eq. (13.8) is sometimes written as

$$B = \mu_r \mu_o H \tag{13.9}$$

where $\mu_r = \mu/\mu_o$ and is called *relative permeability*. For ferromagnetic materials, $\mu_r \gg 1$.

Based on an analogy between a magnetic circuit and a dc resistive circuit, Table 13.1 summarizes the corresponding quantities. In this table, l is the length and A is the cross-sectional area of the path for the current in the electric circuit or for the flux in the magnetic circuit. Based on the above analogy, the laws of resistances in series or parallel also hold for reluctances.

TABLE 13.1

DC resistive circuit	Magnetic circuit
Current, I	Flux, ϕ (in Wb)
Voltage, V	Magnetomotive force, $\mathcal{F}$ (in At)
Conductivity, σ	Permeability, μ (in H/m)
Ohm's law, $I = V/R$	$\phi = \mathcal{F}/\mathcal{R}$
Resistance, $R = l/\sigma A$	Reluctance, $\mathcal{R} = l/\mu A$ (in H^{-1})
Conductance, $G = 1/R$	Permeance, $\mathcal{P} = 1/\mathcal{R}$ (in H)

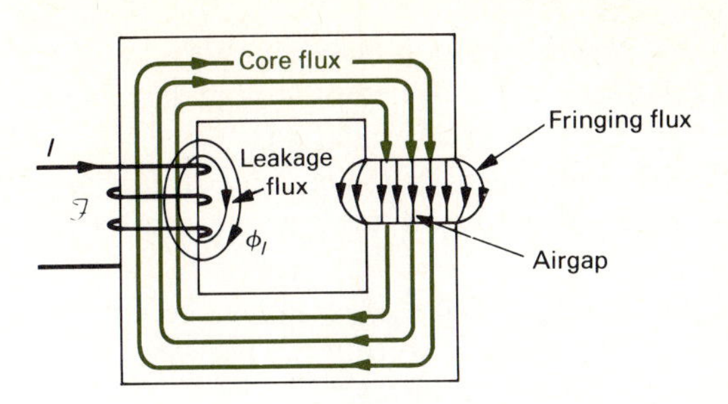

FIGURE 13.3 Leakage flux and fringing flux.

The differences between a dc resistive circuit and a magnetic circuit are that (1) we have an I^2R loss in a resistance but do not have a $\phi^2\mathcal{R}$ loss in a reluctance, (2) magnetic fluxes take *leakage* paths (as ϕ_l in Fig. 13.3) but electric currents (flowing through resistances) do not, and (3) in magnetic circuits with air gaps we encounter *fringing* of flux lines (Fig. 13.3) but do not have fringing of currents in electric circuits. Fringing increases with the length of the air gap and increases the effective area of the air gap.

If the mmf acting in a magnetic circuit is ac, then the B–H curve takes the form shown in Fig. 13.4. The loop shown is known as a hysteresis loop, and the area within the loop is proportional to the energy loss (as heat) per cycle. This energy loss is known as *hysteresis loss*. *Eddy-current loss*, the loss due to the eddy currents induced in the core material of a magnetic circuit excited by an ac mmf, is another feature of an ac-operated magnetic circuit. The power losses due to hysteresis and eddy currents—collectively known as *core losses* or *iron losses*—are approximately given by

eddy-current loss:

$$P_e = k_e f^2 B_m^2 \qquad \text{W/kg} \tag{13.10}$$

hysteresis loss

$$P_h = k_h f B_m^{1.5 \text{ to } 2.5} \qquad \text{W/kg} \tag{13.11}$$

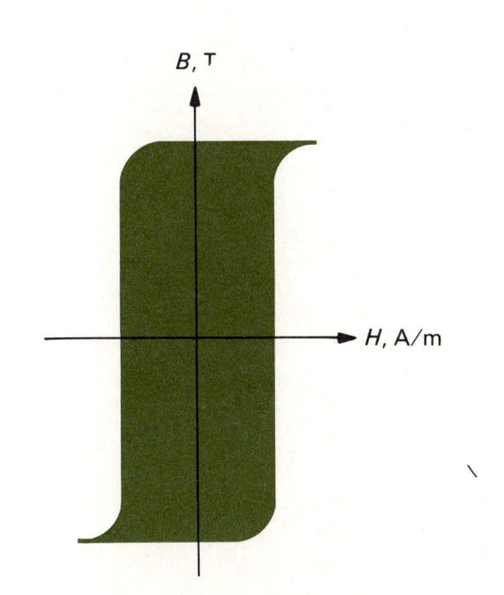

FIGURE 13.4 A hysteresis loop of a core material.

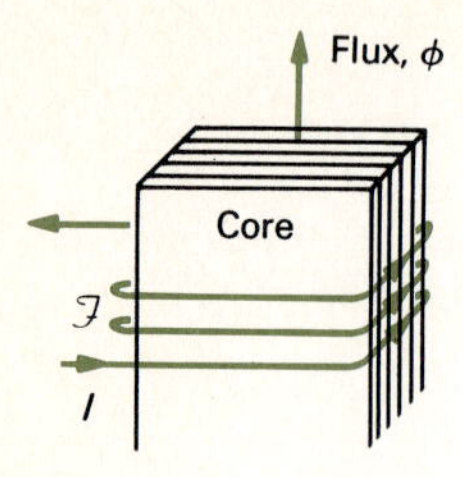

FIGURE 13.5
A laminated core.

where k_e is a constant depending upon material conductivity and thickness, k_h is another constant depending upon the hysteresis loop of the material, B_m is the maximum core flux density, f is the frequency of excitation.

The hysteresis loss component of the core loss in a magnetic circuit is reduced by using "good" quality electrical steel (having a narrow hysteresis loop) for the core material.

The eddy-current loss is reduced by making the core of laminations, or thin sheets, with very thin layers of insulation alternating with the laminations. The laminations are oriented parallel to the direction of flux (Fig. 13.5). Laminating a core increases its cross-sectional area and hence its volume. The ratio of the volume actually occupied by the magnetic material to the total volume of the core is called the *stacking factor*. Table 13.2 gives some values of stacking factors.

TABLE 13.2

Lamination thickness, mm	Stacking factor
0.0127	0.50
0.0254	0.75
0.0508	0.85
0.10 to 0.25	0.90
0.27 to 0.36	0.95

Typical magnetic characteristics of certain core materials are given in Fig. 13.6.

The magnetic circuit concepts developed so far are now illustrated by the following examples.

FIGURE 13.6 E1

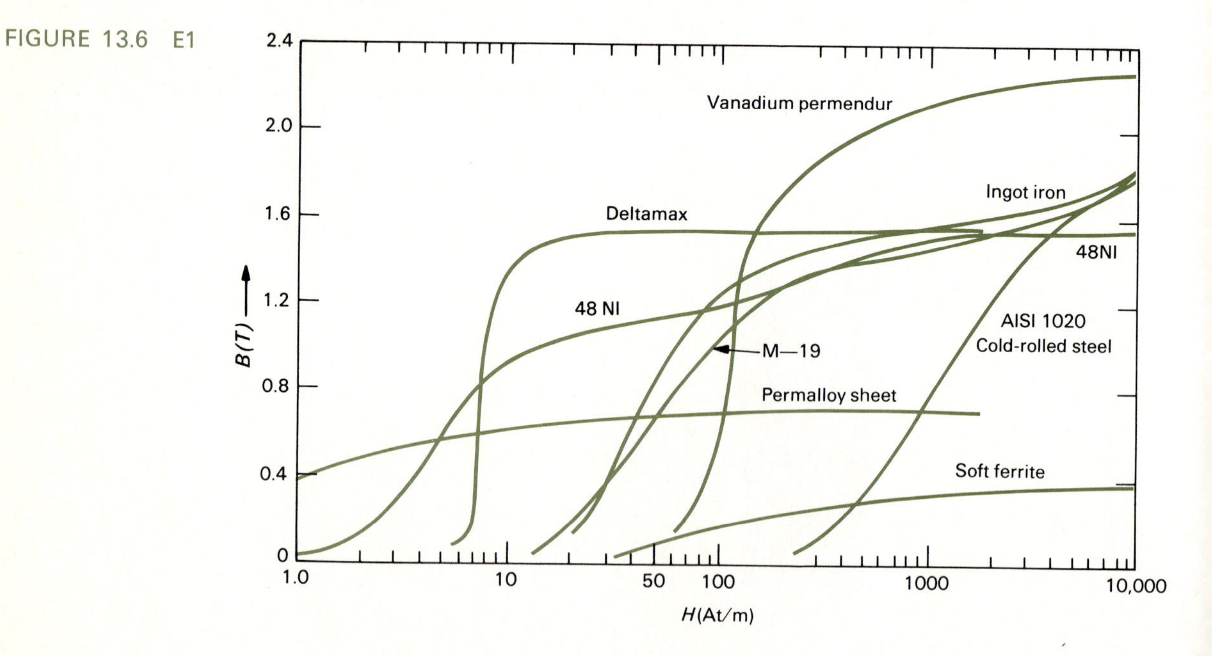

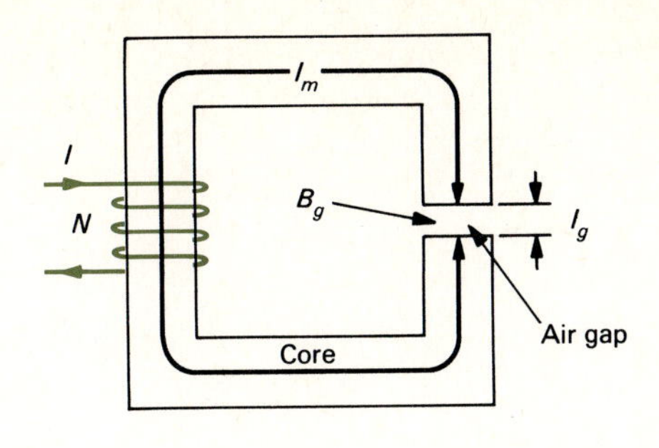

FIGURE 13.7
Magnetic circuit.

Example 13.1 The magnetic circuit shown in Fig. 13.7 carries a 50-turn coil on its core; the total core is made of 0.15-mm-thick laminations of M-19 (Fig. 13.6); the total core length is 10 cm and the air-gap length is 0.1 mm. Calculate the coil current to establish an air-gap flux density of 1 T. Thus, what is the core flux if the core's cross section is 2.5 by 2.5 cm? Neglect fringing and leakage.

Solution For the air gap,

$$H_g = \frac{B_g}{\mu_0}$$

$$= \frac{1.0}{4\pi \times 10^{-7}}$$

$$= 7.95 \times 10^5 \text{ A/m}$$

$$\mathscr{F}_g = H_g l_g$$

$$= (7.95 \times 10^5)(10^{-4})$$

$$= 79.5 \text{ At}$$

For the core, given from Table 13.2 that the stacking factor for a 0.15-mm-thick lamination = 0.9,

$$B_c = \frac{B_g}{0.9}$$

$$= \frac{1}{0.9}$$

$$= 1.11 \text{ T}$$

From Fig. 13.6, for M-19 at 1.11 T we have

$$H_c = 130 \text{ A/m}$$

$$\mathscr{F}_c = H_c l_c$$

$$= 130 \times 0.1$$

$$= 13 \text{ At}$$

For the entire magnetic circuit,

$$NI = \mathscr{F}_{\text{total}}$$
$$= \mathscr{F}_g + \mathscr{F}_c$$
$$= 79.5 + 13$$
$$= 92.5 \text{ At}$$

from which

$$I = \frac{92.5}{50}$$
$$= 1.85 \text{ A}$$

The core flux is thus

$$B_c A_c = 1.11 \times (2.5)^2 \times 10^{-4}$$
$$= 0.694 \text{ milliweber (mWb)}$$

Example 13.2 Define flux linkage λ by $\lambda = N\phi$, where ϕ is the magnetic flux threading an N-turn coil; also define inductance L by $L = \lambda/i$ (or flux linkage per ampere), i being the current in the coil to produce the flux ϕ. Hence, determine the inductance of the coil of Example 13.1.

In addition, evaluate the energy stored in the coil. By computing the energy in the air gap, calculate the magnetic energy stored in the core of the magnetic circuit, and hence show that the energy density in the air gap is much greater than the energy density in the core.

Solution From Example 13.1, $\phi = 6.94 \times 10^{-4}$ Wb and $N = 50$. Thus, for the inductance of the coil we have

$$\lambda = N\phi$$
$$= 50 \times 6.94 \times 10^{-4}$$
$$= 3.47 \times 10^{-2} \text{ Wb-turns}$$

$$L = \frac{\lambda}{i}$$
$$= \frac{3.47}{1.85} \times 10^{-2}$$
$$= 18.76 \text{ mH}$$

Now evaluating the stored energy [in joule (J)], for the energy stored in L we have

$$W_L = \tfrac{1}{2}Li^2$$
$$= \tfrac{1}{2} \times 18.76 \times 10^{-3} \times 1.85^2$$
$$= 0.0321 \text{ J}$$

The energy stored in the air gap is

$$W_g = \frac{B_g^2}{2\mu_o} \times \text{volume}_g$$

$$= \frac{1}{2 \times 4\pi \times 10^{-7}} \times 0.1 \times 10^{-3} \times 2.5^2 \times 10^{-4}$$

$$= 0.0249 \text{ J}$$

and the energy stored in the core is

$$W_c = W_L - W_g$$

$$= 0.0321 - 0.0249$$

$$= 0.0072 \text{ J}$$

and hence

$$\frac{\text{Energy density in air}}{\text{Energy density in core}} = \frac{W_g/\text{volume}_g}{W_c/\text{volume}_c}$$

$$= \frac{0.0249/0.1 \times 10^{-3} \times 2.5^2 \times 10^{-4}}{0.0072/10 \times 10^{-2} \times 2.5^2 \times 10^{-4}}$$

$$= 345.8$$

Example 13.3 Using the definition of inductance given in Example 13.2, obtain an expression for inductance in terms of reluctance $\mathscr{R}$ and number of turns N. Hence, determine the reluctance of the entire magnetic circuit of Example 13.1.

Solution

$$L = \frac{\lambda}{i} = \frac{N\phi}{i} = \frac{N}{i}\frac{\mathscr{F}}{\mathscr{R}} = \frac{N}{i}\frac{Ni}{\mathscr{R}} = \frac{N^2}{\mathscr{R}}$$

Since

$$\frac{N^2}{\mathscr{R}} = 18.76 \times 10^{-3}$$

(from Example 13.2), we obtain

$$\mathscr{R} = \frac{50 \times 50}{18.76 \times 10^{-3}}$$

$$= 1.33 \times 10^5 \text{ H}^{-1}$$

13.2 PRINCIPLE OF OPERATION OF A TRANSFORMER

A transformer is an electromagnetic device having two or more mutually coupled windings. Figure 13.8 shows a two-winding ideal transformer. The transformer is ideal in the sense that its core is lossless, is infinitely permeable, and has no leakage fluxes and that its windings have no losses.

In Fig. 13.8 the basic components are the core, primary winding N_1, and the secondary winding N_2. If ϕ is the mutual (or core) flux linking N_1 and N_2, then, according to Faraday's law of electromagnetic induction, emf's e_1 and e_2 are induced in N_1 and N_2 owing to a finite rate of change of ϕ such that

$$e_1 = N_1 \frac{d\phi}{dt} \tag{13.12}$$

and

$$e_2 = N_2 \frac{d\phi}{dt} \tag{13.13}$$

The direction of e_1 is such as to produce a current which opposes the flux change, according to Lenz's law. The transformer being ideal, $e_1 = v_1$ (Fig. 13.8). From Eqs. (13.12) and (13.13)

$$\frac{e_1}{e_2} = \frac{N_1}{N_2}$$

which may also be written in terms of RMS values as

$$\frac{E_1}{E_2} = \frac{N_1}{N_2} = a \tag{13.14}$$

where a is known as the turns ratio.

Since $e_1 = v_1$ (and $e_2 = v_2$), the flux and voltage are related by

$$\phi = \frac{1}{N_1} \int v_1 \, dt = \frac{1}{N_2} \int v_2 \, dt \tag{13.15}$$

If the flux varies sinusoidally such that

$$\phi = \phi_m \sin \omega t$$

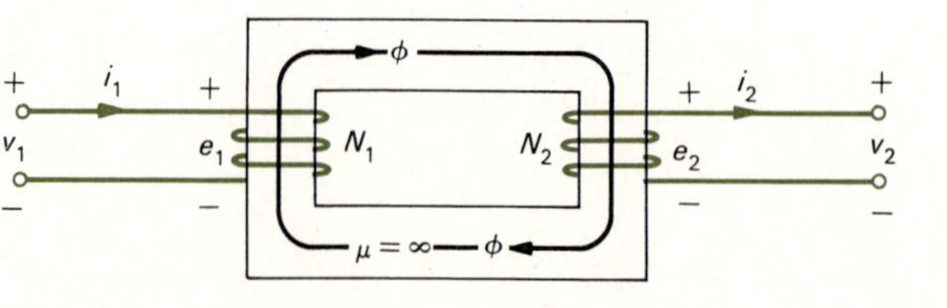

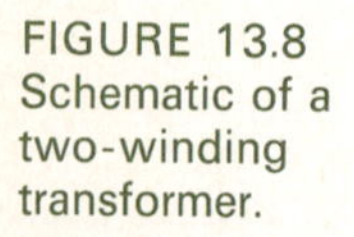
FIGURE 13.8 Schematic of a two-winding transformer.

then the corresponding induced voltage e linking an N-turn winding is given by

$$e = \omega N \phi_m \cos \omega t \tag{13.16}$$

From Eq. (13.16), the RMS value of the induced voltage is

$$\begin{aligned} E &= \frac{\omega N \phi_m}{\sqrt{2}} \\ &= 4.44 f N \phi_m \end{aligned} \tag{13.17}$$

which is known as the emf equation. In Eq. (13.17), $f = \omega/2\pi$ is the frequency in hertz.

13.3 VOLTAGE, CURRENT, AND IMPEDANCE TRANSFORMATIONS

Major applications of transformers are in voltage, current, and impedance transformations and in providing isolation (that is, eliminating direct connections between electrical circuits). The voltage transformation property of an ideal transformer (mentioned in Sec. 13.2) is expressed as

$$\frac{V_1}{V_2} = \frac{E_1}{E_2} = a \tag{13.18}$$

where the subscripts 1 and 2 correspond to the primary and secondary sides, respectively. This property of a transformer enables us to interconnect transmission and distribution systems of different voltage levels in an electric power system.

For an ideal transformer, the net mmf around its magnetic circuit must be zero, implying that

$$N_1 I_1 - N_2 I_2 = 0 \tag{13.19}$$

where I_1 and I_2 are the primary and secondary currents, respectively. From Eqs. (13.14) and (13.19) we get

$$\frac{I_2}{I_1} = \frac{N_1}{N_2} = a \tag{13.20}$$

From Eqs. (13.18) and (13.20) it can be shown that if an impedance Z_2 is connected to the secondary, the impedance Z_1 seen at the primary satisfies

$$\frac{Z_1}{Z_2} = \left(\frac{N_1}{N_2}\right)^2 \equiv a^2 \tag{13.21}$$

Example 13.4 How many turns must the primary and secondary windings of a 220/110-V 60-Hz ideal transformer have if the core flux is not allowed to exceed 5 mWb?

Solution From the emf equation, Eq. (13.17), we have

$$N = \frac{E}{4.44 f \phi_m}$$

Consequently,

$$N_1 = \frac{220}{4.44 \times 60 \times 5 \times 10^{-3}}$$

$$\simeq 166 \text{ turns}$$

$$N_2 = \tfrac{1}{2} N_1$$

$$= 83 \text{ turns}$$

Example 13.5 A 220/110-V 10-kVA transformer has a primary winding resistance of 0.25 Ω and a secondary winding resistance of 0.06 Ω. Determine the primary and secondary currents on rated load, the total winding resistance referred to the primary, and the total winding resistance referred to the secondary.

Solution Given the transformation ratio $a = 220/110 = 2$, the primary current

$$I_1 = \frac{10 \times 10^3}{220}$$

$$= 45.45 \text{ A}$$

and the secondary current

$$I_2 = aI_1$$

$$= 2 \times 45.45$$

$$= 90.9 \text{ A}$$

From Eq. (13.21), the secondary winding resistance referred to the primary $= a^2 R_2 = 2^2 \times 0.06 = 0.24\ \Omega$. The total resistance referred to the primary is thus

$$R_e' = R_1 + a^2 R_2$$

$$= 0.25 + 0.24$$

$$= 0.49\ \Omega$$

The primary winding resistance referred to the secondary = $R_1/a^2 = 0.25/4 = 0.0625\ \Omega$. The total resistance referred to the secondary is thus

$$R_e'' = \frac{R_1}{a^2} + R_2$$

$$= 0.0625 + 0.06$$

$$= 0.1225\ \Omega$$

Example 13.6

Determine the I^2R loss in each winding of the transformer of Example 13.5, and thus find the total I^2R loss in the two windings. Verify that the same result can be obtained by using the equivalent resistance referred to the primary winding.

Solution

$$I_1^2R_1 \text{ loss} = (45.45)^2 \times 0.25$$

$$= 516.425 \text{ W}$$

$$I_2^2R_2 \text{ loss} = (90.9)^2 \times 0.06$$

$$= 495.768 \text{ W}$$

$$\text{Total } I^2R \text{ loss} = 1012.19 \text{ W}$$

For the equivalent resistance,

$$I_1^2R_e' = (45.45)^2 \times 0.49$$

$$= 1012.19 \text{ W}$$

which is consistent with the preceding result.

13.4 THE NONIDEAL TRANSFORMER AND ITS EQUIVALENT CIRCUITS

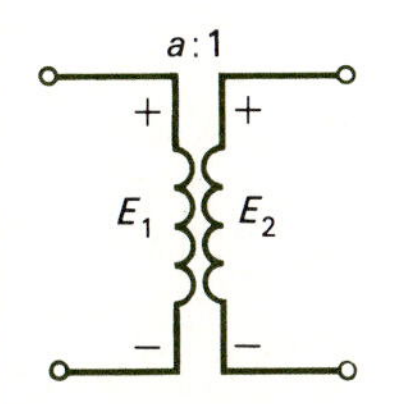

FIGURE 13.9 An ideal transformer.

In contrast to an ideal transformer, a nonideal (or actual) transformer has hysteresis and eddy-current losses (core losses) and has resistive (I^2R) losses in its primary and secondary windings. Furthermore, the core of a nonideal transformer is not perfectly permeable and thus requires a finite mmf for its magnetization. Also, because of leakage, not all fluxes link with the primary and the secondary windings simultaneously in a nonideal transformer.

An equivalent circuit of an ideal transformer is shown in Fig. 13.9. When the nonideal effects of winding resistances, leakage reactances, magnetizing reactance, and core losses are included, the circuit of Fig. 13.9 is modified to that of Fig. 13.10, where the primary and the secondary are coupled by an ideal transformer. By using Eqs. (13.18), (13.20), and (13.21), we may remove the ideal transformer from Fig. 13.10 and refer the entire equivalent circuit either to the primary, as shown in Fig. 13.11, or to the secondary, as shown in Fig. 13.12.

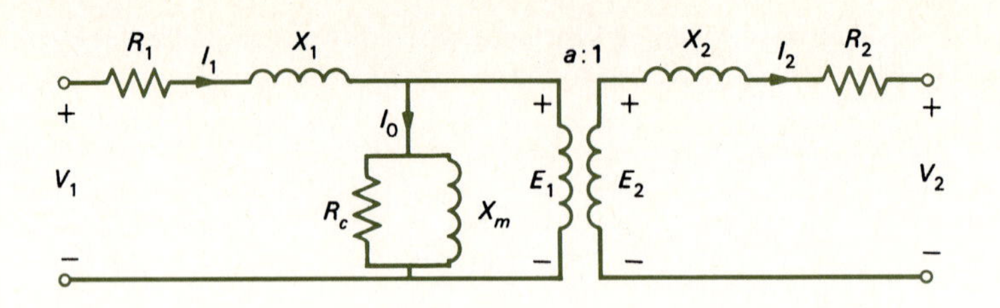

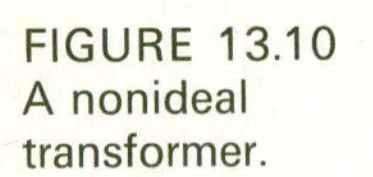
FIGURE 13.10
A nonideal transformer.

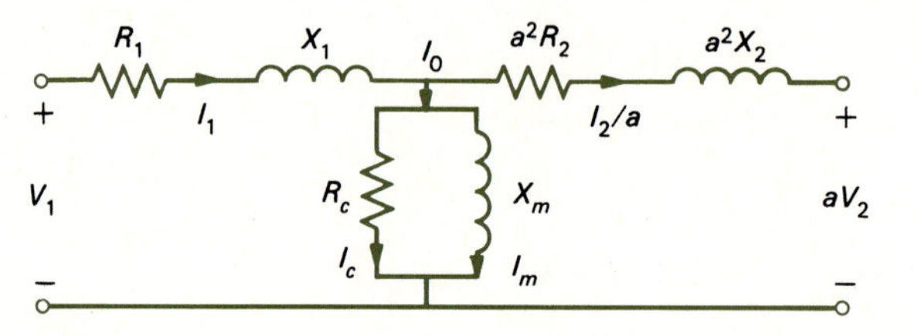

FIGURE 13.11
Equivalent circuit referred to primary.

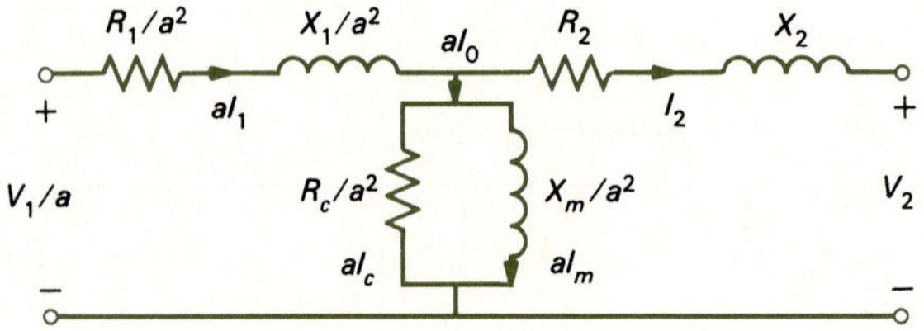

FIGURE 13.12
Equivalent circuit referred to secondary.

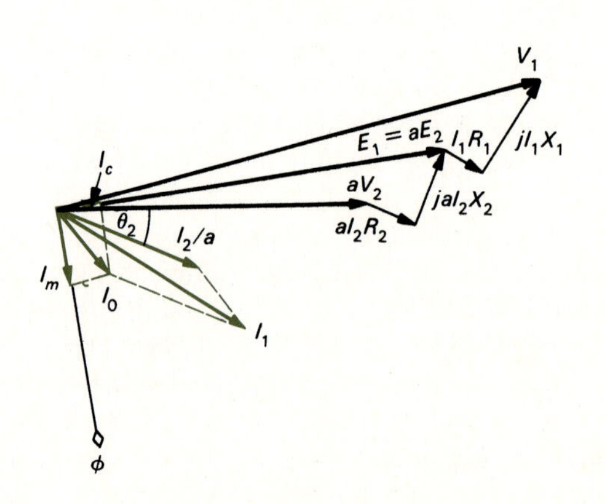

FIGURE 13.13
Phasor diagram corresponding to Fig. 10.

A phasor diagram for the circuit of Fig. 13.11, for lagging power factor, is shown in Fig. 13.13. In Figs. 13.9 to 13.13 the various symbols are

$a \equiv$ turns ratio

$E_1 \equiv$ primary induced voltage

$E_2 \equiv$ secondary induced voltage

$V_1 \equiv$ primary terminal voltage

$V_2 \equiv$ secondary terminal voltage

$I_1 \equiv$ primary current

$I_2 \equiv$ secondary current

$I_0 \equiv$ no-load (primary) current

$R_1 \equiv$ resistance of the primary winding

$R_2 \equiv$ resistance of the secondary winding

$X_1 \equiv$ primary leakage reactance

$X_2 \equiv$ secondary leakage reactance

$I_m, X_m \equiv$ magnetizing current and reactance

$I_c, R_c \equiv$ current and resistance accounting for the core losses

The constants of the equivalent circuit can be found from certain tests (discussed in Sec. 13.5). The major use of the equivalent circuit of a transformer is for determining its characteristics. The characteristics of most interest to power engineers are voltage regulation and efficiency. *Voltage regulation* is a measure of the change in the terminal voltage of the transformer with load. From Fig. 13.13 it is clear that the terminal voltage V_1 is load-dependent. Specifically, we define voltage regulation as

$$\text{Percent regulation} = \frac{V_{\text{no-load}} - V_{\text{load}}}{V_{\text{load}}} \times 100 \tag{13.22}$$

With reference to Fig. 13.13, we may rewrite Eq. (13.22) as

$$\text{Percent regulation} = \frac{V_1 - aV_2}{aV_2} \tag{13.23}$$

for a given load.

There are two kinds of transformer efficiencies of interest to us; they are known as *power efficiency* and *energy efficiency*. These are defined as follows:

$$\text{Power efficiency} = \frac{\text{output power}}{\text{input power}} \tag{13.24}$$

$$\text{Energy efficiency} = \frac{\text{output energy for a given period}}{\text{input energy for the same period}} \tag{13.25}$$

Generally, energy efficiency is taken over a 24-h period and is called *all-day efficiency*. In such a case, Eq. (13.25) becomes

$$\text{All-day efficiency} = \frac{\text{output for 24 h}}{\text{input for 24 h}} \tag{13.26}$$

We will use the term *efficiency* to mean power efficiency from now on. The output power is less than the input power because of losses: the I^2R losses in the windings and the hysteresis and eddy-current losses in the core. Thus, in terms of these losses, Eq. (13.24) may be more meaningfully expressed as

$$\text{Efficiency} = \frac{\text{input power} - \text{losses}}{\text{input power}}$$

$$= \frac{\text{output power}}{\text{output power} + \text{losses}}$$

$$= \frac{\text{output power}}{\text{output power} + I^2R \text{ loss} + \text{core loss}} \tag{13.27}$$

Obviously, I^2R loss is load-dependent, whereas the core loss is constant and independent of the load on the transformer. The next examples show the voltage regulation and efficiency calculations for a transformer.

Example 13.7 A 150-kVA 2400/240-V transformer has the following parameters: $R_1 = 0.2\ \Omega$, $R_2 = 0.002\ \Omega$, $X_1 = 0.45\ \Omega$, $X_2 = 0.0045\ \Omega$, $R_c = 10{,}000\ \Omega$, and $X_m = 1550\ \Omega$, where the symbols are shown in Fig. 13.10. Refer the circuit to the primary. From this circuit, calculate the voltage regulation of the transformer at rated load with 0.8 lagging power factor.

Solution The circuit referred to the primary is shown in Fig. 13.11. From the given data we have $V_2 = 240$ V, $a = 10$, and $\theta_2 = -\cos^{-1} 0.8 = -36.8°$.

$$I_2 = \frac{150 \times 10^3}{240}$$

$$= 625 \text{ A}$$

$$\frac{\mathbf{I}_2}{a} = \frac{I_2}{a} \underline{/\theta_2}$$

$$= 62.5\underline{/-36.8°}$$

$$= 50 - j37.5 \quad \text{A}$$

$$a\mathbf{V}_2 = 2400\underline{/0°}$$

$$= 2400 + j0 \quad \text{V}$$

$$a^2R_2 = 0.2\ \Omega \quad \text{and} \quad a^2X_2 = 0.45\ \Omega$$

Hence,

$$\mathbf{E}_1 = (2400 + j0) + (50 - j37.5)(0.2 + j0.45)$$
$$= 2427 + j15$$
$$= 2427\underline{/0.35^\circ} \text{ V}$$

$$\mathbf{I}_m = \frac{2427\underline{/0.35^\circ}}{1550\underline{/90^\circ}}$$
$$= 1.56\underline{/-89.65^\circ}$$
$$= 0.0095 - j1.56 \quad \text{A}$$

$$\mathbf{I}_c = \frac{2427 + j15}{10{,}000}$$
$$\simeq 0.2427 + j0 \quad \text{A}$$

$$\mathbf{I}_0 = \mathbf{I}_c + \mathbf{I}_m$$
$$= 0.25 - j1.56 \quad \text{A}$$

$$\mathbf{I}_1 = \mathbf{I}_0 + \frac{\mathbf{I}_2}{a}$$
$$= 50.25 - j39.06$$
$$= 63.65\underline{/-37.85^\circ} \quad \text{A}$$

$$\mathbf{V}_1 = (2427 + j15) + (50.25 - j39.06)(0.2 + j0.45)$$
$$= 2455 + j30$$
$$= 2455\underline{/0.7^\circ} \text{ V}$$

$$\text{Percent regulation} = \frac{V_1 - aV_2}{aV_2} \times 100$$
$$= \frac{2455 - 2400}{2400} \times 100$$
$$= 2.3\%$$

Example 13.8 Determine the efficiency of the transformer of Example 13.7, operating on rated load and 0.8 lagging power factor.

Solution

$$\text{Output} = 150 \times 0.8 = 120 \text{ kW}$$

$$\text{Losses} = I_1^2 R_1 + I_c^2 R_c + I_2^2 R_2$$
$$= (63.65)^2 \times 0.2 + (0.2427)^2 \times 10{,}000 + (625)^2 \times 0.002$$
$$= 2.18 \text{ kW}$$

$$\text{Input} = 120 + 2.18$$
$$= 122.18 \text{ kW}$$

$$\text{Efficiency} = \frac{120}{122.18}$$
$$= 98.2\%$$

Example 13.9 The transformer of Example 13.7 operates on full-load 0.8 lagging power factor for 12 h, on no-load for 4 h, and on half-full-load 0.8 lagging power factor for 8 h. Calculate the all-day efficiency.

Solution The output for 24 h is

$$(150 \times 0.8 \times 12) + (0 \times 4) + (150 \times \tfrac{1}{2} \times 8) = 2040 \text{ kWh}$$

The losses for 24 h are as follows.

The core loss is

$$(0.2427)^2 \times 10{,}000 \times 24 = 14.14 \text{ kWh}$$

The I^2R loss on full load for 12 h is

$$12[(63.65)^2 \times 0.2 + (625)^2 \times 0.002] = 19.1 \text{ kWh}$$

The I^2R loss on half-full load for 8 h is

$$8\left[\left(\frac{63.65}{2}\right)^2 \times 0.2 + \left(\frac{625}{2}\right)^2 \times 0.002\right] = 3.18 \text{ kWh}$$

The total losses for 24 h are

$$14.14 + 19.1 + 3.18 = 36.42 \text{ kWh}$$

The input for 24 h is

$$2040 + 36.42 = 2076.42 \text{ kWh}$$

The all-day efficiency is

$$\frac{2040}{2076.42} = 98.2\%$$

13.5 TESTS ON TRANSFORMERS

Transformer performance characteristics can be obtained from the equivalent circuits of Sec. 13.4. The circuit parameters are determined either from design data or from test data. The two common tests are as follows.

Open-Circuit (or No-Load) Test Here one winding is open-circuited and voltage—usually, rated voltage at rated frequency—is applied to the other winding. The voltage, current, and power at the terminals of this winding are measured. The open-circuit voltage of the second winding is also measured, and from this measurement a check on the turns ratio can be obtained. It is usually convenient to apply the test voltage to the winding that has a voltage rating equal to that of the available power source. In step-up voltage transformers, this means that the open-circuit voltage of the second winding will be higher than the applied voltage, sometimes much higher. Care must be exercised in guarding the terminals of this winding to ensure safety for test personnel and to prevent these terminals from getting close to other electrical circuits, instrumentation, grounds, and so forth.

In presenting the no-load parameters obtainable from test data, it is assumed that voltage is applied to the primary and that the secondary is open-circuited. The instrumentation is shown in Fig. 13.14. The no-load power loss is equal to the wattmeter reading in this test; the core loss is found by subtracting the ohmic loss in the primary, which is usually small and may be neglected in some cases. Thus, if P_0, I_0, and V_0 are the input power, current, and voltage, then the core loss is given by

$$P_c = P_0 - I_0^2 R_1 \tag{13.28}$$

The primary induced voltage is given in phasor form by

$$\mathbf{E}_1 = V_0\underline{/0^\circ} - (I_0\underline{/\theta_0})(R_1 + jX_1) \tag{13.29}$$

where $\theta_0 \equiv$ no-load power-factor angle $= \cos^{-1}(P_0/V_0 I_0) < 0$. (*Note*: The determination of R and X is discussed under "Short-Circuit Test," which follows. See also Examples 13.10 and 13.11.)

$$R_c = \frac{E_1^2}{P_c} \tag{13.30}$$

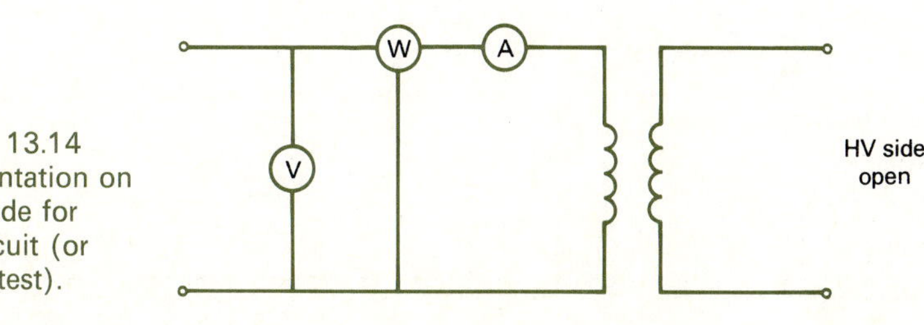

FIGURE 13.14 Instrumentation on the LV side for open circuit (or no-load test).

$$I_c = \frac{E_1}{R_c} \tag{13.31}$$

$$I_m = \sqrt{I_0^2 - I_c^2} \tag{13.32}$$

$$X_m = \frac{E_1}{I_m} \tag{13.33}$$

$$a \approx \frac{V_0}{E_2} \tag{13.34}$$

Short-Circuit Test In this test, one winding is short-circuited across its terminals, and a reduced voltage is applied to the other windings. This reduced voltage is of such a magnitude as to cause a specific value of current—usually, rated current—to flow in the short-circuited winding. Again, the choice of the winding to be short-circuited is usually determined by the measuring equipment available for use in the test. However, care must be taken to note which winding is short-circuited, for this determines the reference winding for expressing the impedance components obtained by this test. Let the secondary be short-circuited and the reduced voltage be applied to the primary, as shown in Fig. 13.15.

With a very low voltage applied to the primary winding, the core-loss current and magnetizing current become very small, and the equivalent circuit reduces to that of Fig. 13.16. Thus, if P_s, I_s, and V_s are the input power, current, and voltage under short circuit, then, referred to the primary,

$$Z_s = \frac{V_s}{I_s} \tag{13.35}$$

$$R_1 + a^2R_2 \equiv R_s = \frac{P_s}{I_s^2} \tag{13.36}$$

$$X_1 + a^2X_2 \equiv X_s = \sqrt{Z_s^2 - R_s^2} \tag{13.37}$$

The primary resistance R_1 can be measured directly and, knowing a, R_2 can be found from Eq. (13.36). In Eq. (13.37) it is usually assumed that the leakage reactance is divided equally between the primary and the secondary; that is,

$$X_1 = a^2X_2 = \tfrac{1}{2}X_s$$

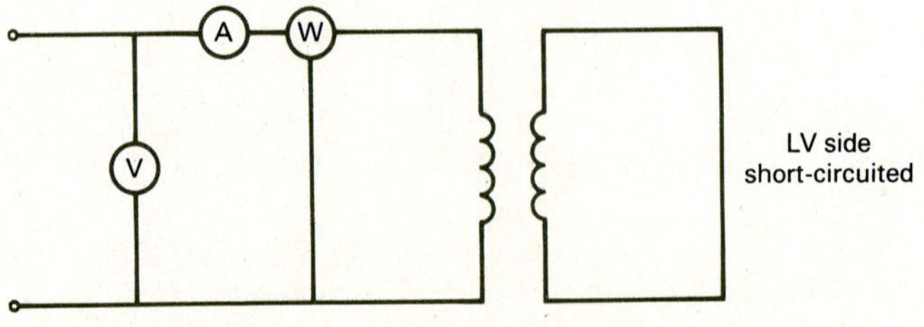

A W V

LV side short-circuited

FIGURE 13.15 Instrumentation on the HV side for short-circuit test.

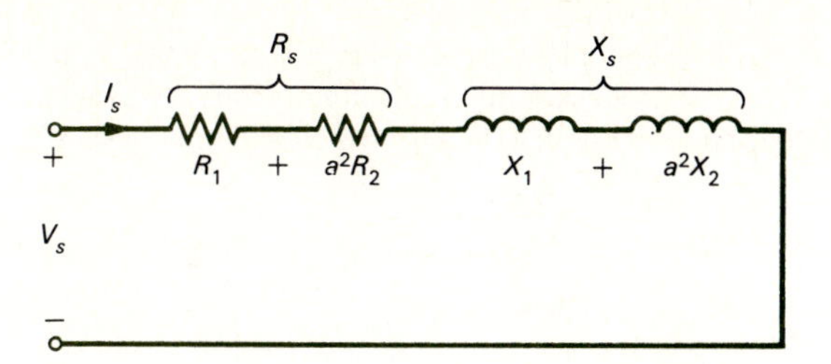

FIGURE 13.16 Equivalent circuit test for short circuit.

Example 13.10 A certain transformer, with its secondary open, takes 80 W of power at 120 V and 1.4 A. The primary winding resistance is 0.25 Ω and the leakage reactance is 1.2 Ω. Evaluate the magnetizing reactance X_m and the core-loss equivalent resistance R_c.

Solution For the no-load test, we have the power factor angle

$$\theta_0 = \cos^{-1}\frac{80}{1.4 \times 120}$$
$$= -61.6°$$

and the primary induced voltage is therefore

$$\mathbf{E}_1 = 120\underline{/0°} - 1.4\underline{/-61.6°}(0.25 + j1.25)$$
$$\simeq 118.29 \text{ V}\underline{/0°}$$

Thus, from Eqs. (13.28)-(13.30)

$$R_c = \frac{(118.29)^2}{79.5}$$
$$= 176\ \Omega$$

$$I_c = \frac{118.29}{176}$$
$$= 0.672 \text{ A}$$

$$I_m = \sqrt{(1.4)^2 - (0.672)^2}$$
$$= 1.228 \text{ A}$$

$$X_m = \frac{118.29}{1.228}$$
$$= 96.3\ \Omega$$

Example 13.11 The results of open-circuit and short-circuit tests on a 25-kVA 440/220-V 60-Hz transformer are as follows:

Open-circuit test Primary open-circuited, with instrumentation on the low-voltage side. Input voltage, 200 V; input current, 9.6 A; input power, 710 W.

Short-circuit test Secondary short-circuited, with instrumentation on the high-voltage side. Input voltage, 42 V; input current, 57 A; input power, 1030 W. Obtain the parameters of the exact equivalent circuit (Fig. 13.11), referred to the high-voltage side. Assume that $R_1 = a^2R_2$ and $X_1 = a^2X_2$.

Solution From the short-circuit test:

$$Z_{s1} = \frac{42}{57}$$

$$= 0.737\ \Omega$$

$$R_{s1} = \frac{1030}{(57)^2}$$

$$= 0.317\ \Omega$$

$$X_{s1} = \sqrt{(0.737)^2 - (0.317)^2}$$

$$= 0.665\ \Omega$$

where the subscript s corresponds to the short-circuit condition and the subscript 1 implies that the quantities are referred to the primary. Consequently,

$$R_1 = a^2R_2$$

$$= 0.158\ \Omega$$

$$R_2 = 0.0395\ \Omega$$

$$X_1 = a^2X_2$$

$$= 0.333\ \Omega$$

$$X_2 = 0.0832\ \Omega$$

From the open-circuit test:

$$\theta_0 = \cos^{-1}\frac{710}{9.6 \times 220}$$

$$= \cos^{-1} 0.336$$

$$= -70^\circ$$

$$\mathbf{E}_2 = 220\underline{/0^\circ} - (9.6\underline{/-70^\circ})(0.0395 + j0.0832)$$

$$\approx 219\underline{/0^\circ}\ \text{V}$$

$$P_{c2} = 710 - (9.6)^2(0.0395)$$

$$\approx 706\ \text{W}$$

$$R_{c2} = \frac{(219)^2}{706}$$

$$= 67.9\ \Omega$$

$$I_{c2} = \frac{219}{67.9}$$

$$= 3.22\ \text{A}$$

$$I_{m2} = \sqrt{(9.6)^2 - (3.22)^2}$$

$$= 9.04\ \text{A}$$

$$X_{m2} = \frac{219}{9.04}$$

$$= 24.22\ \Omega$$

$$X_{m1} = a^2 X_{m2}$$

$$= 96.9\ \Omega$$

$$R_{c1} = a^2 R_{c2}$$

$$= 271.6\ \Omega$$

where the subscript 2 implies that the quantities are referred to the secondary.

13.6 TRANSFORMER POLARITY

Polarities of a transformer identify the relative directions of induced voltages in the two windings. The polarities result from the relative directions in which the two windings are wound on the core. For operating transformers in parallel, it is necessary that we know the relative polarities. Polarities can be checked by a simple test requiring only voltage measurements with the transformer on no-load. In this test, rated voltage

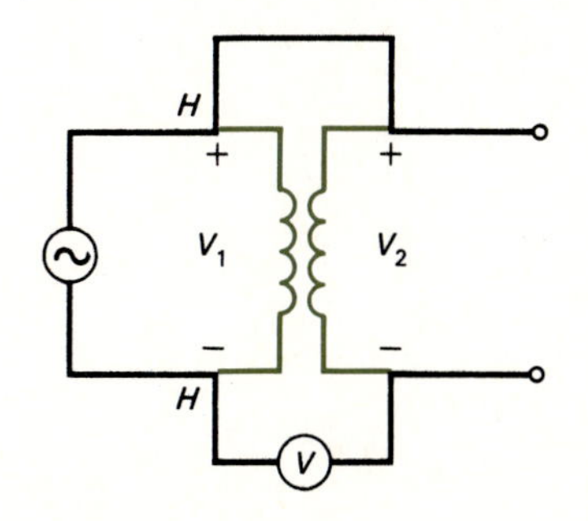

FIGURE 13.17 Polarity test on a transformer.

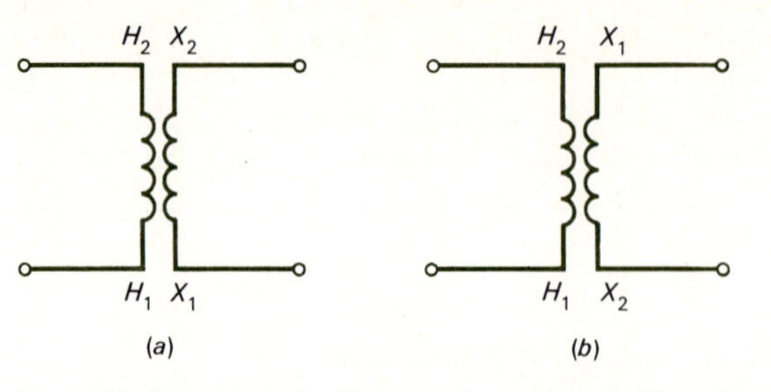

FIGURE 13.18 (*a*) Subtractive and (*b*) additive polarities of a transformer.

is applied to one winding, and an electrical connection is made between one terminal from one winding and one from the other, as shown in Fig. 13.17. The voltage across the two remaining terminals (one from each winding) is then measured. If this measured voltage is larger than the input test voltage, the polarity is *additive*; if smaller, the polarity is *subtractive*.

A standard method of marking transformer terminals is as follows: The high-voltage terminals are marked $H_1, H_2, H_3, \ldots$, with H_1 being on the right-hand side of the case when facing the high-voltage side. The low-voltage terminals are designated $X_1, X_2, X_3, \ldots$, and X_1 may be on either side, adjacent to H_1 or diagonally opposite. The two possible locations of X_1 with respect to H_1 for additive and subtractive polarities are shown in Fig. 13.18. The numbers must be so arranged that the voltage difference between any two leads of the same set, taken in order from smaller to larger numbers, must be of the same sign as that between any other pair of the set taken in the same order. Furthermore, when the voltage is directed from H_1 to H_2, it must simultaneously be directed from X_1 to X_2. Additive polarities are required by the American National Standards Institute (ANSI) in large ($>$ 200-kVA) high-voltage ($>$ 8660-V) power transformers. Small transformers have subtractive polarities (which reduce voltage stress between adjacent leads).

Knowing the polarities of a transformer aids in obtaining a combination of several voltages from the transformer. Certain single-phase connections of transformers are illustrated in the next example.

Example 13.12 Two 1150/115-V transformers are given. Using appropriate polarity markings, show the interconnections of these transformers for 2300/230-V operation and for 1150/230-V operation.

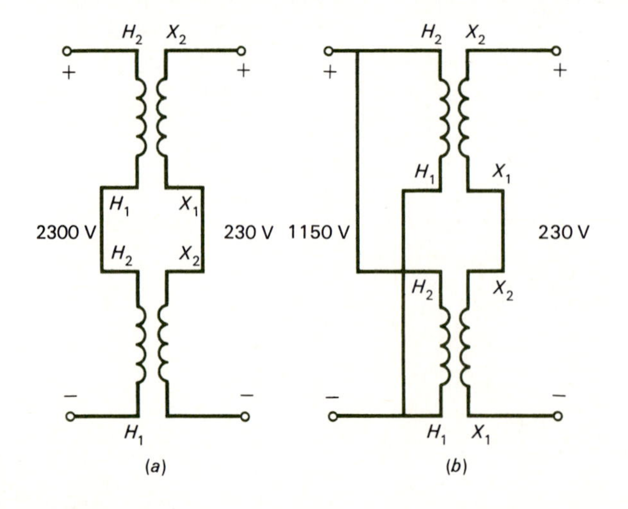

FIGURE 13.19 Example 13.12.

Solution We mark the high-voltage terminals by H_1 and H_2 and the low-voltage terminals by X_1 and X_2 (as in Figs. 13.17 and 13.18). According to the ANSI convention, H_1 is marked on the terminal on the right-hand side of the transformer case (or housing) when facing the high-voltage side. With this nomenclature, the desired connections are shown in Fig. 13.19*a* and *b*.

13.7 AUTOTRANSFORMERS

In contrast to the two-winding transformers considered so far, the autotransformer is a single-winding transformer having a tap brought out at an intermediate point. Thus, as shown in Fig. 13.20, *ac* is the single winding (wound on a laminated core) and *b* is the intermediate point where the tap is brought out. The autotransformer may be used either as a step-up or a step-down operation, as with a two-winding transformer. Considering a step-down arrangement, let the primary applied (terminal) voltage be V_1, resulting in a magnetizing current and a core flux ϕ_m. Let the secondary be open-circuited. Then the primary and secondary voltages obey the same rules as in a two-winding transformer, and we have

$$\frac{V_1}{V_2} = \frac{E_1}{E_2} = \frac{N_1}{N_2} = a$$

with $a > 1$ for step-down.

Furthermore, ideally,

$$V_1 I_1 = V_2 I_2$$

and

$$\frac{V_1}{V_2} = \frac{I_2}{I_1} = a$$

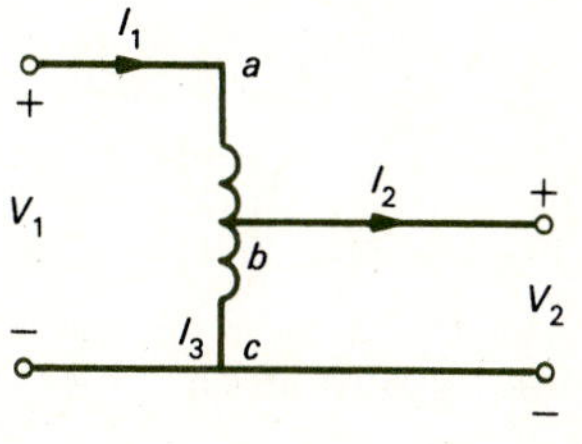

FIGURE 13.20
A step-down autotransformer.

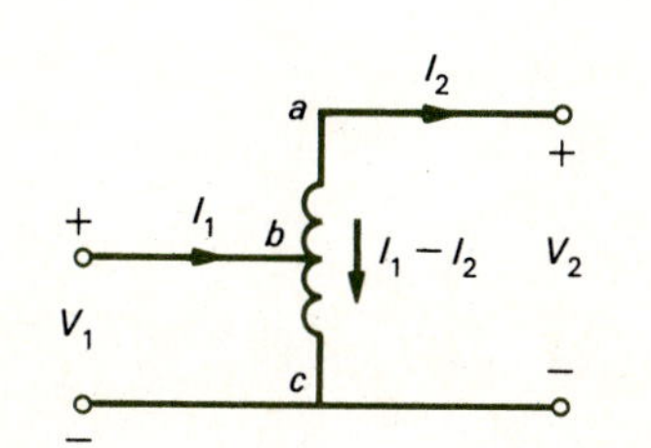

FIGURE 13.21
A three-winding transformer.

Neglecting the magnetizing current, we must have the mmf balance equation as

$$N_2 I_3 = (N_1 - N_2) I_1$$

or

$$I_3 = \frac{N_1 - N_2}{N_2} I_1 = (a - 1) I_1 = I_2 - I_1$$

The apparent power delivered to the load may be written as

$$P = V_2 I_2 = V_2 I_1 + V_2 (I_2 - I_1) \qquad (13.38)$$

In Eq. (13.38) the power is considered to consist of two parts:

1 conductively transferred power through *bc*

$$V_2 I_1 \equiv P_c$$

2 inductively transferred power 2.34 through *ab*

$$V_2 (I_2 - I_1) \equiv P_i$$

These powers are related to the total power by

$$\frac{P_i}{P} = \frac{I_2 - I_1}{I_2} = \frac{a - 1}{a} \qquad (13.39)$$

and

$$\frac{P_c}{P} = \frac{I_1}{I_2} = \frac{1}{a} \qquad (13.40)$$

where $a > 1$.

It may be shown that for a step-up transformer the power ratios are obtained as follows:

$$P = V_1 I_1 = V_1 I_2 + V_1 (I_1 - I_2)$$

implying that the total apparent power consists of two parts:

$$P_c = V_1 I_2 = \text{conductively transferred power}$$

$$P_i = V_1 (I_1 - I_2) = \text{inductively transferred power}$$

Hence,

$$\frac{P_i}{P} = \frac{I_1 - I_2}{I_1} = 1 - a \tag{13.41}$$

and

$$\frac{P_c}{P} = \frac{I_2}{I_1} = a \tag{13.42}$$

where $a < 1$.

Example 13.13 A two-winding 10-kVA 440/110-V transformer is reconnected as a step-down 550/440-V autotransformer. Compare the voltampere rating of the autotransformer with that of the original two-winding transformer, and calculate P_i and P_c.

Solution Refer to Fig. 13.20. The rated current in the 110-V winding (or in *ab*) is

$$I_1 = \frac{10{,}000}{110}$$

$$= 90.91 \text{ A}$$

The current in the 440-V winding (or in *bc*) is

$$I_3 = I_2 - I_1$$

$$= \frac{10{,}000}{440}$$

$$= 22.73 \text{ A}$$

which is the rated current of the winding *bc*. The load current is

$$I_2 = I_1 + I_3$$

$$= 90.91 + 22.73$$

$$= 113.64 \text{ A}$$

Check: For the autotransformer

$$a = \frac{550}{440}$$

$$= 1.25$$

and

$$I_2 = aI_1$$

$$= 1.25 \times \frac{10{,}000}{110}$$

$$= 113.64 \text{ A}$$

which agrees with I_2 calculated above. Hence, the rating of the autotransformer is

$$P_{\text{auto}} = V_1 I_1$$

$$= 550 \times \frac{10{,}000}{110}$$

$$= 50 \text{ kVA}$$

Inductively supplied apparent power is

$$P_i = V_2(I_2 - I_1)$$

$$= \frac{a-1}{a} P$$

$$= \frac{1.25 - 1}{1.25} \times 50$$

$$= 10 \text{ kVA}$$

which is the voltampere rating of the two-winding transformer. Conductively supplied power is

$$P_c = \frac{P}{a}$$

$$= \frac{50}{1.25}$$

$$= 40 \text{ kVA}$$

Example 13.14 Repeat the problem of Example 13.13 for a 440/550-V step-up connection.

Solution The step-up connection is shown in Fig. 2.35. The rating of the winding *ab* is 110 V and the load current I_2 flows through *ab*. Hence,

$$I_2 = \frac{10{,}000}{110}$$

$$= 90.91 \text{ A}$$

The output voltage is

$$V_2 = 550 \text{ V}$$

Thus, the voltampere rating of the autotransformer is

$$V_2 I_2 = 550 \times \frac{10{,}000}{110}$$

$$= 50 \text{ kVA}$$

which is the same as in the last example.

The power transferred conductively is

$$V_1 I_2 = 440 \times 90.91 = 40 \text{ kVA}$$

and the power transferred inductively is

$$50 - 40 = 10 \text{ kVA}$$

PROBLEMS

13.1 A 220/110-V 60-Hz ideal transformer is rated as 5 kVA. Calculate (*a*) the turns ratio and (*b*) the primary and secondary currents at full load and at a 3-kW load at 0.8 lagging.

13.2 A 10-Ω resistance is connected across the secondary of a 220/110-V 60-Hz ideal transformer. The secondary voltage is 110 V. Calculate (*a*) the primary current and (*b*) the equivalent resistance referred to the primary, and (*c*) determine the power dissipated in this equivalent resistance if the primary current calculated in (*a*) flows (through this resistance).

13.3 The secondary winding of a 60-Hz transformer has 50 turns. If the induced voltage in this winding is 220 V, determine the maximum value of the core flux.

13.4 On no-load 60-Hz transformer having a 300-turn primary takes 60 W of power and a current of 0.8 A at an input voltage of 110 V. The resistance of the winding is 0.2 Ω. Calculate (*a*) the core loss, (*b*) the magnetizing reactance X_m, and (*c*) the core-loss equivalent resistance R_c. Neglect leakage reactance.

13.5 Repeat the calculations of Example 13.4, using the circuit referred to the secondary.

13.6 Refer to the data of the transformer of Example 13.4. Using the equivalent circuit referred to the secondary, calculate the efficiency of the transformer at full load and 0.8 leading power factor.

13.7 Verify that the circuit shown in Fig. 13.11 is a valid equivalent circuit of a transformer referred to its secondary side.

13.8 Two 1150/115-V transformers are given. Show the interconnections of these transformers for (*a*) 2300/115-V operation and (*b*) 1150/115-V operation. Indicate the polarity markings.

13.9 A 25-kVA 440/220-V 60-Hz transformer has a constant core loss of 710 W. The equivalent resistance and reactance referred to the primary are each 0.16 Ω. Plot the following curves: (*a*) core loss versus primary current, (*b*) I^2R loss versus primary current, and (*c*) efficiency versus primary current—the primary current to range from 0 to 100 A. What is the I^2R loss when the efficiency of the transformer is maximum? (*Note*: This is a special case of the general rule that the efficiency of a transformer is maximum when the total I^2R losses in the two windings equal the core losses.)

13.10 A 75-kVA transformer has an iron loss of 1 kW and a full-load copper loss of 1 kW. Calculate the transformer efficiency at unity power factor at (*a*) full load and (*b*) half-full load.

13.11 The transformer of Prob. 13.10 operates on full load at unity power factor for 8 h, on no-load for 8 h, and on one-half load at unity power factor for 8 h, during 1 day. Determine the all-day efficiency.

13.12 The maximum efficiency of a 100-kVA transformer is 98.4 percent and occurs at 90 percent of full load. Calculate the efficiency of the transformer at unity power factor at (*a*) full load and (*b*) 50 percent of full load.

13.13 The results of open-circuit and short-circuit tests on a 10-kVA 440/220-V 60-Hz transformer are as follows:

Open-circuit test, HV side open:

$V_o = 220$ V

$I_o = 1.2$ A

$W_o = 150$ W

Short-circuit test, LV side short-circuited:

$V_s = 20.5$ V

$I_s = 42$ A

$W_s = 140$ W

Determine the parameters of the approximate equivalent circuit from the above data. Refer the circuit to the HV side.

13.14 Calculate the primary voltage of the transformer of Prob. 13.13 when the secondary is supplying full load at 0.8 lagging power factor. What is the efficiency of the transformer at this load?

13.15 A 1000-kVA transformer has a 94 percent efficiency at full load and at 50 percent of full load. The power factor is unity in both cases. (*a*) Segregate the losses and (*b*) determine the efficiency of the transformer for unity power factor and 75 percent of full load.

13.16 The primary and secondary voltages of an autotransformer are 440 and 360 V, respectively. If the secondary current is 80 A, determine the primary current.

13.17 A 5-kVA 220/220 V 60-Hz two-winding transformer has a full-load efficiency of 96 percent at unity power factor. The iron loss at 60 Hz is 60 W. This transformer is next connected as an autotransformer and delivers a 5-kVA unity power-factor load at 220 V. The primary of the autotransformer is connected to a 440-V source. Calculate (*a*) the primary current and (*b*) the efficiency of the autotransformer.

REFERENCES

13.1 S. A. Nasar, *Schaum's Outline of Theory and Problems in Electric Machines and Electromechanics*, McGraw-Hill, New York, 1981.

13.2 J. J. Cathey and S. A. Nasar, *Schaum's Outline of Theory and Problems in Basic Electrical Engineering*, McGraw-Hill, New York, 1983.

13.3 S. A. Nasar and L. E. Unnewehr, *Electromechanics and Electric Machines*, 2d ed., Wiley, New York, 1983.

13.4 S. A. Nasar, *Electric Machines and Transformers*, Macmillan, New York, 1983.

Chapter

14

DC Machines

In Chap. 13 we studied the transformer, which is in a sense an *energy transfer* device: energy is transferred from the primary to the secondary. During the energy transfer process, the form of energy remains unchanged; that is, the energy at the input (primary) and output (secondary) terminals is electrical. In a rotating electric machine, on the other hand, electrical energy is converted into mechanical form and vice versa, depending on the mode of operation of the machine. An electric *motor* converts electrical energy into mechanical energy, whereas an electric *generator* converts mechanical energy into electrical energy. For this reason, electric machines are also called electromechanical energy converters.

In essence, the process of electromechanical energy conversion can be expressed as

$$\text{Electrical energy} \underset{\text{generator}}{\overset{\text{motor}}{\rightleftarrows}} \text{mechanical energy}$$

Figure 14.1 schematically represents an ideal electric machine, for which we have, over a certain time interval Δt,

$$vi\,\Delta t = T_e\omega_m\,\Delta t$$

$$vi = T_e\omega_m \qquad (14.1)$$

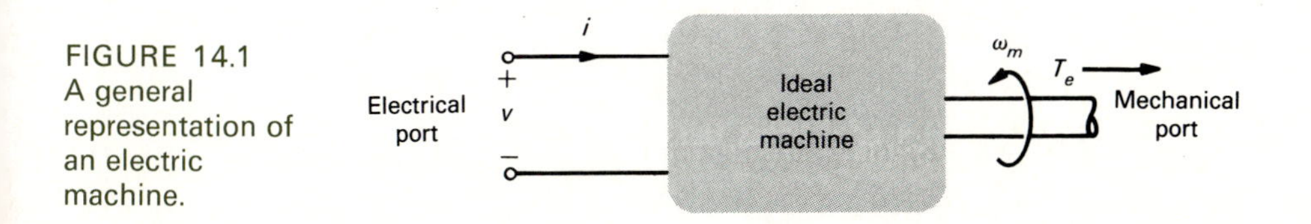

FIGURE 14.1 A general representation of an electric machine.

where, respectively, v and i are the voltage and the current at the electrical port, and T_e and ω_m are the torque [in newton-meters (N · m)] and the angular rotational velocity [in radians per second (rad/s)] at the mechanical port. We wish to reiterate that Eq. (14.1) is valid for an ideal machine, in that the machine is lossless (in Sec. 14.10, we will have more to say about losses in machines).

The first machine devised for electromechanical energy conversion was the dc machine—in the form of the Faraday disk.

14.1 THE FARADAY DISK AND FARADAY'S LAW

Applying his law of electromagnetic induction, Michael Faraday demonstrated in 1832 that if a copper disk with sliding contacts, or brushes, mounted at the rim and at the center of the disk is rotated in an axially directed magnetic field (produced by a permanent magnet), then a voltage will be available at the brushes. Such a machine is shown in Fig. 14.2 and is commonly known as the Faraday disk, or homopolar generator (another name for it is the *acyclic* machine); this device is also capable of operating as a motor. It is called a *homopolar* machine because the conductors (which may be thought of as spokes in the disk) are under the influence of the magnetic field of one polarity at all times. In contrast, the moving conductors in *heteropolar* machines are under the influence of magnetic fields of opposite polarities in an alternating fashion, as we shall soon see.

Faraday's law of electromagnetic induction states that an emf is induced in a circuit placed in a magnetic field if either (1) the magnetic flux linking the circuit is time-varying or (2) there is a relative motion between the circuit and the magnetic field such that the conductors comprising the circuit cut across the magnetic flux lines. The first form of the law, stated as (1) above, is the basis of operation of transformers. The second form, stated as (2) above, forms the basic principle of operation of electric generators, including the homopolar generator.

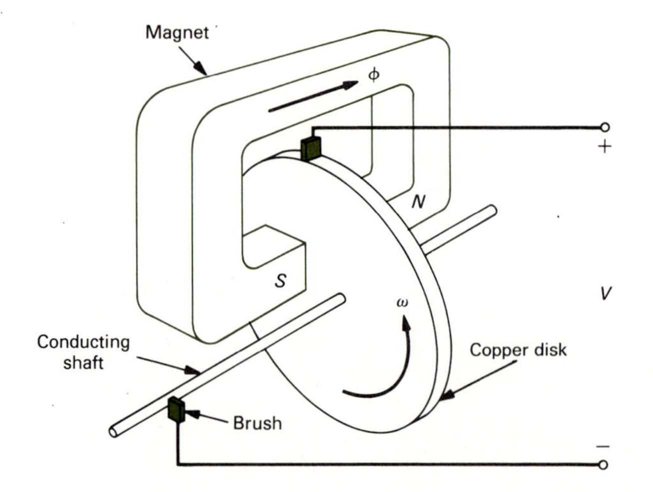

FIGURE 14.2 Faraday disk, or homopolar machine.

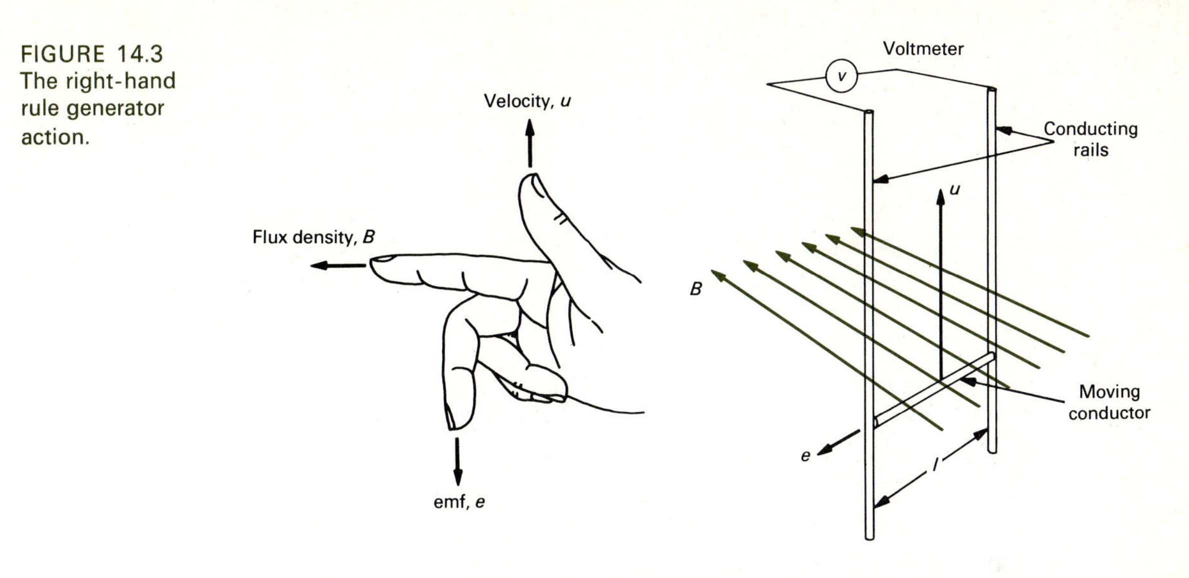

FIGURE 14.3 The right-hand rule generator action.

Consider a conductor of length l located at right angles to a uniform magnetic field of flux density B, as shown in Fig. 14.3. Let the conductor be connected with fixed external connections to form a closed circuit. These external connections are shown in the form of conducting rails in Fig. 14.3. The conductor can slide on the rails, and thereby the flux linking the circuit changes. Hence, according to Faraday's law, an emf will be induced in the circuit and a voltage v will be measured by the voltmeter. If the conductor moves with a velocity u [in meters per second (m/s)] in a direction at right angles to B and l both, then the area swept by the conductor in 1 second is lu. The flux in this area is Blu, which is also the flux linkage (since, in effect, we have a single-turn coil formed by the conductor, the rails, and the voltmeter). In other words, flux linkage per unit time is Blu, which is thus the induced emf e. We write this in equation form as

$$e = Blu \tag{14.2}$$

This form of Faraday's law is also known as the *flux-cutting rule*. Stated in words, an emf e, as given by Eq. (14.2), is induced in a conductor of length l if it "cuts" magnetic flux lines of density B by moving at right angles to B at a velocity u (at right angles to l). The mutual relationships between e, B, l, and u are given by the *right-hand rule*, as shown in Fig. 14.3.

Example 14.1

A conducting disk of 0.5-m radius rotates at 1200 revolutions per minute (r/min) in a uniform magnetic field of 0.4 T. Calculate the voltage available between the rim and the center of the disk.

Solution Referring to Eq. (14.2), in this problem we have $B = 0.4$ T, $l = 0.5$ m, and $u = r\omega$, where $\omega = 2\pi n/60$ radians per second is the angular velocity of the disk. Substituting

$n = 1200$ r/min and $r = 0.5$ m, we obtain

$$u = 0.5 \times 2\pi \times \frac{1200}{60}$$

$$= 20\pi \text{ m/s}$$

Hence, from Eq. (14.2)

$$e = 0.4 \times 0.5 \times 20\pi$$

$$= 12.76 \text{ V}$$

Let us now summarize the salient points of this section:

1 The Faraday disk operates on the principle of electromagnetic induction enunciated by Faraday.

2 There are two forms of expression of Faraday's law—time rate of change of flux linkage (with a circuit) or flux cut (by a conductor) results in an induced emf, but the latter is contained in the former.

3 Electric generators operate on the flux-cutting principle.

4 Example 14.1 shows that a rather small voltage (12.76 V) is induced in a disk of large diameter (1 m) rotating at a reasonably high speed (1200 r/min).

Thus, the Faraday disk in its primitive form is not a practical type of dc generator. Indeed, present-day dc generators have little resemblance to the Faraday disk.

14.2 THE HETEROPOLAR, OR CONVENTIONAL, DC MACHINE

Consider an N-turn coil rotating at a constant angular velocity ω in a uniform magnetic field of flux density B, as shown in Fig. 14.4*a*. Let l be the axial length of the coil and r its radius. The emf induced in the coil can be found by an application of Eq. (14.2). However, we must be careful in determining the u of Eq. (14.2). Recall from the last section that u is the velocity perpendicular to B. In the system under consideration, we have a rotating coil. Thus, u in Eq. (14.2) corresponds to that component of velocity which is perpendicular to B. We illustrate the components of velocities in Fig. 14.4*b*, where the tangential velocity $u_t = r\omega$ has been resolved into a component u_1 along the direction of the flux and another component u_2 across (or perpendicular) to the flux. This latter component cuts the magnetic flux and is the component responsible for the emf induced in the coil. The u in Eq. (14.2) should thus be replaced by u_2. The next term in Eq. (14.2) that needs careful consideration is l_1, the effective length of the conductor. For the N-turn coil of Fig. 14.4*a*, we effectively have $2N$ conductors in series (since each coil side has N conductors and there are two coil sides). If l_1 is the length of each conductor, then the total effective length of the N-turn coil is $2Nl_1$,

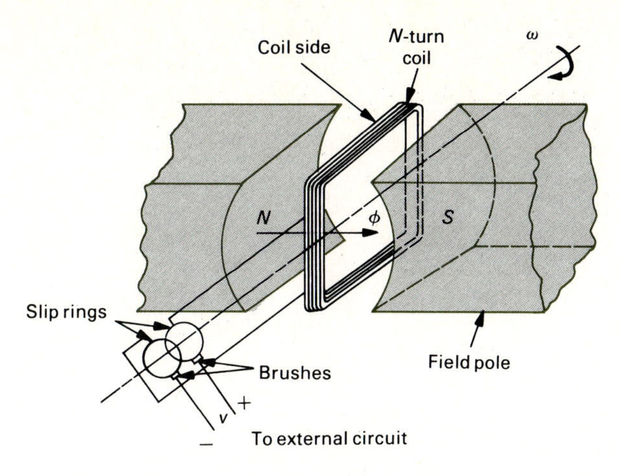

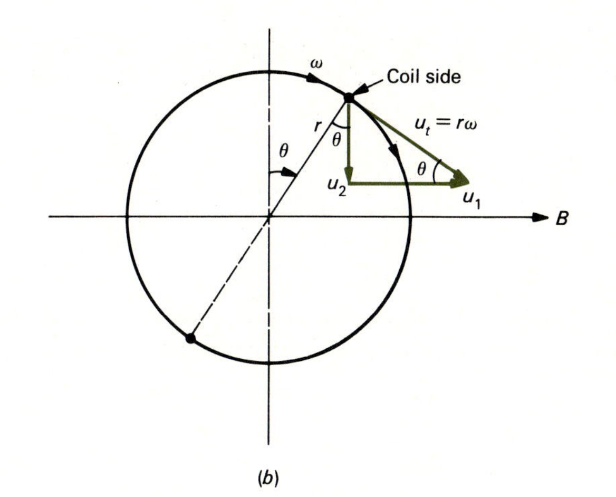

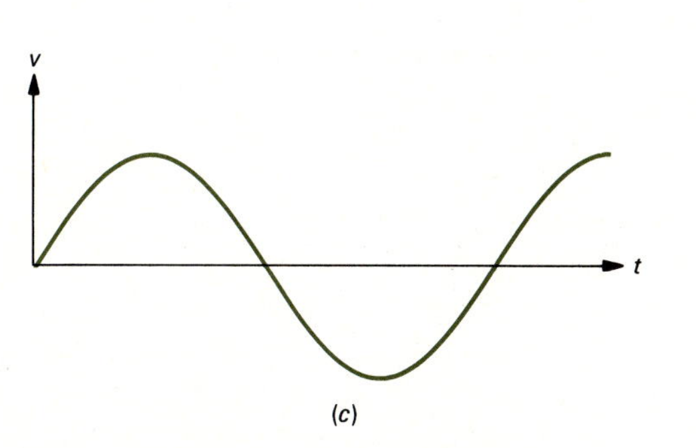

FIGURE 14.4 (*a*) An elementary generator. (*b*) Resolution of velocities into parallel and perpendicular components. (*c*) Voltage waveform at the brushes.

which should be substituted for l in Eq. (14.2). The form of Eq. (14.2) for the N-turn coil thus becomes

$$e = B(2Nl_1)u_2 \tag{14.3}$$

From Fig. 14.4*b* we have

$$u_2 = u_t \sin\theta = r\omega \sin\theta \tag{14.4}$$

If the coil rotates at a constant angular velocity ω, then

$$\theta = \omega t \tag{14.5}$$

Consequently, Eqs. (14.3) to (14.5) yield

$$e = 2BNl_1 r\omega \sin\omega t = E_m \sin\omega t \tag{14.6}$$

where $E_m = 2BNl_1 r\omega$. A plot of Eq. (14.6) is shown in Fig. 14.4*c*. The conclusion is that a sinusoidally varying voltage will be available at the slip rings, or brushes, of the rotating coil shown in Fig. 14.4*a*. The brushes reverse polarities periodically.

In order to obtain a unidirectional voltage at the coil terminals, we replace the slip rings of Fig. 14.4*a* with the commutator segments shown in Fig. 14.5*a*. It can be readily verified, by applying the right-hand rule, that in the arrangement shown in Fig. 14.5*a* the brushes will maintain their polarities regardless of the position of the coil. In

FIGURE 14.5
(*a*) An elementary dc generator.
(*b*) Output voltage at the brushes.

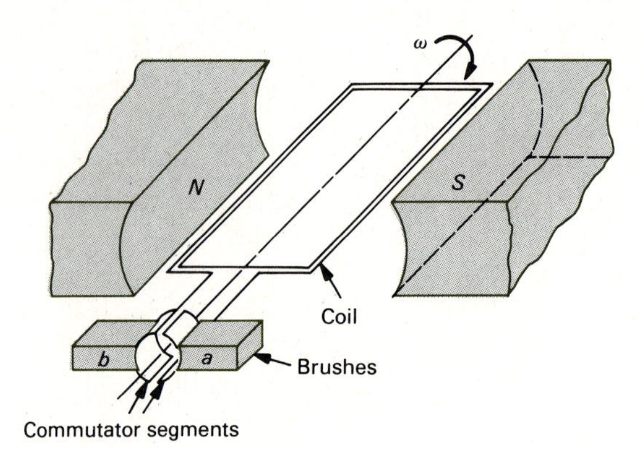

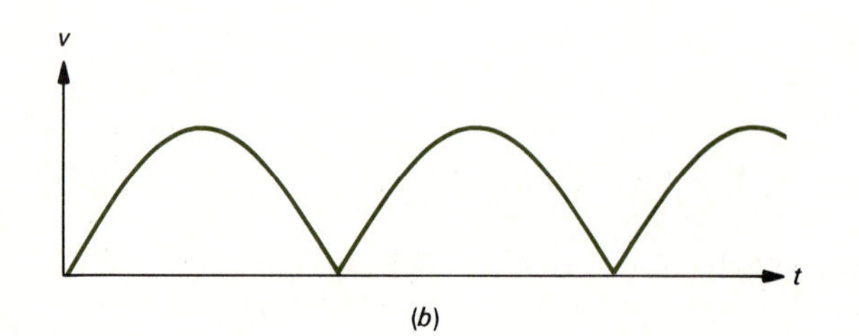

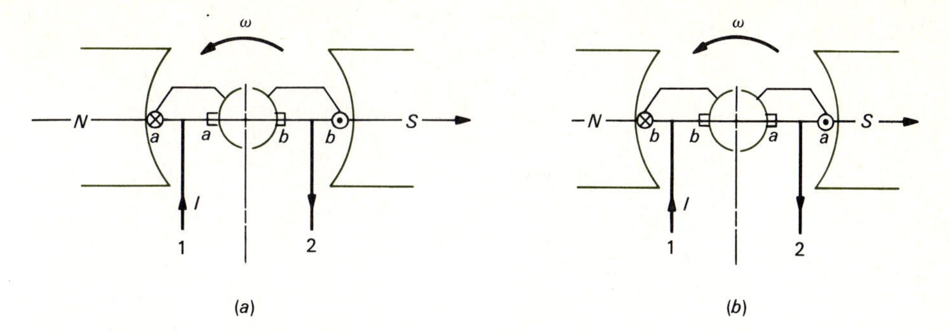

FIGURE 14.6 Production of a unidirectional torque and operation of an elementary dc motor: (*a*) position of conductor *a* under *N* pole; (*b*) position of conductor *a* under *S* pole.

other words, brush *a* will always be positive and brush *b* always negative for the given relative polarities of the flux and the direction of rotation. Thus, the arrangement shown in Fig. 14.5 forms an elementary heteropolar dc machine. The main characteristic of a heteropolar dc machine is that the emf induced in a conductor, or the current flowing through it, has its direction reversed as it passes from a north-pole to a south-pole region. This reversal process is known as *commutation* and is accomplished by the commutator-brush mechanism, which also serves as a connection to the external circuit. The voltage available at the brushes will be of the form shown in Fig. 14.5*b*.

The natural question to ask at this point is "What have we gained, compared to the Faraday disk, by introducing the complications of a commutator?" The answer lies in the fact that a full range of ratings can be realized from a heteropolar cylindrical configuration. Recall from Example 14.1 that we could get only 13 V at the terminals of a homopolar machine. Obtaining much higher voltages is no problem with a heteropolar machine, as we shall presently see. As a matter of convention, we will term the dc heteropolar commutator machine simply a dc machine (unless otherwise stated).

Before leaving the subject of the homopolar and elementary heteropolar machines, let us recall Eq. (14.1), according to which these machines must also be able to operate as motors. The production of a unidirectional torque—and hence a machine's operation as a dc motor—can be verified by referring to Fig. 14.6.

14.3 CONSTRUCTIONAL DETAILS

In order that we may subsequently consider some conventional dc machines, it is best that we briefly study the constructional features of their various parts and discuss the usefulness of these parts.

We observe from the last section that the basic elements of a dc machine are the rotating coil, a means for the production of flux, and the commutator-brush arrangement. In a practical dc machine the coil is replaced by the *armature winding* mounted on a cylindrical magnetic structure. The flux is provided by the *field winding* wound on field poles. Generally, the armature winding is placed on the rotating member, the

FIGURE 14.7
A dc machine. (Courtesy of General Electric Co.)

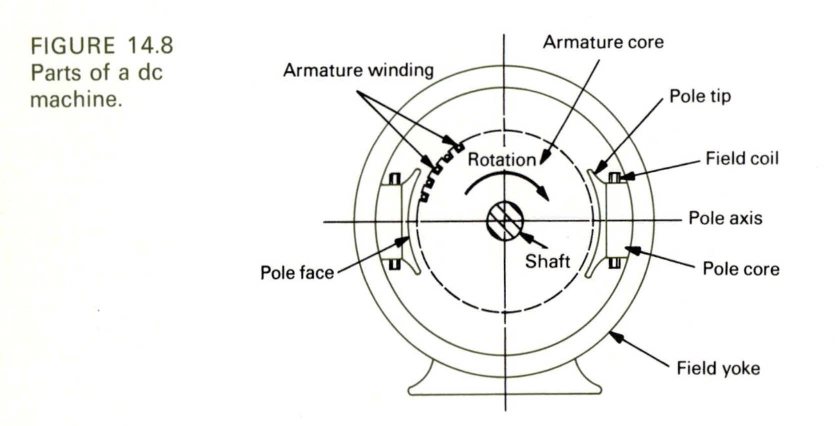

FIGURE 14.8
Parts of a dc machine.

rotor, and the field winding is on the stationary member, the *stator*, of the dc machine. A common, large dc machine is shown in Fig. 14.7, which, along with the schematic of Fig. 14.8, shows most of the important parts of a dc machine.

The field poles, mounted on the stator, carry the field windings. Some machines carry more than one separate field winding on the same core. The cores and faces of the poles are built of sheet-steel laminations. The pole faces have to be laminated because of their proximity to the armature windings, but because the field windings carry direct current, it is not necessary to have the pole cores laminated; nevertheless, the use of laminations for the cores as well as for the pole faces facilitates assembly.

The rotor (or armature) core, which carries the rotor or armature windings, is generally made of sheet-steel laminations. These laminations are stacked together to

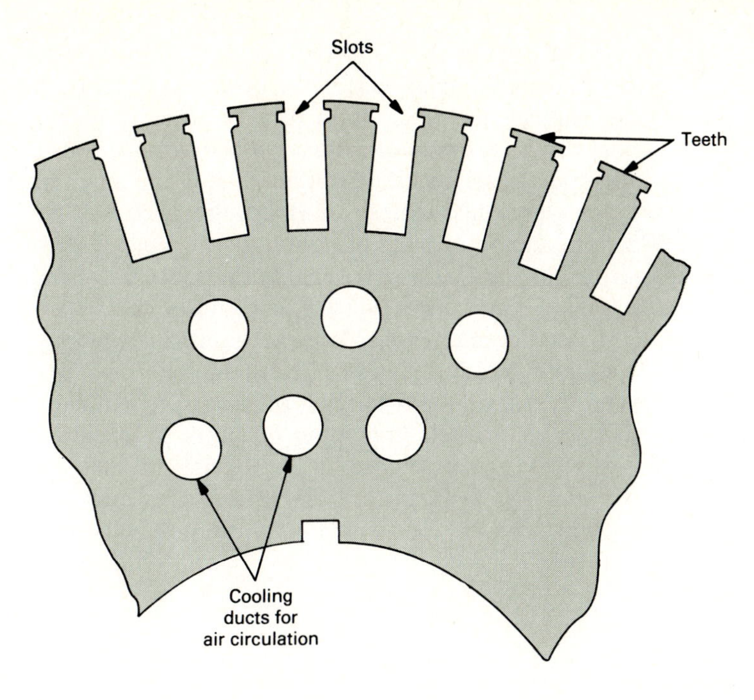

FIGURE 14.9
Portion of an armature lamination of a dc machine, showing slots and teeth.

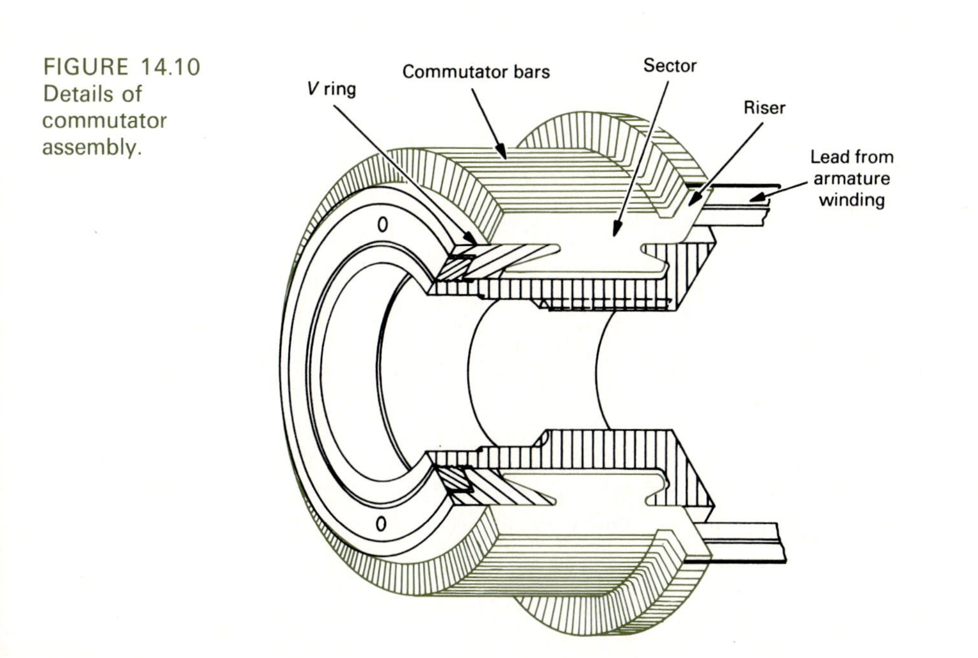

FIGURE 14.10
Details of commutator assembly.

form a cylindrical structure. On its outer periphery, the armature (rotor) has the *slots* in which the armature coils that make up the armature winding are located. For mechanical support, protection from abrasion, and greater electrical insulation, non-conducting slot liners are often wedged in between the coils and the slot walls. The projections of magnetic material between the slots are called *teeth.* A typical slot/teeth geometry for a large dc machine is shown in Fig. 14.9.

The commutator is made of hard-drawn copper segments insulated from one another by mica. The details of the commutator assembly are given in Fig. 14.10. The armature windings are connected to the commutator segments or bars, over which the carbon brushes slide and serve as leads for electrical connection.

The armature winding may be a lap winding (Fig. 14.11*a*) or a wave winding (Fig. 14.11*b*), and the various coils forming the armature winding may be connected in a series-parallel combination. In practice, the armature winding is housed as two layers in the slots of the armature core. In large machines the coils are preformed in the shapes shown in Fig. 14.12 and are interconnected to form an armature winding. The coils span approximately a pole pitch—the distance between two consecutive poles.

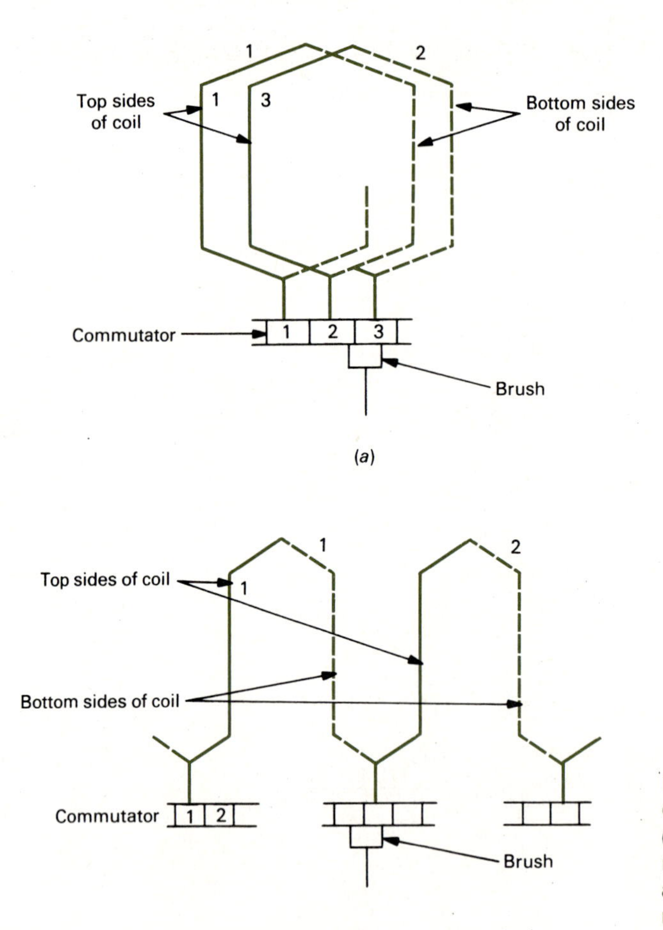

FIGURE 14.11 Elements of (*a*) a lap winding; (*b*) a wave winding. Odd-numbered conductors are at the top and even-numbered conductors are at the bottom of the slots.

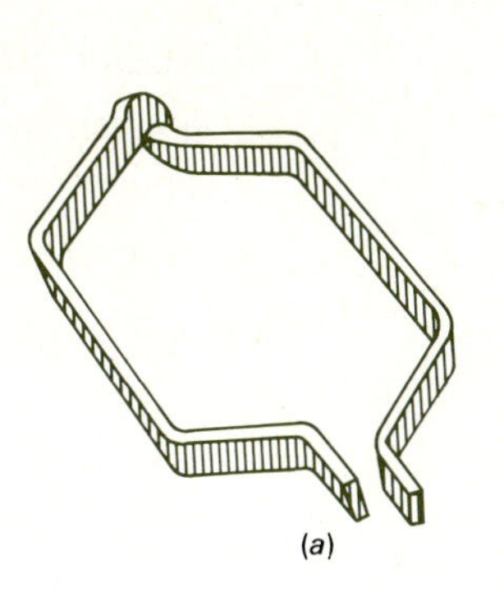
(a)

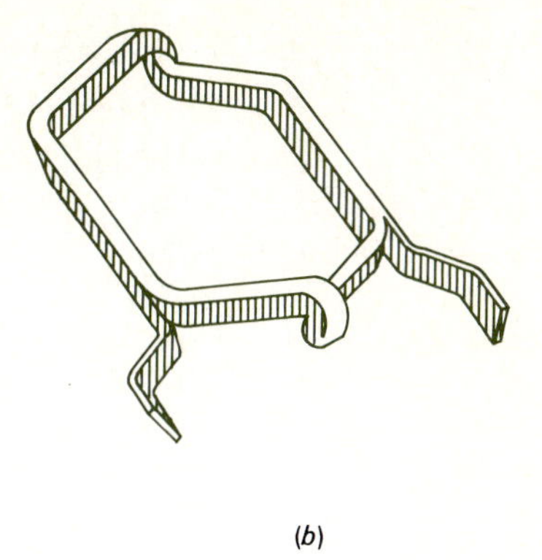
(b)

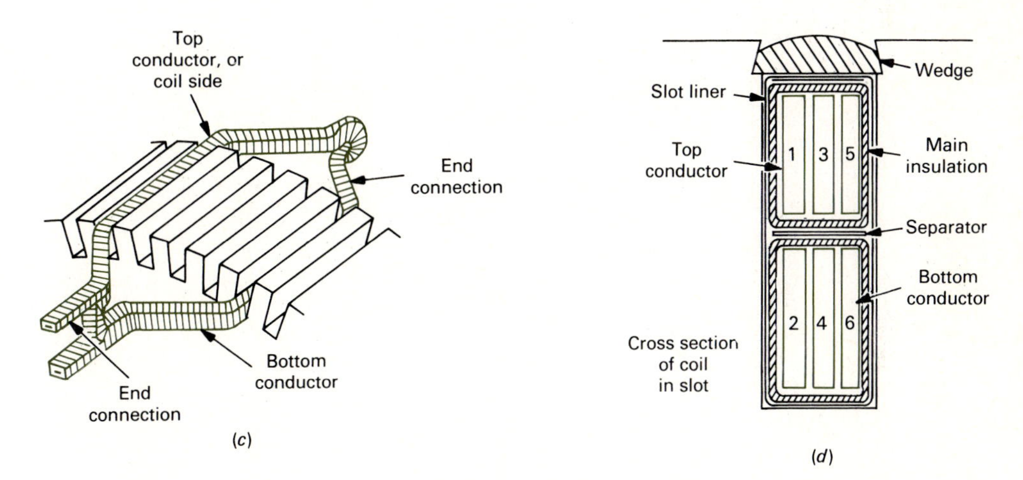

FIGURE 14.12 (*a*) Coil for a lap winding, made of a single bar. (*b*) Multiturn coil for a wave winding. (*c*) A coil in a slot. (*d*) Slot details, showing several coils arranged in two layers.

14.4 CLASSIFICATION ACCORDING TO FORMS OF EXCITATION

Conventional dc machines having a set of field windings and armature windings can be classified, on the basis of mutual electrical connections between the field and armature windings, as follows:

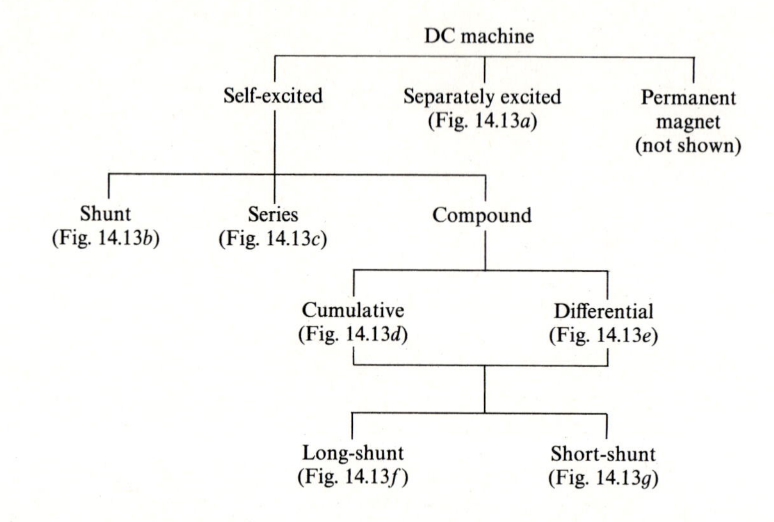

These interconnections of field and armature windings essentially determine the machine's operating characteristics.

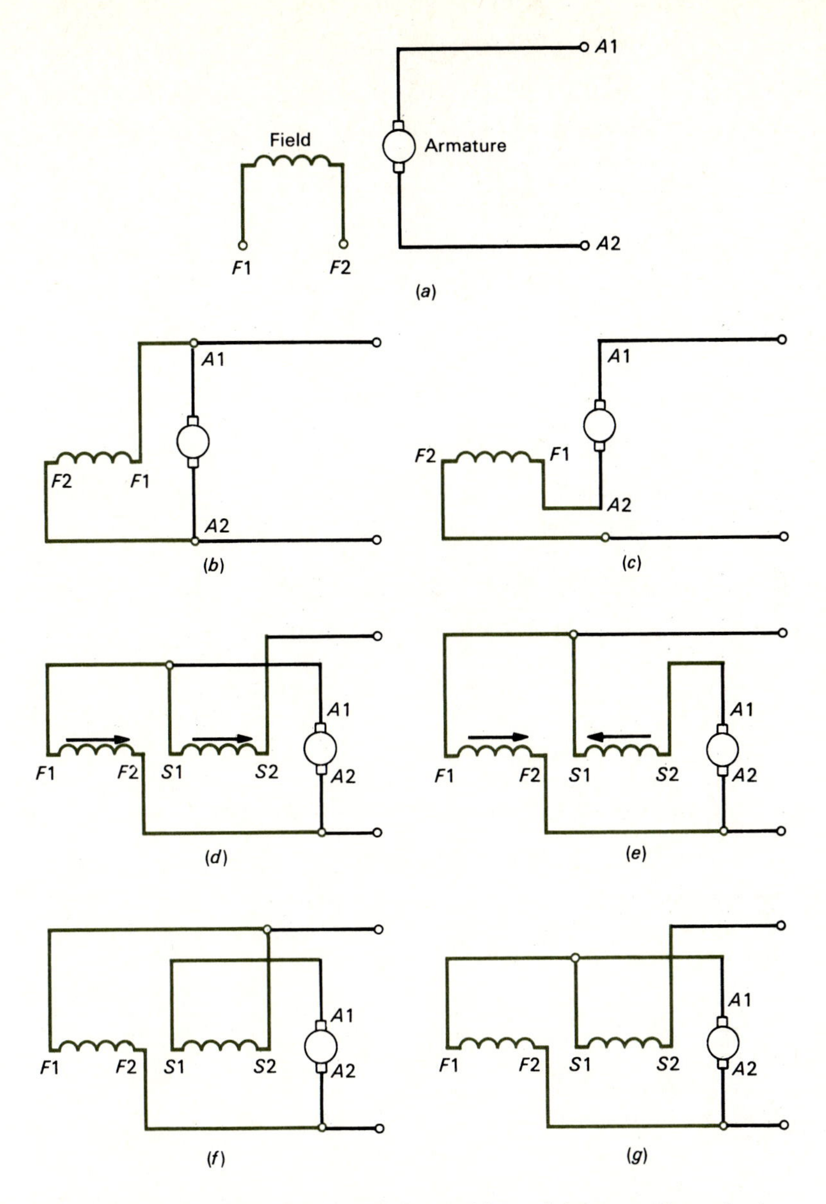

FIGURE 14.13 Classification of dc machines: (*a*) separately excited; (*b*) shunt; (*c*) series; (*d*) cumulative compound; (*e*) differential compound; (*f*) long-shunt; (*g*) short-shunt.

14.5 PERFORMANCE EQUATIONS

The three quantities of greatest interest in evaluating the performance of a dc machine are (1) the induced emf, (2) the electromagnetic torque developed by the machine, and (3) the speed corresponding to (1) and/or (2). We will now derive the equations which enable us to determine the above quantities.

EMF Equation The emf equation yields the emf induced in the armature of a dc machine. The derivation follows directly from Faraday's law of electromagnetic induction, according to which the emf induced in a moving conductor is the flux cut by the conductor per unit time.

Let us define the following symbols:

Z = number of active conductors on the armature

a = number of parallel paths in the armature winding

p = number of field poles

ϕ = flux per pole

n = speed of rotation of the armature, r/min

Thus, with reference to Fig. 14.14,

Flux cut by one conductor in 1 rotation $= \phi p$

Flux cut by one conductor in n rotation $= \phi np$

Flux cut per second by one conductor $= \phi np/60$

Number of conductors in series $= Z/a$

Flux cut per second by Z/a conductors $= \phi np/60\,(Z/a)$

Hence, the emf induced in the armature winding is

$$E = \frac{\phi n Z}{60}\frac{p}{a} \tag{14.7}$$

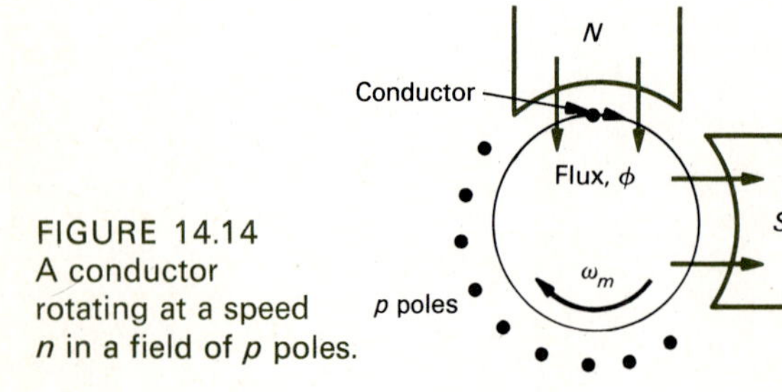

FIGURE 14.14
A conductor rotating at a speed n in a field of p poles.

Equation (14.7) is known as the *emf equation* of a dc machine. In Eq. (14.7) we have separated p/a from the rest of the terms because p and a are related to each other for the two types of windings. For a lap winding $p = a$, and $a = 2$ in a wave winding.

Example 14.2 Determine the voltage induced in the armature of a dc machine running at 1750 r/min and having four poles. The flux per pole is 25 mWb, and the armature is lap-wound with 728 conductors.

Solution Since the armature is lap-wound, $p = a$, and Eq. (14.7) becomes

$$E = \frac{\phi n Z}{60}$$

$$= \frac{25 \times 10^{-3} \times 1750 \times 728}{60}$$

$$= 530.8 \text{ V}$$

Torque Equation The mechanism of torque production in a dc machine has been considered earlier (see Fig. 14.6), and the electromagnetic torque developed by the armature can be evaluated either from the BlI rule or from Eq. (14.1). Here, we choose the latter approach. Regardless of the approach, it is clear that for torque production we must have a current through the armature, as this current interacts with the flux produced by the field winding. Let I_a be the armature current and E the voltage induced in the armature. Thus, the power at the armature electrical port (see Fig. 14.1) is EI_a. Assuming that this entire electric power is transformed into mechanical form, we rewrite Eq. (14.1) as

$$EI_a = T_e \omega_m \tag{14.8}$$

where T_e is the electromagnetic torque developed by the armature and ω_m is its angular velocity in rad/s. The speed n (in r/min) and ω_m (in rad/s) are related by

$$\omega_m = \frac{2\pi n}{60} \tag{14.9}$$

Hence, from Eqs. (14.7) to (14.9) we obtain

$$\frac{\omega_m}{2\pi} \phi Z \frac{p}{a} I_a = T_e \omega_m$$

which simplifies to

$$T_e = \frac{Zp}{2\pi a} \phi I_a \tag{14.10}$$

which is known as the *torque equation.* An application of Eq. (14.10) is illustrated by the next example.

Example 14.3 A lap-wound armature has 576 conductors and carries an armature current of 123.5 A. If the flux per pole is 20 mWb, calculate the electromagnetic torque developed by the armature.

Solution For the lap winding, we have $p = a$. Substituting this and the other given numerical values into Eq. (14.10) yields

$$T_e = \frac{576}{2\pi} \times 0.02 \times 123.5$$

$$= 226.4 \text{ N} \cdot \text{m}$$

Example 14.4 If the armature of Example 14.3 rotates at an angular velocity of 123.5 rad/s, what is the induced emf in the armature?

Solution To solve this problem, we use Eq. (14.8) rather than Eq. (14.7). From Example 14.3 we have $I_a = 123.5$ A and $T_e = 226.4$ N · m. Hence, Eq. (14.8) gives

$$E \times 123.5 = 226.4 \times 123.5$$

or

$$E = 226.4 \text{ V}$$

Speed Equation and Back EMF The emf and torque equations discussed above indicate that the armature of a dc machine, whether operating as a generator or as a motor, will have an emf induced while rotating in a magnetic field and will develop a torque if the armature carries a current. In a generator, the induced emf is the internal voltage available from the generator; when the generator supplies a load, the armature carried a current and develops a torque. This torque opposes the prime-mover (such as a diesel engine) torque.

In motor operation, the developed torque of the armature supplies the load connected to the shaft of the motor, and the emf induced in the armature is termed the *back emf.* This emf opposes the terminal voltage of the motor.

Referring to Fig. 14.15, which shows the equivalent circuit of a separately excited dc motor running at speed n while taking an armature current I_a at a voltage V_t, we

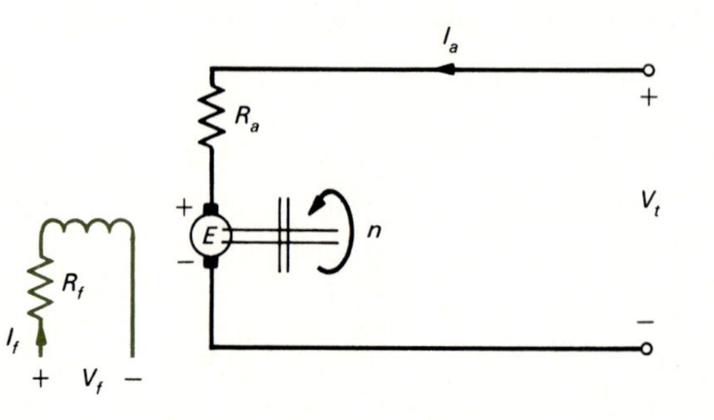

FIGURE 14.15 Equivalent circuit of a separately excited dc motor.

have from this circuit

$$V_t = E + I_a R_a \tag{14.11}$$

Substituting Eq. (14.7) in Eq. (14.11), putting $k_1 = Zp/60a$, and solving for n, we get

$$n = \frac{V_t - I_a R_a}{k_1 \phi} \tag{14.12}$$

This equation is known as the *speed equation*, as it contains all the factors which affect the speed of a motor. In Sec. 14.9 we shall consider the influence of these factors on the speed of dc motors. For the present, let us focus our attention on k_1 and ϕ. The term k_1 replacing $Zp/60a$ is a design constant in the sense that Z, p, and a cannot be altered once the machine has been built. The magnetic flux ϕ is primarily controlled by the field current I_f (Fig. 14.15). If the magnetic circuit is unsaturated, then ϕ is directly proportional to I_f. Thus, we may write

$$\phi = k_f I_f \tag{14.13}$$

where k_f is a constant. We may combine Eqs. (14.12) and (14.13) to obtain

$$n = \frac{V_t - I_a R_a}{k I_f} \tag{14.14}$$

where $k = k_1 k_f$ is a constant. This form of the speed equation is more meaningful because all of the quantities in Eq. (14.14) can be conveniently measured; in contrast, it is very difficult to measure ϕ in Eq. (14.12).

Example 14.5 A 250-V shunt motor has an armature resistance of 0.25 Ω and a field resistance of 125 Ω. At no-load, the motor takes a line current of 5.0 A while running at 1200 r/min. If the line current at full load is 52.0 A, what is the full-load speed?

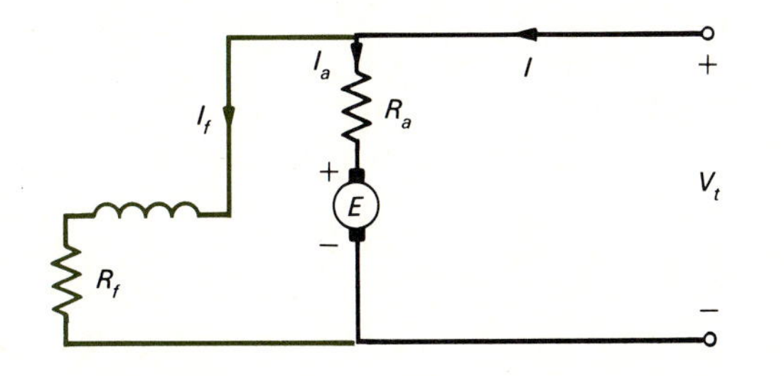

FIGURE 14.16 Equivalent circuit of a shunt motor.

Solution The motor equivalent circuit is shown in Fig. 14.16. The field current is

$$I_f = \frac{250}{125} = 2.0 \text{ A}$$

At no load, the armature current is

$$I_a = 5.0 - 2.0$$
$$= 3.0 \text{ A}$$

the back emf is

$$E_1 = V_t - I_a R_a$$
$$= 250 - (3 \times 0.25)$$
$$= 249.25 \text{ V}$$

and the speed is

$$N_1 = 1200 \text{ r/min} \qquad \text{given}$$

At full load, the armature current is

$$I_a = 52.0 - 2.0$$
$$= 50.0 \text{ A}$$

the back emf is

$$E_2 = 250 - (50 \times 0.25)$$
$$= 237.5 \text{ V}$$

and the speed is

$$N_2 = \text{unknown}$$

Now, since

$$\frac{N_2}{N_1} = \frac{E_2}{E_1}$$
$$= \frac{237.5}{249.25}$$

thus

$$N_2 = \frac{237.5}{249.25} \times 1200$$
$$= 1143 \text{ r/min}$$

14.6 EFFECTS OF SATURATION—BUILDUP OF VOLTAGE IN A SHUNT GENERATOR

Saturation plays a very important role in governing the behavior of dc machines. It is extremely difficult, however, to take the effects of saturation quantitatively. For the time being, then, let us consider qualitatively the consequences of saturation on the operation of a self-excited shunt generator.

A self-excited shunt machine is shown in Fig. 14.16. Writing the steady-state equations for the operation of the machine as a generator from the circuit of Fig. 14.16, we have

$$V_t = R_f I_f$$

and

$$E = V_t + I_a R_a = I_f R_f + I_a R_a$$

These equations are represented by the straight lines shown in Fig. 14.17*a*. Notice that the voltages V_t and E will keep building up and that no equilibrium point can be

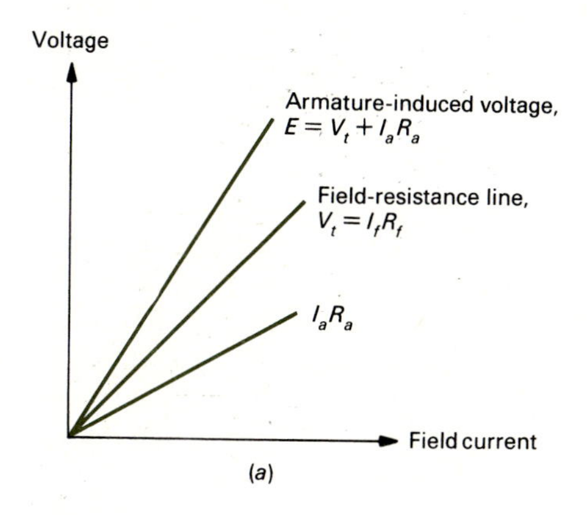

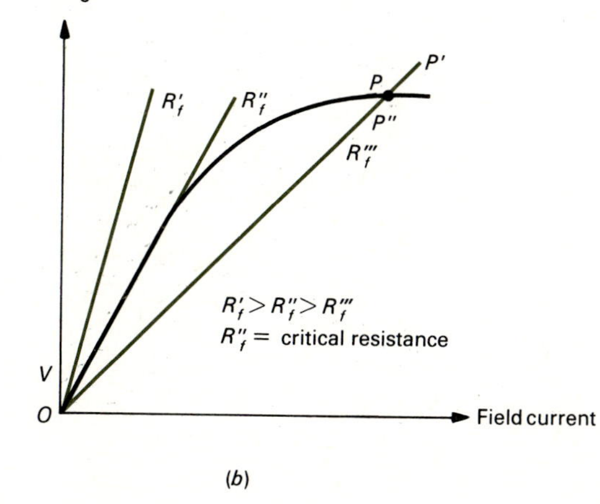

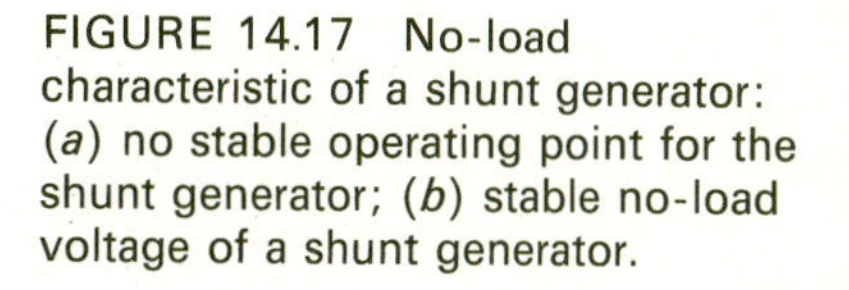
FIGURE 14.17 No-load characteristic of a shunt generator: (*a*) no stable operating point for the shunt generator; (*b*) stable no-load voltage of a shunt generator.

reached. On the other hand, if we include the effect of saturation, as in Fig. 14.17*b*, the point P defines the equilibrium, because at this point the field-circuit resistance line intersects the saturation curve. A deviation from P to P' or P'' would immediately show that at P' the voltage drop across the field is greater than the induced voltage, which is not possible, and at P'' the induced voltage is greater than the field-circuit voltage drop.

Figure 14.17*b* shows some residual magnetism, as measured by the small voltage OV. Evidently, without it the shunt generator will not build up any voltage. Also shown in Fig. 14.17*b* is the critical resistance. A field-circuit resistance greater than the critical resistance (for a given speed) would not let the shunt generator build up any appreciable voltage. Finally, we should ascertain that the polarity of the field winding is such that a current through it produces a flux which aids the residual flux. If it does not, the two fluxes tend to neutralize and the machine voltage will not build up. To summarize, the conditions for the buildup of a voltage in a shunt generator are: the presence of residual flux, a field-circuit resistance less than the critical resistance, and an appropriate polarity of the field winding.

14.7 GENERATOR CHARACTERISTICS

The no-load and load characteristics of dc generators are usually of interest in determining their potential applications. Between the two, load characteristics are of greater importance. As the names imply, no-load and load characteristics respectively correspond to the behavior of the machine when it is supplying no power (open-circuited, in case of a generator) and when it is supplying power to an external circuit.

The only no-load (or open-circuit) characteristics which are meaningful are those of the shunt and separately excited generators. In the last section we discussed the no-load characteristic of a shunt generator as a voltage buildup process. For the separately excited generator, the no-load characteristic corresponds to the magnetization, or saturation, characteristic—the variation of E (or V_t) under open-circuit condition as a function of the field current I_f. This characteristic is illustrated in Fig. 14.18.

The load characteristics of self-excited generators are shown in Fig. 14.19. The shunt generator has a characteristic similar to that of a separately excited generator, except for the cumulative effect mentioned in Sec. 14.6. If the shunt generator is loaded beyond a certain point, it breaks down, in that the terminal voltage collapses. In a series generator, the load current flows through the field winding; this implies that the field flux, and hence the induced emf, increases with the load until the core begins to saturate magnetically. Thus, a load beyond a certain point would result in a collapse of the terminal voltage of the series generator, too. Compound generators have the combined characteristics of shunt and series generators. In a differential compound generator, the shunt and series fields are in opposition; hence, the terminal voltage drops very rapidly with the load. On the other hand, cumulative compound generators have shunt and series fields aiding each other. The two field mmf's may be adjusted such that the terminal voltage on full load is less than the no-load voltage, as

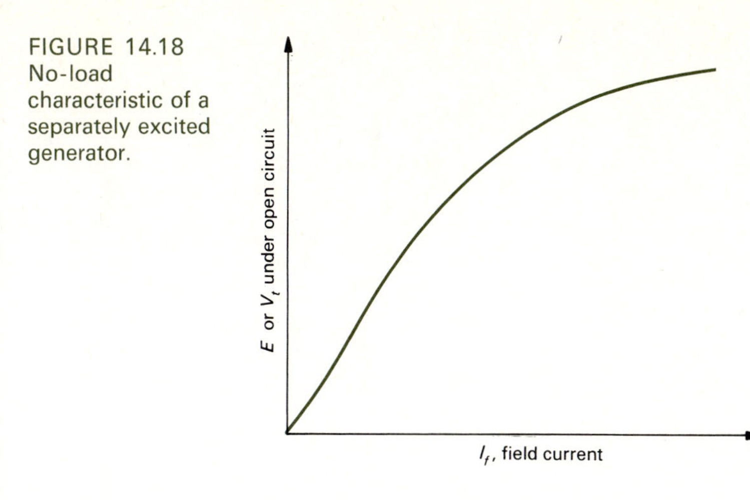

FIGURE 14.18 No-load characteristic of a separately excited generator.

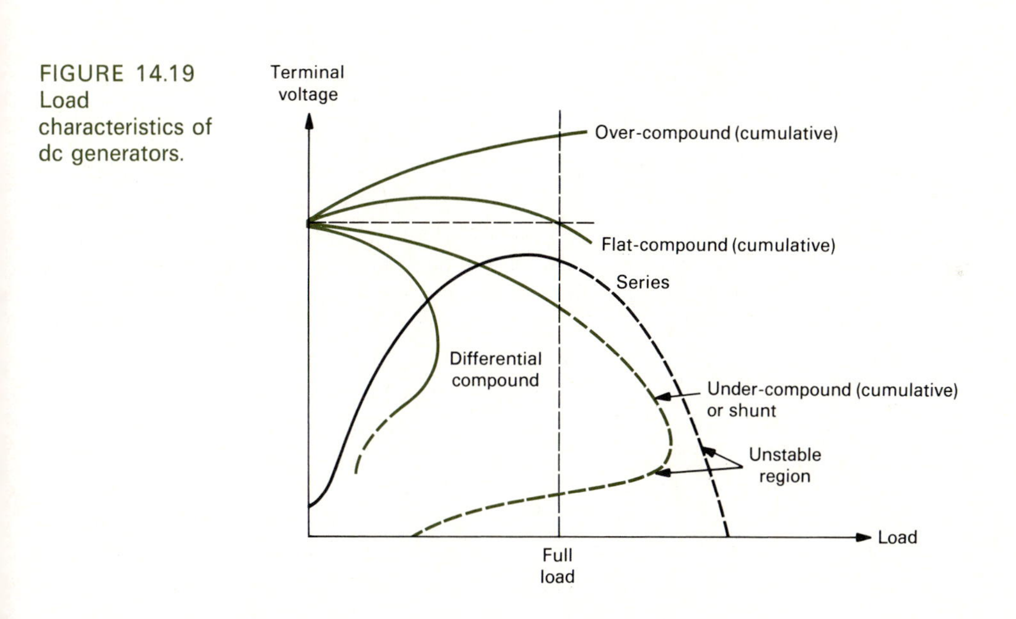

FIGURE 14.19 Load characteristics of dc generators.

in an under-compound generator; or the full-load voltage may be equal to the no-load voltage, as in a flat-compound generator. Finally, the terminal voltage on full load may be greater than the no-load voltage, as in an over-compound generator.

Example 14.6 A 50-kW 250-V short-shunt compound generator has the following data: $R_a = 0.06\ \Omega$, $R_{se} = 0.04\ \Omega$ and $R_f = 125\ \Omega$. Calculate the induced armature emf at rated load and terminal voltage. Take 2 V as the total brush-contact drop.

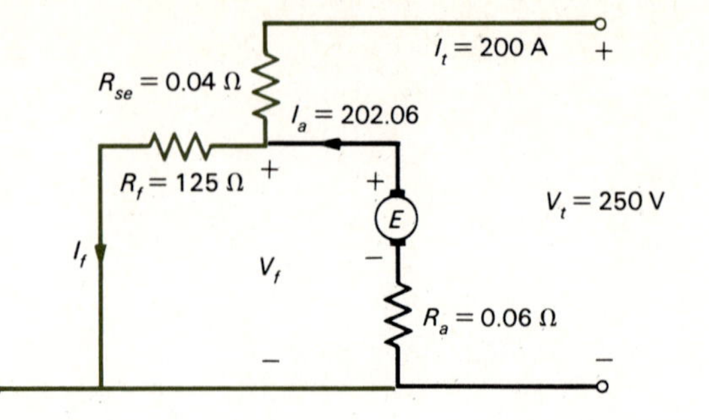

FIGURE 14.20
Example 14.6.

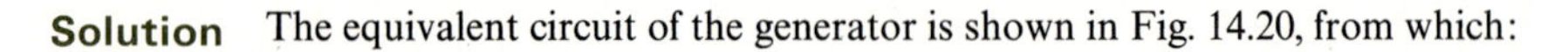

Solution The equivalent circuit of the generator is shown in Fig. 14.20, from which:

$$I_t = \frac{50 \times 10^3}{250}$$

$$= 200 \text{ A}$$

$$I_t R_{se} = 200 \times 0.04$$

$$= 8 \text{ V}$$

$$V_f = 250 + 8$$

$$= 258 \text{ V}$$

$$I_f = \frac{258}{125}$$

$$= 2.06 \text{ A}$$

$$I_a = 200 + 2.06$$

$$= 202.06 \text{ A}$$

$$I_a R_a = 202.06 \times 0.06$$

$$= 12.12 \text{ V}$$

$$E = 250 + 12.12 + 8 + 2$$

$$= 272.12 \text{ V}$$

14.8 MOTOR CHARACTERISTICS

Among the various characteristics of dc motors, their torque-speed characteristics are most important from a practical standpoint. The torque and speed equations derived earlier govern the motor characteristics. From these equations (and after accounting for magnetic saturation) it follows that the shunt, series, and compound motors have the torque-speed characteristics of the forms shown in Fig. 14.21. The governing equations also yield the motor speed-current characteristics of Fig. 14.22.

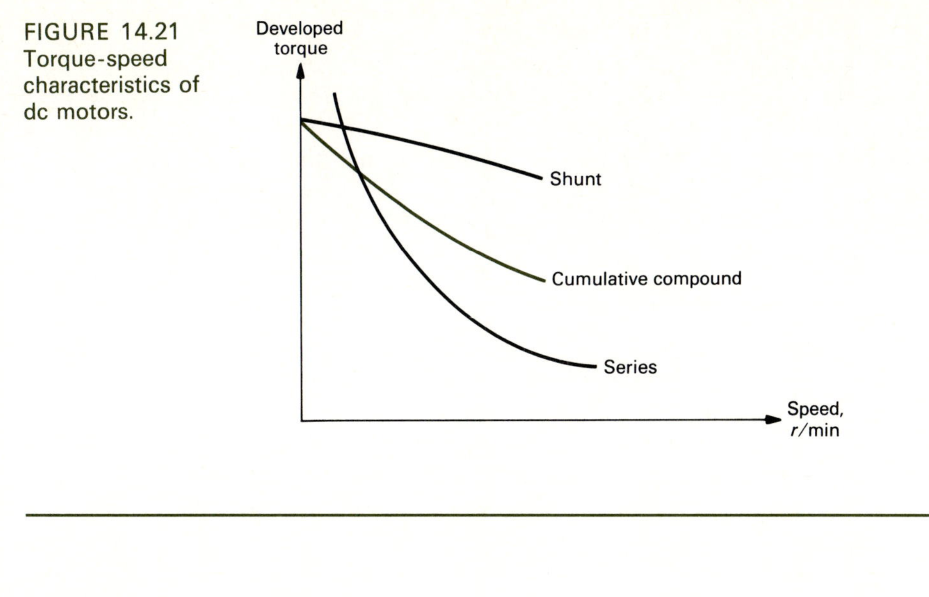

FIGURE 14.21 Torque-speed characteristics of dc motors.

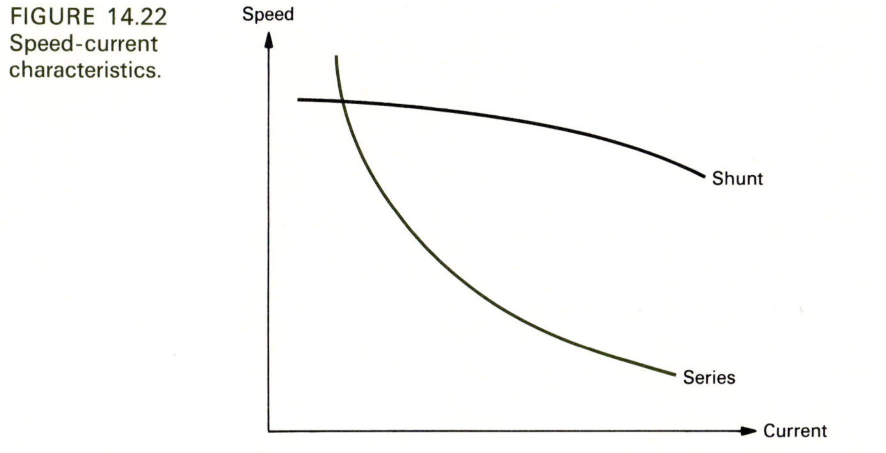

FIGURE 14.22 Speed-current characteristics.

In Fig. 14.21, we have also shown the developed torque versus speed for shunt and series motors. As we recall from Sec. 14.1, developed power is simply the product of developed torque (in N · m) and angular speed (in rad/s).

14.9 STARTING AND SPEED CONTROL OF MOTORS

In addition to certain operational conveniences, the basic requirements for satisfactory starting of a dc motor are (1) sufficient starting torque and (2) armature current, within safe limits, for successful commutation and for preventing the armature from overheating. The second requirement is obvious from the speed equation, according to which the armature current I_a is given by $I_a = V_t/R_a$ when the motor is at rest (n or $\omega_m = 0$). A typical 50-hp 230-V motor having an armature resistance of 0.05 Ω, if connected across 230 V, shall draw a 4600-A current. This current is evidently too large for the motor, which might be rated to take 180 A on full load. Commonly, double the full-load current is allowed to flow through the armature at the time of starting. For the motor under consideration, therefore, an external resistance $R_x = [230/(2 \times 180)] - 0.05 = 0.59\ \Omega$ must be inserted in series with the armature to limit I_a within double the rated value.

In practice, the necessary starting resistance is provided by means of a push-button starter.

From the speed equation of a dc motor, it follows that the speed of the motor can be varied by varying (1) the field-circuit resistance, to control I_f and hence the field flux; (2) the armature circuit resistance; and (3) the terminal voltage. Method (1) is commonly used to increase the speed of the motor, as illustrated by Example 14.7. Method (2) is used to decrease the speed of the motor, but is a wasteful method, as shown by Example 14.8. Method (3) is used either to increase or decrease the speed of the motor, but is feasible only if a variable voltage source is available. Method (3) in conjunction with method (1) constitutes the *Ward-Leonard system*, which gives a wide variation of speed. This method is presented in a simplified form in Example 14.9.

Example 14.7 A 230-V shunt motor has an armature resistance of 0.05 Ω and a field resistance of 75 Ω. The motor draws a 7-A line current while running light at 1120 r/min. The line current at a certain load is 46 A. (*a*) What is the motor speed at this load? (*b*) At this load, if the field-circuit resistance is increased to 100 Ω, what is the new speed of the motor? Assume that the line current remains unchanged.

Solution *a* On no-load

$$n_o = 1120 \text{ r/min} \qquad \text{given}$$

$$I_f = \frac{230}{75}$$

$$= 3.07 \text{ A}$$

$$I_a = 7 - 3.07$$

$$= 3.93 \text{ A}$$

The speed equation gives

$$1120 = \frac{230 - (3.93 \times 0.05)}{3.07k}$$

or

$$k = 0.0668$$

On load (with $R_f = 75\ \Omega$)

$$I_f = 3.07\ \text{A}$$

$$I_a = 46 - 3.07$$

$$= 42.93\ \text{A}$$

$$n = \frac{230 - (42.93 \times 0.05)}{3.07 \times 0.0668}$$

$$= 1111\ \text{r/min}$$

b On load (with $R_f = 100\ \Omega$)

$$I_f = \frac{230}{100} = 2.3\ \text{A}$$

$$I_a = 46 - 2.3$$

$$= 43.7\ \text{A}$$

$$n = \frac{230 - (43.7 \times 0.05)}{2.3 \times 0.0668}$$

$$= 1483\ \text{r/min}$$

Example 14.8 Refer to part *a* of the solution to Example 14.7. The no-load conditions remain unchanged. On load, the line current remains at 46 A, but a 0.1-Ω resistance is inserted in the armature. Determine the speed of the motor.

Solution In this case we have (from Example 14.7)

$$I_f = 3.07\ \text{A}$$

$$I_a = 42.93\ \text{A}$$

$$k = 0.0668$$

Thus (with $R_a = 0.05 + 0.1 = 0.15\ \Omega$)

$$n = \frac{230 - (42.93 \times 0.15)}{3.07 \times 0.0688}$$

$$= 1058\ \text{r/min}$$

Example 14.9 The system shown in Fig. 14.23 is called the Ward-Leonard system for controlling the speed of a dc motor. Discuss the effects of varying R_{fg} and R_{fm} on the motor speed. Subscripts g and m correspond to generator and motor respectively.

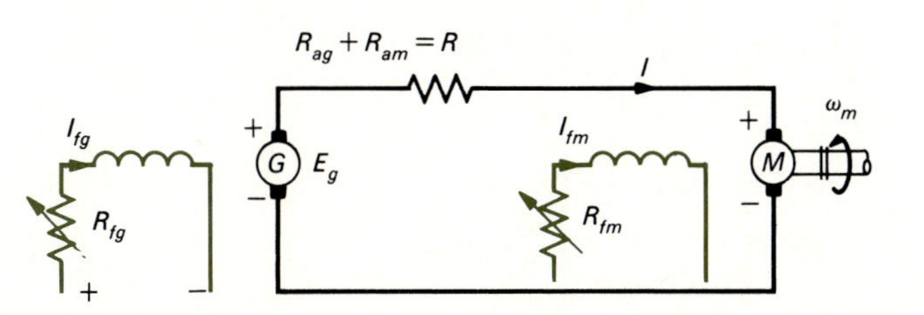

FIGURE 14.23 The Ward-Leonard system.

Solution Increasing R_{fg} decreases I_{fg} and hence E_g. Thus, the motor speed will decrease. The opposite will be true if R_{fg} is decreased.

Increasing R_{fm} will increase the speed of the motor, as shown in Example 14.7. Decreasing R_{fm} will result in a decrease of the speed.

14.10 LOSSES AND EFFICIENCY

Besides the voltamperage and speed-torque characteristics, the performance of a dc machine is measured by its efficiency:

$$\text{Efficiency} \equiv \frac{\text{power output}}{\text{power input}} = \frac{\text{power output}}{\text{power output} + \text{losses}} \tag{14.15}$$

Efficiency may therefore be determined either from load tests or by determination of losses. The various losses are classified as follows:

1 Electrical. These are the copper losses in various windings, such as the armature winding and different field windings, and the loss due to the contact resistance of the brush with the commutator.

2 Magnetic. These are the iron losses; they include the hysteresis and eddy-current losses in the various magnetic circuits, primarily in the armature core and pole faces.

3 Mechanical. These include the bearing-friction, windage, and brush-friction losses.

4 Stray-load. These are other load losses not covered above. They are taken as 1 percent of the output (as a rule of thumb). Another recommendation often followed is to take the stray-load loss as 28 percent of the core loss at the rated output.

The power flow in a dc generator or motor is represented in Fig. 14.24 in which the symbols are as follows:

E = back or induced emf in the armature, V

I_a = armature current, A

I_f = current through the field winding, A

P_{elec} = electric power, W

P_{mag} = magnetic loss, W

P_{mech} = rotational mechanical loss, W

P_{SL} = stray-load loss, W

T_e = electromagnetic torque developed by the armature, $\text{N} \cdot \text{m}$

T_s = torque available at the shaft, $\text{N} \cdot \text{m}$

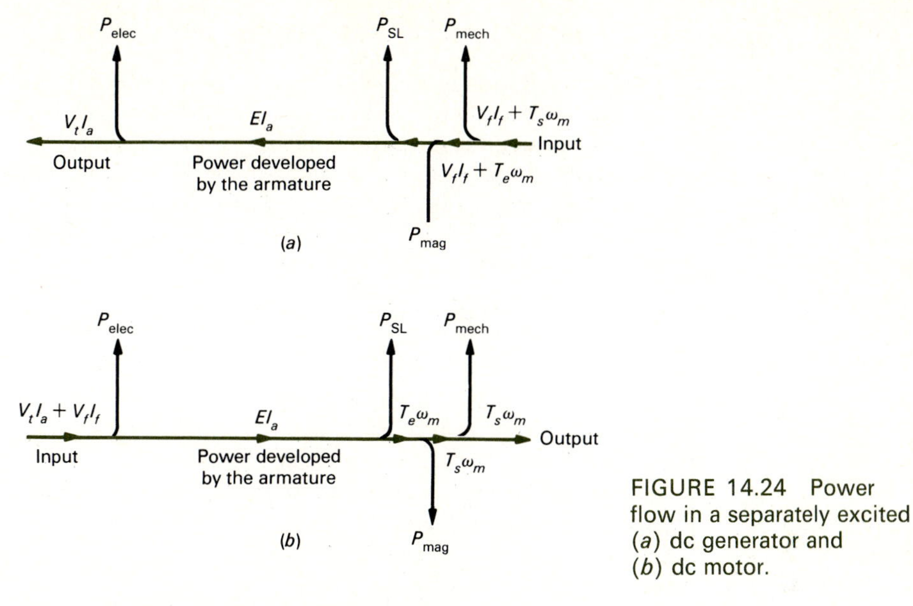

FIGURE 14.24 Power flow in a separately excited (*a*) dc generator and (*b*) dc motor.

V_f = voltage across the field winding, V

V_t = terminal voltage, V

ω_m = mechanical angular velocity, rad/s

Example 14.10 A 230-V 10-hp shunt motor takes a full-load line current of 40 A. The armature and field resistances are 0.25 and 230 Ω, respectively. The total brush-contact drop is 2 V, and the core and friction losses are 380 W. Calculate the efficiency of the motor. Assume that the stray-load loss is 1 percent of the rated output.

Solution

Input: $40 \times 230 = 9200$ W

Field-resistance loss: $\dfrac{230^2}{230} = 230$ W

Armature-resistance loss: $(40 - 1)^2 0.25 = 380$ W

Core loss and friction loss: $= 380$ W

Brush-contact loss: $2 \times 39 = 78$ W

Stray-load loss: $\dfrac{10}{100} \times 746 = 75$ W

Total losses: $= 1143$ W

Power output: $9200 - 1143 = 8057$ W

Efficiency: $\dfrac{8057}{9200} = 87.6\%$

14.11 CERTAIN APPLICATIONS

DC machines find applications in the following: electric traction and diesel-electric locomotives; large rolling mills; electrochemical plants and metal-refining plants; earth-moving equipment; battery charging; ship, train, and aircraft auxiliaries; isolated experimental stations; exciters for synchronous machines (see next chapter); automatic control systems; wind-generating systems; etc.

The choice of a dc motor for a specific application depends upon the nature of the load. For instance, permanent magnet dc motors are preferred for actuators requiring high-peak and steady power and fast response; they are basic drives in aircraft control systems. For traction loads, dc series motors are ideally suited; series motors are used to drive cranes, hoists, and high-inertia loads. In contrast, the shunt motor is essentially a constant-speed motor. Its speed can be easily controlled by adjusting the field current or armature voltage. Hence, it has numerous industrial applications, such as in driving pumps, compressors, and punch presses.

DC generators are used in the chemical industry. Differential compound generators find application in welding.

PROBLEMS

14.1 The armature of a four-pole dc machine has 32 conductors. Draw the winding layout for a double-layer lap winding, and verify that the winding has four parallel paths.

14.2 The armature of a four-pole dc machine has 30 conductors. Draw the winding layout for a two-layer wave winding. Verify that the winding has two parallel paths.

14.3 The flux per pole of a generator is 75 mWb. The generator runs at 900 r/min. Determine the induced voltage (*a*) if the armature has 32 conductors connected as lap winding and (*b*) if the armature has 30 conductors connected as wave winding.

14.4 Determine the flux per pole of a six-pole generator required to generate 240 V at 500 r/min. The armature has 120 slots, with 8 conductors per slot, and is lap-connected.

14.5 If the armature current in the generator of Prob. 14.4 is 25 A, what is the developed electromagnetic torque?

14.6 The armature and field resistances of a 240-V shunt generator are 0.2 and 200 Ω, respectively. The generator supplies a 9600-W load. Determine the induced voltage, taking 2 V as the total brush-contact voltage drop.

14.7 The armature, shunt-field, and series-field resistances of a compound generator are 0.02, 80, and 0.03 Ω, respectively. The generator-induced voltage is 510 V and the terminal voltage is 500 V. Calculate the power supplied to the load (at 500 V) if the generator has (*a*) a long-shunt connection and (*b*) a short-shunt connection.

14.8 A four-pole shunt-connected generator has a lap-connected armature with 728 conductors. The flux per pole is 25 mWb. If the generator supplies two hundred 110-V 75-W bulbs, determine the speed of the generator. The field and armature resistances are 110 and 0.075 Ω, respectively.

14.9 A shunt generator delivers a 50-A current to a load at 110 V, at an efficiency of 85 percent. The total constant losses are 480 W, and the shunt-field resistance is 65 Ω. Calculate the armature resistance.

14.10 For the generator of Prob. 14.9, plot a curve for efficiency versus armature current. At what armature current is the efficiency maximum?

14.11 A separately excited six-pole generator has

a 30-mWb flux per pole. The armature is lap-wound and has 534 conductors. This supplies a certain load at 250 V while running at 1000 r/min. At this load, the armature copper loss is 640 W. Calculate the load supplied by the generator. Take 2 V as the total brush-contact drop.

14.12 In a dc machine, the hysteresis and eddy-current losses at 1000 r/min are 10,000 W at a field current of 7.8 A. At 750-r/min speed and 7.8-A field current, the total iron losses become 6,000 W. Assuming that the hysteresis loss is directly proportional to the speed and that the eddy-current loss is proportional to the square of the speed, determine the hysteresis and eddy-current losses at 500 r/min.

14.13 The saturation characteristic of a dc shunt generator is as follows:

Field current, A	1	2	3	4	5	6	7
Open-circuit voltage, V	53	106	150	192	227	252	270

The generator speed is 900 r/min. At this speed, what is the maximum field-circuit resistance such that the self-excited shunt generator would not fail to build up?

14.14 To what value will the no-load voltage of the generator of Prob. 14.13 build up, at 900 r/min, for a field-circuit resistance of 42 Ω?

14.15 A 240-V separately excited dc machine has an armature resistance of 0.25 Ω. The armature current is 56 A. Calculate the induced voltage for (*a*) generator operation and (*b*) motor operation.

14.16 The field- and armature-winding resistances of a 400-V dc shunt machine are 120 and 0.12 Ω, respectively. Calcuate the power developed by the armature (*a*) if the machine takes 50 kW while running as a motor and (*b*) if the machine delivers 50 kW while running as a generator.

14.17 The field and armature resistances of a 220-V series motor are 0.2 and 0.1 Ω, respectively. The motor takes a 30-A current while running at 700 r/min. If the total iron and friction losses are 350 W, determine the motor efficiency.

14.18 A 400-V shunt motor delivers 15 kW of power at the shaft at 1200 r/min while drawing a line current of 62 A. The field and armature resistances are 200 and 0.05 Ω, respectively. Assuming a 1-V contact drop per brush, calculate (*a*) the torque developed by the motor and (*b*) the motor efficiency.

14.19 A 400-V series motor, having an armature circuit resistance of 0.5 Ω, takes a 44-A current while running at 650 r/min. What is the motor speed for a line current of 36 A?

14.20 A 220-V shunt motor, having an armature resistance of 0.2 Ω and a field resistance of 110 Ω, takes a 4-A line current while running on no-load. When loaded, the motor runs at 100 r/min while taking a 42-A current. Calculate the no-load speed.

14.21 The machine of Prob. 14.20 is driven as a shunt generator to deliver a 44-kW load at 220 V. If the machine takes 44 kW while running as a motor, what is its speed?

14.22 A 220-V shunt motor, having an armature resistance of 0.2 Ω and a field resistance of 110 Ω, takes a 4-A line current while running at 1200 r/min on no-load. On load, the input to the motor is 15 kW. Calculate (*a*) the speed, (*b*) developed torque, and (*c*) efficiency at this load.

14.23 A 400-V series motor has a field resistance of 0.2 Ω and an armature resistance of 0.1 Ω. The motor takes a 30-A current at 1000 r/min while developing a torque T. Determine the motor speed if the developed torque is $0.6T$.

14.24 A shunt machine, while running as a generator, has an induced voltage of 260 V at 1200 r/min. Its armature and field resistances are 0.2 and 110 Ω respectively. If the machine is run as a shunt motor, it takes 4 A at 220 V. At a certain load the motor takes 30 A at 220 V. On load, however, armature reaction weakens the field by 3 percent. Calculate the motor speed and efficiency at the specified load.

14.25 The machine of Prob. 14.24 is run as a motor. It takes a 25-A current at 800 r/min. What resistance must be inserted in the field circuit to increase the motor speed to 100 r/min? The torque on the motor for the two speeds remains unchanged.

14.26 The motor of Prob. 14.24 runs at 600 r/min while taking 40 A at a certain load. If a 0.8-Ω resistance is inserted in the armature circuit, determine the motor speed, provided that the torque on the motor remains constant.

14.27 A 220-V shunt motor delivers 40 hp on full load at 950 r/min and has an efficiency of 88 percent. The armature and field resistances are 0.2 and 110 Ω, respectively. Determine (*a*) the starting resistance, such that the starting line current does not exceed 1.6 times the full-load current, and (*b*) the starting torque.

14.28 A 220-V series motor runs at 750 r/min while taking a 15-A current. The total resistance in the armature circuit, including the field resistance, is 0.4 Ω. What is the motor speed if it takes a 10-A current? The torque on the motor is such that it increases as the square of the speed.

REFERENCES

14.1 M. G. Say and E. O. Taylor, *Direct Current Machines*, Halsted Press, New York, 1980.

14.2 S. A. Nasar, *Schaum's Outline of Theory and Problems in Electric Machines and Electromechanics*, McGraw-Hill, New York, 1981.

14.3 S. A. Nasar, *Electric Machines and Transformers*, Macmillan, New York, 1983.

14.4 A. E. Fitzgerald, C. Kinsley, Jr., and S. D. Umans, *Electric Machinery*, 4th ed., McGraw-Hill, New York, 1983.

Chapter

15

Synchronous Machines

The bulk of electric power for everyday use is produced by polyphase synchronous generators, which are the largest single-unit electric machines in production. For instance, synchronous generators with power ratings of several hundred megavolt-amperes (MVA) are fairly common, and it is expected that machines of several thousand megavoltamperes will be in use in the near future. These are called synchronous machines because they operate at constant speeds and constant frequencies under steady-state conditions.

Like most rotating machines, synchronous machines are capable of operating both as motors and as generators. They are used as motors in constant-speed drives; where a variable-speed drive is required, a synchronous motor is used with an appropriate frequency changer. As generators, several synchronous machines often operate in parallel, as in a power station. While operating in parallel, the generators share the load with each other. At a given time, one of the generators may not carry any load; in such a case, instead of shutting down the generator, it is allowed to "float" on the line as a synchronous motor on no-load.

The operation of a synchronous generator is based on Faraday's law of electromagnetic induction. Thus, an ac synchronous generator works very much like a dc generator, in which emf is generated by the relative motion of conductors and magnetic flux (Fig. 14.4 illustrated a synchronous generator in its elementary form). Unlike a dc generator, however, a synchronous generator does not have a commutator. The two basic parts of a synchronous machine are the magnetic field structure, carrying a dc-excited winding, and the armature; the armature often has a three-phase winding in which the ac emf is generated. Almost all modern synchronous machines have stationary armatures and rotating field structures. The dc winding on the rotating field structure is connected to an external source through slip rings and brushes. Some field structures do not have brushes but, instead, have brushless excitation by rotating diodes. In some respects the stator carrying the armature windings is similar to the stator of a polyphase induction motor (discussed in the next chapter).

15.1 SOME CONSTRUCTION DETAILS

Some of the factors that dictate the form of construction of a synchronous machine are as follows:

1 *Form of excitation.* Notice from the preceding remarks that the field structure is usually the rotating member of a synchronous machine and is supplied with a dc-excited winding to produce the magnetic flux. This dc excitation may be provided by a self-excited dc generator mounted on the rotor shaft of the synchronous machine; such a generator is known as the *exciter.* The direct current thus generated is fed to the synchronous machine field winding. In slow-speed machines with large ratings, such as hydroelectric generators, the exciter may not be self-excited; instead, a pilot exciter, which may be self-excited or may have a permanent magnet, activates the exciter. A hydroelectric generator and its rotor and exciters are shown in Fig. 15.1. The maintenance problems of direct-coupled dc generators impose a limit on this form of excitation at a rating of about 100 megawatts (MW). An alternative form of excitation is provided by silicon diodes and thyristors, which do not present excitation problems for large synchronous machines; the two types of solid-state excitation systems are:

   a Static systems that have stationary diodes or thyristors, in which the current is fed to the rotor through slip rings.

   b Brushless systems that have shaft-mounted rectifiers that rotate with the rotor, thus avoiding the need for brushes and slip rings. Figure 15.2 shows a brushless excitation system.

2 *Field structure and speed of machine.* We have already mentioned that the synchronous machine is a constant-speed machine. This speed is known as synchronous speed. For instance, a two-pole 60-Hz synchronous machine must run at 3600 r/min, whereas the synchronous speed of a 12-pole 60-Hz machine is only 600 r/min. The rotor field structure consequently depends on the speed rating of the machine. Turbogenerators, which are high-speed machines, have *round cylindrical rotors,* as shown in Figs. 15.3 and 15.4. Hydroelectric and diesel-electric generators are low-speed machines and have *salient-pole rotors,* as shown in Figs. 15.1, 15.2, and 15.5. Such rotors are less expensive to fabricate than round rotors; they are not suitable for large, high-speed machines, however, because of the excessive centrifugal forces and mechanical stresses that develop at speeds around 3600 r/min.

3 *Stator.* The stator of synchronous machine carries the armature, or load, winding. We recall from Chap. 14 that the armature of a dc machine has a winding distributed around its periphery and that this armature winding consists of slot-embedded conductors which cover the entire surface of the armature and are interconnected in a predetermined manner. Likewise, in a synchronous machine the armature winding is formed by interconnnecting the various conductors in the slots spread over the periphery of the stator of the machine. Often, more than one independent winding is on the stator. When constructing a three-phase machine, for instance, the engineers displace the three phases from each other in space as

FIGURE 15.1
A hydroelectric generator.

they distribute the windings in the slots over the entire periphery of the stator; generally, each slot contains two coil sides. Such a winding is known as a *double-layer* winding. The mounting of stator conductors in the slots of one stator half to make the armature winding of a large synchronous machine is shown in Fig.15.6.

4 *Cooling*. Because synchronous machines are often built in extremely large sizes, they are designed to carry very large currents. A typical armature current density may be of the order of 10 A/mm^2 in a well-designed machine. Also, the magnetic loading of the core is such that it reaches saturation in many regions. These severe electric and magnetic loadings in a synchronous machine produce heat that must be appropriately dissipated. Thus, the manner in which the active parts of a

FIGURE 15.2
Rotor of a 3360 kVA 6kV brushless synchronous generator, with rotating diodes.

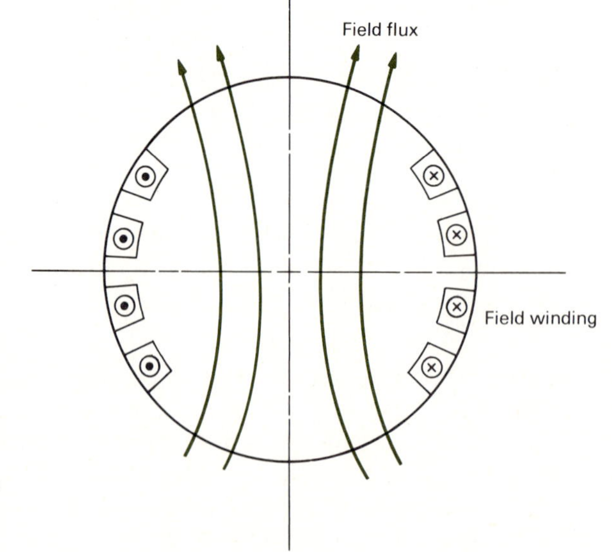

FIGURE 15.3
Field winding on a round or cylindrical rotor.

FIGURE 15.4
Turbine rotor with direct water cooling during the mounting off damper hollow bars.

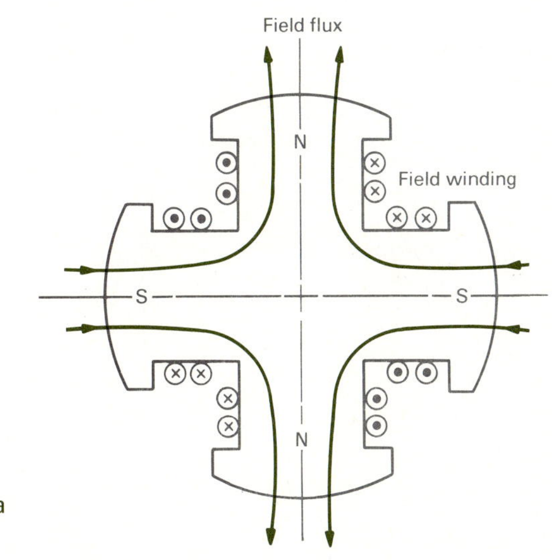

FIGURE 15.5
Field winding on a salient rotor.

FIGURE 15.6 Mounting stator conductors in slots of one stator half of a synchronous motor.

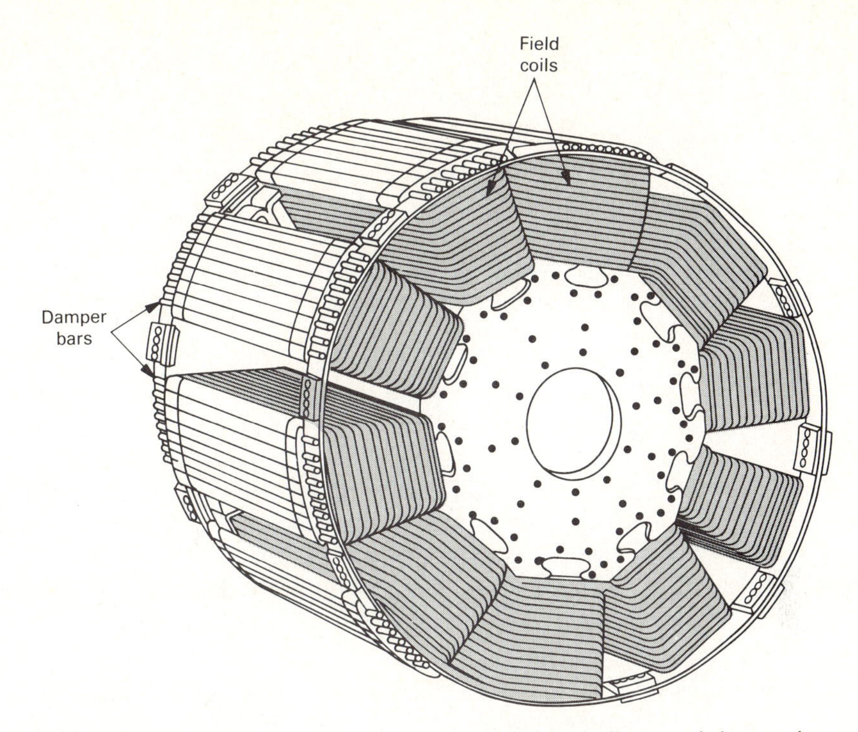

FIGURE 15.7 A salient rotor, showing the field windings and damper bars (shaft not shown).

machine are cooled determines its overall physical structures. In addition to air, some of the coolants used in synchronous machines include water, hydrogen, and helium.

5 *Damper bars.* So far we have mentioned only two electrical windings of a synchronous machine: the three-phase armature winding and the field winding. We have also pointed out that, under steady-state, the machine runs at a constant speed, that is, at the synchronous speed. However, like other electric machines, a synchronous machine undergoes transients during starting and abnormal conditions. During transients, the rotor may undergo mechanical oscillations and its speed deviate from the synchronous speed, which is an undesirable phenomenon. To overcome this, a winding (resembling the cage of an induction motor) is mounted on the rotor; this set constitutes the damper windings. When the rotor speed is different from the synchronous speed, currents are induced in the damper winding. The damper winding acts like the cage rotor of an induction motor, producing a torque to restore the synchronous speed. Also, the damper bars provide a means of starting the machine, which is otherwise not self-starting. Figures 15.4 and 15.7 show the damper bars on round and salient rotors, respectively.

15.2 MAGNETOMOTIVE FORCES (MMFs) AND FLUXES DUE TO ARMATURE AND FIELD WINDINGS

In general, we may say that the behavior of an electric machine depends on the interaction between the magnetic fields (or fluxes) produced by various mmfs acting on the magnetic circuit of the machines. For instance, in a dc motor the torque is produced by the interaction of the flux produced by the field winding and the flux produced by the current-carrying armature conductors. In a dc generator, too, we must consider the effect of the interaction between the field and armature mmfs (as discussed in Sec. 14.2).

As mentioned in Sec. 15.1, the main sources of fluxes in a synchronous machine are the armature and field mmfs. In contrast to a dc machine, in which the flux due to the armature mmf is stationary in space, the fluxes due to each phase of the armature mmf pulsate in time and the resultant flux rotates in space, as will be demonstrated in Eq. (15.1) and in Sec. 16.1. For the present, we shall consider the mmfs produced by a single full-pitch coil having N-turns where the slot opening is negligible and the machine has two poles, as shown in Fig.15.8*a*. The mmf has a constant value of Ni between the coil sides, as shown in Fig.15.8*b*. Traditionally, the magnetic effects of a winding in an electric machine are considered on a per pole basis. Thus, if i is the

FIGURE 15.8 Flux and mmf produced by a concentrated winding: (*a*) flux lines produced by an N-turn coil; (*b*) mmf produced by the N-turn coil; (*c*) mmf per pole.

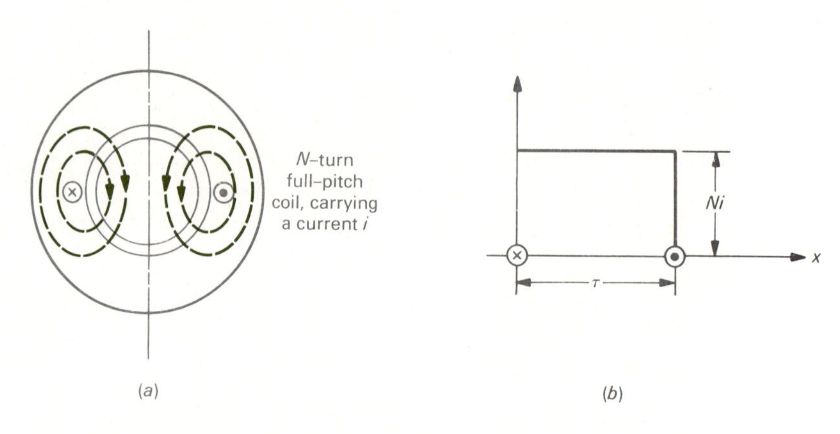

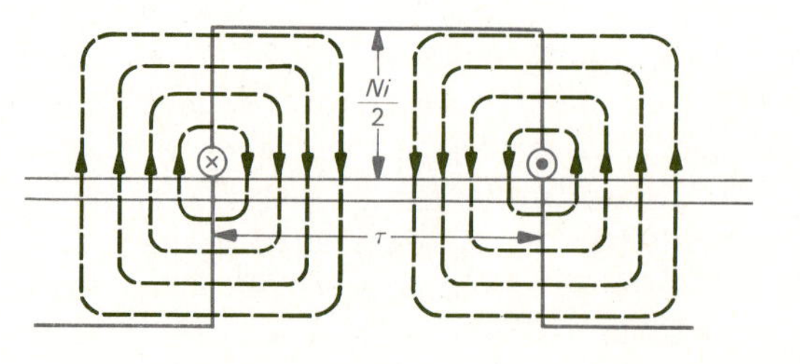

current in the coil, the mmf per pole is $Ni/2$; this is plotted in Fig.15.8*c*. The reason for such a representation is that Fig. 15.8*c* also represents a flux density distribution, but to a different scale. Obviously, the flux density over one pole (say the north pole) must be opposite to that over the other (south) pole, thus keeping the flux entering the rotor equal to that leaving the rotor surface. Comparing Fig. 15.8*b* and *c*, we notice that the representation of the mmf curve with positive and negative areas (Fig. 15.8*c*) has the advantage that it gives the flux density distribution, which must contain positive and negative areas. The mmf distribution shown in Fig. 15.8*c* may be resolved into its harmonic components. In practice, we may assume harmonics to be absent in properly designed armature windings, and the resultant mmf of each phase may be ideally taken as sinusoidal.

Let the three-phase stator (or armature) winding of a synchronous machine be excited by three phase currents. As a result, the mmf's produced by the three phases are displaced from each other by 120° in time and space. If we assume the mmf distribution in space to be sinusoidal, we may write for the three mmf's:

$$\mathscr{F}_a = F_m \sin \omega t \cos p\theta$$
$$\mathscr{F}_b = F_m \sin (\omega t - 120^\circ) \cos (p\theta - 120^\circ)$$
$$\mathscr{F}_c = F_m \sin (\omega t + 120^\circ) \cos (p\theta + 120^\circ)$$

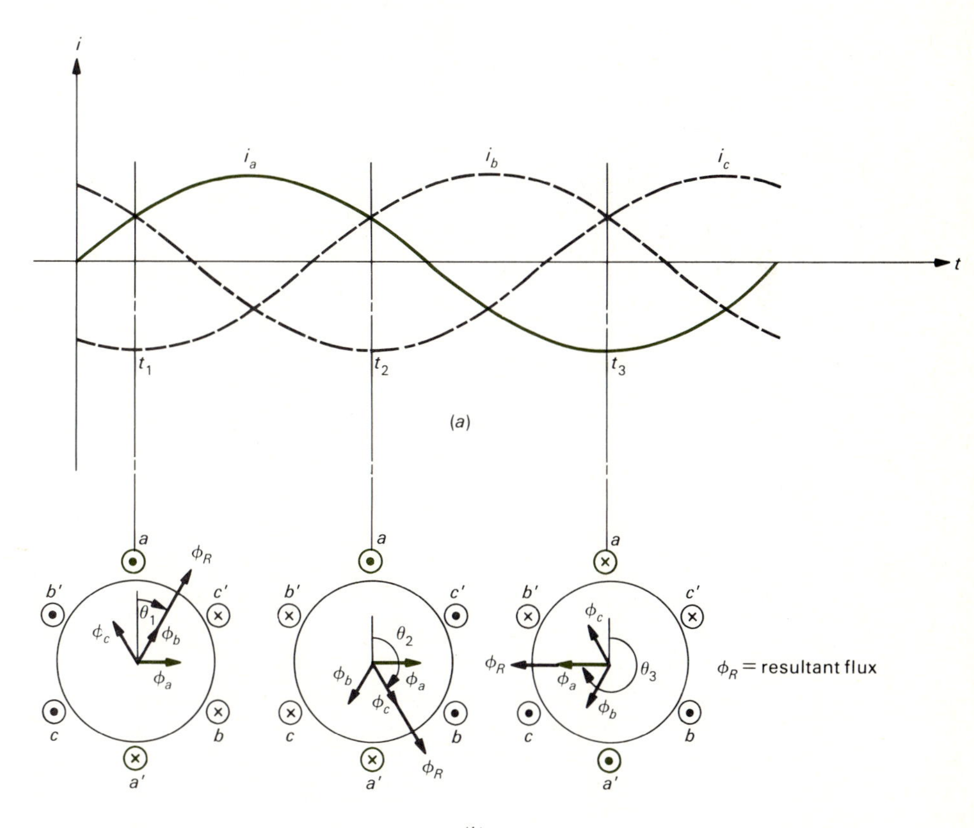

FIGURE 15.9 Production of a rotating magnetic field by a three-phase excitation: (*a*) time diagram; (*b*) space diagram.

where F_m is the amplitude of the mmf and ω and p are defined in Sec. 15.3. The space and time variations of the resultant mmf is then the sum of the above three mmf's. Observing that $\sin A \cos B = \frac{1}{2}\sin(A - B) + \frac{1}{2}\sin(A + B)$ and adding $\mathscr{F}_a$, $\mathscr{F}_b$, and $\mathscr{F}_c$, we obtain the resultant mmf as

$$\mathscr{F}(\theta, t) = 1.5F_m \sin(\omega t - p\theta) \tag{15.1}$$

The magnetic field resulting from the mmf of Eq. (15.1) is a *rotating magnetic field.* Graphically, the production of the rotating field is illustrated in Fig.15.9. The existence of the rotating magnetic field is essential to the operation of a synchronous motor.

15.3 THE SYNCHRONOUS SPEED

To determine the velocity of the rotating field given by Eq. (15.1) imagine an observer traveling with the mmf wave from a certain point. To this observer, the magnitude of the mmf wave will remain constant (independent of time), implying that the right side of Eq. (15.1) would appear constant. Expressed mathematically, this would mean that

$$\sin(\omega t - p\theta) = \text{constant}$$

or

$$\omega t - p\theta = \text{constant}$$

Differentiating both sides with respect to t, we obtain

$$\omega - p\dot{\theta} = 0$$

or

$$\omega_m \equiv \dot{\theta} = \frac{\omega}{p} \tag{15.2}$$

This speed is known as the *synchronous speed.*

In Eqs. (15.1) and (15.2), ω is the frequency of the stator mmf's. What is the significance of p? Notice that the given mmf's vary in space as $\cos p\theta$, indicating that for one complete travel around the stator periphery, the mmf undergoes p cyclic changes. Thus, p may be considered as the order of harmonics, or the number of *pole pairs* in the mmf wave. If P is the number of *poles* then obviously $P = 2p$. Writing ω_m in terms of speed n_s in r/min and ω in terms of the frequency f, we have

$$\omega_m = \frac{2\pi n_s}{60}$$

and

$$\omega = 2\pi f$$

Substituting for P, ω_m, and ω in Eq. (15.2) yields

$$n_s = \frac{120f}{P} \tag{15.3}$$

which is the synchronous speed in r/min.

An alternative form of Eq. (15.3) is

$$f = \frac{Pn_s}{120} \tag{15.4}$$

which implies that the frequency of the voltage induced in a synchronous generator having P poles and running at n_s r/min is f Hz. We could reach the same conclusion from Figs. 14.4 and 14.14; namely, in a 2-pole machine, one cycle is generated in one rotation. Thus, in a P-pole machine $P/2$ cycles are generated in one rotation; and, in n_s rotation, $Pn_s/2$ cycles are generated. Since n_s rotations take 60 s, in 1 s $Pn_s/2 \times 60 = Pn_s/120$ cycles are generated, which is the frequency f.

Example 15.1 For a 60-Hz generator, list four possible combinations of the number of poles and the speed.

Solution From Eq. (15.4) we must have $Pn_s = 7200 = 120 \times 60$. Hence, we obtain the following table:

Number of poles	Speed, r/min
2	3600
4	1800
6	1200
8	900

15.4 SYNCHRONOUS GENERATOR OPERATION

Like the dc generator, a synchronous generator functions on the basis of Faraday's law. If the flux linking the coil changes in time, a voltage is induced in the coil. Stated in alternative form, a voltage is induced in a conductor if it cuts magnetic flux lines (recall Fig. 14.4). Consider the machine shown in Fig. 15.10; assuming that the flux density in the air gap is uniform, we see that sinusoidally varying voltages will be induced in the three coils aa', bb', and cc' if the rotor, carrying dc, rotates at a constant

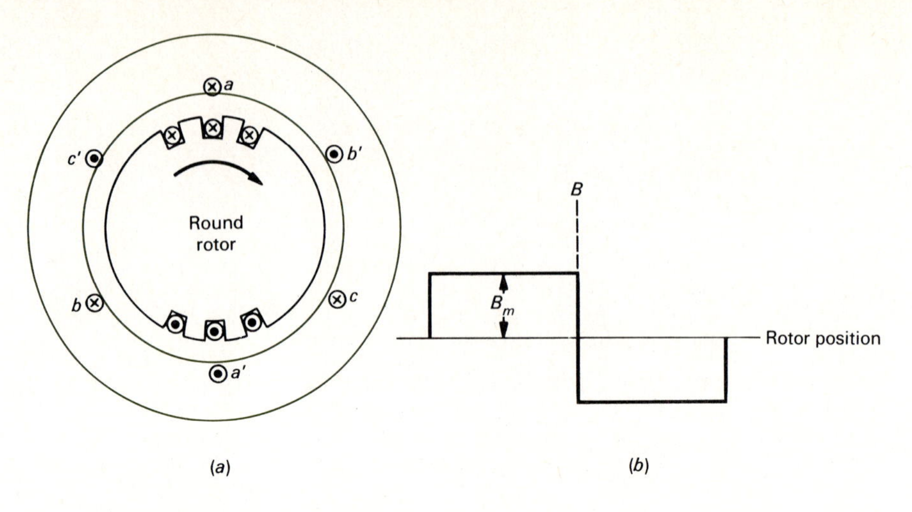

FIGURE 15.10 (*a*) A three-phase round-rotor machine. (*b*) Flux density distribution produced by the rotor excitation.

speed n_s. Recall from Chap. 14, Eq. (14.6), that if ϕ is the flux per pole, ω is the angular frequency, and N is the number of turns in phase a (coil aa'), then the voltage induced in phase a is given by

$$\begin{aligned} e_a &= \omega N \phi \sin \omega t \\ &= E_m \sin \omega t \end{aligned} \tag{15.5}$$

where $E_m = 2\pi f N \phi$ and $f = \omega/2\pi$ is the frequency of the induced voltage. Because phases b and c are displaced from phase a by $\pm 120°$ (Fig. 15.10), the corresponding voltages may be written as

$$e_b = E_m \sin (\omega t - 120°)$$

$$e_c = E_m \sin (\omega t + 120°)$$

These voltages are sketched in Fig. 15. 11 and correspond to the voltages from a three-phase generator.

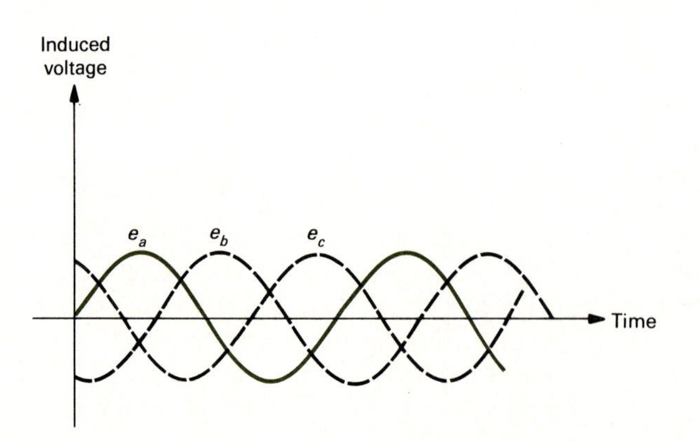

FIGURE 15.11 A three-phase voltage produced by 2 three-phase synchronous generators.

15.5 PERFORMANCE OF A ROUND-ROTOR SYNCHRONOUS GENERATOR

At the outset we wish to point out that we will study the machine on a per phase basis, implying a balanced operation. Thus, let us consider a round-rotor machine operating as a generator on no-load. Variation of V_o with I_f is shown in Fig.15.12 and is known as the open-circuit characteristic of a synchronous generator. Let the open-circuit phase voltage be V_o for a certain field current I_f. Here, V_o is the internal voltage of the generator. We assume that I_f is such that the machine is operating under unsaturated condition. Now let us short-circuit the armature at the terminals, keeping the field current unchanged (at I_f), and measure the armature phase current I_a. In this case, the entire internal voltage V_o is dropped across the internal impedance of the machine. In mathematical terms,

$$\mathbf{V}_o = \mathbf{I}_a \mathbf{Z}_s$$

where Z_s is known as the *synchronous impedance.* One portion of Z_s is R_a, the armature resistance per phase, and the other is a reactance X_s, which is known as *synchronous reactance*; that is,

$$\mathbf{Z}_s = R_a + jX_s \tag{15.6}$$

If the generator operates at a terminal voltage V_t while supplying a load corresponding to an armature current I_a, then

$$\mathbf{V}_o = \mathbf{V}_t + \mathbf{I}_a(R_a + jX_s) \tag{15.7}$$

where X_s is the synchronous reactance.

In an actual synchronous machine (expect in very small ones) we almost always have $X_s \gg R_a$, in which case $Z_s \simeq jX_s$. We will use this restriction in most of our analyses. Among the steady-state characteristics of a synchronous generator, its voltage regulation and power-angle characteristics are the most important ones. As for a

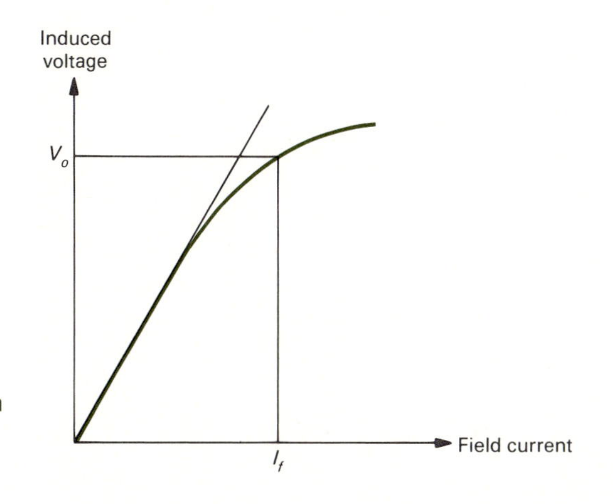

FIGURE 15.12 Open-circuit characteristics of a synchronous generator.

transformer and a dc generator, we define the voltage regulation of a synchronous generator at a given load as

$$\text{Percent voltage regulation} = \frac{V_o - V_t}{V_t} \times 100 \tag{15.8}$$

where V_t is the terminal voltage on load and V_0 is the no-load terminal voltage. Clearly, for a given V_t, we can find V_o from Eq. (15.7) and hence the voltage regulation, as illustrated by the following examples.

Example 15.2 Calculate the percent voltage regulation for a three-phase wye-conected 2500-kVA 6600-V turboalternator operating at full load and 0.8 lagging power factor. The per-phase synchronous reactance and the armature resistance are 10.4 and 0.071 Ω, respectively.

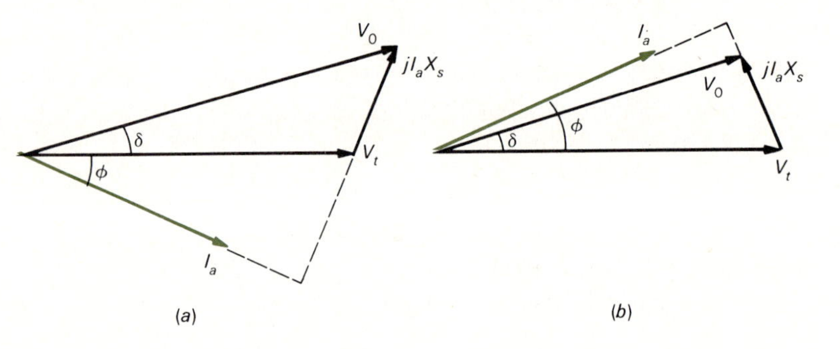

FIGURE 15.13 Phasor diagrams: (*a*) lagging power factor; (*b*) leading power factor.

Solution Here we have $X_s \gg R_a$. The phasor diagram for the lagging power factor, neglecting the effect of R_a, is shown in Fig. 15.13*a*. The numerical values are as follows:

$$V_t = \frac{6600}{\sqrt{3}}$$

$$= 3810 \text{ V}$$

$$\mathbf{I}_a = \frac{2500 \times 1000}{\sqrt{3} \times 6600} \angle -36.87^\circ$$

$$= 218.7 \text{ A} \angle -36.87^\circ$$

From Eq. (15.7) we have

$$\mathbf{V}_0 = 3810 + 218.7(0.8 - j0.6)j10.4$$

$$= 5485\angle 19.3^\circ$$

and from Eq. (15.8) the percent regulation is

$$\frac{5485 - 3810}{3810} \times 100 = 44\%$$

Example 15.3 Repeat the preceding calculations with 0.8 leading power factor.

Solution In this case we have the phasor diagram shown in Fig.15.13*b*, from which we get

$$\mathbf{V}_0 = 3810 + 218.7(0.8 + j0.6)j10.4$$
$$= 3048\underline{/36.6^\circ}$$

and the percent voltage regulation is

$$\frac{3048 - 3810}{3810} \times 100 = -20\%$$

We observe from these examples that the voltage regulation is dependent on the power factor of the load. Unlike what happens in a dc generator, the voltage regulation for a synchronous generator may even become negative. The angle between V_0 and V_t is defined as the *power angle* δ. To justify this definition, we reconsider Fig. 15.13*a*, from which we obtain

$$I_a X_s \cos\phi = V_0 \sin\delta \tag{15.9}$$

Now, from the approximate equivalent circuit (assuming $X_s \gg R_a$) the power delivered by the generator = power developed: $P_d = V_t I_a \cos\phi$ (which also follows from Fig. 15.13*a*). Hence, in conjunction with Eq. (15.9), we get

$$P_d = \frac{V_0 V_t}{X_s} \sin\delta \tag{15.10}$$

which shows that the internal power of the machine is proportional to sin δ. Equation (15.10) is often said to represent the *power-angle characteristic* of a synchronous machine. A plot of Eq. (15.10) is shown in Fig. 15.14*b*, which shows that for a negative δ the machine will operate as a motor, as discussed in the next section.

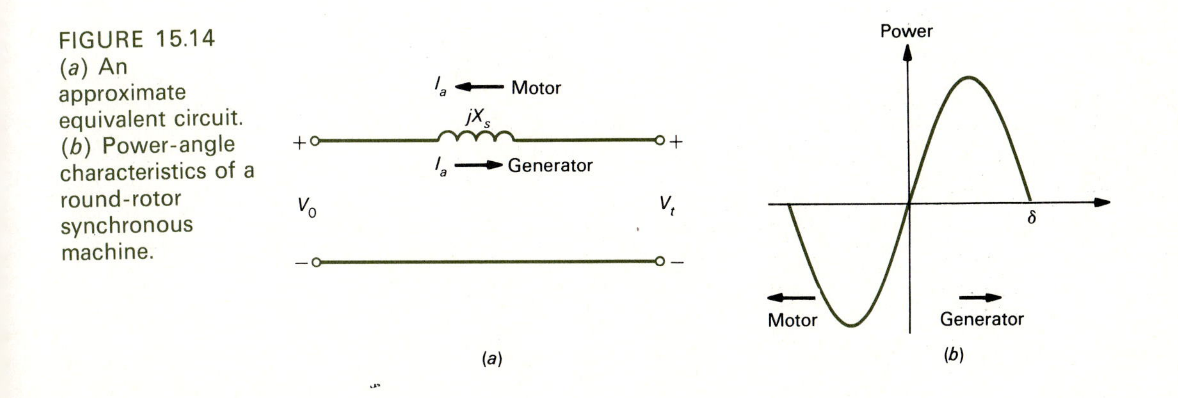

FIGURE 15.14 (*a*) An approximate equivalent circuit. (*b*) Power-angle characteristics of a round-rotor synchronous machine.

15.6 SYNCHRONOUS MOTOR OPERATION

We know from Sec. 15.2 that the stator of a three-phase synchronous machine carrying a three-phase excitation produces a rotating magnetic field in the air gap of the machine. Referring to Fig. 15.15, we will have a rotating magnetic field in the air gap of the salients-pole machine when its stator (or armature) windings are fed from a three-phase source.

Now let the rotor (or field) winding of this machine be unexcited. The rotor will have a tendency to align with the rotating field at all times in order to present the path of least reluctance. Thus, if the field is rotating, the rotor will tend to rotate with the field. A round rotor, on the other hand, will not tend to follow the rotating magnetic field; we see from Fig. 15.10*a* that this is because the uniform air gap presents the same reluctance all around the air gap, and thus the rotor does not have any preferred direction of alignment with the magnetic field. The torque which we have in the salient-rotor machine of Fig. 15.15 but not in the round-rotor machine of Fig. 15.10*a* is called the *reluctance torque*. It is present by virtue of the variation of the reluctance around the periphery of the machine.

Now let the field winding of either machine (Fig 15.10*a* or 15.15) be fed by a dc source that produces a rotor magnetic field of definite polarities, and the rotor will tend to align with the stator field and to rotate with the rotating magnetic field. We observe that both a round rotor and a salient rotor, when excited, will tend to rotate with the rotating magnetic field, although the salient rotor will have an additional reluctance torque because of the saliency, whether excited or unexcited. In Sec. 15.8 we will derive expressions for the electromagnetic power in a synchronous machine attributable to field excitation and to saliency.

So far, we have indicated the mechanism of torque production in a round-rotor and in a salient-rotor machine. To recapitulate, we might say that the stator rotating magnetic field has a tendency to "drag" the rotor along, as if a north pole on the stator "locks in" with a south pole of the rotor. However, if the rotor is at a standstill, the stator poles will tend to make the rotor rotate in one direction and then in the other as they rotate and sweep across the rotor poles. Therefore, a synchronous motor is not

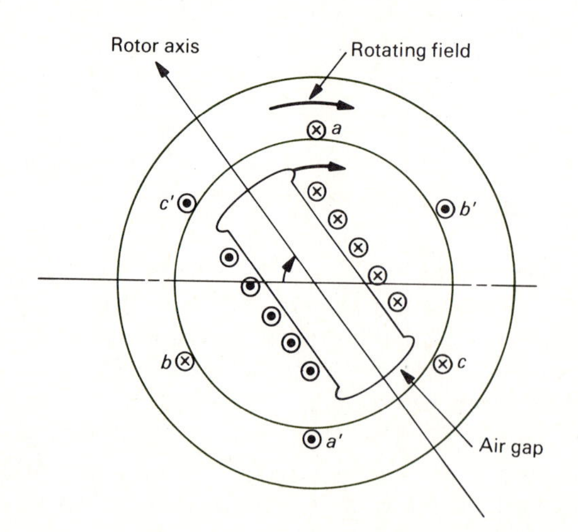

FIGURE 15.15 A salient-rotor machine.

self-starting. In practice, as mentioned in Sec. 15.1, the rotor carries damper bars that act like the cage of an induction motor and thereby provide a starting torque. The mechanism of torque production by the damper bars is similar to the torque production in an induction motor (discussed in the next chapter). Once the rotor starts running and almost reaches the synchronous speed, it locks into position with the stator poles. The rotor pulls into step with the rotating magnetic field and runs at the synchronous speed; the damper bars go out of action. Any departure from the synchronous speed results in induced currents in the damper bars, which tend to restore the synchronous speed. Machines without damper bars—or very large machines with damper bars—may be started by an auxiliary motor. We discuss the operating characteristics of synchronous motors in the next section.

15.7 PERFORMANCE OF A ROUND-ROTOR SYNCHRONOUS MOTOR

Except for some precise calculations, we may neglect the armature resistance as compared to the synchronous reactance. Therefore, the steady-state per phase equivalent circuit of a synchronous machine simplifies to the one shown in Fig. 15.14*a*. Notice that this circuit is similar to that of a dc machine, where the dc armature resistance has been replaced by the synchronous reactance. In Fig. 15.14*a* we have shown the terminal voltage V_t, the internal excitation voltage V_o, and the armature current I_a going into the machine or out of it, depending on the mode of operation—"into" for motor and "out of" for generator. With the help of this circuit and Eq. (15.10), we will study some of the steady-state operating characteristics of a synchronous motor. In Fig. 15.14*b* we show the power-angle characteristics as given by Eq. (15.10). Here, positive power and positive δ imply generator operation, while a negative δ corresponds to motor operation. Because δ is the angle between $\mathbf{V}_o$ and $\mathbf{V}_t$, $\mathbf{V}_o$ is ahead of $\mathbf{V}_t$ in a generator, whereas in a motor $\mathbf{V}_t$ is ahead of $\mathbf{V}_o$. The voltage-balance equation for a motor is, from Fig. 15.14*a*,

$$\mathbf{V}_t = \mathbf{V}_0 + jI_a X_s$$

If the motor operates at a constant power, then Eqs. (15.9) and (15.10) require that

$$V_0 \sin \delta = I_a X_s \cos \phi = \text{constant} \tag{15.11}$$

We recall from Fig. 15.12 that V_o depends on the field current I_f. Consider two cases: (1) when I_f is adjusted so that $V_o < V_t$ and the machine is underexcited and (2) when I_f is increased to a point that $V_o > V_t$ and the machine becomes overexcited. The voltage-current relationships for the two cases are shown in Fig. 15.16*a*. For $V_o > V_t$ at constant power, δ is greater than the δ for $V_o < V_t$, as governed by Eq. (15.11). Notice that an underexcited motor operates at a lagging power factor ($\mathbf{I}_a$ lagging $\mathbf{V}_t$), whereas an overexcited motor operates at a leading power factor. In both cases the terminal voltage and the load on the motor are the same. Thus, we observe that the operating

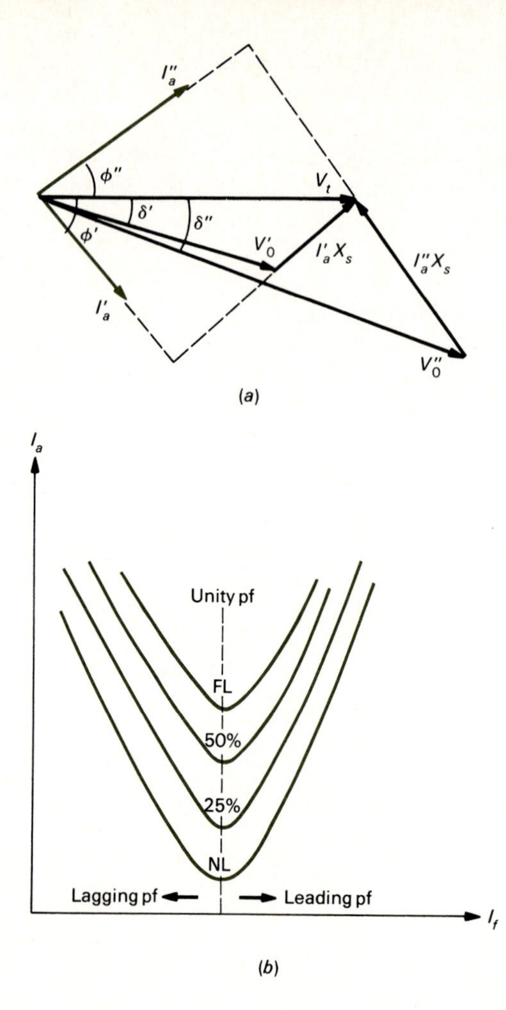

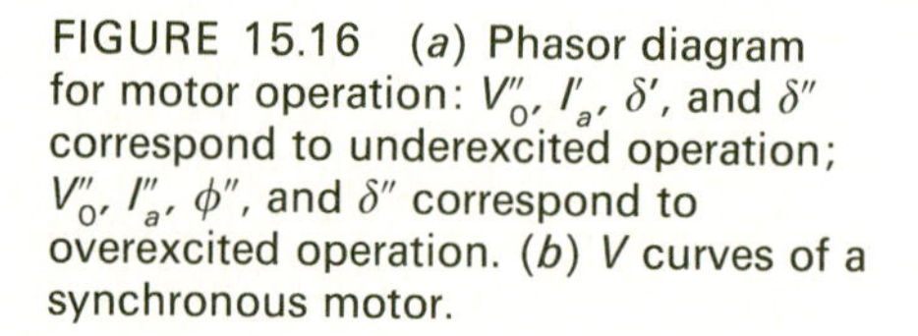

FIGURE 15.16 (*a*) Phasor diagram for motor operation: V''_0, I'_a, δ', and δ'' correspond to underexcited operation; V''_0, I''_a, ϕ'', and δ'' correspond to overexcited operation. (*b*) *V* curves of a synchronous motor.

power factor of the motor is controlled by varying the field excitation, hence altering V_o. This is a very important property of synchronous motors. The locus of the armature current at a constant load, as given by Eq. (15.11), for varying field current is also shown in Fig. 15.16*a*. From this we can obtain the variations of the armature current I_a with the field current I_f (corresponding to V_o), and this can be done for different loads, as shown in Fig. 15.16*b*. These curves are known as the *V curves* of the synchronous motor. One of the applications of a synchronous motor is in power-factor correction, as demonstrated by the following examples.

Example 15.4

A three-phase wye-connected load takes a 50-A current at 0.707 lagging power factor at 220 V between the lines. A three-phase wye-connected round-rotor synchronous motor, having a synchronous reactance of 1.27 Ω per phase, is connected in parallel with the load. The power developed by the motor is 33 kW at a power angle of 30°. Neglecting the armature resistance, calculate the reactive kilovoltamperes of the motor and the overall power factor of the motor and the load.

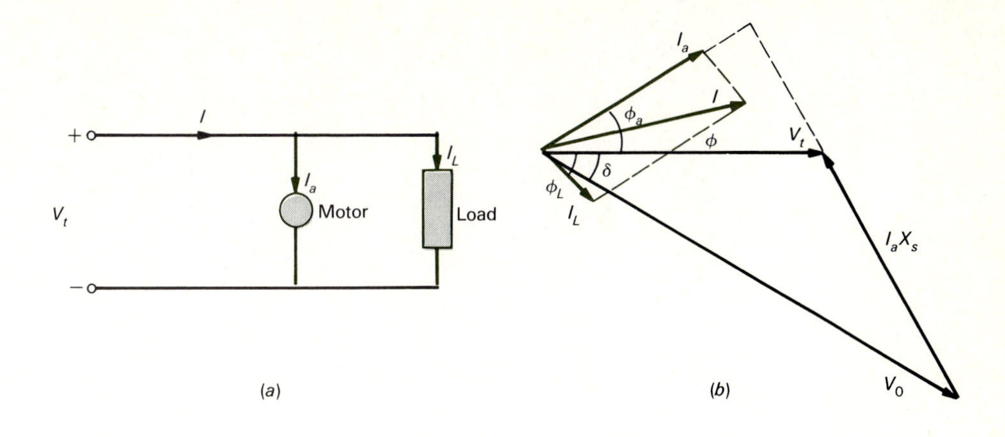

FIGURE 15.17 (*a*) Circuit diagram. (*b*) Phasor diagram.

Solution The circuit and the phasor diagram on a per phase basis are shown in Fig. 15.17. From Eq. (15.10) we have

$$P_d = \tfrac{1}{3} \times 33{,}000$$

$$= \frac{220}{\sqrt{3}} \frac{V_0}{1.27} \sin 30^\circ$$

which yields $V_0 = 220$ V. From the phasor diagram, $I_aX_s = 127$ or $I_a = 127/1.27 = 100$ A and $\phi_a = 30^\circ$. The reactive kilovoltamperes (kVAr) of the motor $= \sqrt{3}\ V_tI_a \sin\phi_a = \sqrt{3} \times 220/1000 \times 100 \times \sin 30 = 19$ kVAr.

The overall power-factor angle is given by

$$\tan\phi = \frac{I_a \sin\phi_a - I_L \sin\phi_L}{I_a \cos\phi_a + I_L \cos\phi_L} = 0.122$$

or $\phi = 7^\circ$ and $\cos\phi = 0.992$ leading.

Example 15.5 For the generator of Example 15.4, calculate the power factor for zero voltage regulation on full load.

Solution Let ϕ be the power factor angle. Then

$$\mathbf{I}_a\mathbf{Z}_s = 218.7 \times 10.4\underline{/\phi + 89.6}$$

$$= 2274.48\underline{/\phi + 89.6} \quad \text{V}$$

For voltage regulation to be zero, $|V_0| = |V_t|$. Hence

$$|3810| = |3810 + 2274.48[\cos(\phi + 89.6) + j \sin(\phi + 89.6)]|.$$

$$3810^2 = [3810 + 2274.48 \cos(\phi + 89.6)]^2 + [2274.48 \sin(\phi + 89.6)]^2$$

from which

$$\phi = 17.76^\circ \qquad \text{and} \qquad \cos\phi = 0.95 \text{ leading}$$

15.8 SALIENT-POLE SYNCHRONOUS MACHINES

In the preceding discussion we have analyzed the round-rotor machine and have made extensive use of the machine parameter, which we defined as synchronous reactance. Because of saliency, the reactance measured at the terminals of a salient-rotor machine will vary as a function of the rotor position, but this is not so in a round-rotor machine. Thus, a simple definition of the synchronous reactance for a salient-rotor machine is not immediately forthcoming.

To overcome this difficulty, we use the two-reaction theory proposed by André Blondel. This theory proposes to resolve the given armature mmf's into two mutually perpendicular components, with one located along the axis of the rotor salient pole, known as the direct (or *d*) axis, and with the other in quadrature, known as the quadrature (or *q*) axis. Correspondingly, we may define the *d*-axis and *q*-axis synchronous reactances X_d and X_q for a salient-pole synchronous machine. Thus, for generator operation, we draw the phasor diagram of Fig. 15.18. Notice that $\mathbf{I}_a$ has been resolved into its *d*- and *q*-axis (fictitious) components I_d and I_q. With the help of this phasor diagram, we obtain

$$I_d = I_a \sin(\delta + \phi)$$

$$I_q = I_a \cos(\delta + \phi)$$

$$V_t \sin\delta = I_q X_q = I_a X_q \cos(\delta + \phi)$$

From these we get (after some manipulation)

$$\tan\delta = \frac{I_a X_q \cos\phi}{V_t + I_a X_q \sin\phi} \tag{15.12}$$

With δ known (in terms of ϕ), the voltage regulation may be computed from

$$V_0 = V_t \cos\delta + I_d X_d$$

$$\text{Percent regulation} = \frac{V_0 - V_t}{V_t} \times 100\%$$

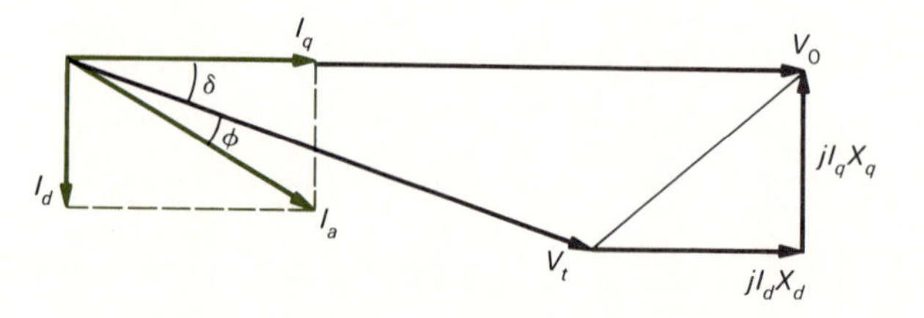

FIGURE 15.18 Phasor diagram of a salient-pole generator.

In fact, the phasor diagram depicts the complete performance characteristics of the machine under steady-state.

Let us now use Fig. 15.18 to derive the power-angle characteristics of a salient-pole generator. If armature resistance is neglected, then $P_d = V_t I_a \cos \phi$. Now, from Fig. 15.18, the projection of I_a on V_t is

$$\frac{P_d}{V_t} = I_a \cos \phi$$

$$= I_q \cos \delta + I_d \sin \delta \tag{15.13}$$

Solving

$$I_q X_q = V_t \sin \delta \qquad \text{and} \qquad I_d X_d = V_0 - V_t \cos \delta$$

for I_q and I_d, and substituting in Eq. (15.10), gives

$$P_d = \frac{V_0 V_t}{X_d} \sin \delta + \frac{V_t^2}{2}\left(\frac{1}{X_q} - \frac{1}{X_d}\right) \sin 2\delta \tag{15.14}$$

Equation (15.14) can also be established for a salient-pole motor ($\delta < 0$); the graph of Eq. (15.14) is given in Fig. 15.19. Observe that for $X_d = X_q = X_s$, Eq. (15.14) reduces to the round-rotor equation, Eq. (15.10).

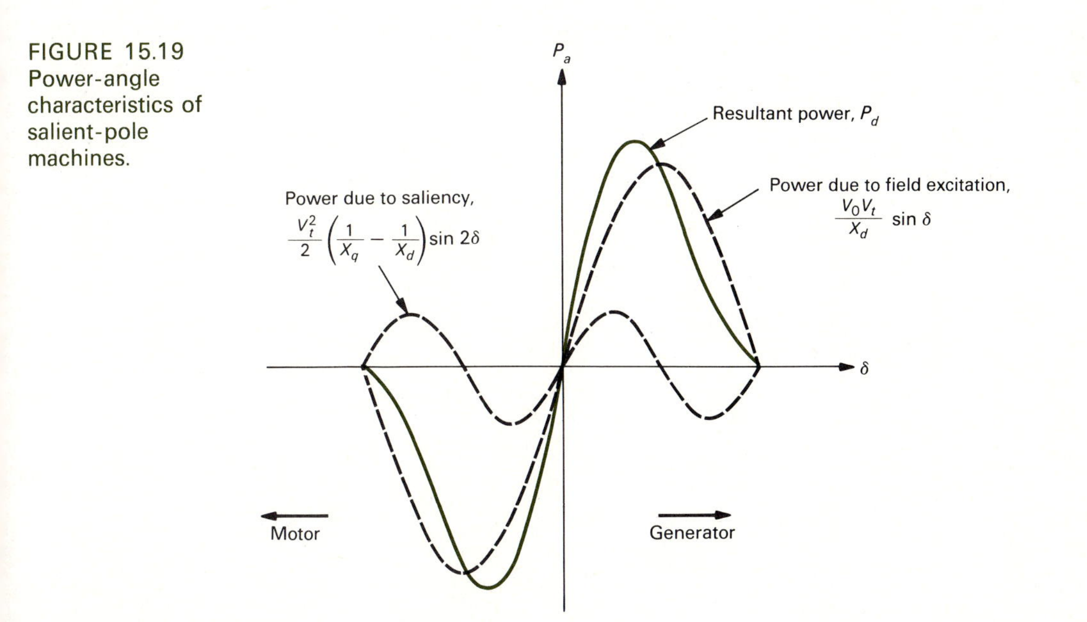

FIGURE 15.19 Power-angle characteristics of salient-pole machines.

Example 15.6 A 20-kVA 220-V 60-Hz wye-connected three-phase salient-pole synchronous generator supplies rated load at 0.707 lagging power factor. The per phase constants of the machine are $R_a = 0.5\ \Omega$ and $X_d = 2X_q = 4.0\ \Omega$. Calculate the voltage regulation at the specified load.

Solution

$$V_t = \frac{220}{\sqrt{3}}$$

$$= 127 \quad \text{V}$$

$$I_a = \frac{20{,}000}{\sqrt{3} \times 120}$$

$$= 52.5 \quad \text{A}$$

$$\phi = \cos^{-1} 0.0707$$

$$= 45^\circ$$

Neglecting R_a, from Eq. (15.12)

$$\tan \delta = \frac{I_a X_q \cos \phi}{V_t + I_a X_q \sin \phi}$$

$$= \frac{52.5 \times 2 \times 0.707}{127 + 52.5 \times 2 \times 0.707}$$

$$= 0.37$$

or

$$\delta = 20.6^\circ$$

$$I_d = 52.5 \sin (20.6 + 45)$$

$$= 47.5 \quad \text{A}$$

$$I_d X_d = 47.5 \times 4$$

$$= 190.0 \quad \text{A}$$

$$V_0 = V_t \cos \delta + I_d X_d$$

$$= 127 \cos 20.6 + 190$$

$$= 308 \quad \text{V}$$

and the percent regulation is

$$\frac{V_0 - V_t}{V_t} \times 100\% = \frac{308 - 127}{127} \times 100$$

$$= 142\%$$

PROBLEMS

15.1 At what speed must a six-pole synchronous generator run to generate a 50-Hz voltage?

15.2 The open-circuit voltage of a 60-Hz generator is 11,000 V at a field current of 5 A. Calculate the open-circuit voltage at 50 Hz and a 2.5-A field current. Neglect saturation.

15.3 A 60-kVA thre-phase wye connected 440-V 60-Hz synchronous generator has a resistance of 0.15 Ω and a synchronous reactance of 3.5 Ω per phase. At rated load and unity power factor, calculate the percent voltage regulation.

15.4 A 1000 kVA 11-kV three-phase wye-connected synchronous generator supplies a 600-kW 0.8 leading power factor load. The synchronous reactance is 24 Ω per phase and the armature resistance is negligible. Calculate (*a*) the power angle and (*b*) the voltage regulation.

15.5 A 1000 kVA 11-kV three-phase wye connected synchronous generator has an armature resistance of 0.5 Ω and a synchronous reactance of 5 Ω. At a certain field current the generator delivers rated load at 0.9 lagging power factor at 11 kV. For the same excitation, what is the terminal voltage at 0.9 leading power factor full load?

15.6 An 11-kV three-phase wye-connected generator has a synchronous impedance of 6 Ω per phase and negligible armature resistance. For a given field current, the open-circuit voltage is 12 kV. Calculate the maximum power developed by the generator. Determine the armature current and power factor for the maximum power condition.

15.7 A 400-V three-phase wye-connected synchronous motor delivers 12 hp at the shaft and operates at 0.866 lagging power factor while taking current. If the armature resistance is 0.75 Ω per phase, determine the efficiency of the motor.

15.8 The motor of Prob. 15.17 has a synchronous reactance of 6 Ω per phase and operates at 0.9 leading power factor while taking an armature current of 20 A. Calculate the induced voltage. Neglect armature resistance.

15.9 A 1000-kVA 11-kV three-phase wye-connected synchronous motor has a 10-Ω synchronous reactance and a negligible armature resistance. Calculate the induced voltage for (*a*) 0.8 lagging power factor, (*b*) unity power factor, and (*c*) 0.8 leading power factor when, in each case, the motor takes 1000 kVA.

15.10 The per-phase induced voltage of a synchronous motor is 2500 V. It lags behind the terminal voltage by 30°. If the terminal voltage is 2200 V per phase, determine the operating power factor. The per-phase sychronous reactance is 6 Ω. Neglect armature resistance.

15.11 The per-phase synchronous reactance of a synchronous motor is 8 Ω, and its armature resistance is negligible. The per-phase input is 400 kW and the induced voltage is 5200 V per phase. If the terminal voltage is 3800 V per phase, determine (*a*) the power factor and (*b*) the armature current.

15.12 A 2200-V three-phase 60-Hz four-pole wye-connected synchronous motor has a synchronous reactance of 4 Ω and a negligible armature resistance. The excitation is so adjusted that the induced voltage is 2200 V (line-to-line). If the line current is 220 A at a certain load, calculate (*a*) the input power, (*b*) the developed torque, and (*c*) the power angle.

15.13 An overexcited synchronous motor is connected across a 150-kVA inductive load of 0.7 lagging power factor. The motor takes 12 kW while running on no-load. Calculate the kVA rating of the motor if it is desired to bring the overall power factor of the motor-inductive load combination to unity.

15.14 Repeat Prob.15.13 if the synchronous motor is used to supply a 100-hp load at an efficiency of 90 percent.

REFERENCES

15.1 A. E. Fitzgerald, D. Higginbotham, and A. Gabel, *Basic Electrical Engineering*, 5th ed., McGraw-Hill, New York, 1981.

15.2 S. A. Nasar, *Electric Machines and Transformers*, Macmillan, New York, 1983.

Chapter

16

Induction Machines

The induction motor is the most commonly used electric motor. It is considered to be the workhorse of the industry. Like the dc machine and the synchronous machine, an induction machine consists of a stator and a rotor. The rotor is mounted on bearings and separated from the stator by an air gap. Electromagnetically, the stator consists of a core made up of punchings (or laminations) carrying slot-embedded conductors. These conductors are interconnected in a predetermined fashion and constitute the armature windings, which are similar to the windings of synchronous machines (Chap. 15).

Alternating current is supplied to the stator windings, and the currents in the rotor windings are induced by the stator currents. The rotor of the induction machine is cylindrical and carries either (1) conducting bars short-circuited at both ends, as in a *cage-type* machine, or (2) a polyphase winding with terminals brought out to slip rings for external connections, as in a *wound-rotor* machine. A wound-rotor winding is similar to that of the stator. Sometimes the cage-type machine is also called a *brushless* machine and the wound-rotor machine a *slip-ring* machine. The stator and the rotor, in its three different stages of production, are shown in Fig. 16.1. The motor is rated at 2500 kW, 3 kV, 575 A, two-pole, and 400 Hz. A finished cage-type rotor of a 3400-kW 6-kV motor is shown in Fig. 16.2, and Fig. 16.3 shows the wound rotor of a three-phase slip-ring 15,200-kW 2.4-kV four-pole induction motor. A cutaway view of a completely assembled motor with a cage-type rotor is shown in Fig. 16.4. The rotor is housed within the stator and is free to rotate therein.

An induction machine operates on the basis of the interaction of the induced rotor currents and the air-gap fields. If the rotor is allowed to run under the torque developed by this interaction, the machine will operate as a motor. On the other hand, when the rotor is driven by an external source beyond a certain speed, the machine begins to deliver electric power and operates as an induction generator (instead of as an induction motor, which absorbs electric power). Thus, we see that the induction machine is capable of functioning either as a motor or as a generator. In practice, its application as a generator is less common than its application as a motor. We will first study the motor operation, then develop the equivalent circuit of an induction motor, and subsequently show that the complete characteristics of an induction machine, operating either as a motor or as a generator, are obtainable from the equivalent circuit.

FIGURE 16.1
Rotor for a 2500-kW 3-kV two-pole 400-Hz induction motor in different stages of production.

FIGURE 16.2
Complete rotor of a 3400-kW 6-kV 990-r/min induction motor.

FIGURE 16.3
Rotor of a 15,200-kW 2.4-kV three-phase slip-ring induction motor.

FIGURE 16.4
Cutaway view of a three-phase cage-type induction motor.

16.1 OPERATION OF A THREE-PHASE INDUCTION MOTOR

The key to the operation of an induction motor is the production of the rotating magnetic field. We have established in the last chapter (Sec. 15.3) that a three-phase stator excitation produces a rotating magnetic field in the air gap of the machine and that the field rotates at a synchronous speed given by Eq. (15.3). As the magnetic field rotates, it cuts the rotor conductors. By this process, voltages are induced in the conductors. The induced voltages give rise to rotor currents, which interact with the air-gap field to produce a torque. The torque is maintained as long as the rotating magnetic field and the induced rotor current exist. Consequently, the rotor starts rotating in the direction of the rotating field. The rotor will achieve a steady-state speed n such that $n < n_s$. When $n = n_s$, there will be no induced currents and hence no torque. The condition $n > n_s$ corresponds to the generator mode, as we shall see in Sec. 16.9.

An alternative approach to explaining the operation of the polyphase induction motor is to consider the interaction of the (excited) stator magnetic field with the (induced) rotor magnetic field. The stator excitation produces a rotating magnetic field, which rotates in the air gap at a synchronous speed. The field induces polyphase currents in the rotor, thereby giving rise to another rotating magnetic field, which also rotates at the same synchronous speed as that of the stator and with respect to the stator. Thus, we have two rotating magnetic fields, both rotating at a synchronous speed with respect to the stator but stationary with respect to each other. Consequently, according to the principle of alignment of magnetic fields, the rotor experiences a torque. The rotor rotates in the direction of the rotating field of the stator.

16.2 SLIP

The actual mechanical speed n of the rotor is often expressed as a fraction of the synchronous speed n_s as related by *slip s*, defined as

$$s = \frac{n_s - n}{n_s} \tag{16.1}$$

where n_s is given by Eq. (15.3), which is repeated below for convenience

$$n_s = \frac{120f}{P} \tag{15.3}$$

The slip may also be expressed as percent slip as follows:

$$\text{Percent slip} = \frac{n_s - n}{n_s} \times 100 \tag{16.2}$$

At standstill, the rotating magnetic field produced by the stator has the same relative speed with respect to the rotor windings as with respect to the stator windings. Thus, the frequency of rotor currents f_r is the same as the frequency of stator currents f. At synchronous speed, there is no relative motion between the rotating field and the rotor, and the frequency of rotor current is zero. At other speeds, the rotor frequency is proportional to the slip s; that is,

$$f_r = sf \tag{16.3}$$

where f_r is the frequency of rotor currents and f is the frequency of stator input current (or voltage).

Example 16.1 A six-pole three-phase 60-Hz induction motor runs at 4 percent slip at a certain load. Determine the synchronous speed, the rotor speed, the frequency of rotor currents, the speed of the rotor rotating field with respect to the stator, and the speed of the rotor rotating field with respect to the stator rotating field.

Solution From Eq. (15.3), the synchronous speed is

$$\begin{aligned} n_s &= \frac{120 \times 60}{6} \\ &= 1200 \text{ r/min} \end{aligned}$$

From Eq. (16.1), the rotor speed is

$$\begin{aligned} n &= (1 - s)n_s \\ &= (1 - 0.04) \times 1200 \\ &= 1152 \text{ r/min} \end{aligned}$$

From Eq. (16.3), the frequency of rotor currents is

$$\begin{aligned} f_r &= 0.04 \times 60 \\ &= 2.4 \text{ Hz} \end{aligned}$$

The six poles on the stator induce six poles on the rotor. The rotating field produced by the rotor rotates at a corresponding synchronous speed n_r relative to the rotor such that

$$n_r = \frac{120f_r}{P} = \frac{120f}{P} s = sn_s$$

But the speed of the rotor with respect to the stator is

$$n = (1 - s)n_s$$

Hence, the speed of the rotor field with respect to the stator is

$$\begin{aligned} n_s' &= n_r + n \\ &= sn_s + (1 - s)n_s \\ &= 1200 \text{ r/min} \end{aligned}$$

The speed of the rotor field with respect to the stator field is

$$n_s' - n_s = n_s - n_s = 0$$

16.3 DEVELOPMENT OF EQUIVALENT CIRCUITS

In order to develop an equivalent circuit of an induction motor, we consider the similarities between a transformer and an induction motor (on a per phase basis). If we consider the primary of the transformer to be similar to the stator of the induction motor, then its rotor corresponds to the secondary of the transformer. From this analogy, it follows that the stator and the rotor have their own respective resistances and leakage reactances. Because the stator and the rotor are magnetically coupled, we must have a magnetizing reactance, just as in a transformer. The air gap in an induction motor makes its magnetic circuit relatively poor, and thus the corresponding magnetizing reactance will be relatively smaller than that of a transformer. The hysteresis and eddy-current losses in an induction motor can be represented by a shunt resistance, as was done for the transformer. Up to this point, we have mentioned the similarities between a transformer and an induction motor. A major difference between the two, however, is introduced because of the rotation of the rotor. Consequently, the frequency of rotor currents is different from the frequency of stator currents; see Eq. (16.3). Keeping these facts in mind, we now proceed to represent a three-phase induction motor by a stationary equivalent circuit.

Considering the rotor first and recognizing that the frequency of rotor currents is the slip frequency, we may express the per phase rotor leakage reactance x_2 at a slip s in terms of the standstill per phase reactance X_2:

$$x_2 = sX_2 \tag{16.4}$$

Next we observe that the magnitude of the voltage induced in the rotor circuit is also proportional to the slip.

A justification of this statement follows from transformer theory because we may view the induction motor at standstill as a transformer with an air gap. For the transformer, we know that the induced voltage, say E_2, is given by

$$E_2 = 4.44 f N \phi_m \tag{16.5}$$

But at a slip s, the frequency becomes sf. Substituting this value of frequency into Eq. (16.5) yields the voltage e_2 at a slip s as

$$e_2 = 4.44 sf N \phi_m = sE_2$$

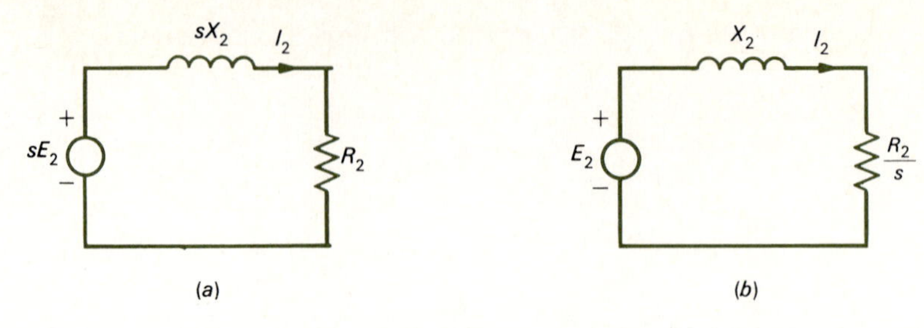

FIGURE 16.5
Two forms of rotor equivalent circuit.

We conclude, therefore, that if E_2 is the per-phase voltage induced in the rotor at standstill, then the voltage e_2 at a slip s is given by

$$e_2 = sE_2 \tag{16.6}$$

Using Eqs. (16.4) and (16.6), we obtain the rotor equivalent circuit shown in Fig. 16.5*a*. The rotor current I_2 is given by

$$I_2 = \frac{sE_2}{\sqrt{R_2^2 + (sX_2)^2}}$$

which may be rewritten as

$$I_2 = \frac{E_2}{\sqrt{(R_2/s)^2 + X_2^2}} \tag{16.7}$$

resulting in the alternative form of the equivalent circuit shown in Fig. 16.5*b*. Notice that these circuits are drawn on a per phase basis. To this circuit we may now add the per phase stator equivalent circuit to obtain the complete equivalent circuit of the induction motor.

In an induction motor, only the stator is connected to the ac source. The rotor is not generally connected to an external source, and rotor voltage and current are produced by induction. In this regard, as mentioned earlier, the induction motor may be viewed as a transformer with an air gap, having a variable resistance in the secondary. Thus, we may consider that the primary of the transformer corresponds to the stator of the induction motor, whereas the secondary corresponds to the rotor on a per phase basis. Because of the air gap, however, the value of the magnetizing reactance X_m tends to be relatively low, compared to that of a transformer. As in a transformer, we have a mutual flux linking both the stator and the rotor, represented by the magnetizing reactance and various leakage fluxes. For instance, the total rotor leakage flux is denoted by X_2 in Fig. 16.5. Now considering that the rotor is coupled to the stator as the secondary of a transformer is coupled to its primary, we may draw the circuit shown in Fig. 16.6. To develop this circuit further, we need to express the rotor quantities as referred to the stator. The pertinent details are cumbersome and are not considered here. However, having referred the rotor quantities to the stator, we obtain from the circuit given in Fig. 16.6 the exact equivalent circuit (per phase) shown in Fig. 16.7.

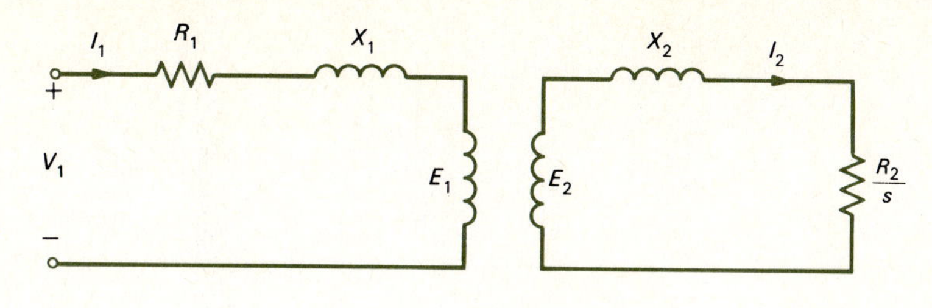

FIGURE 16.6
Stator and rotor as coupled circuits.

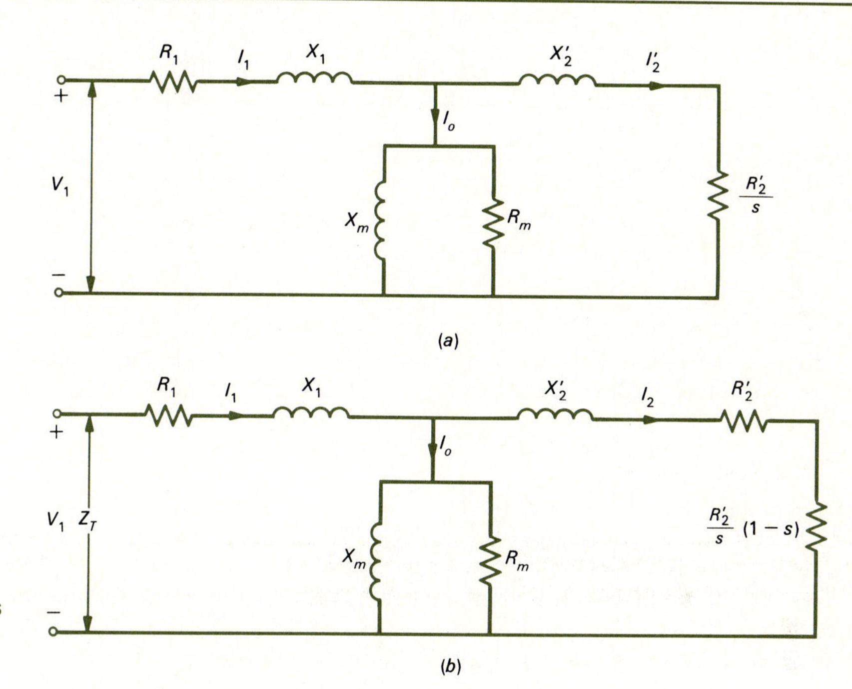

FIGURE 16.7
Two forms of equivalent circuits of an induction motor.

For reasons that will become immediately clear, we split R_2'/s as

$$\frac{R_2'}{s} = R_2' + \frac{R_2'}{s}(1-s)$$

to obtain the circuit shown in Fig. 16.7. Here, R_2' is simply the perphase standstill rotor resistance referred to the stator, and $R_2'(1-s)/s$ is a dynamic resistance that depends on the rotor speed and corresponds to the load on the motor. Notice that all the parameters shown in Fig. 16.7 are standstill values and that the circuit is the per-phase exact equivalent circuit referred to the stator.

16.4 PERFORMANCE CALCULATIONS

We will now show the usefulness of the equivalent circuit in determining motor performance. To illustrate the procedure, we refer to Fig. 16.7. We redraw this circuit in Fig.

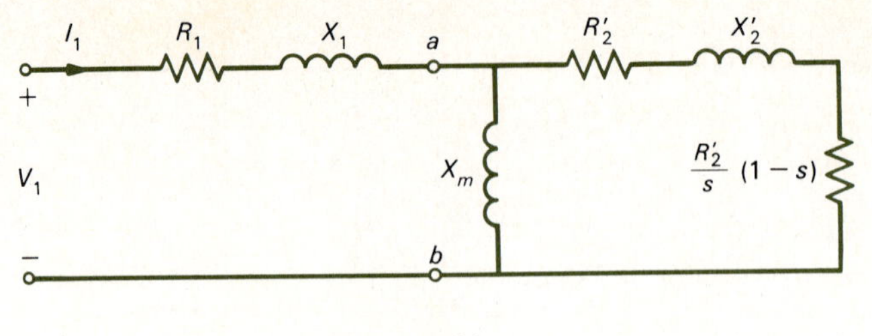

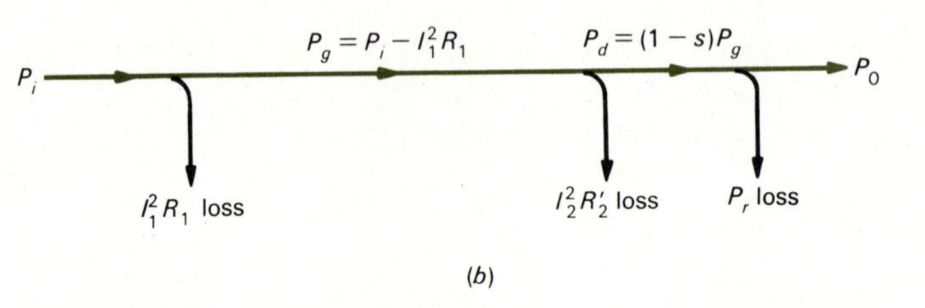

FIGURE 16.8 (*a*) An approximate equivalent circuit of an induction motor. (*b*) Power flow in an induction motor.

16.8, where we also show approximately the power flow and various power losses in one phase of the machine. From Fig. 16.8 we obtain the following relationships on a per phase basis:

$$\text{Input power} = P_i = V_1 I_1 \cos\phi \tag{16.8a}$$

$$\text{Stator } I^2R_1 \text{ loss} = I_1^2 R_1 \tag{16.8b}$$

$$\text{Power crossing the air gap} = P_g = P_i - I_1^2 R_1 \tag{16.8c}$$

Since this power P_g is dissipated in R_2'/s (of Fig. 16.7*a*), we also have

$$P_g = \frac{I_2^2 R_2'}{s} \tag{16.9}$$

Subtracting $I_2^2R_2'$ loss from P_g yields the developed electromagnetic power P_d. Thus

$$P_d = P_g - I_2^2 R_2' \tag{16.10}$$

From Eqs. (16.9) and (16.10) we get

$$P_d = (1-s)P_g \tag{16.11}$$

0.2   j0.5   a   3.77   j0.95

+ $I_1 = 30$ A

$V_1 = 127$ V

−   b

FIGURE 16.9 Example 16.2.

This power appears across the resistance $R_2'(1 - s)/s$ (of Fig. 16.7*b*), which corresponds to the load. Subtracting the mechanical rotational power P_r from P_d gives the output power P_o. Hence,

$$P_o = P_d - P_r \tag{16.12}$$

and

$$\text{Efficiency} = \frac{\text{output power}}{\text{input power}} = \frac{P_o}{P_i} \tag{16.13}$$

Torque calculations can be made from the power calculations. Thus, to determine the electromagnetic torque T_e developed by the motor at a speed ω_m(rad/s), we write

$$T_e \omega_m = P_d \tag{16.14}$$

But

$$\omega_m = (1 - s)\omega_s \tag{16.15}$$

where ω_s is the synchronous speed in rad/s. From Eqs. (16.11), (16.14), and (16.15) we obtain

$$T_e = \frac{P_g}{\omega_s} \tag{16.16}$$

which gives the torque at a slip s. At standstill $s = 1$; hence, the standstill torque developed by the motor is given by

$$T_{e,\,\text{standstill}} = \frac{P_{gs}}{\omega_s} \tag{16.17}$$

Notice that we have neglected the core losses, most of which are in the stator. We will include core losses only in efficiency calculations. The reason for this simplification is to reduce the amount of the complex arithmetic required in numerical computations. The following example illustrates the calculation details.

Example 16.2

The parameters of the equivalent circuit in Fig. 16.8 for a 220-V three-phase four-pole wye-connected 60-Hz induction motor are

$R_1 = 0.2\ \Omega \qquad R_2' = 0.1\ \Omega$

$X_1 = 0.5\ \Omega \qquad X_2' = 0.2\ \Omega$

$X_m = 20.0\ \Omega$

The total iron and mechanical losses are 350 W. For a slip of 2.5 percent, calculate the input current, output power, output torque, and efficiency.

Solution From Fig. 16.8 the total impedance is

$$\mathbf{Z}_1 = R_1 + jX_1 + \frac{jX_m(R_2'/s + jX_2')}{R_2'/s + j(X_m + X_2')}$$

$$= 0.2 + j0.5 + \frac{j20(4 + j0.2)}{4 + j(20 + 0.2)}$$

$$= (0.2 + j0.5) + (3.77 + j0.95)$$

$$= 4.23\underline{/20^\circ}\ \Omega$$

the phase voltage is

$$V_1 = \frac{220}{\sqrt{3}}$$

$$= 127\ \text{V}$$

the input current is

$$I_1 = \frac{127}{4.23}$$

$$= 30\ \text{A}$$

the power factor is

$$\cos\phi = \cos 20^\circ$$

$$= 0.94$$

and the total input power is

$$3V_1I_1\cos\phi = \sqrt{3} \times 220 \times 30 \times 0.94$$

$$= 10.75\ \text{kW}$$

From the equivalence of Figs. 16.8*a* and 16.9, we obtain the total power across the air gap:

$$P_g = 3 \times 30^2 \times 3.77$$

$$= 10.18\ \text{kW}$$

Notice that 3.37 Ω is the resistance between the terminals *a-b* in the two circuits. The total power developed is

$$P_d = (1 - s)P_g$$

$$= 0.975 \times 10.18$$

$$= 9.93\ \text{kW}$$

the total output power is

$$P_d - P_{\text{core}} = 9.93 - 0.35$$
$$= 9.58 \text{ kW}$$

and the total output torque is

$$\frac{\text{Output power}}{\omega_m} = \frac{9.58}{184} \times 1000$$
$$= 52 \text{ N} \cdot \text{m}$$

where $\omega_m = 0.975 \times 60 \times \pi = 184$ rad/s. Thus

$$\text{Efficiency} = \frac{\text{output power}}{\text{input power}}$$
$$= \frac{9.58}{10.75}$$
$$= 89.1\%$$

Using this procedure, we can calculate the performance of the motor at other values of the slip, ranging from 0 to 1. The characteristics thus calculated are shown in Fig. 16.10.

FIGURE 16.10 Characteristics of an induction motor. T_m = maximum torque; T = starting torque.

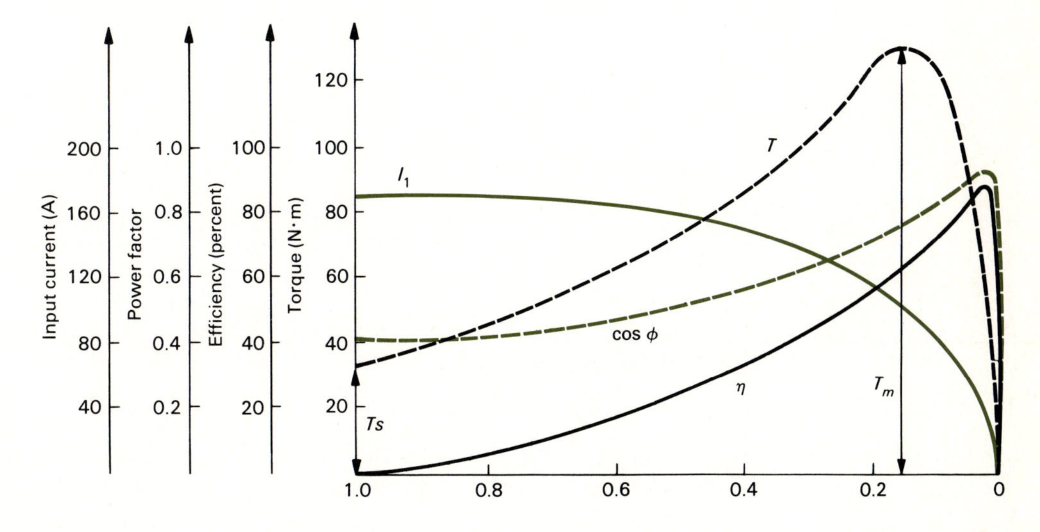

Example 16.3 A two-pole three-phase 60-Hz induction motor develops 25 kW of electromagnetic power at a certain speed. The rotational mechanical loss at this speed is 400 W. If the power crossing the air gap is 27 kW, calculate the slip and the output torque.

Solution From Eq. (16.10) we have

$$1 - s = \frac{P_d}{P_g}$$

or

$$1 - s = \frac{25}{27}$$

Thus,

$$s = 0.074 \qquad \text{or} \qquad 7.4\%$$

The developed torque is given by Eq. (16.16). Substituting $\omega_s = 2\pi n_s/60 = 2\pi \times 3600/60$ and $P_g = 27$ kW (given) in Eq. (16.16) we have

$$T_e = \frac{27{,}000}{2\pi \times 3600/60}$$

$$= 71.62 \text{ N} \cdot \text{m}$$

The torque lost due to mechanical rotation is found from

$$T_{\text{loss}} = \frac{P_{\text{loss}}}{\omega_m}$$

$$= \frac{400}{(1 - 0.074)2\pi \times 3600/60}$$

$$= 1.15 \text{ N} \cdot \text{m}$$

Hence, the output torque is

$$T_e - T_{\text{loss}} = 71.62 - 1.15$$

$$= 70.47 \text{ N} \cdot \text{m}$$

16.5 APPROXIMATE EQUIVALENT CIRCUIT FROM TEST DATA

The preceding examples illustrate the usefulness of equivalent circuits of induction motors. For most purposes, an approximate equivalent circuit is adequate. One such

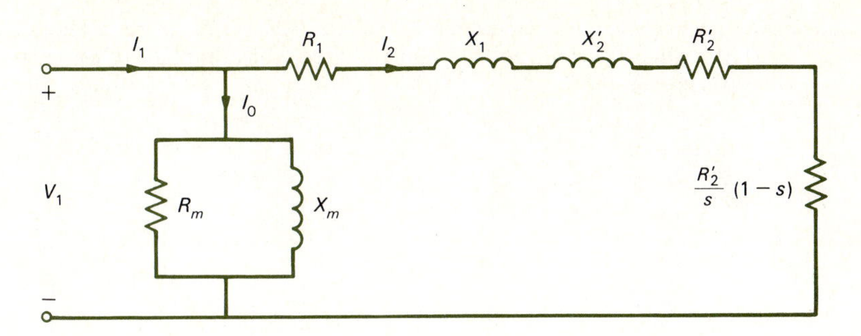

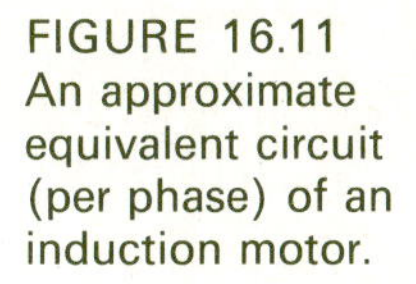
FIGURE 16.11
An approximate equivalent circuit (per phase) of an induction motor.

circuit is shown in Fig. 16.11. Obviously, in order to use this circuit for calculations, its parameters must be known. The parameters of the circuit shown in Fig. 16.11 can be obtained from the following tests.

No-Load Test In this test, rated voltage is applied to the machine and it is allowed to run on no-load. The input power (corrected for friction and windage loss), voltage, and current are measured: these, reduced to per phase values, are denoted by P_0, V_0, and I_0, respectively. When the machine runs on no-load, the slip is close to 0 and the circuit in Fig. 16.11 to the right of the shunt branch is taken to be an open circuit. Thus, the parameters R_m and X_m are found from

$$R_m = \frac{V_0^2}{P_0} \tag{16.18}$$

$$X_m = \frac{V_0^2}{\sqrt{V_0^2 I_0^2 - P_0^2}} \tag{16.19}$$

Blocked-Rotor Test In this test, the rotor of the machine is blocked ($s = 1$), and a reduced voltage is applied to the machine so that the rated current flows through the stator windings. The input power, voltage, and current are recorded to per phase values; these are denoted by P_s, V_s, and I_s, respectively. In this test, the iron losses are assumed to be negligible and the shunt branch of the circuit shown in Fig. 16.11 is considered to be absent. The parameters are thus found from

$$R_e = R_1 + a^2 R_2 = \frac{P_s}{I_s^2} \tag{16.20}$$

$$X_e = X_1 + a^2 X_2 = \frac{\sqrt{V_s^2 I_s^2 - P_s^2}}{I_s^2} \tag{16.21}$$

In Eqs. (16.20) and (16.21) the constant a is the turns ratio. The stator resistance per phase, R_1, can be directly measured, and knowing R_e from Eq. (16.20), we can determine $R_2' = a^2 R_2$, the rotor resistance referred to the stator. There is no simple method for determining X_1 and $X_2' = a^2 X_2$ separately. The total value given by Eq. (16.21) is sometime equally divided between X_1 and X_2'.

Example 16.4 The results of no-load and blocked-rotor tests on a three-phase wye-connected induction motor are as follows:

No-load test:

line-to-line voltage: 400 V

input power: 1770 W

input current: 18.5 A

friction and windage loss: 600 W

Blocked-rotor test:

line-to-line voltage: 45 V

input power: 2700 W

input current: 63 A

Determine the parameters of the approximate equivalent circuit (Fig. 16.11).

Solution From no-load test data:

$$V_0 = \frac{400}{\sqrt{3}} = 231 \text{ V}$$

$$P_0 = \tfrac{1}{3}(1770 - 600) = 390 \text{ W}$$

$$I_0 = 18.5 \text{ A}$$

Then, by Eqs. (16.18) and (16.19),

$$R_m = \frac{(231)^2}{390}$$

$$= 136.8 \ \Omega$$

$$X_m = \frac{(231)^2}{\sqrt{(231)^2(18.5)^2 - (390)^2}}$$

$$= 12.5 \ \Omega$$

From blocked-rotor test data:

$$V_s = \frac{45}{\sqrt{3}} = 25.98 \text{ V}$$

$$P_s = \frac{2700}{3} = 900 \text{ W}$$

$$I_s = 63 \text{ A}$$

Then, by Eqs. (16.20) and (16.21),

$$R_e = R_1 + a^2 R_2$$

$$= \frac{900}{(63)^2}$$

$$= 0.23\ \Omega$$

$$X_e = X_1 + a^2 X_2$$

$$= \frac{\sqrt{(25.98)^2(63)^2 - (900)^2}}{(63)^2}$$

$$= 0.34\ \Omega$$

16.6 PERFORMANCE CRITERIA OF INDUCTION MOTORS

Example 16.2 shows the usefulness of the equivalent circuit in calculating the performance of the motor. The performance of an induction motor may be characterized by the following major factors:

1 Efficiency
2 Power factor
3 Starting torque
4 Starting current
5 Pullout (or maximum) torque

Notice that these characteristics are shown in Fig. 16.10. In design considerations, the heating because of I^2R losses and core losses and the means of heat dissipation must be included. It is not within the scope of this book to present a detailed discussion of the effects of design changes and consequently of parameter variations on each performance characteristic. Here we summarize the results qualitatively. For example, the efficiency is approximately proportional to $(1 - s)$; thus, the motor would be most compatible with a load running at the highest possible speed. Because the efficiency is clearly dependent on I^2R losses, R_2' and R_1 must be small for a given load. To reduce core losses, the working flux density B must be small. But this imposes a conflicting requirement on the load current I_2' because the torque is dependent on the product of B and I_2'. In other words, an attempt to decrease the core losses beyond a limit would result in an increase in the I^2R losses for a given load.

It may be seen from the equivalent circuits (developed in Sec. 16.4) that the power factor can be improved by decreasing the leakage reactances and increasing the magnetizing reactance. However, it is not wise to reduce the leakage reactances to a mini-

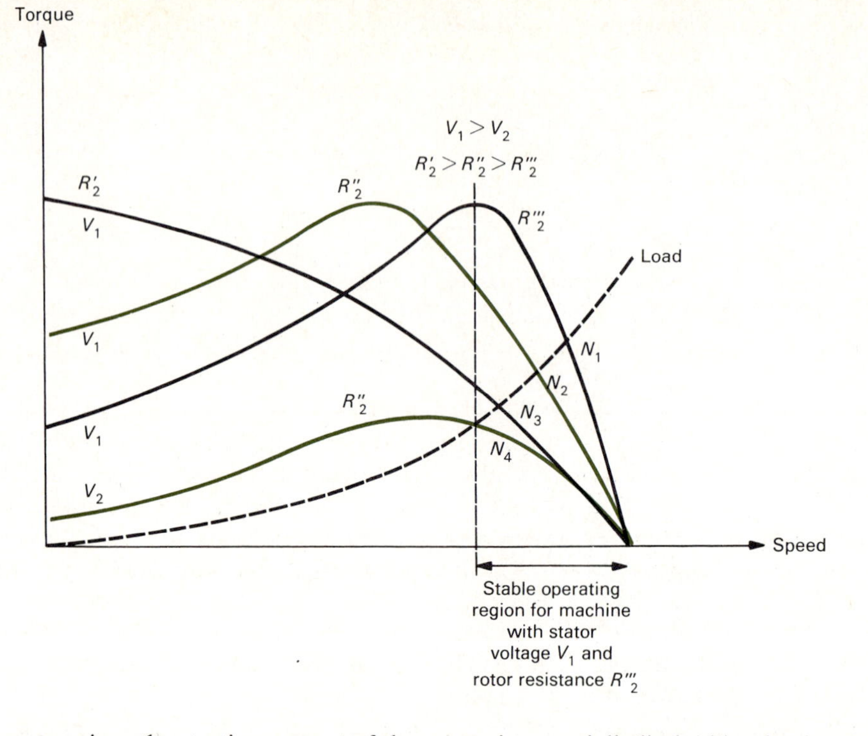

FIGURE 16.12 Effect of rotor resistance on torque-speed characteristics.

mum, since the starting current of the motor is essentially limited by these reactances. Again, we notice the conflicting conditions for a high power factor and a low starting current. Also, the pullout torque would be higher for lower leakage reactances.

A high starting torque is produced by a high R_2'; that is, the higher the rotor resistance, the higher would be the starting torque. A high R_2' is in conflict with a high efficiency requirement. The effect of varying rotor resistance on the motor torque-speed characteristics is shown in Fig. 16.12, which also shows three different steady-state operating speeds for three values of the rotor resistance and two stator voltages V_1 and V_2.

16.7 SPEED CONTROL OF INDUCTION MOTORS

Because of its simplicity and ruggedness, the induction motor finds numerous applications. However, it suffers from the drawback that, in contrast to dc motors, its speed cannot be easily and efficiently varied continuously over a wide range of operating conditions. We will briefly review the various possible methods by which the speed of the induction motor can be varied either continuously or in discrete steps. We will not consider all these methods in detail here; certain details are given in Chap. 17.

The speed of the induction motor can be varied (1) by varying the synchronous speed of the rotating field or (2) by varying the slip. Because the efficiency of the induction motor is approximately proportional to $(1 - s)$, any method of speed control that depends on the variation of slip is inherently inefficient. On the other hand, if the supply frequency is constant, varying the speed by changing the synchronous speed results only in discrete changes in the speed of the motor. We will now consider these methods of speed control in some detail.

Recall from Eq. (15.3) that the synchronous speed n_s of the traveling field in a rotating induction machine is given by

$$n_s = \frac{120f}{P}$$

where P = number of poles and f = supply frequency, which indicates that n_s can be varied (1) by changing the number of poles P or (2) by changing the frequency f. Both methods have found applications, and we consider here the pertinent qualitative details.

Pole-Changing Method In this method the stator winding of the motor is so designed that, by changing the connections of the various coils (the terminals of which are brought out), the number of poles of the winding can be changed in the ratio of 2:1. Accordingly, two synchronous speeds result. We observe that only two speeds of operation are possible. If more independent windings (e.g., two) are provided—each arranged for pole-changing—more synchronous speeds (e.g., four) can be obtained; however, the fact remains that only discrete changes in the speed of the motor can be obtained by this technique. The method has the advantage of being efficient and reliable because the motor has a squirrel-cage rotor and no brushes.

Another method of pole-changing is by means of pole-amplitude modulation; single-winding squirrel-cage motors are reported to have been developed that yield three operating speeds. Another method (based on pole-changing) which produced three or five speeds, has been termed *phase-modulated pole-changing.* Like the simplest pole-changing method, the pole-amplitude modulation and the phase-modulated pole-changing methods give discrete variations in the synchronous speed of the motor.

Variable-Frequency Method We recall that the synchronous speed is directly proportional to the supply frequency. If it is practicable to vary the supply frequency, the synchronous speed of the motor can also be varied. The variation in speed is continuous or discrete according to continuous or discrete variation of the supply frequency. However, the maximum torque developed by the motor is inversely proportional to the synchronous speed. If we desire a constant maximum torque, both the supply voltage and the supply frequency should be increased if we wish to increase the synchronous speed of the motor. The inherent difficulty in the application of this method is that the supply frequency, which is commonly available, is fixed. Thus, the method is applicable only if a variable-frequency supply is available. Various schemes have been proposed to obtain a variable-frequency supply. With the advent of solid-state devices with comparatively large power ratings, it is now possible to use static inverters to drive the induction motor for a variable-speed operation.

Variable-Slip Method Controlling the speed of an induction motor by changing its slip may be understood by reference to Fig. 16.12. The dotted curve shows the speed-torque characteristic of the load. The curves with solid lines are the speed-torque characteristics of the induction motor under various conditions, such as different rotor resistances (R_2', R_2'', R_2''') or different stator voltages (V_1 and V_2). We have four different torque-speed curves, and therefore the motor can run at any one of four speeds (N_1, N_2, N_3, and N_4) for the given load. Note that to the right of the peak torque is the stable operating region of the motor. In practice, the slip of the motor can be changed by one of the following methods.

Variable-Stator-Voltage Methods Since the electromagnetic torque developed by the machine is proportional to the square of the applied voltage, we obtain different torque-speed curves for different voltages applied to the motor. For a given rotor resistance R_2, two such curves are shown in Fig. 16.12 for two applied voltages V_1 and V_2. Thus, the motor can run at speeds N_2 or N_4. If the voltage can be varied continuously from V_1 to V_2, the speed of the motor can also be varied continuously between N_2 and N_4 for the given load. This method is applicable to cage-type as well as to wound-rotor-type induction motors.

Variable-Rotor-Resistance Method This method is applicable only to the wound-rotor motor. The effect on the speed-torque curves of inserting external resistances in the rotor circuit is shown in Fig. 16.12 for three different rotor resistances R_2', R_2'', and R_2'''. For the given load, three speeds of operation are possible. Of course, by continuous variation of the rotor resistance, continuous variation of the speed is possible.

Control by Solid-State Switching Other than the inverter-driven motor, the speed of the wound-rotor motor can be controlled by inserting the inverter in the rotor circuit or by controlling the stator voltage by means of solid-state switching devices, such as silicon-controlled rectifiers (SCRs, or thyristors). The output from the SCR feeding the motor is controlled by adjusting its firing angle. The method of doing this is similar to the variable-voltage method outlined earlier. However, it has been found that control by an SCR gives a wider range of operation and is more efficient than other slip-control methods. For details, see Chap. 17.

16.8 STARTING OF INDUCTION MOTORS

Most induction motors—large and small—are rugged enough that they can be started across the line without incurring any damage to the motor windings, even though about 5 to 7 times the rated current flows through the stator at rated voltage at standstill. In large induction motors, however, large starting currents are objectionable in two respects. First, the mains supplying the induction motor may not be of a sufficiently large capacity. Second, a large starting current may cause excessive voltage drops in the lines, resulting in a reduced voltage across the motor. Because the torque varies approximately as the square of the voltage, the starting torque may become so small at the reduced line voltage that the motor might not even start on load. Thus, we formulate the basic requirement for starting: The line current should be limited by the capacity of the mains, but only to the extent that the motor can develop sufficient torque to start (on load, if necessary).

Example 16.5 An induction motor is designed to run at 5 percent slip on full load. If the motor draws 6 times the full-load current at starting at the rated voltage, estimate the ratio of the starting torque to the full-load torque.

Solution The torque at a slip s is given by Eq. (16.16), which in conjunction with Eq. (16.9) becomes

$$T_e = \frac{I_2^2 R_2'}{s\omega_s}$$

At full load, with $I_2 = I_{2f}$, the torque is

$$T_{ef} = \frac{I_{2f}^2 R_2'}{0.05\omega_s}$$

At starting, $I_{2s} = 6I_{2f}$ and $s = 1$, so that

$$T_{es} = \frac{(6I_{2f})^2 R_2'}{\omega_s}$$

Hence,

$$\frac{T_{es}}{T_{ef}} = \frac{(6I_{2f})^2 R_2'}{\omega_s} \frac{0.05\omega_s}{I_{2f}^2 R_2'} = 1.8$$

Example 16.6 If the motor of Example 16.5 is started at a reduced voltage to limit the line current to 3 times the full-load current, what is the ratio of the starting torque to the full-load torque?

Solution In this case we have

$$\frac{T_{es}}{T_{ef}} = 3^2 \times 0.05 = 0.45$$

Notice that the starting torque has been reduced by a factor of 4, relative to the case of full-voltage starting. In many practical cases, the line current is limited to 6 times the full-load current and the starting torque is desired to be about 1.5 times the full-load torque.

There are numerous types of push-button starters for induction motors now commercially available. In the following, however, we will briefly consider only the principles of the two commonly used methods. We consider the current limitation first. Some of the common methods of limiting the stator current while starting are:

Reduced-Voltage Starting A reduced voltage is applied to the stator at the time of starting, and the voltage is increased to the rated value when the motor is within 25

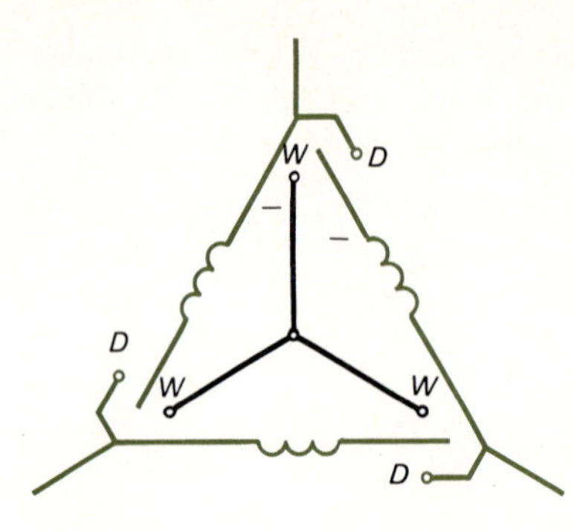

FIGURE 16.13 Wye-delta starting. Switches on *W* correspond to the wye connection, and switches on *D* correspond to the delta connection.

percent of its final speed. This method has the obvious limitation that a variable-voltage source is needed and the starting torque drops substantially. The so-called wye-delta method of starting is a reduced-voltage starting method. If the stator is normally connected in delta, reconnection to wye reduces the phase voltage, resulting in less current at starting; for example, if the line current at starting is about 5 times the full-load current in a delta-connected stator, the current in the wye connection will be less than 2 times the full-load value. But, at the same time, the starting torque for a wye connection would be about one-third its value for a delta connection. One advantage of wye-delta starting is that it is inexpensive and requires only a three-pole (or three single-pole) double-throw switch or switches, as shown in Fig. 16.13.

Current Limiting by Series Resistance Series resistances inserted in the three lines sometimes are used to limit the starting current. These resistances are shorted out, once the motor has gained speed. Because of the extra losses in the external resistances, this method has the obvious disadvantage of being inefficient.

Turning now to the starting torque, we recall from the last section that the starting torque is dependent on the rotor resistance. Thus, a high rotor resistance results in a high starting torque. Therefore, in a wound-rotor machine external resistance in the rotor circuit may be conveniently used (see Fig. 16.14). In a cage rotor, deep slots are used, where the slot depth is 2 or 3 times greater than the slot width (see Fig. 16.15). Rotor bars embedded in deep slots provide a high effective resistance and a large torque at starting. Under normal running conditions with low slips, however, the rotor resistance becomes lower and the efficiency high. This characteristic of rotor bar resistance is a consequence of *skin effect*. Because of skin effect, the current will have a tendency to concentrate at the top of the bars at starting, when the frequency of rotor currents is high; at this point, the frequency of rotor currents will be the same as the stator input frequency (e.g., 60 Hz). While running, the frequency of rotor currents (=slip frequency = 3 Hz at 5 percent slip and 60 Hz) is much lower; at this level of operation, skin effect is negligible and the current is almost uniformly distributed throughout the entire bar cross section.

Skin effect is used in an alternative form in a *double-cage* rotor (Fig. 16.16), where the inner cage is deeply embedded in iron and has low-resistance bars. The outer cage has relatively high-resistance bars close to the stator. At starting, because of skin effect, the influence of the outer cage dominates, thus producing a high starting torque. While the motor is running, the current penetrates to full depth into the lower cage

—because of insignificant skin effect—which results in an efficient steady-state operation. Notice that under normal running conditions both cages carry current, thus somewhat increasing the rating of the motor. The rotor equivalent circuit of a double-cage rotor then becomes as shown in Fig. 16.17.

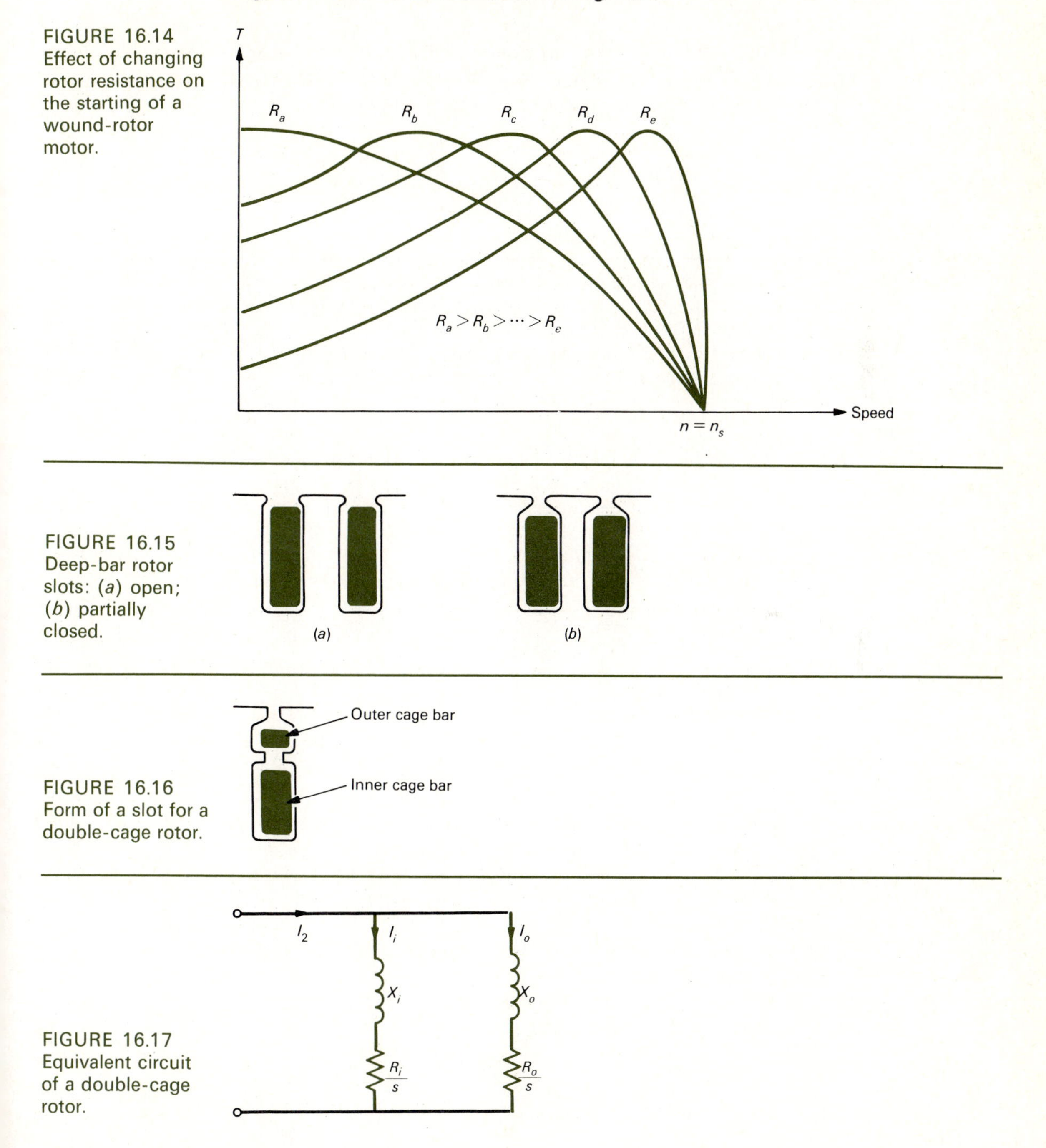

FIGURE 16.14 Effect of changing rotor resistance on the starting of a wound-rotor motor.

FIGURE 16.15 Deep-bar rotor slots: (*a*) open; (*b*) partially closed.

FIGURE 16.16 Form of a slot for a double-cage rotor.

FIGURE 16.17 Equivalent circuit of a double-cage rotor.

Example 16.7 A motor employs a wye-delta starter which connects the motor phases in wye at the time of starting and in delta when the motor is running. The full-load slip is 4 percent and the motor draws 9 times the full-load current if started directly from the mains. Determine the ratio of starting torque T_s to full-load torque T_{FL}.

Solution When the phases are switched to delta, the phase voltage, and hence the full-load current, is increased by a factor of $\sqrt{3}$ over the value it would have had in a wye connection. Then it follows from the last equation in Example 16.5 that

$$\frac{T_s}{T_{FL}} = \left(\frac{9}{\sqrt{3}}\right)^2 (0.04)$$

$$= 1.08$$

Example 16.8 To obtain a high starting torque in a cage-type motor, a double-cage rotor is used. The forms of a slot and of the bars of the two cages are shown in Fig. 16.16. The outer cage has a higher resistance than the inner cage. At starting, because of the skin effect, the influence of the outer cage dominates, thus producing a high starting torque. An approximate equivalent circuit for such a rotor is given in Fig. 16.17. Suppose that, for a certain motor, we have the per phase values

$$R_i = 0.1\ \Omega \qquad R_o = 1.2\ \Omega \qquad X_i = 2\ \Omega \qquad X_o = 1\ \Omega$$

Determine the ratio of the torques provided by the two cages at starting and at 2 percent slip.

Solution From Fig. 16.17, at $s = 1$,

$$Z_i^2 = (0.1)^2 + (2)^2 = 4.01\ \Omega^2$$

$$Z_o^2 = (1.2)^2 + (1)^2 = 2.44\ \Omega^2$$

the power input to the inner cage is

$$P_{ii} = I_i^2 R_i = 0.1 I_i^2$$

the power input to the outer cage is

$$P_{io} = I_o^2 R_o = 1.2 I_o^2$$

and the ratio between the torque due to inner cage and the torque due to the outer cage is

$$\frac{T_i}{T_o} = \frac{P_{ii}}{P_{io}} = \frac{0.1}{1.2}\left(\frac{I_i}{I_o}\right)^2 = \frac{0.1}{1.2}\left(\frac{Z_o}{Z_i}\right)^2$$

$$= \frac{0.1}{1.2}\left(\frac{2.44}{4.01}\right) = 0.05$$

Similarly, at $s = 0.02$,

$$Z_i^2 = \left(\frac{0.1}{0.02}\right)^2 + (2)^2 = 29\ \Omega^2$$

$$Z_o^2 = \left(\frac{1.2}{0.02}\right)^2 + (1)^2 = 3601\ \Omega^2$$

$$\frac{T_i}{T_o} = \frac{0.1}{1.2}\left(\frac{3601}{29}\right) = 10.34$$

16.9 INDUCTION GENERATORS

Up to this point, we have studied the behavior of an induction machine operating as a motor. We recall from the preceding discussions that for motor operation the slip lies between 0 and 1, and for this case we have a conversion of electric power into mechanical power. If the rotor of an induction machine is driven by an auxiliary means such that the rotor speed n becomes greater than the synchronous speed n_s, then from Eq. (16.1) we have a negative slip. A negative slip implies that the induction machine is now operating as an induction generator. Alternatively, we may refer to the rotor portion of the equivalent circuit, such as that of Fig. 16.11. If the slip is negative, the resistance representing the load becomes $R_2'[1 - (-s)]/(-s)$, which results in a negative value of the resistance. Because a positive resistance absorbs electric power, a negative resistance may be considered as a source of power. Hence, a negative slip corresponds to a generator operation.

To understand the generator operation, we consider a three-phase induction machine to which a prime mover is coupled mechanically. When the stator is excited, a synchronously rotating magnetic field is produced and the rotor begins to run, as in an induction motor, while drawing electric power from the supply. The prime mover is then turned on (to rotate the rotor in the direction of the rotating field). When the rotor speed exceeds synchronous speed, the direction of electric power reverses. The power begins to flow into the supply as the machine begins to operate as a generator. The rotating magnetic field is produced by the magnetizing current supplied to the stator winding from the three-phase source. This supply of the magnetizing current must be available as the machine operates as an induction generator. Stated differently, an induction generator is not self-exciting. It must operate in parallel with a source capable of supplying it with the necessary exciting current at a fixed frequency.

The operating characteristics of an induction machine's motor and generator modes are shown in Fig. 16.18. Unlike in a synchronous generator, for a given load the output current and the power factor are determined by the generator parameters. Therefore, when an induction generator delivers a certain power, it also supplies a certain in-phase current and a certain quadrature current. However, the quadrature component of the current generally does not have a definite relationship to the quadrature component of the load current. The quadrature current must be supplied by the synchronous generators operating in parallel with the induction generator.

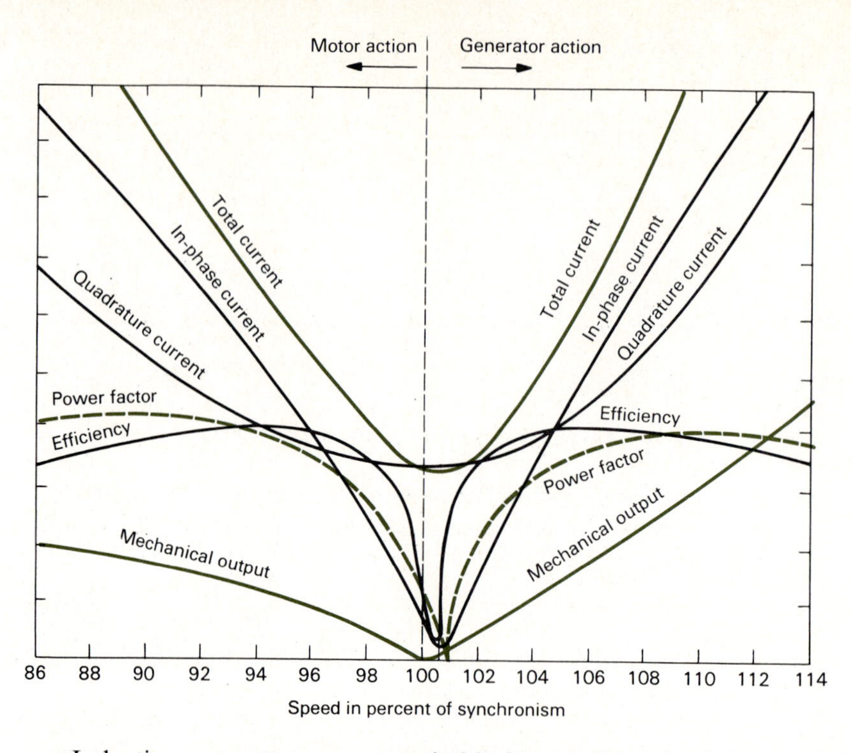

FIGURE 16.18 Motor and generator characteristics of an induction machine.

Induction generators are not suitable for supplying loads having low lagging power factors. In the past, induction generators have been used in variable-speed constant-frequency generating systems. Large induction generators have found applications in hydroelectric power stations. Induction generators are promising for windmill applications.

PROBLEMS

16.1 A six-pole 60-Hz induction motor runs at 1152 r/min. Determine the synchronous speed and the percent slip.

16.2 A six-pole induction motor is supplied by a synchronous generator having four poles and running at 1500 r/min. If the speed of the induction motor is 750 r/min, what is the frequency of rotor current ?

16.3 A three-phase four-pole 60-Hz induction motor develops a maximum torque of 180 N · m at 800 r/min. If the rotor resistance is 0.2 Ω per phase, determine the developed torque at 1000 r/min.

16.4 A 400-V four-pole three-phase 60-Hz induction motor has a wye-connected rotor having an impedance of $(0.1 + j0.5)$ Ω per phase. How much additional resistance must be inserted in the rotor circuit for the motor to develop the maximum starting torque? The effective stator-to-rotor turns ratio is 1.

16.5 For the motor of Prob. 16.4, what is the motor speed corresponding to the maximum developed torque without any external resistance in the rotor circuit?

16.6 A two-pole 60-Hz induction motor develops a maximum torque of twice the full-load torque. The starting torque is equal to the full-load torque. Determine the full-load speed.

16.7 The input to the rotor circuit of a six-pole 60-Hz induction motor running at 1000 r/min is 3 kW. What is the rotor copper loss?

16.8 The stator current of a 400-V three-phase wye-connected four-pole 60-Hz induction motor running at a 6 percent slip is 60 A at 0.866 power factor. The stator copper loss is 2700 W and the total iron and rotational losses are 3600 W. Calculate the motor efficiency.

16.9 An induction motor has an output of 30 kW at 86 percent efficiency. For this operating condition, stator copper loss = rotor copper loss = core losses = mechanical rotational losses. Determine the slip.

16.10 A four-pole 60-Hz three-phase wye-connected induction motor has a mechanical rotational loss of 500 W. At 5 percent slip, the motor delivers 30 HP at the shaft. Calculate the (*a*) rotor input, (*b*) output torque, and (*c*) developed torque.

16.11 A wound-rotor six-pole 60-Hz induction motor has a rotor resistance of 0.8 Ω and runs at 1150 r/min at a given load. The load on the motor is such that the torque remains constant at all speeds. How much resistance must be inserted in the rotor circuit to bring the motor speed down to 950 r/min? Neglect the rotor leakage reactance.

16.12 A 400-V three-phase wye-connected induction motor has a stator impedance of $(0.6 + j1.2)$ Ω per phase. The rotor impedance referred to the stator is $(0.5 + j1.3)$ Ω per phase. Using the approximate equivalent circuit, determine the maximum electromagnetic power developed by the motor.

16.13 The motor of Prob. 16.12 has a magnetizing reactance of 35 Ω. Neglecting the iron losses, at 3 percent slip calculate (*a*) the input current and (*b*) the power factor. Use the approximate equivalent circuit.

16.14 On no-load, a three-phase delta-connected induction motor takes 6.8 A and 390 W at 220 V. The stator resistance is 0.1 Ω per phase. The friction and windage loss is 120 W. Determine the values of the parameters X_m and R_m of the equivalent circuit of the motor.

16.15 On blocked-rotor, the motor of Prob. 16.14 takes 30 A and 480 W at 36 V. Using the data of Prob. 16.14, determine the complete exact equivalent circuit of the motor. Assume that the per phase stator and rotor leakage reactances are equal.

16.16 Determine the parameters of the approximate equivalent circuit of a three-phase induction motor from the following data:

No-load test:

applied voltage: 440 V

input current: 10 A

input power: 7600 W

Blocked-rotor test:

applied voltage: 180 V

input current: 40 A

input power: 6240 W

The stator resistance between any two leads is 0.8 Ω and the no-load friction and windage loss is 420 W.

16.17 A four-pole 400-V three-phase 60-Hz induction motor takes a 150-A current at starting and 25 A while running at full-load. The starting torque is 1.8 times the torque at full-load at 400 V. If it is desired that the starting torque be the same as the full-load torque, determine (*a*) the applied voltage and (*b*) the corresponding line current.

16.18 An induction motor is started by a wye-delta switch. Determine the ratio of the starting torque to the full-load torque if the starting current is 5 times the full-load current and the full-load slip is 5 percent.

16.19 An induction motor is started at a reduced voltage. The starting current is not to exceed 4 times the full-load current, and the full-load torque is 4 times the starting torque. What is the full-load slip? Calculate the factor by which the motor terminal voltage must be reduced at starting.

16.20 The per-phase parameters of the equivalent circuit of a double-cage rotor are: $R_0 = 0.4$ Ω, $X_o = 0.2$ Ω, $R_i = 0.04$ Ω, and $X_i = 0.8$ Ω. At starting, determine the ratio of the torques provided by the outer and inner cages.

16.21 At what slip will the torques contributed by the outer and inner cages of the rotor of Prob. 16.20 be equal?

REFERENCES

16.1 S. A. Nasar and L. E. Unnewehr, *Electromechanics and Electric Machines*, 2d ed., Wiley, New York, 1983.

16.2 A. E. Fitzgerald, C. Kinsley, Jr., and S. D. Umans, *Electric Machinery*, 4th ed., McGraw-Hill, New York, 1983.

16.3 S. A. Nasar, *Electric Machines and Transformers*, Macmillan, New York, 1983.

Chapter

Control of Electric Motors

We have discussed the operating principles of dc, synchronous, and induction motors in Chaps. 14 to 16, in which we also examined basic methods for controlling such motor characteristics as speed and torque. In this chapter we will examine more advanced means of control. The devices and techniques developed in recent years to make the control of speed, torque, output power, and other motor output parameters more precise have led to greatly increased operational efficiency.

To illustrate the importance of choosing the most efficient—and hence the most cost-effective—means of control, let us consider a variable-discharge pump. We could choose to use a nonadjustable drive for the pump and to control the flow by throttling, or we could use an adjustable drive and no throttling. The results of the two choices in terms of their power requirements are shown in Fig. 17.1, which clearly

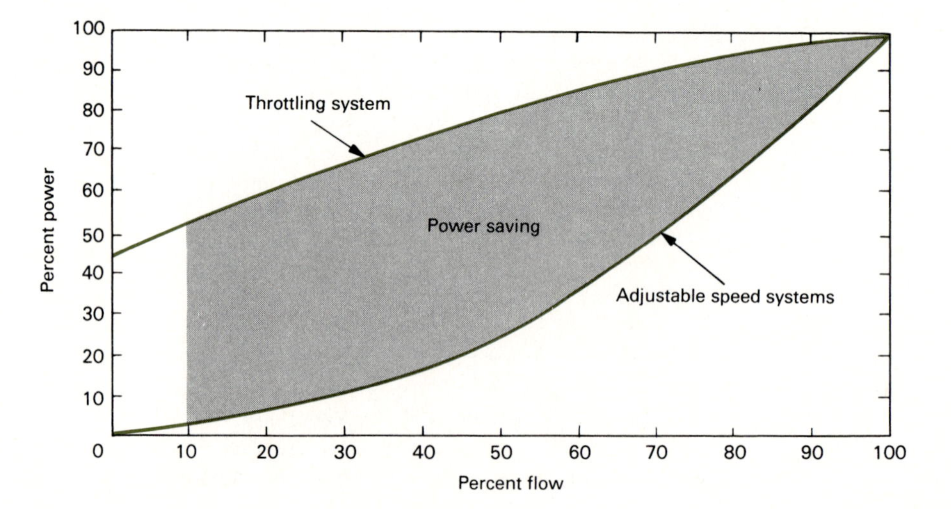

FIGURE 17.1 Comparison of percent power required for percent flow—hydroviscous adjustable-speed drive versus throttling system.

shows that the best choice is an adjustable-speed drive that can rotate the pump at the correct speed to deliver the exact flow and pressure required. We see that when the pump flow is reduced to 20 percent of maximum pump flow, an adjustable-speed system (a hydroviscous drive in this example) has approximately 57 percent less wasted power than a throttling system. We also see that even at 70 percent of maximum pump flow, the throttling system wastes 42 percent more power than the adjustable-speed drive. These figures indicate that the power saved by an adjustable-speed drive would pay for the drive in a few months.

The preceding is a simple example of where an adjustable-speed drive is most desirable. Some of the applications requiring adjustable drives and motor control are in glass industry, food industry, machine tool, material handling, petrochemicals, water and wastewater treatment, paper and paper converting, test stands, and textiles and synthetic fibers. The range of applications of adjustable drives is thus very broad. Because much of present-day motor control is accomplished by solid-state electronic devices, we will begin with a brief review of the devices used in power electronics and motor control.

17.1 POWER SEMICONDUCTORS

Many types of semiconductors are available for electronic motor control applications. The type to be used in a specific application will depend primarily upon the power, voltage, and current requirements of the motor to be controlled. Other factors include the ambient temperature, the control modes to be used, and overall system cost considerations.

The power semiconductors and their associated circuitry are often referred to as the *power circuit* of a motor control system. Table 17.1 lists the principal semiconductor devices used in motor control systems, shows the standard symbols for these devices in circuit diagrams, and gives the state-of-the-art maximum voltage, current, and time-response capabilities of each class of device. The time (or speed) parameter in the right column generally refers to the minimum turnoff time. The device symbols are A, anode; K, cathode; G, gate; E, emitter; C, collector; and B, base. Although the Zener diode is not a true power-controlling device, it is included in Table 17.1 because it is widely used as a voltage control and sensing device in many motor controllers. Of the devices listed in Table 17.1, only the silicon rectifier, silicon-controlled rectifier (SCR, or thyristor), and power transistor are considered in detail here.

Silicon Rectifier Symbolic representations and the ideal $v - i$ characteristic of a silicon rectifier are shown in Fig. 17.2, where it is assumed that the diode acts like an ideal switch. In practice, however, the silicon rectifier is not a perfect conductor; it has a forward voltage drop of approximately 1 V at all currents within the rating of the rectifier.

The principal parameters of a silicon rectifier are a repetitive peak reverse voltage (PRV), or blocking voltage, an average forward current, and a maximum operating junction temperature (which is 125°C for most silicon devices). Another important

TABLE 17.1 Power Semiconductors

Power semiconductor	Symbol	Maximum capabilities: Voltage, V	RMS Current, A	Speed, μs
Silicon rectifier	A	5000	7500	—
Silicon-controlled rectifier (SCR)	G, A, K	5000	3000	1
Bidirectional switch TRIode thyristor for AC control (TRIAC)	G, A, K	1000	2000	1
Gate-turnoff (GTO) SCR Gate-controlled switch (GCS)	G, A, K	400 1000	— 200	0.2 2
Power transistor	E, C, (PNP) B	3000	500	0.2
Darlington	B, C, E	1000	200	1
Zener diode	A, K	500	—	100

characteristic in certain applications is the recovery time of the rectifier. The recovery time is the time following the end of forward conduction before the rectifier assumes its full reverse blocking capability.

In motor control systems, silicon rectifiers commonly function as "freewheeling" diodes, providing a path for continuation of motor current following the switching off of another power device between the motor terminals and the power source.

Silicon-Controlled Rectifier (SCR) This device is also frequently termed a *thyristor*. A symbolic representation of an SCR and its ideal $v - i$ characteristic are shown in Fig. 17.3. The SCR has three terminals: anode, cathode, and gate. A gate signal is used to turn on the SCR. In the reverse, or blocking, direction, the SCR functions very much like the silicon rectifier. In the forward direction, conduction can be controlled to a limited extent by the action of the gate, which is connected to low-signal circuitry that can be electrically isolated from the power circuitry connected to the anode and cathode. Control is limited, and only momentary, because the gate can turn on the device (that is, initiate the conditions for forward current flow) but cannot stop the current flow. Thus, when current begins to flow through the anode, the SCR

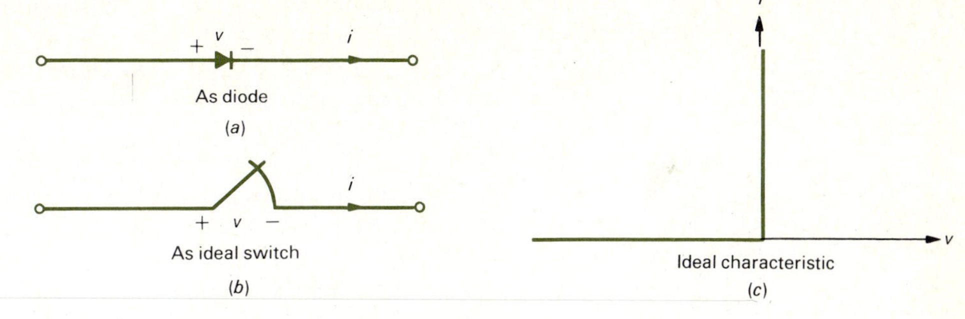

FIGURE 17.2 An ideal diode and its v-i characteristic.

latches on and remains in the conductive state until turned off by external means, in a process known as *commutation*.

The principal parameters for applying SCRs to motor controllers are: repetitive PRV; maximum value of average on-state current, which determines the heating within the SCR; maximum value of on-state current, which is the rating of the metal conductor portions of the device; peak one-cycle on-state current, which is the surge current limit; critical rate of rise of forward blocking voltage, which has two values —initial (when the SCR is first turned on) and reapplied (following commutation) turnoff time, which is the off time required following commutation before forward voltage can be reapplied; maximum rate of rise of anode current during turn-on; and maximum operating junction temperature. Although not included above, thermal management of the SCR—design of its heat sink, mounting method, and cooling—is extremely critical in all applications.

For motor controller use, there are two broad classes of SCRs: *Inverter-type* SCRs are applicable in inverters, cycloconverters, and brushless dc motor systems; *chopper-type* SCRs apply in choppers, phase-controlled rectifiers, regulators, and the like. The primary difference between the two types is in respect to time of response: another difference is that inverter types are generally more costly than chopper types. There are further ways to classify SCRs, especially in the smaller ratings used in control and communications applications. Many SCRs rated at 35 A (RMS) or below can be packaged in plastic cases, which makes them cheaper than metallic SCRs of similar ratings. Another classification relates to the gate signal. A very useful class of SCRs for position sensing in motor control systems is the light-activated SCR (LASCR). In place of energy injection by electric current, the LASCR is triggered on by photon

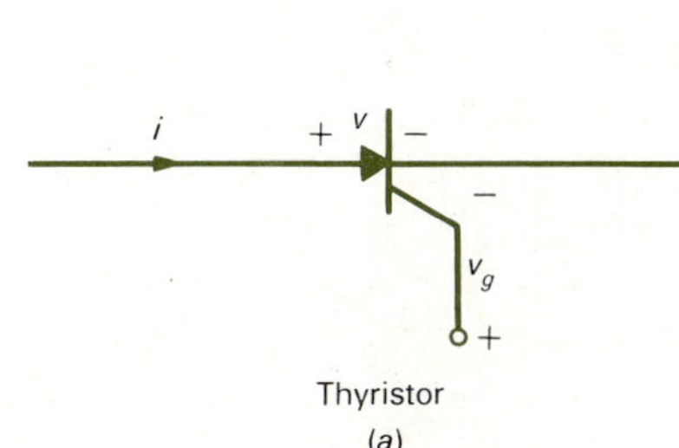

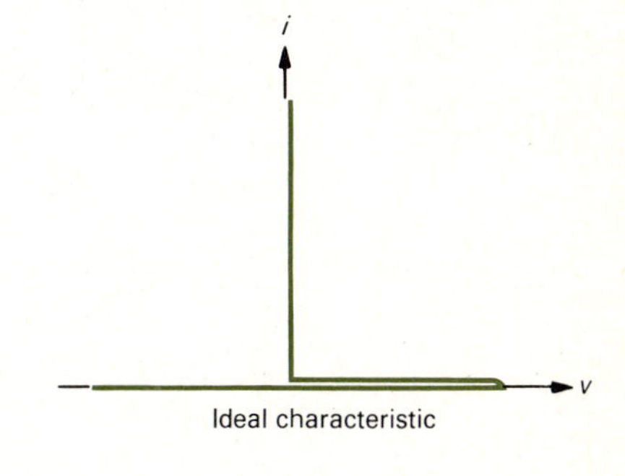

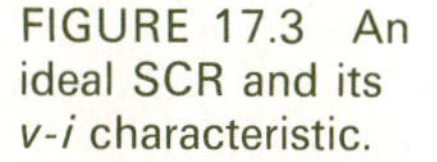
FIGURE 17.3 An ideal SCR and its v-i characteristic.

energy. Light activation is possible in all the other three- and four-terminal devices listed in Table 17.1.

The forward voltage drop of an SCR varies considerably during anode current conduction and may be very large during the first few microseconds after turn-on. The average value is 1.5 to 2.0 V.

Power Transistor When used in motor control circuits, power transistors are almost always operated in a switching mode. The transistor is driven into saturation and the linear gain characteristics are not used. The common-emitter configuration is the most common, because of the high power gain in this connection. The collector-emitter saturation voltage $V_{CE,\,sat}$ for typical power transistors is from 0.2 to 0.8 V. This range is considerably lower than the on-state anode-cathode voltage drop of an SCR; therefore, the average power loss in a power transistor is lower than that in an SCR of equivalent power rating. The switching times of power transistors are also generally faster than those of SCRs, and the problems associated with turning off or commutating an SCR are almost nonexistent in transistors. However, a power transistor is more expensive than an SCR of equivalent power capability. In addition, the voltage and current ratings of power transistors available are much lower than those of existing SCRs. It has already been stated that the maximum ratings listed in Table 17.1 are generally unobtainable concurrently in a single device. This is particularly true of power transistors. Devices with voltage ratings of 1000 V or above have limited current ratings of 10 A or less. Similarly, the devices with high current ratings, 50 A and above, have voltage ratings of 200 V or less. There has been relatively little operating

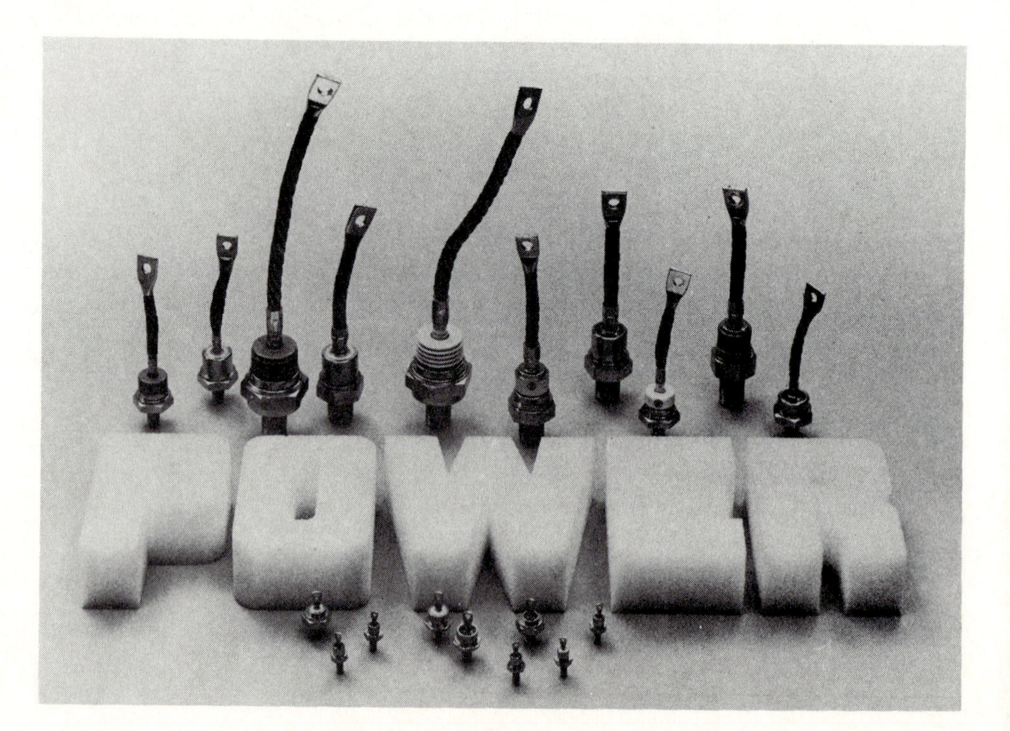

FIGURE 17.4 A family of power semiconductors.

FIGURE 17.5
Disk-type devices and mounting with heat sink.

FIGURE 17.6
Heat sinks and other components for power semiconductor circuits.

experience in practical motor control circuits with transistors whose voltage or current capabilities approach the upper limits listed in Table 17.1. For handling motor control requiring large current ratings at 200 V or below, it has been common to parallel transistors of lower current rating. This requires great care to assure equal sharing of collector currents and proper synchronization of base currents among the paralleled devices.

Figure 17.4 shows a family of power semiconductors—rectifiers, SCRs, and transistors. Disk-type (or hockey-puck) devices—unmounted and with heat sink—are shown in Fig. 17.5. Heat sinks and various components used in power electronics are shown in Fig. 17.6. Having acquired a basic knowledge of power semiconductor devices, we are now ready to study their applications to electric drives.

17.2 CONTROLLERS FOR dc MOTORS

The torque and speed characteristics (under steady-state) of dc motors are essentially governed by the equations derived in Chap. 14. For convenience we repeat these pertinent equations in a slightly modified form as follows:

$$E = \frac{\phi n Z}{60}\frac{P}{a} = k_a \phi \omega_m \tag{17.1}$$

$$T_e = k_a \phi I_a \tag{17.2}$$

$$\omega_m = \frac{V_t - I_a R_a}{k_a \phi} \tag{17.3}$$

where E = voltage induced in the armature (in V)

I_a = armature current (in A)

ϕ = flux per pole (in W)

Z = number of armature conductors

a = number of parallel paths

P = number of poles

ω_m = armature speed (in rad/s)

n = armature speed (in r/min)

V_t = armature terminal voltage (in V)

and

$$k_a = \frac{ZP}{2\pi a} \text{ (a constant)}$$

These equations indicate the great flexibility we have in controlling the motor. For instance, the speed of a dc motor may be varied by varying V_t, R_a, or ϕ (i.e., the field current). Control of dc motors is governed by Eqs. (17.1) to (17.3), and the various practical schemes are manifestations of these equations in one form or another.

From the governing equations, Eqs. (17.2) and (17.3), it is clear that the motor torque and speed can be controlled by controlling ϕ (i.e., the field current), V_t, and R_a, and changes in these quantities can be accomplished as follows. In essence, the method of control involves field control, armature control, or a combination of the two.

Chopper Control Resistance control, either in the field or armature of a dc motor, results in a poor efficiency. High-power solid-state controllers, on the other hand, offer the most practical, reliable, and efficient method of motor control. The most commonly used solid-state devices in motor control are power transistors and thyristors (or SCRs). The main differences between the two are that the transistor requires a continuous driving signal during conduction, whereas the thyristor requires only a pulse to initiate conduction; and the transistor switches off when the driving signal is removed, whereas the thyristor turns off when the load current is reduced to zero or when a reverse-polarity voltage is applied.

Utilizing power semiconductors, the dc chopper is the most common electronic controller used in electric drives. In principle, a chopper is an on-off switch connecting the load to and disconnecting it from the battery (or dc source), thus producing a chopped voltage across the load. Symbolically, a chopper as a switch is represented in

FIGURE 17.7 (*a*) Symbolic representation of a chopper switch, and output waveforms. (*b*) Basic circuits of a thyristor, and output waveforms.

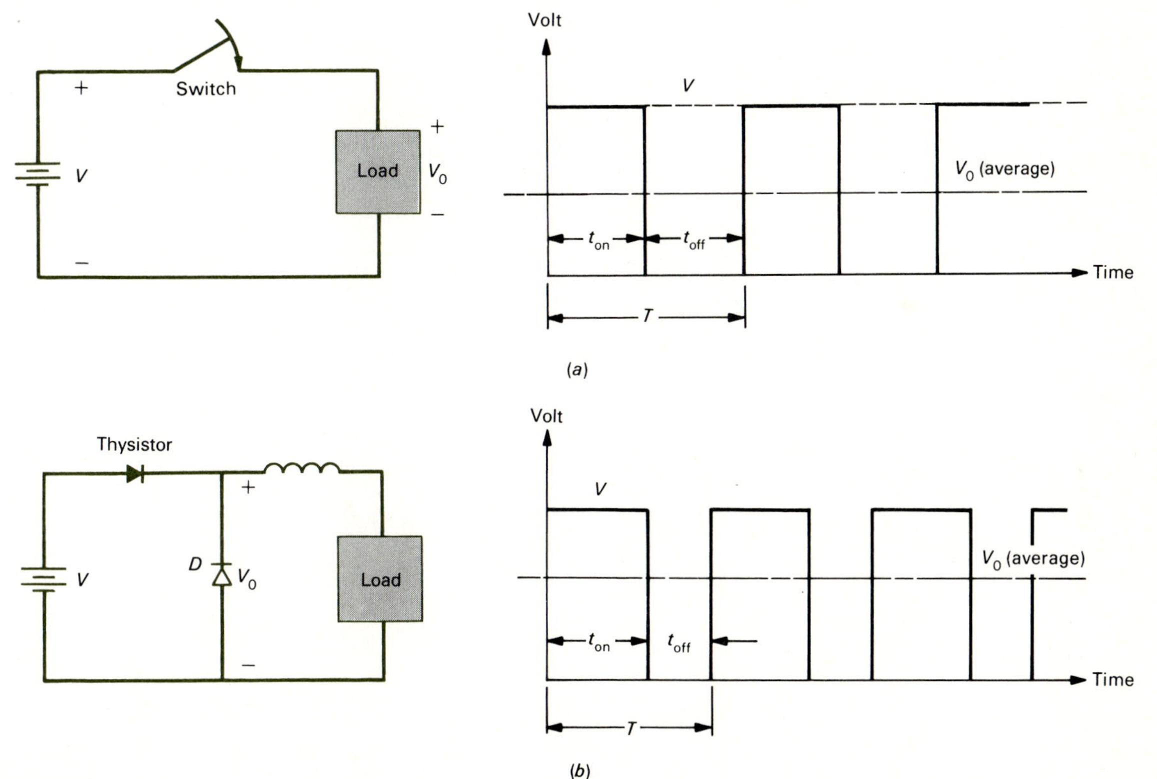

Fig. 17.7*a*, and a basic chopper circuit is shown in Fig. 17.7*b*, when the thyristor does not conduct, the load current flows through the freewheeling diode *D*. From Fig. 17.7*b* it is clear that the average voltage across the load V_0 is given by

$$V_0 = \frac{t_{on}}{t_{on} + t_{off}} V = \frac{t_{on}}{T} V = \alpha V \tag{17.4}$$

where the various times are shown in Fig. 17.7, T is known as the chopping period, and $\alpha = t_{on}/T$ is called the *duty cycle*. Thus, the voltage across the load varies with the duty cycle.

There are three ways in which the chopper output voltage can be varied, and these are illustrated in Fig. 17.8. In the first method, the chopping frequency is kept constant and the pulse width (or on-time t_{on}) is varied; this method is known as *pulse-width modulation*. The second method, called *frequency modulation*, has either t_{on} or

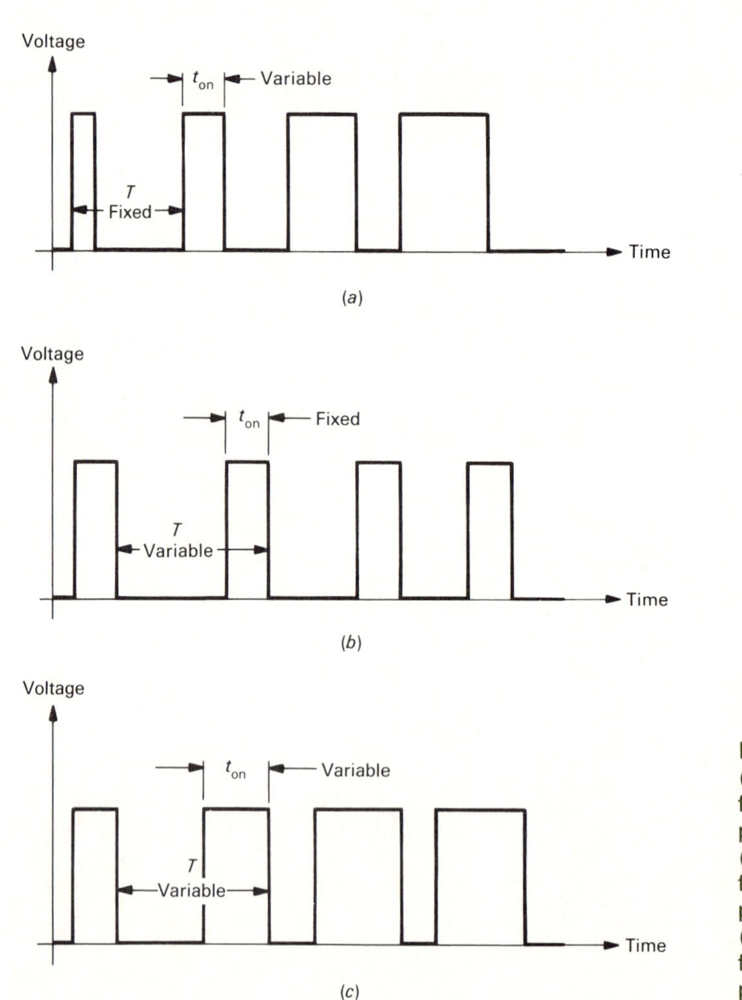

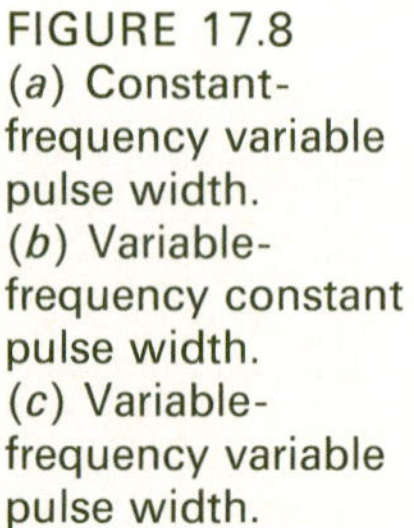
FIGURE 17.8
(*a*) Constant-frequency variable pulse width.
(*b*) Variable-frequency constant pulse width.
(*c*) Variable-frequency variable pulse width.

t_{off} fixed, and a variable chopping period, as indicated in Fig. 17.8*b*. The third method combines the preceding two methods to obtain pulse-width and frequency modulation, shown in Fig. 17.8*c*, which is used in current limit control. In a method involving frequency modulation, the frequency must not be decreased to a value that may cause a pulsating effect or a discontinuous armature current, and the frequency should not be increased to such a high value as to result in excessive switching losses; the switching frequency of most choppers for electric drives range from 100 to 1000 pulses per second. The drawback of a high-frequency chopper is that the current interruption generates a high-frequency noise.

A chopper circuit in a simplified form is shown in Fig. 17.9*a*, where the chopper is shown to supply a dc series motor. The circuit is a pulse-width modulation circuit, where t_{on} and t_{off} determine the average voltage across the motor, as given by Eq. (17.4). As mentioned in Sec. 17.1, the SCR cannot turn itself off, once it begins to conduct. So, to turn the SCR off, we require a commutating circuit which impresses a negative voltage on the SCR for a very short time (of the order of microseconds). The

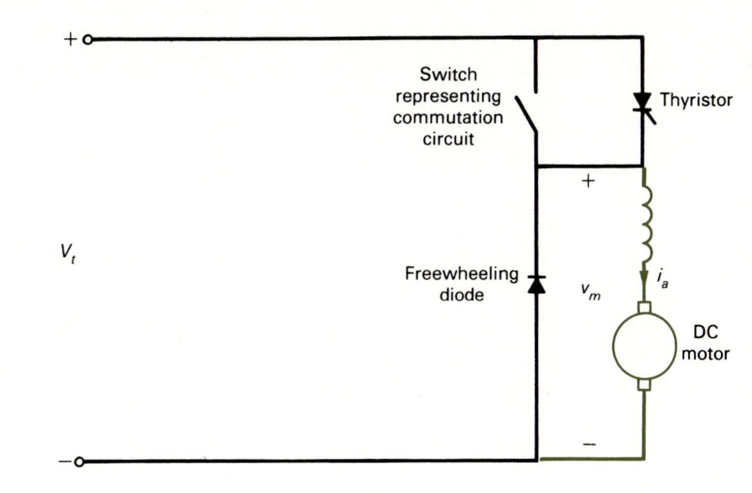

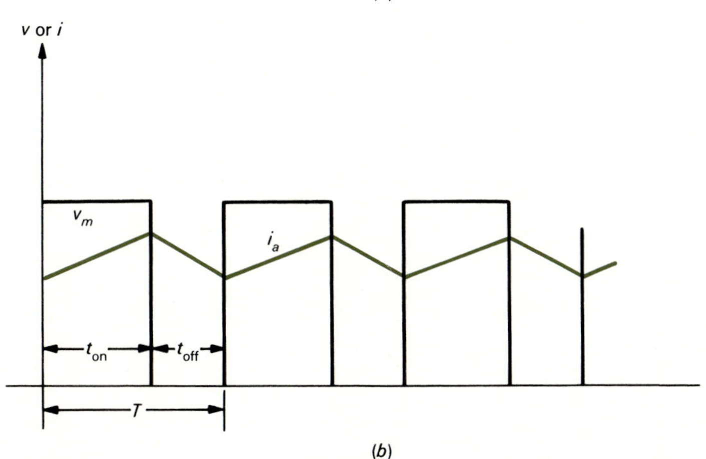

FIGURE 17.9 (*a*) A dc motor driven by a chopper. (*b*) Motor voltage and current waveforms.

circuitry for commutation is often quite involved. For our purposes, we denote the commutating circuit by a switch in Fig. 17.9*a*.

The motor current and voltage waveforms are shown in Fig. 17.9*b*. The SCR is turned on by a gating signal at $t = 0$. The armature current i_a builds up and the motor starts and picks up speed. After the time t_{on}, the SCR is turned off, and it remains off for a period t_{off}. During this period, the armature current continues to drop through the freewheeling diode circuit. Again, at the end of t_{off}, the SCR is turned on and the on-off cycle continues. The chopper thus acts as a variable-voltage source.

Figure 17.10 shows the Jones chopper circuit, which works as follows: S_1 is turned on, resulting in current flow through S_1-L_2-motor; L_2 and L_1 are magnetically coupled (both windings are usually wound on the same magnetic core). Therefore, current flow through L_2 causes a proportional current through L_1-D_1-C, charging C, with the lower plate positive. The pulse through S_1 is ended by turning on S_2, which reverse-biases S_1 and causes the typical sinusoidal pulse through the path C-S_2-L_2 motor. Shortly after this current pulse reaches its maximum value, D_2 is forward-biased and begins to conduct and S_2 is reverse-biased and turns off. During the next period of operation, beginning with the turning on of S_1, C reverses its voltage through the path C-S_1-L_1-D_1, which also contributes to the load current through L_2. When the voltage on C is reversed, further charging may continue through D_1 by the coupling action of load current in L_2 until D_1 is reverse-biased. When D_1 is turned off by this reverse-biased condition and C is charged, with the lower plate positive, the circuit is in the condition to repeat the previously described sequence of events.

The disadvantages of the Jones circuit are the size and weight of the coupled inductances L_1 and L_2. However, this circuit has resulted in low manufacturing costs, partly because of the use of a commutating capacitor of smaller size. Other advantages of the Jones chopper circuit are that (1) the circuit can be utilized for shunt or series

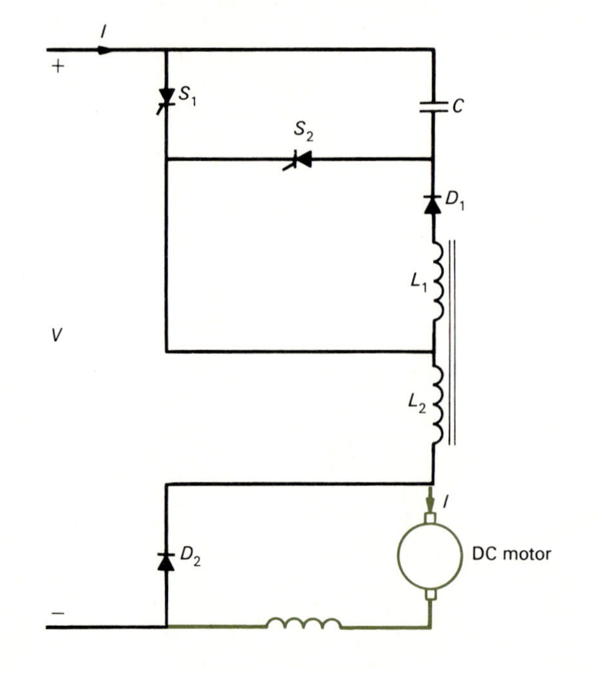

FIGURE 17.10 Jones chopper circuit.

motors and (2) the output voltage may be varied either by varying the chopper frequency or the pulse width.

The equations governing the design of the Jones chopper are as follows:

$$t_{\text{off}} = \frac{CV}{I} \tag{17.5}$$

$$\tfrac{1}{2}CV_c^2 = \tfrac{1}{2}L_1 I^2 \tag{17.6}$$

When S_1 is turned on, Eq. (17.6) implies that the energy stored in the capacitance is transferred to the inductance L_1. In Eqs. (17.5) and (17.6), I is the rated current of the circuit and V_c is the initial charge across the capacitance. The optimum frequency for a minimum commutation loss and smaller commutation loss requires that

$$C \geq \frac{\pi I t_{\text{off}}}{2V} \tag{17.7}$$

and that

$$L_1 = L_2 \tag{17.8}$$

where the symbols are defined in Fig. 17.10.

Example 17.1 For a dc series motor driven by a chopper, the following data are given: supply voltage = 440 V; duty cycle = 30 percent; motor circuit inductance = 0.04 H; maximum allowable change in the armature current = 8 A. Determine the chopper frequency.

Solution From Eq. (17.4) the average voltage across the motor is

$$0.3 \times 440 = 132 \text{ V}$$

The voltage across the motor circuit inductance is

$$V_L = 440 - 132$$
$$= 308 \text{ V}$$

This voltage is also given by

$$V_L = L\frac{\Delta I}{\Delta t}$$

where

$$V_L = 308 \text{ V}$$
$$L = 0.04 \text{ H}$$
$$\Delta I = 8 \text{ A}$$
$$\Delta t = t_{\text{on}}$$

Hence,

$$t_{on} = \frac{0.04 \times 8}{308}$$

$$= 1.04 \text{ ms}$$

But

$$\alpha = 0.3$$

$$= \frac{t_{on}}{t_{on} + t_{off}}$$

$$= \frac{t_{on}}{T}$$

or

$$T = \frac{1.04}{0.3}$$

$$= 3.46 \text{ ms}$$

Hence, the chopper frequency is

$$\frac{1}{T} = 289 \text{ pulses per second}$$

Example 17.2 If the chopper of Example 17.1 is to be designed for a current rating of 50 A, determine the values of C, L_1, and L_2 of the circuit shown in Fig. 17.10.

Solution From Example 17.1

$$t_{off} = T - t_{on}$$

$$= 3.46 - 1.04$$

$$= 2.42 \text{ ms}$$

From Eq. (17.7) we have

$$C \geq \frac{\pi I}{2V} t_{off}$$

$$\geq \frac{\pi \times 50}{2 \times 440} \times 2.42 \times 10^{-3}$$

$$\geq 432 \; \mu\text{F}$$

or

$$C = 450 \; \mu\text{F}$$

Now

$$t_{\text{off}} = \sqrt{L_1 C}$$

from Eqs. (17.5) and (17.6), or

$$L_1 = \frac{t_{\text{off}}^2}{C}$$

$$= \frac{(2.42)^2 \times 10^{-6}}{450 \times 10^{-6}}$$

$$= 13 \text{ mH}$$

and

$$L_1 = L_2$$

$$= 13 \text{ mH}$$

Converters A converter changes an ac input voltage to a controllable dc voltage. Converters have an advantage over choppers in that *natural, or line, commutation* is possible in converters, and thereby no complex commutation circuitry is required. Controlled converters use SCRs and operate either on single-phase or three-phase ac. The four types of phase-controlled converters commonly used for dc motor control are:

1. Half-wave converters
2. Semiconverters
3. Full converters
4. Dual converters

Each of the above could be either a single-phase or a three-phase converter. Semiconverters are one-quadrant converters in that the polarities of voltage and current at the dc terminals do not reverse. In a full converter, the polarity of the voltage reverses, but the current is unidirectional. In this sense, a full converter is a two-quadrant converter. Dual converters are four-quadrant converters.

A half-wave converter and its quadrant operation are shown in Fig. 17.11. This type of a converter is used for motors with ratings up to $\frac{1}{2}$ hp. Other types of single-phase converters have been used in drives rated up to 100 hp.

The performance of converters is measured in terms of the following parameters:

$$\text{Input power factor} = \frac{\text{mean input power}}{\text{RMS input voltamperage}} \tag{17.9}$$

$$\text{Input displacement factor} = \cos \phi_1 \tag{17.10}$$

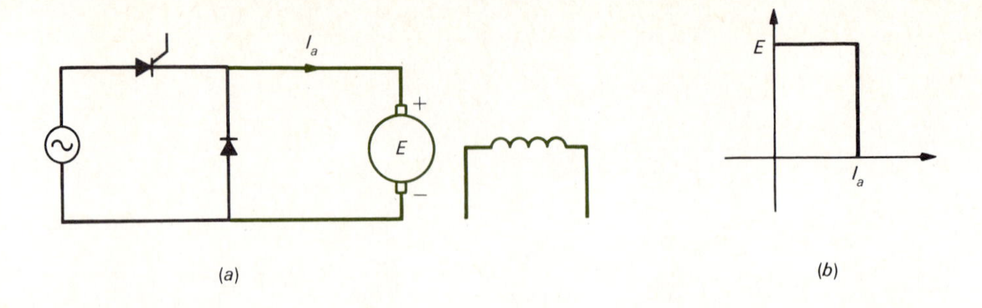

FIGURE 17.11
(*a*) A half-wave converter and (*b*) its quadrant operation.

where ϕ_1 is the angle between the supply voltage and the fundamental component of the supply current.

$$\text{Harmonic factor} = \frac{\text{RMS value of the } n\text{th harmonic current}}{\text{fundamental component of the supply current}} \tag{17.11}$$

To understand the operation of a converter-fed dc motor, we consider the simplest converter—the single-phase half-wave converter. Figure 17.12*a* shows such a system. The waveforms of motor current and voltage are given in Fig. 17.12*b*. By controlling the firing angle α of the SCR, we control the voltage across the motor armature and hence the motor speed. Notice, however, that the armature current does not begin to flow immediately after the SCR is turned on. Only when the line voltage v_t becomes greater than the motor voltage v_m does the armature current begin to flow. The current continues to flow for the period γ, as shown in Fig. 17.12*b*. This period is also known as the conduction angle; it is determined by the equality of the shaded areas A_1 and A_2 in Fig. 17.12*b*. The area A_1 corresponds to the energy stored in the inductance L of the motor circuit while the armature current is building up. This stored energy is returned to the source during the period that the armature current decreases. When the armature current becomes zero, the SCR then turns off until it is turned on again. The motor speed is controlled by the firing angle α.

A converter-driven motor is analyzed on the basis of averaging over the period of the line voltage, in which case Eqs. (17.1) to (17.3) are applicable.

Example 17.3 A 1-hp 240-V dc motor is designed to run at 500 r/min when supplied from a dc source. The motor armature resistance is 7.56 Ω. The torque constant k is 4.23 N·m/A and the back-emf constant is 4.23 V/(rad·s). This motor is driven by a half-wave converter at 200 V 50 Hz ac and draws a 2.0-A average current at a certain load. For the period during which the SCR conducts, the average motor voltage V_m is 120 V. Determine the torque developed by the motor, the motor speed, and the supply power factor.

Solution The torque is

$$kI_a = 4.23 \times 2$$

$$= 8.46 \text{ N} \cdot \text{m}$$

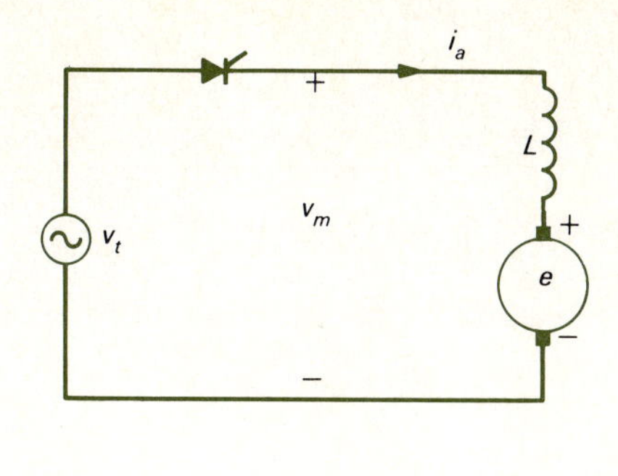

(a)

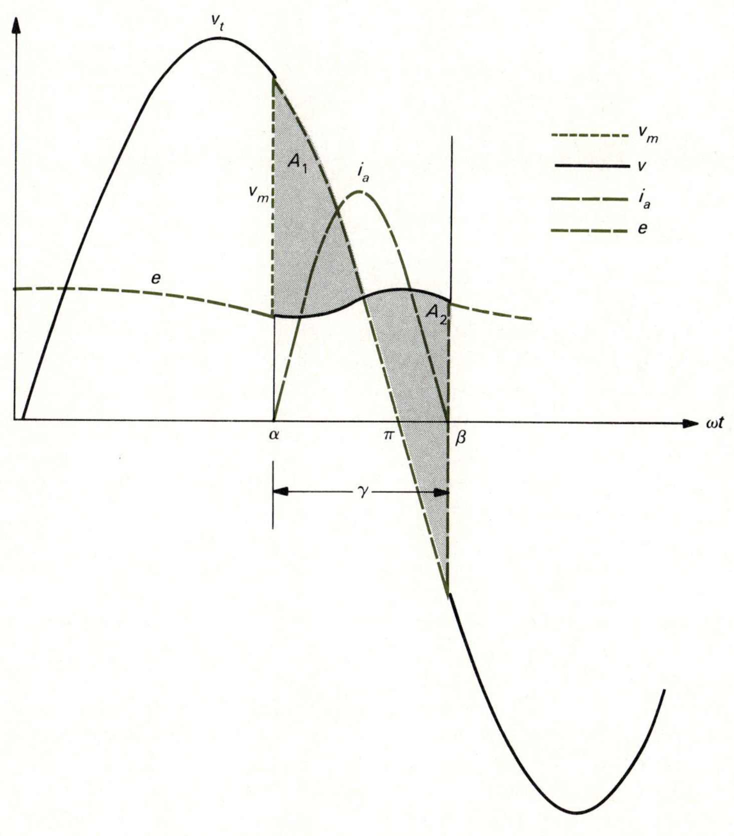

(b)

FIGURE 17.12 (a) A half-wave SCR drive and (b) its v, e, i_a, and v_m waveforms.

The back emf is

$$E = 120 - 2.0 \times 7.56$$
$$= 104.88$$

But

$$E = k\Omega_m$$

or

$$104.88 = 4.23\Omega_m$$
$$= \frac{4.23 \times 2\pi N}{60}$$

Hence,

$$N = \frac{60 \times 104.88}{2\pi \times 4.23}$$
$$= 237 \text{ r/min}$$

The supply voltamperage is

$$200 \times 2 = 400 \text{ VA}$$

and the power taken by the motor is

$$V_m I_a = 120 \times 2$$
$$= 240 \text{ W}$$

The power factor is thus

$$\frac{240}{400} = 0.6$$

17.3 CONTROLLERS FOR AC MOTORS

A list of the different types of ac motors and of the various techniques for speed and torque control of these drives is given in Fig. 17.13. A control scheme (in the form of a block diagram) for an induction motor is illustrated in Fig. 17.14, in which the power-conditioning unit (PCU) includes the energy source (which may even be a dc source), a means of producing an ac variable-frequency source (the inverter) from the available dc source, and some means of controlling the output voltage [the adjustable voltage inverter (AVI) or the pulse-width modulated (PWM) inverter]. In fact, it would be desirable to keep the voltage-to-frequency (V/f) ratio fixed, as we shall see later.

Inverters Basically, an inverter has a dc input and an ac output. We see from the preceding paragraph that the inverter is the backbone of an ac drive system. There is a wide variety of inverter circuits that may be used for a drive motor. Four common inverter types are:

1 AC transistor inverter
2 AC SCR McMurray inverter
3 AC SCR load-commutated inverter
4 AC SCR current inverter (ASCI)

For the present, however, we will consider the full-bridge inverter circuit shown in Fig. 17.15, which also shows the voltage and current waveforms. When T_1 and T_3 are conducting, the battery voltage appears across the load with the polarities shown in

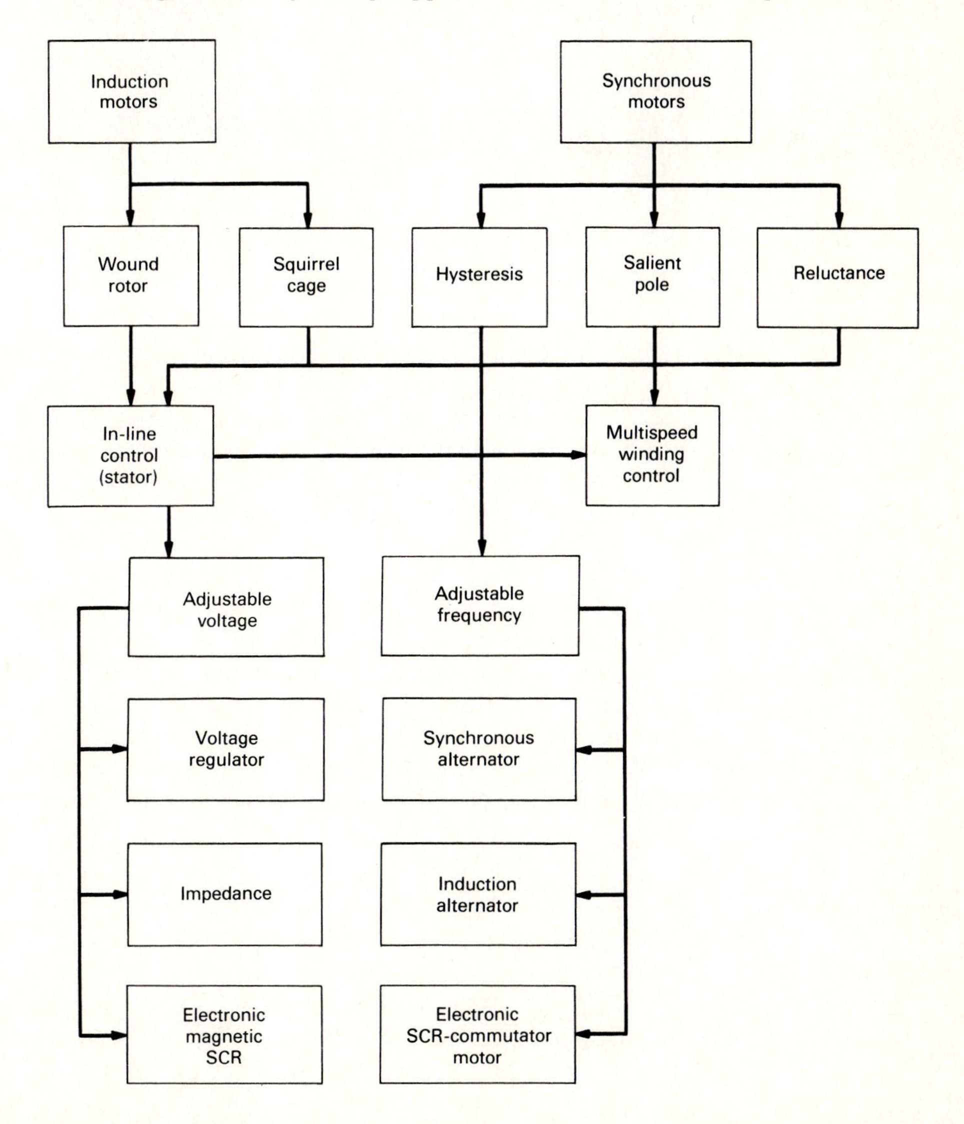

FIGURE 17.13 AC motors and controls.

FIGURE 17.14 Block diagram for induction motor control.

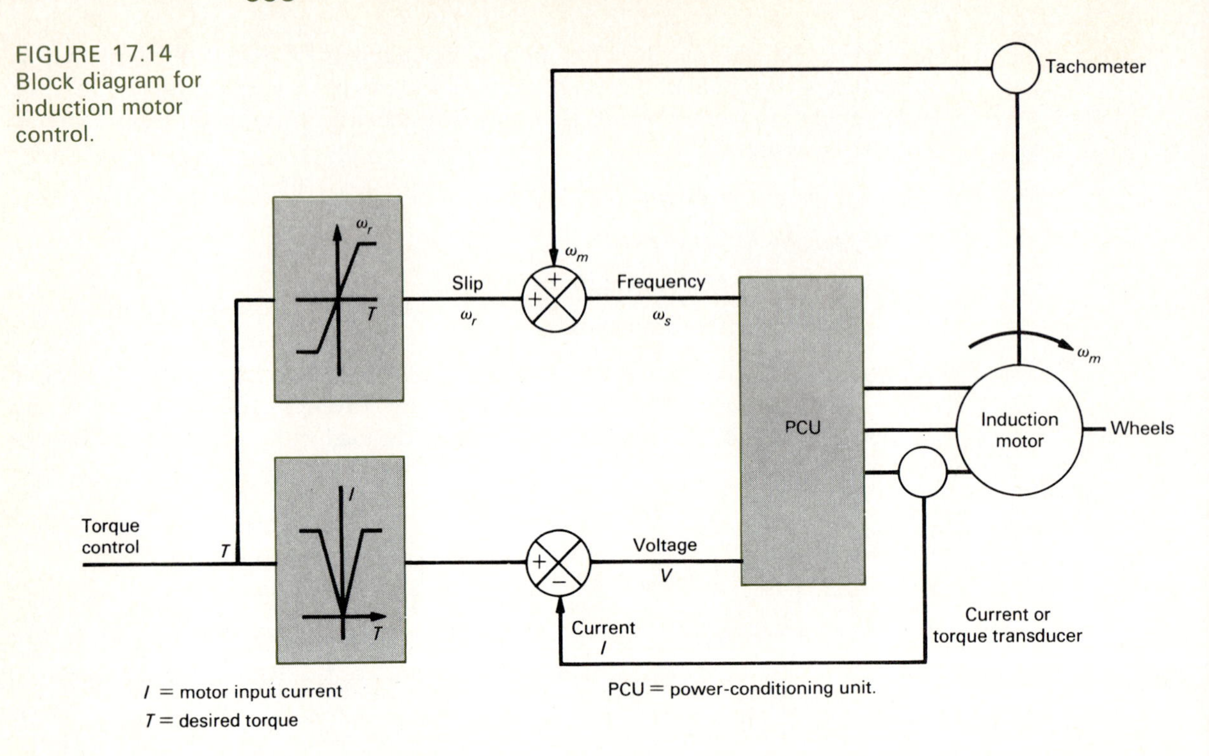

Fig. 17.15*b*. But when T_2 and T_4 are conducting, the polarities across the load are reversed. Thus, we get a square-wave voltage across the load, and the frequency of this wave can be varied by varying the frequency of the gating signals. If the load is not purely resistive, the load current will not reverse instantaneously with the voltage. The antiparallel-connected diodes, shown in Fig. 17.15*a*, allow for the load current to flow after the voltage reversal. The principle of the bridge inverter can be extended to form the three-phase bridge inverter.

Adjustable Voltage Inverter (AVI) In an AVI, the output voltage and frequency can both be varied. The voltage is controlled by including a chopper between the battery and the inverter, whereas the frequency is varied by the frequency of operation of the gating signals. In an AVI, the amplitude of the output decreases with the output frequency and the V/f ratio essentially remains constant over the entire operating range. The AVI output waveform does not contain second, third, fourth, sixth, eighth, and tenth harmonics. Other harmonic contents as a fraction of the total RMS output voltage are:

Fundamental: 0.965

Fifth harmonic: 0.1944

Seventh harmonic: 0.138

Eleventh harmonic: 0.087

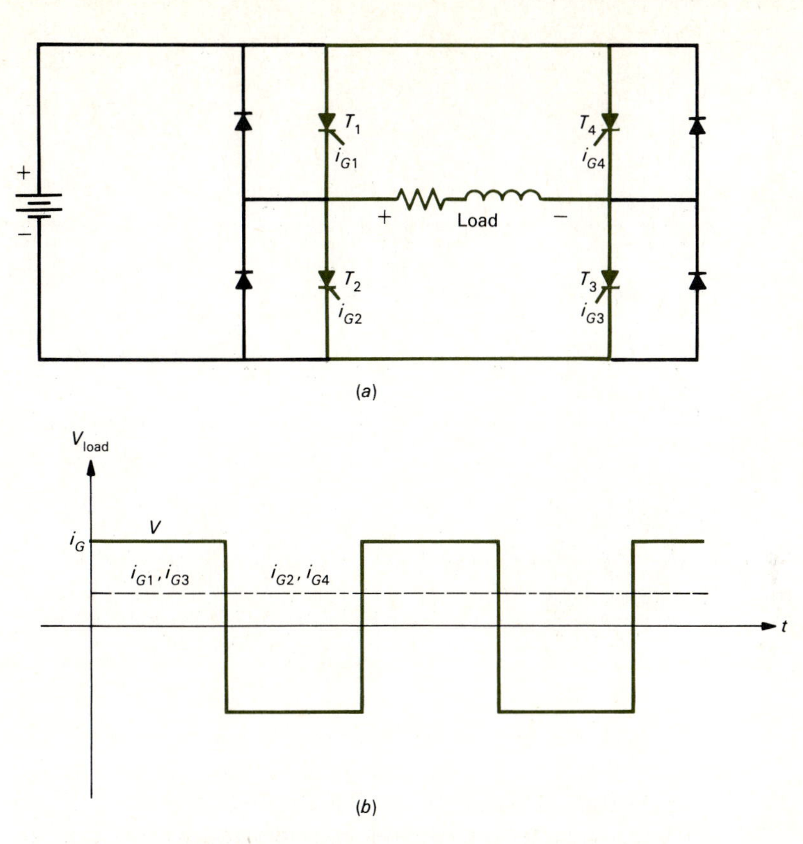

FIGURE 17.15 (*a*) Single-phase full-bridge inverter and (*b*) its load voltage waveform.

The losses produced by these harmonics tend to heat the motor, and thereby decrease the motor efficiency by about 5 percent compared to a motor driven by a purely sinusoidal voltage.

Pulse-Width-Modulated (PWM) and Pulse-Frequency-Modulated (PFM) Inverters The voltage control in PWM and PFM inverters is obtained in a manner similar to that for a chopper, as was shown in Fig. 17.8. In a PWM inverter, the

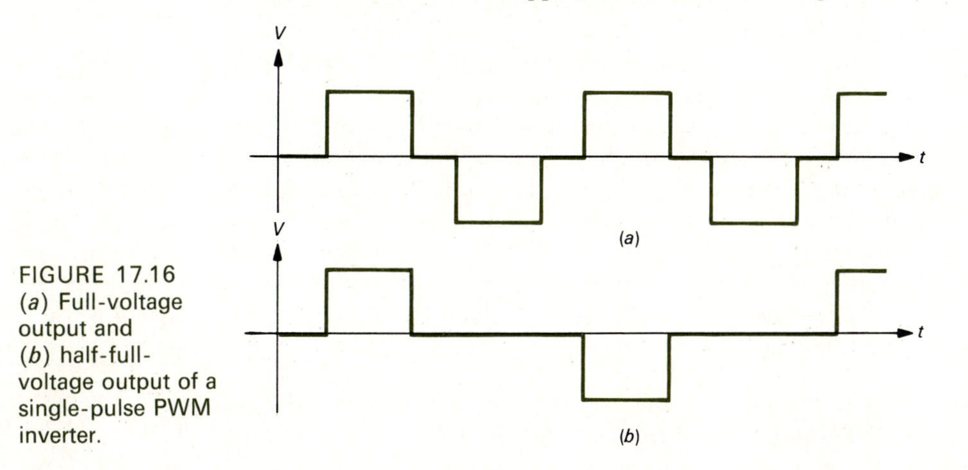

FIGURE 17.16 (*a*) Full-voltage output and (*b*) half-full-voltage output of a single-pulse PWM inverter.

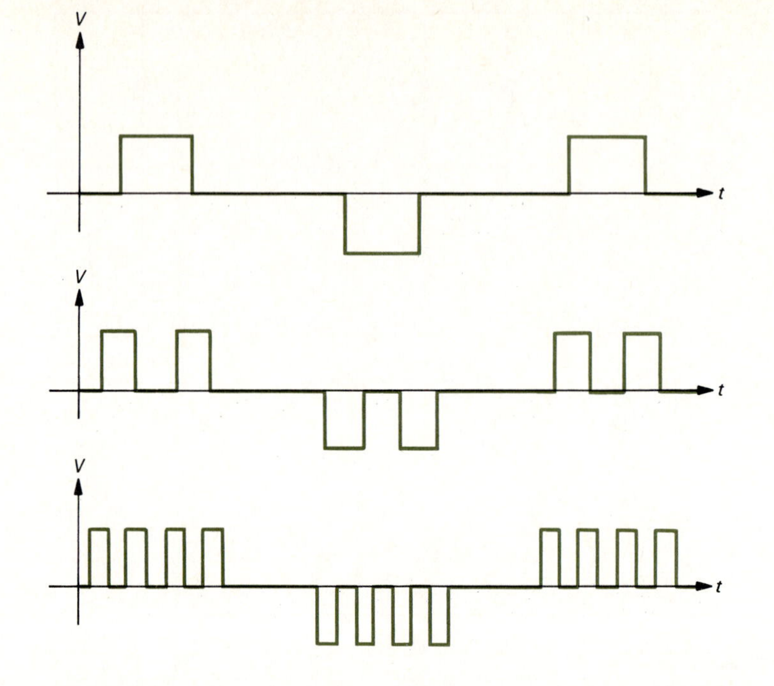

FIGURE 17.17 Outputs from a PFM inverter.

output-voltage amplitude is fixed and equal to the battery voltage. The voltage is varied by varying the width of the pulse "on time" relative to the fundamental half-cycle period, as illustrated in Figs. 17.16*a* and *b*. As the voltage is reduced (Fig. 17.16*b*), the harmonics vary rapidly in magnitude.

Low-frequency harmonics can be reduced by PFM, in which the number and width of pulses within the half-cycle period are varied. PFM waveforms for three different cases are given in Fig. 17.17. Harmonics from the output of a PFM inverter can be reduced by increasing the number of pulses per half cycle. But this requires a reduction in the pulse width, which is essentially limited by the thyristor turn-off time and the switching losses of the thyristor. Furthermore, an increase in the number of pulses increases the complexity of the logic system and thereby increases the overall cost of the PFM inverter system.

In comparing an AVI system with a PWM (or PFM) inverter system, we observe that whereas the AVI system is efficient but expensive, the PWM inverter is relatively inexpensive but inefficient. Both types (AVI and PFM) of inverter systems are suitable for induction motors. However, for the control of a synchronous motor, a thyristor inverter requires a motor voltage greater than the dc link voltage in order to turn off the thyristors. Also, no diodes are required in the ac portion of the circuit, thus avoiding uncontrolled currents. If a synchronous motor is controlled by a transistor inverter, then the motor internal voltage must be less than the battery voltage; otherwise, the diodes in the inverter circuit will conduct, resulting in uncontrollable currents and high losses.

Variable-Frequency Operation of Induction Motors The envelope illustrated in Fig. 17.18*a* shows the torque-speed characteristics of an induction motor for several frequencies. The dashed-line envelope defines two distinct operating regions of con-

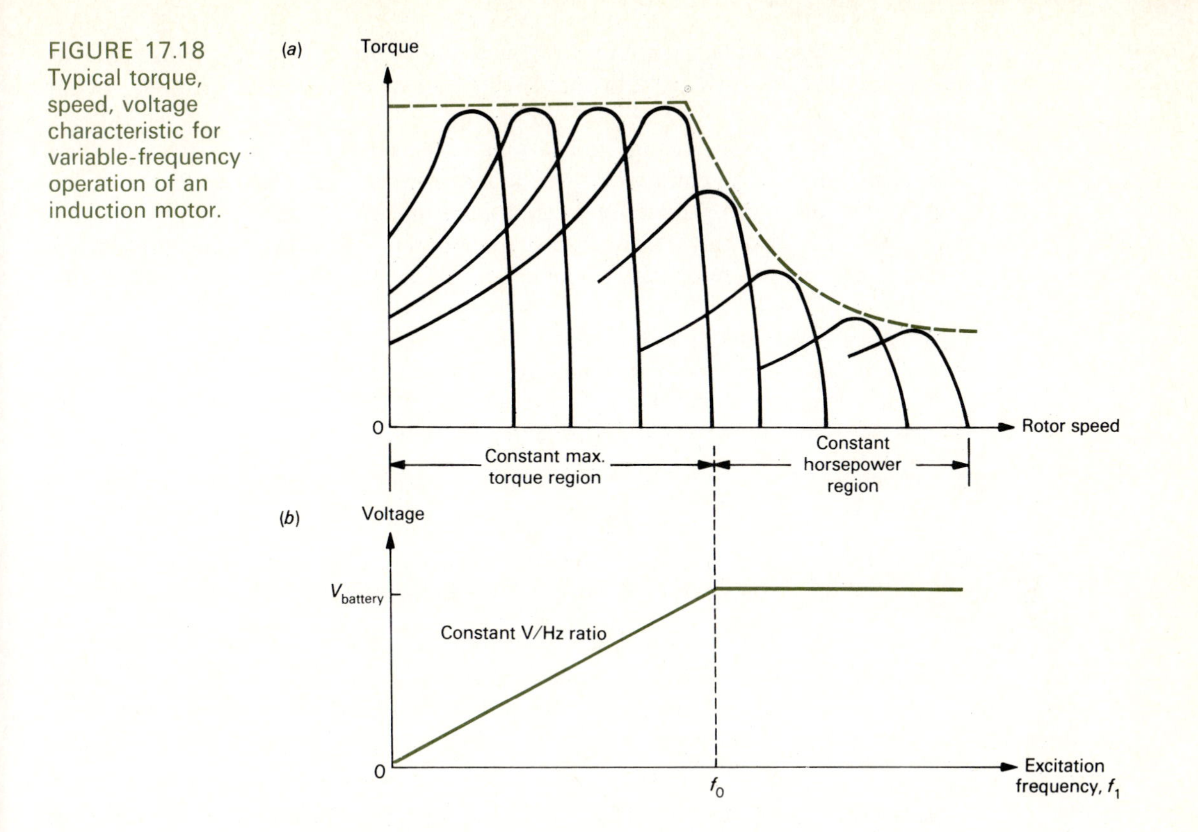

FIGURE 17.18 Typical torque, speed, voltage characteristic for variable-frequency operation of an induction motor.

stant maximum torque and constant horsepower. In the constant maximum torque region, the ratio of applied motor voltage to supply frequency (in volts per hertz) is held constant by increasing motor voltage directly with frequency, as shown in Fig. 17.18*b*.

The volts/hertz ratio defines the air-gap magnetic flux in the motor and can be held constant for constant torque within the excitation frequency range and corresponding rotor speeds, extending to frequency f_0. Beyond frequency f_0, the motor voltage cannot be increased to maintain a constant volts/hertz ratio due to the limitation of a finite dc voltage. For motor speeds in the range f_1 is greater than f_0, the supply voltage is held constant at the battery voltage, and the supply frequency is increased to provide the motor speed demand. This is the constant maximum horsepower region of operation, since the maximum developed torque decreases nonlinearly with speed.

The envelope illustrated in Fig. 17.18*a* characterizes induction motor operation in a typical electric vehicle drive system. Any speed-torque combination within the envelope can be provided by appropriate voltage and frequency control. For heavy loads, such as vehicle acceleration, the motor is operated in the constant torque region to provide the torque demanded. For high-speed operation at vehicle cruising speeds, the motor operates in the constant horsepower region at a frequency to satisfy the demanded speed. Voltage control is usually accomplished in the constant torque region by pulse-width (duty cycle) modulation. The motor operates with a fixed-voltage variable-frequency square wave in the constant horsepower region. Voltage and frequency control are used in both the driving and braking operating modes of the vehicle.

The inverter output voltage and current waveforms are rich in harmonics. These harmonics have detrimental effects on motor performance. Among the most important effects are the production of additional losses and harmonic torques. The additional losses that may occur in a cage induction motor owing to the harmonics in the input current are summarized below:

1 Primary I^2R losses: The harmonic currents contribute to the total RMS input current. Skin effect in the primary conductors may be neglected in small wire-wound machines, but it should be taken into account in motor analysis when the primary conductor depth is appreciable.

2 Secondary I^2R losses: When calculating the additional secondary I^2R losses, we must take skin effect into account for all sizes of motor.

3 Core losses due to harmonic main fluxes: These core losses occur at high frequencies, but the fluxes are highly damped by induced secondary currents.

4 Losses due to skew-leakage fluxes: These losses occur if there is a relative skew between the rotor and stator conductors. At 60 Hz the loss is usually small, but it may be appreciable at harmonic frequencies. Since the time-harmonic mmf's rotate relative to both the primary and the secondary, skew-leakage losses are produced in both members.

5 Losses due to end-leakage fluxes: As in the case of skew-leakage losses, these losses occur in the end regions of both the primary and the secondary and are a function of harmonic frequency.

6 Space-harmonic mmf losses excited by time-harmonic currents: These correspond to the losses that, in the case of the fundamental current component, are termed high-frequency stray-load losses.

In addition to these losses, and harmonic torques, the harmonics act as sources of magnetic noise in the motor.

PROBLEMS

17.1 A dc motor is connected to a 96-V battery through a chopper. If the duty cycle is 45 percent, what is the average voltage applied to the motor?

17.2 A chopper is on for 20 ms and off for the next 50 s. Determine the duty cycle.

17.3 A pulse-width-modulated output waveform from a chopper is shown in Fig. P17.3. This chopper is connected to a dc motor having an armature resistance of 0.25 Ω and running at 350 r/min. If the motor back-emf constant is 0.12 V/(r · min), calculate the average armature current.

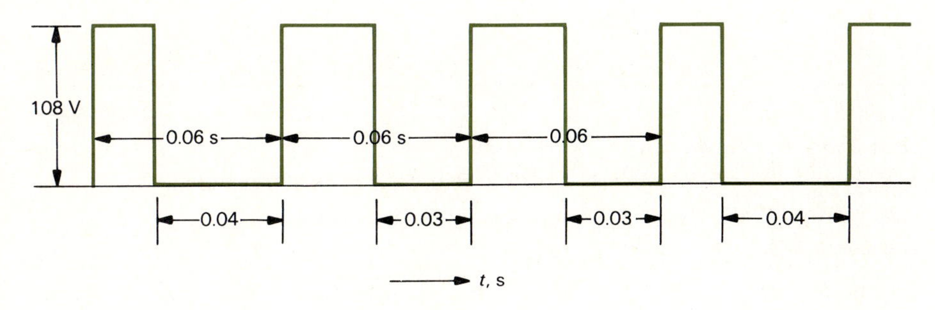

P17.3

17.4 Repeat Prob. 17.3 if the motor is fed from the frequency-modulated voltage waveform shown in Fig. P17.4.

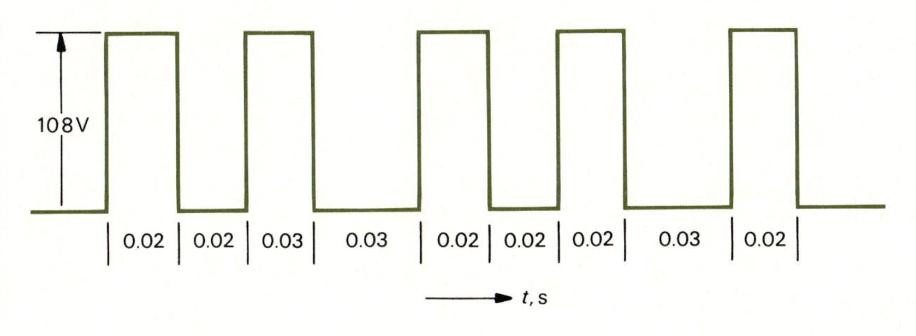

P17.4

17.5 A chopper having the following data drives a dc motor: supply voltage = 400 V, duty cycle = 50 percent, and armature circuit inductance = 20 mH. If the chopper frequency is 270 pulses per second, what is the change in the armature current?

17.6 For the motor of Prob. 17.5, if the allowable change in the armature current is 5 A, determine the chopper frequency.

17.7 If the chopper of Prob. 17.5 is of the form of Fig. 17.10 and can carry a 40-A current, what are the values of the parameters C, L, and L_2? Assume $L_1 = L_2$.

17.8 A 220-V dc shunt motor runs at 750 r/min when supplied from a dc source. When supplied by a half-wave converter at 110 V 60 Hz ac, the motor draws a 1.8-A average armature current. During the conduction period, the average motor voltage is 60 V. The motor armature resistance is 6.4 Ω and the electromechanical energy conversion constant is 3.5 V/(rad · s) (or N · m/A). Calculate the motor torque and speed.

17.9 A six-pole 480-V 60-Hz induction motor is driven by an inverter. At 4 percent slip the motor develops a 320-N · m torque at rated voltage and frequency. Determine the motor speed (at 4 percent slip) if the motor voltage is reduced to 240 V and the frequency to 30 Hz. Also estimate the developed torque.

REFERENCES

17.1 S. A. Nasar and L. E. Unnewehr, *Electromechanics and Electric Machines*, 2d ed., Wiley, New York, 1983.

17.2 S. A. Nasar, *Electric Machines and Transformers*, Macmillan, New York, 1983.

17.3 P. C. Sen, *Thyristor DC Drives*, Wiley, New York, 1981.

17.4 S. B. Dewan, G. R. Slemon, and A. Straughen, *Power Semiconductor Drives*, Wiley, New York, 1984.

Chapter

Electric Power Systems

In the preceding chapters, we have discussed various types of electric machines, which are electromechanical energy converters; that is, electric machines converting electrical energy into mechanical form (as in motors) and vice versa (as in generators). We treated these machines as isolated entities. In practice power generating stations all over the country are interconnected with each other. Such an interconnection of generators, transformers, transmission lines, and diverse other components constitutes an *electric power system*, commonly known as power system.

The concept of *central station* electric power service was first applied at Thomas Edison's Pearl Street Station in New York City in 1882, when the first distribution line was strung along a few city blocks to provide lights in a few homes. Since that time, the electric power industry in the United States has grown at a phenomenal rate, so that it now provides the main driving force of the greatest industrial nation on earth. All of this has taken place in the equivalent of one human life span. The high standard of living enjoyed in this country is closely tied to electric labor-saving appliances and tools and to the high productivity made possible by the electric machinery of industry.

Central station service, as opposed to individual generators in each home, possessed all of the inherent advantages to make it a huge success. Convenience, economies of scale, and relative continuity all combined to promote the growth of Edison's idea at a rapid rate. The first limiting factors encountered were the voltage drop and the resistance losses on the low-voltage dc distribution circuits. Because the problem of voltage regulation became more pronounced out toward the end of the line, the distance of the customer from the generating station was severely limited. The solution to these problems came with the introduction of the *ac transformer* by George Westinghouse in 1885. An ac distribution circuit was placed in service at Great Barrington, Massachusetts, by William Stanley in 1886, proving the feasibility of the technique. Power could be generated at low voltage levels by the simpler ac generator and stepped up to higher voltage for sending over long distances. Because the current required to deliver power at a given voltage is inversely proportional to the voltage, the current requirements and consequently the conductor size could be kept within practical limits and still deliver large amounts of power to distant areas. The first ac transmission line was put into service at Portland, Oregon, in 1890, carrying power 13 miles from a hydrogenerating station on the Willamette River.

During the next decade, two-phase and three-phase motors and generators were developed and were demonstrated to be superior to single-phase machines from the standpoints of size, weight, and efficiency. Three-phase transmission was shown to possess inherent advantages over single-phase transmission in terms of conductor requirements and losses for a given power need. Consequently, by the turn of the century it was apparent that three-phase ac transmission systems would become standard. Three frequencies—25, 50, and 60-Hz—battled for dominance, and it was several decades before the conversion to 60 Hz was complete.

18.1 COMPONENTS OF A POWER SYSTEM

The major electrical components of a power system are as follows:

1 *Generators* These are generally synchronous generators, already discussed in Chap. 15.

2 *Transformers* These are power transmission and distribution transformers, discussed in Chap. 12. The transformer at the generator end is used to step the voltage up to a much higher level than generated so that power can be transmitted up to hundreds of kilometers while the size of the transmission line conductors and the corresponding losses are kept down within practical limits. Physically, a power transformer bank may consist of three single-phase transformers with electrical connections external to the cases, or it may consist of a three-phase unit contained in a single tank, as shown in Fig. 18.1. For economic reasons, the latter is predominant in the larger power ratings. The windings of power transformers are usually immersed in a special-purpose oil for insulation and cooling.

3 *Transmission lines* A transmission line usually consists of three conductors (generally stranded or bundled) and one or more *neutral* conductors, although it is possible sometimes to omit the neutral conductor since it carries only the unbalanced return portion of the line current. A three-phase circuit with perfectly balanced phase currents has no neutral return current at all. In most instances the current is accurately balanced among the phases at transmission voltages, so that the neutral conductor may be much smaller. In some locations the soil conditions permit an effective neutral return current through the earth. The neutral conductors have another equally important function; they are installed above the phase conductors and provide an effective electrostatic shield against lightning (Fig. 18.2).

Manufacturers of high-voltage equipment tend to standardize as much as possible on a few *nominal voltage classes.* The most common transmission voltage classes in use with the United States are 138, 230, 345, 500, and 765 kV (phase-to-

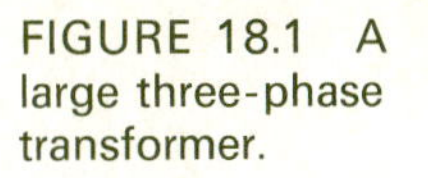
FIGURE 18.1 A large three-phase transformer.

phase). Development work is being done for utilizing voltages up to 2000 kV. Costs of line construction, switchgear, and transformers rise exponentially with voltage, leading to the use of the lowest voltage class capable of carrying the anticipated load.

4 *Circuit breakers* Circuit breakers are large three-pole switches located at each end of every transmission line section and on either side of large transformers. The primary function of a circuit breaker is to open under the control of automatic *protective relays* in the event of a fault or short circuit in the protected equipment. The relays indicate the severity and probable location of the fault; they may contain sufficient electromechanical or solid-state logic circuitry to decide whether the line or transformer could be reenergized safely and to initiate reclosure of its circuit breakers. In very critical extra high-voltage (EHV) switchyards or sub-

FIGURE 18.2 A three-phase transmission line with neutral conductors.

stations, a small digital computer may be used to analyze fault conditions and perform logical control functions. If the fault is a transient one from which the system may recover, such as a lightning stroke, the integrity of the network is best served by restoring equipment to service automatically, preferably within a few cycles. If the fault is persistent, such as a conductor on the ground, the relays and circuit breakers will isolate the faulted section and allow the remainder of the system to continue in normal operation. The secondary function of a circuit breaker is that of a switch to be operated manually by a local or remote operator to deenergize an element of the network for maintenance. When a circuit breaker's contacts open under load, there is a strong tendency to arc across the contact gap as it separates. Various methods are used to suppress the arc, including submersion of the contact mechanism in oil or in a gas such as sulfur hexafluoride (SF_6). The highest voltage classes use a powerful air blast to quench the arc in open air, using several interrupters or contact sets in series for each phase (Fig. 18.3).

5 *Voltage regulators* When electric power has been transmitted into the area where it is to be used, it is necessary to transform it back down to a voltage which can be utilized locally. The step-down transformer bank may be very similar to the step-up bank at the generating station, but of a size to fill the needs only of the immediate area. In order to provide a constant voltage to the customer, a *voltage regulator* is usually connected to the output side of the step-down transformer. It is a special type of 1:1 transformer, with several discrete taps of a fractional percent each over a voltage range of ± 10 percent. A voltage-sensing device and an

FIGURE 18.3 An air-blast circuit breaker.

automatic control circuit will position the tap contacts automatically to compensate the low side voltage for variations in transmission voltage. In many cases the same effect is accomplished by incorporating the regulator and its control circuitry into the step-down transformer, resulting in a combination device called a *load tap-changer* (LTC).

6 *Subtransmission* Some systems have certain intermediate voltage classes known as subtransmission. At the time it was installed, such a voltage class was probably considered to be transmission, but with rapid system growth and a subsequent overlay of higher-voltage transmission circuits, the earlier lines were tapped at intervals to serve more load centers and become local feeders. In most systems 23, 34.5, and 69 kV are considered to be subtransmission, and on some larger systems 138 kV may also be included in that category, depending upon the application. As the frontiers of higher voltages are pushed back inexorably, succeedingly higher voltage classes may be relegated to subtransmission service.

7 *Distribution systems* A low-voltage distribution system is necessary for the practical distribution of power to numerous customers in a local area. A distribution system resembles a transmission system in miniature—having lines, circuit breakers, and transformers, but on a much smaller scale of voltage and power. The electrical theory and analytical methods are identical for both, since the distinction is purely arbitrary. Distribution voltages range from 2.3 to 35 kV, with 12.5 and 14.14 kV predominant. Such voltage levels are sometimes referred to as *primary* voltages with reference to the 240/120 V at which most customers are served. Single-phase distribution circuits are supplied from three-phase transformer banks, balancing the total load on each phase as nearly as possible. Three-phase distribution circuits are erected only to serve large industrial or motor loads. The transformer which ultimately steps the voltage down from the distribution level to the customer service level may be mounted on a pole for overhead distribution systems or mounted on a pad or in a vault for underground distribution. Such transformers are usually protected by *fuses* or *fused cutouts.*

8 *Loads* Countless volumes have been written about the systems and techniques necessary for the protection and delivery of electric power, but very little has been written about loads, for which all the other components exist. Perhaps the main reason is that loads are so varied in nature that they defy comprehensive classification. Simply stated, any device which utilizes electric power can be said to impose a load on the system. Viewed from the source, all loads can be classed as resistive, inductive, capacitive, or some combination of these. Loads may also be time-variant, ranging from a slow random swing to the rapid cyclic pulses which cause a distracting flicker in the lights of nearby customers. The composite load on a system has a predominant resistive component and a small net inductive component. Inductive loads (such as induction motors) are far more prevalent than capacitive loads. Consequently, in order to keep the resultant current as small as possible, *capacitors* are usually installed in quantities adequate to balance most of the inductive current. It has been found from experience that the power consumed by the composite load on a power system varies with system frequency. This effect is imperceptible to the customer in the range of normal operating frequencies (± 0.02 Hz) but can make an important contribution to the control of systems operating in synchronism. System load also varies through daily and annual cycles, creating difficult operating problems.

9 *Capacitors* When applied on a power system for the reduction of inductive current (power-factor correction), capacitors can be grouped into either a transmission or a distribution class. In either case, for maximum effectiveness they should be installed electrically as near to the load as possible. When properly applied, capacitors balance out most of the inductive component of current to the load, leaving essentially a unity power-factor load. The result is a reduction in the size of the conductor required to serve a given load and a reduction in I^2R losses. *Static capacitors* may be used at any voltage, but practical considerations impose an upper limit of a few kilovolts per unit; therefore, high-voltage banks must be composed of many units connected in series and parallel. High-capacity transmission capacitor banks should be protected by a high-side circuit breaker and its associated protective relays. Small distribution capacitors may be vault or pole-top mounted and protected by fuses.

Industrial loads occasionally require very large amounts of power-factor correction, varying with time and the industrial process cycle. The *synchronous condenser* (an overexcited synchronous motor running light) is ideally suited to such an application. Its contribution of either capacitive or inductive current can be controlled very rapidly over a wide range by using automatic controls to vary the excitation current. Physically, it is very similar to a synchronous generator operating at a leading power factor, except that it has no prime mover. The synchronous condenser is started as a motor and has its losses supplied by the system to which it supplies capacitive current.

18.2 REPRESENTATION OF AN ELECTRIC POWER SYSTEM

In the preceding section we have identified the basic components of an electric power system. We now consider the representation of these components interconnected to constitute a power system. First, we will review the graphical representation and the one-line diagram of a power system. This will be followed by the impedance diagrams obtained from the most commonly used equivalent circuits of the components. Finally, because the components have different voltage and kVA ratings, we will introduce the per unit quantities as a common basis for analyzing the interconnected components and systems.

18.2.1 Graphical Representation and the One-Line Diagram

Figure 18.4 shows the symbols used to represent the typical components of a power system. Using these symbols, Fig. 18.5 shows a one-line diagram of a system consisting of two generating stations interconnected by a transmission line. Even from such an elementary network as this, it is easy to imagine the confusion which would result in making diagrams showing all three phases. The advantage of such a one-line representation is rather obvious in that a complicated system can be represented concisely. A concerted effort is made to keep the currents equal in each phase; consequently, on a balanced system, one phase can represent all three by proper mathematical treatment. An impedance or reactance diagram can also be conveniently developed from the one-line diagram as shown in Sec. 18.2.2. A further advantage of the one-line diagram is in power-flow studies. The one-line diagram becomes rather second nature to the power system engineer.

18.2.2 Equivalent Circuits and Reactance Diagrams

Equivalent circuits of components—generators, transformers, transmission lines, and loads—can be interconnected to obtain a circuit representation for the entire system. In other words, the one-line diagram may be replaced by an *impedance diagram* or a *reactance diagram* (if resistances are neglected). Thus, corresponding to Fig. 18.5, the impedance and reactance diagrams are shown in Fig. 18.6*a* and *b* on a per phase basis.

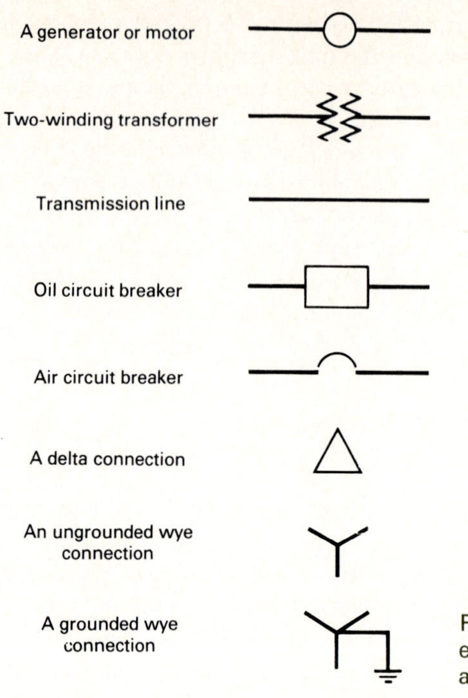

FIGURE 18.4 Symbolic representation of elements of a power system. Other legends are also used.

In the equivalent circuits of the components in Fig. 18.6*a*, we have made the following assumptions: (1) A generator can be represented by a voltage source in series with an inductive reactance; the internal resistance of the generator is negligible as compared to the reactance. (2) The loads are inductive. (3) The transformer core is assumed to be ideal and can be represented by a reactance. (4) The transmission line is of medium length and can be denoted by a T circuit (as we will see in Sec. 18.3.2). (5) The Δ/Y-connected transformer T_1 is replaced by an equivalent Y/Y-connected transformer (by a Δ-to-Y transformation) so that the impedance diagram may be drawn on a per phase basis.

The reactance diagram (Fig. 18.6*b*) is drawn by neglecting all the resistances, loads, and capacitances of the transmission line. The reactance diagram is generally used for short-circuit calculations, whereas the impedance diagram is used for load-flow studies.

18.2.3 Per Unit Representation

When making computations on a power system network having two or more voltage classes, it is very cumbersome to convert currents to a different voltage level at each point where they flow through a transformer, the change in current being inversely proportional to the transformer turns ratio. A much simplified system has been devised whereby a set of *base quantities* is assumed for each voltage class, and each parameter is expressed as a decimal fraction of its respective base.

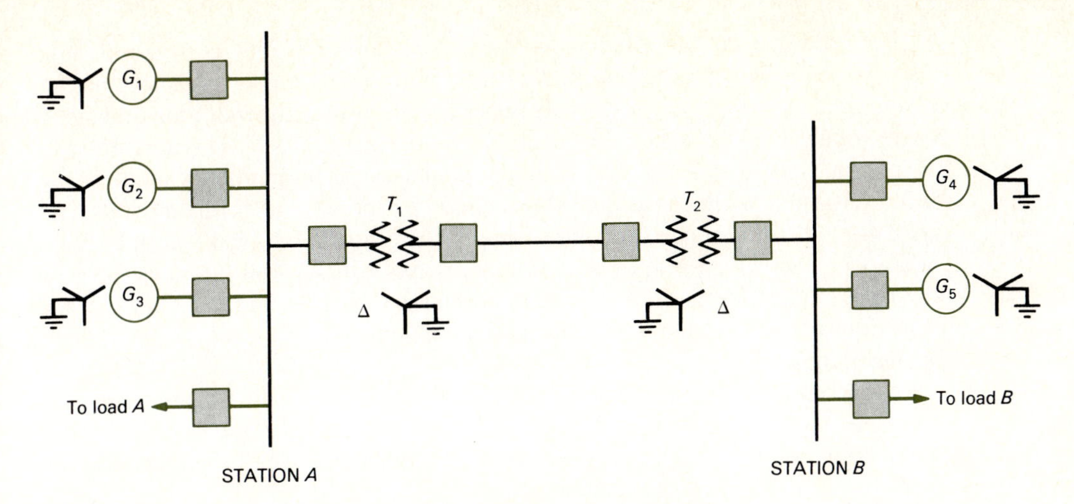

FIGURE 18.5 One-line diagram of a portion of a power system.

FIGURE 18.6 (*a*) Impedance diagram for Fig. 18.5. (*b*) Corresponding reactance diagram, neglecting loads and resistances.

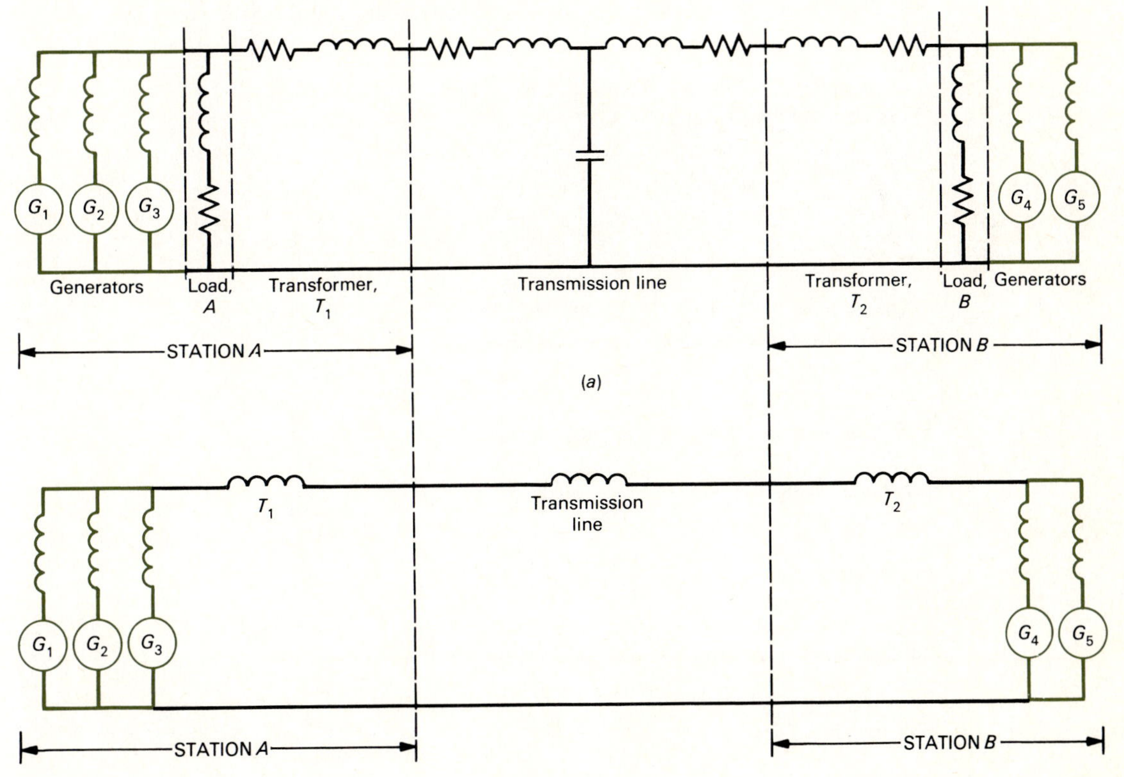

For convenience, base quantities are chosen such that they correspond rather closely to the range of values normally expected in the parameter. For instance, if a voltage base of 345 kV is chosen and under certain operating conditions the actual system voltage is 334 kV, the ratio of actual to base voltage is 0.97. This is expressed as 0.97 *per unit* (*pu*) volts. An equally common practice is to automatically multiply each result by 100, under which the preceding would be expressed as 97 *percent* volts. With experience, this technique can give an excellent grasp or "feel" of the system conditions, whether normal or abnormal. It works equally well for single-phase or polyphase circuits. Per unit or percent quantities and their bases must obey the same relationships (such as Ohm's law and Kirchhoff's laws) to each other as with other systems of units.

A minimum of four base quantities is required to completely define a per unit system: voltage, current, power, and impedance (or admittance). If two of these are chosen arbitrarily (voltage and power, for example), then the others are fixed, as illustrated by the following example.

Example 18.1 A 10-kVA 0.8 lagging power-factor (pf) 1800-V load is connected to a generator by a transmission line having an impedance of $(12 + j20)\ \Omega$. Assume 10 kVA and 2000 V as base quantities, and calculate the per unit voltage at the load, the per unit current, and the per unit impedance of the line.

The following table identifies the per unit (pu) values and the base quantities:

Quantities	pu value	Unit value or base quantity
Voltamperage	1.0	10 kVA (assumed)
Voltage	1.0	2000 V (assumed)
Current	1.0	10,000/2000 = 5 A
Impedance	1.0	2000/5 = 400 Ω
Admittance	1.0	5/2000 = 0.0025 Ω

Solution Using these base values, we have

pu voltage at the load: $\dfrac{1800}{2000} = 0.9$

pu voltamperage: $\dfrac{10{,}000}{10{,}000} = 1.0$

pu impedance of the line: $\dfrac{12 + j20}{400} = 0.03 + j0.05$

pu current: $\dfrac{1.0}{0.9} = 1.111$

From the preceding example the following general equations may be written:

$$\text{Base current} = \frac{\text{base voltampere}}{\text{base voltage}} \quad \text{A} \tag{18.1}$$

$$\text{Base impedance} = \frac{\text{base voltage}}{\text{base current}} \quad \Omega \tag{18.2}$$

$$\text{Per unit voltage} = \frac{\text{actual voltage}}{\text{base voltage}} \quad \text{pu} \tag{18.3}$$

$$\text{Per unit current} = \frac{\text{actual current}}{\text{base current}} \quad \text{pu} \tag{18.4}$$

$$\text{Per unit impedance} = \frac{\text{actual impedance}}{\text{base impedance}} \quad \text{pu} \tag{18.5}$$

We have stated earlier that a power system consists of generators, transformers, transmission lines, and loads. The per unit impedances of generators and transformers, as supplied from tests by manufacturers, are generally based on their own ratings. However, these per-unit values could be referred to a new voltampere base according to the following equation:

$$(\text{pu impedance})_{\text{new base}} = \frac{\text{new base voltamperage}}{\text{old base voltamperage}} (\text{pu impedance})_{\text{old base}} \tag{18.6}$$

The impedances of transmission lines are expressed in ohms, which can be easily converted to the pu value on a given voltamperage base.

Example 18.2 A one-line diagram of a portion of a power system is shown in Fig. 18.7. The transformer bank consists of three transformers, each rated at $\frac{1}{3}$ of 10 MVA at 66/13.2 kV. Notice that a transformer must have two voltage bases. The impedance of each transformer is $(0.01 + j0.05)$ pu on its own base. If the load on the transformer is 8 MVA 1.0 pf at 13 kV, calculate the input voltage and current to the transformer. Neglect no-load losses.

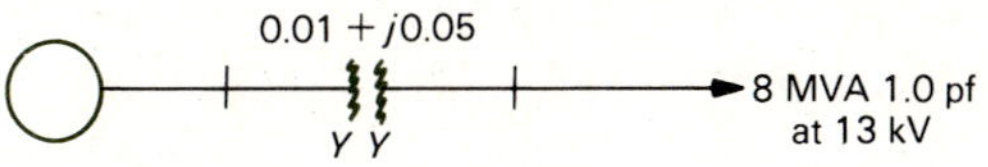

FIGURE 18.7 Example 18.2

Solution

$$\text{Base voltamperage} = 10{,}000 \text{ kVA}$$

$$\text{Base voltage} = \frac{66}{13.2} \quad \text{kV}$$

$$\text{Base current} = \frac{87.5}{437.5} \quad \text{A}$$

$$\text{pu output voltage} = \frac{13}{13.2} = 0.985$$

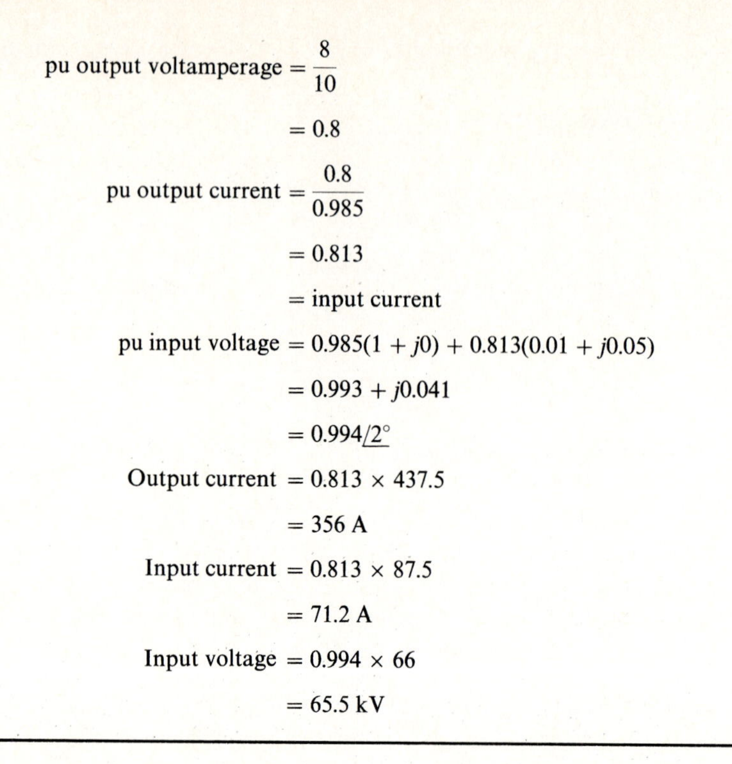

$$\text{pu output voltamperage} = \frac{8}{10} = 0.8$$

$$\text{pu output current} = \frac{0.8}{0.985} = 0.813 = \text{input current}$$

$$\text{pu input voltage} = 0.985(1 + j0) + 0.813(0.01 + j0.05) = 0.993 + j0.041 = 0.994\underline{/2^\circ}$$

$$\text{Output current} = 0.813 \times 437.5 = 356 \text{ A}$$

$$\text{Input current} = 0.813 \times 87.5 = 71.2 \text{ A}$$

$$\text{Input voltage} = 0.994 \times 66 = 65.5 \text{ kV}$$

The per unit system gives us a better feeling of the relative magnitudes of various quantities, such as voltage, current, power, and impedance. We will have a better appreciation of the usefulness of the per unit concept later, particularly in fault calculations (Sec. 18.5). But now we will proceed to study transmission lines, the final major component of an electric power system.

18.3 TRANSMISSION OF ELECTRIC POWER

Transmission lines physically integrate the output of generating plants and the requirements of customers by providing pathways for the flow of electric power between various circuits in the system. These circuits include those between generating units, between utilities, and between generating units and substations (or load centers).

For our purposes, we will consider a transmission line to have a sending end and a receiving end, and we will consider its primary parameters to be a series resistance and inductance and a shunt capacitance and conductance. We will classify transmission lines as short, medium, and long. In a short line, the shunt effects (conductance and capacitance) are negligible; often, this approximation is valid for lines up to 80 kilometers (km) long. In a medium line, the shunt capacitances are lumped at a few predetermined locations along the line; a medium line may be anywhere between 80 and 240 km long. Lines longer than 240 km are considered to be long lines which are represented by (uniformly) distributed parameters.

The operating voltage of a transmission line often depends upon the length of the line. Some of the common operating voltages for transmission lines are 115, 138, 230, 345, 500, and 765 kV. It is estimated that by 1990 the United States will have a total length of 76,974 km [46,184.4 miles (mi)] of transmission lines, of which the 345-kV line 27,320 km (16,391.9 mi) long will be predominant.

Most transmission lines operate on ac, but with the advent of high-power static (solid-state) conversion equipment, high-voltage dc transmission lines are being seriously considered for long distances [exceeding 600 km (375 mi)].

18.3.1 Transmission Line Parameters

The first basic parameter in a transmission line is the resistance of the conductors constituting the line. Resistance is the cause of I^2R loss in the line and also results in an IR-type voltage drop. The dc resistance R of a conductor of length l and of cross-sectional area A is given by

$$R = \rho \frac{l}{A} \quad \Omega \tag{18.7}$$

where ρ is the *resistivity* of the conductor in ohm-meters ($\Omega \cdot m$). The dc resistance is affected only by the operating temperature of the conductor, linearly increasing with the temperature. When the line is operating on ac, however, the current-density distribution across the conductor cross section becomes nonuniform, and is a function of the ac frequency; this is the phenomenon known as skin effect. As a consequence of skin effect, the ac resistance is higher than the dc resistance. At approximately 60 Hz, the ac resistance of a transmission line conductor may be 15 to 25 percent higher than its dc resistance.

The temperature dependence of a resistance is given by

$$R_2 = R_1[1 + \alpha(T_2 - T_1)] \tag{18.8}$$

where R_1 and R_2 are the resistances at temperatures T_1 and T_2, respectively, and where α is defined as the temperature coefficient of resistance. The resistivities and temperature coefficients of certain materials are given in Table 18.1.

TABLE 18.1 Values of ρ and α

Material	Resistivity ρ at 20°C, $\mu\Omega$·cm	Temperature coefficient α at 20°C
Aluminum	2.83	0.0039
Brass	6.4–8.4	0.0020
Copper		
Hard-drawn	1.77	0.00382
Annealed	1.72	0.00393
Iron	10.0	0.0050
Silver	1.59	0.0038
Steel	12–88	0.001–0.005

In practice, transmission line conductors consist of a stranded steel core for mechanical strength. The core is then surrounded by aluminum. Such a conductor is designated as ACSR (aluminum conductor steel reinforced). Since it is difficult to calculate the exact ac resistance of such a conductor, tabulated values of the resistances of various types of stranded conductors are available in the literature.

The next basic parameter is the transmission line inductance. Considering the two-wire single-phase line first, it can be shown that its inductance is given by

$$L = \frac{\mu_o}{4\pi}\left(1 + 4\ln\frac{d-r}{r}\right) \qquad \text{H/m} \tag{18.9}$$

where $\mu_o = 4\pi \times 10^{-7}$ H/m (the permeability of free space), d is the distance between the centers of the conductors, and r is the radius of the conductors. For a three-phase line, it can further be shown that the per phase, or line-to-neutral, inductance of a line with equilaterally spaced conductors is

$$L = \frac{\mu_o}{8\pi}\left(1 + 4\ln\frac{d-r}{r}\right) \qquad \text{H/m} \tag{18.10}$$

where the symbols are the same as in Eq. (18.9). In practice, the three conductors of a three-phase line are seldom spaced equilaterally. Such an unsymmetrical spacing results in unequal inductances in the three phases, leading to unequal voltage drops and an imbalance in the line. To offset this difficulty, the positions of the conductors are interchanged at regular intervals along the line. This practice is known as *transposition* and is illustrated in Fig. 18.8, which also shows the unequal spacings between the conductors. The average per phase inductance for such a case is still given by Eq. (18.10), except that the spacing d in Eq. (18.10) is replaced by the equivalent spacing d_e obtained from

$$d_e = (d_{ab} d_{bc} d_{ca})^{1/3} \tag{18.11}$$

where the distances d_{ab}, etc., are shown in Fig. 18.8.

The final basic parameter is the capacitance of the transmission line. The capacitance per unit length of a single-phase two-wire line is given by

$$C = \frac{\pi\varepsilon_o}{\ln[(d-r)/r]} \qquad \text{F/m} \tag{18.12}$$

FIGURE 18.8 Transposition of unequally spaced three-phase transmission line conductors.

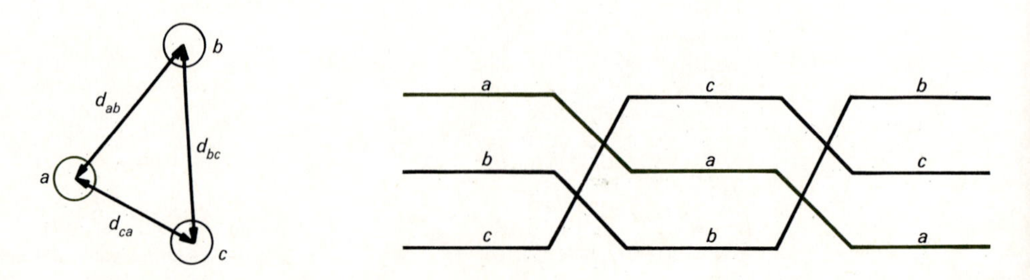

where ε_o is the permittivity of free space and the other symbols mean the same as in Eq. (18.9). For a three-phase line having equilaterally spaced conductors, the per phase (or line-to-neutral) capacitance is

$$C = \frac{2\pi\varepsilon_o}{\ln\,[(d - r)/r]} \tag{18.13}$$

For unequal spacings between the conductors, d in Eq. (18.13) is replaced by a d_e of Eq. (18.11), as was done for the case of inductance.

Example 18.3 Determine the resistance of a 10-km-long solid cylindrical aluminum conductor, having a diameter of 250 mils, at 20°C and at 120°C.

Solution Since 250 mils = 0.25 in (0.635 cm), the area of the cross section = $(\pi/4)(0.635)^2 = 0.317\ \text{cm}^2$. From Table 18.1, $\rho = 2.83\ \mu\Omega\cdot\text{cm}$ at 20°C. Therefore, from Eq. (18.7), at 20°C

$$R_{20} = 2.83 \times 10^{-6} \times \frac{10 \times 10^3}{0.317}$$

$$= 0.0893\ \Omega$$

From Eq. (18.8) and Table 18.1, at 120°C we obtain

$$R_{120} = R_{20}[1 + \alpha(120 - 20)]$$

$$= 0.0893(1 + 0.0039 \times 100)$$

$$= 0.124\ \Omega$$

Example 18.4 A single-circuit three-phase 60-Hz transmission line consists of three conductors arranged as shown in Fig. 18.9. If the conductors are the same as in Example 18.3, find the inductive and capacitive reactances per kilometer per phase.

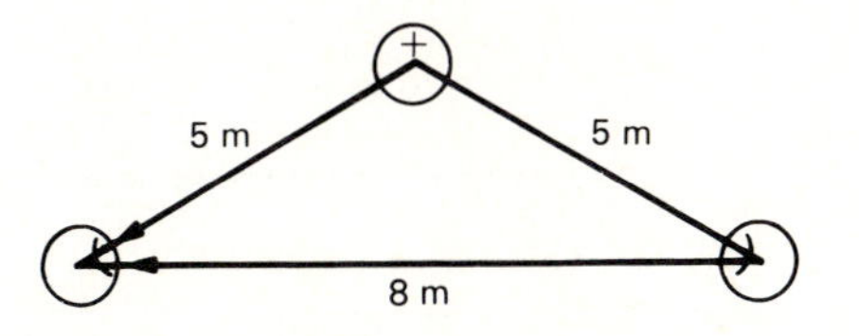

FIGURE 18.9
Example 18.4

Solution From Eq. (18.11)

$$d_e = (5 \times 5 \times 8)^{1/3}$$

$$= 5.848\ \text{m}$$

From Example 18.3

$$r = \tfrac{1}{2} \times 0.635 \times 10^{-2}\ \text{m}$$

Thus

$$\frac{d_e - r}{r} \simeq \frac{d_e}{r} = \frac{5.858 \times 2 \times 10^2}{0.635}$$

$$= 1841.9$$

and

$$\ln[(d_e - r)/r] = 7.52$$

Hence, from Eq. (18.10) (with $\mu_o = 4\pi \times 10^{-7}$ H/m)

$$L = \frac{4\pi \times 10^{-7}}{8\pi}(1 + 4 \times 7.52) \times 10^3$$

$$= 1.554 \text{ mH/km}$$

and the inductive reactance per kilometer is

$$X_L = \omega L$$

$$= 377 \times 1.554 \times 10^{-3}$$

$$= 0.5858\ \Omega$$

From Eq. (18.13) (with $\varepsilon_o = 10^{-9}/36\pi$ F/m)

$$C = \frac{2\pi \times 10^{-9}/36\pi}{7.52} \times 10^3$$

$$= 7.387 \times 10^{-9} \text{ F/km}$$

Hence, the capacitive reactance per kilometer is

$$X_c = \frac{1}{\omega C}$$

$$= \frac{10^9}{377 \times 7.387}$$

$$= 0.36 \times 10^6\ \Omega$$

18.3.2 Transmission Line Representation

In the preceding section we have discussed the three basic parameters of transmission lines. A fourth parameter—the leakage resistance or conductance to ground, which represents the combined effect of various currents from the line to ground—has not been considered because it is usually negligible for most calculations. However, it has been estimated that on 132-kV transmission lines the leakage losses vary between 0.3 and 1.0 kW/mi.

Returning to the three basic parameters, as calculated in Examples 18.3 and 18.4, we observe that the capacitance is much too low for a 1-km-long line. In fact (as mentioned earlier), for a transmission line up to 80 km long, the shunt effects due to

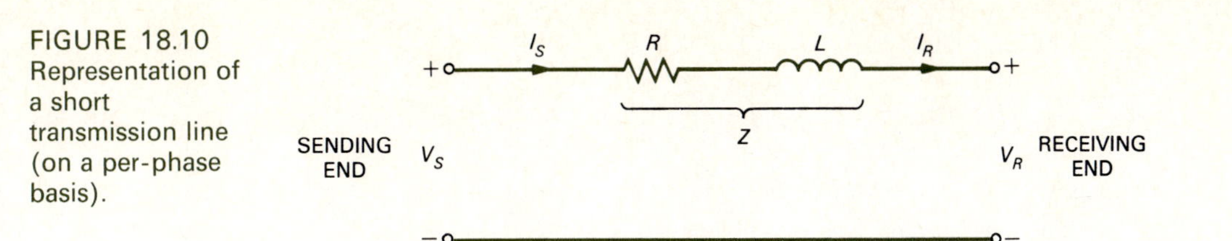

FIGURE 18.10 Representation of a short transmission line (on a per-phase basis).

capacitance and leakage resistance are negligible. Such a line is known as the *short transmission line* and is represented by the lumped parameters R and L, as shown in Fig. 18.10. Notice that R is the resistance (per phase) and L is the inductance (per phase) of the entire line, even though we have introduced the transmission line parameters in terms of per unit length of the line. The line is shown to have two ends—the sending end at the generator end and the receiving end at the load end. The problems of significance to be solved here are the determination of voltage regulation and of efficiency of transmission. These quantities are defined as follows:

$$\text{Percent voltage regulation} = \frac{|V_{R,\,\text{no-load}}| - |V_{R,\,\text{load}}|}{|V_{R,\,\text{load}}|} \times 100 \tag{18.14}$$

$$\text{Efficiency of transmission} = \frac{\text{power at the receiving end}}{\text{power at the sending end}} \tag{18.15}$$

Often, Eqs. (18.14) and (18.15) are evaluated at full-load values. The calculation procedure is illustrated in the following example.

Example 18.5 Let the transmission line of Example 18.4 be 60 km long. The line supplies a 3-phase wye-connected 100-MW 0.9 lagging power-factor load at a 215-kV line-to-line voltage. If the operating temperature of the line is 60°C, determine the regulation and efficiency of transmission in percent.

Solution The line resistance R (see Example 18.4) is given by

$$R = R_{60}\{0.00893[1 + 0.0039(60 - 20)]\} \times 60$$
$$= 0.62\ \Omega$$

The inductance L (from Example 18.4) is

$$L = 1.554 \times 10^{-3} \times 60$$
$$= 93.24\ \text{mH}$$

and

$$X_L = \omega L$$
$$= 377 \times 93.24 \times 10^{-3}$$
$$= 35.15\ \Omega$$

The line current is

$$I = I_S = I_R = \frac{100 \times 10^6}{\sqrt{3} \times 215 \times 0.9}$$

$$= 298.37 \text{ A}$$

The phase voltage at the receiving end is

$$V_R = \frac{215 \times 10^3}{\sqrt{3}}$$

$$= 124.13 \text{ kV}$$

The phasor diagram illustrating the operating conditions is shown in Fig. 18.11, from which

$$\mathbf{V}_S = \mathbf{V}_R + \mathbf{I}(R + jX_L)$$

$$= 124.13 \times 10^3\underline{/0^\circ} + 298.37\underline{/-25.8^\circ}(0.62 + j35.15)$$

$$\simeq 124.13 \times 10^3\underline{/0^\circ} + 298.37\underline{/-25.8^\circ} \times 35.15\underline{/90^\circ}$$

$$= (128.69 + j9.44) \quad \text{kV}$$

$$\simeq 129.04\underline{/4.2^\circ} \text{ kV}$$

Hence, for the percent voltage regulation we have

$$\frac{129.04 - 124.13}{124.13} \times 100 = 3.955\%$$

To calculate the efficiency, we determine the loss in the line as:

$$\text{Line loss} = 3 \times 298.37^2 \times 0.62$$

$$= 0.166 \text{ MW}$$

$$\text{Power received} = 100 \text{ MW} \qquad \text{given}$$

$$\text{Power sent} = 100 + 0.166$$

$$= 100.166 \text{ MW}$$

$$\text{Efficiency} = \frac{100}{100.166}$$

$$= 99.83\%$$

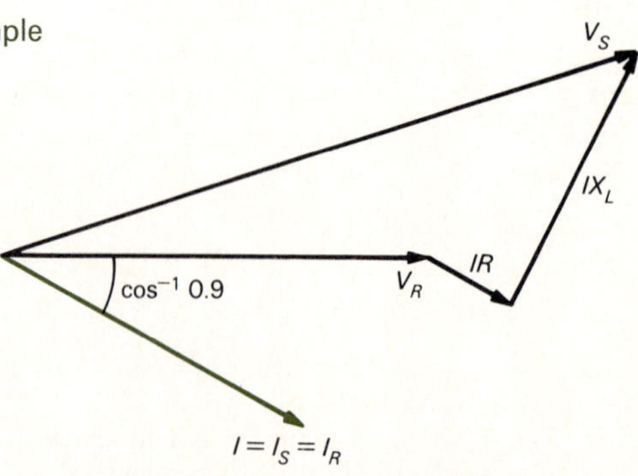

FIGURE 18.11 Example 18.5.

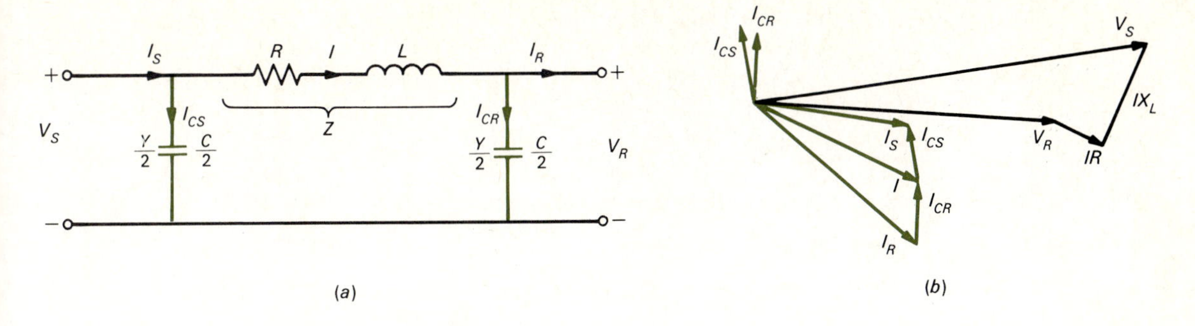

FIGURE 18.12 (*a*) A nominal-π circuit. (*b*) Corresponding phasor diagram.

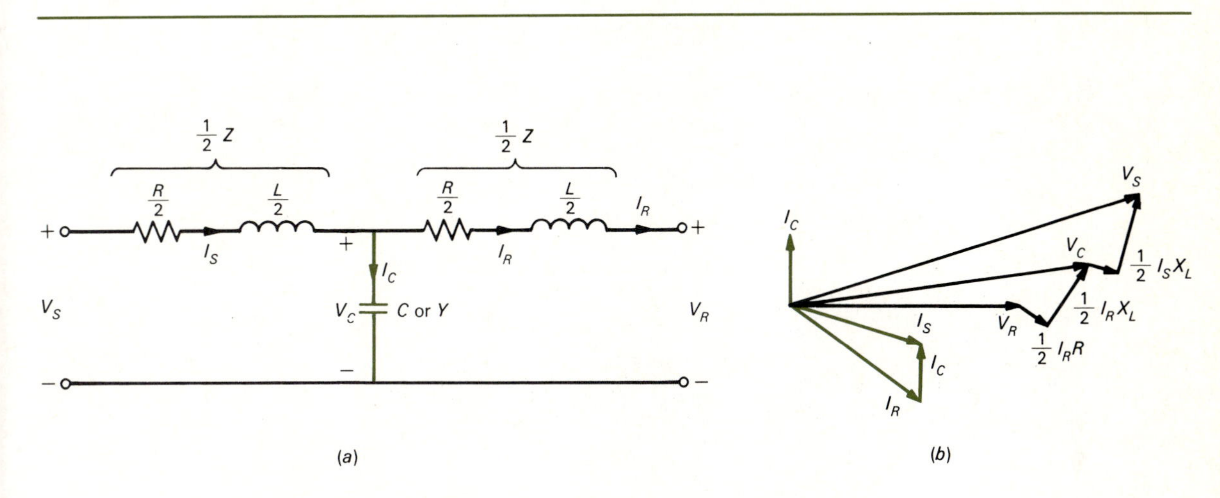

FIGURE 18.13 (*a*) A nominal-T circuit. (*b*) Corresponding phasor diagram.

The *medium-length transmission line* is considered to be up to 240 km long. In such a line the shunt effect due to the line capacitance is not negligible. Two representations for the medium-length line are shown in Figs. 18.12 and 18.13. These are respectively known as the nominal-π circuit and the nominal-T circuit of the transmission line. In Figs. 18.12 and 18.13 we also show the respective phasor diagrams for lagging load conditions. These diagrams aid in understanding the mutual relationships between currents and voltages at certain places on the line. The following examples illustrate the applications of the nominal-π and nominal-T circuits in calculating the performances of medium-length transmission lines.

Example 18.6 Consider the transmission line of Example 18.5, but let the line be 200 km long. Determine the voltage regulation of the line for the operating conditions given in Example 18.5, using the nominal-π circuit and the nominal-T circuit.

Solution From the data of Examples 18.4 and 18.5, we already know that

$$R = 2.07\ \Omega$$

$$L = 310.8\ \text{mH}$$

$$C = 1.4774\ \mu\text{F}$$

$$\mathbf{V}_R = 124.13\ \text{kV}$$

$$\mathbf{I}_R = 298.37\underline{/-25.8^\circ}\ \text{A}$$

For the nominal-π circuit, using the nomenclature of Fig. 18.12, we have

$$\mathbf{I}_{CR} = \frac{\mathbf{V}_R}{\mathbf{X}_{c/2}}$$

$$= \frac{124.13 \times 10^3\underline{/0^\circ}}{1/(377 \times 0.5 \times 1.4774 \times 10^{-6})\underline{/90^\circ}}$$

$$= 34.57\underline{/90^\circ}\ \text{A}$$

$$\mathbf{I} = \mathbf{I}_R + \mathbf{I}_{CR}$$

$$= 298.37\underline{/-25.8^\circ} + 34.57\underline{/90^\circ}$$

$$= 285\underline{/-19.5^\circ}\ \text{A}$$

$$R + jX_L = 2.07 + j377 \times 0.3108$$

$$\simeq 117.19\underline{/88.98}\ \Omega$$

$$\mathbf{I}(R + jX_L) = 285\underline{/-19.5^\circ} \times 117.19\underline{/88.98^\circ}$$

$$= 33.34\underline{/69.48^\circ}\ \text{kV}$$

$$\mathbf{V}_S = \mathbf{V}_R + \mathbf{I}(R + jX_L)$$

$$= 124.13\underline{/0^\circ} + 33.4\underline{/69.48^\circ}$$

$$= 139.39\ \text{kV}$$

Thus, the percent regulation is

$$\frac{139.39 - 124.13}{124.13} \times 100 = 12.3\%$$

For the nominal-T circuit, using the nomenclature of Fig. 18.13, we have

$$\mathbf{V}_C = \mathbf{V}_R + \tfrac{1}{2}\mathbf{I}_R(R + jX_L)$$

$$= 124.13\underline{/0^\circ} + \frac{10^{-3}}{2} \times 298.37\underline{/-25.8^\circ} \times 117.19\underline{/88.98}$$

$$= 132.92\underline{/67.4^\circ}$$

$$= 132 + j15.6 \qquad \text{kV}$$

$$\mathbf{I}_C = \frac{\mathbf{V}_C}{\mathbf{X}_C}$$

$$= \frac{132.92 \times 10^3\underline{/67.4}}{1/(377 \times 1.4774 \times 10^{-6})\underline{/90^\circ}}$$

$$= 74\underline{/157.4^\circ}\ \text{A}$$

$$\mathbf{I}_S = \mathbf{I}_R + \mathbf{I}_C$$

$$= 298.37\underline{/-25.8^\circ} + 74\underline{/157.4}$$

$$= 287.14\underline{/-20.68^\circ}\ \text{A}$$

$$\mathbf{V}_S = \mathbf{V}_C + \tfrac{1}{2}\mathbf{I}_S(R + jX_L)$$

$$= 132.92\underline{/67.4^\circ} + \frac{10^{-3}}{2} \times 287.14\underline{/-20.68^\circ} \times 117.19\underline{/88.98^\circ}$$

$$= 140.0\ \text{kV}$$

Thus, the percent regulation is

$$\frac{140 - 124.13}{124.13} \times 100 = 12.78\%$$

The long-line calculations are rather specialized and are not considered here.

18.4 FAULT ANALYSIS

Under normal conditions, a power system operates as a balanced three-phase ac system. A significant departure from this condition is often caused by a fault. A fault may occur on a power system for a number of reasons, some of the common ones being lightning, high winds, snow, ice, and frost. Faults give rise to abnormal operating conditions, such as excessive currents and voltages at certain points on the system. Protective equipment is used on the system to guard against abnormal conditions; for example, the magnitudes of fault currents determine the ratings of the circuit breakers and the settings of the protective relays. Faults may occur within a generator or at the terminals of a transformer, but here we will be mostly concerned with faults on transmission lines.

A fault study includes the following:

1. Determination of maximum and minimum three-phase short-circuit currents
2. Determination of unsymmetrical fault currents, as in single line-to-ground, double line-to-ground, line-to-line, and open-circuit faults
3. Determination of the ratings of circuit breakers
4. Investigating the operations of protective relays
5. Determination of voltage levels at strategic points during the fault

Of the items listed above, we will consider only the first one here. Per unit values and system representation, as discussed in Sec. 18.2, should be reviewed at this point.

Balanced Three-Phase Short Circuit Balanced three-phase fault calculations can be carried out on a per phase basis, so that only single-phase equivalent circuits are used. Invariably, the circuit constants are expressed in per unit, and all calculations are made on a per unit basis. In short-circuit calculations, we often evaluate the short-circuit MVA which is equal to $\sqrt{3}V_l I_f 10^6$, where V_l is the nominal line voltage and I_f is the fault current. We illustrate the procedure in the following examples.

Example 18.7 An interconnected generator-reactor system is shown in Fig. 18.14*a*. The generator and reactor ratings are as shown. The values of the corresponding reactances in percent, with the ratings of the equipment as base values, are also shown. A three-phase short circuit occurs at *A*. Determine the fault current and the fault kVA if the busbar line-to-line voltage is 11 kV.

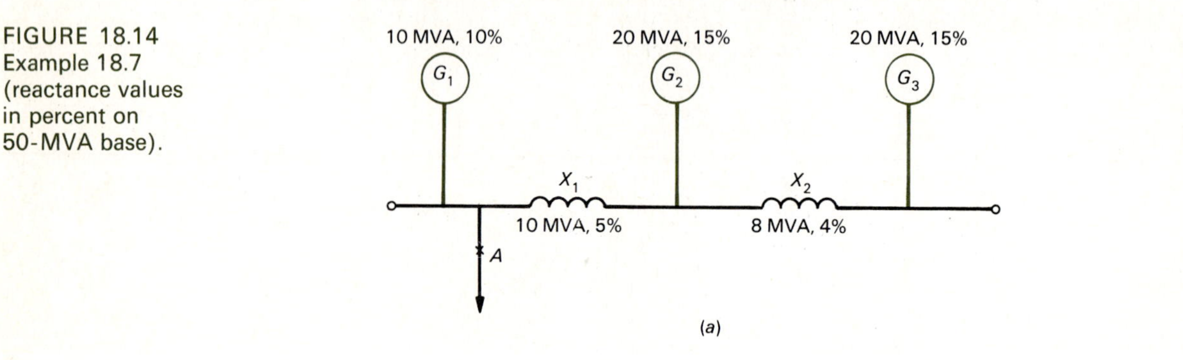

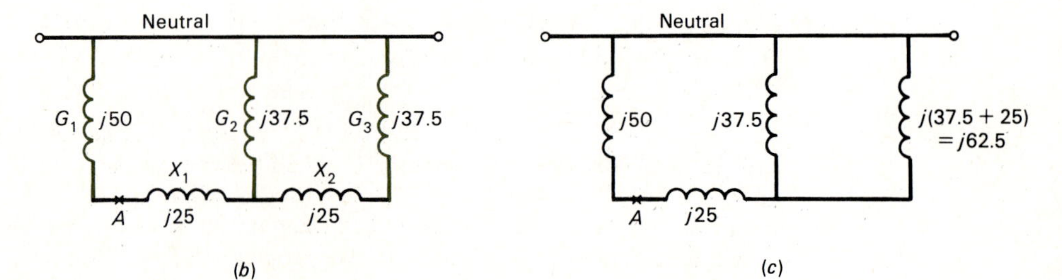

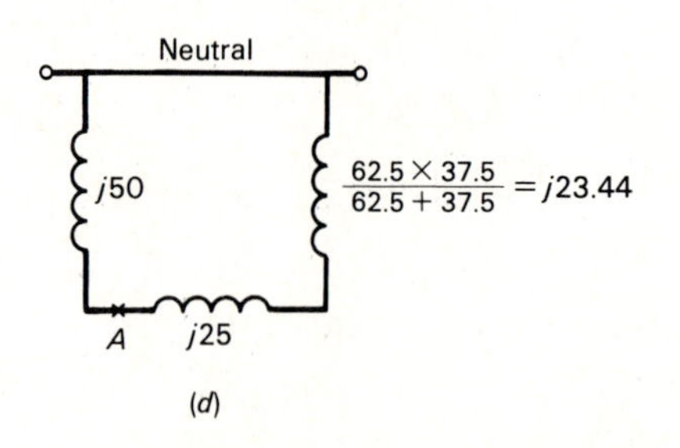

FIGURE 18.14 Example 18.7 (reactance values in percent on 50-MVA base).

Solution First we choose a base MVA for the system and express the reactances in percent for this base value. Let 50 MVA be the base MVA. On this base,

$$\text{Reactance of generator } G_1 = \tfrac{50}{10} \times 10$$
$$= 50\%$$

$$\text{Reactance of generator } G_2 = \tfrac{50}{20} \times 15$$
$$= 37.5\%$$

$$\text{Reactance of generator } G_3 = \tfrac{50}{20} \times 15 = 37.5\%$$

$$\text{Reactance } X_1 = \tfrac{50}{10} \times 5 = 25\%$$

$$\text{Reactance } X_2 = \tfrac{50}{8} \times 4 = 25\%$$

With these values of reactances, we draw a reactance diagram for the system, as shown in Fig. 18.14*b*. This diagram is drawn per phase and does not contain any sources. The reduction of the reactance diagram is illustrated in Figs. 18.14*c* and *d*. Finally, the total reactance from the neutral to the fault (at A) is

$$\text{Percent } X = j\frac{50 \times (23.44 + 25)}{50 + (23.44 + 25)} = j24.6\%$$

$$\text{fault MVA} = \frac{50}{24.6} \times 100 = 203.25 \text{ MVA}$$

$$\text{fault current} = \frac{203.25 \times 10^6}{\sqrt{3} \times 11 \times 10^3} = 10{,}668 \text{ A}$$

Example 18.8 A three-phase short-circuit fault occurs at point A on the system shown in Fig. 18.15*a*. The ratings, reactances, and impedance values are as shown. Calculate the fault current.

Solution Let the base MVA be 30 MVA and 33 kV the base voltage. Then, on this base we have the following values of reactance and impedance:

$$\text{Reactance of generator } G_1 = \tfrac{30}{20} \times 15 = 22.5\%$$

$$\text{Reactance of generator } G_2 = \tfrac{30}{10} \times 10 = 30\%$$

$$\text{Reactance of transformer} = \tfrac{30}{30} \times 5 = 5\%$$

$$\text{Impedance of line} = (3 + j15)\frac{30}{(33)^2} \times 100 = (8.26 + j41.32)\%$$

These values in percent are shown on the diagram of Fig. 18.15*b*, which is then reduced to Fig. 18.15*c*. Finally, the total impedance from the generator neutral to the fault is (from Fig. 18.15c)

$$\text{Percent } Z = 8.26 + j59.18 = 59.75\%$$

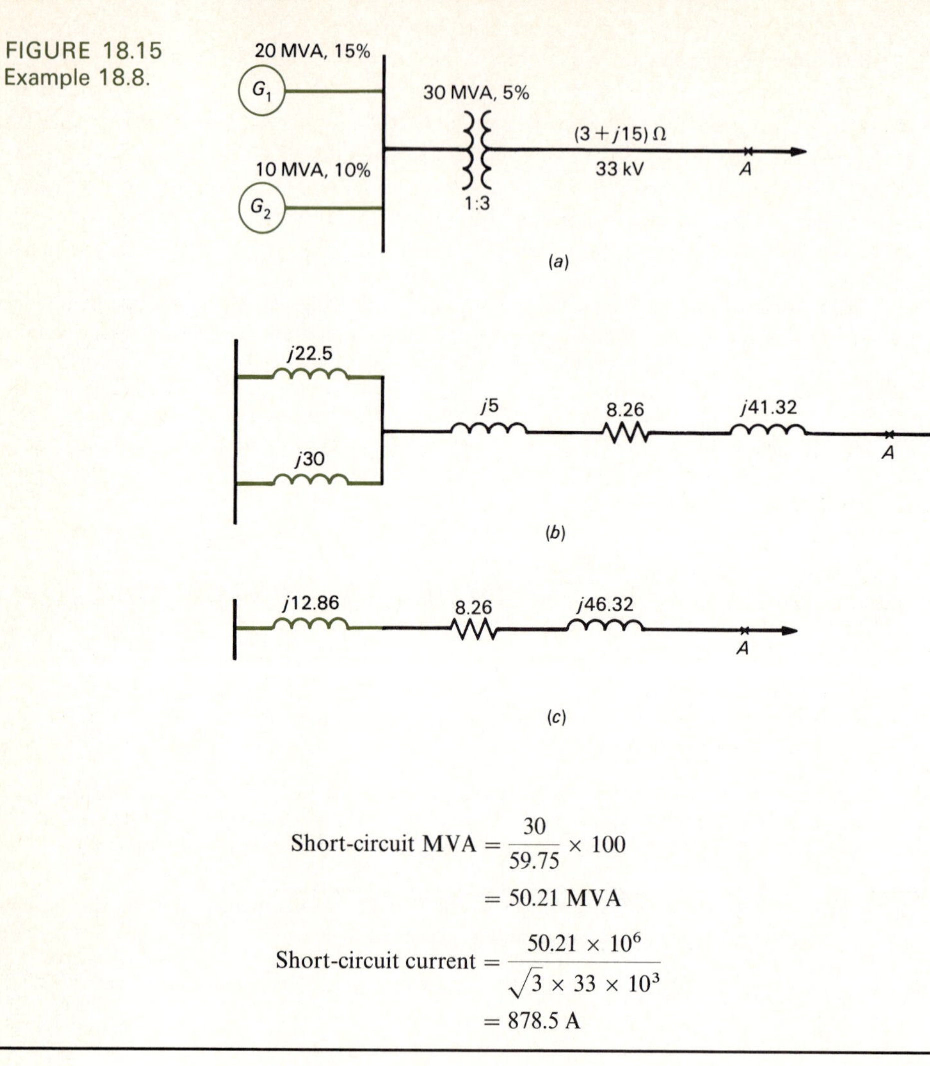

FIGURE 18.15
Example 18.8.

$$\text{Short-circuit MVA} = \frac{30}{59.75} \times 100$$

$$= 50.21 \text{ MVA}$$

$$\text{Short-circuit current} = \frac{50.21 \times 10^6}{\sqrt{3} \times 33 \times 10^3}$$

$$= 878.5 \text{ A}$$

18.5 POWER-FLOW STUDY

Power-flow studies, also known as load-flow studies, are extremely important in evaluating the operations of power systems, in controlling them, and in planning for future expansions. Basically, a power-flow study yields the real and reactive power and phasor voltage at each bus on the system, although a wealth of information is available from the printout of a digital computer solution of a typical power-flow study conducted by a power company. As a consequence of a power-flow study, we can optimize the system operation with regard to system losses and load distribution. The effect of temporary loss of generation capacity or of transmission circuits can also be investigated via a power-flow study.

Whereas the principles of a power-flow study are straightforward, a realistic study relating to a power system can only be carried out with the digital computer. In such a case, numerical computations are carried out in a systematic manner by an iterative procedure. Two of the commonly used numerical methods are the Gauss-Seidel method and the Newton-Raphson method. For details pertaining to these methods, any of the references cited at the end of this chapter may be consulted. In the following, we will consider only simple examples to illustrate the principles and certain procedures.

In order to develop a feel for the parameters which control power flow, let us first consider the power relationships for a transmission line.

Example 18.9 A short transmission line, shown in Fig. 18.16*a*, has negligible resistance and a series reactance of $jX\ \Omega$ per phase. If the per phase sending- and receiving-end voltages are V_s and V_R, respectively, determine the real and reactive powers at the sending end and at the receiving end. Assume that V_s leads V_R by an angle δ.

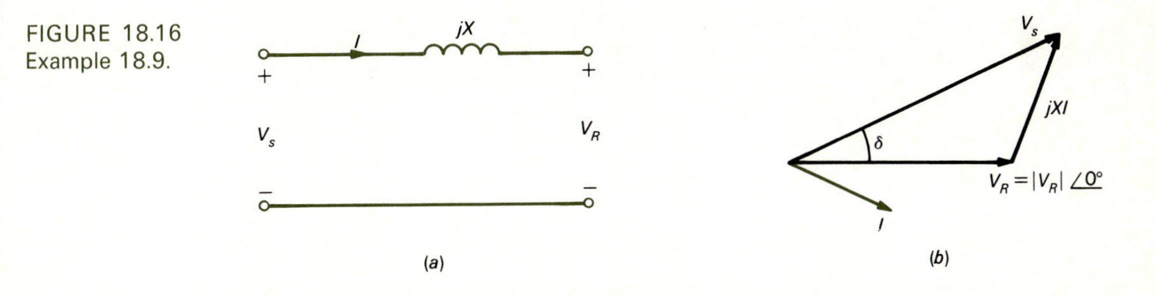

FIGURE 18.16 Example 18.9.

Solution The complex power **S**, in voltamperes, is given in general by

$$\mathbf{S} = P + jQ = \mathbf{V}\mathbf{I}^* \qquad \text{VA} \tag{18.16}$$

Where $\mathbf{I}^*$ is the complex-conjugate of $\mathbf{I}$. Thus, on a per phase basis, at the sending end we have

$$\mathbf{S}_s = P_s + jQ_s = \mathbf{V}_s\mathbf{I}^* \qquad \text{VA} \tag{18.17}$$

From Fig. 18.16*a*, **I** is given by

$$\mathbf{I} = \frac{1}{jX}(\mathbf{V}_s - \mathbf{V}_R)$$

and

$$\mathbf{I}^* = \frac{1}{-jX}(\mathbf{V}_s^* - \mathbf{V}_R^*) \tag{18.18}$$

Substituting Eq. (18.18) in Eq. (18.17) yields

$$S_s = \frac{\mathbf{V}_s}{-jX}(\mathbf{V}_s^* - \mathbf{V}_R^*) \tag{18.19}$$

From the phasor diagram of Fig. 18.16*b* we have

$$\mathbf{V}_R = |\mathbf{V}_R|\underline{/0^\circ} \qquad \mathbf{V}_R = \mathbf{V}_R^*$$

and

$$\mathbf{V}_S = |\mathbf{V}_S|\underline{/\delta}$$

Hence, Eq. (18.19) becomes

$$\mathbf{S}_s = \frac{|\mathbf{V}_s|^2 - |\mathbf{V}_R||\mathbf{V}_s|e^{j\delta}}{-jX} = P_s + jQ_s$$

$$= \frac{|\mathbf{V}_s||\mathbf{V}_R|}{X} \sin \delta + j\frac{1}{X}(|\mathbf{V}_s|^2 - |\mathbf{V}_s||\mathbf{V}_R| \cos \delta)$$

Finally,

$$P_s = \frac{1}{X}(|\mathbf{V}_s||\mathbf{V}_R| \sin \delta) \qquad \text{W} \tag{18.20}$$

and

$$Q_s = \frac{1}{X}(|\mathbf{V}_s|^2 - |\mathbf{V}_s||\mathbf{V}_R| \cos \delta) \qquad \text{VA} \tag{18.21}$$

Similarly, for the receiving end we have

$$\mathbf{S}_R = P_R + jQ_R = \mathbf{V}_R\mathbf{I}^*$$

Proceeding as above finally yields

$$P_R = \frac{1}{X}(|\mathbf{V}_s||\mathbf{V}_R| \sin \delta) \qquad \text{W} \tag{18.22}$$

$$Q_R = \frac{1}{X}(|\mathbf{V}_s||\mathbf{V}_R| \cos \delta - |\mathbf{V}_R|^2) \qquad \text{VA} \tag{18.23}$$

From this simple example, a number of significant conclusions may be derived. First, the angle δ is known as the *power angle*. The transfer of real power depends on δ alone, and not on the relative magnitudes of the sending- and receiving-end voltages (unlike in a dc system). The transmitted power varies approximately as the square of the voltage level. The maximum power transfer occurs when $\delta = 90°$ and

$$(P_R)_{\max} = (P_s)_{\max} = \frac{|\mathbf{V}_s||\mathbf{V}_R|}{X} \tag{18.24}$$

Finally, from Eqs. (18.21) and (18.23) it is clear that reactive-power flow will be in the direction of the lower voltage. If the system operates with $\delta \simeq 0$, then the average reactive-power flow over the line is given by

$$Q_{\text{av}} = \tfrac{1}{2}(Q_s + Q_R) = \frac{1}{2X}(|\mathbf{V}_s|^2 - |\mathbf{V}_R|^2) \qquad \text{VAr} \tag{18.25}$$

Equation (18.25) shows the strong dependence of the reactive-power flow on the difference of the voltages.

Up to this point we have neglected the I^2R loss in the line. If R is the resistance of the line per phase, then the line loss is given by

$$P_{\text{line}} = |\mathbf{I}|^2R \qquad \text{W} \tag{18.26}$$

From Eq. (18.16) we have

$$\mathbf{I}^* = \frac{P + jQ}{\mathbf{V}}$$

and

$$\mathbf{I} = \frac{P - jQ}{\mathbf{V}^*}$$

Thus

$$\mathbf{I}\mathbf{I}^* = |\mathbf{I}|^2 = \frac{P^2 + Q^2}{|\mathbf{V}|^2}$$

and Eq. (18.26) becomes

$$P_{\text{line}} = \frac{(P^2 + Q^2)R}{|\mathbf{V}|^2} \quad \text{W} \tag{18.27}$$

indicating that real and reactive powers both contribute to the line losses. Thus, it is important to reduce reactive-power flow to reduce the line losses. Later, we will see that reactive power is required for the voltage control on a bus. This is accomplished by injecting reactive power locally by installing shunt capacitors at the buses.

The preceding remarks give us some idea of real- and reactive-power flow over transmission lines. In particular, we observe that to raise the voltage level at a given bus, we must supply reactive power at the bus. We also observe that reactive power flows from higher to lower voltages and that the direction of flow of real power depends not on the relative magnitudes of the voltages, but on their phase displacements. In essence, a power-flow analysis yields the voltage levels and the flow of real and reactive powers at various buses of the system. However, explicit analytical solutions are not forthcoming because of load fluctuations on the buses and because the receiving-end voltage may not be known. In such cases, numerical methods are used to solve the problem, as mentioned at the beginning of this section. An iterative procedure may be used to handle simple problems such as the following, since solving them by a noniterative procedure is very cumbersome.

Example 18.10 A two-bus system is shown in Fig. 18.17. The load on bus 2 requires 1.0 pu real power and 0.6 pu reactive power per phase. The line impedance per phase is $(0.05 + j0.02)$ pu. The voltage on bus 1 is $1\underline{/0^\circ}$ pu. Determine on a per phase basis the voltage on bus 2 and the real and reactive power on bus 1.

Solution In Fig. 18.17 we show the real power by solid arrows and the reactive power by dashed arrows. The governing equations for the system are (on a per phase basis)

$$\mathbf{S}_2 = \mathbf{V}_2\mathbf{I}^*$$

$$\mathbf{V}_1 = \mathbf{V}_2 + \mathbf{Z}_l\mathbf{I}$$

FIGURE 18.17
Example 18.10.

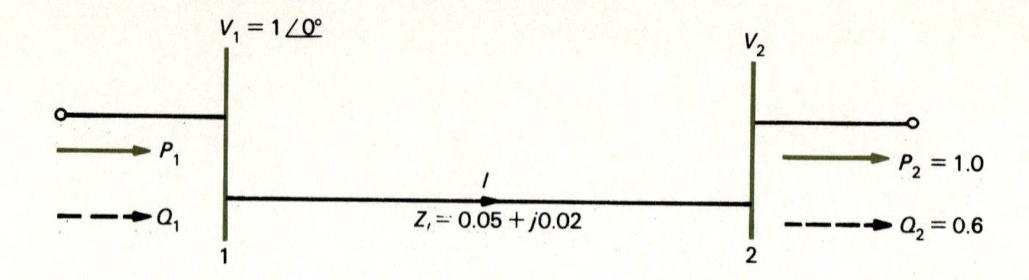

where the symbols are defined in Fig. 18.17. Solving for $\mathbf{V}_2$ and eliminating $\mathbf{I}$ from these equations yields

$$\mathbf{V}_2 = \mathbf{V}_1 - \mathbf{Z}_l\mathbf{I} = \mathbf{V}_1 - \mathbf{Z}_l \frac{\mathbf{S}_2^*}{\mathbf{V}_2^*} \tag{18.28}$$

To solve Eq. (18.28) iteratively, we assume a value for $\mathbf{V}_2$ and call it $\mathbf{V}_2^{(0)}$. We substitute this in the right-hand side of Eq. (18.28) and solve for $\mathbf{V}_2$, calling this value of V_2 (as the value of $\mathbf{V}_2$ after the first iteration) $\mathbf{V}_2^{(1)}$. We substitute $\mathbf{V}_2^{(1)}$ in the right-hand side of Eq. (18.28) and obtain a $\mathbf{V}_2^{(2)}$. This procedure is continued until convergence is achieved. The iterative procedure is thus given by the general equation, or algorithm, as

$$\mathbf{V}_2^{(k)} = \mathbf{V}_1 - \frac{\mathbf{Z}_l\mathbf{S}_2^*}{\mathbf{V}_2^{(k-1)*}} \tag{18.29}$$

For the given numerical values, we first assume that $V_2 = 1\underline{/0^\circ}$ and then use Eq. (18.29) in succession to obtain the following table:

Iteration No.	$V_2 = 1$ pu
0	$1.0 + j0$
1	$0.962 - j0.05$
2	$0.9630 - j0.054$
3	$0.9635 - j0.054$
4	$0.9635 - j0.054$

Notice that convergence is achieved in just four iterations. For other data, such as a greater load, it may take more iterations to converge to the solution. Of course, in some cases convergence may not be achieved because a solution may not exist or the starting point of the iteration process may not be appropriate.

To pursue the problem further, we have

$$\mathbf{V}_2 = 0.9635 - j0.054 \quad \text{pu}$$

Hence,

$$\mathbf{I} = \frac{\mathbf{S}_2^*}{\mathbf{V}_2^*}$$

$$= \frac{1.0 - j0.6}{0.9635 + j0.054}$$

$$= 1.208\underline{/-27.75^\circ} \text{ pu}$$

or

$$\mathbf{I}^* = 1.208\underline{/27.75^\circ}$$

and

$$\mathbf{V}_1 = 1\underline{/0^\circ} \qquad \text{given}$$

Thus,

$$\begin{aligned} P_{1+j}Q_1 &= \mathbf{S}_1 \\ &= \mathbf{V}_1\mathbf{I}^* \\ &= (1.208\underline{/27.75})(1\underline{/0^\circ} \\ &= 1.069 + j0.5625 \end{aligned}$$

The real and reactive powers on bus 1 are therefore

$$P_1 = 1.069 \text{ pu}$$

$$Q_1 = 0.5625 \text{ pu}$$

Example 18.11 For the system presented in Example 18.10, it is desired to have $|\mathbf{V}_1| = |\mathbf{V}_2| = 1.0$ pu by supplying reactive power at bus 2. Determine the value of the reactive power.

Solution From Eq. (18.16) we have

$$\mathbf{I} = \frac{\mathbf{S}^* + jQ_2}{\mathbf{V}_2^*}$$

which, when substituted in Eq. (18.28), yields

$$\mathbf{V}_1 = \mathbf{V}_2 + \frac{\mathbf{Z}_l}{\mathbf{V}_2^*}(\mathbf{S}_2^* + jQ_2) \tag{18.30}$$

We now substitute the following numerical values in Eq. (18.30):

$$|\mathbf{V}_1| = 1 \qquad \mathbf{Z}_l = 0.05 + j0.02$$

$$\mathbf{V}_2 = 1\underline{/0^\circ} \qquad \mathbf{S}_2^* = 1 - j0.6$$

We thus obtain

$$1 = |1 + (0.05 + j0.02)[1 + j(Q_2 - 0.6)]|$$

Hence,

$$Q_2 = 4.02 \text{ pu}$$

As already mentioned, these examples merely illustrate the utility of some aspects of a power-flow study. Detailed calculations for an actual power system must be carried out on a digital computer.

PROBLEMS

18.1 A three-phase 60-Hz 25-km transmission line is made of hard-drawn copper conductor 500 mils in diameter. What is the per phase resistance of the line at 60°C? Refer to Table 18.1 for the values of constants.

18.2 The conductors of the line of Prob. 18.1 are arranged in the form of an equilateral triangle, the sides of which are 6 m. Evaluate the per phase inductive and capacitive reactances of the line.

18.3 A three-phase wye-connected 15-MW 0.866 lagging power-factor load is supplied by the line of Prob. 18.1 at a 115-kV line-to-line voltage. The operating temperature of the line is 60°C. Calculate the line losses and the percent voltage regulation.

18.4 A 138-kV three-phase short transmission line has a per phase impedance of $(2 + j4)\ \Omega$. If the line supplies a 25-MW 0.8 lagging power-factor load, calculate (*a*) the efficiency of transmission and (*b*) the sending-end voltage and power factor.

18.5 Consider the data of Prob. 18.4. If the sending-end voltage is 152 kV (line-to-line), determine (*a*) the receiving-end voltage, (*b*) the efficiency of transmission, and (*c*) the angle between the sending-end and receiving-end voltages.

18.6 A three-phase short transmission line having a per phase impedance of $(2 + j4)\ \Omega$ has equal line-to-line receiving-end and sending-end voltages of 115 kV (at both ends) while supplying a load at a 0.8 leading power factor. Calculate the power supplied by the line.

18.7 A three-phase wye-connected 20-MW 0.866 power-factor load is to be supplied by a transmission line at 138 kV. It is desired that the line losses do not exceed 5 percent of the load. If the per phase resistance of the line is $0.7\ \Omega$, what is the maximum length of the line?

18.8 The impedance of a single-phase transmission line is $\mathbf{Z} = R + jX$. The sending-end and receiving-end voltages are $\mathbf{V}_s$ and $\mathbf{V}_R$, respectively. Show that the maximum power which can be transmitted by the line is given by

$$P_{\max} = \frac{|\mathbf{V}_s||\mathbf{V}_R|}{|\mathbf{Z}|} - \frac{R|\mathbf{V}_R|^2}{|\mathbf{Z}|^2}$$

18.9 The per phase constants of a 345-kV three-phase 150-km-long transmission line are

$$\text{Resistance} = 0.1\ \Omega/\text{km}$$

$$\text{Inductance} = 1.1\ \text{mH/km}$$

$$\text{Capacitance} = 0.02\ \mu\text{F/km}$$

The line supplies a 180-MW 0.9 lagging power-factor load. Using the nominal-π circuit, determine the sending-end voltage.

18.10 Repeat Prob. 18.9, using the nominal-T circuit.

18.11 Determine the efficiency of transmission of the line of Prob. 9.8, using (*a*) the nominal-π and (*b*) the nominal-T circuit.

18.12 A portion of a power system is shown in Fig. P18.12, which also shows the ratings of the generators and the transformer and their respective percent reactances. A symmetrical short circuit occurs on a feeder at F. Find the value of the reactance X (in percent) such that the short-circuit MVA does not exceed 300.

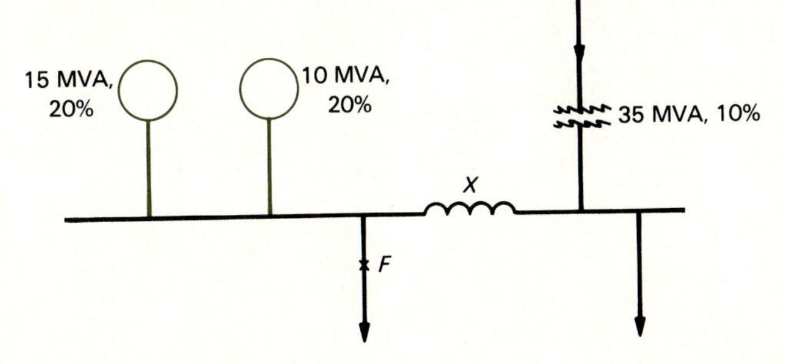

FIGURE P18.12

18.13 Three generators, each rated at 10 MVA and having a reactance of 10 percent, are connected to common busbars and supply the load through two 15-kVA step-up transformers. Each transformer has 7 percent reactance. Determine the maximum fault MVA (*a*) on the high-voltage side and (*b*) on the low-voltage side.

18.14 For the system shown in Fig. P18.14, calculate the short-circuit MVA at A and at B.

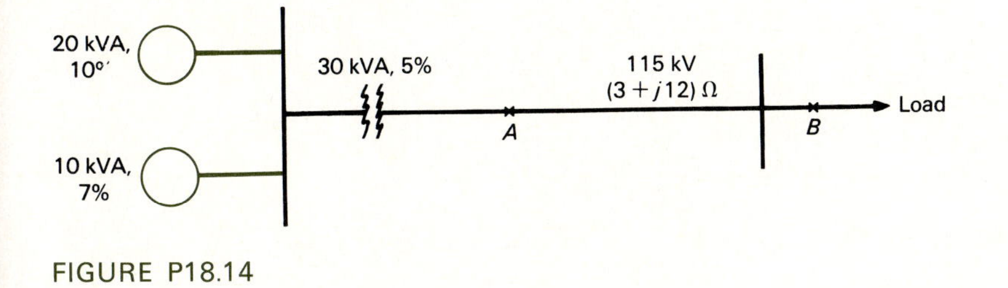

FIGURE P18.14

18.15 For the system shown in Fig. P18.15, it is desired that $|\mathbf{V}_1| = |\mathbf{V}_2| = 1$ pu. The loads are as shown in Fig. P18.15 and are

$$\mathbf{S}_1 = 6 + j10 \quad \text{pu}$$

$$\mathbf{S}_2 = 14 + j8 \quad \text{pu}$$

The line impedance is $j0.05$ pu. If each generator supplies 10 pu real power, calculate the power and the power factors at the two ends.

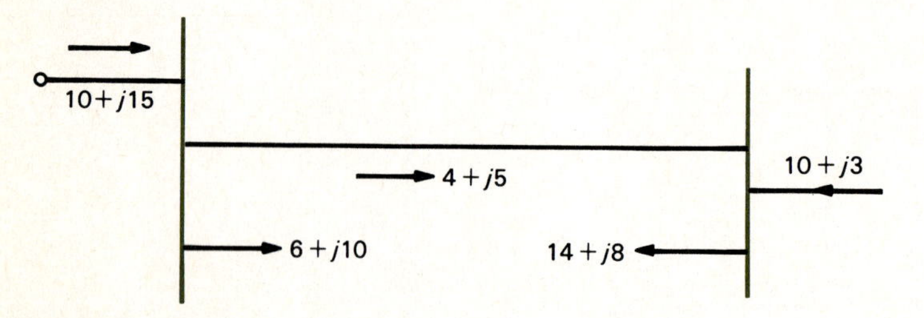

FIGURE P18.15

18.16 At a certain leading power-factor load the sending- and receiving-end voltages of a short transmission line of impedance $(R + jX)$ are equal. Determine the ratio X/R so that maximum power is transmitted over the line.

REFERENCES

18.1 W. D. Stevenson, Jr., *Elements of Power System Analysis*, 4th ed., McGraw-Hill, New York, 1982.

18.2 S. A. Nasar, *Electric Energy Conversion and Transmission*, Macmillan, New York, 1985.

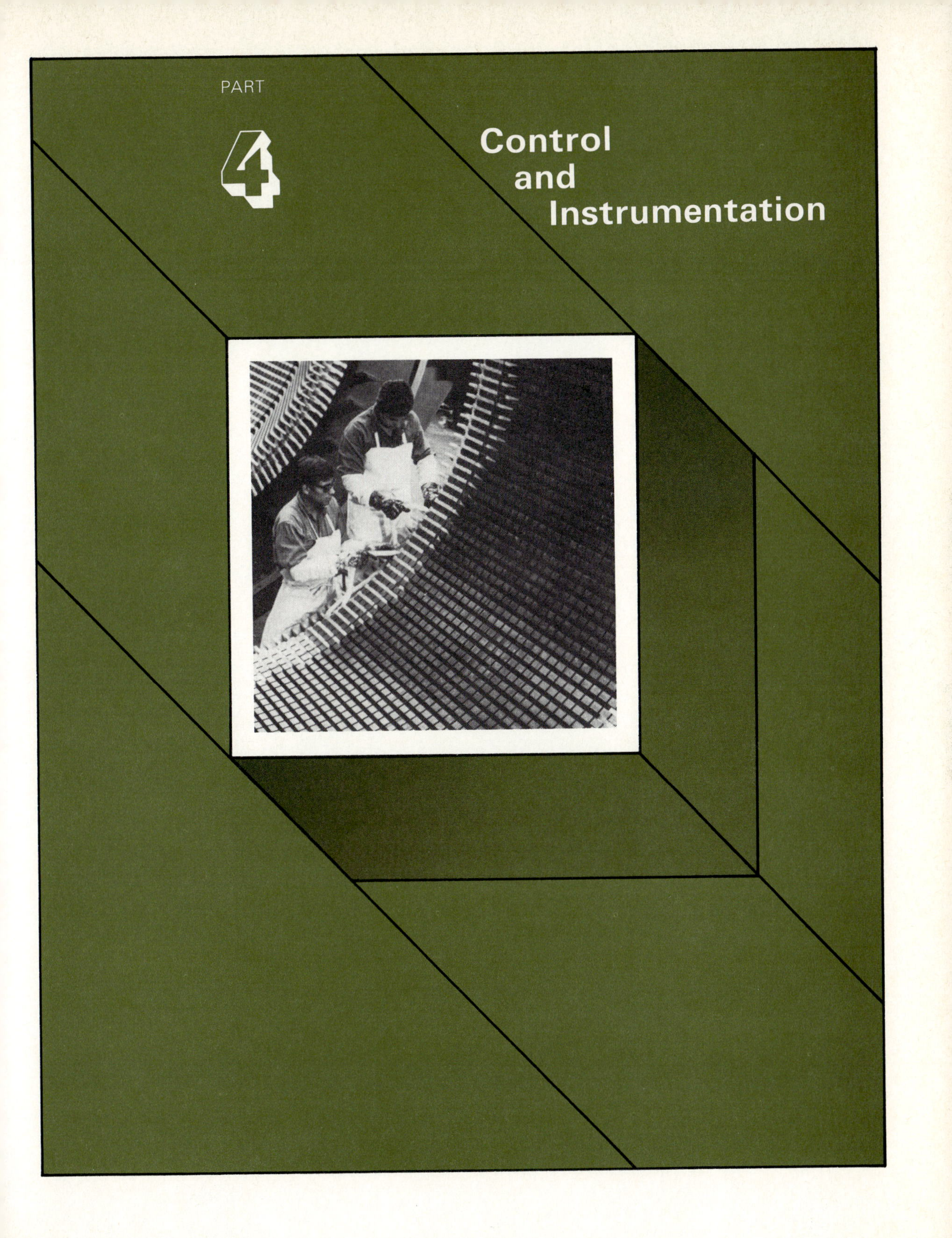
PART
4
Control
and
Instrumentation

Chapter

Feedback Control Systems

The title's three words—feedback, control, and systems—describe the key themes of this chapter. In reverse order, these words lead us from generalized to fairly specific ideas and concepts having to do with the control of physical systems, which is the subject of this chapter.

Systems, one of the catchwords of our age, is a term used in many fields of study —ecology, economics, and the social sciences, as well as the physical sciences. We use the term to describe an assemblage of components, subsystems, and interfaces arranged or existing in such a manner as to perform a function or functions or to achieve a goal. *Control* refers to the function or purpose of the system we wish to discuss; it does not refer to such functions as power, motion, sensing, or communicating, which are the functions of most of the components and circuits previously discussed. A chief distinction is that control is almost always realized not by a single component (such as a transistor, resistor, or motor) but by an entire system of components and interfaces. And finally, we use the term *feedback* in this chapter to mean a specific system configuration—as distinct from the more generalized meaning in common usage today that associates the term with human communications and interrelationships.

Much of the meaning and significance of these three words, which have become so ingrained in our culture, actually developed out of the early studies of physical control systems in the 1930s and 1940s. From these early studies, first known as *linear control theory*, has developed a vast knowledge that is the foundation for much of modern technology—from large-scale computer development to the many satellites and space vehicles that permeate outer space, and from the bionic arm to the automated assembly line.

19.1 BASIC CONCEPTS OF CONTROL

Block diagrams, which serve as an invaluable method for illustrating configurations and interfaces in all types of systems, have long been an important tool in control systems analysis. We will use this tool to define some basic control systems concepts and to develop some of the mathematical reductions of more complex systems.

Block diagrams can be used with an almost infinite variety of generality. As used in economic and social systems, for instance, the diagrams tend to be quite general, illustrating with relatively little mathematical rigor the interactions and interfaces of major subsystems. In the control systems context of this text, however, the elements of the block diagram take on a definite mathematical meaning—a useful merit of this graphical tool.

Figure 19.1 illustrates a basic feedback control system in block diagram format and introduces some notation which will be used throughout this chapter. Each block of the diagram in Fig. 19.1 has a specific meaning. For example, relationships between an input function and an output function are defined by the equation

$$A_n(s) = \frac{V_{no}(s)}{V_{ni}(s)} \tag{19.1}$$

where $A_n(s)$ is defined as the transfer function, $V_{no}(s)$ is the Laplace transform of an output function, and $V_{ni}(s)$ is the Laplace transform of an input function.

The use of the argument (s) in the above equation indicates Laplace-transformed variables. Such a formulation results in the simplification of the "algebra of block diagrams," plus many useful theorems related to stability and response which will be discussed subsequently. For a discussion of Laplace transforms, see App. D.

Other definitions associated with Fig. 19.1 are related to topography. The upper branch of the diagram is known as the *forward loop*. The output of this branch is called the *controlled variable* $C(s)$, or sometimes the *output variable*. The input is the *error variable* $E(s)$. The lower branch is termed the *feedback loop*, since it is a path for feeding back the output variable to the input. The circle at the left represents an *error detector*, or *comparator*. The incoming signal on the left is called the *reference variable* $R(s)$, or sometimes the *input variable*. The symbols V, C, E, and R are used in Fig. 19.1 and throughout this chapter to represent generalized variables; these may be state variables—current, voltage, velocity, position, force, etc.—or variables related to the state variable. The algebraic symbols in the error-detector circle are such as to indi-

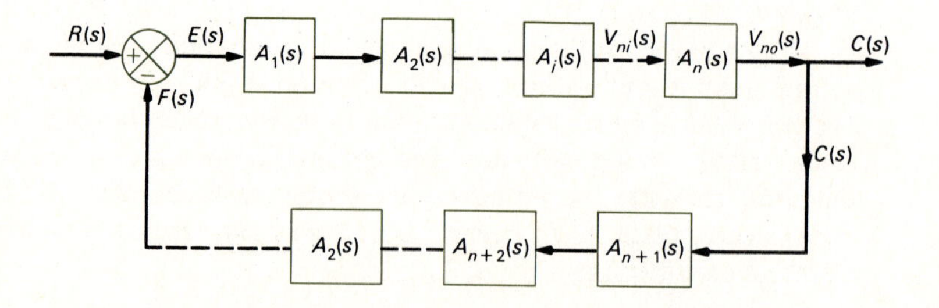

FIGURE 19.1 Generalized feedback control system.

cate *negative feedback*; i.e., the *feedback signal* $F(s)$ is algebraically subtracted from the reference signal $R(s)$. If the sign within the lower quadrant of the circle were positive, *positive feedback* would be indicated. We will now attempt to give mathematical and physical significance to these definitions.

19.1.1 Block Diagrams for Linear Control Systems

The linearity of a control system influences not only its characteristics and performance, but also the means of mathematical representation and analysis. A linear system is one that can be described by linear differential equations. In a physical sense, it is also useful to consider linearity in terms of the physical coefficients or circuit constants that compose a given system. For example, a linear resistance is defined by the familiar Ohm's law relationship $r(t) = v(t)/i(t) = R$ (a constant). If in a certain control system the voltage across this resistance is considered an output variable and the current through this resistance the input variable, then the resistance function could be described by the Laplace-transformed equation

$$V(s) = RI(s) \tag{19.2}$$

and the transfer function

$$A(s) = \frac{V(s)}{I(s)} = R \tag{19.3}$$

A nonlinear resistance might be described by the equation $r(t) = R_o[i(t)]^2$, indicative of the variation of resistance with current magnitude (or temperature), or by the equation $r(t) = f[v(t)]^a$, typical of the varistor and other metal-oxide resistors. Note that the resistor defined by the relationship $r(t) = tR_o$ would be considered a linear resistor with time-varying coefficients and could be treated by means of linear time-variant differential equations. In conclusion, a linear coefficient or circuit constant is one that is independent of the dependent variables. In general, most circuit constants in electrical systems contain some degree of nonlinearity: amplifiers saturate at some level of the dependent variables, most circuits containing the common magnetic steels and alloys show many nonlinear characteristics (saturation, hysteresis, etc.), and electric motor torque is a function of the *product* of state variables. Many mechanical systems also exhibit nonlinearities, such as stiction-type friction, dead band or "windup" in gearing, and fluid heating in hydraulic systems. Although there are some very elegant and interesting mathematical methods for treating nonlinear control systems analytically and/or graphically, such systems are better and more efficiently treated by means of computer simulation techniques. However, many systems containing nonlinearities may be treated as piecewise-linear systems by considering variations over a relatively small variation of the dependent variables or within a region in which the nonlinearity is insignificant (such as occurs when the degree of magnetization of a transformer or motor is limited to magnetic intensities below the knee of the saturation characteristic). In the remaining sections, linear systems are assumed.

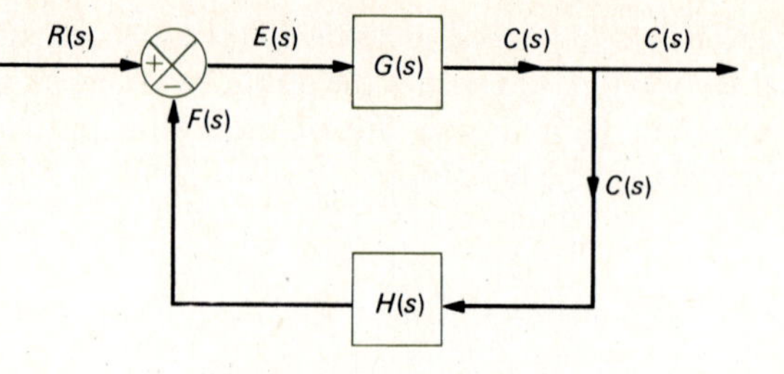

FIGURE 19.2 Classical linear control theory block diagram.

The generalized block diagram of Fig. 19.1 may be conveniently represented by means of the simplified diagram shown in Fig. 19.2. In this diagram, the symbol $G(s)$ represents the product of all forward-loop transfer functions;

$$G(s) = \prod_{j=1}^{n} A_j(s) \tag{19.4}$$

where the symbol $\prod$ indicates the multiplication, or product, of the n terms. Likewise, the feedback branch is simplified by a similar product term $H(s)$;

$$H(s) = \prod_{j=n+1}^{Z} A_j(s) \tag{19.5}$$

Let us look at the algebra associated with Fig. 19.2. Note that it represents negative feedback. The relationship between the controlled variable and the error variable is

$$C(s) = G(s)E(s) \tag{19.6}$$

Other relationships include:

$$F(s) = H(s)C(s) \tag{19.7}$$

$$E(s) = R(s) - F(s) \tag{19.8}$$

$$F(s) = G(s)H(s)E(s) \tag{19.9}$$

By combining Eqs. (19.6) to (19.8),

$$\frac{C(s)}{R(s)} = \frac{G(s)}{1 + G(s)H(s)} \tag{19.10}$$

Equation (19.9) is known as the *open-loop* or return function, and Eq. (19.10) is called the *closed-loop* transfer function.

The individual transfer function terms, or $A_j(s)$, consist of numerator and denominator functions of polynomials in the variable s:

$$A_j(s) = \frac{A_{jn}(s)}{A_{jd}(s)} \tag{19.11}$$

In linear systems, both the numerator and denominator will consist of products of one or more of four types of polynomials: constants, first degree polynomials

$$s \pm a \tag{19.12}$$

nth degree polynomials

$$(s \pm a)^n \tag{19.13}$$

and complex pairs

$$(s^2 \pm bs + \omega^2)^n \tag{19.14}$$

It is seen that the forward-loop transfer function $G(s)$ and the feedback-loop transfer function $H(s)$ consist of numerators and denominators of products of similar terms. When multiplied out, the numerators and denominators are said to be expressed as polynomials of the nth degree in s (or $j\omega$ or p):

$$A_{jn}(s), A_{jd}(s) = s^n \pm c_1 s^{n-1} \pm c_2 s^{n-2} \pm \cdots \pm c_i s^{n-i} \pm \cdots \pm c_{n-1} s \pm c_n \tag{19.15}$$

These various forms of the transfer function all have mathematical significance, as will be illustrated in subsequent sections of this chapter.

We have now talked about the concept of feedback and have illustrated this concept with Figs. 19.1 and 19.2 and Eqs. (19.6) to (19.10). There are many useful and significant methods of controlling physical processes that do not involve feedback, but this discussion is limited to systems in which feedback *is* involved. Notice that feedback is involved *naturally* in many physical systems—particularly in biological and ecological systems—and is commonplace in many manufactured systems, such as the dc motor. Whereas natural or inherent feedback is most significant in many aspects of our lives and our society, and will be illustrated in some examples below, we are particularly concerned with systems in which feedback has been purposefully added to the system. The fact that feedback is purposefully added to systems raises the question "Why?" There are a number of reasons for the use of feedback, and usually several of these reasons are involved in choosing this mode of control:

1 Precise control around a given set point or reference variable
2 Control of impedances in electronic systems
3 Improvement of the stability of a system
4 Reduction of the sensitivity of the system characteristics to changes in one or more parameters
5 Improvement of system response time and reducing overshoot

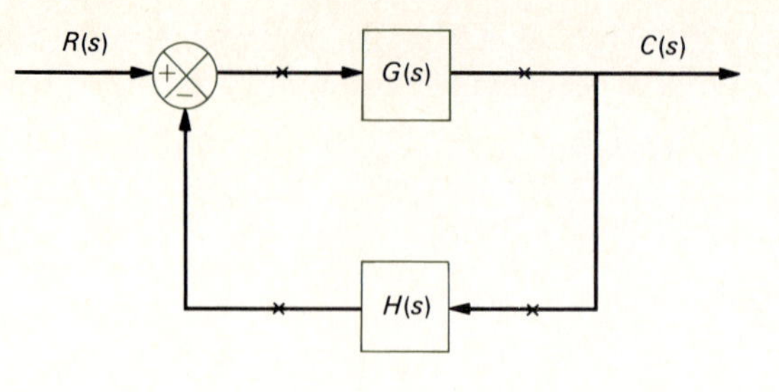

FIGURE 19.3 Classical block diagram with Xs marking "cuts" for evaluating return differences.

Related to the concept of impedance changes in electrical systems is the concept of *return difference*. To describe this concept, it is necessary to return to the feedback circuit of Fig. 19.2. This circuit is redrawn as Fig. 19.3, which shows several breaks, or "cuts," in the circuit. The reader should verify that if an output signal of 1.0 is applied in a clockwise direction at any cut, the difference between the input signal and the returned signal will be

$$\text{Return difference} = 1 - G(s)H(s) \tag{19.16}$$

This value, as expressed in Eq. (19.16), is known as the return difference. Note that this same result occurs regardless of where the break in the closed-loop circuit is made in Fig. 19.3.

The form of the transfer function mathematical expression depends upon the nature of the physical system being analyzed. There is a broad dichotomy in control systems based upon the nature of the signals that occur in the system; these signals may be *continuous* or *discontinuous*, and these two classes of systems are often rather loosely called *analog* and *digital*, respectively. Although these terms are not in any sense rigorous definitions, they are very helpful in describing the types of components or hardware used in constructing a system and also, surprisingly, the type of software or mathematics used to analyze a system. Much of classical control theory has been developed for continuous systems, but it can also be applied to many digital or discontinuous systems by using average or mean signals, or by applying classical theory to individual discrete time intervals of a discontinuous system. These concepts will be illustrated in examples and discussions in later sections of this chapter. With the tremendous developments in power semiconductors and integrated circuits in recent years, digital and discontinuous control systems are becoming widely used in many applications. The ultimate in digital control—microprocessor control—will be discussed in Chap. 22.

Transfer functions are obtained by means of the mathematical and physical theory associated with the particular device or system. To obtain the frequency-response form of a transfer function, the first step is to write the differential equation describing the device or system; the second is to obtain the Laplace transform of the differential equations, setting all initial conditions to zero; and the third is to obtain the output/input ratio of the device or system. A few examples will illustrate this process.

Example 19.1 Figure 19.4 illustrates a simple RC network. Determine the transfer function $E_o(s)/E_i(s)$ for this circuit.

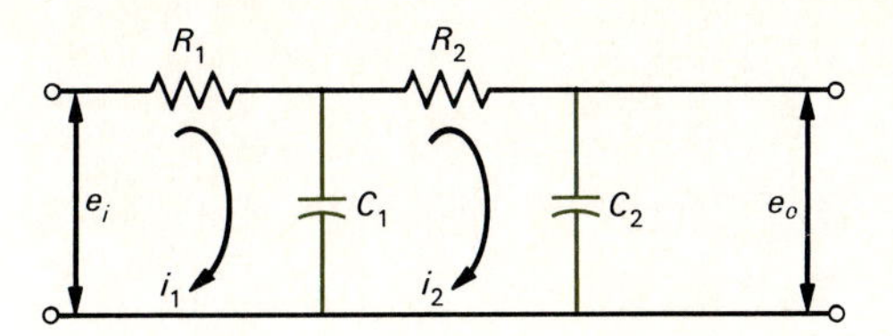

FIGURE 19.4 Example 19.1: a simple RC network.

Solution The differential equations along the current paths shown are

$$e_i = i_1 R_1 + \frac{1}{C_1}\int i_1\,dt - \frac{1}{C_1}\int i_2\,dt$$

$$0 = -\frac{1}{C_1}\int i_1\,dt + R_2 i_2 + \frac{1}{C_1}\int i_2\,dt + \frac{1}{C_2}\int i_2\,dt \tag{19.17}$$

The output voltage is (assuming infinite load impedance)

$$e_o = \frac{1}{C_2}\int i_2 \tag{19.18}$$

Transforming these equations and eliminating i_1 gives

$$E_i(s) = I_2(s)\left[C_1 s\left(R_2 + \frac{1}{C_1 s} + \frac{1}{C_2 s}\right)\left(R_1 + \frac{1}{C_1 s}\right) - \frac{1}{C_1 s}\right] \tag{19.19}$$

$$E_o(s) = \frac{1}{C_2 s} I_2(s) \tag{19.20}$$

The ratio of $E_o(s)$ to $E_i(s)$ is

$$A_1(s) = \frac{E_o(s)}{E_i(s)} = \frac{1}{R_1 R_2 C_1 C_2 s^2 + (R_1 C_1 + R_2 C_2 + R_1 C_2)s + 1} \tag{19.21}$$

Example 19.2 A common analog control element is the dc tachometer. This is basically a permanent magnet generator and is shown schematically in Fig. 19.5. For a tachometer, the input is speed and the output is voltage. Determine the transfer function of this device.

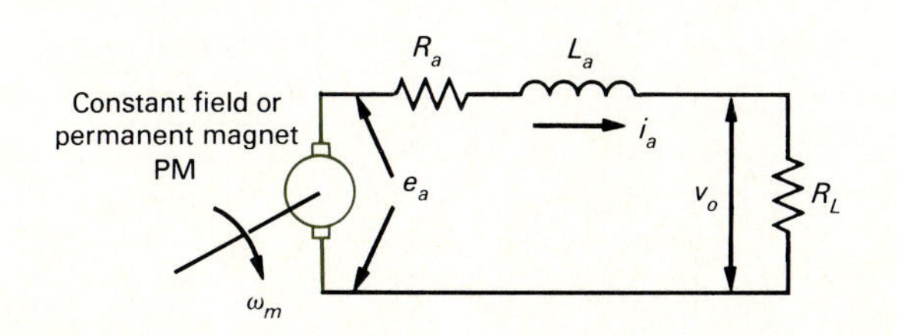

FIGURE 19.5 Example 19.2: the dc tachometer.

Solution The tachometer input is mechanical speed ω_m. The electromagnetically induced armature voltage is

$$e_a = K_a\omega_m = i_a(R_a + R_L) + L_a\frac{di_a}{dt} \tag{19.22}$$

The output of this transfer function is the voltage across an instrument, usually a dc millivoltmeter, and is

$$v_o = i_a R_L \tag{19.23}$$

Taking the Laplace transform of Eqs. (19.22) and (19.23) and solving for the transfer function gives

$$A_2(s) = \frac{V_o(s)}{\Omega_m(s)} = \frac{R_L K_a}{(R_a + R_L) + sL_a} \tag{19.24}$$

Note that, if an electronic instrument with very high impedance ($R_L \to \infty$) were used to sense the generator output, the transfer function would be

$$\begin{aligned} &A_2(s) \to k_a \\ &R_L \to \infty \end{aligned} \tag{19.25}$$

Example 19.3 The dc servomotor is another common analog element in control systems: it is shown schematically in Fig. 19.6. Again the constant field configuration is assumed, and this is frequently realized by means of permanent magnet excitation. Linear elements are assumed.

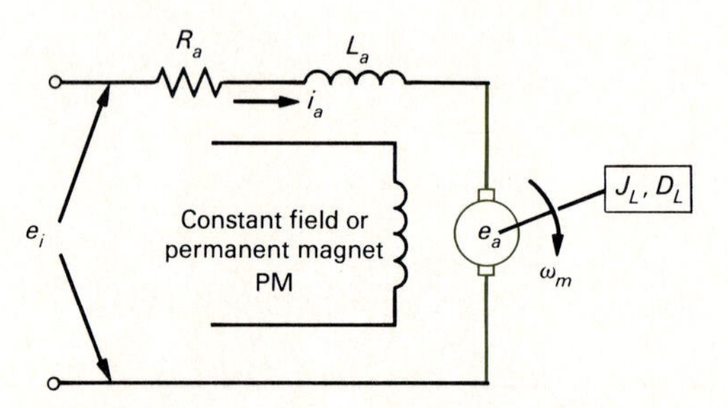

FIGURE 19.6. Example 19.3: the dc servomotor.

The differential equations describing system performance are

$$\begin{aligned} e_i &= i_a R_a + L_a\frac{di_a}{dt} + e_a \\ e_a &= K_a\omega_m \\ T_m &= K_a i_a = D_L\omega_m + J_L\frac{d\omega_m}{dt} \end{aligned} \tag{19.26}$$

in which T_m is used to represent the motor electromagnetic torque, ω_m is the motor output speed, and J_L and D_L are the inertia and viscous friction coefficients of the load, respectively. Transforming the equation set of Eq. (19.26) and solving for the transfer function gives

$$A_3(s) = \frac{\Omega_m(s)}{E_i(s)} = \frac{K_a}{(D_L + sJ_L)(R_a + sL_a) + K_a^2} \tag{19.27}$$

Example 19.4

This example will illustrate a more complex device and also introduce the concept of *on-off control*, or digital control. Figure 19.7 illustrates a basic semiconductor power controller known as a chopper. A chopper-controlled motor is frequently used in control systems, and the differential equations and transfer functions of such systems must be known. A chopper has two distinct states—the on-state, when the switch or semiconductor is fully conductive, and the off-state, when the switch or semiconductor is essentially open. For control systems analysis, it is usually adequate to assume that these two states can be represented by two idealized equivalent circuits: a short circuit for the on-state, and an open circuit for the off-state. This assumption will be used in the following analysis.

Figure 19.7*a* illustrates schematically the equivalent resistance R_a, leakage inductance L_a, and back emf or speed voltage E_a of the armature of a separately excited motor or of the

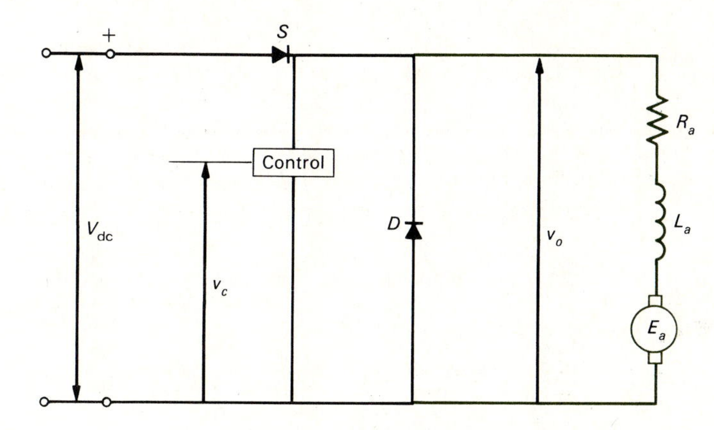

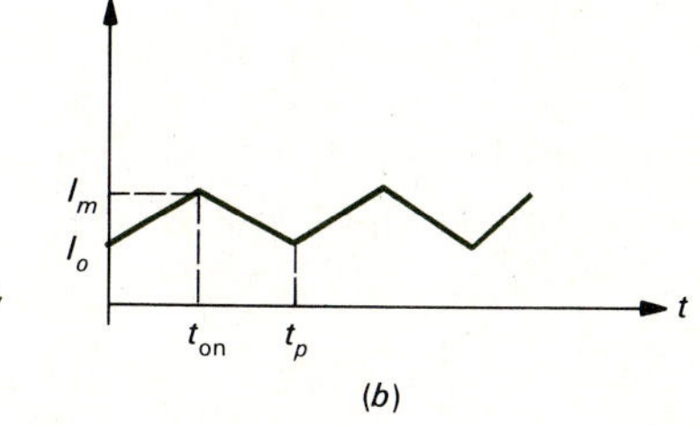

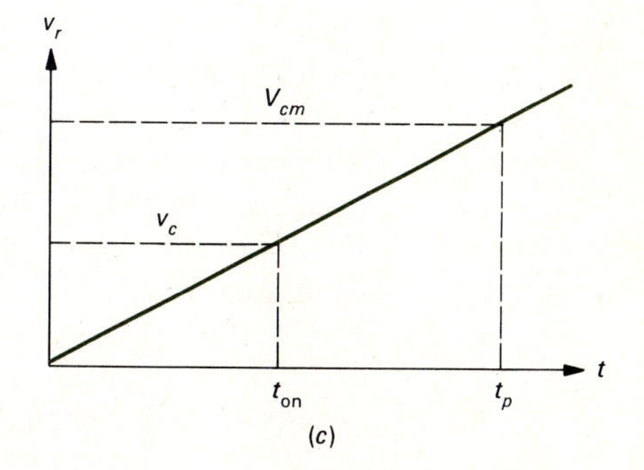

FIGURE 19.7
(*a*) A chopper controller, (*b*) its current waveforms, and (*c*) its control characteristic.

armature plus field of a series motor; it also shows a freewheeling diode D, the chopper S (neglecting commutation circuitry), and the dc source voltage E_b. Figure 19.7*b* illustrates the steady-states waveform of the motor current and an assumed chopper control characteristic, which is the familiar sawtooth capacitor charge characteristic. Note that *continuous current* in the motor is assumed; the case for discontinuous current is left as a problem at the end of this chapter. The chopper period is the time t_p; for time t_{on}, the SCR S is the on-state; for time $t_p - t_{on}$, the SCR is off and current circulates through the motor and freewheeling diode circuit. S is commutated off when the reference voltage v_r equals the capacitor voltage v_c. Determine the chopper transfer function $V_o(s)/V_r(s)$.

Solution From the geometry of Fig. 19.7*b*, note that the average motor current is

$$I_{AV} = \tfrac{1}{2}(I_m + I_o) \tag{19.28}$$

$$v_r = v_c = V_{cm}\frac{t_{on}}{t_p} \tag{19.29}$$

The average voltage across the motor during the on-time (which is also the voltage across D_{FW}) is

$$V_m = I_{AV}R_m + E_m = V_{dc} \tag{19.30}$$

since the *average* voltage across the inductance over a full period is zero. The average voltage across the motor (or D_{FW}) during the off-time is zero, neglecting the diode voltage drop. Therefore, the average motor or diode voltage (which is also v_o in Fig. 19.7) for a full period is

$$v_o = V_{dc}\frac{t_{on}}{t_p} \tag{19.31}$$

Therefore, the chopper transfer function is

$$A_4(s) = \frac{V_o(s)}{V_r(s)} = \frac{V_{dc}}{V_{cm}} \tag{19.32}$$

This is often called the chopper gain; the dynamic terms or time constants associated with the chopper itself are almost always very small compared to the time constants of other portions of the motor control system and can be ignored.

19.1.2 Multiple-Loop and Multiple-Input Systems

The block diagram algebra introduced in Eqs. (19.6) to (19.10) can be extended as a means for simplifying block diagram topography and for treating systems with more than one reference or input function. First note that for a single-loop system of the configuration of Fig. 19.2 but with *positive feedback*, Eq. (19.10) becomes

$$\frac{C(s)}{R(s)} = \frac{G(s)}{1 - G(s)H(s)} \tag{19.33}$$

Positive feedback generally results in instability, as we will soon show. However, it is frequently employed at low magnitudes or in portions of a large control system, such as in an inner loop.

Figure 19.8 illustrates a multiple-loop feedback control system. In the transfer function notation used in this figure, the argument s is omitted for simplicity. Three basic loops are shown in Fig. 19.8*a*. The positive feedback loop 2 can be eliminated by use of Eq. (19.33) and redrawn as a single transfer function (as shown in the subsequent figures) as

$$G_4' = \frac{G_2}{1 - G_2 H_2} \tag{19.34}$$

The takeoff point of a feedback loop can be relocated by including in the relocated feedback path the forward-loop transfer functions that have been bypassed. This process is illustrated by the relocation of the takeoff point for loop 3 in Fig. 19.8*c*. An inverse process is also illustrated in the relocation of loop 1 in Fig. 19.8*d*. Loop 3 in

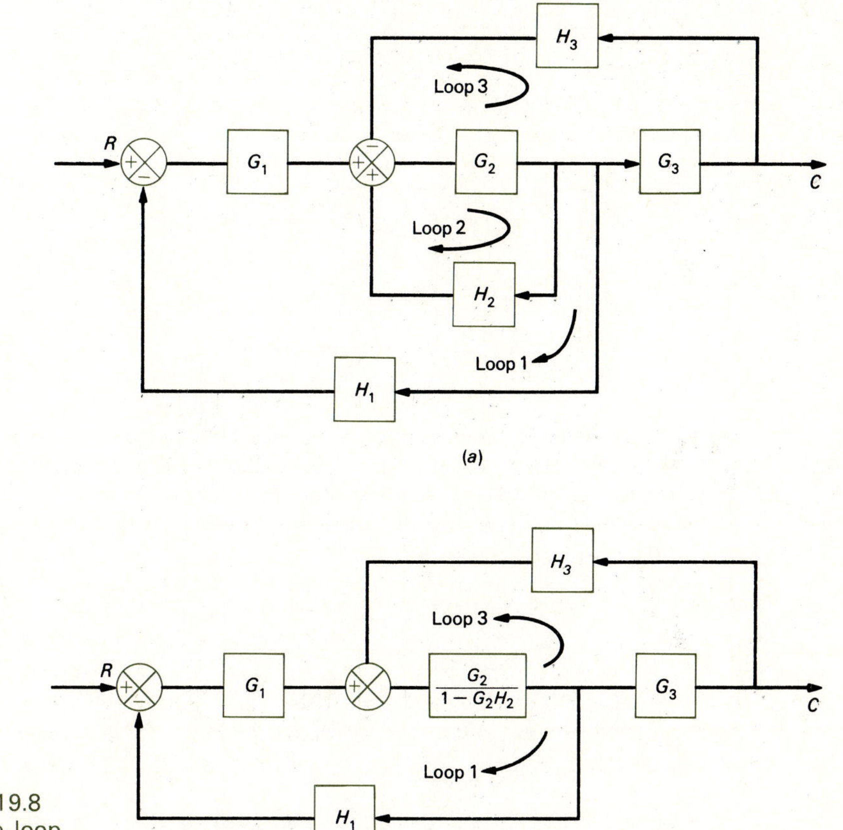

FIGURE 19.8
A multiple-loop feedback control system.

FIGURE 19.8
(*Continued*)

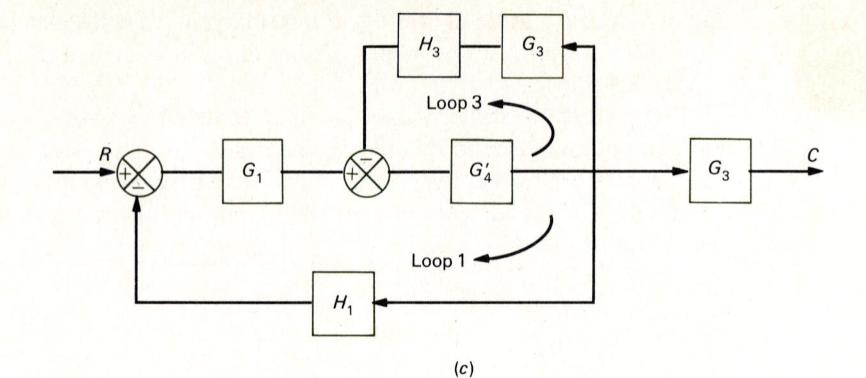

(*c*)

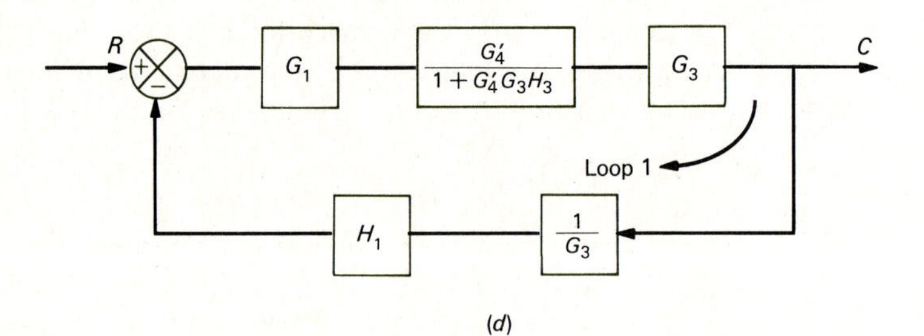

(*d*)

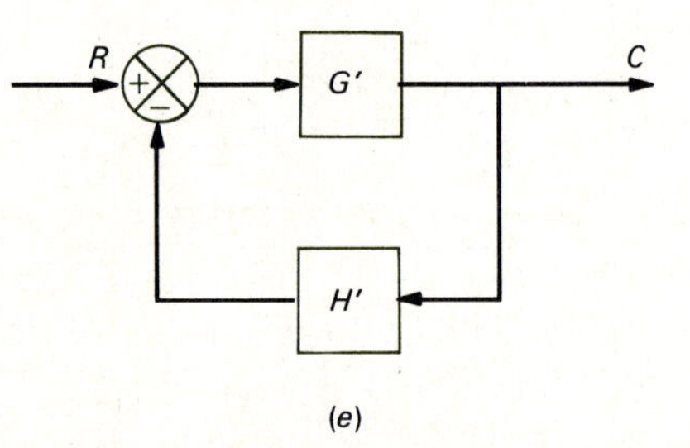

(*e*)

Fig. 19.8*c* is eliminated and expressed as a single transfer function in Fig. 19.8*d* by use of Eq. (19.10). Finally, a single forward and feedback path representation can be made as shown in Fig. 19.8*e*, where

$$G' = G_1 G_3 \frac{G_4'}{1 + G_4' G_3 H_3} \tag{19.35}$$

$$H' = \frac{H_1}{G_3} \tag{19.36}$$

Some additional loop topography that is relatively common in control systems is illustrated in Fig. 19.9*a*. The upper loop is a feed-forward loop (in contrast to a feed-back loop), quite common in digital control systems. The lower loop with parallel

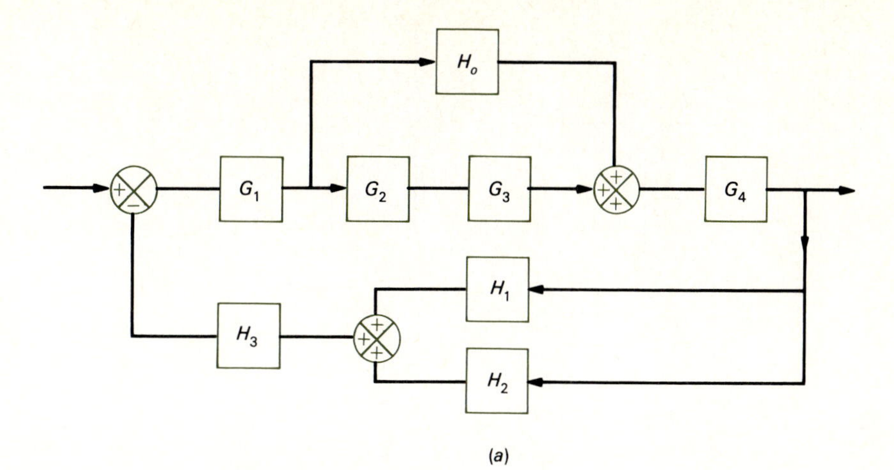

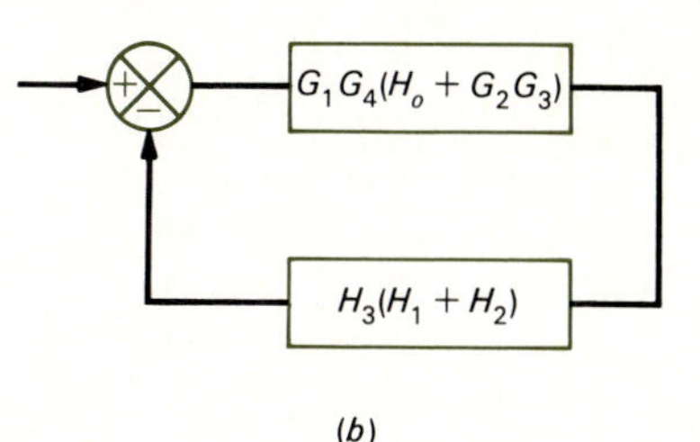

FIGURE 19.9
(*a*) Typical, additional loop topography and (*b*) its single-loop representation.

branches represents the combination of velocity and acceleration feedback that is used in many position control systems. The algebra developed in Sec. 19.1.1 is generally adequate for simplifying the system topography of Fig. 19.9*a*, as well as of many other configurations. The single-loop representation for Fig. 19.9*a* is given in Fig. 19.9*b*. The details of obtaining the simplified form are left for a problem at the end of this chapter.

In many types of feedback control systems there is more than one input or reference function. In some systems, the multiple inputs may be designed into the systems as part of the control philosophy; in others, one or more of the inputs may be reactions produced by the physical elements which compose the control system. A common example of the latter type of input is the load torque of a motor used as a control element. This type of input is often called a *load disturbance*. Examples of the former type of multiple input systems (i.e., where all input functions are designed or planned) generally result from systems in which the reference function, $R(s)$ in Fig. 19.1, is varying in response to other references. For example, a positioning system may have a reference that is moving as a function of another reference position. A classic example of such a system is the air-fuel ratio control of an internal combustion engine in which the reference for the magnitude of the air-fuel ratio is a function of engine speed, vehicle speed, engine torque, and engine temperature.

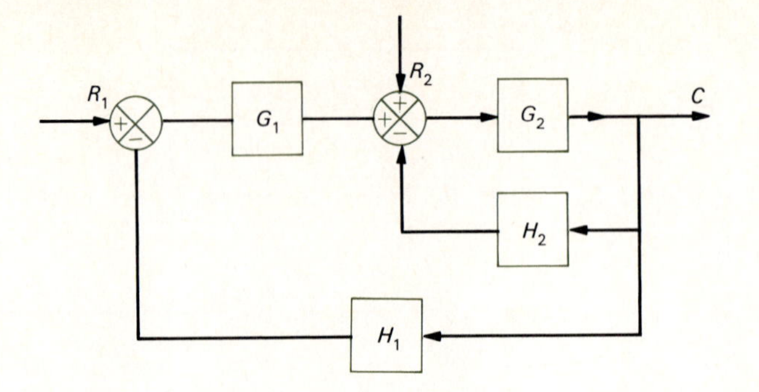

FIGURE 19.10 Multiple-loop system with two input functions, R_1 and R_2.

A generalized 2-input system is shown in Fig. 19.10. Again, we have omitted the argument s in all functions for simplicity. In *linear control systems*, multiple inputs are treated by means of the *principle of superposition*, which states that the response of a system with multiple inputs is the sum of the responses of the system to each input individually. To evaluate the system of Fig. 19.10, we will first simplify the inner loop system and replace it with a single transfer function:

$$G_2' = \frac{G_2}{1 + G_2 H_2} \tag{19.37}$$

The response to reference function R_1 is

$$C_1 = \frac{G_1 G_2'}{1 + G_1 G_2' H_1} R_1 \tag{19.38}$$

The response to R_2 is

$$C_2 = \frac{G_2'}{1 + G_1 G_2' H_1} R_2 \tag{19.39}$$

The system response is

$$C = C_1 + C_2 = \frac{G_1 G_2' R_1 + G_2' R_2}{1 + G_1 G_2' H_1} \tag{19.40}$$

The following examples will illustrate how a load torque on a dc motor may be treated as an input function.

Example 19.5 Add load torque to the load on the servomotor of Fig. 19.6, and from the resulting equations develop a block diagram with motor speed as reference and load torque as a second input or load disturbance.

Solution A voltage signal will serve as a speed reference; also, we will illustrate the inherent feedback due to the motor back emf by rearranging the equations developed in Example 19.3.

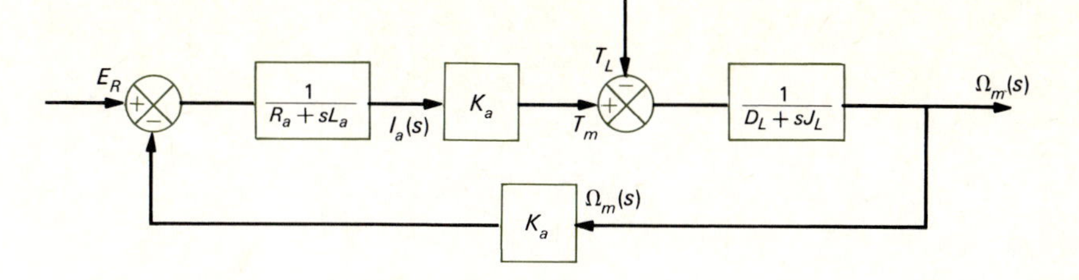

FIGURE 19.11 Example 19.5: block diagram of dc servomotor with load torque.

The only change in these equations describing motor performance is the addition of the load torque T_L to the right side of the third equation of Eq. (19.26), giving

$$T_m = K_a i_a = D_L \omega_m + J_L \frac{d\omega_m}{dt} + T_L \tag{19.26a}$$

To show the inner feedback path, we will maintain the second equation of Eq. (19.26); also, to introduce the load disturbance input, we will maintain the summation shown by Eq. (19.26*a*). Transforming Eq. (19.26*a*) and the first two equations of Eq. (19.26) and adding the reference signal $E_R(s)$ gives the configuration of Fig. 19.11.

It is important to remember that the multiple-loop and multiple-input techniques discussed above are valid only in linear systems. In general, there are no similar reduction or superposition methods for nonlinear systems, and computer modeling is generally required.

19.2 SOME ELEMENTARY FEEDBACK SYSTEMS

A few of the basic electrical or electronic feedback systems will be described in this section by means of examples. In developing the differential equations and transformed equations for these systems, we will use lowercase letters for time functions and uppercase letters for frequency-response or Laplace-transformed functions. Also, circuit constants (resistances, amplifier gains, etc.) will be represented by uppercase letters.

Example 19.6: Feedback Amplifier

The feedback, or operational, amplifier has been described in detail earlier. The purpose of this example is to relate control theory concepts and nomenclature to the feedback amplifier. The operational amplifier is one of the most widely used components in control systems and is the cornerstone of many analog control systems. It is a means of simulating many transfer functions in control systems as well as in analog computers. Recall from Chap. 9 that the basic operational amplifier has a tremendously large gain and that the input current to the amplifier can be considered zero. Figure 19.12*a* illustrates an amplifier with feedback and input impedances Z_{fb} and Z_i, respectively. It was shown in Chap. 9 that the transfer function for this system is

$$G(s) = \frac{E_o(s)}{E_i(s)} = -\frac{Z_{fb}}{Z_i} \tag{19.41}$$

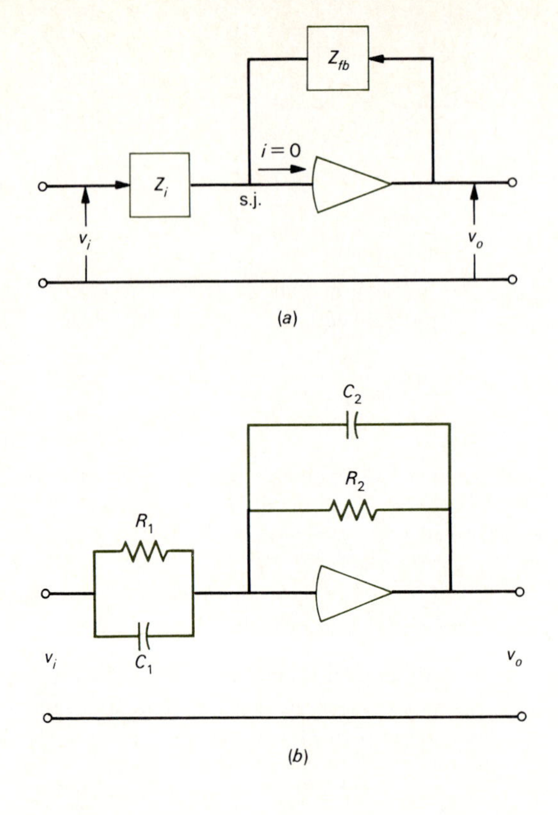

FIGURE 19.12 (*a*) Feedback amplifier with input impedance (s.j. = summing junction). (*b*) Feedback amplifier with *RC* networks.

With the two impedances taking on the *RC* network configurations as shown in Fig. 19.12*b*, the transfer function becomes

$$G(s) = -\frac{R_2}{R_1}\frac{1 + R_1C_1S}{1 + R_2C_2S} \tag{19.42}$$

This function is frequently used in control systems to modify phase characteristics.

Example 19.7: Wien-bridge Oscillator

This device is primarily used as an oscillator or variable-frequency source rather than as a control system, but it is a good example of the use of feedback in an electronic circuit—in this case, positive feedback, a characteristic of electronic oscillators. The basic Wien-bridge oscillator circuit is shown in Fig. 19.13*a*. This figure illustrates the source of the name of this circuit by showing the circuit elements in the form of the familiar Wien bridge, a circuit long in use as a means of accurate measurement of capacitance. A bridge balance (when used as a capacitance-measuring circuit) is achieved when the voltage $v_o = 0$. Note that this condition corresponds to the assumption of infinite gain in the operational amplifier, as noted in Chap. 9 and the previous example, and is a condition for oscillator operation in the Wien-bridge oscillator. The bridge form of the circuit may be redrawn to illustrate the feedback characteristics, as shown in Fig. 19.13*b*. Note the existence of positive feedback. The forward-loop transfer function is

$$G(s) = \frac{R_2 + R_1}{R_2} \tag{19.43}$$

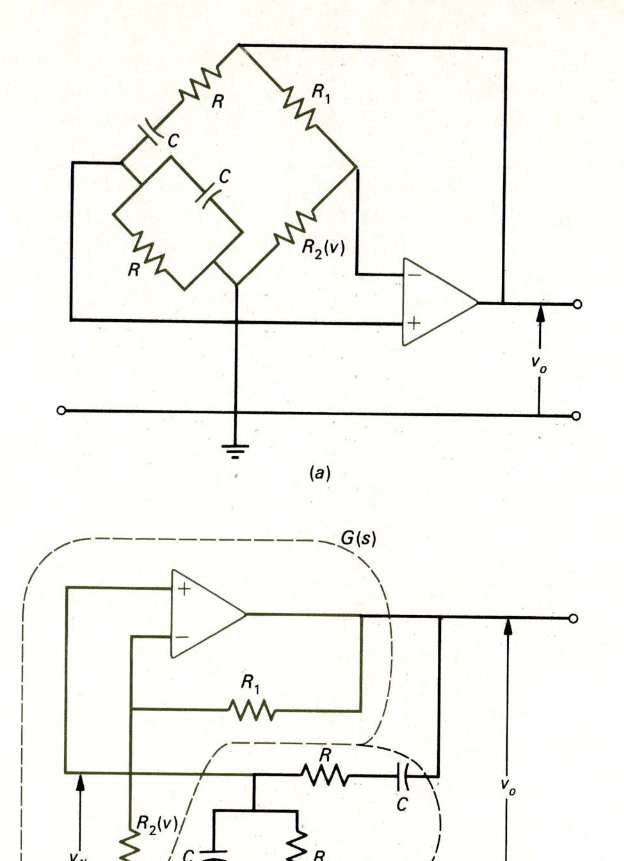

FIGURE 19.13
(*a*) Wien-bridge oscillator.
(*b*) Wein-bridge oscillator, illustrating positive feedback.

The feedback transfer function is

$$H(s) = \frac{1}{3 + RCS + 1/RCS} \tag{19.44}$$

Oscillation occurs when the following conditions are satisfied:

$$f_o = \frac{1}{2\pi RC} \tag{19.45}$$

$$R_2 = 2R_1 \tag{19.46}$$

The resistances R_1 and R_2 are nonlinear, the former consisting of an incandescent lamp or other impedance whose resistance increases with temperature, and the latter consisting of a voltage-sensitive resistance whose resistance decreases with voltage, such as a field-effect transistor.

Stable operation exists when the voltages across the two resistive arms are equal as a result of variations in the nonlinear resistances in these arms, which result in satisfying Eq. (19.46). The conditions for oscillations will be further amplified in the analysis section which follows.

Example 19.8: Ward-Leonard Control

This means of very precise control of huge electromagnetic machines has been in use for many years and is one of the early examples of feedback control applications. The Ward-Leonard system is used in the control of large dc motors used in rolling mills and similar applications. With feedback, this system is capable of such astonishing feats as reversing a 10,000-hp motor in less than 1 s while carrying full load. A simplified schematic circuit is shown in Fig. 19.14. For a

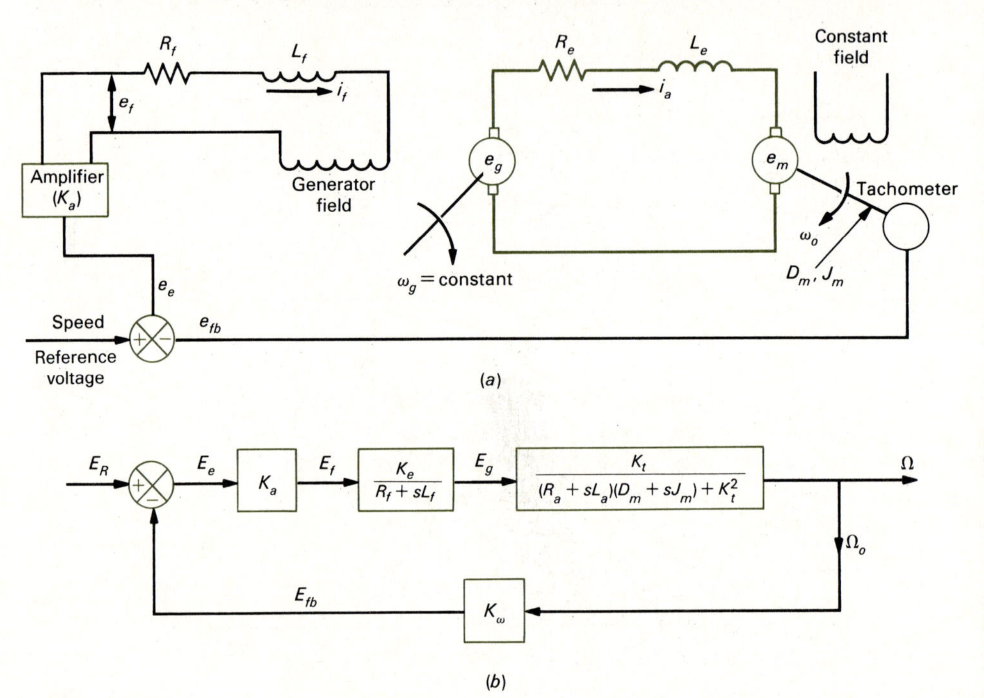

FIGURE 19.14 (*a*) Schematic representation of Ward-Leonard dc motor control. (*b*) Block diagram of the Ward-Leonard system.

given load and base speed condition, it is frequently possible to assume that the field current (and hence machine magnetic excitation) is held constant. Therefore, the simplified constant-field motor and generator (tachometer) transfer functions developed in Examples 19.3 and 19.2, respectively, are used here. We will also ignore load torque in the development, but it may be included in the analysis by the methods discussed in Sec. 19.1.2. Likewise, it can be assumed that the main generator speed is held constant under all conditions of operation. In Fig. 19.14*a*, the tachometer, error detector, and amplifier are represented symbolically rather than by electric circuit connections. The parameters R_e and L_e represent the equivalent resistance and inductance, respectively, of the generator and motor armature circuits, which effectively are in series. Let us determine the transfer functions and block diagram for this system.

First the power generator will be developed. The differential equations describing this element are (referring to Fig. 19.14*a*)

$$e_f = i_f R_f + L_f \frac{di_f}{dt} \tag{19.47}$$

$$e_g = K_e i_f \qquad \text{for } \omega_g = \text{constant} \tag{19.48}$$

Transforming and rearranging gives the transfer function for the generator operated at constant speed:

$$\frac{E_g}{E_f} = \frac{K_e}{R_f + sL_f} \tag{19.49}$$

The differential equation in the armature circuits is

$$e_g = i_a R_e + L_e \frac{di_a}{dt} + e_m \tag{19.50}$$

The motor differential equations are

$$e_m = K_t \omega_o \qquad \text{for constant field current} \tag{19.51}$$

$$T_d = K_t i_a = D_m \omega_o + J_m \frac{d\omega_o}{dt} \qquad \text{neglecting load torque} \tag{19.52}$$

Transforming Eqs. (19.50) to (19.52) and rearranging gives the armature circuit and motor transfer function:

$$\frac{\Omega_o}{E_g} = \frac{K_t}{(R_a + sL_a)(D_m + sJ_m) + K_t^2} \tag{19.53}$$

The remaining component in the forward loop is the amplifier. The transfer function of electronic amplifiers (such as the operational amplifiers described in Chap. 9) can generally be approximated by a gain term with no dynamic or time constant terms, since the time constants of these devices are extremely small compared to electric power or electromechanical time constants. With this assumption, the forward-loop transfer function of the Ward-Leonard system is

$$G(s) = \frac{\Omega_o}{E_e} = \frac{K_a K_e K_t}{(R_f + sL_f)[(R_a + sL_a)(D_m + sJ_m) + K_t^2]} \tag{19.54}$$

All symbols are defined in Fig. 19.14*a*. The feedback loop consists of the tachometer. This could be described in terms of the dc generator with constant-field excitation transfer function developed in Example 19.2. However, due to the small physical size of the typical dc or ac tachometer, the dynamic terms shown in Eq. (19.24) can be ignored. The typical tachometer is generally treated as a speed-to-volts transducer with a transfer function K_ω which is a gain term. The feedback-loop transfer function is

$$H(s) = K_\omega \tag{19.55}$$

The block diagram of the Ward-Leonard system is shown in Fig. 19.14*b*.

Example 19.9: Voltage Regulator

Figure 19.15*a* illustrates the simplified circuit diagram of the buck regulator circuit, one of many types of circuits used to control the output voltage of electronic power supplies. The control mechanism used to control output voltage in this—and many other—circuits is pulse-width modulation (PWM), a technique for controlling *average* voltage by varying the amount

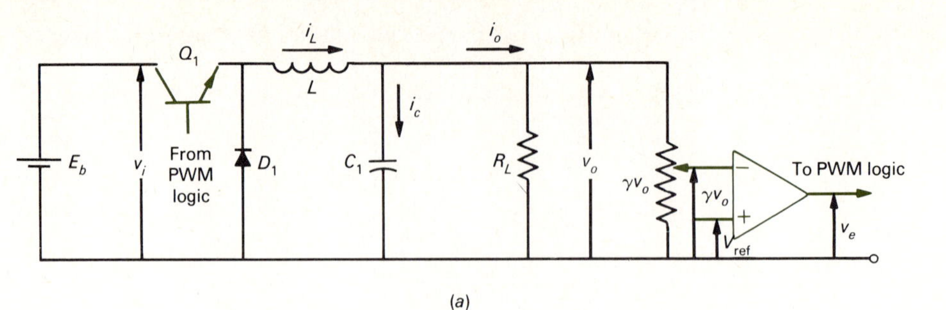

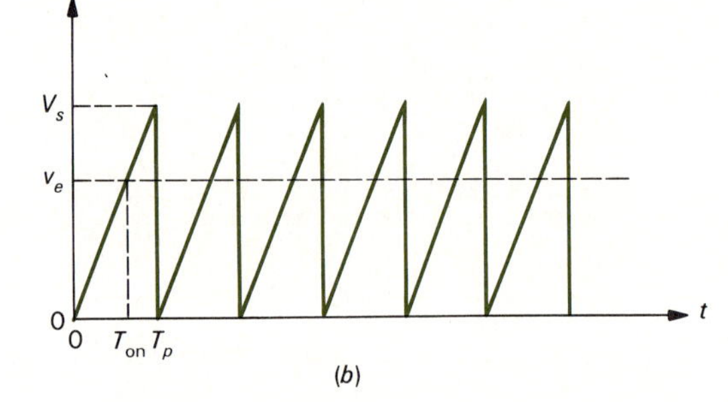

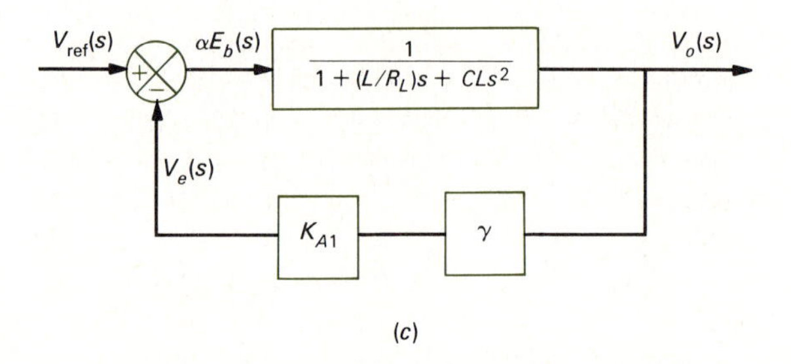

FIGURE 19.15 (*a*) Schematic circuit for electronic voltage regulator. (*b*) Logic relationships for PWM control. (*c*) Block diagram showing control signal flow for electronic regulator.

of time the input voltage v_i is connected to the circuit by means of transistor Q_1. The output voltage is controlled by the width of the "on" pulse of v_i, as shown in Fig. 19.15*b*. Also shown in Fig. 19.15*b* is one of the common means of controlling the pulse period—which is to compare the output of the error-detector amplifier A_1 with the voltage of a standard capacitor charging voltage (sawtooth) wave and to turn on Q_1 when these two signals are equal. Note that this gives a limited range of control determined by the sawtooth peak voltage V_s, which is one example of the nonlinearity known as *saturation* that occurs in most electronic circuits. We will determine the transfer functions and block diagram for this circuit.

Neglecting the internal impedances of the voltage source E_b, the average voltage applied to the regulator during the period T_p is

$$V_i = \frac{T_p - T_{\text{on}}}{T_p} E_b = \alpha E_b \tag{19.56}$$

Since continuous current is maintained by the freewheeling diode D_1, it can be assumed that this average voltage is applied to the smoothing inductor and output capacitor each period, resulting in the differential equations

$$v_i = L\frac{di_L}{dt} + v_o$$

$$i_c = C\frac{dv_o}{dt} \tag{19.57}$$

$$i_L = i_o + i_c = \frac{v_o}{R_L} + i_c$$

Transforming Eq. (19.57) and solving for the ratio of output to input voltage, we have

$$G(s) = \frac{V_o(s)}{V_i(s)} = \frac{1}{1 + (L/R_L)s + CLs^2} \tag{19.58}$$

The block diagram is shown in Fig. 19.15*c*

19.3 LINEAR CONTROL SYSTEMS ANALYSIS

Classical linear control theory—which has fostered the rapid growth of control systems in industrial societies—includes a large body of theorems and analytical techniques. Portions of this theory encompass and define some of the fundamental concepts that underlie the nature of systems operations, including the basic concepts of electrical systems theory. Much of linear control theory is based upon the frequency-response formulation of the systems equations, and a large body of quasi-graphical and algebraic techniques has been developed to analyze and design linear control systems based upon frequency-response methods. An introduction to several of these techniques will be given in this section; for more complete discussion of frequency-response methods, a great many articles and texts are available, including Refs. 19.1 to 19.3 at the end of this chapter.

With the development and widespread use of discrete (digital) control systems and the advent of relatively low-cost digital computers, time-response methods have become both more necessary and more available. Time-response systems analysis may be divided into two broad methodologies: (1) the actual simulation or modeling of the system differential equations by means of either analog or digital computers and (2) the state-variable formulation of the system state equations and their solution by means of a digital computer. A good basic reference for the state-variable formulation (among very many) is given as Ref. 19.4. State-variable methods are probably the most general approach to system analysis and are useful in the solution of both linear and nonlinear systems equations. The brief introduction to multiple-loop and multiple-input systems given in Sec. 19.1.2 may have already alerted the reader to the conclusion that this approach—using signal-flow block diagrams and frequency-response formulation—is very limited, generally to not more than three inputs and probably the same number of inner loops. The matrix formulations associated with state-variable techniques have largely supplanted the block diagram formulations as illustrated in Sec. 19.1.2, and computer software for solving a great variety of state equation formulations is available on most computer systems today.

However, it should be noted that much of the physical reality of any system is lost in the state-variable formulation, including the relationships between system response and system parameters. Furthermore, a great many of the concepts and design parameters used in many practical control systems of industry and aerospace today are based upon both the nomenclature and theory of the frequency-response methods. Therefore, although frequency-response techniques are limited to relatively simple systems—and, in the rigorous mathematical sense, to linear systems—they are still most useful in system design, stability analysis of practical systems, and can give a great deal of information about the relationships between system parameters (time constants, gains, etc.) and system response. Under certain conditions, frequency-response methods can be adapted to nonlinear systems.[19.1,19.5] We will now examine the basic concepts of frequency response and introduce a few of these valuable frequency-response methods for designing and analyzing control systems.

19.3.1 Frequency-Response Formulation

Most of the equations presented so far in this chapter have been in the *frequency-response formulation.* This term refers to the Laplace transformed form of the differential equations describing systems performance. In electric circuits and machines, the frequency-response formulation can also be developed (with care) by expressing the steady-state ac equations in terms of the frequency variable $j\omega$. The differences between these two variables, the Laplacian s and the Fourier $j\omega$, are very meaningful and should be understood by the reader. However, *as a methodology for developing frequency-response transfer functions and in total disregard of mathematical rigor,* either formulation is acceptable and should be used when appropriate for a given problem, and the interchange of variables.

$$s \leftrightarrow j\omega \tag{19.59}$$

is generally acceptable. We will illustrate this approach by means of a return look at Example 19.1:

Example 19.10 Develop the transfer function of the RC network shown in Fig. 19.4 be means of steady-state ac network analysis.

Solution Assuming that the input function e_i can be represented by an RMS voltage function E_i, the network equations are:

$$\begin{aligned} E_i &= (R_1 - jX_{C1})I_1 + jX_{C1}I_2 \\ 0 &= jX_{C1}I_1 + [R_2 - j(X_{C1} + X_{C2})]I_2 \\ E_o &= -jX_{C2}I_2 \end{aligned} \tag{19.60}$$

where X has the meaning $1/\omega C$.

Solving for the ratio of output voltage over input voltage gives

$$A_1(j\omega) = \frac{E_o(j\omega)}{E_i(j\omega)} = \frac{1}{\dfrac{R_1R_2}{j^2X_{C1}X_{C2}} + 1 + j\left(\dfrac{R_1}{X_{C2}} + \dfrac{R_2}{X_{C2}} + \dfrac{R_1}{X_{C1}}\right)} \tag{19.61}$$

It is seen that interchanging the frequency variables as in Eq. (19.59) makes Eq. (19.61) equivalent to Eq. (19.21). Therefore, the use of steady-state ac network theory as presented earlier can be most useful in deriving transfer functions for electrical and electromechanical networks and systems.

However, frequency-response expressions are generally formulated in terms of the Laplace transform variable s. Generalized forms of frequency-response expressions have already been presented in Eqs. (19.11) to (19.15). An additional standard form for the complex second-order expression, Eq. (19.14), is given by

$$(s_2 \pm 2\zeta\omega_o s + \omega_o^2) \tag{19.62}$$

where ω_o is called the natural frequency, in s^{-1} and

ζ is called the damping ratio, dimensionless

The solutions for s in Eqs. (19.12) to (19.14) and (19.62), obtained by setting the polynomial equal to zero, are known as roots of the polynomial. The roots of Eqs. (19.12) and (19.13) are obviously

$$s = \pm a \tag{19.63}$$

The complex roots of Eqs. (19.62) are

$$s = \pm\zeta\omega_o \pm j\omega_o\sqrt{1-\zeta^2} \tag{19.64}$$

Roots of polynomials are frequently illustrated graphically by being plotted on the complex plane of complex variable theory known as the S plane, a concept useful in frequency-response analysis.

Two polynomial expressions of significance in the analysis of linear feedback control systems are the *open-loop response*, given by Eq. (19.9), and the *closed-loop response*, given by Eq. (19.10). A number of examples given previously have illustrated the methods for obtaining the forward-loop transfer function $G(s)$ and the feedback-loop transfer function $H(s)$ for various types of electrical and electromechanical systems. The next step in the analysis is to rearrange these equations into the standard polynomial forms given by Eqs. (19.11) to (19.15). Let us perform this step for the Ward-Leonard system of Example 19.8, with $G(s)$ given by Eq. (19.54) and $H(s)$ given by Eq. (19.55). Taking the product of G and H and factoring through by the coefficient of the term with highest order in s gives, for the open-loop function.

$$G(s)H(s) = \frac{(K_aK_tK_\omega K_eL_aL_fJ_m)}{\left\{s^3 + \left[\dfrac{R_a}{L_a} + \dfrac{R_f}{L_f} + \dfrac{D_m}{J_m}\right]s^2 + \left[\dfrac{R_f}{L_f}\left(\dfrac{D_m}{J_m} + \dfrac{R_a}{L_a}\right) + \dfrac{(R_aD_m + K_t^2)}{L_aJ_m}\right]s + \dfrac{R_f(R_aD_m + K_t^2)}{L_aL_fJ_m}\right\}} \tag{19.65}$$

Note that this is the unfactored form of the polynomial Eq. (19.15), which is the form usually obtained from determination of the transfer functions, except in the simplest configurations. This type of expression is often written in the form

$$G(s)H(s) = \frac{K_{GH}}{s^3 + C_1 s^2 + C_2 s + C_0} \tag{19.66}$$

From basic Laplace transform theory, the reader should note that the *steady-state gain* of the open-loop system when excited by a unit step function is

$$\lim_{s \to 0} [sG(s)H(s)] = \lim_{s \to 0} \frac{sK_{GH}}{s(s^3 + C_1 s^2 + C_2 s + C_0)} = \frac{K_{GH}}{C_0} \tag{19.67}$$

The closed-loop response is found by inserting Eqs. (19.65) and (19.53) into the right-hand side of Eq. (19.10). This is a most tedious process to perform by hand for even the relatively simple third-order function of the Ward-Leonard example; using the simplified notation of Eq. (19.66), we obtain

$$\frac{C(s)}{R(s)} = \frac{G(s)}{1 + G(s)H(s)} = \frac{K_{GH}/K_\omega}{s^3 + C_1 s^2 + C_2 s + C_0 + K_{GH}} \tag{19.68}$$

Again we have ended up with an expression in the unfactored form of Eq. (19.15).

Now, let us look once again at Eqs. (19.9) and (19.10). The left-hand side of both equations is a ratio of a transformed output function to a transformed input function. Cross multiplying by the input function gives the transformed differential equation, from which the output function or output response can be determined. We will repeat these equations in this form:

$$F(s) = G(s)H(s)E(s) \qquad \text{open loop} \tag{19.69}$$

$$C(s) = \frac{G(s)R(s)}{1 + G(s)H(s)} \qquad \text{closed loop} \tag{19.70}$$

To help understand these equations, refer to the system diagrams (Figs. 19.1 to 19.3). Also note that the denominators of the open-loop and closed-loop transfer functions, examples of which are given in Eqs. (19.65) to (19.68), represent the characteristic equations of Eqs. (19.69) and (19.70) from which the form of the system response or output function is determined. The transfer functions result from the transformed homogeneous differential equations describing the system, which, it is recalled, determines the particular solution (often termed the transient or impulsive response) of the differential equations. The concept of impulsive response (i.e., the response to the unit impulse input) is particularly useful in the evaluation of linear control systems since the Laplace transform of the unit impulse is unity.

Frequency-response methods are basically, therefore, an evaluation to determine as much about system response as possible from the frequency-response formulations given in this section. Ideally, if the unfactored form of the characteristic equation could be factored into real and complex-pair factors, the system response could be determined exactly by taking the inverse transform of Eqs. (19.69) and (19.70). There are two cautions to this approach, however: first, the factoring of the form, Eq. (19.5), requires a sizeable computer capability for all but the simplest of systems; second, even if the exact solution is obtained by this method, the results are usually difficult to interpret as far as the quality of response is concerned—except by brute-force plotting of the time-response function based upon the inverse transforms, and this likewise requires a large-scale computer system with considerable graphics capability. Since frequency-response methods were developed before the advent of low-cost large-capacity computing systems, they attempt to relate the nature and quality of closed-loop response to the nature of an open-loop characteristic equation without determining the exact system response. They consist of a number of numerical and graphical techniques, generally most useful for linear continuous systems with fairly simple topography and a low number of inputs. Although some of the basic theory of system operation has been formulated in terms of frequency response, as noted above, the principal function of frequency-response techniques today is as a first-stage or preliminary design tool and as a means of relating stability and other system response functions to circuit parameters.

19.3.2. The S Plane; Poles and Zeros

Much of this material should be familiar to the reader and is inserted here primarily for completeness and as a reference for subsequent developments.

Briefly, the S plane is merely the two-dimensional graph on which the roots of the transformed differential equation are plotted. The roots of differential equations, such as Eqs. (19.69) and (19.70), are classified as poles or zeros depending upon their location in the denominator or numerator, respectively, of the equation. The name "zero" is obvious, because when the transform variable s is set equal to the root of the numerator polynomial, the output function becomes zero. The name "pole" is less obvious; when the transform variable is set equal to the root of the polynomial in a denominator term, the transformed equation "blows up"; that is, it mathematically takes on the value of infinity. From the methods of solving linear differential equations by means of Laplace transform, note that the nature of the denominator roots, or poles, determines the form of the time response. Simple poles of the form $(s + a)$ result in the time response e^{-at}; multiple poles $(s + a)^n$ result in the time response $t^{(n-1)}e^{-at}/n!$; the time response resulting from the complex pairs given in Eqs. (19.62) and (19.63) result in damped sinusoidal functions. Note the inverse relationship between the sign of the pole and the sign of the damping coefficient in the exponential terms. Negative real roots or negative real parts of complex roots result in exponential terms with negative damping coefficients ($s = -a$, or $s = -\zeta \pm j\omega_o$). From the final value theorem noted in the previous section and by observation of the form of the time response exponential terms, it is seen that the final value (steady-state value) of the time-response term resulting from negative real poles or from complex pair poles with negative real parts is zero. Time-response terms resulting from poles with positive real parts become infinite mathematically.

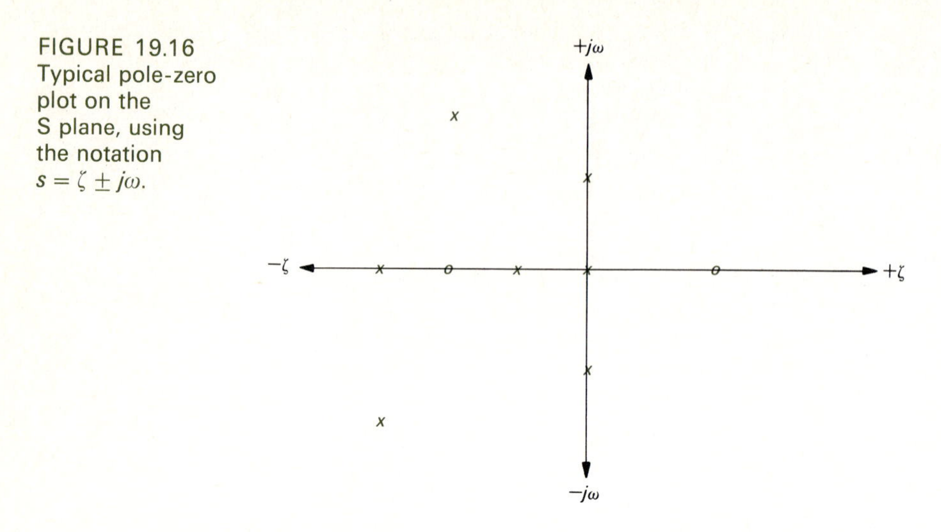

FIGURE 19.16 Typical pole-zero plot on the S plane, using the notation $s = \zeta \pm j\omega$.

Figure 19.16 illustrates a typical S-plane graph of the poles and zeros of a transformed equation. Poles are designated by the symbol x and zeros by the symbol o. Complex-pair roots always occur symmetrically about the horizontal axis on the S plane. Real roots lie on the horizontal axis. Purely imaginary roots (complex pairs with zero real part) lie on the vertical axis. The origin represents the location of the real root $s = 0$, which in electrical systems represents a response equivalent to the steady-state response of a circuit excited by direct current. In regard to control system response, the vertical axis represents a very significant dividing line: roots with negative real parts lie to the *left* of this line, and this portion of the plane is, not surprisingly, called the left-hand S plane. The significance of topography will now be formalized.

19.3.3 Control System Stability

The systems concept of stability refers to the ability of a system to arrive at a stable condition of operation—what we have been referring to as the *steady-state*—in which input and output functions are related by numerical constants derived from system coefficients in the complementary solution of the system differential equation. Mathematically, the concept of stability in a linear system can be precisely defined: the system characteristic homogeneous equation must contain no poles in the right-half S plane, that is, no poles with positive real parts. Note that this mathematical criterion is an either-or criterion; the system is either stable or it blows up, goes to an infinite output. In a practical system this condition of *infinite response* seldom if ever occurs, for it is usually prevented by the onset of nonlinearities in system parameters—such as saturation in electronic amplifiers, stress in a structural member beyond the yield strength, or actual fracturing of a structural member. However, the tendency towards infinite response until something saturates or breaks is very real in linear systems and must be eliminated in order for a system to operate properly.

The linear stability of negative feedback systems constructed from basically stable components and subsystems is primarily a function of system or "loop" gain. This is the parameter that has already been defined as steady-state gain in Eq. (19.67) and is the product of gain terms in the forward and feedback loops. The word "gain" is here used somewhat loosely and actually refers to the magnitude terms in the open-loop transfer function; the system gain may, in actuality, be a number less than 1.0, depending upon the choice of system units. Because of the importance of the gain or magnitude function upon system stability—as well as upon other important aspects of system response—the typical open-loop transfer function, Eq. (19.66), is often written in the form

$$G(s)H(s) = KG'(s) = \frac{K_{GH}/C_0}{s^3/C_0 + s^2 C_1/C_0 + sC_2/C_0 + 1} \tag{19.71}$$

where K is the open-loop or steady-state gain

$G'(s)$ contains all terms involving s and has a final value of 1.0

A negative feedback system that is found to be unstable at a given value of K can usually be made stable by reducing the value of K. In electrical systems, this is normally accomplished by adjustment of the gain of an electronic amplifier in the forward loop, such as K_a in the Ward-Leonard system of Fig. 19.14. By contrast, in a system exhibiting nonlinear instability, stability can be achieved in many cases by *increasing* the forward-loop gain. In linear systems (and most nonlinear systems) the reduction of system gain always results in some compromises or tradeoffs with other aspects of system response. In linear systems, the speed of response or response time generally deteriorates as the system gain is reduced; that is, the system responds more slowly. Also, there may be some deterioration in the accuracy of the steady-state response as the system gain is reduced. Therefore, the principal problem of control systems analysis is to achieve stable operation and the required or specified response time and steady-state accuracy.

The either-or stability criterion for linear systems can be most easily evaluated by means of an analysis of the coefficients (including algebraic sign) of the transformed characteristic equation, i.e., the denominators of such expressions as illustrated in Eqs. (19.65) to (19.71). We have been calling the form of these expressions the *unfactored form*; that is, the roots are not known. A general form for the unfactored polynomial in the transformed variable s is given by the right-hand side of Eq. (19.15). Observe that the polynomial must be arranged in decreasing powers of s. In the following discussions, the algebraic sign in front of the coefficents c_i is assumed to be included as part of the coefficient. Linear instability is indicated by the presence of negative real roots in the *characteristic*, or homogeneous, equation of the form

$$s^n \pm c_1 s^{n-1} \pm c_2 s^{n-2} \pm c_i s^{n-i} \pm \cdots \pm c_{n-1} s \pm c_n = 0 \tag{19.72}$$

Visual examination of the form of Eq. (19.72) can give considerable information on the presence of roots with negative real parts, which will be indicated by

1 A negative c_i

2 The absence of a c_i [other than c_n itself, which means that the polynomial can be factored to the $(n-1)$ form]

Further analysis when all c_i's are positive and nonzero is made by means of a coefficient manipulation technique known as Routh's criterion.[19.1,19.2] Routh's criterion is as follows: With a polynomial arranged in decreasing powers of s, as in Eq. (19.72), set up an array of coefficients and derived terms. For example, with $n = 6$, Eq. (19.72) becomes

$$s^6 + c_1 s^5 + c_2 s^4 + c_3 s^3 + c_4 s^2 + c_5 s + c_6 = 0 \tag{19.72a}$$

The array is formed by arranging the c_n's as follows:

$$\begin{array}{llll l} 1 & c_2 & c_4 & c_6 & n \text{ even} \\ c_1 & c_3 & c_5 & & n \text{ odd} \\ d_1 & d_3 & d_5 & & \\ e_1 & e_3 & & & \\ f_1 & f_3 & & & \\ g_1 & & & & \\ h_1 & & & & \end{array} \tag{19.73}$$

where $$d_1 = \frac{\begin{matrix} 1 & c_2 \\ c_1 & c_3 \end{matrix}}{c_1} = \frac{c_1 c_2 - c_3}{c_1}$$

$$d_3 = \frac{\begin{matrix} 1 & c_4 \\ c_1 & c_5 \end{matrix}}{c_1} = \frac{c_1 c_4 - c_5}{c_1}$$

$$d_5 = \frac{\begin{matrix} 1 & c_6 \\ c_1 & \end{matrix}}{c_1} = c_6$$

$$e_1 = \frac{\begin{matrix} c_1 & c_3 \\ d_1 & d_3 \end{matrix}}{d_1} = \frac{d_1 c_3 - c_1 d_3}{d_1}$$

$$e_3 = \frac{\begin{matrix} c_1 & c_5 \\ d_1 & d_5 \end{matrix}}{d_1} = \frac{d_1 c_5 - c_1 d_5}{d_1}$$

$\vdots$

$$g_1 = \frac{\begin{matrix} e_1 & e_3 \\ f_1 & f_3 \end{matrix}}{f_1} = \frac{f_1 e_3 - e_1 f_3}{f_1}$$

$$h_1 = \frac{\begin{matrix} f_1 & f_3 \\ g_1 & \end{matrix}}{g_1} = f_3$$

The process of deriving new coefficients continues until only zeros are produced for additional coefficients. Routh's criterion states

> The presence of roots with negative real parts is indicated by a sign change in the left-most column of [Eq. (19.73)]; the *number* of roots with negative real coefficients is indicated by the *number of sign changes.*

Equation (19.73) is known as the *Routh array.*

Example 19.11 Determine the number of roots with positive real parts in the characteristic equation

$$s^4 + 5s^3 + 16s^2 + 31s + 18 = 0$$

Solution The Routh array is

$$\begin{array}{lll} 1 & 16 & 18 \\ 5 & 31 & \\ 9.8 & 18 & \\ 21.8 & & \\ 18 & & \end{array} \qquad \begin{aligned} d_1 &= \frac{(5 \times 16) - (1 \times 31)}{5} = 9.8 \\ d_3 &= 18 \\ e_1 &= \frac{(9.8 \times 31) - (5 \times 18)}{9.8} = 21.8 \\ f_1 &= 18 \end{aligned}$$

Since there are no sign changes in the left column, there are no roots with a positive real part.

Example 19.12 Determine the number of roots with positive real parts in the characteristic equation

$$s^5 + s^4 + 2s^3 + 5s^2 + 3s + 8 = 0$$

Solution The Routh array is

$$\begin{array}{lll} 1 & 2 & 3 \\ 1 & 5 & 8 \\ -3 & -5 & \\ 3.33 & 8 & \\ 2.2 & & \\ 8 & & \end{array}$$

Since there are two sign changes in the left column, the presence of two roots with positive real parts is indicated.

Several other characteristics of the Routh array are of interest:

1 If the first element in the first column of one of the derived rows is zero but at least one other element in that same row is not zero, the first column zero may be replaced by an arbitrarily small number and the array formation process can be continued and the same criterion applied.

2 If *all elements of a row* are zero, the presence of a purely imaginary pair of roots (roots with zero real part) is indicated.

The Routh array and criterion are convenient means of quickly appraising either denominator or numerator roots (poles or zeros) in linear control systems transfer functions. When applied to a denominator polynomial to determine the presence of poles with positive real parts, the Routh criterion becomes a simple means of determining *linear stability*.

19.3.4 Root Loci

Several frequency-response methods and theories have been widely used in the design and analysis of linear control systems, including the Nyquist stability criterion, Nichols charts, Bode or frequency-response logarithmic plots, and root locus diagrams. Whereas these graphical methods have been largely supplanted by more accurate computer simulation techniques and state equation solutions, the Bode and root locus diagrams are still most useful as quick and reasonably accurate methods of relating system characteristics, particularly linear stability, to system parameters, especially in the early stages of control system design. Also since a considerable amount of knowledge about system performance can be gained from the nature of the system's poles and zeros, the root locus methods can be a simple aid in understanding much about the nature of control systems, electric circuits, and many other systems. In this section, we will give an introduction to the root locus method of determining the response of a linear feedback control system without getting into much of the mathematical rigor behind this method. For more theory and applications of the root locus method, consult Refs. 19.1 to 19.3.

Frequency-response methods in general and the root locus in particular are methods for evaluating *closed-loop response* from the *open-loop transfer function*. The nomenclature and definitions for frequency-response formulations have been given in Sec. 19.3.1 by Eqs. (19.65) to (19.70), and the S plane has been defined in Sec. 19.3.2. The poles and zeros of the open-loop transfer function $G(s)H(s)$ must be known in order to develop a root locus. Poles and zeros come from what has been called the *factored form* of the transfer function. When the open-loop transfer function is formulated by developing the functional blocks of individual components, as shown by Fig. 19.14*b*, the factored form generally results naturally, as shown by Eq. (19.54). However, in many cases, as illustrated in Eqs. (19.65) and (19.66), the *unfactored* form must be factored before proceeding with the root locus method. The forms of these factors have been discussed several times and are illustrated by Eqs. (19.12) to (19.14) and

(19.62). For the following analysis, we will generalize the factored form, using p_i to represent a pole and z_i to represent a zero, as follows:

$$G(s)H(s) = \frac{K(s + z_1)(s + z_2) \cdots (s + z_i) \cdots}{(s + p_1)(s + p_2) \cdots (s + p_i)} \tag{19.74}$$

where K represents the system gain. From Eq. (19.70) the characteristic equation of the closed-loop response is

$$1 + G(s)H(s) = 0$$

or (19.75)

$$G(s)H(s) = -1$$

The lower form of Eq. (19.75) brings out the two criteria for the loci of the roots of the closed-loop characteristic equation: the open-loop transfer function $G(s)H(s)$ must have a magnitude of 1.0 and a phase of 180°. Values of the complex variable s which satisfy these criteria describe the loci of roots of the closed-loop characteristic equation. These loci are usually formed by letting the system gain K in Eq. (19.74) vary from zero to infinity. The location of the loci on the S plane indicates some of the performance characteristics of the closed-loop system as a function of system gain. The most obvious performance parameter indicated by the loci is linear stability, and portions of loci located in the right-half S plane represent values of gain for which the system is unstable.

Let us describe some of the features of root loci be means of a very simple open-loop transfer function:

$$G_1H_1 = \frac{K(s + z_1)}{(s + p_1)(s + p_2)} \tag{19.76}$$

Assume that the two poles p_1 and p_2 and the zero z_1 are located on the S plane, as shown in Fig. 19.17. First let us substitute Eq. (19.76) into Eq. (19.75) and cross multiply, giving

$$K(s + z_1) = -(s + p_1)(s + p_2) \tag{19.77}$$

The start of the root locus is made by setting $K = 0$; note that the roots of the closed-loop equation for this condition are equal to the open-loop poles. Next we divide both sides of Eq. (19.77) by K and let $K \rightarrow \infty$. Note that the roots of the closed-loop equation for this condition are the open-loop zeros (one is assumed at infinity). Without any mathematical justification, we will generalize from this simple example to state that the root loci start at the open-loop poles and end at the open-loop zeros (or infinity), which is a fundamental property of the root loci. The loci are shown by the heavy lines segments in Fig. 19.17. The arrows on these segments indicate the movement on the loci for *increasing* gain K. In this example, both the open-loop roots (p_1, p_2, z_1) and the closed-loop roots are real, as evidenced by the locus remaining on

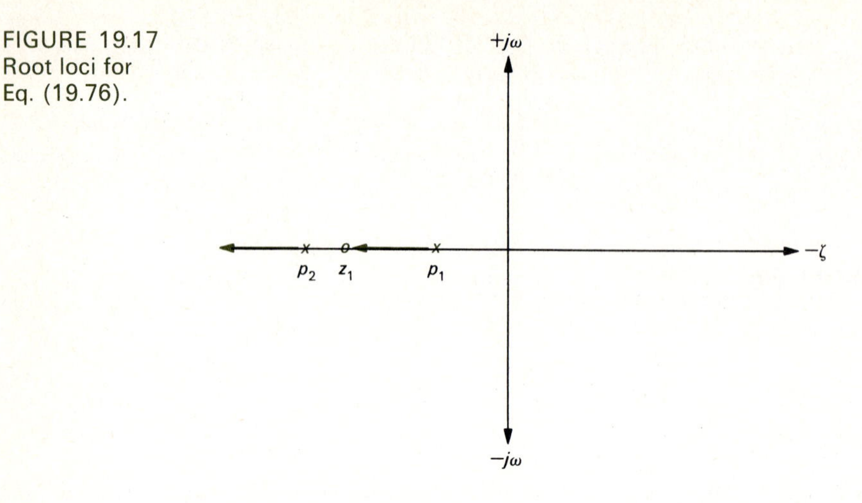

FIGURE 19.17 Root loci for Eq. (19.76).

the negative real axis. It is seen that the system is stable for all values of gain K, which could also have been observed from the form of the open-loop transfer function of Eq. (19.76).

There are several other attributes of typical open-loop transfer functions, Eq. (19.74), which are of significance in drawing the root loci: (1) Generally, the number of zeros is less than the number of poles. (2) Since the open-loop function represents the product of component transfer functions, it is usually—though not necessarily—true that these components are linearly stable; therefore, there will generally be no right-hand plane poles in the open-loop function. (3) There may be right-hand plane zeros in the open-loop function, as noted previously, which characterize nonminimum phase functions.

Next we add a third pole to the open-loop function given by Eq. (19.76)—a pole at the origin, as shown in Fig. 19.18. The entire character of the resulting loci of closed-loop poles is altered by this addition. Complex poles may now exist, but the system remains stable for all K. Again, the arrows represent directions on the loci for increas-

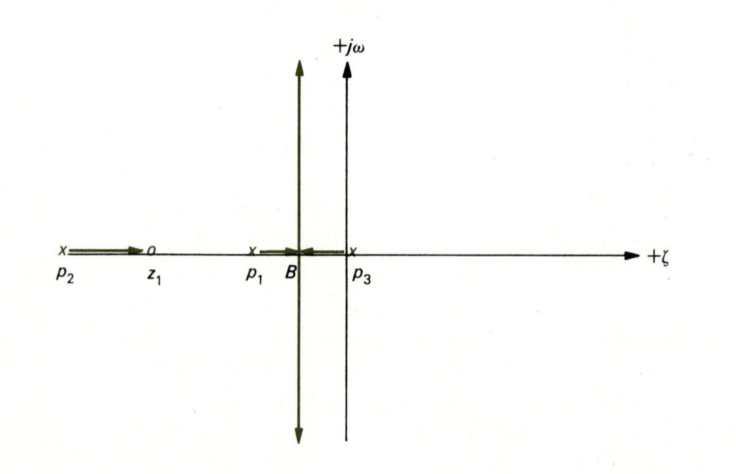

FIGURE 19.18 Root loci with added pole on Eq. (19.76).

TABLE 19.1 Guidelines for Drawing Root Loci[19.3]

1 A point on the real (horizontal) axis is on a root locus if, and only if, that point lies to the left of an odd number of poles and zeros. This principle can be observed rather easily by applying the phase criterion given in Eq. (19.75).

2 The root loci are symmetrical with respect to the real axis.

3 The loci start at an open-loop pole ($K = 0$) and end on an open-loop zero ($K = \infty$).

4 The angle by which the root locus leaves an open-loop pole or approaches an open loop is found by applying Eq. (19.75) and can be expressed as: $180^\circ\, q$ minus the sum of the angles of vectors from the remaining poles and zeros to the pole or zero in question, where q is any odd integer. In the summation, pole arguments are given a negative sign, zero arguments a positive sign.

5 The point at which the locus leaves the real axis is known as the *breakaway point*, point B in Fig. 19.18. This point is found by trial and error or computer iterative techniques by equating the reciprocals of the distances from the poles and zeros *on the real axis* to zero.

6 The direction of the asymptotic lines (see Fig. 19.18) to the horizontal axis is given by

$$\beta = \frac{\pm 180^\circ q}{n - m} \tag{19.78}$$

where n = number of open-loop poles

m = number of open-loop zeros

q = any odd integer

For the open-loop function given by Fig. 19.18, it is seen that β is 90°.

7 The point at which the asymptotes cross the real axis is known as the *centroid*. Its location is given by

$$\frac{\sum \text{real parts of poles} - \sum \text{real parts of zeros}}{n - m} \tag{19.79}$$

ing gain. We will now give some general rules for drawing root loci and then illustrate these rules with some practical examples. The theory and mathematical justification for these rules can be found in Refs. 19.1 to 19.3.

Example 19.13 Sketch the root locus as a function of open-loop gain for

$$GH(s) = \frac{K}{s(s + 3)(s + 15)}$$

Solution There are three poles in this transfer function at $s = 0$, $s = -3$, and $s = -15$, all on the horizontal axis of the S plane. These are shown in Fig. 19.19. The root locus on the horizontal axis is found from guideline 1 in Table 19.1 and is also shown by the dark line in Fig. 19.19. The breakaway angle β is seen (from guideline 6) to be $\pm 60^\circ$. The centroid (guideline 7) is $(0 + 3 + 15)/3 = 6$. The breakaway point B is determined as follows: Assume that the breakaway point B is at a distance b from the S-plane origin: therefore,

$$-\frac{1}{b} + \frac{1}{3 - b} + \frac{1}{15 - b} = 0 \tag{19.80}$$

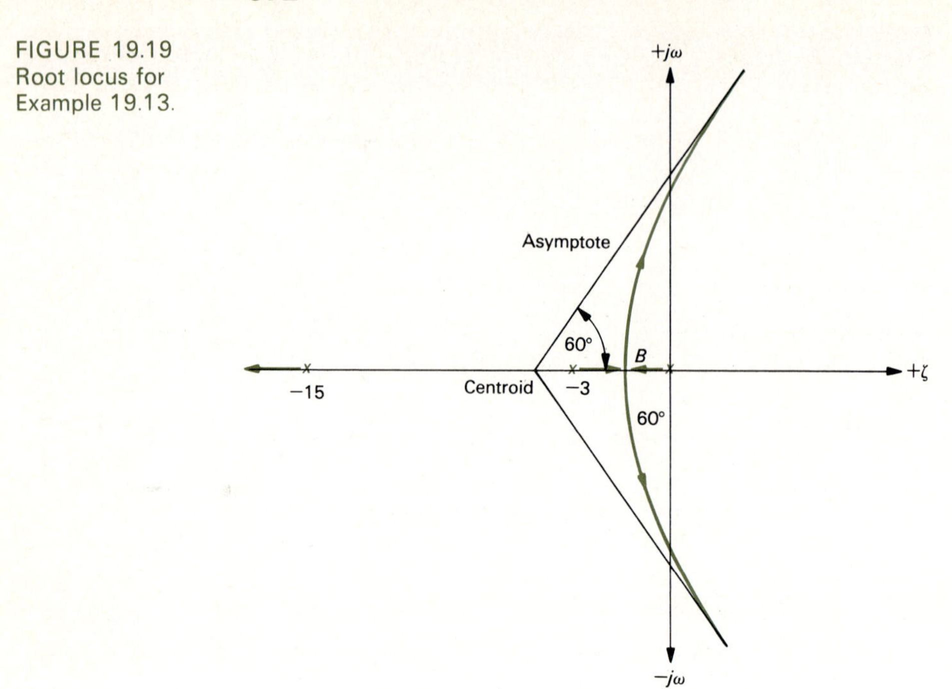

FIGURE 19.19 Root locus for Example 19.13.

Solving for b gives $b = 1.42$ or 10.6; the former value is the breakaway point and is shown in Fig. 19.19. The remainder of the loci are sketched as shown. It is seen that unstable operation is possible for some values of K. The points at which the loci cross into the right half of the S plane will be illustrated in the next example.

Example 19.14 Determine the point at which the root locus of Example 19.13 crosses the imaginary axis and the value of gain K at this point. This value of K is the maximum gain that gives linearly stable performance.

Solution The value of K that gives purely imaginary roots ($s = \pm j\omega$) can be found by forming the Routh array of the *closed-loop* characteristic equation. The closed-loop characteristic equation is

$$1 + GH(s) = 0 = 1 + \frac{K}{s(s+3)(s+15)}$$

or

$$s^3 + 18s^2 + 45s + K = 0$$

The Routh array is

$$\begin{array}{cc} 1 & 45 \\ 18 & K \\ \dfrac{810 - K}{18} & 0 \\ K & \end{array}$$

As noted in Sec. 19.3.3, the existence of purely imaginary roots is indicated when all elements of one row are zero. By making the first element of the third row above equal to zero, this condition can be realized. The first element will be zero by making $K = 810$. The value of $s = \pm j\omega$ at $K = 810$ can be found by forming what is called the *subsidiary equation* of a two-element row. Using the second row,

$$18s^2 + 810 = 0$$

$$s = \frac{\sqrt{810}}{18} = \pm j6.7$$

Example 19.15 Sketch the root locus for the open-loop function

$$GH(s) = \frac{K(s + 0.4)}{(s + 0.2)(s + 0.8)(s^2 + 2s + 10)}$$

Solution Factoring the last term in the denominator shows a complex pair of roots: $s = -1 \pm j3$. The four poles and one zero are plotted in Fig. 19.20.

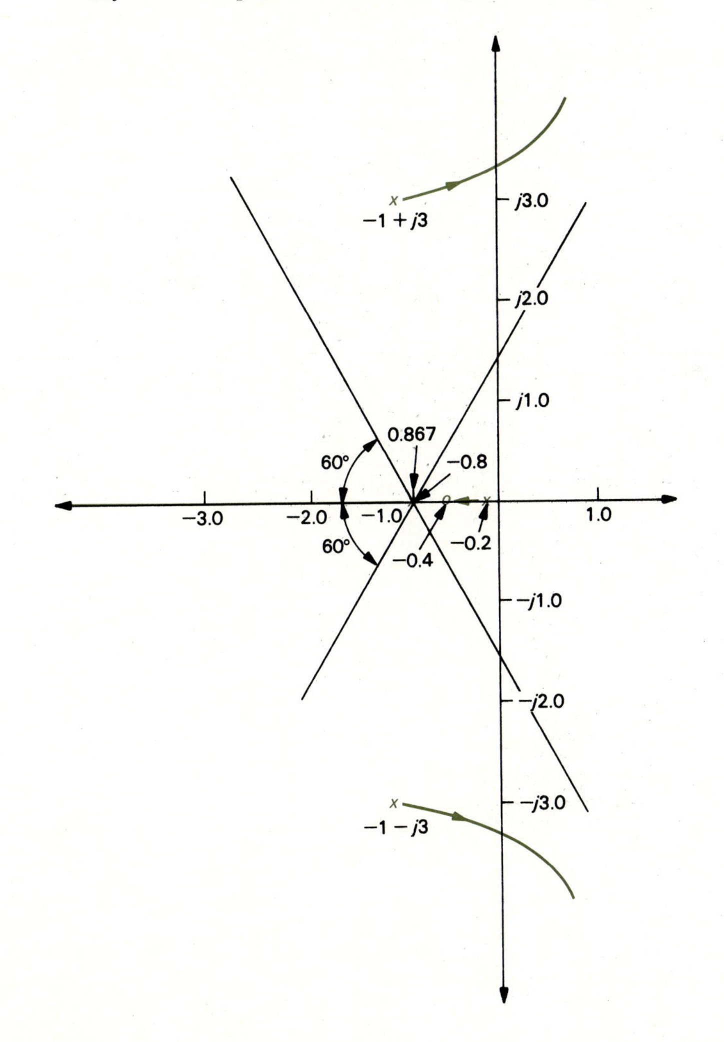

FIGURE 19.20 Root loci for Example 19.15.

The root locus along the real axis is found by application of guideline 1 of Table 19.1 and is shown in Fig. 19.20. The asymptotic angles are

$$\beta = \frac{\pm 180°q}{4-1} = \pm 60°, \pm 180°, \pm 300°$$

The centroid is

$$\frac{-2 - 0.8 - 0.2 - (-0.4)}{4-1} = -0.867$$

The locus to the left of the pole $(s + 0.8)$ approaches a hypothetical zero along the 180° asymptote. The remaining sections of the locus leave the two complex poles. The angle at which the locus leaves the upper pole $(s + 1 - j3)$ is found from guideline 4, giving

$$180° - (-104.9° + 101.3° - 93.8° - 90°) = 7.4°$$

This angle would indicate that the locus would approach the 60° asymptote. If this portion of the locus does indeed approach the 60° asymptote, it would eventually enter the right-half plane, indicating that at some value of K the closed-loop system would be unstable. Therefore, a further check on the nature of this portion of the locus is desirable. This can be done by checking for the presence of imaginary poles (where the locus crosses the imaginary axis) by the method described in Example 19.14, using the Routh array. In the present example, the Routh array will indicate that there is a value of K, K_j, that will cause all elements of one row to become zero. Therefore, the locus crosses the imaginary axis and, for values of K above K_j, unstable operation will exist. The locus from the lower complex pole is symmetrical about the horizontal axis. The details of this Routh array evaluation are left for Prob. 19.7.

The root locus analysis is a quick method for evaluating closed-loop system stability without the need for complex computer simulation. It is invaluable as a means for observing the effect of open-loop poles and zeros on closed-loop stability and is thus a useful design tool. With experience, one can usually sketch the locus quickly from the guidelines listed in Table 19.1. Also, the plot of the locus gives much more information than merely the effect of gain on stability. For example, if portions of the locus are very near the imaginary axis (indicating closed-loop poles with small real parts), a highly oscillatory closed-loop response is indicated.

PROBLEMS

19.1 Determine the transfer functions $G(s) = v_o(s)/v_i(s)$ of the networks shown in Fig. P19.1.

19.2 Determine the transfer function $G(s)$ for the platform position $Y(s)$ for the spring-mounted platform shown in Fig. P19.2. Consider only the translations motion in a vertical plane for both table and platform.

19.3 The network shown in Fig. P19.3 is an example of a nonminimum phase transfer function. Derive the transfer function $G(s) = V_o(s)/V_i(s)$ and show that zeros exist in the right half of the S plane.

19.4 Figure P19.4 illustrates a practical op-amp circuit used for differentiation in analog computers. Assuming an ideal op amp, determine the transfer function $G(s) = V_o(s)/V_i(s)$.

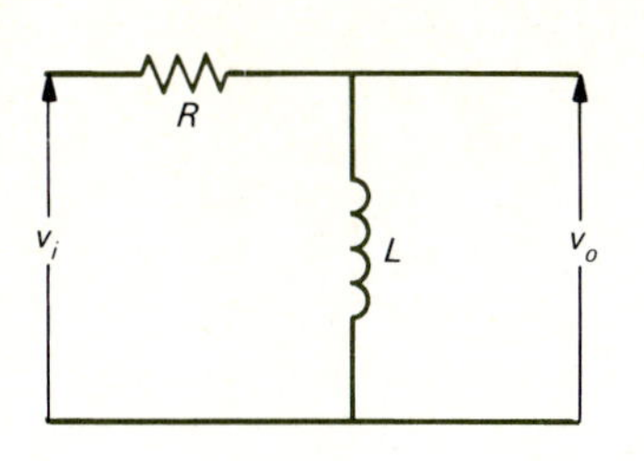

(a)

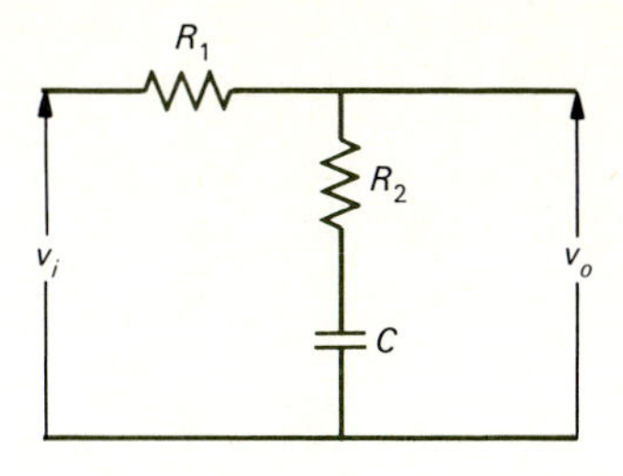

(b)

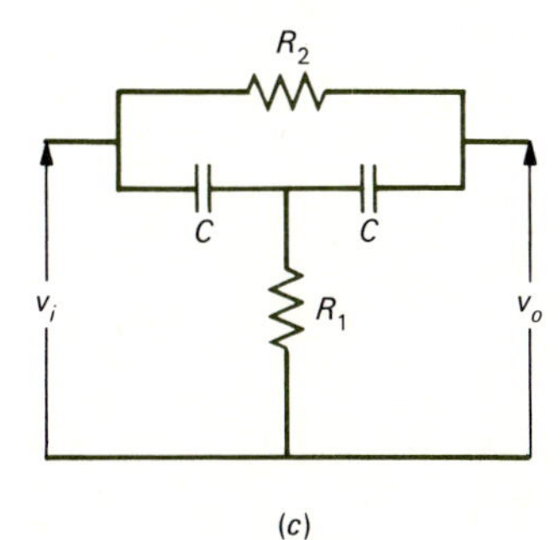

(c)

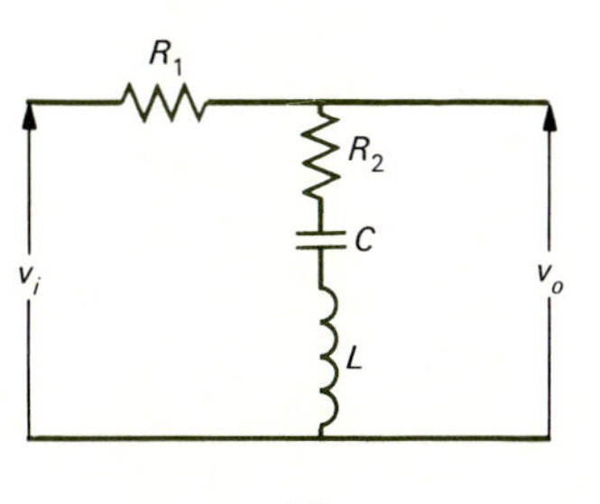

(d)

FIGURE P19.1

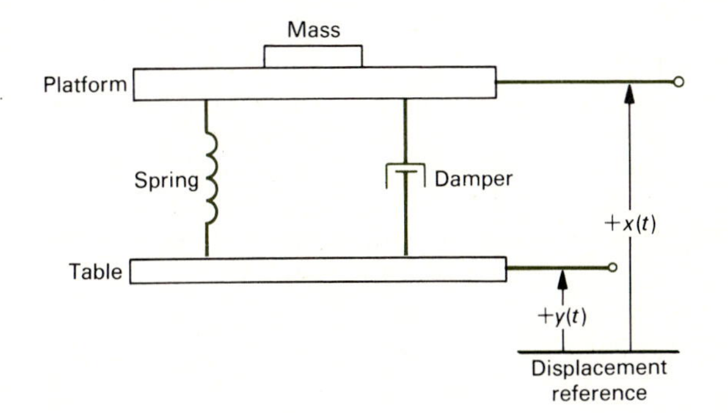

FIGURE P19.2

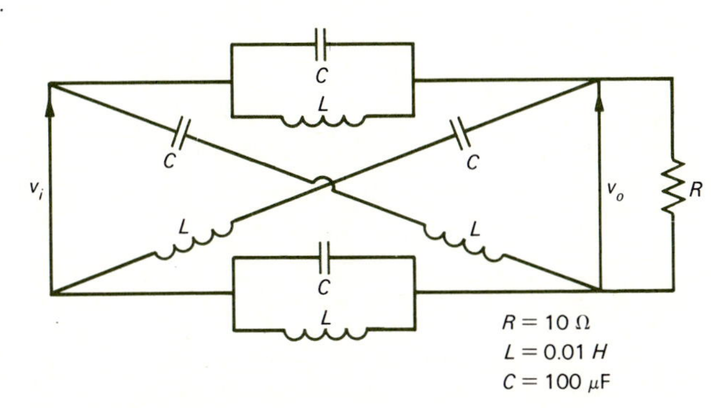

FIGURE P19.3

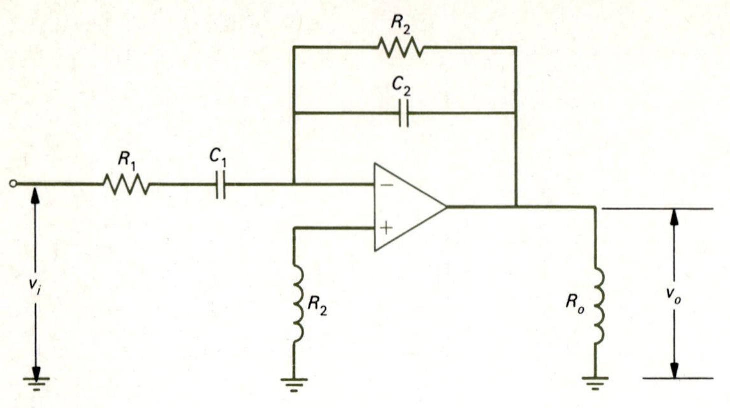

FIGURE P19.4

19.5 Repeat the analysis of Example 19.4 (Fig. 19.7) for the on-off controller, or chopper, and derive the transfer function for the case of discontinuous load current flow.

19.6 Figure P19.6 shows a common-emitter three-stage transistor amplifier. The small-signal transistors Q are similar to the 2N718 described in Chap. 10, and the dc bias voltages are such that the transistors are operated in the linear range of operation with $h_{FE} = 90$. Other circuit values are as follows: $C_1 = 10\ \mu f$, $C_2 = 50\ \mu f$, $R_1 = 20\ k\Omega$, $R_2 = 6400\ \Omega$, $R_3 = 2000\ \Omega$, $R_4 = 1000\ \Omega$, $R_L = 600\ \Omega$. Determine the transfer function for ac operation of the amplifier $V_o(s)/V_i(s)$, including the effects of the coupling capacitors. The small-signal equivalent common-emitter circuit developed in Chap. 10 should be used; neglect h_{ie} and all bias currents and voltages.

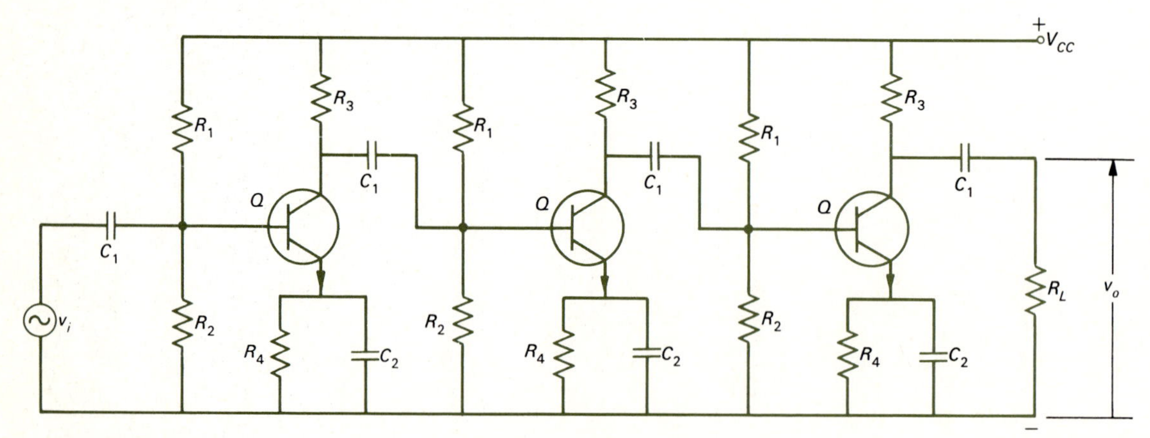

FIGURE P19.6

19.7 By means of the Routh criterion, determine the number of roots that lie in the right half of the S plane in the following characteristic equations.

(*a*) $s^4 + 3.2s^3 + 13.76s^2 + 21.1s + 18.72 = 0$

(*b*) $s^4 + s^3 + s^2 + s + 2 = 0$

(*c*) $0.29s^5 + 2.59s^4 + 6.86s^3 + 9.45s^2 + 10.28s + 4.64 = 0$

(*d*) $s^6 + 3s^5 + s^4 + 7.5s^3 + s^2 + 11s + 40 = 0$

(*e*) The closed-loop characteristic equation of Example 19.15

19.8 Figure P19.8*a* represents the block diagram of a motor speed control system with a load disturbance *T*. Show that the block diagram of Fig. P19.8*a* can be reduced to the form shown in Fig. P19.8*b* and determine the transfer functions for each block in Fig. P19.8*b*.

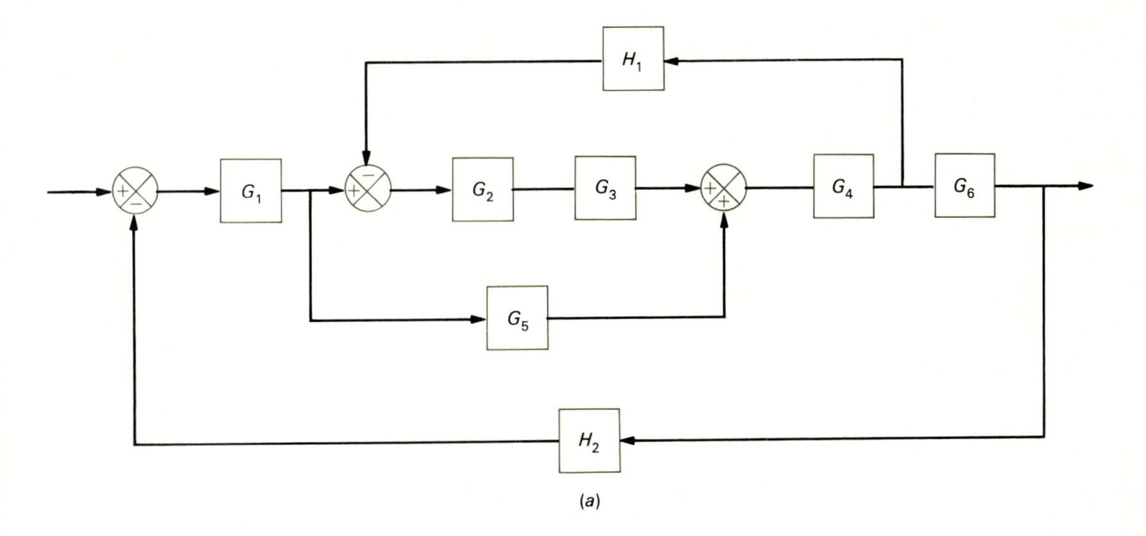

(*a*)

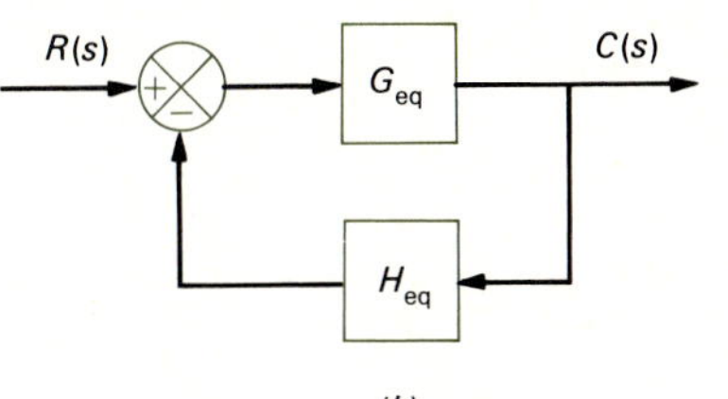

(*b*)

FIGURE P19.8

19.9 Determine the characteristic equation for the *closed-loop* response $C(s)/R(s)$ for the system shown in Fig. P19.9.

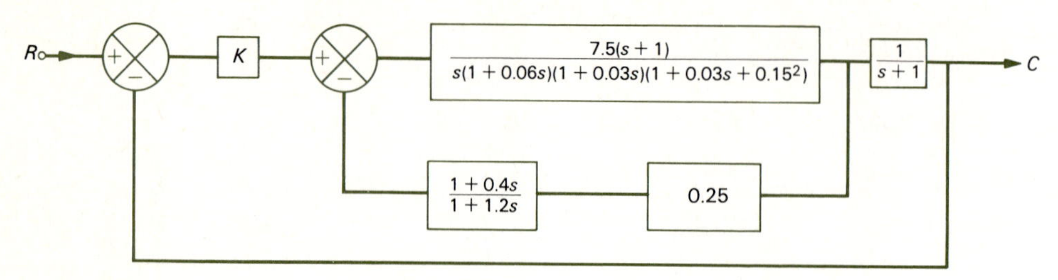

FIGURE P19.9

19.10 Determine the open-loop transfer function $G(s)$ for the system shown in Prob. 19.9. Plot the root locus diagram as a function of K for this system.

19.11 Plot the root locus diagram for a system having the open-loop transfer function

$$GH = \frac{K(s + 0.4)}{(s + 0.02)(s + 0.8)^2(s^2 + 16s + 1088)}$$

19.12 Derive the simplified block diagram of Fig. 19.9*b* from the original multiple-loop diagram shown in Fig. 19.9*a*.

19.13 Verify root locus guideline 1 of Table 19.1.

19.14 Plot the root loci for a system with the open-loop transfer function

$$GH = \frac{K}{s(s + 1)(s + 2)}$$

and determine the maximum value of K for which the system is stable.

REFERENCES

19.1 H. Chestnut and R. W. Mayer, *Servomechanisms and Regulating System Design*, 2d ed., vol. 1, Wiley, New York, 1959.

19.2 R. C. Dorf, *Modern Control Systems*, 3d ed., Addison-Wesley, Reading, Mass., 1980.

19.3 J. J. D'Azzo and C. H. Houpis, *Linear Control System Analysis and Design*, 2d ed., McGraw-Hill, New York, 1981.

19.4 P. M. DeRusso, R. J. Roy, and C. M. Close, *State Variables for Engineers*, Wiley, New York, 1965.

19.5 G. J. Thaler and M. P. Pastel, *Analysis and Design of Nonlinear Feedback Control Systems*, McGraw-Hill, New York, 1962.

19.6 B. C. Kuo, *Digital Control Systems*, Holt, Rinehart & Winston, New York, 1980.

19.7 G. H. Hostetter, C. J. Savant, Jr., and R. T. Stefani, *Design of Feedback Control Systems*, Holt, Rinehart & Winston, New York, 1982.

Chapter

Electrical Instrumentation

The measurement of signals, parameters, and dimensions is one of the most significant branches of science and engineering. The activity of measuring tells each of us much about our universe, our environment, our economy, and ourselves. Much of the world's industry, commerce, transportation, and even human relationships is based upon accurate measurements. Moreover, the basis for all scientific principles and engineering concepts rests, ultimately, on their verification by test and measurement. The theory and principles that have been presented so far in this text have little meaning except as verified by measured test data—and since this is a relatively basic text, the reader can be assured that almost all of the theories and concepts presented here do rest on sound experimental evidence. Therefore, our aim in this chapter is not to explore the methods of verification of the fundamental theorems of science or philosophy, but rather to describe the usefulness of accurate measurement techniques in "everyday" engineering, to explore some of the limitations of measurements, especially those of electrical parameters, and to describe some of the more common electrical measuring devices.

20.1 SOME BASIC CONCEPTS OF ELECTRICAL MEASUREMENTS

The measurement of physical parameters can be broadly divided into two categories—analog and digital. This classification is perhaps more appropriately defined in terms of the nature of the signals used in the measurement process—continuous and discrete. Both sets of nomenclature developed out of the analog and digital computer fields and have become convenient means of distinguishing two basic approaches to many scientific and engineering developments. Analog instruments and analog computers generally share common hardware (or circuit components), common performance characteristics, and common methods of mathematical analysis. The same can also be said of digital instruments and digital computers. In fact, instrumentation of both types is often a link in a computing system or a control system. For a great many electrical measurement requirements today, both analog and digital instruments are available. Therefore, it is often necessary to choose between these two different classes of instruments for a specific application; the following gives a few general guidelines when such decisions are required.

Digital meters are generally more accurate and can be equipped with more scales and broader ranges than analog meters, and they frequently have faster time response and greater frequency response. On the other hand, analog meters are generally less expensive and less susceptible to electromagnetic noise. A more subtle difference is the information gleaned by the person reading the meter. In digital meters, one sees only one specific reading or number; in analog meters, this reading or number is seen in light of an entire range or scale of reading, which is often very informative. When the reading is changing, one has no idea where a digital meter is going, whereas an analog meter shows an increasing or decreasing reading towards some other numerical value; this is especially helpful when one is trying to set a parameter, such as motor speed, at a specific value. Digital meters can also be confusing or at least annoying when the reading is oscillating around a certain value for one reason or another. However, in general, more skill and experience are required to read an analog meter, especially if precision is required, since interpolation between scale markings is generally required and since the angle at which the reader looks at the instrument may cause parallax effects. Therefore, if an unequivocal, accurate reading is required—and especially one by an unskilled reader, as is the case in many assembly-line measurements—digital meters are usually preferred.

So far, we have intimated that the purpose of electrical instrumentation is to provide information to the reader or test personnel. An equally important function is as a means of *sensing* the value of a parameter for use in a control or computational system. Often associated with the task of sensing is the task of providing an appropriate signal proportional to the numerical value of the parameter being sensed. Sensors are an important branch of the electrical instrumentation field and are briefly discussed in several sections of this text. We will confine our discussion in this chapter to the electrical meters used for measurements of electrical parameters.

Finally, there are certain types of electrical instrumentation that use both analog and digital circuitry, probably the best known of which are the analog-to-digital or digital-to-analog converters. These and similar types of instruments are often classified as *hybrid* instruments.

20.1.1 Accuracy, Precision, and Errors

Accuracy is perhaps the first characteristic that one thinks of when selecting or applying an electrical meter. Accuracy is defined as "freedom from mistake or error" or "conformity to truth or a standard." It is the second of these definitions that most closely defines the meaning of the term *accuracy* as applied to meters and instrumentation. Conformity to truth is the fundamental concept of accuracy, and a meter of perfect accuracy is perfectly truthful in indicating the value of the quantity being measured. In the strict sense of this meaning, however, there is no such thing as a "perfect" meter, even though certain measurements of time and distance do approach perfection. Therefore, the concept of conformity to a *standard of measure* is closer to defining the function of a meter. The degree to which a meter or instrumentation system conforms to a standard of measure is known as the accuracy of the meter. Standards of the basic MKS or SI units are maintained in various national standards laboratories. The standard for the secondary unit upon which most electrical units rest, the ampere, is maintained in the French National Laboratory.

Accuracy is often confused with or interchanged with the concept of precision, which may be simply defined as the "quality of being sharply or concisely defined." Related to meters and instruments, precision describes the degree to which a scale, dial, or gauge can be read. For example, on a simple ruler marked with $\frac{1}{8}$-in markings, a distance can be read to a precision of only $\frac{1}{8}$ in. Or we often say, "To the nearest $\frac{1}{8}$ in." This does not mean that the reading of distance using this scale is *accurate* to $\frac{1}{8}$ in, but only that the meter (in this case, a ruler) can be read with a *precision* of $\frac{1}{8}$ in. This author once had a summer job between college terms as a surveyor working the "nonreading" end of a surveyor's tape, the finest scale markings of which were 0.1 in. On the opposite "reading" end of the tape was an old-time employee who would consistently call out such readings as "43 ft 6.578 in" and would defend them all most passionately to the young doubting engineering student. Most of us made similar faux pas at one time or another, but in the recording or reporting of technical data this is a trap to be avoided.

Related to the concept of precision in measurement and meter reading is the concept of the number of significant figures in a numerical value of a meter or instrument reading, a concept probably quite familiar to the reader. When measured data are used in calculations or data processing, it is important to maintain only the significant figures that can be derived from the basic precision with which the measurements were made. In this era of performing most data manipulation on calculators or computers, it is often too easy to present an answer using the 8- or 10-digit accuracy of the calculating device based upon meter readings that have only three significant figures. This is another trap to be avoided in the presentation of data based upon measured values.

Precision in digital meters is much more straightforward and is equal to the number of digits in the readout. Many digital meters have the number of readout digits equal to an integer plus one-half, such as a "$3\frac{1}{2}$-digit readout." In this example, the three digits can register a count of zero to nine; the $\frac{1}{2}$ refers to the fourth digit, which is always the left-most digit and which can register zero or one, as well as the sign of the reading (+ or −). Thus, a $3\frac{1}{2}$-digit readout can register a maximum count of 1999.

All practical meters have a certain degree of inaccuracy, owing to the presence of several types of errors that may occur while an operator is reading and recording an indicated value from a meter. A very general summary of measurement errors is given in Table 20.1.

Most of these errors are obvious from their definition. The first class of error is called "instrumental" rather than "instrument," since it may result from an improper

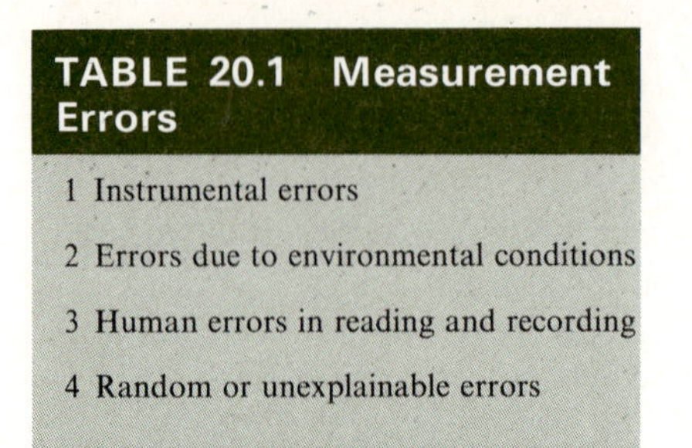
TABLE 20.1 Measurement Errors

1 Instrumental errors

2 Errors due to environmental conditions

3 Human errors in reading and recording

4 Random or unexplainable errors

use of an instrument or an improper planning of a test procedure as well as from the inherent error of the instrument itself. Environmental errors result from the effects of temperature, pressure, humidity, etc., and corrections for such errors can usually be made if the environmental conditions are known. Random or erratic errors frequently occur in the best-designed schemes of measurement and are often unexplainable; allowances for such errors should be made in all measurement planning. Instrument errors as specified by the instrument manufacturer usually take into consideration this possibility of random or unexplainable errors. In other words, manufacturer-specified instrument errors are generally conservative. Instrument errors are generally specified (1) as a percent of full scale reading, (2) as a percent of the actual reading, or in some cases (3) as a fixed error. Also, errors due to temperature deviations are often specified.

20.1.2 Errors in Digital and Analog Meters

Analog meter error is generally stated as a percent of full scale. Digital meter error is generally specified as a combination of errors, including a percent of full scale. However, since this latter is generally very small, the comparison of errors in the two types of instruments may be somewhat misleading. As an example, a good-quality analog voltmeter may be specified as having an instrument error of 0.5 percent of full scale and 0.002 percent of reading per degree Celsius. A digital meter of equivalent rating may have the following error specifications: 0.05 percent of reading, 0.05 percent of scale, 0.005 percent of reading per degree Celsius, and 0.005 percent of scale per degree Celsius. Let us assume that both meters have a full scale of 200 V and that a reading of voltage is being made at 35.0 V in an environment consisting of a temperature of 40°C above ambient. Note that the above error specifications are maximum or guaranteed maximum errors; actual errors of instruments manufactured by reliable dealers are generally far below these values. Continuing with the above example, the analog voltmeter could have an error of $0.005 \times 200 + 0.00002 \times 35 \times 40 = 1.0 + 0.028 = 1.028$ V. The error as a percent of reading is $1.028 \times 100/35 = 2.94$ percent. The digital meter will have an error of $0.0005 \times 200 + 0.0005 \times 35 + 0.00005 \times 200 \times 40 + 0.00005 \times 35 \times 40 = 0.5875$ V. As a percent of reading, this is an error of 1.68 percent. Figure 20.1 gives a picture of the relative differences between analog and digital meter accuracy. It is seen that readings on the low end of a meter scale will contain a large percent error.

Most meter manufacturers use the word "accuracy" to describe what has been properly termed "error" in the above paragraph—probably to avoid the use of the

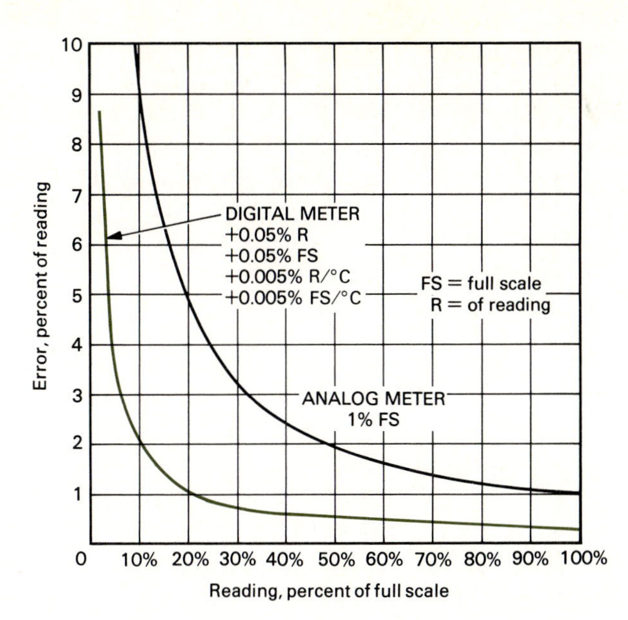

FIGURE 20.1. Comparison of digital and analog meter accuracies.

word "error" in describing their product. Also, it is assumed that the algebraic sign of the error may be either positive or negative. Thus, the analog meter discussed in the previous paragraph might be specified by the manufacturer as having an "accuracy" of ± 0.5 percent of its full scale value. This specification is often termed a *known* instrument error in contrast to a *probable* error, which has a considerably different meaning. By this specification, the manufacturer guarantees that the meter reading will be within an increment around the true value, the increment being equal to ± 0.5 percent of full scale. Note that the increment (when specified as a function of full scale) is constant over the entire scale. The fact that there may be an error in every measured value (with maximum possible value equal to known error) must be considered when performing mathematical manipulation on measured values. For example, if a measured voltage is to be multiplied by a measured current to obtain power or apparent power, the actual multiplication should be

$$P = (V + e_v)(I + e_i) = VI(1 + e_v + e_i + e_v e_i) \tag{20.1}$$

Since the latter term in Eq. (20.1) is very small, it may be dropped, and the difference between the true power (VI) and the measured power (that is, the error in the measured power) is seen to be

$$e_p = e_v + e_i \tag{20.2}$$

Similar relationships can be derived for the other mathematical manipulations.

Instruments should be checked periodically to ensure that indicated readings are within this error (or accuracy) increment. This process is known as *calibration.* A meter is calibrated by first comparing its indicated value with that of a standard meter and then making ajustments, if necessary, to bring the meter's reading within the error increment at major divisions of each scale. A *standard meter* is one of high accuracy which is used *only* for calibration purposes (that is, not for general measurements), which itself is frequently calibrated against primary standard meters maintained by national standards laboratories, and which is usually kept within a controlled environment. As a rule of thumb, the standard meter should have an error increment, or accuracy, one-tenth that of the meter being calibrated. Certain classes of instruments, such as voltmeters and frequency counters, are often calibrated against standard *sources* of the parameter being measured.

20.2 MEASUREMENT OF VOLTAGE, CURRENT, AND POWER

A wide variety of instruments is available for measuring voltage and current over a broad spectrum of frequencies, from dc to the gigahertz range. Digital meters have gradually become more popular for this function, although analog meters are still widely used, especially in the dc and power frequency range.

In selecting a *voltmeter*, the following parameters must be considered: ranges of voltage to be measured, accuracy, scale precision, effect of current drain (which is a function of the meter's impedance), frequency of the signals to be measured, waveform of the signals to be measured, power required to operate the meter, location of measurements (panel or portable), available accessories, and cost. Accuracy and precision have been discussed in Sec. 20.1 and the ranges required for the signals to be measured are usually obvious, keeping in mind that most instruments give greater accuracy if the readings are near full scale. The other parameters deserve some further discussion:

1 *Current drain:* Electronic meters generally have very high internal resistance, usually in the megohm range, and current drain is therefore negligible in most applications; Table 20.2 lists the specifications of a typical digital multimeter

TABLE 20.2 Specifications of a Typical DMM (*Courtesy of Keithley Instrument Co.*)

	Model					
	130	169	178	179 (179-20A)	177	191
Counts	1999	1999	19999	19999	19999	199999
Digits	$3\frac{1}{2}$	$3\frac{1}{2}$	$4\frac{1}{2}$	$4\frac{1}{2}$	$4\frac{1}{2}$	$5\frac{1}{2}$
DC volts						
Maximum sensitivity	100 μV	100 μV	100 μV	10 μV	1 μV	1 μV
Maximum reading	1000 V	1000 V	1200 V	1200 V	1200 V	1200 V
Basic long-term accuracy	0.5%	0.25%	0.04%	0.04%	0.03%	0.007%

C volts						
Method	Average	Average	Average	TRMS	TRMS	Average
Maximum sensitivity	100 μV	100 μV	100 μV	10 μV	10 μV	10 μV
Maximum reading	750 V	1000 V	1000 V	1000 V	1000 V	1000 V
Basic long-term accuracy	1%	0.75%	0.3%	0.5%	0.5%	0.1%
hms						
Special features	Diode test on 20-kΩ range	Two low voltage ranges	—	Hi-lo switch selectable	Front panel lead compensation	Auto 2/4 terminal
Maximum sensitivity	100 mΩ	100 mΩ	100 mΩ	100 mΩ	1 mΩ	1 mΩ
Maximum reading	20 MΩ	20 MΩ	20 MΩ	20 MΩ	20 MΩ	20 MΩ
Basic long-term accuracy	0.5%	0.2%	0.04%	0.04%	0.04%	0.012%
C amps						
Maximum sensitivity	1 μA	100 nA	—	10 nA	1 nA	—
Maximum reading	10 A	2 A	—	2 A (20 A)	2 A	—
Basic long-term accuracy	1%	0.75%	—	0.2%	0.2%	—
C amps						
Method	Average	Average	—	TRMS	TRMS	—
Maximum sensitivity	1 μA	100 nA	—	10 nA	10 nA	—
Maximum reading	10 A	2 A	—	2A (20 A)	2 A	—
Basic long-term accuracy	2%	1.5%	—	1.0%	0.8%	—
Ranging	Manual	Manual	Manual	Manual	Manual	Manual
Outputs	—	—	—	BCD opt IEEE opt	Analog BCD opt IEEE opt	—
Battery operation	Standard	Standard	Optional	Optional	Optional	—
Capsule comment	Pocket size; 0.6-digits; rugged case	Function and range annunciators; 0.6-in digits; 1-year battery; float up to 1.4 kV	Float up to 1.4 kV	1-kV protection on Ω; float up to 1.4 kV; 20-A ac and dc current as 179-20A.	1-μV, 1-nA, 1-mΩ sensitivity	μP control allows up to four conversions per second; digital filter; pushbutton null

(DMM), probably the most common type of lab meter in use today, and the impedance of the voltmeter portions of this meter are seen to be very high on both ac and dc signals. Analog meters, on the other hand, may have relatively low internal impedance, often as low as 100 Ω/V. Therefore, current drain by the meter may be significant and the voltmeter may have a significant effect upon the circuit being measured. Meter capacitance must also be considered when measuring signals of high frequency or short pulses.

2 *Frequency range:* The typical multimeter maintains good accuracy well into the upper audio ranges. *Dynamometer* and *iron vane* analog meters are generally restricted to measurements at power frequencies, i.e., below 1000 Hz. *Thermocouple* analog meters are suitable for frequency ranges well into the upper audio ranges and are probably the most versatile type of meter available for measuring signals with pulses and highly irregular waveforms. *Rectifier* analog and special electronic instruments are available for measuring high-frequency signals above the audio-frequency range. Most electronic instruments have digital circuitry and digital readouts. Analog meters use continuous circuits and have analog scale readouts. Some of these meters are *true analogs* of the signals being measured. For example, the dynamometer meter is a miniature "motor" which models the square of the root-mean-square (RMS) expression

$$v^2 = \frac{1}{\tau}\int_0^t v^2\,dt \tag{20.3}$$

The natural form of the scale on a dynamometer meter is therefore a "square-law" scale, which on a typical meter is made into a quasi-linear scale by the choice of scale markings; because of this square-law action, dynamometer meters measure both dc and ac. Obviously, a dynamometer meter has very poor accuracy at low-scale readings. The thermocouple is also a true analog of RMS voltage (and current), since equivalent heating effect is the very basis for the definition of an RMS parameter. Thermocouple meters are actuated by the signal from a thermocouple as a function of the heat generated by a current (or current proportional to a voltage) and hence measure the *true RMS value* of the signal; this will be discussed further below. The principal analog meter used to measure dc signals is known as the d'Arsenvol meter and, like the dynamometer meter, is based upon the motor action of a current-carrying conductor in a magnetic field produced by a permanent magnet. The force on the conductor is basically a function of the instantaneous current in the coil, although, due to the relatively high inertia of the moving coil system in this instrument, the meter reading is normally the average of the instantaneous forces on the coil except for low-frequency variations; hence, this meter indicates the average or dc values of the signal. Table 20.3 summarizes some of the characteristics of the common analog instruments.

3 *Waveform:* The waveform of the signal being measured may have a profound influence on the numerical value of the meter reading. Many electronic instruments use sampling techniques for obtaining the numerical value of the signal. The

TABLE 20.3 Analog Voltmeter Characteristics

Type	Frequency range, Hz	Resistance, Ω/V	Accuracy, %
D'Arsenvol	dc	100–5000	0.01–0.5
Iron vane	25–125	75–200	0.75–2.0
Dynamometer	0–3200	10–100	0.1–1.5
Thermocouple	0–50,000	100–500	0.1–1.0
Rectifier	20–50 × 10^6	1000–5000	1.0–5.0
Electrostatic	0–10,000	> 10,000	0.5–2.0

Note: Analog ammeters generally have frequency and accuracy capabilities similar to their voltmeter counterparts.

sampled values—values measured at each 10 μs, for example—are combined mathematically by means of digital circuitry to give a meter reading of the average or RMS value of the signal. To perform this mathematical manipulation, a fixed waveform must be assumed, and for ac measurements this is almost always a sinusoidal waveform; therefore, such instruments read *true* RMS only for sinusoidal signals. Other electronic voltmeters actually perform the root-mean-squaring process by analog or digital multiplication and square-rooting and can therefore indicate true RMS regardless of waveform. Figure 20.2 illustrates a typical "true RMS" multimeter. One feature of electronic voltmeters that is not available on analog meters is the ability to measure and hold *peak* values of a signal; the meter shown in Figure 20.2 can measure and hold both the instantaneous peak and the short-time peak, such as occurs during motor starting. Most

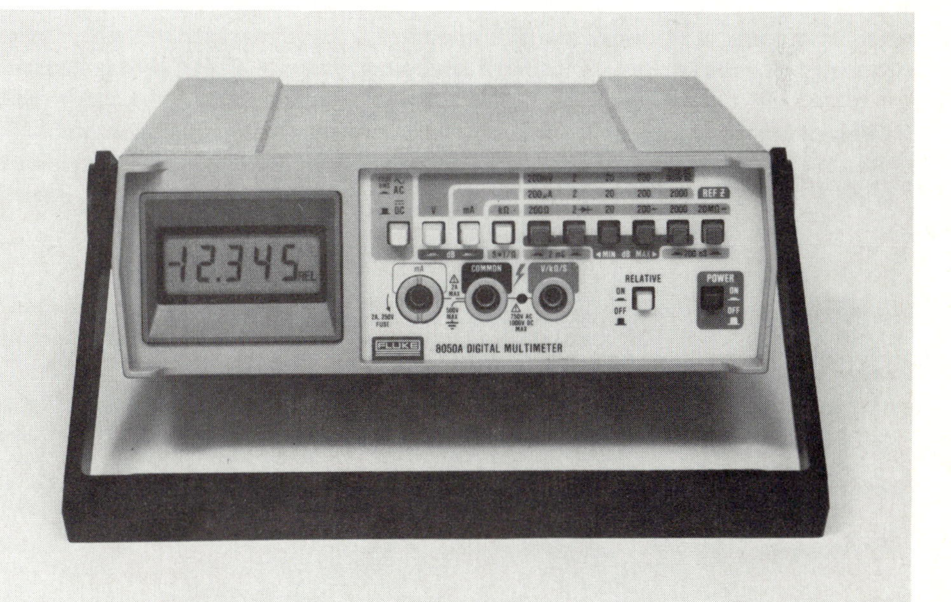

FIGURE 20.2 Typical digital multimeter, with true RMS readout. (With permission from and copyright © 1983 by John Fluke Mfg. Co., Inc.)

TABLE 20.4* Specifications for the Voltmeter and Ohmmeter Sections of a Typical Digital Multimeter (DMM) (*Courtesy of John Fluke Mfg. Co., Inc.*)

DC VOLTS†

Range	Resolution	Accuracy for 1 year
±200 mV	100 μV	±(0.1% of reading + 1 digit)
±2 V	1 mV	
±20 V	10 mV	
±200 V	100 mV	
±1000 V	1 V	

AC volts (true RMS responding):‡

		Accuracy for 1 Year			
Range	Resolution	45 Hz to 1 kHz	45 Hz to 10 kHz	10 kHz to 20 kHz	20 kHz to 50 kHz
200 mV	100 μV	±(0.5% of reading + 2 digits)		±(1.0% of reading +2 digits)	±(5% of reading +3 digits)
2 V	1 mV				
20 V	10 mV				
200 V	0.1 V				
750 V	1 V	±(0.5% of reading +2 digits)			

Resistance§

Range	Resolution	Accuracy: for 1 Year	Full-scale voltage	Maximum test current
200 Ω	0.1 Ω	±(0.2% of reading +1 digit)	0.25 V	1.3 mA
2 kΩ	1 Ω		1.0 V	1.3 mA
20 kΩ	10 Ω		<0.25 V	10 μA
200 kΩ	100 Ω		1.0 V	35 μA
2000 kΩ	1 kΩ	±(0.5% of reading +1 digit)	<0.25 V	0.10 μA
20 MΩ	10 kΩ		1.5 V	0.35 μA

analog meters are *inherently true RMS meters*; this has already been noted for the thermocouple meter, and it is also true of the dynamometer analog meters. However, the frequency response of any circuit, instrument, or machine is primarily a function of its inductance and capacitance. Dynamometer meters have relatively high inductance, as compared to thermocouple and rectifier instruments, and hence their frequency response is limited to the low-audio frequencies, generally to less than 1000 Hz; their ability to respond accurately to signals containing sharply rising wavefronts is thus very limited. The iron vane meter is likewise limited to near sinusoidal signals at power frequencies.

4 *Required power:* Analog meters generally require no external power for operation, but all electronic meters do. Portable electronic meters are battery-operated, which can be a nuisance in some applications. It is important that electronic instruments have some means of indicating when the battery voltage level is below that required for proper meter operation.

5 *Meter location:* All types of meters may be generally classified as either portable or panel types. Panel instruments are used in a great variety of industrial applications, and in terms of numbers of meters sold per year, they probably far outsell the portable type, although the latter is probably much more familiar to the student reader. In general, the differences between these two classes are structural, and most of the comments concerning other parameters, such as accuracy or frequency range, apply equally to panel or portable instruments. Vibration may be of concern in many applications of analog panel instruments and must be considered in such applications.

6 *Available accessories:* Electronic meters generally have many alternative functions which may or may not be included in the price of the "standard package" meter. For example, the typical electronic multimeter (or DMM, as it is commonly called) contains a milliammeter and ohmmeter as well as ac and dc voltmeter functions. Many DMMs, such as the one shown in Fig. 20.2, and Table 20.4 also

---

* The electrical specifications given assume an operating temperature of 18°C to 28°C, humidity up to 90%, and a 1-year calibration cycle. The functions are dc volts, ac volts, dc current, resistance and conductance.

† Input impedance 10 MΩ, all ranges

Normal mode rejection ratio >60 dB at 60 Hz (at 50 Hz on 50 Hz option)

‡ Volt-hertz product 10^7 max (200 V max @ 50 kHz)

Extended frequency response Typically ±3 dB at 200 kHz

Common-mode noise rejection ratio (1 kΩ unbalance) >60 dB at 50 Hz and 60 Hz

Crest fractor range 1.0 to 3.0

Input impedance 10 MΩ in parallel with <100 pF

§ Overload protection 300 V dc/ac RMS on all ranges

Open-circuit voltage Less than 3.5 V on all ranges

Response time 1 second, all ranges except 2000 kΩ and 20 MΩ ranges; 4 seconds these two ranges

Diode test These three ranges have enough voltage to turn on silicon junctions to check for proper forward-to-back resistance. The 2-kΩ range is preferred and is marked with a larger diode symbol on the front panel of the instrument. The three nondiode test ranges will not turn on silicon junctions, so in-circuit resistance measurements can be made with these three ranges.

can be used to measure temperature with the purchase of the required thermocouple. Most electronic meters have facilities for providing an output signal proportional to the signal being measured for use in data acquisition and/or manipulation in a digital computer.

7 *Cost:* For pure voltmeter applications, electronic and DMM instruments offer many advantages over analog instruments, the main advantages being their inherent high accuracy, multirange, and ac/dc capability. Analog meters of comparable accuracy are generally very expensive and are of limited voltage and frequency range. However, where accuracy is not important and the range of voltages to be measured is rather limited, such as on many panel meter applications, an analog voltmeter will usually be much less costly than an equivalent electronic or digital meter. The differences in how a meter is read must also be considered in panel meter applications, as discussed in the previous section.

The above comments on voltmeters *generally apply to ammeters*, too. There is one major difference related to measuring current, however, and this is that electronic or digital ammeters are generally limited in range to milliamperes or a few amperes. Also, the resistance that an electronic ammeter places in series with the current being measured is generally much higher than that of an equivalent analog meter. Therefore, in contrast to the situation in voltage measurement, an electronic ammeter may have more effect on the circuit being measured than an analog ammeter does. The lower-frequency meters, such as the dc d'Arsenvol, iron vane, and dynamometer, generally have negligible effect (i.e., negligible series resistance) on the circuit being measured. The thermocouple and rectifier instruments have somewhat higher series resistance, yet nevertheless a much lower one than that of an equivalent electronic ammeter.

Analog meters can be constructed to measure current magnitudes of very large values, into the thousands of amperes. At some point, however, it becomes desirable from the technical and economic standpoints to limit the range of an ammeter and use a *current transformer* to achieve the higher magnitude capabilities. Current transformers also permit the use of the low-scale capabilities of electronic ammeters and DMMs. Current transformers, briefly discussed in Chap. 13, are basically two-winding transformers connected in series with the current being measured. There is an error associated with a current transformer (an error in both the magnitude and phase angle of the current being measured) which must be considered in determining the overall error of a measuring system.

One of the most useful instruments for measuring currents in the ampere range is the *clip-on ammeter* which combines the current transformer (with a 1-turn primary) and the measurement functions into one device. Figure 20.3 illustrates a typical clip-on ammeter.

20.2.1 Power Measurements

Power is perhaps the most difficult electrical parameter to measure with any degree of accuracy. Instantaneous power is defined as the product of voltage across a given element and the current through this element, or

$$p = ei \tag{20.4}$$

FIGURE 20.3
Clip-on ammeter with associated voltage- and resistance-measuring circuits. (Courtesy of The Triplett Corp., Blufton, Ohio)

Average power is defined as

$$P = \frac{1}{\tau}\int_0^\tau p\,dt \tag{20.5}$$

where τ is a specific time period, such as the period of a sinusoidal signal.

The dynamometer analog instrument is an excellent power-measuring device since, as has been noted in Eq. (20.3), the motor action of the dynamometer performs the product function required by Eq. (20.4) and the inertia of the movement and the proper design of the meter's scale perform the averaging function indicated in Eq. (20.5). However, given the high inductance of the moving and stationary coils of this instrument, its frequency response is limited to power and low-audio frequencies. The maximum frequency for which dynamometer wattmeters have been adapted is approximately 3200 Hz.

Hall device wattmeters have been developed to provide a means of measuring power in the frequency range above 1000 Hz. The Hall phenomenon is utilized in obtaining the product described in Eq. (20.4). A number of commercial Hall wattmeters are available today. However, the frequency response of Hall wattmeters is limited to the midaudio frequency range. For power measurement at frequencies (or waveforms having frequency components) above this range, other circuits are required.

Electronic multiplier wattmeters extend the frequency response of power measurements to approximately 50 kHz. Figure 20.4 shows the block diagram for an electronic multiplier wattmeter designed to measure both instantaneous and average power. This type of meter has been applied successfully in the measurement of the nonsinusoidal low-power-factor power associated with power electronic circuits, such as the power loss in thyristor commutating circuits (Ref. 20.1) and core losses in motors excited from power semiconductor choppers.

The accurate measurement of power at low power factors is particularly difficult even with low-frequency sinusoidal signals. For sinusoidal signals, Eq. (20.5) becomes

$$\begin{aligned} P &= \frac{1}{\pi}\int_0^{\pi} (E_m \sin \omega t)[I_m \sin (\omega t - \theta)]\, d(\omega t) \\ &= \tfrac{1}{2}E_m I_m \cos \theta \\ &= EI \cos \theta \end{aligned} \tag{20.6}$$

Average power is thus the product of three parameters—RMS voltage, RMS current, and $\cos \theta$ or power factor (pf) (in sinusoidal circuits)—and, according to the

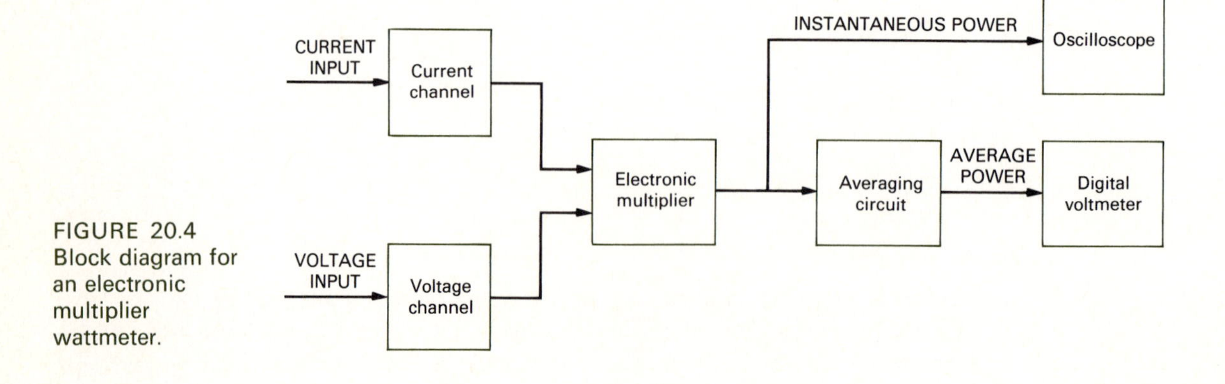

FIGURE 20.4 Block diagram for an electronic multiplier wattmeter.

relationships derived in Sec. 20.1.2, the total wattmeter error is the *sum* of the errors in the measurement of all three parameters. At low power factors, θ approaches 90° and $\cos\theta$ approaches zero. At a power-factor angle of 80°, which is not untypical of many reactive circuits and motor and transformer circuits at light load conditions, an error in θ of 0.5° results in an error in $\cos\theta$ of almost 5 percent. Therefore, care must be taken in the design of both the current and the voltage circuits in any type of wattmeter to ensure minimum magnitude and angle errors. In electronic multiplier and Hall wattmeters, a large number of scales can be made available by means of variable gain circuitry, and power even at very low power factors can be read accurately. However, on standard dynamometer wattmeters, scales are designed for unity power-factor measurements, which means that full scale in watts occurs at rated or full scale in the voltage and current circuits. When measuring power at a low pf, this will require reading the meter in the lower—and hence the less accurate—portion of the watts scale. Therefore, wattmeters with *low-power-factor scales* (full-scale watts occurs at rated current, rated voltage, and 20 percent power factor) are available and should be used when measuring power in reactive circuits. In many power electronic circuits, in which the signals are usually nonsinusoidal, the simple concept of the power factor as the cosine of the angle between the voltage and current signals, as shown in Eq. (20.6), has little meaning. The more general definition of *power factor* is

$$\text{pf} = \frac{\text{average power}}{(\text{RMS volts})(\text{RMS amps})} \tag{20.7}$$

Example 20.1 A certain low-power-factor wattmeter is rated at 200 V 10 A, and the full scale on the watts scale is 400 W. The known errors are ± 0.5 percent of full-scale volts and ± 0.5 percent of full-scale amperes, and there is an angle error of 0.5°. The wattmeter is used to measure power in a sinusoidal system in which the voltage is 175 V, the current is 5 A, and the power factor is 0.15. Assuming errors at the maximum value, what is the maximum possible error in the power reading?

Solution The power factor of 0.15 implies a power-factor angle of 81.37°. The maximum possible power reading is therefore

$$(175 - 1)(5 - 0.05)\cos 81.87° = 122.56 \text{ W}$$

The power is truly equal to

$$175 \times 5 \times 0.15 = 131.25$$

The error is therefore

$$131.25 - 122.56 = 8.69 \text{ W}$$

or

$$8.69 \times \frac{100}{131.25} = 6.6\%$$

Proper connection of the two circuits on a wattmeter is essential for preventing damage to the meter and for ensuring accurate readings. Most commercial wattmeters have the terminals distinguished by polarity markings, usually the ($\pm$) marking. Obviously, it is most essential to connect the voltage and current circuits in the proper orientation since the current circuit on most wattmeters is a very low-resistance delicate circuit and easily damaged by excessive current. But it is also important to connect the two circuits with the proper polarity connections in order to ensure an upscale reading on the meter and to minimize stray capacitive coupling between the coils or electronic circuits that make up the voltage and current channels of the meter. Figure 20.5 illustrates the correct connection of a wattmeter. Frequently, it is necessary to measure voltage and current as well as power, and so the correct location of the voltmeter and ammeter are also shown in Fig. 20.5. Note that either the voltmeter current drain or the ammeter voltage drop will result in a slight error in the wattmeter reading no matter where these instruments are located in the circuit. It is usually desirable to eliminate the voltmeter error and accept the ammeter error, as shown in Fig. 20.5, since the latter is usually much smaller than the former.

The connection of wattmeters to measure power in three-phase circuits can become rather cumbersome, since so many circuits are involved. In four-wire three-phase circuits, three wattmeters must be used, each connected as a single-phase meter between one line and neutral, using the polarity markings as shown in Fig. 20.5. In three-wire three-phase circuits, only two wattmeters are required. The two-wattmeter method of measuring three-phase power is shown in Fig. 20.6.

20.3 MEASUREMENT OF IMPEDANCE

Electrical impedance is composed of real and reactive components, as has been described in earlier chapters of this text. The concept of impedance generally implies

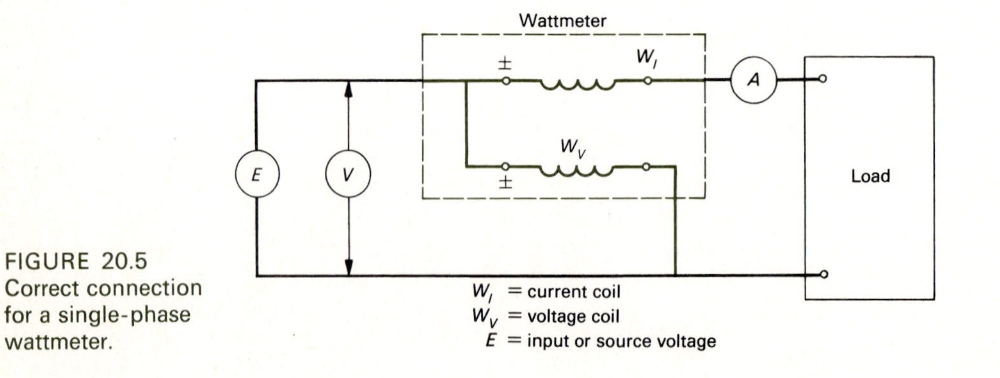

FIGURE 20.5 Correct connection for a single-phase wattmeter.

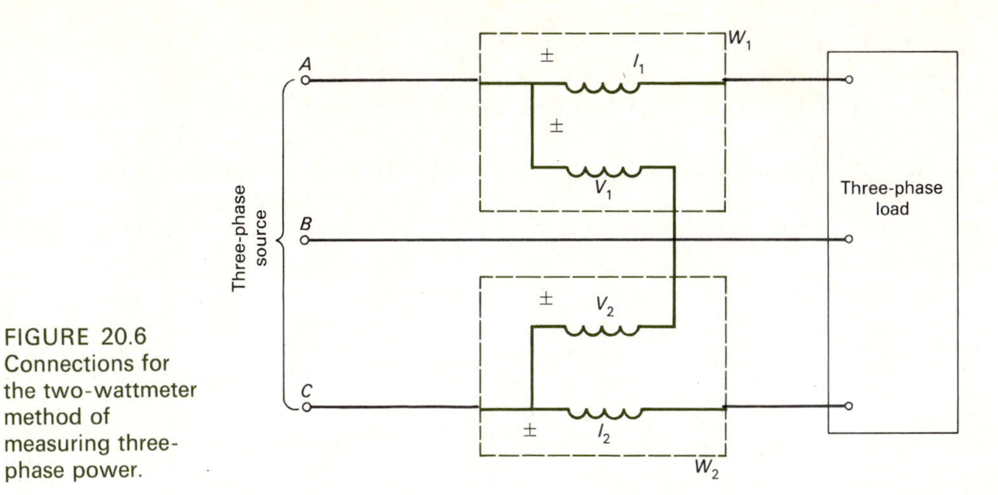

FIGURE 20.6 Connections for the two-wattmeter method of measuring three-phase power.

sinusoidal excitation of the electric circuit. With the advent and widespread use of semiconductors and other switching and nonlinear elements in a great many types of circuits, single-frequency sine-wave impedance or reactance may not be adequate in some applications. It is generally more important to know the value of the circuit elements—that is, resistance, inductance, and capacitance (or their reciprocals)—rather than the reactance value (which is a product of the circuit element value and the frequency). However, much of the nomenclature of the sinusoidally based era carries over into most of the measuring devices used to measure circuit element values, and we have therefore entitled this section "Measurement of Impedance." Most impedance-measuring devices actually measure the impedance of the circuit element at a specific frequency, but read out the measured value as inductance or capacitance rather than as reactance. Impedance over a range of frequency is usually required in electronic and pulse-circuit applications.

20.3.1 Resistance Measurements

Resistance, of course, is theoretically unaffected by the frequency of the applied signal, and therefore many of the above comments would be expected to be inappropriate in relation to resistance measurements. In actuality, however, resistance is probably the most elusive of the three electric circuit parameters. The resistances of almost all practical resistors are frequency-dependent and often include major components of the other circuit elements—inductance and capacitance. For example, the typical wire-wound resistor which will appear purely resistive at power frequencies (below $\simeq 1$ kHz) may appear as primarily inductive at audio frequencies, and it will almost certainly be primarily capacitive at frequencies in the RF range and above. Most resistances are measured at zero (dc) or power frequencies. In general, this type of measurement is not even adequate for evaluating resistances in power circuits, due to the phenomenon known as skin effect (described in Chap. 5). Therefore, if at all possible, resistance should be measured under the excitation that will be used in the ultimate application of this resistance.

However, notwithstanding the above comments, a number of so-called resistance-measuring devices are among the most useful instruments available for many types of experimental work. Among these is the *ohmmeter*, which is often one element in vacuum-tube voltmeters and digital multimeters (DMMs). One function of an ohmmeter is to show circuit continuity (i.e., lack of an open circuit), and in this application it is invaluable. For measurement of the high-resistance values found in many electronic circuits, it gives good accuracy. For low-resistance measurements, its accuracy may be poor, and ohmmeter scales are often difficult to read with precision. Table 20.2 gives the specifications of a typical ohmmeter circuit in a DMM. In terms of measurement, the ohmmeter should never be used below resistance values of about 1.0 Ω. The ohmmeter is basically a dc device measuring the quotient of voltage (at a set value) and current.

More accurate resistance measurements are performed by means of resistance bridges. By far the most common is known as a *Wheatstone bridge*, the circuit diagram of which is shown in Fig. 20.7. In the Wheatstone bridge, as in all other bridge circuits, the measurement is made by means of a comparison with known values. In the case of Fig. 20.7, the measurement of the unknown resistance R_x is performed by balancing the variable resistances R_A and R_B until zero current flows through meter A. At this condition,

$$R_x = \frac{R_A}{R_B} R_S \tag{20.8}$$

The Wheatstone bridge is generally accurate for resistance measurements of 1.0 Ω and above; in most bridges, the accuracy of the measurement is equal to the accuracy to which the standard resistance R_S is known.

To measure resistances below 1.0 Ω (which includes the resistance of a great many rotating machine and transformer circuits used for power and control), the *Kelvin*, or *double, bridge* is required. When measuring resistances of very low magnitude, the voltage drop across the contacts connecting the resistance to the external circuit may be appreciable and cause an error in the resistance measurement. To eliminate contact voltage drop, a four-terminal resistance is used: two outer

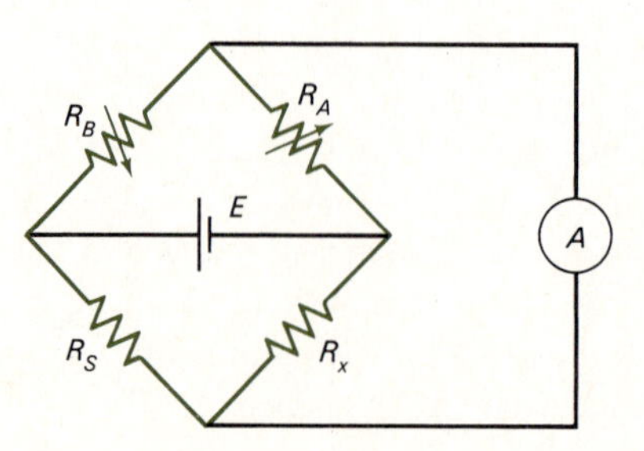

FIGURE 20.7 Elementary Wheatstone-bridge circuit for measurement of resistance.

terminals for connection to the external circuit (i.e., the terminals carrying high current), and two inner terminals connecting to the instrumentation circuit. Thus, the contact potential of the current-carrying terminals is eliminated from the measuring circuit. This principle is used in the Kelvin bridge and also in instrument shunts.

20.3.2 Impedance Measurements

Bridge circuits are also used to measure the two reactive circuit elements, inductance and capacitance. The two most common bridge circuits—which are used to measure inductance in the common impedance, or *universal*, bridge—are the Maxwell and Hay bridge circuits, shown in Fig. 20.8.

Both of these circuits measure inductance in terms of a standard capacitance C_1. At balance—the condition of zero current in the sensing meter A—the unknown resistance and inductance are found in the Maxwell bridge to be

$$L_x = R_2 R_3 C_1 \qquad R_x = \frac{R_2 R_3}{R_1} \tag{20.9}$$

The Maxwell bridge is suited for the measurement of coils of relatively low-quality factor $Q = \omega L_x / R_x$. When it is used to measure coils with higher Q, a very large value of R_1 is required and the resistive balance (to obtain R_x) will not be well defined. For high-Q coils, the Hay circuit is preferred, although its balance equations [similar to Eq. (20.9)] are more complicated and are a function of the frequency of the signal source e.

In general, both the inductance and the resistance of a coil may vary significantly with frequency, and therefore the frequency at which the measurement is performed should be stated. The magnitude of the signal source in all commercial bridges, that is, the value of e in Fig. 20.8, is very low. This introduces another problem when measuring the inductance of an iron-core coil, which is that the core will be magnetized at its level of *initial* magnetization near the origin of its magnetic characteristic, or B-H curve (see Chap. 17). The slope of the B-H curve, which

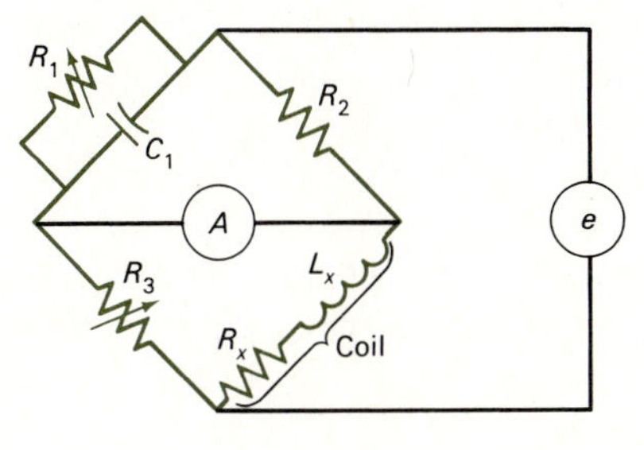

Maxwell bridge

(a)

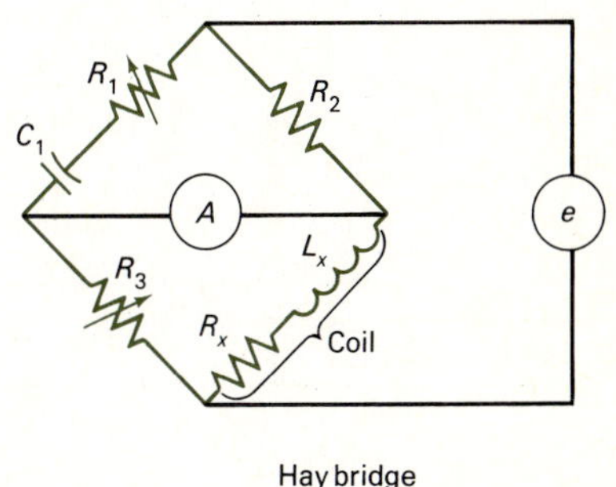

Hay bridge

(b)

FIGURE 20.8 Inductance-measuring circuits.

determines inductance, is usually small at the origin and much different from the slope in the linear operating region of the curve; therefore, the inductance of an iron-core coil as measured on an inductance bridge will almost always be too low. The inductance of iron-core coils should be measured by means of a voltmeter, ammeter, and wattmeter with the coil exciting at the frequency and current level at which it is to be used.

Capacitance is measured by means of bridge circuits similar to the inductance bridge circuits shown in Fig. 20.8. Two most common circuits are the Wien bridge circuit and the capacitance comparison bridge. Capacitance is generally the most constant and stable of the three circuit elements and can be measured with a high degree of accuracy. It is also probably the most "pure" circuit element in that the typical capacitor contains very little resistance and no inductance. For this reason, it is used as a standard of measurement in most bridge circuits, including those of Fig. 20.8.

20.4 OSCILLOSCOPES

The electronic oscilloscope is probably the most versatile instrument in the electronics lab, and it also serves many useful functions on the production line, in the automotive repair shop, in the TV and radio repair shop, and in the households of many engineers and "tinkerers." The oscilloscope is not only a means of *measuring* signals—and with generally excellent accuracy—but also a means of giving a picture of signals, which is perhaps its greatest asset.

The basic element of an oscilloscope is the cathode-ray tube (CRT), although in recent years liquid-crystal displays (LCDs) have been used; LCDs result in a much smaller and more compact scope package. The CRT consists of three major elements: the electron source, often called an electron "gun"; a means of focusing the ray of electrons emitted by the gun, using both electric and magnetic fields; and a screen on which the traverse of the electron beam is exhibited. The CRT screen is constructed of a phosphor layer which emits visible light with a very short time constant upon being excited by the electron beam. The yellow-green portion of the visible light spectrum is the most commonly used phosphor on CRTs since the human eye is most sensitive to this frequency range; however, phosphors emitting many colors can be used, and of course are used in the sophisticated relatives of the CRT screen: the color TV screen, CAD/CAM computer terminal screens, etc. The electron beam is controlled at any given instant by two inputs: the *horizontal control*, which even in complex oscilloscopes is basically the field resulting from the simple triangular waveform produced by the charging of a resistive-capacitive (RC) series circuit; and the *vertical control*, which is determined by the instantaneous magnitude of the signal being observed through the activation of magnetic and/or electric fields as a function of this signal. Secondary control is also exerted by various fields that focus and improve the quality of the beam and by other signals that modify the horizontal control to synchronize it with various external processes or signals, a process generally referred to as *triggering* the oscilloscope. Many oscilloscopes have provision for a wide variety of modifications of the incoming signal being measured; these include a wide range of attenuators

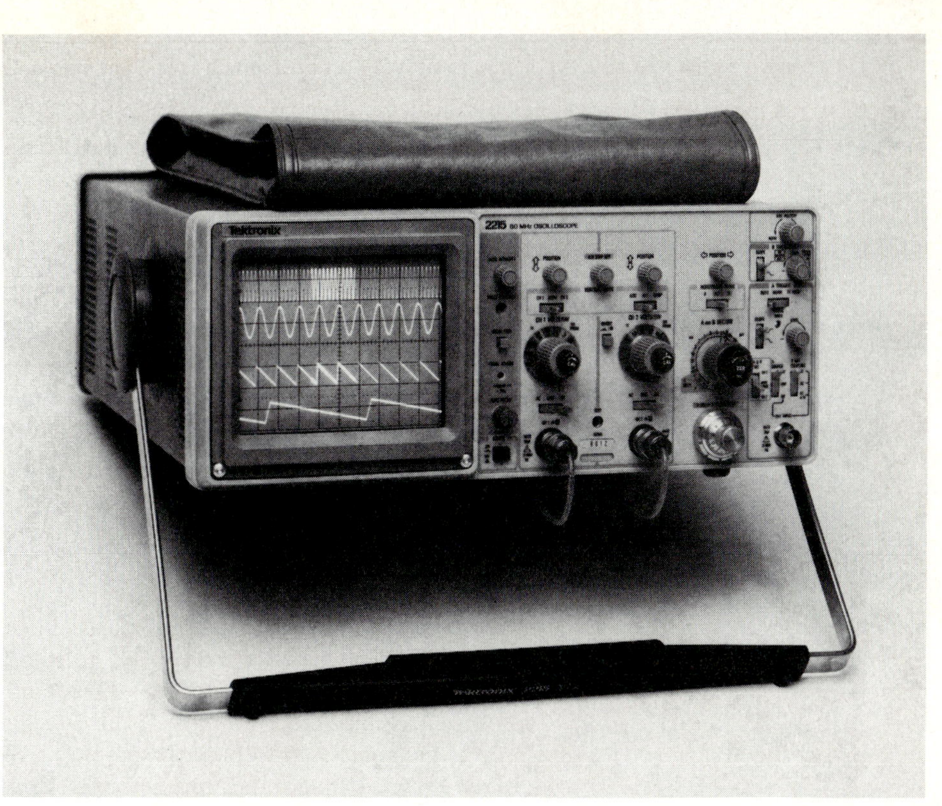

FIGURE 20.9. Typical monolithic oscilloscope for general lab applications. (Courtesy of Tektronix, Inc., Beaverton, Oregon).

for adjusting the input-voltage level, current probes (actually current transformers) for converting current signals to voltage signals of the proper magnitude, circuits for resolving the incoming signal into its Fourier components, and filtering circuits. Thus, a total oscilloscope system is a versatile system capable of reproducing almost any type of signal, providing that the oscilloscope has been designed to handle the frequency, voltage, and rise-time range of the signal.

Figure 20.9 illustrates a typical oscilloscope. Oscilloscopes are basically divided into two classes in terms of construction and configuration: *modular* (often called *plug-in*) oscilloscopes and integrated, or *monolithic* oscilloscopes. The former cover the many options—including those mentioned above—which are realized by means of interchanging plug-in electronic circuits on the oscilloscope chassis. But monolithic oscilloscopes will not accept plug-in circuits; they are designed to perform the basic oscilloscope functions and control capabilities. Figure 20.9 illustrates a monolithic oscilloscope; the specifications of such an oscilloscope, which indicate its capabilities, are given in Table 20.5.

The two most important parameters in the process of choosing an oscilloscope are generally *bandwidth* and *rise time*—factors which determine the frequency response of the scope. These two parameters are defined for an oscilloscope in the same manner as the general definitions given in Chap. 9. Various constraints in scope

TABLE 20.5 Oscilloscope Specifications (*Courtesy of Leader Instruments Corp.*)

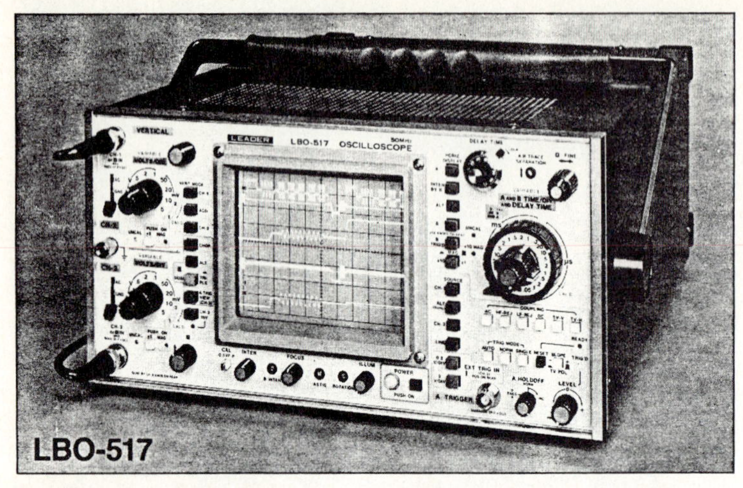

The LBO-517 is a high performance 50-MHz oscilloscope with 1 mV sensitivity up to 10 MHz and 5 mV sensitivity up to 50 MHz. Dual time bases permit detailed observations and accurate time interval measurements. The two time bases may be alternately displayed for simultaneous viewing of both the main time base (with the delayed portion intensified) and the delayed time base for both input channels. Composite triggering permits stable triggering on two asynchronous signals. A trigger viewing function also displays the trigger waveforms for both time bases. The LBO-517 uses a new dome mesh CRT with 20 kV accelerating potential for bright, clearly defined displays, even with very low repetition rates.

SPECIFICATIONS

VERTICAL DEFLECTION

Bandwidth (−3 dB, 8 div.)
dc: 0 Hz to 50 MHz.
ac: 10 Hz to 50 MHz.

Rise Time
7 ns.

Deflection Coefficients
5 mV/cm to 5 V/cm in 10 steps, 1-2-5 sequence, continuously variable between steps, uncalibrated warning lights, x5 multiplier provides 1 mV/cm sensitivity up to 10 MHz.

Accuracy
± 3% (0-40°C), ± 5% with x5 mag.

Imput Impedance
1 MΩ, ± 2%, 35 pF ± 3 pF.

Maximum Input
600 V (dc plus ac peak).

Signal Delay
Leading edge can be observed.

Display Modes
CH-1, CH-2, alternate, chop, add, subtract (CH-2 invert), triple, quad.

Common Mode Rejection Ratio
26 dB at 1 KHz.

Output
CH-1 output on rear panel of 0.1 mV/cm deflection.

EXTERNAL HORIZONTAL DEFLECTION (X-Y MODE)

Input
Via CH-1 vertical amplifier.

Bandwidth (−3 dB, 10 cm)
dc: 0 Hz to 1 MHz.
ac: 10 Hz to 1 MHz.

Rise Time
350 ns.

Phase Shift
<3° at 100 KHz.
All other external horizontal deflection specifications are identical to vertical deflection.

INTERNAL HORIZONTAL DEFLECTION (SWEEP MODE)

Display Modes
Main time base, main time base intensified by delayed time base, alternate main and delayed time base, delayed time base.

Main Time Base
0.05 μS/cm to 0.5 S/cm in 22 steps, 1-2-5 sequence, continuously variable between steps, uncalibrated warning light.

Delayed Time Base
0.05 μS/cm to 0.1 S/cm in 20 steps, 1-2-5 sequence.

Magnifier
Times 10 magnifier extends maximum sweep rate to 5 nS/cm.

Accuracy
± 3% (± 5% with magnifier).

MAIN TIME BASE TRIGGERING (CH-3)

Sources
Internal CH-1, CH-2, Alt., Line.
External (÷1 or ÷10).

Modes
Auto (≥20 Hz).
Normal.
Single.

Coupling
ac, dc, lf reject, hf reject, TV vertical, TV horizontal

Slope
+ or −.

Sensitivity
Internal: 1.5 cm (0.5 cm, 30 Hz-10 MHz)
External: 0.2 V p-p or 2 V p-p switchable.

External Input
Impedance: 1 MΩ, 30 pF.
Maximum Level: 600 V (dc plus ac peak).

Hold-off
Variable sweep hold-off control with B-ends-A switch.

DELAYED TIME BASE TRIGGERING (CH-4)

Modes
Immediate: delayed time base begins immediately after delay time.
Triggered: delayed time base begins on the first trigger after the delay time.

Delay Time Jitter
0.01% (1 part in 10,000) of 10 times the main time base (A TIME/DIV) setting. All other delayed time base specifications are identical to main time base specifications.

Z-AXIS (INTENSITY) MODULATION

Input Level
TTL compatible, DC coupled.

Maximum Input
50 V p-p

INTERNAL CALIBRATOR

Output
0.5 V p-p, ± 2%.

Wave Shape
Square wave, 1 KHz nominal.

CRT DISPLAY

Phosphor
P 31 (P 7 optional).

Graticule
Internal, illuminated 8 x 10 div (1 div = 1 cm).

Accelerating Potential
20 kV.

Trace Alignment
Front panel trace rotation control.

POWER REQUIREMENTS
100, 117, 200, 217, 234 Vac, ± 13%, 50 to 60 Hz.

PHYSICAL

Size (W x H x D)
11¼ x 6¼ x 14¾ in. (290 x 160 x 375 mm).

Weight
25.5 lbs.,
11.5 kg.

ENVIRONMENTAL

Temperature (Operating)
0-40°C.

Vibration
2 mm p-p displacement at 12 to 33 Hz.

Shock
30 g.

SUPPLIED ACCESSORIES
Instruction Manual.
Two (2) Type LP-011 X10 Probes.

AVAILABLE ACCESSORIES
(See pages 16 & 17)
LP-2010 Probe Pouch.
LC-2009 Protective Front Cover.
LRA-517 Rack Mounting Adapter.

construction and design make these two parameters interrelated. As a rule of thumb, the product of bandwidth (MHz) and rise time (ns) should be approximately 0.35. Also, the rise-time rating or capability of an oscillosope should be approximately one-fifth that of the fastest signal to be measured. Using these two approximate relationships gives a desired relationship between the two important oscilloscope parameters of[20.2]

$$\text{Minimum bandwidth (MHz)} \simeq \frac{1.7}{\text{fastest rise time (ns)}} \tag{20.10}$$

An oscilloscope should have a flat response over its rated bandwidth and a smooth roll-off at its upper frequency.

Other important scope parameters are *sensitivity* and *sweep rate*. In an oscilloscope, sensitivity refers to the input voltage required to produce a certain deflection of the spot on the CRT. Most oscilloscopes have a very wide range of voltage sensitivities, but for special applications, other ranges must be considered. Sensitivity is the vertical calibration on an oscilloscope. Most oscilloscsopes have a built-in sawtooth sweep generator that causes the spot on the CRT to move in the horizontal direction at a certain rate. In modern oscilloscopes, the sweep rate is calibrated in the unit of time, usually s/div on the horizontal scale. The sweep rate should be adequate to properly represent the highest frequency signal (or fastest rise-time signal) to be displayed on the oscilloscope. As a rule of thumb, a sweep is considered adequate if it is capable of displaying one cycle across the full horizontal scale.[20.2] Most modern oscilloscopes are equipped with means for calibrating the sweep rate in terms of a direct unit of time; hence, they are generally very accurate in measuring frequency and rise time. A signal trace on an oscilloscope is initiated by the triggering process mentioned earlier. Triggering can be caused by the signal that is being observed, by the power line supplying the oscilloscope, or by external, independent signals.

Many oscilloscopes are capable of displaying several signals concurrently. One technique used for this purpose is to electronically switch the deflection system among two or more input signals; since the switching is done at a very high rate, two or more input signals can virtually be displayed simultaneously. This electronic switching among two input signals is known as a *dual-trace* process. *Dual-beam* oscilloscopes contain two independent beams and deflection systems and, in general, can be operated as two independent oscilloscopes, although some of the triggering options may be shared by the two systems. A dual-beam scope with dual-trace switching can display four signals simultaneously. Dual-trace oscilloscopes can operate in the $X - Y$ mode to present displays of a function of two variables. This mode is similar to that of the familiar $X - Y$ plotter; the variable signal X is inputted to one channel and Y to the other channel. On some oscilloscopes a third variable input, called the Z axis, is available. The Z axis controls the beam intensity, permitting shading of the $X - Y$ display.

Another useful feature of certain oscilloscopes is that of the long-persistent screen, often called a *storage* oscilloscope. In long-persistent screen oscilloscopes, the image displayed on the screen remains for relatively long periods of time (10 s to 2 min). A storage oscilloscope is the extension of this time period to an indefinitely long period. This capacity is excellent for observing and storing all sorts of transient signals, especially those that occur randomly or which are not repeatable.

Recently, storage oscilloscopes have also been constructed on the basis of an entirely different principle, resulting in what is known as the *digital* oscilloscope. In digital storage, the incoming signal is first digitized by a sampling process and then quantized—converted to a binary number which can be stored in data memory, just as computer signals are stored in memory. This process is discussed in more detail in Chap. 22. Since the data memory is permanent, there is no fading or distortion of the trace signal, as may occur over time with the persistent screen storage oscilloscopes. The power requirement of a digital storage oscilloscope is considerably less than that of the persistent screen oscilloscope, since relatively high power is required to drive the persistent screen. Also, the trace that has been stored digitally can be modified in several ways by means of electronic controls contained in the digital oscilloscope; for example, a short portion of the stored trace can be expanded and examined in detail on the CRT.

PROBLEMS

20.1 A voltmeter has a full scale of 500 V and a probable error of 0.1 percent of full scale. When using this voltmeter to measure a signal of 75 V, what percent probable error can exist?

20.2 Compare the percent error in two 10 A ammeters, one digital and one analog, when measuring (*a*) a current of 5 A. The error specifications on the digital meter are 0.07 percent of the reading, 0.05 percent of full scale, 0.005 percent of reading per degree Celsius, and 0.002 percent of full scale per degree Celsius; the analog meter error specifications are 0.5 percent of full scale and 0.001 percent of the reading per degree Celsius; assume that the temperature at the time of the measurement is 20°C above the ambient. Repeat this comparison when measuring (*b*) a current of 2 A at the same temperature.

20.3 The ohmic (I^2R) loss in a power resistor is to be measured. The current is measured with an analog ammeter having a probable error of ± 0.5 percent of full scale of 100 A. The resistance is measured with an ohmmeter having a probable error of ± 1 percent of a 10 Ω scale. The current is measured to be 63 A and the resistance 1.2 Ω. What is the maximum probable error (in watts and in percent or measured value) in the loss measurement?

20.4 Figure 20.6 shows the two-wattmeter connection for making power measurements on balanced three-phase loads. Derive the expression for the reading of each wattmeter in terms of the line voltage V_1, the line current I_1, and the functions of the load power-factor angle θ. Show that the sum of the two wattmeter readings equals $\sqrt{3}V_1 I_1 \cos\theta$. At what power factor does one wattmeter indicate a zero scale reading?

20.5 A standard wattmeter has a current coil rated at 20 A, a voltage coil rated at 200 V, and a full scale of 4000 W. This meter is to be used to measure the no-load losses on an induction motor; the losses are expected to be about 1500 W at a 25 percent power factor. If the measurement is made at 200 V_1, will the current coil be operated over its rating? If the meter's probable error is 0.75 percent of the scale, what is the percent probable error in the reading?

20.6 A certain DMM reads true RMS values of current. If the *peak* value of each of the following periodic current waves is 1.0 A, what will the meter reading be for (*a*) sine wave, (*b*) square wave, (*c*) triangular wave, (*d*) sawtooth wave?

20.7 In the electronic multiplier-type wattmeter, Fig. 20.4, the output of the multiplier is $p = e \cdot i$, instantaneous power. Derive an *RC*-op amp network for the block labeled "Averaging circuit" in Fig. 20.4 to convert this signal to a signal proportional to average power, as given in Eq. (20.5).

REFERENCES

20.1 D. R. Hamburg and L. E. Unnewehr, "An Electronic Wattmeter for Nonsinusoidal Low Power Factor Measurements," *IEEE Transactions on Magnetics*, vol. MAG-7, no 3, September 1971.

20.2 *Tektronix Reference Manual*, Tektronics, Inc., Beaverton, Oregon, 1981.

20.3 T. G. Beckwith, N. L. Buck, and R. D. Marangoni, *Mechanical Measurements*, 3d ed., Addison-Wesley, Reading, Mass., 1982.

20.4 A. Desa, *Principles of Electronic Instrumentation*, Halsted Press, New York, 1981.

Chapter

Digital Logic Circuits

The concepts broadly categorized by the words "digital" and "analog" have been introduced and discussed in several previous chapters. A brief insight into analog computation was given in Chap. 9, and analog and digital instrumentation were discussed and compared in Chap 20.

The word "analog" has actually taken on a meaning in engineering systems that is considerably removed from its basic definition, which is "something that is analogous or similar to something else."[21.1] This definition does apply to the analog computer and, from concepts employed in analog computation, the more general meaning of "analog" has arisen to include circuits, processes, and instruments in which *continuous or infinitely variable signals are present*, as in an analog computer. The word "digital," on the other hand, is much closer to its basic root origin in the word "digit": as defined for engineering purposes, digital means "of or relating to calculations by numerical methods or by discrete units."[21.1] In this text, we make full use of these broader meanings of the words—*analog* to refer to circuits, systems, or mathematical processes in which the signals are continuous and can take on *all values* between certain limiting values—and *digital* to refer to circuits, systems, or mathematical processes in which the signal can take on only *certain discrete values* between the limiting values.

Digital circuits of any type (electric, mechanical, hydraulic, acoustical, etc.) can operate properly in only *two* discrete states, or values. These are generally called one (1) and zero (0). A digital circuit is either in one of these states or it is malfunctioning. The state of 1, called *logic* 1, means that the system is in the higher of two signal levels. For example, in many electrical logic systems, this state may have a voltage level in the range of 2 to 10 V. It is the "on" state, or high state. Conversely, logic 0 is the "off" state, or low state. Thus, the concept of logic 1 and logic 0 also means on-off. Much of the nomenclature in digital circuits has come from an earlier field of study known as *switching theory* and is related to concepts used to describe the operation of on-off devices such as relays and bang-bang controllers.

Digital circuits were originally developed for the on-off controllers mentioned above but, of course, they became much more significant with the development and tremendous use of the digital computer. Mainly as a result of the many by-products of computer technology, digital circuitry has been applied in almost every type of engineering system today, particularly in control and instrumentation. Even the analog computer has been largely displaced—as a purely computational device—by digital modeling programs on digital computers. It can be stated that in almost all types of electrical systems today—with the exception of power production. distribution, and transmission and in high-power motors—there are competing digital and analog systems available to the user.

Many of the comments made in the previous chapter to compare digital and analog instrumentation can also be applied to the general analog-digital comparison for other types of systems. However, as we will see, down at the fundamental level there is little difference between analog and digital circuits or systems, and the basic laws and theorems of electricity presented in earlier chapters of this text apply equally to either type of circuit. It is in the final package—that is, in the discrete bundles or continuous signals—that a difference exists; therefore, in order to analyze this final package, we will have to introduce some different mathematics and circuit-analysis techniques.

21.1 BINARY BITS

The output of a digital circuit is a single binary digit, either logic 1 or logic 0. These two output states are also frequently termed *true* (1) and *false* (0). The abbreviation for binary digit is *bit*.

Two advantages to this type of output are that it is not necessary to measure the output precisely—it is either on or off, 1 or 0, true or false—and that electromagnetic interference (emi) is generally less detrimental to digital circuits than to analog circuits. Obviously, the single bit is a good means of answering questions that have a yes/no answer or of operating a control function that is either on or off. However, for more complex questions and more sophisticated control schemes, the single bit is inadequate; thus, almost all practical circuits require a multitude of bits. Because digital circuits are most commonly used in informational systems, such as computers, calculators, and control systems, the question naturally arises as to the relationship between the informational content and the number of bits required. A general theory[21.2] states that

> The minimum number of bits required to express n different things is k, where k is the smallest positive integer such that

$$2^k \geqslant n \tag{21.1}$$

The base 2, or binary, system of numbers is thus obviously well-suited to the mathematics required for digital systems.

21.1.1 Decimal to Binary Conversion

It is assumed that the reader is familiar with both binary and decimal systems of numbers. The following rules and examples explain how to convert from one system to another. The theory behind these conversions, not included here, can be found in Refs. 21.3 and 21.4.

Let N refer to a specific whole decimal number:

1 Determine if N is odd or even.
 a If odd, write 1 and subtract 1 from N; go to step 2.
 b If even, write 0; go to step 2.

2 Divide the N obtained from step 1 by 2.
 a If $N > 1$, go back to step 1 and repeat.
 b If $N = 1$, write 1.

The number written by this process is the binary (or base 2) equivalent of the original decimal number. The number written first is the least significant bit; the number written last is the most significant bit.

Example 21.1 Find the binary equivalent of the decimal number 332.

Solution The decimal number 332 is even, so we write a 0; dividing 332 by 2 gives 166, which is also even; write another 0; dividing 166 by 2 gives 83, which is odd; write a 1 and subtract 1 from 83, giving 82; divide 82 by 2, giving 41, which is odd; write a 1 and subtract 1 from 41, giving 40; divide 40 by 2, giving 20, which is even; write a 0; divide 20 by 2, giving 10, which is even; write a 0; divide 10 by 2, giving 5, which is odd; write a 1 and subtract 1 from 5, giving 4; divide 4 by 2, giving 2, which is even; write a 0; divide 2 by 2, giving 1; write a 1. The final result is 101001100.

A decimal fraction may be converted to binary by converting the numerator and denominator to binary by means of the method described above. Decimals may not have an exact binary equivalent, and the decimal-to-binary conversion often results in a repetitive sequence of binary bits.[21.3] The method for directly converting decimals to binary equivalent is as follows, with the decimal equal to N:

1 Double N
 a If this gives a value greater than 1, write a 1 and subtract 1 from the doubled value of N and go back to step 1.
 b If the doubled value of N is less than 1, write a 0 and go back to step 1.

2 Repeat step 1 until the number of significant figures of the binary fraction is sufficient for the application at hand; note that the first number obtained from step 1 is the first number to the right of the decimal point, i.e., the most significant digit.

Example 21.2 Find the binary equivalent of the decimal 0.654.

Solution Doubling N gives 1.308, which is greater than 1; write a 1 and subtract 1 from 1.308, giving 0.308; double 0.308, giving 0.616, which is less than 1; write a 0 and double 0.616, giving 1.232, which is greater than 1; write a 1 and subtract 1, giving 0.232; double, giving 0.464; write a zero; double, giving 0.928; write a zero; double, giving 1.856; write a 1 and subtract 1, giving 0.856; double, giving 1.712; write a 1 and subtract 1, giving 0.712. As of this point, our binary equivalent is 0.1010011; the process could be continued to give the desired accuracy of the decimal equivalent. There is no assurance that the binary process will end with a firm 1, and therefore the above process might continue indefinitely.

It should be noted that binary equivalents can be checked out in a manner similar to that used in the decimal system from the basic theory of numbers. This check-out process is, of course, applicable to numbers expressed in any base. For example, the binary number derived in Example 21.1 can be checked out as follows:

$$
\begin{aligned}
0 \times 2^0 &= 0 \\
0 \times 2^1 &= 0 \\
1 \times 2^2 &= 4 \\
1 \times 2^3 &= 8 \\
0 \times 2^4 &= 0 \\
0 \times 2^5 &= 0 \\
1 \times 2^6 &= 64 \\
0 \times 2^7 &= 0 \\
1 \times 2^8 &= \underline{256} \\
& \quad 332
\end{aligned}
$$

Likewise, checking out the decimal conversion of Example 21.2 gives

$$
\begin{aligned}
1 \times 2^{-1} &= 0.5 \\
0 \times 2^{-2} &= 0.0 \\
1 \times 2^{-3} &= 0.125 \\
0 \times 2^{-4} &= 0.0 \\
0 \times 2^{-5} &= 0.0 \\
1 \times 2^{-6} &= 0.015625 \\
1 \times 2^{-7} &= \underline{0.0078125} \\
& \quad 0.6484375
\end{aligned}
$$

21.2 INTRODUCTION TO BOOLEAN ALGEBRA

Boolean algebra is a useful tool in the analysis and design of digital circuits, and we will now introduce a few of this methodology's basic concepts. Boolean algebra is the mathematical expression of the logic and decision-making aspects of digital circuits. Thus, it is useful in both the design and analysis stages of developing digital circuitry; in a more general sense, it is quite useful in any analysis of logic processes, decision making, relay circuits, etc.

Boolean algebra is a system of mathematics which is abstract and independent of its applications. However, it has been largely associated with the mathematics of logic circuits, and our brief presentation will be aimed at this application. In boolean algebra, variables have only two possible values, 1 and 0. This must be remembered in studying the summary of basic boolean relationships listed below, for some of these relationships will be in marked contrast to the common algebra associated with real and complex continuous variables. We present these relationships as "practical tools" in the simplification of logic and relay circuits. These mathematical relationships are also presented to help the reader understand their electronic circuit realization by practical electronic components and circuits; this implementation of boolean logic will be presented in the subsequent sections of this chapter. For further study of boolean theory, consult Refs. 21.2 to 21.7.

21.2.1 Basic Laws of Boolean Algebra

Let x, y, and z be boolean variables in the set $T\colon\{0, 1\}$.

1. The symmetric law: If $x = y$, then $y = x$.
2. The transitive law: If $x = y$ and $y = z$, then $x = z$; and x may be substituted for y or z in any formula involving y or z.
3. Joining, the OR function: The boolean sum expression $x + y$ has the meaning of "joining x and y," the "union of x and y," or, more specifically in logic circuits, x *or* y.
4. Multiplication, the AND function: The boolean expression xy has the meaning x *and* y.
5. Commutative laws:

$$x + y = y + x \tag{21.1a}$$

$$xy = yx \tag{21.1b}$$

6. Distributive laws:

$$x(y + z) = xy + xz \tag{21.2a}$$

$$x + yz = (x + y)(x + z) \tag{21.2b}$$

Note that the second of the distributive laws has no counterpart in conventional algebra.

7 Associative laws:

$$(x + y) + z = x + (y + z) \tag{21.3a}$$

$$(xy)z = x(yz) \tag{21.3b}$$

8 Idempotent laws:

$$x + x = x \tag{21.4a}$$

$$xx = x \tag{21.4b}$$

There is no equivalent to these relationships in conventional algebra.

9 Multiplication and addition by 1 or 0:

$$0 + x = x \tag{21.5a}$$

$$1 + x = 1 \tag{21.5b}$$

$$0 \cdot x = 0 \tag{21.5c}$$

$$1 \cdot x = x \tag{21.5d}$$

10 The NOT function: This is often called the complementary function; it refers to the complement (or opposite) of the listed function. The complement of 1 is, of course, 0, and vice versa. This is a most important concept in the development of electronic logic circuits, the physical embodiment of NOT being the logic inverter. NOT is further described by the laws of complementarity:

$$x\bar{x} = 0 \tag{21.6a}$$

$$x + \bar{x} = 1 \tag{21.6b}$$

where $\bar{x}$ is the complement of x. The logic inverter may be represented in logic diagrams as a separate entity, as in the third and fourth diagram of Fig. 21.1, or by means of a circle on a logic symbol—often called a "bubble."

11 De Morgan's theorem: Two important relationships between functions and their inverse, or complements, are useful in reducing logic expressions and simplifying logic hardware:

$$(x + y) = \bar{x}\bar{y} \tag{21.7a}$$

$$(\overline{xy}) = \bar{x} + \bar{y} \tag{21.7b}$$

It should be noted that other symbols for the OR and AND functions are frequently found in the literature: v and + to represent OR, and $\cdot$ to represent AND. There are several other mathematical functions not included in the above boolean formulas that are of importance in electronic logic circuits. These are:

12 Exclusive OR: This has the meaning "x or y, but not both" and is designated by the symbol $x \oplus y$

13 NAND: This has the meaning "NOT (x and y)" and is designated by the formula

$$\overline{x \cdot y} = z \tag{21.8}$$

21.2.2 Truth Tables

A simple graphical means of describing the logic associated with the boolean expressions given above is the *truth table*. The adjective "truth" arises from the frequently used true/false notation for the logic variables 1/0. For example, the OR function is defined by the truth table of Table 21.1. The exclusive OR (XOR) function is likewise defined by the truth table of Table 21.2. The product, or AND, function is defined by the truth table of Table 21.3. Truth tables are also sometimes called *connectives*. Note in Tables 21.1 to 21.3 that each truth table has four rows. The four rows are combinations of the two states of the boolean variable, 1and 0. From Eq. (21.1) there should be

TABLE 21.1 Truth Table for OR Logic Function

x	y	$x + y$
0	0	0
0	1	1
1	0	1
1	1	1

TABLE 21.2 Truth Table for Exclusive OR Function

x	y	$x \oplus y$
0	0	0
0	1	1
1	0	1
1	1	0

TABLE 21.3 Truth Table for AND Function

x	y	xy
0	0	0
0	1	0
1	0	0
1	1	1

TABLE 21.4

Connective	Truth table	Meaning
0	0000	Totally false
1	0001	xy
2	0010	$x\bar{y}$
3	0011	x
4	0100	$\bar{x}y$
5	0101	y
6	0110	$(x\bar{y} + \bar{x}y) = x \oplus y$
7	0111	$x + y$
8	1000	$(\overline{xy})$
9	1001	$(xy + \overline{xy})$
10	1010	$\bar{y}$
11	1011	$(\bar{y} + x)$
12	1100	$\bar{x}$
13	1101	$(\bar{x} + y)$
14	1110	$(\bar{x} + y) = \overline{xy}$
15	1111	Totally true

$2^4 = 16$ possible ways to fill out a column of truth tables (consult Ref. 21.4), and this is illustrated in Table 21.4.

In this table, the columns of the previous three tables are represented by rows. For example, the truth table of connective 6 in Table 21.4 is the exclusive OR of the right column of Table 21.2; the truth table for x (connective 3) is seen to be the left column of Tables 21.1, 21.2, and 21.3; and so forth.

Example 21.3 Prepare a truth table for the boolean expression $A = x + \bar{y}z$.

Solution This equation is read, "A equals x OR (complement of y AND z)." The truth table is as follows:

x	y	z	$\bar{y}$	$\bar{y}z$	A
0	0	0	1	0	0
0	0	1	1	1	1
0	1	0	0	0	0
0	1	1	0	0	0
1	0	0	1	0	1
1	0	1	1	1	1
1	1	0	0	0	1
1	1	1	0	0	1

This truth table has been prepared in more detail than is usually necessary in order to illustrate the individual steps in solving a boolean equation.

Example 21.4 Determine $\bar{A}$ of A in Example 21.3 in equation form from De Morgan's theorems and from the truth table of Example 21.3.

Solution In words, De Morgan's theorems [Eqs. (21.7*a*) and (21.7*b*)] state that the complement of a function A can be found by replacing each variable by its complement and by interchanging all AND and OR signs. Therefore,

$$\bar{A} = \overline{x + \bar{y}z} = \bar{x}(y + \bar{z})$$

The truth table for $\bar{A}$ can be found by replacing the last column in the table of Example 21.3 by the complement of the individual element. Shown as rows rather than columns (to conserve space), this is

A	0	1	0	0	1	1	1	1
$\bar{A}$	1	0	1	1	0	0	0	0

Example 21.5 Simplify the expression

$$B = (x + \bar{y})(\overline{y + z\alpha}) + (\beta y + \alpha\bar{x})(\overline{y + z})$$

Note that the complement sign over an entire parenthetical expression refers to the complement of the entire expression, not to the complements of the individual elements.

Solution In the analysis of this problem, we show at the right (in parentheses) the numbers of the boolean laws from Sec. 21.2.1.

$$\overline{(y + z\alpha)} = \bar{y}(\bar{z} + \bar{\alpha})$$

De Morgan's theorem (11)

$$\overline{(y + z)} = \bar{y}\bar{z}$$

Substituting,

$$\begin{aligned} B &= (x + \bar{y})\bar{y}(\bar{z} + \bar{\alpha}) + (\beta y + \alpha\bar{x})\bar{y}\bar{z} \\ &= (x\bar{y} + \bar{y})(\bar{z} + \bar{\alpha}) + (\beta y + \alpha\bar{x})\bar{y}\bar{z} && \text{(6 and 8)} \\ &= (x\bar{y} + \bar{y})(\bar{z} + \bar{\alpha}) + \alpha\bar{x}\bar{y}\bar{z} && \text{(10)} \\ &= \bar{y}(\bar{z} + \bar{\alpha}) + \alpha\bar{x}\bar{y}\bar{z} && \text{(6 and 9)} \\ &= \bar{y}\bar{z} + \bar{y}\bar{\alpha}(1 + \bar{x}\bar{z}) && \text{(6)} \\ &= \bar{y}\bar{z} + \bar{y}\bar{\alpha} && \text{(9)} \end{aligned}$$

21.3 IMPLEMENTATION OF BOOLEAN LOGIC

We have been discussing mainly theoretical concepts of logical decision making so far. The arithmetic of these processes is based on the base 2, or binary, number system. However, the physical realization of this system of mathematics is one of the most exciting developments in modern culture, influencing almost every aspect of industrialized society. We are primarily concerned in this text and in this chapter with electronic and electric implementation of Boolean logic; but before we enter into discussion of electric hardware and software, we should note that similar developments in the realization of these logic concepts have been made in many other areas of physics, including mechanical systems, hydraulic systems, and pneumatic systems.

The three elementary logic elements, often called *gates*, are items 1, 7, and 14 in Table 21.4—the AND, OR, and NAND gates, respectively. The NAND function read as "not (x and y)"; it is the inverted, or complementary, AND function and is represented by the AND symbol with a circle, or bubble, on its output. Likewise, the NOR function is the inverted OR function, defined by the OR symbol with a bubble on its output. There are many other important and widely used logic circuits available, some of which are not listed in Table 21.4.

There have been various symbols for the hardware implementation of logic gates and logic functions used through the years, just as there has been some variation in the mathematical symbols used to express boolean functions. Recently, the hardware symbols have been standardized by the American National Standards Institute (ANSI), the Institute of Electrical and Electronics Engineers (IEEE), and other standards agencies. Table 21.5 (in Sec. 21.4) lists the ANSI standard symbols for a number of logic functions, along with the meaning of these functions.[21.5] The group of functions designated FF in Table 21.5 are flip-flop functions—switching functions, each associated with various types of logic—which will be discussed in Sec. 21.4. Figure 21.1 illustrates the use of these hardware symbols in developing logic circuits; it shows models of several simple boolean equations. The triangle in Fig. 21.1 represents the electronic inverter—the complementary, or NOT, function described in Sec. 21.2.1.

21.3.1 Standard Forms of Boolean Equations

Figure 21.1 introduces the concept of boolean equations or boolean functions which are useful in the analysis, design, and simplification of digital electronic circuits. A general form for a boolean equation can be given in terms of the fourth function shown in Fig. 21.1:

$$f(A, B, C) = D = A + BA + C \tag{21.9}$$

In Eq. (21.9), A, B, and C are variables. Symbols representing variables (A, B, $\bar{A}$, etc.) are called *literals*, a term often used in boolean algebra. We have been using the terms x, y, and z as variables in most of the previous expressions, as in conventional algebra, but almost all letters of the Latin and Greek alphabets seem to be acceptable for boolean expressions.

There are two forms of expression which occur frequently in electronic logic circuits: sum of products (SOP) and product of sums (POS). A good example of the latter form is given in the second circuit shown in Fig. 21.1; an example of the former (sum of products) results by multiplying out the output of the first circuit of Fig 21.1, giving $D = AC + BC$. Product terms and sum terms are in *standard form* if they contain one literal from every variable in the domain. For example, the POS given in the

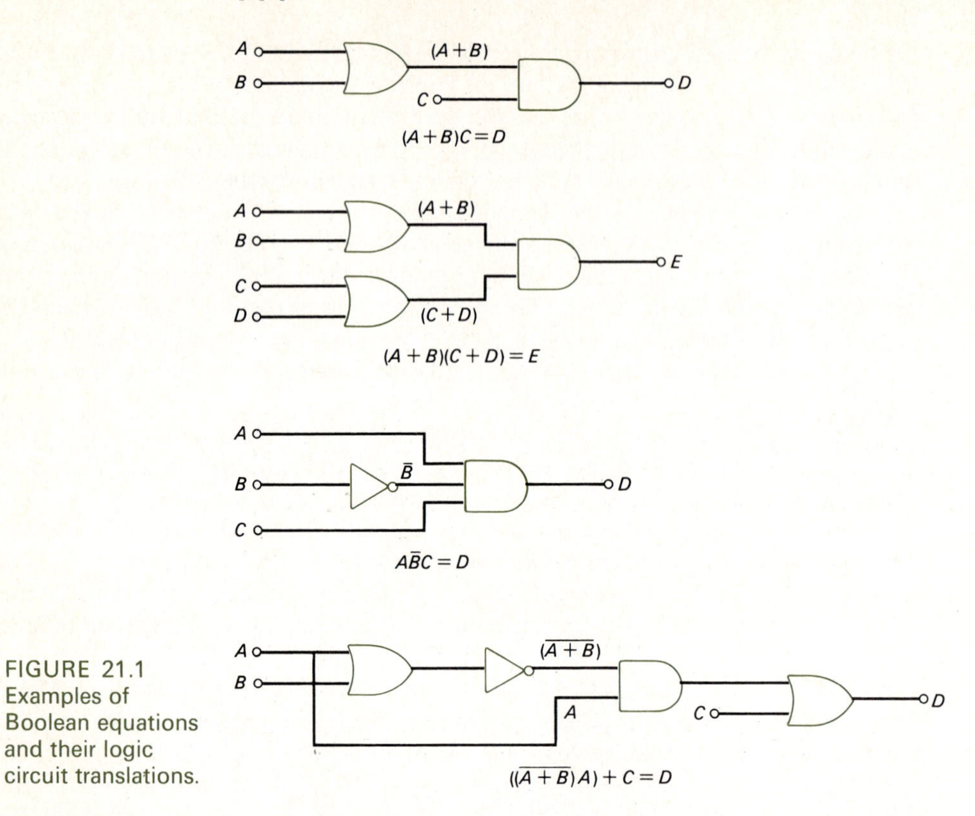

FIGURE 21.1 Examples of Boolean equations and their logic circuit translations.

second circuit of Fig. 21.1 indicates that the output E is a function of A, B, C, and D. These variables define the *domain* of E, and this is indicated by writing E as $E(A, B, C, D)$. Obviously, the second equation in Fig. 21.1 is *not* in standard form by the above definition, since neither sum term contains all four literals. Likewise, the output of the first circuit of Fig. 21.1 would be written as $D(A, B, C) = AC + BC$, and it is seen that this is *not* in standard SOP form since neither sum term contains A, B, and C. For many mathematical manipulations, including the important task of minimizing the number of logic functions required, it is important to describe logic outputs in standard forms. This can usually be done by means of some of the boolean identities listed in the previous section without altering the value of the output. We will illustrate this procedure by means of several examples.

Example 21.6 Convert the form of the output of the first circuit in Fig. 21.1 to standard SOP.

Solution This output is $D(A, B, C) = AC + BC$; it is seen that B is missing from the first product and A from the second. The first term can be multiplied by $(B + \bar{B})$, which equals 1 [from Eq. (21.6*b*)], giving $ABC + A\bar{B}C$. The second term is altered by multiplying it by $(A + \bar{A})$, giving $ABC + \bar{A}BC$. The standard form of this equation is therefore

$$D(A, B, C) = ABC + A\bar{B}C + \bar{A}BC$$

Note that although the term ABC appears twice from this expansion, it is written only once in the standard form. This expression is standard SOP since each variable appears in each product term.

Example 21.7 Convert the form of the output of the second circuit in Fig 21.1 to standard POS.

Solution The output of this circuit is given as $E(A, B, C, D) = (A + B)(C + D)$. C and D are missing from the first product term. It can be shown that the missing terms can be introduced by the following product terms:

$$(A + B + C + D)(A + B + C + \bar{D})(A + B + \bar{C} + \bar{D})(A + B + \bar{C} + D)$$

The reader should ascertain that this substitution is indeed an identity by multiplying out the terms and using the equations in Sec. 21.2.1. The second term in the original POS is missing the literals A and B. This can be introduced by means of a substitution similar to that used for the first term. The result in standard form (with all literals in each product term) is

$$E(A, B, C, D) = (A + B + C + D)(A + B + C + \bar{D})(A + B + \bar{C} + \bar{D})(A + B + \bar{C} + D)$$
$$(A + \bar{B} + C + D)(\bar{A} + \bar{B} + C + D)(\bar{A} + B + C + D)$$

Again note that the term $(A + B + C + D)$ appears twice in the expansion but is written only once.

Example 21.8 Convert the expression $f(a, b, c, d) = a\bar{b} + ac\bar{d}$ to standard SOP form.

Solution The literals c and d are missing from the first term; they can be introduced by multiplying first by $(c + \bar{c})$ and then by $(d + \bar{d})$, as explained in Example 21.6. The second term is missing the literal b and can be put in standard form by multiplying this term by $(b + \bar{b})$. Performing these expansion operations gives

$$f(a, b, c, d) = a\bar{b}\bar{c}\bar{d} + a\bar{b}\bar{c}d + a\bar{b}c\bar{d} + a\bar{b}cd + abc\bar{d}$$

The term $a\bar{b}c\bar{d}$ appeared twice as a result of the above expansion operations but needs to be written only once in the standard form.

21.3.2 Numerical Representation of Standard Forms

In many operations with boolean algebra, it is useful to represent the standard forms by the sum ($\sum$) of a group of numbers, as will be illustrated in the minimization process of Sec. 21.3.4. Without getting into the theory behind this process, we will now illustrate the technique for converting the standard forms into numerical representations.

Standard SOP In converting this form to numerical representation, literals are given the value of 1 and complements of literals are represented by 0. The resultant sums represent a series of binary numbers which are then converted to decimal numbers and expressed as a summation. For example, the standard SOP derived in Example 21.6 is converted to numeric form as follows:

$$\begin{aligned} D(A, B, C) &= ABC + A\bar{B}C + \bar{A}BC \\ &= 111 + 101 + 011 \\ &= \sum(7, 5, 3,) \end{aligned}$$

The standard SOP of Example 21.8 is converted as

$$\begin{aligned} f(a, b, c, d) &= a\bar{b}\bar{c}\bar{d} + a\bar{b}\bar{c}d + a\bar{b}c\bar{d} + a\bar{b}cd + abc\bar{d} \\ &= 1000 + 1001 + 1010 + 1011 + 1110 \\ &= \sum(8, 9, 10, 11, 14) \end{aligned}$$

Standard POS In this conversion, the numeric value of 0 is assigned to literals and the value 1 to complements of literals— just the opposite of the conversion used for standard SOPs. The resulting product of binary numbers is expressed as a product of decimal numbers. For the standard POS of Example 21.7,

$$\begin{aligned} E(A, B, C, D) &= (0000)(0001)(0011)(0010)(0100)(1100)(1000) \\ &= \prod(0, 1, 3, 2, 4, 12, 8) \end{aligned}$$

It is possible to go to the numeric form directly from the original form of a logic equation by inserting a symbol, say X, for the missing literals, and letting X take on the values of 1 and 0. For example, the original form of the POS equation of Example 21.7 is $(A + B)(C + D)$. The first term can be written as $(A + B + X + X)$; letting X take on both a 0 and a 1 gives 0000, 0001, 0010, 0011. The second term can be written as $(X + X + C + D)$; letting X take on 0 and 1 gives 0000, 1000, 1100, 0100. Using only one of the 0000 terms, it is seen that this gives the same set of seven terms in the product as was obtained from the standard SOP form used above. The same can be done in converting standard SOP terms to numeric value. For example, the original equation in Example 21.8 is $a\bar{b} + ac\bar{d}$. Expressing the first term as $a\bar{b}XX$ gives 1000, 1001, 1010, 1011; the second term can be written as $aXc\bar{d}$, which results in 1010 and 1110. Using the term 1010 only once, it is seen that the same five terms are present as were in the conversion using standard SOP above.

The numerical representation is a good shorthand method of writing a function and can easily be converted to the original equation (in standard form) by the reverse of the process described above.

Example 21.9 Write the equation in standard POS for the numerical representation $f(x, y, z) = \prod(0, 2, 4, 5, 7)$.

Solution First express the numbers in binary, giving 000, 010, 100, 101, 111. Then use the reverse of the rule for converting literals to numerals. For POS, a 0 becomes a literal and a 1 becomes a complement. Therefore, 000 becomes $(x + y + z)$, 010 becomes $(x + \bar{y} + x)$, and so on, giving

$$f(x, y, z) = (x + y + z)(x + \bar{y} + z)(\bar{x} + y + z)(\bar{x} + y + \bar{z})(\bar{x} + \bar{y} + \bar{z})$$

Example 21.10 Write the equation in standard form for $f(w, x, y, z) = \sum (1, 5, 13, 15)$.

Solution The binary form of the numerics is 0001,0101,1101,1111. For SOP, a 1 becomes a variable and a 0 becomes a complement. Therefore,

$$f(w, x, y, z) = \bar{w}\bar{x}\bar{y}z + \bar{w}x\bar{y}z + wx\bar{y}z + wxyz$$

21.3.3 Karnaugh Maps

Logic functions which are expressed in standard form can be displayed graphically by a technique known as Karnaugh mapping (see Refs. 21.2, 21.4, and 21.8).

Logic circuits are initially developed from a verbal description of the functions or processes to be performed; from this description, logic diagrams and their associated equations are developed, simple examples of which are shown in Fig. 21.1. But there is generally no way of ascertaining during this early development stage that the resulting logic circuit is the best possible circuit for performing the desired functions. It is desirable from both cost and reliability standpoints to use the minimum number of logic functions to perform the required functions—and circuits can indeed be simplified on a somewhat hit-or-miss basis by use of some of the equations listed in Sec. 21.2 or from past experience. The Karnaugh mapping technique, on the other hand, is a systematic circuit-reduction technique that generally gives the minimum-component system. It is applicable to domains with 3, 4, and 5 variables. Other techniques are available for larger domains, but these are more cumbersome and time-consuming.

For a three-variable domain, the Karnaugh map is drawn in three stages, as shown in Fig. 21.2. The map layout is illustrated in Fig. 21.2*a*; it is a 2×4 matrix. The three variables are separated into two groups—the most significant variable (usually the first listed: x in this case) and the remaining two variables. Along the 4-square side are listed the four possible values of the two remaining values; on the 2-square side are listed the two possible values of the significant variable. The three binary numbers represented at each square (in the sequence of the most significant variable's number first and then the other two) represent an equivalent decimal number, which is then listed in each square, as shown in Fig. 21.2*b*. Note that the order of the two binary number sequence for y and z has been altered, with 11 coming before 10; this is done so that no coordinate differs from its adjacent coordinate by more than one bit position. The next step is to convert the logic expression to standard form and then to obtain the numerical representation of this standard form; taking the SOP used in Example 21.6, for instance, the numerical representation of this expression was found in Sec. 21.3.2 to be $\sum (7, 5, 3)$. The final step in preparing the Karnaugh map is to place a 1 in the squares represented by the decimal numbers of the numerical representa-

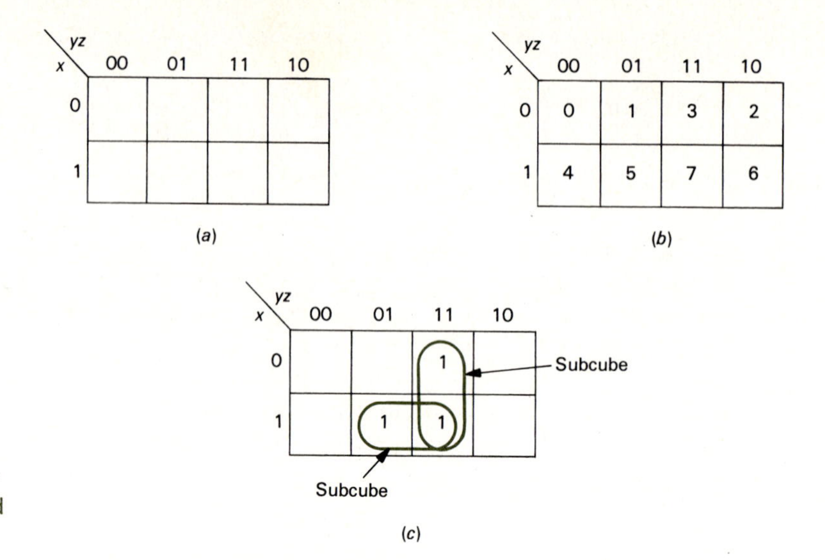

FIGURE 21.2 Karnaugh mapping: (*a*) matrix for 3-variable domain; (*b*) matrix with decimal equivalent; (*c*) subcubes for SOP example discussed in Sec. 21.3.3.

tion, in this case 7, 5, and 3. This is shown in Fig. 21.2*c* and represents the Karnaugh map for the domain (with x, y, z replacing A, B, C),

$$D(x, y, z) = xyz + x\bar{y}z + \bar{x}yz$$
$$= \sum (7, 5, 3)$$

Using the Karnaugh map, we reduce and simplify a logic equation by the following three steps:

1 Form subcubes that cover all 1s on the Karnaugh map. Subcubes are groups of 1s that are adjacent to each other; squares located along a diagonal are not considered to be adjacent. In Fig. 21.2*c* there are two subcubes, which are indicated by the rounded rectangles.

2 For 2-square subcubes, it can be shown that one variable changes value and that the other two have a common value: From our example in Fig. 21.2*c* the binary values of the two adjacent squares in the horizontal subcube are 101 and 111; it is the middle variable, y, that changes. For the vertical subcube, the two adjacent binary values are 011 and 111; it is the first variable, x, that changes.

3 Write the resulting logic expression. The SOP expression for a subcube consists of the variables that do not change. In our example, the two resulting variables that do not change in the horizontal subcube are x and z, and both have values of 1; therefore, the literals appear in the equation. For the vertical subcube, y and z do not change and both have values of 1. Therefore, the resulting equation is $D(x, y, z) = xz + yz$, which is identical to the equation written in the nomenclature of Example 21.6 as $D(A, B, C) = AC + BC$. This represents the minimum-component system, which we might have expected due to the simplicity of the first circuit in Fig. 21.1. When a square containing a 1 is all alone (i.e., not adjacent to any

other square containing a 1), the literal for each variable appears in the resulting equation. For example, a 1 in the upper left-most square of Fig. 21.2c would add the term $\bar{x}\bar{y}\bar{z}$ to the equation since all three variables have the value of 0 in this square.

We have gone through the above example and accompanying explanation to give the general method of preparing a Karnaugh map for logic equations. This method is applicable to most systems with a domain size of 3, 4, and 5. However, the end result was not very meaningful in this case since the system was already a minimum-component system and there was only one possible result. In the general case, there will be a choice of subcubes in step 1 above. We also still have to discuss the handling of POS forms. Therefore, let us return to step 1 above and try to gain some insight into choosing the best subcubes that result in the most economical equation—and hence in the most economical logic system.

21.3.4 Minimization of Logic Functions

It has been stated that the main purpose of Karnaugh mapping is to derive the "best" or "most economical" or "minimum" logic system to accomplish the given function. All of the above terms in quotation marks are used to describe the goal of Karnaugh mapping, and the first two terms are introduced here to remind us that this is not a minimization process in the rigorous mathematical sense but one that requires some judgment and selection. Proceeding on this basis, we will introduce some more interesting Karnaugh maps.

Figure 21.3 illustrates another 3-domain system with a number of possible subcubes. *Subcubes must be a power of 2, i.e., 2, 4, 8, 16, etc.* Subcubes with three adjacent squares are not allowed and must be broken up into two 2-square subcubes. In Fig. 21.3 there are five possible 2-square subcubes and one 4-Square subcube that are obvious. There are also more subcubes that are not obvious until we point out that a Karnaugh map is a continuous surface—that is, the 3-domain map "wraps around" on itself, the right edge of the column headed by $yz = 10$ also being the left edge of the column headed by $yz = 00$. Therefore, there is also a 4-square subcube composed of the squares in the right-most and left-most columns, and two more 2-square subcubes making up the big square subcube. We have listed nine subcubes in this 3-domain map. In higher-order domains, the situation is generally even more complex. What is the best choice of subcubes, and is there a best choice? The answer to the latter part of

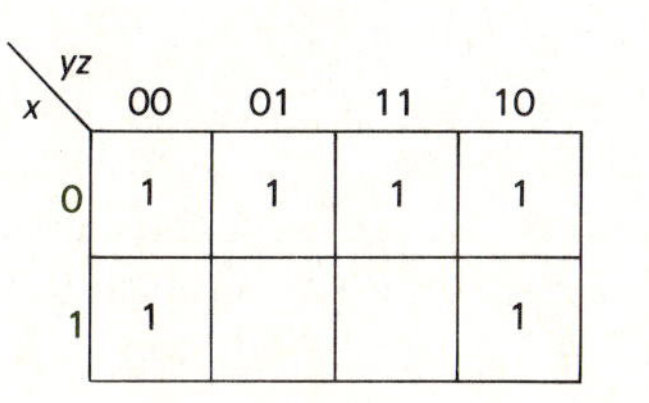

FIGURE 21.3 Karnaugh map with many subcubes (as discussed in Sec. 21.3.4).

this question is "sometimes yes." To the first part of the question, the following steps are added to step 2 in the preceding section for determining the best SOP equation:

2a All 1s must be covered by subcubes, including the "loner" 1s which will introduce the product of all variables into the equation.

2b Start the process of covering 1s with subcubes by finding the *essential subcube*: the subcube that covers a certain 1 that no other subcube can cover.

2c All subcubes should be as large as possible.

2d There should be as few subcubes as possible.

2e Sometimes additional factoring, using the equations and techniques of Sec. 21.2 will further reduce the system after the system equations have been written (step 3 above).

In Fig. 21.3 there are two 4-square subcubes that cover all of the 1s; the choice of these subcubes satisfies steps 2*c* and 2*d* above, and it is these two steps that are the greatest aid in finding the system with the *minimum* number of components. The first 4-square subcube is the obvious one and contains the entire row $x = 0$. In this subcube, y and z vary and $x = 0$ throughout; therefore, its equation is $\bar{x}$. The second 4-square subcube is the square composed of the two squares at each end of the matrix of Fig. 21.3; it results when the two ends are wrapped around together. The binary values in the four subsquares of this subcube are 000, 010, 100, and 110. Note that the third digit, representing z, is 0 in all four squares and that x and y vary. Therefore, the equation from this subcube is $\bar{z}$. The resulting logic equation is

$$f(x, y, z) = \bar{x} + \bar{z} \tag{21.10}$$

The 4-variable domain is handled in the same manner as the 3-variable one and the same rules generally apply, although there are usually many more choices of subcubes. The decimal representation of squares for a 4-variable matrix is shown in Fig. 21.4*a*; this has been obtained by the same means as was Fig. 21.2*b* for the 3-variable matrix. The 1s from Example 21.8 are plotted in Fig. 21.4*b*, with the general variables w, x, y, z replacing the variables used in the example; the numerical representation for

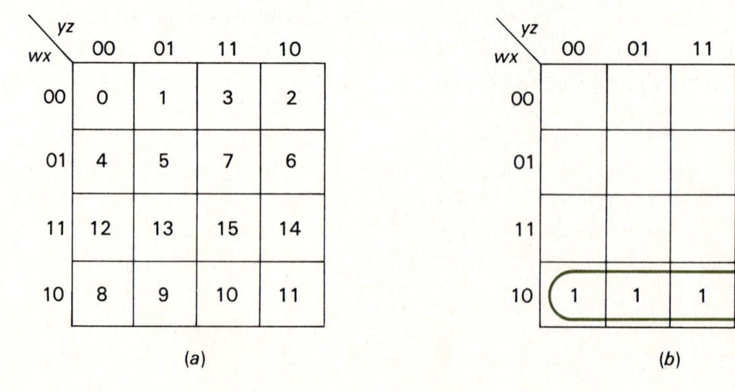

FIGURE 21.4 Karnaugh map for 4-variable domain: (*a*) decimal representation; (*b*) Karnaugh map for example discussed in Sec. 21.3.4.

this example was derived in the second paragraph (Standard SOP) of Sec. 21.3.2. Here there are only two possible subcubes, and these are encircled in Fig. 21.4*b*. For the 4-square subcube, both w and x are constant, giving the product $w\bar{x}$; for the 2-square subcube, w, y, and z are constant with the values 1, 1, and 0, respectively. Therefore, the resulting equation is composed of the SOP

$$f(w, x, y, z) = w\bar{x} + wy\bar{z}$$

which agrees with the original equation of Example 21.8.

Example 21.11 Determine the minimum equation for

$$f(A, B, C, D) = \sum (1, 4, 5, 6, 8, 12, 13, 15)$$

This is somewhat of a contrived example to illustrate steps 2*a* to 2*e* above.

Solution The Karnaugh map is obtained by plotting 1s in the cubes of Fig. 21.4*a*. Represented by the numbers in the $\sum$ above. This is shown in Fig. 21.5*a*. By step 2*c*, one might start with the 4-square subcube at the left center of Fig. 21.5*a*. This gives the equation $B\bar{C}$, since $B = 1$ and $C = 0$. However, this leaves a lot of 1s dangling around the 4-square subcube and

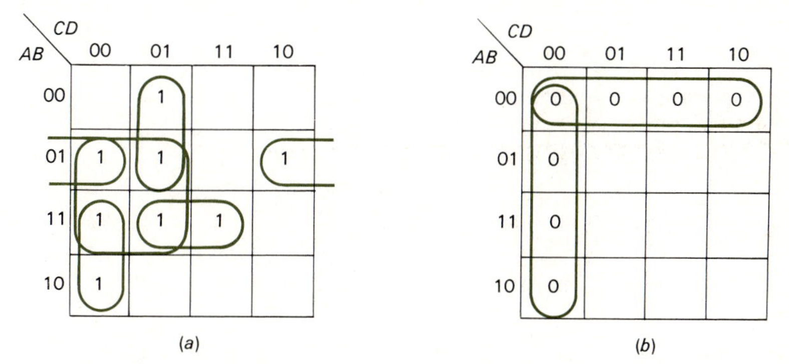

FIGURE 21.5 (*a*) Karnaugh map for Example 21.11. (*b*) Karnaugh map for Example 21.12.

these also must be covered by 2-square subcubes. Note that the 4-variable map wraps around on its edges to give a continuous surface (both vertical and horizontal edges) just as does the 3-variable map. Thus, the 1 in square 0110 is joined with that in square 0100 to give a 2-square subcube, as shown. Adding the equations of the four 2-square subcubes gives a final equation

$$f(A, B, C, D) = B\bar{C} + \bar{A}B\bar{D} + \bar{A}\bar{C}D + A\bar{C}\bar{D} + ABD$$

which has five terms and requires five functions to implement. However, note that all four 2-square subcubes are essential—that is, they cover a 1 that no other subcube can cover. Also, the 4-square subcube is *not* essential. Therefore, starting with the essential subcubes (step 2*b*), we find that one term can be eliminated (since the 4-square subcube will not be needed) and the minimum equation is

$$f(A, B, C, D) = \bar{A}B\bar{D} + \bar{A}\bar{C}D + A\bar{C}\bar{D} + ABD$$

21.3.5 Karnaugh Map for POS Expressions

For plotting POS functions, a 0 is placed in the squares for which the function has a numerical value. Keep in mind that POS functions are represented in an inverse manner to SOP functions; i.e., a variable is represented by a 0 and a complement of a variable is represented by a 1. We will illustrate this process with an example.

Example 21.12 Verify the equation for the logic circuit of Fig. 21.1, second figure; the standard form and numerical representation of this figure have been evaluated in Example 21.7 and in the third paragraph (Standard POS) of Sec. 21.3.2, respectively.

Solution The numerical representation of the second circuit of Fig. 21.1 is

$$f(A, B, C, D) = \Pi(0, 1, 2, 3, 4, 8, 12)$$

To obtain the Karnaugh map, we place a 0 on the squares of Fig. 21.4*a* that are designated in the product above. This is shown in Fig. 21.5*b*. It is seen that there are two 4-square subcubes. For the horizontal subcube, A and B are constant and equal to 0. This gives the equation $(A + B)$. The vertical subcube supplies the equation $(C + d)$. The resulting equation is

$$f(A, B, C, D) = (A + B)(C + D)$$

which agrees with that of the second diagram of Fig. 21.1.

21.3.6 Don't Care Logic State

To complete our discussion of electronic logic and boolean algebra, we will note a third logic state called, very simply, *don't care*. This will aid the reader in evaluating the logic circuit specifications found in most manufacturers' catalogs.

Many types of logic devices have certain combinations of input variables which give outputs that are not important or meaningful. This may be due to a variety of reasons: for instance, these combinations never occur in the proposed application, the outputs have no effect on the output circuitry, or the outputs "should not exist." An example of the latter is a binary logic circuit designed to identify decimal numbers 0 through 9; since 4 bits are needed for this and a 4-bit system can identify 0 through 15, the outputs associated with the decimal numbers 10 through 15 are actually not wanted. Unwanted outputs are designated by the symbol d and called don't care logic states. The same term is also often used to designate input signals—i.e., input signals whose state (0 or 1) is not important for a specific output condition—and is usually designated by the symbol x. The logic circuit truth table shown in Figs. 21.7 and 21.10 illustrates the use of don't care input states x.

Don't care outputs are plotted on the Karnaugh map along with the 1s for SOP or 0s for POS. They can be used to form subcubes *if their use simplifies the resulting*

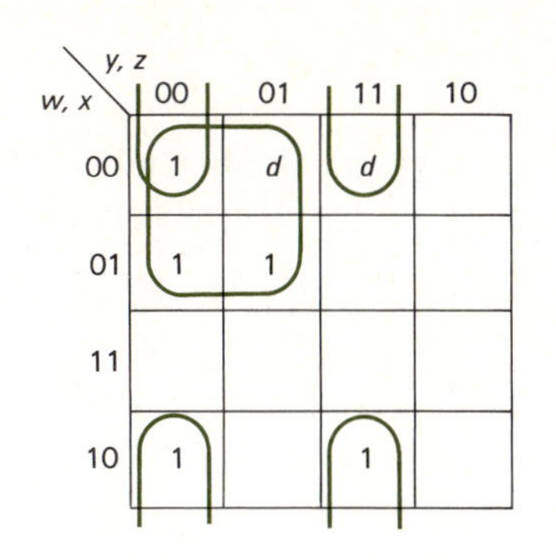

FIGURE 21.6 SOP Karnaugh map with don't care logic states designated by the symbol *d*.

equation; otherwise, they may be ignored. Usually, don't care outputs help in the simplifying process by creating *larger* subcubes (step 2c above). An output function containing don't cares can be expressed in numeric form as

$$F(w, x, y, z) = \sum (0, 4, 5, 8, 11) + d(1, 3)$$

The associated Karnaugh map and the three resulting subcubes are shown as Fig. 21.6. The resulting logic equation is

$$F(w, x, y, z) = \bar{w}\bar{y} + \bar{x}yz + \bar{x}\bar{y}\bar{z}$$

21.4 SEQUENTIAL LOGIC

Thus far we have been concerned with a class of logic circuits often called *combinational logic*. In this class of logic, the device output is a function of the device input only. A second broad category of devices is known as *sequential logic*. In this class of devices, device output may be a function of past history, or of the sequence in which certain signals change logic state, or, in the broadest sense, of time itself.

There are many examples of sequential logic in many devices and activities associated with our daily lives. Combination locks are a simple example in which the *sequence* of setting the numbers is as important as the value of the numbers. Most card games, chess, checkers, and many other games depend upon the sequences of plays. A few years ago, certain automobiles could not be started—at least without the sound of a bothersome buzzer—unless a certain sequence of activities by the operator occurred: sitting in or depressing the driver's seat, closing the left front door, fastening the seat belt, putting the transmission in neutral or park, and finally turning the ignition key. And the automotive assembly line is a sequential system on a grandiose scale.

In most computer and control applications, sequential functions are required along with the gating functions described in the previous section of this chapter. We also include in the sequential category many timing or clock functions and memory functions.

21.4.1 Flip-Flops

Flip-flops are basic sequential logic devices and are shown in Table 21.5. A flip-flop is a *bistable* device; that is, it has two stable modes of operation, which again correspond to the two binary values 0 or 1, on or off, etc.

The *RS* flip-flop is perhaps the basic or elementary flip-flop. The symbols *R* and *S* signify "reset" and "set," respectively, and these two words actually describe the basic function of a flip-flop. In the truth table given in Table 21.5 for the *RS* flip-flop, the symbol *h* generally stands for the symbol 0 as used elsewhere in this text, that is, the low state or low output; the symbol n.c. stands for "no change from the previous condition of inputs." It is seen that an active (or 1) signal on the input of the *RS* flip-flop ($S = 1$) causes the output to SET, or cause Q to be equal to 1. Once SET, the flip-flop remains in SET even after the signal on S is removed. It remains in SET ($Q = 1$) until an active signal of $R = 1$ is applied. $\bar{Q}$ is always the inverse or complement of Q. Likewise, the flip-flop remains in the reset condition ($Q = 0$ and $\bar{Q} = 1$) until the next SET signal is applied to the S terminal. The states called "indeterminate" in Table 21.5 would be included in the don't care states discussed above. The logic shown for this flip-flop is realized by two cross-coupled NAND gates, some embodiments of which will be discussed in later sections. When constructed of NOR gates, the logic for the $R = S = 0$ and $R = S = 1$ states will be the inverse of that shown in Table 21.5.

The *JK* flip-flop can be constructed from two NAND devices; this eliminates some of the undefined states of the *RS* flip-flop. The basic characteristics are shown in the truth table of Table 21.5. The logic of the upper *JK* diagram in Table 21.5 can be described as follows:

1 When $J = K = 1$, the flip-flop will not change state; that is, logic state $Q_{n+1} =$ logic state Q_n.

2 If $J = 1$ and $K = 0$, the flip-flop will set on the next $(n + 1)$ S pulse; $Q_{n+1} = 1$.

3 If $J = 0$ and $K = 1$, the flip-flop will reset (or clear) the R pulse; $Q_{n+1} = 0$.

4 If $J = K = 1$, the flip-flop "toggles"; that is, it changes state in response to each successive R or S pulse.

The second and third versions of the *JK* flip-flop in Table 21.5 are gated by a clock signal (to be described below), which is a useful feature in many computer applications.

21.4.2 Clocks and Registers

Timing in logic circuits is performed by a wide variety of circuits and systems known, not suprisingly, as *clocks*. The simplest clock circuits operate as a function of transient rise times in *RC* networks. Examples of such timers are those used in automotive turn signals, some sweep circuits used in simple oscilloscopes, and highway flashers. Such

TABLE 21.5 Logic Symbol Formats (ANSI)*

Symbol	Function	Description
	Amplifier	Output active only when the input is active (can be used with polarity or logic indicator at input or output to signify inversion.)
or	AND	Output assumes indicated active state only when all its inputs assume their indicated active levels.
1 or	OR	Output assumes its indicated active state only when any of its inputs assume their indicated active levels.
1 or	Exclusive OR	Output assumes its indicated active level if, and only if, only one of the inputs assumes its indicated active level.
	NAND	Inverted AND
	NOR	Inverted OR
	Bilateral switch	A binary-controlled circuit that acts as an on-off switch to analog or binary signals flowing in both directions.
$\geqslant m$	Logic threshold	Output will assume its active state if m or more inputs are active.
$= m$	M and only M	Output will be active when m and only m inputs are active (for example, exclusive OR).
$>n/2$	Majority function	Output will be active only if more than half the inputs are active.
mod 2	Odd function	Output is active only if an odd number of inputs are active.
x/y	Even function	Output is active only if an even number of inputs are active.
00	Signal-level converter	Input levels are different from output levels.

Symbol	Function	Description
FF, R, S	RS	R S | Q $\overline{Q}$ 1 1 | n.c. n.c. 1 h | h 1 h 1 | 1 h h h | undetermined
FF, T	T	Toggling occurs with every clock pulse
FF, C, D, R, S	D	Data output follows data input; input is gated by C
FF, J, K, R, S	JK	J K | Q $\overline{Q}$ 1 1 | n.c. n.c. 1 h | 1 h h 1 | h 1 h h | toggles
FF, J_G, G, K_G, R, S	JK (gated)	J and K inputs are gated by C
FF, J_G, G, K_G, R, S	JK (master-slave)	Outputs are dependent on the negative-going edge of the clock

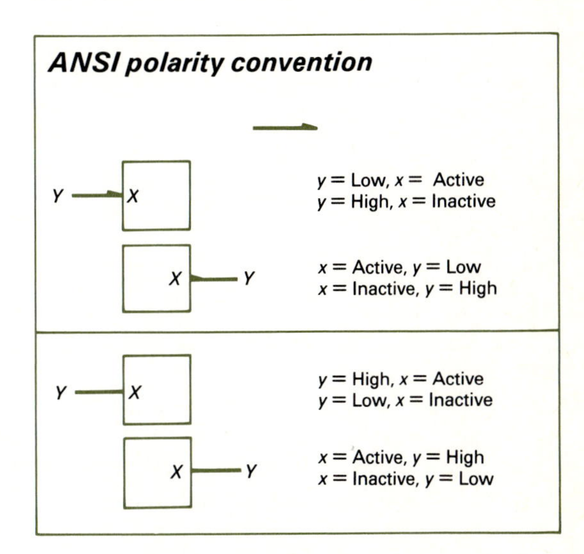

* Based on *Machine Design* [Ref. 21.5]. Used by permission.

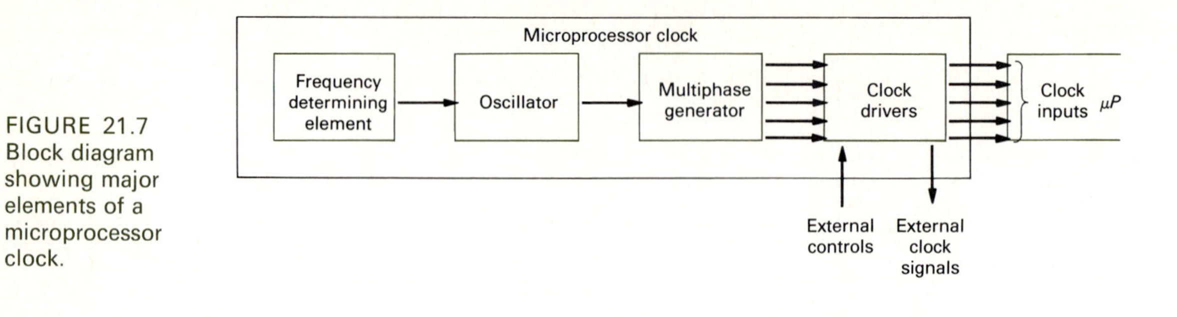

FIGURE 21.7 Block diagram showing major elements of a microprocessor clock.

timing circuits are, of course, susceptible to probable errors in the value of R and C and are subject to temperature variations, only portions of which can be compensated.

A much greater accuracy is required for most logic circuit applications, and this is generally supplied by temperature-compensated crystal-controlled RC oscillators which are constructed of an integrated circuit. Since most microprocessors and computer circuits require many clock signals, the output of a *master clock* is divided by means of a multiphase generator to provide many output timing signals. The block diagram for a clock in a typical microprocessor is shown in Fig. 21.7. The output signal from a clock used in logic timing applications appears generally as shown in Fig. 21.8 and has a square-output waveform. When the value of the clock signal increases to its high (or positive) value, that rising portion of the waveform is called the leading edge, or *positive edge*, of the clock output; the decreasing portion of each pulse, when the clock signal returns to zero or negative or low value, is called the lagging edge, or *negative edge*. Logic devices which are triggered by clock pulses may be designed to operate on either the positive or negative edge of the clock pulses, although the latter is generally more common. These are edge-triggered logic devices. Level sensitive logic devices are designed to operate on either the high or low level of the clock pulses; one such device is the polarity-hold SRL (shift-register latch) described in Ref. 21.14. Note the several versions of the JK flip-flop shown in Table 21.5. The rise and fall times of a clock pulse are of the order of 50 ns or less. Typical frequencies of the master clock are of the order of 20 MHz.

Another clock-triggered flip-flop is the *type-D* flip-flop, also shown in Table 21.5. The complete truth table for the type-D flip-flop is shown in Table 21.6, using the terminal designation as shown in Table 21.5. The upward arrow under the C (clock)

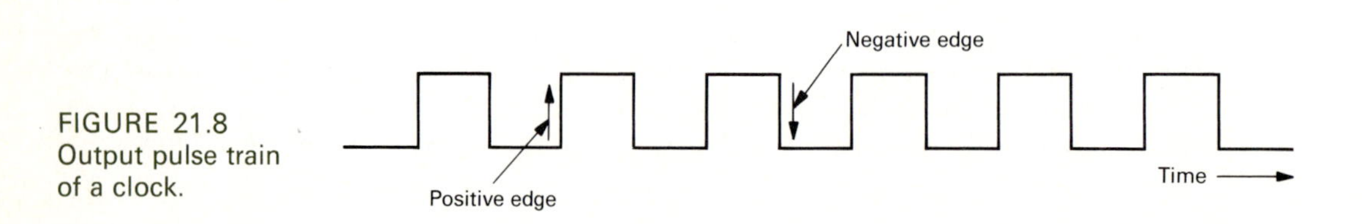

FIGURE 21.8 Output pulse train of a clock.

TABLE 21.6 Truth Table for Type D Flip-Flop

Inputs				Outputs		
C	D	R	S	Q	$\bar{Q}$	
↑	0	0	0	0	1	
↑	1	0	0	1	0	
↓	×	0	0	Q_n	$\bar{Q}_n$	(No change)
×	×	1	0	0	1	
×	×	0	1	1	0	
×	×	1	1	1	1	

input terminal column indicates the positive edge of the clock pulse; the downward arrow indicates the negative edge. Also note the use of don't care inputs designated by the symbol X. The D input channel is called the *data input*.

One of the principal uses of the type-D flip-flop is as a temporary storage or memory element in digital control and computer systems. A grouping of type-D flip-flops used for temporary storage is known as a *register* or sometimes a *shift register*. A grouping of N flip-flops is called an N-bit register and is used for storing N logic bits as defined in Sec. 21.1. Other flip-flops and other hardware are commonly used for this purpose, but the grouping of four type-D flip-flops shown in Fig. 21.9 will serve to illustrate the general concept of register operation.

Figure 21.9*a* illustrates the circuit diagram of the basic elements of a 4-bit register using type-D flip-flops. Figure 21.9*b* describes a typical timing sequence that might go along with this type of register. Note that this timing diagram shows the actual finite rise- and fall-time characteristic of a practical timer. Its operation is as follows: Data, which in this case is the value of a bit (1 or 0), comes in on the D_n ($n = 1 - 4$) channels at the rate shown on the upper trace of Fig. 21.9*b*. This particular register is triggered on the positive edge of the clock pulse train. Therefore, when the clock signal is changing from its low to its high state, D_n is transferred to the Q_n output port and held there until changed either to an opposite value of D_n or a reset signal. The reset (R) signal *asynchronously* (i.e., at any time and not in synchronism with the clock pulse train) resets Q_n to zero or low whenever R = low. The output $\bar{Q}_n$ is always at the complement state of Q_n. This type of register operation is known as the *parallel mode*.

The same circuit shown in Fig. 21.9 can also be caused to operate in a *series mode*, in which mode it is known as a *shift register*. In this mode, the bit in one flip-flop is shifted to an adjacent flip-flop, either to the right or to the left. This serial shifting process can be continued sequentially to the last flip-flop in the sequence, or it can be halted at another flip-flop, depending upon the system logic. Figure 21.10 is a copy of the data sheet for a Motorola shift register, MC14194B. The packaging and construction of this device will be discussed in later sections. The MC14194B can be operated

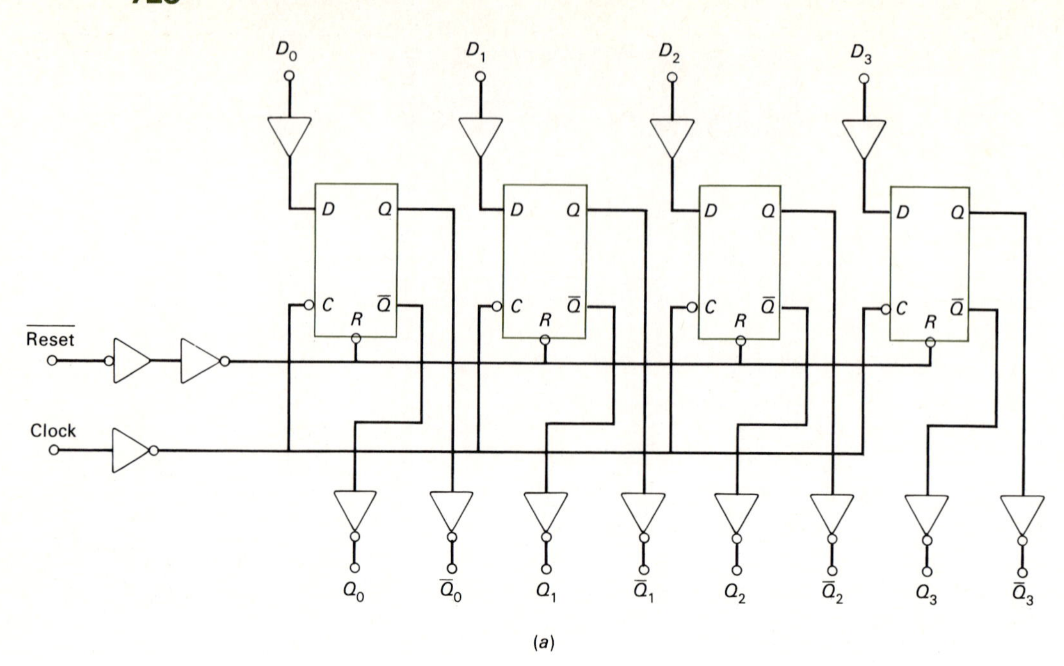

(*a*)

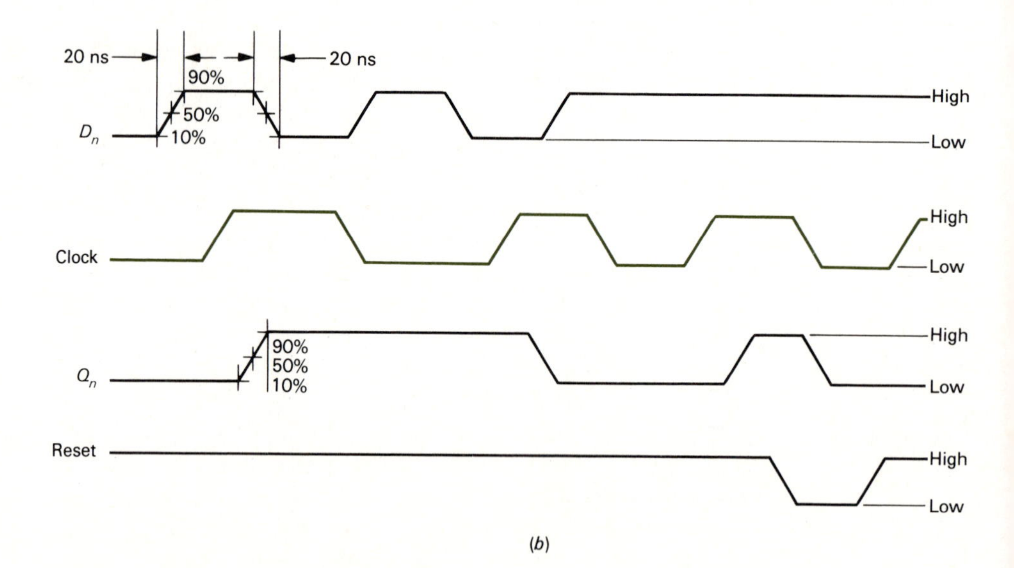

(*b*)

FIGURE 21.9 Four-bit register using type-*D* flip-flops and inverters: (*a*) circuit diagram; (*b*) timing diagram.

MC54/74HC194

Advance Information

4-BIT BIDIRECTIONAL UNIVERSAL SHIFT REGISTER

The MC54/74HC194 is identical in pinout to the LS194 and the MC14194B metal gate CMOS device. The device inputs are compatible with standard CMOS outputs; with pull-up resistors, they are compatible with LSTTL outputs.

This static shift register features parallel load, serial load (shift right and shift left), hold, and reset modes of operation. These modes are tabulated in the Function Table, and further explanation can be found in the Pin Description section.

- Low Power Consumption Characteristic of CMOS Devices
- Output Drive Capability: 10 LSTTL Loads Minimum
- Operating Speeds Similar to LSTTL
- Wide Operating Voltage Range: 2 to 6 Volts
- Low Input Current: 1 μA Maximum
- Low Quiescent Current: 80 μA Maximum (74HC Series)
- High Noise Immunity Characteristic of CMOS Devices
- Diode Protection on All Inputs

HIGH-PERFORMANCE
CMOS
LOW-POWER COMPLEMENTARY MOS
SILICON-GATE

4-BIT BIDIRECTIONAL UNIVERSAL SHIFT REGISTER

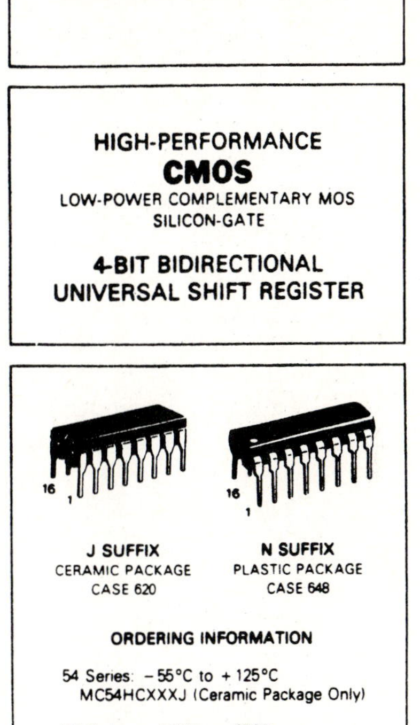

J SUFFIX
CERAMIC PACKAGE
CASE 620

N SUFFIX
PLASTIC PACKAGE
CASE 648

ORDERING INFORMATION

54 Series: −55°C to +125°C
MC54HCXXXJ (Ceramic Package Only)

74 Series: −40°C to +85°C
MC74HCXXXN (Plastic Package)
MC74HCXXXJ (Ceramic Package)

BLOCK DIAGRAM

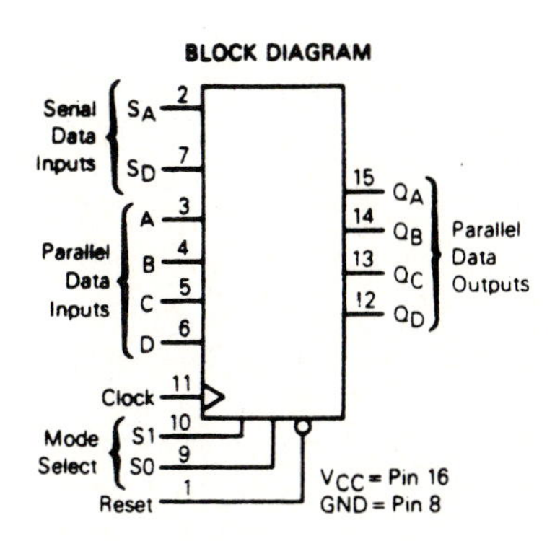

FUNCTION TABLE

Inputs										Outputs				Operating Mode
	Mode Select			Serial Data		Parallel Data								
Reset	S1	S0	Clock	S_D	S_A	A	B	C	D	Q_A	Q_B	Q_C	Q_D	
L	X	X	X	X	X	X	X	X	X	L	L	L	L	Reset
H	H	H	↑	X	X	a	b	c	d	a	b	c	d	Parallel Load
H	L	H	↑	X	H	X	X	X	X	H	Q_{An}	Q_{Bn}	Q_{Cn}	Shift Right
H	L	H	↑	X	L	X	X	X	X	L	Q_{An}	Q_{Bn}	Q_{Cn}	
H	H	L	↑	H	X	X	X	X	X	Q_{Bn}	Q_{Cn}	Q_{Dn}	H	Shift Left
H	H	L	↑	L	X	X	X	X	X	Q_{Bn}	Q_{Cn}	Q_{Dn}	L	
H	L	L	X	X	X	X	X	X	X	no change				Hold
H	X	X	L	X	X	X	X	X	X	no change				
H	X	X	H	X	X	X	X	X	X	no change				

H = high level (steady state)
L = low level (steady state)
X = don't care
↑ = transition from low to high level.

a, b, c, d = the level of steady-state input at inputs A, B, C, or D, respectively.
Q_{An}, Q_{Bn}, Q_{Cn}, Q_{Dn} = the level of Q_A, Q_B, Q_C, or Q_D, respectively, before the most-recent ↑ transition of the clock.

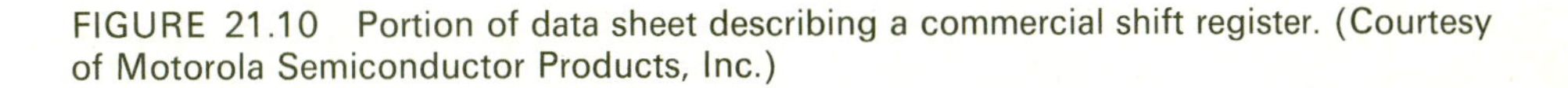
FIGURE 21.10 Portion of data sheet describing a commercial shift register. (Courtesy of Motorola Semiconductor Products, Inc.)

in the parallel mode, two series modes, and the *hold mode*. In hold, the outputs are fixed at their values at the instant S_1 and S_2 are set at 0 until reset by $R = 0$.

21.4.3 Counters

Counting is a digital (in contrast to an analog) function, and the logic circuits that have been discussed so far are ideally suited for this application. Digital counters are used to count pulses, events (such as the operation of a particular gate), frequency, etc., and are widely used in both control and computing applications. Counters are sequential circuits, as is the process of counting itself. The size of a counter—that is, the highest count that can be reached—is called the counter, *modulus* and is designated by N. This number also equals the total number of states of the counter logic. Counting is almost universally performed in binary logic starting with the least significant bit. The modulus N depends upon the number of bits contained in the counter circuitry. A 1-bit counter can obviously count the numbers 0 and 1, and $N = 1$; a 2-bit counter can count from binary 0 up to 11, or decimal 3, and $N = 3$; a 3-bit counter has an N of binary 111 or decimal 7. In general, when a counter has counted up to the value of its modulus, the contents of the counter are reset or cleared, and the counting process starts over again. Certain classes of counters, known as *up-down* counters, effectively double the modulus by counting up to N and down to 0 before the reset process is applied.

Figure 21.11 illustrates a simple 3-bit counter using the type JK flip-flop described above and defined in Table 21.5. For this counter application, the J and K inputs are set at $J = K = 1$, or high. In Fig. 21.11*a*, the flip-flop on the left with output of Q_1 is the least significant bit, and the bit significance increases in each successive flip-flop to the right. Some of the waveforms of the counting process are shown in Fig. 21.11*b*. These waveforms assume an initial reset or cleared condition of the counter in which all three flip-flops are set at 0, giving the bit count of 000. The input is applied to the flip-flop representing the least significant bit, the flip-flop on the left of Fig. 21.11*a*. Each flip-flop changes state on the negative edge of the input pulse train—which are the pulses to be counted. The first input pulse is counted when Q_1 changes from 0 to 1 on the negative edge of the first input pulse. Q_1 returns to 0 on the negative edge of the second pulse; however, since Q_1 is applied to the input of the second flip-flop and this change from 1 to 0 represents a negative edge, Q_2 becomes 1. This gives a binary count (reading from right to left, i.e., $Q_3Q_2Q_1$) of 010, which is decimal 2, which correctly indicates a count of two pulses. The next input pulse causes Q_1 to change to 1, giving binary 011, or decimal 3, which is correct. The next, or fourth, input pulse causes Q_1 to return to 0, which inputs a negative edge to the second flip-flop, causing Q_2 to change to 0—which, in turn, inputs a negative edge to the third flip-flop, causing it to change state to a 1, giving a binary count of 100, or 4. This process is continued until the binary count is 111, or decimal 7. The next pulse causes the counter to revert back to 000, and the process of counting up to 7 (actually a count of 8) is repeated.

Counters can be constructed from many other logic devices and circuit configurations and may have logic considerably different from that illustrated above. The *down* counter has already been mentioned and can be simply derived from the up

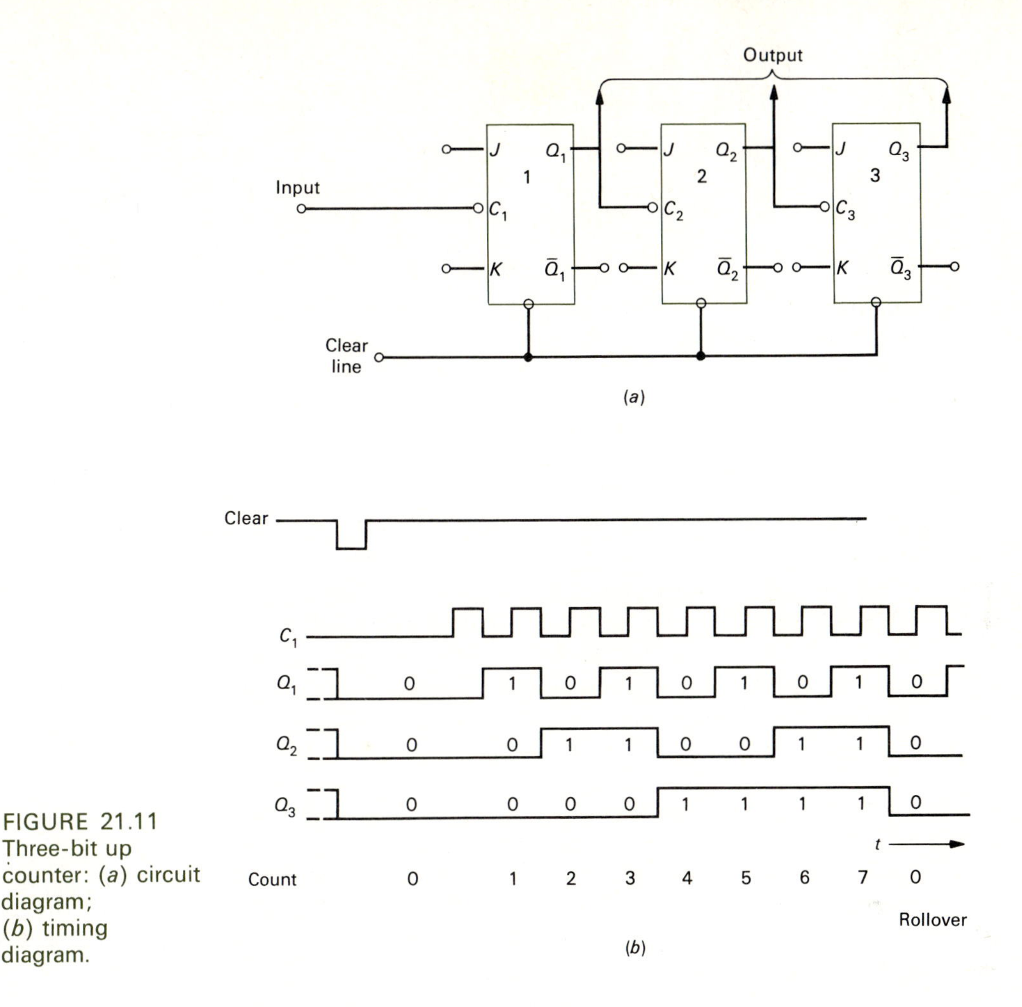

FIGURE 21.11 Three-bit up counter: (*a*) circuit diagram; (*b*) timing diagram.

counter of Fig. 21.11 by using the $\bar{Q}$ outputs instead of the Q outputs. *Synchronous* counters are controlled by a clock such that all flip-flops change with clock pulse, regardless of the number of stages. Divide-by-N counters are useful in relating the count to the time period of the count and hence in determining pulses per second, or frequency. *Johnson* counters consist of D flip-flops arranged in a ring; the input pulses are applied to all clock terminals as in synchronous counters, but the complemented output of the last stage goes back into the D terminal of the first stage. The 5-stage Johnson counter is called a divide-by-10 counter since the waveforms complete one full cycle for every 10 input pulses.

Example 21.13 Design a counter that can count up to 32.

Solution Since $N = 32 = 2^5$, a 5-bit counter is required. This can be realized by adding two more flip-flops to the circuit shown in Fig. 21.11*a*. With five flip-flops the maximum binary count is 11111, which equals 31, giving a count of 32 when 0 is included.

21.5 HARDWARE IMPLEMENTATION OF LOGIC CIRCUITS

Thus far we have dealt primarily with theoretical and mathematical concepts in describing the properties of digital electronic logic. Now we will look at the means of realizing these properties by physical circuits and devices.

We have introduced the term *hardware* here because it has come into common usage in the engineering lexicon and has led to the formulation of a relatively new antonym, *software*. Software generally describes the nonhardware components of a computer, in particular the programs needed to made computers perform their intended tasks[21.10] The first software systems were libraries of mathematical routines. Software thus includes the mathematical formulation of boolean algebra, truth tables, and many other mathematical operations, as well as "housekeeping" functions for proper operations of the computer or control system and the input and output functions, as will be discussed in the next chapter. The means of realizing all of these functions and operations in a practical physical sense is commonly called hardware in today's lexicon.

Hardware refers to physical devices, circuits, and systems. Associated with this noun are a number of adjectives—developmental, meaning still in the research or development stage; state-of-the-art, meaning available today but at no specified cost; and commercial, meaning available from conventional markets today—with many other nuances defining these basic three divisions of marketplace reality. At the time of this writing, electronic logic hardware costs are in a stage of rapid decrease and are lower than are software costs in many computer and processor applications. As a result of these cost relationships, electronic logic is dominant in almost all aspects of computer and control applications, from space-war games and numerical control of machine tools to all types of engineering and scientific mathematical calculations. However, many of the concepts of boolean algebra and the mathematics of logic circuits arose from the use of other types of hardware, and before we proceed into electronic hardware discussions, a brief mention of the other systems may be of interest.

Much of the early interest in boolean algebra resulted from the widespread use of electromagnetic relay systems in power systems and motor protection schemes; this type of hardware is still in common use in many applications, and there is generally a direct analogy between relay logic and electronic logic circuits, as discussed in Ref. 21.6. The early computer circuits were basically mechanical computers of various types, and some of the earliest computer languages that led to such widely used electronic computer languages as FORTRAN (formula translation) were originally developed for mechanical computers.[21.10]. Mechanical counters are still in very common use. Fluidic computers and controllers also generally preceded electronic systems and are still used in many applications. Superconducting computers have been proposed, and a few prototypes have been constructed. And of course one of the most ancient of counters and calculators, the abacus, still has its adherents. Thus, the mathematical logic or software that has been presented in this chapter goes back to very ancient beginnings—in many cases, the operation of the hardware was understood and used long before any formal mathematics were developed. However, the advent of the integrated-circuit (IC) semiconductor "makes ancient good uncouth," to paraphrase the poet James Russell Lowell, and these circuits and devices have burst into almost every aspect of modern industrial culture.

21.6 IC LOGIC FAMILIES

The first thing one encounters when trying to use or understand electronic logic devices and systems is an unholy array of acronyms. We will try to alleviate this alphabetic stumbling block by listing some of the more common acronyms, although this is somewhat of a dangerous step, since IC logic systems come and go and get outdated in a matter of months. Some of the following may be ancient history by the time you read this.

IC—integrated circuit—an assemblage consisting of a number of discrete devices (transistors, diodes, etc.), circuit constants (resistors, capacitors), and interconnected leads integrated to form a complete system in a small region of semiconductor material; usually called a *chip*

DIP—dual in-line package—a common means of "packaging" semiconductor chips (see Fig. 21.12)

SMT—surface mount technology

SOP—small-outline package

LCC—leadless chip carriers

FET—field-effect transistor

MOS—metal-oxide semiconductor

MOSFET—metal-oxide-semiconductor field-effect transistor

PC—printed circuit

LSI—large-scale integration (a chip containing densely packed functions, or the process of developing such ICs)

The principal logic families are:

CMOS—Complementary MOS

HMOS—high-speed CMOS

ECL—emitter-coupled logic

I^2L—integrated injection logic

NMOS—*n*-channel MOS

PMOS—*p*-channel MOS

RTL—resistor-transistor logic

TTL (T^2L)—transistor-transistor logic

The purpose of the remaining sections of this chapter is to present the external characteristics and some very brief looks at the internal circuitry of the electronic logic used to achieve the mathematical functions described previously in the chapter. The semiconductor physics and electronic circuit theory associated with ICs have been briefly discussed in earlier chapters of this text. For further study of IC design, structure, and performance, consult Refs. 21.2, 21.7, and 21.11 to 21.17.

ICs are used in most logic applications today because of their low cost, high-volume density, and suitability for many computer and control systems. The internal structure of ICs, that is, the electrical circuitry, more or less delineates what are called *logic families.* Many other "families" are in use today besides those listed above, and new families are continuously being introduced as the state of the art of semiconductor manufacture improves. The choice of a particular family largely depends upon the associated components to be used in the system—the power supplies (and associated bus voltages), output devices, frequency response, type of PC board or other mounting structure, etc. The volume, or packing, density of all types of ICs is generally very high as compared with mechanical or electromagnetic logic, so this is generally not an issue in the choice of logic family. ICs are available in a number of physical forms, although the DIP (dual in-line package) is probably the most popular. We use the term *package* to describe the physical and structural members used to contain, support, and protect the actual semiconductor logic member. Figures 21.12 and 21.13 illustrate several typical IC packages in common use.[21.5] The DIP is also illustrated in Figs. 21.10 and 21.15. SMT packages are gaining acceptance in semiconductor applications. The principal advantages of SMT are reduced package size, improved thermal characteristics, simplification of automatic mounting on PC boards, and elimination of hole-drilling steps required for conventional DIP package. Common

FIGURE 21.12 A comparison of SMT and DIP packaging of ICs. The package on the left is a small-outline package (SOP), housing the same IC as the DIP on the left. (Courtesy of North American Phillips SMD Technology, Inc.)

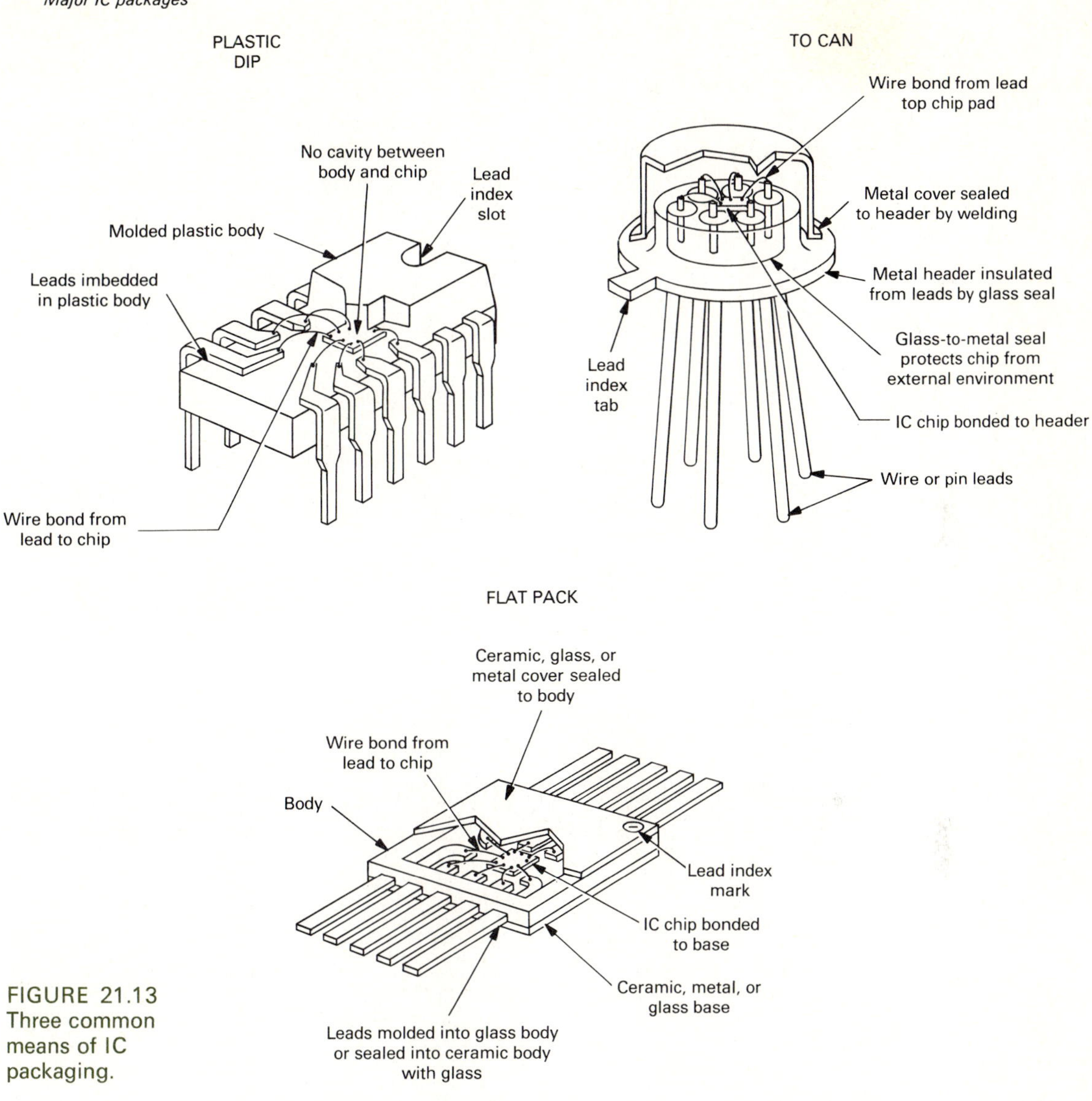

FIGURE 21.13 Three common means of IC packaging.

SMT packages are known as leadless chip carriers (LCC), tape-automated bonding (TAB), and small-outline package (SOP). See Fig. 21.12.

21.6.1 CMOS Logic

The acronym CMOS refers to complementary-symmetry metal-oxide semiconductor. The letter C contrasts this semiconductor's structure and internal circuitry with the earlier *n*MOS and *p*MOS structures. It should be noted that the symbols for logic functions, such as those shown in Fig. 21.1 and Table 21.5, are common for *all* logic families, as are the mathematical relationships developed in the earlier pages of this chapter.

The CMOS (complementary MOS) is constructed of metal-oxide semiconductors (MOS), and the basic structure of CMOS logic makes use of both *n*-channel and *p*-channel enhancement-mode transistors. Figure 21.14 illustrates the cross-sectional structure of recent CMOS devices and clearly shows their size reduction. The advantages of CMOS logic are low power dissipation and a wide range of power-supply voltages. The circuit diagram of a logic circuit using CMOS is shown in Fig. 21.15.[21.11] This figure is also an example of a commercial NOR gate, which was defined in Sec. 21.3. Figure 21.15 describes a typical commercial logic device (circa 1981), shows the logic symbol, defines power-supply voltage range (V_{DD}), illustrates the use of *buffering* to improve noise immunity and to increase the device's capacity to handle load current, and gives other pertinent information for the user. CMOS switching circuits operate at switching rates up to 15 MHz, and CMOS memories are used up to 50 MHz. The disadvantages of CMOS are that this logic system is generally more costly than other systems, that it does not offer all of the circuits available in other systems, particularly TTL logic, and that it does not offer the frequency response of TTL.

We have devoted much of our study of logic algebra in this chapter to systems involving the direct gates, the AND and OR gates, since the concepts and operation of

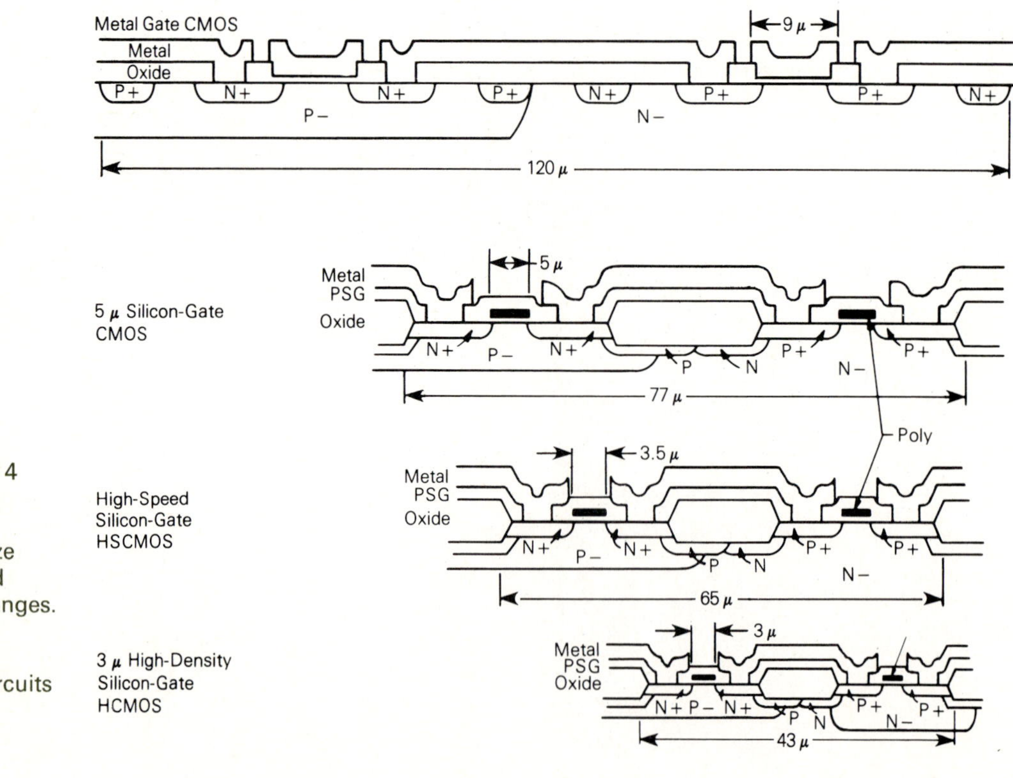

FIGURE 21.14 Evolution of CMOS chips, illustrating size reduction and structural changes. (Courtesy of Motorola Integrated Circuits Div., Austin, Texas.)

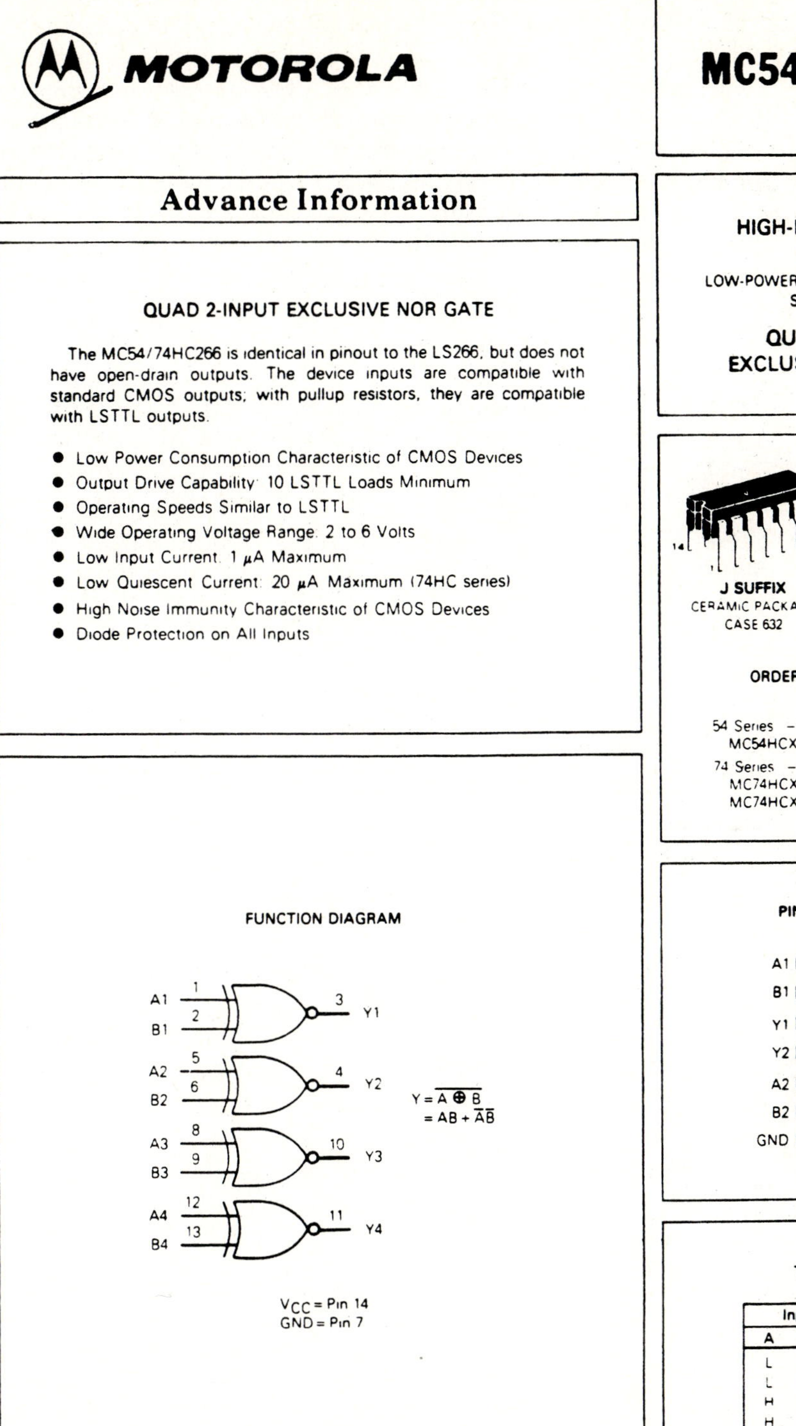

MOTOROLA

MC54/74HC266

Advance Information

QUAD 2-INPUT EXCLUSIVE NOR GATE

The MC54/74HC266 is identical in pinout to the LS266, but does not have open-drain outputs. The device inputs are compatible with standard CMOS outputs; with pullup resistors, they are compatible with LSTTL outputs.

- Low Power Consumption Characteristic of CMOS Devices
- Output Drive Capability: 10 LSTTL Loads Minimum
- Operating Speeds Similar to LSTTL
- Wide Operating Voltage Range: 2 to 6 Volts
- Low Input Current: 1 μA Maximum
- Low Quiescent Current: 20 μA Maximum (74HC series)
- High Noise Immunity Characteristic of CMOS Devices
- Diode Protection on All Inputs

HIGH-PERFORMANCE
CMOS
LOW-POWER COMPLEMENTARY MOS
SILICON-GATE

QUAD 2-INPUT
EXCLUSIVE NOR GATE

J SUFFIX
CERAMIC PACKAGE
CASE 632

N SUFFIX
PLASTIC PACKAGE
CASE 646

ORDERING INFORMATION

54 Series: −55°C to +125°C
MC54HCXXJ (Ceramic Package Only)
74 Series: −40°C to +85°C
MC74HCXXN (Plastic Package)
MC74HCXXJ (Ceramic Package)

FUNCTION DIAGRAM

PIN ASSIGNMENT

Signal	Pin	Pin	Signal
A1	1	14	V_{CC}
B1	2	13	B4
Y1	3	12	A4
Y2	4	11	Y4
A2	5	10	Y3
B2	6	9	B3
GND	7	8	A3

TRUTH TABLE

Inputs		Outputs
A	B	Y
L	L	H
L	H	L
H	L	L
H	H	H

FIGURE 21.15 Data sheet for CMOS NOR gate. (Courtesy of Motorola Semiconductor Products Div.)

of these gates are, in general, easily grasped. The complementary gates, NAND and NOR, have been introduced in several examples and are implied from De Morgan's theorems (Sec. 21.2.1). In practice, the complementary gates are more widely used than the direct gates. This is due to the inherent inversion property of many logic circuits as well as other electronic circuits and devices, such as operational amplifiers. The standard symbol to indicate inversion is the circle (bubble) at the output terminal of the NOR gates, as shown in the logic diagram of Fig. 21.15. In this same diagram, referring now to the numbered terminals of the dual NOR gate, the logic states that the output at terminal 1 is the *inverse* of the results of the OR process at the input terminals 2, 3, 4, and 5. Calling the output Y and the inputs A, this may be stated in boolean form as

$$Y_1 = \overline{(A_2 + A_3 + \overline{A}_4 + A_5)} \tag{21.11}$$

From the first De Morgan theorem, Eq. (21.7*a*), it is seen that Eq. (21.11) can also be written as

$$Y_1 = \overline{A}_2\overline{A}_3\overline{A}_4\overline{A}_5 \tag{21.12}$$

which reads, "Y_1 equals not A_2 and not A_3 and not A_4 and not A_5." Pause for a moment or two to be sure that you agree that these two forms are equivalent—that they are stating the same logic.

There is a similar relationship for the NAND gate, which can be expressed by two equations analogous to the above (for a 4-input gate):

$$Y_1 = \overline{A_1A_2A_3A_4} \tag{21.13}$$

or

$$Y_1 = \overline{A}_1 + \overline{A}_2 + \overline{A}_3 + \overline{A}_4 \tag{21.14}$$

Symbolically, the two forms are represented by (1) changing the bubbles from input to output, or vice versa, and (2) changing AND to OR, or vice versa.

Example 21.14 Determine the alternative form for the NOR gate of Fig. 21.15 and develop the truth table.

Solution The alternative form is an AND gate with complementary (NOT) inputs, that is, with the bubbles on the input leads; this form corresponds to Eq. (21.12) and is shown in Fig. 21.16. The truth table follows from *either* logic form Eq. (21.11) or Eq. (21.12). In this table, 1 and 0 are used symbolically to represent high and low logic states, respectively.

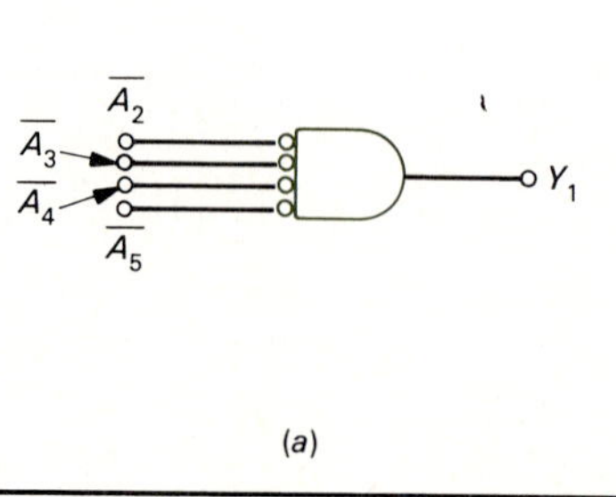

A_2	A_3	A_4	A_5	Y_1
0	0	0	0	1
1	0	0	0	0
0	1	0	0	0
0	0	1	0	0
0	0	0	1	0
1	1	0	0	0
1	1	1	0	0
1	1	1	1	0

(*b*)

FIGURE 21.16 Example 21.14: alternative form and truth table for the 4-input NOR gate of Fig. 20.15: (*a*) alternative form; (*b*) truth table.

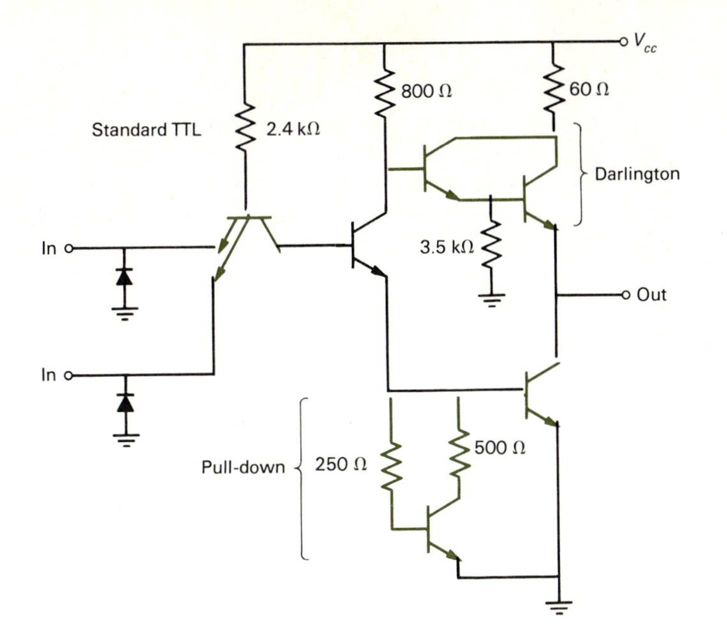

FIGURE 21.17 Conventional circuitry for a 2-input NAND gate in the TTL logic family.

21.6.2 Transistor-Transistor Logic

Transistor-transistor logic (TTL, or T^2L) is a very popular type of logic. It is generally lower in cost and faster in switching speeds than CMOS logic, and it is still the most widely available logic technology. Figure 21.17 illustrates the circuit for a simple 2-input NAND gate. Both inputs, A and B, come into the input transistor Q_1. The output voltage Y is derived from a Darlington circuit, which is often called a "totem pole" arrangement. The Darlington circuit provides a low output impedance for increased on-off switching speed, and hence a higher operation frequency. In the basic TTL gates, transistors are driven into saturation, and speed limitations result primarily from storage time delays associated with transistor turnoff. TTL gates operate at frequencies up to approximately 40 MHz.

21.6.3 Emitter-Coupled Logic

Emitter-coupled logic (ECL) is faster than TTL and is used in applications where very high switching speed is required. ECL has high switching speeds since the transistors are operated in a differential mode as emitter followers and are not driven to saturation. Switching speeds up to 100 MHz are possible with ECL. Because of these high frequencies, the ancillary leads and circuit connections associated with ECL systems must be designed with extreme care in order to minimize circuit inductance. Figure 21.18 illustrates a basic ECL circuit which can be used as either an OR or NOR gate or both. Figure 21.19 illustrates a recent development by Motorola to improve switching speed by means of series-coupled logic.[21.13] This figure also describes the exclusive

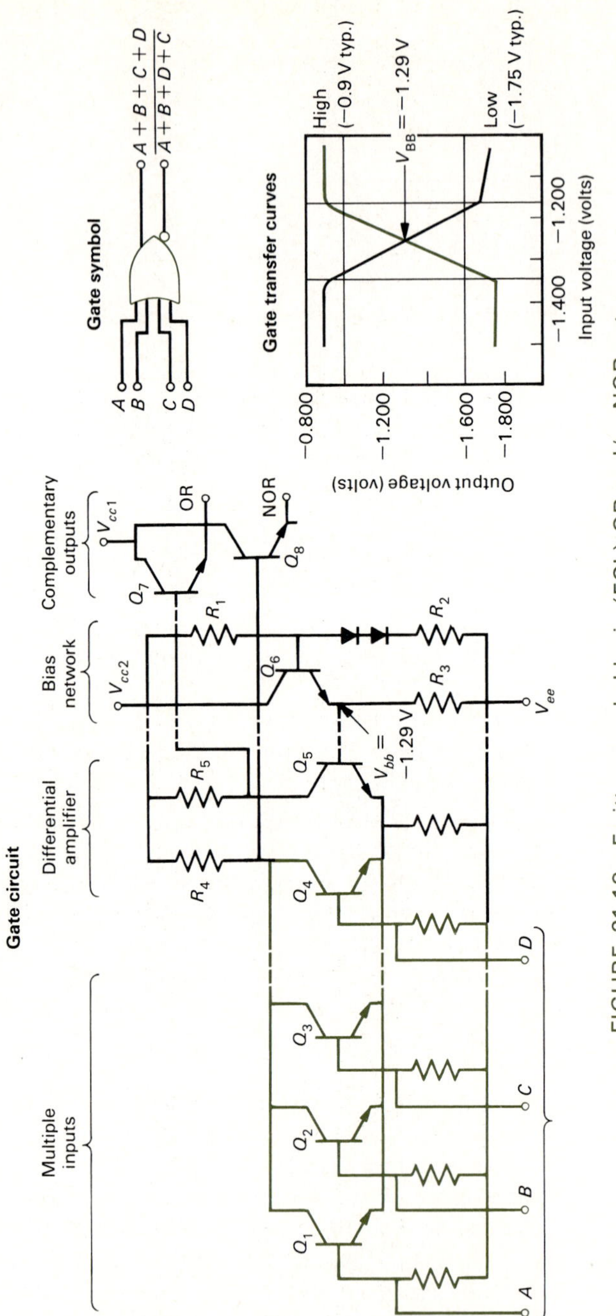

FIGURE 21.18 Emitter-coupled logic (ECL) OR and/or NOR gate.

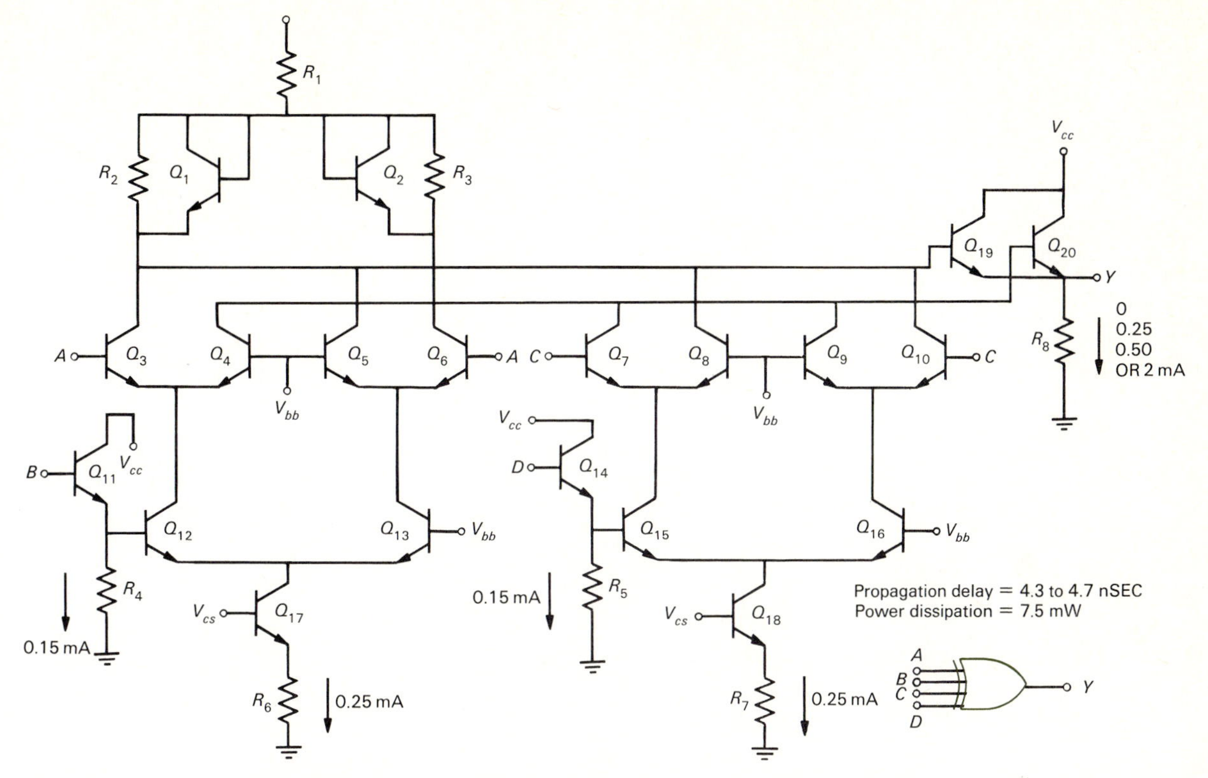

FIGURE 21.19 A 4-input XOR gate using series-coupled logic to improve response time. (Courtesy of Motorola Semiconductor Products Div.).

OR gate, which has some very interesting characteristics and will be our last sortie into the realm of boolean algebra.

The 2-input exclusive OR circuit has been defined in Sec. 21.2.1, Table 21.2. It differs from the regular OR circuit by eliminating from the conditions resulting in a high or true or 1 output that condition when *all* inputs are high or true or 1. The 4-input exclusive OR gate (often called XOR gate) shown in Fig. 21.19 has the equation

$$Y = A\bar{B}CD + \bar{A}BCD + AB\bar{C}D + ABC\bar{D} + A\bar{B}\bar{C}\bar{D} + \bar{A}B\bar{C}\bar{D} + \bar{A}\bar{B}C\bar{D} + \bar{A}\bar{B}\bar{C}D \tag{21.15}$$

In the symbolic form suggested in Sec. 21.2.1, this could also be described by the equation

$$Y(A, B, C, D) = A \oplus B \oplus C \oplus D \tag{21.16}$$

where the $\oplus$ symbol means exclusive OR. The numerical representation of the 4-input XOR gate can be found by the methods of Sec. 21.3.1 [note that Eq. (21.15) is in standard SOP form] as

$$Y(A, B, C, D) = \sum (1, 2, 4, 7, 8, 11, 13, 14) \tag{21.17}$$

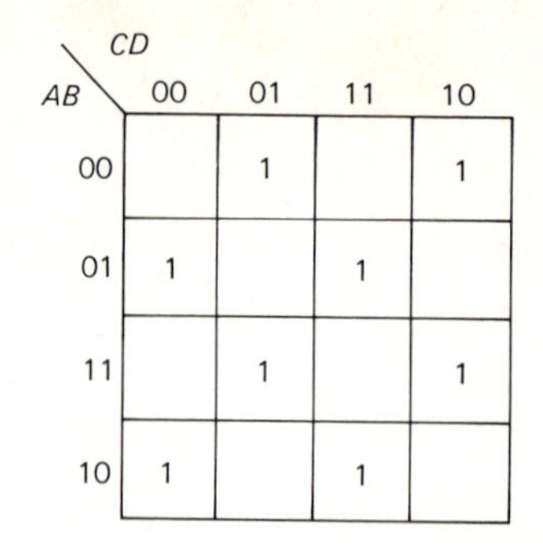

FIGURE 21.20 Karnaugh map for a 4-input XOR gate.

The Karnaugh map for this function is shown in Fig. 21.20. This gives an interesting checkerboard type of map, and no 1s can be combined into subcubes. From Eq. (21.12) it would appear that eight NAND circuits would be required to achieve this logic equation. However, from Eq. (21.16) it can be seen that three 2-terminal XOR circuits can also simulate this logic, which is what is done in Fig. 21.19.

21.7 SUMMARY AND CONCLUSIONS

This chapter has introduced a very important technology that influences almost every aspect of modern industrial society. The basis for understanding and applying this technology is the mathematics of logic systems, the foundation of which is boolean algebra. The basic concepts of this algebra have been presented. The hardware for implementing logic systems has been discussed, and particular emphasis has been placed on integrated circuits and electronic logic. The many families of electronic logic have been defined, three of them—CMOS, TTL, and ECL—have been discussed in some detail, and circuit diagrams and manufacturers' data sheets describing practical logic have been presented. Some applications of this technology will be discussed in the following chapter.

PROBLEMS

21.1 Find the binary equivalent of the following decimal numbers:

(*a*) 76 (*b*) 1218 (*c*) 750
(*d*) 0.061 (*e*) 0.89 (*f*) 0.175

21.2 Verify that the binary numbers derived in Prob. 21.1 are equivalent to the decimal numbers from which they were derived.

21.3 Develop truth tables for the
(*a*) 2-input NAND function
(*b*) 4-input NOR function
(*c*) fourth logic circuit of Fig. 21.1

21.4 Write the logic equation $Y = f(A, B, C)$ and prepare a truth table for the circuit shown in Fig. P21.4.

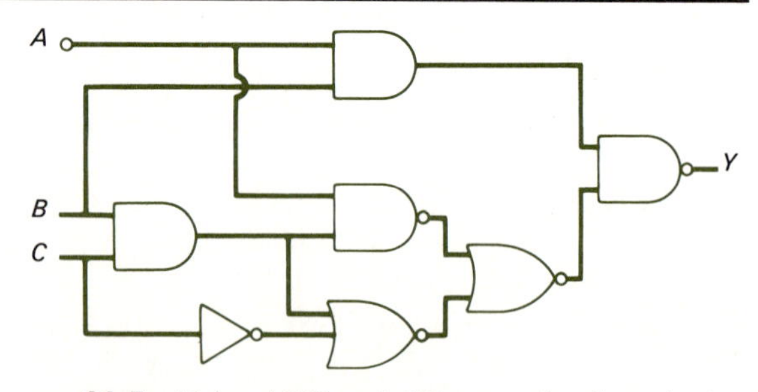

21.5 Using AND and OR gates, develop a logic circuit to express the logic equation

$$X = A + \overline{B}\overline{D} + BD + CD$$

with A, B, C, AND D being the input variables.

21.6 Does $X + WZ = WX + \bar{W}X\bar{Y} + WX\bar{Z} + \bar{W}Y$?

21.7 Does $BC + ABD + A\bar{C} = BC + A\bar{C}$?

21.8 Convert the following expressions to standard SOP form:

(*a*) $f(ac\,b, c) = a\bar{b} + c$
(*b*) $f(a, b, c, d) = b + \bar{a}\bar{c} + a\bar{b}cd$
(*c*) $f(W, X, Y, Z) = W\bar{X} + XYZ + \bar{W}\bar{Y}Z$

21.9 Determine the numerical representation ($\sum$ of a group of numbers) for the three equations of Prob. 21.9.

21.10 Convert the three expressions in Prob. 21.9 to standard POS form.

21.11 Determine the numerical representation (Π of a group of numbers) for the three resulting equations in Prob. 21.11.

21.12 Prepare the Karnaugh maps for the three logic equations of Probs. 21.9, 21.10, and 21.11, and find the minimum expression.

21.13 Find the minimum expression for

$$f(A, B, C, D) = \sum (0, 2, 3, 4, 8, 9, 10, 11, 15)$$

21.14 Find the minimum expression for the system defined by the Karnaugh map of Fig. P21.14.

wx \ yz	00	01	11	10
00	1		1	1
01	1		1	1
11	1	1	1	
10	1	1	1	

FIGURE P21.14

21.15 Find the minimum expression for the system defined by the Karnaugh map of Fig. P21.15.

wx \ yz	00	01	11	10
00	0		0	0
01			0	
11				
10	0		0	0

FIGURE P21.15

REFERENCES

21.1 *Webster's Ninth New Collegiate Dictionary*, Merriam-Webster, Springfield, Mass.,1984.

21.2 J. D. Greenfield, *Practical Digital Design Using ICs*, 2d ed., Wiley, New York, 1983.

21.3 A. Barna and D. I. Porat, *Integrated Circuits in Digital Electronics*, Wiley, New York, 1973.

21.4 F. E. Hohn, *Applied Boolean Algebra—Elementary Introduction*, Macmillan, New York, 1960.

21.5 1983 Electrical and Electronics Reference Issue, "Digital Logic," *Machine Design*, May 1983.

21.6 J. Prioste and T. Balph, "Relay to IC Conversion," *Machine Design*, May 28, 1970.

21.7 D. G. Fink and D. Christiansen, *Electronics Engineer's Handbook*, 2d ed., McGraw-Hill, New York, 1982.

21.8 M. Karnaugh, "The Map Method for Synthesis of Combinational Logic Circuits," *AIEE Transactions*, vol. 72, pt. 1, 1953.

21.9 A. Rappaport, "Automated Design and Simulation Aids Speed Semicustom IC Development," *EDN*, vol. 27, no. 15, August 4, 1982.

21.10 J. W. Hunt, "Programming Languages," *Computer* (IEEE), April 1982.

21.11 Motorola CMOS Handbook: *CMOS Integrated Circuits*, Motorola Semiconductor Products, Austin, Texas 78721, 1980.

21.12 J. Kasper and S. Feller, *Digital Integrated Circuits*, Prentice-Hall, Englewood Cliffs, New Jersey, 1982.

21.13 L. Teschler, "Update of Electronic Logic," *Machine Design*, August 20, 1981.

21.14 E. B. Eichelberger, "A Logic Design Structure for LSI Testability," *14th Design Automation Conference*, June 1977, pp. 462–468.

21.15 Anthony Ralston (ed.), *Encyclopedia of Computer Science and Engineering*, 2d ed., Van Nostrand Reinhold, New York, 1982.

21.16 A. B. Williams, *Designers Handbook of Integrated Circuits*, McGraw-Hill, New York, 1984.

21.17 D. G. Ong, *Modern MOS Technology*, McGraw-Hill, New York, 1984.

Chapter

Digital Systems

In several previous chapters it has been noted that there is a distinct dichotomy in the nature of electric circuits and devices. This dichotomy encompasses two broad circuit classifications which have gradually become known as analog and digital, based upon the means of energy control—either in a continuous manner (analog) or in discrete bundles (digital). Circuits and systems in which circuit control is exclusively of an analog nature and composed exclusively of analog circuits and components are known as *analog* systems. Where control components are exclusively digital, the system is classified as *digital*. A great many circuits and systems are composed of both digital and analog circuits or components and these are referred to as *hybrid* systems. This chapter is concerned primarily with digital and hybrid circuits and systems.

The basic building blocks of digital circuits—gates, relays, switches, etc.—have been described in Chap. 21. Obviously, there are countless hosts of additional digital components, circuits, and devices, as witnessed by the countless number of very thick circuit description books provided by each electronic circuit supplier. There are also many interesting electromechanical, mechanical and hydraulic digital devices, such as the electric stepper motor, which are mated with the digital circuits discussed here. Digital circuits are generally thought of as low-signal low-power circuits used in control and information applications, such as calculators, computers, and radar, but digital circuitry is also widely used in many high-power motor, machine tool, and energy conversion applications, usually as part of a hybrid system. An example of this type of application, in the control of an electric motor, is shown in Fig. 22.1. In the control

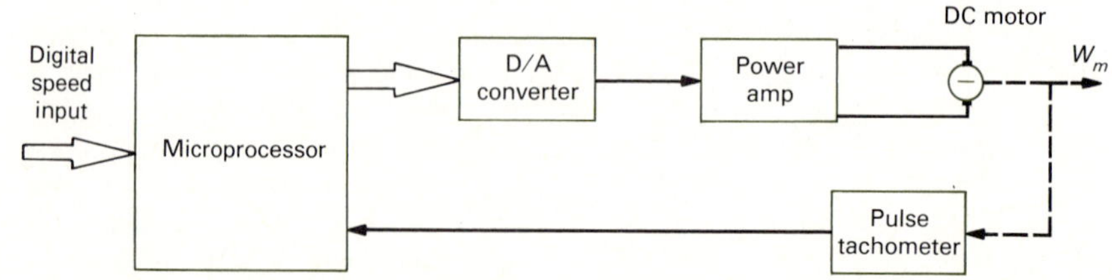

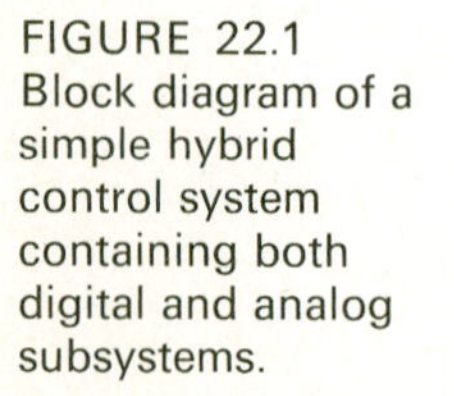
FIGURE 22.1 Block diagram of a simple hybrid control system containing both digital and analog subsystems.

system of Fig. 22.1, two main subsystems, the microprocessor and the digital tachometer used in the feedback branch, are digital, whereas the power amplifier and motor are analog components. The link between these two types of circuits is known as a digital-to-analog (D/A) converter. The inverse conversion, analog-to-digital (A/D), is equally important in hybrid systems. Both systems will also be discussed in this chapter.

Digital circuits may be assembled or manufactured by wiring together a number of discrete elements (transistors, relays, resistors, capacitors, etc.) or by integrating the entire assembly of devices and associated interconnections into a small semiconductor chip known as an integrated circuit (IC), which has been defined in the previous chapter. It is this latter form of digital circuit assemblage that has fostered the tremendous growth of digital circuit use in control and computational applications. One of the features of IC development that has contributed to this dazzling growth of digital circuit usage is the rapid increase during recent years in the IC packing density—that is, the number of digital functions (OR gates, NAND gates, transistors, etc.) that can be integrated into a given volume. Figure 22.2*a* illustrates this particular feature of IC development.[22.1] The dots on this graph are based upon the past history of two of the logic families discussed in the preceding chapter, MOS, (CMOS, NMOS, etc) and bipolar TTL. Figure 22.2*b* and *c* gives further illustration of the past and projected increase in chip density and complexity.

A major difference between analog and digital electric circuits is the nature of the waveforms of the electrical signals in the circuits. The signals (voltage or current) in analog systems are continuous and have classically been analyzed on the basis of either sinusoidal or continuous dc waveforms; much of classical electric circuit theory is based upon the assumption of one or the other of these waveforms. In contrast, waveforms in digital systems are composed of a series of pulses separated by periods of

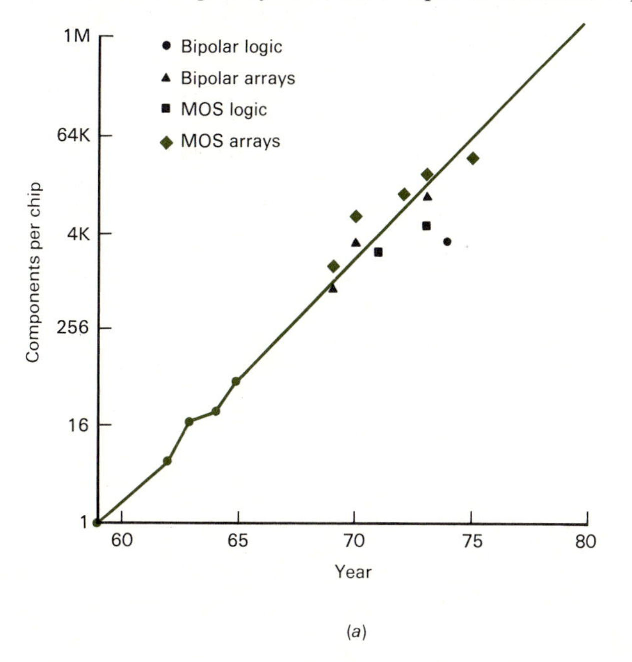

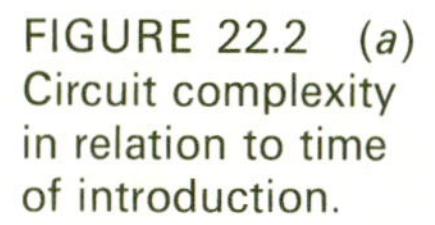
FIGURE 22.2 (*a*) Circuit complexity in relation to time of introduction.

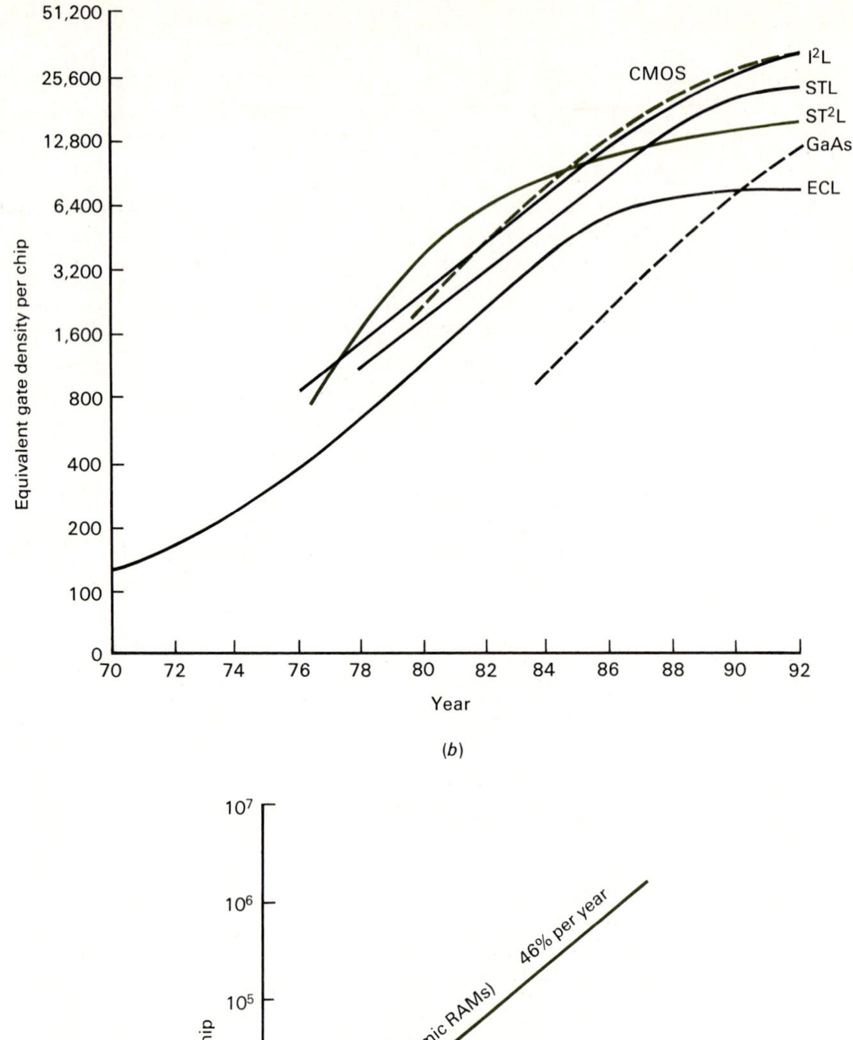

(*b*)

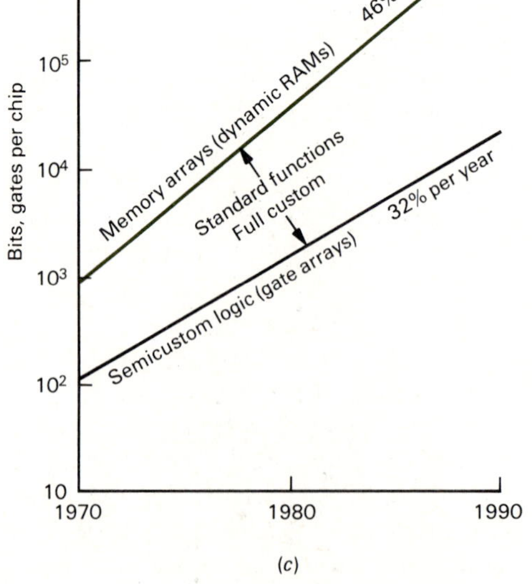

(*c*)

FIGURE 22.2 (*b*) Gate array chip complexity. (*c*) Very large-scale integration (VLSI) chip complexity trends.

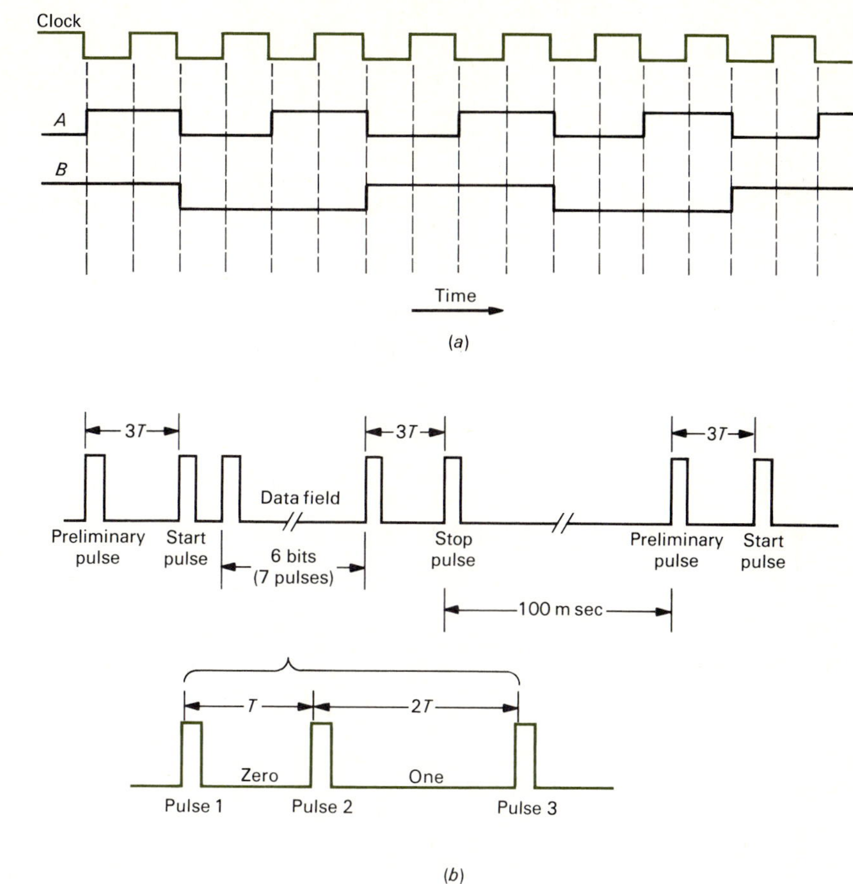

FIGURE 22.3 Examples of digital circuit waveforms: (*a*) clock and derived pulses; (*b*) an example of pulse-position modulation (PPM) (see Sec. 22.1).

zero signal level, or off time. In most cases the pulses can be represented as square or rectangular in shape. Typical digital signals are shown in Fig. 22.3. These signals are unipolar, but digital signals, in general, may be either unipolar or bipolar.

The control of analog signals is accomplished by varying the magnitude or frequency of the signal. Information transferred by means of analog signals is accomplished by means of superimposing magnitude or frequency variation upon sinusoidal signals, a process known as *modulation.* The processes of control and information transfer in digital systems are roughly analogous to those used in analog systems, although there is a greater flexibility of control and modulation in digital systems due to the nature of digital waveforms.

Another difference is that digital systems generally have a greater immunity from extraneous signals, or *noise*, than do analog systems. This again is due primarily to the nature of the digital waveforms. Figure 22.4 is an attempt to illustrate in a very simple manner the noise immunity of digital signals. In this figure, the square pulse represents an original digital pulse that represents a binary bit of information. The second pulse shape, v_2, represents the change in the original pulse due to transmission and transformer impedances, induced noise caused by switching of adjacent circuits, noise due

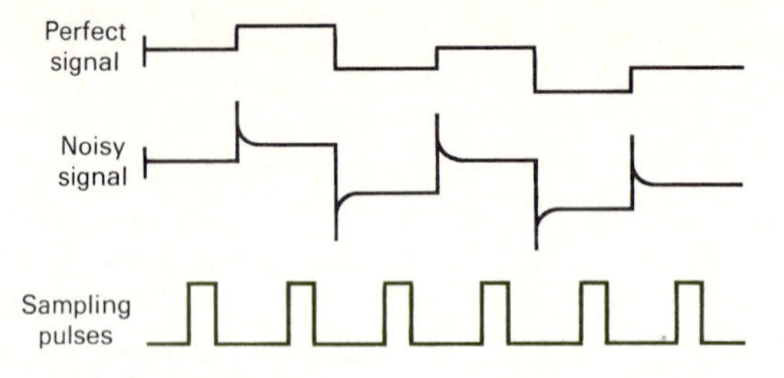

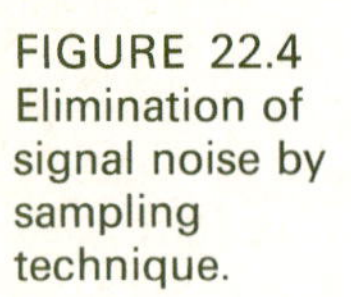
FIGURE 22.4 Elimination of signal noise by sampling technique.

to imperfect grounding (ground loops, etc.), and other effects along the paths of a digital signal between its source and its sink. As has been noted, binary signals are detected either by their absence or presence (1 or 0, true or false, etc.); this detection is often performed by a process called *sampling*, which will be discussed in Sec. 22.1.1. We see from the third pulse of Fig. 22.4 that the distortion of the waveshape caused by system impedances and noise effects will have little effect upon detecting the existence of the pulse, except for a massive shift in the pulse location (frequency). Further illustrations of the effects of noise upon digital signals will be given in the following discussion of modulation.

22.1 PULSE MODULATION AND SAMPLING

Pulse modulation in the most general sense refers to the control of one or more of the parameters associated with a digital pulse. The term has most frequently been applied to the process of encoding information from digital pulse trains, but more recently, similar techniques have been applied to control the power associated with digital signals, such as the average or RMS value of pulse currents or voltages.

The three principal parameters of the pulse train used in modulation and demodulation schemes are the pulse amplitude (or magnitude), the pulse width, and the frequency of repetition of the pulse. All three parameters and several combinations of them are used in control and communications applications. Power applications have generally been based upon only pulse-width modulation (PWM).

Modulation refers to the process of superimposing a signal containing intelligence upon a second signal known as the carrier signal—or sometimes known in pulse modulation as the pulse train. The reader is doubtless familiar with the well-known processes—used in radio and telephone communications, amplitude modulation (AM) and frequency modulation (FM)—of modulating sinusoidal carriers with a spectrum of sinusoidal signals, such as the spectrum resulting from a human voice. The intent of pulse modulation is similar to that of AM and FM, but the mechanism of the modulating techniques and the resulting waveforms are much different. Again, as we have often done on previous pages, we might use the terms *analog* and *digital* to contrast the two groups of modulating techniques; this terminology is in fact used in many practical applications, such as in distinguishing digital and analog recording methods. Demodulation is, of course, the reverse of the modulation process; that is, demodulation is the extracting of the intelligent signal from the modulated signal.

Pulse modulation schemes are generally named according to the pulse parameter which is being controlled or modulated; pulse-amplitude modulation (PAM); pulse-time modulation (PTM), which is so named because modulation is achieved by operating on the pulse's time parameters, and which includes the techniques with more familiar names, pulse-width modulation (PWM) and pulse-position modulation (PPM); pulse-frequency modulation (PFM); and pulse-code modulation (PCM), which includes such subgroupings as delta modulation (DM). Before discussing a few of these techniques in more detail, it is necessary to introduce a process required in all pulse modulation schemes, the process of sampling.

22.1.1 Sampling

Sampling is the process of evaluating one signal at discrete time intervals for the purpose of deriving another signal. Sampling is used in many areas of technology, such as in electrical measurements and digital instrumentation, in reliability and failure-rate analysis, and even in many socioeconomic areas, such as in public opinion polls. Our primary concern in this discussion, however, is the role of sampling in pulse modulation as used in digital systems of all types.

In this context, sampling generally refers to the evaluation of an analog, or continuous, signal at discrete time intervals. The ideal sampler is an *impulse sampler*, which multiplies a continuous signal $f(t)$ with a train of unit impulses $P(t)$ of period T. Such a process is illustrated in Fig. 22.5. The train of unit impulses has been converted to a train of impulses of varying magnitude, or amplitude, varying according to the analog function $f(t)$. Mathematically, the process of taking one sample of $f(t)$ is described by

$$f(t) = \int_{-\infty}^{\infty} f(\lambda)\, \delta(t - \lambda)\, d\lambda \tag{22.1}$$

where $\delta(t - \lambda)$ is the unit impulse at $t = \lambda$. This integral is known as the *sampling property* of the unit impulse and is related to the *impulsive response* of a system, a concept used in circuit and control theory. The amount of information contained in the resulting pulse train of Fig. 22.5 (the sequence of vertical arrows under the analog function) depends upon the *sampling rate* (or its reciprocal, the sampling period, T). A large body of knowledge has developed around this concept of the relationship between the information and the technical signals representing or transmitting that information, and this body of knowledge is known as *information theory*. A key theorem of information theory, known as Shannon's sampling theorem, addresses the relationship between the sampling period and the informational content.[22.2] If the frequency spectrum of the sampled function $f(t)$ contains no frequencies above f_c, then Shannon's sampling theorem states that the continuous signal $f(t)$ can be reproduced if and only if $1/T > 2f_c$. The minimum sampling frequency $2f_c$ is known as the *Nyquist rate*. In practice, due to noise and other effects, a signal may not be completely limited to the frequency band, so the sampling frequency $f_s = 1/T$ is usually made greater than twice f_c.[22.3]

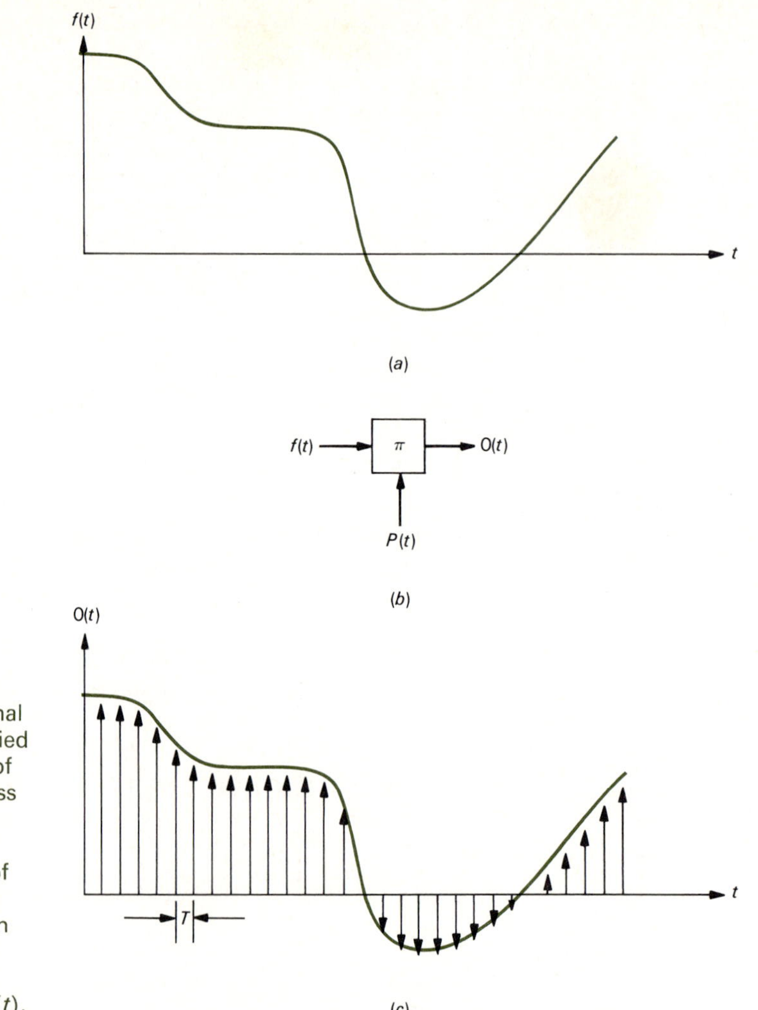

FIGURE 22.5 Ideal impulse sampler: (*a*) analog or modulating signal $f(t)$; (*b*) simplified representation of sampling process where $P(t)$ represents an impulse strain of hint magnitude and of repetition period T; (*c*) modulated impulse train $O(t)$.

Also in the practical case, the ideal unit impulse (of zero pulse width) is not realizable and narrow but finite-width pulses represent the practical sampling pulse train. Sampling pulses may be unipolar or bipolar, or flat-topped or conformal-topped, depending upon the required fidelity, the cost restraints, and the frequency spectrum of the analog signal. Figure 22.6 illustrates two different types of sampling pulses used in a PAM system. The pulse width of the sampling pulse τ must be small compared to the sampling period T.

A basic component of information and control systems using sampling techniques is the *sample-and-hold circuit*. The function of this type of circuit is to take a sample of an analog function at a given instant of time and to hold the amplitude of this function for a short interval of time. An obvious purpose of such a device is to keep the value of a sample available for a longer period of time than the sampling period T to permit processing of this value by the circuit or system. As will be seen,

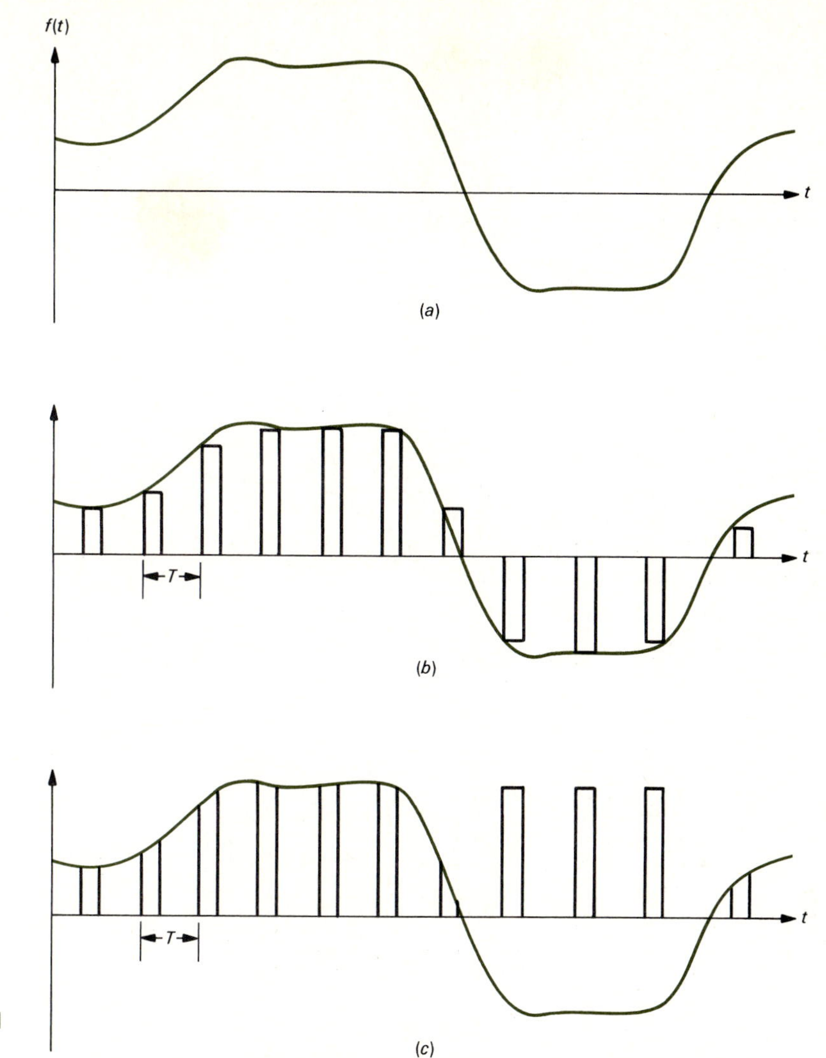

FIGURE 22.6 Pulse-amplitude modulation: (*a*) modulating signal; (*b*) pulses from bipolar square-topped PAM; (*c*) pulses from unipolar conformal PAM.

various sample-and-hold circuits are useful in the demodulation process. Figure 22.7 illustrates a simple electronic sample-and-hold circuit and its associated functioning on a typical analog signal $f(t) = v_s(t)$. This type of hold is known as *zero-order hold*, indicating that the value of $f(t)$ held at each sampling instant is the value of $f(t)$ at that instant. As seen from Fig. 22.7, the zero-order hold tends to convert a continuously varying analog signal into a staircase-type signal.

22.1.2 Pulse-Amplitude Modulation (PAM)

In this scheme, the modulating process consists of deriving a train of pulses, the amplitudes of which vary with the amplitude of the analog signal $f(t)$. This process is shown in Fig. 22.6. PAM is analogous to AM in the continuous-signal case and has the same

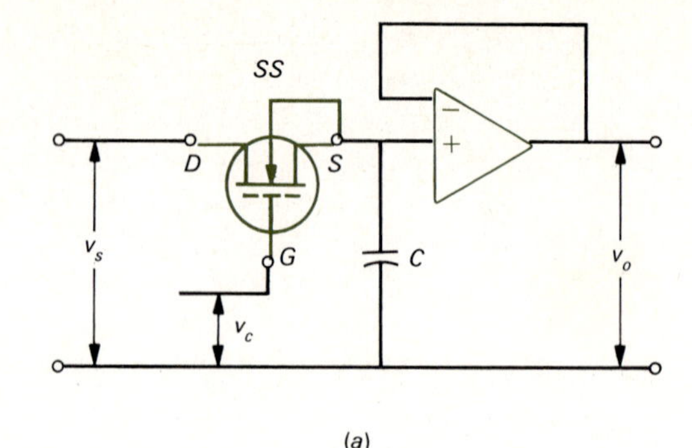

(a)

f(t)
v_o
t
v_c
t
τ_s
T

(b)

FIGURE 22.7 Sample-and-hold circuit: (*a*) typical circuit diagram using MOSFET and operational amplifier; (*b*) waveforms of voltage signals.

relatively low signal-to-noise ratio as does AM. Neither method is generally applicable to wide-band (f_c very large) data transmission. PAM is frequently used in multiplex systems, with the sampling of the individual channels made consistent with the sampling theorem given in the previous section. The pulse train in PAM is analogous to the carrier in AM. The hardware for performing PAM, called an *encoder*, is based upon an electronic multiplication system (as shown in Fig. 22.5), with filtering, various auxiliary subsystems, and synchronizing circuits, depending upon the specific type of PAM—unipolar or bipolar, square or conformal-topped, etc. In *square-topped PAM* the pulse has a flat top, the magnitude of which represents the average value of the analog signal during the pulse period τ. In conformal or *top-modulated PAM* the pulse conforms to the changing amplitude of the analog signal during the pulse period, as shown in Fig. 22.6.

An important class of control system based on PAM is known as *sample-data control*. In sample-data control, a sampling of the system error signal (or signal related to system error) of any waveform similar to those shown in Fig. 22.6 is fed back to the reference signal rather than a continuous function proportional to system error. This permits the processing of the error signal and the development of feedback signals to be performed by means of a digital system rather than an analog, with the associated reduction in the effects of system noise and often with reduced system costs. A transformation calculus known as Z transforms has been developed for the analysis and design of sample-data systems.

22.1.3 Pulse-Width Modulation (PWM)

This scheme is a specific form of the more general pulse-time modulation (PTM) methods. Information or control is transmitted by means of the time width of pulses having a constant amplitude and (usually) having a constant frequency of repetition. Pulse-width modulation is also known as pulse-duration modulation (PDM) or pulse-length modulation (PLM). There are two basic methods of PWM generation—the *uniform sampling* method in which the pulse-frequency or pulse-repetition rate is fixed, as in the case of PAM, and the *natural sampling* method in which there is some slight modification (causing a small amount of distortion of the output signal) of the time at which the samples are actually taken.

A basic block diagram of the uniformly sampled PWM is shown in Fig. 22.8, and a typical waveform of PWM is shown in Fig. 22.9. This waveform results from what is called *leading edge synchronization* and can be described as follows: Each pulse in the pulse train (Fig. 22.9*c*) is initiated by the leading edge of the triangular wave (Fig. 22.9*b*), which is the sum of a sawtooth wave of fixed frequency and the staircase wave created from a sample-and-hold circuit operating on the analog signal $f(t)$ (Fig. 22.9*a*); each pulse is ended each time the sawtooth goes through its vertical wavefront; thus, the pulse width is determined by the time between the leading edge of the sawtooth (i.e., the instant at which the triangular wavefront crosses the horizontal) and the time the vertical wavefront crosses the horizontal—that is, each pulse width is proportional to the width of the sawtooth at the horizontal axis of Fig. 22.9*b*. The importance of synchronizing the sample-and-hold circuit with the sawtooth wave generator is obvious from the pulse development shown in Fig. 22.9. The resulting pulse train has pulses of constant amplitude and constant frequency of repetition. Information is contained in the relative widths of the individual pulses. Note that zero on modulating signal is well defined by the pulse of mean width. There are many variations in methods of setting pulse widths proportional to the amplitude of the modulating signal $f(t)$, which results in minor differences in the frequency spectrum of the

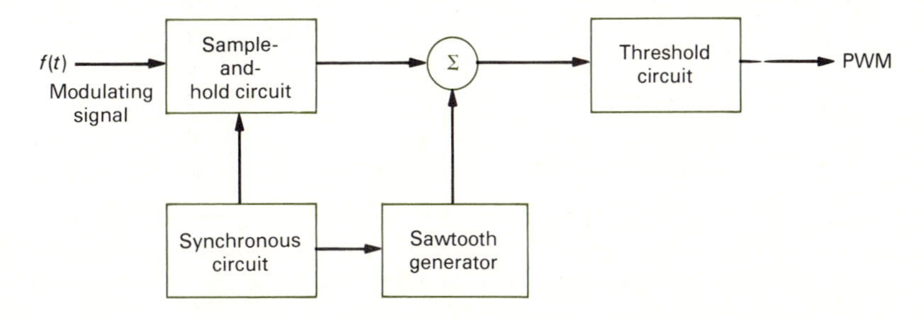

FIGURE 22.8 Block diagram of uniform-sampling type of pulse-width modulation.

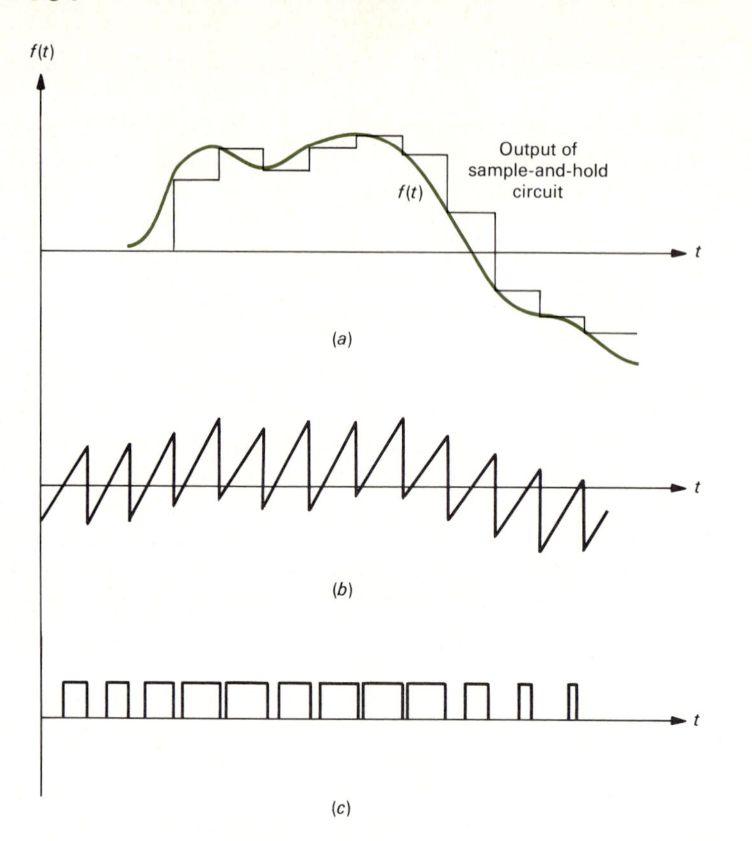

FIGURE 22.9 Waveforms associated with uniform sampling PWM: (*a*) modulating function and sample-and-hold output; (*b*) sum of sample-and-hold output and sawtooth wave; (*c*) generated pulses using leading edge modulation.

pulse train, the signal-to-noise ratio, the demodulation technique, etc., but the basic concepts of PWM are similar to those shown in Figs. 22.8 and 22.9.

PWM, through the use of constant-amplitude pulse trains, allows the circuitry creating the pulses to operate at its saturated level, which is usually the condition of maximum power efficiency. This is especially important where modulation of high-power signals is required as has been noted. PWM is used to control motors and other power devices through modulation of power semiconductors. Power applications of PWM are described elsewhere in the text. The minimum off interval between pulses, known as the *guard interval*, must be nonzero even though the sampling theorem is satisfied with a zero interval, owing to finite pulse rise times. Overlapping of pulses will result in a distortion of the demodulated signal and in cross talk in multichannel systems.

22.1.4 Pulse-Code Modulation (PCM)

PCM is markedly different from the two modulation techniques previously discussed in that it converts the modulating signal into a group of pulses, each group of which represents the numerical value of the modulating signal at the sampling instant. The translation of the quantized analog signal at the sampling instant into a "code" of pulses is the basis for the naming of this modulation technique. It has become more feasible in recent years owing to the great improvements in technical characteristics and to the reduced cost of integrated circuits. PCM lends itself readily to time multiplexing (the transmitting of several signals over the same medium at distinct time intervals in a given time period) and generally has a high signal-to-noise ratio. PCM signals are readily amplified by regenerative repeater stations for long-distance transmission with a minimum reduction as compared with most other modulation techniques. Error-detecting and -correcting processes are generally more easily accomplished in PCM systems.

Figure 22.10 illustrates waveforms at various stages in a PCM system. The analog, or modulating, signal is quantized by the use of sample-and-hold circuitry, giving a distinct amplitude of the analog signal at each sampling instant. This amplitude is then converted into a group of pulses by a specific instruction code. A wide variety of codes is possible, and the resulting pulses may be binary (consisting of magnitude levels 1 and 0), tertiary (1, 0, and -1), or, for that matter, n-ary. Binary pulses are the most common. The *number* of pulses in each group is related to the frequency spectrum desired for the transmission of the analog signal. In Fig. 22.10, three pulses per group have been used, which permits the encoding of eight levels of the analog signal in binary representation. Using an n-ary code with m pulses per group permits transmission of n^m values of the analog, or modulating, signal. Keeping in mind the conclusions of the sampling theorem in Sec. 22.2.1, we see that signals using PCM require a

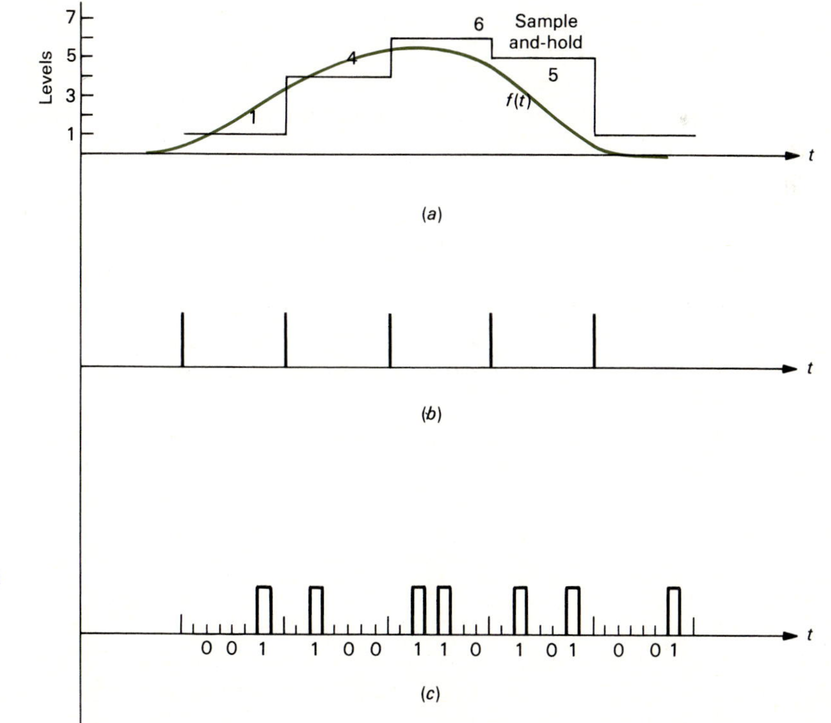

FIGURE 22.10 Pulse-code modulation (PCM) signals: (*a*) modulating signal; (*b*) synchronizing pulse; (*c*) output-coded pulses.

relatively wide-band frequency spectrum. The pulse capacity in bits per second for pulse-modulated systems can be shown to be an n-ary, m-pulse/group code

$$C = mf_{sN} \log_2 n^2 \qquad \text{bits/s} \tag{22.2}$$

where f_{sN} is a sampling rate based upon the twice critical frequency criterion (the Nyquist rate, as defined in Sec. 22.2.1). Shannon has shown that the maximum possible rate of binary pulses is

$$C = B \log_2 \left(1 + \frac{S}{N}\right) \qquad \text{bits/s} \tag{22.3}$$

where S/N is the ratio of signal power to noise power in dB and B is the bandwidth in Hz. The *bandwidth* of PCM is mf_{sN}, as seen from Eq. (22.2).

In PCM, there is an inherent error in the modulation process due to the quantization of the modulating signal, as shown in Fig. 22.10*a*. This error is often referred to as *quantization noise* and is the major noise component of N in the above equations. Quantization noise is expressed as a function of the noise at the demodulator Ku, where u is the RMS noise at the demodulator input and $N = u^2$. Using this notation, it can be shown that the pulse capacity of PCM is

$$C = B \log_2 \left(1 + \frac{12S}{K^2 N}\right) \tag{22.4}$$

In PCM there is a fairly optimum S/N ratio beyond which the quantization error decreases less rapidly. This ratio is about 9.2, or 20 dB. Using this ratio in Eq. (22.4) and typical values of K, it can be shown that the power (S) required to transmit a given bit capacity is about 7 times that of the ideal binary system described in Eq. (22.3). However, this is still a more efficient means of pulse data transmission than PAM, PWM, or PFM.

Delta modulation (DM) is a simplified version of PCM in which $n = 1$. The one digit (in contrast to binary or tertiary, etc.) is used to transmit information about the differences of successive samples of the modulating signal. At the receiver, or demodulator, the pulses are integrated to obtain the original signal. DM circuitry is very simple and low in cost, but DM requires a sampling rate much higher than the Nyquist rate of $2f_c$ and a wider bandwidth than binary PCM.

22.1.5 Signal Processing, Demodulation

Digital systems used for the transmission and processing of intelligence signals generally include a great number of subsystems with many diverse functions. These functions come under the general classification of signal processing and handling. Some of these functions are similar to their analog system counterparts, such as data transmission from one location in space to another and amplification at various intervals during transmission, although there will obviously be differences in the required frequency ranges and in other details of the hardware implementation. Other subsystems devoted to signal shaping and filtering are generally totally different in analog and digital systems. Included in this latter category are the processes of signal modulation and demodulation.

The purpose of demodulation, or decoding a digital signal, is to recover the analog, or modulating, signal which has modulated the carrier pulse train. In PAM and PTM, demodulation generally consists of low-frequency filtering, which filters out the higher frequencies associated with the carrier pulse train and passes the low-frequency analog signal carrying the intelligence. Low-frequency or low-pass filtering is generally acceptable where the interpulse period is relatively large. In many cases, however, it is necessary to modulate on a pulse-by-pulse basis. Figure 22.11*a* illustrates the block diagram of a PWM demodulation system in which each pulse is processed, first with an integration stage. Signals at various stages in the demodulation process are shown in Fig. 22.11*b*. The sample-and-hold circuitry sums the pulse integrals and resets the integrator in synchronization with the PWM waveform. The smoothing filter smooths the staircase-type waveform of the hold circuit to form an analog signal.

PCM demodulation, usually called decoding, is almost exactly the inverse of the modulation process. Each group of pulses arriving at the receiver or decoder

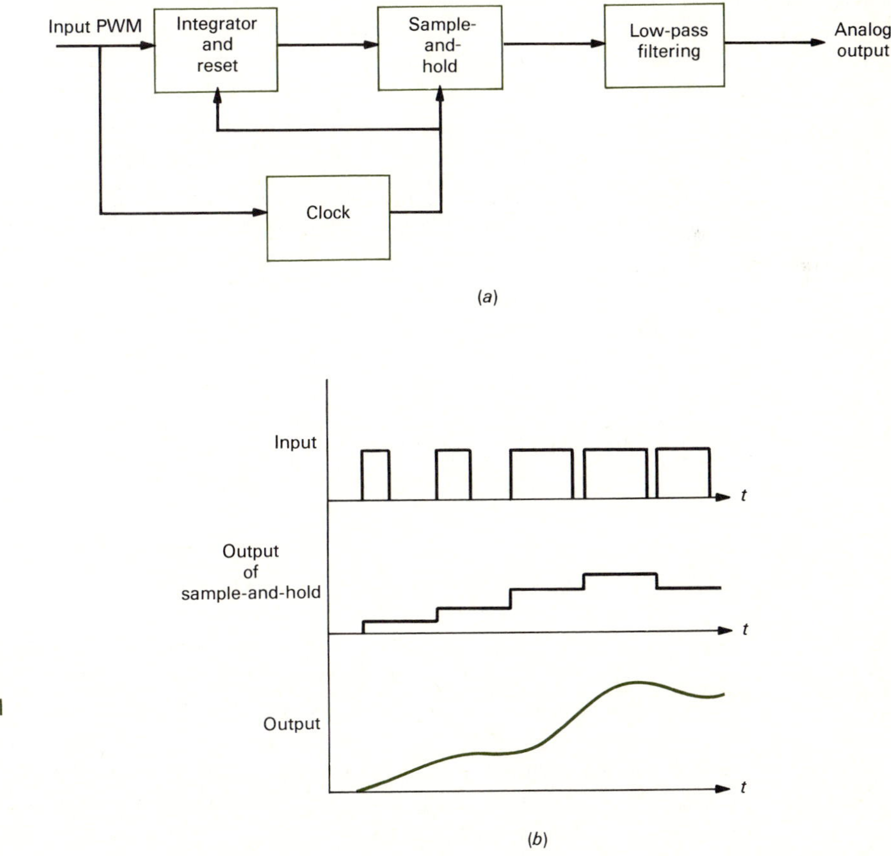

FIGURE 22.11 (*a*) Block diagram for PWM demodulations. (*b*) Signal waveforms for PWM demodulator.

represents an analog level. These pulses are decoded—for example, by means of a counter—and an analog signal proportional to the count is produced. The demodulation process must be exactly synchronized with the pulse-coding process.

The modulation-demodulation process begins and ends with a signal, generally analog, containing or representing some form of information or intelligence. The use of digital transmission and signal processing is advantageous because of generally faster response times, better immunity from noise, compatibility with all-digital systems of microprocessors and computers, and generally lower hardware costs as compared to digital systems.

22.2 SIGNAL CONVERSION

We have seen in previous sections of this chapter the need for conversion of signals from analog-to-digital or from digital-to-analog forms. These processes should not be confused with the modulation-demodulation process discussed in the previous section, in which a signal of one form—usually analog—is used to vary the parameters of the other form—digital, in the cases discussed in the previous section. However, in many applications and in many subsystems, it is desirable to convert a signal in one form to the other form. This was discussed, for example, in Chap. 20 relating to electrical measurements. In many types of measurements, the measured parameter (such as motor speed in revolutions per unit time) is counted, which is a purely digital process, but it is desired to present a readout or display in analog form, or a signal proportional to motor speed is to be used in an analog type control system. In such cases, the count of revolutions per unit time would be converted to a continuous signal, a process known as digital-to-analog conversion, or DAC. However, a more common DAC application is the conversion of a binary number to an analog voltage or other form of continuous signal. The reverse process, analog-to-digital conversion, or ADC, is generally more common than is DAC, since more and more signal and data processing is being performed by digital circuits, whereas a great many sensors of data, including most human processes, are of a generally analog nature. The point at which there is a conversion between analog and digital signals is an example of a system *interface*. This term, interface, is of course equally applicable to other types of interconnections among subsystems making up a larger system in which digital or analog signals exist on both sides of the interconnection.

22.2.1 Digital-to-Analog Conversion (DAC)

DAC is generally a less complex process than its inverse, ADC, and consists of developing a specific, continuous signal level related to a specific number of incoming discrete digital signals. Most commonly, the digital input signals are in binary form, having the discrete values 1 and 0, or high and low, etc., as discussed in Chap. 21.

The majority of DACs available today are based upon the use of *weighted networks*. A schematic arrangement of a weighted network for DAC application is illustrated in Fig. 22.12*a*. This is a summing network in which the analog output (shown

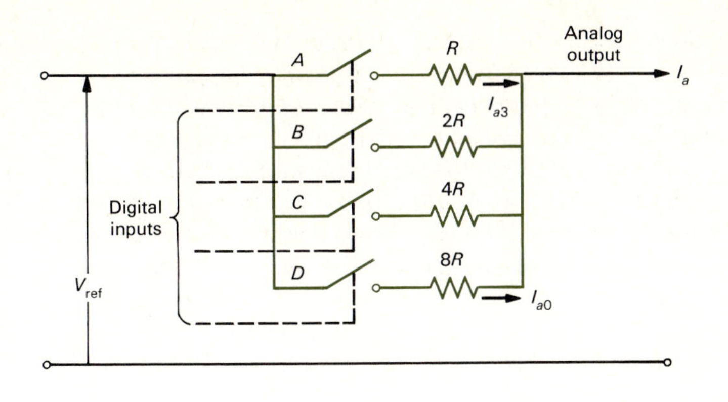

(a)

Decimal	Gate or binary column A	B	C	D	I_a, pu
0	0	0	0	0	0
1	0	0	0	1	0.125
2	0	0	1	0	0.25
3	0	0	1	1	0.375
4	0	1	0	0	0.5
5	0	1	0	1	0.625
6	0	1	1	0	0.75
7	0	1	1	1	0.875
8	1	0	0	0	1.0
9	1	0	0	1	1.125
10	1	0	1	0	1.25
11	1	0	1	1	1.375
12	1	1	0	0	1.5
13	1	1	0	1	1.625
14	1	1	1	0	1.75
15	1	1	1	1	1.875

(b)

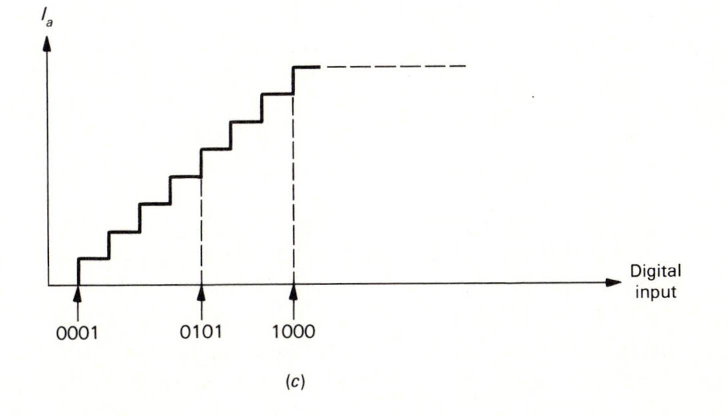

(c)

FIGURE 22.12 Relationships in a simple, weighted network DAC: (a) schematic circuit diagrams; (b) per-unit (pu) values of analog output ($1 \text{ pu} = V_{ref}/R$); (c) analog output waveform.

simply as a current here, but usually the voltage across a fixed resistor) is the sum of a number of branch currents—each of which represents a signal proportional to a column in a binary number. This relationship is illustrated in Fig. 22.12*b*. Figure 22.12*c* illustrates the staircase shape of the analog voltage output of a DAC. The operation is as follows: Each binary input channel (or binary number column) is controlled by logic gates (shown in Fig. 22.12*a* as on-off switches), which can be of any of the types discussed in Chap. 21. When a gate is in the on state, the reference voltage is applied to a *weighted resistor* associated with that gate. For example, the analog output current resulting when the gate representing the most significant digit (MSD, the left-most digit of the binary number) is energized or is in the on state is

$$I_{a3} = \frac{V_{ref}}{R} \tag{22.5}$$

When the gate for the least significant digit is energized, the output is

$$I_{a0} = \frac{V_{ref}}{8R} \tag{22.6}$$

The general analog output is the sum of these currents (or a voltage proportional to the sum):

$$I_a = \sum_{i=0}^{n} I_{ai} \tag{22.7}$$

where it is recognized that each of the currents I_{ai} may have one of two discrete values, 0 or $V_{ref}/(i + 1)R$. It is seen that for the case depicted in Fig. 22.12, $n = 3$ and represents a binary number of maximum length of four digits, or a decimal number of value up to 15. The decimal number is also the measure of the analog output signal. Note that this signal is a function of the reference voltage V_{ref}, which is usually determined by the voltage level for which the particular logic family is designed. Eight volts is a common logic family voltage level and is shown in Fig. 22.12*b*; 5-, 10-, and 15-V levels are also commonly used in various logic families.

Example 22.1 Prepare a weighted network diagram and determine the levels of analog voltage output for a 6-bit DAC using a 15-V logic family.

Solution It is seen from Fig. 22.12 and from Eqs. (22.5) to (22.7) that the multiplier for the resistance in the least significant digit path is

$$R_0 = 2^n R \tag{22.8}$$

Therefore, for 6-bit conversion, the largest resistor is 2^5 or $32R$, and the others are in descending binary multiples of R. This is shown in Fig. 22.13. The voltage levels can be reasoned as follows: Starting from zero output voltage, the next step on the staircase must be one thirty-second of

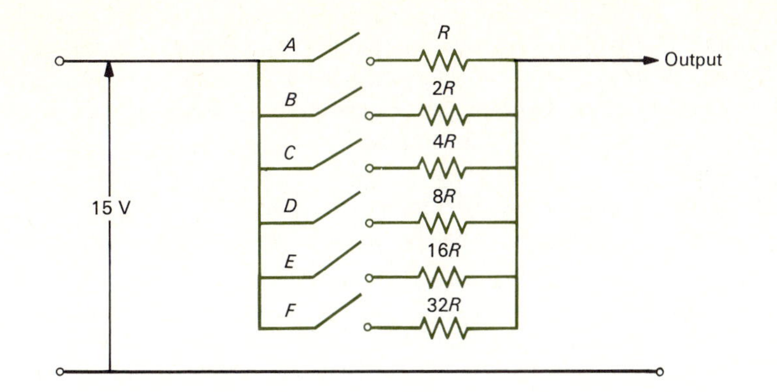

FIGURE 22.13 Example 22.1: 6-bit resistor network.

the maximum possible voltage, and continue in increments of $V_{ref}/32$; therefore, the voltage levels of the output function are 0, 0.46875, 0.9375, ... 15.0 V.

The disadvantage of the weighted network scheme is the large number of precision resistors of many different values required which are costly even where the network is part of an IC chip. The resistance values must be relatively small and the output impedance must be very large in order to minimize the value of the current components I_{ai} and the associated voltage drops through leads and connections. Also, variations due to temperature changes are relatively large in a multivalue resistance network.

A resistor system that achieves the same weighting characteristic as the networks of Figs. 22.12 and 22.13 without requiring so many different resistance values is a *ladder network*, an example of which is shown in Fig. 22.14. In ladder DACs, precision resistors of only two different values—values in the ratio of 2:1—are required. However, it is now necessary to use two different reference values, and a larger number of resistors are required. One of the two reference voltages is zero (in contrast to "NOT V_{ref}," as in the case of Fig. 22.12). This zero reference value is electrically equivalent to

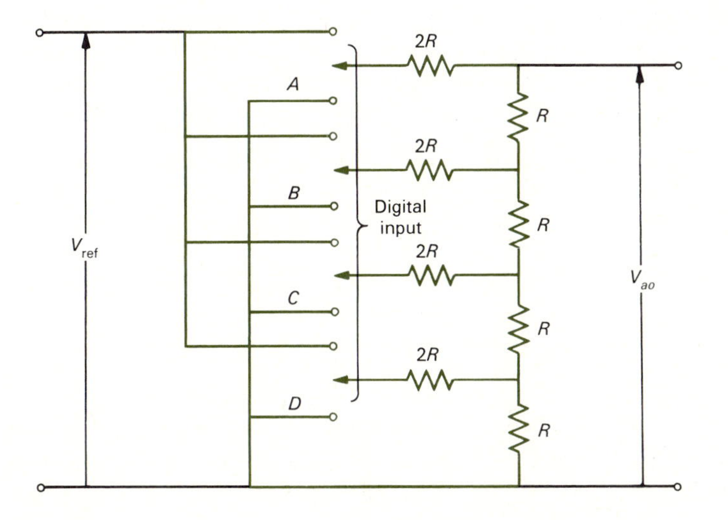

FIGURE 22.14 Ladder network DAC. The digital input controls the operation of switches *A, B, C, D*; V_{ao} is analog output.

grounding the gates associated with the specific binary channels. While this simplifies the hardware associated with achieving the DAC, it does complicate the analysis of the electric circuit. A few problems related to the DAC ladder network realization are included at the end of this chapter for those interested in the analysis of resistive networks.

Now, with a simple example based upon a 2-bit ($n = 1$) DAC, we will illustrate how such two-value resistor networks achieve binary conversion.

Example 22.2

Determine the relative (or per unit) output voltage of the 2-bit ladder network shown in Fig. 22.15 as a function of the digital switch (gate) settings. Switch *A* represents the most significant digit (MSD), *B* the least significant. To produce an analog output voltage V_a for a single digital channel, the switch for that channel is in the up position of Figs. 22.14 and 22.15; all other switches are in the down position, which connects those circuits to the system ground.

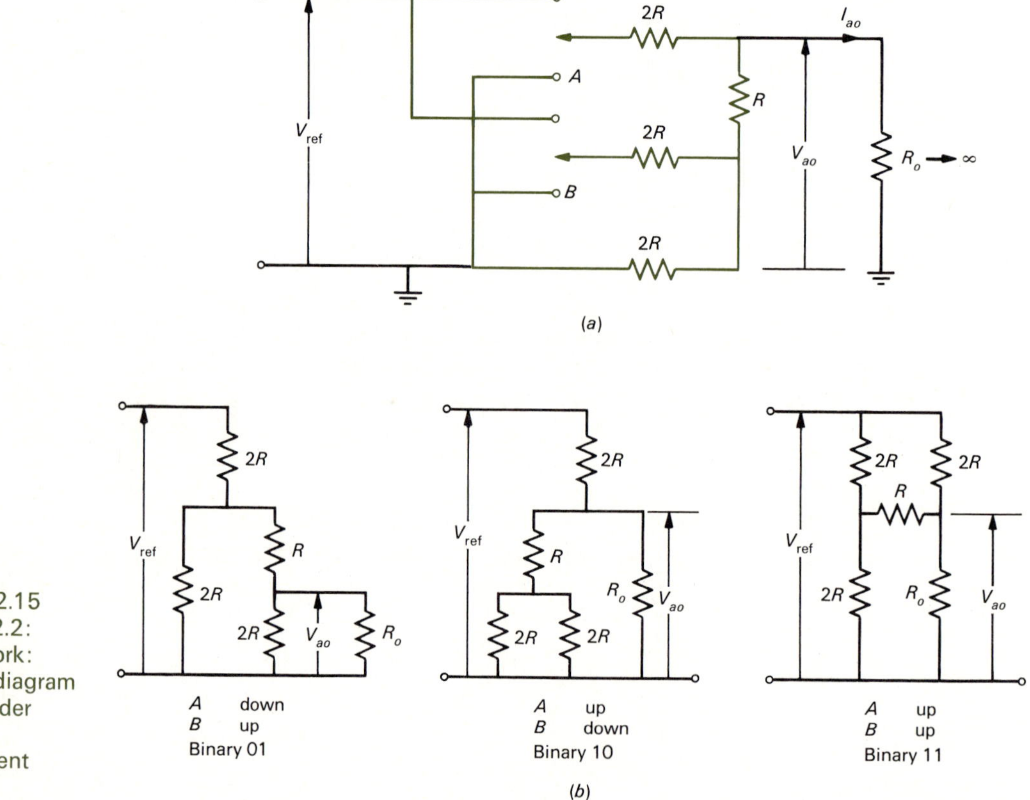

FIGURE 22.15 Example 22.2: 2-bit network: (*a*) circuit diagram of 2-bit ladder network; (*b*) equivalent circuits.

Solution From the general relations given above, a 2-bit or two-channel DAC can provide three different analog output signal levels, representing the binary numbers 01, 10, and 11 and the decimal numbers 1, 2, and 3. The first of these is with *A* down and *B* up; the second number, binary 10 or decimal 2, results from *A* up and *B* down; the third, binary 11 and decimal 3, is with both A and B up. The three networks that result from these three switch settings are shown in Fig. 22.15*b*. From simple network analysis, it is seen that the three resulting values of V_a are, respectively, $V_{ref}/4$, $V_{ref}/2$, and $3V_{ref}/4$. These are in the proper decimal relationship of 1–2–3 and represent the first three steps of the staircase analog output signal.

Since only two values of resistance are required in the ladder DAC, the system characteristics are more stable with temperature variations. Also, the ladder configuration has a constant output resistance, the proof of which is left for a problem at the end of this chapter. Present-day DAC technology includes many configurations built on a single IC chip and thick-film hybrid circuits. The more precise units are constructed with thick-film resistive networks located on the surface of a ceramic substrate. The digital input circuits of a DAC are often impedance-matched to the external logic circuits by means of operational amplifiers or RC networks or time-matched by means of a register, a process called *buffering*. A typical 16-bit DAC with CMOS logic will have a temperature variation of about 1 ppm/0°C, require about 4 mw of power, and have an output-current settling time of 1 to 3 ms.

22.2.2 Analog-to-Digital Conversion (ADC)

Analog-to-digital conversion is generally more complex than DAC and there are a wider variety of circuit configurations and techniques, each with its own distinct merits. The most common as well as one of the simplest configurations is known as the *method of successive approximations.* Figure 22.16 depicts a block diagram representation of this ADC configuration. This method requires a comparator, a DAC, and a clock-controlled counter. The inputs to the comparator are the analog voltage and the output of the DAC. When the analog input V_a equals the output of the DAC (that is, when the two comparator inputs are equal), the comparator stops the clock and the counter. At this instant, the input to the DAC is the digital output of the counter, which is the digital equivalent of V_a. This is the output of the ADC system and represents the digital value of the analog input V_a. A simple up counter is acceptable, but because it has to count through the full range of counts for every negative increment of the input V_a, it is therefore slow to track decreasing analog signals. An up-down counter is preferred; this type of counter can operate at clock rates of 1 MHz and above in practical ADCs. To further illustrate the operation of this type of ADC we will introduce some numerical relationships in the following example.

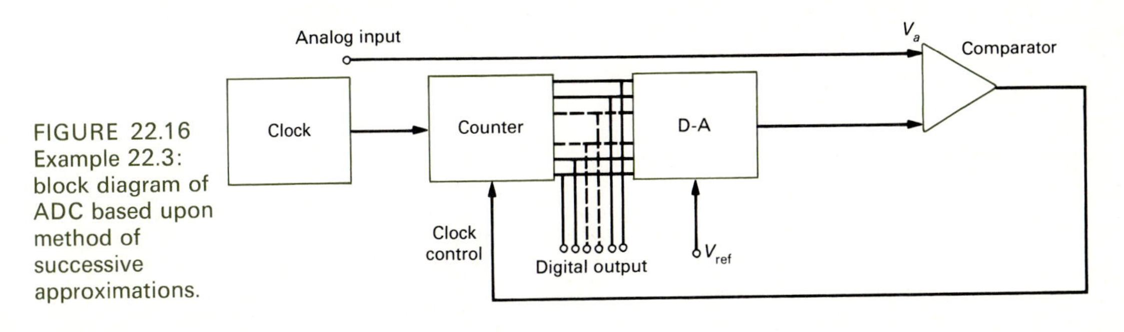

FIGURE 22.16 Example 22.3: block diagram of ADC based upon method of successive approximations.

Example 22.3 An ADC based upon the circuit shown in Fig. 22.16 has an analog input that has a value of +6 V at a given instant; the counter is a binary 6-bit counter (see Chap. 21) which develops a value of 0.175 V per count (often called the counter *resolution*). Determine the digital output for this 6-V input and the time required for the counter to arrive at this input value if the clock frequency is 1 MHz.

Solution The clock will stop when the output of the DAC equals the analog input V_a, which is 6 V in this example. This will occur after $6/0.175 = 34.286$, or 35, counts. Therefore, the output (which is actually the input to the DAC) is the binary equivalent of 35, or 100011. The time required for the count up to 35 is $35 \times 1 \times 10^{-6}$, or 35 μs. Note that the accuracy in this particular example is 34.286/35, or 98 percent. Using counters with smaller resolution and amplification of the digital output, accuracies of 99.9 percent are realizable.

Another circuit configuration for ADC is illustrated in Fig. 22.17. This configuration uses op amps A_1 and A_2, a comparator, logic switches S_1 to S_6, and control logic. Operation is initiated by sampling the analog input V_a, and this value is compared with V_{ref}. The difference is used to develop a digital equivalent of the analog input. The accuracy of this configuration is about the same as that of the system in Fig. 22.16. Practical embodiments of both systems are available as single chips in present technology. For further information on DAC and ADC, consult Refs. 22.4 to 22.6.

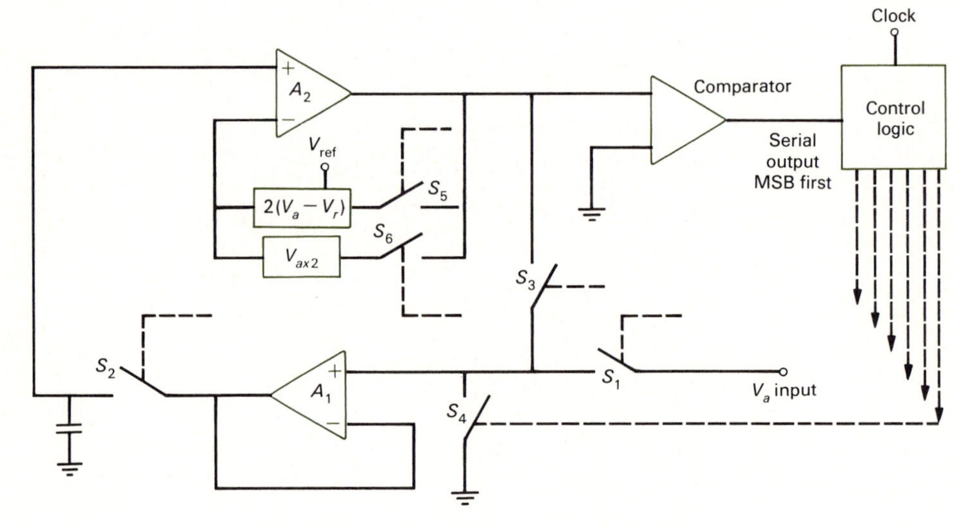

FIGURE 22.17 Block diagram of a recirculating-type ADC; MSB stands for most significant bit. (Courtesy of Siliconix, Inc., Santa Clara, California.)

22.3 DATA ACQUISITION

A major application of the vast array of digital circuitry now available is that of acquiring and processing measurements made on practically all types of technical processes. Digital circuitry and instrumentation are available for evaluation of almost any measurement conceivable, whether it be in outer space or within a vital human organ or in very mundane environments such as steam electric power plants. Both the measuring function and the data-processing function can be accomplished by a wide variety of both analog and digital circuitry at the present time. However, digital circuitry appears to offer more advantages—particularly economic advantages—owing to the rapid development of IC chips, as has been noted frequently in previous chapters. Also, the significant developments in digital display techniques and the proliferation of mini- or microcomputers to almost every corner of industrialized society has further fostered the use of digital circuits for data acquisition and processing. However, the availability of a wide variety of low-cost ADC chips, such as discussed in the previous section, has resulted in a continuation of analog measurements in a great variety of applications. As was noted in Chap. 20, there are many arguments favoring analog sensing and measurement of many parameters and there is also "human prejudice" toward digital instrumentation in certain functions. Therefore, many data acquisition systems tend to be of a hybrid nature.

Figure 22.18 illustrates a very simplified block diagram of a hybrid data acquisition system, with analog sensing of the measured signals and conversion to digital form for data processing, storage, and possible use in control and further computer manipulation. The measured signals include temperatures, speeds, voltages, currents, powers, and such mechanical parameters as strain, pressure, velocity, and position, as well as most other types of measurements (chemical, optical, acoustical, etc.). It is often desirable to display the measurements in their "raw" analog form, as shown at the lower left of Fig. 22.18. For transmission to remote locations, the signals are not only amplified but frequently require multiplexing (as shown). They can be transmitted at either the analog or digital stage, and many transmission systems can handle either type of signal, but usually at some stage the signals are converted to digital form by means of ADCs. Figure 22.19 illustrates a single-chip 8-bit ADC with 16-channel multiplexer. The function of the multiplexer is to *time-share* the incoming analog signals—that is, to sample an incoming analog signal for a short time period (usually a few microseconds or less), apply this signal to the ADC, and then return to the next analog signal and repeat the process. The timesharing at the multiplexer stage permits the processing of a large quantity of information (in this case, 16 signals representing 16 different measurements) with a high degree of accuracy. Most present-day data

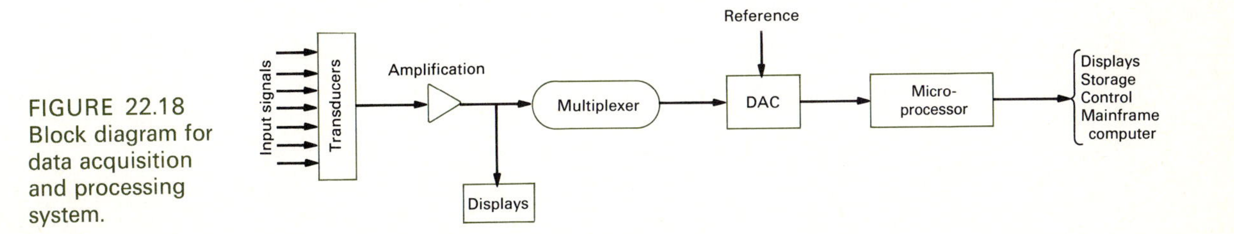

FIGURE 22.18 Block diagram for data acquisition and processing system.

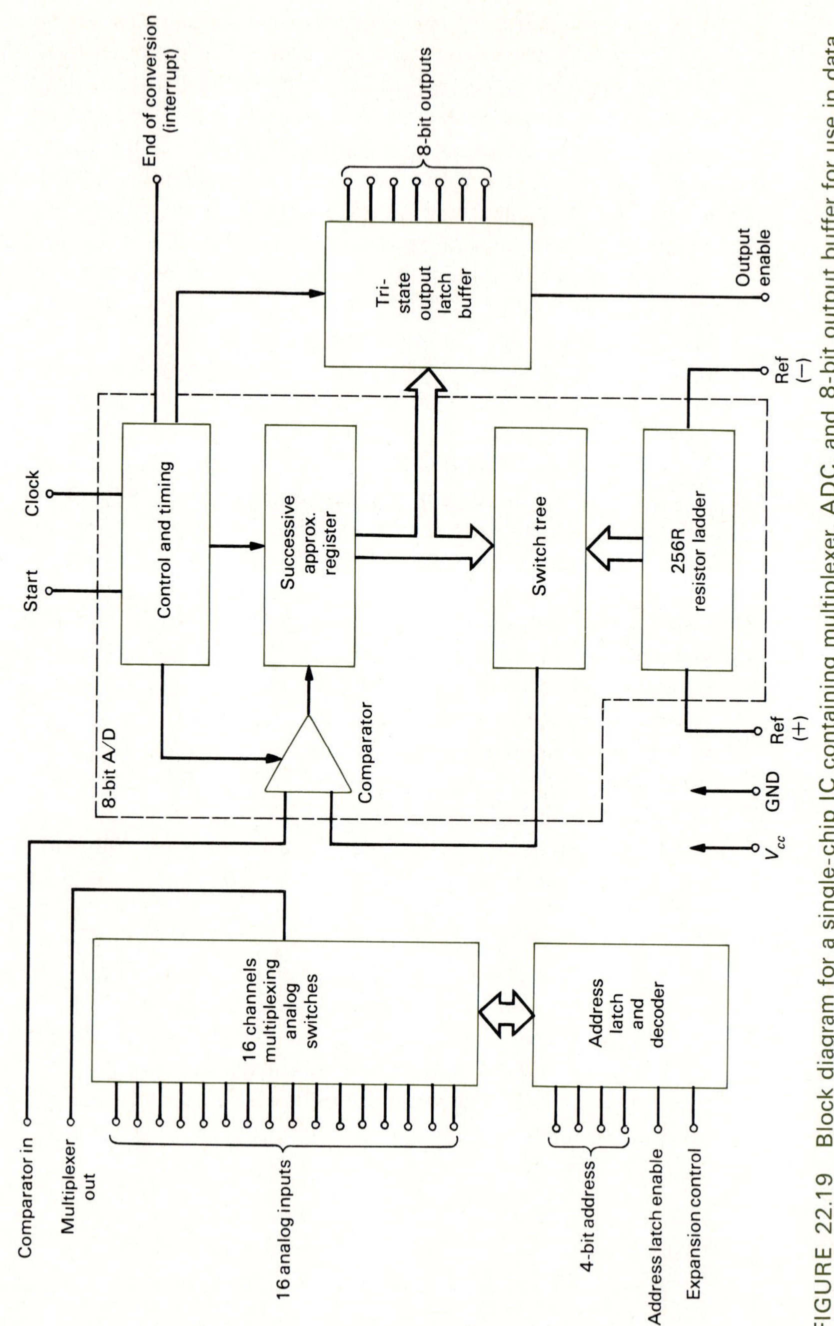

FIGURE 22.19 Block diagram for a single-chip IC containing multiplexer, ADC, and 8-bit output buffer for use in data acquisition. (Courtesy of National Semiconductor Corp.)

acquisition systems are designed to be compatible with microprocessors or other digital systems for processing the measured data. The output of the system shown in Fig. 22.19 is compatible with an 8-digit microprocessor.

The output of the ADC stage may be used in a variety of applications. The most obvious are those that truly process the information obtained. For example, assume that a data acquisition system such as illustrated in Fig. 22.18 is used to measure the parameters of electric motors being manufactured on a major motor manufacturer's assembly line. The parameters measured would as a rule be those specified by the

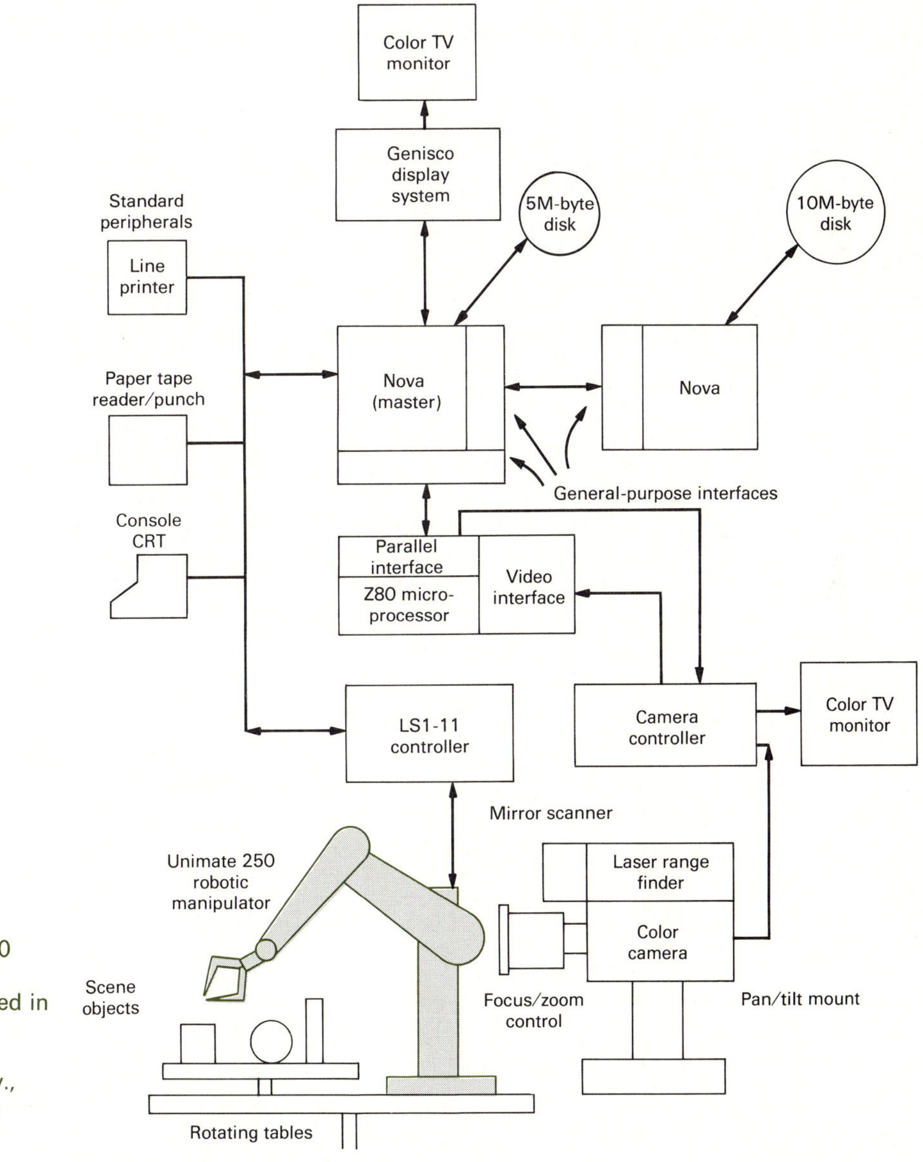

FIGURE 22.20 Visual data acquisition used in robot control. (Courtesy of Unimation Div., Westinghouse Corp.)

standards agencies, such as the National Electrical Manufacturers Association (NEMA). Thus, for an induction motor, the measurements might include the voltage, current, and power in all phase windings, the speed, the temperature at selected motor regions, the output torque or power, the response to high-potential insulation tests, and perhaps the mechanical vibration at the no-load and full-load conditions. Many of these measurements might be displayed at the instant of measurement on analog or digital meters. All measurement data would be sent on through the amplification, multiplexing, and ADC stages to a microprocessor. Typical ways in which the microprocessor would process these data would be to calculate such parameters as average voltages and currents, RMS voltages and currents, average power and torque and efficiencies and slip; to compare the readings of a given motor with previous motors, establish trends, and note unusual changes in any parameters; and to record and store details about the machine rating, the various verifications of the manufacturing process made along each motor's assembly stages, and perhaps even such marketing data as cost and type of customer. As indicated in Fig. 22.18, the output of the microprocessor can be used in a variety of ways. The obvious output is some form of data storage—a printout of all of the data processing calculation results, for example, or storage on magnetic tape or magnetic disks. But in many cases the microprocessor output is used in a more dynamic mode, such as to alert manufacturing personnel about a deterioration in certain specific materials, tolerances, or other manufacturing processes—perhaps even to control modifications and changes in materials or manufacturing processes in an adaptive manner—or to transmit all of the measured data to a larger computer system for subsequent analysis of long-term corporate trends in costs, sales potential, energy improvement, and so forth.

One of the most interesting applications of data acquisition systems is where the output of the microprocessor in Fig. 22.18 is fed into a robot or other type of automatic processor. The sensors or transducers for robot systems tend to be of the visual type—sensing position, distance, three-dimensional location, color, etc.—rather than of the more prosaic type that indicates such parameters as voltage, current, or temperature. Figure 22.20 illustrates a very generalized block diagram of a robot information acquisition system.

Perhaps the most sophisticated of all applications are those being carried on in outer space. The environmental and reliability constraints of outer space greatly exceed those on earth, but the concepts of data acquisition, processing, control, and storage remain generally the same as we have presented here.

22.4 ARITHMETIC DIGITAL SYSTEMS

One of the earliest and still one of the principal functions of digital circuits is to perform arithmetic calculations. Computer arithmetic is performed in the binary system or in alternative binary codes related to the binary system, a few of which will be discussed in Sec. 22.4.5. The purpose of the following discussion is to illustrate the use of digital circuits to perform arithmetic operations. The reader may wish to review binary arithmetic before proceeding with this topic.

22.4.1 Addition

Addition in any number system results in two classes of sums: one with a *carry* to the next column, such as 6 + 8 in decimal, and one without a carry. In binary, there is only one type in each class: (1 + 1) which has a carry of 1 and a *sum* of 0, and (0 + 1) which has no carry and a sum of 1. Two digital circuit configurations are required to perform these two classes of summations; these configurations are known as the *full adder* and the *half-adder*, respectively. The complete binary adder for adding two n-bit numbers consists of one half-adder and $(n - 1)$ full adders. Figure 22.21 shows the truth table and basic schematic circuit diagram for a half-adder; Fig. 22.22 illustrates the same for the full adder. It is seen that the half-adder accepts as inputs only the two numbers to be added, X and Y, whereas the inputs to the full adder are X, Y, and C_{in}, where C_{in} represents the carry input. To perform binary addition, the half-adder is used to add the *least significant digits* of X and Y since there will never be a carry input with these digits. We will denote the least significant digits by X_0 and Y_0. The carry output of the half-adder is fed into a full-adder stage which also has the inputs X_1 and Y_1, that is, the digit in the second column from the right in each number; the carry from this stage is fed into the third stage, a full adder, which also has the digits X_2 and Y_2 as inputs; and so on. This process is illustrated in block diagram form in Fig. 22.23, where each block represents the complete circuit as shown in either Fig. 22.21 or Fig. 22.22. The sum output of each stage is read out and represents the arithmetic sum for that column of digits.

Inputs		Outputs	
X	Y	S	C
0	0	0	0
0	1	1	0
1	0	1	0
1	1	0	1

(a)

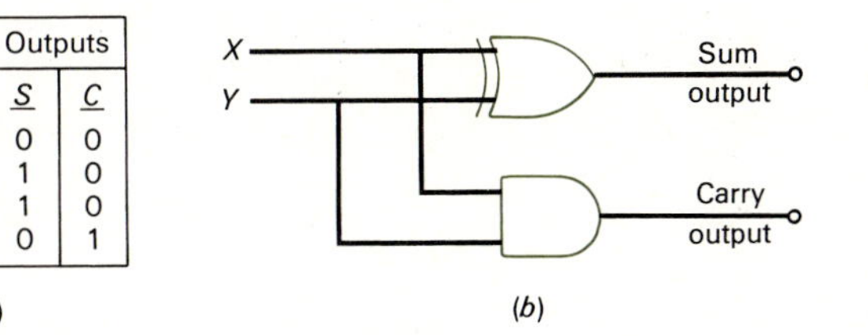

(b)

FIGURE 22.21 Binary half-adder: (a) truth table; (b) circuit diagram.

Inputs			Outputs	
C_{in}	X	Y	S	C_{out}
0	0	0	0	0
0	0	1	1	0
0	1	0	1	0
0	1	1	0	1
1	0	0	1	0
1	0	1	0	1
1	1	0	0	1
1	1	1	1	1

(a)

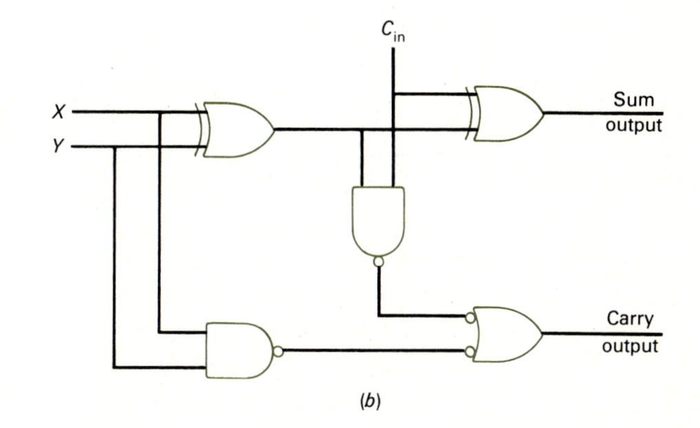

(b)

FIGURE 22.22 Binary full adder: (a) truth table; (b) circuit diagram.

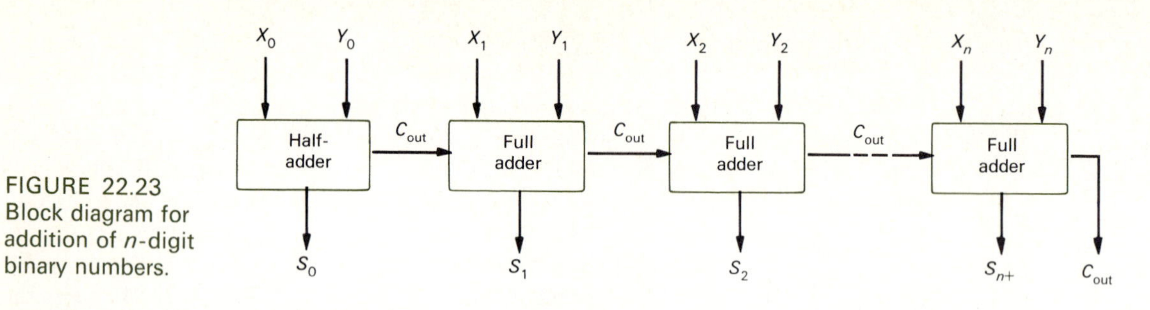

FIGURE 22.23 Block diagram for addition of n-digit binary numbers.

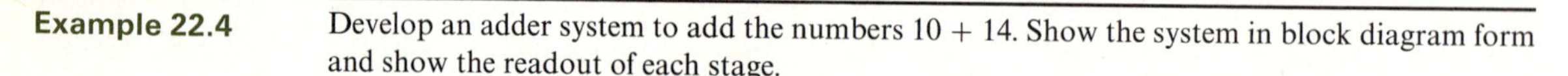

Example 22.4 Develop an adder system to add the numbers 10 + 14. Show the system in block diagram form and show the readout of each stage.

Solution Converting 10 and 14 to binary numbers gives 1010 and 1110, respectively. These are 4-bit numbers and will require four stages of adders, one half-adder and three full adders. These are shown in Fig. 22.24 along with the inputs, sums, and carrys of each stage.

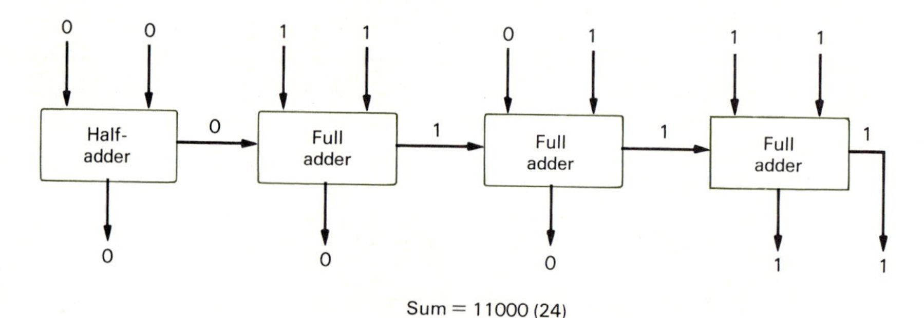

FIGURE 22.24 Example 22.4: block diagram and numerical values.

By developing Karnaugh maps from the truth table of the full adder (Fig. 22.22), we can find the boolean equations expressing the sum and carry outputs. These are, respectively,

$$S = X \oplus Y \oplus C_{in} \tag{22.9}$$

$$C_{out} = XY + C_{in}(X + Y) \tag{22.10}$$

22.4.2 Subtraction

The process for finding the difference of two binary numbers X and Y is quite similar to that for finding their sum. Two similar digital circuits are used, the half-subtracter for the least significant digit and the full subtracter for the remaining digits of the binary numbers. The circuit configuration for the half-subtracter is shown in Fig. 22.25; the truth table and circuit for the full subtracter are shown in Fig. 22.26. In these figures, B stands for *borrow* and B_{out} represents a number borrowed from the column to the left when required. The truth table of Fig. 22.26 is for the difference $X - Y$. The least significant digit difference, and hence the half-adder, requires no B_{in}. Note that when borrowing is required in the binary system, the difference performed is that of $10 - 1$, giving a difference D of 1 and a borrow out B_{out} of 1.

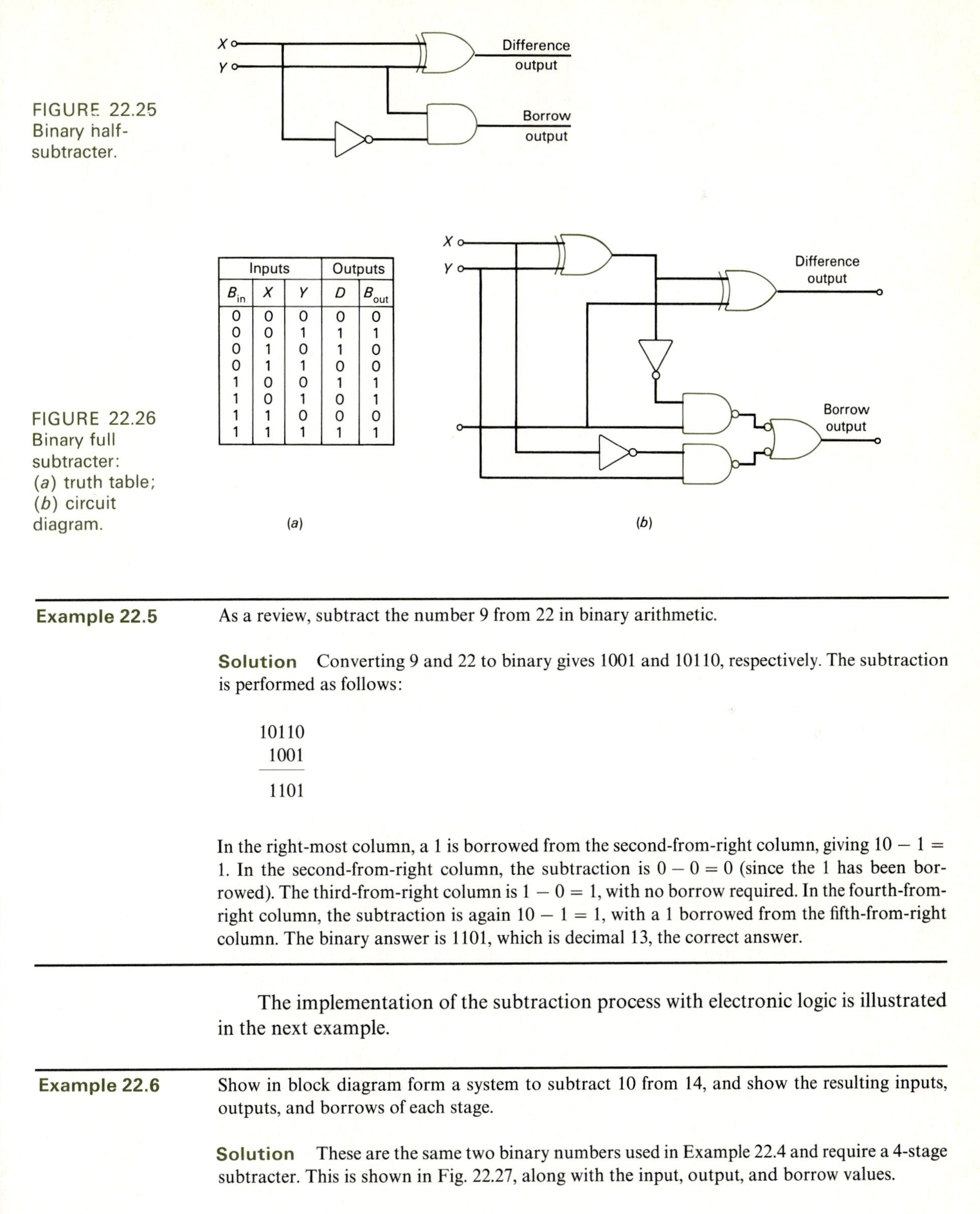

FIGURE 22.25 Binary half-subtracter.

Inputs			Outputs	
B_{in}	X	Y	D	B_{out}
0	0	0	0	0
0	0	1	1	1
0	1	0	1	0
0	1	1	0	0
1	0	0	1	1
1	0	1	0	1
1	1	0	0	0
1	1	1	1	1

FIGURE 22.26 Binary full subtracter: (*a*) truth table; (*b*) circuit diagram.

Example 22.5 As a review, subtract the number 9 from 22 in binary arithmetic.

Solution Converting 9 and 22 to binary gives 1001 and 10110, respectively. The subtraction is performed as follows:

```
10110
 1001
-----
 1101
```

In the right-most column, a 1 is borrowed from the second-from-right column, giving 10 − 1 = 1. In the second-from-right column, the subtraction is 0 − 0 = 0 (since the 1 has been borrowed). The third-from-right column is 1 − 0 = 1, with no borrow required. In the fourth-from-right column, the subtraction is again 10 − 1 = 1, with a 1 borrowed from the fifth-from-right column. The binary answer is 1101, which is decimal 13, the correct answer.

The implementation of the subtraction process with electronic logic is illustrated in the next example.

Example 22.6 Show in block diagram form a system to subtract 10 from 14, and show the resulting inputs, outputs, and borrows of each stage.

Solution These are the same two binary numbers used in Example 22.4 and require a 4-stage subtracter. This is shown in Fig. 22.27, along with the input, output, and borrow values.

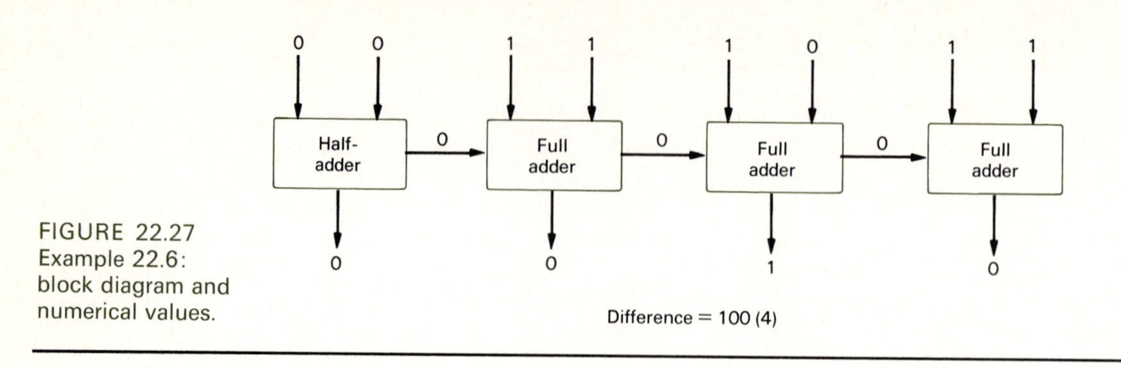

FIGURE 22.27
Example 22.6: block diagram and numerical values.

22.4.3 Signed Integers and Word Length

The process of subtraction introduces the concept of negative numbers and how they can be handled in digital circuits. Obviously, it is important to be able to perform addition, subtraction, multiplication, and division using negative numbers or mixtures of negative and positive numbers in calculators, computers, and similar digital systems. For instance, if the number Y were subtracted from the number X in Example 22.6, we would end up with a negative number, in this case -4. But it is seen from examination of the block diagram of Fig. 22.27 that if X and Y were interchanged, there would be a borrow output from the right subtracter, which has no meaning, and that the number represented by the four difference (D) signals would be binary 1100, or decimal 12, which is also meaningless. One method used on some computers to solve this dilemma is to use another digit to represent the sign of the number. For the case we have just discussed, where we want to find the difference of $10 - 14$, the subtraction process would be performed just as in Example 22.6 and Fig. 22.27, but another digit would be carried along with both numbers (X and Y) and with the difference to indicate their sign, usually 0 for positive and 1 for negative.

In computer-type systems, the range of numbers that can be handled in arithmetic calculations and other mathematical manipulation and processing is a function of what is known as *word length.* The word length is the number of digits or bits making up the maximum binary number that can be processed with the hardware from which the computing system is constructed. Specifically, the temporary storage devices or registers in a computing system are sized exactly according to word length. Thus, an 8-bit word length requires an 8-bit register for storing 8 binary digits. An n-digit binary number can represent up to 2^n decimal numbers; an 8-bit word length can therefore process positive decimal numbers up to $256 - 1$ (since 0 is one of the numbers), or 255. It is customary to use the most significant bit (MSB) of the word to represent the sign of a number, a process called the *signed modulus* method for representing negative numbers. Thus, in the previous examples the number X, decimal 14 or binary 1110, would be stored in an 8-bit register as 00001110; the decimal number -14 would be stored as 10001110.

There are several attributes of the signed modulus method of sign representation that are not desirable. First of all, using one digit to represent sign reduces the number of decimal numbers considerably (from 255 to 127 in the case of 8-bit word length), although the range of numbers is essentially the same (from 6 −127 to +127). Second, a signed modulus positive number is not the additive inverse of the negative number of the same value; that is, the two do not add up to zero. Also, the signed modulus numbering system results in the rather strange situation of having two zeros—a negative zero and a positive zero. A method which overcomes some of these disadvantages of signed modulus notation and which is more widely used in computer systems is known as *2s complement notation.*

The 2s complement method of forming signed binary numbers is somewhat analogous in methodology to the 9s complement method used in decimal number subtraction. To form a decimal 9s complement, each digit in the decimal number is replaced by a number found by subtracting the digit from 9. To form a 2s complement in binary, each digit of the binary number is replaced by a number found by subtracting the digit from 1. This has the effect of replacing all 0s by 1s and all 1s by 0s. The replacement number is called the *complement* of the original number. *Negative numbers in 2s complement notation are found by adding 1 to the complement of the positive number.* Thus, to determine the binary representation of decimal −14, the normal binary representation of 14 is first found. For an 8-bit word length, this is 00001110; the complement is 11110001; adding 1 gives 11110010, which is the 2s complement notation for decimal −14. One representation for zero is used. The list below shows 2s complement notation for decimal numbers between −10 and +10 in an 8-bit word length system:

TABLE 22.1 2s Complement Notation

Binary	Decimal equivalent
00001010	10
00001001	9
00001000	8
00000111	7
00000110	6
00000101	5
00000100	4
00000011	3
00000010	2
00000001	1
00000000	0
11111111	−1
11111110	−2
11111101	−3
11111100	−4
11111011	−5
11111010	−6
11111001	−7
11111000	−8
11110111	−9
11110110	−10

It is seen that the negative number is the additive inverse of the positive number (the carry from the MSB is thrown away when adding inverses). The range of number representation is about the same as for the signed modulus method, $2^{(n-1)} - 1$ on the positive side and $2^{(n-1)}$ on the negative side. For an 8-bit word length, this gives a range of from +127 to −128, which is 256 when zero is included.

Example 22.7 Find the 2s complement representation of −31, assuming an 8-bit word length.

Solution Positive 31 has a binary representation of 00011111. The complement is 11100000. Adding 1 gives 11100001.

Example 22.8 Subtract 14 from 10 using 2s complement notation for an 8-bit word length.

Solution This is equivalent to adding −14 to 10. Negative 14 was found to be 11110010 above; adding this to positive 10 gives

$$\begin{array}{r} 11110010 \\ 00001010 \\ \hline 11111100 \end{array}$$

which is seen to be −4 in Table 22.1.

Obviously, with finite word lengths in a computing system, the processes of adding or subtracting can result in numbers larger than can be represented by the word length. This causes a condition known as *overflow* when the number is larger than the positive range of allowable numbers and *underflow* when the number is smaller than the negative range. For example, in a system with an 8-bit word length, adding the decimal numbers 100 and 50 gives decimal 150, which is beyond the allowable range of positive numbers. In 2s complement notation, this would result in 01100100 + 00110010 = 10010110. This is equivalent to the decimal number −106, which is of course a ridiculous answer. To detect such situations, the computing system must contain circuitry that indicates when an overflow or underflow exists. This circuitry is based upon the logic that goes along with the mathematical process. In the process of addition, for example:

1. If the two numbers being added are of opposite sign, there can be no overflow.
2. If two positive numbers are being added and the sum is negative, an overflow condition exists (as illustrated above).
3. If two negative numbers are being added and the sum is positive, an underflow condition exists.

Thus, overflow-underflow detection requires detection of the MSB (MSB is 1) (MSB is 0) of the sum and the sign of the numbers to be added. The design of a logic circuit to detect these conditions is given as a problem at the end of this chapter. Overflow-underflow can occur during subtraction, too, and similar logic relations can be developed for such conditions.

22.4.4 Additional Add/Subtract Circuits

The adders that are shown in Figs. 22.21 to 22.24 are parallel adders since the inputs to each stage are processed nearly simultaneously. However, a finite time is required for each carry to propagate from one stage to the next. An adder circuit known as the *look-ahead carry* anticipates whether there is to be a carry or not across all of the stages. The actual addition is performed by the same hardware as shown in Figs. 22.21 and 22.22. The carry anticipation circuitry adds considerable circuit complexity to an adder and increases its cost, but it greatly increases the response time of the adder.

Figure 22.28 illustrates the data sheet of a CMOS 4-bit adder with the look-ahead carry circuitry; note the 160-ns response time. Separate look-ahead carry ICs are also available to add to existing adders to improve their response time. The carry anticipation circuitry is based upon an implementation of Eq. (22.10); the carry in the ith stage can be written

$$C_i = G_i + P_i C_{i-1} \tag{22.11}$$

where $G_i = X_i Y_i$ = the input to the ith stage, called the *generate term*. In Fig. 22.28, $P_i = A_i B_i$ and is called the *propagate term*; $C_{i-1} = C_{\text{in}}$ for the ith stage. The carry anticipation circuitry requires the inputs from each stage and the initial carry C_o (as shown) and from this information determines almost instantaneously what the carry from each stage to the next will be. The truth table of Fig. 22.28 further defines the operation of this circuit.

It can be seen from Figs. 22.21 to 22.27 that the hardware implementation of adders and subtracters is very similar. It is possible to use the basic full-adder circuit to perform both addition and subtraction operations with very little addition of hardware. Such combined circuits are called, not surprisingly, adder/subtracters. Figure 22.29 illustrates the basic elements of a 4-bit adder/subtracter circuit. The input to the C_o terminal determines the arithmetic mode of the circuit. When C_o is zero, the circuit performs as an adder since a zero on the left terminals of the XORs causes the XORs to operate as normal or gates and the inputs on the Y terminals are unchanged. The initial carry is 0 in this case. When the input to the C_o terminal is 1, the circuit operates in the subtraction mode as follows: Placing a 1 on the left terminals of the XORs causes the complement of the digits of Y to be inputted to the adders; a 1 from C_o is added to the complement, giving the 2s complement notation negative value of Y. Thus, negative Y is added to X, and this addition is of course the difference $X - Y$. The digits shown in Fig. 22.29 are for the values $X = 1001$ (9) and $Y = 0101$ (5). The addition process forms the sum $1001 + 0101 = 1110$ (14). The subtraction process first forms the 2s complement negative of Y, which is $1010 + 1 = 1011$; this is added to X, giving 10100; the left digit of this number is the carry out and appears on the terminal $C4$ in Fig. 22.29 and is ignored. The remaining four digits appear on the S terminals and are 0100; this is binary 4 and is the correct answer.

MOTOROLA

DESCRIPTION — The SN54LS/74LS83A is a high-speed 4-Bit Binary Full Adder with internal carry lookahead. It accepts two 4-bit binary words (A_1 — A_4, B_1 — B_4) and a Carry Input (C_0). It generates the binary Sum outputs Σ_1 — Σ_4) and the Carry Output (C_4) from the most significant bit. The LS83A operates with either active HIGH or active LOW operands (positive or negative logic). The SN54LS/74LS283 is recommended for new designs since it is identical in function with this device and features standard corner power pins.

4-BIT BINARY FULL ADDER WITH FAST CARRY

LOW POWER SCHOTTKY

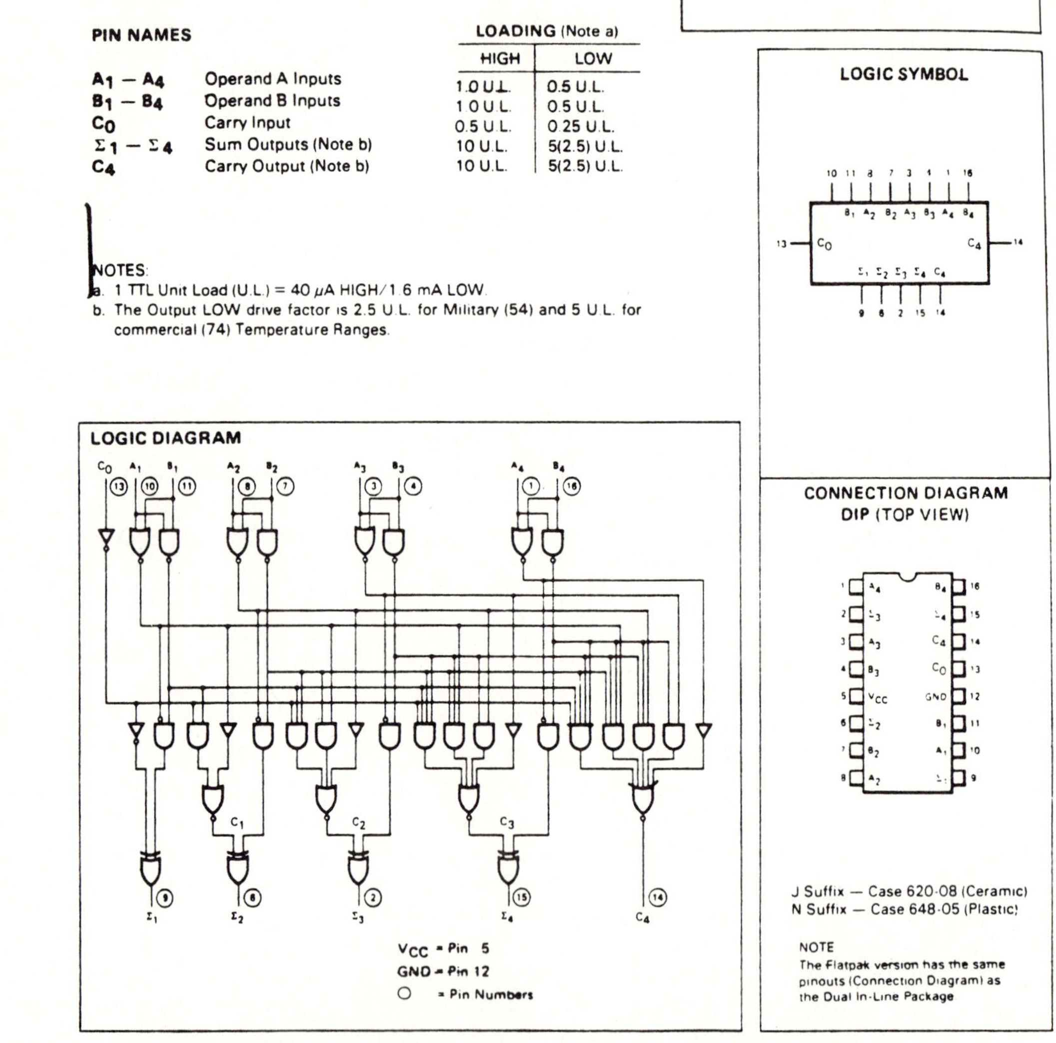

PIN NAMES

		LOADING (Note a) HIGH	LOW
A_1 — A_4	Operand A Inputs	1.0 U.L.	0.5 U.L.
B_1 — B_4	Operand B Inputs	1.0 U.L.	0.5 U.L.
C_0	Carry Input	0.5 U.L.	0.25 U.L.
Σ_1 — Σ_4	Sum Outputs (Note b)	10 U.L.	5(2.5) U.L.
C_4	Carry Output (Note b)	10 U.L.	5(2.5) U.L.

NOTES:
a. 1 TTL Unit Load (U.L.) = 40 μA HIGH/1.6 mA LOW.
b. The Output LOW drive factor is 2.5 U.L. for Military (54) and 5 U.L. for commercial (74) Temperature Ranges.

FIGURE 22.28 Portions of data sheets for a typical CMOS 4-bit full adder. (Courtesy of Motorola Semiconductor Products.)

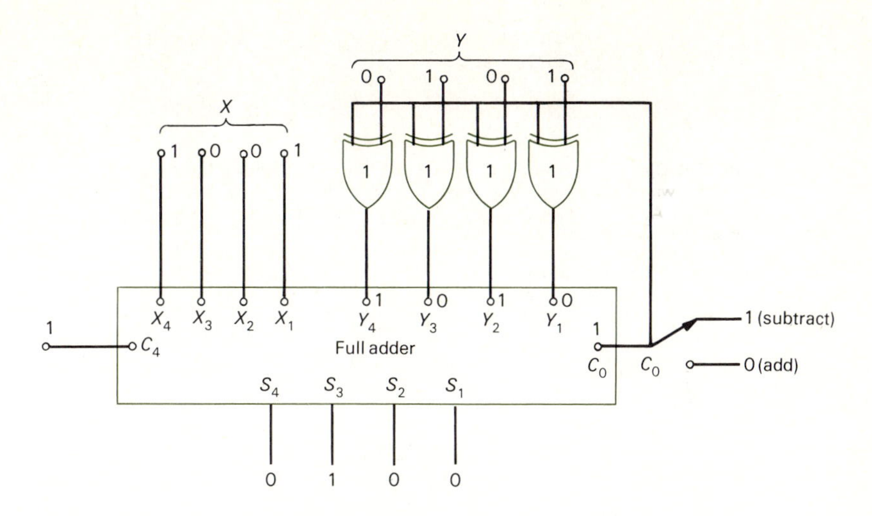

FIGURE 22.29
Block diagram for adder/subtracter.

22.4.5 Alternative Binary Codes

In various types of digital circuits, alternative representations of binary numbers are used. These *binary codes* are usually used in order to simplify hardware or software or both, or to make more efficient use of the word length limitations. Codes are also used to improve or simplify the human-computer interface, since binary representation is not the most suitable form for writing a computer program or feeding in data to a computer.

The *Gray code*, used in many computer systems and digital circuits (DCs), is listed here primarily for references purposes. In this code, an alternative to conventional binary counting, only one digit (or bit) changes between successive numbers. This feature simplifies the logic and hence the electronic circuitry in some applications.

The *octal code* (*base 8*) is used to reduce the problems of dealing with long strings of binary 1s and 0s. It is a shorthand for replacing three binary digits with a single octal number from 0 to 7. For example, a 6-bit binary word would be split into two groups of three; thus, 100 010 would be represented in octal as 42 and would be written as 42_8. The octal number 576_8 is equal to 101111110 in binary. Octal numbering follows the rules for conventional numbering in base 8 and is not listed here.

Hexadecimal coding is widely used in computer systems with 8- and 16-bit word lengths. Hexadecimal numbering, shown in Table 22.2, uses the decimal digits plus the first six letters of the alphabet. In hexadecimal coding, binary numbers are divided into groups of four binary digits, and each group of four is assigned a hexadecimal digit. For example, binary 1100 0101 becomes (from Table 22.2) *C*5 and is often written $C5_{16}$ because it is essentially a base 16 numbering system with some new definitions of numbers from 10 to 15. Conversely, $A2F_{16}$ is converted to binary as 1010 0010 1111.

The *binary-coded decimal* (*BCD*) *system* codes each digit of *decimal* numbers with its *binary* equivalent. For example, the BCD equivalent of 65 is 0110 0101. Although

TABLE 22.2 The Gray Code and Hexadecimal

Decimal number	Binary number	Gray code number	Hexadecimal
0	0000	0000	0
1	0001	0001	1
2	0010	0011	2
3	0011	0010	3
4	0100	0110	4
5	0101	0111	5
6	0110	0101	6
7	0111	0100	7
8	1000	1100	8
9	1001	1101	9
10	1010	1111	A
11	1011	1110	B
12	1100	1010	C
13	1101	1011	D
14	1110	1001	E
15	1111	1000	F

this is a type of binary notation for decimal numbers, it is not the same as conventional binary counting, as seen by converting decimal 65 to the conventional binary number—which is 1000001. The decimal number 472 becomes 0100 0111 0010 in BCD. Likewise, BCD 100101100101 equals 965 in decimal. A number in BCD code is much easier to produce than the pure binary equivalent of a decimal number, which requires the tedious process of continuous divide-by-2. The reverse process is also much simpler; for example, the super-long binary string 0110001110000011 0110 is converted to BCD by splitting it up into groups of four binary digits (starting on the right) and assigning the decimal equivalent number to each group, which gives 5 3 8 3 6. On the other hand, it can be seen that BCD "wastes" a lot of digits since four binary digits can count up to decimal 15, whereas BCD codes only the decimal digits from 0 to 9. Thus, as we have seen, conventional binary 8-bit words cover a range of decimal numbers from 0 to 255, whereas an 8-bit word in BCD can cover a range of only 0 to 99. However, BCD is widely used in many computing systems, particularly calculators. Note that this waste of digits does not occur in octal and hexadecimal coding.

BCD adders and subtracters must differ in circuitry from conventional binary systems (such as illustrated in Figs. 22.21 to 22.29) owing to the differences in the meaning of digit magnitude in the two systems. For example, 110111 in conventional binary represents decimal 55; 00110111 in BCD represents decimal 37. An adder/subtracter must be able to detect this difference in the meaning of the digits. The basic BCD adder can be constructed from the same circuit elements as used in the basic binary adder, but logic must be added to the BCD adder to cause the sum to be presented in BCD format rather than conventional binary. This is illustrated by the 4-bit BCD adder in Fig. 22.30. The full-adder blocks in this diagram are the conven-

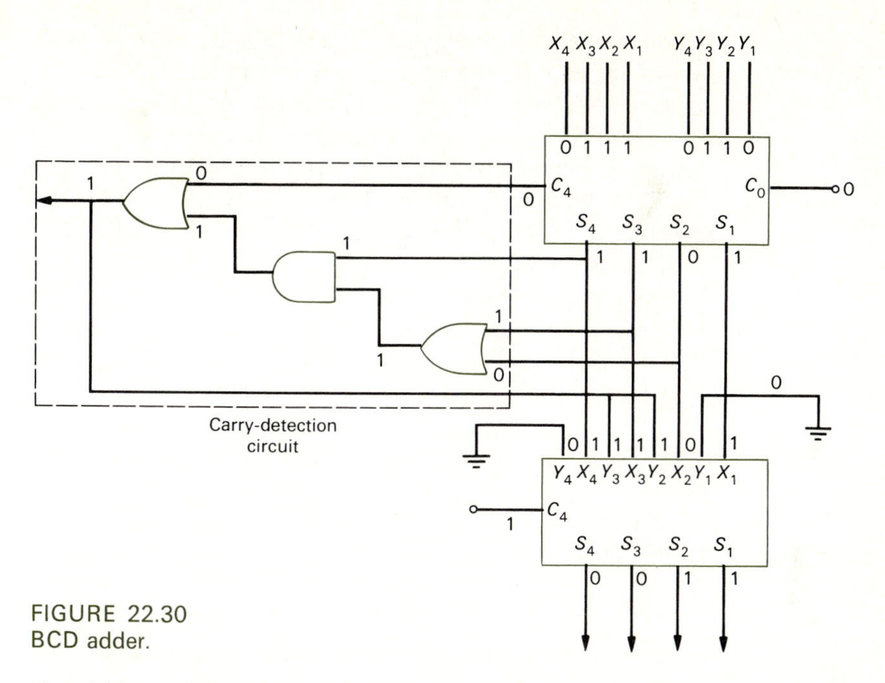

FIGURE 22.30
BCD adder.

tional binary full adders of Fig. 22.22. The function of the carry-detection circuit in Fig. 22.30 is to detect when a sum of the upper full adder is greater than decimal 9 or binary 1001. Thus, the operation of a BCD basic adder is as follows: One full adder accepts the sum of two BCD numbers and a carry from the addition of BCD digits of lower significance; the output from this adder of the three most significant digits is supplied to the carry-detection circuitry; if the values of these three digits indicate a sum greater than decimal 9 (note that the least significant digit doesn't matter in this decision), the lower full adder is modified to give the correct BCD summation; note that the lower-adder carry C_4 represents the left digit of summation (1 or 0001 in Fig. 22.30) and is applied to the next stage of BCD digit addition, if such exists. If the summation from the upper full adder is decimal 9 or less, there is no modification of the lower full adder and the summing process is equivalent to conventional binary summation. In Fig. 22.30, the BCD representations of decimal 7 (0111) and decimal 6 (0110) are being added. A conventional binary adder would give the sum 1101 (13), which of course is correct. But the BCD representation of decimal 13 is 0001 0011. The carry-detection circuitry of a BCD adder accomplishes this change in output, as shown in Fig. 22.30.

22.4.6 Multiplication

There are many different circuit configurations used to perform binary multiplication. These are generally based on bit-by-bit multiplication and storing with the proper column alignment in a shift register. Some computing systems perform multiplication by means of software rather than hardware, and multiplication can of course be performed by means of repetitive summations, using the adder circuits described earlier in this chapter. However, with the decreasing cost profile of ICs, the use of hardware multiplication has become very inexpensive and has resulted in the replacement of some analog multipliers by binary multipliers in certain applications.

Figure 22.31*a* illustrates the circuit configuration for a 1-bit, or "bit-by-bit," multiplier. This circuit, which is composed of CMOS AND, OR, and XOR gates, which should be quite familiar to the reader by now, is the core of the IC chip forming the 2-bit multiplier and adder shown in Fig. 22.31*b*. The mathematical equations performed by the IC are listed in Fig. 22.31*c*, and a partial data sheet listing some of the characteristics and input-output parameters are shown in Fig. 22.31*d*. This chip and many

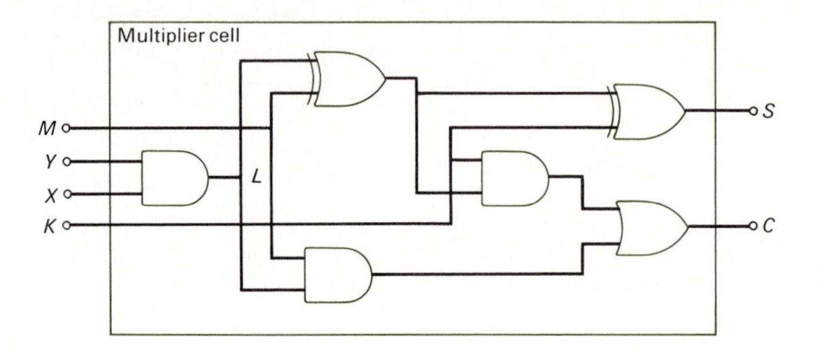

(*a*)

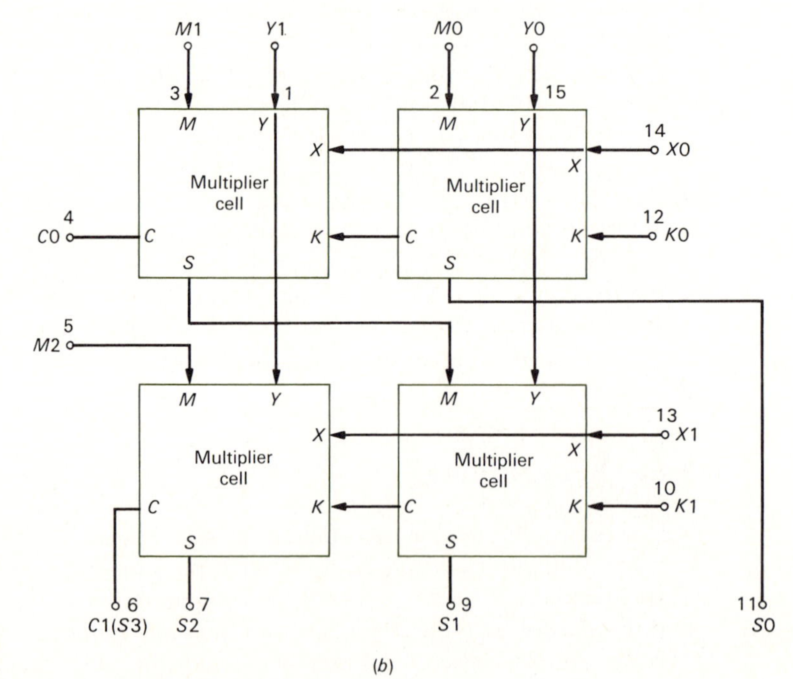

(*b*)

FIGURE 22.31 CMOS multiplier circuit and specifications: (*a*) circuit diagram of basic 1-bit multiplier; (*b*) 2-bit multiplier using basic circuit of (*a*).

EQUATIONS

$S = (X + Y) + K + M$

Where:

$\times$ Means Arithmetic Times.

$+$ Means Arithmetic Plus.

$S = S3\ S2\ S1\ S0$, $X = X1\ X0$, $Y = Y1\ Y0$.

$K = K1\ K0$, $M = M1\ M0$ (binary numbers).

Example:

Given: $X = 2(10)$, $Y = 3(11)$
$K = 1(01)$, $M = 2(10)$

Then: $S = (2 \times 3) + 1 + 2 = 9$
$S = (10 \times 11) + 01 + 10 = 1001$

Note: $C0$ connected to $M2$ for this size multiplier.
See general expansion diagram for other size multipliers.

(*c*)

2-BIT BY 2-BIT PARALLEL BINARY MULTIPLIER

The MC14554B 2 x 2-bit parallel binary multiplier is constructed with complementary MOS (CMOS) enhancement-mode devices. The multiplier can perform the multiplication of two binary numbers and simultaneously add two other binary numbers to the product. The MC14554B has two multiplicand inputs ($X0$ and $X1$), two multiplier inputs ($Y0$ and $Y1$), five cascading or adding inputs ($K0$, $K1$, $M0$, $M1$, and $M2$), and five sum and carry outputs ($S0$, $S1$, $S2$, $C1$ [$S3$], and $C0$). The basic multiplier can be expanded into a straightforward m-bit-by-n-bit parallel multiplier without additional logic elements.

Application areas include arithmetic processing (multiplying/adding, obtaining square roots, polynomial evaluation, obtaining reciprocals, and dividing), fast Fourier transform processing, digital filtering, communications (convolution and correlation), and process and machine controls.

- Diode protection on all inputs
- All outputs buffered
- Quiescent current = 5.0 nA typical @ 5 V dc
- Straightforward m-bit-by-n-bit expansion
- No additional logic elements needed for expansion
- Multiplies and adds simultaneously
- Positive logic design
- Supply voltage range = 3.0 to 18 V dc
- Capable of driving two low-power TTL loads, one low-power Schottky TTL load, or two HTL loads over the rated temperature range

MAXIMUM RATINGS (voltages referenced to V_{SS})

Rating	Symbol	Value	Unit
DC supply voltage	V_{DD}	−0.5 to +18	V dc
Input voltage, all inputs	V_{in}	−0.5 to V_{DD} +0.5	V dc
DC current drain per pin	I	10	mA dc
Operating temperature range—AL device CL/LP device	T_A	−55 to +125 −40 to +85	°C
Storage temperature range	T_{stg}	−65 to +150	°C

(*d*)

FIGURE 22.31 CMOS multiplier circuit and specifications: (*c*) mathematical processes performed by multiplier; (*d*) data specifications.

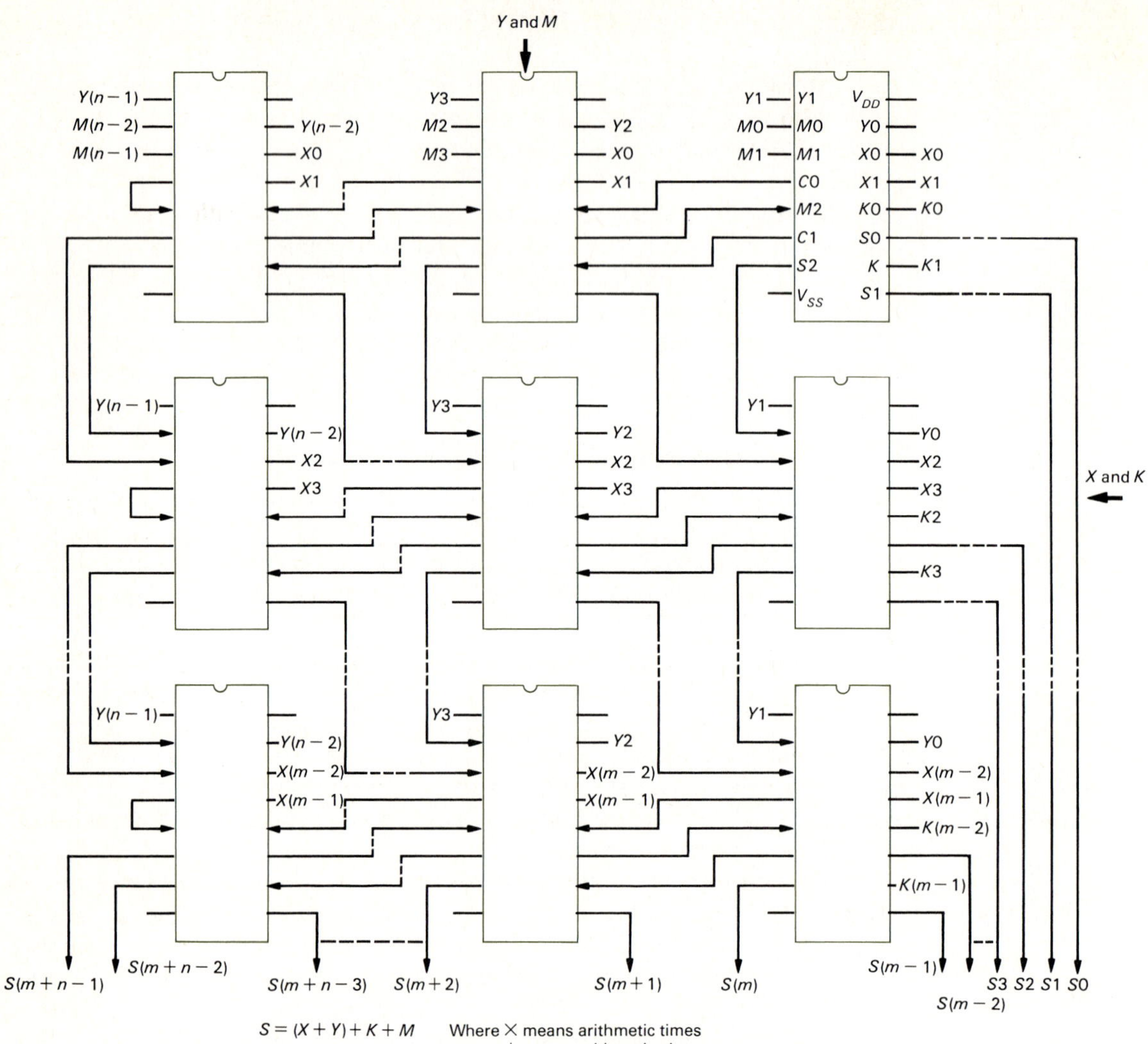

$S = (X \times Y) + K + M$ Where $\times$ means arithmetic times
$+$ means arithmetic plus

$S = S(m+n-1)S(m+n-2) \cdots S2\,S1\,S0$
$X = X(m-1)X(m-2) \cdots X2\,X1\,X0$, $Y = Y(n-1)Y(n-2) \cdots Y2\,Y1\,Y0$
$K = K(m-1)K(m-2) \cdots K2\,K1\,K0$ and $M = M(n-1)M(n-2) \cdots M2\,M1\,M0$
(Binary numbers)

Number of output binary digits $= m + n$
Number of packages $= mxn/4$ (for m or n or both odd, select next highest even number)

FIGURE 22.32 An *m*-bit–by–*n*-bit multiplier using the basic 2-bit multiplier of Fig. 22.31*b*. (Courtesy of Motorola Semiconductor Products Div.)

similar chips by other manufacturers perform the multiplying function in many microprocessors, calculators, and computers. The reader should work through the circuit of Fig. 22.31*a* and *b* to further understand the mechanism of binary multiplication and to observe how the most elemental binary multiplication—1×0 and 1×1—is performed. The outputs of the 2-bit multiplier—$S0$, $S1$, and $S3$—are the digits of the product and are generally stored in a register of some sort.

The basic 2-bit multiplier is commonly used in tandem with many other similar circuits to perform the multiplication of large binary numbers. Figure 22.32 shows the interconnections among the basic multipliers for an *m*-bit-by-*n*-bit multiplier 9.

Note that the basic 2-bit multiplier of Fig. 22.31 can be used for a variety of mathematical processes—division, obtaining square roots, etc. For further study of these more complex binary mathematical embodiments, consult Refs. 22.7 to 22.9.

22.5 MICROPROCESSORS

A microprocessor, also known as a *microcontroller*, is an assemblage of digital circuits used to perform a specific function. (Most of these circuits have been described in previous sections of this text.) A microprocessor consists of three basic elements: an arithmetic logic unit (ALU), short-term memory registers and accumulators, and a control block. A *microcomputer* is, in the simplest form, a microprocessor to which have been added on-chip (or on-board) memory location input-output (I/O) ports (or terminals), interrupt functions, and additional ROM (read-only memory). There are also secondary elements in microprocessors: buffers, used to accept input data and control signals; buses, which may be a common set of physical connections or a channel along which data can be sent; a small amount of memory or long-term data storage, usually for the purpose of storing the processor program or set of instructions; and a clock or timing control device.

The microprocessor is actually a miniaturization of the basic computing element of a computer, often called the *central processing unit* (CPU). With the addition of permanent data storage (memory) and facilities for communication with peripherals such as printers, video terminals, disk storage, and tape decks, the microprocessor becomes a minicomputer or microcomputer. A microprocessor is often called the "guts" or the "heart" of a computer system. Each microprocessor is supplied with an "instruction set," the software that contains the total list of instructions that can be executed by the microprocessor.

However, a microprocessor is also a stand-alone system and has found widespread use in almost every area of system control. A great many automobiles today have engines controlled by microprocessors. These microprocessors accept inputs from sensors which sense such parameters as vehicle speed, engine r/min, engine torque (or manifold air pressure), exhaust vacuum, and exhaust temperature and determine the best mode of engine (and sometimes transmission) operation to minimize exhaust pollutants and maintain good vehicle efficiency at each condition of vehicle speed and torque. In the chemical and petroleum industries, as well as in similar process control industries, microprocessors control many refining, mixing, and other processes. Microprocessors are also widely used in operations as diverse as the control of steel- or aluminum-rolling processes, robotic control (as seen in Fig. 22.20), food processing, and many, many others.

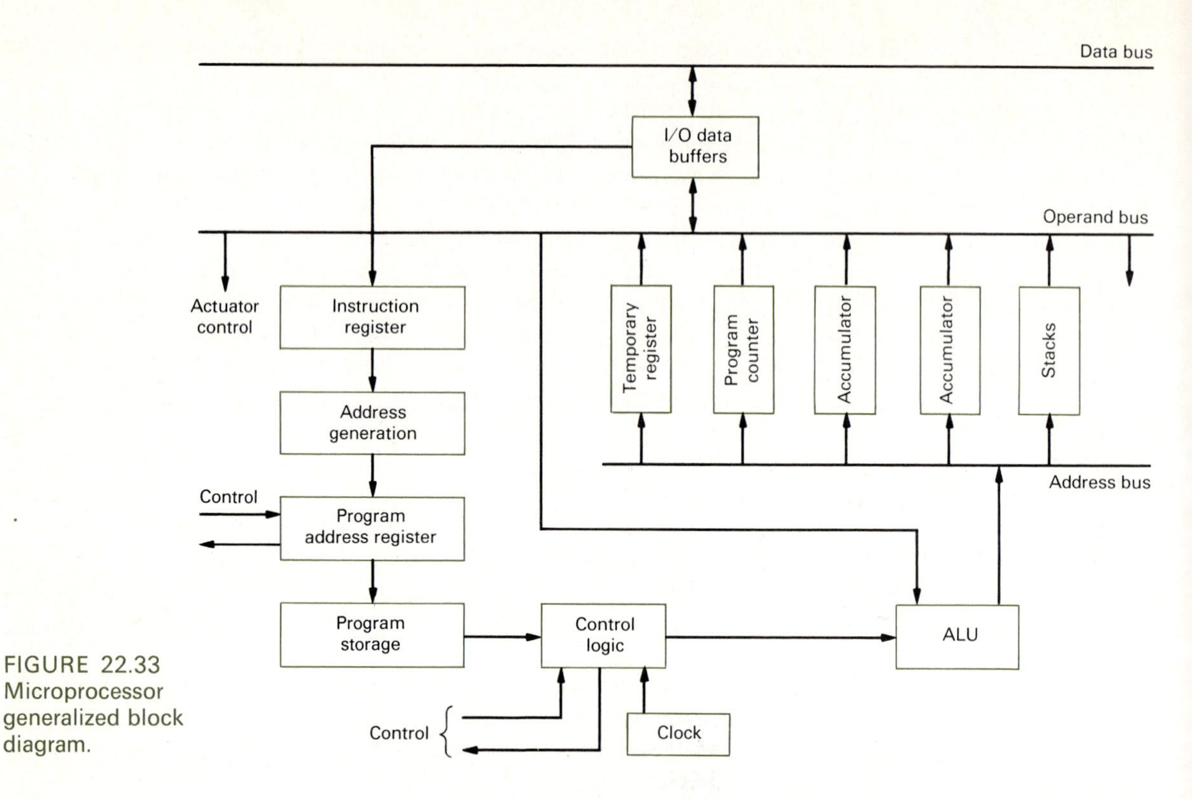

FIGURE 22.33 Microprocessor generalized block diagram.

One example of a block diagram representation of a microprocessor is shown in Fig. 22.33. There are virtually countless variations to microprocessor circuit configurations (often called system *architecture*), which is an indication of the great diversity of control and computing functions that can be performed by microprocessors. Most of the blocks in Fig. 22.33 should by now be familiar to the reader; the remainder may be described as follows:

The *arithmetic logic unit* (ALU) is basically an assemblage of adder/subtracter circuits (described in Sec. 22.4.4) plus some logic circuitry. The ALU carries out the fundamental arithmetic operations of adding and subtracting and the logical operations of AND, OR, or their complements. In some microprocessors, multiplication is also included in the ALU. The ALU accepts data from the data bus, processes it according to fixed instructions (from program storage) and/or to external control signals, and feeds the results into temporary storage, from which various external control and actuation functions are performed.

The *accumulators* are parallel storage registers (see Sec. 21.4.2) used for processing the work in progress. Addresses and data may be stored in them temporarily, and some are used for housekeeping functions.

The *stacks* provide temporary data storage in a sequential order—the last-in first-out (LIFO) order. The stacks are used mainly during the execution of subroutines and eliminate the need to read and write temporary data into memory.

The *program counter* is a register/counter which holds the *next* instruction address. It is also used in program control.

Microprocessors have instruction sets ranging from around 20 to several hundred instructions. These instructions—called *microprograms*—consist of a series of operations of both the arithmetic and logical types, and they include instructions for fetching and transferring data. The microprograms are stored in ROM and initiate the routines of the microprocessor. In our discussions of ROMs, we will be using these acronyms:

ROM—read-only memory

PROM—programmable ROM

EPROM—erasable PROM

EEPROM, or E^2—electronically erasable PROM

The E^2 PROM is the latest development in this sequence and gives the IC designer a much greater flexibility than the other three do, since it can be programmed and erased in situ (i.e., in the circuit in which it is operating) and hence as a portion of a chip. The other three devices must be removed from the circuit for one or both of these operations.

Microprocessors are categorized by the *word size* in bits; 1-, 4-, 8-, and 16-bit microprocessors are common. Usually, the larger the word size, the more powerful the processor. Another term related to word size is the *byte*, which is defined as 8 bits.

22.5.1 Microprocessor Buses and Internal Connections

The term *bus* has essentially the same meaning in microprocessing as in electric power systems—namely, a location where common electrical connections are made. In power systems, the word also connotates a point of fixed electric potential, which is not necessarily true of microprocessor buses. In power systems, buses are frequently constructed of long, parallel electrical conductors, often called *busbars*. A microprocessor bus is also constructed of parallel electrical conductors (although of vastly different current-carrying ability), with each conductor representing a digit in the word length. A bus is used to facilitate communication between more than two devices. A microprocessor system bus consists of three physical buses: the address bus, the data bus, and the control bus. In a power system, all incoming circuits to a bus are generally energized simultaneously. The types of circuits connected to microprocessor buses are registers, accumulators, or buffer circuits between the bus and the external memory or input-output circuits. It is required that these many incoming circuits to a bus be electrically connected to the bus individually, one at a time—a process called READ if the signal is incoming to the bus, and WRITE if it is outgoing from the bus. Every time that the system clock generates a READ or WRITE pulse, this causes the registers to feed signals into or out of a bus; at all other times, the registers normally

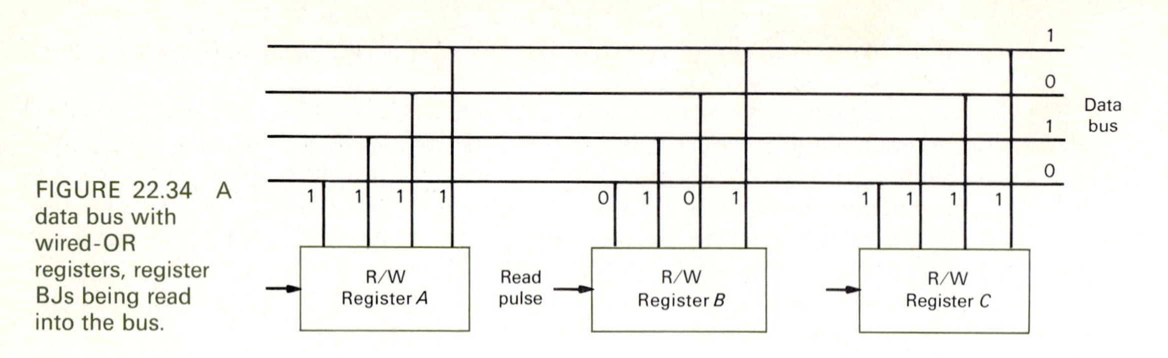

FIGURE 22.34 A data bus with wired-OR registers, register BJs being read into the bus.

read a binary 1 into all conductors of the bus. Therefore, if the selected register is to read in some binary 0s, there will be a conflict on the lines to which 0s are being conflict:

1. Wired-OR connection: In this technique, the registers and buses are designed such that a binary 0 always overrides a binary 1. This is illustrated in Fig. 22.34.
2. Tri-state output: This method adds a third operating state to the register. This is a state of very high impedance, which more or less disconnects the register from the bus when it is not being read.
3. Multiplexing: This is probably the most obvious solution, which is to connect each register to the bus on a time-shared basis only when it is being read.

There are several types of buses in most microprocessors: the *address bus*, which consists of up to 16 parallel lines into which binary codes can be fed; the *data bus*, with a number of lines equal to the word length; and sometimes an *operating bus*, used to transfer various internal operations and commands. The address bus is the path by which locations in memory or input-output peripheral devices are called or selected, a process somewhat analogous to dialing a telephone number. The data bus carries data between the microprocessor and external memory or input-output peripherals. The connections between the buses and internal components shown in Fig. 22.33 are by no means unique or the best or even the most desirable connections, and many other configurations are possible. In many cases, these so-called connections are not permanently wired connections but are formed for the program software stored in the system's ROM memory.

22.5.2 Addressing

Addressing is the process of locating bits of information in memory. Large microprocessors generally have an address bus of 16 lines or 16 bits. As we have seen in Sec. 22.4.3, n-binary bits can represent up to 2^n decimal numbers; therefore, a 16-bit address bus can address up to $2^{16} = 65{,}536$ memory locations, an 8-bit bus can address 256, and so on.

In many microprocessors and computers, addressing is performed with the hexadecimal code in order to avoid the use of long strings of binary numbers. The relationship between binary and hexadecimal is given in Table 22.2. Groups of hexadecimal (or binary) numbers which represent alphanumeric characters (numerals, letters, and other symbols) are called a *character code.* The most common code in microprocessors is the American Standard Code for Information Interchange (ASCII), which is a 7-bit code and is shown in Table 22.3. The relationship between the bit location, binary values, and hexadecimal values for the ASCII character ESC is:

	Left	Right
Binary	001	1011
Hexadecimal	1	B

ASCII words are frequently stored as 8-bit words. It is common to use a 1 or 0 for the left-most bit (MSB). Another practice is to use this extra bit as a means of checking that a word is unaltered as it is moved around the processor, in which case the extra bit is called the *parity bit.*

TABLE 22.3 ASCII 7-Bit Character Code*

Right		Left							
		Binary 000	001	010	011	100	101	110	111
Binary	Hexadecimal	Hexadecimal 0	1	2	3	4	5	6	7
0000	0	NUL	DLE	SP	0	@	P	/	P
0001	1	SOH	DC1	!	1	A	Q	a	q
0010	2	STX	DC2	"	2	B	R	b	r
0011	3	EXT	DC3	#	3	C	S	c	s
0100	4	EOT	DC4	S	4	D	T	d	t
0101	5	ENQ	NAK	%	5	E	U	e	u
0110	6	ACK	SYN	&	6	F	V	f	v
0111	7	BEL	ETB	.	7	G	W	g	w
1000	8	BS	CAN	(	8	H	X	h	x
1001	9	HT	EM	)	9	I	Y	i	y
1010	A	LF	SUB	*	:	J	Z	j	ż
1011	B	VT	ESC	+	;	K	[	k	{
1100	C	FF	FS	,	<	L	/	1	/
1101	D	CR	GS	—	=	M	]	m	}
1110	E	SO	RS	0	>	N	↑	n	~
1111	F	SI	US	/	?	O	←	o	DEL

* *Right* refers to the four right-most, or least significant, digits; *Left* to the left three, or most significant, digits.

22.5.3 Microprocessor Instructions

Instructions generally consist of 1, 2, or 3 bytes. The first byte contains the type of operation to be performed and is known as the *operating code*. The second and third bytes would contain an address and possibly some data. The three main types of instructions in a microprocessor are:

1 Data transfer—moving data in and out of registers from memory or inputs.

2 Arithmetic and logic—the normal mathematical and logic functions that we associate with calculators and computers.

3 Test and branch—the word or data is tested to verify its correctness and also to determine the next step in the program.

Symbols in common use for describing some of the instruction processes are shown in Table 22.4. Instructions are designated by a code name called a *mnemonic*. This code name is not an abbreviation; rather, it is a group of letters used to describe the instruction when programming a microprocessor in *assembler language*, which is a level of programming between the higher level languages, such as BASIC (beginner's all-purpose symbolic instruction code) or FORTRAN (formula translation), and the machine code itself. For example, in the Motorola 6800 microprocessor an instruction is described as follows:

ABA: Add accumulator *A* to accumulator *B*

$A \leftarrow (A) + (B)$

Binary code: 0001 1011

Hexadecimal code: 1*B*

The upper line is the instructions mnemonic and its description, the second line is the symbolic representation of what the instruction will do, and the third and fourth lines are the binary and hexadecimal codes for the instruction. The complete meaning of this particular instruction *ABA* is "Add the contents of accumulator *A* to the contents of accumulator *B*, placing the sum in *A* and leaving the contents of *B* unchanged." For example, if the contents of *A* were 01100010 and the contents of *B* were 00001111

TABLE 22.4 Symbols in Common Use

Symbol	Meaning
$\leftarrow$	Is transferred to
( ) [ ]	Contents of register or memory location named in the parenthesis
$\leftrightarrow$	Is exchanged with
$[A]'$, $(A)'$	Contents of register *A* are complemented

when this instruction command was made, the following changes in the two accumulators would take place:

	A	B
Before	01100010	00001111
After	01110001	00001111

A complete list of instructions in a small 1-bit microprocessor is given in Fig. 22.36 and will be discussed in the next section.

Each instruction of a microprocessor causes the microprocessor to go through a number of routines and operations—that is, through what we have described as a microprogram. A microprogram has two basic sets of routines known as *fetch* and *execute*. In the fetch portion of a microprogram, the instructions to be implemented are retrieved from memory and stored in registers of the microprocessor; the operating code of the instruction is always stored in a register known as the *index register*. In the execute portion, the mathematical, logical, transfer, or cycling processes called for in the instructions are executed to completion. A collection of microprograms arranged to perform one or more functions is known as a microprocessor program. For further information on microprocessor programming, consult Refs. 22.8 to 22.11.

22.5.4 Commercial Microprocessor Example

Figure 22.35 gives the data sheet and specifications for a commercially available microprocessor IC. This is a relatively small but powerful (as noted by its capabilities listed on the data sheet) 1-bit microprocessor used in industrial control applications. Data are fed into the microprocessor on a single binary channel by means of a pulse train, and information is contained in the modulation of the pulse train. In the block diagram shown in Fig. 22.35, the component designated as LU is actually the arithmetic logic unit (ALU) in the notation we have been using. The instruction set is contained in a 4-bit address, giving a maximum of $2^4 = 16$ instructions (shown at the bottom right of the figure). It is seen that this instruction set is aimed toward control applications rather than mathematical processing. Instructions are called by the oscillator signal $X1$. Signal levels at certain points in the microprocessor during the process of retrieving instructions are shown in Fig. 22.37. Groups of instructions are called in response to an external program or control algorithm stored in external ROM.

The Motorola MC14500B described in Figs. 22.35 and 22.36 is a 16-pin DIP. It has very low current drain, 1.5 mA during operation and only 5 nA in the quiescent or standby state. It is relatively fast, performing bitwise decision making in 1 μs or less, and is used in a variety of relay-ladder logic applications (such as discussed in Chap. 21) and in programmable controllers for industrial control. It can operate from supply voltages between 3 and 18 V, which is an advantage when operating from unregulated power supplies. Table 22.5 summarizes the data on the MC14500B and several other microprocessors in general use.

MOTOROLA

MC14500B

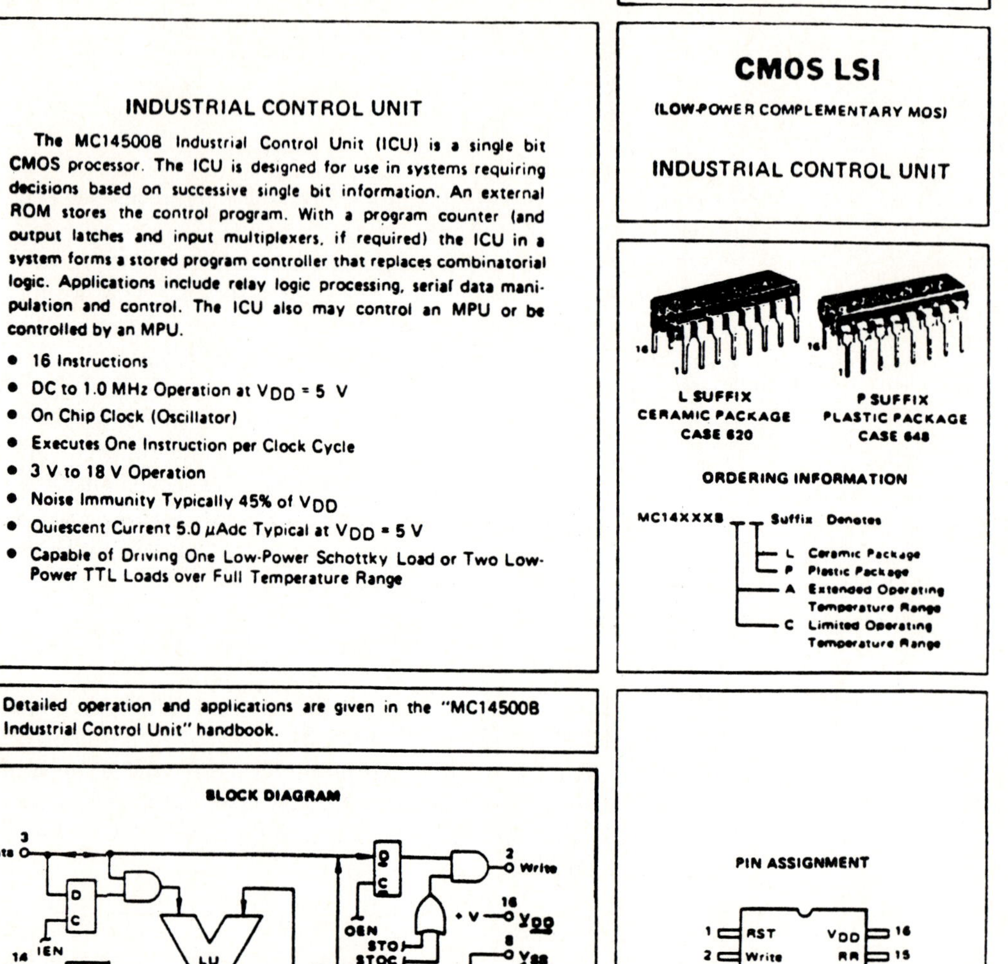

CMOS LSI

(LOW-POWER COMPLEMENTARY MOS)

INDUSTRIAL CONTROL UNIT

INDUSTRIAL CONTROL UNIT

The MC14500B Industrial Control Unit (ICU) is a single bit CMOS processor. The ICU is designed for use in systems requiring decisions based on successive single bit information. An external ROM stores the control program. With a program counter (and output latches and input multiplexers, if required) the ICU in a system forms a stored program controller that replaces combinatorial logic. Applications include relay logic processing, serial data manipulation and control. The ICU also may control an MPU or be controlled by an MPU.

- 16 Instructions
- DC to 1.0 MHz Operation at V_{DD} = 5 V
- On Chip Clock (Oscillator)
- Executes One Instruction per Clock Cycle
- 3 V to 18 V Operation
- Noise Immunity Typically 45% of V_{DD}
- Quiescent Current 5.0 μAdc Typical at V_{DD} = 5 V
- Capable of Driving One Low-Power Schottky Load or Two Low-Power TTL Loads over Full Temperature Range

L SUFFIX CERAMIC PACKAGE CASE 620

P SUFFIX PLASTIC PACKAGE CASE 648

ORDERING INFORMATION

MC14XXXB — Suffix Denotes

- L Ceramic Package
- P Plastic Package
- A Extended Operating Temperature Range
- C Limited Operating Temperature Range

Detailed operation and applications are given in the "MC14500B Industrial Control Unit" handbook.

BLOCK DIAGRAM

PIN ASSIGNMENT

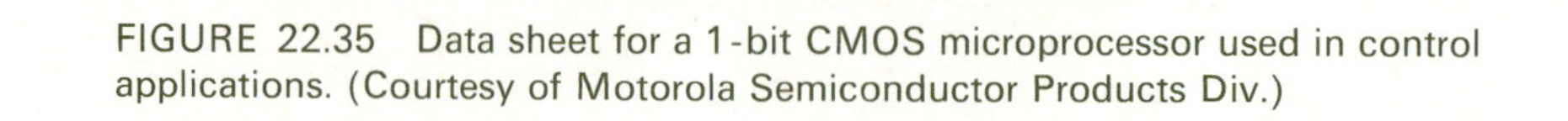

FIGURE 22.35 Data sheet for a 1-bit CMOS microprocessor used in control applications. (Courtesy of Motorola Semiconductor Products Div.)

Instruction Code		Mnemonic	Action
0	0000	NOPO	No change in registers. RR → RR, Flag O → ⎍
1	0001	LD	Load result register. Data → RR
2	0010	LDC	Load complement. $\overline{\text{Data}}$ → RR
3	0011	AND	Logical AND. RR·Data → RR
4	0100	ANDC	Logical AND complement. RR·$\overline{\text{Data}}$ → RR
5	0101	OR	Logical OR. RR + Data → RR
6	0110	ORC	Logical OR complement. RR + $\overline{\text{Data}}$ → RR
7	0111	XNOR	Exclusive NOR. If RR = Data, RR → 1
8	1000	STO	Store. RR → Data Pin, Write → ⎍
9	1001	STOC	Store complement. $\overline{\text{RR}}$ → Data Pin, Write → ⎍
A	1010	IEN	Input enable. Data → IEN Register
B	1011	OEN	Output enable. Data → OEN Register
C	1100	JMP	Jump. JMP Flag → ⎍
D	1101	RTN	Return. RTN Flag → ⎍ and skip next instruction
E	1110	SKZ	Skip next instruction if RR = 0
F	1111	NOPF	No change in registers. RR → RR, Flag F → ⎍

FIGURE 22.36 Instruction set for the MC14500B microprocessor. (Courtesy of Motorola Semiconductor Products Div.)

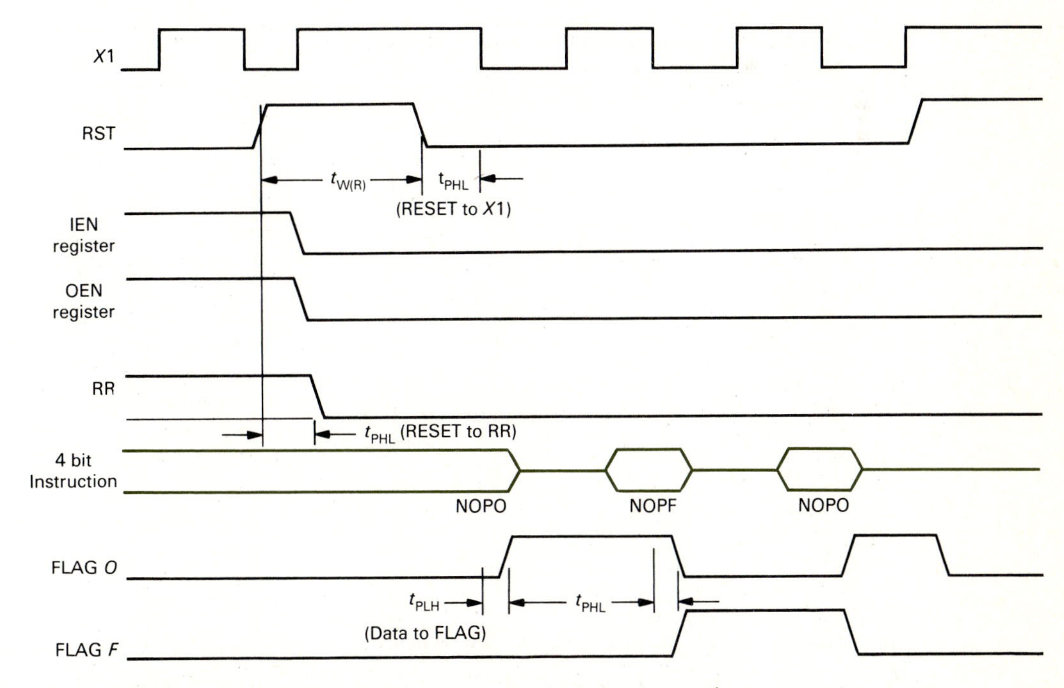

FIGURE 22.37 Fundamental parts of an instruction cycle.

TABLE 22.5 Characteristics of Selected Microprocessors

Designation	Manufacturer	Technology	Maximum power dissipation, w	Minimum instruction time, μs	Word length data/instr, bits	On-board clock/pulse rate, MHz	Direct address range, words
Mc14500	Motorola	CMOS		1.0	1/4	Yes/1.0	0
COP 402	National	NMOS	0.75	4.0	4/8	Yes/1.0	1K
8 × 305	Signetics	Bipolar	1.65	0.25	8/8	No/10.0	8K
8080A	Intel	NMOS		1.5	8/8	No/2.0	64K
80188	Intel	NMOS	3.0	0.2	8/16	Yes/8.0	1M
80286	Intel	NMOS	3.0	0.25	16/16	No/8.0	1G (Virtual)
MC68020	Motorolo	HCMOS	1.5	0.13	32/32	/16.67	16M
Micro/J-11	D.E.C.	CMOS	1.0	0.2	32/16	Yes/20	4M

22.5.5 Digital Control Systems

Digital control circuits are gradually supplanting analog circuits in many control applications, given the rapid decrease in digital IC costs and the small size and great diversity of these components. Microprocessors and small computers give a totally new and expanded perspective to the precision and sophistication possible in the control of physical processes, which was hardly conceivable when viewing control from an analog perspective. Microprocessors have led to the development of "intelligent" control systems, some present applications of which include the automated assembly line, robots, optimum internal combustion engine control, and human body member replacements controlled by human nerve signals.

Microprocessors are also used in the relatively old-fashioned types of control applications to add further precision and more decision-making ability to the control process. Figure 22.38 illustrates a digital circuit (DC) motor armature control system,

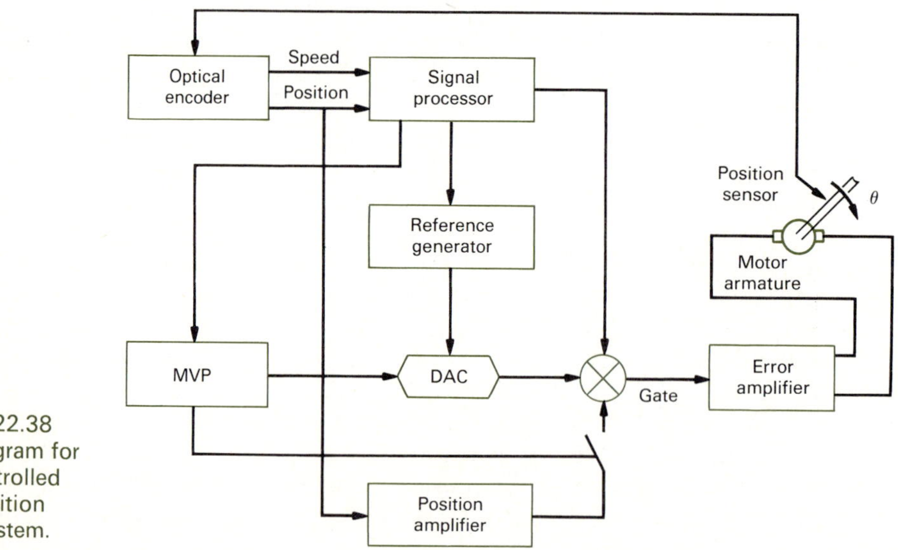

FIGURE 22.38 Block diagram for MVP-controlled motor position control system.

Number of basic instructions	Number of general-purpose registers	Floating-point arithmetic?	Interrupt levels	Package type/pins	High-level language?	Number of Stack registers	RAM,bits
16	1	No	1	DIP/16	No	0	—
49	64	Yes	3	DIP/40	No	RAM	64 × 4
8	8	No	—	DIP/50	No	0	
78	8	Yes	1	DIP/40	Yes	RAM	128 × 8
88	8	Yes	4	SMT/68	Yes	4 × 16	
92	8	Yes	1	SMT/68	Yes	4 × 16	
61	16	Yes	7		Yes	3 × 32	
112	12	Yes	4	DIP/60	Yes		

one serving the same basic function as some of the analog motor control systems discussed in Chap. 19. Figure 22.38 represents an all-digital control system using a microprocessor (or at least nearly all-digital, since the final control—the output of the D/A converter—is in analog form). This is a position control system and the input is the desired final position of the motor shaft. The initial, or present-time, speed is also inputted to the microprocessor, which calculates an optimum speed profile for the motor to reach its final position at zero speed in minimum time and with the required accuracy of position.

22.6 MICROCOMPUTERS

As we have noted, a microprocessor is actually the heart of a full computing system. It is essentially what gives a computer its intelligence. Any computer, whether a microcomputer, a minicomputer, or a mainframe computer, consists of four principal parts: the central processing unit (CPU), which is basically a microprocessor unit (MPU); storage (memory); input components; and output components. Whereas an MPU is generally dedicated to performing fairly specific tasks, the CPU of a computer must have much more general and diverse capabilities. Therefore, it generally has an expanded control section, many more instructions, and additional buses.

In addition to the ROM used to contain instructions and other basic computer housekeeping functions, computer storage is generally of two types: main memory and offline memory. Main (or core) memory is the working memory of a computer. Programs permanently stored on disks (or other types of offline memory, such as drums, tape decks, and optical storage) are brought into the working memory and partitioned and stored temporarily in such a way as to make the operation of the program efficient. Both the working and offline memories are of the type known as random-access memory (RAM); as its name implies, any bit of information stored in RAM can be accessed within a very short time.

The working memory is constructed of semiconductor memory elements, although magnetic bubble memories are becoming more common. The offline memory most often consists of rotating disks, although rotating drums are also common. Disks and drums consist of surfaces coated with a thin magnetic oxide film upon which

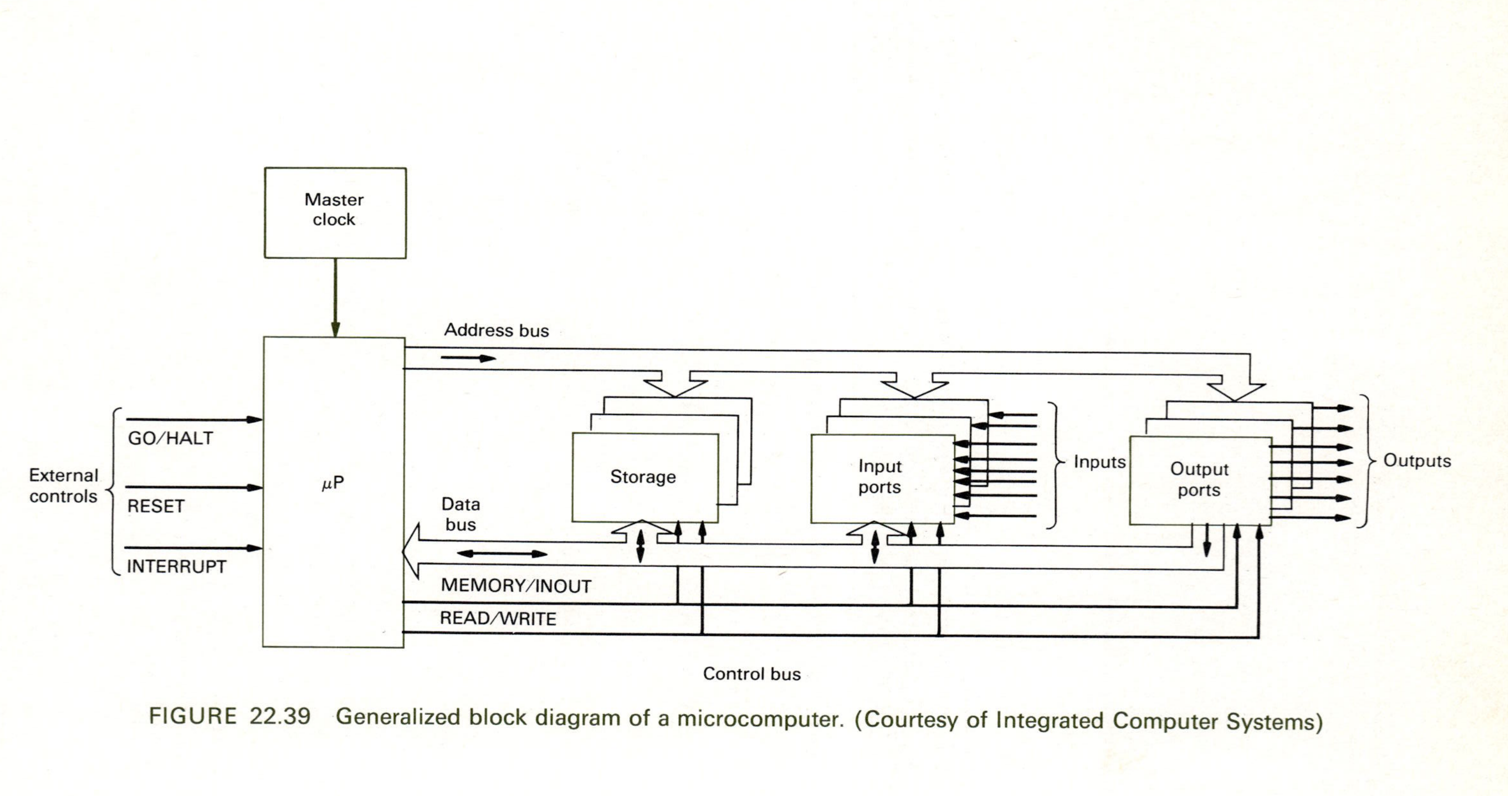

FIGURE 22.39 Generalized block diagram of a microcomputer. (Courtesy of Integrated Computer Systems)

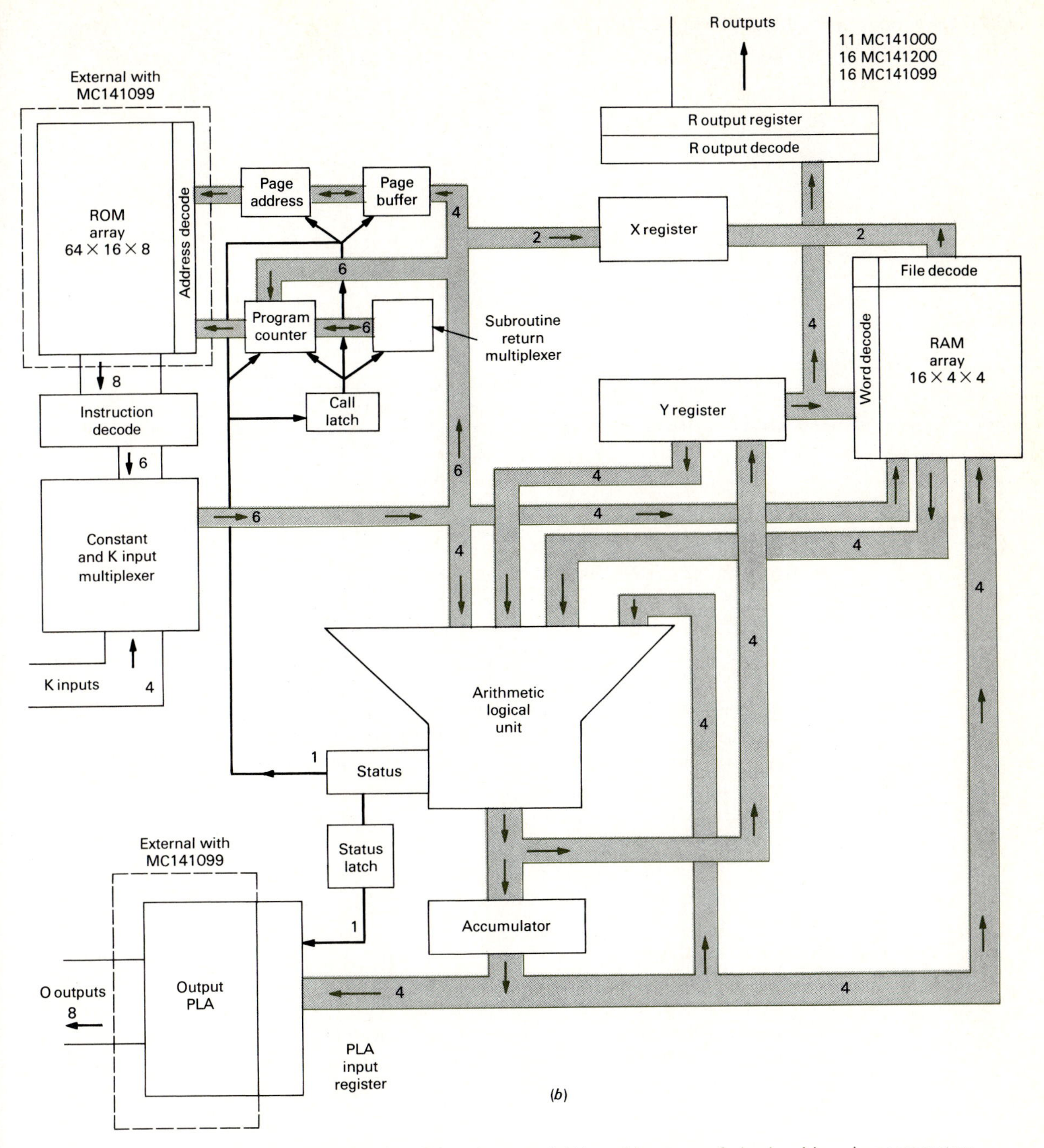

FIGURE 22.40 (*a*) Data sheet and (*b*) architecture of single-chip microcomputers. (Courtesy of Motorola Semiconductor Products Div.)

TABLE 22.6 Characteristics of Selected Microcomputers

Designation	Manufacturer	Technology	Maximum power dissipation, w	Word length data/instruction bits	On-board clock/pulse rate, MHz	Minimum instruction time, μs	Number of basic instructions	On-board RAM (b)
COP420	National	NMOS	0.75	4/8	Yes/2.0	4.0	26	64 × 4
PPS-4/1	Rockwell	PMOS	1.0	4/8	Yes/0.2	5.0	50	48 × 4
Z8	Zilog	NMOS		8/8	Yes/4.0	1.5	47	144 × 8
MC6801	Motorola	NMOS	1.2	8/8	Yes/4.0	2.0	82	128 × 8
MC6801U4	Motorola	HNMOS	1.2	8/8	Yes/4.0		98	192 × 8
MC6805	Motorola	HNMOS	0.7	8/8	Yes/4.0	2.0	61	96 × 8
MC68HC11	Motorola	HCMOS	0.11	8/8	Yes/8A		139	256 × 8
8022	Intel	NMOS		8/8	Yes/3.6	8.4	70	64 × 8
8048AH	Intel	HNMOS	1.5	8/8	Yes/6.0	2.5	96	64 × 8
8748	Intel	CMOS	1.	8/8	Yes/6.0	1.36	96	64 × 8
80C51	Intel	CHMOS	2.0	8/8	Yes/12	1.0	111	128 × 8
8751	Intel							
8396	Intel	NMOS	1.5	16/8	Yes/12	1.0	83	232 × 8
68200	MOSTEK	NMOS	1.0	16/16	Yes/6.0	1.0		128 × 16

small regions of magnetization represent a binary 1 or 0. The packing density (bits per square inch) is very high and access by means of a high-speed pickup head is very rapid. Memory size is one means of quantizing the size of a computer. Memory size is designated by rounding off the memory locations discussed in Sec. 22.5.2. A 16-bit memory size which can store 2^{16} (or 65,536) bits is designated as a 64K memory; likewise, a 2^{10}-bit (or 1024-bit) memory is designated as a 1K memory. A computer with a core memory of 64K and above is generally classified as a minicomputer. Disk memory sizes vary from a few K bits in the very small floppy-disk memories to many megabytes in the large steel disks used in minicomputers and mainframe computers.

Disks, tape decks, and printers are the principal input and output devices associated with microcomputers and minicomputers and are jointly termed *peripherals.* Figure 22.39 illustrates the basic organization of a microcomputer. A microprocessor is the arithmetic and control section of the microcomputer, to which have been added memory, input-output (I/O) ports, additional clocks and/or timing circuits, and interrupts. The function of an interrupt is to halt the normal computer program in response to certain conditions, such as a logic decision in the computer program, a flag (or warning) about a program system malfunction, or a request for higher-priority operations. After the interrupt, the normal program may or may not be resumed, depending upon the reason for the interrupt and the program philosophy. In Fig. 22.39, the extent of the microcomputer is within the heavy-lined box. The power supply and interfaces to peripheral devices are external to the microcomputer.

Table 22.6 summarizes some of the characteristics of several microcomputers in general use. The reader is probably well aware that the computing systems designated by the prefixes "mini" and "micro" are often more powerful and capable than many of the large mainframe computers of just a few years ago. Figure 22.40 illustrates the data sheet and architecture of single-chip CMOS 4-bit microcomputers.

On-board ROM, KB	On-board EPROM, Bhts	I/O ports Ser	Par	Interrupt levels	Package/pins	Floating-point arithmetic	Number of General-purpose registers	On-board timers	On-board ADC	Maximum temperature °C Op.	Ston
1 × 8	0	3	20	1	DIP/28	No	4		—	70	150
2 × 8	0	3	28	2	DIP/40	Yes	1		—		
2/4 × 8	0	4	28	6	DIP/40	Yes	124		—		
2 × 8	0	1	29	2	DIP/40	Yes	2	1	—		
4 × 8	0	1	29	2	DIP/40	Yes		1	—		
3.6 × 8	128	1	32	1	DIP/40	Yes		2	—		
8 × 8	512	2	5	2	SMT/52	Yes	7	1	Yes	85	150
2 × 8	0	3	24	1	DIP/28	Yes	8	1	Yes	85	150
1 × 8	0	3	24	1	DIP/40	Yes	7	1	—	70	150
—	1024	3	24	1	DIP/40	Yes	7	1	—	70	150
4 × 8	0	2	32	2	DIP/40	Yes	32	2	—		
8 × 8	0	2	5	8	SMT/68	Yes		4	Yes	70	150
2 × 16		1	6		DIP/48	Yes	8	3	—		

PROBLEMS

22.1 Determine the first three terms of the Fourier series for a square wave with a period of 1 ms. Assuming these three terms adequately represent the square wave, what is the minimum sampling frequency, or Nyquist rate?

22.2 In the PWM scheme illustrated in Fig. 22.9, it is desired to cover a range of numerical values from 0 to 20 by means of the square pulse widths shown in Fig. 22.9*c*. If the half period of the analog function $f(t)$ is 2 ms, suggest a magnitude and frequency for the sawtooth wave (Fig. 22.9*b*) to accomplish this and sketch the resulting waveforms. What is the average value (over a full period) of the modulated square pulse when representing a magnitude of 2? of 5? of 20?

22.3 A certain signal is represented by the equation $f(t) = 4 \cos \omega t = \cos 3\, \omega t$, where $\omega = 377$. A PAM scheme is to sample at 4 times the Nyquist rate. Sketch the resulting pulses for (*a*) bipole square-topped PAM and (*b*) unipolar conformal PAM. (*c*) What is the pulse period T?

22.4 What is the pulse capacity C of a binary PCM scheme with five pulses per group? How many values of the analog signal can be transmitted?

22.5. Prepare a weighted network diagram (Fig. 22.12) and determine the levels of analog voltage output for an 8-bit DAC using a 12-V logic family.

22.6 A DAC has a maximum output of 10 V and accepts 6 binary bits as inputs. How many volts does each analog step represent?

22.7 Repeat Example 22.2 for a 3-bit ladder network.

22.8 Prove that the ladder networks of the type shown in Figs. 22.14 and 22.15 have constant output impedance.

22.9 Repeat Example 22.3 when the analog signal has a value of 10.

22.10 Develop an adder system to add the numbers 18 and 20, and show the system block diagram and the readout.

22.11 Repeat Prob. 22.10 for subtracting 18 from 20.

22.12 Find the 2s complement representation of −25. Subtract 25 from 10 using 2s complement notation. Assume an 8-bit word length.

REFERENCES

22.1 R. N. Noyce, "From Relays to MPUs," *Computer* (IEEE), vol. 9, no. 12, September 1977.

22.2 C. E. Shannon, "Communication in the Presence of Noise," *Proceedings IRE*, vol. 37, no. 1, January 1979.

22.3 Mischa Schwartz, *Information Transmission, Modulation, and Noise: A Unified Approach*, 3d ed., McGraw-Hill, New York, 1980.

22.4 *Analog Switches and Their Applications*, Siliconix, Inc., Santa Clara, California, 1980.

22.5 D. F. Hoeschele, Jr., *Analog-to-Digital, Digital-to-Analog Conversion Techniques*, Wiley, New York, 1968.

22.6 H. Schmid, "Electronic Design Practical Guide to D/A and A/D Conversion," *Electronic Design*, October 1968.

22.7 J. D. Greenfield, *Practical Digital Design Using ICs*, 2d ed., Wiley, New York, 1983.

22.8 D. E. Heffer, G. A. King, and D. Keith, *Basic Principles and Practices of Microprocessors*, Halsted Press, New York, 1981.

22.9 E. A. Lamagna, "Fast Computer Algebra," *Computer* (IEEE), vol. 15, no. 9, September 1982, pp.

22.10 Ron Bishop, *Basic Microprocessors and the Sixty-Eight Hundred*, Hayden Book, Rochelle Park, New Jersey, 1979.

22.11 I. Flores and C. Terry, *Microcomputer Systems*, Van Nostrand Reinhold, New York, 1982.

22.12 M. D. Freedman and L. B. Evans, *Designing Systems with Microcomputers: A Systematic Approach*, Prentice-Hall, Englewood Cliffs, New Jersey, 1983.

22.13 D. F. Stout, *Microprocessor Application Handbook*, McGraw-Hill, New York, 1982.

22.14 P. Lister (ed), *Single-Chip Microcomputers*, McGraw-Hill, New York, 1984.

Appendix

Unit Conversion

Symbol	Description	One (SI unit)	Is equal to (USGS unit)
B	Magnetic flux density	tesla (T)(= 1 Wb/m^2)	6.452×10^4 lines/in^2
H	Magnetic field intensity	ampere per meter (A/m)	0.0254 A/in
ϕ	Magnetic flux	weber (Wb)	10^8 lines
D	Viscous damping coefficient	newton-meter-second (N · m · s)	0.73756 lb · ft · s
F	Force	newton (N)	0.2248 lb
J	Inertia	kilogram-square meter	23.73 lb · ft^2
T	Torque	newton-meter (N · m)	0.73756 ft · lb
W	Energy	joule (J)	1 W · s

Appendix

Table of Certain Constants

Permeability of free space = $4\pi \times 10^{-7}$ H/m
Permittivity of free space = 8.85×10^{-12} F/m
Resistivity of copper = $1.724 \times 10^{-8}\ \Omega \cdot m$
Resistivity of aluminum = $2.83 \times 10^{-8}\ \Omega \cdot m$
Charge of an electron = -1.6×10^{-19} C
Electron mass = 9.11×10^{-31} kg
Acceleration of gravity = $9.807\ m/s^2$

Appendix

C

Fourier Series

Fourier series is a representation of periodic functions that fulfill certain conditions in terms of their harmonics. A function is said to be periodic if it repeats at certain intervals. Explicitly, for periodicity,

$$F(t) = F(t \pm nT) \tag{C.1}$$

where T is the period and n is an integer.

Stated formally, if $F(t)$ is a single-valued periodic function, which is finite, has a finite number of discontinuities, and has a finite number of maxima and minima over a period T, then $F(t)$ may be represented over a complete period by a series of simple harmonic functions, the frequencies of which are integral multiples of the fundamental frequency. This series is known as *Fourier series*. Expressed mathematically, we have

$$F(t) = \frac{A_0}{2} + \sum_{n=1}^{\infty} (A_n \cos n\omega t + B_n \sin n\omega t) \tag{C.2}$$

where $\omega = 2\pi/T$ and the coefficients A_n and B_n are given by

$$A_n = \frac{2}{T}\int_0^T F(t) \cos n\omega t \, dt \tag{C.3}$$

and

$$B_n = \frac{2}{T}\int_0^T F(t) \sin n\omega t \, dt$$

The term $A_0/2$ in Eq. (2) is introduced so that the general result of Eq. (3) is applicable when $n = 0$, that is, for A_0 as well. It may be seen from (2) that $A_0/2$ is the average value of the function over the period. Using (3) for A_0 we have

$$A_0 = \frac{2}{T}\int_0^T F(t) \, dt \tag{C.5}$$

Appendix

Laplace Transforms

We define the Laplace transform of a function $f(t)$, where $f(t) \equiv 0$ for $t \leq 0$, as

$$\mathscr{L}\{f(t)\} \equiv F(s) = \int_0^\infty e^{-st} f(t)\, dt \tag{D.1}$$

In Eq. (D.1) s is a complex variable, $s = \sigma + j\omega$, with σ chosen large enough to make the infinite integral converge. Given $f(t) = e^{-at}$ $(a > 0)$, we find $F(s)$.

Substituting in Eq. (D.1) yields

$$F(s) = \int_0^\infty e^{-st} e^{-at}\, dt = \left. \frac{e^{-(s+a)t}}{-(s + a)} \right|_0^\infty = \frac{1}{s + a}$$

where, to make the exponential vanish at the upper limit, it has been assumed that $\sigma > -a$. However, this restriction may be dropped, and the result

$$F(s) = \frac{1}{s + a}$$

may be considered valid over the entire complex S plane.

Hence, the Laplace transform pairs in Table D.1 are developed.

TABLE D.1 Laplace Transform Pairs

	$f(t)$		$F(s)$
1	$\delta(t)$		1
2	a		$\frac{a}{s}$
3	t		$\frac{1}{s^2}$
4	e^{-at}		$\frac{1}{s+a}$
5	te^{-at}		$\frac{1}{(s+a)^2}$
6	$\sin \omega t$		$\frac{\omega}{s^2+\omega^2}$
7	$\cos \omega t$		$\frac{s}{s^2+\omega^2}$
8	$e^{-at} \sin \omega t$		$\frac{\omega}{(s+a)^2+\omega^2}$
9	$e^{-at} \cos \omega t$		$\frac{s+a}{(s+a)^2+\omega^2}$
10	$\frac{d}{dt}[f(t)]$		$sF(s) - f(0+)$
11	$\frac{d^2}{dt^2}[f(t)]$		$s^2F(s) - sf(0+) - f'(0+)$
12	$\frac{d^n}{dt^n}[f(t)]$		$s^nF(s) - s^{n-1}f(0+) - s^{n-2}f'(0+) - \cdots - f^{n-1}(0+)$
13	$\int_0^t f(\tau)\, d\tau$		$\frac{1}{s} F(s)$
14	$af(t) + bg(t)$		$aF(s) + bG(s)$
15	$\int_0^t f(\tau)g(t-\tau)\, d\tau$		$F(s)G(s)$
16	$f(\infty)$	final value	$\lim_{s \to 0} sF(s)$
17	$f(0+)$	initial value	$\lim_{\substack{s \to \infty \\ s \text{ real}}} sF(s)$

Bibliography

Analog Switches and Their Applications, Siliconix, Inc., Santa Clara, California, 1980.

Barna, A., and D. I. Porat: *Integrated Circuits in Digital Electronics*, Wiley, New York, 1973.

Beckwith, T. G., N. L. Buck, and R. D. Marangoni: *Mechanical Measurements*, 3d ed., Addison-Wesley, Reading, Mass., 1982.

Bishop, Ron: *Basic Microprocessors and the Sixty-Eight Hundred*, Hayden Book, Rochelle Park, New Jersey, 1979.

Boylestad, R. L.: *Introductory Circuit Analysis*, 4th ed., Charles E. Merrill, Columbus, Ohio, 1982.

______ and L. Nashelsky: *Electronic Devices and Circuit Theory*, 3d ed., Prentice-Hall, Englewood Cliffs, New Jersey, 1982.

Carlson, A. B., and D. G. Gisser: *Electrical Engineering: Concepts and Applications*, Addison-Wesley, Reading, Mass., 1981.

Cathey, J. J., and S. A. Nasar: *Schaum's Outline of Theory and Problems in Basic Electrical Engineering*, McGraw-Hill, New York, 1983.

Chestnut, H., and R. W. Mayer: *Servomechanisms and Regulating System Design*, 2d ed., vol. 1, Wiley, New York, 1959.

D'Azzo, J. J., and C. H. Houpis: *Linear Control System Analysis and Design*, 2d ed., McGraw-Hill, New York, 1981.

DeRusso, P. M., R. J. Roy, and C. M. Close: *State Variables for Engineers*, Wiley, New York, 1965.

Desa, A.: *Principles of Electronic Instrumentation*, Halsted Press, New York, 1981.

Dewan, S. B., G. R. Slemon, and A. Straughen: *Power Semiconductor Drives*, Wiley, New York, 1984.

Dorf, R. C.: *Modern Control Systems*, 3d ed., Addison-Wesley, Reading, Mass., 1980.

Eichelberger, E. B.: "A Logic Design Structure for LSI Testability," *14th Design Automation Conference*, June 1977, pp. 462–468.

Edminster, J. A.: *Schaum's Outline of Electric Circuits*, 2d ed., McGraw-Hill, New York, 1983.

Fink, D. G., and D. Christiansen: *Electronics Engineer's Handbook*, 2d ed., McGraw-Hill, New York, 1982.

Fitzgerald, A. E., D. Higginbotham, and A. Grabel: *Basic Electrical Engineering*, 5th ed., McGraw-Hill, New York, 1981.

______, C. Kingsley, Jr., and S. D. Umans: *Electric Machinery*, 4th ed., McGraw-Hill, New York, 1983.

Flores, I., and C. Terry: *Microcomputer Systems*, Van Nostrand Reinhold, New York, 1982.

Freedman, M. D., and L. B. Evans: *Designing Systems with Microcomputers: A Systematic Approach*, Prentice-Hall, Englewood Cliffs, New Jersey, 1983.

Greenfield, J. D.: *Practical Digital Design Using ICs*, 2d ed., Wiley, New York, 1983.

Hamburg, D. R., and L. E. Unnewehr: "An Electronic Wattmeter for Nonsinusoidal Low Power Factor Measurements," *IEEE Transactions on Magnetics*, vol. MAG-7, no. 3, September 1971.

Hayt, W. H., Jr., and J. E. Kemmerly: *Engineering Circuit Analysis*, 3d ed., McGraw-Hill, New York, 1978.

Heffer, D. E., G. A. King, and D. Keith: *Basic Principles and Practices of Microprocessors*, Halsted Press, New York, 1981.

Hoeschele, D. F., Jr.: *Analog-to-Digital, Digital-to-Analog Conversion Techniques*, Wiley, New York, 1968.

Hohn, F. E.: *Applied Boolean Algebra—An Elementary Introduction*, Macmillan, New York, 1960.

Hostetter, G. H., C. J. Savant, Jr., and R. T. Stefani: *Design of Feedback Control Systems*, Holt, Rinehart & Winston, New York, 1982.

Hubert, C. I.: *Electric Circuits AC–DC: An Integrated Approach*, McGraw-Hill, New York, 1982.

Hunt, J. W.: "Programming Languages," *Computer* (IEEE), April 1982.

Irwin, J. D.: *Basic Engineering Circuit Analysis*, Macmillan, New York, 1983.

Karnaugh, M.: "The Map Method for Synthesis of Combinational Logic Circuits," *AIEE Transactions*, vol. 72, pt. 1, 1953.

Kasper, J., and S. Feller: *Digital Integrated Circuits*, Prentice-Hall, Englewood Cliffs, New Jersey, 1982.

Kuo, B. C.: *Digital Control Systems*, Holt, Rinehart & Winston, New York, 1980.

Lamagna, E. A.: "Fast Computer Algebra," *Computer*, vol. 15, no. 9, September 1982.

Lister, P. (ed.): *Single-Chip Microcomputers*, McGraw-Hill, New York, 1984.

Lowenberg, E. C.: *Electronic Circuits*, McGraw-Hill, New York, 1967.

Mix, D. F., and N. M. Schmitt: *Circuit Analysis for Engineers*, Wiley, New York, 1985.

Motorola CMOS Handbook: *CMOS Integrated Circuits*, Motorola Semiconductor Products, Austin, Texas 78721, 1980.

Nasar, S. A.: *Electric Energy Conversion and Transmission*, Macmillan, New York, 1985.

______: *Electric Machines and Transformers*, Macmillan, New York, 1983.

______: *Schaum's Outline of Theory and Problems in Electric Machines and Electromechanics*, McGraw-Hill, New York, 1981.

______ and L. E. Unnewehr: *Electromechanics and Electric Machines*, 2d ed., Wiley, New York, 1983.

Noyce, R. N.: "From Relays to MPUs," *Computer*, vol. 9, no. 12, September 1977.

Oldham, W. G., and S. E. Schwartz: *An Introduction to Electronics*, Holt, Rinehart & Winston, New York, 1972.

Ong, D. G.: *Modern MOS Technology*, McGraw-Hill, New York, 1984.

Prioste, J., and T. Balph: "Relay to IC Conversion," *Machine Design*, May 28, 1970.

Ralston, Anthony (ed.): *Encyclopedia of Computer Science and Engineering*, 2d ed., Van Nostrand Reinhold, New York, 1982.

Rappaport, A.: "Automated Design and Simulation Aids Speed Semicustom IC Development," *EDN*, vol. 27, no. 15, August 4, 1982.

Say, M. G., and E. O. Taylor: *Direct Current Machines*, Halsted Press, New York, 1980.

Schilling, D. L., and C. Belove: *Electrical Circuits: Discrete and Integrated*, 2d ed., McGraw-Hill, New York, 1979.

Schmid, H.: "Electronic Design Practical Guide to D/A and A/D Conversion," *Electronic Design*, October 1968.

Schwartz, Mischa: *Information Transmission, Modulation, and Noise: A Unified Approach*, 3d ed., McGraw-Hill, New York, 1980.

Schwartz, S. E., and W. G. Oldham: *Electrical Engineering: An Introduction*, Holt, Rinehart & Winston, New York, 1984.

Sen, P. C.: *Thyristor DC Drives*, Wiley, New York, 1981.

Shannon, C. E.: "Communication in the Presence of Noise," *Proceedings IRE*, vol. 37, no. 1, January 1979.

Smith, Ralph J.: *Circuits, Devices and Systems: A First Course in Electrical Engineering*, 4th ed., Wiley, New York, 1984.

Stevenson, W. D., Jr.: *Elements of Power System Analysis*, 4th ed., McGraw-Hill, New York, 1982.

Stout, D. F.: *Microprocessor Application Handbook*, McGraw-Hill, New York, 1982.

Tektronix Reference Manual, Tektronics, Inc., Beaverton, Oregon, 1981.

Teschler, L.: "Update on Electronic Logic," *Machine Design*, August 20, 1981.

Thaler, G. J., and M. P. Pastel: *Analysis and Design of Nonlinear Feedback Control Systems*, McGraw-Hill, New York, 1962.

Webster's Ninth New Collegiate Dictionary, Merriam-Webster, Springfield, Mass., 1984.

Williams, A. B.: *Designers Handbook of Integrated Circuits*, McGraw-Hill, New York, 1984.

1983 Electrical and Electronics Reference Issue: "Digital Logic," *Machine Design*, May 1983.

Answers to Problems

CHAPTER 2

2.1 6.25×10^{18} electrons. **2.2** 20.43×10^{-3} N.
2.3 (*a*) 3.6×10^{-2} N; (*b*) 0. **2.4** -0.375 J; -7.5×10^4 V; 0.525 J; 7.5×10^4 V.
2.5 120 C. **2.6** 1.0 C **2.8** 4 V; 2 V; -3 A; -4 A. **2.10** 4 V; 4 V; -5 A.
2.11 5×10^6 J. **2.12** 36 W. **2.13** 0.5 A; 1 V. **2.15** 7/30 W.
2.16 (*a*) 2.7 W; (*b*) 2.7 W.

CHAPTER 3

3.1 (*a*) 1 Ω; (*b*) 4 Ω; (*c*) 2 Ω. **3.2** (*a*) 5 Ω; (*b*) 2 Ω; (*c*) 3 Ω; (*d*) 1 Ω. **3.3** 0.25 A; 1.5 V
3.4 2.5 A; 20 V. **3.5** 1 A. **3.6** 1 A. **3.7** 0.5 V. **3.8** 16/3 A. **3.9** 5 A.
3.10 4 V. **3.11** -5 A. **3.12** 3 A. **3.13** $V_{TH} = 1$ V; $R_{TH} = 252$.
3.14 $I_{SC} = 0.5$ A. **3.15** 2 A. **3.16** 0.75 V. **3.17** 7/11 V. **3.18** $-21/23$ V.
3.19 $-4/3$ A. **3.20** 2.5 W (total). **3.21** 152/3 W. **3.22** 14 W. **3.23** 1 Ω.
3.24 0 Ω.

CHAPTER 4

4.1 44,526.51 in. **4.4** -5×10^{-3} C. **4.5** 1.5 μF; 5 V.
4.8 3 mH; 3 A. **4.11** 4.5×10^{-2} J. **4.12** 10^{-2} J. **4.13** 3 V; 0 V; 2 A.
4.14 11/2 V; 11/4 A; 1/2 V; 1/4 A.

CHAPTER 5

5.1 $A = 10$, $\omega = 1.29$, $\theta = -53.13$. **5.2** $6.36 \sin(2t - 45°)$. **5.3** $3 \sin t$ A.
5.4 (*a*) $3.16/18.43°$; (*b*) $3.16/71.57°$; (*c*) $2/90° = j2$; (*d*) $4/0°$; (*e*) $4.24/-45°$.
5.7 (*a*) $2.24/63.43°$; $2.24/-26.57°$; (*b*) $3.16/18.43°$; (*c*) $3.16/108.43°$; (*d*) $1.41/-135°$; (*e*) $-j$; (*f*) $5/36.87°$. **5.8** (*d*) $3.06/-66.7$; (*e*) $6.23/-1.2°$.
5.9 (*a*) $56.69/105.2°$; (*b*) $6.46 + j1 = 6.54/8.8°$. **5.10** $6.36 \sin(2t - 45°)$.
5.11 $3 \sin t$. **5.12** $0.24 \sin(2t - 71.57°)$. **5.13** $6.36 \sin(2t - 45°)$.
5.14 $3 \sin t$. **5.15** $0.24 \sin(2t - 71.5°)$. **5.16** $0.18 \cos(3t - 135°)$.
5.17 $0.38 \cos(3t + 10.19°)$. **5.18** $1.34 \sin(2t + 13.43°) + 1.41 \cos(t + 55°)$.
5.19 $2.12 \sin(t + 5°) + 1.41 \cos(t + 55°)$. **5.20** 1.47 W; 9.41 W. **5.21** $2/-90°$.
5.22 0.89; 4 Ω; 1 H. **5.23** 0.62; 8.62 μF. **5.24** 0.25 W. **5.25** 77.97 Hz.
5.26 711.76 Hz; 30 Hz; 149.

CHAPTER 6

6.10 $0.03e^{-10^4 t} - 0.01e^{-3 \times 10^4 t}$. **6.11** $0.02e^{-2 \times 10^4 t} + 400te^{-2 \times 10^{-4} t}$.
6.12 $(0.02 \cos 10^4 t + 2 \sin 10^4 t)e^{-2 \times 10^4 t}$. **6.13** $-\frac{50}{3}(e^{-t} - e^{-9t}) + 5$.
6.14 $0.75(e^{-6000t} - e^{-2000t})$. **6.15** 3 A. **6.16** 10 V. **6.17** 7.5 V; 10^{-3} s.
6.18 2 s; 0 V. **6.20** $Ae^{-t} + \beta e^{-9t} + 5$.

CHAPTER 7

7.1 $10^{-9}(e^{v/24.84 \text{ mV}} - 1)$. **7.2** 0.9 mA; 0.477 mW. **7.3** 1.8 mA; 1.04 mW.
7.4 4.5 mA; 0.61 V; 2.745 mW. **7.5** 3.3 mA; 0.6 V; 78.6 mW.
7.6 0.9 mA; 0.53 V; 0.477 mW. **7.7** 10 Ω; 0.58 V. **7.8** 8.9 Ω; 53.48 Ω.
7.9 $(4.5 + 1.684 \sin \omega t)$ mA. **7.13** 0 for all t. **7.16** 50 to 500 Ω.

CHAPTER 8

8.1 10^3; 100. **8.3** 9.6 V; 5.2 mA; −2 V; 3.6.
8.4 5.2 V; 5 mA; −2 V; A_v not defined. **8.5** 11.4 V; 4.35 mA; 3.6. **8.7** 1.4
8.8 70; 0.99. **8.9** 0.58 V; 0.04 mA. **8.10** 0.58 V; 0.03 mA. **8.11** 0.6 V; 0.06 mA.
8.12 0.58 V; 0.04 mA. **8.13** $V_{BE} = 0.58$ V; $I_B = 0.0404$ mA; $V_{CE} = 6.4$ V; $I_C = 2.4$ mA.
8.14 1709 Ω; 135.5 kΩ. **8.16** $R_{B1} = 11.01$ kΩ; $R_{B2} = 109.1$ kΩ.

CHAPTER 9

9.1 375; 100; 200 Ω; 750 Ω. **9.2** 300 Ω. **9.3** 8.8 mA.
9.4 1.475×10^{-3} ℧; 5×10^{-5}. **9.5** −10; −50; 50 kΩ; 10 kΩ.
9.6 −2.5; −12.5; 50 kΩ; 10 kΩ. **9.7** 0.91; 4.55; 50 kΩ; 476 Ω.
9.8 10; 0.46; 455 Ω; 10 kΩ. **9.9** 17.195 kΩ. **9.11** 0.5 **9.12** 16.67. **9.13** 20 MHz.
9.14 416.7 Ω; 72.5; 1.2×10^{-9} ℧. **9.15** −100; −47.6; 476 Ω; 1 kΩ.
9.16 −0.49; −4.4; 9103 Ω; 1 kΩ. **9.17** 1000; 0.5; 5 Ω; 10 kΩ.
9.18 0.98; 25.19; 12853 Ω; 7.84 Ω. **9.20** −3.21; 32.55; 500 Ω; 50 Ω. **9.21** 574,836 Hz.

CHAPTER 10

10.1 −2 V. **10.2** 2 mV. **10.3** 1 mV. **10.4** $10^{-3}R$ V. **10.5** $n_2/n_1(v_2 - v_1)$.
10.8 n_1. **10.9** −10 kΩ. **10.10** 159 ρF. **10.11** 3183 Ω; 1592 Ω. **10.14** 0.1 μF.

CHAPTER 11

11.7 43 kΩ; 2400 Ω. **11.8** 3 kΩ; 400 Ω. **11.13** 763 Ω.

CHAPTER 12

12.1 8.33 $\angle 15°$ A; 14.43 $\angle 45°$ A; 2414 W; 2500 VA.
12.2 46.19 $\angle -53.13°$ A; 191.7 kW; 319.52 kVA.
12.3 14.5 $\angle -160°$ A; 10.77 $\angle 51.8°$ A; 7.81 $\angle -26.3°$ A.
12.4 2.88 $\angle -30°$ A; 5.77 $\angle -120°$ A; 2.31 $\angle 90°$ A.
12.5 14.33 $\angle -89°$ A; 10.77 $\angle 111.8°$ A; 6.07 $\angle 45.5°$ A. **12.6** 1766 W.
12.7 2.05 $\angle 79°$ A. **12.8** 604.6 W; 1095.5 VA. **12.9** 64 kW.
12.10 $(17.77 + j18.15)$Ω. **12.11** 1913 W; 2734 VA.

CHAPTER 13

13.1 (*a*) 2; (*b*) (*i*) 22.73 A, 45.45 A; (*ii*) 10.91 A, 21.82 A.
13.2 (*a*) 5.5 A; (*b*) 40 Ω; (*c*) 1210 W. **13.3** 16.5 mWb.
13.4 $X_m = 187.6$ Ω; $R_c = 202.1$ Ω. **13.6** 98.4 percent.
13.10 (*a*) 97.4 percent; (*b*) 96.8 percent. **13.11** 96.4 percent.
13.12 (*a*) 98.39 percent; (*b*) 98.12 percent.
13.13 $X_m = 891.15$ Ω; $R'_e = 0.0794$ Ω; $X'_e = 0.0398$ Ω. **13.14** 450.9 V; 97.58 percent.
13.15 94.31 percent. **13.16** 65.45 A. **13.17** 98.09 percent; 23.17 A.

CHAPTER 14

14.3 (*a*) 36 V; (*b*) 67.5 V. **14.4** 0.03 Wb. **14.5** 114.6 N·m. **14.6** 250.24 V.
14.7 (*a*) 96,875 W; (*b*) 98,009 W. **14.8** 396.6 rpm. **14.9** 0.1215 Ω. **14.11** 11.428 W.
14.12 1200 W. **14.15** (*a*) 254 V; (*b*) 226 V. **14.16** (*a*) 46.89 kW.
14.17 90.6 percent. **14.18** (*a*) 188.6 N·m; (*b*) 93.17 percent. **14.19** 802.8 rpm.
14.20 1035.8 rpm. **14.21** 692.8 rpm.
14.22 (*a*) 1129 rpm; (*b*) 108.9 N·m; (*c*) 88.3 percent. **14.23** 1297.6 rpm.
14.24 1020 rpm; 84.3 percent. **14.25** 226.52 Ω. **14.26** 514 rpm.
14.27 (*a*) 0.704 Ω; (*b*) 5117.5 N·m. **14.28** 753.5 rpm.

CHAPTER 15

15.1 1000 rpm. **15.2** 0.966; 0.966. **15.3** 1841.4 V. **15.4** 15 mWb.
15.6 50.72 percent. **15.7** 7.44°; −8.16 percent. **15.8** 11.54 kV.
15.9 (*a*) 22 MW; (*b*) 1566.4 A; (*c*) 0.737. **15.10** 93.2 percent. **15.11** 260.5 V/phase.
15.12 (*a*) 10.48 kV; (*b*) 11.04 kV; (*c*) 11.56 kV. **15.13** $\cos \phi = 1$.
15.14 $\cos \phi = 0.53$; 196.9 A. **15.15** (*a*) 786.4 kW; (*b*) 6260 N·m; (*c*) 40.5°.
15.16 107.67 kVA. **15.17** 150.64 kVa.

CHAPTER 16

16.1 4 percent. **16.2** 25 percent. **16.3** 144.35 N·m. **16.4** 0.4 Ω. **16.5** 960 rpm.
16.6 3341 rpm. **16.7** 500 W. **16.8** 76.95 percent. **16.9** 3.27 percent.
16.10 (*a*) 24,084 W; (*b*) 187.5 N·m; (*c*) 191.7 N·m. **16.11** 3.2 Ω. **16.12** 20,888 W.
16.13 (*a*) 14.61 A; (*b*) 0.84. **16.14** 547.14 Ω; 56.4 Ω.
16.15 $R_2 = 0.433\ \Omega$; $X_1 = X_2 = 1$. **16.17** (*a*) 298.14 V; (*b*) 111.8 A.
16.18 41.67 percent of full-load torque.
16.19 (*a*) 4.69 percent; (*b*) $V_{st} = 0.5\ V_f$. **16.20** 32.

CHAPTER 17

17.1 43.2 V. **17.2** 0.286. **17.5** 18.52 A. **17.6** 1000 pulses/s.
17.7 290.9 μF; 11.8 mH. **17.8** 239 rpm. **17.9.** 576 rpm; 80 N·m.

CHAPTER 18

18.1 $40.17 \times 10^{-3}\ \Omega$. **18.2** $L = 35.5$ mH; $C = 0.203\ \mu$F.
18.3 Power loss = 907.4 W; $V_R = 66.4$ kV; $V_s = 66988$ V; regulation = 0.88 percent.

18.4 (*a*) $\eta = 98.78$ percent; (*b*) $V_s = 80.57/0.32$ kV.
18.5 (*a*) Hence, $V_R = 87.29$ kV/phase, or 151.18 kV line-to-line; (*b*) 99.66 percent.
18.6 839.2 MW. **18.7** 51 KM. **18.9** 202.56 kV/phase, or 350.8 kV line-to-line.
18.10 207.46 kV/phase, or 359.3 kV line-to-line. **18.11** (*a*) 97.3 percent; (*b*) 97 percent.
18.12 $X = 30$ percent. **18.13** (*a*) 44 percent; 68.18 MVA; (*b*) 30 percent; 100 MVA.
18.14 The short-circuit MVA at $B \approx$ short circuit MVA at $A = 0.218$ MVA.

CHAPTER 19

19.1 (*b*) $(1 + T_s s)/(1 + T_2 s)$, $T_1 = R_2 C$, $T_2 = (R_1 + R_2)C$;
(*c*) $[R_2(1 + T_1 s)/(1 + T_1 s + T_1 T_2 s^2)]$, $T_1 = 2R_1 C$, $T_2 = R_2 C/2$.
19.2 $(1 + k + Ds)/(k + Ds + Ms^2)$.
19.7 (*a*) Stable; (*b*) stable; (*c*) stable; (*d*) unstable; (*e*) stable for $K < 25.16$.
19.9 $C/R = KGx/(s + 1 + KGx)$. **19.14** $K = 6$.

CHAPTER 20

20.1 $\pm$ 0.67 percent. **20.2** (*a*) Digital: 0.27 percent; analog: 1.02 percent;
(*b*) digital: 0.168 percent; analog: 2.52 percent.
20.3 2.42 percent; 115 W. **20.5** $\pm$ 2 percent.
20.6 (*a*) 0.707 A; (*b*) 1.0 A; (*c*) 0.577 A; (*d*) 0.577 A.

CHAPTER 21

21.1 (*a*) 1001100; (*c*) 1011101110; (*e*) 0.1110001. **21.4** $Y = ABC$.
21.6 $Y = (S2)(S1 + CR1)$. **21.7** No. **21.8** Yes.
21.11 (*b*) $(a + b + \bar{c} + \bar{d})(a + b + \bar{c} + d)$.
21.12 (*a*) $\Pi(6,2,0)$; (*b*) $\Pi(10,9,8,3,2)$.
21.13 (*a*) $a\bar{b} + c$; (*b*) $b + \overline{ac} + a\bar{b}cd$. **21.14** $A\bar{B} + ACD + \overline{ACD} + \overline{\bar{A}BC}$.
21.15 $\overline{yz} + wz + \bar{w}y$. **21.16** $(x + \bar{y})(x + y + z)(w + \bar{y} + \bar{z})$.

INDEX